BRITISH BASIDIOMYCETAE

British Basidiomycetae

A HANDBOOK TO THE LARGER BRITISH FUNGI

BY

CARLETON REA, B.C.L., M.A.

Hon. Member of the British Mycological Society and the Cryptogamic Society of Scotland, Membre titulaire de la Société Mycologique de France, etc.

PUBLISHED UNDER THE AUSPICES OF THE BRITISH MYCOLOGICAL SOCIETY

CAMBRIDGE
AT THE UNIVERSITY PRESS
1922

CAMBRIDGE
UNIVERSITY PRESS

University Printing House, Cambridge CB2 8BS, United Kingdom

Cambridge University Press is part of the University of Cambridge.

It furthers the University's mission by disseminating knowledge in the pursuit of education, learning and research at the highest international levels of excellence.

www.cambridge.org
Information on this title: www.cambridge.org/9781107487369

Published August 1, 1922
First published 1922
First paperback edition 2015

A catalogue record for this publication is available from the British Library

ISBN 978-1-107-48736-9 Paperback

PREFACE

AFTER thirty years of study of our British Basidiomycetae I have ventured to write this text-book at the request of many mycological friends. The well-known classification of Fries is now insufficient for modern requirements and must be altered to embody the results obtained from a more detailed anatomical and microscopical examination of these plants. The present work is based chiefly on the excellent system set forth by N. Patouillard in his *Essai taxonomique sur les familles et les genres des Hyméno-mycètes*, published in 1900. Since this date several alterations and additions to this scheme have been made, due to the investigations of the eminent mycologists J. Bresadola, E. A. Burt, H. Bourdot and A. Galzin, F. von Hoehnel and V. Litschauer, and René Maire. My very best thanks are due to Mr J. Ramsbottom for his invaluable technical advice and assistance in the preparation of the manuscript and for kindly correcting the proof-sheets of this work, and to Miss E. M. Wakefield for valuable information respecting the Thelephoraceae. I am also very greatly indebted to Messrs A. D. Cotton, C. H. Grinling, A. A. Pearson and J. Ramsbottom for obtaining the large subsidy for the publication of this work, to the generous contributors to the same, and to The Royal Society for a grant-in-aid of £10.

CARLETON REA.

19*th June*, 1922.

CONTENTS

INTRODUCTION

FUNGI are non-chlorophyllous cryptogams reproduced by spores. In the larger fungi these spores are borne either to the exterior of a cell called the *basidium* or are contained within a sac-like cell called the *ascus*. Fungi with basidia constitute the BASIDIOMYCETAE, whilst those with asci are known as the ASCOMYCETAE. The present work deals only with the former group.

The Basidiomycetae were, until quite recently, divided into two main groups, HYMENOMYCETAE and GASTEROMYCETAE: in the former the spores are freely exposed to the air at maturity, whilst in the latter they remain inclosed within the body of the fungus. The spores on germination either give rise to threads or hyphae (collectively known as the *mycelium* or spawn) or they produce secondary spores termed *sporidiola* which on germination develop the mycelium. This mycelium constitutes the vegetative portion of the fungus and consists of septate hyphae, in some cases with lateral outgrowths known as *clamp connections* which arise near the septa and connect two adjacent cells. The mycelium is either filamentous or forms cord-like strands. At their extremities the hyphae give rise to other hyphae which terminate in basidia, sterile *paraphyses* which act as spacing elements and in some cases other sterile cells termed *cystidia*. These elements together constitute the *hymenium* and generally form a homogeneous layer but in some of the Gasteromycetae the basidia are irregularly distributed as in Scleroderma. The tissue between the hymenial layers is known as the *trama*. The basidium may consist either of a continuous cell, or be longitudinally divided, or transversely septate. It is usually surmounted at the apex by short stalks, the *sterigmata*, which bear the spores; sometimes these are lateral or arise from the division of the transversely septate basidia: sometimes the spores are borne directly on the hyphae. In some genera long vesicular hyphae are present which traverse the tissue in various directions and these are often filled with a granular coloured or colourless latex. The vegetative hyphae constitute the main portion of the sporophore except in the small effused forms. The hymenium may be spread over radiating gills or plates as in the mushroom, line the interior of tubes or pores as in Polyporus, cover teeth-like projections as in Hydnum, be immersed in a gelatinous mass as in Tremella, or be borne on variously shaped structures known as *receptacles* as in the Phalloids.

The sporophore is of different shapes, *e.g.* globose as in puff-balls, sessile with outstretched flaps as in Stereum, erect, clavate, coralloid or dendroid as in Clavaria, or differentiated into distinct stem, pileus or cap and hymenium-bearing surface as in the mushroom. When young the sporophore is often completely surrounded by a *universal* veil or membrane termed the *volva* which is ruptured by the growth of the stem and in many species portions or traces of this remain permanently at the base of the stem. A second membrane or *partial veil* consisting of either interwoven or arachnoid threads often protects the hymenial surface before it is mature and extends from the stem to the margin of the pileus: it either disappears completely or remains as a *ring* on the stem, or in appendiculate fragments at the margin of the pileus. The wall surrounding the Gasteromycetae is termed the *peridium* and consists of one, two, or more layers, the *exoperidium* to the exterior and the *endoperidium* to the interior. The peridium sometimes incloses separate hymenium-bearing bodies, the *peridiola*, which are either free, attached by a cord (*funiculus*) or form a compound structure. The contents of the peridium are collectively known as the *gleba*; in addition to spores there are often certain threads, the *capillitium*, which arise either directly from the base, from the walls of the peridium or are attached to a denser central portion, the *columella*.

In some cases the external walls of the hyphae deliquesce and this gives a jelly-like consistency to the whole fungus as in Tremella, whilst in other cases the deliquescence is confined to certain areas and constitutes a valuable diagnostic feature. In the genus Coprinus the edge of the gill gradually undergoes a process of autodigestion which ensures the economical dispersion of the spores. The tramal plates of many Gasteromycetae finally dissolve and disappear. All fungi since they have no chlorophyll and are thus unable to form carbohydrates are either *saprophytes* obtaining their nourishment from dead organic matter or are *parasites* dependent on a living host. Sometimes the mycelium forms a somewhat dense stratum (*stroma*) on which the receptacle is seated or immersed; at other times it forms hard compact masses known as *sclerotia* which often lie dormant for a considerable length of time.

In more recent classifications the BASIDIOMYCETAE are primarily divided into two main divisions, the HOMOBASIDIAE and the HETEROBASIDIAE. In the former, the basidium is an undivided cell usually clavate in shape: the spores on germination give rise to a mycelium which reproduces the sporophore. In the latter, the basidia are either transversely, longitudinally, or vertically septate, or sometimes continuous, but the spores on germination give rise to sporidiola which germinate in their turn to form the mycelium which reproduces the fruit body.

The HOMOBASIDIAE are divided into two main subdivisions based upon their parasitic or saprophytic *habit*. The parasitic forms constitute the subdivision EXOBASIDIINEAE, the saprophytic forms the subdivision EU-HOMOBASIDIINEAE. The latter are divided into the three orders **Gasteromycetales, Agaricales** and **Aphyllophorales.**

The **Gasteromycetales** include the species having the hymenium still surrounded at maturity by a peridium: the **Agaricales** have the hymenium originally protected by a volva or a ring but at maturity fully exposed, whilst the **Aphyllophorales** have the hymenium exposed from the first.

The HETEROBASIDIAE are divided into four orders according to the nature of the basidium, viz. **Auriculariales**, **Tremellales**, **Tulasnellales** and **Calocerales**. In the **Auriculariales** the basidia are transversely septate; in the **Tremellales** the basidia are longitudinally, cruciately divided; the **Tulasnellales** have simple basidia but the sterigmata are at first of such a wide diameter that they were formerly considered to be spores, whilst the **Calocerales** have cylindrical basidia terminated by two pointed, usually long, sterigmata. All these orders are in general saprophytic and have the hymenium fully exposed from the first but the Auriculariales include the three parasitic suborders Pucciniineae, Coleosporiineae and Ustilagineae which are not dealt with in the present work and one suborder the Ecchynineae which has the hymenium inclosed within a peridium at maturity.

LIST OF ABBREVIATIONS

cm., centimetre.

mm., millimetre.

μ, one-thousandth part of a millimetre.

P., Pileus.

Pe., Peridium.

R., Receptacle.

st., stem.

v.v., Living specimens seen by the author.

Spores ferruginous, rough; general veil persistent	*Rozites.*
Spores ochraceous, or ferruginous, generally smooth; general veil none, or fugacious	*Pholiota.*
Spores purple, or fuscous	*Stropharia.*

**With an arachnoid or filamentous general veil.

1. Veil forming an arachnoid, fugacious ring on the stem. Spores ochraceous, or ferruginous	*Cortinarius.*
2. Veil concrete with the epidermis of the pileus.	
Spores ochraceous, or ferruginous, elliptical, smooth	*Inocybe.*
Spores ochraceous, or ferruginous, irregular, angular, echinulate, or verrucose	*Astrosporina.*

***Gills sinuate.

Spores white	*Tricholoma.*
Spores pink	*Entoloma.*
Spores ochraceous, or ferruginous	*Hebeloma.*
Spores purple, or fuscous	*Hypholoma.*

****Gills decurrent, or adnato-decurrent by a tooth.

Spores white; hymenium not waxy, nor pulverulent	*Clitocybe.*
Spores white; hymenium pulverulent	*Laccaria.*
Spores white; hymenium waxy	*Hygrophorus.*
Spores pink	*Clitopilus.*
Spores ochraceous, or ferruginous	*Flammula.*
Spores greenish fuscous, or blackish; gills mucilaginous	*Gomphidius.*

C. Pileus confluent with, but heterogeneous from, the cartilaginous stem.

*Gills adnate, or sinuato-adnate.

†Margin of pileus at first incurved, or exceeding the gills.

Spores white	*Collybia.*
Spores pink	*Leptonia.*
Spores ochraceous, or ferruginous	*Naucoria.*
Spores purple, or fuscous	*Psilocybe*
Spores black, or blackish	*Panaeolus.*

††Margin of pileus straight, at first adpressed to the stem.

Spores white — *Mycena.*

Spores pink — *Nolanea.*

Spores ochraceous, or ferruginous — *Galera.*

Spores purple, or fuscous — *Psathyra.*

Spores black, or blackish — *Psathyrella.*

**Gills decurrent.

Spores white — *Omphalia.*

Spores pink — *Eccilia.*

Spores ochraceous, or ferruginous — *Tubaria.*

D. Pileus confluent with the excentric, or lateral stem, dimidiate, sessile, or resupinate.

Spores white, gill edge entire — *Pleurotus.*

Spores white, gill edge longitudinally split — *Schizophyllum.*

Spores pink — *Claudopus.*

Spores ochraceous. — *Crepidotus.*

II. Receptacle fleshy, trama vesiculose and traversed by lacticiferous vessels. Spores white, or yellow.

Latex watery, uncoloured — *Russula.*

Latex milk-white, or coloured, rarely like serum — *Lactarius.*

III. Receptacle membranaceous, or fleshy membranaceous, fragile, rapidly putrescent, or shrivelling up.

Spores ochraceous, or ferruginous — *Bolbitius.*

Spores black, or blackish fuscous. Gills auto-digested from below upwards — *Coprinus.*

IV. Receptacle membranaceous, tough, reviving with moisture, not putrescent.

Spores white.

*Pileus with a thin, unspecialized cellular pellicle — *Marasmius.*

**Pileus with a thick, cellular pellicle.

Cells of the pellicle upright, echinulate, or verrucose — *Androsaceus.* (*Marasmius p.p.*)

Cells of the pellicle decumbent, very long, fibrillose — *Crinipellis.*

V. Receptacle coriaceous, fleshy coriaceous, or woody. Spores white.

Pileus fleshy coriaceous, gills somewhat soft	*Panus.*
Pileus membranaceous coriaceous, gills coriaceous, branched, obtuse	*Xerotus.*
Pileus coriaceous, or woody, pliant; gills firm, often toothed	*Lentinus.*

CANTHARELLINEAE.

Same characters as the suborder	CANTHARELLACEAE.

CANTHARELLACEAE.

*Spores white.

Receptacle fleshy, stipitate; gills simple. Parasitic on other Agarics	*Nyctalis.*
Receptacle fleshy, stipitate; gills forked	*Cantharellus.*
Receptacle fleshy, membranaceous, funnel-shaped or umbilicate. Hymenium veined, or smooth	*Craterellus.*
Receptacle membranaceous, spathulate, or cup-shaped, pendant. Hymenium veined, or smooth	*Dictyolus.* (*Cantharellus p.p.*)

**Spores ochraceous.

Receptacle fleshy coriaceous, stipitate. Hymenium fold-like	*Neurophyllum.*

BOLETINEAE.

Same characters as the suborder	BOLETACEAE.

BOLETACEAE.

1. Hymenium spread over gills, which anastomose by veins, and form irregular pores, especially at the apex of the stem. Spores white, ochraceous, or ferruginous	*Paxillus.*

2. Hymenium lining the inside of fleshy tubes.

Spores white, or pale yellowish	*Gyroporus.* (*Boletus p.p.*)
Spores pink	*Tylopilus.* (*Boletus p.p.*)

Spores purple	*Phaeoporus.* (*Boletus p.p.*)
Spores blackish, or fuscous. Pileus covered with imbricate scales	*Strobilomyces.*
Spores ochraceous, ferruginous, or olivaceous.	
Tubes short, alveolar, decurrent	*Boletinus.*
Tubes very short, gyroso-plicate	*Gyrodon.*
Tubes long	*Boletus.*

APHYLLOPHORALES.

I.	Receptacle pileate, stipitate, sessile, or resupinate; hymenium inferior	POROHYDNINEAE.
II.	Receptacle erect, dendroid, coralloid, simple, or branched, never pileate; hymenium more or less amphigenous	CLAVARIINEAE.

I. POROHYDNINEAE.

1.	Hymenium lining tubes coherent throughout their length, forming a layer distinct from the substance of the pileus, sometimes becoming torn into teeth, or gill-like plates, and separated by dissepiments sterile on the edge	POLYPORACEAE.
2.	Hymenium lining tubes, or covering gills, or teeth, homogeneous with the substance of the pileus, not forming a distinct layer, sterile on the edge	POLYSTICTACEAE.
3.	Hymenium spread over veins, anastomosing pores, or quite smooth; edge of veins or pores fertile	MERULIACEAE.
4.	Hymenium inferior, lining free and separate tubes	FISTULINACEAE.
5.	Hymenium spread over the surface of spines, granules, warts, or other protuberances, or quite smooth, intervening spaces fertile. Receptacle fleshy, coriaceous, waxy, crustaceous, or floccose, rarely none	HYDNACEAE.

6. Hymenium spread over a smooth, rugose, or ribbed surface, either resting upon an intermediate layer of hyphae running longitudinally between it and the mycelium, or seated directly upon the mycelium	THELEPHORACEAE.
7. Hymenium covering the whole of the interior of cup-shaped, urceolate, or cylindrical receptacles, smooth, or veined	CYPHELLACEAE.

POLYPORACEAE.

Receptacle stipitate, or sessile, fleshy, cheesy, or coriaceous. Tubes homogeneous, or heterogeneous, dissepiments entire, or toothed. Spores white, or coloured	*Polyporus.*
Receptacle stipitate, fleshy. Tubes becoming torn into teeth, or gill-like plates, and anastomosing at the base. Spores white	*Sistotrema.*
Receptacle sessile, hard, woody, or corky. Pileus often concentrically zoned, covered with a hard crust, or villose. Tubes homogeneous, or heterogeneous, often stratose. Spores white, or coloured	*Fomes.*
Receptacle stipitate, or sessile, corky. Pileus covered with a rigid, laccate, shining crust. Tubes often stratose. Spores coloured, oval, truncate at the base	*Ganoderma.*
Receptacle resupinate, membranaceous, soft, coriaceous, or corky. Tubes often inserted directly on the mycelium, round, or angular. Spores white, or coloured	*Poria.*

POLYSTICTACEAE.

Receptacle sessile, thin, coriaceous, or membranaceous. Tubes homogeneous, developing from the centre outwards. Spores white	*Polystictus.*
Receptacle sessile, or resupinate, membranaceous, or coriaceous. Tubes alveolar, becoming torn, or toothed. Spores white	*Irpex.*

Receptacle sessile, corky, or coriaceous. Hymenium spread over gills, which anastomose at the base, homogeneous with the substance of the pileus, and not forming a distinct layer. Spores white	*Lenzites.*
Receptacle sessile, corky. Tubes homogeneous with the substance of the pileus, not forming a distinct layer, regular, round, or oblong. Spores white, rarely yellowish	*Trametes.*
Receptacle stipitate, or sessile, spongy, or corky. Tubes homogeneous with the substance of the pileus, not forming a distinct layer, irregular, sinuous, or labyrinthiform, often becoming torn, or toothed. Spores white.	*Daedalea.*

MERULIACEAE.

Receptacle sessile, or resupinate, more or less gelatinous. Veins anastomosing to form irregular pores. Spores white or coloured	*Merulius.*
Receptacle erect, or resupinate, waxy, firm. Veins radial. Spores white	*Phlebia.*
Receptacle sessile, spongy coriaceous. Veins gill-like, crisped. Spores white	*Plicatura.* (*Trogia.*)
Receptacle resupinate, waxy. Hymenium granular, or smooth. Spores coloured, smooth. No cystidia	*Coniophora.*
Like *Coniophora*, but with cystidia	*Coniophorella.*

FISTULINACEAE.

Same characters as the family	*Fistulina.*

HYDNACEAE.

Receptacle none. Spines simple, cylindrical, acute, seated directly on the fugacious mycelium. Spores white	*Mucronella.*
Receptacle simple, or branched, stipitate, sessile, or dimidiate, fleshy, coriaceous, or corky. Spines subulate. Spores white, or coloured	*Hydnum.*
Receptacle resupinate, or reflexed, membranaceous coriaceous. Spines subulate, apex hispid. Spores white, oval, or oblong. Cystidia present.	*Mycoleptodon.* (*Hydnum p.p.*)

Receptacle resupinate, thin, waxy, inseparable. Tubercles or spines obtuse, often deformed, irregularly scattered, or confluent. Spores white, or coloured. Cystidia none, cystidioles (sterile basidia) sometimes present	*Radulum.*
Receptacle resupinate, thin, waxy. Spines thin, subulate, generally entire, distinct, or connate at the base. Spores white. Cystidia none, cystidioles very thin, or absent	*Acia.* (*Hydnum p.p.*)
Receptacle resupinate, thin, membranaceous, pelliculose, or crustaceous. Tubercles or spines obtuse, or pointed, entire. Spores white, or coloured. Cystidia none	*Grandinia.*
Receptacle resupinate, thin, membranaceous, waxy, crustaceous or mealy. Spines conical, ciliate, or penicillate at the apex. Spores white. Cystidia present	*Odontia.*
Receptacle resupinate, subgelatinous. Spines very minute, sterile. Spores white	*Kneiffia.*
Like *Odontia*, but spores coloured, smooth	*Hydnopsis.*
Receptacle resupinate, soft, floccose. Spines soft, conical, villose, fimbriate at the apex. Flesh coloured. Spores coloured, verrucose, or echinulate	*Caldesiella.*
Receptacle pileate, stipitate, sessile, or resupinate, without a distinct pellicle, coriaceous. Hymenium smooth, granular, or faintly ribbed. Flesh coloured. Spores coloured, angular, echinulate, or verrucose	*Phylacteria.* (*Thelephora p.p.*)
Receptacle resupinate, soft, floccose. Hymenium granular, or smooth, floccose. Flesh coloured. Spores coloured, echinulate, or angular	*Hypochnus.* (*Tomentella.*)
Like *Hypochnus*, but spores violet, smooth	*Hypochnella.*
Receptacle resupinate, effused, flocculose-pulverulent, *Hypochnus*-like. Spores straw coloured, subelliptical, hyaline-appendiculate	*Jaapia.*
Receptacle resupinate, subgelatinous, then cartilaginous. Hymenium smooth. Spores olive, elliptical, smooth	*Aldrigea.*
[Irregular abnormal growths the conidial forms of Porohydnineae]	[*Ptychogaster.*]

THELEPHORACEAE.

1. Hymenium separated from the mycelium by an intermediate layer of hyphae.

Description	Genus
Receptacle erect, much branched, branches flattened in a lamellar, or plate-like manner, fleshy. Hymenium smooth. Spores white	*Sparassis.*
Receptacle simple, or branched, stipitate, sessile, effuso-reflexed, rarely resupinate, coriaceous. Hymenium smooth. Flesh pale. Spores white. Cystidia hyaline	*Stereum*[1]
Like *Stereum*, but cystidia, or setae coloured. Flesh coloured. Spores white, or coloured	*Hymenochaete.*
[Like *Stereum*, but hymenium smooth, granular, or faintly ribbed. Flesh coloured. Spores coloured, angular, echinulate, or verrucose]	[*Phylacteria.*]
Receptacle sessile, or produced behind into a stem-like base, coriaceous, or woody. Hymenium with fan-like folds, or radiating woody, branched ribs, or veins. Spores white	*Cladoderris.*

2. Hymenium seated directly on the mycelium.

Description	Genus
Receptacle resupinate, waxy, or floccose. Hymenium smooth, with scattered protuberances caused by the breaking through of fasciculate, sterile, mycelial hyphae. Spores white. Cystidia none	*Epithele.*
Receptacle saucer-shaped with a free margin, or resupinate and adnate, floccose, or crustaceous, becoming coriaceous. Hymenium smooth, pulverulent, with much granular, or crystalline matter. Spores white, large; basidia large, sterile basidia or paraphyses moniliform, or racemose	*Aleurodiscus.*
Receptacle resupinate, waxy, crustaceous, or floccose. Hymenium waxy, smooth, or tubercular, continuous, often cracked. Spores white, rarely faintly coloured, smooth. No cystidia; sterile basidia (cystidioles) sometimes emergent	*Corticium*[2].
Like *Corticium* but the hyphae and hymenium traversed by long, cystidia-like bodies, whose walls are never thickened, nor incrusted with crystalline deposits (gloeocystidia)	*Corticium.* (Sub-genus *Gloeocystidium.*)

[1] Cf. *Eichleriella.* [2] Cf. *Sebacina.*

Receptacle incrusting, variously branched, lobed, or effused, fibrillosely floccose, soft. Spores white, echinulate. Growing on fallen twigs, and mosses	*Cristella.* (*Thelephora p.p.*)
[Receptacle resupinate, soft, floccose. Hymenium granular, or smooth, floccose. Flesh coloured. Spores coloured, echinulate, or angular]	[*Hypochnus.* (*Tomentella.*)]
[Like *Hypochnus*, but spores violet, smooth]	[*Hypochnella.*]
[Receptacle resupinate, effused, flocculose-pulverulent, *Hypochnus*-like. Spores straw coloured, subelliptical, hyaline-appendiculate]	[*Jaapia.*]
[Receptacle resupinate, waxy. Hymenium granular, or smooth. Spores coloured, smooth. No cystidia]	[*Coniophora.*]
Like *Corticium*, but with prominent, hyaline, or subhyaline cystidia in the hymenium or subhymenial tissues, which are generally thick walled, or incrusted with crystalline deposits. Spores white, rarely slightly coloured	*Peniophora.*
[Like *Coniophora*, but with cystidia]	[*Coniophorella.*]

CYPHELLACEAE.

Receptacles sessile, scattered, crowded, or confluent, coriaceous-gelatinous. Hymenium smooth, becoming wrinkled, or veined. Spores white, or pale	*Cytidia.* (*Auriculariopsis.*)
Receptacles stipitate, or sessile, scattered, or crowded, membranaceous, or waxy. Hymenium smooth, or veined. Spores white	*Cyphella.*
Receptacles sessile, seated on a superficial, felt-like, then floccose and fugacious mycelium, gregarious, or fasciculate. Hymenium smooth. Spores white	*Solenia.*
Receptacles sessile, more or less crowded, distinct, seated on, or immersed in an effused, membranaceous, or floccose stroma. Spores white	*Porothelium.*
Like *Cyphella*, but spores coloured, smooth, or echinulate	*Phaeocyphella.*

II. CLAVARIINEAE.

Same characters as the suborder — CLAVARIACEAE.

II. CLAVARIACEAE.

Receptacle erect, simple, or branched, branches cylindrical, smooth, or longitudinally striate, fleshy, or subcoriaceous, generally putrescent. Spores white, or ochraceous, smooth, or rough. Growing on the ground, or on wood	*Clavaria.*
Receptacle erect, simple, very rarely branched, cylindrically-clavate, with a long, thin stem, often springing from a sclerotium, fleshy, waxy, or tough. Spores white. Growing on fallen twigs and dead leaves	*Typhula*
Receptacle erect, simple, very rarely forked, club shaped, with a short, thick, glabrous, or villose stem, fleshy, or waxy. Spores white. Growing on herbaceous plants	*Pistillaria.*
Receptacle filiform, simple, or branched, firm, tough. Spores white. Growing on the ground, or on wood	*Pterula.*

**EXOBASIDIINEAE.

EXOBASIDIALES.

Same characters as the order — EXOBASIDIACEAE.

EXOBASIDIACEAE.

Mycelium vegetating in the interior of the living host, and giving rise, on the exterior, to basidia	*Exobasidium.*

HETEROBASIDIAE.

AURICULARIALES.

1 Parasites, with, or without, probasidia	PUCCINIINEAE[1]. COLEOSPORIINEAE[1]. USTILAGINEAE[1].
2. Saprophytes, without probasidia	
(*a*) Hymenium fully exposed from the first	AURICULARIINEAE.
(*b*) Hymenium inclosed within a peridium	ECCHYNINEAE.

[1] Not dealt with in the present work.

AURICULARIINEAE.

Same characters as suborder — AURICULARIACEAE.

AURICULARIACEAE.

Receptacle effused, incrusting, membranaceous, soft, floccose. Hymenium smooth. Basidia more or less incurved, transversely septate; sterigmata subulate, unilateral. Spores white	*Helicobasidium.*
Receptacle effused, or upright, thin, waxy, or gelatinous. Hymenium smooth. Basidia cylindrical, straight, transversely septate. Spores white	*Platygloea.*
Receptacle dimidiate, cup-shaped, sessile, or substipitate, gelatinous coriaceous, then cartilaginous. Hymenium smooth, reticulate, or ribbed. Basidia cylindrical, transversely 3-septate. Spores white, cylindrical, or subreniform	*Auricularia.*
Receptacle erect, filiform, or subclavate. Hymenium smooth. Basidia cylindrical, transversely 3-septate. Spores white	*Eocronartium.* (*Clavaria p.p.* = *Helicobasidium* sec. Patouillard.)
Receptacle erect, globose, stipitate. Hymenium consisting of branched threads terminated by a basidium. Basidia short, pear-shaped, transversely 1-septate. Spores white, elliptical. Growing on dead wood	*Stilbum.*

ECCHYNINEAE.

Same characters as suborder — ECCHYNACEAE.

ECCHYNACEAE.

Peridium globose, stipitate, or substipitate, thin, fugacious. Threads of gleba bearing the basidia on their lower portion, either in tufts, or scattered. Basidia transversely 3-septate, bearing the spores either sessile, or on very short sterigmata. Spores brown	*Ecchyna.*

TREMELLALES.

Same characters as the order	TREMELLACEAE.

TREMELLACEAE.

Receptacle foliaceous, brain-like, or tubercular, gelatinous, soft, fertile over the whole surface, very rarely papillose, sometimes with an irregular nucleus formed by mineral concretions. Spores white, globose or elliptical. Growing on dead wood, rarely on the ground	*Tremella.*
Like *Tremella,* but spores coloured	*Phaeotremella.*
Receptacle erect, ear-shaped, or spathulate, substipitate, or sessile, gelatinous, firm. Hymenium inferior, smooth, or indistinctly veined. Spores white. Growing on the ground, or on rotten wood	*Guepinia.*
Receptacle cupulate, discoid, foliaceous, or effused, marginate, pendant, sterile on upper surface, gelatinous, soft, pellucid. Hymenium inferior, smooth, reticulately veined, or foliaceous, often papillose. Spores white, allantoid. Growing on wood	*Exidia.*
Receptacle dimidiate, substipitate, or sessile, gelatinous. Hymenium with fertile spines or teeth. Spores white. Growing on wood	*Tremellodon.*
Like *Odontia,* but with subgelatinous teeth, and longitudinally septate basidia. Spores white. Growing on dead wood	*Protodontia.*
Receptacle effused, incrusting, like *Corticium,* coriaceous, gelatinous, or waxy. Hymenium smooth. Spores white. Growing on the ground, or on wood	*Sebacina.* (*Thelephora* and *Corticium p.p.*)
Like *Sebacina,* but hymenium possessing true cystidia	*Sebacina,* subg. *Heterochaetella.*
Like *Sebacina,* but hymenium possessing gloeocystidia filled with a coloured juice	*Sebacina,* subg. *Bourdotia.*
Receptacle cup-shaped, or resupinate with the margin free, or reflexed, membranaceous, waxy, or coriaceous, soft. Hymenium smooth, rugulose, or tubercular. Spores white. Growing on dead branches	*Eichleriella.*

TULASNELLALES.

Same characters as the order — TULASNELLACEAE.

TULASNELLACEAE.

Receptacle effused, fleshy membranaceous, or gelatinous, then cartilaginous. Hymenium smooth, exposed from the first. Spores white, producing sporidiola on germination; sterigmata very thick and stout. Growing on dead wood, and fallen pine needles	*Tulasnella.*

CALOCERALES.

Same characters as the order — CALOCERACEAE.

CALOCERACEAE.

Receptacle more or less tubercular, or cup-shaped, entirely gelatinous. Hymenium smooth, or plicate. Spores white, simple, septate, or muriform. Growing on dead wood	*Dacryomyces.*
Receptacle cup-shaped, or lobed, stipitate, or substipitate, gelatinous, or cartilaginous; stem firm, indurated. Hymenium smooth. Spores white, elliptic-oblong, becoming 1–3-septate. Growing on dead wood	*Ditiola.*
Receptacle erumpent, convex, then plane, sessile, gelatinous, or floccose. Hymenium smooth, becoming plicate. Spores yellowish, oblong, becoming 8–10-, or more, septate. Growing on dead, rarely living, wood	*Femsjonia.*
Receptacle upright, cylindrical, apex globose, or elongate, stipitate, gelatinous, firm. Hymenium smooth, or rugosely plicate, confined to the upper portion of the receptacle. Spores white. Growing on dead wood	*Dacryomitra.*
Receptacle upright, cylindrical, simple, or branched, gelatinous-coriaceous, cartilaginous when dry. Hymenium smooth, amphigenous. Spores white. Growing on wood	*Calocera.*

BASIDIOMYCETAE.

Fungi reproduced by spores borne on basidia.

HOMOBASIDIAE.

Basidia simple; spores on germination giving rise to a mycelium.

*EU-HOMOBASIDIINEAE.

Saprophytes.

GASTEROMYCETALES.

Hymenium inclosed at maturity within a peridium.

PHALLINEAE.

Peridium globose, consisting of three layers, the middle one gelatinous, at length ruptured, with the lower portion forming a volva at the base. Gleba at length mucilaginous, consisting of labyrinthiform cells, attached to the variously shaped receptacle, and finally borne upwards.

CLATHRACEAE.

Receptacle *trellised*, or dividing into arms, or branches at the apex, sessile, or stipitate. Gleba internal, or between the arms or branches.

Clathrus (Micheli) Pers.

(κλεῖθρον, lattice.)

Peridium globose, becoming torn into irregular lobes at the apex. Receptacle forming an obovate, or globose, hollow *lattice*, covered on the inside with the mucilaginous gleba. Basidia bearing 4–8, sessile, or subsessile, smooth, colourless, cylindrical spores. Growing on the ground.

1. **C. ruber** (Mich.) Pers. (= *Clathrus cancellatus* (Tourn.) Fr.) Rolland, Champ. t. 108, no. 245, as *Clathrus cancellatus*.

Ruber, red.

Volva 5–10 cm., *white*, globose, becoming torn into irregular lobes at the apex, attached at the base by a cord-like mycelium. Receptacle *vermilion*, or *pinkish red*, obovate, or globose, sessile, forming a hollow, pentagonal net-work, perforated in lattice-, or trellis-fashion, flattened on the outer surface, torn, and irregular on the inner side, and covered with olive brown mucus. Spores white, cylindrical, 5–6 × 2μ. Smell extremely foetid. Woods, plantations, gardens and stoves. Sept.—Nov. Rare. (*v.v.*)

Lysurus Fr.

(λύσις, loosing; οὐρά, tail.)

Peridium globose, becoming torn at the apex into irregular lobes. Receptacle stipitate, dividing at the apex *into free arms, or lobes,* distinct from the stem, bearing the mucilaginous gleba. Basidia with 4–6, sessile, or subsessile, cylindrical, or oblong, coloured spores. Growing on the ground.

2. **L. australiensis** Cke. & Mass. (= *Lysurus borealis* (Burt) P. Henn.[1]) Trans. Brit. Myc. Soc. II, t. 3.

Australiensis, belonging to Australia.

Volva 4–5 cm., *white*, globose, becoming torn above into irregular lobes, attached to the soil around the base by numerous white, cord-like mycelial strands. Receptacle 6 × 2 cm., *whitish*, cylindrical, attenuated at the base, hollow, cellular, dividing at the apex into six arms. Arms *deep reddish brown*, mucilaginous on the inside, 15–20 mm. long, 4–5 mm. wide at the base, attenuated at the apex, with a longitudinal groove down the centre and transversely ribbed, not cellular, differing in texture from the receptacle, erect, slightly incurved at the apex. Spores reddish brown, oblong elliptical, $3 \times 1{\cdot}5\mu$. Pasture where refuse of sacks had been emptied out, and amongst stable refuse. Sept.—Nov. Rare. (*v.v.*)

Aseroe La Billard.

(ἀσηρός, disgusting.)

Peridium globose, becoming torn at the apex into irregular lobes. Receptacle stipitate, crowned at the apex by *a disc, from which the arms radiate*; arms covered with the gleba. Growing on the ground.

3. **A. rubra** La Billard. Engl. & Prantl. Nat. Pflanz. Fam. I**, t. 137, figs. A—C.

Rubra, red.

Receptacle stipitate, *red, or pale rose*, sheathed by the volva at the base, pervious at the apex, which is expanded into a *bright red disc*, furnished at the margin with from five to eight *bifid rays*. Spores "hyaline, oblong, $6–10 \times 1{\cdot}5–2\mu$" Petch. On soil brought from Australia. Rare.

PHALLACEAE.

Receptacle hollow, *cylindrical, or fusiform*, with, or without, a campanulate pileus at the apex. Gleba external.

[1] Recorded as a distinct British species by Wakefield in Kew Bulletin of Miscel. Inf. no. 7 (1918), 231.

Cynophallus (Fr.) Cda.

(**Mutinus** Fr.)

(*κύων*, dog; *φαλλός*, penis.)

Peridium oval, or oblong, becoming split at the apex into two or three lobes. Receptacle hollow, cylindrical, or fusiform. Pileus apical, *adnate*, covered on the outside with the mucilaginous gleba. Basidia with 4–6, sessile, oblong, or cylindrical, pale yellowish spores. Growing on the ground.

4. **C. caninus** (Huds.) Fr. Sow. Engl. Fung. t. 330, as *Phallus inodorus*. *Caninus*, pertaining to a dog.

Volva 1–2 cm., *white, or yellowish*, oval, or oblong, splitting into two or three lobes at the apex, springing from a white, cord-like mycelium at the base. Receptacle 6–9 × 1 cm., *white, or rosy*, subfusiform, apex perforate or imperforate, hollow, cellular. Pileus *red*, 2 cm. long, adnate to the apex of the receptacle, acutely digitaliform, covered at first with *green mucus*. Spores pale yellowish, oblong, 3–5 × 2μ. Smell slight. Amongst dead leaves, and on old stumps, especially in mixed woods. June—Dec. Common. (*v.v.*)

5. **C. bambusinus** (Zoll.) Rea. Engl. & Prantl. Nat. Pflanz. Fam. I**, t. 142, figs. G—I, as *Mutinus bambusinus* Zoll. *Bambusinus*, pertaining to bamboos.

Receptacle 10–12 × 1 cm., *bright red, or pinkish*, the upper half sporiferous and *tapering into an acute point, purplish red*, covered at first *with green mucus*. Spores cylindrical, 4 × 1·5μ. Smell very foetid. Probably introduced with plants from Java. Rare.

Phallus (Micheli) Pers.

(*φαλλός*, penis.)

Peridium globose, becoming torn into irregular lobes at the apex. Receptacle hollow, cylindrical, or fusiform. Pileus reticulated, apical, *attached only by the apex*, covered on the outside with the mucilaginous gleba. Basidia with 4–8, sessile, oblong, pale yellowish spores. Growing on the ground.

6. **P. impudicus** (Linn.) Pers. Grev. Scot. Crypt. Fl. figs. 213–214, as *Phallus foetidus*. *Impudicus*, shameless.

Volva 3–5 cm., *white, or yellowish*, globose, then oval, splitting into irregular lobes at the apex, springing at the base from dense masses of *white*, cord-like mycelium. Receptacle *white*, cylindrical, attenuated at both ends, 10–30 × 1–3 cm., cellular, hollow, perforate at the apex. Pileus 3–5 cm. long, cylindrical, *white*, at first covered with *green mucus*, attached at the apex to the receptacle by a *narrow disc*, *reticulated* externally. Spores pale yellowish, oblong, 3–5 × 2μ. Smell strong, very foetid. Woods, plantations, and gardens, especially under conifers. May—Nov. Common. (*v.v.*)

var. **togatus** (Kalchbr.) Cost. & Duf. *Togatus*, cloaked.

Differs from the type in having a *white, reticulately pierced, pendant veil, attached to the base of the pileus.* Woods. Sept.—Oct. Rare.

var. **iosmos** (Berk.) Cke. Curt. Brit. Ent. x, t. 469.
ἴον, violet; ὀσμή, scent.

Differs from the type in its *pale reddish grey colour, the strongly toothed borders of the reticulations on the conical pileus and the sweet smell of violets when fresh.* Sandhills. Rare.

7. **P. imperialis** Schulz. Kalchbr. Icon. t. 40, fig. 1.
Imperialis, imperial.

Volva 2·5–7 cm., *pink on the outside, white inside*, pear-shaped, splitting at the apex into several lobes, springing at the base from a *pinkish, or pale blue*, cord-like mycelium. Receptacle *white, slightly pinkish at the extreme base*, 10–25 × 2–3 cm., cylindrical, attenuated at both ends, hollow, cellular. Pileus 3–5 cm. long, *white*, at first covered with *dark green mucus*, campanulate, attached at the apex by a *broad*, circular disc which often becomes *yellowish and crenate*, reticulated on the outside. Spores hyaline, 3–4 × 1·5–2μ. Smell *pleasant*, like that of *Glycyrrhiza* (Liquorice). Micaceous sandy soil. Oct. Rare. (*v.v.*)

HYMENOGASTRINEAE.

Peridium globose, consisting of one layer, indehiscent. Gleba formed of cells lined by the hymenium. Cystidia often present.

Hysterangiaceae.

Gleba cells *radially arranged on the sterile basal hyphae.* Spores olivaceous, oblong, or oblong elliptical. Subterranean.

Hysterangium Vitt.

(ὑστέρα, the womb; ἀγγεῖον, a vessel.)

Peridium globose, separating from the gleba at maturity. Gleba cartilaginous, or mucilaginous, cells at first empty. Basidia sometimes bearing eight spores. Spores olivaceous, oblong, or oblong elliptical. Subterranean.

8. **H. nephriticum** Berk. νεφρός, the kidneys.

Pe. 1–2·5 cm., *white*, globose, or globoso-depressed, springing from a much branched, white mycelium, tomentose, peridium rather thick, elastic. Gleba *pinkish, then pale blue, or grey, and finally greenish*, cells minute, *radiating from the base*, contracting into a very small space when dried. Spores greenish olivaceous in the mass, drab colour under the microscope, oblong elliptical, pointed at both ends, or blunt at the one end, 10–12 × 4μ. Smell at first like that of *Helianthus tuberosus*, then disagreeable. Gregarious, sometimes confluent. Buried in the ground. Woods. May—Feb. Uncommon. (*v.v.*)

9. **H. Thwaitesii** B. & Br. G. H. K. Thwaites.

Pe. 2 cm., *white, becoming rufous when touched*, subglobose, or slightly irregular, slightly silky, peridium membranaceous. Gleba *brownish olive*. Spores pale olive, oblong, apiculate, 25–30 × 7–9μ. Buried in the ground. Woods. Aug.—Oct. Rare.

HYMENOGASTRACEAE.

Surface of peridium sometimes traversed by mycelial strands. Gleba cells *arising from the peridium*, sometimes empty at first, with, or without, a sterile base. Spores coloured, elliptical, fusiform, globose, smooth, or echinulate. Subterranean or superficial.

Hymenogaster (Vitt.) Tul.

(ὑμήν, a membrane; γαστήρ, belly.)

Gleba cells empty at first, sterile base well-developed. Basidia generally with two sterigmata. Spores coloured, *elliptical to fusiform, with a prominent papilla*. Subterranean, or superficial.

10. **H. Klotzschii** Tul. Tul. Fung. Hypog. t. 10, fig. 12. J. F. Klotzsch.

Pe. 1–1·5 cm., *dirty white*, obovate, or subglobose, adpressedly tomentose, base fibrillose. Gleba *pallid, becoming rufous ochre*. Spores pale brown, minutely tuberculose, broadly elliptic, ends obtuse, 18–20 × 11–13μ. Pot in greenhouse, and sandy soil. Dec. Rare.

11. **H. muticus** Berk. Tul. Fung. Hypog. t. 10, fig. 7.
Muticus, curtailed.

Pe. 1·5–2·5 cm., *white, then tinged with brown*, globose, scarcely lobed, at length much cracked. Gleba *pale yellow-brown*, cells loose, small. Spores pale brown, obovate, oblong, very obtuse (figured by Massee as apiculate at both ends), 18–21 × 10–12μ. Smell slight. Under trees. Nov. Rare.

12. **H. luteus** Vitt. Vitt. Mon. Tub. t. 3, fig. 9. *Luteus*, yellow.

Pe. 2–3 cm., *white, then brownish*, subglobose, soft, silky, peridium very thin. Gleba *bright yellow*, cells small, flexuose. Spores yellowish, oval, or elliptical, 24–28 × 10μ. Smell pleasant, of "musk" Quél, of "strawberry" Vitt., sometimes "powerfully foetid" Berk. Woods. Sept.—March. Not uncommon.

13. **H. decorus** Tul. Tul. Fung. Hypog. t. 10, fig. 9. *Decorus*, graceful.

Pe. 2·5–5 cm., *dirty white, becoming yellowish in places*, roundish. Gleba *lilac brown, then blackish*, sterile base almost obsolete. Spores ochraceous, then brown, rugulose, broadly elliptical, obtuse, or obtusely apiculate, 24–28 × 13–15μ; basidia long, slender, sometimes flexuose, monosporous rarely bisporous. Wood and tan pits. Oct.—Nov. Rare.

14. **H. lycoperdineus** Vitt. Vitt. Mon. Tub. t. 2, fig. 5. *Lycoperdon*, a puff-ball.

Pe. 2–5 cm., *white, then brownish*, subglobose, somewhat deformed, plicate at the base, smooth, silky. Gleba *whitish, then fuliginous*, cells large, irregular. Spores "brownish yellow, oblong or elliptical, somewhat uneven, 19–23 × 9–11μ" Rabenh. Smell strong, of onion. Gregarious. In earth, and clay. Nov. Rare. (*v.v.*)

15. **H. vulgaris** Tul. Tul. Fung. Hypog. t. 10, fig. 13. *Vulgaris*, common.

Pe. 2–3 cm., *whitish, becoming discoloured*, subglobose, regular, or variously lobed, or sulcate, soft. Gleba *dirty white, then dark brown*, cells rather large, irregular, sterile base minute. Spores blackish brown, rugulose, oblong, or oblong-lanceolate, acute, attenuated at the base, 34–40 × 12–14μ. Gregarious, sometimes subcaespitose. Underground. July—Oct. Rare.

16. **H. pallidus** B. & Br. *Pallidus*, pale.

Pe. 6–12 mm., *white, then dirty tan colour*, round, depressed, nearly smooth. Gleba *white, then yellow* and finally *pale brown*, sterile base obsolete. Spores brown, rather rough, lanceolate, acute, shortly pedicellate, 30–36 × 12–14μ. Underground, under firs. Oct. Rare.

17. **H. citrinus** Vitt. Tul. Fung. Hypog. t. 1, fig. 1; t. 10, fig. 3. *Citrinus*, lemon yellow.

Pe. 2–4 cm., *lemon, or golden yellow, then rufous black*, rotundato-gibbous, shining as if silky. Gleba *lemon yellow, then brown*, cells small, tramal plates yellow. Spores reddish brown, rugulose, lanceolate, apiculate, 40 × 17–20μ. Smell cheesy. Underground. May—Dec. Uncommon.

echinulate, elliptic-oblong, 12–15 × 9–10μ, with a large central gutta. Often somewhat superficial. Woods and downs under trees. Aug.—Jan. Uncommon. (*v.v.*)

27. **H. carneum** Wallr. Boud. Icon. t. 192. *Carneum*, flesh colour.

Pe. 1·5–3 cm., *flesh colour*, subglobose, or irregular, slightly tomentose, then smooth, and somewht marbled, attached at the base to the soil. Gleba *concolorous*, cells small, irregular. Spores yellowish in the mass, hyaline under the microscope, with long acute spines, globose, 13–18μ. Subterranean, or somewhat superficial. About the roots of *Eucalyptus*. Oct.—Dec. Rare.

Rhizopogon Fr.

(ῥίζα, root; πώγων, beard.)

Peridium globose, or oblong, *covered on the surface with mycelial strands*. Gleba cells arising from the peridium, empty at first. Basidia bearing 2–8 sessile spores. Spores coloured, oblong elliptical. Subsuperficial.

28. **R. rubescens** Tul. Tul. Fung. Hypog. t. 11, fig. 4; t. 2, fig. 1. *Rubescens*, becoming red.

Pe. 2–6 cm., *white, becoming reddish when exposed to the air, then yellow or olive*, ovate, or globose, silky, *covered with numerous strands of the mycelium which become reddish when touched.* Gleba *yellowish then brownish*, cells small, irregular. Spores pale ochraceous, oblong elliptical, 7–8 × 3μ, 1–3-guttulate; basidia with 2–8-sterigmata. Smell somewhat acid, then unpleasant. Somewhat superficial. Sandy fir woods. Sept.—Dec. Uncommon. (*v.v.*)

29. **R. luteolus** Fr. Tul. Fung. Hypog. t. 1, fig. 5; t. 11, fig. 5. *Luteolus*, yellowish.

Pe. 2–5 cm., *whitish, becoming dirty yellow, then olive brown*, globose, or oblong ovate, clothed with numerous free, or adnate, mycelial strands, peridium thick, subcoriaceous. Gleba *olivaceous*, tramal plates *whitish*, cells minute, rounded. Spores olivaceous, oblong elliptical, 6–7 × 3μ, 2-guttulate. Smell slight, then strong. Somewhat superficial. Sandy fir woods. Sept.—Nov. Not uncommon. (*v.v.*)

LYCOPERDINEAE.

Peridium globose, or variously shaped, consisting of two or more layers; dehiscing by an apical aperture, or by the gradual falling away of the upper peridial walls. Gleba consisting of cells lined by the hymenium, finally breaking down into a powdery mass, consisting of spores, and capillitium threads, attached to the endoperidial walls or springing from a central columella, or entirely free with, or without, a sterile base. Basidia bearing 4–8 sessile, or stipitate spores. Spores coloured, smooth, verrucose, or echinulate, globose, subglobose, or elliptical, sometimes with the sterigma remaining attached. Superficial.

Lycoperdaceae.

Same characters as the suborder.

Lycoperdon (Tournef.) Pers.

(λύκος, a wolf; πέρδομαι, I break wind.)

Peridium globose, or variously shaped; exoperidium pseudoparenchymatous, fleshy, or membranaceous, spinulose, warted, granular, or smooth, fugacious; endoperidium membranaceous, or papyraceous, thin, dehiscing by an apical aperture, or by the gradual falling away of the upper portion. Gleba with, or without, a sterile base. Capillitium threads long, branched, not consisting of a distinct stem and branches, attached to the peridium or to a central columella. Spores coloured, echinulate, verrucose, or smooth, globose, or elliptical. Superficial.

I. Peridium dehiscing by the upper portion gradually falling away in pieces. Capillitium very long, and much branched. Sterile base persistent.

30. **L. giganteum** (Batsch) Pers. (= *Lycoperdon Bovista* (Linn.) Fr.) Boud. Icon. t. 188–189, as *Lycoperdon Bovista* Linn.

γίγας, giant.

Pe. 15–16 cm., *white, then yellowish, or olivaceous,* globose, or depressed, oval, pumpkin-shaped, often more or less plicate at the base, sessile, attached by a cord-like mycelium; exoperidium at first subtomentose, then becoming smooth like a kid glove, fragile, ultimately splitting up and falling away in pieces from the endoperidium, which is also very thin, brittle and evanescent above. Gleba *white, then yellowish and finally olivaceous,* compact. Sterile base very thin, or almost absent. Spores olivaceous, or brownish, verrucose, globose, sometimes pedicellate, 4–5μ. Capillitium brown, very long, branched, septate, 3–5μ in diam., persistent. Edible. Pastures, gardens and roadsides. May—Nov. Common. (*v.v.*)

31. **L. caelatum** (Bull.) Fr. (= *Lycoperdon favosum* (Rostk.) Bonord.) Berk. Outl. Brit. Fung. t. 20, fig. 7. *Caelatum*, engraved.

Pe. 7–12 cm., *white, then ochraceous, and finally tinged brownish,* subglobose, oval or depressed, contracted below into a more or less stem-like base with thick mycelium; exoperidium floccose, covered with large, distant warts, and *cracking into net-like areolae*; warts evanescent above, and separating in patches from the endoperidium; endoperidium thick, fragile, thinner in the upper half and finally falling away in pieces, leaving only the cup-like sterile base *with its diaphragm.* Gleba *white, then yellowish, and finally olivaceous,* compact. Sterile base, thick, persistent, forming nearly half the peridium, separated from the fertile portion by a *distinct, membranaceous diaphragm.* Spores dark olivaceous, globose, rarely very shortly pedicellate, 4–5μ. Capillitium yellowish, very long, flexuose, branched, brittle, 6–7μ in diam. Edible. Woods, heaths and pastures. May—Nov. Common. (*v.v.*)

32. **L. saccatum** (Vahl.) Fr. Krombh. Icon. t. 30, figs. 11–12.
σάκκος, a bag.

Pe. 7–18 cm. high, 3–12 cm. wide, *whitish, or greyish,* becoming tinged brownish with age, *clavate,* or *pestle-like,* rounded above, obtuse, plicato-lacunose below and continued into a long stem-like base, 2·5–6 cm. wide, cylindrical, or subventricose, often scrobiculate, exoperidium consisting of small fugacious, spinulose warts, and granules which soon disappear from the upper portion—the warts split at the base and coalesce in a fine point at the apex; endoperidium *concolorous, very thin,* fragile, falling away in patches. Gleba *white, then yellow, and finally olivaceous,* compact. Sterile base reaching to the apex of the stem-like portion of the peridium, convex, cellular, firm. Spores olivaceous, verrucose, globose, 4–5μ. Capillitium pale yellowish, very long, branched, 3–5μ in diam. Edible. Woods, heaths and pastures. Aug.—Nov. Not uncommon. (*v.v.*)

33. **L. excipuliforme** (Scop.) Pers. Fr. Sverig. Svamp. t. 73.
Excipula, a vessel; *forma,* shape.

Pe. 5–13 cm. high, 4–11 cm. wide, *greyish, becoming tinged with yellow or brown,* globose, often compressed, plicate on the underside and continued into a short, or fairly long, stout, broad, stem-like base; exoperidium consisting of long, delicate, floccose spines, separate at their base but confluent at their apices, becoming smaller downwards, wearing away with age and weathering; endoperidium *floccose, thick,* firm, only gradually wearing away and disappearing in the upper portion. Gleba white, then yellowish, and finally brownish olivaceous. Sterile base whitish, becoming yellowish or greenish, cellular, *concave,* extending to the apex of the stem-like base of the peridium. Spores fuscous olivaceous, *echinulate,* globose, 3–5μ. Capillitium olivaceous, becoming hyaline, flexuose, rarely branched, 3–5μ in diam. Edible. Woods and pastures. April—Nov. Not uncommon. (*v.v.*)

var. **flavescens** Quél. *Flavescens,* becoming yellowish.

Differs from the type in its *smaller size, its club-shaped peridium and its brighter yellow colour.* Heaths and pastures. Sept.—Oct. Uncommon. (*v.v.*)

II. Peridium dehiscing by an apical mouth, followed by the upper portion falling away in pieces. Sterile base persistent, separated from the gleba by a distinct diaphragm.

34. **L. depressum** Bon. (= *Lycoperdon hyemale* (Pers.) Vitt. sec. Hollós, *Lycoperdon pratense* Pers. sec. Lloyd.) Trans. Brit. Myc. Soc. II, t. 9. *Depressum*, depressed.

Pe. 2–5 cm., *yellowish white, then greyish yellow, and finally brownish*, obconic, at first rounded at both ends, then flattened on the top, often compressed at the sides, more or less contracted at the base and plicate; exoperidium consisting of whitish spines united at the apex, intermixed with minute, simple spines and furfuraceous granules, all of which disappear with age and weathering; endoperidium *concolorous*, thin above, dehiscing by a well-defined apical mouth which soon extends until the whole of the upper portion of the peridium disappears. Gleba white, then yellowish, and finally fuscous olivaceous, *separated from the sterile base by a distinct membranaceous diaphragm*. Sterile base with large cells, often forming one half of the peridium. Spores olivaceous, globose, 4μ. Capillitium colourless, branched, flexuose, rough, $4–6\mu$ in diam. Heaths, pastures and hillsides. Aug.—March. Common. (*v.v.*)

35. **L. candidum** Pers. (= *Lycoperdon papillatum* (Schaeff.) Hollós.) Lloyd, The Genus Lycop. in Eur. t. 51, as *Lycoperdon cruciatum*. *Candidum*, shining white.

Pe. 2–5 cm., *white, then yellowish, and finally pale darkish brown*, globose, or usually depressed, often plicate beneath and continued into a stem-like base attached to the white cord-like mycelium; exoperidium consisting of *white*, blunt cruciate spines *which adhere together and peel off in patches*; endoperidium *yellowish, then pale darkish brown*, minutely furfuraceous, thin. Gleba olive, then dark brown, *with a distinct diaphragm* separating it from the sterile base. Sterile base with large cells, about a quarter to a third of the peridium, rarely very small. Spores dark brown, globose, often pedicellate, $3{\cdot}5–4\mu$. Capillitium coloured, sparingly branched, $5–7\mu$ in diam. Pastures and heaths. Sept.—Oct. Uncommon. (*v.v.*)

III. Peridium dehiscing by an apical mouth. Sterile base not separated from the gleba by a diaphragm.

A. Spores strongly echinulate or verrucose.

36. **L. echinatum** Pers. Rolland, Champ. t. 110, no. 251. ἐχῖνος, a hedgehog.

Pe. 2–6 cm., *white, then ochraceous and finally brown*, obovate, or subglobose, often compressed, sometimes rather attenuated at the base, attached by a long, white, cord-like mycelium; exoperidium consisting of long, conical warts, separate at the base and often coalescent at their apices, white, then ochraceous, and finally brownish, surrounded at the base of the warts by a ring of minute, mealy warts; the warts on the upper portion of the peridium disappear with age and weathering and then the *pale brown* inner peridium presents a *net-like appearance* from the persistent rings of darker brown, mealy warts. Mouth simple, apical, torn. Gleba olivaceous, then *violet, or brownish purple*, compact. Sterile base about one-third of the peridium, sometimes very small, *cellular*. Spores purple umber, *echinulate*, globose, 4–6μ. Capillitium purplish, much branched, branches pointed, 3–4μ in diam. Woods and plantations, especially beech. March—Nov. Uncommon. (*v.v.*)

37. **L. Hoylei** Berk. Hoyle.

Pe. 3–4 cm., *brownish*, subglobose, ovate, or subpyriform; exoperidium consisting of long, pyramidal warts, separate at the base, coalescent at the apices, ochraceous at first, then brownish, at the base of the larger warts surrounded by a ring of minute, dark brown warts, that give a *net-like appearance* to the *paler* inner peridium when the larger warts fall away. Mouth small, irregularly torn. Gleba olivaceous, *then purplish*, compact. Sterile base, *bright olive, compact*. Spores purple, *verrucose*, globose, 5μ. Capillitium yellowish, sparingly branched, flexuose, uneven, 4–5μ in diam. Amongst leaves in woods. Oct.—Dec. Rare. (*v.v.*)

38. **L. atropurpureum** Vitt. Vitt. Mon. Lyc. t. 2, fig. 6.
Ater, black; *purpureum*, purple.

Pe. 2·5–6 cm., *greyish, or brownish, yellowish towards* the base, subglobose, or pyriform, sessile, or attenuated into a stem-like base, often plicate below, thin, soft, flexible; exoperidium consisting of long, thin, brownish spines, often coalescent at their apices, becoming shorter towards the base, brittle, falling away and exposing the *smooth, somewhat shining, light brown, or purplish endoperidium*. Mouth small, irregular. Gleba olivaceous, then brownish and finally dark purple. Sterile base cellular, shallow, rarely reaching a third of the peridium. Spores dark purple, strongly verrucose, globose, sometimes pedicellate, 5–7μ. Columella globose. Capillitium branched, 4–6μ in diam. Oak woods and heaths. Sept.—Nov. Uncommon. (*v.v.*)

39. **L. umbrinum** Pers. Pers. Icon. Pictae, t. 18, fig. 3.
Umbrinum, umber colour.

Pe. 2·5–5 cm., *umber*, obovate, or pear-shaped; exoperidium densely covered with long, brown, slender spines, that are generally connivent by twos at their acute apices, simple at the base and arising from the *pale brown* endoperidium, somewhat brittle and deciduous. Mouth small, round, or toothed, apical. Gleba dark umber in the centre, paler towards the periphery and more lax. Sterile base olivaceous, about one-third of the peridium. Spores reddish brown, verrucose, globose, 4μ. Capillitium pale yellowish, branched, flexuose, uneven, forming a small pseudo-columella. Woods and heaths. July—Dec. Common. (*v.v.*)

40. **L. velatum** Vitt. Trans. Brit. Myc. Soc. II, t. 3. *Velatum*, veiled.

Pe. 3–6 cm., *snow white, then flesh colour, and finally greyish, or yellowish*, subglobose, or pyriform, often slightly umbonate, attached by a white cord-like mycelium at the base; exoperidium *white, then slightly yellowish*, tomentose, breaking up into evanescent, *star-shaped rosettes* and often forming a ring-like zone at the apex of the sterile basal stratum, finally disappearing almost completely; endoperidium *concolorous*, furfuraceous, minutely spinulose. Mouth small, apical, irregular. Gleba white, then fulvous, and finally ash colour, or purplish. Sterile base whitish, cellular, reaching to the apex of the stem-like portion of the peridium. Spores yellow, obtusely verrucose, globose, 4–5μ. Capillitium *yellow*, with darker walls, 3–4μ in diam. Woods and heaths. Sept.—Oct. Uncommon. (*v.v.*)

B. Spores smooth or only minutely verrucose, or punctate.

*Sterile base with large cells.

41. **L. perlatum** Pers. (= *Lycoperdon gemmatum* Auct. pl.) Rolland, Champ. t. 109, no. 247, as *Lycoperdon gemmatum*.
Perlatum, very wide-spread.

Pe. 2·5–5 cm., *snow white, then yellowish, and finally brownish*, especially above, turbinate, or subglobose with an elongated, cylindrical stem-like base, rarely subglobose, or depressed and nearly sessile, *always umbonate*, generally plicate and lacunose below, and attached, often in pairs, to a white, cord-like mycelium; exoperidium consisting of acute, or obtuse spines, each surrounded by a ring of smaller, obtuse warts, which give a net-like appearance to the endoperidium when the large spines are rubbed off or fall away. Mouth small, at the apex of the umbo. Gleba *white, then greenish yellow, and finally olivaceous*. Sterile base convex, cellular, reaching to the apex of the stem-like base. Spores olivaceous, smooth, or minutely punctate, globose, 4μ. Columella *prominent*, elliptical, loose. Capillitium olivaceous, simple, sparingly branched, 3–6μ in diam. Woods and pastures. July—Nov. Common. (*v.v.*)

var. **lacunosum** (Bull.). Bull. Hist. Champ. Fr. t. 52.
Lacunosum, full of hollows.

Differs from the type in the *lacunose, scrobiculate, depressed pits on the stem-like base of the peridium*. Heaths. Oct. Uncommon.

42. **L. molle** Pers. *Molle*, soft.

Pe. 1–3 cm., *white, then yellowish, or tan colour*, turbinate, or globose, depressed above, and abruptly attenuated into a short, thick, stem-like base, and attached by a white, fibrous mycelium; exoperidium consisting of fugacious, furfuraceous spines and granules; endoperidium *olive brown*, thin, papyraceous, collapsing, shining. Mouth small, irregular. Gleba greenish yellow, then brownish olivaceous. Sterile base paler, cellular, one-third of the peridium. Spores ochraceous olive, very minutely warted, globose, often shortly pedicellate, 3–4μ. Capillitium yellow, branched, 4–6μ in diam. Woods, especially oak. Sept.—Oct. Uncommon. (*v.v.*)

43. **L. nigrescens** Pers. Lloyd, The Genus Lycop. in Eur. t. 123. *Nigrescens*, becoming black.

Pe. 3–5 cm. high and wide, *brown*, subglobose, depressed above, attenuated downwards into a stem-like base; exoperidium consisting of long, *stiff, brown spines*, connivent at their apices and surrounded by a circle of minute, brown warts, or granules, which, when the larger spines fall away, give a net-like appearance to the endoperidium; endoperidium *paler*, thin, smooth. Gleba olive umber, somewhat lax. Sterile base of large cells, filling the stem-like portion of the peridium. Spores olivaceous umber, globose, very minutely verrucose, with caducous pedicels 4–5μ. Capillitium olivaceous, rarely branched, 4–6μ in diam. Woods. Sept.—Nov. Uncommon. (*v.v.*)

44. **L. pyriforme** (Schaeff.) Pers. Grev. Scot. Crypt. Fl. t. 304. *Pyrus*, pear; *forma*, shape.

Pe. 2·5–10 cm. high, 1–3 cm. wide, *white, grey, or brownish*, pyriform, or subglobose, subumbonate, attached at the base by long, white, cord-like mycelial strands, thin, flaccid; exoperidium consisting of minute, fugacious, pointed spines and granules; endoperidium *concolorous*, smooth. Mouth small, apical, torn. Gleba white, then greenish yellow, and finally brownish. Sterile base white, becoming discoloured, of rather small cells, forming the stem-like portion of the peridium. Spores olivaceous, globose, 4μ. Columella *distinct*, subglobose. Capillitium olivaceous, branched, long, 4–5μ in diam. Generally caespitose. Stumps, logs and buried *débris* of wood. May—Feb. Common. (*v.v.*)

var. **serotinum** (Bon.) Hollós. Lloyd, The Genus Lycop. in Eur. t. 50, figs. 1–2, as *Lycoperdon serotinum*. *Serotinum*, late.

Differs from the type in the peridium becoming *broken up into areolae*. Stumps and logs. Oct.—Dec. Not uncommon. (*v.v.*)

var. **excipuliforme** Desmaz. Lloyd, The Genus Lycop. in Eur. t. 49, as *Lycoperdon Desmazieres*. *Excipula*, a vessel; *forma*, shape.

Differs from the type in *the peridium being contracted abruptly into a long, slender stem.* Stumps and logs. Sept.—Nov. Not uncommon. (*v.v.*)

var. **tessellatum** Pers. Lloyd, The Lycop. Unit. St. t. 50, figs. 3–6. *Tessellatum*, checkered.

Differs from the type in the *reddish brown exoperidium becoming broken up into indurated areolae.* Stumps. Oct.—Nov. Uncommon. (*v.v.*)

45. **L. spadiceum** Pers. (= *Lycoperdon Cookei* Massee sec. Hollós.) Lloyd, The Genus Lycop. in Eur. t. 54. *Spadiceum*, date brown.

Pe. 1–2 cm., *bluish grey, soon yellowish, and finally light brown*, obovate and flattened below, or globose, abruptly contracted into a stem-like base, and somewhat pear-shaped, *whitish, becoming yellowish towards the base*; exoperidium consisting of minute, nodular, granular or subfurfuraceous spines; endoperidium *concolorous*, thin, often covered with lime granules. Mouth apical, small, irregular. Gleba olive, then brown. Sterile base whitish, then yellowish, and finally umber brown, fairly large celled, convex, reaching to a third of the peridium. Spores yellow, then olivaceous, globose, sometimes pedicellate, 4μ. Capillitium yellowish, simple, rarely branched, 4–6μ in diam. Gregarious. Sandy soil on heaths and lawns. Sept.—Oct. Uncommon. (*v.v.*)

**Sterile base with minute cells.

46. **L. polymorphum** Vitt. (= *Lycoperdon furfuraceum* (Schaeff.) Sacc.) Lloyd, The Genus Lycop. in Eur. t. 34 and 52. *πολύς*, many; *μορφή*, shape.

Pe. ·5–3 cm., *white, then dirty yellow, or greyish brown, and finally yellowish brown, somewhat reddish at the base when quite mature*, round, often depressed, sometimes pear-shaped, or attenuated into a stem-like short base, thin, membranaceous; exoperidium consisting of minute, fugacious, furfuraceous spines and granules; endoperidium thin, smooth and shining. Mouth apical, small, becoming torn. Gleba yellowish, then olivaceous brown. Sterile base, *very compact*, consisting of cells only perceptible under a lens, concolorous, reaching to the apex of the stem-like base of the peridium. Spores yellowish, very minutely warted, globose, sometimes with a wart-like basal apiculus the remains of the sterigma, 3–4μ. Capillitium yellowish, or yellowish brown, branched, 4–6μ in diam. Sandy pastures and heaths. Aug.—Nov. Not uncommon. (*v.v.*)

var. **cepaeforme** (Bull.) Lloyd. Morgan, N. Amer. Fung. in Journ. Cincinnati Soc. Nat. Hist. XIV, t. 2, fig. 9.
Cepa, onion; *forma*, shape.

Differs from the type in its *constant subglobose shape, and in the very scanty sterile base.* Sandy soil on heaths. Sept.—Oct. Uncommon. (*v.v.*)

***Sterile base absent.

47. **L. pusillum** (Batsch) Pers. Lloyd, The Genus Lycop. in Eur. t. 53, figs. 9–11. *Pusillum*, very small.

Pe. 9–20 mm., *white, then yellowish*, globose, attenuated at the base into a tapering root ending in the white mycelial strands, membranaceous, flaccid; exoperidium consisting of minute, adpressed, fugacious, mealy squamules; endoperidium smooth, shining, thin. Mouth apical, small, irregular. Gleba white, then yellowish, or greenish yellow, and finally brownish olivaceous. Sterile base *absent.* Spores olivaceous ochre, very minutely warted, globose, sometimes pedicellate, 3·5–4μ. Capillitium yellow, much branched, tapering at the ends, 3·5–4μ in diam. Sandy soil on heaths. Sept.—Nov. Not uncommon. (*v.v.*)

Bovistella Morgan.

(Diminutive of *Bovista*, a puff-ball.)

Peridium subglobose; exoperidium thick, or thin, floccose, or smooth, fugacious; endoperidium membranaceous, thin, dehiscing by an apical aperture. Gleba with a *well-developed sterile base.* Capillitium threads *free, consisting of a thick stem, and dichotomous, pointed branches.* Spores coloured, globose, or oval, smooth, pedicellate. Superficial.

48. **B. paludosa** (Lév.) Lloyd. (= *Bovista paludosa* Lév.) Trans. Brit. Myc. Soc. III, t. 8. *Paludosa*, of marshes.

Pe. 3 cm. high and wide, *pale yellow, tinged with reddish brown*, subglobose, plicate below and abruptly attenuated into a well-developed stem-like base; exoperidium *pale ochraceous* (like a coat of whitewash), gradually disappearing, very thin; endoperidium *concolorous, becoming somewhat brownish with age*, thin, flexible. Mouth apical, minute. Gleba dark olive. Sterile base *well developed*, reaching to the apex of the stem-like portion of the peridium. Spores olive, globose, 4–5μ, with long, hyaline, slender pedicels 9–10μ long. Capillitium yellowish, thick walls deeper coloured, consisting of separate, branched threads tapering to a point; branches 3–4μ in diam., main stem 9–12μ in diam. Moors. Aug. Rare. (*v.v.*)

49. **B. ammophila** (Lév.) Lloyd. (= *Bovista ammophila* Lév.) Lloyd, Myc. Writings, II, t. 87, figs. 5–6[1]. *ἄμμος*, sand; *φίλος*, loving.

Pe. 3 cm., *whitish, then pallid,* broadly obovate, plicate below and attenuated into a long, slender, taproot-like base, thin, brittle, rigid, hard; exoperidium *whitish* broken up into tomentose warts; endoperidium *pallid,* thin. Mouth small, apical, irregularly torn. Gleba dark brown. Sterile base of large cells, very firm, rigid, about one-third of the peridium. Spores olive in the mass, pale under the microscope, oval, 4–5μ with slender, tapering pedicels. Capillitium olive, thick walled, consisting of separate, short, branched threads. Sandy places. Sept. Rare.

Bovista (Dill.) Morgan.

(*Bofist,* a puff-ball.)

Peridium subglobose; exoperidium fleshy, smooth, fugacious, sometimes persistent at the base; endoperidium membranaceous, becoming papyraceous, thin, soft, dehiscing by an apical aperture, or opening irregularly. Gleba *without a sterile base.* Capillitium threads *free, consisting of a thick stem, and dichotomous, long pointed branches.* Spores coloured, globose, oval, or elliptical, smooth, pedicellate. Superficial.

50. **B. nigrescens** Pers. Berk. Outl. Brit. Fung. t. 20, fig. 5. *Nigrescens,* becoming black.

Pe. 2·5–6 cm., *whitish, then pale dark brown, or umber brown, and finally blackish umber,* globose; exoperidium *whitish,* papyraceous, soon breaking away; endoperidium *concolorous,* thin, tough, *shining,* smooth. Mouth apical, irregular, torn. Gleba white, then ochraceous, or olivaceous, and finally purple, soft, loose. Spores umber purple, globose, or slightly oval, 5–6μ, with long, hyaline pedicels. Capillitium dark brown, thick walled, bent, branched, branches pointed at the ends, 12–18μ in diam. Pastures and heaths. Jan.—Dec. Common. (*v.v.*)

51. **B. plumbea** Fr. (= *Bovista ammophila* Lév. ex Massee in Journ. of Bot. (1883), 133.) Berk. Outl. Brit. Fung. t. 20, fig. 6. *Plumbea,* lead colour.

Pe. 1–3 cm., *whitish, then lead colour,* globose, or depressed; exoperidium *white,* thin, smooth, soon peeling off, sometimes leaving a persistent portion near the base; endoperidium *lead colour,* thin, tough, *opaque.* Mouth apical, round, oval, or irregular. Gleba white, then ochraceous, or olive, and finally purplish brown, soft, loose.

[1] Lloyd states, l.c. II, 262, that the British record rests on an erroneous determination.

Spores brown, subglobose, or oval, 6–7 × 5–6μ, with long, hyaline pedicels. Capillitium brown, thick walled, branched, branches pointed at the ends, 12–16μ in diam. Pastures and heaths. Jan.—Dec. Common. (*v.v.*)

52. **B. olivacea** Cke. & Massee. *Olivacea*, olive colour.

Pe. 3–5 cm., *white, or ochraceous*, globose; exoperidium very thin, fugacious; endoperidium *concolorous*, thick, soft, becoming brittle and breaking away in patches upwards. Gleba citron, then olive, dense. Spores pale yellow, globose, 5μ, sometimes pedicellate. Capillitium pale, thin, flaccid. Pastures. Sept. Rare.

53. **B. ovalispora** Cke. & Massee. *Ovalis*, oval; σπορά, seed.

Pe. 5–6 cm., *whitish, or ochraceous*, subglobose; exoperidium breaking away in patches above, subpersistent towards the base; endoperidium *dull lead colour*, thin, flaccid, smooth. Mouth apical, irregular. Gleba umber. Spores brownish umber, *with a narrow hyaline border, elliptical*, 6 × 4–5μ, with long, stout, hyaline pedicels. Capillitium umber, thick walled, much and vaguely branched, tapering to long slender tips, 12–16μ in diam. Lawns. Rare.

Myriostoma Desv.

(μυρίος, countless; στόμα, mouth.)

Peridium subglobose; exoperidium consisting of two layers, a fibrous, or mycelial layer, and a pseudo-parenchymatous layer, thick, fleshy-coriaceous, *splitting at maturity from the apex downwards into several star-like lobes which become reflexed*; endoperidium membranaceous, then papyraceous, thin, *supported on several short stems, dehiscing by many apertures, or mouths*. Capillitium threads *simple*, rarely branched, tapering at the end. Spores coloured, minutely verrucose, globose. Superficial.

54. **M. coliforme** (Dicks.) Cda. (= *Geastrum coliforme* (Dicks.) Pers.)
Dicks. Pl. Crypt. Brit. t. 3, fig. 4, as *Lycoperdon coliforme*.
Colum, a strainer; *forma*, shape.

Exoperidium 7–10 cm., *ochraceous*, round, covered with large, angular *dark brown* scales, splitting into 4–7 sharp pointed lobes, divided almost up to the middle, reflexed, seldom inflexed; endoperidia *lead colour, or brownish*, round, compressed, with a silvery sheen, minutely warted, supported on numerous, slender, angular, or cylindrical, sometimes branched pedicels, mouths numerous, ciliated. Spores umber brown, verrucose, globose, 4–6μ. Columellas numerous, filamentous, branched, or unbranched. Capillitium pale brown, simple, flexuose, thick walled, pointed at the ends, rarely branched, 3–4μ in diam. Sandy soil. Oct. Rare. (*v.v.*)

Geaster (Micheli) Fr.

(γῆ, earth; ἀστήρ, star.)

Peridium subglobose, rarely ovate acuminate; exoperidium consisting of two layers, a fibrous, or mycelial layer, and a pseudoparenchymatous layer, thick, fleshy-coriaceous, at first closely investing the endoperidium but distinct *splitting at maturity from the apex downwards into several, star-like lobes*, which often become reflexed; endoperidium membranaceous, then papyraceous, thin, *shortly stipitate, or sessile, dehiscing by a single aperture or mouth.* Capillitium threads *simple*, long, slender, tapering at each end, attached to the peridium, or a central columella, the other end free. Basidia bearing 4–8 spores. Spores coloured, minutely verrucose, globose. Half buried at first, then superficial.

I. Exoperidium not splitting up into two portions when expanded.

*Peristome sulcate.

†Endoperidium stipitate.

55. **G. Bryantii** Berk. Trans. Brit. Myc. Soc. III, t. 18. Charles Bryant.

Exoperidium 2–6 cm., *snow white, then pale ochraceous, and brownish*, globose, coriaceous, splitting up into 8–10 unequal, acute lobes, divided almost to the middle, expanded, then recurved, brownish inside, fleshy, then cracked, the collenchyma layer finally disappearing with the exception of a circular ring at the base of the stem; endoperidium ·5–2 cm., *snow white, then ochraceous, becoming blackish blue when weathered*, mealy, then smooth, subglobose, or pear-shaped, compressed above, pedicellate, *with a distinct, permanent groove round the apex of the stem.* Peristome long, conical, deeply furrowed, striate. Stem 5–10 × 2 mm., *whitish, or brownish*, cylindrical, or compressed, slightly enlarged at the apex. Spores fuscous, obtusely warted, globose, 4–5μ. Columella globose, broad at the base. Capillitium brownish, subfusiform, or subcylindrical, rarely slightly branched towards the ends, 4–6μ in diam. Amongst leaves in woods and hedgerows. Jan.—Dec. Not uncommon. (*v.v.*)

var. **minor** Berk. *Minor*, smaller.

Differs from the type in its *smaller size.*

56. **G. pectinatus** (Pers.) Lloyd. (= *Geaster Schmideli* Vitt.) Lloyd, The Geastrae, figs. 19–22. *Pectinatus*, with teeth like a comb.

Exoperidium 3–6 cm., *white, then ochraceous*, globose, splitting up into 5–10, subequal, acute lobes, divided up to about the middle, revolute, *whitish, or ochraceous* inside, fleshy, the flesh cracking and falling away; endoperidium 1–2·5 cm., *brown, or lead colour,* subglobose, mealy, attenuated into the stem and *striate at the base.* Peristome prominent, long, conical, deeply sulcate, apex fimbriate. Stem 6–8 × 2–3 mm., *whitish, or concolorous,* cylindrical. Spores blackish umber, verrucose, globose, 4–6μ. Columella thick, half as high as the endoperidium. Capillitium brown, fusiform, simple, 4–7μ in diam. Pine woods and under conifers. Rare.

57. **G. Berkeleyi** Massee. Trans. Brit. Myc. Soc. III, t. 18, as *Geaster asper* Lloyd.

Rev. Miles Joseph Berkeley, the father of British mycology.

Exoperidium 6–9 cm., *ochraceous, then brownish,* globose, splitting up into 7–9, unequal, acute lobes, divided to the middle, expanded, then slightly recurved, hard, firm, *brown inside,* becoming slightly cracked, even; endoperidium 2–3 cm., *brown, becoming paler,* broadly ovate, *coarsely papillose, or granular,* pedicellate. Peristome long, prominent, conical, sulcato-striate, surrounded by a smooth, depressed, silky zone. Stem 3–5 × 6–8 mm., *pale,* compressed. Spores umber, acutely warted, globose, 5–6μ. Columella globose, short. Capillitium brown, cylindrical, 9–10μ in diam. Under trees and amongst fir leaves. Oct. Rare. (*v.v.*)

††Endoperidium sessile.

58. **G. umbilicatus** Fr. (= *Geaster striatus* DC. ex W. G. Smith, Grevillea, II, t. 16, fig. 1, sec. Hollós, *Geaster Smithii* Lloyd.)

Umbilicatus, having a navel.

Exoperidium 2–4 cm., *whitish, or tan colour, then brown,* globose, splitting into 4–12, unequal lobes, divided almost to the middle, convex at the base and reflexed, the tips incurved when dry, the outer mycelial layer thin, usually adnate with adhering sand, the inner fleshy layer *brownish,* adnate, thin when dry; endoperidium 5–15 mm., *whitish grey,* opaque, roundish, or oval, *sessile,* appearing slightly pedicellate when dried. Peristome *flattened* (or when old conical), *seated on a depressed area, regularly sulcato-striate.* Spores blackish fuscous, slightly verrucose, apiculate, globose, 4–6μ. Capillitium 4–6μ in diam. Sandy places and coniferous woods. Nov. Rare.

**Peristome not sulcate.

†Endoperidium stipitate.

59. **G. limbatus** Fr. Trans. Brit. Myc. Soc. III, t. 18.

Limbatus, fringed.

Exoperidium 3·5–9 cm., *blackish, or dark brown,* globose, splitting into 7–10, unequal, acute lobes, divided nearly to the middle, expanded, or recurved, leathery, flexible, fibrillose, *dark brown, or grey* inside, fleshy, smooth or cracked; endoperidium 1–3 cm., *grey, sometimes light or dark brown,* globose, or subpyriform, compressed, sometimes swollen at the base near the apex of the stem. Peristome depressed, conical, subacute, fimbriato-ciliate, often surrounded by a pale silky circle. Stem 3–5 × 4–10 mm., *concolorous, or paler,* compressed. Spores blackish purple, acutely warted, globose, 4–5μ. Columella almost wanting. Capillitium brownish, fusiform, 5–7μ in diam. Woods, hedgebanks, amongst firs and leaves. Sept.—Nov. Uncommon. (*v.v.*)

††Endoperidium sessile.

(*a*) Exoperidium strongly incurved when dry.

60. **G. mammosus** Chev. Sow. Eng. Fung. t. 401, as *Lycoperdon recolligens.* *Mammosus,* full-breasted.

Exoperidium 2·5–5 cm., *ochraceous, variegated with white, or silvery white,* globose, splitting into 7–10, acute, somewhat narrow lobes, divided nearly to the base, *very hygroscopic, strongly inrolled when dry,* often umbilicate at the base; *chestnut brown* inside, smooth; endoperidium 8–15 mm., *yellowish,* or *light brown,* globose, *sessile,* smooth. Peristome conical, acute, fimbriato-ciliate, surrounded by a pale narrow silky circle. Spores dark brown, verrucose, globose, 3–6μ. Columella dark brown with a purplish tinge, short, cylindrical, conical, broad at the base. Capillitium hyaline, simple, cylindrical, blunt at the ends, 4–6μ in diam. Sandy woods and fields. Feb.—Dec. Uncommon. (*v.v.*)

G. hygrometricus Pers. = **Astraeus hygrometricus** (Pers.) Morgan.

(*b*) Exoperidium not incurved when dry.

a. Unexpanded plants globose.

61. **G. fimbriatus** Fr. Trans. Brit. Myc. Soc. III, t. 19. *Fimbriatus,* fringed.

Exoperidium 2·5–6 cm., *yellowish,* globose, splitting into 5–15, unequal, pointed lobes, divided to the middle or a little deeper, and strongly recurved below forming a convex cushion at the base of the sessile endoperidium, outer layer membranaceous, *deep ochraceous* inside, fleshy, soon cracked, and often peeling off; endoperidium 1–2 cm., *concolorous,* globose, *sessile,* smooth. Mouth indeterminate, piloso-fimbriate. Spores blackish umber, minutely verrucose, globose, 3–4μ. Columella obovate, slender. Capillitium yellowish brown, simple, cylindrical, 3–6μ. in diam. Coniferous and beech woods, and on heaths. May—Dec. Common. (*v.v.*)

62. **G. saccatus** Fr. Grevillea, II, t. 20. σάκκος, a bag.

Exoperidium 2–5 cm., *yellowish,* globose, splitting into 6–9, thin, equal, acute lobes, divided to the middle, *deeply saccate at the base,* recurved, becoming incurved when dry, soft, flaccid, *densely floccose outside*, becoming smooth; inside fleshy layer thin, adnate; endoperidium 1–1·5 cm., *yellowish*, globose, *sessile*, smooth. Mouth acute, silky, surrounded by a broad, depressed zone. Spores minutely verrucose, globose, 3–4μ. Capillitium light brown, 4–5μ in diam. Sandy ground in hedgerows. Aug. Rare.

β. Unexpanded plants ovate acuminate.

63. **G. lageniformis** Vitt. Grevillea, II, t. 14, fig. 1.
λάγηνος, a flagon; *forma*, shape.

Exoperidium 4–8 cm., yellowish, *ovate acuminate*, splitting into 6–9, very long, pointed, nearly equal lobes, divided beyond the middle, usually saccate but also recurved, with whitish mycelial strands at the base, the mycelial layer closely adnate, often separating and splitting into parallel lines; *ochraceous* inside, *becoming brown*, fleshy layer soft, disappearing; endoperidium 1–2·5 cm., *ochraceous*, or *brownish*, subglobose, *sessile*, soft, membranaceous. Mouth planoconic, silky, striate, *surrounded by an orbicular silky zone*. Spores yellowish brown, minutely verrucose, globose, 3–4μ. Columella small, clavate. Capillitium pale brownish, fusiform, simple, or slightly branched towards the ends, 6–8μ in diam. Sandy soil. Woods and hedgerows. April—Nov. Rare. (*v.v.*)

64. **G. triplex** Jungh. (= *Geaster Michelianus* W. G. Sm.) Lloyd, The Geastrae, figs. 47–49. *Triplex*, three-fold.

Exoperidium 5–10 cm., *brownish olivaceous*, *ovate-acuminate*, splitting into 4–7, subequal, broad, acute lobes, divided to the middle, *often much cracked up into areolae on the outside*; *brownish* inside, fleshy layer *very thick*, cracking and peeling off with the exception of a disc-like portion which forms *a cup* at the base of the endoperidium; endoperidium 1·5–3·5 cm., *pale brownish*, subglobose, compressed, *sessile*, membranaceous. Mouth paler, broadly conical, fibrillose. Spores brown, verrucose, globose, 4–5μ. Columella pale brown, clavate, long. Capillitium light brown, simple, fusiform, 6–7μ in diam. Woods and pastures. Sept.—Oct. Uncommon. (*v.v.*)

†††Endoperidium sessile, or substipitate.

65. **G. rufescens** Pers. Fr. Trans. Brit. Myc. Soc. III, t. 19.
Rufescens, becoming reddish.

Exoperidium 4–8 cm., *yellowish*, globose, splitting into 6–10, broad, acute lobes, divided to the middle or beyond, expanded, then recurved, rigid, thick, firm; *ochraceous* inside, becoming *rufescent*, fleshy layer thick, soon cracking, and often peeling off; endoperidium 1·5–3 cm., *yellowish, or pale brownish*, globose, or subovate, *sessile or substipitate*. Mouth fibrillose, indefinite, frequently torn. Spores brownish olivaceous, echinulate, globose, 4μ. Columella brownish olivaceous, short, globose. Capillitium olivaceous, fusiform, simple, 6–7μ in diam. Woods and pastures. May—Dec. Not uncommon. (*v.v.*)

var. **minor** Pers. *Minor*, smaller.

Differs from the type in its *smaller size*. Woods. Sept.—Oct. Uncommon. (*v.v.*)

II. Exoperidium splitting up into two portions, the lower portion forming a hollow sphere at the base.

66. **G. coronatus** (Schaeff.) Lloyd. Lloyd, The Geastrae, figs. 58–61. *Coronatus*, crowned.

Exoperidium 2·5–5 cm., *yellowish, then brown*, globose, splitting up into four rarely more, equal, pointed lobes, divided nearly to the middle, lobes attached by their apex nearly perpendicularly to the mycelial layer which remains in the ground and forms a hollow cup, *brown* inside, fleshy layer becoming cracked and finally peeling off; endoperidium 4–10 mm., *bluish grey, sometimes whitish, or brownish*, oval, oblong, or pear-shaped, with an apophysis above its attachment to the stem, apex pale yellow with a sharply defined zone bordering the base of the fibrous, *projecting peristome*, surface often rough with white crystals. Stem 2–3 × 2–6 mm., *whitish*, often compressed. Spores brown, verrucose, globose, 4–5μ. Columella brown with a purplish tinge, slender, elliptical. Capillitium brown, cylindrical, 5–7μ in diam. Densely gregarious. Amongst coniferous needles. Sept.—Oct. Locally common. (*v.v.*)

67. **G. fornicatus** (Huds.) Fr. Trans. Brit. Myc. Soc. III, t. 17. *Fornicatus*, arched.

Exoperidium 4–9 cm., *yellowish*, globose, both the outer and the inner layers splitting up into 4–5 lobes, the outer layer remaining sunk in the ground and forming a hollow cup, whilst the inner lobes, divided beyond the middle, stand perpendicularly (erect) on the end of the lobes attached to the tips of the outer layer, hard, leathery, thick, *dark brown* inside, the fleshy layer cracking and peeling off in places; endoperidium 1·5–3·5 cm., *rust colour, or dark brown*, globose, depressed, or urn-shaped, with a ring-like apophysis above its attachment to the stem, somewhat downy. Peristome conical, then tubular, scarcely furrowed, ciliate. Stem 2–5 × 4–15 mm., *whitish*, cylindrical, often compressed. Spores *purplish*, echinulate, globose, 3–4μ. Columella brown tinged with purple, slender, clavate. Capillitium brown, fusiform, 10–12μ in diam. Meadows, pastures, heaths and amongst firs. March—Nov. Uncommon. (*v.v.*)

NIDULARIINEAE.

Peridium campanulate, cylindrical, or cup-shaped, consisting of one to three layers, *inclosing several peridiola*, and sometimes covered at the apex by a membranaceous epiphragm. Peridiola lenticular, attached, or not, to the peridium by a funiculus, consisting of two layers and lined on the inside with the basidia and paraphyses. Basidia bearing 2–4 stipitate, or sessile spores. Spores white, elliptical, oval, or subglobose, smooth. Growing on dead wood and twigs, more rarely on the ground.

NIDULARIACEAE.

Same characters as the suborder.

Nidularia (Fr.) Tul.

(*Nidulus*, a little nest.)

Peridium subglobose, sessile, *consisting of a single layer, without an epiphragm*, dehiscing irregularly, or in a circumscissile manner, by the rupture of the upper portion. Peridiola lenticular, biconvex, or compressed, numerous, *without a funiculus* at maturity, and involved in mucus. Spores white, elliptical, or subglobose, smooth. Growing on the ground, wood, or leaves.

68. **N. pisiformis** (Roth) Tul. Massee, Mon. Brit. Gastromyc. t. 37.
Pisum, pea; *forma*, shape.

Pe. 4–10 mm., *whitish, then cinnamon, or brownish*, subglobose, seated on a broad base, sessile, minutely tomentose, dehiscing in a circumscissile manner. Peridiola *brown*, 2 mm. across, subrotund, biconvex, shining. Spores white, broadly elliptical, or subglobose, 7–8 × 6–7μ. Cystidia "large, fusiform" Massee. Gregarious, or solitary. Dead branches. May—Oct. Uncommon. (*v.v.*)

var. **Broomei** Massee. C. E. Broome, the eminent mycologist.

Differs from the type in the *narrowly elliptical spores, with a thick hyaline epispore.* Pine wood. Rare.

69. **N. Berkeleyii** Massee. (= *Nidularia pisiformis* (Roth) Tul. sec. Lloyd.) Massee, Mon. Brit. Gastromyc. t. 38.
Rev. Miles Joseph Berkeley, the father of British mycology.

Pe. 5–7 mm. broad and high, *bright cinnamon*, subglobose, becoming broadly open, thick, felt-like, hirto-tomentose; *bright cinnamon* inside, velvety. Peridiola *bright brown*, about 2 mm. in diam., numerous (40–50), circular, biconvex, smooth, shining, much wrinkled when dry, firmly agglutinated together by mucus. Spores colourless, then *becoming pale brownish olive*, elliptical, 9–10 × 5–6μ. Solitary, or two or three together. On wood, twigs, etc. Sept. Rare.

70. **N. confluens** Fr. (=*Nidularia pisiformis* (Roth) Tul. sec. Lloyd.)
Confluens, crowded together.

Pe. 4–11 mm., *whitish*, subglobose, compressed, villose, dehiscing irregularly, peridium thin. Peridiola *deep chestnut colour*, orbicular, compressed, 1·5–2 mm. across, shining. Spores white, elliptical, 6–7 × 4–5μ. Crowded. On the ground, twigs and amongst leaves. Aug.—Oct. Uncommon. (*v.v.*)

N. dentata With. = **Sphaerobolus dentatus** (With.) W. G. Sm.

Crucibulum Tul.

(*Crucibulum*, a crucible.)

Peridium globose, then campanulate, or shortly cylindrical, sessile, *consisting of two layers*, and *closed by a membranaceous epiphragm* at the apex, which is finally ruptured. Peridiola lenticular, compressed, biconvex, numerous, *attached by a papilla to the funiculus*. Spores white, oblong-elliptical, smooth. Growing on wood, twigs, and dead herbaceous stems.

71. **C. vulgare** Tul. Berk. Outl. Brit. Fung. t. 2, fig. 2.
Vulgare, common.

Pe. 5–8 mm. high, 6 mm. across, *greyish, or dirty cinnamon*, globose, then bell-shaped, or shortly cylindrical, at first closed by a fugacious epiphragm, then broadly open, minutely tomentose on the outside, soon becoming smooth, inside *whitish*, smooth, and shining. Peridiola *pale*, 1·5–2 mm. across, circular, biconvex, attached by a nipple-like tubercle to the funiculus. Spores white, oblong-elliptical, 8–12× 4–6μ. Gregarious, or crowded. Wood, twigs and dead fern stems. Sept.—March. Common. (*v.v.*)

Cyathus Haller.

(*κύαθος*, a cup.)

Peridium cylindrical, then broadly campanulate, sessile, *consisting of three layers*, and *closed* at the apex *by a membranaceous epiphragm* which finally becomes ruptured and disappears. Peridiola lenticular, compressed, *umbilicate*, numerous, *attached by a funiculus*. Spores white, elliptical, smooth. Growing on wood, more rarely on the ground.

72. **C. striatus** (Huds.) Pers. Rolland, Champ. t. 109, no. 246.
Striatus, furrowed.

Pe. 10–15 mm. high, 8–10 mm across, *reddish brown*, or *ferruginous* and *strigosely hairy* on the outside, obconic, or bell-shaped, truncate at the base, apex at first incurved, and the interior closed with a pale, fugacious epiphragm, then opening out and disclosing the *lead coloured*, *shining, fluted* inner surface of the peridium. Peridiola *whitish*, subcircular, compressed, umbilicate, and attached to the funiculus, 2 mm. thick. Spores white, oblong elliptical, 18–22 × 10μ. Fasciculate. On stumps, wood, twigs, and fir-cones. Jan.—Dec. Common. (*v.v.*)

73. **C. olla** (Batsch) Pers. (= *Cyathus vernicosus* (Bull.) DC.) Berk. Outl. Brit. Fung. t. 21, fig. 1, as *Cyathus vernicosus*.

Olla, a pot.

Pe. 10–15 mm. high, 8–15 mm. across, *greyish, or ochraceous* on the outside, and *minutely silky, then smooth*, broadly bell-shaped, mouth broadly open, undulate, tapering downwards to a narrow base, *lead coloured, or brownish* inside, *smooth*. Peridiola *blackish, or greyish*, shining, circular, biconvex, 3–4 mm. across, umbilicate, and attached to the white funiculus. Spores white, broadly elliptical, 10–14 × 8μ. On bare soil, rotten wood, sticks, and in flower-pots. Feb.—Nov. Common. (*v.v.*)

var. **agrestis** Pers. *Agrestis*, pertaining to the fields.

Differs from the type in its *smaller size, subhemispherical shape, and erect margin*. Stubble fields. Aug.—Sept. Common. (*v.v.*)

PLECTOBASIDIINEAE.
(SCLERODERMINEAE.)

Peridium subglobose, obovate, clavate, or variously shaped, sessile, stipitate, or prolonged into a stem-like base, consisting of one or more layers, dehiscing by the gradual falling away of portions of the peridial walls, by the rupture of the exoperidium or endoperidium in an irregular, or circumscissile manner, or by a well-defined apical aperture. Gleba *not divided up into cells*, with, or without, a sterile base, and traversed, or not, by sterile veins, which rarely break up into peridiola, finally breaking down into a powdery mass, rarely becoming slimy. Basidia clavate to pear-shaped, bearing 2–12, pedicellate, or sessile, apical, or lateral spores, *irregularly arranged*, rarely tufted, *and not forming a distinct hymenium*, intermixed, or not, with capillitium threads. Spores white, or coloured, subglobose, or elliptical, echinulate, verrucose, reticulate, or smooth. Subterranean, or superficial.

SCLERODERMATACEAE.

Peridium subglobose, turbinate, or irregularly pear-shaped, sessile, or prolonged into a stem-like base, consisting of one or more layers, fleshy, leathery, or membranaceous, dehiscing in an irregular manner. Gleba *traversed by sterile veins*. Capillitium *rudimentary*. Spores coloured, globose, or elliptical, echinulate, verrucose, reticulate, or smooth. Subterranean, or superficial.

Melanogaster Cda.

(μέλας, black; γαστήρ, belly.)

Peridium subglobose, or elliptical, *with branched mycelial strands springing from every part of the surface,* fleshy, firm, *not sharply separated from the gleba.* Basidia pear-shaped, or elliptical to club-shaped, bearing 3–4, apical, or sublateral, sessile spores. Spores coloured, elliptical, or obovate, smooth, or papillate. *Subterranean,* or half buried.

74. **M. variegatus** (Vitt.) Tul. Tul. Fung. Hypog. t. 11, fig. 4, and t. 12, fig. 6. *Variegatus,* of different colours.

Pe. 2–3 cm., *ochraceous, or clear yellow, then reddish ferruginous,* irregularly globose, adpressedly tomentose, and ornamented with the brown, fibrous, cord-like anastomosing mycelium. Gleba *fuliginous, then black,* tramal plates *whitish, then bright orange.* Spores brown, elliptic oblong, 10 × 5μ. Smell pleasant, aromatic. Amongst leaves, and twigs. June—Nov. Uncommon.

var. **Broomeianus** (Berk.) Tul. Sow. Eng. Fung. t. 426, as *Tuber moschatum.* C. E. Broome, the eminent mycologist.

Differs from the type in the *tramal plates never being bright yellow, or orange.* In tufts of five or six, under beech, and Lombardy poplars. June—Nov. Not uncommon.

75. **M. ambiguus** (Vitt.) Tul. Tul. Fung. Hypog. t. 2, fig. 5, and t. 11, fig. 5. *Ambiguus,* changeable.

Pe. 2·5–3·5 cm., *pale olive, becoming brownish when exposed to the air,* globose, or elliptical. Gleba *jet black,* tramal plates *white,* unchangeable, "becoming reddish" Berk. Spores brown, obovate, or elliptical, apex acute, or obtuse and papillate, 13–15 × 7–8μ. Smell very foetid. Under fir, deodar, beech, poplar, and oak. April—Oct. Rare. (*v.v.*)

var. **intermedius** Tul. *Intermedius,* lying between.

Differs from the type in the *obovate, obtuse, very rarely slightly papillate spores, and in the yellowish tramal plates becoming red when dried.* Rare.

Scleroderma Pers.

(σκληρός, tough; δέρμα, skin.)

Peridium subglobose, obovate, or turbinate, sessile, or prolonged into a stem-like base, consisting of one, or two layers, firm, *leathery or corky,* warted, scaly, granular, or smooth, dehiscing irregularly, or by the exoperidium splitting at the apex in a star-like manner, *sharply separated from the gleba.* Gleba at length becoming *pulverulent.* Basidia pear-shaped, to clavate, bearing 2–5, shortly pedicellate spores. Capillitium *rudimentary,* the remains of the sterile veins. Spores coloured, globose, echinulate, verrucose, or reticulate. *Superficial.*

*Spores reticulate.

76. **S. aurantium** Pers. (= *Scleroderma vulgare* (Hornem.) Fr.) Berk. Outl. Brit. Fung. t. 15, fig. 4, as *Scleroderma vulgare* Fr. in text.
Aurantium, golden.

Pe. 5–15 cm., *whitish,* or *yellowish,* often becoming pink when cut, globose, subsessile, or substipitate, often depressed, verrucose, or broken up into minute, rigid scales; peridium *thick,* tough, often attached by a dense mass of cord-like mycelium at the base. Gleba *greyish white, then blackish with a purple tinge,* tramal plates *white.* Dehiscing by an irregular mouth. Spores blackish with a purple tinge, reticulate with small meshes, verrucose, globose, 8–12μ. Smell often somewhat unpleasant. Woods, heaths, and under trees. July—Jan. Common. (*v.v.*)

var. **laevigatum** (Fuck.) W. G. Sm. *Laevigatum,* made smooth.

Differs from the type in the *even, smooth peridium.* Woods and heaths. Sept.—Oct. Not uncommon. (*v.v.*)

var. **aurantiacum** (Bull.) W. G. Sm. *Aurantiacum,* golden.

Differs from the type in its *brassy yellow colour.* Woods. Sept.—Oct. Not uncommon. (*v.v.*)

var. **spadiceum** (Pers.) W. G. Sm. *Spadiceum,* date brown.

Differs from the type in its *smooth, date brown peridium.* Beech woods. Sept. Uncommon. (*v.v.*)

var. **cervinum** (Pers.) W. G. Sm. *Cervinum,* fawn colour.

Differs from the type in its *small size, and granular surface.* Coniferous woods. Uncommon.

77. **S. Bovista** Fr. Hollós, Gasteromyc. Ung. t. 23, figs. 16–20.
Bofist, puff-ball.

Pe. 2–5 cm., *yellowish,* subsessile, obovate, often irregular, *thin,* pliant, smooth, or slightly scaly, sometimes breaking away in patches, rarely substipitate, springing from a dense mass of cord-like mycelium at the base. Gleba *olive brown*; tramal plates *yellow, floccose,* the flocci with clamp connections. Dehiscing by an irregular mouth. Spores olive brown, reticulate with large meshes, verrucose, surrounded by an irregular, transparent border, globose, 10–13μ. Sandy soil in woods, and on heaths. Sept.—Nov. Uncommon. (*v.v.*)

78. **S. Geaster** Fr. Boud. Icon. t. 186. *Geaster*, the genus Geaster.

Pe. 5–15 cm., *greyish ochraceous, or yellow*, globose, or turbinate, sessile, minutely tomentose, *granular*, exoperidium *very thick, splitting at the apex in a stellate manner into recurved subequal lobes.* Gleba *purple umber*, floccose. Spores purple umber, reticulate, obtusely verrucose, globose, 12–15μ, 1-pluri-guttulate. Sandy soil. Aug.—Nov. Uncommon. (*v.v.*)

**Spores echinulate.

79. **S. verrucosum** (Vaill.) Pers. Grev. Scot. Crypt. Fl. t. 48. *Verrucosum*, warted.

Pe. 2·5–8 cm. wide, 2–10 cm. high, *ochraceous, or dingy brown*, subglobose, continued below into a more or less elongated stem-like base, sometimes almost sessile, covered with minute, darker warts, rarely almost smooth; peridium *thin above, fragile.* Gleba *umber brown*, tramal plates *whitish.* Dehiscing by an irregular mouth. Spores dark brown, *bluntly* echinulate, globose, 10–14μ. Sandy soil in woods and on heaths. July—Nov. Common. (*v.v.*)

80. **S. cepa** (Vaill.) Pers. Hollós, Gasteromyc. Ung. t. 23, figs. 3–7. *Cepa*, onion.

Pe. 1–5 cm., *reddish brown, or bay*, globose, or bulbous, compressed, sessile, or with a very short stem, smooth, or rough with very small warts on the upper part; peridium *very thick* when fresh, thinner and somewhat leathery and wrinkled when dry. Gleba *white, becoming sooty black tinged with lilac*, tramal plates *whitish, then greyish tinged darker or lilac.* Dehiscing by an irregular mouth. Spores blackish, *acutely* echinulate, globose, 9–10μ. Woods. Aug.—Oct. Uncommon. (*v.v.*)

Pisolithus A. & S. (**Polysaccum** DC.)

(πίσος, peas; λίθος, stone.)

Peridium irregularly globose, attenuated downwards into a stem-like base, thin, *membranaceous*, dehiscing by the falling away of the upper portion. Gleba *forming round, or polygonal peridiola*, that finally become free and separate from each other. Basidia pear-shaped, bearing 2–6, almost sessile spores. Spores coloured, globose, verrucose. Capillitium *rudimentary.* Half buried in the ground.

81. **P. arenarius** A. & S. (= *Polysaccum pisocarpium* (Nees) Fr.) Sow. Eng. Fung. t. 425, as *Lycoperdon capsuliferum*. *Arenarius*, pertaining to sand.

Pe. 2·5–8 cm., *ochraceous, then olivaceous brown*, irregularly globose, or pear-shaped, attenuated downwards into a stem-like base which is sunk in the ground, smooth, or rough; very fragile. Gleba *consisting of many peridiola*; peridiola *sulphur yellow, then brown*, irregularly angular. Spores reddish brown, warted, globose, 9–10μ. Sandy soil. May—Oct. Rare. (*v.v.*)

Calostomataceae.

Peridium globose, consisting of several layers. Gleba *traversed by sterile veins*. Basidia pear-shaped or clavate, bearing sessile, lateral spores. Capillitium *well developed*, springing from the inside of the endoperidium. Spores coloured, globose or elliptical, verrucose, or smooth. Superficial or half buried in the ground.

Astraeus Morgan.

(ἄστρον, a star.)

Peridium globose; exoperidium consisting of three layers, the outer composed of thin, irregularly interwoven hyphae, the middle layer of a corky consistency and the inner cartilaginous and collenchymatous, at first concrete with the endoperidium, then at maturity *splitting at the apex in a star-like manner* into several lobes and separating from the endoperidium; endoperidium thin, membranaceous or papyraceous, sessile, dehiscing by an apical aperture. Capillitium *well developed*, springing from the inner surface of the endoperidium, threads long, much branched and interwoven. Basidia pear-shaped, bearing 4, sessile, lateral spores. Spores coloured, globose, minutely verrucose. Half buried in the ground.

82. **A. hygrometricus** (Pers.) Morgan. (= *Geastrum hygrometricum* Pers.) Trans. Brit. Myc. Soc. III, t. 17.

ὑγρός, wet; μέτρον, a measure.

Exoperidium 4–8 cm., *grey, or greyish brown* outside, *brownish* inside and becoming deeply cracked, globose, tough, leathery, cartilaginous, splitting up into 7–20, acute lobes, divided up almost to the base, strongly incurved and depressed over the apex of the endoperidium when dry, reflexed and standing up on the apices of the lobes when moist; endoperidium 1·5–2·5 cm., *grey, or brown*, sessile, globose, depressed, smooth, or subreticulate, dehiscing by an irregular, small apical mouth. Spores brown, minutely verrucose, globose, 8–11μ. Capillitium hyaline, thick walled, branched, 6–7μ in diam. Woods, and under trees. Feb.—Dec. Uncommon. (*v.v.*)

Tulostomataceae.

Peridium subglobose, *stipitate, or prolonged into a stem-like base*, consisting of a thin, fugacious exoperidium, and a thin, membranaceous endoperidium, which is raised upwards by the firm, fibrous basal portion. Gleba without cells. Basidia club-shaped, irregularly scattered on the hyphae, bearing the spores laterally at various levels. Capillitium well developed, attached to the endoperidium. Spores coloured, globose, verrucose. Subterranean, or half buried in the ground, then superficial.

Tulostoma Pers.

(τύλος, a knob; στόμα, mouth.)

Peridium depressed globose; exoperidium, thin, fugacious; endoperidium membranaceous, thin, *dehiscing by an apical aperture*; stem elongate, inserted into a socket at the base of the peridium. Capillitium well developed, threads hyaline, very long, much branched, attached to the endoperidium, interwoven. Spores coloured, globose, verrucose. At first subterranean, then superficial.

83. **T. brumale** Pers. (= *Tulostoma mammosum* (Mich.) Fr.) Sow. Eng. Fung. t. 406, as *Lycoperdon pedunculatum*.

Brumale, pertaining to the winter.

Pe. 5–10 mm., *whitish, then yellowish*, globose, or somewhat depressed; outer peridium friable, inner peridium smooth, thin, membranaceous, papyraceous; mouth small, slightly prominent, scarcely toothed, entire. Stem 2–5 cm. × 2–3 mm., *reddish brown*, equal, slightly attenuated upwards to the base of the peridium, and surrounded by a ball of mycelium at the base, smooth, or more or less fibrillose, concolorous inside, stuffed. Spores pinkish, acutely warted, globose, 4–5μ. Capillitium threads hyaline, branched, thick walled, *nodose septate*, 4–6μ in diam. Amongst sand, and on old walls. May—Feb. Uncommon. (*v.v.*)

Queletia Fr.

(Dr Lucien Quélet, the eminent French mycologist.)

Peridium subglobose, then prolonged into a stem-like basal portion; exoperidium thin, fugacious; endoperidium firm, hard, breaking away from the stem-like base and *dehiscing by the gradual falling away of the peridial walls from the base upwards.* Capillitium sparse. Basidia with one to three, apical, or lateral, shortly pedicellate spores. Spores coloured, globose, verrucose. Subterranean, then superficial.

84. **Q. mirabilis** Fr. Bull. Soc. Myc. Fr. 29, t. 28.

Mirabilis, wonderful.

Pe. 3–7 cm., *whitish*, subglobose, at first rounded above and slightly conical at the base, which subsequently developes in a stem-like manner; exoperidium thin, brittle, breaking up into fugacious granules; endoperidium *yellowish, becoming brownish,* firm, hard, and finally breaking away from the stem-like base, dehiscing by the falling away of portions from the base upwards. Stem-like base 4–15 × 1·5–5 cm., *whitish, then concolorous,* torn up into revolute, squarrose fibrils, and finally breaking away from the peridium at maturity and exposing the rusty brown gleba. Spores rusty brown, coarsely warted, globose, often shortly pedicellate, 6–8μ; basidia 1–3-spored. Capillitium almost colourless, rarely branched or septate, 8–10μ in diam. Amongst rotten leaves, and spent tan. Sept. Rare. (*v.v.*)

Battarrea Pers.

(J. A. Battarra, author of Fungorum Agri Ariminensis Historia.)

Peridium globose, exoperidium becoming irregularly torn at the apex, leaving a volva-like basal portion; stem elongated, hollow, becoming torn into fibrous scales; endoperidium hemispherical, plane, or concave underneath, dehiscent by a circular fissure beneath the margin, the upper part coming off like a lid. Capillitium of simple, or branched threads, *with spiral, or annular thickenings.* Spores coloured, globose, verrucose. Subterranean, then superficial.

85. **B. phalloides** (Dicks.) Pers. Sow. Eng. Fung. t. 390, as *Lycoperdon phalloides* Dicks. φαλλός, penis; εἶδος, like.

Volva 2–4 cm., *white* outside, parchment-like, pitted, filled when young with colourless mucus, globose, splitting above into four or five more or less pointed lobes, *rust colour* inside, fibrous, woody, often fringed. Stem 10–30 × ·5–2 cm., *rusty brown,* attenuated at both ends, slightly ventricose in the middle, woody, firm, covered with long, twisted fibres, whitish inside, and stuffed with long, transparent threads. Inner peridium *rusty brown,* campanulate, somewhat flattened, fibrous, splitting horizontally and filled with the yellowish brown capillitium and spores. Spores rust colour, obtusely verrucose, globose, often with a hyaline apiculus, 6μ. Capillitium threads *pale rust colour,* 63–80 × 8μ, with spiral thickenings. Sandy places, and in hollow trees. Sept.—Dec. Rare. (*v.v.*)

Glischrodermataceae.

Peridium globose, consisting of a *single layer, seated on a ring-like mass of mycelium* surrounding the sessile base, dehiscing by a well-defined apical aperture. Capillitium attached to the inner wall of the peridium. Spores coloured, globose, minutely warted. Superficial.

Glischroderma (Fuck.) Rea.

(γλίσχρος, clammy; δέρμα, skin.)

With the same characters as the family.

86. **G. cinctum** (Fuck.) Rea. Trans. Brit. Myc. Soc. IV, t. 2. *Cinctum*, girdled.

Pe. 5–15 mm., *pale grey, becoming darker with age*, globose, slightly sticky, then scurfy, dehiscing by a well-defined apical pore which becomes larger, *seated on a ring-like mass of white mycelium* ·5–2 *mm. wide at the base*. Spores pale pink, minutely warted, showing 4–5 warts in a row across the hemisphere, globose, 4μ. Capillitium hyaline, thick walled, septate, 5–10μ in diam., attached to the inner wall of the peridium. Charcoal heaps in woods. Sept.—Oct. Uncommon. (*v.v.*)

SPHAEROBOLACEAE.

Peridium subglobose, sessile; exoperidium consisting of three layers, an outer gelatinous layer, a middle pseudo-parenchymatous layer, and an inner fibrous layer, splitting in a star-like manner into pointed lobes when mature. Gleba consisting of a peripheral layer of upright palisade cells and a central fertile portion, finally *becoming slimy* and ejecting the whole mass. Basidia pear-shaped bearing 5–8, sessile spores. Spores white, elliptical, or oblong elliptical, smooth. Growing on wood and leaves.

Sphaerobolus (Tode) Pers.

(σφαῖρα, a ball; βόλος, a throw.)

With the same characters as the family.

87. **S. stellatus** (Tode) Pers. Berk. Outl. Brit. Fung. t. 21, fig. 2. *Stellatus*, set with stars.

Pe. 2 mm. high and broad, *whitish, or pale yellow*, globose, then oval, seated on an arachnoid mycelium, splitting above in a stellate manner into 6–8, acute teeth, tomentose, then smooth. Gleba at *first whitish* and transparent, *then brown*, broadly elliptical. Spores white, broadly elliptical, 10–11 × 5–6μ. Crowded. On wood, twigs, sawdust, and leaves. Jan.—Dec. Common. (*v.v.*)

88. **S. dentatus** (With.) W. G. Sm. (= *Sphaerobolus stellatus* (Tode) Pers. sec. Lloyd.) *Dentatus*, toothed.

Pe. 1·5 mm., *pale livid buff, to vinous brownish, pale brown, or brown*, springing from scanty brownish mycelium, finely pilose with erect hairs, opening above in a 4–7-stellate manner, the rays clad with long, white hairs, which at first converge over the opening, *white, ivory, white-greysih, or faint olive ivory* inside, ejecting a *reddish brown* gleba. Scattered. Dead elder. Aug. Rare.

89. **S. terrestris** (A. & S. non Tode) W. G. Sm. Brit. Basidiomyc. fig. 139. *Terrestris*, pertaining to the earth.

Pe. ·5–2 mm., *saffron yellow, or sienna*, at first hemispherical, urceolato-ventricose, seated on a dense tomentose subiculum of buff white, here and there brownish, mycelium. Gleba *concolorous*, solitary, spherical. Spores white, elliptic-oblong, 10–12 × 5–6μ. Crowded, in troops. Rotten wood, and running over leaves. March—Nov. Rare. (*v.v.*)

Thelebolus Tode = A genus of the Ascobolaceae.

AGARICALES.

Hymenium strictly defined from the first, covering the exterior of gills, or lining the interior of fleshy tubes, or pores, or spreading over a smooth surface; at first protected by the universal, or partial veil (the volva, or ring), then finally, at maturity, fully exposed.

AGARICINEAE.

Hymenium inseparable from the pileus and spread over the surface of gills radiating from a stem, or central point. Receptacle fleshy, membranaceous, or coriaceous, fragile, firm, or tough, putrescent, or not putrescent, reviving with moisture or not, and sometimes containing lacticiferous vessels; consisting of a pileus with, or without, a central or lateral stem, and sometimes with a volva or a ring, and broad gills acute at the margin. Growing on the ground, or on wood.

AGARICACEAE.

Same characters as the suborder.

I. Receptacle fleshy, trama not vesiculose, nor traversed by lacticiferous vessels, not membranaceous, not rapidly putrescent, nor tough and woody.

A. Pileus distinct, and easily separable from the fleshy stem.

*Without a ring, or a volva.

Spores white.

Schulzeria Bres.

(Stephan Schulzer.)

Pileus fleshy, regular; margin at first incurved. Stem central. Gills free from the stem. Spores white, oval, or clavate, wall continuous. Growing on the ground.

90. **S. lycoperdoides** Cke. & Massee.

Lycoperdon εἶδος, resembling a Lycoperdon.

P. 3–4 cm., *tan coloured, beset with darker pyramidal warts resembling those of some species of Lycoperdon, and often splitting at the base*, convex, then expanded; margin appendiculate. St. 5 × 1 cm., *whitish*, equal, minutely fibrillose. Gills *white*, free, rather crowded. Flesh *white, brownish under the cuticle.* Spores white, oval, 5 × 4μ. Under cedars. Sept.—Oct. Rare.

91. **S. Grangei** Eyre. Trans. Brit. Myc. Soc. II, t. 5.

Grange Park, Hants.

P. 2·5–4 cm., *dark green, cracking into fibrous scales on a white ground,* fleshy, convex, flatly umbonate. St. 5–6 cm. × 5–7 mm., *brownish, squamose, squamules tipped with the same colour as the p.*, equal, base attenuated. Gills *yellowish*, free, broader in front, *minutely denticulate.* Flesh *white, reddish in the stem.* Spores white, club-shaped, 11–13 × 3–5μ, multi-guttulate. Caespitose. On soil among beech leaves. Nov. Uncommon. (*v.v.*)

S. Eyrei Massee = **Glaucospora Eyrei** (Massee) Rea.

Spores pink.

Pluteus Fr.

(*Pluteus*, a movable pent-house.)

Pileus fleshy, regular. Stem central. Gills free from the stem, rounded behind. Spores pink, rarely pale yellowish, globose, subglobose, or elliptical, wall continuous. Cystidia on edge of gills pear-shaped, or inflated clavate; on the sides of the gill fusiform, or bottle-shaped, and hooked at the apex. Growing on wood, more rarely on the ground.

*Cuticle of the pileus separating into fibrils or flocci.

92. **P. cervinus** (Schaeff.) Fr. *Cervinus*, pertaining to deer.

P. 4–10 cm., *fuliginous, becoming paler, and broken up into fibrils, or squamules,* and often streaked, fleshy, somewhat fragile, *campanulate, then expanded, viscid.* St. 7–10 cm. × 5–15 mm., *white, covered with black fibrils,* firm, equal, often bulbous at the base. Gills *white, then flesh colour,* free, rounded behind, crowded, ventricose, somewhat crenulated. Flesh white, soft. Spores pink, elliptical, 8–10 × 4–5μ. Cystidia on gill edge pear-shaped, 25 × 22μ, on gill surface fusiform, hooked at the apex, 55–75 × 15–18μ. On fallen trunks, stumps, sawdust heaps. Jan.—Dec. Common. (*v.v.*)

var. **Bullii** Berk. Cke. Illus. no. 304, t. 357.

Dr H. G. Bull of Hereford, the originator of the Woolhope Club fungus forays.

P. 10–16 cm., *pallid, disc darker,* convex, then expanded and gibbous. St. 8–16 × 2–3 cm., *pale brown, darkest at the swollen base,* fibrillose. Gills *white, then pink,* free, rounded behind, *very broad,* crowded. Flesh *white,* thick. Spores elliptical, 7–8 × 5μ. Stumps and rotten wood. Sept.—Oct. Uncommon. (*v.v.*)

var. **rigens** Pers. *Rigens,* stiff.

Differs from the type in the *cinereous pileus, with black fibrils, or squamules, and the glabrous, somewhat shining stem.*

93. **P. eximius** Saund. & Sm. Cke. Illus. no. 303, t. 302. *Eximius,* distinguished.

P. 6–20 cm., *rufescent umber, when young darker, and tinged with carmine round the margin,* campanulate, then convex and expanded, viscid pellicle separable. St. 10–14 × 4 cm., *pallid, at length becoming blackish,* nearly equal, *sulcate upwards,* fibrillose. Gills *white, then pale rose,* becoming rufous when bruised, *very broad,* crowded, free, but very close to the stem. Flesh *yellowish, cartilaginous.* Spores pink, spherical, but somewhat irregular, 7 × 5μ. On sawdust. Nov. Rare.

94. **P. patricius** Schulz. (= *Pluteus cervinus* (Schaeff.) Fr. sec. Quél.) Boud. Icon. t. 87. *Patricius,* noble.

P. 6–15 cm., *white, or greyish, disc covered with brown, hairy, pointed squamules,* convex, then expanded, more or less silky; margin lobed, and often split. St. 5–15 × 1–3 cm., *white, covered with small whitish squamules* that become brownish at the base, equal. Gills *white, then flesh colour, free, very broad.* Flesh *white,* firm. Spores pink, elliptical, 5–8 × 3–5μ. On dead logs, stumps. June—Sept. Uncommon. (*v.v.*)

95. **P. petasatus** (Fr.) Karst. Cke. Illus. no. 305, t. 303. πέτασος, a travelling hat with a broad brim.

P. 8–15 cm., *whitish cinereous,* at length somewhat date brown, disc fleshy, campanulate, then expanded, *umbonate,* viscid pellicle separable, *at length striate to the middle,* margin membranaceous. St. 10–20 × 1–1·5 cm., *pallid, at length becoming tawny,* rigid, fibrillosely-striate, attenuated upwards from the silky base. Gills *white, then reddish,* at length *tawny at the edge, very broad,* very crowded, drying up. Flesh *white,* soft. Spores pink, "broadly oval, 7·5–9 × 4·5–5μ. Cystidia fusoid-bottle-shaped, 11–14μ broad, with a few hooks" Lange. On heaps of straw and dung, sawdust. Uncommon. (*v.v.*)

96. **P. sororiata** Karst. *Soror,* a sister.

P. 3–6 cm., *yellow,* somewhat fleshy, campanulate, then expanded, *floccosely-squamulose*; margin somewhat striate, often repand. St. 6 × 1 cm., *pallid, becoming yellowish, squamulose* at the enlarged base. Gills *flesh colour, margin at first yellow,* rounded behind, free, crowded, oblong. Spores hyaline, or becoming very pale yellowish, broadly elliptical, 7–8 × 6μ. On burnt and rotten wood, and rotting branches. Sept.—Oct. Uncommon.

97. **P. umbrosus** (Pers.) Fr. (= *Entoloma nigrocinnamomeum* Schulz. sec. Quél.) Boud. Icon. t. 88. *Umbrosus*, shady.

P. 6–10 cm., *umber*, campanulate, then expanded, subumbonate, *more or less rugulose, covered with adpressed fibrils*, then squamulose; margin more or less lobed, *ciliato-fimbriate*. St. 5–10 × ·5–3 cm., *pale, covered with villose, brown squamules*, base white, floccose. Gills *white, then rosy, fuliginous-fimbriate at the margin*, free, broad. Flesh *white*. Spores pink, elliptical, 6–7 × 4–5μ. "Cystidia on edge fusoid bladder-shaped, 15–25μ broad, content yellowish-brown" Lange. Smell of radish, taste slightly bitter. On stumps and fallen logs. Aug.—Oct. Uncommon. (*v.v.*)

98. **P. ephebeus** Fr. (=*Pluteus villosus* (Bull.) Quél.) *ἔφηβος*, arrived at man's estate.

P. 5–7 cm., *violaceous bistre*, fleshy, convex, then plane, *at first villose, finally floccose*. St. 3–4 cm. × 5–8 mm., *white*, base *swollen, violaceous bistre*, rigid, *striate*. Gills *white, then rosy*, free, *very ventricose*, margin unequal. Flesh *white, brownish under the cuticle*. Spores pink, globose, 6–7μ, with a large central gutta. On rotten wood, and trunks. Sept. Uncommon. (*v.v.*)

99. **P. argenteo-griseus** Rea. Trans. Brit. Myc. Soc. v, t. 4. *Argenteus*, silvery; *griseus*, grey.

P. 3·5–4·5 cm. wide, 2·5 cm. high, *snow white, then becoming smoky grey*, campanulate, obtusely umbonate, *floccosely-silky, atomate*; *margin splitting, exceeding the gills*. St. 6–7 cm. × 5–6 mm., *concolorous*, curved, striate, base bulbous. Gills *white, then pink*, 11–12 mm. broad, free, ventricose, crowded. Flesh *white*. Spores pink, globose, or elliptical, 6 × 6μ, or 7–8 × 6μ, 1-many-guttulate. Cystidia none. Dead wood. Sept. Uncommon. (*v.v.*)

100. **P. violarius** Massee. Cke. Illus. no. 311, t. 518, fig. B. *Violarius*, a dyer of violet colour.

P. 1·5–2·5 cm., *dark purple, darkest at the disc*, hemispherical, then nearly plane, *minutely velvety*; margin undulate. St. 2–3 cm. × 2–3 mm., *pale umber, sprinkled with delicate, black fibrils below, whitish above*, attenuated upwards, silky. Gills *whitish, then bright flesh colour*, free, crowded, *margin serrulate*. Flesh *grey*, thickish. Spores pink, subglobose and apiculate, 5–6μ. On stumps, and rotten wood. Rare.

101. **P. salicinus** (Pers.) Fr. Cke. Illus. no. 1157, t. 1169, fig. A.
Salicinus, pertaining to a willow.

P. 2–3 cm., *bluish-grey, then cinereous*, disc darker, slightly fleshy, *convex, then plane*, subumbonate, *flocculoso-rugulose*. St. 3–5 cm. × 2–6 mm., *white-azure-blue*, or sometimes becoming green, equal, often thicker at the base, *fibrillose*, fragile. Gills *white, then rose-colour*, free, ventricose. Flesh *white, tinged with green*. Spores pink, elliptical, 8–9 × 6–7μ, with a large central gutta. "Cystidia on edge inflated clavate, 16–18μ broad, on sides fusoid bottle-shaped, with hooks" Lange. On willow, and alder trunks and branches. Feb.—Dec. Not uncommon. (*v.v.*)

var. **beryllus** (Pers.) Fr. βήρυλλος, a jewel of sea-green colour.

Differs from the type in the *pileus being streaked with green, the ash-coloured rugose disc, and the whitish stem with greenish fibrils*. On alder. Sept.—Oct. Uncommon. (*v.v.*)

var. **floccosus** Karst. *Floccus*, a flock of wool.

Differs from the type in the *floccosely squamulose pileus*. Rare.

102. **P. hispidulus** Fr. Fr. Icon. t. 90, fig. 2.
Hispidulus, somewhat hairy.

P. 1–2 cm., *grey*, thin, *convex, then plane*, obtuse, *silky, or slightly pilose*; margin at length *slightly striate*. St. 3–4 cm. × 2–4 mm., *silvery white*, equal, fragile, curved-ascending. Gills *white, then rose colour*, free, broadest in front. Flesh *white, grey under the pellicle* of the pileus. Spores pink, globose, 6–7μ. "Cystidia inflated-club-shaped, 13–16μ broad" Lange. On beech stumps, and rotten wood. Sept.—Oct. Uncommon. (*v.v.*)

103. **P. pellitus** (Pers.) Fr. Quél. Jur. et Vosg. t. 5, fig. 4.
Pellitus, covered with skins.

Entirely white. P. 2–5 cm., fleshy, convex, then plane, subumbonate, *silky*. St. 4–5 cm. × 4–6 mm., equal, slightly thickened at the base, *shining*, fragile. Gills *white, then flesh colour*, free, rounded behind, crowded, ventricose, *margin slightly toothed*. Flesh white, soft, thin. Spores pink, "broadly ovate, 6–7 × 4–5μ. Cystidia on sides, subfusoid with hooks, on edge, inflated obtuse" Lange. On and near trunks. July—Sept. Uncommon.

var. **punctillifer** Quél. *Punctillum*, a little dot; *fero*, I bear.

Differs from the type in the *disc of the pileus being tinged fuscous and covered with minute umber scales, and in the base of the stem being covered with minute umber scales*. Spores pink, globose, 6μ, 1–3-guttulate. On the ground. Oct. Uncommon. (*v.v.*)

**P. pruinate, somewhat pulverulent.

104. **P. nanus** (Pers.) Fr. Cke. Illus. no. 309, t. 305, fig. A.
νάννος, a dwarf.

P. 1–5 cm., *umber, sprinkled with fuscous, pulverulent sootiness,* disc often darker, *covered with veined, radiating, umber wrinkles,* thin, convex, then flattened. St. 3–4 cm. × 2–4 mm., *whitish,* rigid, equal, or attenuated downwards, *slightly striate.* Gills *white, then flesh colour,* free, ventricose. Flesh *white.* Spores pink, broadly elliptical, 5–6 × 4–5μ, 1-guttulate. "Cystidia cylindric bladder-shaped" Lange. On fallen sticks, especially birch and beech, and on sawdust. May—Oct. Not uncommon. (*v.v.*)

var. **lutescens** Fr. Cke. Illus. no. 309, t. 305, fig. B.
Lutescens, becoming yellow.

Differs from the type in the *stem and often the gills and flesh being yellow.* Spores pink, globose, 5–6μ. On stumps. May—Oct. Not uncommon. (*v.v.*)

var. **major** Mass. Cke. Illus. no. 309, t. 305, fig. C. *Major,* larger.

Differs from the type in being *larger, with an even greyish pileus.* Fallen sticks. Rare.

105. **P. melanodon** (Secr.) Fr. μέλας, black; ὀδούς, a tooth.

P. 2·5 cm., *dull yellow,* thin, convex, then plane, conically umbonate, *pulverulent; margin slightly striate.* St. 6–7 cm. × 1–2 mm., *yellowish brown, apex white, swollen base blackish,* firm, *polished.* Gills *whitish, tinged with rose colour,* free, ventricose, *margin black, denticulate.* Flesh *white.* Rotten wood in beech woods. Sept. Rare.

106. **P. spilopus** B. & Br. (= *Pluteus spodopileus* Sacc.) Cke. Illus. no. 310, t. 325. σπῖλος, a spot; πούς, foot.

P. 2–4 cm., *brown, or dark fawn colour,* fleshy, convex, then expanded, subumbonate, often depressed round the umbo, *radiately rugulose.* St. 4–5 cm. × 4–8 mm., *whitish,* or *faintly tinged with fawn colour, punctate with scattered black points,* equal, incurved. Gills *white, then pink,* free. Flesh *white.* Spores pink, globose, 7–8μ. On stumps. July—Oct. Uncommon. (*v.v.*)

107. **P. semibulbosus** (Lasch) Fr. Boud. Icon. t. 89.
Semi, half; *bulbosus,* bulbous.

P. 1–4 cm., *pale ashy ochraceous, becoming rosy and pale,* submembranaceous, thin, somewhat diaphanous, convex, then plane, *pulverulently pruinose, deeply striate.* St. 3–6 cm. × 4–6 mm., *white, delicately pubescent, and velvety, minutely striate; base bulbous, velvety.*

Gills *white, then flesh colour, or slightly yellowish,* free, ventricose. Flesh *white,* somewhat filamentous in the stem. Spores pink, broadly elliptical, 7–8 × 5–7μ. "Cystidia obtuse, elongated, cylindric, very prominent, 13–14μ broad, entire length, 75–115μ" Lange. On poplar trunks, sawdust and rotten wood. Aug.—Sept. Uncommon.

***P. glabrous.

108. **P. leoninus** (Schaeff.) Fr. (= *Pluteus leoninus* var. *coccineus* Massee.) Cke. Illus. no. 313, t. 421, figs. A, B.

Leoninus, pertaining to a lion.

P. 3–7 cm., *yellow, lemon yellow, crimson orange, or vermilion,* thin, *fragile, campanulate, then expanded,* margin striate. St. 4–7 × 1 cm., *whitish light yellow, often vermilion at the thickened base,* fragile, *striate,* fibrillose. Gills *white, then flesh colour, margin often light yellow,* free. Flesh *white, or yellowish, reddish under the cuticle* in the vermilion specimens. Spores pink, elliptical, or globose, 5–7 × 5μ, multi-guttulate. Cystidia "on gill surface flask-shaped to fusiform 60–110 × 24–33μ apex rounded, untoothed, on edge of gill clavate to bottle-shaped, 60–75 × 15–27μ" Rick. On old willows, and stumps. Sept.—Jan. Uncommon. (*v.v.*)

109. **P. roseo-albus** Fr. *Roseus,* rosy; *albus,* white.

P. 7–8 cm., *rosy,* thin, convex, then expanded. St. 4–6 × 1 cm., *white, pruinose,* curved. Gills *white, then flesh colour,* free. Spores pink. Poplar trunks. Sept. Rare.

110. **P. chrysophaeus** (Schaeff.) Fr. Boud. Icon. t. 91.

χρυσός, gold; *φαιός,* dusky.

P. 2–4 cm., *dark cinnamon,* or *pale umber, often streaked with black,* submembranaceous, convex, then expanded; margin striate. St. 4–6 cm. × 3–9 mm., *yellow,* equal, *striate, fibrillose.* Gills *white, then flesh colour,* free. Flesh *yellowish,* deeper coloured in the stem. Spores pink, subglobose, 6–7μ, multi-guttulate. "Cystidia obtusely fusiform, subventricose, inflated" Lange. On twigs and stumps. May—Nov. Common. (*v.v.*)

111. **P. phlebophorus** (Ditm.) Fr. Cke. Illus. no. 315, t. 422, fig. A.

φλέψ, a vein; *φέρω,* I bear.

P. 3–5 cm., *umber,* slightly fleshy, convex, then expanded, *wrinkled with veins.* St. 3–6 cm. × 5–6 mm., *white, shining,* somewhat incurved; base swollen, floccose. Gills *white, then flesh colour,* free. Flesh *white,* fragile. Spores pink, subglobose, 5–9 × 5–8μ, 1-guttulate. Cystidia "bladder-shaped on edge of gill, 30–40 × 15–18μ" Rick. On rotten wood, dead twigs, and sawdust. June—Oct. Uncommon. (*v.v.*)

var. **albo-farinosus** Rea. *Albus*, white; *farinosus*, mealy.

Differs from the type in the *white mealy apex of the stem.* On rotten wood in woods. Oct. Rare. (*v.v.*)

var. *reticulatus* Cke. = **Pleurotus palmatus** (Bull.) Quél.

var. **marginatus** Quél. Quél. As. Fr. (1884), t. 8, fig. 4.

Marginatus, bordered.

Differs from the type in *the chestnut brown peridium, and the bistre black, crenulate edge of the gills.*

112. **P. umbrinellus** (Sommerf.) Fr. *Umbrinellus*, brownish.

P. 15 mm., *bistre*, convex, *tough*; margin *paler*, *fimbriate*. St. 5–7 cm. × 3–4 mm., *white, shining, tough, rooting.* Gills *white, then flesh colour*, free. Spores "subglobose or elliptical 8–9 × 6–7μ or 7–8 × 6μ; cystidia 60–90 × 18–30μ" Sacc. In coppices, and gardens. June—Oct. Uncommon.

Spores green.

Glaucospora Rea[1].

(γλαυκός, green; σπορά, seed.)

Pileus fleshy, regular; margin at first incurved. Gills free from the stem. Spores bluish green, elliptical, wall continuous. Growing on the ground.

113. **G. Eyrei** (Massee) Rea. (= *Schulzeria Eyrei* Massee.) Grevillea, XXII (1894), t. 185, fig. 1, as *Schulzeria Eyrei* Massee.

Rev. W. L. W. Eyre, the ardent mycologist of Swarraton.

P. 2–4 cm., *pallid, disc ochraceous*, minutely granular, campanulate, then expanded and plane, *broadly umbonate*; margin incurved, and appendiculate with the membranaceous veil when young, often split, and revolute when old. St. 5 cm. × 2–3 mm., *pallid, becoming ochraceous*, flexuose, subequal, apex minutely mealy. Gills *pale green, then deep bluish green, free*, narrowed, thin. Flesh *white, yellowish in the stem.* Spores *bluish green*, elliptical, 4·5 × 2·5–3μ. Under spruce firs, and in pastures. Aug.—Oct. Uncommon. (*v.v.*)

Spores ochraceous.

Pluteolus Fr.

(*Pluteolus*, a little pent-house.)

Pileus fleshy, very thin, viscid; margin at first straight, adpressed to the stem. Stem central, subcartilaginous. Gills free, rounded behind. Spores ochraceous, ferruginous, or ochraceous brown, elliptical, smooth. Cystidia obpyriform, or ventricose. Growing on wood.

[1] The name *Chlorospora* proposed by Massee for this genus cannot stand, as Spegazzini had previously used it for a genus of the Peronosporaceae.

114. **P. reticulatus** (Pers.) Fr. (= *Pluteolus aleuriatus* Fr. sec. Quél.) Cke. Illus. no. 516, t. 495. *Reticulatus*, netted.

P. 4–5 cm., *delicate bistre, becoming violaceous, fuscous, or livid grey*, fleshy, campanulate, then expanded, umbonate when mature, subrepand, *viscid at first*, and *covered with a network of anastomosing veins*, becoming more even, or slightly pitted; margin slightly striate. St. 4–5 cm. × 4–6 mm., *white*, equal, fragile, *fibrillose*, the cuticle becoming polished, even, subcartilaginous, apex mealy. Gills *dingy cinnamon, distinctly free*, ventricose, crowded, arid. Flesh *white*, thin. Spores ochraceous ferruginous, broadly elliptical, often slightly depressed on one side, 9–10 × 5–6μ, 2–4-guttulate. Cystidia obpyriform, or ventricose and apiculate, 15 × 8μ. Dead wood. Sept.—Oct. Uncommon. (*v.v.*)

115. **P. aleuriatus** Fr. (= *Pluteolus reticulatus* (Pers.) Fr. sec. Quél.) Fr. Icon. t. 126, fig. 5. *ἄλευρον*, wheaten flour.

P. 1–2·5 cm., *bluish grey, livid*, or *rose colour*, submembranaceous, conical, then convexo-plane, viscid, *striate to the disc*. St. 2·5–4 cm. × 2–3 mm., *white*, slightly attenuated upwards, straight, or incurved, *pulverulent*. Gills *saffron ochraceous, then cinnamon*, free, ventricose, 2 mm. broad, thin. Flesh *white*, very thin. Spores ferruginous, oblong elliptical, 6–10 × 4μ, 1-guttulate. Rotten sticks, and stumps. Sept.—Nov. Not uncommon. (*v.v.*)

116. **P. Mulgravensis** Massee & Crossl.
Mulgravensis, belonging to the Mulgrave woods, near Whitby.

P. 5–6 cm., *grey*, somewhat fleshy, convex, then expanded, *umbonate, flocculose, becoming broken up into squamules, striate*. St. 4 cm. × 3–4 mm., *whitish*, subequal, *base subclavate, smooth*. Gills *white, then cinnamon*, free, crowded, broad. Spores ochraceous brown, elliptical, 9–10 × 4–5μ. On wood. Sept. Rare.

Spores purple, or fuscous.

Pilosace Fr.

(*πῖλος*, a cap; *σάκος*, a shield.)

Pileus fleshy, regular. Stem central, stout. Gills free from the stem. Spores bay purple, globose, smooth, with a germ-pore. Growing on the ground.

117. **P. Algeriensis** Fr. in Quél. (? = *Stropharia epimyces* (Peck) Atk. sec. Harper.) *Algeriensis*, Algerian.

P. 10 cm., *snow white, then reddish, or bistre*, fleshy, convex, then plane, smooth, shining like a kid glove. St. 4–5 × 4–5 cm., *white, incrassated at the base*, silky. Gills *rosy flesh colour*, then *bistre violaceous*, free, horizontal, *narrow*. Flesh *white*, compact, soft. Spores "bay purple, globose, 8μ" Quél. Smell and taste pleasant. Edible. On the ground amongst ferns. Aug. Rare.

**With a ring on the stem.

Spores white.

Lepiota (Pers.) Fr.

(λεπίς, a scale; οὖς, ear.)

Pileus fleshy, regular. Stem central. Ring membranaceous, free, or adnate, persistent, or fugacious, *always manifest in the adult stage.* Gills free, adnate, or sinuato-adnate, often attached to a collar. Spores white, rarely pinkish, or ochraceous, oval, elliptical, pip-shaped, fusiform, subreniform, deltoid, or projectile shaped; continuous, or with a germ-pore. Cystidia rare. Growing on the ground, rarely on wood.

A. EPIDERMIS DRY.

*P. squamulose, or becoming broken up into scales.

(*a*) Ring movable, distinct from the volva; apex of stem surrounded by a cartilaginous collar.

118. **L. procera** (Scop.) Fr. Rolland, Champ. t. 11, no. 15.
Procera, tall.

P. 10–25 cm., *whitish, the brownish cuticle breaking up into thick, separable scales*, ovato-acorn-shaped, then campanulate, and flattened, *with a broad, obtuse, prominent umbo*; margin fimbriate, fibrillose. St. 15–30 × 1·5–2 cm., *brownish*, breaking up into *snake-like markings*, due to the slower growth of the external hyphae, cylindrical, base bulbous. Ring *white* above, *brownish on the exterior*, thick, movable, persistent, cartilaginous near the stem, fibrillose at the margin. Gills *whitish, often becoming fuscous at the edge, remote from the stem, and separated by a cartilaginous collar*, ventricose, crowded, broader in front, soft, crowded. Flesh *white, floccose*. Spores white, elliptical, 15–18 × 10μ, multi-guttulate, with a germ-pore. Cystidia "on edge of gill bottle-shaped to clavate, 40–50 × 15–20μ" Rick. Taste and smell pleasant. Edible. Woods, heaths, and pastures. July—Nov. Common. (*v.v.*)

119. **L. prominens** Fr. Viv. Ital. t. 12. *Prominens*, prominent.

P. 5–10 cm., *ochraceous, disc covered with brownish-ochre, imbricate scales*, which are more sparse towards the fibrillose margin, hemispherical, then flattened, *with a prominent, darker umbo*. St. 7–18 × 1–2 cm., *brownish, or ochraceous, with a few scattered adnate squamules*, equal, *base abruptly bulbous*. Ring *whitish*, movable, fimbriate at the margin. Gills *white*, free, separated by a cartilaginous collar. Flesh *white, floccose*. Spores white, elliptical, 14–16 × 8–9μ. Taste and smell pleasant. Edible. Woods, heaths, and upland downs. Sept.—Oct. Uncommon. (*v.v.*)

120. **L. rhacodes** (Vitt.) Fr. Boud. Icon. t. 10. ῥάκος, ragged.

P. 7–18 cm., *greyish ochre, covered with large, thick, angular, ragged, yellowish scales, which become darker at their margin, disc flat, and deeper in colour*, very fleshy, globose, then flattened, or depressed. St. 7–25 × 1·5–2 cm., *white, bruising reddish*, conical, then elongated, and attenuated upwards, base *large, marginately bulbous, smooth*. Ring *white*, or *brownish*, clothed on the outside with one or two zones of scales, fimbriate at the margin. Gills *whitish, or reddish*, free, separated by a cartilaginous collar, lanceolate or ventricose, crowded. Flesh white, *reddening on exposure to the air, especially in the stem*. Spores white, elliptical, 12–15 × 6–8μ, 1–2-guttulate, with a germ-pore. Cystidia "on edge of gill coloured, ventricose-bottle-shaped, 30–36 × 12–15μ" Rick. Smell and taste pleasant. Edible. Under trees in pastures, and in coniferous woods. July—Nov. Common. (*v.v.*)

121. **L. puellaris** (Fr.) Rea. *Puellaris*, girlish.

P. 5–8 cm., *white, disc gibbous and ochraceous*, campanulate, then convex, *surface breaking up into delicate, floccose scales*. St. 9–12 × 1 cm., *white*, equal, *slightly mealy above the ring*, base sub-bulbous. Ring *white*, movable, narrow. Gills *white*, free, separated by a narrow, cartilaginous collar, narrowed behind, crowded. Flesh *white, or faintly tinted reddish*. Spores white, oblong-elliptical, 12–18 × 7–8μ, 1–3-guttulate. "Cystidia obovate-bottle-shaped, 16μ broad, occasionally with a somewhat protruding apex" Lange. Taste pleasant. Edible. In pastures, generally under oaks. Aug.—Oct. Uncommon. (*v.v.*)

122. **L. permixta** Barla. Barla, Champ. Alp. Marit. t. 10, figs. 1–4. *Permixta*, mixed up.

P. 12–15 cm., disc *brown cinnamon*, the paler ground colour elsewhere *covered with cinnamon brown patches* of the cuticle, campanulato-convex, then expanded, *subumbonate*; margin *whitish, torn*. St. 12–15 × 1·5–2 cm., *white, covered with small, irregular brownish scales*, base bulbous. Ring *tawny brown*, membranaceous, large, margin fimbriate. Gills *yellowish white, or flesh colour*, remote from the stem, and separated by a cartilaginous collar, narrowed behind, crowded. Flesh white, *becoming reddish*, floccose. Spores "oval, elliptical or almond shape, 12–20 × 8–12μ, hyaline, surrounded by a golden ring" Sacc. Taste pleasant. Edible. Woods, and pastures. Sept.—Nov. Uncommon. (*v.v.*)

123. **L. excoriata** (Schaeff.) Fr. Krombh. t. 24, figs. 27–28.
Excoriata, peeled.

P. 6–10 cm., *whitish, disc often brown, gibbous,* fleshy, globose, then expanded and plane, the *very thin cuticle breaking up into large patches and appearing as if it had been drawn inwards from the fimbriate margin.* St. 4–7·5 × ·5–1 cm., *white, or tinged greyish, equal,* base bulbous. Ring concolorous, firm, movable. Gills *white,* remote from the stem, and separated by a cartilaginous collar, soft, crowded. Flesh *white.* Spores white, elliptical, 14–15 × 9–11μ, with an apical germ-pore. "Cystidia obtusely fusiform, 50 × 10μ" Lange. Taste and smell pleasant. Edible. Heaths, and pastures, rarely in woods. May—Nov. Common. (*v.v.*)

124. **L. gracilenta** (Krombh.) Fr. *Gracilenta*, slender.

P. 7–12 cm., *whitish, the fuscous cuticle breaking up into closely adnate scales,* ovate, then campanulate, and at length flattened, *umbonate*; margin *deprived of its cuticle.* St. 12–15 × ·5–1 cm., *whitish, covered with small, distinct, yellowish scales,* attenuated upwards, base subbulbous. Ring *white, floccose,* very laxly woven, movable, *fugacious.* Gills *white, often dingy at the edge,* remote from the stem and separated by a broad, cartilaginous collar, very crowded. Flesh *white.* Spores white, pip-shaped, 12–13 × 7–8μ, with a large central gutta, and an apical germ-pore. "Cystidia on edge of gill ventricose-bottle-shaped, 30–36 × 12–15μ" Rick. Taste and smell pleasant. Edible. Woods, heaths, and pastures. Aug.—Nov. Common. (*v.v.*)

125. **L. mastoidea** Fr. Cke. Illus. no. 23, t. 24.
μαστός εἶδος, breast-like.

P. 3–6 cm., *whitish, the fuscous cuticle becoming broken up into adpressed scales,* campanulate, then convex, *acutely umbonate.* St. 7–10 cm. × 3–4 mm., *whitish, or bistre,* obsoletely squamulose, tough, flexible, *attenuated at the apex,* base bulbous. Ring *white, margin brownish,* entire, movable. Gills *white, or cream colour, very remote* from the stem, and separated by a cartilaginous collar, lanceolate, soft, very crowded. Flesh *white.* Spores white, elliptical, 15 × 9–10μ. Taste and smell pleasant. Edible. Heaths, pastures, and clearings in woods. Aug.—Oct. Uncommon. (*v.v.*)

126. **L. nympharum** Kalchbr. Kalchbr. Icon. t. 2, fig. 1.
Nympha, a bride.

P. 3–10 cm., *white, covered with white, concentric, squarrulose, torn scales,* that become somewhat ochraceous at their margin with age, disc *ochraceous,* campanulate, then conico-convex. St. 7·5–10 × ·5–1 cm., *white,* attenuated upwards, base bulbous, *apex mealy.* Ring *white,* distant. Gills *white,* remote from the stem, and separated by a cartilaginous collar, attenuated behind. Flesh *white, becoming pinkish under the cuticle of the pileus and at the base of the st.* Spores white, subglobose, or elliptical, 7 × 6μ, or 7–10 × 6μ, with a large central gutta. Taste pleasant. Edible. Heaths and hedgerows. Oct. —Dec. Uncommon. (*v.v.*)

(*b*) Ring fixed, homogeneous with the universal veil which clothes the st.; apex of st. without a cartilaginous collar; p. torn into scales, or flocci.

127. **L. acutesquamosa** (Weinm.) Fr. (= *Lepiota aspera* (Pers.) Quél.) Rolland, Champ. t. 13, no. 20. *Acutus,* sharp; *squamosa,* scaled.

P. 3–12 cm., *pale ferruginous, covered with small, rigid, apiculate, fuscous, deciduous warts, which leave areolate scars,* fleshy, hemispherical, then expanded, convex, very obtuse, *tomentose.* St. 7·5–10 × 1–2·5 cm., *white, becoming ferruginous downwards with the fibrils and spirally arranged scales* (the remains of the universal veil) attenuated upwards, base subbulbous. Ring *white, becoming yellow, margin sprinkled with rust coloured warts on the underside,* large, soft, pendulous from the apex of the stem. Gills *white,* free, very crowded, lanceolate, *often branched.* Flesh *white,* thick. Spores white, elliptical, or globose, 3–6 × 3–4μ. Cystidia "obovate-subrotund" Lange. Taste slightly bitter, smell strong. Woods, pastures, and bare soil. Sept.—Nov. Not uncommon. (*v.v.*)

128. **L. Friesii** (Lasch) Fr. Cke. Illus. no. 1105, t. 941.
Elias Fries, the prince of mycologists.

P. 9–10 cm., *ferruginous fuscous, covered with adpressed, tomentose, reddish brown scales,* very fleshy, campanulate, then convex, soft. St. 8–11 × 1·5–2 cm., *concolorous,* cylindrical, or subbulbous, scaly at the base. Ring *white,* superior, pendulous. Gills *white,* linear, free, often veined, branched. Flesh *white, becoming yellowish,* thick at the disc. Spores white, elliptical, 7 × 3–4μ, 1-guttulate. Cystidia "on edge of gill vesiculose, 15–18 × 10–13μ" Rick. Smell strong, taste unpleasant. On bare soil in gardens, and in oak and beech woods. Sept.—Oct. Uncommon. (*v.v.*)

129. **L. hispida** (Lasch) Fr. Fr. Icon. t. 14. *Hispida,* rough.

P. 3–7 cm., *fuscous umber,* fleshy, soft, hemispherical, then expanded, *umbonate, tomentose,* then *breaking up into thin, pointed, fugacious papillae, or scales.* St. 7·5 × ·5–1 cm., *fuscous umber,* attenuated upwards, *densely floccosely scaly below the ring.* Ring whitish, superior, membranaceous, reflexed, floccose. Gills white, remote from the stem, with a prominent collar encircling the stem, crowded, ventricose. Flesh *white*, thin. Spores white, elliptical, 6–7 × 4μ. Cystidia none. Smell of radish. Shady beech woods, and coniferous woods. Aug.—Oct. Rare. (*v.v.*)

130. **L. Badhami** B. & Br. Boud. Icon. t. 11. Dr. C. D. Badham.

Whole plant becoming saffron-red when touched or wounded, then finally blackish. P. 5–12 cm., *greyish*, campanulate, obtuse, at length expanded, often depressed and umbonate, hispid, with minute, velvety, fuliginous scales, but sometimes entirely fuliginous without any distinct scales. St. 5–18 cm. × 6–12 mm., *white*, silky, or floccoso-scaly, attenuated above, base bulbous. Ring *white*, firm, erect, and deflexed, more or less movable, often clothed with dingy granules on the outside. Gills *white*, remote from the stem. Flesh *white, instantly becoming red* when cut, and *finally blackish*. Spores white, elliptical, or pip-shaped, 6–7 × 3–4μ, 1–2-guttulate. Smell rather disagreeable. Under oaks, Spanish chestnuts, yews, and in hedgerows. Sept.—Nov. Uncommon. (*v.v.*)

131. **L. meleagris** (Sow.) Fr. Cke. Illus. no. 26, t. 26.
Meleagris, a guinea-fowl.

P. 2–5 cm., *fawn colour, covered with minute blackish scales*, fleshy, thin, ovate, or hemispherical, very obtuse, minutely tomentose and warty, then expanded, somewhat campanulate. St. 4–7·5 cm. × 5–8 mm., *concolorous, here and there tinged with yellow, minutely squamulose below the ring, fusiform*, or attenuated upwards from the bulbous base. Ring *white, often covered with minute blackish scales on the outside*, torn, very fugacious. Gills *white*, then *rose colour*, rarely *lemon colour, becoming reddish by rubbing*, remote from the stem and separated by a collar, rounded behind, sometimes connected, ventricose. Flesh *turning red*, as does the whole plant when dried. Spores "elliptical, 6–7 × 4μ" Massee. Taste not disagreeable. Plantations, hedgerows, hot beds, and spent tan. May—Oct. Rare.

132. **L. emplastrum** Cke. & Massee. Cke. Illus. no. 1106, t. 1164.
ἔμπλαστρον, a plaster.

P. 5–7·5 cm., *pallid, covered with a smooth, membranaceous, dark brown cuticle, which becomes broken up into large, persistent patches*, convex, then expanded, silky below the cuticle. St. 7·5 × 1–1·5 cm., *pallid*, equal, base slightly thickened, more or less striate. Ring *whitish, externally brown at the margin*, rather distant, erect. Gills *whitish*, remote from the stem, crowded, narrowed behind. Flesh *white, becoming pink, or reddish when cut.* Spores white, elliptical, obliquely apiculate, 18–20 × 10–12μ. Under trees. Oct. Rare.

133. **L. biornata** B. & Br. Cke. Illus. no. 27, t. 37.
Bis ornata, doubly adorned.

P. 2·5–5 cm., *white, or yellowish, sprinkled with scattered, minute, dark red scales,* fleshy, convex, broadly campanulate. St. 10 cm. × 8 mm., *whitish spotted with red,* attenuated at the base, rooting. Ring *white,* spotted at the edge like the pileus, descending. Gills *white,* approximate, ventricose, 4 mm. broad. Flesh *white, or yellow, reddish in the stem.* Spores white, elliptical, 10 × 8μ. The whole plant becomes blackish when dry. Melon, and cucumber frames. July. Rare.

134. **L. clypeolaria** (Bull.) Fr. (= *Lepiota metulaespora* B. & Br. of many British authors.) Cke. Illus. no. 28, t. 27, as *Lepiota hispida* Lasch. *Clypeus*, a shield.

P. 3–7·5 cm., *very variable in colour, at first covered with a yellow, or brownish, dense felt, which breaks up into floccose, torn patches,* fleshy, campanulate, then convex, and flattened, disc gibbous; margin *appendiculate* with the remains of the ring. St. 6–8 cm. × 4–10 mm., *concolorous,* equal, or slightly thickened at the base, *fragile, clothed with the same felt-like covering below the ring.* Ring *concolorous, floccose, fugacious.* Gills *white or becoming yellow,* free, 6 mm. broad, soft, somewhat crowded. Flesh white, floccose. Spores white, *fusiform,* 14–15 × 6μ, 1–many-guttulate. Smell and taste pleasant. Edible. Woods. Sept.—Oct. Not uncommon. (*v.v.*)

135. **L. clypeolarioides** Rea (= *Lepiota clypeolaria* Auct. plur. non Quél.) Cke. Illus. no. 29, t. 38, as *Lepiota clypeolaria* Bull.
Clypeolaria εἶδος, resembling *L. clypeolaria.*

P. 3–5 cm., *tan colour, covered with small, adpressed reddish brown scales and fibrils,* fleshy, convex, obtusely umbonate, then plane and depressed. St. 7·5–10 cm. × 4–8 mm., *concolorous,* scaly below the ring, slightly attenuated upwards. Ring *concolorous, narrow,* distant. Gills *white, becoming yellowish,* free, crowded. Flesh *white.* Spores white, elliptical, 6–8 × 3–4μ, or 8 × 5μ, 1-guttulate. Woods, and hedgerows. Sept.—Nov. Uncommon. (*v.v.*)

136. **L. pratensis** (Fr.) Rea. *Pratensis*, growing in meadows.

P. 2–5 cm., *yellowish tawny, disc darker, margin paler,* convex, then expanded, obtusely umbonate, *almost smooth.* St. 6–7 cm. × 6–10 mm., *yellowish, densely clothed with erect, white flocci up to the ring.* Ring *white, floccose,* somewhat fugacious. Gills *white,* free, attenuated at both ends, crowded. Flesh *white, brownish under the epidermis and at the base of the stem.* Spores white, *fusiform,* 12–14 × 4–5μ, 1–2-guttulate. Smell and taste pleasant. Edible. Heaths, hillsides, and pastures. Aug.—Nov. Not uncommon. (*v.v.*)

137. **L. alba** (Bres.) Sacc. Bres. Fung. Trid. t. 16, fig. 1, as *Lepiota clypeolaria* Bull. var. *alba* Bres. *Alba*, white.

P. 3–7 cm., *whitish, becoming yellowish with age*, fleshy, convex, then expanded, broadly umbonate, disc glabrous; margin fibrillosely flocculose, then squamulose. St. 4–6 cm. × 8–10 mm., *white, somewhat fuscous at the base*, equal, or attenuated downwards, *white floccose and spotted below the ring*, often forming a spurious second ring, *finally becoming glabrous*. Ring *white*, floccose and yellow on the outside, striate inside, distant, fugacious. Gills *white, then ochraceous*, free, crowded, 5–7 mm. broad. Flesh *white, somewhat yellowish in the stem*. Spores white, obovate-oblong, rounded at the one end, apiculate at the other, 11–14 × 6–7μ, 1-guttulate. Edible. Heaths, and hillsides. Sept.—Oct. Uncommon. (*v.v.*)

138. **L. gracilis** (Quél.) Rea. *Gracilis*, slender.

P. 2–3 cm., *whitish, disc brown, becoming broken up into brownish or rufous squamules*, convex, then plane, floccose; margin *white*, becoming torn. St. 3–4 cm. × 3–5 mm., *whitish, becoming discoloured*, equal, *smooth*. Ring *white, with a few scattered brownish squamules*, median, silky, floccose, fugacious. Gills *white*, free, crowded. Flesh *white*, thin, floccose. Spores white, pip-shaped, or elliptical with an oblique basal apiculus, 9–11 × 5–6μ. Cystidia none. Under beeches. Sept.—Oct. Uncommon. (*v.v.*)

139. **L. fulvella** Rea. Trans. Brit. Myc. Soc. VI, t. 2, fig. 2. *Fulvella*, somewhat tawny.

P. 3–5 cm., *somewhat tawny*, fleshy, convexo-campanulate, then expanded and subumbonate, *covered with closely adpressed, darker squamules*; margin thin. St. 3–6 cm. × 3–6 mm., *concolorous*, equal, or attenuated downwards, hollow, smooth. Ring *whitish*, inferior, fugacious. Gills *pallid, then ochraceous*, 4–6 mm. broad, free, rounded behind, crowded. Flesh *whitish*, watery. Spores white, oblong, angular, truncate at the base, acute, or acutely angular at the apex, 9–10 × 3·5–4μ, 1–2-guttulate; basidia clavate, 4-sterigmata. Cystidia subglobose, or pyriform, 14–18 × 8–12μ. Smell and taste none. Bare soil in woods. Sept.—Oct. Uncommon. (*v.v.*)

L. metulaespora B. & Br. The records of this as British are erroneous; they should be referred to *Lepiota clypeolaria* (Bull.) Fr. as defined above.

140. **L. helveola** Bres. Trans. Brit. Myc. Soc. IV, t. 8. *Helveola*, pale yellowish.

P. 1·5–3 cm., *madder brown*, somewhat fleshy, convex, then expanded, *subumbonate, scaly*. St. 2–4 cm. × 3–7 mm., *concolorous*, equal, fibrilloso-tomentose. Ring *whitish*, distant, fugacious. Gills *creamy white*, free, crowded, 4–5 mm. wide, ventricose, edge fimbriate. Flesh *white, becoming reddish when dry*. Spores white, elliptical, or subreniform, 6–10 × 4–6μ, granular. Poisonous. Amongst short grass. Sept.—Oct. Rare. (*v.v.*)

141. **L. felina** (Pers.) Fr. Cke. Illus. no. 1108, t. 943, fig. A.
Felina, cat-like.

P. 2–3 cm., *whitish, covered with concentric, small, blackish bistre scales, disc blackish*, convex, subumbonate. St. 3–5 cm. × 4–5 mm., *white, often sprinkled with blackish bistre scales near the base*, equal, or subbulbous. Ring *white, often sprinkled with blackish bistre scales*, superior, membranaceous, fugacious. Gills *white, or yellowish*, free, ventricose. Flesh *white*, thin. Spores white, elliptical, 8–9 × 4μ, 1-guttulate. Cystidia "on edge of gill clavate-vesiculose, 33–36 × 8–10μ" Rick. Under conifers. Sept.—Oct. Not uncommon. (*v.v.*)

142. **L. micropholis** B. & Br. Cke. Illus. no. 1108, t. 943, fig. B.
μικρός, small; φολίς, a scale.

P. 1–1·5 cm., *white, covered with minute, concentric, dark grey, or blackish scales*, conical, then plane; margin slightly striate. St. 2–2·5 cm. × 2 mm., *white*, curved, base *minutely bulbous*. Ring *white, blackish on the under surface*, distant, spreading. Gills *white*, free, 2 mm. broad, crowded, ventricose. Flesh white, very thin. Spores white, elliptical, 5–6 × 3–4μ, with a large central gutta. On coconut fibre in stoves, and on soil in pots. Aug. Rare. (*v.v.*)

143. **L. nigromarginata** Massee. *Niger*, black; *marginata*, margined.

P. 3–5 cm., *pale sienna-ochre, covered with small, concentric, umber scales*, campanulate, soon expanded, subumbonate. St. 5–6 cm. × 3 mm., *buff*, peronate below the ring, attenuated upwards. Ring *white*, membranaceous, distant, persistent. Gills *whitish, edge bordered with dark umber*, free, broader in front, narrow. Flesh whitish, thin. Spores white, elliptical, 6–7 × 4μ, 1-guttulate. Amongst grass. Rare.

144. **L. cristata** (A. & S.) Fr. Cke. Illus. no. 31, t. 29.
Cristata, crested.

P. 2–7 cm., *whitish, disc brown, covered with reddish brown scales*, slightly fleshy, campanulate, then expanded, often umbonate, *silky*. St. 4–6 cm. × 3–8 mm., *white, yellowish, or rufescent, equal*, silky, fragile. Ring *white, often tinged reddish*, distant, membranaceous, narrow, fugacious. Gills *white*, free, very crowded, plane. Flesh *white, often tinged reddish*, thin. Spores white, elliptical, 6–8 × 3–4μ. Cystidia "on edge of gill bottle-shaped, 30–36 × 8–12μ" Rick. Smell strong, often of radish, taste unpleasant. Woods, pastures, and lawns. July—Nov. Common. (*v.v.*)

145. **L. castanea** Quél. Quél. As. Fr. (1880), t. 8, fig. 1.
κάσταvov, the chestnut tree.

P. 1–3 cm., *reddish brown*, campanulate, often umbonate, *tomentose, then shaggy*. St. 3–4 cm. × 3–4 mm., *white, becoming concolorous with the tawny fibrils*, firm, base *bulbous*. Ring *white*, narrow, thin, membranaceous, silky, *tawny on the outside*, fugacious. Gills *cream colour, often tinged reddish when old*, free, ventricose. Flesh *cream colour, becoming reddish in the stem and occasionally in the pileus*. Spores white, oblong elliptical, or projectile-shaped, *often with a spine-like appendage on the one side*, 10–11 × 3·5–4·5μ, 1–2-guttulate. Cystidia "hair-shaped, rather broad and obtuse" Lange. Smell pleasant. Poisonous. Woods, and roadsides. Sept. Uncommon. (*v.v.*)

146. **L. scobinella** Fr. Trans. Brit. Myc. Soc. VI, t. 7.
Scobinella, fine sawdust.

P. 3–6 cm., *mouse grey, disc darker*, convex, then plane, umbonate, *pellicle breaking up into minute, separable, bistre scales*; margin *whitish*, smooth, silky. St. 4–6 cm. × 4–7 mm., *white*, stuffed, equal, slightly attenuated at the apex and base, *covered with white squamules that become tinged with bistre below the ring, striate above*. Ring *whitish, becoming tinged with bistre at the edge*, membranaceous, superior, often fugacious. Gills *white, becoming yellowish*, 3–4 mm. wide, ventricose, free, crowded. Flesh white, often tinged with fulvous at the base of the stem, thick at the disc, very thin at the margin of the pileus, floccose. Spores white, elliptical, 6–7 × 3–4μ, contents granular. Cystidia hyaline, clavato-cylindrical, 28–30 × 6μ, sparse. Woods and pastures. Sept.—Oct. Uncommon. (*v.v.*)

147. **L. citrophylla** B. & Br. Boud. Bull. Soc. Myc. Fr. IX (1893), t. II, fig. 1. κίτρov, lemon; φύλλov, gill.

P. 1·5–2 cm., *lemon yellow, covered with rufous scales*, convex, then expanded, obtuse, or broadly umbonate, at length depressed. St. 2–4 cm. × 2–4 mm., *lemon yellow*, equal, squamulose. Ring fugacious. Gills *lemon yellow*, free, rounded behind, or attenuated, minutely serrate. Flesh *white*, thin. Spores white, elliptical, 7–8 × 4μ. On the ground. Oct. Rare.

**P. not, or rarely squamulose, often granular, mealy or pruinose.

(*a*) Ring superior, fixed, subpersistent; universal veil adnate to the p. Collar wanting, or similar in texture to the flesh of the p.

L. Vittadinii (Moretti) Fr. = **L. Amanita Vittadinii** (Moretti) Vitt.

148. **L. naucina** Fr. (= *Lepiota pudica* (Bull.) Quél; *Psaliota cretacea* Fr.) Rolland, Champ. t. 12, no. 17. *Nucinus*, nutty.

P. 5–10 cm., *white, often pinkish or yellowish*, the thin cuticle *breaking up into evanescent granules*, fleshy, soft, globose, then expanded, gibbous, or obtusely umbonate. St. 4–8 × 1–1·5 cm., *white*, fibrillose, attenuated upwards from the *swollen base*. Ring *white*, membranaceous, superior, thick, *fimbriate at the margin*, often finally fugacious. Gills *white*, free, separated by a collar, ventricose, soft, crowded. Flesh *white*, thick. Spores white, broadly ovoid, 8–9 × 5·5μ, with a large central gutta. "Cystidia club-shaped, 55 × 10–11μ" Lange. Smell and taste pleasant. Edible. Pastures and gardens. July—Oct. Not uncommon. (*v.v.*)

var. **leucothites** (Vitt.) Fr. Vitt. Fung. Mang. t. 40. λευκός, white.

Differs from the type in the *p. breaking up into squamules especially near the margin, and in the gills becoming pink with age*. Spores white, elliptical, 7–8 × 5μ, or 9 × 7μ. Edible. Pastures and heaths. Sept. —Oct. Not uncommon. (*v.v.*)

149. **L. holosericea** Fr. Cke. Illus. no. 34, t. 41.
ὅλος, wholly; σηρικός, silky.

P. 5–10 cm., *white, or yellowish*, fleshy, soft, convex, then expanded, obtuse, *fibrillosely silky*. St. 6–10 × 1–1·5 cm., *whitish*, soft, fragile, base *bulbous*, silky-fibrillose. Ring *whitish*, membranaceous, *superior*, large, soft, pendulous. Gills *white, becoming cream colour*, free, ventricose, broad, crowded. Flesh *white*, soft. Spores white, elliptical, 8–9 × 4–5μ. Edible. Bare ground in arable fields, hopyards, and gardens. Sept.—Oct. Not uncommon. (*v.v.*)

150. **L. erminea** Fr. Cke. Illus. no. 32, t. 40. *Erminea*, white.

Entirely white with the exception of the ochraceous disc of the p. P. 3–6 cm., campanulate, then flattened, slightly gibbous at the prominent disc, *becoming fibrillosely silky towards the margin*. St. 5–7·5 cm. × 3–6 mm., *equal, very fragile*, silky. Ring membranaceous, *distant*, narrow, at length torn and fugacious. Gills sinuate, then free, somewhat crowded. Spores white, oblong elliptical, 9–11 × 3–4μ. "Cystidia on edge of gill vesiculose-pyriform, 36–40 × 12–16μ" Rick. Smell and taste of radish. Roadsides, and hilly pastures. Sept.—Nov. Uncommon. (*v.v.*)

151. **L. constricta** (Fr.) Quél. (= *Armillaria constricta* Fr.) Fr. Icon. t. 18. *Constricta*, compressed.

Entirely white, becoming ochraceous when bruised. P. 3–5 cm., fleshy, convex, then plane, obtuse, *pruinose*, then *silky*; margin at first involute, and *villous*. St. 4–5 cm. × 6–9 mm., equal, or thickened at the base, slightly rooting, *fibrillose*, or squamulose. Ring *superior*, *narrow*, adhering obliquely, at length fugacious. Gills emarginate, then free, very crowded, unequal. Flesh compact. Spores white, elliptical, granular, 7–8 × 4–5μ. Smell of new meal. Pastures, especially where the grass is scorched by urine, and amongst short grass under conifers. Aug.—Oct. Uncommon. (*v.v.*)

152. **L. cepaestipes** (Sow.) Fr. (= *Leucocoprinus cepaestipes* (Sow.) Pat.) Gillet, Champ. Fr. t. 414. *Cepa*, onion; *stipes*, stem.

Caespitose. P. 2·5–6 cm., *white, or yellowish, covered with floccose, fugacious, yellowish scales*, disc deeper coloured, umbonate, *membranaceous*, obtusely conical, then campanulate; margin *striate.* St. 7·5–10 cm. × 5–8 mm., *white, attenuated upwards from the bulbous base, covered with fugacious, delicate flocci.* Ring *white*, narrow, membranaceous, separating-free, fugacious. Gills *white, then flesh colour*, free, at length remote, attenuated at both ends, 4 mm. broad, very crowded. Flesh *white, then pinkish*, thin. Spores white, elliptical, 6–7 × 4–5μ, 1-guttulate, with an apical germ-pore. Taste bitter. On tan in hot-houses, greenhouses, and in frames. March—Oct. Not uncommon. (*v.v.*)

var. **cretacea** (Bull.) Fr. Grev. Scot. Crypt. Fl. t. 333. *Cretacea*, chalk-like.

Differs from the type in its *chalk white colour and darker scales.* Spores white, elliptical, 7–9 × 6–7μ, 1-guttulate. Greenhouses, and on spent tan. May—Sept. Not uncommon. (*v.v.*)

153. **L. lutea** (Bolt.) Quél. (= *Agaricus flos sulfuris* Schnitz.) Boud. Icon. t. 19. *Lutea*, yellow.

Entirely sulphur colour. P. 1–4 cm., campanulate, thin, *deeply striate, covered with concolorous flocci.* St. 5–12 cm. × 3–4 mm., *covered with concolorous flocci*, apex smooth, base bulbous. Ring membranaceous. Gills free, remote, narrow. Flesh *concolorous.* Spores white, oval, 8–10 × 5–7μ, with a large central gutta. On coconut fibre in greenhouses. Jan.—Dec. Not uncommon. (*v.v.*)

154. **L. medioflava** Boud. Bull. Soc. Myc. Fr. x (1894), t. I, fig. 1. *Medius*, middle; *flava*, yellow.

P. 2–3 cm., *white*, soon expanded, and depressed, *umbo prominent, becoming light yellow, deeply striate*, minutely tomentose. St. 4–7 cm. × 2–3 mm., *white, minutely mealy* above the ring, tomentose, *often becoming light yellow* at the bulbous base. Ring *white*, median, reflexed. Gills *white*, free, crowded, rounded behind. Flesh *white.* Spores white, ovate, obtuse, 5–6 × 3μ, 1-guttulate. On decaying coconut fibre, and soil in greenhouses. June—Oct. Uncommon.

155. **L. pseudo-licmophora** Rea. (= *Lepiota licmophora* auct. non B. & Br. and Petch.)

ψευδής, false; λικμός, a winnowing fan; φέρω, I bear.

P. 2·5–5 cm., *lemon-yellow, sometimes wholly sulphur-white*, membranaceous, plane, *depressed, deeply sulcate up to the central disc, glabrous*; margin crenate. St. 7·5–10 cm. × 2–3 mm., *lemon-yellow*, attenuated upwards, base tomentose. Ring *median*. Gills *lemon-yellow*, remote, narrow, 2 mm. broad, slightly arched, distant, *interstices veined*. Spores white, elliptical, 9–10 × 5μ. In greenhouses. Aug.—Oct. Uncommon. (*v.v.*)

156. **L. serena** Fr. Cke. Illus. no. 57, t. 47, as *Armillaria subcava* Schum. fide Boudier. *Serena*, clear.

P. 2–4 cm., *white, becoming yellowish with age, fragile*, campanulate, thin, expanded, becoming silky, margin *slightly striate*. St. 4–7 cm. × 4–5 mm., *white, becoming greyish*, equal, base *subbulbous*. Ring *white*, membranaceous, median, thin, narrow, erect, fugacious. Gills *white*, free, ventricose. Spores white, elliptical, 5–6 × 4μ, multiguttulate. Amongst grass, and larch needles. Sept.—Nov. Uncommon. (*v.v.*)

(*b*) Universal veil sheathing the st., at first extending continuously from the st. to the p., at length ruptured and forming an inferior ring. P. granular or warted, consisting chiefly of globose cells.

L. pyrenaea Quél. = **Pholiota aurea** (Mattusch) Fr. fide R. Maire.

157. **L. granulosa** (Batsch) Fr. Cke. Illus. no. 39, t. 18, upper figs. only. *Granulosa*, granular.

P. 3–5 cm., *ferruginous, or rusty-brown, becoming pale-hoary when dry*, fleshy, convex then flattened, *obtusely umbonate*, furfuraceo-granular, often wrinkled; margin appendiculate with the veil. St. 5–9 cm. × 4–9 mm., *white at the apex, covered below with fine, brownish granules*, equal. Ring *concolorous*, membranaceous, inferior, torn. Gills *whitish, or cream colour, slightly adnexed, or sinuato-adnate*. Flesh *yellowish, becoming reddish in the lower portion of the st.* Spores white, elliptical, 7–8 × 5μ. "Cystidia hair-shaped, acute, small, 2–3μ broad" Lange. Taste pleasant. Edible. Heaths, and hilly woods. July—Oct. Not uncommon. (*v.v.*)

var. **rufescens** B. & Br. Cke. Illus. no. 40, t. 213, upper figs. *Rufescens*, becoming reddish.

Differs from the type in its *smaller size, and the pure white p. and st. partially turning red with age or when bruised*. Spores oval, 3–4 × 2μ. Amongst beech leaves. Sept.—Oct. Uncommon. (*v.v.*)

158. **L. amianthina** (Scop.) Fr. Cke. Illus. no. 40, t. 213, lower figs. ἀμίαντος, unspotted.

P. 3–5 cm., *ochraceous*, somewhat fleshy, convex, then plane, subumbonate, furfuraceo-granulose, often wrinkled. St. 3–5 cm. × 4–6 mm., *whitish at the apex, covered with ochraceous granules below the ring*, equal. Ring *concolorous*, granular on the outside, inferior, fugacious. Gills *cream colour*, *adnate*, crowded. Flesh *yellow*. Spores white, elliptical, 4–5 × 3μ. Cystidia none. Taste pleasant. Edible. Coniferous woods, heaths, and lawns. Aug.—Nov. Common. (*v.v.*)

var. **Broadwoodiae** B. & Br. Miss S. Broadwood.

Differs from the type in the *delicately tomentose p.*, *and inflexed margin*. Woods. Rare.

var. **alba** René Maire. *Alba*, white.

Differs from the type in being *entirely white*. Woods. Oct. Rare. (*v.v.*)

159. **L. cinnabarina** (A. & S.) Fr. Cke. Illus. no. 38, t. 43.

κιννάβαρι, dragon's blood.

P. 5–8 cm., *cinnabar-colour*, fleshy, convex, then plane, obtuse, granuloso-furfuraceous; margin *fimbriate*. St. 4–7 × 1–2 cm., *concolorous, covered with reddish granules below the ring, subbulbous*. Ring *concolorous*, thin, narrow, inferior, fugacious. Gills *white*, free, lanceolate. Flesh *ochraceous, reddish under the cuticle of the p. and st.* Spores white, elliptical, obtuse, 4 × 2·5–3μ, 1-guttulate. "Cystidia hair-shaped, acute" Lange. Taste pleasant. Edible. Coniferous woods. Sept.—Oct. Rare. (*v.v.*)

var. **Terreyi** B. & Br. Saund. & Sm. t. 35, figs. 1–5.

Michael Terrey.

P. 2·5–5 cm., *bright tawny*, somewhat hemispherical, pulverulent, roughened with minute warts. St. somewhat equal, often cylindrical, covered below the ring *with furfuraceous scales of the same colour as the p.* Ring at length torn into fragments. Gills *white*, remote, narrow, not branched. Spores white, elliptical, 5 × 4μ. Sandy ground. Rare.

160. **L. carcharias** (Pers.) Fr. Cke. Illus. no. 37, t. 42.

κάρχαρος, sharp-pointed.

P. 2–5 cm., *flesh coloured*, fleshy, convex, then plane, *often umbonate*, covered with minute granules. St. 3–6 cm. × 4–8 mm., *concolorous*, and *covered with minute, pointed warts below the ring, apex white*, subbulbous, or equal and attenuated upwards. Ring *concolorous, covered on the outside with the same minute, pointed warts*. Gills *white, adnate*. Flesh *whitish*, or *ochraceous*. Spores white, elliptical, obtuse, 4–5 × 2–3μ, 1–3-guttulate. Taste disagreeable, smell unpleasant. Coniferous woods, and amongst short grass. May—Nov. Common. (*v.v.*)

161. **L. rosea** Rea. Trans. Brit. Myc. Soc. VI, t. 2, fig. 1.

Rosea, rose colour.

P. 2–3·5 cm., *bright rose colour*, somewhat fleshy, convex, then expanded, *densely granular*, or mealy, consisting of globose cells, 45–50μ in diam.; margin thin. St. 5–6 cm. × 3–5 mm., *whitish, becoming concolorous*, equal, hollow, smooth. Ring *concolorous*, membranaceous, medial, narrow, soon fugacious. Gills *whitish, then ochraceous*, free, rounded behind, crowded, 4–5 mm. broad. Flesh *whitish, becoming reddish* especially in the stem. Spores white, elliptical, 5 × 3μ, 1-guttulate; basidia clavate, with 4-sterigmata. Cystidia none. Bare soil in moist, shady woods. Sept.—Oct. Uncommon. (*v.v.*)

162. **L. atrocrocea** W. G. Sm. *Ater*, black; *crocea*, saffron.

P. 3–4 cm., *bright salmon orange, more or less covered with purple brown, almost black, granular flocci*, expanded, then slightly depressed. St. *bright salmon orange*, attenuated upwards, *covered with salmon brown squamules*. Ring fugacious. Gills *salmon white, broadly adnate*. Flesh *salmon orange brown*, thin. Oct. Rare.

163. **L. haematosperma** (Bull.) Boud. (= *Lepiota echinata* (Roth) Boud.) Boud. Icon. t. 12. *αἷμα*, blood; *σπέρμα*, seed.

P. 2–3·5 cm., *blackish grey, more or less olivaceous*, convex, then plane, *very thin, fragile, finely granular*; margin paler, appendiculate with the veil. St. 3–6 cm. × 2–3 mm., *dark, becoming vinous*, equal, or *slightly bulbous, base covered with pulverulent, fugacious granules*. Ring *reddish*, floccose, granular on the outside, inferior, fugacious. Gills *blood red*, free, ventricose, rounded behind. Flesh *whitish, becoming reddish under the epidermis and in the stem*. Spores *pale ochraceous in the mass, becoming reddish with age*, elliptical, 4–5 × 3μ. Cystidia none. Smell strong. Hedgerows, gardens, and occasionally in woods. July—Oct. Uncommon. (*v.v.*)

164. **L. polysticta** Berk. Cke. Illus. no. 41, t. 30.
πολύς, many; *στικτός*, spotted.

P. 2–5 cm., *ochraceous, covered on the obtusely umbonate disc with minute red brown scales* from the breaking up of the cuticle, fleshy, firm, tough, convex, then expanded; margin often appendiculate with the veil. St. 3–5 cm. × 5–9 mm., *white at the apex, densely clothed with reddish, ferruginous scales below the ring*, equal, or attenuated downwards. Ring *concolorous*, very narrow, inferior, very fugacious. Gills *white, then yellowish*, free, rounded before and behind, *broad*, ventricose, crowded. Flesh, *whitish, slightly reddish under the epidermis of the st. and p.* Spores white, sub-globose, 4 × 3μ, 1–4-guttulate. Open pastures, and amongst short grass in woods. Sept.—Nov. Uncommon. (*v.v.*)

(*c*) Smaller, slender. P. dry, cuticle entire, not scaly nor granular.

165. **L. parvannulata** (Lasch) Fr. Fr. Icon. t. 16, fig. 3.
Parvus, small; *annulata*, ringed.

P. 1–2 cm., *white, becoming yellowish when dry*, thin, ovato-campanulate, then plane, *pruinose, then silky*. St. 1–2 cm. × 2–3 mm., *white, equal, fibrillose below the ring.* Ring white, *very small,* distant, entire. Gills *cream colour,* free, crowded. Flesh *white,* very thin. Spores white, elliptical, 4–5 × 3μ. Cystidia none. Amongst mosses and short grass in woods, and pastures. Sept.—Oct. Uncommon. (*v.v.*)

166. **L. sistrata** Fr. Cke. Illus. no. 42, t. 85, fig. A.

Sistrum, a rattle.

P. 1–2 cm., *whitish, becoming light yellowish, or flesh colour, disc often darker,* slightly fleshy, campanulate, then expanded and obsoletely umbonate, *pruinate with shining atoms,* margin often appendiculate with the veil. St. 2·5–5 cm. × 1–4 mm., *white, or flesh colour,* equal, *fibrillosely silky and pruinose* below the ring. Ring *concolorous,* fibrillose, fugacious. Gills *white,* free, reaching the stem, crowded, 4 mm. broad, ventricose. Flesh *white, often pinkish in the stem,* fragile. Spores white, elliptical, 3–4 × 1·5–2μ. On bare ground in woods, and by roadsides. Sept.—Nov. Not uncommon. (*v.v.*)

167. **L. seminuda** (Lasch) Fr. Cke. Illus. no. 43, t. 19, fig. a.

Semi, half; *nuda,* naked.

P. 1–2 cm., *whitish, or flesh colour, becoming yellowish,* very thin, campanulato-expanded, *umbonate, at first covered with fugacious, floccose meal,* margin appendiculate with the veil. St. 2–3 cm. × 1–2 mm., *whitish, or flesh colour,* equal, *mealy* below the ring. Ring *concolorous, mealy,* torn, fugacious. Gills *white, then cream colour,* free, but reaching the st., thin, crowded, ventricose. Flesh *white, often pinkish in the st.* Spores white, elliptical, 4 × 2μ. Smell pleasant. Amongst moss in woods. Aug.—Oct. Uncommon. (*v.v.*)

168. **L. Bucknallii** B. & Br. (= *Lepiota lilacina* Quél.) Boud. Bull. Soc. Myc. Fr. IX (1893), t. II, as *Lepiota lilacina* Quél.

Cedric Bucknall.

P. 1·5–3 cm., *white,* fleshy, campanulate, then convex, *minutely mealy, and becoming tinged with lilac.* St. 5–7 cm. × 3–5 mm., *white,* gradually attenuated upwards, *densely mealy, and becoming deep lilac* below the ring *with age or bruising.* Ring *concolorous, mealy,* fugacious. Gills *yellowish,* free, not crowded. Flesh *white, becoming deep lilac in the lower two-thirds of the st.* Spores white, boat-shaped, 7–8 × 3μ, 3-guttulate. Smell *strong of gas-tar.* Amongst grass. Oct. Rare. (*v.v.*)

169. **L. mesomorpha** (Bull.) Fr. μέσος, middle; μορφή, form.

P. 1·5–2·5 cm., *whitish, or yellowish flesh colour, umbo deeper coloured,* slightly fleshy, very thin, campanulate, then expanded, at first pubescent, then becoming smooth, or minutely granular. St. 5–7·5 cm.

× 2 mm., *concolorous*, slightly attenuated upwards, silky. Ring membranaceous, floccose, superior, entire, spreading, fugacious. Gills *whitish, or cream colour*, free, crowded, ventricose. Flesh *white*, thin. Spores white, elliptical, 6–8 × 3–4μ, guttulate. Woods. Sept.—Nov. Uncommon.

170. **L. ianthina** Cke. Cke. Illus. no. 1112, t. 944, fig. A.
ἰάνθινος, coloured violet.

P. 2 cm., *whitish, covered with violet, radiating, hair-like squamules, umbonate disc dark violet, fibrillose*, thin, campanulate, then expanded. St. 2–3 cm. × 2–3 mm., *whitish*, subequal, somewhat flexuose. Ring distant, *narrow*, fugacious. Gills whitish, free, 2 mm. broad, lanceolate, scarcely crowded. Flesh *white*, thin. Stoves. March. Rare.

171. **L. martialis** Cke. & Massee. Cke. Illus. no. 1112, t. 944, fig. B.
Martialis, belonging to Mars.

P. 2–3 cm., *clear deep pink, disc darker, becoming yellowish with age*, thin, campanulate, then plane, *minutely silky*; margin *striate*. St. 2·5–4 cm. × 3 mm., *pale ochraceous at the apex, pinkish red below the ring*, slightly attenuated upwards. Ring *white*, broad, pendulous, rather distant, persistent. Gills *whitish*, free, 2 mm. broad, somewhat lanceolate, rather crowded. Flesh *white*, thin. Spores white, elliptical, 8 × 4μ. On the trunk of a tree fern. March. Rare.

172. **L. submarasmioides** Speg.
Sub, somewhat; *Marasmius* εἶδος, like a Marasmius.

P. 2–5 cm., *pale buff, umbo tawny*, convex, then expanded, *floccosely wrinkled* towards the margin. St. 5 cm. × 3 mm., *white*, equal, *slightly striate*. Ring *whitish*, superior, fugacious. Gills *ochraceous*, free, 3 mm. wide, rounded at both ends, crowded. Flesh *white, reddish under the epidermis of the p. and in the st.*, tough. Spores *pale ochraceous, deltoid, or pyramidal*, 5–6 × 3μ. On bare ground, Malvern Hills. Sept. Rare. (*v.v.*)

B. EPIDERMIS VISCID.

173. **L. medullata** Fr. Fr. Icon. t. 16. *Medullata*, pithy.

P. 4–7 cm., *white, often greyish at the disc*, slightly fleshy, convexo-plane, *umbonate*, viscid; margin appendiculate with the veil. St. 7·5 cm. × 6 mm., *white*, dry, equal, *silky and squamulose below the ring*, apex striate, *stuffed with a distinct separable pith*. Ring white, *incomplete, torn*. Gills *white*, free, broader in front, ventricose, crowded. Flesh *white*, soft, watery. Smell strong of radish. Coniferous woods. Aug.—Oct. Uncommon. (*v.v.*)

174. **L. arida** (Fr.) Gillet. (= *Amanita arida* Fr.) Fr. Icon. t. 12, as *Amanita arida* Fr. *Arida*, dry.

P. 5–7 cm., *greyish*, thin, convex, then plane, obtuse, silky; margin *whitish*, sulcato-striate. St. 6–9 × 1 cm., *white*, glabrous, *floccose at the incrassated base*. Ring *concolorous*, distant. Gills *white, then flesh colour*, attenuato-adnate. Flesh *white*, soft. Spores white, elliptical, 9–10 × 7–7·5μ. Birch and fir woods. Sept.—Oct. Uncommon.

175. **L. lenticularis** (Lasch) Cke. (= *Amanita lenticularis* (Lasch) Fr., *Lepiota guttata* (Pers.) Quél.) Fr. Icon. t. 13, as *Amanita lenticularis* Lasch. *Lenticula*, a lentil.

P. 4–10 cm., *pinkish tan colour*, fleshy, globose, then campanulato-convex; margin *paler, slightly glutinous*. St. 8–10 × 1–2 cm., *white, or cream colour, apex marked with dark green, watery drops in very wet weather, which on drying become dingy*, equal, or subbulbous, floccose, or smooth below the ring. Ring *concolorous, often spotted like the apex of the stem, large*, superior. Gills *whitish, sometimes inclining to olivaceous*, free, approximate, ventricose, broader in front, very crowded, sometimes forked. Flesh *white, reddish at the base of the st.* Spores white, pip-shaped, or elliptical, 7–8 × 4–5μ. Smell mouldy. Edible. Deciduous woods, and heaths. Sept.—Oct. Not uncommon. (*v.v.*)

var. **megalodactylus** (B. & Br.) Rea. Cke. Illus. no. 15, t. 11, as *Amanita megalodactylus* B. *μέγας*, large; *δάκτυλος*, finger.

Differs from the type in being *thinner, and paler in colour*. Woods. Oct.—Nov. Uncommon.

176. **L. irrorata** Quél. Trans. Brit. Myc. Soc. III, t. 8. *Irrorata*, bedewed.

P. 2·5–5 cm., *yellowish, then straw colour*, firm, convex, *covered like the stem with dew-like transparent drops*. St. 3–4 cm. × 7–10 mm., *white*, satiny above the ring, silky and *variegated with small yellow, or brownish squamules below*, equal. Ring *concolorous*, membranaceous, narrow. Gills *white, then cream colour*, free, emarginate, 4 mm. wide, ventricose. Flesh *white*. Spores white, ovoid, 4–5 × 4μ, punctate. Pastures, and clearings in woods. June—Oct. Uncommon. (*v.v.*)

177. **L. illinita** Fr. Fr. Icon. t. 16. *Illinitus*, besmeared.

P. 4–9 cm., *white, or yellowish*, fleshy, globose, then convex, umbonate, umbo *often becoming fuscous, viscid*; margin *slightly striate*, sometimes fimbriate. St. 5–8 × ·5–1 cm., *white, very viscid*, equal, or subbulbous, fragile. Ring *white*, membranaceous, thin. Gills *white*, free, at length remote, crowded, soft, somewhat connected by veins. Flesh *white*, floccose, thin. Spores white, broadly elliptical, or subglobose, 6 × 4–5μ, 1–2-guttulate. Smell pleasant. Plantations. Aug.—Sept. Rare. (*v.v.*)

178. **L. glioderma** Fr. (= *Armillaria glioderma* (Fr.) Quél.) Fr. Icon. t. 15. *γλοιόν*, viscid; *δέρμα*, skin.

P. 4 cm., *brownish red*, slightly fleshy, campanulate, then convex, broadly gibbous, or obtuse, *glutinous*. St. 7·5 cm. × 4–6 mm., *whitish, or rufescent*, equal, fragile, *dry*, floccosely scaly up to the ring. Ring *white above, rufescent squamulose on the outside*, fibrillose, silky, torn. Gills *white, or cream colour*, free, approximate, ventricose, broad, crowded. Flesh *white, then pinkish*, soft, thin. Spores white, subglobose, 5μ. Fir woods. July—Oct. Uncommon. (*v.v.*)

L. delicata Fr. = **Armillaria delicata** (Fr.) Boud.

179. **L. Georginae** W. G. Sm. Cke. Illus. no. 47, t. 132. Miss Georgina E. Johnstone.

Entirely white, turning crimson everywhere when touched, and finally becoming brown when dry. P. 1–3 cm., slightly fleshy, fragile, campanulate, then expanded, *covered with a minute, dense, viscid pruinosity*; margin *at length striate*. St. 2·5–5 cm. × 3–4 mm., slightly attenuated upwards, *viscido-pruinose*. Ring fugacious. Gills free, very thin, moderately distant, somewhat ventricose, 3 mm. broad. Spores pinkish in the mass, pip-shaped, 6–8 × 4μ, 1- rarely 2-guttulate. Pine woods and amongst mosses in a cool fernery. May—Nov. Rare. (*v.v.*)

Hiatula Fr.

(*Hio*, I gape.)

Pileus slightly fleshy at the disc, campanulate. Stem central. Ring *very fugacious, not manifest in the adult stage.* Gills free, or adnate. Spores white, subglobose, smooth, with a germ-pore. Growing on wood.

180. **H. Wynniae** B. & Br. Cke. Illus. no. 676, t. 688. Mrs Lloyd Wynne.

Entirely shining white. P. 3–4 cm., *very thin*, campanulate, then plane, with a trace of an umbo, *striate, pulverulent, disc darker*. St. 2·5 cm. × 1–2 mm., equal, striate. Gills free, or very slightly adnexed at first, subdistant, 2 mm. broad, *scarious*. Spores white, subglobose, 5 × 4μ, with a germ-pore. On wood in stoves. Phosphorescent. A native of Queensland.

Spores pink.

Annularia Schulz.

(*Annularia*, pertaining to a signet-ring.)

Pileus fleshy, regular. Stem central. Ring large, free, or adnate. Gills free. Spores pink, globose, or oval, smooth, continuous. Cystidia ventricose. Growing on the ground, or on wood.

181. **A. laevis** (Krombh.) Schulz. (= *Lepiota pudica* (Bull.) Quél.) Krombh. Icon. t. 26, figs. 16 and 17, as *Agaricus laevis* Krombh

Laevis, smooth.

P. 5–12 cm., *white, disc brownish, or yellowish*, convex, expanded, obtuse, or subumbonate, sometimes minutely squamulose; margin often cracked, appendiculate with the remains of the ring. St. 4–12 × ·5–1 cm., *white*, slightly attenuated upwards, base bulbous, *silky*. Ring *white*, large, *free*, subdistant. Gills *white, then flesh colour*, free, narrowed behind, somewhat crowded, thin. Flesh *white*, firm. Spores pink, "subglobose, 7–8μ" Massee. Bushy places amongst grass. Aug.—Oct. Uncommon.

182. **A. transilvanica** Schulz.

Transilvanica, belonging to Transylvania.

P. *whitish, disc darker*, campanulate, striate to the vertex; margin lobed. St. *paler than the p.*, flocculose, hollow. Ring membranaceous, complete. Gills crowded, unequal.

Spores purple, or fuscous.

Psaliota Fr.

(ψάλιον, a ring.)

Pileus more or less fleshy, regular. Stem central. Ring membranaceous, adnate, persistent, rarely fugacious. Gills free. Spores fuscous purple, reddish purple, blackish purple, or fuscous, elliptical, oval, globose, or obovate, with an apical germ-pore. Cystidia present, or absent. Growing on the ground.

*Large, fleshy.

183. **P. augusta** Fr. (= *Psaliota Elvensis* B. & Br. sec. Quél.) Bres. Fung. Trid. t. 60, as *Psalliota villatica* Brond.

Augusta, majestic.

P. 10–30 cm., *whitish, fuscous citron, or dark straw colour*, fleshy, globose-hemispherical, then expanded, very obtuse, silky, *soon breaking up into adpressed squamules*; margin exceeding the gills, tomentosely toothed. St. 6–20 × 1·5–5 cm., *white, becoming yellowish when bruised, then brownish*, very firm, attenuated upwards from the base which is sunk in the earth, smooth, flocculose just under the ring. Ring *white and smooth above, yellowish and areolately floccose on the under side, very wide*, adnate to the st. for 2–3 cm., then free and pendulous. Gills *pallid, then fuscous*, free, *separated from the st. by a broad collar*, narrow at first, becoming wider, simple, thin, crowded. Flesh *whitish, becoming yellowish or brownish in the st. when broken*, soft, floccose like that of *Lepiota procera*. Spores brownish purple, elliptical, 7–13 × 5–6μ. Smell pleasant like anise, sometimes disagreeable. Taste pleasant, often like almonds. Edible. Woods, and pastures, often near ant hills. Aug.—Sept. Uncommon. (*v.v.*)

184. **P. peronata** Massee non Roz. et Rich. (= *Psaliota augusta* Fr. sec. René Maire.) *Peronata*, booted.

P. 10–12·5 cm., *pale dull ochraceous, densely covered with small, brown, silky scales,* that become larger towards the margin, fleshy, hemispherical, then expanded. St. 12·5–15 × 1·5 cm., *white,* equal, *marginately bulbous at the base, covered with large, white, upward pointing, squarrose scales below the ring,* smooth above the ring. Ring yellowish, large, spreading. Gills *pink, then pale purple brown,* very distant from the st., 6–7 mm. broad, crowded. Flesh *white, becoming brownish in the st. when cut.* Spores purple-brown, obliquely elliptical, 6 × 4μ. Taste pleasant. Edible. Pine woods. Oct. Rare.

185. **P. Elvensis** B. & Br. Boud. Icon. t. 134.
Elvensis, belonging to the river Elwy.

P. 8–15 cm., *purplish brown,* fleshy, subglobose, then hemispherical, at length often flattened and slightly depressed at the areolate disc, fibrillose, *breaking up into large, persistent, floccose, pointed, somewhat revolute darker scales*; margin very obtuse, thick, covered with pyramidal warts. St. 7–15 × 2·5–5 cm., *concolorous and fibrillose below the ring,* apex paler, equal, becoming swollen in the centre, and attenuated at the base. Ring *concolorous,* membranaceous, thick, deflexed, broken here and there, more or less floccose at the margin and on the underside which is often areolate. Gills *brownish flesh colour, then brownish purple,* free, 6–8 mm. broad. Flesh *turning red when cut, then becoming brownish,* thick, firm. Spores brownish purple, globose, or subglobose, 6 × 4–5μ, with a large central gutta. Smell and taste pleasant. Edible. Often caespitose. Under oaks, beeches, firs, and on roadsides. Aug.—Oct. Uncommon. (*v.v.*)

186. **P. Bernardii** Quél. Trans. Brit. Myc. Soc. III, t. 14.
G. Bernard.

P. 10–20 cm., *white, then becoming ferruginous at the apex of the warts,* fleshy, convex, then expanded, firm, the tomentose surface soon *breaking up into thick, angular warts.* St. 6–7 × 4–5 cm., *white, becoming reddish brown with age,* attenuated upwards from the bulbous base, apex striate. Ring *white,* membranaceous, soon disappearing, striate on the upper surface. Gills *greyish flesh colour, then blackish purple,* free, attenuated at both ends, 8–12 mm. broad. Flesh *white, then tinged with purple, and finally stained with reddish brown,* firm. Spores blackish purple, ovoid elliptical, 9–11 × 6–7μ, 1-guttulate, with an apical germ-pore. Smell unpleasant. Taste disagreeable. Pastures near the sea. Sept.—Oct. Uncommon. (*v.v.*)

187. **P. arvensis** (Schaeff.) Fr. Rolland, Champ. t. 71, no. 159.
Arvensis, belonging to cultivated fields.

P. 7–20 cm., *whitish, becoming stained with yellow*, fleshy, globoso-campanulate, then flattened, obtuse, *flocculoso-mealy when young*, then slightly silky even or squamulose, dry. St. 7–12 × 2·5–3 cm., *white, often stained with yellow*, thickened at the base, obsoletely marginato-bulbous when young, villose. Ring *white, superior*, large, pendulous, *formed as it were of two growing together*, the interior one membranaceous, uniform, the *exterior one thicker and shorter, somewhat free at the circumference, often appendiculate at the margin of the p., radiately split*. Gills *white, at length reddish fuscous*, free, *approximate*, ventricose, broader in front, always arid. Flesh *white, or tinged with yellow*, compact, firm, juicy, at length softer. Spores brownish purple, elliptical, 8–10 × 5–6μ. Smell pleasant, often like new meal. Taste mild. Edible. Often forming large rings. Pastures, and woods. May—Dec. Common. (*v.v.*)

var. **purpurascens** Cke. Cke. Illus. no. 541, t. 584.
Purpurascens, becoming purple.

Differs from the type in its *smaller size and in the p. becoming tinged with purple*. Spores reddish brown, broadly elliptical, 4–5 × 3μ, 1-guttulate. Woods, pastures and under trees. July—Oct. Uncommon. (*v.v.*)

var. **albosquamosa** W. G. Sm. Field and cultivated mushrooms, fig. 8.
Albus, white; *squamosa*, scaly.

Differs from the type in the *snow-white patches on the fawn-coloured p.—the remains of the universal veil*. Artificially made mushroom beds. Not uncommon.

var. **vaporaria** (Otto) W. G. Sm. (= *Psaliota campestris* (Linn.) Fr. var. *vaporaria* (Otto) Fr.) *Vaporaria*, of hothouses.

Differs from the type in the *smooth, pilose brown p., and the white st., reddish at the apex, and brownish, or reddish at the base*. A cultivated form in mushroom beds. Common. (*v.v.*)

var. **hortensis** (Cke.) W. G. Sm. (= *Psaliota campestris* (Linn.) Fr. var. *hortensis* Cke.) Cke. Illus. no. 545, t. 527, as *Psaliota campestris* Linn. var. *hortensis*. *Hortensis*, of gardens.

Differs from the type in the *fibrillose, or squamulose brownish p.* Artificially made mushroom beds. Common. (*v.v.*)

var. **Buchananii** (Berk.) W. G. Sm. (= *Psaliota campestris* (Linn.) Fr. var. *Buchanani* Berk.) Field and cultivated mushrooms, fig. 10.
Buchanan, a gardener.

Differs from the type in the *white, opaque, nearly smooth, depressed p.* Artificially made mushroom beds.

var. **cryptarum** (Letell.) W. G. Sm. (= *Psaliota campestris* (Linn.) Fr. var. *cryptarum* (Letell.) Fr.) κρύπτη, a cave.

Differs from the type in the *greyish white, or brownish white p.* A cultivated form in caves.

var. **intermedia** W. G. Sm. Field and cultivated mushrooms, fig. 5. *Intermedia*, intermediate.

Differs from the type in the *pale, livid brassy-yellow p., with small rusty-yellowish spots at the middle, in the very short, obese, dull whitish, stained pale rusty st., and in the white flesh, becoming pale vinous-brown.* Amongst rank grass in fields, often near trees.

var. **epileata** W. G. Sm. Field and cultivated mushrooms, fig. 16. *E*, without; *pileatus*, having a cap.

Differs from the type in the *almost or quite obsolete p. and in the white, ringless st. inflated below.* It is really an aborted form and unworthy of a varietal name. Artificially prepared mushroom beds. Common. (*v.v.*)

var. **obesa** W. G. Sm. Field and cultivated mushrooms, fig. 15. *Obesa*, stout.

Differs from the type in the *ventricose st. equalling, or exceeding the width of the p., in the very narrow gills, and in the white flesh sometimes changing to deep mahogany brown.* This is a monstrous form and unworthy of a varietal name. Artificially prepared mushroom beds. Often common. (*v.v.*)

188. **P. xanthoderma** Genev. (= *Pratella cretacea* Quél. sec. Maire.) Roze et Richon, t. 17, figs. 5–8. ξανθός, yellow; δέρμα, skin.

P. 8–12 cm., *white, then somewhat tawny, becoming stained with yellow, especially when touched or rubbed*, fleshy, campanulate, then convex, at length expanded, silky. St. 8–12 × 2–3 cm., *white, becoming yellow where touched or bruised*, attenuated at the apex, more or less bulbous at the base, silky. Ring *white, often stained with yellow at the margin.* Gills *white, then pink, cinereous, violet, or brownish*, free, crowded. Flesh *white, becoming yellow especially* at the base of the st., and under the cuticle of the p. and st. Spores brownish purple, pip-shaped, 6 × 4μ. Smell and taste *unpleasant, almost foetid.* Poisonous for some persons. Woods, pastures, and hedgerows. July—Nov. Not uncommon. (*v.v.*)

var. **lepiotoides** René Maire. Cke. Illus. no. 542, t. 524, as *Psaliota cretacea* Fr. *Lepiota*, the genus Lepiota; εἶδος, like.

Differs from the type in the *p. greyish white at first, then covered with greyish brown squamules, larger and denser at the disc, separated by whitish cracks, and finally becoming tinged with reddish purple.* Pastures, and stoves. Feb.—Oct. Rare.

189. **P. flavescens** Gillet. Trans. Brit. Myc. Soc. III, t. 16.
Flavescens, becoming yellow.

P. 5–12 cm., *white, at once turning saffron colour, then finally light brown when touched or bruised*, campanulate, then expanded, smooth, dry, *shining with a satin-like sheen*, pellicle easily separable. St. 10–14 × 1·5–2 cm., *white, with a satiny sheen*, tinged reddish yellow at the base on one side, cylindrical. Ring *dirty white, yellow on the outside, and more deeply coloured at the margin*, membranaceous, soon disappearing. Gills *pale pink, then darker, and finally brownish*, free, crowded. Flesh *white, turning instantly bright saffron yellow when fresh, and reddish yellow when drier*, especially near the cuticle of the p. and at the base of the st. Spores reddish brown, oval, 5–6 × 4–5μ, 1-guttulate, with an apical germ-pore. Smell none. Taste not disagreeable. Poisonous for some persons. Solitary, or in rings. Pastures, and fir woods. July—Nov. Not uncommon. (*v.v.*)

P. cretacea Fr. = **Lepiota naucina** Fr.

190. **P. perrara** Schulz. (=*Psaliota augusta* Fr. sec. Maire.) Bres. Fung. Trid. t. 89. *Perrara*, very uncommon.

P. 6–14 cm., *yellow, covered with dense, imbricate,* Lepiota-*like, fulvous scales*, fleshy, campanulate, then expanded. St. 9–11 × 1·5–3 cm., *whitish, covered below the ring with evanescent, fulvous scales*, incrassated at the base. Ring *white, becoming discoloured*, squamosely floccose on the under side, large, superior, reflexed. Gills *white, then rosy, and at length fuscous*, free, often very remote, equally attenuated at both ends, 5 mm. broad, crowded. Flesh *white, becoming yellowish in the st. when broken*, soft. Spores purplish fuscous, obovate, 7–9 × 4–5μ, 1-guttulate. Oak woods, and under oaks. Aug.—Oct. Uncommon. (*v.v.*)

191. **P. pratensis** (Schaeff.) Fr. Cke. Illus. no. 543, t. 525.
Pratensis, growing in meadows.

P. 5–9 cm., *whitish, becoming cinereous*, fleshy, ovoid, then expanded, obtuse, silky, villous under a lens, *becoming rimosely squamulose*, dry. St. 5–8 × 1 cm., *white*, equal, slightly incrassated at the base, firm, smooth. Ring *white*, membranaceous, median, deciduous. Gills *cinereous, then fuscous*, free, approximate, *rounded behind, acutely attenuated in front*. Flesh *white*, thick, firm. Spores brown, elliptical, or pip-shaped, 5–6 × 3μ, 1-guttulate. Smell and taste pleasant. Edible. Woods, pastures, and hedgerows. July—Oct. Uncommon. (*v.v.*)

192. **P. campestris** (Linn.) Fr. *Campestris*, belonging to a plain.

P. 5–12 cm., *white, or rufescent*, fleshy, lens-shaped-convex, then flattened, obtuse, dry, *silky-even, or squamulose*. St. 4–8 × 2–4 cm., *white*, firm, bulbous when young, then somewhat equal, even, or squamulose. Ring *white*, membranaceous, rarely in the form of a cortina, median, or more strictly sheathed to the middle, spreading, or reflexed, torn, often fugacious. Gills *whitish, then soon flesh coloured*, and at *length umber-fuscous*, free, approximate, ventricose, *equally attenuated at both ends*, crowded, often deliquescent. Flesh *white, becoming reddish, or sometimes fuscous*, thick, soft. Spores brownish purple, broadly elliptical, 6–7 × 5–5·5μ. Smell and taste pleasant. Edible. Pastures, and heaths, rarely in woods. May—Dec. Common. (*v.v.*)

var. **alba** Viv. Berk. Outl. t. 10, fig. 2. *Alba*, white.

Differs from the type in the *white, silky pileus and short st.* Hardly worthy of a varietal name. Generally found along with the type. (*v.v.*)

var. **praticola** (Vitt.) Fr. Vitt. t. 7. *Praticola*, living in meadows.

Differs from the type in the *rufous-scaly p., and in the flesh becoming immediately rufescent.*

var. **subvolvacea** W. G. Sm. Field and cultivated mushrooms, fig. 13. *Sub*, somewhat; *volvacea*, having a volva.

Differs from the type in the *pale brown p. breaking up into dark umber scales, and in the long pale brownish st. furnished with a thin brown volva at the base.* Fields, and artificially made mushroom beds.

var. **rufescens** Berk. Berk. Outl. t. 10, fig. 3. *Rufescens*, becoming reddish.

Differs from the type in the *rufous, minutely squamulose p., the elongated st., and in the bright rose, sometimes crimson flesh when cut.*

var. **umbrina** (Vitt.) Fr. Vitt. t. 8. *Umbrina*, umber.

Differs from the type in the *umber p. becoming even, and in the stout, squamulose st.*

var. **fulvaster** Viv. Viv. t. 45, upper fig. *Fulvaster*, yellowish.

Differs from the type in the *ochraceous tawny p., and in the rose coloured gills becoming blackish.*

var. **costata** (Viv.) Fr. Cke. Illus. no. 546, t. 528, fig. A. *Costata*, ribbed.

Differs from the type in the *sulcate, repand p.* Woods. Rare.

var. **elongata** Berk. Field and cultivated mushrooms, fig. 3.
Elongata, elongated.

Differs from the type in the *even, shining white p., in the margin permanently appendiculate with the veil, and in the long, bulbous st.* Scarcely worthy of a varietal name. Pastures under trees. Uncommon. (*v.v.*)

var. **exannulata** Cke. Cke. Illus. no. 546, t. 528, fig. B.
Exannulata, without a ring.

Differs from the type in the *evanescent, or obsolete ring*. Scarcely worthy of a varietal name. Pastures. Occasionally. (*v.v.*)

193. **P. sylvicola** (Vitt.) Fr. Cke. Illus. no. 547, t. 529, as *Psaliota campestris* Linn. var. *sylvicola* Vitt.
Sylvicola, inhabiting woods.

P. 7–11 cm., *white, or yellowish*, fleshy, globose, then convexo-expanded, silky, *becoming even, shining*; margin often appendiculate with the partial veil. St. 10–15 × 1–1·5 cm., *concolorous*, slightly attenuated upwards from the *subbulbous base*, smooth. Ring *concolorous*, membranaceous, large, reflexed. Gills *whitish, then slowly becoming fuscous*, free, acute behind. Flesh *whitish, at length becoming brownish*, thin at the margin. Smell and taste pleasant. Edible. Woods and shrubberies. Aug.—Nov. Common. (*v.v.*)

194. **P. exserta** (Viv.) Rea. Trans. Brit. Myc. Soc. III, t. 15.
Exserta, thrust out.

P. 6–18 cm., *white, becoming yellowish ochraceous and broken up into minute adpressed scales*, fleshy, campanulate, then convexo-expanded. St. 10–15 × 3–6 cm., *white*, either slightly attenuated upwards from the base, or ventricose at the middle, *bleeding when cut or wounded*, almost smooth. Ring *white, covered on the underside with yellowish, fugacious warts*, membranaceous, large, thick, *double*, made up of two layers that split apart. Gills *whitish, then pinkish, and finally fuscous*, free, 5–10 mm. broad, somewhat crowded. Flesh *white, immediately turning bright red when bruised, cut, or wounded, and exuding a bright red juice which finally stains the part affected deep brown*. Spores deep ochre when deposited in the mass, subglobose, 5–6 × 4–5μ, 1-guttulate, with an apical germ-pore. Smell and taste pleasant. Edible. Solitary, or in rings. Pastures. May—Nov. Uncommon. (*v.v.*)

195. **P. villatica** (Brond.) Magn. Cke. Illus. no. 538, t. 521, as *Psaliota augusta* Fr. *Villa*, a country house.

P. 10–40 cm., *pale brown, sometimes with a yellowish tinge*, fleshy, globose, then expanded, very obtuse, disc even, *minutely fibrillose*, *adpressedly silky, squamose towards the paler margin*, the scales appearing as if they had been pressed down with a hot iron. St. 10–20 × 3–5 cm., *white, becoming tinged with brown especially at the base*, slightly attenuated upwards from the incrassated base, becoming smooth. Ring *white above, yellowish and floccosely scaly on the under side*, membranaceous, soft, median, thick, reflexed. Gills *pallid, then fuscous cinereous*, free, attenuated behind, 10–15 mm. broad, crowded. Flesh *white, becoming reddish brown when cut, especially under the cuticle of the pileus and at the base of the st.*, compact, thick at the disc, thin at the margin. Spores rich brown, elliptical, 7–9 × 5–6μ, with an apical germ-pore. Smell unpleasant. Taste mild. Edible. Pastures and gardens. July—Oct. Uncommon. (*v.v.*)

196. **P. sylvatica** (Schaeff.) Fr. Bres. Fung. Trid. t. 90.
Sylvatica, of woods.

P. 7·5–11 cm., *subferruginous, scales rufescent, or becoming fuscous*, fleshy, oval, then campanulate and flattened, *often somewhat umbonate, the whole surface floccose, torn into squamules*, disc often remaining continuous, and at length denuded of scales; margin often rimosely incised. St. 6–9 × 1–1·5 cm., *dingy white*, at first stuffed with a cylindrical, separate, white pith, *equal*, or bulbous at the base, the bulb sometimes marginate, fibrillose below the ring, smooth above. Ring *white, distant*, floccose on the underside, sometimes wide, thin and membranaceous, sometimes narrow, incomplete, fugacious. Gills *white, then reddish, at length cinnamon fuscous, or umber fuscous*, free, ventricose, *equally attenuated at both ends*, thin, arid, crowded. Flesh *white, generally rufescent, often yellowish at the apex of the st.*, thin, fragile. Spores tawny flesh colour, elliptical, 6–7 × 3·5–4μ. Smell pleasant, or strong. Taste mild. Edible. Woods, and under cedars. July—Sept. Not uncommon.

197. **P. haemorrhoidaria** Kalchbr. (= *Pratella sylvatica* Schaeff. sec. Quél.) Kalchbr. Icon. t. 18, fig. 1. *αἱμορροΐδες*, hemorrhoids.

P. 5–12·5 cm., *rufous fuscous, or brownish*, fleshy, ovate, then expanded, *covered with broad, adpressed, darker scales*; *margin at first incurved*. St. 8–12 × 2–3 cm., *white, becoming blood red when bruised*, equal, often more or less bulbous at the base, silky, fibrillose. Ring *white, becoming discoloured*, large, persistent, superior, membranaceous. Gills *rosy flesh colour, then purple umber*, free, approximate, 6–12 mm. broad, crowded. Flesh *white, immediately turning blood red when broken*, thick. Spores purple-brown, elliptical, 6–7 × 3–4μ. Smell and taste pleasant. Edible. Woods, especially coniferous woods, pastures, and under conifers. Aug.—Jan. Common. (*v.v.*)

198. **P. setigera** Fr.[1] (= *Pratella sylvatica* Schaeff. sec. Quél.) Paul. t. 132, figs. 3–4. *Setigera*, having coarse hairs.

P. *pale umber*, fleshy, convex, then plane, obtuse, smooth, silky. St. *covered with pale umber, pilose squamules*, equal. Ring thin, fugacious. Gills *fuscous umber*, free. Woods.

199. **P. rubella** (Gillet) Rea (= *Pratella sylvatica* Schaeff. sec. Quél.) Gill. Hym. Fr. t. 102, as *Pratella rubella* Gillet. *Rubella*, reddish.

P. 4–8 cm., *entirely covered with red fibrils, or scales, disc red brown, paler towards the margin*, convex, or obtusely umbonate. St. 5 cm., *white, or whitish, becoming stained with blood red like the p., slightly subbulbous at the base*, cartilaginous, fibrillose. Ring fugacious. Gills *rosy flesh colour, then brownish purple*, free, slightly ventricose, crowded. Flesh *white, becoming blood red*, firm. Spores "4·75–6 × 3–4, generally 5 × 3·5" Sacc. Under conifers. Sept.—Oct. Rare.

**Smaller, p. thinly fleshy.

200. **P. comtula** Fr. Fr. Icon. t. 130, fig. 1. *Comtula*, adorned.

P. 3–5 cm., *yellowish white, disc often tawny*, fleshy, convex, then plane, obtuse, *adpressedly fibrillosely silky*. St. 3–5 cm. × 4–6 mm., *white, becoming somewhat light yellow, somewhat attenuated at the base, apex striate, satiny*. Ring *concolorous*, membranaceous, median, torn, fugacious, very thin. Gills *flesh colour, then rose, and finally fuscous flesh colour*, free, rounded behind, broader in front, crowded. Flesh *white, or slightly yellowish*, thin, soft. Spores purple fuscous, elliptical, 5 × 3μ, 1-guttulate. Smell and taste strong of anise. Edible. Woods, heaths and pastures. May—Oct. Common. (*v.v.*)

201. **P. amethystina** Quél. Roz. & Rich. t. 18, figs. 1–5. *Amethystina*, amethyst colour.

P. 3–5 cm., *white, becoming either rose, lilac, or amethyst coloured from the centre outwards*, fleshy, convex, then expanded, umbonate, villose, or fibrillose. St. 3–6 cm. × 5–10 mm., *white*, subbulbous at the base, fragile, glabrous. Ring *white*, thin, satiny. Gills *light grey, then bay brown*, free, ventricose, 6–7 mm. broad, crowded. Flesh *white*, thin. Spores brownish purple, roundish oblong, 5–7 × 4μ, 1–2-guttulate. Smell like that of *Psaliota sylvicola*. Taste pleasant. Edible. Woods, and pastures. Aug.—Oct. Uncommon. (*v.v.*)

202. **P. subgibbosa** Fr. *Sub*, somewhat; *gibbosa*, humpbacked.

P. 2–2·5 cm., yellowish, fleshy, convexo-plane, subumbonate, smooth; margin silky fibrillose. St. 2·5 cm. × 2–4 mm., pallid. Ring cortinate, fugacious. Gills *white, then cinereous fuscous*, free, remote. Spores fuscous. Fir woods. Rare.

[1] This is listed as British by Massee in his Eur. Fung. Fl. Agar. 207.

203. **P. sagata** Fr. Cke. Illus. no. 1177, t. 968.
Sagata, clothed in a mantle.

P. 3–5 cm., *yellowish tawny, or reddish brown*, fleshy, convex, then plane, at length revolute, obtuse, *smooth, subpelliculose*, shining. St. 5 cm. × 5–6 mm., *yellowish*, equal, at length compressed, fragile, smooth. Ring *white, distant, entire*, persistent. Gills *pinkish, then umber*, free, ventricose, 6 mm. broad, crowded. Flesh *white*, thin. Spores purplish umber, elliptical, 6 × 3–4μ. Cystidia "on edge of gill basidia-like, 36–40 × 8–12μ" Rick. Grassy places, and under beeches. Oct. Rare.

204. **P. dulcidula** Schulz. Kalchbr. Icon. t. 17, fig. 1.
Dulcidula, sweetish.

P. 2·5–5 cm., *lurid white, or ochraceous, disc subfuscous, or dirty violaceous*, fleshy, convex, then plane, slightly gibbous, somewhat smooth, *dry*. St. 2·5–4 cm. × 4–6 mm., *concolorous*, fragile, base incurved, subbulbous, almost smooth. Ring *concolorous*, membranaceous, *median, erect, persistent*. Gills *pallid greyish, then black*, free, widest in front, 4 mm. broad, rather crowded. Flesh *white*, thin at the margin. Smell pleasant. Taste sweetish. Under oaks, and on heaths. Oct. Rare.

205. **P. rusiophylla** (Lasch) Fr. *Russus*, red; *φύλλον*, a leaf.

P. 2–3·5 cm., *flesh colour, or ruddy, becoming pale*, fleshy, campanulate, then expanded, *umbonate*, fibrillose; margin somewhat appendiculate with the partial veil. St. 2·5–4 cm. × 2–3 mm., *white*, slightly attenuated from the thickened base, fibrillose. Ring superior, reflexed, persistent. Gills *rosy, then fuscous*, free, crowded. Flesh *pallid*, thin at the margin. Spores reddish, elliptical, 5 × 3μ. Frondose woods, and parks. Oct. Rare.

P. haematosperma (Bull.) Fr. = **Lepiota haematosperma** (Bull.) Boud.

P. echinata (Roth) Fr. = **Lepiota haematosperma** (Bull.) Boud.

Spores black, or blackish fuscous.

Anellaria Karst.

(*Anellus*, a little ring.)

Pileus fleshy, campanulate. Stem central. Ring membranaceous, persistent, or fugacious. Gills adnate, or often almost free. Spores black, or blackish fuscous, pip-shaped, or elliptical, smooth, with an apical germ-pore. Growing on dung, or on the ground.

206. **A. separata** (Linn.) Karst. (= *Panaeolus separatus* (Linn.) Fr.) Cke. Illus. no. 623, t. 623, as *Panaeolus separatus* Fr.
Separata, distinct.

P. 2–6 cm., *clay whitish, or yellowish,* fleshy, ovato-campanulate, 2·5–4·5 cm. high, not expanding, obtuse, *viscid,* smooth, often wrinkled, or cracked when old; margin often appendiculate with the veil. St. 5–20 cm. × 4–8 mm., *whitish,* tense and straight, *rigid,* gradually attenuated upwards from the *thickened base,* striate under a lens, smooth. Ring *white,* membranaceous, *distant, entire,* narrow, persistent, often striate. Gills *whitish, then cinereous black,* adnate, but almost separating, ascending, 4–8 mm. broad, *edge often whitish.* Flesh *whitish, yellowish under the cuticle, and towards the base of the st.,* thick at the disc. Spores black, pip-shaped, 16–20 × 10–12μ. Cystidia bottle-shaped, 30–40 × 8–14μ × 5–8μ at apex. On dung, especially that of horse. Woods, fields, and gardens. April—Dec. Common. (*v.v.*)

207. **A. fimiputris** (Bull.) Karst. (= *Panaeolus fimiputris* (Bull.) Fr.) Cke. Illus. no. 626, t. 626, as *Panaeolus phalenarum* Bull.

Fimus, dung; *putris,* rotten.

P. 2–3 cm., *fuliginous-cinereous, or livid,* fleshy, conical, then expanded, somewhat gibbous, *viscid,* smooth, generally beaded with the veil. St. 5–10 cm. × 2 mm., *pallid, equal,* smooth, *girt with an annular zone above the middle.* Gills *livid blackish,* adfixed. Flesh *whitish,* thin. Spores black, "elliptical, apiculate, 9–10 × 6μ" Massee. On dung. Fields, roadsides, and gardens. April—Nov. Common. (*v.v.*)

208. **A. scitula** Massee. Cke. Illus. no. 625, t. 927, fig. B, as *Panaeolus scitulus* Massee. *Scitula,* pretty.

P. 1–1·5 cm., *dirty pale ochre,* fleshy, campanulate, obtuse, smooth, *viscid,* margin exceeding the gills. St. 2–3 cm. × 1–2 mm., *white,* equal, shining, *base peronate, sheath ending in a persistent ring below the middle of the st.* Gills *becoming ashy grey, speckled with the black spores,* almost free, narrow, crowded. Flesh *white,* thin. Spores black, opaque, with a colourless hilum, elliptical, 12–13 × 4μ. On soil in a flower-pot. Rare.

***With a volva at the base of the stem.

Spores white.

Amanitopsis Roze.

(*Amanita,* the genus; ὄψις, like.)

Pileus fleshy, regular. Stem central, fleshy. Volva membranaceous, free, lax, sheathing. Gills free, or adnate. Spores white, globose, subglobose, oblong elliptic, smooth, continuous. Growing on the ground.

209. **A. vaginata** (Bull.) Roze. Gonn. & Rabenh. t. 7, fig. 1.

Vagina, a sheath.

P. 3–10 cm., *livid, or mouse grey, covered with large, white, or grey, fugacious patches of the fragments of the volva,* slightly fleshy, campanulate, then flattened, obtuse, slightly viscid at first; margin *deeply striate.* St. 12–15 × 1–1·5 cm., *white, or grey,* floccose, slightly attenuated upwards, surrounded at the base by a *large, free, lax, often lobed, white, or grey membranaceous volva,* often inclosing a ring-like mark around the stem. Gills *white, or greyish,* free, ventricose. Flesh *white,* thin. Spores globose, 10–12μ, with a large central gutta. Taste pleasant. Edible. Woods, heaths, and pastures. June—Nov. Common. (*v.v.*)

210. **A. fulva** (Schaeff.) W. G. Sm. Boud. Icon. t. 7. *Fulvus,* tawny.

P. 4–10 cm., *tawny,* disc deeper coloured, campanulate, then flattened, *umbonate,* slightly viscid, *covered with a few, fugacious patches of the yellowish volva;* margin *striate.* St. 7–20 × ·5–1 cm., *paler tawny,* squamulose, base *surrounded by the upright, lax, free yellowish, membranaceous volva.* Gills *white, tinged with yellow,* free. Flesh *white, yellow under the epidermis.* Spores globose, 9–14μ, multi-guttulate. Taste pleasant. Edible. Woods, heaths, etc., especially under birch trees. May—Nov. Common. (*v.v.*)

211. **A. nivalis** (Grev.) Rea. Grev. Scot. Crypt. Fl. t. 18. *Nivalis,* snow coloured.

P. 5–9 cm., *white, disc pale ochraceous, covered at first with the very fugacious, white fragments of the volva,* campanulate, then convex and plane, or slightly umbonate; margin striate. St. 7–13 × 1 cm., *white, the subbulbous base surrounded by a white, lax, free, membranaceous volva.* Gills *white,* free, broader in front. Flesh *white,* thin. Spores oblong elliptic, 11–12 × 9μ, with a large central gutta. Taste pleasant. Edible. Woods, and heaths. Aug.—Oct. Uncommon. (*v.v.*)

212. **A. strangulata** (Fr.) Roze. Boud. Icon. t. 9, as *Amanitopsis inaurata* (Secr.) Boud. *Strangulata,* choked.

P. 8–15 cm., *bright tawny, or tawny brown, covered with numerous large, grey, patches of the fragments of the volva,* convex, then plane, slightly viscid; margin deeply striate. St. 12–30 × 3–4 cm., *greyish white,* stout, attenuated upwards, *encircled by one to three greyish rings* on the lower half—the *remnants of the friable volva which disintegrates at the base.* Gills *white, or tinged yellowish,* adnate, crowded, ventricose. Flesh *white, tinged slightly yellowish under the cuticle.* Spores globose, 8–13μ, multi-guttulate. Taste pleasant. Edible. Woods, and pastures, chiefly on the chalk and limestone. May—Oct. Not uncommon. (*v.v.*)

213. **A. adnata** (W. G. Sm.) Sacc. (= *Amanita junquillea* Quél.) Saund. & Sm. t. 20. *Adnata,* adnate.

P. 6–7 cm., *pale yellowish buff, covered with white, woolly patches of the volva,* fleshy, very firm, convex, then expanded; margin *exceeding the gills.* St. 5–10 × 1·5 cm., *pale buff,* fibrillose, base slightly swollen and covered by the *adnate volva*, which has only a small, free, lax margin, sometimes almost obsolete. Ring generally absent. Gills white, truly adnate, crowded. Flesh *white, buff beneath the epidermis*. Spores subglobose, 7–9 × 6–7μ, with a large central gutta. Woods, and heaths. July—Oct. Uncommon. (*v.v.*[1])

Spores pink.

Volvaria Fr.

(*Volvaria*, having a volva, or wrapper.)

Pileus fleshy, regular. Stem central. Volva membranaceous, free, sheathing. Gills free. Spores pink, elliptical, or subglobose, smooth, continuous. Growing on the ground, or on wood.

*P. dry, silky, or fibrillose.

214. **V. bombycina** (Schaeff.) Fr. (= *Volvaria Loveiana* Berk. sec. Barb.) *Bombycina*, silky.

P. 7–20 cm., *white, becoming yellowish*, fleshy, soft, globose, then campanulate and convex, subumbonate, *silky, becoming squamuloso-villose*, disc rarely becoming smooth. St. 7–15 × 1–2 cm., *white*, attenuated upwards, *base bulbous*, often curved. Volva *whitish*, becoming discoloured, soon torn asunder, ample, 3–8 cm. broad, membranaceous, lax, laciniate, somewhat viscid, persistent. Gills *white, then flesh colour*, free, very crowded, ventricose, becoming toothed. Flesh *white, becoming yellowish*. Spores pink, elliptical, 6–7 × 4–5μ. Cystidia sparse, clavate, often slightly constricted near the apex, 55–60 × 18μ. Smell and taste pleasant. On decayed wood, stumps and sawdust. June—Sept. Uncommon. (*v.v.*)

215. **V. volvacea** (Bull.) Fr. (= *Volvaria Taylori* Berk. sec. Quél.) Cke. Illus. no. 294, t. 294. *Volvacea*, having a volva.

P. 7–10 cm., *cinereous, black-streaked with adpressed fibrils*, campanulate, then expanded, obtuse. St. 5–12·5 × 1 cm., *white, villose*, somewhat equal. Volva *whitish, fuscous at the apex*, oval, membranaceous, lax, often adpressed to the stem. Gills *white, then flesh colour*, free, ventricose. Flesh *white*, floccose. Spores pink, elliptical, 6–7 × 4–5μ, 1-guttulate. Gardens, in stoves, roadsides. July—Oct. Uncommon. (*v.v.*)

216. **V. Loveiana** Berk. (= *Volvaria plumulosa* (Lasch) Quél.; *Volvaria hypopitys* Fr. sec. Quél.; *Volvaria bombycina* Schaeff. sec. Barb.) Berk. Outl. t. 7, fig. 2. Rev. R. T. Lowe

P. 5–7 cm., *white, with a very slight shade of pink, or cinereous*, subtruncato-globose, then convex, or slightly expanded, *beautifully silky*;

[1] At Sandringham, Norfolk, on the 30th October, 1899, specimens were found both with a well-defined membranaceous ring and without any trace of a ring. C. R.

margin involute. St. 5 × ·5–1 cm., *pure white*, bulbose, attenuated upwards, *closely fibrillose, with a little matted down*, very juicy. Volva *pure white*, with a little downy prominence within round the base of the stem. Gills *white, becoming gradually pale pink*, free, broad in front, subdeliquescent. Flesh *white, becoming yellowish*. Spores pink, elliptical, 5–6 × 3–4μ, 1-guttulate. On *Clitocybe nebularis* (Batsch) Fr. Oct. Rare. (*v.v.*)

217. **V. Taylori** Berk. (= *Volvaria volvacea* Bull. sec. Quél.) Cke. Illus. no. 296, t. 296. M. A. Taylor.

P. 2–5 cm., *livid, conico-campanulate*, obtuse, *striato-rimose* from the apex, thin; margin lobed, sinuate. St. 3–6 cm. × 3–5 mm., *pallid*, nearly equal, slightly bulbose at the base. Volva *date-brown*, lobed, somewhat lax, small. Gills *rose colour*, free, broad in front, *very much attenuated behind*, uneven, *edge floccose, white*. Spores pink, elliptical, 7–8 × 5μ, 2-guttulate. Gardens. July—Oct. Uncommon. (*v.v.*)

218. **V. temperata** B. & Br. Cke. Illus. no. 300, t. 300, upper fig. *Temperata*, moderate.

P. 6 mm., *whitish, disc tinged tawny*, convex, then expanded, umbonate, *pulverulent*, striate. St. 1·5–2·5 cm. × 1–2 mm., *whitish*, pellucid. Volva *white*, ample. Gills *pinkish*, free. Spores pink, elliptical, 4 × 2·5μ. Greenhouses. Feb. Rare.

**P. more or less viscid, smooth.

219. **V. speciosa** Fr. (= *Volvaria gloiocephala* DC. sec. Dumée.) Boud. Icon. t. 84. *Speciosa*, handsome.

P. 7–13 cm., *whitish, subumbonate disc grey, or umber*, fleshy, globose, then campanulate, at length plane, *viscid*. St. 10–20 × 1–2·5 cm., *white*, firm, slightly attenuated upwards, *base white-villose when young*. Volva *white*, membranaceous, bulbous, free, variously torn into loops, *externally tomentose*. Gills *white, then flesh colour*, free, ventricose. Flesh *white*, floccose. Spores pink, elliptical, 15–16 × 8–10μ, 1–2-guttulate. Cystidia "vesiculose-pyriform, 60–70 × 20–36μ, sometimes pointed" Rick. Smell none, or somewhat strong. Edible sec. Maire. Dunghills, roadsides, and occasionally in woods. June—Oct. Uncommon. (*v.v.*)

220. **V. gloiocephala** (DC.) Fr. (= *Volvaria speciosa* Fr. sec. Dumée.) Cke. Illus. no. 298, t. 298. γλοιός, sticky; κεφαλή, head.

P. 7–11 cm., *fuliginous*, fleshy, campanulate, then expanded, *umbonate, glutinous*; margin striate. St. 8–18 × 1–2 cm., *white, becoming fuscous, or tawny*, attenuated upwards, base subbulbose and villose. Volva *whitish, grey, or fuscous*, circularly split, lobed, *villose*, often adpressed to the stem. Gills *white, then reddish*, free, broad, especially in front, attenuated behind, margin slightly toothed. Flesh *white*, *fuscous under the cuticle of the pileus.* Spores pink, elliptical, 12 × 7μ. Smell and taste unpleasant. Poisonous. On the ground. June—Nov. Uncommon. (*v.v.*)

221. **V. viperina** Fr.[1] (= ?*Volvaria conica* (Pico.) Quél.; *Volvaria speciosa* Fr. sec. Maire.) *Viperina*, of a snake.

P. 3–4 cm., *grey, or cinereous,* fleshy, *persistently conical, acute,* viscid, silky shining when dry. St. *white,* equal, subflexuose. Volva *thin, entire, closely sheathing.* Gills *tinged yellowish then flesh colour.* Spores pink, "6–8 × 4–4·5μ" Herpell.

222. **V. media** (Schum.) Fr. Cke. Illus. no. 299, t. 299. *Medius*, middle.

P. 3–5 cm., *white, disc brownish, or yellowish,* slightly fleshy, convexo-plane, obtuse, *viscid, silky when dry and shining.* St. 4–7 cm. × 4–6 mm., *white, subbulbose,* equal. Volva *white,* membranaceous, *sheathing,* lobed. Gills *white, then rosy flesh colour,* free, broad in front, attenuated behind. Flesh *white.* Spores pink, elliptical, 5–6 × 4μ. Woods, and pastures. Aug.—Oct. Not uncommon. (*v.v.*)

223. **V. parvula** (Weinm.) Fr. (= *Volvaria pusilla* (Pers.) Quél.) Boud. Icon. t. 86. *Parvulus*, very small.

P. 1–3 cm., *whitish, disc yellowish,* slightly fleshy, conical, then campanulate, at length rather plane and *umbonate,* at first slightly viscid, soon dry *silky.* St. 2·5–4 cm. × 2–4 mm., *white,* equal, *silky,* base villose. Volva *white,* membranaceous, *free,* lobed, *minutely tomentose on the outside.* Gills *white, then flesh colour,* free, broad in front. Flesh *white.* Spores pink, elliptical, 5 × 3μ, 1–2-guttulate. Pastures, gardens, and woods. May—Oct. Common. (*v.v.*)

var. **biloba** Massee (= *Volvaria parvula, forma B.* Fries Monogr.). *Bi*, two; λοβός, the lobe of the ear.

Entirely white when young. P. conical, 6–8 mm. high, dry, *sometimes floccosely squamulose.* St. 2·5 cm. × 1–2 mm., equal, *pubescent.* Volva *bilobed, sheathing, externally adpressedly silky.* Pastures. July—Sept. Not uncommon. (*v.v.*)

Spores ochraceous.

Locellina Gill.

(**Acetabularia** Berk.)

(*Locellus*, a casket.)

Pileus fleshy, regular. Stem central, thin. Volva membranaceous, sheathing. Gills free, or adnate. Spores ochraceous, or somewhat fuscous, oval, or oblong, smooth, continuous. Cystidia ventricose, pointed. Growing on the ground.

[1] This is listed as British by Massee in his Eur. Fung. Fl. Agar. 120.

224. **L. Alexandri** Gillet. Alexandre.

P. 2–3 cm., *yellowish-tan, disc darker*, convex, umbonate, *viscid*, pellicle easily separable; margin *appendiculate with the cinnamon fibrils of the arachnoid veil*. St. *white, or whitish*, equal, or slightly thickened at the base, flexuose, striate, *covered with a fibrillose, cinnamon veil* up to 1–2 cm. of the apex. Volva *white*, or *whitish, becoming reddish when handled*, irregular, torn at the edge. Gills *reddish flesh colour, paler at the edge*, adnato-decurrent, crowded. Flesh *whitish*. Spores "becoming fuscous, oblong" Big. & Guill. Woods, at the base of beeches. Oct. Rare.

225. **L. acetabulosa** (Sow.) Sacc. (= *Acetabularia acetabulosa* Berk.; *Pluteus semibulbosus* Lasch sec. Boud.) Sow. Eng. Fung. t. 303, as *Agaricus acetabulosus* Sow. *Acetabulum*, a vinegar cup.

P. 2–3 cm., *tan colour*, convex; margin plicate, or deeply striate. St. 4–5 cm. × 3 mm., *white*, equal. Volva *discoid, socket-like*. Gills *tawny*, free, lanceolate, 3 mm. broad. Flesh very thin. River bank, near high-water mark. May. Rare.

Spores purple, or fuscous.

Clarkeinda O. Kuntz.

(**Chitonia** Fr.)

(C. B. Clarke, 'Ινδός, pertaining to India.)

Pileus fleshy, regular. Stem central, thin. Volva membranaceous, sheathing. Gills free. Spores brownish purple, elliptical, smooth, with an apical germ-pore. Growing on the ground.

226. **C. rubriceps** (Cke. & Massee) Rea. Cke. Illus. no. 1176, t. 967, as *Chitonia rubriceps* Cke. & Massee. *Ruber*, red; *caput*, head.

P. 1·5–2·5 cm., *testaceous*, fleshy, campanulate, then expanded, umbonate, often becoming depressed round the umbo, smooth; margin arched, faintly striate. St. 7·5 cm. × 3–4 mm., *paler than the p.*, equal, smooth, rooting. Volva *whitish*, sheathing, saccate, torn at the margin. Gills *purplish brown*, free, lanceolate, narrow, rather crowded. Flesh *white*, fairly thick at the disc. Spores brownish purple, elliptical, a little attenuated at both ends, 12 × 6μ. On soil in Aroid house. Dec. Rare.

****With a ring on the stem, and a volva at the base of the stem.

Spores white.

Amanita (Pers.) Fr.

(Probably from Mount Amanus in Cilicia.)

Pileus fleshy, regular. Stem central, fleshy. Ring membranaceous, adnate, persistent, rarely fugacious. Volva membranaceous, free, or adnate, persistent, or friable. Gills free, subadnate, or decurrent by a tooth, intermediate gills *cut squarely behind.* Spores *white,* rarely tinged greenish, globose, subglobose, oval, or elliptical, smooth, very rarely verrucose. Cystidia subglobose, or cylindrical ventricose. Growing on the ground.

(*a*) Margin of volva free, persistent. P. generally naked.

227. **A. verna** (Lam.) Fr. Syst. (= *Amanita virosa* Fr. Hym. Eur.) Cke. Illus. no. 1, t. 1, as *Amanita virosa* Fr. *Ver*, spring.

Entirely white. P. 5–8 cm., fleshy, *conical, acute,* then campanulate, expanded, and *subumbonate, glutinous,* shining when dry; margin often *unequal, repand,* inflexed. St. 8–12 × 1·5–2 cm., cylindrical from the bulbose base, often compressed at the apex, split up into longitudinal fibrils, *floccosely squamulose.* Ring apical, lax, silky, splitting up into floccose fragments. Volva *thick, lax, wide.* Gills free, thin, linear-lanceolate, a little broader in front, crowded, edge often floccose. Spores white, globose, 7–8μ. Smell foetid. Taste unpleasant. Poisonous. Moist woods. Aug.—Oct. Uncommon. (*v.v.*)

var. **grisea** Massee. *Grisea*, grey.

Differs from the type in the *p. being shaded with grey.* Woods. Rare.

228. **A. phalloides** (Vaill.) Fr. (= *Amanita virescens* (Vaill.) Quél.) Rolland, Champ. t. 3, no. 3. φαλλός, Phallus; εἶδος, like.

P. 7–10 cm., *greenish, or yellowish olive, streaked with dark, innate fibrils,* fleshy, *ovato-campanulate,* then expanded, *obtuse, viscid,* rarely covered with one or two fragments of the volva. St. 8–12 × 1·5–2 cm., *white,* rarely besprinkled with olive or pale yellowish olive, adpressed squamules, smooth, or floccose, attenuated upwards, base bulbous. Ring *white,* superior, reflexed, slightly striate, swollen, generally entire. Volva free for half its depth, generally splitting up into three or four, more or less acute segments. Gills *white,* free, ventricose, 8 mm. broad. Flesh white. Spores white, subglobose, 8–11 × 7–9μ, with a large central gutta. Smell foetid when old. Taste unpleasant. Poisonous. Woods, and adjoining pastures. July—Nov. Common. (*v.v.*)

var. **verna** (Bull.) Fr.[1] Boud. Icon. t. 2, as *Amanita verna* (Bull.) Fr. *Verna*, occurring in spring.

Differs from the type in being entirely white. Spores white, globose, 7–8μ, with a large central gutta. July—Aug. Uncommon. (*v.v.*)

var. **umbrina** (Ferry) Maire.

Differs from the type in the *brownish umber p.,* and in the *fuscous, adpressed squamules on the st.* (*v.v.*)

[1] Boudier describes this as a distinct species with oval spores 10–14 × 7–9μ.

229. **A. porphyria** (A. & S.) Fr. Barla, Champ. t. 3, figs. 5–6. *πορφύρα*, purple.

P. 3–6 cm., *greyish bistre with a purplish tinge*, campanulate, then expanded, moist; margin rarely slightly striate. St. 7–9 × 1 cm., *white tinted with grey*, base bulbous. Ring *white, becoming fuscous*, distant. Volva *erect, white becoming fuscous*. Gills *white*, adnexed, crowded, thin. Flesh *white*. Spores white, globose, 9μ, multi-guttulate. Smell strong. Poisonous. Pine woods. July—Oct. Uncommon. (*v.v.*)

230. **A. lutea** Otth. *Lutea*, yellow

P. 4–7·5 cm., *yellow, or yellowish ochre, conical*, then expanded, viscid, *disc papillose, usually with broad scattered scales*; margin involute, *striate*. St. 7–8 cm., bulbous, rather narrowed upwards. Ring *white*, thin. Volva membranaceous. Gills *white*, crowded. Woods. Rare.

(*b*) Volva circumscissile, or fugacious. P. generally covered with fragments of the volva.

231. **A. recutita** Fr. Gonn. & Rabenh. I and II, t. 2, as *Amanita Secretanii* Rabenh. *Recutita*, circumcised.

P. 6–9 cm., *fuliginous, bistre, or brown, but without any tinge of purple, dry*, convex, then plane, generally covered with fragments of the volva, silky. St. 8–10 × 1·5–3 cm., *white*, silky, attenuated upwards, base bulbous. Ring *white*, distant. Volva *greyish bistre, closely sheathing, ending abruptly*. Gills *white*, adnexed with a decurrent line. Flesh *white*. Spores white, globose, 8–9μ. Smell slightly foetid. Poisonous. Pine, and birch woods. July—Oct. Uncommon. (*v.v.*)

A. junquillea Quél. = **Amanitopsis adnata** (W. G. Sm.) Sacc.

232. **A. mappa** (Batsch) Fr. (= *Amanita citrina* (Schaeff.) Quél.) Cke. Illus. no. 4, t. 4. *Mappa*, a napkin.

P. 6–9 cm., *white, or becoming yellow*, covered with patch-like fragments of the volva, slightly fleshy, *dry*, convexo-plane, obtuse, or depressed, orbicular. St. 5–8 × 1–1·5 cm., *white*, equal, base bulbous. Ring *white, yellowish on the exterior*, superior, soft, lax, minutely striate. Volva *yellowish*, or *fuliginous, obtuse, the friable upper portion disappearing and leaving a distinct groove round the base of the st.* Gills *white*, adnexed, crowded, narrow, edge often *yellowish*. Flesh *white, yellowish under the cuticle*. Spores white, subglobose, apiculate at the base, 8–10 × 7–9μ, with a large central gutta. Smell foetid. Poisonous. Woods, and heaths. July—Oct. Common. (*v.v.*)

var. **citrina** (Gonn. & Rabenh.) Rea. Gonn. & Rabenh. I and II, t. 4. *Citrina*, lemon yellow.

P. 8–12 cm., *bright yellow with white patches of the fragments of the volva*, convex, obtuse. St. 10–12 × 2 cm., *white*, stout. Volva imperfect. Spores white, "globose, *warted*, 6–7 μ" Massee. Rare.

var. **alba** (Gillet) Rea. Cke. Illus. no. 3, t. 3. *Alba*, white.

Differs from the type in being *white, and then becoming discoloured*. Woods. Sept.—Oct. Not uncommon. (*v.v.*)

(*c*) Volva floccose, or friable. P. floccose, or verrucose with the fragments of the volva, rarely naked.

233. **A. muscaria** (Linn.) Fr. Cke. Illus. no. 5, t. 117. *Musca*, a fly.

P. 10–20 cm., *scarlet, or orange*, covered with white, or yellowish fragments of the volva, fleshy, viscid, globose, then convex, and at length flattened; margin *slightly striate* when mature. St. 10–22 × 2·5 cm., *white*, or *yellowish*, firm, often torn into scales, apex striate, base *bulbous, encircled by several concentric rings* formed from the fragments of the volva. Ring *white, yellowish on the exterior*, superior, very soft, torn, somewhat striate. Gills *white, rarely becoming yellow*, free, but reaching the stem, crowded, thick, broader in front, minutely denticulate. Flesh *white, yellow under the epidermis*. Spores white, elliptical, apiculate, 8–10 × 6–7 μ. Taste mild. Poisonous. Birch, and coniferous woods, and under birches and conifers. July—Dec. Common. (*v.v.*)

var. **regalis** Fr. *Regalis*, royal.

Differs from the type in the *very glutinous, liver coloured p., and in the st. becoming light yellow internally*. Beech woods. Uncommon. (*v.v.*)

var. **formosa** Fr. Gonn. & Rabenh. I and II, t. 10, fig. 2. *Formosa*, handsome.

Differs from the type in the *lemon yellow p. covered with lax, mealy yellowish, fugacious fragments of the veil, and in the st. and ring often becoming yellow*. Beech woods. Uncommon. (*v.v.*)

var. **umbrina** Fr. *Umbrina*, umber.

Differs from the type in being *thinner, and more slender, and in the umber, or livid p., fuscous at the disc*. Woods. Rare.

var. **puella** (Batsch) Cda. Gonn. & Rabenh. I and II, t. 7, fig. 2. *Puella*, a girl.

Differs from the type in its *smaller size, and in the p. being destitute of any fragments of the volva*. Woods. Rare. (*v.v.*)

var. **aureola** (Kalchbr.) Quél. Kalchbr. Icon. t. 1, fig. 1.
Aureola, golden.

Differs from the type in the *erect, membranaceous volva*. Under birches. Uncommon. (*v.v.*)

234. **A. Emilii** Riel. Bull. Soc. Myc. Fr. XXIII (1907), t. 1.
Emile Boudier, the eminent French mycologist.

P. 13–17 cm., *yellowish butter colour, becoming tawny purplish*, disc finally *dark fuscous*, covered with *cream coloured fragments of the volva*, fleshy, viscid, convex, then hemispherical, and finally expanded, and depressed; margin *paler*, finally striate. St. 12–20 × 1·5–3 cm., *white*, bulbous. Ring *white*, thick, especially at the margin, and covered with the fragments of the cream coloured volva, crenulate, torn. Volva friable, forming three to four concentric rings round the apex of the globose, rarely fusiform base of the stem. Gills *whitish, or pale rose colour*, attenuated or rounded near the stem, broad, somewhat crowded, edge denticulate, floccose. Flesh *white, pale rose red under the epidermis*. Spores white, subglobose, 9–10μ, 1-guttulate. Taste *nutty*. Poisonous. Deciduous woods. Sept.—Oct. Uncommon.

235. **A. solitaria** (Bull.) Fr. (= *Amanita strobiliformis* (Vitt.) sec. Quél.) Boud. Icon. t. 3. *Solitaria*, lonely.

P. 8–12 cm., *white, then pearl grey, covered with moderately thick, angular, wart-like fragments of the volva*, which are at first *plate-like*, floccose, white, and easily separable, then becoming greyish and *hardened*, very fleshy, moist, convex then expanded; margin *appendiculate with the veil*. St. 10–20 × 3 cm., *white, covered with thick, floccose, imbricate scales*; base bulbous, *prolonged into a root-like point*. Ring *cream colour*, floccose, often torn, and finally disappearing, striate. Volva *white*, or *greyish*, very friable. Gills *snow white*, free, decurrent by a tooth, ventricose, minutely crenulate. Flesh white. Spores white, elliptical, 13–15 × 8–10μ. Smell and taste pleasant. Edible. Clearings in woods, and adjacent pastures. July—Oct. Rare. (*v.v.*)

236. **A. strobiliformis** (Paul.) Quél. Cke. Illus. no. 9, t. 277.
Strobilus, a pine cone.

P. 6–30 cm., *grey, covered with very thick, somewhat separable, angular, pyramidal, wart-like, grey fragments of the volva*, very fleshy, hemispherical, then plane. St. 15–22 × 3–5 cm., *whitish, clothed with grey flocci*; base bulbous, *immersed in the soil and surrounded by two or three circles* formed by the remains of the volva. Ring *white*, apical, torn, dependent, wide, striate. Volva *greyish*, friable. Gills *white*, free, decurrent by a tooth. Flesh *white*. Spores white, elliptical,

10–11 × 7μ. Smell and taste pleasant. Edible. Downs, and woods especially on the chalk. July—Oct. Locally common, rare elsewhere. (*v.v.*)

237. **A. aculeata** Quél. Quél. Champ. Jur. et Vosg. I, t. 1, fig. 1, as *Amanita strobiliformis* Fr. *Aculeata*, prickly.

P. 5–10 cm., *white, becoming greyish*, fleshy, convex, then plane, *densely covered with erect, slender, pointed, angular, firm, adnate, whitish or greyish warts, that become tinged with bistre with age*; margin *white*, smooth. St. 5–12 cm. × 2–5 cm., *whitish*, solid, equal, floccosely scaly; base bulbous, often attenuated downwards, *surrounded by several concentric crenulate zones*, the remains of the volva. Ring *white*, superior, thin, torn, striate, often becoming fugacious. Gills *white*, becoming yellowish with age, 5–15 mm. wide, sinuate behind, crowded. Flesh *white, then tinged with yellow*, thick, soft. Spores white, broadly elliptical, or subglobose, with a basal apiculus, 10–11 × 8–9μ, contents granular. Smell and taste pleasant. Amongst beech leaves, in woods. Oct. Rare. (*v.v.*)

238. **A. Vittadinii** (Moretti) Vitt. (= *Amanita umbella* (Paul.) Quél.) Krombh. t. 27. Vittadini, an Italian mycologist.

P. 6–12 cm., *white, densely covered with small, erect, wart-like* fragments of the volva, convex, then plane, silky. St. 15–20 × 2–2·5 cm., *white, becoming tinged with greenish*, floccosely scaly, base often somewhat bulbous. Ring *white*, superior, large, flexuose, often double. Volva *white*, or *grey*, friable. Gills *cream colour, finally becoming greenish*, decurrent by a tooth, ventricose, *thick*. Flesh *white*, then *tinged greenish*. Spores white ("greenish" Quél.), elliptical, 6–9 × 6μ. Smell and taste unpleasant. Poisonous. Downs, and woods on the chalk and limestone. July—Oct. Rare. (*v.v.*)

239. **A. echinocephala** Vitt. Cke. Illus. no. 1102, t. 939, fide Boudier, as *Amanita solitaria* Bull. ἐχῖνος, hedgehog; κεφαλή, head.

P. 6–8 cm., *white, or greyish*, covered with *thin, pointed, wart-like* fragments of the volva, convex, then flattened and depressed at the disc; margin floccose. St. 8–14 × 2–4 cm., *white*, clothed with revolute squamules, base *napiform and rooting*. Ring *white, distant, persistent*, slightly striate. Volva *greyish*, friable. Gills *greenish yellow*, free, broad. Flesh *white, yellowish at the base of the stem*. Spores "white, ovoid, 11–13 × 7–10μ" Boud. Smell and taste unpleasant. Poisonous. Limestone pastures, and woods. July—Sept. Rare.

240. **A. excelsa** Fr. (= *Amanita ampla* (Pers.) Quél.) Rolland, Champ. t. 6, no. 8, as *Amanita ampla*. *Excelsa*, tall.

P. 9–15 cm., *reddish grey, or brownish grey*, covered with mealy, fugacious patches of the volva, *streaked with innate, blackish bistre fibrils*, globose, then plane, *viscid*, rugose, uneven; margin often finally striate. St. 12–20 × 2–3 cm., *greyish*, equal, or bulbous at the base, *villose, concentrically scaly below the ring from the breaking up of the epidermis*. Ring *white*, large, superior, *dependent*, torn, often fugacious. Volva *whitish grey*, friable. Gills *white*, free, very broad, 12–15 mm., ventricose. Flesh *white, soft, fragile*. Spores white, subglobose, 9 × 7–8μ, multi-guttulate. Cystidia "on edge of gill, globular, 20–35μ in diam." Lange. Taste pleasant, smell unpleasant. Poisonous. Deciduous woods. July—Oct. Not uncommon. (*v.v.*)

241. **A. pantherina** (DC.) Fr. Rolland, Champ. t. 7, no. 10.
Pantherina, deceitful.

P. 6–10 cm., *olivaceous umber, fuliginous, or greyish olive, rarely whitish, covered with numerous small, white, moderately persistent fragments of the volva*, fleshy, convex, then flattened, or subdepressed, *viscid*, shining when dry; *margin striate*. St. 7–9 × 1 cm., *white*, equal, or attenuated upwards, base bulbous. Ring *white, distant*, thin, *striate*, adhering obliquely, somewhat fugacious. Volva *white, forming one or two concentric rings* at the apex of the globose base of the stem. Gills *white*, free, reaching the stem, broader in front, 6–8 mm. broad. Flesh *white*. Spores white, elliptical, 11–12 × 7–9μ. Cystidia "mostly cylindric-vesiculose about 12μ in diam." Lange. Taste insipid, smell unpleasant. Poisonous. Woods, heaths, and pastures. July—Oct. Not uncommon. (*v.v.*)

242. **A. cariosa** Fr. Gonn. & Rabenh. t. 9, fig. 2, as *Amanita aspera*.
Cariosa, rotten.

P. 6–12 cm., *umber, or dark cinereous, covered with white, mealy fragments of the volva*, convex, then plane, often hemispherical, *tender*; margin often striate. St. 12–14 × 2–4 cm., *white, fragile*, attenuated upwards, *not bulbous*, villose, mealy. Ring *white*, superior, broad, fugacious. Volva *white*, friable. Gills *white*, adnate, becoming free. Flesh *white*, fragile. Spores white, ovoid, 11–13 × 7–10μ. Taste acid. Poisonous. Woods. Sept. Rare.

243. **A. spissa** Fr. Rolland, Champ. t. 6, no. 9. *Spissa*, thick.

P. 8–15 cm., *umber fuliginous, or grey*, fleshy, compact, convexo-plane, obtuse, *covered with whitish, or greyish, fugacious patches of the volva*; margin often fibrillose. St. 10–11 × 3–4 cm., *white, clothed with concentric squamules below the ring*, base *bulbous*, somewhat rooting. Ring *white*, superior, large, *striate*. Volva *whitish*, or *greyish*, friable. Gills *white, slightly striato-decurrent*, broad, crowded. Flesh *white*, firm. Spores white, subglobose, 9–10 × 8–9μ. Cystidia "on edge of gills globular, 18–30μ in diam." Lange. Taste insipid, or slightly biting. Edible. Deciduous woods. July—Oct. Not uncommon. (*v.v.*)

244. **A. rubescens** (Pers.) Fr. (= *Amanita rubens* (Scop.) Quél.) Rolland, Champ. t. 9, no. 13. *Rubescens*, becoming red.

P. 8–12 cm., *reddish brown, or dingy reddish brown, sometimes pale, covered with large, grey (sometimes white, or yellowish) mealy patches* of the volva, fleshy, convex then plane, obtuse, moist; margin slightly striate when old. St. 7–12 × 3–4 cm., *reddish white*, deeper in colour at the bulbous base, *squamulose*, attenuated upwards. Ring *white*, superior, large, membranaceous, soft, *striate*. Volva evanescent. Gills *white, then spotted with red*, decurrent by a tooth, attenuated behind, thin, crowded, soft. Flesh *white, becoming reddish when broken. The whole plant becomes reddish with injury, or handling.* Spores white, ovoid, or elliptical, 8–10 × 7μ, 1–2-guttulate. Taste sweet, then acrid. Edible. Woods, heaths, pastures, etc. Common. (*v.v.*)

var. **magnifica** (Fl. Dan.) Rea. Cke. Illus. no. 14, t. 34, as *Amanita magnifica* Fr. *Magnifica*, splendid.

Differs from the type in the *smooth pileus, equal stem, and fugacious ring.* Woods. Uncommon. (*v.v.*)

var. **alba** W. G. Smith. *Alba*, white

Differs from the type in being *entirely white.* Woods. Rare.

var. **annulo-sulphurea** Gillet.
Annulus, a ring; *sulphurea*, sulphur-yellow.

Differs from the type in having *a persistent, sulphur coloured ring.* Woods, and heaths. Not uncommon. (*v.v.*)

var. **gracilis** Cat. de S. et L. *Gracilis*, thin.

Differs from the type in being *thinner, and smaller in all its parts.* Heaths. Uncommon. (*v.v.*)

245. **A. nitida** Fr. *Nitida*, shining.

P. 6–10 cm., *white, or yellowish, shining, covered with large, angular, thick fragments of the volva, which become fuscous*, convex, then plane, fleshy. St. 6–7 × 1–2·5 cm., *white, firm*, slightly attenuated upwards, *squamulose* below the ring, base bulbous. Ring *white*, superior, thin, torn, slightly striate, villous outside, at length fugacious. Volva *whitish, becoming fuscous*, evanescent. Gills *white*, free, crowded, very broad, 8–12 mm., ventricose. Flesh *white*. Spores white, elliptical, 6–9 × 4–5μ. Taste sweet, or slightly acrid. Poisonous. Deciduous woods. Aug.—Sept. Uncommon. (*v.v.*)

246. **A. aspera** (Fr.) Quél. *Asper*, rough.

P. 5–8 cm., *straw colour, grey, olive, or bistre, covered with small, pointed, floccose, sulphur coloured, persistent fragments of the volva,* which become whitish or brownish in dry weather, convex, then plane. St. 5–8 × 1 cm., *white*, attenuated upwards, slightly floccose, base bulbous, *surmounted by sulphur coloured flocci that become brownish.* Ring *white*, distant, *margin sprinkled with sulphur coloured flocci.* Volva *sulphur coloured*, friable. Gills *white, or tinged sulphur colour*, rounded-free, ventricose. Flesh *white, yellowish, or brownish under the epidermis.* Spores white, ovoid, 7–8 × 6–7 μ, 1-guttulate. Cystidia sparse, vesiculose, 20–40 × 18–25 μ. Smell and taste pleasant. Poisonous. Beech woods. July—Oct. Uncommon. (*v.v.*)

A. magnifica (Fl. Dan.) Fr. = **Amanita rubescens** (Pers.) Fr. var. **magnifica** (Fl. Dan.) Rea.

A. arida Fr. = **Lepiota arida** (Fr.) Gillet.

A. lenticularis (Lasch) Fr. = **Lepiota lenticularis** (Lasch) Cke.

A. megalodactylus Berk. & Br. = **Lepiota lenticularis** (Lasch) Cke. var. **megalodactylus** (B. & Br.) Rea.

B. Pileus confluent, and homogeneous with the fleshy stem.

*With a membranaceous ring on the stem.

Spores white.

Armillaria Fr.

(*Armilla*, a ring.)

Pileus fleshy, regular. Stem central, fleshy. Ring membranaceous, or subarachnoid, adnate, persistent, or fugacious. Gills sinuato-adnexed, decurrent, or adnate. Spores white, elliptical, oval, or globose, smooth. Growing on the ground, and on wood, sometimes caespitose.

(*a*) Gills sinuato-adnexed.

247. **A. bulbigera** (A. & S.) Fr. *Bulbus*, a bulb; *gero*, I bear.

P. 7·5–10 cm., *pale yellowish brick colour*, fleshy, not compact, convexo-flattened, obtuse, moist; margin *paler, squamuloso-fibrillose from the fragments of the veil.* St. 5–7·5 × 1·5–2 cm., *white*, equal, *floccose with the remains of the arachnoid veil up to the ring,*—sometimes the separable cuticle is marked longitudinally with blackish fibrils, *base marginately bulbous.* Ring *white, arachnoid*, silky, fugacious. Gills *white, then cream colour, or reddish*, broadly emarginate, ventricose, broad. Flesh *white, reddish under the cuticle, and above the base of the gills.* Spores white, elliptical, 6–7 × 4–5 μ. In pine woods. Sept.—Nov. Uncommon. (*v.v.*)

Exactly like a white-spored *Cortinarius*.

248. **A. rufa** (Batt.) Quél. (= *Agaricus causetta* Barla sec. Quél.; *Armillaria focalis* Fr. sec. Quél.) Cke. Illus. no. 51, t. 33, as *Armillaria aurantia* Fr. fide Boudier. *Rufa*, red.

P. 8–12 cm., *chestnut, or brownish tawny*, convex, then plane; margin *fibrillose*, torn. St. 6–9 × 3 cm., *white, covered with reddish tawny squamules up to the ring*, apex glabrous, equal, *attenuated and rooting at the base*. Ring *reddish*, squamulose. Gills *greenish white*, sinuate. Flesh *white*, compact. Spores white, globose, 3–4μ, punctate. Taste pleasant. Edible. Sandy coniferous woods. Sept.—Oct. Uncommon.

249. **A. focalis** Fr. (= *Armillaria rufa* (Batt.) Quél.) *Focale*, a neck-cloth.

P. 10–12·5 cm., *reddish tawny*, fleshy, convex, then flattened, obtuse, slightly shining, *silky-fibrillose*. St. 7·5–9 × 2·5 cm., *whitish, becoming tawny, equal, fibrillose*. Ring *concolorous*, median, oblique. Gills *white, then pale*, emarginato-free, crowded. Flesh *pale tawny*. Spores white, "4–5 × 3μ" Rick. Pine woods, and under old laurel trees. Aug.—Oct. Rare.

var. **Goliath** Fr. The giant Goliath.

Differs from the type in its *larger size, in the revolute torn margin of the p., in the st. becoming tawny fibrillose downwards, in the fugacious ring, and in the thinner flesh at the margin of the p.*

250. **A. robusta** (A. & S.) Fr. (= *Agaricus caligatus* Viv. sec. Quél.) Boud. Icon. t. 22. *Robusta*, strong.

P. 5–15 cm., *bay brown rufescent, margin paler*, very fleshy, convex, then expanded, obtuse; margin *scaly-fibrillose*. St. 4–7 × 2–3 cm., *white, covered with rufescent squamules up to the ring, fusiform*, apex mealy. Ring *white, streaked with rufescent fibrils*, large, subpersistent. Gills *whitish, or cream colour*, broadly emarginate, almost free, 10–12 mm. broad, crowded, often transversely veined. Flesh *white, reddish under the cuticle of the p.* Spores white, globose, 4–5μ. Taste and smell pleasant. Edible. Coniferous woods. Sept.—Oct. Common. (*v.v.*)

var. **minor** Fr. Quél. Jur. et Vosg. part III, t. 1, fig. 4, as *Armillaria subannulata* Batsch. *Minor*, smaller.

Differs from the type in the *smooth p. and in very narrow gills and ring*. Woods. Sept.—Oct. Uncommon. (*v.v.*)

251. **A. caligata** (Viv.) Fr. (= *Armillaria robusta* A. & S. sec. Quél.) Boud. Icon. t. 21. *Caliga*, a soldier's shoe.

P. 6–12 cm., *brownish chestnut, somewhat purplish, covered with adpressed, denticulate, darker squamules on the disc*, firm, convex, depressed at the centre; margin *white, incurved, appendiculate with the veil.* St. 5–8 × 1–2·5 cm., *white, and mealy at the apex, covered with large, denticulate, dark chestnut squamules below the ring,* attenuated at the base. Ring *white inside, torn, covered on the outside with similar squamules,* ascending. Gills *white,* adnate, slightly decurrent, broad. Flesh white, firm. Spores white, elliptical, 5–6 × 4μ. Smell of pear. Taste bitter. Edible. Pine woods. Sept.—Oct. Uncommon. (*v.v.*)

A. aurantia (Schaeff.) Fr. = **Tricholoma aurantium** (Schaeff.) Fr. Syst.

252. **A. colossa** (Fr.) Boud. (=*Tricholoma colossum* Fr.) Boud. Icon. t. 20. *κολοσσός*, a gigantic statue.

P. 10–22 cm., *reddish tawny, darker at the centre,* globose, then expanded, plano-convex, then depressed, always very obtuse, repand, becoming broken up into scales; margin *whitish, incurved, slightly viscid, cottony.* St. 7·5–10 × 6–10 cm., *concolorous, or more deeply coloured than the p. below the ring,* apex *white, floccose,* base bulbous. Ring *white, then becoming reddish,* membranaceous, *soon fugacious.* Gills *white, then pale brick red,* rounded, sinuate, very wide, 12 mm. broad, fragile, torn. Flesh *white, then pale brick-red, very hard,* thick. Spores white, globose, 6–7 × 5–7μ, with a large central gutta. Taste nutty, then slightly bitter. Edible. Pine woods. June—Oct. Rare. (*v.v.*)

253. **A. ramentacea** (Bull.) Fr. (= *Tricholoma ramentaceum* (Bull.) Quél.) Cke. Illus. no. 53, t. 71. *Ramenta*, shavings.

P. 5–7·5 cm., *whitish, or greyish, covered with adpressed, floccose, dark grey, or bistre scales,* convex, then plane, obtuse, or gibbous, at length depressed, and revolute. St. 2·5–5 cm. × 6–12 mm., *white, covered with adpressed, fuscous, or bistre squamules below the ring,* firm, unequal, often thickened at the base. Ring *white above, greyish and scaly on the outside,* often stained yellowish, *membranaceous,* fugacious. Gills *white, often stained with yellow,* emarginato-adnexed, separating free, 6–8 mm., broad, crowded, then subdistant, thin. Flesh *white, yellowish under the epidermis.* Spores white, elliptical, 5–6 × 4μ, with a large central gutta. Taste sweet. Smell unpleasant. Under pines. Sept.—Oct. Uncommon. (*v.v.*)

A. constricta Fr. = **Lepiota constricta** (Fr.) Quél.

A. glioderma (Fr.) Quél. = **Lepiota glioderma** Fr.

254. **A. delicata** (Fr.) Boud. (= *Lepiota delicata* Fr.) Boud. Icon. t. 23. *Delicata*, tender.

P. 1–3 cm., *rufescent, pale rose, yellowish, or brownish,* convex, then plane, depressed at the centre, often slightly umbonate, *viscid*; margin *faintly sulcate.* St. 2·5–5 cm. × 4–6 mm., *whitish, tinted with the colour of the p., and floccoso-scaly or tomentose below the ring,* equal. Ring *concolorous*, membranaceous, *densely floccoso-scaly*. Gills *white, becoming tinted with the colour of the p. when old*, almost free, crowded, thin, ventricose. Flesh *white, or yellowish*, thin. Spores white, globose, 5–6μ. Coniferous woods, hothouses, and about old stumps. June—Sept. Rare in woods.

255. **A. haematites** B. & Br. Cke. Illus. no. 54, t. 45.

αἱματίτης, bloody.

P. 2–4 cm., *red liver colour*, hemispherical, then somewhat flattened, or depressed at the centre, thin, *slightly hispid*, becoming smooth. St. 4–6 cm. × 3–5 mm., *concolorous below the ring, whitish above*, equal; base *thickened*, white floccose. Ring *whitish, then concolorous*, submembranaceous, narrow, inferior, scaly beneath, torn, often fugacious. Gills *white, then whitish tinged with rose*, and becoming rose colour when rubbed, sinuato-adnate, or shortly decurrent, scarcely crowded, narrow, 3 mm. broad. Flesh *pale liver colour, slightly yellowish in the st.* Spores white, ovoid-ellipsoid, 4 × 3μ. Among fir leaves. Nov. Rare.

256. **A. Jasonis** Cke. & Massee. (= *Lepiota amianthina* (Scop.) Fr. sec. Boud. Cke. Illus. no. 1113, t. 955.

Jason and the golden fleece.

P. 2·5–7·5 cm., *golden yellow, disc tawny*, fleshy, campanulate, then expanded, *with a distinct rounded umbo, granularly papillate, granules innate*; margin appendiculate with the fibrous veil. St. 5–7·5 cm. × 6–9 mm., *concolorous*, equal, or slightly thickened at the base, *squamulose below the ring*. Ring *concolorous*, distant, squarrose, torn. Gills *white, then pallid*, adnate, scarcely crowded, thin. Flesh *reddish*. Spores white, elliptical, 8 × 5μ. Smell strong. Caespitose. On stumps. Sept. Rare.

(*b*) Gills more or less decurrent.

257. **A. mellea** (Vahl.) Fr. Grev. Scot. Crypt. Fl. t. 332.

Mel, honey.

P. 5–15 cm., *ochraceous yellow, tawny, or bistre, covered with olivaceous, or brownish hairy squamules*, fleshy, convex, then flattened, and depressed in the centre; margin paler, striate. St. 7·5–15 × ·5–1 cm., *yellow, tawny, or bistre, often covered with olivaceous down below the ring*, becoming blackish with age, equal, or subbulbous at the base, *elastic*, fibrillose, apex *striate*. *Ring white, becoming discoloured*, apical, silky, membranaceous, thick, swollen at the margin. *Gills whitish flesh colour, then rufescent*, adnate, decurrent by a tooth, subdistant. Flesh *white, becoming discoloured*, floccose. Spores white, elliptical, 8–9 × 5–6μ. Cystidia "on edge of gill basidia-like, 40–60 × 8–12μ" Rick. Taste acrid. Edible. Caespitose. On old stumps, and buried fragments of wood. July—Dec. Very common. (*v.v.*)

var. **sulphurea** (Weinm.) Fr. *Sulphurea*, sulphur colour.

Differs from the type in the *yellow, or yellowish flesh coloured p., and the sulphur coloured gills.* Woods. Uncommon. (*v.v.*)

var. **minor** Barla. Barla, Champ. Alp. Marit. t. 21, figs. 3–4. *Minor*, smaller.

Differs from the type in its *smaller size, and thinner flesh.* Woods (*v.v.*)

var. **maxima** Barla. Barla, Champ. Alp. Marit. t. 22, figs. 1–2. *Maxima*, very large.

Differs from the type in the *very large p.* 20 *cm. or more, st.* 15×4 *cm., ventricose, attenuated at the base, and very wide ring, tawny on the outside.* Woods. Uncommon. (*v.v.*)

var. **obscura** Gillet. *Obscura*, dark.

Differs from the type in the *brownish p., covered with numerous black scales.* Woods. Uncommon. (*v.v.*)

var. **glabra** Gillet. *Glabra*, smooth.

Differs from the type in *the smooth pileus.* Woods. Common. (*v.v.*)

var. **bulbosa** Barla. Barla, Champ. Alp. Marit. t. 22, figs. 3–7. *Bulbosa*, bulbous.

Differs from the type in the *reddish, bulbous stem, and ochraceous, or bright bistre ring.* Woods. Not uncommon. (*v.v.*)

var. **viridi-flava** Barla. Barla, Champ. Prov. Nice, t. 11, figs. 1–3. *Viridis*, green; *flava*, yellow

Differs from the type in the *greenish p., covered with yellow scales, or fibrils, in the bright yellow, or sulphur coloured st., the citron yellow ring, and the yellowish gills.* Stumps in hedgerows. Uncommon. (*v.v.*)

var. **laricina** (Bolt.) Fr. Barla, Champ. Alp. Marit. t. 21, figs. 5–6. *Larix*, larch.

Differs from the type in the *flesh coloured, glabrous, not striate p., and the white, narrow gills.* Woods. Uncommon. (*v.v.*)

var. **versicolor** W. G. Sm.[1] *Verto*, I turn; *color*, colour.

Differs from the type in the *bulbous, white, then brown st.*, and *the yellow white, then deep red brown gills.*

[1] W. G. Smith probably referred this wrongly to *Agaricus versicolor* With.

var. **tabescens** (Scop.) Rea. (= *Clitocybe tabescens* (Scop.) Fr.; *Agaricus gymnopodius* Bull. sec. Quél.) Boud. Icon. t. 51, cited in text under t. 61, as *Clitocybe tabescens* (Scop.) Fr.

Tabesco, I waste away.

Differs from the type in the *complete absence of the ring*. Woods, and hedgerows. Uncommon. (*v.v.*)

258. **A. denigrata** Fr. (=*Pholiota erebia* Fr. sec. Lange). Fr. Icon. t. 20.

Denigro, I colour very black.

P. 3–6 cm., *dark brown*, convex, then plane, obtuse, *slightly viscid, looking as if covered with minute drops of water*, owing to the presence of *elevated warts*. St. 5–6 × 1–1·5 cm., *pallid fuscous, brown at the base*, equal, or ventricose and attenuated, elastic, *fibrillosely striate*. Ring *paler*, superior, membranaceous, *narrow*, entire, fugacious. Gills *pale brown, then darker*, sinuato-decurrent, *narrow*. Flesh *bistre*, firm. Solitary, or caespitose. At the base of old trees, and in garden humus. Sept.—Oct. Uncommon.

259. **A. citri** (Inzenga) Fr. *Citrus*, orange.

P. 5 cm., *sulphur yellow*, convex, then plane, subumbonate, fleshy; margin becoming white, *crenulate*. St. 5–7·5 cm. × 1–2 mm., *whitish, base rufescent*, equal, apex white floccose. Gills whitish, adnate, crowded. Spores white, "subglobose, 5 × 4μ" Massee. Smell of new meal. Caespitose. On stumps. Rare.

(*c*) Gills equal behind, st. externally subcartilaginous.

260. **A. subcava** (Schum.) Fr. *Sub*, somewhat; *cava*, hollow.

P. 2–5 cm., *white, umbo umber*, submembranaceous, convexo-plane, *viscid, striate to the middle*. St. 7–9 cm. × 6 mm., *white, fistulose upwards*, equal, *slightly dotted below the ring*. Ring *white*, inferior, torn. Gills white, decurrent, plane. Flesh white, thick. Fir woods. July—Nov. Rare.

261. **A. mucida** (Schrad.) Fr. Cke. Illus. no. 58, t. 16.

Mucida, slimy.

P. 3–8 cm., *white, or grey*, thin, *almost diaphanous*, hemispherical, then expanded, obtuse, *more or less radiato-wrinkled, glutinous*; margin striate when thinner. St. 4–7·5 cm. × 4–15 mm., *white, base thickened and fuliginously scaly*, rigid, striate above the ring. Ring *white, becoming fuscous* from the dried gluten, superior, dependent, often sulcate. Gills *white, then yellowish*, rounded behind, decurrent by a tooth, distant, broad, lax. Flesh *white*, mucilaginous. Spores white, globose, 15–17μ, multi-guttulate. Cystidia none. Taste mild. Edible. On beeches, rarely on oaks and birches. Aug.—Nov. Common. (*v.v.*)

Spores ferruginous, rough; general veil persistent.

Rozites Karst.

(E. Roze, a French mycologist.)

Pileus fleshy, regular, white pruinose with the thin *general veil*. Stem central, fleshy. Ring membranaceous. Gills adnate. Spores ferruginous, pip-shaped, *rough*, with an apical germ-pore. Growing on the ground.

262. **R. caperatus** (Pers.) Karst. (= *Pholiota caperata* (Pers.) Fr.) Rolland, Champ. t. 59, no. 132, as *Pholiota caperata*.

Caperatus, wrinkled.

P. 4–13 cm., *more or less intensely yellow*, campanulate, then expanded, obtuse, viscid only when moist and not truly so, *incrusted with the floccose-mealy universal veil*, which is crowded on the even disc, and *squamulose and fugacious towards the thin, lacunoso-wrinkled, sulcate, splitting margin*. St. 8–17 × 2·5–3·5 cm., *white, becoming tinged with yellow*, stout, fibrillose, striate, equal, base often tuberous, and *the universal veil often cohering in the form of a volva*, squamulose above the ring. Ring *white, becoming yellowish*, membranaceous, *striate*, distant, often oblique and torn. Gills *clay-cinnamon*, adnate, crowded, thin, *denticulate*. Flesh *whitish, becoming yellowish*. Spores ferruginous, pip-shaped, 11–12 × 8μ, rough, 1-guttulate. Cystidia "on edge of gill clavate, 45–50 × 8–10μ" Rick. Smell and taste pleasant. Edible. Woods. Aug.—Dec. Not uncommon. (*v.v.*)

Spores ochraceous, or ferruginous, generally smooth; general veil none, or fugacious.

Pholiota Fr.

(*φολίς*, a scale; *οὖς*, the ear.)

Pileus fleshy, regular. Stem central. Ring membranaceous, persistent, or fugacious, superior, or inferior. Gills adnate, or decurrent by a tooth. Spores ochraceous, or ferruginous, rarely fuscous, elliptical, oval, obovate, subreniform or oblong elliptical, generally *smooth*, continuous, or with a germ-pore. Cystidia variable. Growing on the ground, or on wood, often caespitose.

I. Growing on the ground, not adnate to mosses, rarely caespitose.

263. **P. aurea** (Mattusch) Fr. (= *Lepiota pyrenaea* Quél. sec. Maire; *Pholiota spectabilis* Fr. sec. Quél.) Fr. Icon. t. 101.

Aurea, golden.

Entirely golden-tawny. P. 4–25 cm., fleshy, convex, obtuse, soft, *at first velvety*, then torn into innate, hairy squamules. St. 6–28 × 1–3·5 cm., somewhat equal, becoming pale, *sprinkled below the ring with a separating, ferruginous scurf*, apex flocculose. Ring membranaceous, *externally flocculose and ferruginous-furfuraceous*, internally *golden-tawny*, about 2·5 cm. distant from the p., properly inferior, but appearing to be medial, at first erect, then spreading, sometimes small, often however wide, laciniate. Gills *pallid ferruginous*, adnexed, then free, attenuated at both ends, ventricose, crowded, connected by veins. Flesh *white, becoming yellow*. Spores fuscous, elliptical, 9–10 × 4–5μ, 1-guttulate. Cystidia none. Subcaespitose. On the ground, and on sawdust heaps. Aug.—Nov. Uncommon. (*v.v.*)

var. **Vahlii** (Schum.) Fr. Fl. Dan. t. 1498. M. Vahl in Flora Danica.

Differs from the type in the *smooth pileus, and somewhat free gills*.

var. **Herefordiensis** Renny. Cke. Illus. no. 374, t. 347.
Herefordiensis, belonging to Hereford.

Differs from the type in the *granulate, tuberculate stem*.

264. **P. terrigena** Fr. (= *Pholiota Cookei* Fr. sec. Massee.) Fr. Icon. t. 103, fig. 1. *Terra*, earth; *γίγνομαι*, to be born.

P. 3–8 cm., *dingy yellow*, fleshy, convex, or lens-shaped, then flattened, obtuse, *adpressedly silky with fibrils, fibrillosely scaly towards the margin*. St. 4–5 cm. × 4–12 mm., *concolorous*, equal, fleshy-fibrous, *covered with floccose, squarrose squamules that become ferruginous*. Ring thin, torn. Gills *pallid light yellow*, then *olivaceous-ferruginous*, adnate, decurrent with a tooth, scarcely crowded, 4 mm. broad. Flesh *yellow*. Spores ferruginous, "elliptical, 5–8 × 2–3μ" Karst. Woods, hedgerows, and old earthy stumps. Aug.—Nov. Uncommon. (*v.v.*)

265. **P. erebia** Fr. Cke. Illus. no. 377, t. 358.
ἔρεβος, a place of nether darkness.

P. 2–5 cm., *lurid*, or *becoming ferruginous-lurid when moist, becoming pale* (*ochraceous clay*) *when dry*, slightly fleshy, convex, then flattened, *almost viscid, rugulose*; *margin striate when dry*. St. 2·5–5 cm. × 3–6 mm., *fuliginous, becoming pale*, equal, often cohering at the base, fibrillose, striate. Ring membranaceous, *white, becoming discoloured*, sulcate, *superior*. Gills *pallid*, then *dingy cinnamon*, adnate, subdistant. Flesh *pale brownish*. Spores ferruginous, pip-shaped, 10–12 × 5–6μ, 1–2-guttulate. Woods, pastures, and heaths. Aug.—Oct. Not uncommon. (*v.v.*)

266. **P. ombrophila** Fr. *ὄμβρος*, a storm of rain; *φίλος*, loving.

P. 4–8 cm., *pale ferruginous when moist, clay colour when dry*, convex, then plane, *gibbous*, here and there repand, *almost viscid, very hygrophanous*; *margin striate when moist*. St. 5–8 cm. × 4–8 mm., *pallid*, fragile, equal, obsoletely fibrillose, or slightly striate. Ring membranaceous, *white, distant, entire,* reflexed. Gills *pallid, then watery ferruginous,* adfixed, *then separating* almost free, *ventricose,* crowded. Flesh *becoming white,* thin, soft. Spores ochraceous, elliptical, or pip-shaped, 8–10 × 4–6μ, 1-guttulate. Cystidia "on edge of gill lanceolate-capitate, 50–60 × 12–15μ" Rick. Heaths, and pastures. Sept.—Oct. Not uncommon. (*v.v.*)

var. **brunneola** Fr. Fr. Icon. t. 103, fig. 2. *Brunneola,* brownish.

Differs from the type in its *smaller size, its obtuse, brown pileus, and its narrower gills.*

267. **P. molliscorium** Cke. & Massee. Cke. Illus. no. 1161, t. 1171. *Mollis,* soft; χόριον, skin.

P. 5–7·5 cm., *tawny yellow, disc darker, margin paler,* fleshy, convex, then plane, obtuse, at length depressed, soft like kid leather, shining; margin acute, thin. St. 7·5 cm. × 6–10 mm., *pale yellow,* equal, erect, *silky, apex punctately squamulose.* Ring *yellow,* distant, broad, deciduous. Gills *ferruginous,* narrowly adnate, crowded, thin, ventricose, 4 mm. broad. Flesh *yellow,* thin. Spores ferruginous, elliptical, 12 × 5–6μ. Gregarious. On the ground. June. Rare.

268. **P. togularis** (Bull.) Fr. Boud. Icon. t. 101. *Togula,* a little cloak.

P. 2–4 cm., *pallid ochraceous, disc darker,* campanulate, then expanded, obtuse, orbicular, sometimes striate. St. 3–6 cm. × 2–3 mm., *whitish, becoming fuscous at the thickened base,* equal, fibrillosely striate, apex mealy. Ring *white,* membranaceous, medial, entire, spreading, reflexed. Gills *yellow, at length pallid ferruginous,* adnato-separating, attenuated at both ends, crowded. Flesh *whitish, becoming yellow.* Spores ferruginous, oblong, 7–8 × 3–4μ, 1-guttulate, "with flattened germ-pore. Cystidia on edge of gill fusiform, 25–36 × 6–8μ" Rick. Woods, and pastures. May—Nov. Not uncommon. (*v.v.*)

var. **filaris** Fr. Fr. Icon. t. 104, fig. 4. *Filum,* a thread.

Differs from the type in being *two to three times smaller with a yellowish stem.* Lawns, and pastures. July. Uncommon. (*v.v.*)

269. **P. blattaria** Fr. *Blattarius,* like a cockroach.

P. 1·5–2·5 cm., *ferruginous, becoming pale,* hygrophanous, thin, convex, or umbonate, soon flattened; margin *paler, striate.* St. 2·5–5 cm. × 2 mm., *white,* equal, silky, straight. Ring *white,* membranaceous, distant, silky, entire. Gills *watery-cinnamon,* rounded behind, *free,* ventricose. Flesh *concolorous,* very thin. Spores ferruginous, elliptical, 8–10 × 4–5μ, 1–3-guttulate. Cystidia "on edge of gill fusiform-subulate, 30–36 × 7–9μ" Rick. Lawns, arable fields, and gardens. Aug.—Sept. Rare. (*v.v.*)

270. **P. dura** (Bolt.) Fr. *Durus*, hard.

P. 5–9 cm., *tawny tan colour, becoming fuscous*, fleshy, *somewhat compact*, convexo-plane, obtuse, *becoming cracked into patches*. St. 5–8 cm. × 10–15 mm., *yellowish*, hard, becoming silky-even, then longitudinally cracked when dry, apex *thickened, mealy*, sometimes ventricose and irregularly shaped, furnished with fibrillose rootlets at the base. Ring *white*, membranaceous, thin, apical, *often torn*, fugacious. Gills *white*, then *livid, or fuscous ferruginous, adnate*, striato-decurrent with a tooth, ventricose, 7–12 mm. broad. Flesh *whitish*. Spores ferruginous, elliptical, 9–12 × 6–8μ, with a germ-pore. Cystidia "vesiculose-clavate, on edge of gill almost flask-shaped, 36–50 × 12–18μ" Rick. Smell strong. Fields, and gardens. May—Oct. Not uncommon. (*v.v.*)

var. **xanthophylla** Bres. Bres. Fung. Trid. t. 159.
ξανθός, yellow; *φύλλον*, a leaf.

Differs from the type in the *bright sulphur yellow gills*.

271. **P. praecox** (Pers.) Fr. Cke. Illus. no. 381, t. 360.
Praecox, early.

P. 3–8 cm., *whitish, then tan colour, becoming pale*, fleshy, soft, convex, soon plane, obtuse, moist. St. 4–9 cm. × 6–10 mm., *white, becoming yellowish*, equal, *at first mealy with white flocci*, then somewhat naked, base white floccose. Ring *whitish*, membranaceous, entire, reflexed, striate above. Gills *whitish, then rust coloured, rounded-adnexed*, 4–10 mm. broad, crowded. Flesh *white, yellowish in the stem*, soft. Spores ferruginous, oblong elliptical, 10–13 × 6–8μ, 1-guttulate. Cystidia sack-shaped, often slightly constricted towards the apex, 30–40 × 13–15μ. Taste sweet. "Edible," Quélet. Woods, pastures, and roadsides. May—Oct. Common. (*v.v.*)

var. **minor** (Batt.) Fr. *Minor*, smaller.

Differs from the type in its *smaller size* (*scarcely* 2·5 *cm. broad*), *and in the torn, appendiculate ring*. Pastures. Uncommon. (*v.v.*)

272. **P. sphaleromorpha** (Bull.) Fr.
σφαλερός, deceptive; *μορφή*, form.

P. 2–5 cm., *light yellow*, fleshy, *thin*, convexo-plane, obtuse. St. 7–10 cm. × 4–8 mm., *yellowish*, attenuated upwards, silky, *base incrassated*, villose. Ring *whitish*, membranaceous, very thin, median, ample, spreading, lax, entire. Gills *yellowish, then ferruginous-tan*, equally broad, *truly decurrent*, linear, 3 mm. broad. Flesh *whitish*, hygrophanous. Spores "almost colourless under the microscope, cylindrical-elliptical, 6–7 × 3–4μ, smooth. Cystidia ventricose-fusiform, 60–75 × 10–12μ" Rick. Leaf soil, heaths, and peat bogs. July—Oct. Uncommon.

II. Growing on wood, or epiphytal, most frequently caespitose.

*P. naked not scaly, but here and there rimoso-rivulose. Gills pallid, then rufescent, or becoming fuscous.

273. **P. radicosa** (Bull.) Fr. Cke. Illus. no. 382, t. 361.
Radicosus, having a root.

P. 5–13 cm., *clay-coloured, then spotted rufous*, fleshy, convexo-plane, *viscid*, becoming dry. St. 7·5–24 × 1–2·5 cm., *white*, firm, *thickened at the base and fusiform rooted, concentrically scaly below the ring, the floccose, erect scales becoming rufous, apex mealy, pruinose.* Ring *white*, membranaceous, distant, rather erect, entire, scaly. Gills *pallid, then rufescent ferruginous*, rounded behind, *somewhat free*, very crowded, 6 mm. broad. Flesh *whitish*, moderately thick. Spores ochraceous, elliptical, 8–9 × 5μ, "rough. Cystidia on edge of gill filamentous-clavate, 36–40 × 6–8μ, thin walled" Rick. Smell pleasant, like cherry laurel, or bitter almonds. Taste pleasant. Solitary or gregarious. Woods about stumps. Aug.—Nov. Not uncommon. (*v.v.*)

274. **P. pudica** (Bull.) Fr. *Pudica*, modest.

P. 5–10 cm., *whitish, or slightly tawny, disc darker*, globose, or oval, then convex, and expanded, obtuse, *umbonate, dry*. St. 3–6 cm. × 8–10 mm., *whitish*, straight, or curved at the base, equal, or attenuated upwards, sometimes excentric, *fibrous*. Ring *white*, membranaceous, large, spreading, persistent. Gills *whitish, then tawny*, adnato-decurrent, wide, ventricose. Spores "ferruginous 8 × 5–6μ" Sacc. Often solitary. In woods on old trunks, at the base of trees, and on elder trunks. May—Nov. Uncommon.

275. **P. leochroma** Cke. Cke. Illus. no. 384, t. 363.
λέων, a lion; χρῶμα, colour.

P. 5–10 cm., *bright tawny, whitish at the margin*, fleshy, convexo-plane, at length depressed, soft, generally rivulose from the cracking of the cuticle. St. 7–12 cm. × 8–13 mm., *paler than the pileus, white above*, nearly equal, fibrous. Ring *tawny*, persistent, membranaceous. Gills *pallid, then cinnamon*, rounded, adnate, slightly ventricose. Flesh *yellowish, somewhat tawny under the cuticle of the p., and at the base of the st.* Spores ferruginous, elliptical, 7–8 × 5μ, 1–2-guttulate. Smell and taste pleasant. Edible. Caespitose. Elm stumps. July—Oct. Not uncommon. (*v.v.*)

276. **P. aegerita** (Porta) Fr. (= *Pholiota capistrata* Cke., *Pholiota luxurians* (Batt.) Fr.) Cke. Illus. nos. 385, 386, t. 364, as *Pholiota capistrata* Cke. and t. 453. *αἴγειρος*, the black poplar.

P. 3–12 cm., *tawny, becoming pale white towards the margin,* fleshy, convex then plane, *rivuloso-wrinkled,* slightly viscid when moist, silky when dry, *disc often areolately cracked*; margin incurved, scalloped. St. 8–15 × 1·5–3 cm., *white, becoming stained with yellow,* attenuated downwards, fibrillose, often striate. Ring *white,* membranaceous, large, superior, reflexed. Gills *pallid, then fuscous,* adnate, decurrent with a small tooth, crowded. Flesh *white, brownish under the cuticle of the p. and at the base of the st.* Spores ferruginous, elliptical, 9–10 × 5–7μ. Cystidia broadly clavate or pear-shaped, 30–36 × 12–15μ. Smell and taste pleasant. Edible. Caespitose. Stumps, especially elm. May—Nov. Common. (*v.v.*)

277. **P. Junonia** Fr. Cke. Illus. no. 397, t. 369.
Junonia, belonging to Juno.

P. 4–8 cm., *rich yellow, or tawny yellow,* fleshy, firm, convexo-plane, obtuse. St. 4–9 cm. × 6–14 mm., *tawny, yellow and mealy above the ring,* equal, firm, incurved, often excentric. Ring *concolorous,* membranaceous, inferior, reflexed. Gills *yellow, then tawny, adnate,* crowded, *broad.* Flesh *pale yellow,* compact. Spores ferruginous, elliptical, 8–10 × 6–7μ. Usually solitary. Trunks. Oct. Uncommon. (*v.v.*)

**P. scaly, gills changing colour. P. not hygrophanous.

a. Gills pallid, then becoming fuscous, olivaceous, clay coloured, not truly ferruginous.

278. **P. destruens** (Brond.) Fr. (= *Pholiota comosa* Fr. sec. Quél., *Pholiota heteroclita* Fr. sec. Bres.) Bres. Fung. Trid. t. 84.
Destruens, destructive.

P. 6–20 cm., *yellowish white, disc becoming tawny, covered with white, woolly, fugacious scales, fleshy, somewhat viscid,* convex, then flattened, sometimes gibbose, or broadly umbonate; margin at first involute, fibrillose. St. 5–17 × 2–3 cm., *concolorous, covered with white, fugacious squamules,* becoming smooth, attenuated at the apex, *base bulbous and rooting.* Ring *white,* floccose, fugacious. Gills *white, then becoming umber cinnamon,* rounded behind, adnexed, or plano-adnate and striato-decurrent, crowded. Flesh *white, fulvous cinnamon in the base of the st.* Spores fuscous ferruginous, elliptical, 8 × 5μ. Cystidia "on edge of gill cylindrical, or clavate-capitate, 40–60 × 8–12μ" Rick. Smell unpleasant, taste bitter, then sweet. Poplar, birch, beech, and willow trunks. Sept.—Nov. Uncommon. (*v.v.*)

279. **P. heteroclita** Fr. (= *Pholiota destruens* (Brond.) Fr. sec. Bres.) Cke. Illus. no. 389, t. 366. *ἕτερος*, one side; *κλίνω*, I lean.

P. 5–15 cm., *whitish, or yellowish, tawny when old, generally broken up into broad, scattered, innate, adpressed, spot-like (darker) scales,* fleshy, compact, hemispherical, then flattened, very obtuse, sometimes viscid when old and wet; margin often appendiculate with the remains of the ring. St. 5 × 1–2·5 cm., *white,* commonly curved-ascending, base bulbous and bluntly rooting, hard, fibrillose. Ring floccose, fugacious, cortinate, encircling the stem with an annular zone. Gills *pallid, at length dirty ferruginous,* rounded behind, slightly adnexed, very broad, crowded. Flesh *white, rhubarb coloured at the base of the stem.* Spores ferruginous, "8–10 × 5–6μ" Karst. Smell strong, pungent, almost that of horse-radish. Solitary. Trunks of poplar, birch, and willow. Sept.—Nov. Uncommon.

280. **P. aurivella** (Batsch) Fr. *Aurum,* gold; *vellus,* fleece.

P. 5–17 cm., *yellow, or ferruginous yellow,* fleshy, campanulate, then convex, compact at the disc, gibbous when expanded, moist, *with darker adpressed spot-like scales*; margin involute, sprinkled with floccose scales. St. 7–10 × 1–1·5 cm., *yellowish, clothed with adpressed, flocсoso-fibrillose, fuscous ferruginous scales up to the ring,* at length naked, fibrillose, equal, somewhat rooting. Ring floccoso-fibrillose, superior. Gills *whitish, then straw coloured, at length ferruginous date-brown, or somewhat fuscous, sinuato-adnexed,* broad, crowded. Flesh *white, becoming yellowish.* Spores ferruginous, elliptical, 6–7 × 4–5μ, 1-2-guttulate. Cystidia "on edge of gill clavate-fusiform, 30–45 × 6–9μ, contents becoming yellow" Rick. Caespitose, or solitary. On old trunks, and stumps. Sept.—Nov. Uncommon. (*v.v.*)

var. **filamentosa** (Schaeff.) Fr. *Filum,* a thread.

Differs from the type in the *smaller, fulvous pileus, with subconcentric, adnate scales, in the filamentous stem, and the floccoso-radiate ring.* Pine woods.

281. **P. squarrosa** (Müll.) Fr. Cke. Illus. no. 391, t. 367. *Squarrosa,* scaly.

P. 3–10 cm., *saffron ferruginous, or ochraceous,* fleshy, campanulato-convex, then flattened, obtusely umbonate, or gibbose, *squarrose with innate, crowded, revolute, darker (becoming fuscous) persistent scales.* St. 6–20 × 1–2·5 cm., *concolorous,* attenuated downwards, often incrassated at the base, *squarrose up to the ring with crowded, revolute, darker scales.* Ring *of the same colour as the scales,* fibrillose, laciniate, superior. Gills *pallid olivaceous, then ferruginous,* adnate with a decurrent tooth, crowded, narrow. Flesh *light yellow.* Spores ferruginous, elliptical, 7-9 × 4–5μ. Cystidia "clavate, pointed, 30–

45 × 10–12μ, contents olive-yellow" Rick. Smell unpleasant, like rotting wood, sometimes none. Densely caespitose. On and near trunks and stumps, especially ash and apple, more rarely under conifers. July—Dec. Common. (*v.v.*)

var. **Mulleri** Fr. Cke. Illus. no. 392, t. 471.
O. F. Muller, the Danish mycologist.

Differs from the type in the *obtuse, moist, pallid p., with darker adpressed scales, in the equal stem, the entire ring, and in the gills becoming fuscous.* Spores ferruginous, elliptical, 7–9 × 4–5μ, 1-2-guttulate. On beech stumps. Sept.—Oct. Not uncommon. (*v.v.*)

var. **verruculosa** (Lasch) Fr. Cke. Illus. no. 398, t. 370, upper figs., as *Pholiota tuberculosa* (Schaeff.) Fr. sec. Boud.
Verruculosa, full of warts.

Differs from the type in the *compact, obtuse, yellow p., with crowded cinnamon scales and papillae, and in the villose-scaly stem.* On maple trunks. Sept.—Oct. Uncommon.

var. **reflexa** (Schaeff.) Fr. Schaeff. Icon. t. 80. *Reflexa*, bent back.

Differs from the type in the *thinner, cuspidately umbonate, pilosely-squamulose p., the long, equal stem, and the membranaceous ring.* At the base of oak, and beech trees. Uncommon.

282. **P. subsquarrosa** Fr. Fr. Icon. t. 103, fig. 3.
Sub, somewhat; *squarrosa*, scaly.

P. 5–6 cm., *brown-ferruginous, with darker, adpressed, floccose scales,* fleshy, convex, obtuse, or gibbous, *viscid.* St. 5–8 cm. × 8–10 mm., *yellow-ferruginous, clothed with darker, adpressed scales,* equal, *furnished with an annular zone at the apex.* Gills *pale, then dingy yellow,* deeply sinuate, emarginate, *almost free,* arcuate, crowded. Flesh of stem *becoming yellow-ferruginous.* Spores ochraceous, oblong-elliptical, 4·5–5 × 2–2·5μ. Cystidia ochraceous, fusiform, tapering into a long exserted point, 25–30 × 6–8μ, thick walled, contents yellowish, granular. Subcaespitose. On fir stumps and at the base of trunks. Sept. Uncommon. (*v.v.*)

283. **P. grandis** Rea. *Grandis*, large.

P. 20–30 cm., *fulvous tawny,* fleshy, convex, then expanded and broadly gibbous, *covered with innate, deep tawny squamules at the circumference, which become revolute at the disc.* St. 25–30 × 6–7 cm., *tawny* below the ring, *paler and deeply striate* for 5 cm. above, fusiform, slightly squamulose. Ring distant, *almost fugacious.* Gills *pallid, then fuscous, deeply sinuato-decurrent, very broad,* 1·5–2 cm. wide, attenuated in front, somewhat crowded. Flesh *light yellow, ferruginous in the stem.* Spores fuscous, oblong, 6 × 3μ, 1-guttulate. Smell and taste very pleasant. Caespitose. At base of oak, and ash trees. Aug.—Oct. Not uncommon. (*v.v.*)

β. Gills yellow then truly ferruginous, or tawny.

284. **P. spectabilis** Fr. Fr. Icon. t. 102. *Spectabilis*, remarkable.

P. 5–13 cm., *tawny, or golden yellow, then becoming pale,* fleshy, compact, convex, obtuse, *shining as if varnished* in dry weather, *torn into adpressed, innate, pilose squamules of the same colour,* continued into the veil at the inflexed margin. St. 6–13 × 2–3 cm., *sulphur yellow,* hard, more or less *ventricose, extended into a fusiform root, sheathed with the veil,* sometimes squamulose, sometimes smooth, shining, apex *mealy.* Ring *yellowish,* becoming discoloured, inferior, persistent, spreading. Gills *pure yellow, becoming ferruginous,* adnate, most frequently with a small decurrent tooth, very crowded. Flesh *sulphur yellow, reddening when touched,* thick, hard. Spores ferruginous, elliptical, 8–9 × 5μ, 1-guttulate. Smell pleasant. Taste bitter-aromatic. Caespitose. Stumps, and at the base of trees, especially ash and apple. Aug.—Dec. Common. (*v.v.*)

285. **P. adiposa** Fr. Cke. Illus. no. 395, t. 353. *Adiposa*, fat.

P. 3–17·5 cm., *yellow, covered with superficial, floccose, subconcentric, fugacious, ferruginous scales,* fleshy, convex, then plane, somewhat gibbous, *very viscid.* St. 6–15 × 1–2·5 cm., *whitish, then light yellow, clothed with squarroso-reflexed, separating, ferruginous scales,* thickened downwards, *viscid.* Ring *yellow, at length ferruginous,* floccoso-radiate, cortinate. Gills *pallid light yellow, then ferruginous, adnate,* slightly rounded, broad. Flesh *yellowish, tawny at the base of the stem.* Spores fuscous ferruginous, elliptical, 6–7 × 3–4μ. Beech, ash, and birch trunks. July—Oct. Not uncommon. (*v.v.*)

286. **P. lucifera** (Lasch) Fr. Bres. Fung. Trid. t. 85.
Lucifera, light-bringing.

P. 3–5 cm., *yellow, covered with minute, adpressed, fugacious, tawny scales,* fleshy, convexo-plane, at length umbonate, sometimes gibbous, *viscid,* becoming smooth with age. St. 2–5 cm. × 4–8 mm., *yellow, covered with floccose, fugacious, ferruginous scales, pale yellow above the ring,* equal, or attenuated downwards. Ring *ferruginous,* apical, floccose, fugacious. Gills *yellow, then cinnamon, or ferruginous,* sinuato-adnate, crowded, margin *crenulate, pubescent, white.* Flesh *yellow, ferruginous in the stem.* Spores *yellow* under the microscope, obovate, or subreniform, 7–8 × 5–6μ. Cystidia "on edge of gill clavate-subulate, 30–35 × 6–8μ" Rick. Trunks, branches, straw, and burnt earth. Sept.—Oct. Rare. (*v.v.*)

287. **P. flammans** Fr. Fr. Icon. t. 104, fig. 1. *Flammans*, flaming.

P. 2–10 cm., *yellow tawny, sprinkled with superficial, pilose, somewhat concentric, paler, or sulphur yellow, squarrose, or curly scales*, convex then plane, subumbonate. St. 5–8 cm. × 4–10 mm., *very light yellow, as are also the crowded squarrose scales*, equal, often flexuose. Ring *concolorous*, floccose, apical. Gills *bright sulphur yellow, then ferruginous, adnate*, somewhat thin, crowded. Flesh *light yellow*, thin. Spores ferruginous, elliptical, 8 × 4μ, 1-guttulate. Cystidia "clavate-bottle-shaped, rarely pointed, 30–33 × 6–8μ, contents becoming yellow" Rick. Gregarious, or solitary. Pine woods. July—Oct. Common. (*v.v.*)

288. **P. tuberculosa** (Schaeff.) Fr. Fr. Icon. t. 104, fig. 2, as *Pholiota tuberculata*. *Tuberculosa*, having swellings.

P. 3–5 cm., *tawny yellow, compactly fleshy*, convexo-plane, obtuse, sometimes depressed, smooth, *then broken up into innate, broad, adpressed scales*. St. 2·5–4 cm. × 4–6 mm., *bright light yellow*, incurved, *base bulbous, rooting, fibrillose*, somewhat scaly, often excentric. Ring *concolorous*, floccose, reflexed, fugacious. Gills *light yellow, then tawny, or spotted ferruginous, emarginate*, crowded, broad, plane, *edge serrulated, white floccose*. Flesh *becoming yellow*, not very thick. Spores ochrey-ferruginous, elliptical, 5–8 × 3μ. Solitary, or caespitose. Beech and birch trunks, and on sawdust. Oct. Uncommon.

289. **P. curvipes** Fr. Fr. Icon. t. 104, fig. 3. *Curvus*, bent; *pes*, foot.

P. 3–5 cm., *tawny yellow, or orange*, fleshy, thin, but slightly *firm and tough*, convex, then plane, obtuse, *wholly innato-flocculose, then torn into minute scales*. St. 2·5–5 cm. × 2–6 mm., *light yellow*, equal, incurved, tough, *fibrillose, or delicately squamulose*, sometimes attenuated downwards. Ring floccose, fugacious, rarely manifest. Gills *light yellowish, at length tawny, adnate*, crowded, 4–6 mm. broad, *edge white, at length floccoso-crenate*. Flesh *yellowish, darker in the stem*. Spores slightly tawny, "6–7 × 3–4μ" Karst. Smell and taste pleasant. On sawdust, and fallen trunks, especially poplar, birch, and rose bushes. Oct. Rare.

290. **P. muricata** Fr. *Muricata*, pointed.

P. 2–3 cm., *yellow, covered with innate, fasciculate, granular, or needle-shaped, tawny flocci, that are either erect, or convergent like the warts of a Lycoperdon*, convex, then expanded, obtuse, or umbilicate. St. 2·5–5 cm. × 2–4 mm., *pale yellow*, becoming fuscous, *covered up to the ring with floccose, subsquarrose, or peronate, brownish tawny squamules*, equal, or curved. Ring inferior, small, or incomplete, squamulose, or floccosely radiating, fugacious, like that of *Lepiota granulosa*. Gills *whitish, or yellow, at length pale cinnamon*, adnexed, separating, thin, broad, *edge yellowish*. Flesh *whitish*, *or yellowish*, thin. Spores fuscous, elliptical, "7–8 × 4–5μ" Harper. Gregarious. Beech trunks. Sept. Rare.

291. **P. erinacea** (Fr.) Quél. (= *Naucoria erinacea* Fr.) Cke. Illus. no. 513, t. 480, as *Naucoria erinacea* Fr. *Erinaceus*, a hedgehog.

P. 5–15 mm., *ferruginous-umber, scaly and prickly with very dense, crowded, fasciculate, squarrose flocci*, slightly fleshy, convex, *umbilicate*; margin at first involute, *appendiculate with the fibrillose remains of the ring*. St. 1–1·5 cm. × 2–3 mm., *concolorous*, tough, equal, incurved, *everywhere shaggy* with strigose hairs. Ring *yellowish, fibrillose, ferruginous on the under side*. Gills *concolorous*, adnate, subdistant, ventricose. Flesh *reddish*, dry, tough. Spores ferruginous, elliptical, 9–11 × 6–7μ, 1–2-guttulate. Dead branches. Jan.—Dec. Not uncommon. (*v.v.*)

292. **P. cruentata** Cke. & Sm. Cke. Illus. no. 399, t. 502. *Cruentata*, stained with blood.

P. 4–5 cm., *yellow, then turning red, breaking up into darker, adpressed scales*, fleshy, convex, then expanded, obtuse; margin incurved. St. 3–5 cm. × 6–8 mm., *concolorous*, curved, attenuated at the base and *rooting*, *dark red brown* and sparsely squamulose below. Ring fibrillose. Gills *yellow, then clay colour*, emarginate, rather distant, 4–6 mm. broad, finally separating from the stem. Flesh *pale yellow, at length changing to cinnabar colour*. Taste insipid. Oak stumps, and burnt ground. Aug.—Sept. Uncommon.

***Hygrophanous. Gills cinnamon (not at first light yellow).

293. **P. paxillus** Fr. *Paxillus*; like the species *Paxillus involutus*.

Entirely cinnamon. P. 7–15 cm., subcompact, convexo-gibbous, then expanded, moist, rather repand, becoming pale, somewhat silky when dry. St. 8–16 × 1·5–2 cm., firm, slightly attenuated upwards. Ring *white*, membranaceous, narrow, spreading. Gills decurrent, crowded, broad. Flesh *concolorous*, compact. Trunks. Sept.—Oct. Rare.

294. **P. dissimulans** B. & Br. Cke. Illus. no. 400, t. 371. *Dissimulans*, deceiving.

P. 2–3 cm., *lurid, becoming pale, or whitish*, campanulate, very obtuse, at length flattened, scarcely viscid, hygrophanous; margin involute. St. 3–5 cm. × 3–4 mm., *shining white*, thickened downwards, *fistulose with transverse dissepiments*, base cottony. Ring erect, mostly persistent. Gills *pallid clay colour*, sinuato-adnate, at length decurrent. Flesh *white*, thin. Spores fuscous, elliptical, 7 × 4–4·5μ. Sloe and hawthorn sticks. Oct.—Nov. Rare.

295. **P. sublutea** (Fl. Dan.) Fr. *Sub*, somewhat; *lutea*, yellow.

P. 6–7·5 cm., *yellow*, fleshy, thin, expanded, umbonate, moist, *covered with darker squamules*; margin striate when moist. St. 10–12 × ·5–1 cm., *yellow*, base thickened. Ring narrow, spreading. Gills *cinnamon*, decurrent, crowded. Grassy ground. Oct. Rare.

296. **P. phalerata** Fr. Fr. Icon. t. 105, fig. 1. *Phalerata*, decorated.

P. 5 cm., *yellow, covered with superficial, fugacious, pale scales*, fleshy, thin, convex, then plane, obtuse, moist; margin involute, appendiculate. St. 7–10 cm. × 6 mm., *concolorous, base becoming ferruginous*, equal, *everywhere fibrillose*, or fasciculately squamosely pilose, *adpressedly flocculose above the ring.* Ring *white, distant*, entire, reflexed. Gills *yellow, then cinnamon*, adnato-decurrent, 2 mm. broad, attenuated at the margin. Flesh *concolorous.* Spores "elliptical, 6–7 × 3–4μ, smooth. Cystidia on edge of gill fusiform-filamentous, 30–36 × 5–6μ" Rick. Pine woods, on twigs, pine needles, and the ground. Rare.

297. **P. confragosa** Fr. Fr. Icon. t. 105, figs. 2, 3. *Confragosa*, rough.

P. 3–4 cm., *brick-red, or cinnamon rufous when moist, tawny when dry*, hygrophanous, slightly fleshy, convexo-plane, very obtuse, *densely flocculoso-furfuraceous*, becoming smooth with age. St. 2·5–7·5 × 2–4 mm., *pale ferruginous*, fragile, equal, flexuose, or incurved, *fibrillosely peronate with the veil*, or naked, *striate above the ring.* Ring *white*, membranaceous, spreading, fibrillose, persistent, or fugacious, at length reflexed. Gills *rufous, or cinnamon rufous*, sometimes tinged with purple, adnate, scarcely decurrent, very thin, crowded, linear, 2 mm. broad, *edge* under a lens *unequal, crenulate.* Flesh *concolorous, becoming pale.* Spores "ferruginous, elliptic-oblong, 8 × 4μ" Massee. Subcaespitose. Beech, and fir trunks, and on old fallen elm. Oct. Uncommon.

298. **P. mutabilis** (Schaeff.) Fr. Cke. Illus. no. 402, t. 355. *Mutabilis*, changeable.

P. 3–6 cm., *cinnamon when moist, becoming pale when dry*, hygrophanous, slightly fleshy, convex, then flattened, commonly obtusely umbonate, sometimes depressed, sometimes squamulose when young. St. 4–8 × ·5–1 cm., *ferruginous blackish, or umber downwards, paler upwards, rigid*, equal, or attenuated downwards, *squarrosely scaly up to the ring.* Ring *concolorous*, membranaceous, externally squamulose. Gills *pallid*, then cinnamon, *adnato-decurrent*, crowded, rather broad. Flesh *white, tinged brownish under the cuticle of the p. and in the st.* Spores deep ochraceous, elliptical, or subglobose, 7–8 × 4–5μ, 1-guttulate. Cystidia on edge of gill clavate, or cylindrical, flexuose, obtuse, 23–29 × 4–5μ. Caespitose. Stumps and trunks, especially *Tilia cordata*. April—Dec. Common. (*v.v.*)

299. **P. marginata** (Batsch) Fr. Cke. Illus. no. 403, t. 372.
Marginata, furnished with a border.

P. 2–5 cm., *honey coloured when moist, tan when dry*, hygrophanous, slightly fleshy, convex, then expanded, *obtuse*, margin *striate*. St. 3–9 cm. × 2–8 mm., *concolorous*, equal, *fibrillose, striate, but becoming fuscous, and commonly white velvety at the base.* Ring *pale yellow*, membranaceous, distant, often cortinate and fugacious. Gills *pallid, then darker cinnamon, adnate*, crowded, thin, narrow. Flesh *concolorous, becoming paler*. Spores deep ochraceous, fusiform-elliptical, 8–10 × 5μ, 1-2-guttulate, "almost punctate" Rick. Cystidia "ventricose-fusiform, 50–60 × 10–12μ" Rick. On twigs, and on the ground, especially in coniferous woods. Aug.—Dec. Common. (*v.v.*)

300. **P. sororia** Karst. *Sororia*, sisterly.

P. 2·5–4 cm., *tawny cinnamon*, convex, then expanded, *slightly striate, squamulose*. St. 5 cm. × 4–6 mm., *concolorous, then paler, variegated with white squamules*, equal, wavy, apex scurfy. Gills pallid, then cinnamon, sinuato-adnate, crowded. Spores 6–7 × 3–4μ. On chips, and twigs. Rare.

301. **P. mustelina** Fr. Cke. Illus. no. 404, t. 356.
Mustelina, belonging to a weasel.

P. 1–1·5 cm., *yellow, ochraceous*, slightly fleshy, campanulate, convex. St. 2–2·5 cm. × 1–2 mm., *pallid*, equal, *base thickened and white villose, white mealy above the ring*. Ring *brown*, membranaceous, reflexed. Gills *tawny cinnamon*, adnate, subdistant, *edge white, crenulate*. Stumps, especially pine. Sept. Uncommon. (*v.v.*)

302. **P. unicolor** (Fl. Dan.) Fr. *Unicolor*, of one colour.

P. 6–20 mm., *bay brown, then ochraceous*, hygrophanous, fleshy, *campanulate, then convex, umbonate*, at length striate. St. 3–4 cm. × 2 mm., *concolorous*, or *bistre brown at the base*, equal, apex *mealy*. Ring *yellowish*, membranaceous, thin, *entire*, distant, *persistent*. Gills *pallid-ochrey, slightly adnexed, ventricose, broad, edge white*. Flesh *concolorous*, thin. Spores deep ochraceous, elliptical, 7–8 × 4–5μ, 1-guttulate, "almost punctate. Cystidia on edge of gill filamentous, ventricose at base, 50–60 × 8–10μ, apex 5μ in diam." Rick. Trunks and branches of larch, and on the ground. Sept.—Nov. Not uncommon. (*v.v.*)

III. Growing amongst mosses. Like ringed *Galerae*, hygrophanous.

303. **P. pumila** Fr. Fr. Icon. t. 105, fig. 4. *Pumila*, dwarf.

P. 8–15 mm., *ochraceous*, somewhat fleshy for its size, campanulate, then *hemispherical*, obtuse. St. 2·5–4 cm. × 2–3 mm., *concolorous*, equal, *lax*. Ring distinct, *floccoso-woven in the form of a zone like that of the Cortinarii*. Gills *pallid, wholly adnate*, at first ascending, then plane, 4–6 mm. broad, almost triangular, crowded, stopping short of the acute margin. Flesh *concolorous*, thin, watery. Spores ferruginous, pip-shaped, 9 × 4–5μ. Amongst moss in pastures, on twigs, sawdust, and old walls. Aug.—Jan. Common. (*v.v.*)

304. **P. mycenoides** Fr. (= *Galera mycenoides* (Fr.) Quél.) Boud. Icon. t. 102. *Mycena*, the genus Mycena; εἶδος, shape.

P. 1–2·5 cm., *ferruginous, disc darker, transparent* when moist; *becoming tawny, or pale when dry*, membranaceous, hygrophanous, *striate*, campanulate, soon hemispherical, somewhat obtuse. St. 4–10 cm. × 1–3 mm., *tawny ferruginous, darker than the pileus*, filiform, apex *furfuraceous*, paler, then concolorous. Ring *white, membranaceous, entire*, large, persistent. Gills *yellowish, then ferruginous, adnate*, with a small decurrent tooth, *subdistant*, at length plane. Flesh *concolorous, darker in the stem*. Spores ferruginous, broadly elliptical, 10–11 × 6–7μ, 1–3-guttulate. Amongst moss in pastures, bogs, and on lawns. Sept.—Dec. Not uncommon. (*v.v.*)

305. **P. rufidula** Kalchbr. Kalchbr. Icon. t. 37, fig. 3. *Rufus*, red.

P. 6–10 mm., *rufous brick-red, clay colour when dry, often becoming livid round the umbo*, somewhat fleshy, convex, then somewhat plane, *disc always depressed, white-flocculose at the margin* from the white veil, rarely appendiculate. St. 3–4 cm. × 2–3 mm., *watery rufescent, sprinkled with white fugacious fibrils*, floccose at the white base. Ring *white*, floccose, somewhat persistent, subapical, narrow, spreading, reflexed. Gills *reddish*, adnate, then decurrent, linear, branched, of different lengths, *subdistant*. Flesh *concolorous*. Spores ochraceous ferruginous, "8–10 × 4–6μ," Massee. Pastures, and lawns. April. Rare.

Spores purple, or fuscous.

Stropharia Fr.

(στρόφος, a belt.)

Pileus fleshy, regular. Stem central. Ring membranaceous, adnate, persistent, or fugacious. Gills more or less adnate. Spores purple, or fuscous, rarely blackish, elliptical, elliptic-oblong or pip-shaped, smooth, with an apical germ-pore. Cystidia variable. Growing on the ground, on dung, and on wood, sometimes caespitose, subcaespitose, or fasciculate.

A. Pellicle of the p. even, or scaly, most frequently viscid.

*Not growing on dung.

306. **S. depilata** (Pers.) Fr. *Depilata*, plucked.

P. 3–12 cm., *yellowish-livid, then tan*, fleshy, convex, then plane, obtuse, smooth, *viscid*. St. 4–15 cm. × 5–13 mm., *white*, equal, *clothed with white, revolute, squarrose scales below the ring*. Ring *white*, large, distant. Gills *white, then blackish*, adnato-decurrent, 3–8 mm. broad. Flesh *white*, compact. Spores brownish purple, elliptical, 7–8 × 4–5μ. Amongst straw refuse, and in pine woods. Nov. Rare. (*v.v.*)

307. **S. Percevalii** B. & Br. Cke. Illus. no. 554, t. 550. Cecil H. Spencer Perceval.

P. 3–6 cm., *ochraceous*, fleshy, umbonate, then flattened, slightly viscid, *white floccose here and there especially at the margin*, flocci at length coming off. St. 5–7·5 cm. × 6 mm., *pallid upwards*, attenuated upwards from the thickened base, rooting, *transversely scaly*. Ring narrow, more or less persistent. Gills *white, then somewhat cinereous*, at *length pallid umber*, adfixed, 8 mm. broad, *very distant*. Flesh at length *dull umber*. Spores fuscous, oblong elliptic, 16–17 × 7μ. Sawdust, and rotten wood. Sept.—Nov. Uncommon. (*v.v.*)

308. **S. versicolor** (With.) Fr. (= *Armillaria mellea* (Vahl.) Fr. var. *versicolor* W. G. Sm.) *Versicolor*, of various colours.

P. 2·5–10 cm., *becoming greenish brown*, fleshy, convexo-plane, scaly, the scales of the disc crowded. St. 5 cm., as thick as a swan-quill, *whitish fuscous*, spongy-stuffed, *bulbous*. Ring persistent. Gills *pallid, then rufous brown*, decurrent. On the ground.

309. **S. aeruginosa** (Curt.) Fr. Cke. Illus. no. 555, t. 551. *Aeruginosa*, full of copper rust.

P. 3–8 cm., *verdigris green from the azure blue slime, becoming pale and yellowish* as the slime separates, fleshy, campanulato-convex, then flattened, subumbonate, obtuse when larger, *viscoso-pelliculose, often covered with white, fugacious squamules*. St. 4–10 cm. × 4–12 mm., *concolorous*, equal, *viscid, covered with white, fugacious squamules below the ring*, smooth above. Ring *whitish above, concolorous on the underside*, distant, floccose, fugacious. Gills *white, then fuscous, at length somewhat purple*, adnate, soft, plane, 4–8 mm. broad, not crowded, edge often white. Flesh *bluish, becoming whitish*, not compact, thin at the margin. Spores brownish purple, elliptical, 7–10 × 5μ. Cystidia "on surface of gill, clavate, often with a short point, 30–33 × 8–10, contents becoming yellow, on edge of gill filamentous-clavate, 40–75 × 8–10μ" Rick. Smell none, or somewhat strong. Poisonous. Woods, pastures, heaths, gardens, and thatched roofs. May—Nov. Common. (*v.v.*)

310. **S. squamulosa** Massee. *Squamulosa*, squamulose.

P. 4–6 cm., *very deep verdigris-green, margin whitish, disc becoming ochraceous with age*, fleshy, subglobose, then expanded, and slightly depressed, margin drooping, often appendiculate with the veil, *dry and silky from the first, soon becoming broken up into adpressed silky scales*. St. 5–7 × 2 cm., *paler green than the p.*, slightly constricted at the apex, fibrillosely striate, clothed with white patches of the broken up ring, base white. Gills *brown*, sinuately adnate, ventricose, rather broad, crowded, thin, dry. Flesh *of p. white, tinged with green in the st.*, rather thin. Spores pale brown, elliptic-oblong, 8–9 × 5μ, obliquely apiculate. Amongst stones, and in woods. Sept.—Oct. Uncommon. (*v.v.*)

311. **S. albocyanea** (Desm.) Fr. Cke. Illus. no. 556, t. 552.
Albus, white; *κύανος*, dark blue.

P. 1–3 cm., *verdigris green, becoming whitish*, sometimes white and shining when young, fleshy convex, then plane, *viscid with a colourless gluten*, smooth, *naked*. St. 5-8 cm. × 6–8 mm., *whitish, or tinged green*, equal, ascending, or flexuose, *fragile, not viscid*, smooth, pruinose about the ring. Ring *white, becoming stained fuscous with the spores*, narrow, distant, often incomplete. Gills *whitish, becoming fuscous*, sinuato-adnate, 3–4 mm. broad, thin, scarcely crowded. Flesh *white*, soft, watery. Spores purple, elliptical, 8–9 × 4–5μ, with a large central gutta. Woods, heaths, pastures, and ditches. Aug.—Nov. Not uncommon. (*v.v.*)

312. **S. inuncta** Fr. Cke. Illus. no. 557, t. 534. *Inuncta*, anointed.

P. 2·5–5 cm., *pallid light yellow, becoming livid-purple with the dense gluten with which it is at first besmeared*, fleshy, convexo-plane, subumbonate, pelliculose, smooth; margin slightly striate. St. 4–7·5 cm. × 3–4 mm., *shining white*, equal, very flexuose, often decumbent, very soft, *dry*, silky fibrillose below the ring, pruinose above. Ring *white*, median, *distant*, very thin, fugacious. Gills *whitish, then fuscous* when bruised, whitish at the sides, adnate, *with a decurrent* tooth, 6 mm. broad, scarcely crowded. Flesh *white*, thin, soft. Spores fuscous purple, elliptical, 8 × 5μ, 1–2-guttulate. Smell and taste often disagreeable. Heaths, and pastures. Sept.—Nov. Common. (*v.v.*)

var. **Lundensis** Fr. *Lundensis*, appertaining to Lund, Sweden.

Differs from the type in the *campanulate, then expanded p., and the stuffed st.*

var. **Upsaliensis** Fr. *Upsaliensis*, appertaining to Upsala.

Differs from the type in the *convex then plane p. being distinctly umbonate, and in the hollow st.* Woods, and pastures. Sept.—Oct. Uncommon. (*v.v.*)

var. **pallida** B. & Br. *Pallida*, pallid.

Differs from the type in being *very pale*.

313. **S. coronilla** (Bull.) Fr. (= *Stropharia melasperma* Fr. sec. Quél.) Quél. Jur. et Vosg. t. 14, fig. 7. *κορώνη*, a garland.

P. 2–5 cm., *ochraceous tawny*, fleshy, hemispherical, then expanded, smooth; *margin white floccose*. St. 3–4 cm. × 4–10 mm., *white, becoming yellowish with age or when touched*, equal, attenuated at the base. Ring *white*, narrow, median, *striate, then violaceous, with the edge white*, sinuato-adnate, crowded. Flesh *white*, firm. Spores purple, elliptical, 9–10 × 5μ. Cystidia "on surface of gill clavate, 30–40 × 8–10μ, on edge of gill filamentous-capitate, 36–45 × 6–10μ" Rick. Taste mild. Heaths, and pastures. May—Oct. Not uncommon. (*v.v.*)

314. **S. obturata** Fr. (= *Stropharia coronilla* (Bull.) Fr. sec. Quél.) *Obturata*, stopped up.

P. 2–3 cm., *light yellow*, fleshy, convex, then plane, obtuse, obsoletely viscid, *commonly dry*, smooth, *often rimosely squamulose*. St. 2·5–4 cm. × 6–8 mm., *white*, firm, *attenuated downwards*, smooth. Ring *white*, superior, deflexed. Gills *white, then purple umber*, adnate, plane, crowded. Flesh *shining white*, thick, *compact*. Spores purple brown, 7–9 × 4–6μ. Pastures. Sept.—Oct. Rare.

315. **S. melasperma** (Bull.) Quél. non Fr. Bres. Fung. Trid. t. 61. *μέλας*, black; *σπέρμα*, seed.

P. 3–6 cm., *whitish, disc straw coloured*, fleshy, convex, then plane and somewhat depressed, soft, smooth, slightly viscid, soon dry, often broken up into patches. St. 4–6 cm. × 6–7 mm., *white*, subequal, silky fibrillose, becoming smooth, apex striate. Ring *white*, adnate to the st., striately sulcate to the middle on the upper-side, smooth on the under-side, fugacious. Gills *whitish, then violaceous cinereous, at length becoming blackish cinereous*, sinuato-adnate, ventricose, crowded. Flesh *of p. white, at length becoming somewhat straw colour in the st.* Spores fuscous, 9–10 × 6μ. Heaths, pastures, and woods. Aug.—Oct. Not uncommon. (*v.v.*)

var. **lutescens** Boud. Cke. Illus. no. 558, t. 535, as *Stropharia coronilla* Bull. sec. Boud. *Lutescens*, becoming yellow.

Differs from the type in the *yellow p.*

316. **S. squamosa** (Pers.) Fr. Cke. Illus. no. 560, t. 553.
Squamosa, scaly.

P. 2–7·5 cm., *brownish when moist, then becoming ochraceous, disc tawny*, fleshy, hemispherical, then flattened, more frequently obtuse, or gibbous with an obsolete umbo, viscoso-pelliculose when moist, not viscid when dry, *sprinkled with superficial, fugacious, piloso-fasciculate, concentric scales*. St. 6–12·5 cm. × 4–6 mm., *pallid, becoming ferruginous at the base*, equal, tough, pulverulent above the ring, *either squarrose with fibrillose reflexed scales, or covered over with dense, strigose down below the ring*. Ring *pallid*, membranaceous, thin, *distant*. Gills *cinereous, then blackish, edge white*, adnate, ventricose, 10–12 mm. broad. Flesh *pallid, often reddish when moist*, thin, watery. Spores brownish purple, broadly elliptical, often depressed on one side, 14–15 × 7–8μ, with an apical germ-pore. Cystidia "on edge of gill, filamentous-clavate, 50–70 × 4–7μ" Rick. In troops. Woods and heaths. Sept.—Nov. Common. (*v.v.*)

var. **aurantiaca** Cke. Cke. Illus. no. 562, t. 555, as *Stropharia thrausta* Kalchbr. var. *aurantiaca*. *Aurantiaca*, orange.

Differs from the type in the *orange, or brick-red p*. Woods. Sept.—Oct. Uncommon. (*v.v.*)

var. **thrausta** (Kalchbr.) Cke. (= *Stropharia luteo-nitens* (Fl. Dan.) Fr. sec. Quél.) Kalchbr. Icon. t. 15, fig. 2. *θραύω*, I break.

Differs from the type in being *more slender, and in the p. being soon denuded of its scales*. Woods. Sept.—Oct. Not uncommon. (*v.v.*)

317. **S. Worthingtonii** Fr. Cke. Illus. no. 563, t. 556.
Worthington G. Smith, the eminent mycologist.

P. 2–3 cm., *yellow*, fleshy, campanulate, *smooth*, viscid?. St. 4–7 cm. × 1–2 mm., *dark blue*, flexuose, *smooth*. Ring *incomplete*, distant. Gills *brown cinnamon*, adnate, broad. Flesh of p. *pale sulphur yellow, pale bright blue in the st.*, thin. Spores brown, elliptical, 7 × 4μ. Pastures and woods. Nov.—Dec. Uncommon.

**Growing on dung. Ring often incomplete.

318. **S. luteo-nitens** (Fl. Dan.) Fr. *Luteus*, yellow; *nitens*, shining.

P. 2·5–5 cm., *yellow*, fleshy, conico-hemispherical, *umbonate*, smooth, covered with *pallid, superficial, fugacious squamules towards the margin, viscid* when moist, *shining* when dry. St. 5 cm. × 4 mm., *pallid*, equal, *somewhat firm, minutely silky fibrillose*, apex pruinose. Ring *white*, distant, membranaceous, entire, spreading. Gills *cinereous, becoming blackish*, subadnate, truly ventricose, broad, plane. Flesh *white*, thin. Spores "at first violet, then olive yellow under the microscope, elliptical,

15–19 × 9–11μ. Cystidia only on the edge of the gill filamentous-capitate, 25–27 × 3–6μ" Rick. On dung in pastures. Sept. Uncommon.

319. **S. merdaria** Fr. (= *Psilocybe merdaria* (Fr.) Rick.) Cke. Illus. no. 565, t. 537. *Merdaria*, of dung.

P. 2–5 cm., *somewhat cinnamon when moist, then yellow, and at length ochraceous,* fleshy, obtusely campanulate, then convexo-plane, gibbous, *smooth,* pelliculose, *moist, hygrophanous, slightly viscid*; margin thin, deflexed, even, at length somewhat striate. St. 5–7·5 cm. × 4–6 mm., *straw white,* tough, equal, subflexuose, *stuffed with a pith, flocculosely villous, and slightly silky,* dry, apex striate, base white villous. Ring *concolorous, incomplete,* torn, for the most part *commonly adhering to the margin of the p.* Gills *pallid, somewhat isabelline, at length brown fuscous, adnato-decurrent,* plane, 6 mm. broad, very broad behind, somewhat crowded, soft, *edge white.* Flesh *whitish,* thick at the disc, *sometimes fuscous in the st. when old.* Spores black fuscous, broadly elliptical, 13–15 × 8–9μ, with a large central gutta, and flattened germ-pore. Cystidia "on edge of gill cylindrical-filamentous, 25–30 × 3–5μ" Rick. Gregarious, or subcaespitose. On horse dung. Aug.—Oct. Not uncommon. (*v.v.*)

var. **major** Fr. (= *Stropharia ventricosa* Massee.) Fr. Icon. t. 130, fig. 3. *Major*, larger.

Differs from the type in its much *larger size and in the ventricose, rooting st.* On horse dung. Sept.—Oct. Not uncommon. (*v.v.*)

320. **S. stercoraria** Fr. Cke. Illus. no. 566, t. 538. *Stercoraria*, of dung.

P. 2–3 cm., *yellow,* fleshy, hemispherical, then expanded, obtuse, orbicular, *pelliculoso-viscid,* smooth; margin sometimes striate. St. 7–10 cm. × 4–6 mm., *yellow,* equal, *stuffed with a separable, fibrous pith, clothed below the ring with the viscid flocculose veil* so that it appears as if smooth. Ring *viscid,* distant, thin, narrow, spreading, floccose. Gills *white, then umber fuscous, or olivaceous fuscous,* adnate, 4–8 mm. broad, *very broad behind,* somewhat crowded. Flesh *pallid,* thin at the margin. Spores fuscous purple, oblong elliptical, 18–20 × 8–10μ, with a flattened germ-pore. Cystidia "lanceolate, 50–70 × 12–18μ" Rick. Quélet says that the st. often rises from a sclerotium. Dry dung. Pastures. May—Nov. Common. (*v.v.*)

321. **S. semiglobata** (Batsch) Fr. Cke. Illus. no. 567, t. 539. *Semiglobata*, hemispherical.

P. 1–3 cm., *light yellow,* fleshy, *hemispherical,* very obtuse, smooth, *viscid.* St. 6–10 cm. × 2–3 mm., *becoming yellow, apex paler,* equal, *tense and straight, smooth, smeared with the glutinous veil.* Ring *viscid,* incomplete, distant, thin. Gills *clouded with black*, adnate, *very broad*, 8–10 mm., plane. Flesh *pallid*, thin. Spores fuscous purple, elliptical, 15–17 × 9–10μ. Cystidia *only on edge of gill*, filamentous, apex obtuse, 5–6μ in diam., base ventricose, 50–60 × 7–11μ. On dung, especially horse. Woods, and pastures. April—Nov. Common. (*v.v.*)

B. P. without a pellicle, but innato-fibrillose, not viscid.

322. **S. caput-Medusae** Fr. (= *Hypholoma caput-Medusae* (Fr.) Rick.) Cke. Illus. no. 568, t. 540.

Caput, head; *Medusa*, a monster with snakes instead of hair.

P. 3–7·5 cm., *disc umber, paler, somewhat tan colour towards the margin*, fleshy, ovate, then convexo-expanded, obtuse, or obtusely umbonate, *dry, when young very densely, scaly-squarrose with the fuscous veil*, soon becoming smooth especially at the disc, which becomes rimosely warty, or granulose; margin thin, splitting. St. 5–12 × 1–1·5 cm., *whitish*, equally attenuated upwards, somewhat fragile, *covered below the ring with fuscous, crowded, imbricate, squarrose scales*, white mealy above. Ring *white*, superior, membranaceous, pendulous, margin very much swollen, floccose, fuscous. Gills *clay white, then pale umber, fuscous spotted in appearance*, adnate, ventricose, lanceolate, or semiovate, 4–6 mm. broad, not very crowded, fragile. Flesh *pallid*, thick at the disc. Spores fuscous purple, pip-shaped, 8–9 × 4μ, 1-guttulate. Cystidia "vesiculose-flask-shaped, 45–60 × 12–20μ, rounded above" Rick. Subcaespitose, or fasciculate. On and near pine stumps. Sept.—Oct. Rare. (*v.v.*)

323. **S. scobinacea** Fr. (= *Stropharia versicolor* (With.) Quél., *Hypholoma scobinacea* (Fr.) Rick.) Cke. Illus. no. 1179, t. 1189.

Scobinacea, powdered.

P. 3–5 cm., *fuscous, disc livid, becoming yellow, circumference grey-violaceous*, fleshy, hemispherical, then expanded, *gibbous, slightly sulcate, covered with crowded, adpressed, separating, fugacious, blackish squamules*. St. 6–9 cm. × 6–8 mm., *white*, attenuated from the thickened base, fragile, fibrillose, apex mealy. Ring *white*, superior, fugacious. Gills *whitish flesh colour, then purple*, adnate, crowded, crenulated. Flesh *whitish, pinkish when exposed to the air*, thin. Spores brownish purple, elliptical, 7–8 × 3μ. Caespitose. On and near stumps, especially ash. April—Oct. Uncommon. (*v.v.*)

S. cotonea Quél. = **Hypholoma lacrymabundum** Fr.

324. **S. Battarrae** Fr. Quél. Jur. et Vosg. t. 22, fig. 4, as *Stropharia aculeata* Quél.

J. A. Battarra, author of Fungorum Agri Ariminensis Historia.

P. 5–7 cm., *whitish, grey, fuscous, or olivaceous*, fleshy, hemispherical, then plane, obtuse, *covered with darker, adpressed, fibrillose scales*, *which are erect on the disc*; margin appendiculate with the partial veil. St. 2·5–5 cm. × 5–6 mm., *whitish*, incurved, thickened at the base, *covered below the ring with imbricate, erect, fibrillose, fuscous, or olivaceous scales, which often become rufescent*, apex pruinose. Ring *white*, membranaceous, thin, fugacious. Gills *white, then rosy, and at length fuscous*, sinuato-adnate, crowded, *edge white*. Flesh *white*, thin, fragile. Spores "brownish purple, elliptical, 10μ" Quél. Poplars, and poplar stumps. Rare.

325. **S. Jerdonii** B. & Br. Cke. Illus. no. 569, t. 541. A. Jerdon.

P. 4–5 cm., *ochraceous, brown when dry*, fleshy, campanulate, obtuse, with a broad umbo, *minutely rivulose, adorned with superficial, fugacious, white scales*; cuticle not peeling off. St. 5–7 cm. × 4–6 mm., *snow white*, cylindrical, apex pulverulent, *brownish with silky transverse scales below*. Ring superior, deflexed. Gills *pallid, then brown*, adnate, sending a line down the st. but not truly decurrent, *transversely* striate. Flesh *white, brownish towards the base of the st.*, thick at the disc. Spores dark brown, "10 × 5μ" Massee. Caespitose. Fir stumps. Sept.—Nov. Rare.

326. **S. spintrigera** Fr. (= *Hypholoma appendiculatum* Bull. sec. Quél.) Fr. Icon. t. 132, fig. 1.
Spinther, a bracelet; *gero*, I wear.

P. 2–10 cm., *brownish, or pinkish tan, becoming pale*, fleshy, *fragile*, ovate, then expanded, *smooth*, soft. St. 5–15 cm. × 4–10 mm., *shining white*, equal, or slightly ventricose, base often subbulbous, *floccosely squamose, or fibrillose*, apex naked. Ring *white, very distant*, thin, fugacious. Gills *brownish fuscous*, adnate, linear, 2–8 mm. wide, joined behind, subdeliquescent. Flesh *white*, thin, almost none at the margin. Spores fuscous cinereous, elliptical, 7–8 × 4–5μ. Caespitose. On trunks and stumps. Sept.—Oct. Uncommon. (*v.v.*)

327. **S. punctulata** (Kalchbr.) Fr. (= *Flammula gummosa* (Lasch) Fr. sec. Quél.) Kalchbr. Icon. t. 14, fig. 2, as *Pholiota punctulata* Kalchbr. *Punctulata*, minutely dotted.

P. 2·5–3 cm., *pallid, tinged slightly yellowish or fuscous*, fleshy, convex, obtuse, disc somewhat depressed, dry, *minutely squamulosely punctate from the veil*, at length smooth; margin involute. St. 2·5–5 cm. × 4–6 mm., *pallid*, equal, *or bulbous at the base, clothed below the ring with fibrillose, punctiform squamules*, smooth above. Ring superior, or subapical, thin, *formed of the fibrils of the veil brought together in a zone, fugacious*. Gills *pallid, then pale umber*, sinuato-adnate, decurrent with a tooth, 4–6 mm. broad. Flesh *becoming pale fuscous, tawny at the base of the st.*, thin. Spores fuscous. On buried chips. Sept.—Oct. Rare.

328. **S. hypsipus** Fr. Fr. Icon. t. 132, fig. 2. ὑψί-πους, high-footed.

P. 5–9 cm., *livid fuscous when moist, pallid tan when dry*, somewhat fleshy, campanulate, then convexo-plane, obtuse, *smooth, hygrophanous*; margin somewhat striate when moist. St. 7–15 cm. × 4–8 mm., *white, fragile*, equal, *smooth*. Ring *white*, membranaceous, *median, distant*, persistent. Gills *white, then fuscous*, adnate, at length separating, subdistant. Flesh *white*, very thin. Spores "brown, elliptical, 12–14 × 6–7μ" Massee. Damp places amongst grass, sometimes on twigs. Oct. Rare.

329. **S. cothurnata** Fr. Fr. Icon. t. 132, fig. 3. κόθορνος, a high hunting boot.

P. 1–2·5 cm., *white, submembranaceous*, convex, then plane, obtuse, smooth, silky smooth under a lens. St. 3–5 cm. × 2–3 mm., *white*, equal, soft, *floccosely villose below the ring*, smooth above. Ring *white, median, narrow*. Gills *white, becoming fuscous*, adnexed, ventricose, crowded. Flesh *watery white*, thin at the disc. Shady fir woods. Rare.

S. lacrimabunda (Bull.) Quél. = **Hypholoma velutinum** (Pers.) Fr.

S. pyrotricha (Holmsk.) Quél. = **Hypholoma pyrotrichum** (Holmsk.) Fr.

**With an arachnoid or filamentous general veil.

1. Veil forming an arachnoid, fugacious ring on the stem.

Spores ochraceous, or ferruginous.

Cortinarius Fr.

(*Cortina*, a veil.)

Pileus fleshy, regular. Veil arachnoid, distinct from the pellicle of the pileus, viscid, or dry. Stem central. Gills adnate, sinuate, sinuato-adnate, or decurrent, pulverulent with the spores at maturity. Spores ochraceous, citron yellow, golden, clay colour, cinnamon, ferruginous, tawny, or fuscous; oval, elliptical, pip-shaped, oblong elliptical, almond-shaped, or fusiform; smooth, verrucose, granular, aculeolate, or echinulate, continuous. Cystidia none on the surface of the gills (except 333), rarely on the edge. Growing on the ground, solitary, caespitose, or subcaespitose.

1. Phlegmacium Fr.

(φλέγμα, phlegm.)

Veil *viscid* on the pileus, arachnoid, dry on the stem.

I. Partial veil superior, pendulous from the apex of the clavate, or subequal stem as an imperfect ring.

*Gills pallid, then clay colour.

330. **C. (Phleg.) triumphans** Fr. Cke. Illus. no. 682, t. 692.
Triumphans, triumphal.

P. 6–12 cm., *yellow with a tinge of brown or ochraceous when moist, yellow when dry,* disc sometimes variegated with minute, adpressed, spot-like scales, margin *brighter coloured,* fleshy, convexo-plane, obtuse. St. 6–17 × 1–2·5 cm., *yellowish white, adorned with tawny scales arranged in many circles, or rings,* firm, attenuated upwards from the ovato-bulbous base, striate; partial veil superior, woven, *somewhat ringed.* Gills *whitish,* or *inclining to very pale bluish grey, then clay colour, and somewhat cinnamon,* emarginate, crowded, 6 mm. broad; edge *toothed, white.* Flesh *whitish.* Spores brown, punctate, elliptical, 9–10 × 4–5μ, 1-guttulate. Taste mild. Woods, heaths, and commons, generally under birches. Sept.—Oct. Not uncommon. (*v.v.*)

331. **C. (Phleg.) claricolor** Fr. Fr. Icon. t. 141, fig. 2.
Clarus, bright; *color*, colour.

P. 7–12 cm., *yellow,* fleshy, convexo-plane, then depressed, veiled with superficial, *silky-pruinose, villose down,* at first everywhere, soon only round the margin, then smooth, and *for the most part broken up into scales.* St. 7–9 × 1·5–2 cm., *white, or yellow,* hard, *clothed* up to the superior cortina *with fugacious, white, scaly flocci, or lax down,* either short and bulbous, or elongated and conico-attenuated, or cylindrical. Gills *whitish, then clay colour,* emarginate, almost free, or adnate, crowded, *edge toothed.* Flesh *white,* compact. Spores brown, punctate, pip-shaped, 10–11 × 6–7μ. Taste pleasant. Woods, and heaths, under birches and conifers. Aug.—Nov. Uncommon. (*v.v.*)

332. **C. (Phleg.) turmalis** Fr. Cke. Illus. no. 684, t. 694.
Turmalis, belonging to a troop.

P. 5–9 cm., *yellow-tan,* disc often darker, compact, convex, then plane, very obtuse, *when young veiled with pruinate, very fugacious, villose down,* soon smooth, sometimes obsoletely piloso-virgate. St. 7–15 × 2·5 cm., *shining white when dry,* very hard, rigid, *cylindrical,* attenuated at the base, *when young sheathed with a fugacious, white, woolly veil.* Cortina entirely fibrillose, superior and persistent in the form of a ring, at length ferruginous with the spores. Gills *white, then clay colour,* adnexed, rounded, or emarginate, even decurrent with a tooth, crowded, *serrated.* Flesh *white,* soft. Spores ferruginous, palish fuscous under the microscope, elliptical, 8–9 × 5μ. Densely caespitose. Mixed woods, and especially amongst beech leaves. Sept.—Oct. Uncommon.

333. **C. (Phleg.) crassus** Fr. (= *Hebeloma crassum* (Fr.) Rick.) Fr. Icon. t. 142, fig. 1. *Crassus*, thick.

P. 6–12 cm., *dirty yellow*, opaque, very fleshy, convexo-plane, very obtuse, disc depressed, somewhat viscid, *the circumference broken up into innate fibrils*. St. 5 × 2·5 cm., *whitish*, either truly bulbous, or stout and equal, fibrillose, *apex white-mealy*, base often white-tomentose. Cortina very delicate, fugacious. Gills *pallid-clay colour, then clay colour*, rounded behind, crowded, 6 mm. broad. Flesh *becoming pallid*, pouring out a watery juice when the stem is compressed. Spores "brownish, pale yellow under the microscope, punctate, subfusiform, 6–7 × 4μ. Cystidia very abundant, both on the surface and edge of the gill, cylindrical-fusiform, 40–75 × 6–9μ" Rick. Smell strong. Moist woods, and grassy places. Aug.—Oct. Uncommon.

334. **C. (Phleg.) balteatus** Fr. (= *Cortinarius* (*Phlegmacium*) *nemorensis* Fr. sec. Quél.) Fr. Icon. t. 142, fig. 2. *Balteatus*, girdled.

P. 7–10 cm., *disc tawny-fuliginous, or date colour, beautiful violet, or lilac near the floccose margin* (but this colour often disappears with age, or when the plant is dry), compact, flattened, very obtuse, disc depressed, often unequal, shining when dry, *innately floccose*, fibrillose, and shining towards the margin. St. 4–6 × 2·5–3 cm., *whitish*, stout, very compact, equal, or ovately bulbous, *minutely velvety*, and dingy above the fibrillose, adpressed, rusty veil, longitudinally striate, or reticulate below, but sometimes white tomentose. Gills *pallid, then somewhat tan colour*, emarginate, or rounded behind, or subdecurrent, rather crowded. Flesh *white*, compact. Spores ferruginous, pip-shaped, 8–10 × 4–5μ, multi-guttulate. Mixed woods, and amongst pine needles in pine woods. Sept.—Oct. Uncommon. (*v.v.*)

335. **C. (Phleg.) sebaceus** Fr. Boud. Icon. t. 103. *Sebum*, tallow.

P. 5–12 cm., *whitish ochraceous, or deep ochraceous*, convex, then campanulate, umbonate disc flattened, *at first covered over with a whitish pruinose lustre*; margin appendiculate with the remains of the cortina. St. 7–18 × 1–2·5 cm., *concolorous, or paler*, stout, compact, often twisted and compressed, *fusiform* towards the fibrillose base. Cortina *white*, delicate, fugacious. Gills *white, then cinnamon, paler near the margin*, emarginate, adnate, connected by veins, 8 mm. wide, not crowded. Flesh *white*. Spores golden yellow, attenuated at the one end, oblong, 12–15 × 5–6μ, multi-guttulate. Taste pleasant. Mixed woods. Sept.—Nov. Uncommon. (*v.v.*)

336. **C. (Phleg.) lustratus** Fr. Cke. Illus. no. 688, t. 799.
Lustratus, purified.

Entirely whitish. P. 2·5–5 cm., equally fleshy, convex, then expanded, very obtuse. St. 2·5–5 cm. × 8–10 mm., equal, rarely attenuated at the base, covered with a few fibrils of the *white cortina.* Gills *becoming slightly discoloured,* rounded behind, almost free, very crowded, narrow. Flesh *white,* thick. Spores ochraceous. Amongst grass in sunny places. Oct. Rare.

**Gills violaceous, or purplish, then cinnamon.

337. **C. (Phleg.) crocolitus** Quél. Grevillea, t. 127, fig. 1.
κρόκος, saffron; *litus*, daubed.

P. 10 cm., *bright yellow, disc sprinkled with tender, saffron flocci,* convex. St. 6–10 × 1 cm., *white, then citron yellow,* fragile, swollen at the base, fibrillose, apex *silky, adorned with scales, or woolly zones below the membranaceous, very fragile ring.* Cortina *white,* fugacious. Gills *whitish lilac, then nankeen yellow,* uncinate, uneven, *edge white.* Flesh *white, then citron yellow,* soft. Spores "citron yellow, granular, pruniform, 11–12 × 5–6μ" Bat. Taste becoming bitter. Woods, especially birch. Sept.—Oct. Rare.

338. **C. (Phleg.) varius** (Schaeff.) Fr. Cke. Illus. no. 689, t. 698.
Varius, changeable.

P. 5–9 cm., *bright ferruginous-tawny,* or *yellow tawny,* compact, hemispherico-flattened, very obtuse; margin thin, at first incurved, appendiculate with the cortina. St. 4–8 × 2–3 cm., *shining white,* compact, adpressedly flocculose, the superior veil pendulous, *base bulbous.* Cortina *white,* silky. Gills *violaceous-purplish, then ochraceous-cinnamon with the edge violaceous,* emarginate, thin, somewhat crowded. Flesh *white,* firm. Spores ferruginous, warted, broadly elliptical, 8–10 × 6–8μ, often apiculate at one end. Woods. Sept.—Nov. Not uncommon. (*v.v.*)

339. **C. (Phleg.) cyanopus** (Secr.) Fr. Cke. Illus. no. 690, t. 699.
κύανος, dark blue; *πούς*, a foot.

P. 5–8 cm., *date-brown-livid, then tan, and opaque,* fleshy when unfolded, flattened, obtuse, regular, dry in fine weather. St. 5–10 × 1–2 cm., *violaceous, becoming whitish, the apex remaining violaceous,* firm, ventricose, *base bulbous.* Gills *intensely violaceous, or pallid bluish-grey, soon becoming cinnamon,* adnate, then emarginate, broad, 6–10 mm. wide, *not much crowded.* Flesh *violaceous at the apex of the stem, whitish elsewhere.* Spores ferruginous, punctate, pip-shaped, 9–10 × 5–6μ. Woods and under oaks. Aug.—Nov. Not uncommon. (*v.v.*)

340. **C. (Phleg.) variecolor** (Pers.) Fr. *Varie*, diversely; *color*, colour.

P. 8–15 cm., *date-brown, then fulvous-reddish, disc darker, the tomentose margin violet, rarely entirely violet*, convex, then expanded, obtuse. St. 5–8 × 2·5–3 cm., *white with the apex becoming blue, or blue becoming whitish, hard*, stout, base bulbous, diffused upwards into the p., *at first villose, then flocculose.* Gills *blue, then clay-cinnamon*, emarginate, decurrent, somewhat arcuate, thin, 12 mm. wide, *margin crenulate.* Flesh *violaceous, becoming whitish.* Spores ferruginous, "almond-shaped, 15–18 × 8–9μ, warted-rough" Rick. Smell and taste pleasant. Edible. Pine woods. Sept.—Nov. Uncommon. (*v.v.*)

var. **nemorensis** Fr. (= *Cortinarius* (*Phlegmacium*) *balteatus* Fr. sec. Quél.) Cke. Illus. no. 692, t. 863.

Nemorensis, belonging to a wood.

P. 10–12·5 cm., *bay-brown, then yellowish, margin violet, smooth*, slightly viscid at first, soon dry, opaque, *pilosely rivulose.* St. 7·5 × 2·5 cm., *bluish, becoming white, obclavate, not bulbous, nor villose*, apex *mealy.* Gills rounded, subdecurrent. Flesh *white, bluish at the periphery.* Beech woods. Sept.—Oct. (*v.v.*)

341. **C. (Phleg.) largus** Fr. *Largus*, large.

P. 5–15 cm., *sometimes violet when young, date-brown-tawny*, fleshy, compact at the disc, thin at the circumference, convexo-flattened, very obtuse, only slightly viscid, adpressedly silky-fibrillose when dry, commonly rivuloso-squamulose, sometimes fibrillose towards the margin. St. 6–13 × 2–3 cm., *white, tinted violaceous*, equal, *often curved* and ascending, *wholly fibrillose*, apex pruinose. Cortina *white*, silky, *thick, superior, pendulous.* Gills *bluish-grey-clay-colour, then cinnamon*, adnate, or emarginate, crowded, 10–14 mm. broad, *minutely denticulate.* Flesh *whitish-bluish-grey*, becoming white when exposed to the air, that of the stem sometimes becoming bloody when bruised, wholly fibrous, firm. Spores ferruginous, pip-shaped, 10–11 × 5–6μ, "rough" Rick. Smell and taste pleasant. Edible. Caespitose. Deciduous and pine woods. Sept.—Oct. Not uncommon. (*v.v.*)

342. **C. (Phleg.) Riederi** (Weinm.) Fr. Cke. Illus. no. 694, t. 702.

M. Rieder, of Petrograd.

P. 5–7·5 cm., *ochraceous*, compact, campanulate, then expanded, *umbonate, glutinous*, shining when dry. St. 5–12·5 cm. × 5–12 mm., *white, apex violaceous, or lilac, tawny fibrillose*, clavate. Gills *lilac, then cinnamon*, adnate, rather thick, crowded. Flesh *greyish-white, becoming yellow under the cuticle.* Spores ferruginous, "warted, almond-shaped, 15–17 × 8–10μ" Rick. Pine woods. Sept.—Oct. Rare.

***Gills yellow, cinnamon, or ferruginous.

343. **C. (Phleg.) percomis** Fr. Fr. Icon. t. 143, fig. 2.
Percomis, very friendly.

P. 5–7 cm., *pale yellow*, truly fleshy, compact, convex, obtuse. St. 6–8 × 1–2 cm., *sulphur yellow*, compact, firm, *fusiform*, or *clavate*, fibrillose, apex pruinose. Cortina *citron yellow*. Gills *sulphur yellow*, *becoming fulvous*, broadly emarginate, crowded, 4–6 mm. wide. Flesh *sulphur yellow*, compact. Spores ferruginous in the mass, broadly elliptical, 12–14 × 8–9μ, "warted-punctate" Rick. Smell pleasant, "like lavender" Quél, "like toilet vinegar" Peltereau. Taste pleasant. Edible. Pine woods. Sept.—Oct. Uncommon.

344. **C. (Phleg.) latus** (Pers.) Fr. Bres. Fung. Trid. t. 162.
Latus, broad.

P. 6–10 cm., *tan colour*, *disc darker*, fleshy, convex, then expanded, obtuse, slightly viscid, soon dry, fibrillose, then glabrous. St. 5–7 × 1·5–2 cm., *white*, equal, base ovately bulbous, rarely emarginately bulbous, somewhat squamose, then fibrillose. Cortina *white*, superior, forming a ring, fugacious. Gills *pallid, then clay cinnamon*, emarginate, 6 mm. broad, crowded, distantly dentate. Flesh *white*. Spores ochraceous, punctate, minutely rough, oblong elliptical, 10–13 × 6–7μ. Taste pleasant. Gregarious, or subcaespitose. Coniferous woods. Oct. Rare.

345. **C. (Phleg.) saginus** Fr. *Saginus*, fattened.

P. 10–12·5 cm., *yellow*, fleshy, plano-convex, irregular, *repand*. St. 7·5 × 2·5–3 cm., *light yellowish*, somewhat bulbous, fibrillose, *apex naked*. Cortina fibrillose, *fugacious*, not very conspicuous. Gills *dingy-pallid, then cinnamon, truly decurrent*, 8–10 mm. broad, attenuated at both ends, edge eroded. Flesh *white*, soft. Spores "pale yellow under the microscope, almond-shaped, 10–11 × 6–6·5μ, warted-punctate" Rick. Gregarious, subcaespitose. Mountainous fir woods. Oct. Rare.

346. **C. (Phleg.) russus** Fr. Cke. Illus. no. 696, t. 751. *Russus*, red.

P. 6–10 cm., *unicolorous, rufous*, fleshy, convex, then flattened, obtuse, *innately fibrillose round the margin*. St. 7·5 × 1·5–2 cm., *pale white*, attenuated upwards, often curved-ascending, soft, adpressedly fibrillose, apex delicately pruinose. Cortina *concolorous*, very tender, fugacious. Gills *rufous-ferruginous*, obtusely adnate, 8–10 mm. broad, crowded, connected by veins. Flesh *whitish-flesh-colour, violaceous under the cuticle*. Spores ferruginous, elliptical, 8–12 × 5–7μ, "warted-punctate" Rick. Taste bitter, nauseous. Woods. Sept.—Oct. Uncommon. (*v.v.*)

****Gills olivaceous.

347. **C. (Phleg.) infractus** (Pers.) Fr. (= *Cortinarius* (*Phlegmacium*) *anfractus* Fr. sec. Quélet et Bresadola.) Bres. Fung. Trid. t. 163. *Infractus*, broken.

P. 5–10 cm., *olivaceous-fuliginous, becoming fulvous*, fleshy, convex, then plane, streaked, *often fuscous zoned near the undulate, broken margin.* St. 3–7 × 1–3 cm., *concolorous*, ovato-clavate, or elongate and bulbous, adpressedly fibrillose, *apex often violaceous.* Gills *olivaceous-fuliginous, then umber*, crowded, or somewhat distant, broad, undulate, crisped. Flesh *yellowish white, somewhat violaceous at the apex of the stem.* Spores ferruginous in the mass, somewhat ochraceous under the microscope, subglobose, or broadly elliptical, 6–9 × 4–6μ, verrucose. Smell somewhat nauseous. Taste bitter. Woods. Sept.—Oct. Uncommon. (*v.v.*)

348. **C. (Phleg.) praestans** (Cordier) Sacc. (= *Cortinarius* (*Phlegmacium*) *anfractus* Fr. sec. Berk., *Cortinarius* (*Phlegmacium*) *Berkeleyi* Cke., *Cortinarius torvus* Fr. sec. Kalchbr. and Quél.) Boud. Icon. t. 116, as *Cortinarius torvus* Fr. var. *Berkeleyi* Cke. *Praestans*, pre-eminent.

P. 7·5–20 cm., *fuliginous, or brown, disc darker, often with a tinge of violet at the margin, at first inclosed in a whitish volva which breaks up in patches on the disc*, convex, then expanded, shining when dry, very fleshy, sometimes radiately silky, *becoming paler and rivulose with age.* St. 10–20 × 3–6 cm., *white, covered with the general veil, which is at first violaceous, then pale, often remaining appendiculate at the margin of the pileus, finally becoming ochraceous when old*, base bulbous. Gills *dingy olive, then cinnamon*, adnate, slightly emarginate, broad, scarcely distant. Flesh *pale ochraceous, darker under the pellicle of the pileus.* Spores yellow-brown, fusiform, minutely verrucose, 15–16 × 8–9μ. Forming large circles in woods. Sept.—Oct. Uncommon. (*v.v.*)

II. Bulb depressed, or turbinate, *marginate.* St. fleshy, fibrous; *cortina* commonly *inferior*, arising from the margin of the bulb. P. equally fleshy. Gills somewhat sinuate.

*Gills whitish, then clay-coloured, or pale cinnamon.

349. **C. (Phleg.) multiformis** Fr. (= *Cortinarius rapaceus* Fr. sec. Quél., *Cortinarius talus* Fr. sec. Quél.) Boud. Icon. t. 104. *Multiformis*, many shaped.

P. 4–7 cm., *unicolorous, light yellow, clay yellow, tawny*, etc., fleshy, convex, then flattened, very obtuse, at length depressed, very viscid, or somewhat dry, and sprinkled with the universal *white* veil. St. 5–12 × 1–1·5 cm., *white, then yellowish*, equal, or attenuated upwards, often adpressedly fibrillose, with a somewhat marginate bulb. Cortina *white*, fibrillose, fugacious. Gills *whitish, often tinged with violet, then clay colour*, emarginate, free, or with a small decurrent tooth, very thin, crowded, edge serrulate. Flesh *white, becoming yellowish at the base of the stem*. Spores ochraceous-tawny, verrucose, elliptical, $10 \times 6\mu$. Taste mild. Woods, especially beech. Aug.—Nov. Not uncommon. (*v.v.*)

var. **flavescens** (Cke.). Cke. Illus. no. 702, t. 709.
Flavescens, becoming yellow.

Differs from the type *in the yellow gills, and yellowish flesh.*

350. **C. (Phleg.) napus** (Fr.). Cke. Illus. no. 703, t. 710.
Napus, turnip.

P. 5–8 cm., *fuliginous, then date-brown-tawny*, fleshy, convexo-plane, obtuse, *glutinous, margin abruptly bent inwards*. St. 5×1–2 cm., *white, at length becoming yellow at the base*, equal, ascending, firm, inserted in an obconic, *acutely and obliquely marginate bulb*. Gills *whitish-fuliginous*, emarginate, *somewhat distant, broad, crisped*. Flesh *white, with a horny line at the base of the gills*. Spores brownish, elliptical, $10 \times 5\mu$. Pine woods. Sept.—Oct. Uncommon. (*v.v.*)

351. **C. (Phleg.) allutus** (Secr.) Fr. Cke. Illus. no. 704, t. 752.
Allutus, washed.

P. 2–3 cm., *rufescent*, fleshy, conical, then convex, finally expanded, and sometimes depressed, *margin darker*. St. 2–3 cm. × 3–4 mm., *white, striate with reddish lines below*, equal, apex mealy, *viscid*, base marginately bulbous. Gills *whitish, then rufescent*, adnate, rather crowded, edge crenulate. Flesh *rufescent*, thin. Pine woods. Oct. Rare.

352. **C. (Phleg.) talus** Fr. (= *Cortinarius* (*Phlegmacium*) *multiformis* Fr. sec. Quél.) Fr. Icon. t. 145, fig. 2. *Talus*, the ankle bone.

P. 4–8 cm., of a *yellowish dirty colour, becoming pale, margin somewhat olivaceous, yellowish*, fleshy, thin, convexo-plane. St. 7·5 cm. × 12 mm., *pale*, equal, cylindrical, base marginato-bulbous. Gills *beautiful straw colour, or ochrey-pallid*, emarginate, somewhat crowded. Flesh *dingy pallid whitish*, watery, *with hyaline spots, and variegated with a horny line next the gills*. Spores ferruginous, elliptical, 8–10 × 4–5μ, 1–2-guttulate. Woods. Sept.—Nov. Uncommon. (*v.v.*)

**Gills violaceous, dark blue or purplish, at length cinnamon.

353. **C. (Phleg.) glaucopus** (Schaeff.) Fr. Cke. Illus. no. 706, t. 712.
γλαυκός, pale blue; πούς, a foot.

P. 6–12 cm., *dingy yellow, tan-tawny, or clay colour*, very fleshy, *compact*, convex, then flattened, somewhat repand, *often floccoso-scaly* and marked with a raised fuscous zone round the split margin. St. 6–8 × 1–2 cm., *pale azure-blue, becoming yellowish,* firm, *fibrillose,* striate, base marginately bulbous. Gills *azure-blue, then cinnamon,* rounded behind, emarginate, crowded, sometimes crisped. Flesh *white, or bluish, becoming yellowish.* Spores ferruginous, minutely verrucose, broadly elliptical, 9–10 × 5–6μ, with a hyaline apiculus at the one end. Woods, and pastures. Aug.—Oct. Not uncommon. (*v.v.*)

354. **C. (Phleg.) calochrous** (Pers.) Fr. Cke. Illus. no. 707, t. 713.
καλός, beautiful; χρώς, colour.

P. 4–8 cm., *tawny, yellow round the margin,* compact, convex, then plane, obtuse, guttate, often stained with soil; margin involute, flexuose when expanded. St. 4–6 × 1·5–3 cm., *yellowish,* firm, equal, fibrillose, *marginately bulbous, bulb very depressed.* Cortina *yellow* ("amethyst" Quélet), marginal, fugacious. Gills *dark blue-purple, then ferruginous,* emarginate, crowded, *serrated.* Flesh *white,* firm. Spores ferruginous, elliptical, 10–11 × 6–7μ, minutely punctate. Smell sometimes foetid. Taste mild, sometimes acrid. Pastures, and woods, especially beech. Sept.—Nov. Not uncommon. (*v.v.*)

355. **C. (Phleg.) caerulescens** Fr. Cke. Illus. no. 709, t. 722.
Caerulescens, becoming azure.

P. 5–7 cm., *blue-violaceous, becoming tinged with ochre especially on the disc, sometimes entirely yellow ochraceous,* convex, then convexo-plane, sometimes finally a little depressed at the disc, fleshy, pellicle separable; margin incurved, *pubescent, white, then expanded and violaceous.* St. 4–6 × 1–1·5 cm., *blue-violaceous, or violet-amethyst,* cylindrical, conical, fibrillosely silky, then becoming smooth, *marginately bulbous,* bulb *white.* Cortina *violaceous.* General veil fibrillose, *violaceous,* fugacious, little distinct from the cortina. Gills *violet-amethyst, or blue-violaceous, becoming rust colour, edge remaining violet for a long time, broadly adnate, deeply emarginate,* attenuated in front, rounded behind, *wide,* somewhat crowded. Flesh *pale blue-violaceous,* especially in the stem and under the pellicle of the p., *becoming whitish, finally tinged with ochre where wounded.* Spores ferruginous in the mass, yellow ochraceous under the microscope, elliptical, or somewhat almond-shaped, 12–14 × 7·5μ or 6–6·5 × 5μ, compressed on the side, verrucose. Taste sweet, or slightly bitter. Woods, especially beech, and fir. Sept.—Oct. Not uncommon. (*v.v.*)

356. **C. (Phleg.) caesiocyaneus** Britz. Cke. Illus. no. 708, t. 721, as *Cortinarius (Phlegmacium) caerulescens* Fr.
Caesius, bluish grey; κύανος, dark blue.

P. 5–10 cm., *pale blue-violaceous, more or less washed with yellow ochre at the centre,* fleshy, convex, then convexo-plane, *more or less radially streaked with innate fibrils,* sometimes marked with loose, white patches, the remains of the volva. St. 5–8 cm. × 12–15 mm., *bluish, then becoming pale, and finally whitish, attenuated upwards from the distinctly marginate, bulbous base,* fibrillose; bulb *white from the first,* the margin often forming a ledge, or sheath. Cortina *bluish,* fugacious. Gills *whitish, then bluish white, and finally clay colour and rust colour, narrowly adnate, slightly sinuate, or emarginate,* thin, crowded, *somewhat narrow.* Flesh *yellowish, whitish in the bulb, bluish in the stem.* Spores ferruginous in the mass, yellowish brown under the microscope, almond-shaped, 10–12·5 × 5–6μ ("8–10 × 4–5μ" Britz.), verrucose. Smell faint, like that of *Cortinarius purpurascens.* Taste pleasant. Fir woods. Sept.—Oct. Not uncommon. (*v.v.*)

357. **C. (Phleg.) purpurascens** Fr. Cke. Illus. no. 710, t. 723.
Purpurascens, becoming purple.

P. 6–15 cm., *bay brown, or date brown olivaceous, then tawny olivaceous,* fleshy, convex, obtuse, *glutinous,* opaque when dry, *tiger-spotted,* often depressed round the margin which is at first inflexed, then repand, and *marked with a raised, violet fuscous zone.* St. 5–9 × 1·5–3 cm., *intensely pallid azure-blue, darker when touched,* fibrillose, base bulbous, somewhat marginate. Gills *azure-blue-clay, then cinnamon, violaceous purple when bruised,* broadly emarginate, 6–12 mm. wide, crowded. Flesh *azure-blue.* Spores ferruginous, rough, elliptical, 9-11 × 5μ. Woods, and heaths. Sept.—Nov. Common. (*v.v.*)

var. **subpurpurascens** Fr. Cke. Illus. no. 712, t. 725.
Subpurpurascens, becoming somewhat purple.

Differs from the type in the *thinner, somewhat virgate p., becoming pale, in the somewhat equal, bluish white, somewhat marginately bulbous stem only fibrillose at the base, in the pallid, then cinnamon gills becoming somewhat purplish when rubbed, and in the flesh in young specimens becoming purplish when broken, and finally white.* Woods. Sept.—Oct. Not uncommon. (*v.v.*)

***Gills ferruginous, tawny, or yellow.

358. **C. (Phleg.) dibaphus** Fr. Saund. & Sm. t. 10.
δί-βαφος, twice dyed.

P. 5–10 cm., *purplish, disc yellowish, and at length variegated with lilac,* fleshy, convex, then plane, at length depressed, somewhat repand. St. 6–9 × 1–3 cm., *yellow, shining purplish at the apex,* fibrillose, base marginato-bulbous. Gills *purplish-ferruginous,* adnate, slightly rounded, somewhat crowded, broad ("margin lilac" Quél.). Flesh *white, then yellow, variegated under the pellicle with a violet line.* Spores purplish brown, pip-shaped, 12–14 × 7–8μ, verrucose. Smell and taste mild. Beech, and oak woods. Sept.—Nov. Uncommon. (*v.v.*)

var. **xanthophyllus** Cke. Cke. Illus. no. 713, t. 753.

ξανθός, yellow; *φύλλον*, a leaf.

Differs from the type in the *yellow gills*. Woods. Oct.—Nov. Rare. (*v.v.*)

359. **C. (Phleg.) turbinatus** (Bull.) Fr. Boud. Icon. t. 105.

Turbo, a spinning-top.

P. 5–13 cm., *unicolorous, dingy yellow, or green, becoming pale*, hygrophanous, opaque when dry, fleshy, convex, then flattened, obtuse, at length depressed, orbicular, *covered with adpressed fibrils which are deeper coloured and somewhat squamulose at the disc*. St. 5–7 × 2–3 cm., *concolorous, or paler than the p.*, sometimes tinged with violet at the apex, equal, cylindrical, *springing from a globoso-depressed, distinctly marginate, turbinate bulb*. Gills *pallid light yellowish, sometimes tinted with dark purple, then ferruginous*, uncinately adnate, thin, crowded, broad. Flesh *white*, soft. Spores ferruginous, elliptical, 15 × 7–8μ ("8–10 × 5–6μ" Boud.), verrucose. Taste somewhat bitter. Beech woods, and pastures. Sept.—Oct. Uncommon. (*v.v.*)

var. **lutescens** Rea. *Lutescens*, becoming yellow.

Differs from the type in the *bright yellow colour of the flesh*. Woods, and pastures. Oct. Rare. (*v.v.*)

360. **C. (Phleg.) corrosus** Fr. Cke. Illus. no. 715, t. 715.

Corrosus, gnawed to pieces.

P. 5–8 cm., *clay colour, becoming ferruginous, then pallid*, fleshy, compact, expanded, *umbilicate, opaque when dry, rivulose, flocculose*, only fugaciously viscid. St. 2·5–5 × 1–2·5 cm., *white*, cortinately-fibrillose, equal, base marginately bulbous. Gills *somewhat ferruginous from the first*, emarginate, or rounded behind, *very crowded*, narrow, 4 mm. wide, edge unequal. Flesh *white*, rarely zoned with violet. Pine woods. Sept. Rare.

361. **C. (Phleg.) fulgens** (A. & S.) Fr. Boud. Icon. t. 106.

Fulgens, shining.

P. 5–10 cm., *orange-tawny*, very fleshy, convexo-plane, obtuse, occasionally punctate as if with drops, at length *silky-fibrillose*, or squamulose. St. 5–10 × 1·5–2 cm., *yellow, paler at the apex*, equal, *densely fibrillose with the yellow cortina* which is viscid in wet weather; base acutely marginately bulbous, then depressed and oblique. Gills *bright yellow, then tawny, or ferruginous with the spores*, emarginate, 6–10 mm. broad, somewhat crowded. Flesh *white-yellow, compact, then spongy and tan colour*. Spores ferruginous, verrucose, pip-shaped, 11–12 × 6μ. Smell pleasant, "like fennel" Quél. Woods. Sept.—Nov. Not uncommon. (*v.v.*)

362. **C. (Phleg.) fulmineus** Fr. Cke. Illus. no. 717, t. 717.
Fulmineus, pertaining to lightning.

P. 4–8 cm., *tawny, almost brown, margin orange, variegated with dense, irregular, agglutinated scales,* very fleshy, at first hemispherical, and attached to the bulb, then convex, very viscid; margin involute. St. 2–5 × 1–1·5 cm., *yellow, white cortinate at the apex,* when young inclosed in the bulb, bulb very depressed, marginate, *rooting,* wider than the young p. Gills *golden yellow, at length tawny,* rounded, thin, very crowded. Flesh *white, often yellow at the circumference or wholly yellowish.* Spores ferruginous, elliptical, 13–14 × 7–8μ, verrucose. Deciduous woods. Sept.—Oct. Uncommon. (*v.v.*)

363. **C. (Phleg.) orichalceus** (Batsch) Fr. Cke. Illus. no. 718, t. 754.
ὀρεί-χαλκος, copper ore.

P. 4–13 cm., *reddish copper colour, disc darker, often spotted with scales, bluish green towards the margin,* convex, then flattened, fleshy; margin incurved, pubescent, or white, then expanded and concolorous. St. 5–12 × 1·5–2 cm., *pale greenish yellow,* more or less covered with the fibrils of the cortina, fibrillosely silky, somewhat cylindrical, base marginately bulbous. Cortina *whitish,* or *very light greenish yellow, then rust colour from the spores.* General veil *whitish, often becoming reddish copper colour,* fibrillose, scarcely distinct from the cortina. Gills *yellow tinted greenish, then olive, and finally olive rust colour,* slightly adnate, sinuate, or emarginate, attenuated in front, slightly rounded behind, narrow, 4–6 mm. wide, thin. Flesh *greenish yellow, then citron yellow under the pellicle of the pileus and in the base of the stem, finally becoming reddish brown in the bulb, with a strong bluish grey horny line at the base of the gills.* Spores ferruginous, elliptical, or almond-shaped, 10–11 × 6–7μ, verrucose. Smell strong, "of fennel" Quél. Coniferous woods, and under beeches. Sept.—Nov. Uncommon. (*v.v.*)

364. **C. (Phleg.) elegantior** Fr. *Elegantior*, neater.

P. 7–10 cm., *tawny,* often spotted with drops, fleshy, convex, then plane, margin split. St. 5–7 × 2–3 cm., *becoming yellowish,* stout, fibrillose, base marginately bulbous. Cortina *pale.* Gills *egg-yellow, becoming olivaceous,* sinuate, *thin, crowded, serrulate.* Flesh *becoming yellow.* "Spores sphaeroideo-ellipsoid, dark or yellowish (under the microscope), 11–14 × 7–8μ" Sacc. Woods. Oct. Rare.

365. **C. (Phleg.) testaceus** Cke. (= *Cortinarius rufo-olivaceus* Fr. sec. Maire.) Cke. Illus. no. 1188, t. 1190. *Testaceus*, brick coloured.

P. 7–10 cm., *brick-red, rather vinous, becoming paler,* fleshy, convex, then flattened, umbonate, or depressed. St. 7–9 × 1·5 cm., *whitish above, becoming rufous at the base*, attenuated upwards, longitudinally, fibrously striate below; base submarginate, bulbous. Gills *dusky cinnamon*, adnate, a little emarginate behind, 6–10 mm. wide. Flesh *rather flesh-colour, becoming ruddy at apex and base* of stem. Spores elliptical, 9–11 × 5–6μ, rarely 16 × 8μ, narrowed at each end, verrucose. Woods, amongst leaves. Sept.—Nov. Uncommon. (*v.v.*)

****Gills olivaceous.

366. **C. (Phleg.) prasinus** (Schaeff.) Fr. Boud. Icon. t. 107.
πράσον, a leek.

P. 5–8 cm., *olivaceous, aeruginous, or tawny ferruginous, tiger-spotted* as if scaly, convex, then plane and depressed at the centre, *adpressedly fibrillose*; margin involute. St. 5–8 × 1–2 cm., *concolorous*, equal, or slightly attenuated upwards, fibrillosely silky; base marginately bulbous. Cortina *whitish, or pallid-green.* Gills *yellow-olivaceous*, or *somewhat olivaceous, darker and cinereous olivaceous at the base*, emarginate, undulate. Flesh *dingy white, greenish white in the stem, olivaceous under the pellicle of the p. and at the base of the stem.* Spores ferruginous in the mass, ochraceous under the microscope, elliptical, 12–15 × 6–8μ, verrucose. Smell none, or "of sulphur" Quél. Taste mild. Beech woods. Sept.—Oct. Uncommon. (*v.v.*)

367. **C. (Phleg.) atrovirens** (Kalchbr.) Fr. Kalchbr. Icon. t. 19, fig. 3.
Ater, black; *virens*, green.

P. 5–10 cm., *dark green, or olivaceous green*, compact, convex, obtuse. St. 5–8 × 1–2 cm., *yellow*, firm, equal, fibrillose, except the subturbinate, marginate bulb. Gills *sulphur colour, then greenish, at length cinnamon*, adnate, 6–8 mm. broad, crowded. Flesh *greenish yellow, then darker.* Spores ferruginous, "elliptical, 10 × 6μ" Massee. Mycelium *sulphur colour.* Pine woods. Oct.—Nov. Rare.

368. **C. (Phleg.) scaurus** Fr. Cke. Illus. no. 721, t. 755.
σκαῦρος, with projecting ankles.

P. 5–10 cm., *of a peculiar tawny fuliginous colour, more tawny when dry, tiger-spotted*, fleshy, convex, then plane and depressed; margin *thin, slightly striate when old.* St. 6–8 cm. × 8–10 mm., *azure-blue, or olivaceous, becoming white and also yellowish when old, attenuated upwards*, fibrillosely striate; base marginately bulbous, the bulb sometimes evanescent. Cortina *greenish*, fibrillose. Gills *purplish-olivaceous, olivaceous, or fuliginous*, attenuato-adnexed, rounded, 2–4 mm. broad, thin, *very crowded.* Flesh watery, thin, soft. Spores ferruginous, broadly elliptical, 10–11 × 6–7μ, "punctate-rough" Rick. Taste mild. Woods, and bogs. Sept.—Nov. Uncommon. (*v.v.*)

369. **C. (Phleg.) herpeticus** Fr. ἑρπετόν, a creeping thing.

P. 3–8 cm., *olivaceous, then dirty tan colour, disc becoming pale,* fleshy, convexo-plane, obtuse, somewhat spotted, slightly viscid. St. 5–8 × 1 cm., *pallid,* firm, unequal, somewhat twisted, fibrillose; bulb *napiform,* marginate. Gills *violet-umber, then fuliginous-olive,* slightly emarginate, at first crowded, 4–6 mm. broad. Flesh of the pileus *pale violet when young, then becoming dirty white.* Spores "nearly almond-shaped, punctate-rough, 7–8 × 4–5μ" Rick. Woods. Sept. Rare.

III. Cortina *simple, thin,* fugacious, median, or inferior. St. *at the first exserted,* somewhat thin, rigid-elastic, *externally subcartilaginous, polished, shining.* P. thin, often hygrophanous.

*Gills whitish, then clay coloured, or dirty cinnamon.

370. **C. (Phleg.) cumatilis** Fr. Fr. Icon. t. 146, fig. 2. *κῦμα,* a wave.

P. 4–8 cm., *of a very charming violet, or purple violet,* fleshy, convex, obtuse, often irregular. St. 5–10 × 1–1·5 cm., *white,* often curved, cortinate only at the apex, *the universal veil* (which serves as a pellicle of the p.) *ruptured at the base, and adnate to it as a separable, agglutinated membrane of the same colour as the p.* Gills *white, then clay colour,* attenuato-adnexed, almost free, crowded, narrow, 4–6 mm. broad, with a small decurrent tooth. Flesh *white.* Spores ferruginous, pip-shaped, 9–10 × 4–5μ, verrucose. Taste pleasant. Solitary, or subcaespitose. Fir woods. Sept.—Oct. Uncommon. (*v.v.*)

371. **C. (Phleg.) serarius** Fr. *Serarius,* living on whey.

P. 7–10 cm., *reddish-tan,* fleshy, convex, then plane, obtuse, or broadly gibbous, viscid, *opaque, appearing as if pruinately silky when dry.* St. 10 × 1 cm., *white,* equal, entirely fibrillose, and soft, polished, shining. Cortina *white,* inferior, inconspicuous. Gills *white, then clay colour,* arcuately-adnate with a decurrent tooth, crowded, broad. Flesh *white, with a hyaline line near the base of the gills.* Spores "thin, fusiform, 7–8 × 3μ, almost smooth" Rick. Mixed woods. Sept.—Oct. Rare.

372. **C. (Phleg.) emollitus** Fr. *Emollitus,* softened.

P. 5–8 cm., *tawny, then ochraceous yellow,* fleshy, globose, then campanulato-convex, finally plane, or deformed, often fibrillosely virgate, shining when dry; margin incurved, flexuose. St. 4–8 × 1–1·5 cm., *white, becoming yellowish,* equal, or attenuated downwards, often thickened at the apex, striate, or fibrillose, base sometimes thickened, often compressed, curved, or somewhat twisted. Cortina *white,* fugacious, often appendiculate from the margin of the p. Gills *white, then ochraceous,* adnate, or emarginate, *somewhat distant,* 10–12 mm. *broad,* fragile. Flesh *white, very soft.* Spores ferruginous, elliptical,

6–7 × 4μ. Taste very acrid. Often caespitose. Pastures, and woods, especially beech, and oak. Sept.—Oct. Not uncommon. (*v.v.*)

373. **C. (Phleg.) causticus** Fr. Bull. Soc. Myc. Fr. XXVI, t. 5, figs. 1–4. *καυστικός*, burning.

P. 3–5 cm., *ochraceous nankeen yellow*, almost hemispherical, then convex, and plane, sometimes slightly umbonate, and finally slightly depressed at the centre, pellicle easily separable, at first covered with the white fibrils of the universal veil, *soon white pruinose*, silky towards the margin, *only slightly viscid when young*, soon dry and shining; margin slightly incurved, then straight. St. 5–8 cm. × 3–5 mm., *white*, straight, or flexuose, *firm*, elastic, covered with the fibrillose veil, and *slightly viscid when young*, soon dry, very minutely pruinose at the apex; base equal, or somewhat bulbose, sometimes fusiform and slightly rooting. Cortina *white*, fugacious. Gills *cream colour, then ochraceous rust*, broadly adnate, slightly emarginate, diminishing in width towards the margin, slightly crowded. Flesh *yellowish when young*, becoming whitish when dry. Spores ferruginous in the mass, yellowish brown under the microscope, elliptical, 6·5–7·5 × 4μ, apiculate, very minutely verrucose. Smell rather strong. Taste *of the cuticle of the pileus very bitter*, of the flesh sweet, or very slightly bitter. Pine woods, and under conifers. Sept.—Oct. Not uncommon. (*v.v.*)

374. **C. (Phleg.) crystallinus** Fr. Grevillea, t. 107, fig. 3. *κρυστάλλινος*, crystalline.

P. 1·5–4 cm., *shining silvery white towards the margin, disc watery-pallid, becoming altogether shining white when dry*, fleshy, convex, then plane, *hygrophanous*. St. 5–7 cm. × 6–10 mm., *whitish, then straw colour, fragile*, equal, or attenuated at the base, fibrillose. Gills *clay colour*, emarginate, thin, 6 mm. broad, crowded. Flesh *white*, thin. Spores clay colour, elliptical, 4–5 × 3μ, "7–8 × 4–5μ, faintly punctate" Rick. Taste very acrid. Woods, especially beech. Sept.—Nov. Uncommon. (*v.v.*)

375. **C. (Phleg.) decoloratus** Fr. Cke. Illus. no 726, t. 729. *Decoloratus*, stained.

P. 4–10 cm., *clay colour, disc darker*, thin, equally fleshy, campanulate, then convex, obtuse, soft, soon dry, and *flocculose*, corrugated and stained when old. St. 5–10 cm. × 10–12 mm., *silvery*, equal, thickened at the base, sometimes attenuated downwards, fibrillose. Cortina *white*, fibrillose, inferior. Gills *whitish, or bluish, then clay colour and cinnamon*, emarginate, adnate, or decurrent, not much crowded, 6 mm. broad. Flesh *white*, watery, soft. Spores pale ferruginous, pip-shaped, 11–12 × 5–6μ, verrucose. Taste slightly acrid. Woods, especially beech. Aug.—Oct. Not uncommon. (*v.v.*)

**Gills violaceous, purplish, or flesh coloured.

376. **C. (Phleg.) decolorans** (Pers.) Fr. Cke. Illus. no. 727, t. 730.
Decolorans, discolouring.

P. 3–6 cm., *persistently yellow*, fleshy, convex, then flattened, somewhat gibbous. St. 5–7 cm. × 6–10 mm., *shining white*, equal, attenuated downwards, or slightly thickened at the base. Cortina *white*, persistent, median. Gills *purplish, then soon cinnamon*, sinuato-adnexed, thin, crowded, 6 mm. broad. Flesh *white, thin, firm*. Spores pale ferruginous, subglobose, 7–8 × 7μ; "almond-shaped, 10–12 × 5–6μ, warted" Rick. Coniferous woods, and under birches. Sept.—Oct. Not uncommon. (*v.v.*)

377. **C. (Phleg.) porphyropus** (A. & S.) Fr. Cke. Illus. no. 728, t. 731.
πορφύρεος, purple; πούς, foot.

P. 3–8 cm., *livid-light-yellowish, or clay colour*, fleshy, very thin at the margin, convexo-plane, obtuse, *innately streaked*. St. 5–10 cm. × 6–10 mm., *violaceous-lilac, becoming pale, even whitish, but soon becoming violaceous-lilac when touched*, fragile, somewhat bulbous, or rather equally attenuated from the thickened base, sometimes equal. Cortina *violaceous-lilac*, fibrillose, inferior. Gills *purplish, then watery cinnamon, becoming purple again* when touched, rounded, or emarginate, somewhat crowded, 4–10 mm. broad. Flesh of *pileus whitish, soon becoming purple-lilac when broken, of stem purple-lilac becoming whitish*. Spores pale ferruginous, pip-shaped, 10–11 × 6–7μ, "slightly rough" Rick. Woods, especially beech. Sept.—Oct. Uncommon. (*v.v.*)

378. **C. (Phleg.) croceo-caeruleus** (Pers.) Fr. Cke. Illus. no. 729, t. 732.
Croceus, saffron; *caeruleus*, azure.

P. 2–3 cm., *lilac, or faintly violaceous*, fleshy, thin, convex, then plane, obtuse, or gibbous. St. 5 cm. × 4–6 mm., *whitish, fragile*, somewhat equal, or attenuated downwards. Cortina *white*, fibrillose, fugacious. Gills *lilac, then clay-saffron*, attenuated, or broadly emarginate, with a small, very thin decurrent tooth, somewhat distant. Flesh *pallid, lilac under the pellicle*, watery. Spores ferruginous, pip-shaped, 6–8 × 4–5μ, punctate. Taste "bitter" Pers. Woods, especially under beeches, and hazels. Sept.—Oct. Not uncommon. (*v.v.*)

***Gills pure ochre, tawny, or ferruginous.

379. **C. (Phleg.) coruscans** Fr. Cke. Illus. no. 730, t. 733.
Coruscans, glittering.

P. 10 cm., *yellow-ochraceous, often spotted tawny*, fleshy, soon plane, regular, at length depressed, shining when dry. St. 7–15 × 1 cm., *shining white*, elastic, *equal*, apex enlarged, *fibrillosely-striate*. Cortina *white*, fibrillose, fugacious. Gills *bright ochraceous*, decurrent by a tooth, thin, very narrow, 2–4 mm. wide, very crowded, linear. Flesh *white*, soft. Woods. Sept.—Oct. Rare.

380. **C. (Phleg.) papulosus** Fr. Cke. Illus. no. 731, t. 718.
Papulosus, having pimples.

P. 6–9 cm., *honey-tan colour, disc ferruginous, or fuscous*, and here and there gibbous, fleshy, convex, obtuse, then plane, and at length depressed, *the cuticle breaking up into minute, granular, fuscous patches when dry*. St. 6–7 × 1–1·5 cm., *white*, firm, equal, or thickened at the base, densely fibrillose, *apex naked*. Cortina *white*, inferior, very fugacious. Gills *pallid, soon ochraceous, at length very pale yellow cinnamon*, adnato-decurrent, crowded, slightly joined behind, separating from the stem when old, and connected by a spurious collar. Flesh *white*, thick at the disc, thin at the margin. Spores "sub-elliptical, 8–10 × 5–6μ, very slightly punctate" Rick. Pine woods. Oct.—Nov. Rare.

var. **major** Fr. *Major*, larger.

P. *yellowish, ferruginous, margin much paler*, glutinous, disc truly granular. St. *at length coloured like the gills*, attenuated from the base, filamentous from the inferior veil, apex cortinate. Gills slightly sinuate.

381. **C. (Phleg.) vespertinus** Fr. *Vespertinus*, pertaining to evening.

P. 7–9 cm., *yellowish ochraceous, disc egg-yellow*, fleshy, convex, then plane, glutinous, *wrinkled and folded at the margin*. St. 5–7·5 × 1–1·5 cm., *shining white, firm, elastic*, incrassated at the base, fibrillose. Cortina *pallid*, inferior, fugacious. Gills *bright and intense fulvous-cinnamon*, broadly emarginate, *very broad*, firm, little crowded, *shining*. Flesh *white*, firm. Spores "elliptical, 4–5 × 3–4μ, almost smooth" Rick. Deciduous woods. Oct. Rare.

****Gills olivaceous, fuliginous.

382. **C. (Phleg.) olivascens** (Batsch) Fr. Fr. Icon. t. 147, fig. 2.
Olivascens, becoming olivaceous.

P. 3–5 cm., *somewhat fuliginous, or bistre olivaceous, becoming pale*, somewhat fleshy, convexo-plane, obtuse; margin *substriate*. St. 7–9 × 1 cm., *silvery, becoming pallid* ("*whitish lilac, then silvery at the apex, white in the middle and citron yellow at the base*" Quél.), attenuated upwards, somewhat bulbose, fibrillose, striate. Gills *olivaceous, or clay colour, then cinnamon*, adnate, emarginate, thin, little crowded. Flesh *paler* ("*violaceous, then reddish*" Quél.), thin. Spores "tawny olivaceous, pruniform, 10–12 × 5·5–7μ, punctate" Bat. Taste acrid. Damp woods amongst *Sphagna*. Sept. Rare.

2. **Myxacium** Fr.

(μύξα, mucus.)

General veil glutinous. Stem *viscid.* Pileus slightly fleshy.

†St. flocoso-peronate, the flocci at first covered with gluten.

383. **C.** (**Myx.**) **arvinaceus** Fr. (= *Cortinarius* (*Myxacium*) *mucosus* (Bull.) Quél., (*Myxacium*) *alutipes* (Lasch) Fr. sec. Quél.) Cke. Illus. no. 734, t. 739, as *Cortinarius* (*Myxacium*) *mucosus* Fr. *Arvinaceus*, greasy.

P. 6–10 cm., *orange-tawny*, or *reddish tan*, fleshy, soft, convex, then soon flattened, at length reflexed and undulating, viscid, glistening when dry; margin *slightly striate* when in full vigour. St. 10–20 × 1–1·5 cm., *white*, equal, *silky-viscous*. Cortina soon fibrillose, fugacious. Gills *straw colour, then bright ochraceous, adnato-decurrent, very broad*, 12–18 mm., *somewhat distant, edge crenulate.* Spores ochraceous, "fusiform, 15–17 × 8–9μ, rough" Rick. Beech woods. Oct. Rare.

384. **C.** (**Myx.**) **collinitus** (Sow.) Fr. (= *Cortinarius* (*Myxacium*) *mucifluus* Fr. sec. Quél.) Cke. Illus. no. 735, t. 740, as *Cortinarius* (*Myxacium*) *mucifluus* Fr. *Collinitus*, besmeared.

P. 6–11 cm., *orange-tawny, fleshy*, not compact, convex, with the margin bent inwards, then expanded, obtuse, *covered with persistent orange-tawny gluten*, shining when dry. St. 7–12 × 1–2·5 cm., *violaceous, white*, or *yellowish*, firm, *cylindrical*, at length soft, *covered with a floccose, glutinous veil, which is commonly broken up into concentric scales*, near the apex the gluten is continuous with that of the p. and forms an entirely viscous, fugacious ring. Gills *whitish-bluish-grey, or clay colour then cinnamon*, adnate, somewhat crowded. Flesh *whitish, brownish under the cuticle of the p. and at the base of the st.* Spores ferruginous, pip-shaped, 10–11 × 6μ, rough. Woods. July—Nov. Common. (*v.v.*)

385. **C.** (**Myx.**) **mucosus** (Bull.) (= *Cortinarius* (*Myxacium*) *alutipes* (Lasch) Fr. and (*Myxacium*) *arvinaceus* Fr. sec. Quél.) Boud. Icon. t. 108. *Mucosus*, full of mucus.

P. 4–10 cm., *chestnut*, fleshy, campanulato-convex, then expanded, *covered with chestnut gluten*, margin *paler, striate.* St. 5–15 × 2 cm., *whitish ochre, or ochraceous*, cylindrical, slightly attenuated at the base, fibrillosely tomentose. Cortina *white, glutinous.* Gills *whitish, then cinnamon*, adnate. Flesh *whitish, tinged with chestnut under the cuticle of the p. and at the base of the st.* Spores tawny, verrucose, lemon-shaped, 14–17 × 7–8μ. Pine woods. Aug.—Oct. Not uncommon. (*v.v.*)

386. **C. (Myx.) mucifluus** Fr. (= *Cortinarius* (*Myxacium*) *collinitus* Sow. sec. Quél.) Fr. Icon. t. 148, fig. 1.

Mucus, mucus; *φλύω*, I boil over.

P. 3–9 cm., *livid-clay, tan when dry*, opaque, thin, somewhat fleshy, campanulate, then expanded, at length reflexed and repand, *smeared with separating, hyaline gluten*; margin membranaceous, *striate*. St. *white, or inclining to azure-blue*, spongy, attenuated downwards, viscid with the floccose-scaly fugacious veil. Gills *clay colour, then watery cinnamon*, adnate. Spores ferruginous, almond-shaped, granular, 12 × 7μ. Cystidia "on edge of gill, vesiculose, 30–45 × 18–30μ" Rick. Pine woods. Aug.—Oct. Not uncommon. (*v.v.*)

387. **C. (Myx.) elatior** Fr. Fr. Icon. t. 149, fig. 1. *Elatior*, taller.

P. 6–12 cm., *livid-light-yellow when moist, dingy ochraceous when dry*, sometimes whitish, tan fuscous, date brown, violaceous brown, black, whitish round the margin, or grey with the margin violaceous, slightly fleshy only at the disc, cylindrical, or bullate, then campanulate, afterwards flattened and somewhat reflexed, disc above the stem obtuse, *membranaceous and longitudinally plicato-wrinkled at the sides*, fragile. St. 7–18 × 1–5 cm., *violaceous, lilac, becoming white*, commonly attenuated at both ends, especially at the base, *fibrillosely floccose*. Cortina *concolorous*, viscid, fugacious. Gills *ochraceous, or lilac, then dark brown cinnamon*, adnate, broad, connected by veins or wrinkled at the sides. Flesh *whitish, or pale yellowish*. Spores purplish-ferruginous, almond-shaped, 12–14 × 6μ, verrucose. Cystidia "on edge of gill vesiculose-pyriform, 36–45 × 21–28μ" Rick. Woods. Aug.—Nov. Common. (*v.v.*)

388. **C. (Myx.) grallipes** Fr. Cke. Illus. no. 738, t. 734.

Grallae, stilts; *pes*, foot.

P. 4–8 cm., *ferruginous when moist, ochraceous tan when dry*, opaque, *almost membranaceous* with the exception of the *prominent, often acutely umbonate disc*, campanulate, then flattened, hygrophanous, very viscid when wet. St. 10–15 cm. × 4–6 mm., *yellowish tawny, ochraceous* when dry, tough, equal, flexuose, *fibroso-striate*, viscid. Cortina *pale, whitish brown*, fugacious. Gills *clay colour, then ferruginous*, adnate with a decurrent tooth, 12 mm. broad, attenuated in front, crowded. Flesh *white*. Spores ferruginous, "7–8 × 4–5μ" Herpell. Caespitose. Mixed woods, and under oaks and poplars. Oct. Uncommon.

389. **C. (Myx.) livido-ochraceus** Berk. Cke. Illus. no. 739, t. 767.

Lividus, livid; *ochraceus*, ochre.

P. 2·5–5 cm., *livid-ochre, somewhat membranaceous*, convex, then plane, cuticle thick, subcartilaginous, margin very thin, often with a few, indistinct fragments of the veil. St. 2·5–6 cm. × 7–10 mm., *beautiful violet, ochraceous at the base*, attenuated at both ends, somewhat scaly, striate above the fugacious veil. Gills *pale*, then *cinnamon, margin pale*, somewhat adnexed, broad in front, moderately distant. Flesh *yellowish, livid under the pellicle of the p.* Spores ferruginous, elliptical, 8–10 × 5–6μ, rough, 1-guttulate. Taste "like *Ag. campestris*" Berk. Woods. Sept.—Oct. Not uncommon. (*v.v.*)

††Veil entirely viscid, hence the st. is not floccoso-peronate, but only viscid, acquiring a varnished appearance when dry.

*Gills whitish, then clay colour.

390. **C. (Myx.) nitidus** (Schaeff.) Fr. Cke. Illus. no. 1189, t. 1191.
Nitidus, shining.

P. 4–12 cm., *honey-coloured tan, at length whitish, disc tan colour*, fleshy, convex, then expanded, gibbous, or almost obtuse, glutinous, when dry the cuticle often cracked in streaks, and appearing minutely fuscous punctate. St. 5–10 × 1–1·5 cm., *pallid white* and fibrillose when young, then *becoming yellowish* and naked, base clavate, often curved, tough, elastic, *apex at first white-mealy.* Cortina slightly fibrillose, fugacious. Gills *whitish, soon clay colour, and finally watery cinnamon, truly decurrent, arcuate at first*, crowded, narrow, 4 mm. wide. Flesh *white.* Spores light brown, "broadly pip-shaped, 10–12 × 8μ" Cke. Subcaespitose. Beech woods. Sept.—Oct. Rare.

**Gills at first violaceous, dark blue, or reddish.

391. **C. (Myx.) salor** Fr. Fr. Icon. t. 150, fig. 1. σάλος, the high sea.

P. 4–7 cm., *grey, bright violaceous round the inflexed margin, at length of the same colour*, obtusely conical, or parabolic, soon campanulate, and at length flattened, with a broad umbo on account of the fleshy disc, thin towards the circumference, thinly viscid, fibrillose towards the margin when dried. St. 4–8 × 1–1·5 cm., *white, covered up to the apex with the azure-blue glutinous veil when young*, becoming pale when old, *conico-attenuated from the bulbous base*, gradually elongated. Gills *pale grey, with the edge violaceous, or bluish grey, then grey clay colour, or cinnamon, adnate, distant*, 4–6 mm. broad. Flesh *white, becoming yellow, or faintly azure-blue.* Spores ferruginous, "subglobose, 8–9 × 8μ, granular" Rick. Woods. Oct. Rare.

392. **C. (Myx.) delibutus** Fr. Cke. Illus. no. 741, t. 743.
Delibutus, besmeared.

P. 3–7·5 cm., *light yellowish*, fleshy, thin, especially towards the margin, convex, then flattened, obtuse, at length somewhat depressed, viscid with hyaline gluten, slightly silky fibrillose when the gluten has disappeared. St. 5–10 cm. × 6–8 mm., *yellowish white, apex snow white, equally attenuated* from the slightly bulbose base, or somewhat equal, elastic, viscid up to the *white,* scanty, fibrillose, fugacious cortina. Gills *dark-blue,* or *violaceous dark-blue,* then *clay cinnamon, serrulated, pallid or often crisped at the edge,* adnate, at length rounded, or slightly emarginate, more or less *distant,* 4–6 mm. broad. Flesh *white.* Spores pale ochraceous, "subglobose, 7 × 6–7μ, granular" Karst. Taste watery, then slightly pungent. Grassy, and damp places. Sept.—Oct. Uncommon. (*v.v.*)

var. **elegans** Massee. *Elegans,* neat.

P. and st. very glabrous, *yellow-viscid,* shining when dry, *only* apex of stem *white,* flesh *whitish-yellow,* gills paler, more crowded. Grassy banks of streams. Sept.—Oct. Rare.

393. **C. (Myx.) illibatus** Fr. *Illibatus,* unimpaired.

P. 2·5–5 cm., *yellow, disc darker,* slightly fleshy, campanulate, then convex, at length plane, *subumbonate,* pellicle viscid. St. 7·5 cm. × 4 mm., *white, commonly with reddish dots upwards,* slightly attenuated upwards, viscid. Cortina superior, fibrillose, very fugacious. Gills *flesh-colour, then clay and cinnamon,* adnato-decurrent, arcuate, 4 mm. broad, thin, *crowded.* Flesh *white,* very thin at the circumference. Spores cinnamon, elliptical, "15–16 × 6–7μ, granular" Massee, "subglobose, 7–9μ, granular" Bat. Pine woods. Sept.—Oct. Uncommon.

***Gills at first ochraceous, or cinnamon.

394. **C. (Myx.) stillatitius** Fr. Cke. Illus. no. 742, t. 831. *Stillaticius,* dripping.

P. 4–6 cm., *clothed with azure-blue gluten, fuscous-livid* when the gluten separates in the form of drops, *at length grey-white,* slightly fleshy, convex, then plane, subumbonate; margin *smooth.* St. 5–7 cm. × 6–8 mm., *sheathed with thick azure-blue gluten* which is extended into the cortina, *very soft,* equally attenuated. Gills *dark cinnamon, emarginate,* 6 mm. broad. Flesh watery, soft, hygrophanous. Spores ferruginous, "subglobose, 8 × 6μ, 1-guttulate" Sacc., "almond-shaped, 13–15 × 7–8μ, rough. Cystidia on edge of gill, 30–40 × 12–20μ" Rick. Pastures, and amongst dead leaves. Oct. Uncommon.

395. **C. (Myx.) vibratilis** Fr. Cke. Illus. no. 743, t. 744. *Vibratilis,* quivering.

P. 3–6 cm., *yellow, golden* when dry and very shining, fleshy at the disc, thin elsewhere, convexo-plane, obtuse, very glutinous. St. 4–6 cm. × 4–8 mm., *shining white, conically attenuated,* or ventricose, equal and flexuose amongst mosses, fragile, *very soft.* Cortina glutinous, often forming a median ring. Gills *pallid, then bright ochraceous cinnamon*; rounded, emarginate, or decurrent by a tooth, crowded, thin. Flesh *pallid.* Spores ferruginous, elliptical, 6–7 × 4μ, "punctate" Rick. Smell strong, taste very bitter. Woods. Sept.—Oct. Not uncommon. (*v.v.*)

396. **C. (Myx.) pluvius** Fr. Cke. Illus. no. 744, t. 769. *Pluvius*, rainy.

P. 1·5–3 cm., *pale yellow-tawny* when moist, *ochraceous tan* and opaque when dry, slightly fleshy, somewhat globose, then convex, commonly gibbous, *slightly pellucidly striate when more fully grown, hygrophanous*, viscid, shining in rainy weather. St. 4–6 cm.×4–6 mm., *white, then yellow and concolorous, soft, equal*, or slightly attenuated upwards, slightly viscid, silky. Cortina *white, fibrillose*, slightly viscid, soon fugacious. Gills *light yellowish, or at the first whitish, then ochraceous*, adnexed, separating, ventricose, crowded. Flesh *pale yellowish*, becoming white. Spores deep ochraceous, broadly elliptical, 9–10×7–8μ, 1-guttulate. Taste watery, then acrid and pungent. Woods, especially pine. Sept.—Nov. Not uncommon. (*v.v.*)

3. **Inoloma** Fr.

(ἴς, fibre; λῶμα, fringe.)

Pileus equally fleshy, *dry, at first floccose, fibrillose, velvety, pubescent, or silky*, then becoming somewhat smooth. Veil simple.

*Gills at first white, or pallid.

397. **C. (Ino.) opimus** Fr. Fr. Icon. t. 151, fig. 1. *Opimus*, plump.

P. 7–10 cm., *tan colour*, fleshy, *very thick, very hard*, convex, then plane, deformed, repand, *everywhere covered with short tan coloured tomentum*, then rimoso-rivulose; margin involute, *pruinose, white*, often split. St. 2·5–5 × 2–5 cm., *whitish*, covered with the *white fibrils of the veil*, attenuated at the base and rooting. Gills *whitish, then clay colour, emarginate*, much narrower than the flesh of the p., somewhat crowded, flexuose. Flesh *whitish*, firm. Spores ochraceous, "subglobose, 8–9 × 7–8μ, warted" Rick. Smell and taste pleasant. Woods, especially conifers. Sept.—Oct. Uncommon. (*v.v.*)

var. **fulvobrunneus** Fr. *Fulvus*, tawny; *brunneus*, brown.

P. *tawny brown*, undulated, thinner (margin thin), glabrous, rimoso-rivulose. St. 3·5 × 2·5 cm., attenuated downwards, fibrillosely striate. Gills very broad.

398. **C. (Ino.) argutus** Fr. Fr. Icon. t. 151, fig. 2. *Argutus*, pointed.

P. 7–10 cm., *clay ochraceous, or deep ochraceous*, fleshy, broadly conico-campanulate, soon convex, somewhat gibbous, at length plane, obtuse, *fibrillosely silky, here and there minutely squamulose*, becoming smooth with age, rather rimose, opaque. St. 6–10 × 2–3 cm., *white, floccoso-squamulose, becoming smooth and yellowish*, ovately bulbous, or ventricose at the base, often curved and prolonged below the bulb into a pointed root. Veil *white*, superior, simple, forming a ring when young, rarely noticeable when mature. Gills *white, then clay colour*, adnate, somewhat distant. Flesh *very hard, white* ("becoming red on exposure to the air" Quél.). Spores pale ferruginous, elliptical, 7–8×4μ, verrucose, "almond-shaped, 13–15×8–9μ, coarsely warted" Rick. Deciduous woods. Oct. Uncommon. (*v.v.*)

399. **C. (Ino.) turgidus** Fr. Grevillea, t. 109, fig. 1.

Turgidus, swollen.

P. 5–10 cm., *clay colour, silvery-shining* when full grown, very fleshy, compact, convex, then plane, very obtuse, *hoary*, rarely sprinkled with shining atoms; margin silky and white when young. St. 4–6 × 2 cm., *silvery white, stout*, bulbous base much swollen, *externally cartilaginous*, elastic, longitudinally fibrillose under a lens, and *split up into subreticulate cracks*, often undulate. Cortina *white*, fibrillose, fugacious. Gills *whitish, then clay colour*, emarginate, crowded, 4 mm. broad, *denticulate*. Flesh *whitish, tough*. Spores ferruginous, pip-shaped, 7–9 × 4·5–6μ, 1-guttulate. Taste pleasant. Edible. Woods. Sept.—Oct. Uncommon. (*v.v.*)

400. **C. (Ino.) argentatus** (Pers.) Fr. Cke. Illus. no. 745, t. 745.

Argentatus, silvered.

P. 4–10 cm., *silvery-shining, disc becoming pale*, at first *silky-lilac round the margin, then dun-coloured*, fleshy, convexo-plane, at length broadly gibbous, *silky*. St. 8–10 × 1·5–2 cm., *concolorous*, attenuated from the thickened base. Gills *pallid, then watery cinnamon*, emarginate, crowded; edge slightly serrated, *white*. Flesh *whitish, often with a bluish tinge*. Spores ferruginous, pip-shaped, 8–9 × 5μ, punctate. Smell and taste pleasant. Woods. Sept.—Oct. Not uncommon. (*v.v.*)

var. **pinetorum** Fr. Cke. Illus. no. 746, t. 746.

Pinetorum, of pine woods.

Smaller. P. 5 cm., *at first lilac and silky*. St. 5 cm. Smell weak. Pine woods. Oct. Uncommon. (*v.v.*)

401. **C. (Ino.) fusco-tinctus** Rea. Trans. Brit. Myc. Soc. v, t. 8.

Fuscus, dark; *tinctus*, stained.

P. 2–6 cm., *pale ochraceous, becoming blood red immediately in places where touched, then fuscous especially around the margin*, fleshy, convex, subgibbose, *fibrillosely silky*, disc floccosely squamulose under a lens; margin at first involute, arachnoid with the veil. St. 6–10 cm. × 5–10 mm., *concolorous, becoming reddish when touched, and soon fuscous,* fusiform, often incurved at the base, solid, firm, apex minutely white pruinose. Cortina *white,* manifest, median, at length fugacious. Gills *clay colour, then pale cinnamon,* sinuato-adnate, attenuated in front, 4–6 mm. broad, crowded; edge *white,* unequal. Flesh *white,* unchangeable, compact, firm, with a grey horn colour line at the base of the gills. Spores ferruginous in the mass, pale ferruginous under the microscope, elliptical, 9–10 × 5μ, contents granular. Smell and taste none. The change of colour is present only in the cuticle of the p., and st. Oak woods. Sept. Uncommon. (*v.v.*)

**Gills, as well as the st. and veil, violaceous.

402. **C. (Ino.) violaceus** (Linn.) Fr. Sverig. ätl. Svamp. t. 58.
Violaceus, violet.

P. 7–15 cm., *dark violaceous, sometimes purplish-violet,* fleshy, convex, then flattened, regular, obtuse, *villous, the innate persistent villous down for the most part rimoso-squamulose*; margin at first involute. St. 6–10 × 1·5–2 cm., *dark violaceous,* stout, remarkably bulbous, at first tomentose, then fibrillose. Cortina *azure-blue,* woolly, then ferruginous with the spores. Gills *dark, almost black violaceous,* then coloured ferruginous with the spores and again violaceous when these are rubbed off, somewhat adnate, firm, *distant,* connected by veins, broader than the flesh of the pileus. Flesh *blue, becoming white.* Spores ferruginous, broadly elliptical, 11–13 × 7–8μ, verrucose. Taste pleasant. Edible. Woods, especially under birch, and beech. Aug.—Nov. Uncommon. (*v.v.*)

403. **C. (Ino.) cyanites** Fr. Fr. Icon. t. 152, fig. 1. *κύανος,* dark blue.

P. 6–13 cm., *dark blue, becoming azure-blue, or livid-fuscous,* fleshy, soft, convex, then flattened, obtuse, silky. St. 7–13 × 1–2 cm., *concolorous, very bulbous,* fibrillose. Cortina *azure-blue,* fibrillose. Gills *deep dark blue,* adnate, sinuate, *crowded,* thin, 6 mm. broad. Flesh *blue, reddening on exposure to the air, and when compressed giving out a red juice.* Spores pale ferruginous, elliptical, 9–10 × 4–5μ, "warted" Rick. Woods. Sept.—Oct. Rare. (*v.v.*)

var. **major** Fr. *Major,* larger.

Differs from the type in the *compact stem, in the p. tardily becoming reddish, and in the somewhat distant, cinereous dark blue gills.*

404. **C. (Ino.) muricinus** Fr.
Murex, a mollusc from which the Tyrian purple was obtained.

P. 5–10 cm., *violaceous, becoming reddish,* fleshy, compact, convex, then plane, very obtuse, becoming smooth; margin fibrillose. St.

3–10 × 1·5–2·5 cm., *becoming violaceous*, attenuated upwards from the bulbous base, *villous*. Gills *purplish violet*, at length *reddish liver colour*, emarginate, 12 mm. *broad, somewhat crowded*. Flesh *paler, becoming bluish near the gills*, spongy. Spores ferruginous, "almond-shaped, 13–15 × 7–8μ, warted" Rick. Smell strong, peculiar. Fir, and larch woods. Oct. Uncommon.

405. **C. (Ino.) alboviolaceus** (Pers.) Fr. Fr. Icon. t. 151, fig. 3.
Albus, white; *violaceus*, violet.

P. 5–7·5 cm., *whitish violet*, fleshy, convex, *broadly umbonate*, or *rather gibbous*, dry, beautifully *innately silky*, the fibrils longitudinally adpressed as in *Inocybe geophylla*. St. 5–10 × 1–2·5 cm., *concolorous, becoming whitish*, firm, *clavato-bulbous*, or conico-attenuated, *white villous*, fibrillose above with the cortina, and often *zoned with the white veil* at the middle. Gills *greyish lilac, then grey-cinnamon*, adnate, scarcely emarginate, 4–5 mm. broad, subdistant, *subserrulate*. Flesh *azure blue white*, juicy, thick at the disc. Spores ferruginous, oblong elliptical, 9–12 × 5–6μ, punctate. Woods, especially beech. Aug.—Oct. Common. (*v.v.*)

406. **C. (Ino.) malachius** Fr. μαλάχη, a mallow.

P. 5–10 cm., *pale lilac, then fuscous ferruginous, pale brick colour when dry*, very fleshy, compact, convex, then expanded, obtuse, or slightly gibbous, *hoary with minute, fasciculate down*, or silky towards the margin. St. 7–12 × 2·5 cm., *bluish lilac, becoming whitish*, bulbous base slightly marginate, ventricose, or equal, often deformed, *striate with violaceous fibrils*, very rarely having a white membranaceous ring. Cortina *violaceous*, thin. Gills *purple, becoming pale, at length watery ferruginous*, emarginate, *crowded*. Flesh *violaceous, becoming white*, thick, watery, *soft in the st.* Spores ferruginous, pip-shaped, "10–12 × 6–7μ" Cke., "punctate" Rick. Pine, and fir woods. Sept. Rare.

407. **C. (Ino.) camphoratus** Fr. Fr. Icon. t. 152, fig. 2.
Camphoratus, strong scented.

P. 5–8 cm., *lilac, becoming whitish, or yellowish*, very fleshy, convex, then flattened, obtuse, silky, becoming smooth. St. 7–13 × 1–2·5 cm., bulbous, or obclavate, *peronately woolly when young*. Cortina *blue*, fibrillose. Gills *intense azure blue*, becoming purple, decurrent, or emarginate, arcuate, thin, crowded. Flesh *blue, white at the base of the stem*, thick. Spores ferruginous, "somewhat almond shape, 12–14 × 7–8μ, granular" Cke. Smell foetid, exceedingly penetrating, like fenugreek, or curry-powder. Woods, especially pine. Sept. Rare.

408. **C. (Ino.) hircinus** (Bolt.) Fr. (= *Cortinarius amethystinus* (Schaeff.) Quél.) Bolt. Hist. Fung. t. 52. *Hircinus*, of a goat.

P. 4–5 cm., *violet, disc at length becoming ferruginous*, fleshy, convex, obtusely gibbous, *silky with adpressed, violet fibrils*. St. 4–5 × 1·5 cm., *violet, becoming pallid, yellowish at the bulbous base*, cortinate. Gills *violet, then cinnamon*, emarginate, *broad*, thin, subdistant. Flesh *dingy, becoming yellowish especially at the base of the st.*, thick. Spores ferruginous, "8·5–10 × 4·5–5·5μ, minutely verrucose" Maire. Smell strong like goats, or burnt horn. Pine woods. Sept. Rare.

***Gills or veil cinnamon, red, or ochraceous.

409. **C. (Ino.) traganus** Fr. (= *Cortinarius amethystinus* (Schaeff.) Quél.) Cke. Illus. no. 752, t. 757. *τράγος*, a goat.

P. 4–8 cm., *lilac purplish, becoming pale and finally yellowish*, very fleshy, convex, then flattened, obtuse, dry, silky, becoming smooth. St. 7–12 × 1–2·5 cm., *violaceous, then whitish*, spongy, attenuated upwards, base villous, *very bulbous*, silky, then fibrillose. Cortina *pallid violaceous*, continuous with the silky covering of the p. Gills *saffron-ochraceous, then cinnamon*, emarginate, *very broad, thick, distant*, edge often somewhat crenate. Flesh *yellowish*, thick, *deep saffron-ochraceous in the spongy st.* Spores bright ferruginous, elliptical, 8–10 × 5–6μ, verrucose, 1-guttulate. Smell foetid like goats, or the larvae of *Cossus*. Pine woods. Aug.—Oct. Uncommon. *(v.v.)*

var. **finitimus** Weinm. *Finitimus*, nearly related.

Differs from the type in the *yellowish mottled flesh of the st., and the pleasant smell, like gum just beginning to ferment, or like camphor.*

410. **C. (Ino.) suillus** Fr. Fr. Icon. t. 152, fig. 3.
Suillus, pertaining to swine.

P. 7–10 cm., *dingy, or pallid brick-red*, fleshy, convex, obtuse, at length *floccosely squamulose*, silky towards the margin. St. 7–10 × 1–2·5 cm., *dingy pallid*, clavato-bulbous, attenuated upwards, darker when touched, fibrillose, *apex pale violaceous, fugacious*, base white-woolly. Gills *cinnamon*, opaque, adnate, 10–12 mm. broad, subdistant, fragile, often veined at the base. Flesh *dirty pale brick colour, especially in the st.*, thick at the disc, thin elsewhere. Spores "ellipsoid, obtuse at the ends, 10–12 × 6–8μ" Sacc. Fir, and pine woods. Sept.—Nov. Rare.

411. **C. (Ino.) tophaceus** Fr. Fr. Icon. t. 153, fig. 1. *Tophus*, tufa.

P. 7–10 cm., *golden tawny*, opaque, fleshy, hemispherical, *villosely squamulose*, varying slightly silky, and shining. St. 5–10 × 1·5–2 cm., *tawny*, slightly attenuated upwards from the bulbous base, *villosely squamulose*, often twisted. Gills *concolorous, then tawny cinnamon,* broadly emarginate, 12–15 mm. broad, distant. Flesh *white*, compact at the disc, thin at the margin, soft. Spores "roundish, 8–9 × 7μ, punctate" Rick. Subcaespitose, or solitary. Beech woods. Aug.—Sept. Uncommon. (*v.v.*)

var. **redimitus** Fr. Cke. Illus. no. 754, t. 773.

Redimitus, bound round.

Differs from the type in the *thinner, obtusely umbonate, golden yellow p. streaked with adpressed darker fibrils, the yellowish, fibrillosely striate st. slightly thickened at the base, and the light yellow gills adnate with a small decurrent tooth*. Beech woods. Oct. Rare.

412. **C. (Ino.) callisteus** Fr. *κάλλιστος*, very beautiful.

P. 4–6 cm., *yellow tawny*, fleshy, convexo-expanded, *rather smooth, silky towards the margin*, generally *broken up into minute, innate squamules*. St. 7–12 × 1–1·5 cm., *concolorous, or rhubarb colour*, clavato-bulbous, equally attenuated upwards, fibrillosely striate. Cortina *concolorous*, marginal, fibrillose, fugacious. Gills *concolorous*, adnate, *connected together at the base and to the stem by flocci*, plane, *subdistant*, 8 mm. broad, thin. Flesh *yellowish white, rhubarb colour in the st.*, thin at the margin. Spores pale ferruginous, broadly elliptical, 6–8 × 6μ, 1-guttulate, "punctate" Rick. Pine woods. July—Oct. Not uncommon. (*v.v.*)

413. **C. (Ino.) vinosus** Cke. Cke. Illus. no. 758, t. 759.

Vinosus, wine colour.

P. 5–7·5 cm., *vinous red*, fleshy, semiglobose, then expanded, at length flattened, smooth, shining. St. 5–7·5 × 1 cm., *violet*, cylindrical, abruptly thickened into *a marginately bulbous, reddish base*. Cortina *reddish*. Gills *ferruginous cinnamon*, adnexed, ventricose, scarcely crowded. Flesh *pale violet, reddish in the st.* Spores ferruginous, almond-shaped, 16–18 × 8μ, granular. Under trees, and in woods. Sept.—Oct. Not uncommon. (*v.v.*)

414. **C. (Ino.) Bulliardii** (Pers.) Fr. Boud. Icon. t. 109.

Pierre Bulliard, the eminent French mycologist.

P. 4–10 cm., *dark rufescent, bay brown blood colour, becoming pale*, fleshy, convex, then expanded, obtuse, smooth, or fibrillose. St. 5–12 × 1–1·5 cm., *whitish above, blood red downwards*, and *covered with blood red fibrils near the ovate bulb*, which arises from *a blood red mycelium*. Cortina *whitish*, fugacious. Gills *purplish, then ferruginous*, adnexed with a decurrent tooth, 6 mm. broad, somewhat crowded, *often crenulated at the whitish edge*. Flesh *whitish, brownish under the cuticle*, and *reddish at the base of the st.* Spores ferruginous, elliptical, 7–8 × 4–5μ, verrucose, 1-guttulate. Woods, especially beech. Sept.—Oct. Uncommon. (*v.v.*)

415. **C. (Ino.) bolaris** (Pers.) Fr. Cke. Illus. no. 759, t. 760.
βῶλος, a clod of earth.

P. 3–7 cm., *light yellow red, or pale, variegated with innate, adpressed, spot-like red scales*, fleshy, convexo-plane, obsoletely umbonate. St. 4–8 cm. × 6–10 mm., *pale, variegated with saffron-red, adpressed, fibrillose scales*, sometimes entirely scarlet, *apex white*, firm, equal. Cortina *saffron-red*, fibrillose, fugacious. Gills *cream colour, then dark cinnamon*, decurrent, or adnate, arcuate, crowded. Flesh *white, yellowish in the st.*, firm. Spores pale ferruginous, broadly elliptical, 6–7 × 5μ, minutely punctate. Taste acrid. Woods, especially beech. Aug.—Nov. Not uncommon. (*v.v.*)

****Gills or veil dark, fuscous, or olivaceous.

416. **C. (Ino.) pholideus** Fr. Grevillea, t. 117, fig. 1. *φολίς*, a scale.

P. 5–10 cm., *fawn colour, becoming pale, at length somewhat cinnamon*, fleshy, convex, then flattened, subumbonate and depressed round the umbo, *covered with innate, piloso-fasciculate, crowded, fuscous blackish, squarrose scales*. St. 7–10 cm. × 6–12 mm., *brownish*, attenuated upwards, sometimes shorter and clavato-bulbous, *squarrose with fuliginous, blackish scales up to the cortinate, arachnoid ring*, pale violaceous above. Gills *violaceous, then clay colour, and at length cinnamon*, subemarginate, 4–8 mm. broad, thin, crowded. Flesh *pallid*, thin. Spores ferruginous, broadly elliptical, 6–7 × 4–5μ, "punctate" Rick. Deciduous woods. Aug.—Oct. Common. (*v.v.*)

417. **C. (Ino.) sublanatus** (Sow.) Fr. Boud. Icon. t. 111.
Sub, somewhat; *lanatus*, woolly.

P. 4–10 cm., *fawn colour, or olivaceous fawn, becoming tan fuscous, and at length ferruginous*, slightly fleshy, campanulate, then expanded, umbonate, *clothed with innate, floccose, fuscous squamules*. St. 8–11 × 1·5 cm., *pale ochraceous*, conico-elongated, or clavato-bulbous, *clothed to the middle with fuscous down*, continued into a fibrillose cortina, which does not form a zone, *apex slightly violaceous*, naked. Cortina *yellowish*, arachnoid. Gills *olivaceous yellowish, or ochraceous ferruginous, then cinnamon*, adnate, broader behind, 6 mm. broad, scarcely crowded. Flesh *ochraceous yellow, deeper coloured in the st.*, fairly thick, firm. Spores "ochraceous tawny, subglobose, 8–10μ, apiculate at the one end, verrucose" Boud. Smell of radish. Fir and larch woods. Sept.—Oct. Uncommon.

418. **C. (Ino.) phrygianus** Fr. Fr. Icon. t. 153, fig. 3.
Phrygianus, embroidered.

P. 5–7 cm., *honey colour*, fleshy, convex, obtuse, *densely covered with simple, black, hispid fibrils*. St. 3–8 × 1–1·5 cm., *paler than the p., becoming whitish when dry, reticulately clothed with lax, black fibrils,* equal, base bulbous. Gills *dirty yellow*, rounded behind, 4–8 mm. broad, somewhat crowded. Flesh white, firm. Smell of radish. Damp beech woods. Sept.—Oct. Rare.

419. **C. (Ino.) arenatus** (Pers.) Fr. *Arena*, sand.

P. 3–6 cm., *pale yellowish fuscous, or olivaceous*, fleshy, convex, at first gibbous, *punctate with granular, floccose, brown squamules*. St. 5–7 cm. × 6–10 mm., *brown*, clavato-attenuated, *sheathed up to and beyond the middle with fuscous squamules*, apex naked, *cream colour*. Gills *yellowish, then cinnamon*, emarginate, ventricose, somewhat crowded. Spores "obliquely elliptical, 7 × 5μ" Massee. Fir, and mixed woods. Aug.—Oct. Uncommon.

4. **Dermocybe** Fr.

(δέρμα, skin; κύβη, head.)

Pileus thinly, and equally fleshy, *dry, not hygrophanous*, at first silky with subinnate villose down, then smooth. Veil simple, forming a zone in *C. caninus*.

*Gills at first whitish, or pallid.

420. **C. (Dermo.) ochroleucus** (Schaeff.) Fr. Cke. Illus. no. 764, t. 775.
ὠχρός, pale; λευκός, white.

P. 5–8 cm., *pale white, or yellowish, disc ochraceous*, fleshy, broadly campanulate, then expanded, and somewhat gibbous, *slightly silky*, becoming smooth. St. 4–7 cm. × 8–12 mm., *white*, firm, *ventricose*, naked. Cortina *white, fibrillose*. Gills *clay colour, then ochraceous*, sinuato-adnexed, then free, broader behind, 6 mm. broad, crowded. Flesh *white*, thick at the disc, firm. Spores pale ferruginous, broadly elliptical, 6–8 × 4–5μ, 1-guttulate. Taste bitter. Deciduous woods. Aug.—Nov. Not uncommon. (*v.v.*)

421. **C. (Dermo.) decumbens** (Pers.) Fr. Grevillea, t. 127, fig. 3.
Decumbens, lying down.

P. 2·5–4 cm., *white, or yellowish*, fleshy, firm, convex, then plane, gibbous, then obtuse, *silky-shining*. St. 2·5–5 cm. × 5–6 mm., *shining white, ascending, clavato-bulbous*, smooth, apex mealy. Cortina *white, silky*. Gills *white, then clay colour, at length ochre cinnamon*, adnexed, 4 mm. broad, ventricose, crowded. Flesh *white*, firm. Spores ochraceous, "elliptical, 9–12 × 5–6μ" Rick. Taste slightly bitter. Woods, and grassy places. Sept.—Oct. Uncommon. (*v.v.*)

422. **C. (Dermo.) riculatus** Fr. *Rica*, a head veil.

P. 5–8 cm., *honey colour, but only conspicuously so at the disc, elsewhere clothed with a very thin, floccose, adpressed silkiness*, that makes it appear almost glabrous, fleshy, convexo-plane, slightly gibbous. St. 5–6 cm. × 5–6 mm., *pallid, becoming white*, thickened downwards, smooth. Cortina *pallid*, fibrillose. Gills *clay colour, at length watery ferruginous*, adnate, 4 mm. broad, somewhat crowded. Flesh *pallid*, fairly thick, spongy in the st. Spores "pale brown in the mass, oval, 8–11 × 6–7μ." Herpell. Pine woods. Sept.—Oct. Rare.

423. **C. (Dermo.) tabularis** (Bull.) Fr. *Tabula*, a board.

P. 4–8 cm., *clay or fuscous clay colour, sometimes tawny, becoming pale*, fleshy, convexo-plane, broadly gibbous, at length very flat, *veiled at first with very thin, white flocci*, which rarely in wet weather are collected in a zone at the margin, *becoming smooth*; margin silky. St. 5–8 cm. × 6–12 mm., *white, becoming pale*, tough, elastic, equal, or attenuated upwards, erect, either floccosely scaly, or smooth. Cortina *white*, fugacious. Gills *whitish, then clay colour, subemarginate*, 6 mm. broad, thin, crowded. Flesh *white*, thick. Spores ferruginous, pip-shaped, 9 × 6μ. Woods. Common. (*v.v.*)

424. **C. (Dermo.) camurus** Fr. Fr. Icon. t. 154, fig. 1.
Camurus, crooked.

P. 5–8 cm., *fuscous, often hoary, becoming pale, pallid yellowish, umbo deeper in colour*, fleshy, campanulate, then convex, with a broad, obtuse, often oblique umbo, rimosely incised when dry. St. 4–8 cm. × 7–14 mm., *white*, equal, ascending, flexuose, or twisted, fibrillose, apex *silvery-shining, very fragile*. Gills *grey clay colour, then watery cinnamon, and somewhat fuscous*, adnate, or sinuate, 6 mm. broad, thin, crowded. Flesh *white, fuscous under the cuticle when moist*, thin at the margin, loose. Spores ferruginous, subglobose, 7 × 6μ, multiguttulate. Smell unpleasant. Often caespitose. Woods, especially birch, and beech. Sept.—Oct. Uncommon. (*v.v.*)

425. **C. (Dermo.) diabolicus** Fr. Cke. Illus. no. 765, t. 816, fig. B.
διάβολος, the Devil.

P. 2·5–7 cm., *fuscous with a grey bloom, becoming smooth and fuscous yellow*, fleshy, thin, convex, then plane, hemispherical, obtuse, or umbonate, dry, fragile, often splitting at the margin. St. 4–8 cm. × 4–10 mm., *pale, bluish grey at the apex*, attenuated downwards, smooth. Cortina fugacious. Gills *pale bluish grey, soon becoming white, at length clay colour*, adnate, separating, subemarginate, 4–6 mm. broad, firm, somewhat crowded. Flesh *whitish*, thin. Spores "subglobose, 8–10 × 7–8μ, punctate" Rick. Beech woods. Aug.—Oct. Uncommon. (*v.v.*)

**Gills at first violaceous, becoming purple.

426. **C. (Dermo.) azureus** Fr. Quél. Jur. et. Vosg. t. 24, fig. 4.
Azureus, sky blue.

P. 3–6 cm., *lilac, becoming hoary, then fuscous, and pallid*, fleshy, convex, then plane, obtuse, silky-shining, atomate. St. 6–8 cm. × 8–

10 mm., *sky blue, becoming whitish,* thickened at the base, fragile, *silky,* striate, often twisted. Cortina *concolorous.* Gills *bright bluish violet,* slightly emarginate, then decurrent, rather crowded. Flesh *white, bluish in the st.,* thick at the disc. Spores "subglobose, 7–10 × 7–9μ, punctate" Rick. Deciduous woods, especially beech. Sept.—Oct. Uncommon. (*v.v.*)

427. **C. (Dermo.) caninus** Fr. Cke. Illus. no. 768, t. 765.
Caninus, belonging to a dog.

P. 5–10 cm., *fuscous brown, becoming brick-rufescent or tawny when dry,* fleshy, firm, convex, then plane, obtuse, becoming smooth; margin *at first whitish, silky.* St. 7–12 cm. × 8–12 mm., *pale white, apex violaceous, often ochraceous* at the thickened, somewhat bulbous base, equal, fibrillose, elastic. Cortina *forming a white, or fuscous zone* near the apex of the st., fibrillose. Gills *bluish grey, or purplish,* then *cinnamon,* emarginate, 6–10 mm. broad, subdistant, thin. Flesh *white, becoming yellowish,* thick at the disc, soft. Spores ferruginous, elliptical, 9–10 × 6μ, 1-guttulate. Taste mild. Edible. Deciduous woods, and heaths. Sept.—Nov. Common. (*v.v.*)

428. **C. (Dermo.) anomalus** Fr. Cke. Illus. no. 772, t. 776.
ἀ, not; ὁμαλός, even.

P. 3–6 cm., *fuliginous, then rufescent, becoming hoary with separating fibrils, at length yellowish,* fleshy, thin, very convex, then expanded, and gibbous. St. 5–7·5 cm. × 6–8 mm., *violaceous above, whitish below, at length becoming pale and somewhat yellow,* attenuated from the base, slightly sheathed, *fibrillose, or somewhat scaly.* Gills *more or less violaceous, bluish grey purplish, then cinnamon,* adnate, or emarginate, with a decurrent tooth, crowded, thin. Flesh watery, *becoming white when dry, violaceous at the apex of the st.,* thin, soft. Spores ferruginous, elliptical, 8–9 × 6–7μ, punctate. Taste mild. Edible. Woods, and heaths. Aug.—Oct. Common. (*v.v.*)

429. **C. (Dermo.) lepidopus** Cke. Cke. Illus. no. 773, t. 850.
λεπίς, a scale; πούς, a foot.

P. 1·5–7 cm., *umber, with a tinge of violet near the margin, disc becoming rufescent,* fleshy, convex, then expanded, gibbous, *smooth.* St. 6–12 cm. × 8–12 mm., *violet at the apex, dirty white below,* attenuated upwards, *with concentric, fibrillose, darker bands.* Cortina *whitish, with a tinge of violet.* Gills *violet, then cinnamon,* adnate, rather crowded, thin. Flesh *whitish, tinged lilac at the apex of the stem,* rather thin. Spores pale ferruginous, ovate, sometimes almost globose, 8–9 × 6–7μ, with a basal apiculus. Woods, and heaths. Aug.—Oct. Not uncommon. (*v.v.*)

430. **C. (Dermo.) myrtillinus** Fr. μύρτος, the myrtle.

P. 3–7·5 cm., *fuliginous, tinged with lilac, becoming hoary silky* with the dense white fibrils, fleshy, convex, gibbous, becoming plane. St. 5–7 cm. × 6–12 mm., *whitish, streaked with sparse, lilac fibrils,* tough, slightly bulbous. Cortina *white,* fibrillose. Gills *pure amethyst-azure-blue,* scarcely changing colour, *adnate,* subdistant, edge *whitish, denticulate.* Flesh *fuscous, becoming whitish when dry, violaceous at the apex of the st.,* tough, thin. Spores pale ferruginous, broadly elliptical, 7–8 × 6μ. Woods. Sept.—Oct. Not uncommon. (*v.v.*)

431. **C. (Dermo.) albocyaneus** Fr. *Albus,* white; κύανος, dark blue.

P. 2–3 cm., *white, becoming yellow,* fleshy, convexo-plane, obtuse, *hoary silky,* becoming smooth. St. 6–10 × ·5–1·5 cm., *white,* subclavate, *naked.* Cortina *white,* fugacious. Gills *bluish purple, then somewhat ochraceous,* emarginate, 6–8 mm. broad, *crowded.* Flesh *white,* thick at the disc. Spores pale ferruginous, elliptical, 10 × 6–7μ, punctate. Smell "of apple" Quél. Coniferous, birch, and beech woods. Sept. Uncommon.

432. **C. (Dermo.) spilomeus** Fr. Fr. Icon. t. 154, fig. 3. σπῖλος, a spot.

P. 3–7 cm., *rufescent, or clay colour,* fleshy, convex, then expanded, gibbous, becoming smooth. St. 5–7 × 1 cm., *whitish lilac, covered in the basal half with rufous, or tawny scales,* equal, slightly thickened at the base. Cortina *white,* fibrillose. Gills *bluish grey, or violaceous, becoming pale, at length watery cinnamon,* adnate, or emarginate, crowded, thin. Flesh *cinereous, becoming white,* thick at the disc. Spores ferruginous, broadly elliptical, or subglobose, 6–9 × 6–7μ, apiculate at the base, multi-guttulate, "punctate" Rick. Woods. Sept.—Nov. Uncommon. (*v.v.*)

433. **C. (Dermo.) violaceo-fuscus** (Cke. & Massee) Massee. Cke. Illus. no. 1163, t. 1174, as *Inocybe violaceo-fusca* Cke. & Mass. *Violaceus,* violet; *fuscus,* dark.

P. 2·5–5 cm., *umber, often tinged with violet,* fleshy, more or less convex, then expanded, obtusely umbonate, flocculose, fibrillose, *concentrically scaly,* dry; margin thin, torn, fimbriate. St. 5–6 cm. × 6–8 mm., *violet above, pallid below,* equal, silky. Gills *violet, then umber,* adnexed, rounded behind, or slightly sinuate, 4–6 mm. broad, scarcely crowded, *edge paler, serrulate.* Flesh thin. Spores ferruginous, elliptical, 7–8 × 4μ. Amongst grass in open places. Uncommon.

***Gills bright cinnamon, red, or yellow.

434. **C. (Dermo.) phoeniceus** (Bull.) Maire. (= *Cortinarius miltinus* Quél. non Fr.) Boud. Icon. t. 112, as *Cortinarius miltinus* Fr. φοινίκεος, purple-red.

P. 2–5 cm., *bay brown cinnamon, or dark cinnamon when moist, becoming bright bay when dry*, fleshy, convex, then expanded, broadly gibbous, or umbonate, flexuose, fibrillosely silky. St. 5–7 cm. × 4–9 mm., *pale, reddish fibrillose below*, equal, or slightly thickened at the base, rigid, striate. Cortina *red*, fibrillose. Gills *reddish, then ferruginous*, adnate, 3–5 mm. broad, somewhat crowded, thin. Flesh *fuscous under the cuticle of the p., becoming paler, tinged reddish in the st.*, thin. Spores ferruginous, elliptical, 6–8 × 4–5μ, multi-guttulate. Smell none, or of radish. Woods, especially birch. Sept.—Nov. Common. (*v.v.*)

435. **C. (Dermo.) semisanguineus** (Brig.) Maire. Rolland, Champ. t. 66, no. 146. *Semi*, half; *sanguineus*, bloody.

P. 3–6 cm., *tan, or tawny olivaceous, becoming paler*, convex, then plane, silky. St. 3–6 cm. × 6–8 mm., *paler tawny, or yellowish*, equal, often slightly thickened at the base. Cortina *tawny*, fibrillose. Gills *blood red*, sinuato-adnate, broad, or narrow, crowded, thick. Flesh *fuscous, becoming pale*, thin at the margin. Spores pale ferruginous, elliptical, 6–7 × 3–4μ. Woods, and heaths under birches. Aug.—Nov. Common. (*v.v.*)

436. **C. (Dermo.) cinnabarinus** Fr. Boud. Icon. t. 113. *κιννάβαρι*, dragon's blood.

P. 2–7·5 cm., *scarlet-red*, fleshy, campanulate, then flattened, obtuse, or obtusely umbonate, silky, then *becoming smooth and shining*. St. 3–6 cm. × 6–8 mm., *concolorous*, equal, sometimes bulbous, fibrillose, or striate. Cortina *cinnabar* colour, fibrillose, lax. Gills *concolorous*, dark blood colour when bruised, adnate, subdecurrent, *subdistant*, often connected by veins; edge unequal and darker. Flesh *concolorous, then paler*, firm. Spores ferruginous, almond-shaped, 10–13 × 5–6μ, verrucose. Smell of radish. Woods, especially beech. Sept.—Nov. Not uncommon. (*v.v.*)

437. **C. (Dermo.) sanguineus** (Wulf.) Fr. Grevillea, t. 110, fig. 5. *Sanguineus*, bloody.

P. 2·5–5 cm., *dark blood colour, becoming paler when dry*, fleshy, convex, then plane, obtuse, or slightly umbonate, sometimes depressed, *shaggy, or squamulose*. St. 5–10 cm. × 4–6 mm., *concolorous, or darker*, equal, or slightly attenuated downwards, flexuose, *clothed with concolorous fibrils*, base sometimes white. Cortina *blood red*, arachnoid, fugacious. Gills *concolorous, then rust colour*, adnate, sinuate, *crowded*. Flesh *reddish, paler*, thin, pouring out a *blood red juice* when pressed. Spores pale ferruginous, elliptical, 8–9 × 5–6μ, 1-guttulate, verrucose. Smell of radish, sometimes obsolete. Woods, especially of conifers. Sept.—Nov. Not uncommon. (*v.v.*)

438. **C. (Dermo.) anthracinus** Fr. *ἄνθραξ*, coal.

P. 2–3 cm., *dark chestnut,* or *brown fuscous, often reddish rose colour at the margin,* fleshy, convex, then expanded, umbonate, fibrillose, becoming smooth. St. 4–5 cm. × 3–5 mm., *intense blood colour, fuscous, or yellow at the base,* equal, fibrillose. Gills *deep red, or fiery in colour, becoming blood red when bruised, then rust colour with the edge deep red,* sinuato-adnate, crowded. Flesh *concolorous* (lilac according to Quélet), soft, thick at the disc. Spores ferruginous, elliptical, 7 × 5μ, punctate. Woods. Aug.—Oct. Uncommon. (*v.v.*)

439. **C. (Dermo.) cinnamomeus** (Linn.) Fr. Cke. Illus. no. 777, t. 777. *κιννάμωμον*, cinnamon.

P. 1–10 cm., *somewhat cinnamon, or tawny ochraceous,* fleshy, convex, then expanded, *obtusely umbonate, silky, or squamulose with innate yellowish fibrils,* at length becoming smooth. St. 5–9 × ·5–1 cm., *yellowish, equal,* fibrillose. Cortina *yellowish,* fibrillose. Gills *yellowish, then cinnamon,* adnate, broad, thin, crowded, *shining.* Flesh *yellowish,* thin, scissile. Spores dark ochraceous, elliptical, 6–8 × 4–5μ, 1-guttulate, "faintly punctate" Rick. Coniferous, and deciduous woods. Aug.—Feb. Common. (*v.v.*)

var. **croceus** (Schaeff.) Fr. *κρόκος*, saffron colour.

Differs from the type in its *smaller size, and its bright yellow st., and gills.* Woods. Sept.—Oct. Not uncommon. (*v.v.*)

440. **C. (Dermo.) croceo-conus** Fr. *κρόκος*, saffron; *κῶνος*, a cone.

P. 3–5 cm., *fulvous cinnamon,* conical, then campanulate, *persistently acute, almost glabrous.* St. 7–12 cm. × 4 mm., *yellowish,* flexuose. Gills *cinnamon,* ascending, linear, crowded. Flesh *very thin,* 1 mm. thick. Spores "elliptical, almost smooth, 8–9·5 × 5μ" Kauffm. Subcaespitose. Amongst moss in coniferous woods. Sept.—Oct. Uncommon. (*v.v.*)

441. **C. (Dermo.) uliginosus** Berk. Cke. Illus. no. 781, t. 851. *Uligo*, marshy ground.

P. 3–5 cm., *bright red brown, almost brick-red,* fleshy, campanulato-conical, then expanded, *very strongly umbonate,* silky, sometimes streaked. St. 3–8 cm. × 3–8 mm., *paler than the p.,* flexuose. Gills *yellow, becoming olive, then cinnamon,* adnate with a tooth, *distant.* Flesh *yellow-olive, then cinnamon,* thick at the disc. Spores dark ochraceous, elliptical, 8–9 × 4–5μ. Amongst *Sphagnum* in woods. Sept.—Oct. Uncommon. (*v.v.*)

442. **C. (Dermo.) orellanus** Fr. non Quél. Cke. Illus. no. 776, t. 787, lower figs. *ὄρος*, a mountain.

P. 3–7 cm., *orange tawny,* fleshy, convex, then convexo-plane, more or less undulate, umbonate, *covered with concolorous, or deeper coloured fibrillose squamules.* St. 2·5–9 cm. × 4–20 mm., *tawny,* equal, or attenuated upwards, striato-fibrillose, or smooth. Cortina *tawny,* fibrillose. Gills *tawny, then rust colour,* broadly adnato-sinuate, broad, thick, distant, often veined on the sides. Flesh *concolorous, reddening,* thin at the margin. Spores brownish ferruginous, broadly elliptical, 8–11 × 5–6μ, 1-multi-guttulate, verrucose. Woods, and heaths. Aug.—Oct. Rare. (*v.v.*)

443. **C. (Dermo.) malicorius** Fr. Fr. Icon. t. 155, fig. 1.
Malicorium, the rind of a pomegranate.

P. 3–6 cm., *tawny, disc darker, golden, and floccose at the margin,* fleshy, convex, then plane, obtuse, velvety, or fibrillose. St. 4–5 cm. × 12 mm., *golden, at length fuscous, and olivaceous, covered with golden fibrils.* Cortina *golden,* fibrillose. Gills *golden tawny,* rounded behind, adnexed, crowded, *edge at length floccose and discoloured.* Flesh *yellow, then greenish olive,* rather thick, *scissile.* Spores "elliptical, 8–9 × 4–5μ, faintly punctate" Rick. Taste pleasant. Coniferous woods. Sept.—Oct. Rare.

444. **C. (Dermo.) infucatus** Fr. Fr. Icon. t. 155, fig. 2.
Infucatus, painted.

P. 2·5–4 cm., *bright yellow,* fleshy, convex, obtuse, *silky* when dry. St. 4–7·5 cm. × 4–8 mm., *pale light yellow,* equally *attenuated upwards from the clavate base,* fibrillose. Cortina *yellow,* fibrillose. Gills *tawny, then cinnamon,* adnate, almost linear, 2 mm. broad, crowded, thin. Flesh whitish. Spores "elliptical, 10 × 5μ" Massee. Woods. Sept.—Oct. Rare.

445. **C. (Dermo.) colymbadinus** Fr. Fr. Icon. t. 155, fig. 3.
κολυμβάς, swimming.

P. 5–8 cm., *honey tan colour, becoming yellowish when dry,* somewhat fleshy, convex, then expanded, scarcely umbonate, often repand, *covered with yellow, fugacious fibrils,* then smooth, and shining. St. 5–10 cm. × 6–8 mm., *pallid,* equal, somewhat naked, *fibrillosely-striate,* sometimes twisted. Cortina almost none, very fugacious. Gills *dark ferruginous,* adnate, 4–8 mm. *broad, subdistant, thick, edge white-floccose.* Flesh *pallid, darker at the base of the st.,* scissile. Spores "subglobose, 7–8 × 6–7μ, almost spinose" Rick. Smell very strong of radish. Pine, and beech woods. Sept.—Oct. Rare.

****Olivaceous, veil dingy pallid, or fuscous.
P. not torn into scales.

446. **C. (Dermo.) cotoneus** Fr. *κότινος,* the wild olive.

P. 4–8 cm., *olivaceous,* fleshy, campanulate, then expanded, obtuse, somewhat repand, *innately velvety,* fragile when old. St. 5–9 × 1–

1·5 cm., *pale olivaceous*, bulbous, somewhat fibrillose. Cortina *yellow olivaceous*, persistent, *woven into a fuscous zone* towards the apex of the st. Gills *olivaceous, then cinnamon*, adnate, separating, 4–6 mm. broad, somewhat crowded. Flesh *pale olivaceous*, deeper coloured in the st., thin, lax, soft. Spores ferruginous, subglobose, 8–9 × 8μ, granular. Taste mild. Woods, especially oak. Sept.—Oct. Not uncommon. (*v.v.*)

447. **C. (Dermo.) subnotatus** Fr. Cke. Illus. no. 784, t. 832.
Subnotatus, marked.

P. 6–10 cm., *olivaceous, becoming yellowish, then fuscous*, fleshy, conical, campanulate, then expanded, gibbous, *at first covered with hoary, silky fibrils*, then smooth. St. 7–10 × 1–1·5 cm., *pale olivaceous*, conical, equally attenuated upwards, often curved and flexuose, fibrillose, or *squamulose* with the yellowish cortina, apex naked, *silvery-shining*. Cortina yellowish, fibrillose, inconspicuous. Gills *bright ochraceous, then olivaceous cinnamon*, adnate, 6–10 mm. broad, subdistant, *often connected by veins*. Flesh *yellowish*, very thin at the margin. Spores "elliptical, 6–8 × 5–6μ, granular" Massee. Smell of radishes or none. Beech woods. Sept.—Nov. Uncommon. (*v.v.*)

448. **C. (Dermo.) raphanoides** (Pers.) Fr. Cke. Illus. no. 786, t. 833, fig. A. ῥαφανίς, a radish; εἶδος, like.

P. 2·5–5 cm., *fuscous olivaceous, becoming tawny*, fleshy, *campanulate, then expanded*, obtusely umbonate, often undulate, *silky fibrillose*, then smooth. St. 5–8 × ·5–1 cm., *olivaceous, becoming pallid*, equal, or slightly attenuated upwards from the somewhat thickened base, sometimes twisted, fibrillose. Cortina *pallid olive*, filamentous, often forming a narrow ring-like zone on the st. Gills *subolivaceous, then cinnamon, and subferruginous*, adnate, slightly ventricose, scarcely crowded, *edge often paler*. Flesh *pallid, or ochraceous*, thick at the disc, firm, then soft. Spores pale ferruginous, elliptical, 7–8 × 4–5μ, granular. Smell strong of radish. Taste bitter. Beech, birch, and fir woods. Sept.—Oct. Common. (*v.v.*)

449. **C. (Dermo.) valgus** Fr. Cke. Illus. no. 785, t. 750.
Valgus, bow-legged.

P. 5–8 cm., *yellowish fuscous, becoming paler, somewhat brick red when dry*, fleshy, *fragile*, convex, then expanded and subumbonate, *smooth*; margin *submembranaceous*. St. 6–12 × 1–1·5 cm., *pallid, smooth, shining*, attenuated upwards, often somewhat twisted, apex *lilac and substriate*; base *white-tomentose, bulbous, rooting*. Gills *yellowish, then cinnamon*, adnate, somewhat separating, 4–6 mm. broad, subdistant. Flesh *yellowish*, thick at the disc. Spores pale ferruginous, elliptical, 8 × 5μ. Smell none, or of radish. Amongst moss in coniferous woods. Oct. Uncommon.

450. **C. (Dermo.) venetus** Fr. Fr. Icon. t. 155, fig. 4.
Venetus, sea-coloured.

P. 4–5 cm., *green, then greenish yellow, yellowish when dry*, fleshy, hemispherical, obtusely umbonate, *covered with a persistent, erect, yellow, velvety tomentum*. St. 5–8 cm. × 6–8 mm., *concolorous, or paler*, equal, often curved, firm, *very fibrillosely silky, base often yellow and villous*. Cortina *green, or citron yellow*, fibrillose. Gills *olivaceous, darker than the p., then brownish*, adnate, *very broad*, in the form of a segment, *often connected by veins*, subdistant. Flesh *pale yellowish, or greenish yellow*, thick at the disc, soft. Spores "olivaceous, elliptical, 10μ, echinulate" Bataille, "subglobose, 7–8 × 6–7μ, roughish" Rick. Smell of radish. Taste acrid. Beech, and fir woods. Aug.—Oct. Rare.

5. **Telamonia** Fr.

(τελαμών, a broad linen bandage.)

Pileus thinly fleshy, or abruptly thin at the margin, moist, hygrophanous, smooth or sprinkled with superficial whitish fibres of the veil. Stem cortinate, and *annulate*, hence the veil is somewhat double.

I. Gills very broad, rather thick, more or less distant.
St. spongy, and wholly fibrous.

*St. and cortina white, or whitish.

451. **C. (Tela.) macropus** Fr. Cke. Illus. no. 787, t. 788.
μακρός, long; πούς, a foot.

P. 5–9 cm., *brick colour, at length becoming ferruginous, paler at the margin*, which is at first incurved, fleshy, convex, then flattened, obtuse, dry, *hoary with very small squamules*, becoming smooth. St. 7·5–15 × 1–2·5 cm., *dingy whitish, then concolorous*, subequal, fibrillose. Cortina *white*, forming a distant, inferior, narrow woven ring. Gills *pallid, then watery cinnamon*, adnexed, very broad, 1–2·5 cm., distant, edge sometimes crenate. Flesh *whitish, then cinereous*, thin at the margin, firm, then soft. Spores pale ferruginous, elliptical, 9–10×5μ, minutely punctate. Woods. Sept.—Oct. Uncommon. (*v.v.*)

452. **C. (Tela.) laniger** Fr. Fr. Icon. t. 156, fig. 2.
Laniger, wool bearing.

P. 5–9 cm., *bright or dark tawny, sometimes becoming pale*, fleshy, hemispherical, then expanded, obtuse, *at first floccosely squamulose with whitish flocci*, then becoming smooth, silky towards the margin. St. 5–10 × 2–4 cm., *white*, equal, or bulbous, sometimes ventricose, *more or less distinctly sheathed by the veil*. Cortina *white, forming a very soft, shining white, distinct ring*, very delicate above. Gills *bright saffron cinnamon, then shining tawny*, adnate, or sinuate, at first crowded, then subdistant, sometimes transversely veined. Flesh *white, reddish white in the st. at length becoming tawny at the base,* thick at the disc, thin at the margin. Spores ferruginous, pip-shaped, 9–10 × 6μ, punctate. Smell strong. Coniferous woods. Sept.—Oct. Rare. (*v.v.*)

453. **C. (Tela.) bivelus** Fr. Fr. Icon. t. 156, fig. 1.
Bis, twice; *velum*, a veil.

P. 5–12 cm., *tawny ferruginous,* often spotted, or darker at the disc, fleshy, convexo-plane, always obtuse, bibulous, *smooth,* or slightly silky round the margin, shining, rarely opaque, sometimes rivulose. St. 6–8 × 1–2 cm., *dingy white,* bulbous, or equally attenuated, fibrillosely villous. Cortina *white, sheathing, terminating in a spurious and fugacious ring,* thin and vanishing above. Gills *ochraceous, then bright tawny cinnamon,* adnate, or subemarginate, at first crowded, then subdistant. Flesh *white, becoming somewhat ferruginous in the stem,* thick, spongy in the stem. Spores pale ferruginous, elliptical, often pointed at the base, 9–10 × 6–7μ, 1–2-guttulate, punctate, "almost smooth" Rick. Smell "strong," "pleasant" Quél. Taste mild. Woods, and heaths. Sept.—Oct. Uncommon. (*v.v.*)

454. **C. (Tela.) bulbosus** (Sow.) Fr. Sow. Eng. Fung. t. 130.
Bulbosus, bulbous.

P. 5–7·5 cm., *date brown, becoming fuscous brick colour when dry,* fleshy, *campanulato-expanded,* obtuse, or *broadly gibbous,* even, or *fibrillosely squamulose* towards the margin from the torn epidermis. St. 4–7·5 × 1–1·5 cm., *paler than the p., becoming whitish, tinged saffron-yellow at the bulbous base,* equal. Cortina *white,* sheathing, forming a fugacious ring. Gills *dark, then brown-cinnamon,* adnate, broad, subdistant. Flesh *concolorous and pallid when damp, whitish when dry, tinged with saffron-yellow at the base of the st.,* thick and compact at the disc. Spores pale ferruginous, elliptical, 8–9 × 5–6μ, minutely verrucose. Smell none, or of radish. Woods. Sept.—Oct. Uncommon. (*v.v.*)

455. **C. (Tela.) urbicus** Fr. Grevillea, t. 111, fig. 8.
Urbicus, pertaining to the city.

P. 3–5 cm., *clay-whitish,* fleshy, *convexo-plane, smooth,* pitted when larger. St. 5–7·5 cm. × 12–15 mm., *concolorous,* equal, villous above the ring when young. Cortina *white,* forming a narrow ring above the middle of the st. Gills *watery ferruginous,* emarginate, broad, thin, crowded. Flesh *whitish,* firm. Spores "ochraceous, pruniform, 8μ, punctate" Quél. Grassy places. Sept.—Oct. Rare.

456. **C. (Tela.) licinipes** Fr. *Licinium*, lint; *pes*, foot.

P. 5–7·5 cm., *very pale yellow, tan pallid when dry,* fleshy-membranaceous, campanulate, then convex, and flattened, obtusely umbonate, at length depressed round the umbo, smooth. St. 5–12 cm. × 6–8 mm., *pale white*, at length fragile, equal, often flexuose, base white villous, *clothed with shining, white*, fugacious, *floccoso-plumose scales* below the ring, even above. Cortina *white*, forming a distant, membranaceous ring. Gills *watery cinnamon*, adnate, very broad behind, up to 12 mm., somewhat crowded. Flesh *watery white*, thin. Fir woods, and *Sphagnum* swamps. Oct. Rare.

var. **robustior** Cke. Cke. Illus. no. 792, t. 819. *Robustior*, firmer.

Differs from the type in being *larger and stouter*. Spores 10 × 6–7μ. Damp woods. Oct. Rare.

457. **C. (Tela.) microcyclus** Fr. Cke. Illus. no. 793, t. 865.
μικρός, small; *κύκλος*, a ring.

P. 2–3 cm., *brick-red fuscous, disc darker, becoming paler and opaque when dry, almost membranaceous*, plano-convex, minutely umbonate, smooth. St. 2·5–5 cm. × 3–4 mm., *pallid, then white, attenuated upwards from the subbulbous base.* Cortina *white*, forming a ring-like zone on the st. Gills *lilac, then dark cinnamon*, adnate, very broad, almost ovate, *distant, thin.* Flesh thin. Spores "reddish brown in the mass, elliptical, 5–7 × 4μ" Herpell. Coniferous woods, and under trees. Sept.—Oct. Uncommon.

**St. and gills violaceous. Cortina commonly white-violaceous, universal veil white. Very distinguished.

458. **C. (Tela.) torvus** Fr. Fr. Icon. t. 157, fig. 1. *Torvus*, wild.

P. 4–12 cm., *brick colour, date brown, copper brown*, fleshy, convex, then flattened, obtuse, *sprinkled with hoary squamules and fibrils*, at length becoming smooth. St. 7–12 × 1–1·5 cm., *whitish, becoming discoloured*, short and bulbous, then elongated and subequal, often curved, *sheathed to the middle, and forming a white, membranaceous, persistent ring*, fibrillose and floccosely scaly below the ring; apex *pale violaceous*, silky; base white villous. Cortina *white*, villous, then fibrillose. Gills *violaceous, soon purplish umber, then dark cinnamon*, subadnate, very broad, 6–12 mm., *thick, distant*, fragile, at length sometimes veined at the base. Flesh *dingy, becoming whitish when dry*, thick at the disc, firm. Spores ferruginous, pip-shaped, 9–10 × 5–6μ, 1-guttulate, "warted" Rick. Smell "of camphor" Maire. Woods, especially beech. Aug.—Nov. Common. (*v.v.*)

459. **C. (Tela.) impennis** Fr. Fr. Icon. t. 157, fig. 2.
In, not; *penna*, a feather.

P. 5–10 cm., *umber, then brick colour*, decolouring and dingy, *fleshy, convex, very obtuse, smooth*; margin silky when young, at length cracked. St. 5–10 × 1–2·5 cm., *pale, becoming violet at the apex, cylindrical*, scarcely bulbous, fibrillose, veil forming an incomplete white zone towards the apex. Cortina *white*, fibrillose. Gills *intense bright violaceous, somewhat purplish, soon becoming watery* ferruginous, adnate, then emarginate, distant, rather thick. Flesh *pallid, thick, becoming azure blue at the apex of the st.* Spores ferruginous, 9–10 × 6μ, punctate. Woods, especially pine, and among dead leaves. Sept.—Oct. Uncommon. (*v.v.*)

var. **lucorum** Fr. Cke. Illus. no. 1190, t. 1192, as a species.
Lucus, a wood.

Differs from the type in the *unicolorous, clavato-bulbous stem, in the gills only being tinged with a fugacious violet, and in the firm dark watery flesh becoming isabelline when dry.* Woods. Sept. Rare.

460. **C. (Tela.) plumiger** Fr. Grevillea, t. 112, fig. 1.
Plumiger, feather-bearing.

P. 6–9 cm., *fuscous, somewhat olivaceous when moist, brick tan when dry*, fleshy, *conical*, then campanulate, *with a broad, obtuse, very prominent umbo*, then expanded, often cracked, dry, *clothed with dense, white, floccoso-plumose scales*, which are either erect and squarrose, or adpressed and silky. St. 7·5–10 × 1 cm., *pale, then often tinged with citron yellow, very clavate, apex pubescent, floccosely scaly from the veil, which forms a ring-like zone at the apex.* Cortina *white*, floccose. Gills *violaceous, soon watery then pure cinnamon*, adnate, scarcely crowded, broad; *edge lilac, or clay colour*, often denticulate. Flesh *white, or lilac, then yellowish*, thin, firm. Smell unpleasant, foetid. Spores ferruginous, almond-shaped, 13–15 × 7–8μ, minutely echinulate. Coniferous, and mixed woods. Sept.—Oct. Rare.

461. **C. (Tela.) scutulatus** Fr. Fr. Icon. t. 158, fig. 2.
Scutulatus, diamond- or lozenge-shaped.

P. 2–5 cm., *purple umber, or brick fuliginous*, very hygrophanous, *brick colour when dry*, fleshy, ovato-globose, then campanulato-hemispherical, obtuse, sometimes umbonate, or umbilicate, white silky round the margin, then naked, *rivulose* in *the form of innate squamules*, sometimes lacunoso-wrinkled. St. 5–15 cm. × 4–12 mm., *deep violaceous, at length becoming fuscous*, cylindrical, or bulbous at the extreme base, white villous at the base, *rigid*, somewhat rooting, fibrillosely striate, veil sheathing and *forming a white, narrow, membranaceous ring.* Cortina *white*, floccose. Gills *violaceous, then purple, at length cinnamon*, adnate, rarely emarginate, 6 mm. broad, more or less distant, edge often white and serrate when young. Flesh *violaceous*, firm, thick at the disc. Spores ferruginous, pip-shaped, 7–8 × 4μ, 1–multi-guttulate, "slightly punctate" Rick. Smell "strong, of radish, or of violets" Quél. Woods, and moist places. Aug.—Oct. Uncommon. (*v.v.*)

462. **C. (Tela.) evernius** Fr. Luc. Champ. t. 191.

εὐ-ερνής, flourishing.

P. 3–10 cm., *purple bay brown, brick colour when dry, becoming isabelline-hoary when old*, very hygrophanous, fleshy, conico-campanulate, then flattened, obsoletely umbonate, adpressedly silky, then smooth, *at length* rimosely incised, and *torn into fibrils*, fragile. St. 7–15 × 1–1·5 cm., *violaceous, becoming pale, equal, or attenuated downwards*, substriate, *squamulose and obsoletely zoned with the white veil.* Cortina *white*, fibrillose. Gills *violaceous purple, becoming pale, then cinnamon*, adnate, ventricose, *very broad*, 8–20 mm., distant. Flesh *concolorous in the p., violaceous in the st.*, very thin at the margin. Spores ferruginous, elliptical, 7–8 × 5–6μ, 1–2-guttulate, "faintly warted" Rick. Smell like mushrooms. Deciduous, and pine woods, and damp places. Sept.—Dec. Uncommon. (*v.v.*)

463. **C. (Tela.) quadricolor** (Scop.) Fr. Cke. Illus. no. 799, t. 867.

Quadricolor, four coloured.

P. 4–7·5 cm., *pallid yellow, then somewhat tawny, shining when dry*, fleshy, conical, then flattened, umbonate, smooth, at length spotted; margin radiato-striate. St. 5–7·5 cm. × 4–6 mm., *violaceous, becoming whitish, equal*, flexuose, subrigid, fibrilloso-striate with the adpressed veil, which forms an oblique, fugacious, white ring. Cortina *white*, fibrillose. Gills *dark violaceous, or purplish, then cinnamon*, adnate, 6–8 mm. broad, distant, *white-serrated at the edge.* Flesh *yellowish*, thin. Spores ferruginous, broadly elliptical, 7–8 × 5μ, multi-guttulate, "nearly spinulose" Rick. Woods, especially beech. Sept.—Oct. Not uncommon. (*v.v.*)

***St. and veil reddish or yellow. Gills tawny, or cinnamon, never violaceous, nor becoming brown.

464. **C. (Tela.) armillatus** Fr. (= *Cortinarius haematochelis* (Bull.) Fr.) Fr. Icon. t. 158, fig. 1.

Armillatus, having a bracelet.

P. 4–12 cm., *red- or fuscous-brick colour*, fleshy, cylindrical, then campanulate, at length flattened, often gibbous, smooth, *then innately fibrillose*, or squamulose; margin at first incurved. St. 6–15 × 1–2 cm., *white, becoming brownish with age*, equal, *base bulbous, the red veil forming one to four distant, oblique cinnabar zones*, striate when old, and reddish fibrillose at the base. Cortina *reddish white*, fibrillose. Gills *pallid cinnamon, then dark ferruginous, almost bay brown*, adnate, slightly rounded, *very broad*, 10–15 mm., distant. Flesh *dingy pallid, isabelline in the st.*, thin at the margin. Spores pale ferruginous, elliptical, 9–10 × 5–6μ, multi-guttulate, minutely verrucose. Smell of radish, or none. Woods, and heaths. Aug.—Oct. Common. (*v.v.*)

465. **C. (Tela.) paragaudis** Fr.
Paragaudis, a border worked on a garment.

P. 2·5–7·5 cm., *bay, becoming tawny or yellowish tan colour when dry*, fleshy, conical, then campanulate, and expanded, umbonate, often repand and torn on the surface, fragile. St. 7–15 × 1–1·5 cm., *brick-red, becoming pale, reddish at the base*, equal, or ventricose, curved and somewhat twisted, or undulate and flexuose, *covered with reddish flocci, or squamules*. Cortina *whitish*, fibrillose. Gills *pale, then becoming dark cinnamon*, adnate, separating, ventricose, crowded, or subdistant, edge unequal. Flesh *paler*, thick at the disc. Spores "subelliptical, 8–10 × 4–5μ, punctate" Rick. Damp places under pines. Sept.—Oct. Rare.

var. **praestigiosus** Fr. *Praestigiosus*, delusive.

Differs from the type in the *submembranaceous pileus being striate to the disc, in the thin stem*, 2–3 *mm. thick, and the tawny cinnamon, linear gills*. Under pines, and amongst *Scirpus caespitosus*. Rare.

466. **C. (Tela.) croceo-fulvus** (DC.) Fr. Cke. Illus. no. 1191, t. 1193.
κρόκος, saffron; *fulvus*, tawny.

P. 5–10 cm., *orange-tawny*, fleshy, convex, then expanded, obtusely umbonate, or gibbous, smooth. St. 7–10 cm. × 6–18 mm., *yellow, becoming reddish*, equal, *veil forming a rufous orange zone*, apex pale. Gills *becoming ferruginous*, adnate, slightly sinuate, 6–8 mm. broad, rather distant. Flesh *bright yellow*. Spores obovate, 8–10 × 6μ, rough. Woods. Sept. Rare.

467. **C. (Tela.) limonius** Fr. Fr. Icon. t. 159, fig. 1.
Limonius, lemon-yellow.

P. 5–10 cm., *tawny lemon yellow, ochraceous yellow and opaque when dry*, very hygrophanous, fleshy, convexo-plane, *obtuse, smooth when moist, rimosely incised when dry*. St. 6–8 cm. × 12 mm., *yellow*, equal, *base* attenuated or thickened, and *at length deep saffron, floccosely scaly with the light yellow veil*, which often forms a floccose ring at the apex. Gills *yellow, or light yellow, at length tawny cinnamon*, adnate, rarely emarginate, distant. Flesh *concolorous*, soft. Spores golden tawny, elliptical, 8–9 × 4–5μ, minutely echinulate. Smell slight of radish, or none. Coniferous woods. Sept.—Oct. Uncommon. (*v.v.*)

468. **C. (Tela.) helvolus** Fr. Cke. Illus. no. 802, t. 804, fig. B.
Helvolus, pale yellow.

P. 3–7·5 cm., *dark tawny cinnamon, very pale yellow when dry*, fleshy, *convexo-plane, obtuse*, or obtusely umbonate, smooth; margin *incurved, at first cortinate*. St. 5–20 cm. × 4–8 mm., *concolorous, at length fuscous ferruginous*, equal, either attenuated upwards, or at the base, fibrillose, *girt above with an annular, narrow, oblique, ferruginous, margined zone formed by the woven veil.* Gills *tawny, then dark cinnamon, very emarginate,* 8 mm. broad, distant, thick, often veined at the base, opaque. Flesh *tawny,* firm, *fuscous ferruginous in the st.* Spores ferruginous, "elliptical, 9–10 × 5–6μ, verrucose" Rick. Woods, and wooded pastures. Sept.—Oct. Uncommon.

469. **C. (Tela.) hinnuleus** (Sow.) Fr. Cke. Illus. no. 803, t. 805.

Hinnuleus, a young stag.

P. 3–6 cm., *pallid tawny cinnamon, becoming pale,* shining when dry, fleshy, campanulato-expanded, obtuse, or obtusely umbonate, sometimes depressed at the disc, smooth; margin at first silky and white. St. 2·5–10 cm. × 4–12 mm., *dingy tawny, or fuscous,* equal, or attenuated downwards, rigid, *white-silky with the adpressed silky veil, and white-zoned above with the membranaceous, or fibrillose veil,* which is often oblique, or fugacious. Gills *ochraceous, then tawny ferruginous,* more or less emarginato-adnexed, 8–10 mm. *broad, distant, thin,* often connected by veins. Flesh *concolorous, often reddish in the st.,* thick at the disc, firm. Spores ferruginous, pip-shaped, 9–10 × 6–7μ, granular. Smell strong, slightly of radish, or none. Taste mild, then slightly acrid. Woods, and heaths. Aug.—Nov. Common. (*v.v.*)

470. **C. (Tela.) gentilis** Fr. Fr. Icon. t. 159, fig. 2.

Gentilis, of the same race.

P. 1–4 cm., *tawny cinnamon, yellow when dry,* very hygrophanous, fleshy, conico-expanded, then flattened, *acutely umbonate,* rimosely incised, often somewhat silky. St. 6–9 cm. × 2–8 mm., *concolorous,* equal, or attenuated at the base, often curved, fibrillose, *veil forming one or more oblique, yellow annular zones,* sometimes floccoso-scaly below the ring, base white tomentose. Gills *yellow, then tawny cinnamon,* adnate, *thick, very distant,* often connected by veins. Flesh *concolorous,* thin at the margin. Spores bright ochraceous, elliptical, or pip-shaped, 7–8 × 6μ, granular, 1-guttulate. Gregarious. Woods, especially pines, and heaths. Aug.—Oct. Not uncommon. (*v.v.*)

471. **C. (Tela.) helvelloides** Fr. Fr. Icon. t. 159, fig. 3.

Helvella, the genus Helvella; εἶδος, like.

P. 1–3 cm., *ferruginous, becoming tawny when dry, submembranaceous,* convex, then flattened, umbonate, smooth, rarely fibrillose when young, *substriate when moist,* cracked and squarrose when more fully grown. St. 4–7·5 cm. × 2–4 mm., *subferruginous,* equal, *very undulate and flexuose,* apex *white silky* and glittering, veil *forming a yellow, ring-like zone at the apex.* Gills *violaceous umber,* then ferruginous, adnate, rather broad, *very thick, very distant, edge white-floccose.* Flesh ferruginous *in the st.,* very thin at the disc. Spores ferruginous, "elliptical, 9–10 × 5–5·5μ, verrucose" Rick. Moist woods. Aug.—Oct. Uncommon. (*v.v.*)

472. **C. (Tela.) rubellus** Cke. Cke. Illus. no. 806, t. 835.
Rubellus, reddish.

P. 5–7·5 cm., *rufous orange, darker at the umbo*, fleshy, campanulate, then expanded. St. 7–10 × 1–1·5 cm., *pale above, darker below*, equal, or attenuated upwards, *marked with concentric, dark ferruginous, fibrillose bands*. Gills *pale, then bright ferruginous red*, adnate, sinuate, rather narrow, scarcely crowded. Flesh *reddish ochre*, thick at the disc. Spores ferruginous, pyriform, 8 × 5μ, minutely rough. Swampy places. Sept.—Oct. Rare.

****St. becoming fuscous, veil fuscous, or dirty, gills dark coloured.

473. **C. (Tela.) bovinus** Fr. (= *Cortinarius brunneus* (Pers.) Fr. sec Barbier.) Cke. Illus. no. 807, t. 822. *Bovinus*, pertaining to oxen.

P. 6–12 cm., *watery cinnamon, becoming tawny when dry*, convex, then plane, obtuse, or gibbous, smooth, fragile, opaque, hygrophanous. St. 6–8 × 2–2·5 cm., *dingy pallid, becoming fuscous cinnamon, very bulbous, veil forming a simple, interwoven fuscous zone, apex whitish.* Gills *cinnamon, becoming dark*, adnexed, very broad, 12 mm., distant. Flesh *pallid*, watery, thick at the disc, spongy in the st. Spores pale ferruginous, elliptical, 9–13 × 6–7μ, coarsely verrucose. Pine, and deciduous woods. Sept.—Oct. Uncommon. (*v.v.*)

474. **C. (Tela.) nitrosus** Cke. Cke. Illus. no. 808, t. 837.
Nitrosus, full of natron.

P. 5–7·5 cm., *fawn colour, or tawny, disc darker and brownish*, fleshy, obtuse, convex, then expanded, margin undulate, *soon breaking up into minute, subconcentric darker scales*. St. 5–8 × 1 cm., *ochraceous, base darker*, subequal, *marked below with concentric darker squamose bands*. Gills *violet, then watery cinnamon*, emarginate, rather broad, subdistant. Flesh *pale brown*, thin. Spores pale ferruginous, elliptical, 12 × 4μ. Smell *stinking, nitrous.* Mixed woods. Sept.—Oct. Uncommon.

475. **C. (Tela.) brunneus** (Pers.) Fr. Cke. Illus. no. 810, t. 854.
Brunneus, brown.

P. 5–10 cm., *umber, dirty brick tan colour when dry*, fleshy, campanulate, then expanded, disc obtusely umbonate, smooth, innately fibrillose towards the margin. St. 6–10 cm. × 8–12 mm., *becoming fuscous*, clavate, or attenuated upwards from the thickened base, *elastic, covered with dense, minute white striae*, veil *dingy white*, forming *a brownish white, ring-like zone*. Gills *dark purple cinnamon, then brown, at length umber brown*, adnate, then adnexed, 10–15 mm. broad, *thick, distant*, often transversely veined, broadest in the middle. Flesh *pallid fuscous*, thick only at the umbonate disc. Spores ferruginous, broadly elliptical, 7–8 × 5–6μ, minutely verrucose. Woods, heaths, and swampy places. Sept.—Oct. Common. (*v.v.*)

476. **C. (Tela.) injucundus** (Weinm.) Fr. (= *Cortinarius brunneus* (Pers.) Fr. sec. Barbier.) Cke. Illus. no. 809, t. 823.

Injucundus, unpleasant.

P. 6–10 cm., fuscous cinnamon, fleshy, convex, then plane, obtuse, fibrillose. St. 6–10 × 1–1·5 cm., *concolorous, then tawny yellow*, clavate, attenuated upwards, *covered with fuscous fibrils*, veil *fuscous*. Gills *lilac tan, then cinnamon*, emarginate, very broad, 8–10 mm. Flesh *pale reddish*, compact, firm. Spores ferruginous, elliptical, or pip-shaped, 10–11 × 5–6μ, granular. Smell musty, or pleasant. Fir woods, and under conifers. Oct.—Nov. Uncommon. (*v.v.*)

477. **C. (Tela.) brunneofulvus** Fr. *Brunneus*, brown; *fulvus*, tawny.

P. 5–11 cm., *tawny cinnamon*, scarcely changing colour when dry, fleshy, campanulate, then expanded, obsoletely umbonate, smooth, *minutely fibrilloso-virgate with innate adpressed hairs under a lens*, margin at first *white, fibrillose.* St. 7–10 × 1–2 cm., *concolorous, or paler*, attenuated upwards, *fibrillosely striate*, veil *dingy white*, forming a fugacious zone. Gills *tawny cinnamon*, opaque, adnate, *very broad*, 12–20 mm., *subdistant*, soft. Flesh *pale tawny*, thin. Spores ferruginous, elliptical, 7–8 × 4–5μ, granular. Woods, heaths, and swampy places. Sept.—Oct. Uncommon. (*v.v.*)

478. **C. (Tela.) glandicolor** Fr. Cke. Illus. no. 812, t. 789.

Glans, acorn; *color*, colour.

P. 2–5 cm., *brown, or cinnamon-brown, tan colour or isabelline when dry*, submembranaceous, conical, then expanded, generally obtusely umbonate, soon glabrous; margin striate when moist, *sprinkled with thin, short, white fibrils when dry.* St. 7–12·5 cm. × 4–6 mm., *concolorous, at length date brown fuscous, equal, straight*, sometimes undulate, fibrillosely striate, veil *forming a woven, white, distant, fugacious ring.* Gills *concolorous, or umber*, adnate, rounded in front, *very distant, somewhat thick*, up to 8 mm. broad. Flesh *concolorous*, very thin. Spores "tawny, elliptical, 9–10 × 5–6μ, rough" Bataille. Pine woods. Sept.—Oct. Uncommon.

var. **curtus** Fr. *Curtus*, shortened.

Differs from the type in the *umbo of the p. becoming somewhat black, and in the short* (2·5 *cm.*) *flexuose st., peronate and zoned by the white veil.*

479. **C. (Tela.) punctatus** (Pers.) Fr. Cke. Illus. no. 813, t. 855.

Punctatus, dotted.

P. 1–2 cm., *hoary umber, becoming pale, tan colour when dry*, submembranaceous, conico-convex, umbo scarcely prominent, smooth, at length punctate. St. 5–7·5 cm. × 2–4 mm., *yellow fuscous*, equal, undulated, fibrillose; *girt with a pallid fuscous zone* from the fugacious veil. Gills *brown cinnamon*, adnate, very distant. Flesh *yellowish*, thin, firm. Spores ochraceous, "elliptical, 10–12 × 7–8μ, punctate-warted" Rick. Smell strong. Pine, and beech woods. Sept.—Oct. Uncommon.

II. Gills narrow, thin, more or less crowded. P. thin. St. externally more rigid, subcartilaginous, often attenuated downwards.

*St. whitish, pallid, not floccosely scaly.

480. **C. (Tela.) triformis** Fr. Cke. Illus. no. 814, t. 790, as var. *Schaefferi* Fr. *Triformis*, three formed.

P. 4–8 cm., *fawn colour, brownish, or livid yellowish, then yellowish or honey colour, isabelline, or dingy tan when dry, very hygrophanous*, fleshy, convex, then plane, obtuse, or slightly gibbous, superficially fibrillose, or becoming smooth, at length punctate-dotted, *opaque*. St. 7·5 cm. × 12 mm., *pallid*, subbulbous, fragile, rather smooth, ringed upwards with the woven veil, *ring distant, white*. Gills *watery honey colour, then watery cinnamon*, adnate, subemarginate, ventricose, 8 mm. broad, subdistant, often connected by veins. Flesh *whitish*, thin, spongy in the st. Spores ferruginous, "fusiform-elliptical, 9–10 × 4–5μ" Rick. Woods, especially beech. Oct. Uncommon.

var. **fusco-pallens** Fr. *Fuscus*, dark; *pallens*, pale.

Differs from the type in the *fuscous, umbonate p. becoming pale, and in the narrow* (2–4 *mm.*), *watery white gills*. Coniferous woods.

var. **melleo-pallens** Fr. *Melleus*, honey colour; *pallens*, pale.

Differs from the type in the *moist. isabelline yellow p. becoming yellow, in the striate margin and the pallid yellowish, fragile st.* Pine woods.

481. **C. (Tela.) biformis** Fr. Cke. Illus. no. 815, t. 869. *Biformis*, two formed.

P. 3–8 cm., *dark, or ferruginous brown, pale date brown and shining when dry*, submembranaceous, *conical, then campanulate*, at length expanded, *acutely umbonate*, firm, *smooth*, rarely covered with fugacious fibrils. St. 5–10 cm. × 6–8 mm., *paler than the p., attenuated downwards, distinctly striate, adpressedly fibrillose*, firm. Ring *white*, distinct, oblique, interwoven, sometimes obsolete. Gills *grey, then watery cinnamon*, adnate, or emarginate, attenuated behind, connected by veins, 6 mm. broad, rather crowded, edge often crenulate. Flesh brownish, becoming pale, very thin except at the disc. Spores pale ferruginous, elliptical, 7–8 × 3–4μ, minutely punctate. Pine, and mixed woods. Oct. Rare.

482. **C. (Tela.) fallax** Quél. Grevillea, t. 128, fig. 6.
Fallax, deceptive.

P. 1–1·5 cm., *yellow, then cream ochraceous*, campanulato-convex. St. 4–5 cm. × 2 mm., *whitish cream colour*, flexuose, silky, *lilac* and satiny above the ring. Ring *white*, narrow, fugacious. Gills *cream colour, then ochraceous*, adnate, ventricose. Flesh *white*, thin. Spores straw colour, ovoid pruniform, 8μ, punctate. Woods.

**St. inclining to violet.

483. **C. (Tela.) periscelis** Fr. Cke. Illus. no. 816, t. 838.
περισκελίς, a garter.

P. 2–5 cm., *lilac, tawny at the disc, violaceous at the margin*, fleshy, hygrophanous, campanulate, then convex, umbonate, submembranaceous, *covered with white silky fibrils*. St. 7–10 cm. × 4–8 mm., *concolorous, becoming fuscous when dry*, equal, straight, fibrillose, *the fuscous veil forming several fibrillose zones*, base *white-villous*. Gills *pallid, then dark ferruginous*, adnate, narrow, crowded. Flesh *pale tawny*, thin. Spores ferruginous, elliptical, 8–9 × 4–5μ, 1–2-guttulate. Woods, bogs, and under beeches. Sept.—Nov. Uncommon. (*v.v.*)

484. **C. (Tela.) flexipes** Fr. Cke. Illus. no. 817, t. 824, fig. A.
Flexus, bent; *pes*, foot.

P. 1–3 cm., *dark date-brown fuscous, or inclining to violaceous, becoming pale, very pale yellow when dry, becoming tan when old*, fleshy, at first conical and acute, then expanded and *acutely umbonate*, at length depressed round the umbo, *hoary fibrillose*, finally naked, torn when old. St. 6–10 cm. × 4 mm., *pallid, violaceous throughout, or at the apex*, equal, flexuose, *floccoso-scaly* below the ring; ring *white*, woven, distinct. Gills *purple, or umber violaceous, then cinnamon*, adnate, *subdistant, edge whitish*. Flesh *concolorous*, thin. Spores tawny, pip-shaped, 6–7 × 4–5μ, rough, 1-guttulate. Woods. Sept.—Oct. Not uncommon. (*v.v.*)

485. **C. (Tela.) flabellum** Fr. Cke. Illus. no. 817, t. 824, fig. B.
Flabellum, a small fan.

P. 1·5–3 cm., *olivaceous fuscous, tan when dry*, submembranaceous, conical, then flattened, generally acutely umbonate, *at first covered with white, superficial, separating scales*, silky when dry, at length rimosely incised, torn into fibrils. St. 5–10 cm. × 2–4 mm., *pallid, becoming violet at the apex*, equal, undulated, flexuose, *floccosely scaly*. Veil *white*, inferior, giving rise to the scales on the stem, terminating in a ring which is sometimes perfect and entire, sometimes woven and oblique, and sometimes wanting. Gills *dark violaceous, then cinnamon, and at length ferruginous*, adnate, *linear, narrow, crowded*. Flesh *paler*, very thin. Spores "elliptical, 8–9 × 5–6μ, minutely punctate" Rick.

Smell strong, somewhat of radish. Gregarious. Woods, especially beech, and damp places. Sept.—Oct. Uncommon.

***St. and p. tawny, ferruginous.

486. **C. (Tela.) psammocephalus** Fr. non Bull. Cke. Illus. no. 818, t. 839, fig. A. ψάμμος, sand; κεφαλή, head.

Entirely tawny cinnamon, becoming pale and somewhat golden when dry. P. 2·5–5 cm., fleshy, *convex, then plane,* at length umbonate and revolute, *broken up into minute furfuraceous squamules.* St. 2·5–5 cm. × 4–8 mm., somewhat attenuated downwards, *sheathed with the continuous, squamulose* veil. Cortina fibrillose. Gills *at length darker, umber cinnamon,* sinuato-adnate, 4 mm. broad, crowded. Flesh *concolorous, or yellowish,* thin. Spores ochraceous, elliptical, 9–10 × 5–6μ, 1–2-guttulate, minutely verrucose. Woods, and charcoal heaps. Aug.—Oct. Common. (*v.v.*)

487. **C. (Tela.) incisus** (Pers.) Fr. Fr. Icon. t. 160, fig. 1. *Incisus,* cut into.

P. 1–3 cm., *tawny ferruginous, opaque, more rarely date brown, or olivaceous fuscous,* fleshy, *conico-convex,* then expanded, very acutely or obsoletely umbonate, *naked,* then, especially in dry weather, *torn into fibrils, or scales,* even and shining when scorched by the sun. St. 2·5–10 cm. × 2–6 mm., *tawny or ochraceous,* equal, flexuose, fibrillose, *veil forming a woven, white ring, sometimes obsolete.* Gills cinnamon-ferruginous, adnate, *subdistant.* Flesh *concolorous,* thin. Spores ochraceous, elliptical, 8–9 × 5–6μ, 1-guttulate, rough. Subcaespitose, or in troops. Woods, heaths, and dried up swamps. Sept.—Oct. Not uncommon. (*v.v.*)

488. **C. (Tela.) iliopodius** Fr. ἰλύς, mud; πούς, foot.

P. 2·5–5 cm., *opaque cinnamon, tan when dry,* fleshy, conical, then expanded, generally acutely umbonate, silky with hoary fibrils, then becoming smooth. St. 2·5–10 cm. × 2–4 mm., *tawny, becoming fuscous, subcartilaginous, equal,* flexuose, elastic, *sheathed to the middle by the white veil* which becomes even and silky, cortinately ringed where the sheathing ends, apex naked, fibrillosely striate. Gills *cinnamon,* adnate, *thin, somewhat crowded.* Flesh *of st. saffron cinnamon,* thin. Spores ferruginous, elliptical, 8–9 × 5–6μ, with a large central gutta, punctate. Woods, especially pine and beech. July—Dec. Not uncommon. (*v.v.*)

****St. floccosely scaly, and, as well as the p., fuscous.

489. **C. (Tela.) hemitrichus** Fr. Cke. Illus. no. 820, t. 825. ἥμι, half; θρίξ, hair.

P. 2·5–8 cm., *dark fuscous, fuscous tan when dry, umbo generally persistently dark,* fleshy, *convexo-expanded,* acutely or obtusely umbonate, or wholly obtuse, often umbilicate in large specimens, *covered wholly, or only round the margin, with white, fibrillose, curled, erect, superficial flocci*, then becoming smooth. St. 4–7 cm. × 4–8 mm., *concolorous*, equal, firm, *white flocculose below the ring*. Ring shining white, median, woven, often membranaceous and reflexed. Gills *clay colour* ("*bluish clay*" Quél.), *then cinnamon*, adnate, ventricose at the base, rounded, 6 mm. broad, very crowded. Flesh *concolorous, becoming paler*, thick at the disc. Spores ochraceous, elliptical, 8–9 × 5–6μ, punctate. Taste mild. Woods, heaths, and boggy ground, especially under birches. April—Nov. Common. (*v.v.*)

490. **C. (Tela.) stemmatus** Fr. Fr. Icon. t. 160, fig. 3.

στέμμα, a wreath.

P. 2–5 cm., *date brown, becoming pale when dry*, fleshy, convex, then flattened, obtuse, fragile, *hoary silky round the margin when moist, fibrillose when dry*. St. 5–8 cm. × 4–8 mm., *ferruginous date brown*, equal, or slightly attenuated at the base, often curved, soft, *generally floccosely squamulose with two to four white ring-like zones*, sometimes naked, apex *paler*, becoming silky even. Gills *date brown*, opaque, *narrow*, 4 mm. broad, *very crowded*. Flesh *ferruginous date brown*, thin at the margin. Spores ferruginous, pip-shaped, 9–10 × 5–6μ, punctate. Moist woods, and heaths. Sept.—Oct. Uncommon. (*v.v.*)

491. **C. (Tela.) rigidus** (Scop.) Fr. Cke. Illus. no. 822, t. 791.

Rigidus, stiff.

P. 1–4 cm., *bay cinnamon, fuscous when decaying, pale yellow, or fuscous tan colour when dry*, fleshy, conical, then convex, and expanded, acutely or obtusely umbonate, or quite obtuse, at length depressed round the umbo, *smooth*, becoming broken up into scales when fully grown, margin at length pellucidly striate, *at first silky from the white veil*. St. 5–10 cm. × 4–5 mm., *concolorous, or becoming fuscous, or pale*, equal, straight, or flexuose, adpressedly fibrillose, girt with the squamose, white veil. Ring *white*, floccose, sometimes membranaceous. Gills *cream colour, then cinnamon*, adnate, broad, plane, *somewhat crowded*, often connected by veins. Flesh *concolorous*, thin at the margin. Spores ferruginous, elliptical, 7–8 × 4–5μ, minutely punctate. Smell strong, taste mild. Woods, and heaths, especially under birches. Sept.—Nov. Not uncommon. (*v.v.*)

492. **C. (Tela.) paleaceus** (Weinm.) Fr. Fr. Icon. t. 160, fig. 4.

Paleaceus, chaffy.

P. 1–3 cm., *fuscous, dingy when dry*, very hygrophanous, submembranaceous, *conical*, then expanded, acutely or obtusely umbonate, *silky with white, superficial squamules, the remains of the veil*, becoming smooth, opaque. St. 5–7·5 cm. × 2–3 mm., *concolorous*, paler when young, tough, equal, undulate, *squamulose with white flocci*, base white-villose. Ring *white*, fibrillosely floccose, fugacious. Gills *pallid-whitish, then cinnamon*, adnate, broad, *crowded*. Flesh *concolorous*, very thin at the margin. Spores ferruginous, broadly elliptical, 7–8 × 4–5μ, rough. Smell weak. Woods, especially beech, and birch, also on boggy heaths. Sept.—Nov. Not uncommon. (*v.v.*)

493. **C. (Tela.) penicillatus** (Fr.) Quél. (= *Cortinarius* (*Inoloma*) *penicillatus* Fr.) *Penicillatus*, pencilled.

P. 2–4 cm., *ferruginous fuscous, tawny when dry*, fleshy, convex, minutely umbonate, dry, *densely floccoso-scaly with dark, innate, ferruginous fibrils*. St. 5–7·5 cm. × 3–6 mm., *paler than the p., equal*, fragile, *squamose to the apical ring with adpressed, fuscous, ferruginous, concentric scales*, paler and adpressedly silky above the ring. Gills *ochraceous, then cinnamon*, sinuato-adnate, then separating, plane, 6 mm. broad, somewhat crowded. Flesh *concolorous*, thin. Spores pale ferruginous, pip-shaped, 7–8 × 5μ, minutely rough. Coniferous woods. Sept.—Oct. Uncommon. (*v.v.*)

494. **C. (Tela.) Iris** Massee. ἶρις, the rainbow.

P. 2–3 cm., *pale ochraceous brown*, fleshy, hemispherical, then expanded, acutely umbonate, silky, *densely covered with minute white fibrils*, usually splitting at the margin. St. 5–7 cm. × 4 mm., *orange brown*, conical, *covered with concolorous, pointed, fibrillose squamules below the bright brown, fibrillose ring*, smooth, silky, and *violet, becoming pale above the ring*. Gills *dirty ochraceous, then bright orange brown*, very much cut out behind, slightly attached, moderately broad, rather crowded. Flesh *concolorous*, thin. Spores orange brown, elliptical, obliquely apiculate, 10 × 5μ. Solitary or in clusters of two to four. Woods. Oct. Rare.

495. **C. (Tela.) Cookei** Quél. Cke. Illus. no. 821, t. 840, fig. B. M. C. Cooke, the eminent English mycologist.

P. 1–2 cm., *tawny yellow*, conical, umbonate, *fibrillose, covered with a paler, shining, woolly veil*. St. 3–5 cm. × 2 mm., *concolorous*, equal, flexuose, *girt with several yellowish floccose zones*. Gills *violet, then reddish, at length rust colour*, adnate, 2–3 mm. broad, *edge often floccose, white*. Flesh yellowish, thin. Spores ferruginous, elliptical, 7 × 3·5μ. Damp woods. Rare.

6. **Hydrocybe** Fr.

(ὕδωρ, water; κύβη, head.)

Pileus thinly fleshy, rarely compact, moist, *hygrophanous*, smooth, or covered only with white, superficial fibrils. Stem not sheathed, cortina rarely forming an arachnoid ring.

I. P. somewhat fleshy, convex, or campanulato-convex, then expanded, obtuse, or at length gibbous; margin at first *incurved*. St. for the most part attenuated upwards.

*St. white, cortina of the same colour.

496. **C. (Hydro.) firmus** Fr. Cke. Illus. no. 824, t. 792. *Firmus*, firm.

P. 4–8 cm., *tawny ochraceous*, fleshy, convex, then plane, obtuse, firm, smooth, shining, dry. St. 6–8 × 1–1·5 cm., *shining white*, firm, somewhat elastic, base clavate, subbulbous, rarely equal, fibrillosely striate. Cortina *white*, fibrillose, sparse, fugacious. Gills *almost concolorous*, emarginate, crowded, fairly broad, thin. Flesh *white*, thick, *compact*. Spores ferruginous, "tear-drop shaped, 9μ, minutely echinulate" Quél. Smell of horse-radish. Woods, and grassy places. Sept.—Oct. Uncommon.

497. **C. (Hydro.) subferrugineus** (Batsch) Fr.
Sub, somewhat; *ferrugineus*, rust-colour.

P. 4–8 cm., *ferruginous, or watery cinnamon, either tawny and shining when dry, or becoming pale, more or less hygrophanous*, fleshy, convex, then expanded, obtuse, flexuose, firm. St. 6–8 × 1–1·5 cm., *pallid*, more or less bulbous, attenuated upwards, adpressedly fibrillose, rigid, subcartilaginous. Cortina *white*, fibrillose, marginal, very fugacious. Gills *pallid, soon watery, then dark ferruginous, opaque*, very emarginate, often connected by veins, 6 mm. *broad, more or less crowded*. Flesh *dingy isabelline white, saffron yellow at the base of the st.*, scissile, thick at the disc. Spores ferruginous, elliptical, 7–8 × 4–5μ, 1-guttulate, "warted" Rick. Smell strong, taste unpleasant. Deciduous woods, and amongst rotting pine leaves. Sept.—Oct. Not uncommon. (*v.v.*)

498. **C. (Hydro.) armeniacus** (Schaeff.) Fr. Cke. Illus. no. 826, t. 793.
Armeniacum, the apricot.

P. 5–12 cm., *tawny cinnamon, ochraceous when dry*, fleshy, rigid, campanulate, then convex and flattened, broadly and obtusely umbonate, smooth, here and there slightly striate at the margin. St. 5–8 × 1–1·5 cm., *white*, conico-attenuated, fibrillose, *subcartilaginous, rigid*, elastic. Cortina *white, somewhat sheathing*, collapsing and forming an adpressed, silky zone. Gills *pallid, then tawny cinnamon*, shining, adnate, at length slightly rounded, rather broad, crowded. Flesh *somewhat concolorous*, thin at the margin, scissile. Spores ferruginous, elliptical, 7–9 × 4–5μ, granular. Woods, especially pine. Aug.—Oct. Uncommon. (*v.v.*)

var. **falsarius** Fr. *Falsarius*, deceptive.

Differs from the type in the *light yellowish p. becoming white when dry*.

499. **C. (Hydro.) damascenus** Fr. Cke. Illus. no. 827, t. 856.
Damascenus, a damson.

P. 5–8 cm., *bay cinnamon, disc often darker, becoming brick-red when dry, firm,* convex, then plane, globose, obtuse, or very obtusely umbonate, smooth, generally *rivulosely squamulose when dry.* St. 6–8 × 1–1·5 cm., *white, quite cylindrical,* equal, *firm, elastic,* fibrillose. Cortina *white,* fibrillose, fugacious. Gills *pallid, then pale cinnamon,* adnate, narrower in front, thin, subdistant, opaque. Flesh *white,* firm, thin at the margin. Spores ferruginous, elliptical, 12 × 6μ. Taste acrid. Subcaespitose. Grassy places in woods, and pastures. Sept.—Oct. Uncommon.

500. **C. (Hydro.) privignus** Fr. Cke. Illus. no. 828, t. 827.
Privignus, a step-son.

P. 4–6 cm., *fuscous, becoming hoary-pale with a very thin white film, pallid tan when dry, very fragile,* fleshy, convex, then flattened, often reflexed and undulate, obtusely umbonate, dry, very hygrophanous. St. 5–8 cm. × 6–8 mm., *silvery-pale,* equal, or attenuated upwards, often twisted, white-silky. Cortina *white,* silky. Gills *watery, then opaque cinnamon,* adnate, broad, not crowded, *edge white-fimbriate, serrate.* Flesh *white,* hygrophanous, thin at the margin, fragile. Spores pale ferruginous, pip-shaped, 8–9 × 4–5μ, punctate, 1-guttulate. Smell unpleasant. Taste scarcely acrid. Pine, and oak woods. Sept.—Oct. Uncommon. (*v.v.*)

501. **C. (Hydro.) duracinus** Fr. Cke. Illus. no. 829, t. 809.
Duracinus, hard-berried.

P. 4–7 cm., *watery brick colour, tan when dry, always opaque,* fleshy, campanulate, or convex, then plane, gibbous, or with a broad, obtuse umbo, with *an elevated ridge at the circumference,* caused by the margin being at first sharply and regularly bent inwards to the breadth of 1 mm. and white-silky, then becoming flattened and naked; cuticle hard, rigid, fragile. St. 5–8 × 1–1·5 cm., *pale white,* hard, fusiform, or subbulbous, *rooting at the attenuated base,* smooth, rigid, *with a thick, cartilaginous, rigid, separable cuticle* (the fragments of which when it breaks up become revolute). Cortina *white,* appearing only as a narrow zone round the margin of the p. Gills *whitish, then watery cinnamon,* adnate, 4–6 mm. broad, moderately crowded; edge often white, irregular. Flesh *white, then tinged reddish,* thick at the disc. Spores ferruginous, "elliptical-almond-shaped, 10–11 × 5–6μ, punctate" Rick. Woods, and pastures. Aug.—Oct. Uncommon. (*v.v.*)

502. **C. (Hydro.) illuminus** Fr. Cke. Illus. no. 830, t. 841.
Illuminus, dull.

P. 4–8 cm., *pale brick-red, or tawny cinnamon, brick-red tan when dry,* somewhat fleshy, convex, then plane, gibbous, or obtusely umbonate smooth, *minutely innato-fibrillose and virgate under a lens*, moist. St. 7–10 cm. × 6–10 mm., *pallid, becoming ferruginous or yellowish*, attenuated upwards, sometimes twisted, fibrillosely silky, *base white*. Cortina *white*, fibrillose, evident. Gills *pale reddish tan, then cinnamon*, adnate, scarcely crowded, 4–10 mm. broad, often veined at the base. Flesh *white*, thin at the margin. Spores "ferruginous, subelliptical, 9–10 × 4–5μ, punctate" Rick. Pine, and deciduous woods. Sept. Uncommon.

503. **C. (Hydro.) tortuosus** Fr. Fr. Icon. t. 161, fig. 1.
Tortuosus, twisted.

P. 3–7·5 cm., *ferruginous bay*, somewhat shining, *dull ochraceous when dry, submembranaceous*, campanulate, then expanded, acutely or obtusely umbonate, convex, or revolute, smooth, moist. St. 7–10 cm. × 6–8 mm., *silvery, becoming pale*, apex at first tinged with fugacious lilac, fragile, equal, *generally twisted, naked*, rooting. Gills *shining tawny, then ferruginous, becoming blood red when rubbed*, adnate, separating, 4–8 mm. broad, somewhat crowded. Flesh pallid, thin, fragile. Spores ferruginous, "subelliptical, 8–10 × 5–6μ, spinulose" Rick. Damp places in pine woods. Sept.—Oct. Rare.

504. **C. (Hydro.) dilutus** (Pers.) Fr. Grevillea, t. 85, fig. 2.
Dilutus, diluted.

P. 4–5 cm., *bay brown, or watery brick colour, tan colour when dry, opaque*, fleshy, convex, then expanded, umbonate, umbo thin, vanishing, *silky and white at the margin*, becoming smooth. St. 5–8 cm. × 4–8 mm., *whitish*, opaque, slightly attenuated from the base, *white-silky*, becoming smooth. Cortina white, silky, often collapsing into patches, or forming spurious zones on the st. Gills *ochraceous, then pale cinnamon, deeply emarginate, very ventricose*, 6–8 mm. broad, *crowded*. Flesh *white, becoming reddish*, thin at the margin. Spores brownish, "subglobose, 5–6 × 5μ, punctate" Rick. Deciduous woods. Nov. Rare.

**St. and gills commonly inclining to violet.

505. **C. (Hydro.) saturninus** Fr. Fr. Icon. t. 161, fig. 2.
Saturninus, dull.

P. 5–12 cm., *dark bay-brown, somewhat umber when damp, soon becoming pale brick colour*, changing colour very much, fleshy, campanulate, then expanded, obtuse, even, smooth, *superficially white, silky round the margin when young*. St. 5–8 × 1–2·5 cm., *deep violet, becoming white*, firm, *thickened downwards*, sometimes bulboso-ventricose, *fibrillosely striate*. Cortina *white*, fibrillose, inferior, abundant. Gills *purplish, then watery ferruginous*, rounded-adfixed, very broad, 8 mm., crowded, thin, fragile, *edge often white floccose*. Flesh *violaceous*, *then whitish,* thin at the margin. Spores yellowish ferruginous, "almond shaped, 10–12 × 5–6μ, punctate" Rick. Often subcaespitose. Woods, and pastures. Sept.—Nov. Not uncommon. (*v.v.*)

506. **C. (Hydro.) sciophyllus** Fr. (= *Cortinarius saturninus* Fr. var. *sciophyllus* (Fr.) Quél.) Fr. Icon. t. 161, fig. 3.
σκιά, shade; *φύλλον,* leaf.

P. 2–5 cm., *dark sky blue fuscous, or steel blue,* fleshy, convex, then expanded, obtuse, at first white silky round the margin from the veil. St. 5–7 × 1 cm., *violaceous, becoming ferruginous at the base,* attenuated upwards from the thickened base. Cortina *white,* very abundant, collapsing and *leaving many, Telamonia-like, white zones on the st.* Gills *dark umber,* adnate, *narrow,* 1–2 mm. broad, attenuated from the st. to the margin, crowded. Flesh *pale umber,* thick at the disc. Spores "ochraceous, elliptical, or subglobose, 8–9 × 6–8μ, granular" Bataille. Smell somewhat strong. Gregarious, or subcaespitose. Beech woods. Oct. Uncommon.

507. **C. (Hydro.) imbutus** Fr. (= *Cortinarius bicolor* Cke. sec. Bataille.) Cke. Illus. no. 834, t. 870. *Imbutus,* saturated.

P. 5–10 cm., *toast brown, then pale yellowish,* fleshy, *convex,* obtuse, smooth, obsoletely hoary-fibrillose towards the margin. St. 4–7 × 1–2 cm., *whitish,* equal, sometimes twisted, scarcely fibrillose, *apex pale violaceous.* Cortina *white,* appendiculate at the margin of the p. and on the apex of the st., fugacious. Gills *dark bluish grey, or violaceous cinereous, then watery cinnamon,* rounded, 6 mm. *broad,* with narrower and shorter ones intermixed, *subdistant.* Flesh *dingy, violaceous only at the apex of the st., subequal.* Spores ferruginous, elliptical, 7–8 × 5μ. Woods. Sept. Uncommon.

508. **C. (Hydro.) castaneus** (Bull.) Fr. Boud. Icon. t. 117.
κάστανον, the chestnut tree.

P. 2–5 cm., *fuscous chestnut, becoming pale and silky when dry, shining, umbo becoming black, paler at the slightly scalloped margin,* and often white silky with the cortina, fleshy, firm, *almost pliant,* campanulate, then flattened, obtuse, or obtusely umbonate, rarely umbilicate, often irregular, smooth. St. 4–8 cm. × 4–6 mm., *pallid violaceous, or pallid rufescent,* subequal, rarely thickened at the base and rooting, *cartilaginous,* slightly fibrillose with the veil. Cortina *white,* fibrillose, scanty. Gills *violaceous, then ferruginous,* adnate, or emarginate, 4–6 mm. broad, thin, crowded, edge often whitish. Flesh *violaceous,* darker under the cuticle of the p., thin. Spores ferruginous, elliptical, 7–8 × 4–5μ, minutely verrucose. Taste pleasant. Edible. Gregarious, sometimes caespitose. Woods, pastures, and roadsides. June—Nov. Common. (*v.v.*)

509. **C. (Hydro.) bicolor** Cke. (= *Cortinarius imbutus* Fr. sec. Bataille.) Cke. Illus. no. 836, t. 871. *Bicolor*, two coloured.

P. 2–5 cm., *dingy whitish, with an occasional tinge of lilac*, fleshy, campanulate, then expanded, broadly or acutely umbonate, somewhat fragile, smooth, silky shining. St. 5–8 cm. × 6–10 mm., *pallid violet, becoming whitish*, equal, or attenuated downwards. Cortina *white*, fugacious. Gills *purplish violet, then cinnamon*, adnate with a tooth, subventricose, rather broad, scarcely crowded, slightly eroded at the edge. Flesh *colour of the pileus, or paler, bright purplish at the base of the st., pallid above*, thin. Spores ferruginous, elliptical, a little attenuated towards one or both ends, 9–15 × 5–7μ, minutely verrucose. Woods. Aug.—Oct. Not uncommon. (*v.v.*)

***St. and somewhat obsolete veil yellow or rufous.

510. **C. (Hydro.) balaustinus** Fr. Cke. Illus. no. 837, t. 794. *βαλαύστιον*, the flower of the wild pomegranate.

P. 3–8 cm., *reddish ferruginous, tawny brick-red and shining when dry*, fleshy, convex, then plane, obtuse, moist, *fibrillosely virgate* under a lens. St. 5–8 cm. × 10–12 mm., *pale and streaked with red when young, becoming ferruginous*, often curved, *clavately bulbous*, or attenuated upwards, firm. Cortina *reddish*. Gills *reddish, then ferruginous red*, adnate, broad behind, somewhat crowded, at length subdistant. Flesh *ferruginous in the st.*, thin. Spores ferruginous, "subglobose, 6–7 × 5–6μ, punctate" Rick. Beech woods. Oct. Uncommon.

511. **C. (Hydro.) colus** Fr. Paulet, t. 99. *Colus*, distaff.

P. 2·5–5 cm., *brown rufescent, paler brick colour and shining when dry*, fleshy, campanulate, then convex, obtuse, or obtusely umbonate. St. 8–10 cm. × 4 mm., *paler than the p.*, subbulbous, *equally attenuated upwards, base encircled with the blood red mycelium*, sometimes rooting, stiff, *longitudinally fibrillose with fibrils of the same colour as the p.* Cortina *tawny reddish*, fibrillose, fugacious. Gills *pale, then dark cinnamon*, adnate, scarcely sinuate, 6 mm. broad, plane, firm, tough, rather thick, scarcely crowded, veined at the base. Flesh *concolorous, dingy whitish when dry*, thin. Spores ochrey-cinnamon, "dark under the microscope, 9–10 × 5–6μ, almost spinulose, cystidia on edge of gill vesiculose, 25–36 × 12–15μ" Rick. Coniferous woods. Sept.—Oct. Uncommon.

512. **C. (Hydro.) isabellinus** (Batsch) Fr. Cke. Illus. no. 839, t. 829. *Isabellinus*, dirty linen colour.

P. 3–5 cm., *yellowish, honey colour, yellow and shining when dry*, fleshy, convex, subumbonate, *smooth*. St. 7–10 cm. × 8 mm., *yellowish, equal, very rigid*, firm, striate. Cortina *concolorous*, sparse, very fugacious. Gills *yellow, then clay cinnamon,* adnate, broad, thin, subdistant, *edge often yellowish.* Flesh *cream colour,* firm. Spores "elliptical, 7–9 × 4–5μ, rough" Bataille. Coniferous woods. Sept.—Oct. Uncommon.

513. **C. (Hydro.) renidens** Fr. Fr. Icon. t. 162, fig. 1.
Renidens, glistening.

P. 2–5 cm., *ferruginous tawny, ochraceous when dry, or only becoming pale at the disc,* fleshy, firm, convexo-plane, *obtuse,* or gibbous, rarely umbilicate, very smooth, *shining.* St. 4–8 cm. × 6–8 mm., *pale yellowish, then tawny,* firm, *equal,* subcartilaginous, *splitting up into fibrils of the same colour as the p.* Cortina *yellow,* laxly fibrillose, fugacious. Gills *pallid cinnamon, then tawny,* adnate, separating free, 6 mm. broad, somewhat crowded. Flesh *paler,* thin, scissile. Spores dark ochraceous, "subglobose, 6–7 × 6μ, minutely warted" Rick. Smell weak. Deciduous woods. Sept. Uncommon.

514. **C. (Hydro.) angulosus** Fr. Fr. Icon. t. 162, fig. 2.
Angulosus, full of corners.

P. 4–8 cm., *reddish tawny, somewhat variegated with darker spots, opaque tawny yellow when dry,* fleshy, *convex, then plane, very obtuse, repand,* hygrophanous, fragile, very smooth; margin membranaceous, splitting, flexuose. St. 3–7 cm. × 4–8 mm., *somewhat tawny,* firm, equal, often twisted, *striate.* Cortina *tawny,* very fugacious. Gills *tawny,* adnate, 6–8 mm. broad, *thick, subdistant,* fragile. Flesh *yellowish white, darker in the st.,* thin, firm. Spores ferruginous, "roundish-elliptical, 7–8 × 5–6μ, minutely warted" Rick. Coniferous woods. Aug.—Sept. Uncommon.

var. **gracilescens** Fr. *Gracilescens,* becoming slender.

Differs from the type in the *st. being attenuated at the base.* Pine woods.

****St. inclining to fuscous; cortina pallid, dirty, or white, not yellow, gills dark.

515. **C. (Hydro.) uraceus** Fr. Fr. Icon. t. 162, fig. 3. *Uro,* I burn.

P. 2–5 cm., *umber, or brown, sometimes olivaceous,* somewhat shining, *becoming brick colour when young, commonly tan or isabelline when dry,* fleshy, conical, then campanulate and expanded, umbonate, or obtuse, smooth often becoming subfibrillose. St. 5–10 cm. × 4–8 mm., *fuscous, sometimes olivaceous, apex becoming pale, at length becoming entirely fuscous black,* cylindrical, *quite equal,* firm, *fibrillosely striate with paler striae,* becoming smooth, apex naked. Cortina *fuscous, superior,* fibrillose, rarely noticeable. Gills *cinnamon brown,* adnate, ventricose, 6 mm. broad, distant, firm; edge sometimes white and fimbriately serrated. Flesh *fuscous, darker in the st., thin.* Spores ferruginous, "subelliptical, 12–18 × 7–9μ, verrucose," Rick. Coniferous woods. Oct.—Nov. Uncommon. (*v.v.*)

516. **C. (Hydro.) jubarinus** Fr. *Jubar*, radiance.

P. 3–7·5 cm., *bright tawny cinnamon, shining*, fleshy, campanulato-flattened, obsoletely umbonate, often repand, undulate, at length reflexed, disc depressed when large and old, smooth, becoming innately fibrillose under a lens when old, silky round the margin when young. St. 5–6 cm. × 4–12 mm., *pale tawny, paler at the base and naked at the apex*, firm, equal, *fibrillosely striate.* Cortina *white*, fibrillose, fugacious. Gills *tawny cinnamon*, adnate, 2–6 mm. broad, *subdistant.* Flesh *pale tawny in the st.*, thick at the disc. Spores pale ferruginous, pip-shaped, 7–8 × 4–5μ. Coniferous woods. Sept.—Oct. Uncommon. (*v.v.*)

517. **C. (Hydro.) irregularis** Fr. *Irregularis*, irregular.

P. 3–7·5 cm., *brown ferruginous, tawny ferruginous when dry*, fleshy, convex, then plane, acutely umbonate, *repand*, at length depressed round the umbo, dry, smooth, sometimes deformed, undulate, and rugose, white silky round the membranaceous margin when young. St. 5–10 cm. × 4–6 mm., *brick-red, equal*, rigid, *longitudinally fibrillosely striate*, sometimes twisted, and attenuated downwards. Cortina *white*, silky, sparse, sometimes peronate at the base of the st. Gills *watery, then dark ferruginous, adnate with a decurrent tooth, or arcuato-decurrent*, 4–8 mm. broad, *very crowded.* Flesh pallid, thick at the disc. Spores ferruginous, "almond-shaped-elliptical, 8–10 × 5–6μ, almost spinulose" Rick. Caespitose. Coniferous woods. Rare.

518. **C. (Hydro.) pateriformis** Fr. *Patera*, a saucer; *forma*, shape.

P. 2–4 cm., *fuscous chestnut*, fleshy, orbicular, *plane, then depressed*, very obtuse, rigid, dry, smooth, at first white silky round the margin. St. 5–8 cm. × 2–4 mm., *silvery white, becoming fuscous*, firm, *equal*, or attenuated at the base, *straight*, fibrillose, base *white villose.* Cortina *white*, fibrillose, fugacious. Gills *brick-red, or watery cinnamon*, adnate with a decurrent tooth, plano-convex thin, *crowded.* Flesh *pallid*, thin. Spores brownish ferruginous, elliptical, 9 × 4–5μ, with a large central gutta, "minutely warted" Rick. Grassy places in woods. Sept.—Oct. Uncommon. (*v.v.*)

519. **C. (Hydro.) unimodus** Britz. *Unimodus*, uniform.

P. 4–7 cm., *reddish brown*, campanulate, then convex, umbonate, fibrillose. St. 8–11 cm. × 6 mm., *concolorous*, thickened at the base, fibrillose. Gills *brown*, emarginato-adnate, distant. Spores "pruniform, 10–12 × 8μ, rough" Bataille. Woods, and grassy places. Sept.—Oct. Rare.

II. P. submembranaceous, conical, then expanded, umbonate, umbo acute, or more rarely obtuse and vanishing; margin at first *straight*. St. subequal, or attenuated at the base.

*St. white.

520. **C. (Hydro.) dolabratus** Fr. Cke. Illus. no. 845, t. 811.
Dolabra, a pick-axe.

P. 5–10 cm., *brick-red, tan colour when dry, fleshy-membranaceous*, fragile, *campanulate, then convex* and expanded, smooth, superficially silky near the margin. St. 10–15 × 1–2 cm., shining *white, quite equal, cylindrical*, often curved, smooth. Cortina very fugacious. Gills *tawny cinnamon, entirely adnate, with a decurrent tooth*, widest behind, 10–25 mm., *very broad*, somewhat thick, *distant*. Flesh *whitish*, thin at the margin. Spores ochraceous, broadly elliptical, 9–10 × 7μ. Smell strong, stinking. Coniferous woods, and amongst *Vaccinium*. Sept.—Nov. Uncommon. (*v.v.*)

521. **C. (Hydro.) rigens** (Pers.) Fr. Cke. Illus. no. 846, t. 812.
Rigens, rigid.

P. 2·5–6 cm., *opaque tan clay colour, whitish tan when dry*, fleshy, campanulate, lax, then convex, obtuse, or broadly gibbous, smooth, firm. St. 5–10 cm. × 4–10 mm., *pale, white when dry*, equal, sometimes thickened upwards, sometimes downwards, sometimes fusiform, *rooting, tough*, elastic, *rigid, cortex very cartilaginous*, naked, smooth. Cortina scarcely evident. Gills *watery clay colour, then pallid, cinnamon, adnate, subdecurrent*, very broad, 6–10 mm., distant, often veined on the sides. Flesh *white*, thick, somewhat firm. Spores ferruginous, pruniform, 7·5–9 × 5–5·5μ, granular. Smell strong, like iodoform, or balsam. Taste mild, then unpleasant. Woods, especially coniferous woods. June—Oct. Uncommon. (*v.v.*)

522. **C. (Hydro.) fulvescens** Fr. Grevillea, t. 116, fig. 2.
Fulvescens, becoming tawny.

P. 2–3 cm., *cinnamon, brick-red when dry, prominent umbo darker, submembranaceous*, conical, soon convexo-plane, *often acutely umbonate*, shining, at length fibrillose; margin cortinate, at length striate. St. 6–8 cm. × 3–6 mm., *becoming pale*, equal, or slightly attenuated upwards, soft, flexuose, smooth. Cortina *concolorous*, distinct. Gills *tawny cinnamon*, adnate, plane, subdistant, *thin*. Flesh *white*, thin. Spores ferruginous, "almond shaped, minutely echinulate, 12μ" Quél. Pine woods. Sept.—Oct. Rare.

523. **C. (Hydro.) Krombholzii** Fr. (= *Cortinarius leucopus* (Bull.) Fr. sec. Quél.) Cke. Illus. no. 847, t. 813. J. V. Krombholz.

P. 2·5–5 cm., *pale yellowish tan, disc darker*, fleshy, conico-campanulate, then gibbous, smooth; margin *appendiculate with the membranaceous veil.* St. 7–12 cm. × 6 mm., *whitish,* equal, slightly swollen at the base, naked. Veil *white, membranaceous,* fugacious. Gills *ferruginous, edge yellowish,* rounded behind, slightly adnexed, 6 mm. broad. Flesh *whitish,* thin at the margin. Spores "ferruginous, elliptical, 8 × 4–5μ" Massee. Amongst moss, often caespitose. Rare.

524. **C. (Hydro.) Reedii** Berk. Hussey, Illus. Brit. Myc. II, t. 45.
Miss F. Reed, sister of Mrs Hussey.

P. 2–3 cm., *persistently brown,* fleshy, conical, then expanded, *strongly umbonate,* smooth, shining, *disc areolate;* margin splitting. St. 4–5 cm. × 4 mm., *white, slightly bulbous,* fibrillosely striate. Veil fibrillose, evanescent. Gills *white, or pallid, then cinnamon,* ascending, attenuated behind, free, ventricose broad. Flesh *pallid,* thin at the margin. Spores "7–8 × 4μ" Massee. Amongst moss, and beech mast. May. Rare.

525. **C. (Hydro.) leucopus** (Bull.) Fr. Cke. Illus. no. 848, t. 843, fig. B.
λευκόπους, white footed.

P. 2–3 cm., *very pale yellow, tan colour and shining when dry,* fleshy, conical, then expanded, umbonate, smooth, moist. St. 2·5–5 cm. × 4–8 mm., *shining white,* equal, or slightly attenuated upwards, *soft.* Cortina *white,* median. Gills *pallid, then cinnamon, adnexed,* separating, *ventricose, crowded,* thin. Flesh *pallid,* thin at the margin. Spores ferruginous, elliptical, 8–9 × 5–6μ, granular, "spiny" Rick. Woods. Aug.—Nov. Common. (*v.v.*)

526. **C. (Hydro.) scandens** Fr. Fr. Icon. t. 163, fig. 1.
Scandens, climbing.

P. 1–3 cm., *tawny ferruginous, then honey colour, tan colour when dry, umbo becoming tawny,* submembranaceous, conical, then campanulate, acutely or obtusely umbonate, umbo sometimes obsolete; margin *slightly striate.* St. 6–10 cm. × 4 mm., *yellowish, shining whitish when dry, thickened at the apex, attenuated at the white base,* awl-shaped, *flexuose,* soft, *fibrillosely silky, apex often mealy.* Cortina *white,* superior, thin, fibrillose. Gills *yellowish, then tawny cinnamon, adnate, narrow,* 2–4 mm. broad, attenuated in front, thin, subdistant. Flesh *yellowish,* thick at the umbo. Spores yellow, "elliptical, 6–7 × 4–5μ, minutely warted" Rick. Pine woods. Sept.—Nov. Uncommon. (*v.v.*)

**St. inclining to violet, or reddish.

527. **C. (Hydro.) erythrinus** Fr. Cke. Illus. no. 850, t. 798, fig. A.
ἐρυθρός, red.

P. 2·5–4 cm., *bay brown rufous, tawny when dry,* fleshy, *conical, then convex,* regular, umbonate, the *obtuse or obsolete umbo darker,* smooth. St. 4–8 cm. × 4–6 mm., *shining silvery white, violaceous upwards,* equal, rarely thickened at the base, straight, or ascending, fibrillosely striate, apex often pruinose. Cortina *white*, superior, fibrillose. Gills *pallid, then pale cinnamon, slightly adnexed*, ventricose, thin, *subdistant*. Flesh *concolorous when moist*, thin. Spores pale ferruginous, elliptical, 6 × 4–5μ, 1-guttulate, "almost spinulose" Rick. Woods. Sept.—Oct. Not uncommon. (*v.v.*)

var. **argyropus** Fr. ἀργυρό-πους, with silver feet.

Differs from the type in being *more slender, and in the silvery stem white mealy at the apex*.

528. **C. (Hydro.) decipiens** (Pers.) Fr. Cke. Illus. no. 850, t. 798, fig. B. *Decipiens*, deceiving.

P. 2–3 cm., *bay brown, shining and brick colour when dry, umbo always* darker, fleshy membranaceous, campanulato-expanded, *acutely umbonate*, at length depressed round the umbo, smooth; margin at length striate and torn. St. 5–10 cm. × 2–4 mm., *pallid, pale rufescent, or with brick coloured spots*, quite equal, tense and straight, or flexuose, fibrillose, *covered with a pallid separable cuticle*. Cortina *white*, fibrillose, very fugacious. Gills *brick colour ferruginous*, adnate, 4–6 mm. broad, thin, more or less crowded. Flesh *pale, brick colour in the st.*, thin. Spores pale ferruginous, elliptical, 9 × 5μ, granular. Woods. Sept.—Nov. Common. (*v.v.*)

var. **insignis** Fr. *Insignis*, distinguished.

Differs from the type in the *paler p., flexuose, smooth st., and in the less crowded gills*.

529. **C. (Hydro.) germanus** Fr. Cke. Illus. no. 851, t. 844. *Germanus*, born of the same parents.

P. 2–3 cm., *fuscous, clay colour when dry, very hygrophanous*, opaque, submembranaceous, campanulate, obtusely umbonate when expanded, *fragile, somewhat silky with white fibrils*. St. 6–8 cm. × 2–6 mm., *silvery pale, somewhat lilac, equal*, often twisted, smooth. Cortina *white*, fibrillose, fugacious. Gills *watery cinnamon*, adnate, broad, subdistant. Flesh *concolorous, then whitish*, thin. Spores pale ferruginous, elliptical, 7–8 × 4–5μ, 1-guttulate. Smell disagreeable. Beech, and pine woods. Sept.—Oct. Uncommon. (*v.v.*)

530. **C. (Hydro.) ianthipes** (Secr.) Fr. Grevillea, t. 113, fig. 7. ἰάνθινος, coloured violet; *pes*, foot.

P. 1–2 cm., *brown, or tawny*, fleshy, somewhat firm, conical, then convexo-plane, obtusely umbonate, silky, *shining*; *margin silky white, or becoming yellowish*. St. 2–4 cm. × 2–4 mm., *violaceous, becoming reddish downwards*, equal, somewhat fragile and flexuose, *shining*, silky, base white villose. Ring *white*, silky, floccose. Cortina *tawny*, fugacious. Gills *whitish, then greyish olivaceous*, "*lilac, then brownish violet*" Quél., slightly adnexed, somewhat crowded. Flesh *reddish*, thin. Spores ferruginous, "pruniform, 8μ" Quél. Woods. Sept. Rare.

***St. yellowish, generally becoming pale.

531. **C. (Hydro.) detonsus** Fr. *Detonsus*, sheared off.

P. 2·5–5 cm., *bright yellowish, tan colour when dry, submembranaceous*, conical, then expanded, subumbonate, *striate to the middle when moist, slightly silky when dry*, fragile. St. 5–8 cm. × 4 mm., *yellowish*, equal, or attenuated upwards, *soft*, smooth. Gills *bright yellowish, then brick cinnamon*, adnate, ventricose, subdistant. Flesh *whitish*, thin. Spores ferruginous, "elliptical, 7–8 × 3–4μ, minutely verrucose" Rick. Amongst moss in woods. Sept. Rare.

532. **C. (Hydro.) obtusus** Fr. Fr. Icon. t. 163, fig. 3. *Obtusus*, obtuse.

P. 1–4 cm., *bay brown, ferruginous, soon cinnamon, pale ochraceous, or tan whitish when dry, submembranaceous*, conical, then campanulate, at length expanded and obtusely umbonate, smooth; *margin striate*. St. 5–10 cm. × 4–8 mm., *tan yellowish, becoming whitish when dry, ventricose*, curved, flexuose, often attenuated at the base, fragile, *sprinkled with adpressed, white, silky fibrils*. Cortina *white*, very fugacious. Gills *tawny cinnamon*, adnate, ventricose, *very broad, rather thick, subdistant*, the shorter ones narrower, *connected by veins*, edge often white-fringed. Flesh *yellowish, or reddish*, thick at the disc. Spores brownish ferruginous, elliptical, 8–9 × 5·5–6μ, verrucose. Smell strong. Woods, especially pine. April—Nov. Not uncommon. (*v.v.*)

var. **gracilis** Quél. Grevillea, t. 129, fig. 1. *Gracilis*, slender.

Differs from the type in the *pale yellow p.*

533. **C. (Hydro.) saniosus** Fr. Fr. Icon. t. 163, fig. 2.
Saniosus, full of bloody matter.

P. 1–3 cm., *tawny cinnamon, fuscous tawny, or cinnamon, becoming tawny and shining when dry*, slightly hygrophanous, fleshy, campanulate, then expanded, acutely, or obtusely umbonate, smooth, "*covered with yellow fibrils*" Quél.; margin fibrillosely torn. St. 3–6 cm. × 3–4 mm., *dingy yellow, becoming yellow, rarely tawny or fuscous*, equal, firm, somewhat curved, flexuose, *covered with the yellow fibrils of the cortina*, or fibrillosely silky, sometimes becoming saffron red at the base. Cortina *yellow, fibrillose*. Gills *pale, then dark cinnamon*, adnate, ventricose, somewhat crowded, or subdistant. Flesh *concolorous, sometimes becoming yellow in the st.*, thin. Spores ferruginous, "nearly almond-shaped, 9–10 × 4–5μ, verrucose" Rick. Smell strong. Woods, pastures and swamps. Sept.—Oct. Uncommon.

534. **C. (Hydro.) acutus** (Pers.) Fr. Cke. Illus. no. 852, t. 845, fig. B.
Acutus, pointed.

P. 1–3 cm., *honey colour to very pale yellow, somewhat shining, tan, or white when dry,* with a silky appearance, *submembranaceous,* conical, then campanulate and expanded, *acutely umbonate and concolorous,* at length depressed round the umbo, *striate when moist*; margin at first obsoletely fibrillose, then smooth. St. 4–8 cm. × 2–4 mm., *concolorous,* equal, flexuose, white fibrillose, then smooth. Cortina *white,* fibrillose, adhering to the margin of the p. Gills *ochraceous cinnamon,* adnate, sometimes free, lanceolate, *thin,* crowded, becoming subdistant. Flesh *concolorous,* thin. Spores pale ferruginous, elliptical, 9–11 × 6μ, 1-guttulate. Woods and heaths. Aug.—Nov. Common. (*v.v.*)

****St. inclining to fuscous.

535. **C. (Hydro.) Junghuhnii** Fr. Cke. Illus. no. 853, t. 846, fig. A.
Francis Junghuhn.

P. 2–3 cm., *shining cinnamon, tawny when dry, fleshy, convexo-plane, umbonate with a papilla, striate to the middle when moist, somewhat velvety with thin, sparse, white, persistent fibrils* under a lens. St. 5–8 cm. × 4–5 mm., *pale brick colour,* equal, or attenuated at the whitish base, *shining, adpressedly fuscous fibrillose.* Cortina *white, inclining to fuscous,* sparse. Gills *saffron brick colour,* adnate, ventricose, 4 mm. broad, thin, veined at the base. Flesh *of stem darker brick colour,* thick at the disc. Spores ferruginous, "elliptical, 7–8 × 6μ, verrucose" Rick. Woods amongst moss. Aug. Rare.

536. **C. (Hydro.) depressus** Fr. Fr. Icon. t. 163, fig. 4.
Depressus, depressed.

P. 5–8 cm., *ferruginous fuscous, fuscous tan when dry,* fleshy-membranaceous, *conico-convex,* obtusely umbonate, at length almost plane and depressed round the umbo, *superficially silky* at first; margin striate when moist. St. 3–5 cm. × 4 mm., *reddish, at length becoming blackish, base becoming fuscous,* equal, or attenuated downwards, *rigid, subcartilaginous,* white silky, then smooth. Cortina scarcely any. Gills *saffron, becoming yellowish, then ferruginous,* adnate, broad, plane, thin, more or less crowded. Flesh *concolorous,* very thin. Smell faint, of fish, or cucumber. Damp places in woods. Sept.—Oct. Uncommon.

537. **C. (Hydro.) milvinus** Fr. Cke. Illus. no. 853, t. 846, fig. B.
Milvinus, pertaining to the kite.

P. 1–3 cm., *olivaceous fawn colour, hoary tan and opaque when dry,* membranaceous, conical, then convex, obtusely or obsoletely umbonate, striate to the middle when full grown and *beautifully wreathed with white squamules at the margin,* somewhat silky when dry. St.

5–8 cm. × 4 mm., *pallid fuscous*, equal, curved, silky, spotted with the white cortina. Veil obsolete except at the margin of the p. Gills *olivaceous, becoming ferruginous*, adnate, scarcely ventricose, very thin, subdistant, connected by veins. Flesh *concolorous*, thick at the disc. Spores ferruginous, "elliptical, 8–10 × 5–6μ, verrucose" Rick. Smell strong. Woods. Oct. Uncommon.

538. **C. (Hydro.) fasciatus** Fr. Cke. Illus. no. 855, t. 814.
Fasciatus, arranged in bundles.

P. 2–4 cm., *brick colour, acute umbo blackish, becoming pale and silky when dry, submembranaceous, conical*, then expanded, acutely umbonate, smooth. St. 5–8 cm. × 2–3 mm., *pallid fuscous, then cinnamon fuscous*, equal, straight, flexuose, undulate, *fibrillosely fissile*, smooth. Gills *cinnamon, adnate*, subventricose, 3 mm. broad, thin, distant. Flesh *concolorous*, slightly fleshy at the disc. Spores ferruginous, elliptical, 8–9 × 5–6μ, with a large central gutta, punctate. Pine woods, and under pines. Sept.—Oct. Uncommon. (*v.v.*)

2. Veil concrete with the epidermis of the pileus.

Spores ochraceous, or ferruginous, elliptical, smooth.

Inocybe Fr.

(ἴς, fibre; κύβη, head.)

Pileus fleshy, regular. Veil marginal, fugacious. Stem fleshy, central. Gills adnate, sinuato-adnate, or adnexed. Spores ochraceous, ferruginous, olivaceous, or fuscous, elliptical, elliptic-oblong, pip-shaped, or subreniform, smooth. Cystidia present, or absent, ventricose, clavate, fusiform, or cylindrical. Growing on the ground.

I. Cystidia present.

*Stem whitish, or pallid.

†Gills brownish, ochraceous, or cinnamon.

539. **I. scabra** (Mull.) Fr. (= *Inocybe capucina* Fr. sec. Quél.) Cke. Illus. no. 413, t. 391. *Scabra*, rough.

P. 1·5–4 cm., *becoming pale fuliginous, or pale tan, variegated with adpressed, darker (fuscous), spot-like, fibrous scales*, fleshy, somewhat compact, conical, then convex, obsoletely gibbous. St. 2–4 cm. × 6–10 mm., *whitish*, firm, equal, *cuticle cartilaginous, silky-fibrillose*. Gills *whitish, then somewhat fuliginous*, slightly adnexed, thin, somewhat crowded. Flesh *white*. Spores yellowish brown, pip-shaped, 9–11 × 5–6μ. Cystidia slightly ventricose, 65–75 × 12–16μ, abundant. Coniferous, and mixed woods. June—Oct. Uncommon. (*v.v.*)

var. **firma** Fr. *Firma*, firm.

Differs from the type in *the fuscous-tan p. spotted with fuscous scales, and in the velvety stem.*

540. **I. pyriodora** (Pers.) Fr. Cke. Illus. no. 411, t. 472. *Pyrus*, pear; *odora*, scented.

P. 4–8 cm., *pale ochraceous, often reddish when young, campanulate,* obtuse, fleshy at the darker disc, *everywhere torn into fibrils*; margin often bent in and lacerate, and sometimes repand. St. 5–15 × 1–1·5 cm., *pallid often tinged with red,* fragile, equal, or attenuated at the base, often curved, *fibrillose, apex white-mealy.* Gills *whitish, then somewhat fuliginous,* adnate, sinuate, thin, crowded, *edge whitish.* Flesh *becoming reddish.* Spores tawny, pip-shaped, 8–11 × 5–6μ, 2-many-guttulate. Cystidia ventricose, or clavate, 55–60 × 15–21μ, thick walled. Smell *pleasant, like ripe pears.* Woods. May—Dec. Common. (*v.v.*)

541. **I. rimosa** (Bull.) Fr. Cke. Illus. no. 429, t. 384. *Rimosa*, full of cracks.

P. 2·5–5 cm., *yellowish, varying rufescent, and date brown* (*especially when old*), fleshy, *conico-campanulate,* obtuse, then more flattened, and at length reflexed, umbonate, *somewhat fibrillose, longitudinally cracked*; disc sometimes even, sometimes cracked in a tesselated manner. St. 4–7 cm. × 4–8 mm., *whitish, becoming yellow, or fuscous,* firm, *apex mealy.* Gills *whitish, then becoming fuscous, and ferruginous, very much attenuated behind,* free, or slightly adnexed, somewhat ventricose, *edge serrulated, pallid.* Flesh *white.* Spores ferruginous in the mass, ochraceous under the microscope, elliptical, 9–11 × 5–7μ. Cystidia fusoid, or ventricose, 60–68 × 9–14μ. Smell earthy. Woods, and open ground. June—Nov. Common. (*v.v.*)

542. **I. tomentosa** (Jungh.) Quél. (= *Inocybe eutheles* B. & Br.) Cke. Illus. no. 431, t. 386, as *Inocybe eutheles* B. & Br. *Tomentosa*, downy.

P. 2–5 cm., *pale fawn-colour, campanulate,* then expanded, *strongly umbonate,* thin, *villose, fibrillose*; margin *white,* often appendiculate with the veil. St. 4–8 cm. × 5–8 mm., *pallid, or whitish,* equal, slightly swollen at the base, *fibrillose, slightly striate.* Gills *pallid,* adnate, slightly toothed, *margin white.* Flesh *white.* Spores ochraceous, elliptical, 8–9 × 4–5μ. Cystidia ventricose, 60–63 × 12–13μ, abundant. Smell of new meal. Woods, and among fir-leaves. Aug.—Oct. Not uncommon. (*v.v.*)

543. **I. pallidipes** Ellis & Everh. *Pallidus*, pale; *pes*, foot.

P. 2–3 cm., *light brown,* conico-campanulate, then expanded and umbonate, fibrose-squamose, disc innately scaly, margin subrimose.

St. 2·5–5 cm., *white*, slightly narrowed and mealy above, loosely fibrillose below, base subbulbous, white tomentose. Gills *pale cinnamon, edge paler* and fimbriate, broadly attached, with a strong decurrent tooth, ascending at first, then ventricose, scarcely crowded, rather broad. Spores cinnamon, pip-shaped, 8–9 × 5μ Cystidia, numerous, fusoid, or subventricose, 40–50 × 14–18μ. Woods. Oct. Rare.

544. **I. sambucina** Fr. Fr. Icon. t. 109, fig. 2.
Sambucina, belonging to elder.

Entirely white. P. 4–8 cm., *often becoming pallid-yellow*, fleshy, firm, *convex, then expanded, often repand*, obtuse, *fibrilloso-silky*. St. 2·5–4 × 1–2·5 cm., often curved, equal, or subbulbous at the base, *striate*, obsoletely pruinose at the apex. Gills becoming *dingy ochre*, emarginate, slightly adnexed, ventricose, 4–6 mm. broad. Flesh *white*. Spores ochraceous, elliptical, 9–12 × 6μ. Cystidia ventricose, 50–60 × 12–16μ, scattered. Smell strong. Coniferous woods. Sept.—Nov. Uncommon.

545. **I. Clarkii** B. & Br. Cke. Illus. no. 439, t. 429, lower figs.
J. Aubrey Clark.

P. 2–3 cm., *whitish, campanulate*, obtuse, silky-fibrillose. St. 3–5 cm. × 4 mm., *white*, equal, slightly thickened at the base, *flocculose*. Gills *pallid, margin white*, adnexed, rather distant, broadish. Flesh *pallid*. Spores pale, elliptical, 8–10 × 5–6μ. Cystidia ventricose, 55–65 × 12–16μ, some narrower, scattered. Shady places. Oct. Rare.

546. **I. corydalina** Quél. (= *Inocybe pyriodora* Fr. sec. René Maire.) Trans. Brit. Myc. Soc. II, t. 4. *Corydalis*, the genus Corydalis.

P. 3–6 cm., *whitish, covered with bistre fibrils, green at the umbo*, fleshy, campanulate, then expanded, umbonate. St. 3–5 × ·5–1 cm., *whitish*, curved, *subbulbous*, striate, pruinose. Gills *whitish, then brown*; *edge fimbriate, white*, adnate, emarginate, 5–7 mm. wide. Flesh *white, becoming yellowish with age*. Spores brown, elliptical, 7–9 × 4–5μ. Cystidia fusiform, ventricose, 42–51 × 15–18μ. Smell pleasant, like *Corydalis cava*. Deciduous woods. Aug.—Oct. Not uncommon. (*v.v.*)

547. **I. geophylla** (Sow.) Fr. (= *Inocybe geophila* (Bull.) Quél.) Cke. Illus. no. 440, t. 401. γῆ, earth; φύλλον, leaf.

P. 1·5–3 cm., *white, sometimes tinged yellow when old*, somewhat fleshy, *conical, then expanded, umbonate, silky, then fibrillose*, often cracking. St. 4–8 cm. × 2–6 mm., *white*, equal, base slightly thickened, often flexuose, satiny, *apex white-mealy*. Gills *whitish, then clay-fuscous, and earth colour, almost free*, rather broad, ventricose, crowded. Flesh *white*. Spores ferruginous, elliptical, 7–10 × 4–5μ. Cystidia ventricose, 45–60 × 13–15μ, abundant. Smell earthy, taste slightly acrid. Woods, under trees, and hedgerows. July—Dec. Common. (*v.v.*)

var. **lilacina** Fr. (= *Inocybe geophylla* (Sow.) Fr., var. *violacea* Pat.) Boud. Icon. t. 125. *Lilacina*, lilac colour.

Differs from the type in the *violet colour of the p., the yellow umbo, the paler violet stem, and the ochraceous yellow base.* Woods. Aug.—Nov. Common. (*v.v.*)

var. **lateritia** (Weinm.) Stev. *Lateritia*, made of bricks.

Differs from the type in *the brick-red p.*

var. **fulva** Pat. *Fulva*, tawny.

Differs from the type in the *rufous-ochre p., and the paler margin.*

548. **I. Whitei** B. & Br. Cke. Illus. no. 444, t. 404, fig. A. Dr F. Buchanan White, a Scotch botanist.

P. 1·5–2·5 cm., *tawny, margin white, then wholly pale tawny*, conical, then convex, and at length expanded, *slightly viscid*. St. 3–4 cm. × 3–4 mm., *shining white, then tawny*, nearly equal, base slightly thickened. Gills *shining white, then cinnamon*, adnexed, crowded. Flesh *white*. Spores pale yellow-brown, obliquely elliptical, 9–11 × 4–5μ. Cystidia ventricose, or almost cylindrical, 50–60 × 16–20μ, fairly abundant. Pine woods. Oct. Rare.

549. **I. sindonia** Fr. Cke. Illus. no. 438, t. 400. σινδών, muslin.

P. 3–5 cm., *dingy white*, or *at length becoming yellow*, fleshy, somewhat thin, *campanulate, then convex*, gibbous, *silky-velvety, becoming even*, margin appendiculate when young with fibrils of the cortina. St. 5–7·5 cm. × 6 mm., *white*, stuffed *with a separate pith that disappears*, equal, at first slightly fibrillose with the evanescent, delicate cortina. Gills *becoming whitish fuscous*, attenuato-adnexed, linear-lanceolate, 2 mm. broad. Flesh *white*. Spores reddish brown, elliptical, 8–10 × 5–6μ. Cystidia ventricose, 50–60 × 12–16μ. Mixed woods, damp shady places. Sept.—Oct. Rare.

550. **I. descissa** Fr. Bres. Fung. Trid. t. 122, fig. 1. *Descissa*, split up.

P. 2–3 cm., *greyish white, disc somewhat fuscous*, conico-campanulate, then expanded and umbonate, silky, then fibrillosely split up, subsquamulose around the umbo. St. 3–5 cm. × 3–4 mm., *white*, often slightly tinged with flesh colour, somewhat equal, white fibrillose, apex white pruinose, veil white, evanescent. Gills *white*, then *fuliginous, or fuliginous yellow*, sinuato-adnexed, somewhat crowded, edge fimbriate. Flesh *white, sometimes flesh colour in the stem.* Spores yellow, subreniform, 9–10 × 5–6μ. Cystidia fusiform, ventricose, 55–70 × 14–20μ, thick walled, *somewhat fuscous at the apex*. Pine woods. Sept.—Nov. Rare.

var. **auricoma** (Batsch) Fr. *Auricoma*, golden-haired.

Smaller, and thinner than the type, p. becoming yellow, margin striate, gills adfixed, ventricose, whitish fuscous. Remarkably cracked. Woods, and burnt soil.

551. **I. cervicolor** (Pers.) Quél. Fr. Icon. t. 107, figs. 1, 2, as *Inocybe Bongardii* (Weinm.) Fr. *Cervus*, a deer; *color*, colour.

P. 3–5 cm., *pale brown, or fawn colour, covered with brown, recurved firils*, campanulate, thin. St. 4–10 cm. × 4–6 mm., *whitish, fibrillose with brown, recurved filaments, firm*, flexuose. Gills *cream colour, then rusty brown, margin white, denticulate*, emarginate, ventricose, *thick*. Flesh *white, tinged purplish when cut*. Spores brown, elongate pip-shaped, 10–12 × 6–7μ. Cystidia cylindric-fusoid, 45–50 × 12–14μ, numerous. Smell strong, unpleasant, like a mouldy cask. Woods, and heaths. Aug.—Oct. Common. (*v.v.*)

552. **I. deglubens** Fr. Cke. Illus. no. 420, t. 394. *Deglubens*, peeling off.

P. 2–5 cm., *date-brown-rufescent, then becoming yellow*, slightly fleshy, convexo-plane, obtuse, or obtusely umbonate, *adpressedly torn into fibrils*, peeling off in darker, then separating fibrils. St. 4–7 cm. × 4–5 mm., *pallid*, equal, *adpressedly fibrillose, apex slightly rough with brown points*. Gills *grey, then cinnamon*, obtusely adnate, ventricose, *somewhat distant*. Flesh *white*. Spores pallid brown, pip-shaped, 8–10 × 5–6μ. Cystidia ventricose, 50–60 × 10–15μ, fairly abundant. Smell earthy. Pine woods. Aug.—Sept. Rare.

††Gills with an olive tinge.

553. **I. abjecta** Karst. *Abjecta*, mean.

P. 1–3·5 cm., *brownish, becoming ochraceous-brown when dry, everywhere covered with white fibrils, disc with whitish, subsquarrose squamules*, fleshy, subcampanulate, or convex, then expanded, sometimes umbonate. St. 2·5–4 cm. × 4–8 mm., *pallid, everywhere covered with white, fibrous squamules*, equal, or fusiform, rather tough, flexuose, apex white-pruinose. Gills *pale cinnamon-olive*, adnate, ventricose in front, 6–7 mm. wide, rather distant, margin minutely flocculoso-crenulate at first. Flesh *white*. Spores ferruginous, pip-shaped, 14–16 × 6–7μ, 1-guttulate. Cystidia ventricose, 50–65 × 13–16μ, scanty. Amongst sand. Sept.—Oct. Uncommon. (*v.v.*)

554. **I. Godeyi** Gillet. (= *Agaricus* (*Inocybe*) *hiulcus* (Fr.) Kalchbr. and Cke. Illus. no. 427, t. 337.) Trans. Brit. Myc. Soc. II, t. 8. Dr Godey, an eminent French mycologist.

P. 3–5 cm., *whitish at first, then more or less suffused with rose which is usually accompanied by an ochraceous tinge*, fleshy, campanulate, obtusely umbonate, silky-fibrillose, rimose. St. 4–6 cm. × 5–10 mm., *concolorous*, equal, often bulbous at the base, apex white-pruinose. Gills *whitish, then dusky cinnamon, usually with an olive tinge*, edge white, minutely flocculose, narrowed behind, adnexed, almost free, somewhat crowded. Flesh *white, tinged with pink*. Spores ferruginous, elliptical, slightly curved or subreniform, 7–8 × 4–5μ. Cystidia ventricose, 36–48 × 18–25μ, fairly numerous. Woods, and heaths. Aug.—Oct. Not uncommon. (*v.v.*)

555. **I. lucifuga** Fr. Boud. Icon. t. 123. *Lux*, light; *fuga*, avoiding.

P. 2–7·5 cm., *somewhat olivaceous*, sometimes becoming pale, rarely fawn colour, fleshy, *convexo-campanulate, then plane*, more or less umbonate, *longitudinally fibrillose, or covered with minute adpressed scales*. St. 3–7 × 4–12 mm., *pallid*, equal, *rigid*, often flexuose, apex white-farinose. Gills *whitish, then yellowish, and finally pure olivaceous*, sinuate, ventricose, crowded, broad, crisped in large specimens. Flesh *white, slightly coloured under the cuticle*. Spores olivaceous, elliptical, 8–9 × 4–5μ. Cystidia ventricose, 60–70 × 12–14μ, scattered. Smell strong, like radishes. Woods, roadsides, and under trees. Sept.—Oct. Not uncommon.

†††Gills tinged violet.

556. **I. violaceifolia** Peck. *Violaceus*, of a violet colour; *folia*, leaves.

P. 1–1·5 cm., *grey*, convex, or almost plane, *fibrillose, squamulose*. St. 2·5 cm. × 3–4 mm., *whitish*, firm, slender, *fibrillose*. Gills *pale violet, then brownish cinnamon*, adnexed, crowded. Spores elliptical, 10 × 6μ. Cystidia ventricose, 50–60 × 12–16μ, fairly abundant. Amongst moss. Sept. Rare.

**Stem coloured.

†Gills brown, ochraceous, or cinnamon.

557. **I. caesariata** Fr. Cke. Illus. no. 437, t. 388. *Caesariata*, covered with hairs.

P. 4 cm., *tawny-dirty, or like a smooth sugared cake*, fleshy, convex, then expanded, gibbous, repand, *becoming silky-even*, finally almost smooth at the disc, silky at the margin. St. 4 cm. × 6 mm., *pallid*, unequal, *sometimes twisted*, somewhat thickened at the base, *pruinose*. Gills *whitish fuliginous*, somewhat adnate, thin, crowded, ventricose. Spores reniform, 8–10 × 4–5μ. Cystidia narrowly ventricose, 70–80 × 12–15μ, "often septate" Rick., fairly abundant. Smell disagreeable. Beech woods. Sept.—Oct. Uncommon. (*v.v.*)

var. **fibrillosa** Fr. (= *Inocybe delecta* Karst.) Fr. Icon. t. 109, fig. 3. *Fibrillosa*, full of fibrils.

Differs from the type in the *obsoletely umbonate, fibrillose, sometimes squamulose, ochraceous pileus becoming tawny, in the very ochraceous-*

fibrillose stem, and the pallid ochraceous gills at length turning brown. Deciduous woods, especially beech.

558. **I. obscura** (Pers.) Fr. *Obscura*, dark.

P. 1·5–2·5 cm., *brown, more or less suffused with violet*, somewhat fleshy, campanulato-convex, obtuse, or subumbonate, longitudinally fibrillose, *disc scaly*. St. 3–6 cm. × 4–6 mm., *violaceous, becoming fuscous downwards*, flexuose, fibrillose. Gills *olivaceous, then brown*, uncinato-adnexed, crowded, ventricose, edge often unequal. Flesh *tinged bluish especially at the apex of the stem*. Spores brownish, elliptical, 8–9 × 4μ. Cystidia ventricose, 65–75 × 12–16μ, abundant. Smell strong, of radishes. Amongst pines, and in woods. July—Nov. Not uncommon. (*v.v.*)

var. **rufa** Pat. *Rufus*, red.

Differs from the type in the *reddish-brown, strongly umbonate p., the violet gills, and spores narrowed towards one end.*

559. **I. lacera** Fr. *Lacera*, torn.

P. 2–4 cm., *brownish, then mouse colour, becoming pale*, slightly fleshy, *convex, then expanded*, obtuse, or obtusely umbonate, at first adpressedly fibrillose, *then rimosely scaly and squarrose*. St. 3–7 cm. × 5–8 mm., *paler than the p., covered with brown, fibrillose squamules*, tough, equal, or attenuated at the base. Gills *white flesh, then mouse colour*, attenuato-adnexed, ventricose, broad. Flesh of *stem reddish*. Spores ferruginous, elliptical, 10–12 × 6–7μ. Cystidia ventricose, 56–80 × 13–17μ, numerous. Pine, and mixed woods. July—Oct. Not uncommon. (*v.v.*)

560. **I. carpta** (Scop.) Fr. *Carpta*, torn.

P. 3–5 cm., *tan colour, or dusky brown*, convex, then flattened, usually at length more or less depressed at the disc, *everywhere densely fibrillose and woolly*. St. 3–5 × 1 cm., *paler than the pileus, attenuated downwards, covered with a spreading, fibrillose woolliness like the p.* Gills *ochraceous, then brown fuscous*, adnate, then free, ventricose, broad. Flesh *pale ochraceous, becoming tinged with brown*. Spores elliptical, 9–10 × 4–5μ, 1–2-guttulate; "angular, nearly trapezium shaped, 10–12 × 5–6μ" Rick. Cystidia often slightly curved, ventricose, 60–70 × 13–15μ, abundant. Taste sweet, then bitter. Woods. Aug.—Sept. Uncommon. (*v.v.*)

561. **I. hystrix** Fr. Fr. Icon. t. 106, fig. 1. *Hystrix*, porcupine.

P. 4–9 cm., *dull brown, or mouse colour*, fleshy, firm, convex, then flattened, obtuse, or slightly gibbous, orbicular, *squarrose with revolute, floccose scales*, which are fibrilloso-adpressed towards the margin. St. 5–9 cm. × 6–10 mm., *fuscous mouse colour*, firm, equal, or *attenu-*

ated downwards, or subfusiform, *squarrose with revolute, crowded, floccose scales up to the sharply defined annular zone*, apex pallid. Gills *whitish, becoming fuscous*, adnate, crowded, linear, *margin white*. Flesh *white*. Spores brown, pip-shaped, or wedge-shaped, 9–10 × 4–5μ, 1-guttulate. Cystidia ventricose, 75–80 × 12–14μ, fairly abundant. Smell of new meal. Woods. Aug.—Oct. Uncommon. (*v.v.*)

562. **I. incarnata** Bres. (= *Inocybe pyriodora* (Pers.) Fr. sec. René Maire.) Bres. Fung. Trid. t. 53.

Incarnata, made flesh colour.

P. 5–8 cm., *yellowish rufescent, or flesh colour*, fleshy, convexo-campanulate, then expanded and gibbous, or broadly umbonate, fibrillose, then squamulose, margin villosely fimbriate. St. 6–8 cm. × 6–15 mm., *rosy flesh colour, apex white, furfuraceous*, attenuated downwards, somewhat rooting, somewhat fibrillose. Gills *whitish, then cinereous-cinnamon, at length spotted with red, or from the first becoming red*, sinuato-adnate, crowded, edge fimbriate. Flesh *of pileus white, becoming deep red when broken, that of the stem red from the first*. Spores yellowish-brown, broadly elliptical, 12–15 × 6–8μ. Cystidia fusoid, 50–65 × 15–18μ, apex brownish. Smell pleasant, like pears. Woods. June—Oct. Not uncommon. (*v.v.*)

563. **I. nigrodisca** Peck. *Niger*, black; *discus*, a quoit.

P. 1–2 cm., *blackish-brown, margin greyish*, convex, then plane, or depressed, umbonate, very minutely fibrillose. St. 2·5–3·5 cm. long, *reddish brown*, slender, firm, flexuose, minutely pruinosely downy. Gills *greyish, then rusty-brown*, sometimes tinged yellow, free, or subadnexed, rounded behind, crowded. Spores ferruginous, subelongate, 5·5–6·5 × 4·5–5μ. Cystidia fairly abundant, ventricose, 45–55 × 12–16μ. Among moss and grass. Rare.

564. **I. mutica** Fr. Cke. Illus. no. 418, t. 382. *Mutica*, curtailed.

P. 2·5–5 cm., *whitish, or tinged yellowish with darker adpressed squamules*, fleshy, convex, then plane, *always very obtuse*, at length subdepressed in the centre. St. 3–6 cm. × 5–8 mm., *white straw colour, becoming somewhat fuscous*, often attenuated downwards, sometimes enlarged at the base, fibrillose. Gills *white, then slightly fuscous*, adnate, or attenuated behind and becoming free, thin, crowded. Flesh *white*. Spores elliptical, 8–10 × 5μ, 1-guttulate. Cystidia fusoid, 51–80 × 12–15μ, abundant. Woods, and roadsides. Sept.—Nov. Not common. (*v.v.*)

565. **I. brunnea** Quél. Trans. Brit. Myc. Soc. II, t. 9.

Brunnea, brown.

P. 3–5 cm., *chestnut colour*, fleshy, campanulate, umbonate, fibrillosely silky, then cracked. St. 3–6 cm. × 5–6 mm., *concolorous, or paler*, thickened at the base, *apex white, pruinose, fibrillosely striate. Cortina concolorous, and fugacious.* Gills *cream colour, then umber, edge white, indented*, emarginate, broadest in front, 5 mm. wide. Flesh *white*. Spores bistre, elliptical, 7–12 × 4–6μ. Cystidia ventricose, or fusoid, 50–55 × 12–15μ, fairly abundant. Woods, also amongst grass near pines. Aug.—Oct. Not uncommon. (*v.v.*)

566. **I. haemacta** Berk. & Cke. (= *Inocybe pyriodora* (Pers.) Fr. sec. René Maire.) Cke. Illus. no. 410, t. 390. *αἱμακτός*, bloody.

P. 2·5–5 cm , *umber, margin pallid, often becoming stained with red, or verdigris*, disc darker, subsquamose, fleshy, compact, obtuse, campanulate, then expanded, floccosely fibrillose. St. 4–5 cm. × 6–8 mm., *whitish above, tinged verdigris-green at the base, and often higher up*, scarcely fibrillose, nearly equal. Gills *pallid, then dingy tan*, adnate, slightly rounded behind. Flesh *turning blood red when touched or wounded, verdigris at the base of the stem, and also often elsewhere*. Spores reddish brown, elliptical, 8–9 × 5μ, 1-guttulate. Cystidia ventricose, 66–78 × 14–16μ, fairly numerous. Lawns, and gardens. Aug.—Oct. Uncommon. (*v.v.*)

var. **rubra** Rea. *Rubra*, red.

Differs from the type in *the blood red colour of the p. and st., which is only tinged verdigris at the extreme base, although the flesh is tinged verdigris for some considerable distance upwards.* Bare earth. Sept. Rare. (*v.v.*)

567. **I. conformata** Karst. *Conformata*, shaped.

P. 1–3 cm., *pale fuscous, or tinged rusty*, convex, then expanded, umbonate, fibrillosely rimose, sometimes minutely, adpressedly, floccosely squamulose. St. 3–5 cm. × 3–6 mm., *concolorous, apex at first tinged violet*, equal, often flexuose, solid, minutely fibrillose. Gills *pallid, then brownish*, 4–5 mm. wide, adnexed, ventricose, somewhat crowded; margin *white*, fimbriate. Flesh *white, brownish under the cuticle of the p., bluish at first in the stem*, thick at the disc, very thin at the margin of the p., firm. Smell and taste none. Spores brownish in the mass, oblong elliptical, depressed on one side, 8–11 × 4–5μ. Cystidia hyaline, fusiform ventricose, apex muriculate, 65–75 × 15–19μ. Under oaks in woods. Sept. Uncommon. (*v.v.*)

568. **I. flocculosa** Berk. (=*Inocybe scabella* (Fr.) Bres. sec. Bataille.) Cke. Illus. no. 416, t. 393. *Flocculosa*, woolly.

P. 2·5 cm., *brownish fawn colour*, somewhat fleshy, convex, subcampanulate, umbonate, silky squamulose, margin smoother, veil white, fibrillose, fugacious. St. 4 cm. × 4 mm., *pale fawn*, fibrillose, *brown beneath the fibrillae*, apex minutely squamuloso-pulverulent. Gills *pale fawn, at length dull-ferruginous*, rounded behind, adnate but not broadly so, ventricose, margin white. Spores ferruginous, elliptical, 8–10 × 4–5μ. Cystidia 45–60 × 10–12μ, abundant. Smell of new meal, but nauseous. Bare soil, and amongst grass. Sept.—Nov. Uncommon. (*v.v.*)

††Gills tinged olive.

569. **I. dulcamara** (A. & S.) Fr. Cke. Illus. no. 408, t. 582, fig. B.
Dulcis, sweet; *amara*, bitter.

P. 3–6 cm., *olivaceous-fuscous, becoming paler*, campanulato-convex, then expanded and umbonate, *floccoso-scaly*, silky towards the margin. St. 4–6 cm. × 4 mm., *paler than the pileus*, equal, *cortinato-fibrillose, adpressedly squamulose*, apex mealy. Gills *pallid, then olivaceous*, arcuato-adfixed, ventricose, crowded. Flesh becoming *yellowish white*, thin. Spores brown, almost colourless under the microscope, elliptical, 7–9 × 4–5μ. Cystidia ventricose, 55–65 × 15–18μ, fairly abundant, "on edge of gill only" Rick. Taste slightly bitter. Pine, and mixed woods. Aug.—Oct. Not uncommon. (*v.v.*)

570. **I. relicina** Fr. (= *Astrosporina relicina* (Fr.) Schroet.)
Relicina, curled backwards.

P. 1·5–3 cm., *fuliginous*, fleshy, thin, conical, then expanded, obtuse, *everywhere scaly-squarrose with fasciculate down*. St. 4–5 cm. × 4–6 mm., *fuliginous, apex paler, soft*, equal, *floccoso-scaly*, fibrillose. Gills *yellow, then olivaceous*, adnexed, crowded. Spores olive brown, pip-shaped, 10–12 × 7μ. Cystidia ventricose, 70–85 × 14–16μ, scattered, "only on edge of gill" Rick. Damp pine woods amongst *Sphagnum*, and in bogs. July—Oct. Uncommon.

571. **I. Bongardii** (Weinm.) Massee.
H. G. Bongard, a Russian botanist.

P. 3–7 cm., *whitish with a rufescent, or yellowish tinge, covered with darker fibrillose squamules*, fleshy, campanulate, then expanded, obtusely umbonate. St. 5–8 cm. × 5–12 mm., *concolorous, or slightly paler*, equal, straight, or curved, tough, apex white-mealy. Gills *whitish, then olive-cinnamon*, arcuato-adnate, crowded, ventricose, edge white, eroded. Flesh *reddish*. Spores cinnamon brown, yellow under the microscope, elliptical, 9 × 6μ, 1-guttulate. Cystidia ventricose, 50–65 × 12–16μ, scattered, "only on edge of gill, basidia with olive granular contents" Rick. Smell pleasant, like ripe pears. Woods and pastures. May—Oct. Uncommon. (*v.v.*)

†††Gills tinged violet.

572. **I. cincinnata** Fr. Bres. Fung. Trid. t. 51, fig. 2.
Cincinnata, with curled hair.

P. 1·5–3·5 cm., *mouse colour, or brownish chestnut*, somewhat fleshy, convex, then expanded, obtuse, or obsoletely umbonate, *disc besprinkled with floccose scales*; margin fibrillose. St. 3–5 cm. × 2–3 mm., *fuscous, apex tinged violet at first, then becoming pale*, base subbulbose, *fibrillosely-squamulose*. Gills *fuscous-violaceous, then cinnamon*, adnexed, or attenuated behind, separating, ventricose, *crowded*. Flesh *whitish, bluish at the apex of the stem*. Spores brown, elliptical, 7–8 × 4–5μ, 1–2-guttulate. Cystidia subcylindrical, or ventricose, 60–70 × 9–15μ, "sometimes filled with a dark juice" Rick., fairly abundant. Woods. Aug.—Oct. Not uncommon. (*v.v.*)

II. No cystidia.

*St. whitish, or pallid.

†Gills brownish, ochraceous, or cinnamon.

573. **I. perlata** Cke. Cke. Illus. no. 1162, t. 961.

Perlata, very broad.

P. 6–10 cm., *ochraceous, longitudinally streaked with darker, fuscous fibrils, disc dark bistre nearly black*, fleshy, convex, then expanded, broadly umbonate; margin incurved. St. 6–11 × 1·5 cm., *pallid, darker below*, straight, or curved, sometimes twisted, striate, equal, apex mealy. Gills *pallid, then pale umber*, adnexed, somewhat rounded behind, broad, margin whitish, irregular. Flesh *dingy white*. Spores reddish brown, elliptical, 10 × 6–7μ, with a large central gutta. Woods, and under hornbeam. Aug.—Sept. Uncommon. (*v.v.*)

574. **I. perbrevis** (Weinm.) Fr. Cke. Illus. no. 434, t. 519.

Per-brevis, very short.

P. 1·5–3 cm., *fuscous, or rufous, becoming yellowish*, somewhat fleshy, convex, then plane, obtusely umbonate, often depressed round the umbo, fibrillose, or squamulose; margin fibrillose, often splitting. St. 2–2·5 cm. × 3–5 mm., *pallid, somewhat attenuated downwards, white-fibrillose*, apex pruinose, cortinate. Gills *whitish, becoming fuscous, or clay colour*, adnexed with a tooth, ventricose, rather distant. Flesh *white*. Spores brown, elliptic-oblong, 8–9 × 4·5–5μ, apiculate. "Cystidia only on edge of gill, basidia-like, 48–60 × 12–15μ" Rick. Woods, and shady places. Sept.—Oct. Uncommon.

575. **I. squarrosa** Rea. Trans. Brit. Myc. Soc. v, t. 4.

Squarrosa, rough with scales.

P. 5–13 mm., *mouse grey with an ochraceous tinge*, slightly fleshy, campanulate, or convex, *covered with erect, revolute, floccose, darker grey scales especially at the disc*, floccosely fibrillose elsewhere; margin at first involute, *white*. St. 1·5–3 cm. × 1–2 mm., *white with a pinkish tinge*, apex pruinose, base white, subbulbose. Gills *white, then brownish*, adnato-sinuate, 2 mm. wide, edge *white*. Flesh *white*. Spores greenish-fuscous, elliptical, 9–10 × 5–6μ, 1-guttulate. Under willows and alders in a brick pit. July—Aug. Uncommon. (*v.v.*)

576. **I. vatricosa** Fr. Fr. Icon. t. 110, fig. 3.
Vatricosa, with crooked feet.

P. 1–7 cm., *yellowish, becoming whitish with age*, slightly fleshy, convex, then plane, obtuse, or umbonate, *viscid when moist*, shining when dry; margin obsoletely silky. St. 3–7 cm. × 3–10 mm., *white, or greyish*, equal, slightly thickened at the white villose base, *often curved and contorted, white pulverulent*. Gills *white, then fuscous*, adnexed, broadly emarginate, almost free, broad. Spores brownish, elliptical, 5–6 × 3–3·5μ. Woods, dead stumps, chips, and sawdust. Sept.—Oct. Uncommon.

**St. coloured.

†Gills brownish, ochraceous, or cinnamon.

577. **I. Cookei** Bres. Bres. Fung. Trid. t. 121.
Mordecai Cubitt Cooke, the eminent English mycologist.

P. 3–5 cm., *yellowish straw colour, becoming lurid yellowish*, fleshy, conico-campanulate, then expanded and umbonate, margin at length revolute and split, silky-fibrillose, then rimose. St. 4–7 cm. × 5–7 mm., *concolorous*, equal, silky fibrillose, base marginately bulbose. Gills *whitish cinereous, then yellowish cinnamon*, attenuated behind, adnexed, crowded, somewhat ventricose, margin *white, fimbriate*. Flesh *whitish straw colour*. Spores ochraceous, subreniform, 8–10 × 5–5·5μ. Smell somewhat pleasant when young, becoming somewhat earthy. Fir woods. Sept. Uncommon.

578. **I. mimica** Massee. (= *Inocybe adequata* Britz. sec. Cke.)
μιμικά, a mimic.

P. 6–8 cm., *yellow-brown, everywhere covered with large, adpressed, slightly darker, fibrous scales*, fleshy, campanulate, obtusely umbonate, fibrillose. St. 6–8 × 1 cm., *paler than the p.*, equal, fibrillose. Gills *yellow-brown*, deeply sinuate, attached to the stem by a very narrow portion, broad. Flesh *brownish*. Spores brown, subcylindrical, 14–16 × 6–8μ, with an oblique apiculus. Woods. Sept. Rare.

579. **I. rhodiola** Bres. Bres. Fung. Trid. t. 200, as *Inocybe frumentacea* (Bull.) Bres.
ῥόδον, a rose.

P. 4–8 cm., *rufous-chestnut, or fuscous flesh colour*, fleshy, campanulate, then expanded and umbonate, fibrillosely cracked, centre even. St. 5–8 × 1–1·5 cm., *vinous*, fibrilloso-squamulose, becoming glabrous, *apex pallid, subfloccose*. Gills *white, then yellowish umber, often spotted with brownish umber*, sinuato-uncinate, almost free, crowded, edge fimbriate. Flesh *white, vinous at the base of the stem*. Spores yellowish, subreniform, 12–13 × 6–8μ, or 9–10 × 5–7μ, 1–2-guttulate. Cells on edge of gills, clavate, or subfusoid, 45–60 × 12–15μ. Smell fruity. Woods, fields, and parks. Aug.—Oct. Not uncommon. (*v.v.*)

580. **I. hirsuta** (Lasch) Fr. Bres. Fung. Trid. t. 86, fig. 2.
Hirsuta, bristly.

P. 1–4 cm., *fuscous, or ochraceous fuscous, disc sometimes tinged with green*, slightly fleshy, conico-campanulate, then expanded, and acutely or obtusely umbonate, *squarrose with fasciculato-pilose scales*, margin *fibrillose, fimbriate*. St. 4–9 cm. × 2–6 mm., *concolorous, base verdigris*, somewhat equal, or somewhat thickened at the base, tough, *fibrillose, apex white floccoso-scaly*. Gills *pallid, then fuscous cinnamon*, adnate, crowded, narrow, edge whitish, fimbriate. Flesh *pallid, becoming faintly tinged with red on exposure to the air, greenish in the stem*. Spores ochraceous, oblong pip-shaped, 9–11 × 4–5μ. Woods. Sept.—Oct. Not uncommon. (*v.v.*)

581. **I. calamistrata** Fr. Fr. Icon. t. 106, fig. 2.
Calamistrata, curled with the curling-iron.

P. 2·5–6 cm., *fuscous*, slightly fleshy, campanulate, *obtuse*, margin at first bent inwards, often repand, at length flattened, *everywhere squarrose-scaly*. St. 4–9 cm. × 4–6 mm., *concolorous, base dark azure blue*, rigid, tough, equal, *somewhat rooting, squarrose throughout with rigid, recurved scales*. Gills *white, then ferruginous*, adnexed, separating, crowded, *broad*, edge whitish, somewhat serrated. Flesh *reddish, azure blue at the base of the stem*. Spores ochraceous, elliptical, 10–11 × 6μ. Smell *strong*. Pine, and fir woods. Aug.—Oct. Uncommon. (*v.v.*)

I. echinata (Roth) Cke. = **Lepiota haematosperma** (Bull.) Boud.

††Gills tinged olive.

582. **I. destricta** Fr. Boud. Icon. t. 121. *Destricta*, stripped off.

P. 4–8 cm., *livid violet tinged brownish*, fleshy, campanulate, *covered longitudinally with darker fibrils*, then torn into scales, *rimosely cracked*. St. 5–10 × 1–1·5 cm., *concolorous, covered with darker fibrils*, apex mealy. Gills *olivaceous*, uncinately adnate, narrow, 4 mm. wide. Flesh *whitish violaceous*, deeper coloured under the cuticle, *reddish at the base of the stem*. Spores olivaceous, often slightly curved, oblong, 11–13 × 5·5–7μ, multi-guttulate. Cystidia none[1]. Woods, and under poplars. July—Nov. Uncommon.

583. **I. fastigiata** (Schaeff.) Fr. (= *Inocybe Curreyi* Berk. sec. Massee.) Bres. Fung. Trid. t. 57. *Fastigiata*, having a gable.

P. 3–7 cm., *straw colour, yellow fuscous, sometimes fuscous brown, disc ochraceous, or livid-fulvous*, fleshy, conico-campanulate, obtusely, or acutely umbonate, *longitudinally fibrillose and cracked*, rarely adpressedly scaly; margin sometimes lobed. St. 4–9 cm. × 4–8 mm., *pallid, becoming fuscous*, equal, slightly enlarged downwards, minutely fibrillose, often twisted with age. Gills *yellow, then olivaceous*, sinuato-

[1] Massee and Ricken have a different plant in view as they describe cystidia.

free, ventricose, crowded. Flesh *white*. Spores brownish, pip-shaped, 7–10 × 4–5μ, 1-guttulate. "Cystidia only on the edge of the gills, bottle-shaped-clavate, 50–60 × 15–20μ" Rick. Woods, and pastures. June—Oct. Not uncommon. (*v.v.*)

III. Insufficiently described.

584. **I. mamillaris** Pass.[1] *Mamilla*, a teat.

P. *white*, convex, mammillate, squamulose. St. *white*, hollow, equal, flexuose. Gills emarginato-adnexed. Spores smooth.

585. **I. phaeocephala** (?Bull.) Cke. Cke. Illus. no. 425, t. 396. φαιός, dusky; κεφαλή, head.

P. 5–10 cm., *fuliginous, becoming brownish*, subcampanulate, rarely flattened and umbonate, smooth. St 7–13 cm. × 8–14 mm., *grey, with brownish lines*, base *white* and swollen, naked. Gills *yellowish bistre colour*, free, semilunate, very broad. Spores *bright ferruginous red*, elliptical, 6 × 4μ. Pine, and mixed woods. Oct. Rare

586. **I. schista** Cke. & Sm. Cke. Illus. no. 423, t. 504. σχιστός, split.

P. 5–7·5 cm., *bay brown*, obtusely campanulate, broadly subumbonate, cracking longitudinally, rather fibrillose. St. 5–8 × 1 cm., *paler than the p.*, equal, twisted, solid. Gills *rufescent*, adnate with a decurrent tooth, rather broad; edge *pale*, serrate. Lawns. May. Rare.

I. subrimosa (Karst.) Sacc. = **Astrosporina asterospora** (Quél.) Rea.

I. scabella (Fr.) Bres. = **Inocybe flocculosa** Berk.

I. tricholoma (A. & S.) Fr. = **Flammula tricholoma** (A. & S.) Fr.

I. strigiceps Fr. = **Flammula strigiceps** Fr.

I. Curreyi Berk. = **Inocybe fastigiata** (Schaeff.) Fr.

I. adequata Britz. = **Inocybe mimica** Massee.

I. lanuginosa Fr. sec. Bataille = **Astrosporina sabuletorum** (B. & Curt.) Rea.

I. violaceo-fusca Cke. & Massee = **Cortinarius (Dermocybe) violaceo-fuscus** (Cke. & Massee) Massee.

I. Trinii (Weinm.) Bres. = **Inocybe Godeyi** Gill.

I. hiulca (Fr.) Kalchbr. = **Inocybe Godeyi** Gill.

Spores ochraceous, or ferruginous, irregular, angular, echinulate, or verrucose.

Astrosporina Schroet.

(ἀστήρ, star; σπορά, seed.)

Pileus fleshy, regular. Veil marginal, fugacious. Stem fleshy, central. Gills adnate, sinuato-adnate, or adnexed. Spores ochraceous,

[1] Massee, Eur. Fung. Fl. Agar. p. 156.

ferruginous, olivaceous, or fuscous; angular, nodose, verrucose, echinulate, or irregular in shape. Cystidia present, or absent, fusiform, ventricose, cylindrical, or clavate. Growing on the ground, rarely caespitose.

I. Cystidia present.

*Stem whitish, or pallid.

587. **A. fibrosa** (Sow.) Rea. (= *Inocybe fibrosa* (Sow.) Fr.) Bres. Fung. Trid. t. 56, as *Inocybe fibrosa* Sow. *Fibrosa*, fibrous.

P. 6–10 cm., *white, or straw colour, sometimes spotted yellow*, fleshy, obtusely campanulate, then expanded and gibbous, or broadly umbonate, silky, at length cracking; margin deeply split, often lobed. St. 6–10 × 1·5–2·5 cm., *white*, base at length straw colour, subequal, striate, or somewhat sulcate, apex pruinose. Gills *white, becoming cinereous-cinnamon with age*, attenuated behind, *nearly free*, crowded. Flesh *white*. Spores ochraceous, angular, oblong, 10–13 × 5–7μ. Cystidia fusoid, 70–90 × 10–16μ, thick walled. Smell foetid. Coniferous woods. May—Sept.

588. **A. duriuscula** Rea. (= *Inocybe duriuscula* Rea.) Trans. Brit. Myc. Soc. III, t. 3, as *Inocybe duriuscula* Rea.
Duriuscula, somewhat hard.

P. 6–7 cm., *fulvous ochraceous, disc becoming pale*, fleshy, campanulate, then expanded and gibbous, *floccose*, soon longitudinally fibrillose; margin at length revolute. St. 8 × 1·5 cm., *white*, enlarged at the apex and base, firm, striate, apex *ribbed with the decurrent teeth of the gills*. Gills *white, becoming fuscous, sinuato-adnate with a decurrent tooth*, 6–8 mm. wide, somewhat crowded. Flesh *white, very firm*. Spores umber, strongly nodulose, 9–10 × 7–8μ, 1-guttulate. Cystidia ventricose, 52–60 × 13–15μ, abundant. Woods. Sept.—Oct. Not uncommon. (*v.v.*)

589. **A. proximella** (Karst.) Rea. (= *Inocybe proximella* Karst.) Trans. Brit. Myc. Soc. III, t. 2, as *Inocybe proximella* Karst.
Proximella, somewhat near.

P. 2–5 cm., *pallid yellow, disc, and especially the umbo, becoming rusty brown, or bay*, fleshy, conico-convex, then expanded and umbonate, becoming longitudinally fibrillosely cracked. St. 6–8 × ·5–1 cm., *pallid*, slightly narrowed upwards, often attenuated at the base, sometimes wavy, subfibrillose. Gills *pallid, then tan, finally brown*, adnate, 6–7 mm. wide, crowded, *ventricose*. Flesh *white*. Spores ferruginous, nodulose, irregularly oblong, 7–10 × 5–7μ, 1-guttulate. Cystidia ventricose, 55–70 × 15–24μ, abundant. On the ground. Sept.—Oct. Uncommon. (*v.v.*)

590. **A. praetervisa** (Quél.) Schroet. (= *Inocybe praetervisa* Quél.) Trans. Brit. Myc. Soc. II, t. 5, as *Inocybe praetervisa* Quél.
Praetervisa, overlooked.

P. 2–6 cm., *fawn, generally darker at the margin*, conico-campanulate, then expanded, umbonate, or gibbous, often splitting at the margin with age, *slightly viscid, fibrillose*, disc glabrous, soon *longitudinally virgate*. St. 4–7 cm. × 4–8 mm., *white, then pale straw colour*, equal, *base bulbous*, apex pruinose. Gills *white, then dirty cinnamon*, attenuated behind, adnexed, almost free, narrow, 3 mm. wide, crowded; *edge white, fimbriate*. Flesh *white*. Spores dark brownish in the mass, ochraceous under the microscope, angular, 9–11 × 5–6μ, 1–3-guttulate. Cystidia fusiform, ventricose, 55–75 × 18–30μ, *somewhat fuscous at the apex*, "sometimes with olive granular contents" Rick. On the ground under conifers. Aug.—Oct. Uncommon. (*v.v.*)

591. **A. lanuginella** Schroet. *Lanuginella*, somewhat floccose.

P. 1·5–3 cm., *tawny*, or *greyish brown*, campanulato-convex, then plane, obtusely umbonate, fibrillose, cracked ("fibrils septate, apical cell 35–40 × 8–11μ, with rounded ends" Schroeter). St. 1·5–5 cm. × 1·5–5 mm., *pallid, apex at first delicately tinged with lilac, base brownish*, equal, fibrillose. Gills *pallid, then cinnamon*, 2–3 mm. wide, slightly adnexed, somewhat crowded, edge fimbriate. Flesh *white, tinged reddish under the cuticle of the pileus and stem*, thick at the disc, thin at the margin of the pileus, firm. Spores cinnamon in the mass, oblong, obtusely angular, 8–11 × 5–7μ. Cystidia hyaline, either fusiform, ventricose, obtuse at the apex, muriculate or not, 40–70 × 15–23μ, or acicular and acute. On the ground in oak woods. Aug. Uncommon. (*v.v.*)

592. **A. infida** (Peck) Rea. (= *Agaricus* (*Hebeloma*) *infidus* Peck, *Inocybe umbratica* Quél.) Bres. Fung. Trid. t. 58, fig. 2, as *Inocybe commixta* Bres. *Infida*, unsafe.

Entirely white. P. 2–3 cm., fleshy, conico-campanulate, then expanded and umbonate, silky-fibrillose; margin often split. St. 3–4 cm. × 3–6 mm., minutely pruinose, equal, apex mealy, *base bulbous*. Gills *white, then greyish-cinnamon*, free, very crowded, edge minutely fimbriate. Flesh white, somewhat straw colour at the base of the stem. Spores ochraceous, angular, globose-oblong, 9–10 × 6–7μ. Cystidia fusiform or subventricose, somewhat fuscous at the apex, 40–45 × 12–14μ. Smell earthy, strong. Taste mild. Coniferous woods. Aug.—Oct. Uncommon. (*v.v.*)

593. **A. trechispora** (Berk.) Rea. (= *Hebeloma trechisporum* Berk.) Cke. Illus. no. 443, t. 403, upper figs., as *Inocybe trechispora* Berk. *τραχύς*, rough; *σπορά*, seed.

P. 1·5–2·5 cm., *whitish, or cinereous, umbo tawny*, submembranaceous, convex, *strongly umbonate, viscid at first*, then dry and silky;

margin paler with a slight livid tinge, thin. St. 5 cm. × 4 mm., *white,* equal, often flexuose, with a mass of white mycelium at the base, slightly striate under a lens and mealy. Gills *whitish, then pinkish grey,* emarginate, scarcely adnate, ventricose, margin denticulate. Flesh *white.* Spores bistre-brown, *warted,* angular, 7–8 × 5–6μ. Cystidia fusoid, or subventricose, 40–50 × 12–18μ, fairly abundant. Woods. Aug.—Oct. Uncommon.

**Stem coloured.

594. **A. asterospora** (Quél.) Rea. (= *Inocybe asterospora* Quél., *Inocybe subrimosa* Sacc., *Clypeus subrimosus* Karst.) Cke. Illus. no. 430, t. 385, as *Inocybe asterospora* Quél.

ἀστήρ, star; σπορά, seed.

P. 2–5 cm., *bistre, with brown striae,* convex then expanded and umbonate, becoming rimose. St. 5–8 cm. × 6–8 mm., *reddish, streaked with brown fibrils, pubescent,* equal, *base marginately bulbous,* cuticle separable. Gills *whitish bistre, then cinnamon,* emarginate, ventricose, thin. Flesh *white.* Spores ferruginous, stellately-nodulose, subglobose, 9–12μ. Cystidia ventricose, 33 × 18μ, abundant. Smell mouldy. Woods, and under trees. Aug.—Oct. Not uncommon. (*v.v.*)

595. **A. fasciata** (Cke. & Mass.) Rea. (= *Inocybe fasciata* Cke. & Massee.) Cke. Illus. no. 1164, t. 1173, as *Inocybe fasciata* Cke. & Massee. *Fasciata,* bound together in bundles.

Densely caespitose. P. 5–7·5 cm., *tawny, disc rufous,* campanulato-convex, silky, *clad with minute, darker, squarrose scales.* St. 5–7·5 cm. × 3–10 mm., *pallid, base reddish,* equal, or a little attenuated below, fibrillose. Gills *pallid,* attenuated in front, rounded behind, or slightly sinuate, crowded. Flesh *white, reddish towards the base of the stem.* Spores fuscous, angular, elliptical, 10 × 6μ. Cystidia ventricose, 40–50 × 12–15μ, scanty. Amongst grass. Rare.

596. **A. lanuginosa** (Bull.) Schroet. (= *Inocybe lanuginosa* (Bull.) Bres.) Bres. Fung. Trid. t. 117, as *Inocybe lanuginosa* Bull.

Lanuginosa, woolly.

P. 2·5–4 cm., *umber, at length becoming yellow,* slightly fleshy, hemispherical, or campanulato-convex, then expanded and umbonate, *flocculosely-scaly, disc squarrose with hairy scales.* St. 2·5–4 cm. × 2–3 mm., *somewhat concolorous,* equal, *fibrillosely-squamulose*; cortina greyish white, soon disappearing. Gills *pallid clay colour, then reddish cinnamon,* rounded behind, adnexed, then separating, somewhat crowded; margin *white, floccosely crenulate.* Flesh *pallid.* Spores fuscous ochraceous, angular, 10–15 × 8–9μ. Cystidia cylindrical, or clavate, 50–60 × 20–25μ, *somewhat fuscous at the apex.* Mossy stumps, especially conifers, "never on the ground" Bres. July—Nov. Uncommon. (*v.v.*)

597. **A. calospora** (Quél.) Rea. (= *Inocybe calospora* Quél.) Bres. Fung. Trid. t. 21, as *Inocybe calospora* Quél.
καλός, beautiful; σπορά, seed.

P. 2–3 cm., *fuscous rufescent, becoming somewhat yellowish*, somewhat fleshy, conico-campanulate, then expanded and umbonate, *fibrillosely villose, then covered with adpressed, concolorous squamules*; margin *whitish, fibrillose*. St. 3–5 cm. × 2–4 mm., *livid then rufescent*, apex *white-pruinose*, base subbulbose. Gills *pallid, then tawny cinnamon*, free, ventricose, crowded; *edge white-pruinose*. Flesh *somewhat concolorous*. Spores ferruginous in the mass, bright ochraceous under the microscope, globose, 10–14μ, bluntly warted. Cystidia fusiform, 45–55 × 8–10μ, *somewhat fuscous at the apex*, scanty. Taste slightly acid. Deciduous woods, and under trees. Aug.—Oct. Uncommon. (*v.v.*)

598. **A. Gaillardii** (Gillet) Rea. (= *Inocybe Gaillardi* Gillet.)
A. Gaillard.

P. 1–3 cm., *tawny-yellow, or rust colour*, convex, then plane, umbonate, *floccosely squamulose*; *margin fimbriate*. St. 1·5–3 cm. × 2–4 mm., *concolorous*. Gills *brownish cinnamon*, free, ventricose; *edge whitish*. Flesh *concolorous*, or *paler*. Spores ferruginous, globose, *covered with long, slender, hyaline spines*, 8μ, or 10–12μ, including the spines. Cystidia subcylindrical, 40–45 × 10μ, scanty. Woods, and under trees. Aug.—Oct. Uncommon. (*v.v.*)

599. **A. scabella** (Fr.) Schroet. (= *Inocybe scabella* (Fr.) Quél.) Fr. Icon. t. 110, fig. 1, as *Inocybe scabella* Fr. *Scaber*, rough.

P. 1–2 cm., *fuscous, or fuscous rufescent*, somewhat fleshy, conical, then expanded, umbonate, *silky-fibrillose*, at length *torn into scales* around the even umbo. St. 4 cm. × 1–2 mm., *rufescent, or becoming pale*, apex pruinose, equal, tense and straight, or flexuose. Gills *dingy, becoming fuscous, adnexed*, more or less crowded. Flesh *dingy*. Spores bistre, warted, angular elliptical, 8 × 6μ. Cystidia ventricose fusiform, 55–65 × 12–15μ. Woods, and amongst short grass. Sept.—Nov. Not uncommon. (*v.v.*)

600. **A. Trinii** (Weinm.) Rea. (= *Inocybe Trinii* (Weinm.) Fr.) Cke. Illus. no. 435, t. 428, lower figs., as *Inocybe Trinii* Weinm.
Carl Bernard Trinius, the Russian botanist.

P. 1–2 cm., *whitish with a rufous tinge due to longitudinal rufous fibrils, tawny when dry*, hemispherical, obtuse. St. 4–6 cm. × 2–3 mm., *white, covered with loose reddish, or rufous fibrils*, equal, apex *white-mealy*. Gills *dusky cinnamon*, rounded behind, adnexed, ventricose, edge *white-flocculose*. Flesh *white*. Spores angular, subglobose, or somewhat oblong, 9–10μ, or 9–10 × 6–8μ. Cystidia ventricose, 50–60 × 14–17μ, abundant. Smell pleasant, like clove-pinks. Woods, and amongst grass. Aug.—Sept. Uncommon. (*v.v.*)

601. **A. maritima** (Fr.) Rea. (= *Inocybe maritima* Fr.) Cke. Illus. no. 414, t. 392, as *Inocybe maritima* Fr.

Maritima, pertaining to the sea.

P. 2–3 cm., *fuscous, or mouse colour, becoming hoary when dry, hygrophanous*, fleshy, somewhat soft, *convex, then flattened*, obtuse, or umbonate, flocculosely fibrillose, more or less adpressedly scaly. St. 2–4 cm. × 4–6 mm., *a little paler than the pileus*, equal, *fibrillose*, at first furnished with a cortina. Gills *fuscous-grey, then becoming ferruginous*, rounded, adnexed, somewhat separating, ventricose. Flesh *becoming fuscous-grey*. Spores ferruginous, angular, oblong, 9–10 × 6μ, 1-guttulate. Cystidia ventricose, 45–55 × 12–18μ. Often caespitose. Sea shores, and sandy ground in woods. Sept.—Oct. Not uncommon. (*v.v.*)

602. **A. Rennyi** (B. & Br.) Rea. (= *Inocybe Rennyi* B. & Br.) Cke. Illus. no. 442, t. 520, fig. A, as *Inocybe Rennyi* B. & Br.

J. Renny, a British mycologist.

P. 1·5–2 cm., *pale fawn colour, disc brown*, hemispherical, slightly fibrillose. St. 3–5 cm. × 3–4 mm., *paler than the p.*, attenuated downwards, *fibrillose*. Gills *dingy ochraceous*, rounded behind, almost free. Spores ochraceous, angular, slightly nodulose, oblong, 11–13 × 7–8μ, pointed at one end. Cystidia fusoid, 40–50 × 12–16μ, scanty. On the ground.

var. **major** (Massee) Rea. (= *Inocybe Rennyi* B. & Br. var. *major* Massee.) Cke. Illus. no. 442, t. 520, fig. B, as *Inocybe Rennyi* B. & Br. var. *Major*, larger.

Differs from the type in its *larger size, and slightly nodulose spores*, 13–17 × 10μ. Fir woods. Nov.

603. **A. sabuletorum** (B. & Curt.) Rea. (= *Inocybe sabuletorum* B. & Curt., and *Inocybe lanuginosa* Fr. sec. Bataille.)

Sabuletorum, of sandy places.

P. 1–3 cm., *umber, at length becoming yellowish*, slightly fleshy, convex, then expanded, obtuse, or somewhat umbonate, *velvety, the pile becoming matted together into little squamules, which stand erect at the disc*. St. 2–4 cm. × 6–8 mm., *concolorous*, tough, equal, fibrillosely squamulose, or downy, *apex white mealy*. Gills *clay colour, then ferruginous*, sinuate, or separating free, thin, *ventricose*; *edge white, minutely fimbriate*. Flesh *concolorous*. Spores ferruginous, angular, 9–10 × 6–7μ. Cystidia fusoid, or ventricose, 45–50 × 12–15μ, somewhat scanty. Woods. July—Nov. Not uncommon. (*v.v.*)

604. **A. fulva** Rea. Trans. Brit. Myc. Soc. VI, t. 7. *Fulvus*, tawny.

P. 3–4 cm., *tawny, darker at the disc*, fleshy, convex, then expanded, *longitudinally adpressedly fibrillose*; margin thin. St. 5–6 cm. × 5–

6 mm., *concolorous*, apex *lilac colour becoming pale*, equal, slightly attenuated at the base, *fibrillosely striate*. Gills *white, then ochraceous, margin whitish*, sinuato-adnate, 6–7 mm. broad, somewhat crowded. Flesh *white, becoming reddish in the stem*, thin. Spores deep ochraceous, oblong, angular, 10 × 5–6·5μ. Cystidia hyaline, *bladder-like*, obtuse, 42 × 20μ, thin walled, often slightly granular at the apex. Bare ground in frondose woods. Sept. Uncommon. (*v.v.*)

605. **A. fulvella** (Bres.) Rea. (= *Inocybe fulvella* Bres.) Trans. Brit. Myc. Soc. II, t. 8, as *Inocybe fulvella* Bres. *Fulvus*, tawny.

P. 6–12 mm., *olivaceous honey colour with the umbo somewhat darker, then becoming yellowish, or fuscous olive with the umbo tawny*, conico-campanulate, then expanded, and umbonate, flocculosely silky, glabrous at the centre, somewhat hygrophanous. St. 2–3 cm. × 1–2 mm., *lilac, soon changing to rufescent*, attenuated downwards, *apex* white-pruinose. Gills *lilac, then cinnamon*, rounded behind, and almost free, ventricose, 1–5 mm. wide, subdistant, edge fimbriate. Flesb *yellowish, lilac at the apex of the stem* and *then rufescent*. Spores ferruginous, angular, warted, 7–9 × 5–6μ. Cystidia fusiform, ventricose, 45–60 × 10–18μ, *somewhat fuscous at the apex*. Amongst moss. Aug.—Oct. Uncommon. (*v.v.*)

606. **A. Bucknallii** (Massee) Rea. (= *Inocybe Bucknallii* Massee.) Annals of Bot. XVIII (1904), t. 32, figs. 5, 6, as *Inocybe Bucknalli* Massee. Cedric Bucknall, the Bristol mycologist.

P. 1–2 cm., *brownish*, campanulato-convex, fibrillose, disc with a few squamules. St. 2–4 cm. × 2 mm., *concolorous*, equal, or slightly thickened at the base, fibrillose. Gills *rusty-brown*, adnexed, thick, rather distant, edge minutely fimbriate. Spores irregularly oblong, 15–17 × 8–9μ, angular, apiculate at one end. Basidia clavate, exceptionally large, 70–80 × 16–18μ, 4-spored. Cystidia on edge of gill only, clavate, 75–85 × 15–20μ, thin walled. Under bushes. Autumn. Rare.

607. **A. petiginosa** (Fr.) Rea. (= *Hebeloma petiginosum* Fr., *Inocybe petiginosa* (Fr.) Quél.) Fr. Icon. t. 114, fig. 4, as *Hebeloma petiginosum* Fr. *Petiginosa*, scurfy.

P. 1–2 cm., *fuscous* at the gibbous, naked disc, *hoary-silky with superficial, closely adpressed fibrils at the circumference, when old rufescent, or becoming yellow*, slightly fleshy, *conical, then convex*, somewhat umbonate. St. 2·5–5 cm. × 2 mm., *brick rufescent, becoming fuscous*, tough, equal, or slightly attenuated at the base, *white pulverulent*. Gills *light yellow, then olivaceous-date-brown*, beautifully ciliated under a lens, at first slightly adnexed, soon free, ventricose, crowded. Flesh *cream colour, rufescent in the stem*. Spores olivaceous, angular, warted, oblong, 8–9 × 5–6μ. Cystidia fusiform, 48–65 × 9–10μ. Beech woods. Sept.—Oct. Not uncommon. (*v.v.*)

II. No cystidia.

608. **A. margaritispora** (Berk.) Rea. (= *Inocybe margaritispora* Berk.) Cke. Illus. no. 432, t. 505, as *Inocybe margaritispora*.

μαργαρίτης, a pearl; *σπορά*, seed.

P. 3–5 cm., *fawn colour*, or *pale yellowish-brown*, campanulate, then expanded and broadly umbonate, undulating, silky, clad with adpressed fibrillose scales. St. 6–10 cm. × 5–8 mm., *pallid*, equal, fibrillose. Gills *pallid*, reaching the stem, scarcely adnate. Flesh *yellowish*. Spores ochraceous, coarsely warted, subglobose, 8μ. On the ground. Oct. Rare.

III. No record given in the diagnosis whether cystidia are present or not.

609. **A. plumosa** (Bolt.) Rea. (= *Inocybe plumosa* (Bolt.) Fr.) Boud. Icon. t. 118, as *Inocybe plumosa* (Bolt.) Fr.

Plumosa, feathered.

P. 3–5 cm., *dark fuliginous*, campanulate, then expanded, umbonate, fibrillose, *squamulose with recurved scales especially at the darker disc*. St. 4–6 cm. × 4–8 mm., *concolorous, squamulosely fibrillose*, apex naked. Gills *concolorous*, adnate, fairly wide. Flesh *pallid, concolorous under the pellicle of the pileus and stem*. Spores olivaceous, or fuliginous, angular, 8–12 × 5–7μ. Pine woods. Aug.—Nov. Uncommon.

A. leucocephala (Boud.) Rea. (= *Inocybe leucocephala* Boud.) Massee makes this a synonym for *Astrosporina infida* (Peck) Rea, but this requires confirmation.

A. hiulca (Fr.) Rea. (= *Inocybe hiulca* (Fr.) Bres.) This Astrosporina has erroneously been recorded as British, but the specimens should have been referred to *Inocybe Godeyi* Gill.

***Gills sinuate.

Spores white.

Tricholoma Fr.

(*θρίξ*, hair; *λῶμα*, fringe.)

Pileus fleshy, regular, margin incurved. Stem central, fleshy. Gills sinuate, sinuato-adnate, or decurrent by a tooth. Spores white, rarely pinkish, or yellowish in the mass; elliptical, oval, pip-shaped, globose, subglobose, or oblong; smooth, punctate, verrucose, or echinulate, continuous. Cystidia present, or absent. Growing on the ground, very rarely on wood, sometimes forming large rings.

A. P. viscid, fibrillose, scaly or pubescent.

(*a*) Gills not changing colour.

610. **T. equestre** (Linn.) Fr. Barla, Champ. Alp. Marit. t. 24, figs. 1–12. *Equestre*, belonging to the order of knights.

P. 7·5–12·5 cm., *pale yellowish, sulphur-olive, olivaceous, brick rufescent, disc and innate squamules darker, becoming fuscous*, compactly fleshy, unequal, convex, then plane, very obtuse, flexuoso-repand, viscid. St. 2·5–5 × 1·5–2·5 cm., *sulphur-yellow*, hard, *squamulose*. Gills *sulphur-yellow, or sulphur-olive*, emarginate, or rounded, scarcely adnexed, broad, somewhat ventricose, crowded. Flesh *whitish, yellow under the cuticle of the p.*, thick. Spores white, elliptical, 6 × 3μ. "Cystidia on edge of gill cylindrical-clavate, 30–36 × 10–14μ, filled with yellow juice" Rick. Taste pleasant. Edible. Pine woods. Sept.—Nov. Not uncommon. (*v.v.*)

611. **T. coryphaeum** Fr. Bres. Fung. Trid. t. 76. *κορυφαῖος*, a leader.

P. 5–12 cm., *yellowish, disc darker, punctate with brown squamules*, fleshy, convexo-campanulate, then plane, sometimes broadly umbonate, somewhat viscid, soon dry. St. 5–7 × 1 cm., *white, tinged with yellow in the middle*, apex *white mealy*, base *obclavate, or fusiform and somewhat rooting*. Gills *white with a yellow margin*, often grey at the base, sometimes connected by veins, emarginate, crowded, broad. Flesh *white, citron-yellow under the epidermis*, thick. Spores white, subglobose, 5–6 × 4–5μ, 1-guttulate. Taste somewhat bitter. Beech woods. Sept.—Nov. Rare. (*v.v.*)

612. **T. sejunctum** (Sow.) Fr. Sow. Eng. Fung. t. 126. *Sejunctum*, separated.

P. 7·5–10 cm., *light yellow, streaked with innate fuscous fibrils*, convex, then expanded, gibbous, viscid in wet weather; margin *yellowish, or white, villous*, becoming torn. St. 6–12·5 × 1·5–2·5 cm., *white, tinged with yellow*, ventricose, then elongated, apex *delicately squamulose*. Gills *shining white, sometimes tinged with yellow*, emarginate, broad, *subdistant*. Flesh *white, yellowish under the cuticle of the p. and in the st.*, fragile. Spores white, subglobose, 5–7μ. Smell of new meal. Taste slightly bitter. Mixed woods, especially oak, and pine. Sept.—Nov. Common. (*v.v.*)

613. **T. portentosum** Fr. Fr. Icon. t. 24, upper figs. *Portentosum*, monstrous.

P. 7·5–12·5 cm., *fuliginous, livid, sometimes violaceous, streaked with black, innate fibrils*, fleshy, convexo-plane, subumbonate, unequal, repand, viscid. St. 7·5–15 × 2·5 cm., *white, becoming tinged with sulphur-yellow, or greenish glaucous*, equal, sometimes attenuato-rooted at the villous base, *fibrillosely striate.* Gills *white, becoming straw colour, or glaucous,* rounded, almost free, broad, *distant.* Flesh *white, often tinged with yellow or greenish, very thin at the margin of the p.*, fragile. Spores white, elliptical, 4–5 × 3μ, 1-guttulate. Taste pleasant. Edible. Pine, and fir woods. Sept.—Nov. Uncommon. (*v.v.*)

614. **T. fucatum** Fr. Fr. Icon. t. 24, lower figs. *Fucatum*, painted.

P. 5–10 cm., *lurid yellow, or cinereous light yellow, variegated with bistre, tiger-like spots, disc darker, fleshy,* convexo-plane, obtuse, often irregular, viscid, soon dry. St. 5–7·5 cm. × 8–12 mm., *straw colour, streaked with fibrils that become blackish,* apex *white, pruinose,* fragile, subequal. Gills *whitish,* deeply emarginate, 6–10 mm. broad, *somewhat crowded,* fragile. Flesh *yellowish,* thin, fragile. Spores white, "subglobose, 5–6 × 5μ" Rick. Pine woods. Sept.—Nov. Uncommon.

615. **T. quinquepartitum** Fr. Fr. Icon. t. 25.

Quinque, five; *partitum,* divided.

P. 5–7·5 cm., *pallid light yellow, margin paler,* fleshy, convex, then plane, obtuse, *umbonate,* repand, viscid, fragile. St. 6–10 cm. × 12–15 mm., *white, or whitish,* equal, or attenuated upwards, *striate.* Gills *white,* emarginate, 12 mm. broad, not crowded. Flesh *white,* fragile, thin at the margin. Spores "5–6 × 3–4μ" Sacc. Taste mild. Pine woods. Oct.—Nov. Rare.

616. **T. resplendens** Fr. Fr. Icon. t. 29, upper figs.

Resplendens, shining brightly.

Entirely shining white, becoming yellowish externally and internally. P. 5–10 cm., *silvery shining when dry,* often with hyaline spots, *disc becoming yellow,* fleshy, convex, then flattened, obtuse, viscid; margin *straight,* thin. St. 5–7·5 cm. × 12–15 mm., equal, or subbulbous, sometimes curved, apex *slightly flocculose.* Gills emarginate, somewhat crowded, 6–8 mm. broad, equally attenuated in front. Spores white, elliptical, 8 × 4μ. "Cystidia on edge of gill basidia-like, 30–36 × 8–9μ" Rick. Smell pleasant, taste mild. Beech, and hazel woods. Sept.—Nov. Not uncommon. (*v.v.*)

617. **T. spermaticum** (Paul.) Fr. (= *Tricholoma columbetta* Fr. sec. Quél.) Gillet, Hym. t. 62. *σπέρμα,* semen.

Entirely white. P. 4–9 cm., somewhat fleshy, convex, then flattened, obtuse, repand, viscid, shining when dry, and often becoming yellowish; margin *involute.* St. 5–9 × 1·5–2 cm., *twisted.* Gills emarginate, subdistant, *eroded.* Spores white, elliptical, 6–8 × 5–6μ, verrucose. Smell strong, taste unpleasant. Oak, and fir woods. Sept.—Oct. Uncommon. (*v.v.*)

(*b*) Gills changing colour, generally with reddish spots.

T. colossum Fr. = **Armillaria colossa** (Fr.) Boud.

618. **T. nictitans** Fr. (= *Tricholoma acerbum* Bull. sec. Quél.) Hussey, Illus. Brit. Myc. II, t. 46. *Nictitans*, winking.

P. 5–6 cm., *brownish red, becoming yellow, disc darker*, fleshy, convex, then flattened, obtuse, viscid. St. 7·5 cm. × 12 mm., *pallid light yellow*, equal, or attenuated upwards, base truncate, elastic, *dry, apex squamulose.* Gills *light yellow*, rufescent spotted when old, rounded adnexed, *crowded*, rather broad, subventricose. Flesh *white*, thin. Spores white, "elliptical, 7–8 × 5μ" Massee. Taste pleasant. Mixed woods. Sept.—Oct. Rare.

619. **T. fulvellum** Fr. *Fulvus*, tawny.

P. 2·5–8 cm., *pale yellowish rufescent, or tan colour*, fleshy, convex, then plane, viscid, *disc darker, dotted wrinkled.* St. 4–7·5 × 1 cm., *whitish rufescent, fibrillose*, apex naked. Gills *white, then rufescent*, rounded, then emarginate, crowded. Flesh *often more or less yellow.* Spores white, "subglobose, 4–5μ" Massee. Beech woods. Oct.—Nov. Rare.

620. **T. aurantium** (Schaeff.) Fr. (= *Armillaria aurantia* (Schaeff.) Fr. Hym. Eur.) Fr. Icon. t. 27, upper figs., as *Armillaria aurantia* Schaeff. *Aurantium*, orange coloured.

P. 5–8 cm., *rusty orange colour, disc often darker*, fleshy, convex, then plane, obtuse, viscid, *obsoletely innato-squamulose*, or almost smooth. St. 6–8 × 1–1·5 cm., *concolorous with concentric, orange scales*, apex *white, mealy*, equal, or slightly attenuated upwards. Gills *white, then tinged with rufous*, emarginate, adnexed, crowded, 4 mm. broad. Flesh *white, reddish in the st.* Spores white, elliptical, 6–7 × 4–5μ, 1-guttulate. Smell very pleasant. Taste bitter. Edible. Coniferous woods. Sept.—Nov. Uncommon. (*v.v.*)

621. **T. fulvum** (DC.) Fr. (= *Tricholoma flavo-brunneum* Fr. Hym. Eur.) Fr. Icon. t. 26, upper figs., as *Tricholoma flavo-brunneum* Fr. *Fulvum*, tawny.

P. 7·5–15 cm., *reddish brown, or rufous tawny with a darker disc, fleshy*, conico-convex, then flattened, broadly gibbous, viscid, *streaked with fibrils, or innately squamulose.* St. 7·5–12·5 cm. × 12 mm., *rufescent, or becoming fuscous, streaked with rufous fibrils, attenuated at both ends*, equal only when smaller, *viscid at first*, apex *naked.* Gills *light yellow, spotted rufous when old or touched, emarginate, decurrent by a tooth*, crowded. Flesh *white, sulphur-yellow in the stem.* Spores white, elliptical, 5–6 × 3–4μ, 1-guttulate; "basidia on edge of gill filled with yellowish juice" Rick. Smell of new meal, or rancid. Deciduous woods, and heaths under birches. Sept.—Nov. Common. (*v.v.*)

622. **T. albobrunneum** (Pers.) Fr. (= *Tricholoma striatum* (Schaeff.) Quél.) Barla, Champ. Alp. Marit. t. 27, figs. 7–11.
Albus, white; *brunneum*, brown.

P. 7·5–10 cm., *rich brown, or chestnut*, fleshy, compact only at the *papillose disc*, campanulate, then hemispherical, viscid, *streaked with innate fibrils*; margin incurved, often wrinkled-crenate. St. 4–5 × 2·5–4 cm., *rufescent, constantly white at both ends, equal, dry*, apex *white mealy*. Gills *white, becoming pale, or rufescent*, rounded emarginate, scarcely crowded, very broad, 6 mm. and more wide, firm. Flesh *white, scissile*. Spores white globose, or elliptical, 4–6 × 4μ, 1-guttulate. Taste mild. Edible. Coniferous woods. Sept.—Nov. Common. (*v.v.*)

623. **T. irregulare** Karst. *In*, not; *regulare*, regular.

P. 10 cm., *pale, tinged tawny rufous*, convex, then expanded, *very irregular*, rather viscid, *fibrillosely virgate*. St. 6–9 × 1·5–3 cm., *white*, equal, curved, *apex flocculose*. Gills *white, then reddish, or spotted*, emarginate. Flesh *white*. Spores *white*, subglobose, 3–4 × 2–3μ. Smell strong of meal.

624. **T. ustale** Fr. Gonnerm. & Rabenh. t. 14, fig. 2. *Ustale*, burnt.

P. 3–8 cm., *bay brown rufous, disc darker*, fleshy, *not compact*, hemispherico-expanded, *umbonate at first*, soon plane, obtuse, viscid, *smooth*. St. 5–7 cm. × 12 mm., *rufescent, apex whitish*, equal, somewhat rooted, *dry, fibrillose, often rufous scaly downwards*. Gills *white, at length rufescent*, emarginate, *with a small decurrent tooth, crowded*, rather broad. Flesh *white, becoming reddish in places when broken*. Spores white, subglobose, 5 × 4–5μ, 1-guttulate. Taste bitter. Pine woods, and under larches. Sept.—Oct. Not uncommon. (*v.v.*)

625. **T. pessundatum** Fr. Rolland, Champ. t. 21, no. 39.
Pessundatum, ruined.

P. 6–9 cm., *bay brown, or rufescent, paler or whitish at the margin*, compactly fleshy, convex, then expanded, very obtuse, *granulate, or guttato-spotted*, viscid. St. 5–7·5 × 2·5 cm., *white, covered with small, brownish granules*, bulbous at first, then somewhat equal. Gills *white, then rufescent*, deeply emarginate, somewhat free, crowded. Flesh *white*. Spores white, globose, 5μ, 1-guttulate. Smell of new meal, taste acid, and bitter. Edible. Fir woods, heaths, and pastures. Sept.—Oct. Uncommon. (*v.v.*)

626. **T. stans** Fr. (= *Tricholoma striatum* Schaeff. sec. Quél.) Fr. Icon. t. 28, as *Tricholoma pessundatum* Fr. *Stans*, standing.

P. 7–12 cm., *rufescent*, compact, convex, then flattened, *smooth*, viscid. St. 5–7·5 × 1·5–2 cm., *whitish, tinged rufescent, squamulose*, somewhat equal. Gills *whitish, stained reddish on the margin*, rounded behind, crowded. Flesh *reddish under the cuticle*. Spores white, elliptical, 7–8 × 5μ, 1-guttulate. Smell of new meal. Pine woods. Aug.—Oct. Uncommon. (*v.v.*)

T. russula (Schaeff.) Fr. = **Hygrophorus russula** (Schaeff.) Quél.

T. frumentaceum (Bull.) Fr. = **Hygrophorus russula** (Schaeff.) Quél.

B. P. never viscid; torn into scales, or fibrillose.

(*a*) Gills not changing colour.

627. **T. rutilans** (Schaeff.) Fr. Rolland, Champ. t. 18, no. 32.
Rutilans, becoming *reddish*.

P. 5–20 cm., *yellow, densely covered with granular, purplish downy scales*, fleshy, campanulate, then convex and flattened, often umbonate. St. 6–9 × 1–2·5 cm., *light yellow, besprinkled with purple, squamulose flocci*, bulbous, then ventricose. Gills *sulphur-yellow*, sinuato-adnate, crowded, *broad*; margin thickened, obtuse, *floccose*. Flesh *light yellow, golden when broken*, soft, *thick*. Spores white, globose, or elliptical, 5–8 × 4–7μ. Cystidia "on edge of gill, clavate, or clavate-capitate, rarely fusiform, 60–175 × 20–36μ" Rick. Taste mild. Poisonous. Stumps of conifers. Aug.—Nov. Common. (*v.v.*)

628. **T. variegatum** (Scop.) Fr. Cke. Illus. no. 1117, t. 642.
Variegatum, variegated.

P. 4–6 cm., *yellowish, covered with purple-reddish flocci*, slightly fleshy, fragile, convex, then flattened. St. 5–6 × ·5 cm., *yellowish*, equal, curved, sometimes bulbous, rarely sprinkled with purple reddish flocci, *generally smooth*. Gills *light yellowish white, adnate, narrow*, crowded, *thin*. Flesh *pallid, yellowish* in the st. Spores white, subglobose, 7 × 6μ, with a large central gutta. On rotten wood, and pine stumps. June—Oct. Rare. (*v.v.*)

629. **T. decorum** (Fr.) Quél. (= *Pleurotus decorus* Fr. Hym. Eur., *Clitocybe decora* Fr. Icon. t. 60, upper figs., *Tricholoma rutilans* (Schaeff.) Fr. sec. Dumée). Trans. Brit. Myc. Soc. II, t. 10, as *Pleurotus decorus* Fr. *Decorum*, beautiful.

P. 5–12 cm., *yellow, covered with linear, bistre fibrils that become black with age*, fleshy, brittle, convex, then expanded, or depressed, often excentric. St. 6–10 × 1·5–2 cm., *concolorous, covered with fibrils that blacken with age*, equal, slightly enlarged at the base, often twisted. Gills *golden yellow*, adnate, or sinuato-adnate, 5–10 mm. broad, crowded, often separating from the p. when old, edge uneven. Flesh *pale yellow*, thin. Spores white, elliptical, 6 × 4–5μ, with a large central gutta. Taste bitter. Stumps of conifers. Sept.—Oct. Uncommon. (*v.v.*)

630. **T. centurio** Kalchbr. (= *Tricholoma molybdinum* (Bull.) Quél., *Clitocybe ampla* Pers. sec. Quél.) Kalchbr. Icon. t. 4, fig. 2.
Centurio, captain.

P. 7·5–12·5 cm., *fuscous, then livid*, acorn-shaped, then campanulate, *strongly umbonate*, at length expanded, and somewhat repand, *cuticle breaking up into adpressed fibrils*; *margin splitting*. St. 7–10 × 5 cm., *white*, ventricose, *obsoletely fibrillose*. Gills *pallid*, deeply emarginate, almost free, 6–8 mm. wide, scarcely crowded. Flesh white, thick. Spores white, globose, minute. Taste mild. Pine woods. Sept.—Oct. Rare.

631. **T. luridum** (Schaeff.) Fr. Barla, Champ. Alp. Marit. t. 31, figs. 4–8. *Luridum*, lurid.

P. 10 cm., *lurid, becoming yellow cinereous, sometimes light yellow, disc darker*, convex, then plane, obtuse, somewhat repand, irregularly shaped, *dry*, the cuticle *breaking up into free, brown fibrils*; margin *fimbriate*. St. 5–7·5 × 1–2·5 cm., *whitish or yellowish*, unequal, *fibrillose*. Gills *whitish, or glaucous, then greyish*, emarginate, broad, *very crowded*. Flesh *whitish*, soft, fibrous. Spores white, "elliptical, 12 × 7–8μ, 1-guttulate" Bat. Smell of new meal. Taste mild. Fir woods. Sept.—Oct. Uncommon.

632. **T. guttatum** (Schaeff.) Fr. (= *Tricholoma amarum* (A. & S.) Quél., *Tricholoma gentianeum* Quél.) Barla, Champ. Alp. Marit. t. 31, figs. 9–12. *Guttatum*, spotted.

P. 7·5–12 cm., *cinnamon, or pale yellowish, disc darker*, fleshy, convex, then flattened, *broken up into granular, or floccose squamules*; margin involute at first, *white floccose, remotely sulcate*. St. 5–8 × 1·5 cm., *white*, subequal, mealy. Gills *snow-white*, emarginate, decurrent in the form of lines, very crowded. Flesh *white*. Spores white, elliptical, 4–6 × 3μ. Smell and taste bitter, somewhat acrid. Woods. Oct. Uncommon.

633. **T. truncatum** (Schaeff.) Quél. (= *Hebeloma truncatum* (Schaeff.) Fr.) Boud. Icon. t. 26. *Truncatum*, maimed.

P. 5–10 cm., *somewhat rufous, or yellowish tawny*, convex, then depressed, *undulato-repand, very irregular*, smooth, then minutely floccose; margin *whitish, or flesh colour*, lobed, incurved, *pruinose*. St. 4–9 cm. × 8–10 mm., *white*, equal, firm, *pruinosely tomentose*. Gills *cream colour, then flesh colour*, emarginate with a tooth, *narrow*, often *crisped, branched and anastomosing*. Flesh *white*, firm. Spores *yellowish in the mass*, uncoloured under the microscope, elliptical, 5–8 × 3–4μ. Smell weak. Taste somewhat bitter. Edible. Fir, and deciduous woods. Sept.—Oct. Uncommon. (*v.v.*)

634. **T. psammopum** Kalchbr. Kalchbr. Icon. t. 3, fig. 2. *ψάμμος*, sand; *πούς*, foot.

P. 3–5 cm., *tawny brown*, fleshy, campanulate, then convex, obtuse, or slightly umbonate, *minutely fibrillosely squamulose*, or smooth. St. 5–7 × 1·5 cm., *concolorous*, equal, or ventricose, *attenuated at the base*, *granularly punctate from the breaking up of the epidermis*, apex *white*. Gills *straw, or flesh colour*, emarginate, adnate by a tooth, ventricose, subdistant, 4 mm. broad. Flesh *white, or whitish*, firm. Spores white, subglobose, 4–5 × 3–4μ, 1-guttulate. Taste mild, then slightly bitter. Coniferous woods. Sept.—Oct. Uncommon. (*v.v.*)

635. **T. amarum** (A. & S.) Quél. (= *Clitocybe amara* Fr., *Tricholoma guttatum* Schaeff., and *Tricholoma gentianeum* Quél. sec. Quél.) *Amarum*, bitter.

P. 5–8 cm., *rufescent*, or *becoming fuscous, then becoming pale whitish*, fleshy, pliant, convex, then flattened, obtuse, often unequal, and repand, *fibrillosely-silky*, becoming smooth and *rivulose*; margin *paler, or whitish*, thin, involute at first, *tomentose*. St. 2·5–5 cm. × 12 mm., *white*, equal, attenuated at the white floccose base, tough, elastic, firm, tomentose, often striate. Gills *white*, adnato-decurrent, or sinuate, crowded, thin, 4 mm. broad. Flesh *white*, compact. Spores white, subglobose, 6 × 5μ, 1-guttulate. Smell pleasant. Taste very bitter. Woody places, and hedgerows. Sept.—Nov. Uncommon. (*v.v.*)

636. **T. opiparum** (Fr.) Quél. (= *Clitocybe opipara* Fr.) Fr. Icon. t. 49, upper figs., as *Clitocybe opipara* Fr. var. *major* Fr. *Opiparum*, splendid.

P. 5–10 cm., *yellowish flesh colour, pale yellow, or rosy flesh colour, very fleshy*, convex, then plane, obtuse, delicately flocculose at first, then *very smooth, shining*. St. 4–7 × 1–2 cm., *white, often becoming discoloured*, firm, equal, or slightly attenuated at the base, *smooth*. Gills *white*, adnate, or adnato-decurrent, equally attenuated behind, *crowded*, 6–8 mm. broad, often connected by veins. Flesh *white*, compact. Spores white, elliptical, 7–8 × 3–4μ, 1-guttulate. Smell and taste very pleasant. Edible. Mossy places under trees. Oct.—Nov. Uncommon. (*v.v.*)

637. **T. columbetta** Fr. (= *Hypophyllum spermaticum* Paul. sec. Quél.) Fr. Icon. t. 29, lower figs. *Columba*, a pigeon.

Entirely shining white, occasionally spotted with red or blue. P. 5–10 cm., *disc sometimes greyish*, fleshy, convex, then flattened, obtuse, flexuose, *dry, at first smooth, then silky-fibrillose*, becoming even, or squamulose; margin incurved, *tomentose*. St. 7·5–10 × 2·5 cm., equal, or unequal, sometimes attenuated at the base, *often fibrillosely striate and tinged with bluish green towards the base*. Gills somewhat emarginate, *almost* free, 4–6 mm. broad, often subserrulate. Flesh *white*, thin. Spores white, elliptical, 6–7 × 3–4μ, 1-guttulate. Smell and taste pleasant. Edible. Woods, and pastures. Aug.—Nov. Not uncommon. (*v.v.*)

638. **T. scalpturatum** Fr. (= *Tricholoma argyraceum* (Bull.) Quél.) Bres. Fung. Trid. t. 151. *Scalpturatum*, engraved.

P. 5–7·5 cm., *whitish, or greyish, becoming fuscous with the adpressed scattered, floccose, umber, or rufescent scales*, fleshy, conical, then convex and flattened, often umbonate, *floccose when young*; margin *fibrillosely tomentose*. St. 3–7 cm. × 8–12 mm., *whitish*, equal, either thickened, or attenuated at the base, firm, adpressedly fibrillose. Gills *white, or greyish white, becoming yellow, or spotted with yellow*, emarginate, *almost free*, crowded, ventricose. Flesh *whitish, becoming cinereous*, thin, slightly firm. Spores white, elliptical, 5–7 × 3–4μ, 1–2-guttulate. Fir woods. Sept.—Nov. Uncommon. (*v.v.*)

(*b*) Gills rufescent, or becoming cinereous, edge often spotted red, or black.

639. **T. imbricatum** Fr. Fr. Icon. t. 30.

Imbricatum, covered with tiles.

P. 7·5–10 cm., *rufous umber, torn into squamules except at the disc*, fleshy, *broadly convex, then flattened, and obtuse*, very dry; margin thin, at first inflexed, pubescent, then naked. St. 4–5 × 2·5 cm., *brownish*, ventricose, base either conico-bulbous, or attenuated, *adpressedly fibrillose*, apex *white, mealy*. Flesh *white*, often becoming discoloured, firm, compact, thick. Spores white, subglobose, 4–5 × 3·5μ, with a large central gutta. Taste pleasant. Edible. Coniferous woods. Sept. —Nov. Not uncommon. (*v.v.*)

640. **T. vaccinum** (Pers.) Fr. Cke. Illus. no. 80, t. 60.

Vaccinum, cow-like.

P. 2·5–6 cm., *rufous, torn up into floccose, squarrose, or when smaller, adpressed scales*, fleshy, *campanulate, then expanded, umbonate*, dry; margin *involute, very woolly, tomentose for a short distance on the p. over the base of the gills*. St. 7·5–10 cm. × 8–16 mm., *brownish*, equal, thickened, or attenuated at the base, *fibrillose, cortinate*. Gills *whitish, then spotted rufous, and at length rufescent, slightly sinuate*, almost adnate, subdistant, 6–12 mm. broad. Flesh *white, then reddish*, firm, thick. Spores white, subglobose, 6–8 × 6–7μ. Taste somewhat acrid. Coniferous woods. Aug.—Oct. Common. (*v.v.*)

T. immundum Berk. = **Collybia fumosa** (Pers.) Quél.

641. **T. inodermeum** Fr. Cke. Illus. no. 1120, t. 945.

ἴς, fibre; δέρμα, skin.

P. 2–5 cm., *fuscous becoming reddish, becoming torn up into fibrils, and variegated with radiating scales*, slightly fleshy, *conico-campanulate*, acute, then rather convex and subumbonate. St. 7–8 cm. × 3–5 mm., *whitish, becoming rufescent, firm*, equal, scarcely fibrillose, *apex white mealy*. Gills *white, becoming spotted red when touched*, free, or slightly adnexed, *very broad*, strongly ventricose, almost *semicircular*, *distant*. Flesh *white*, *slightly reddening*, thin, fibrous in the stem. Spores white, elliptical, obtuse at both ends, more rarely with a basal apiculus, 7–9 × 4–5μ, 3–4-guttulate. Coniferous woods. Aug.—Oct. Rare. (*v.v.*)

642. **T. unguentatum** Fr. Fr. Icon. t. 31, upper figs.
Unguentatum, anointed.

P. 5–10 cm., *cinereous*, or *greyish brown*, slightly fleshy, campanulate, then expanded, umbonate, *covered with floccose, viscid squamules*; margin at first inflexed, then spreading, smooth. St. 7–10 × 1–2·5 cm., *grey*, or *white*, equal, firm, laxly fibrillose. Gills *dirty white*, emarginate, crowded, very broad. Flesh *white*, firm. Pine woods. Sept. Rare.

643. **T. gausapatum** Fr. Barla, Champ. Alp. Marit. t. 35, figs. 1–4.
γαυσάπης, a shaggy woollen cloth.

P. 5–7·5 cm., *cinereous grey*, fleshy, somewhat thin, *bullate*, or obtusely campanulate, then expanded, repand, *densely tomentose with superficial, separating, silky, adpressed fibrils*; margin *white*, incurved, *woolly*. St. 5–6 × 1 cm., *shining white*, equal, blunt, *laxly fibrillose*, *cortinate*. Gills *grey*, emarginate, free, crowded, 4 mm. broad, edge often uneven. Flesh *white*, thin. Spores white, elliptical, 6–7 × 4–5μ. Woods, and grassy places. Sept.—Oct. Rare. (*v.v.*)

644. **T. terreum** (Schaeff.) Fr. (= *Tricholoma triste* (Scop.) Quél.)
Terreum, earthy.

P. 4–8 cm., *fuscous, mouse grey, bistre, or blackish*, slightly fleshy, *campanulate, then expanded*, repand when larger, *often umbonate, villous*, for the most part *floccoso-scaly*, sometimes broken up into dark innate, adpressed, fibrils; margin inflexed, naked. St. 5–8 cm. × 8–12 mm., *white, or grey*, equal, *floccosely fibrillose*, apex *white pruinose*. Gills *white, becoming cinereous, emarginate*, subdistant, 6 mm. broad, edge *uneven*. Flesh *whitish grey*, thin, soft, fragile, scissile. Spores white, broadly elliptical, 5–7 × 4–5μ, 1-guttulate. Taste slightly acid. Edible. Coniferous, and beech woods. Jan.—Dec. Very common. (*v.v.*)

var. **atrosquamosum** Chev. Cke. Illus. no. 85, t. 51.
Ater, black; *squama*, a scale.

Differs from the type in the *small black squamules on the p.* and in *the apex of the stem having a few black squamulose points*. Woods. Sept.—Oct. Uncommon. (*v.v.*)

645. **T. argyraceum** (Bull.) Fr. (= *Tricholoma scalpturatum* Fr. sec. Quél.) Barla, Champ. Alp. Marit. t. 36, figs. 14–18.
ἄργυρος, silver.

P. 5–7 cm., *whitish, or pale grey, covered with grey scales and fibrils, disc darker, often speckled with yellow, or bistre*, convex, then plane, subumbonate; margin *white*. St. 4–7 cm. × 8–12 mm., *white, or greyish*, equal, often covered with small, blackish scales. Gills *whitish, often tinged with yellow*, sinuato-adnate, *crowded*. Flesh *greyish*. Spores white, pip-shaped, or almond-shaped, 5–6 × 3·5–4μ, with a large central gutta. Beech, oak, and pine woods. Sept.—Nov. Common. (*v.v.*)

646. **T. chrysites** (Jungh.) Gillet. χρυσίτης, like gold.

P. 4–5 cm., *brownish black, often stained yellowish*, campanulate, then plane, umbonate, covered with floccose scales; margin *yellowish*. St. 4–5 cm. × 8–12 mm., *white, often tinged with yellow*, equal, slightly fibrillose, *base reddish*. Gills *white, stained yellowish, especially at the edge*, sinuato-adnate. Flesh *yellowish*. Spores white, elliptical, 6–7 × 4μ. Pine woods, and under pines. Sept.—Oct. Not uncommon. (*v.v.*)

var. **virescens** Wharton. Cke. Illus. no. 1118, t. 641, as *Tricholoma argyraceum* Fr. var. *virescens* Cke. *Virescens*, becoming green.

Differs from the type in *the p., st., and gills becoming greenish*. Woods. Sept.

647. **T. orirubens** Quél. Quél. Jur. et Vosg. II, t. 1, fig. 2. *Os*, mouth; *rubens*, becoming reddish.

P. 6–8 cm., *grey, disc brownish black*, fragile, convex, then plane, smooth and *pruinose, then shaggy*. St. 5–7 cm. × 8–12 mm., *white, streaked with rose colour, more or less coloured blue, or greenish at the base*, fusiform, fibrillose. Gills *white, edge reddish*, emarginate, distant, undulating. Flesh *white*. Spores white, elliptical, 8 × 6μ. Smell of new meal. Taste unpleasant. Woods. Sept.—Nov. Not uncommon. (*v.v.*)

648. **T. triste** (Scop.) Fr. *Triste*, sad.

P. 3–4 cm., *dark grey, minutely scaly fibrillose*, slightly fleshy, *conico-campanulate, then convex* and umbonate; margin involute, *fibrillose, appendiculate with the white, arachnoid veil*. St. 3–5 cm. × 5–12 mm., *white, with a fuscous ring-like mark, the remains of the arachnoid veil*, floccose, then smooth. Gills *whitish becoming cinereous*, emarginato-adnexed, *narrow, subdistant*. Flesh *whitish*, thin. Spores white, subglobose, 3–4 × 3μ, with a large central gutta. Under beeches. Sept.—Nov. Uncommon. (*v.v.*)

T. ramentaceum (Bull.) Quél. = **Armillaria ramentacea** (Bull.) Fr.

649. **T. murinaceum** (Bull.) Fr. *Murinaceum*, like mice.

P. 5–10 cm., *blackish brown, covered with darker, longitudinal fibres, and cracked*, fleshy, convex, broadly umbonate, or gibbous; margin *dark, incurved*, woolly. St. 5–7 × 2·5 cm., *concolorous, longitudinally streaked with blackish fibrils*, thickened at the base. Gills *white, then grey*, deeply sinuate, *very broad, fragile, often undulated and blackish on the edge, distant.* Flesh *white, then greyish,* thick. Spores white, elliptical, 6–8 × 4–5μ. Smell strong, unpleasant. Taste unpleasant. Mixed woods. Sept.—Nov. Not uncommon. (*v.v.*)

650. **T. squarrulosum** Bres. Trans. Brit. Myc. Soc. II, t. 4.
Squarrulosum, scaly.

P. 4–8 cm., *fuscous, densely covered with squarrose, blackish squamules,* fleshy, convex, then expanded, dry; margin fibrillose, exceeding the gills. St. 4–5 cm. × 5–7 mm., *concolorous,* equal, or incrassated at the base, *densely covered with blackish, fuscous squamules.* Gills *grey, becoming reddish when touched,* sinuato-adnexed, almost free, 6–7 mm. broad, crowded. Flesh *greyish, then white.* Spores white, pip-shaped, 7–8 × 4–5μ, with a large central gutta. In pine, and oak woods. Sept.—Oct. Uncommon. (*v.v.*)

651. **T. horribile** Rea. Trans. Brit. Myc. Soc. II, t. 7.
Horribile, terrible.

P. 10–12 cm., *fuscous,* and *densely covered with dark fuscous squamules, that become squarrose with age especially on the disc, the p. becoming tinged with a pink background at maturity,* fleshy, convex, then expanded; margin thin exceeding the gills. St. 8–9 × 2 cm., *white,* smooth, incrassated at the base. Gills *whitish, then pinkish,* emarginate, *very broad,* 2 cm. wide, crowded, edge uneven. Flesh *white, becoming pinkish, everywhere covered with fuscous spots,* which on the interior of the p. suggest that its squamules have been riveted through the depth of the flesh. Spores white, globose, 5μ. Under beech trees. Oct.—Nov. Uncommon. (*v.v.*)

C. Cuticle of the p. rigid, punctato-granulate, or broken up into glabrous squamules when dry.

(*a*) Gills white, or pallid, not spotted.

652. **T. macrorhizum** (Lasch) Fr. Kalchbr. Icon. t. 3, fig. 1, as *Tricholoma macrocephalum* Schulz. *μακρός*, long; *ῥίζα*, root.

P. 20–30 cm., *ochraceous, then darker,* fleshy, compact, convex, then plane, depressed at the disc, at first smooth, *then broken up in a tesselated manner.* St. 10–15 × 5–6 cm., *whitish, ochraceous downwards,* very minutely granular, *ventricose, prolonged into a thick, blunt, white root, deeply sunk in the soil.* Gills *pallid,* deeply emarginate, almost free, attenuated in front, scarcely crowded, 12–18 mm. broad. Flesh *white,* firm, *becoming light yellow when broken.* Spores white, "irregularly globose, 6μ" Massee. Smell very unpleasant, corpse-like Taste unpleasant. Under oaks in pastures. Sept.—Nov. Rare.

653. **T. compactum** Fr. Fr. Icon. t. 35, upper figs. *Compactum*, compact.

P. 7–15 cm., *cinereous-livid,* fleshy, convex, then plane, smooth, *dry,* glabrous. St. 3–5 × 3 cm., *white,* attenuated upwards, *smooth,* polished. Gills *yellow*, rounded, subdistant. Flesh *white*, spongy, compact, softer in the stem. Spores white, "elliptical, 6–7μ, guttulate" Quél. Woods. Rare.

654. **T. cartilagineum** Fr. non Bull. Fr. Icon. t. 33.

Cartilagineum, cartilaginous.

P. 5–7·5 cm., *blackish, becoming black-dotted from the cuticle breaking up into minute granules*, fleshy, *rigid*, convex then expanded, gibbous, undulated; margin persistently incurved, pubescent at first. St. 2·5–5 × 2·5 cm., *shining white*, firm but fragile, equal, *polished*. Gills *white, then pale grey*, emarginato-sinuate, *crowded*, moderately thin, 4 mm. broad. Flesh *white*, not compact, somewhat fragile. Spores white, elliptical, 5–6 × 4μ, with a large central gutta. Grassy places in woods, pastures, and under pines. Aug.—Nov. Uncommon. (*v.v.*)

655. **T. tenuiceps** Cke. & Massee. Cke. Illus. no. 1121, t. 1166.

Tenuis, thin; *caput*, head.

P. 5–8 cm., *fuliginous*, fleshy at the disc, convex, obtuse, or sometimes slightly gibbous, dry, *granular*. St. 7·5 × 2·5 cm., *ochraceous white*, tough, slightly attenuated upwards, *minutely granular*, base abrupt, *furnished with long, spreading, cord-like mycelium*. Gills *white*, adnexed, rounded behind, attenuated in front, 4 mm. broad, ventricose. Flesh *white*, thin at the margin. Spores white, globose, 6–7μ. Amongst grass under trees. July. Uncommon.

T. loricatum Fr. = **Clitocybe cartilaginea** (Bull.) Bres.

656. **T. atrocinereum** (Pers.) Fr. Fr. Icon. t. 31, lower figs.

Ater, black; *cinereum*, ash-coloured.

P. 3–4 cm., *cinereous, disc prominent, darker*, fleshy, convexo-plane, *smooth*, dry, *becoming rimosely incised and revolute at the margin*. St. 5–7·5 cm. × 8 mm., *whitish*, equal, *slightly striate with longitudinally adpressed fibrils, apex naked*. Gills *hyaline white*, free, or decurrent with a tooth, or arcuato-adnexed, *somewhat ventricose*, thin, crowded. Flesh *hyaline when moist*, hygrophanous, fragile. Spores white, elliptical, 10 × 6–7μ, minutely punctate. Smell of new meal. Grassy ground, and coniferous woods. Aug.—Oct. Uncommon. (*v.v.*)

657. **T. cuneifolium** Fr. Cke. Illus. no. 91, t. 52, fig. B.

Cuneus, a wedge; *folium*, a leaf.

P. 1–2·5 cm., *fuscous, or livid, then grey*, slightly fleshy, convex, then plane, obtuse, at length depressed, *pruinose*. St. 2·5–4 cm. × 4–6 mm., *pallid, attenuated downwards*, rarely equal, *hollow*, somewhat fibrillose, becoming smooth. Gills *white, then greyish, ovate-wedge-shaped*, very much attenuated behind, *very broad in front*, deeply sinuato-adnate, crowded, fragile, often connected by veins. Flesh *white, or greyish,* very thin. Spores white, subglobose, 5–6 × 5μ, minutely punctate, 1-guttulate. Smell of new meal. Taste pleasant. Edible. Pastures, lawns, and heaths. Aug.—Oct. Common. (*v.v.*)

var. **griseo-rimosum** (Batsch) Cke. Cke. Illus. no. 92, t. 261, as var. *cinereo-rimosum* Batsch. *Griseus*, grey; *rimosum*, full of cracks.

Differs from the type in its *larger size, and the concentrically rimose p.* Generally abundant with the type. (*v.v.*)

658. **T. amplum** (Pers.) Rea. (= *Clitocybe ampla* (Pers.) Fr., *Tricholoma molybdinum* (Bull.) Quél., *Tricholoma centurio* Kalchbr. sec. Quél.) Fr. Icon. t. 53, as *Clitocybe ampla* Pers. *Amplum*, large.

P. 10–20 cm., *dark bistre, disc reddish, becoming greyish,* fleshy, campanulate, then expanded, *gibbous,* minutely streaked with fibrils. St. 9–14 × 1·5–2 cm., *white,* equal, thickened at the base, firm, *fibrillosely striate, apex mealy.* Gills *greyish, or yellowish,* sinuato-adnate, 10–12 mm. *broad,* subdistant, margin often toothed. Flesh *white,* compact. Spores white, pip-shaped, 9 × 6μ, 1-guttulate. Caespitose. Woods. Sept.—Nov. Uncommon. (*v.v.*)

(*b*) Gills becoming reddish, or cinereous, or spotted.

659. **T. saponaceum** Fr. Cke. Illus. no. 88. t. 91. *Sapo*, soap.

P. 5–10 cm., *fuscous livid,* fleshy, convex, then flattened, obtuse, smooth, becoming cracked, and broken up into scales in dry weather, somewhat *soapy when moist,* margin thin, inflexed. St. 5–10 × 1·5–2 cm., *pallid, often becoming reddish,* often unequal, curved, base attenuated, often rooting. Gills *white with a glaucous tinge,* becoming spotted with red, uncinato-emarginate, distant, thin. Flesh *white, often becoming reddish,* firm. Spores white, elliptical, 5–6 × 3–4μ, 1-guttulate. Smell soapy. Deciduous, and pine woods. Aug.—Nov. Very common. (*v.v.*)

var. **squamosum** Cke. Cke. Illus. no. 89, t. 216, as var. "*stipite squamuloso.*" *Squamosum*, scaly.

Differs from the type in the *st. being covered with darkish scales.* Woods. (*v.v.*)

var. **sulphurinum** Quél. *Sulphurinum*, sulphur coloured.

Differs from the type in the *canary yellow p.* and *the pale yellow st. and gills.* Woods. Uncommon. (*v.v.*)

var. **atrovirens** (Pers.) Quél. *Ater*, black; *virens*, becoming green.

Differs from the type in the *dark green p. covered with black squamules.* Woods. Sept.—Oct. Uncommon. (*v.v.*)

T. crassifolium Berk. = **Collybia crassifolia** (Berk.) Bres.

660. **T. sudum** Fr. Fr. Icon. t. 34, lower figs. *Sudum*, dry.

P. 6–9 cm., *greyish rufous, or brownish*, fleshy, convexo-plane, then reflexed, obtuse, *often rimosely squamulose*; margin paler. St. 6–8 × 1·5 cm., *whitish, tinged with reddish*, slightly attenuated upwards, *punctate with minute squamules, fibrillosely striate*, apex dilated, base white floccose. Gills *whitish, becoming reddish at the edge*, deeply emarginate with a decurrent tooth, *crowded*. Flesh *white*, firm. Spores white, elliptical, 6 × 4μ, punctate. Grassy places in woods. Sept.—Oct. Rare.

661. **T. tumidum** (Pers.) Fr. Barla, Champ. Alp. Marit. t. 39, figs. 1–5. *Tumidum*, swollen.

P. 7·5–9 cm., *cinereous-livid, disc darker, variegated with tiger-like spots*, subcartilaginous, *irregularly shaped, bullate*, then undulated when expanded, *at length rimosely incised, shining* when dry; margin thin, *lobed*. St. 7·5 × 1·5–2 cm., *shining white*, equal, *sometimes swollen, stout, striate*, often attenuated and rooting at the base. Gills *shining white, then cinereous rufescent*, emarginate, 12 mm. *broad*, thicker at the base, *subdistant*. Flesh *white, often tinged reddish in the st.*, firm, rigid. Spores white, elliptical, 6–7 × 4μ, granular. Smell slight, pleasant. Moist pine woods. Sept.—Nov. Uncommon.

var. **Keithii** Phill. & Plowr. Rev. Dr J. Keith.

Differs from the type in the *cinereo-rufescent p., the dirty white st., with brownish innate fibrils, becoming tinged with red especially near the base, and in often having a strong smell of new meal.* Pine woods. Oct.

662. **T. hordum** Fr. *Hordum*, pregnant.

P. 8 cm., *grey, disc darker, soon cracking, and covered with minute, squarrose scales*, slightly fleshy, campanulato-expanded, then plane, subumbonate, revolute, dry. St. 6–8 × 1–1·5 cm., *whitish, or greyish*, equal, *striate*. Gills *white, becoming cinereous*, emarginate, broad, *subdistant*. Flesh *white*, thin. Spores white, "subglobose, 10μ, with a large central gutta" Quél., "elliptical, 6–7 × 4–5μ" Rick. Taste mild. Beech woods. Sept.—Oct. Uncommon.

663. **T. virgatum** Fr. Fr. Icon. t. 34, upper figs. *Virgatum*, striped.

P. 5–10 cm., *grey-cinereous, umbo often darker*, fleshy, *rigid*, convex, then flattened, subumbonate, very dry, *streaked with fine black innate fibrils, or black squamules*; margin *straight*, naked. St. 7·5–10 × 1–1·5 cm., *whitish, or greyish*, firm, equal, or tuberous at the very base, *striate*, sometimes squamulose. Gills *white, then greyish, becoming hoary*, broadly emarginate, 6–10 mm. broad, crowded. Flesh *greyish white, becoming white in the st.*, thin. Spores white, broadly elliptical, 6–7 × 4–5μ, punctate. Cystidia "on edge of gill clavate, 45 × 10–13μ, filled with darkish juice" Rick. Taste very bitter when young, then mild. Mixed, and pine woods. Aug.—Oct. Not uncommon. (*v.v.*)

664. **T. elytroides** (Scop.) Fr. ἔλυτρον, a cover; εἶδος, like.

P. 6–9 cm., *mouse grey, or brownish black*, slightly fleshy, convex, then plane, obtuse, fragile, very dry, *rough with crowded, erect minute granules*, disc floccosely scaly, becoming hoary when old. St. 7–8 × 1–1·5 cm., *white, becoming cinereous, attenuated downwards, covered with erect, light grey fibrils.* Gills *cinereous, then pruinose,* deeply emarginate, very broad, somewhat thick, fragile, veined at the base. Flesh *white, sometimes becoming reddish,* thick at the disc. Spores white, 6–7 × 4–5μ, or "5–6 × 3·5–4μ" Sacc. Smell weak, of new meal. Grassy places. Oct. Rare.

665. **T. opicum** Fr. *Opicum,* clownish.

P. 2·5–3·5 cm., *grey, or brownish grey,* disc blackish, slightly fleshy, convex, then expanded, *obtusely, or conically umbonate,* at length split, revolute, very dry, smooth, then *minutely squamulose*; margin *often white, or whitish.* St. 4–7·5 × ·5 cm., *pallid, becoming cinereous,* somewhat shining, equal, fibrillose, becoming smooth. Gills *white, or slightly greyish, becoming hoary,* broadly emarginate, ventricose, *somewhat thick, scarcely distant.* Flesh *greyish,* very thin at the margin. Pine woods. Sept.—Nov. Rare.

D. P. at first slightly silky, soon becoming smooth, very dry.

(*a*) Gills broad, rather thick, subdistant.

666. **T. sulphureum** (Bull.) Fr. Cke. Illus. no. 96, t. 62.
Sulphureum, sulphur coloured.

P. 4–8 cm., *sulphur yellow, disc dingy, or rufescent,* fleshy, globose, then convexo-plane, *subumbonate,* at length depressed, unequal, silky, becoming smooth. St. 5–11 × 1 cm., *concolorous,* equal, often curved, *striate.* Gills *sulphur yellow,* adfixed, narrowed behind, arcuato-emarginate, *rather thick, distant.* Flesh *yellow.* Spores white, elliptical, 9–10 × 4–5μ. Smell pleasant, of gas tar. Oak, and mixed woods. Sept.—Nov. Common. (*v.v.*)

667. **T. bufonium** (Pers.) Fr. Kalchbr. Icon. t. 39, fig. 1. *Bufo,* a toad.

P. 3–7 cm., *umber, fuscous tan, or tawny, disc rufescent,* fleshy, convexo-plane, *subumbonate,* silky, becoming smooth, *rugulose.* St. 4–5 cm. × 8–14 mm., *yellow, or tinged with red,* equal, or attenuated at the base, *flocculose.* Gills *yellow tan, pallid,* arcuato-subdecurrent, subdistant, margin often irregular. Flesh *yellow.* Spores white, elliptical, 7–8 × 5μ. Smell weak, of gas tar. Pine woods, and pastures. Sept.—Nov. Uncommon. (*v.v.*)

668. **T. lascivum** Fr. Fr. Icon. t. 38, upper figs. *Lascivum,* wanton.

P. 5–9 cm., *pallid tan, disc darker,* fleshy, convex, then plane, obtuse, at length *somewhat depressed,* delicately silky, then smooth, dry; margin at first involute. St. 7·5–11 × 1 cm., *whitish, becoming discoloured,* rigid, equal, fibrillose, *apex white pruinose*; *base tomentose,* white. Gills *whitish,* arcuato-adnexed, at length arcuato-decurrent, thin, *crowded,* broad. Flesh *white,* compact. Spores white, elliptical, 6–7 × 4–5μ. Smell very pleasant, like that of the flowers of *Syringa vulgaris* according to Quélet. Deciduous woods. Sept.—Nov. Uncommon. (*v.v.*)

var. **robustum** Cke. Cke. Illus. no. 99, t. 217. *Robustus,* strong.

Differs from the type in being *more robust, in the p. being almost white, and in there being scarcely any perceptible odour.* Pastures. May.

669. **T. inamoenum** Fr. Fr. Icon. t. 38, lower figs.

Inamoenum, unpleasant.

P. 3–6 cm., *dingy white,* fleshy, convex, then flattened, *subumbonate,* very dry, slightly silky, then smooth, or cracked. St. 7·5–10 cm. × 6–12 mm., *white,* firm, equal, pruinose, villous. Gills *shining white,* emarginate with a decurrent tooth, or adnato-decurrent, plane, *rather thick,* very broad, *very distant.* Flesh *white,* firm, thick at the disc. Spores white, *elliptical,* 9–10 × 6–7μ, with a large central gutta. Cystidia "clavate, with a prominent point, 33–40 × 9–12μ" Rick. Smell "pleasant, like honeysuckle" Quélet. Fir woods. Sept.—Nov. Uncommon. (*v.v.*)

var. **insigne** Massee. *Insignis,* striking.

Differs from the type in the *decurrent gills.*

670. **T. interveniens** Karst. *Interveniens,* intermediate.

P. 8 cm., *tan, becoming pale,* fleshy, convexo-plane, then depressed, silky, then smooth, somewhat shining; margin involute, *distantly rugulose.* St. 11 × 1 cm., *pallid,* equal, fibrillose, *apex pruinose,* base tomentose, *rooting.* Gills *becoming pale,* arcuato-adnexed, *crowded.* Flesh *watery.* Spores white, oblong, attenuated at both ends, 6–7 × 2–2·5μ. Smell almost none. Mixed woods. Sept. Rare.

(*b*) Gills thin, crowded, narrow. Small, inodorous.

671. **T. cerinum** (Pers.) Fr. Fr. Icon. t. 39, upper figs.

Cerinum, wax coloured.

P. 3–5 cm., *dingy wax colour, or becoming fuscous,* fleshy, convex, then flattened, obtuse, at length depressed, very opaque, very dry, silky, then smooth; *margin often paler.* St. 4–6 cm. × 4–8 mm., *light yellow, often fuscous at the base,* equal, *fibrillosely striate.* Gills *dark yellow, or wax colour,* sinuato-adnexed, separating, horizontal, plane, *very thin, very crowded,* 2 mm. broad. Flesh *white,* thin, firm. Spores white, "2–3 × 2–3μ" Rick. Taste becoming bitter. Fir woods. July—Oct. Rare.

672. **T. fallax** Peck. *Fallax*, deceptive.

P. 2·5 cm., *yellow, disc sometimes rufous*, thin, convex, then expanded, rarely depressed at the centre, moist, smooth. St. 2–5 cm. × 3 mm., *pale yellow*, base sometimes narrowed. Gills *white, then yellowish*, adnexed, rounded behind, crowded. Flesh thin. Spores white, elliptical, 4–5 × 3μ. Under firs. Sept. Rare.

673. **T. onychinum** Fr. Fr. Icon. t. 39, lower figs.
Onychinum, yellowish marble colour.

P. 4–6 cm., *dingy purple, or reddish bay*, fleshy, convexo-plane and umbonate, then expanded and obtuse, very dry, *opaque*; margin *paler*, somewhat silky, narrowly incurved, *striate*. St. 4–6 cm. × 4–8 mm., *pallid, or yellow, apex becoming purple, then lilac, delicately fibrillosely silky*, equal. Gills *dark yellow*, adnexed, *then free*, plane, horizontal, rounded behind, *rather crowded*, 2–4 mm. broad, unequal. Flesh *white, becoming yellow*, thin, firm. Spores white, ovoid, 4–5μ, punctate. Smell and taste pleasant. Under pines, and in fir woods. Sept.—Oct. Rare.

674. **T. ionides** (Bull.) Fr. Boud. Icon. t. 24. *ἴον*, violet; *εἶδος*, like.

P. 3–6 cm., *deep violet, becoming paler*, fleshy, campanulato-convex, then plane, *often umbonate*, becoming smooth; margin incurved, *pruinose*. St. 3–6 cm. × 7–12 mm., *paler violet*, elastic, attenuated downwards, or thickened at the base, *fibrillosely striate*. Gills *white, becoming yellowish*, emarginate with a decurrent tooth, crowded, thin, 6 mm. broad, *edge uneven*. Flesh *white*, tinted violet in the p., and at the base of the st. Spores white, elliptical, 5–6 × 3μ, 1–2-guttulate. Smell and taste pleasant. Edible. Moist woods, especially beech, and open pastures. Aug.—Oct. Uncommon. (*v.v.*)

var. **pravum** (Lasch) Fr. *Pravum*, deformed.

Differs from the type in being *smaller, thinner, and more fragile*, and in *its fuscous-reddish, fuscous-purple, lilac-reddish, or livid colour*. Stoves. July.

675. **T. persicolor** Fr. Boud. Icon. t. 25. *Persicum*, peach; *color*, colour.

P. 1–2 cm., *ochraceous reddish yellow*, convex, then flattened, slightly umbonate, *hygrophanous*, smooth. St. 3–4 cm. × 3–6 mm., *concolorous*, equal, smooth. Gills *whitish, tinged slightly with the colour of the p.*, sinuate, narrow, not crowded. Flesh *concolorous*. Spores white, elliptical, 4–5 × 3–3·5μ. Pastures. Sept. Rare. (*v.v.*)

676. **T. carneum** (Bull.) Fr. (= *Tricholoma carneolum* Fr. sec. Quél.) Fr. Icon. t. 40, fig. 2, as *Tricholoma paeonium*.
Carneum, flesh coloured.

P. 2–3 cm., *red, then flesh colour, becoming pale, and shining*, thin, slightly fleshy, hemispherical, then convex and regular, obtuse, at length flattened and obtuse, often umbonate, *smooth, dry.* St. 2·5 cm. × 4–6 mm., *flesh colour, becoming pale, attenuated downwards, tough, almost cartilaginous,* rigid, apex somewhat pruinose. Gills *shining white, rounded,* somewhat free, horizontal, *very crowded,* wider at the base, 2–3 mm. broad. Flesh *white,* thin. Spores white, elliptical, 5–6 × 2–3μ. Pastures, heaths, and downs. July—Oct. Common. (*v.v.*)

T. carneolum Fr. = **Tricholoma carneum** (Bull.) Fr.

677. **T. caelatum** Fr. Fr. Icon. t. 37, lower figs. *Caelatum,* engraved.

P. 2·5–3 cm., *fuscous, becoming pale grey when dry,* slightly fleshy, convex, *umbilicate,* smooth, *becoming flocculose and rimosely cracked.* St. 2·5–3·5 cm. × 3–5 mm., *concolorous,* tough, elastic, equal, or slightly thickened at the *pruinose apex.* Gills *dingy whitish, or grey,* sinuato-adnate, with a small decurrent tooth, slightly arcuate, crowded. Flesh *subfuscous,* thin. Spores *dingy white,* elliptical, 8 × 4μ. Cystidia "subulate, 40–60 × 5–6μ, filled with olive yellow juice" Rick. Woods, and downs. Uncommon. (*v.v.*)

E. P. fleshy, soft, fragile, spotted, or rivulose. Vernal.

(*a*) Gills whitish.

678. **T. gambosum** Fr. (= *Tricholoma Georgii* (L'Ecluse) Quél.) Barla, Champ. Alp. Marit. t. 41, figs. 1–7. *Gamba,* a hoof.

P. 7·5–11 cm., *ochraceous, or pale tan,* fleshy, *hemispherico-convex, then flattened,* obtuse, undulated, repand, even, smooth, *spotted as with drops, at length widely cracked;* margin at first *involute, tomentose.* St. 5–10 × 1–3 cm., *white, or ochraceous,* firm, *almost equal,* often curved-ascending at the base; *apex white, villous, striate* from the decurrent tooth of the gills. Gills *whitish,* rounded, or emarginato-adnexed, with a subdecurrent tooth, sinuato-decurrent when old, ventricose, 4–6 mm. wide, *crowded.* Flesh *white,* thick, soft, fragile. Spores white, elliptical, 6–7 × 3–4μ, 2-guttulate. Smell and taste strong, of new meal. Edible. Pastures, often forming large rings. April—June. Not uncommon. (*v.v.*)

679. **T. Georgii** (Clus.) Fr. Barla, Champ. Alp. Marit. t. 41, figs. 8–11. *Georgius,* Saint George.

P. 4–6 cm., *ochraceous, then white,* fleshy, convexo-plane, then somewhat repand, often gibbous, dry, *slightly floccose;* margin *naked.* St. 4–6 × 1·5–2 cm., *whitish, or tinged yellowish,* attenuated downwards, fibrillose. Gills *whitish,* attenuato-adnexed, *narrow, crowded.* Flesh *white,* firm, thick. Spores white, elliptical, 5–6 × 3μ. Taste and smell weak, of new meal. Edible. Woods, and open downs. April—June. Not uncommon. (*v.v.*)

680. **T. albellum** Fr. Barla, Champ. Alp. Marit. t. 41, figs. 12–17.
Albellum, whitish.

P. 6–7 cm., *yellowish, then whitish, becoming greyish when dry*, fleshy, *conical, then convex, gibbous* when expanded, moist, *spotted as with scales*; margin thin, naked. St. 5–8 × 1–1·5 cm., *concolorous, fibrillosely striate*, often incrassated at the base. Gills *white, then cream colour*, attenuato-adnexed, *broad in front*, very crowded. Flesh *white*, thick at the disc, soft, floccose. Spores white, elliptical, 5–6 × 3μ, 1-guttulate. Smell and taste weak, of new meal. Edible. Woods, and under yews. April—June. Uncommon. (*v.v.*)

681. **T. boreale** Fr. (= *Tricholoma irinum* (Fr.) Quél.) Fr. Icon. t. 41, upper figs. *Boreale*, northern.

P. 5–7·5 cm., *flesh colour, then whitish tan*, fleshy, very variable in shape, convex, umbonate, *unequal*, often flexuose, smooth, *cracked when dry*; margin thin, involute, naked. St. 5–7·5 × 1 cm., *paler than the p., attenuated downwards*, often twisted and incurved, *apex obsoletely pruinose*. Gills *whitish*, emarginate with a decurrent tooth, thin, crowded. Flesh *white*, soft. Spores white, elliptical, 7–8 × 4–5μ. Smell of new meal. Grassy places in woods. June—July. Uncommon. (*v.v.*)

(*b*) Gills discoloured, rufescent, or smoky.

682. **T. amethystinum** (Scop.) Fr. non Quél.
Amethystinum, amethyst colour.

P. 3–5 cm., *livid, spotted with azure blue*, fleshy, convexo-plane, obtuse, repand, smooth, even, moist; margin paler, wrinkled. St. 3–5 × 1–1·5 cm., *paler than the p.*, attenuated at the base. Gills *white, then rufescent*, subadnate, crowded. Spores white, "ovoid, 7μ long" Sacc. Pine woods. Sept. Rare.

T. tigrinum Fr. Icon. non Schaeff. = **Hygrophorus camarophyllus** (A. & S.) Fr.

683. **T. pes-caprae** Fr. (= *Clitocybe conglobata* (Vitt.) Bres.)
Pes, foot; *capra*, she-goat.

P. 5–10 cm., *grey, becoming fuscous*, variegated, fleshy, *conical*, then expanded, *umbonate*, unequal, *rimosely incised*; margin thin, naked. St. 6–7 × 1 cm., *white*, equal or attenuated downwards, *naked*. Gills *white, becoming cinereous*, emarginate, *very broad*, 12–15 mm., *at length distant*. Flesh thick at the disc. Spores white, "6 × 4" Sacc. Smell of new meal. Edible. Open places under oaks. Oct.—Nov. Rare.

var. **multiforme** (Schaeff.) Cke. Schaeff. Icon. t. 14.
Multiforme, many shaped.

Differs from the type in its *smaller size, and in the irregular thinner p.* Lawns. Nov. Rare.

F. P. compact, then spongy, obtuse, even, smooth, moist, but not hygrophanous.

(*a*) Gills not discoloured.

684. **T. Schumacheri** Fr. Christian Friedrich Schumacher.

P. 5–8 cm., *cinereous livid, disc darker, becoming paler and whitish*, fleshy, convex, then flattened, obtuse, regular, smooth, moist; margin *exceeding the gills*, inflexed. St. 7–10 cm. × 10–12 mm., *white*, equal, villous and sometimes ventricoso-bulbous at the base, *slightly striate*, fibrous. Gills *white, or cream colour*, emarginate, very crowded, plane, 6–8 mm. broad. Flesh *white, pinkish when broken*, spongy, compact. Spores "elliptical, 9 × 6–7μ, minutely punctate, *glaucous*" Quél. Taste mild. Woods, hot-houses. Sept.—Oct. Rare.

685. **T. amicum** Fr. Fr. Icon. t. 36, upper figs. *Amicum*, friendly.

P. 5–9 cm., *fuscous, disc sometimes darker, not becoming paler when dry*, fleshy, convex, then expanded, gibbous, very regular, smooth. St. 7–9 × 1·5 cm., *white*, attenuated upwards from the subbulbous base, fibrillosely striate, elastic. Gills *shining white*, deeply emarginate, almost free, *distant*, 6–8 mm. broad. Flesh *white, brownish under the cuticle*, thick at the disc, rather soft. Spores "elliptical, 6 × 4μ" Massee. Pine woods. Sept.—Oct. Rare.

686. **T. circumtectum** Cke. Cke. Illus. no. 1125, t. 1182.

Circumtectum, covered round.

P. 5–8 cm., *olive, or dusky, disc tawny*, fleshy, convex, very obtusely umbonate, or only obtuse, rarely becoming slightly depressed, cracking slightly when old; margin incurved, *tomentose*, wavy. St. 3–4 × 1–2 cm., *whitish*, firm, obclavate, or sometimes attenuated downwards, *striate*. Gills *white*, slightly sinuate, adnexed, 3–4 mm. broad, scarcely crowded. Flesh white, thick. Spores white, subglobose, 4–5μ. Taste pleasant. Bare ground under trees. Sept. Rare. (*v.v.*)

687. **T. patulum** Fr. Fr. Icon. t. 37, upper figs. *Patulum*, spread out.

P. 4–12 cm., *pallid cinereous*, fleshy, firm, convexo-plane, obtuse, often repand, smooth. St. 4–6 × 1–3 cm., *white*, firm, equal, somewhat elastic, *smooth*. Gills *whitish*, emarginate, almost free, *crowded*, fairly narrow, 4 mm. broad, plane, somewhat veined at the sides. Flesh *white*, thin, not compact. Spores white, elliptical, 8–9 × 4–5μ. Woods and pastures. Sept.—Oct. Uncommon. (*v.v.*)

688. **T. oreinum** Fr. ὀρεινός, hilly.

P. 3–9 cm., *fuscous livid, fuliginous grey, or light bistre, becoming darker*, convex, slightly gibbous, then globose, and plane; margin at first incurved, *prolonged into a narrow membrane along the edge of the gill*. St. 5–7 cm. × 4–12 mm., *white*, rigid, fibrillosely striate, apex *floccosely mealy, base bulbous*. Gills *white*, rounded, then free, crowded, thin, 4–6 mm. broad. Flesh *white, bistre under the cuticle.* Spores white, elliptical, 7–9 × 5–6μ, minutely punctate. Cystidia lanceolate subulate, 25–30 × 4–6μ, apex shaggy, 3μ in diam. Taste pleasant. Edible. Heaths, and downs. Aug.—Oct. Uncommon. (*v.v.*)

689. **T. album** (Schaeff.) Fr. *Album*, white.

Entirely white, or becoming yellowish at the disc. P. 5–12 cm., fleshy, convex, then flattened, becoming plane, obtuse, *very dry*, smooth; margin at first incurved, *floccosely striate over the base of the gills.* St. 7–8 × 1–1·5 cm., *elastic*, attenuated upwards, subbulbous, *apex slightly mealy.* Gills emarginate, somewhat crowded, 8 mm. broad. Flesh *white*, thick, not compact, fibrous in the stem. Spores white, broadly elliptical, 6 × 4–5μ, 1-guttulate. Smell pleasant, taste bitter. Poisonous. Mixed woods. Sept.—Nov. Not uncommon. (*v.v.*)

var. **caesariatum** Fr. *Caesariatum*, covered with hair.

Differs from the type in being *more slender, and in the silky fibrillose p.* Shady beech woods. Sept.—Oct. Uncommon.

690. **T. leucocephalum** Fr. Barla, Champ. Alp. Marit. t. 33, figs. 8–13. λευκός, white; κεφαλή, head.

P. 3–6 cm., *white*, disc sometimes becoming ochraceous, fleshy, thin, tough, convexo-plane, obtuse, or often umbonate, moist, smooth, *when young covered with shining whitish, adpressed silky, at length separating villous down*; margin acute, spreading, smooth. St. 4–6 cm. × 5–10 mm., *white, subcartilaginous*, polished, twisted, smooth, rooting at the attenuated base. Gills *shining white, rounded-free, crowded*, thin. Flesh *watery, becoming whitish*, thin, compact. Spores white, ovoid, 6–7 × 4–6μ, minutely echinulate, 1-guttulate. Smell pleasant, of new meal. Deciduous woods. Aug.—Oct. Uncommon. (*v.v.*)

(*b*) Gills discoloured.

691. **T. acerbum** (Bull.) Fr. (= *Tricholoma nictitans* Fr. sec. Quél.) Barla, Champ. t. 44, figs. 1–5. *Acerbum*, bitter.

P. 7–12 cm., *yellowish buff, becoming rufous at the disc*, fleshy, convexo-expanded, smooth, moist; margin involute, *viscid, tomentosely ribbed over the base of the gills.* St. 5–9 × 2–3 cm., white, becoming yellowish, apex *white, mealy*, base generally bulbous, sometimes attenuated. Gills *pallid*, then rufescent, emarginate, crowded. Flesh *white*, compact, firm. Spores white, globose, 3–4μ. Smell unpleasant, taste bitter. Edible. Woods and downs. Aug.—Oct. Not uncommon. (*v.v.*)

692. **T. luteocitrinum** Rea. Trans. Brit. Myc. Soc. III, t. 8. *Luteus*, yellow; *citrinus*, citron.

P. 2–7 cm., *bright yellow*, fleshy, campanulate, then expanded and gibbous, *floccose, soon breaking up into adpressed, small squamules, which become revolute and fibrillose with age*; margin *citron yellow*, involute. St. 6–7 × 1–2 cm., *whitish, spotted with yellow*, attenuated downwards. Gills *whitish, becoming yellowish*, sinuato-adnate, 4–5 mm. broad, somewhat crowded. Flesh *whitish, citron yellow under the epidermis of the p.*, and dark yellow at the base of the stem. Spores white, elliptical, 6 × 4μ, 1-guttulate. Taste mild. Under larches. Sept.—Oct. Uncommon. (*v.v.*)

693. **T. militare** (Lasch) Fr. Cke. Illus. no. 112, t. 169.
Militare, soldier-like.

P. 10–17 cm., *reddish cinnamon*, fleshy, gibbous, then plane, or depressed, *compact*, flexuous, rather smooth, viscid; margin involute, *white, floccose.* St. 7–12 × 2–2·5 cm., *pallid, becoming stained with red, fibrillosely striate*, base subbulbous. Gills *whitish, or reddish, then livid-spotted*, emarginate, somewhat crowded, torn, margin *eroded.* Flesh *white, brownish under the cuticle of the p. and in the centre of the stem*, thick, compact, firm. Spores white, elliptical, 4–5 × 3μ, 1-guttulate. Smell very strong, taste unpleasant. Woods. Oct. Rare. (*v.v.*)

694. **T. civile** Fr. Fr. Icon. t. 42, upper figs. *Civile*, citizen-like.

P. 6–9 cm., *tawny yellow, becoming paler, disc darker*, fleshy, fragile, convexo-plane, obtuse, very smooth, moist, *pellicle separable.* St. 5–8 × 2–3 cm., *whitish, fragile*, attenuated upwards from the thickened base, *fibrillose, or squamulose.* Gills *whitish, becoming yellowish*, deeply emarginate, *almost free, crowded*, 6 mm. broad, very soft. Flesh *whitish*, soft, fragile, spongy. Spores white, elliptical, or pip-shaped, 5–6 × 3μ. Smell none, or like hay according to Barla. Taste sweet, then astringent, and bitter. Woods. Oct.—Nov. Rare.

695. **T. irinum** Fr. (= *Tricholoma boreale* Fr. sec. Quél.) Trans. Brit. Myc. Soc. IV, t. 8. *ἴρινον*, belonging to the Iris.

P. 5–12 cm., *pale flesh colour*, fleshy, convex, then plane, obtuse, moist, glabrous, sometimes obscurely virgate; margin *white*, incurved, *pruinose.* St. 6–12 × 2–3 cm., *paler than the p.*, subbulbous, *striate, apex whitish, pruinose.* Gills *pale ochre, becoming somewhat lurid with age*, sinuato-adnate, narrow, 5–6 mm. broad, crowded. Flesh *tinted pale pink, then white*, firm, thick, compact. Spores *dirty pink, or yellowish in the mass*, hyaline, elliptical, or pip-shaped, 7–9 × 4–5μ, 1-guttulate. Smell very pleasant, like *Iris*, or *Viola*, according to Quélet. Taste mild. Edible. Pastures, and orchards. Sept.—Oct. Uncommon. (*v.v.*)

696. **T. personatum** Fr. (= *Tricholoma amethystinum* Quél. non Fr.) Cke. Illus. no. 113, t. 66. *Personatum*, masked.

P. 6–15 cm., *tan colour, fuscous, or whitish*, very fleshy, thick, hemispherical, then convex and flattened, very obtuse, regular, at length also repand, smooth, moist in rainy weather, opaque when dry; margin *white*, involute at first, *exceeding the gills*, *pruinose*. St. 5–9 × 2–3 cm., *whitish*, *covered with evanescent*, *blue fibrils*, equal, *base often bulbous*, firm, sometimes very short. Gills *whitish*, *becoming discoloured*, rounded, then free, broad, 10–15 mm., *crowded*. Flesh *greyish when moist*, *white when dry*, compact, then *spongy-soft*, *thick*. Spores white, elliptical, 7 × 5μ, 1-guttulate. Smell and taste pleasant. Edible. Forming large rings in pastures. Sept.—Dec. Common. (*v.v.*)

697. **T. saevum** Gillet. *Saevum*, wild.

P. 6–9 cm., *pale tawny*, *or tan colour*, *sometimes tinted with violet*, compact, fleshy, convex; margin incurved, exceeding the gills. St. 3–5 × 1·5–2 cm., *lilac*, *or violaceous*, *apex paler*, *base white*, slightly squamulose, equal. Gills *dirty white*, *or yellowish white*, emarginate, *narrow*, crowded. Flesh *whitish*, *or slightly tinted with violet*, thick, spongy. Spores white, elliptical, 7 × 5μ. Woods, and pastures. Oct.—Nov. Uncommon. (*v.v.*)

698. **T. glaucocanum** Bres. Bres. Fung. Trid. t. 2.
γλαυκός, bluish grey; *canum*, hoary.

P. 6–9 cm., *bluish grey*, *becoming hoary*, fleshy, somewhat soft, convex, then expanded, smooth, moist; margin involute, *floccosely pruinose*. St. 4–5 × 1·5–2 cm., *concolorous*, *becoming paler*, *base bulbous*, *fibrillosely-striate*, *apex subsquamulose*. Gills *greyish violet*, *becoming hoary*, emarginate, *easily separating from the hymenophore*, *very* crowded. Flesh *pale bluish grey*, soft, thick. Spores white, elliptical, 6–7 × 3μ, 1-guttulate. Smell strong of new meal. Taste pleasant. Edible. Coniferous, and beech woods on calcareous soil. Sept.—Nov. Not uncommon. (*v.v.*)

699. **T. nudum** (Bull.) Fr. Berk. Outl. t. 4, fig. 7. *Nudum*, naked.

P. 7–10 cm., *entirely purple violaceous*, *or with the disc brownish*, *or tawny*, *becoming paler*, fleshy, convexo-plane, then depressed, obtuse, even, moist; margin inflexed, thin, naked. St. 5–9 × 2 cm., *violaceous*, *becoming pale*, *elastic*, *equal*, *apex mealy*. Gills *concolorous*, *becoming pale*, rounded, then decurrent, crowded, *narrow*. Flesh *tinged with violet*, *becoming whitish when dry*, thick at the disc. Spores white, elliptical, 7 × 3–4μ, 1-guttulate. Smell and taste pleasant. Edible. Coniferous, and deciduous woods, rarely in pastures. Sept.—Dec. Common. (*v.v.*)

var. **majus** Cke. *Majus*, larger.

Differs from the type in being *larger*, *and more robust*. Woods. Sept.—Nov. Uncommon. (*v.v.*)

var. **lilaceum** Quél. Quél. Jur. et Vosg. t. 3, fig. 1, as *Tricholoma sordidus* Fr. *Lilaceum*, lilac colour.

Differs from the type in its *small size, and the blue violaceous colour of all its parts.* Spores white, elliptical, 6 × 3μ, 1-guttulate. Woods, and pastures. Sept.—Oct. Uncommon. (*v.v.*)

700. **T. cinerascens** (Bull. non Fr.) Quél. (= *Clitocybe fumosa* Fr. sec. Quél.) Boud. Icon. t. 29. *Cinerascens*, becoming ash-colour.

P. 4–10 cm., *fuliginous fuscous, or livid, becoming grey when dry,* fleshy, *firm*, convex, then plane, gibbous, sometimes slightly umbonate, either circinate and regular, or flexuose and undulated, smooth, somewhat hygrophanous. St. 5–10 cm. × 8–16 mm., *whitish, often stained with brown towards the base*, cylindrical, or attenuated at the base, fibrillose, apex pruinose, often connate at the base, or branched. Gills *greyish white, paler than the p.*, adnate, or emarginate, *crowded.* Flesh *greyish, white when dry*, thick at the disc, firm. Spores white, globose, 5–6μ, punctate. Taste becoming bitter. Woods. Aug.—Oct. Not uncommon. (*v.v.*)

701. **T. panaeolum** Fr. (= *Tricholoma nimbatum* (Batsch) Quél.) Fr. Icon. t. 36, lower figs. *παναίολος*, variegated.

P. 5–10 cm., *cinereous-fuliginous, becoming paler, pruinosely hoary, and spotted*, fleshy, convex, then plane, or here and there depressed, obtuse, often repand and excentric; margin thin, involute, *mealy.* St. 3–7·5 × 1–2 cm., *whitish-grey*, fragile, equal, or attenuated at the base, *fibrillosely striate.* Gills *white, then grey, or dingy rufescent,* emarginate, or rounded, at length decurrent, *very crowded, plane,* 4 mm. broad. Flesh *greyish, becoming whitish, with a horn colour line at the base of the gills*, thick at the disc, soft. Spores *dirty pink* in the mass, elliptical, 4–5 × 3μ. Smell and taste pleasant. Edible. Forming large rings in pastures. Oct.—Nov. Common. (*v.v.*)

var. **caespitosum** Bres. Bres. Fung. Trid. t. 153. *Caespitosum*, caespitose.

Differs from the type in its *caespitose habit, the mealy apex of the st., and the narrower gills.* Hilly pastures, and heaths. Oct. Uncommon. (*v.v.*)

var. **calceolum** (Sterb.) Fr. *Calceolus*, a small shoe.

Differs from the type in the *deformed p., the naked, incised margin, the excentric, short, fusiform st. and the fuliginous gills.* Under hazels. Oct. Rare.

702. **T. cnista** Fr. (? = *Tricholoma melaleucum* (Pers.) Fr. sec. Dumée.) *κνῖσα*, smell of burnt sacrifice.

P. 5–8 cm., *pale tan, or whitish, disc darker*, fleshy, convex, then plane, obtuse, smooth, moist; margin incurved naked. St. 3–5 cm. × 8–10 mm., *white*, tough, equal, smooth. Gills *white, pallid yellow when bruised*, adnexed, rounded behind, inclined to separate from the p., *transversely veined, crisped when dry*. Flesh *white*, rather thick, soft. Spores white, elliptical, 7–10 × 4μ, "roughish" Rick. Smell of cooked flesh. Amongst grass in open places. Sept.—Oct. Rare.

703. **T. duracinum** Cke. Cke. Illus. no. 1126, t. 640.

Durus, hard; *acinum*, berry.

P. 5–7·5 cm., *cinereous*, fleshy, firm, convex, broadly umbonate, dry, smooth, *shining*; margin involute. St. 5–7·5 × 2·5–3 cm., *paler than the p., or greyish white*, attenuated upwards, striate below, *apex reticulately squamose*. Gills *cinereous*, emarginate, arcuate, *narrow*, 2–3 mm. broad, crowded. Flesh *nearly white*, thick, firm. Under cedars. Oct. Rare.

G. P. thin, subumbonate, hygrophanous (*Melanoleuca* Pat.).

(*a*) Gills whitish, unspotted.

704. **T. grammopodium** (Bull.) Fr. (= *Tricholoma melaleucum* (Pers.) Fr. sec. Dumée.) Cke. Illus. no. 118, t. 98. γραμμή, a line; πούς, foot.

P. 7–15 cm., *pallid-livid, or fuscous rufous, whitish when dry*, fleshy, *campanulate, then convex*, at length flattened, *obtusely umbonate*, smooth, moist pellicle separable. St. 7–10 × 1–1·5 cm., *whitish, longitudinally striate with bistre, or brownish fibrils*, elastic, equal, *base thickened*, firm. Gills *whitish, becoming brownish*, arcuato-adnate, or broadly, horizontally emarginate, often acute at both ends, *very crowded*, very many shorter gills, often branched behind. Flesh *bistre when moist, becoming white when dry*, thick at the disc, spongy. Spores white, *elliptical*, 7–8 × 4–6μ, punctate. Smell mouldy. Edible. Forming large rings in pastures and orchards, and solitary, on leaf mould, in deciduous woods. Sept.—Nov. Not uncommon. (*v.v.*)

705. **T. melaleucum** (Pers.) Fr. (= *Tricholoma arcuatum* Fr. sec. Quél.) Fr. Icon. t. 44, upper figs. μέλας, black; λευκόν, white.

P. 4–10 cm., *dark fuliginous when moist, then livid fuscous, paler when dry, umbo blackish*, fleshy, *convex, then flattened*, umbonate, smooth, moist. St. 5–8 cm. × 4–8 mm., *whitish with darker striae, base fuscous*, elastic, thickened at the base, *fibrillosely striate*. Gills *white*, emarginato-adnexed, *horizontal, straight*, broad, more or less ventricose, crowded. Flesh *white, becoming fuliginous*, soft. Spores white, elliptical, 8 × 5μ, warted, apiculate at the base. Cystidia on edge of gill lanceolate subulate, 50–60 × 10–15μ, shaggy at the apex. Taste mild. Edible. Woods, and pastures. Sept.—Nov. Common. (*v.v.*)

var. **adstringens** (Pers.) Quél. *Adstringens*, abridged.

Differs from the type in the *rigid, pitch black p. and in the white gills becoming pinkish*. Pastures, and downs. Sept.—Oct. Not uncommon. (*v.v.*)

706. **T. porphyroleucum** (Bull.) Fr. (= *Tricholoma melaleucum* (Pers.) Fr. sec. Dumée.) Cke. Illus. no. 119, t. 119.

πορφύρα, purple; *λευκόν*, white.

P. 4–6 cm., *fuliginous, or fuscous, becoming rufescent*, with a darker, evanescent umbo, fleshy, firm, convex, then plane and depressed, smooth. St. 3–5 cm. × 6–9 mm., *bistre*, equal, *striate, apex white mealy*. Gills *white, often becoming yellowish*, sinuato-adnate, attenuated in front, *somewhat distant*. Flesh *white, cinereous under the cuticle of the p., bistre in the st.* Spores white, elliptical, 8 × 5μ. Pastures, and open downs. Sept.—Oct. Not uncommon. (*v.v.*)

707. **T. polioleucum** Fr. (= *Tricholoma medium* (Paul.) Quél., *Tricholoma melaleucum* (Pers.) Fr sec. Dumée.) *πολιός*, grey; *λευκόν*, white.

P. 5–7 cm., *grey, umbo darker, whitish at the margin*, slightly fleshy, convex, then plane and depressed, umbo often evanescent, pruinose, margin often scalloped. St. 4–6 × 1 cm., *concolorous*, elastic, *striate*, apex mealy, equal. Gills *whitish, or greyish*, sinuate, decurrent by a tooth, attenuated in front, *edge denticulate*, crowded. Flesh *greyish, becoming white*, thin. Spores white, elliptical, 6–7 × 4μ, rough. Taste pleasant. Edible. Lawns, and hilly pastures. Aug.—Oct. Not uncommon. (*v.v.*)

708. **T. phaeopodium** (Bull.) Quél. (= *Tricholoma melaleucum* (Pers.) Fr. sec. Dumée.) Barla, Champ. Alp. Marit. t. 46, fig. 16.

φαιός, dusky; *πούς*, foot.

P. 3–5 cm., *blackish bistre*, becoming paler, fleshy, convex, then plane, subumbonate, smooth; margin incurved. St. 5–7 cm. × 4–9 mm., *concolorous*, equal, striate; base subbulbous, clothed with the white mycelium. Gills *white*, sinuato-adnate, 7–9 mm. *broad*, crowded. Flesh *dark bistre*, very thin at the margin. Spores white, pip-shaped, 8–9 × 5–6μ, minutely warted. Damp pastures, and woods. Sept.—Oct. Uncommon. (*v.v.*)

709. **T. arcuatum** (Bull.) Quél. (= *Tricholoma arcuatum* var. *cognatum* (Fr.) Quél. and René Maire, *Tricholoma melaleucum* (Pers.) Fr. sec. Dumée.) Gillet, Champ. t. 665. *Arcuatum*, curved.

P. 5–10 cm., *pale brownish, brownish clay colour, or coffee and milk colour*, fleshy, convex, then plane, becoming smooth, moist, hygrophanous; margin at first involute. St. 4–8 cm. × 7–14 mm., *concolorous*, firm, equal, *fibrillose*, base thickened. Gills *yellowish, tinged with pinkish flesh colour*, emarginate, decurrent with a long tooth, broad, moderately crowded. Flesh *white, tinged with flesh colour*, soft, thick at the disc. Spores white, elliptical, 9–10 × 6μ, punctate. Cystidia "on edge of gill lanceolate, 40–60 × 10–12μ" Rick. Taste mild. Heaths, and on twigs, and rotting wood, in coniferous woods. Oct.—Nov. Rare.

710. **T. brevipes** (Bull.) Fr. (= *Tricholoma melaleucum* (Pers.) Fr. sec. Dumée.) Cke. Illus. no. 120, t. 68. *Brevis*, short; *pes*, foot.

P. 5–8 cm., *umber, becoming pale*, fleshy, *convex, then plane*, smooth, moist, opaque when dry; margin at first strongly incurved. St. 1–3 × 1–3 cm., *fuscous, or bistre*, rigid, firm, equal, attenuated downwards, or bulbous, fibrillose, *apex pruinose*. Gills *fuscous, or bistre, becoming whitish*, emarginato-free, ventricose, *crowded*. Flesh *of p. fuscous, becoming white when dry, fuscous in the st., especially at the base*, firm, then soft. Spores white, elliptical, 7–8 × 5–6μ, punctate. Cystidia on edge of gill sparse, lanceolate, 55–65 × 11–14μ, shaggy at the apex. Taste mild. Edible. Pastures, woods, and cinder paths. June—Oct. Uncommon. (*v.v.*)

711. **T. humile** (Pers.) Fr. (= *Tricholoma exscissum* Fr. sec. Quél., *Tricholoma melaleucum* (Pers.) Fr. sec. Dumée.) Cke. Illus. no. 122, t. 263, fig. A. *Humile*, lowly.

P. 5–12 cm., *fuscous-cinereous, mouse grey, or bistre, becoming pale*, fleshy, *convex, soon flattened*, somewhat repand, often umbonate, sometimes depressed, *pruinose, pulverulent*, hygrophanous; margin thin, *exceeding the gills*, often white. St. 4–9 × 1·5–2·5 cm., *white, or becoming greyish*, fragile, *somewhat equal, villosely pulverulent*. Gills *white, then greyish*, rounded-adnexed, decurrent with a tooth, and often arcuato-decurrent, crowded, narrow, 4–6 mm. broad. Flesh *whitish, grey under the cuticle*, soft, thick at the disc. Spores white, elliptical, 9 × 6μ, minutely echinulate. Cystidia "on edge of gill lanceolate, pointed, 55–65 × 10–13μ, shaggy at the apex" Rick. Smell of new meal. Edible. Woods, and pastures. April—Oct. Uncommon. (*v.v.*)

var. *blandum* Berk. = **Tricholoma melaleucum** (Pers.) Fr.

var. **evectum** Grove. *Evectum*, carried out.

P. 7·5–9 cm., *fuscous, becoming pale*, plane, then depressed, or concave, smooth; margin entire, not striate. St. 7·5–8 cm. × 8 mm., *somewhat fuscous*, incrassated at the apex, thickened at the base, fibrous, *punctately squamulose*, striate, *apex white, pulverulent*. Gills *pale ochraceous*, sinuate, crowded, thin, edge entire. Flesh pallid. Spores white, oval, 6–7 × 4·5μ. Amongst heaps of leaves. Sept.—Oct. Uncommon.

712. **T. exscissum** Fr. (= *Tricholoma humile* Pers. sec. Quél.) Fr. Icon. t. 44, lower figs. *Exscissum*, torn out.

P. 3–7 cm., *mouse grey, or fuscous cinereous, becoming paler and often yellowish*, slightly fleshy, *campanulate*, soon plane, *with a prominent umbo*, smooth. St. 2·5–5 cm. × 2–8 mm. *white, then ochraceous, equal, polished*. Gills *white*, emarginate, *linear, narrow*, 2–3 mm.

broad, *edge uneven*. Flesh *white, or yellowish, thin*. Spores white, broadly elliptical, 10 × 6–7μ, punctate. Pastures. May—Sept. Uncommon. (*v.v.*)

713. **T. subpulverulentum** (Pers.) Fr. (= *Tricholoma medium* Paul. sec. Quél.) Hussey, Illus. Brit. Myc. II, t. 39.

Sub, somewhat; *pulverulentum*, dusty.

P. 4–8 cm., *livid, becoming whitish with the innate pruina*, fleshy, convex, then plane, or depressed; margin inflexed, *exceeding the gills*. St. 4–6 × 1 cm., *concolorous*, equal, *smooth, substriate*, apex obsoletely pruinate. Gills *white, becoming darker*, rounded, crowded, narrow. Flesh white, hygrophanous. Spores *very pale ochraceous* in the mass, elliptical, 6–7 × 4μ, minutely punctate. Often forming large rings in pastures, and woods. Aug.—Oct. Uncommon. (*v.v.*)

(*b*) Gills becoming violet, grey, or fuliginous.

714. **T. sordidum** (Schum.) Fr. Barla, Champ. Alp. Marit. t. 47, figs. 10–18. *Sordidum*, dirty.

P. 3–8 cm., *bistre, or livid-lilac, becoming fuscous and pale when old*, somewhat fleshy, campanulato-convex, then plane, or depressed, subumbonate, often undulate, or excentric when old, smooth; margin often slightly striate at maturity. St. 4–6 × 1–2·5 cm., *concolorous*, flexuose, *pliant*, often thickened at the base, *fibrillosely striate*. Gills *violaceous, becoming pale, or fuliginous*, rounded, then sinuato-decurrent, *at length* distant. Flesh *grey, or tinged with lilac*, thin. Spores *pale ochraceous* in the mass, elliptical, 7–8 × 4μ. Taste mild. Edible. Pastures, hedgerows, and manure heaps. July—Nov. Common. (*v.v.*)

715. **T. paedidum** Fr. Fr. Icon. t. 46, upper figs. *Paedidum*, nasty.

P. 3–5 cm., *fuliginous mouse grey*, somewhat fleshy, flaccid, campanulate, then convex, flattened, umbonate, at length *depressed round the conico-prominent umbo*, moist, *radiately streaked with innate fibrils*, becoming smooth. St. 2·5–4 cm. × 4–6 mm., *dingy grey*, subcartilaginous, slightly bulbous at the base, *substriate*. Gills *whitish, then grey*, sinuate with a small decurrent tooth, *narrow*, crowded. Flesh *becoming white*, very thin, very tough. Spores white, "elliptic-fusiform, 10–11 × 5–6μ" Massee. Grassy places in woods. Aug. Rare.

716. **T. lixivium** Fr. (= *Tricholoma arcuatum* (Bull.) Quél.) Fr. Icon. t. 45, lower figs. *Lixivium*, made into lye.

P. 5–10 cm., *cinereous fuscous, then umber*, slightly fleshy, *convex, then plane*, umbonate (the umbo vanishing), sometimes sinuous, smooth; margin *flattened, membranaceous, at length slightly striate*. St. 5–8 × ·5–2 cm., *concolorous, at first white-floccoso-pruinose*, often flexuose, *apex white, fragile*. Gills *grey*, rounded-adnexed, truncato-free, *distant, soft*, 6 mm. broad, sometimes crisped, attenuated from the stem towards the margin. Flesh white, thin at the margin. Spores white, elliptical, 5–6 × 3μ, 1-guttulate. Pine woods, and under conifers. Oct.—Nov. Uncommon. (*v.v.*)

717. **T. putidum** Fr. Barla, Champ. Alp. Marit. t. 47, figs. 19–22. *Putidum*, stinking.

P. 3–6 cm., *greyish bistre, or olivaceous, becoming hoary when dry,* often *sprinkled with white silkiness,* hygrophanous, *slightly fleshy, hemispherical, umbonate,* soft; margin straight. St. 4–9 × ·5–1 cm., *grey,* equal, *covered with a white, thin, evanescent pruina, fibrillosely striate,* somewhat fragile. Gills *cinereous,* sinuato-adnate, ventricose, *crowded,* 4–6 mm. broad. Flesh *bistre,* thin. Spores white, pip-shaped, 9–11 × 4–5μ, with a large central gutta. Smell rancid, of new meal. Fir woods, and amongst pine needles. Oct.—Dec. Uncommon. (*v.v.*)

Spores pink.

Entoloma Fr.

(*ἐντός*, within; *λῶμα*, a fringe.)

Pileus fleshy, regular, or irregular. Stem central, fibrous, or fleshy. Gills sinuate, sinuato-adnate, or adnexed. Spores pink, angular, globose, elliptical, or verrucose. Cystidia rarely present. Growing on the ground, very rarely on wood; solitary, gregarious, or caespitose.

I. Pileus fleshy, smooth, moist or viscid.

718. **E. sinuatum** Fr. (= *Entoloma lividum* (Bull.) Fr. sec. Dumée.) Cke. Illus. no. 316, t. 310. *Sinuatum*, waved.

P. 8–25 cm., *becoming yellowish white,* very fleshy, convex, then expanded, at first gibbous, then depressed; margin repand, sinuate. St. 7–18 × 2·5 cm., *shining white,* firm, equal, compact, *fibrillose,* then smooth. Gills pale *yellowish-rufescent, emarginate,* slightly adnexed, 12–18 mm. broad, *crowded.* Flesh *white.* Spores pink, angular, globose, 8–9μ. Smell strong, pleasant, almost like burnt sugar. Taste pleasant. Poisonous. Gregarious, in mixed woods. July—Oct. Common. (*v.v.*)

719. **E. lividum** (Bull.) Fr. Fr. Icon. t. 90, fig. 3. *Lividus*, livid.

P. 7–15 cm., *livid tan,* becoming pale, fleshy, disc somewhat compact, convex, then plane, somewhat gibbous, *silky,* fibrillose under a lens. St. 7–8 × 2·5–3 cm., *shining white,* equal, *slightly striate,* apex *pruinose.* Gills *whitish, then flesh colour,* rounded, somewhat free, *attenuated in front,* 6–10 mm. broad, *distant.* Flesh *white, brownish under the cuticle of the pileus.* Spores pink, angular, elliptical, 9–11 × 6–8μ, 1–3-guttulate. Smell pleasant, of new meal, then becoming unpleasant. Taste pleasant. Poisonous. Woods, and pastures. April—Oct. Not uncommon. (*v.v.*)

var. **roseum** Cke. Cke. Illus. no. 318, t. 469. *Roseum*, rosy.

Differs from the type in the *rosy disc, and the whitish margin of the p.* On logs.

720. **E. prunuloides** Fr. Fr. Icon. t. 91, fig. 1.
Prunulus, the species *prunulus*; εἶδος, like.

P. 5–7 cm., *whitish, becoming yellow, or livid*, fleshy, *campanulate*, then convex, at length flattened, *subumbonate, viscid*, finally longitudinally rimose; margin at length slightly striate. St. 6–8 cm. × 6–8 mm., *white*, fibrous-fleshy, even, often slightly striate. Gills *white, then flesh colour, somewhat free, emarginate*, rarely rounded, slightly adnexed at first, 6–8 mm. broad, crowded, ventricose. Flesh *white, yellowish in the centre of the stem.* Spores pink, angular, 8–10 × 8μ, 1-multi-guttulate. Smell strong, of new meal. In woods, and pastures. May—Sept. Uncommon. (*v.v.*)

721. **E. porphyrophaeum** Fr. (= *Entoloma phaeocephalum* (Bull.) Quél., *Entoloma placenta* Batsch sec. Quél.) Fr. Icon. t. 93, fig. 1.
πορφύρα, purple; φαιόν, dusky.

P. 4–10 cm., *brownish bistre, becoming paler and mouse colour when dry*, slightly fleshy, campanulate, then expanded, *umbonate*, cracking, *minutely fibrillose.* St. 5–7 cm. × 8–12 mm., *greyish, streaked with violet or lilac fibrils*, base *subbulbose*, white villose, attenuated upwards. Gills *greyish white, then reddish grey*, truncate behind, almost free, ventricose, rather distant, 4–6 mm. broad. Flesh *bluish*, or *brownish, becoming white.* Spores pink, angular globose, 7–8 × 7μ, 1-guttulate. Cystidia "inflated, large, flask-shaped, occasionally with a roundish head" Lange. Woods, and pastures. Aug.—Oct. Uncommon. (*v.v.*)

722. **E. repandum** (Bull.) Fr. Bull. Champ. t. 423, fig. 2.
Repandum, bent backwards.

P. 2·5–5 cm., *whitish, or ochraceous*, fleshy, conical, then expanded, *umbonate, striate with darker spots*, somewhat silky when dry; margin incurved, lobed. St. 4 cm. × 5–6 mm., *white*, silky. Gills *flesh colour*, broader in front. Spores "substellate, 11 × 6 μ, becoming yellow" Sacc. Smell of new meal. Pastures. June—Oct. Rare.

723. **E. erophilum** Fr. ἦρ, spring; φίλον, loving.

P. 3–4 cm., *brownish*, or *greyish*, slightly fleshy, convex, then expanded, obtuse, *venosely striate and virgate.* St. 3–5 cm. × 4–8 mm., *white, becoming greyish*, equal, pruinose. Gills *greyish*, then flesh colour, rounded behind, adnate, wide. Flesh *white, yellowish under the cuticle of the p.* Spores pink, globose, 7–9μ, 1-guttulate. Woods, and hilly pastures. May—Sept. Uncommon. (*v.v.*)

724. **E. placenta** (Batsch) Fr. (= *Entoloma phaeocephalum* (Bull.) Quél.) Cke. Illus. no. 321, t. 314. *Placenta*, a flat cake.

P. 4 cm., *brown*, fleshy, *convex then flattened*, umbonate, orbicular, moist when damp. St. 5–8 cm. × 4–6 mm., *brown*, wholly fibrous, equal, *fibrilloso-striate*. Gills *whitish*, then pallid flesh colour, emarginato-adnexed, crowded, rather thick. Flesh *becoming pale*, thin. Spores pink, globose, 8μ. Taste becoming acrid. Hedgerows, and damp places. April—Oct. Rare.

725. **E. helodes** Fr. Cke. Illus. no. 322, t. 339. ἕλος, a marsh.

P. 3–6 cm., varying *cinereous, fuliginous, and at the same time becoming purple*, slightly fleshy, convex, then *rather plane*, tough, umbonate, often depressed round the umbo, often as if variegated with tiger-spots; margin spreading, sometimes striate. St. 5–7·5 cm. × 6 mm., *cinereous-fibrillose, becoming pallid cinereous*, fragile, equal, or slightly thickened at the base, sometimes twisted. Gills *white*, then flesh colour, *obtusely adnate*, 6 mm. broad, subdistant. Flesh *pallid*, thin. Spores pink, subglobose, 10μ, coarsely warted. Smell of new meal, taste watery. Heathy pastures, and bogs. Sept.—Oct. Uncommon.

726. **E. Batschianum** Fr. Cke. Illus. no. 325, t. 326.
A. J. G. C. Batsch, author of Elenchus Fungorum.

P. 1–4 cm., *dark fuscous*, or *fuliginous black, slightly fleshy, viscid, shining when dry*, slightly convex, scarcely umbonate, then depressed; margin at first manifestly involute. St. 5–7·5 cm. × 4–10 mm., *grey, moderately tough*, wholly fibrous, equal, or attenuated at both ends, slightly striate with adpressed fibrils. Gills *whitish, then cinereous*, or *fuliginous*, narrowed behind, wholly adnexed at the apex, crowded, becoming subdistant. Flesh *yellowish*. Spores pink, angular, subglobose, 6–9 × 6–7μ. On the ground, and among fir needles in damp places. Sept.—Oct. Uncommon. (*v.v.*)

727. **E. Bloxamii** Berk. (= *Entoloma madidum* (Fr.) Quél.) Cke. Illus. no. 326, t. 327. Rev. A. Bloxam, a British mycologist.

P. 2–10 cm., *blackish-blue*, compact, *campanulate*, very obtuse, somewhat lobed, moist, slightly silky. St. 3–8 cm. × 12–15 mm., *concolorous, base white*, obtuse, slightly attenuated upwards, fibrillose. Gills *yellowish*, then pale pink, attenuated behind, or slightly adnexed, moderately broad. Flesh *white, bluish under the cuticle*, very thick in the centre. Spores pink, subglobose, minutely warted, 8–9μ. Heaths, and pastures. Sept.—Nov. Not uncommon. (*v.v.*)

var. **triste** Boud. Boud. Icon. t. 92. *Triste*, gloomy.

Differs from the type in its *blackish grey slate colour*.

728. **E. Farrahii** Massee & Crossland. Naturalist, 1904, t. 1, figs. 1–4. John Farrah, F.L.S.

P. 5–6 cm., *blackish-blue*, cylindrico-ovate, then campanulate and somewhat repand, *umbonate, fibrilloso-silky*; margin pallid. St. 6–7 × 1–1·5 cm., *concolorous*, ventricose, base white. Gills *salmon colour*, sinuato-adnexed, broad, ventricose, somewhat distant. Spores pink, *elliptical*, smooth, 10 × 4–5μ. Cystidia *cuspidate*, ventricose, 50–60 × 12–15μ. Taste mild. Pastures. Sept. Uncommon.

729. **E. ardosiacum** (Bull.) Fr. (= *Entoloma nitidum* Quél.) Cke. Illus. no. 327, t. 328. ἄρδειν, to water.

P. 2–5 cm., *steel-blue-fuscous, becoming blackish when young, and cinereous when older*, slightly fleshy, campanulate, then convex, obtuse, often slightly depressed at the disc. St. 5–8 cm. × 4–6 mm., *steel-blue*, slightly attenuated upwards, easily splitting into fibres. Gills *white, or grey*, then flesh colour, free, broad, ventricose, attenuated behind, crowded, edge uneven. Flesh *white, bluish under the cuticle of the p.* Spores pink, angular, subglobose, 8–10 × 6–8μ, 1-guttulate. Moist meadows, and bogs. July—Sept. Not uncommon. (*v.v.*)

730. **E. madidum** Fr. (= *Entoloma Bloxamii* Berk. sec. Quél.) Fr. Icon. t. 91, fig. 3. *Madidum*, soaked.

P. 2–6 cm., *blackish violet* when young, *fuliginous* when old, fleshy, *campanulate, then convex, viscid* in wet weather, shining when dry; margin inflexed, thin, slightly striate. St. 5–7·5 cm. × 4–10 mm. at the white base, × 4–6 mm. at the apex, *violet, thickened in a clavate manner below*, fleshy fibrous, surface fibrillose, apex naked. Gills *greyish white, slightly adnexed*, almost free, ventricose, watery, soft. Flesh *white, darkish under the cuticle of the pileus*. Spores pink, globose, angular, 6–8μ. Smell *strong, somewhat like that of Russula foetens*. In pastures, and amongst leaves. Sept.—Oct. Uncommon. (*v.v.*)

731. **E. liquescens** Cke. (= *Psilocybe spadiceo-grisea* (Schaeff.) Fr. sec. Boud.) Cke. Illus. no. 328, t. 581. *Liquescens*, dissolving.

P. 5–6 cm., *yellowish ochre, disc reddish brown*, convex, then plane, broadly umbonate, margin thin, flexuose. St. 5–8 cm. × 4–8 mm., *white*, equal, flexible. Gills *white, then pale dirty lilac*, free, crowded, *deliquescent*. Flesh *white*, thin. Spores pink, subglobose, irregular, 7–8μ. On the ground under trees. April.

732. **E. ameides** B. & Br. Cke. Illus. no. 329, t. 341. ἀμειδής, gloomy.

P. 2·5–6 cm., *pale reddish grey, irregular*, broadly campanulate, thin, gibbous, centre polished; margin *white-flocculent*, at length smooth, silky-shining, undulated. St. 3–4 cm. × 5–10 mm., *whitish*, compressed, striate, fibrillose, apex flocculent, base villose. Gills *greyish*, then flesh colour, slightly adnexed, distant, wrinkled. Flesh *becoming reddish*. Spores pink, angular, oblong, 12–13 × 9μ, 1-guttulate. Smell at *first unpleasant*, like a mixture of orange-flower water and starch, or of burnt sugar. Pastures. Aug.—Nov. Not uncommon. (*v.v.*)

E. frumentaceum (Bull.) Berk. = **Hygrophorus russula** (Schaeff.) Quél.

E. Cookei Rich. = **Pleurotus palmatus** (Bull.) Fr.

II. P. absolutely dry, flocculose, or somewhat scaly.

733. **E. Saundersii** Fr. Cke. Illus. no. 331, t. 306.
W. W. Saunders who collaborated with W. G. Smith in Mycological Illustrations.

P. 4–12 cm., *white, becoming fuscous when old*, fleshy, campanulate, then expanded, obtuse, or umbonate, repando-lobed, *adpressedly tomentose*. St. 3–10 × 1–1·5 cm., *white*, equal, silky-fibrous, apex furfuraceous. Gills *reddish*, slightly adnexed, often quite free, *broad*, distant. Flesh *white, yellowish under the cuticle of the pileus*. Spores pink, elliptical, 4–6 × 3–4μ, 1-guttulate ("round, slightly angular, 11–13μ" Boud.). On the ground, river-sand, sawdust. June—Oct. Uncommon. (*v.v.*)

734. **E. fertile** Berk. (= *Entoloma lividum* (Bull.) Fr. sec. Big. & Guill.) Cke. Illus. no. 332, t. 316. *Fertile*, fruitful.

P. 10–15 cm., *pinkish-buff*, fleshy, convex, then plane, obtuse, *pulverulento-squamulose*. St. 6–12 × 2–3 cm., *paler than the p.*, firm, fibrillose, *subsquamulose*, subcompressed, base subbulbose. Gills *flesh colour*, adnexed, nearly free. Flesh *white*, thickest at the disc. Smell pleasant, of new meal. Woods. Aug.—Sept. Rare.

735. **E. Rozei** Quél. Quél. Soc. Bot. XXIII, t. 2, fig. 2.
E. Roze, the eminent French mycologist.

P. 3 cm., *pearl grey, lilac at the margin*, thin, convex, umbonate, then plane, *velvety with thin, very short, white hairs*. St. 5–6 × ·5 cm., *white, somewhat silvery, silky-fibrillose, apex bistre*. Gills *white*, then flesh colour, adnate, emarginate. Spores pink, angular, elliptical, 10μ. Amongst *Sphagnum*, and under pines. July—Oct. Uncommon.

736. **E. jubatum** Fr. Fr. Icon. t. 92, fig. 1.
Jubatum, having a mane.

P. 3–8 cm., *mouse colour*, somewhat fleshy, *campanulate, then expanded* and flattened, umbonate, *floccoso-scaly*, or *fibrillose*. St. 5–12 cm. × 4–10 mm., *concolorous*, fleshy-fibrous, though rigid, fragile, equal, *clothed with fuliginous fibrils*. Gills *dark fuliginous, then purple fuliginous, slightly adnexed*, somewhat emarginate, easily separating, crowded, ventricose. Flesh *white*, thin, easily scissile. Spores pink, angular, oblong, 10–12 × 7–8μ, multi-guttulate. Heaths, and pastures May—Dec. Common. (*v.v.*)

737. **E. resutum** Fr. Fr. Icon. t. 92, fig. 2. *Resutum*, ripped open.

P. 2–3 cm., *becoming fuscous, disc darker*, slightly fleshy, *convex*, somewhat obtuse, *densely floccoso-scaly*, sometimes with darker adpressed scales, sometimes becoming even, longitudinally fibrillose. St. 4–7 cm. × 2–4 mm., *somewhat grey*, wholly fibrous, soft, equal, *polished*, slightly striate. Gills *grey*, at first darker, *adnexed*, very ventricose, almost free, 4 mm. broad, rather crowded, rather thick. Spores pink, "irregular, globose-elliptical, angular, 9–12 × 7–8μ, also subglobose, 7–10μ" Herpell. Woods, and pastures. Oct. Uncommon.

738. **E. griseocyaneum** Fr. Fr. Icon. t. 94, fig. 1.

Griseus, grey; *κύανος*, dark blue.

P. 2–4 cm., *grey, or inclining to lilac*, thin, *campanulate, then convex, obtuse, wholly floccoso-scaly*. St. 4–6 cm. × 4–6 mm., *pallid, then becoming azure-blue*, sometimes white, fibrous, *floccoso-fibrillose*. Gills *whitish*, then flesh colour, *adnexed*, separating-free, ventricose. Flesh *bluish, becoming white*. Spores pink, angular, elliptical, 9–11 × 7–8μ, 1-guttulate. Pastures, downs, and woods. June—Oct. Not uncommon. (*v.v.*)

var. **roseum** Maire. Trans. Brit. Myc. Soc. III, t. 11. *Roseum*, rosy.

Differs from the type in the *pink stem, and the pink pileus covered with darker scales*. Downs, and pastures. Aug.—Oct. Not uncommon. (*v.v.*)

739. **E. Wynnei** B. & Br. Cke. Illus. no. 339, t. 329.

Mrs Lloyd Wynne, of Coed Coch.

P. 3–4 cm., *fuliginous*, convex, then plane, more or less umbonate, often wavy, *velvety, then squamulose*; margin striate, often undulated. St. 3–5 cm. × 3–6 mm., *fuliginous-azure-blue*, often compressed; base cottony, white. Gills *pallid*, then flesh colour, almost free, 4 mm. broad, *transversely ribbed*, edge crenulate. Flesh *white*, very thin except at the disc. Spores "elliptic-oblong, apiculate, coarsely warted, 10–11 × 7–8μ" Massee. Smell *unpleasant, like bugs*. Fir woods. Sept.—Nov. Uncommon.

740. **E. bulbigenum** B. & Br. (= *Entoloma Persoonianum* Phill. & Plowr., *Entoloma Persoonii* Du Port.) Cke. Illus. no. 324, t. 315, as *Entoloma Persoonianum* Du Port.

βολβός, a bulb; *γίγνομαι*, to be born.

P. 1–4 cm., *shining white*, campanulate, or hemispherical, *minutely tomentosely scaly*. St. 6–8 cm. × 3–5 mm., *whitish*, equal, *piloso-*

squamulose, base *bulbous, solid*. Gills *red flesh colour*, slightly adnexed, 3–4 mm. broad. Flesh *white*, thin. Spores pink, angular, 15μ. Grassy places. Feb.—Oct. Uncommon.

741. **E. pulvereum** Rea. Trans. Brit. Myc. Soc. II, t. 14. *Pulvereum*, dusty.

P. 5–30 mm., *fuliginous*, thin, campanulate, then expanded and flattened, *densely covered with very minute scales* which are erect at the centre, striate at first; margin very slightly incurved. St. 2·5–5 cm. × 2–3 mm., *fuliginous*, equal, densely velvety, *covered when young with a reddish, rust coloured meal* which becomes darker with age. Gills *grey*, at length dusted with the pinkish ferruginous spores, adnate with a minute sinus, *veined*, subdistant, *exceeding the margin of the p.*; edge irregular, thick, pale or pinkish at first then deep rose colour. Spores pink, irregular, angular, 12–13 × 6μ, 1-guttulate. Woods, and pastures. Aug.—Nov. Not uncommon. (*v.v.*)

742. **E. dichroum** (Pers.) Fr. Fr. Icon. t. 92, fig. 3. *δίχρουν*, two coloured.

P. 2·5–5 cm., *violet, then livid mouse colour*, somewhat fleshy, campanulate, then expanded and plane, *obtusely umbonate, squamuloso-fibrillose*. St. 3–5 cm. × 2–8 mm., *blue*, becoming paler at the base, wholly fibrous, *fibrillosely-mealy*, and sublacunose. Gills *white*, or *pallid*, then flesh colour, sinuato-adnexed, crowded, 2–3 mm. wide. Flesh *white, tinged with blue*. Spores pink, *becoming fuscous*, "longish, 9–11 × 6–7μ, with a few distinct angles" Rick. Woods, and hilly pastures. June—Sept. Uncommon.

E. sericellum Fr. = **Leptonia sericella** (Fr.) Quél.

743. **E. Thomsonii** B. & Br. Cke. Illus. no. 336, t. 374. Dr Thomson

P. 2·5–3 cm., *grey*, convex, then expanded, more or less umbonate, *tomentose, adorned with raised, radiating ribs, which form reticulations in the centre*. St. 3–5 cm. × 4 mm., *paler than the pileus, fibrillose, tomentose*. Gills *flesh colour*, rounded behind, very slightly adnexed, 4 mm. broad, rather distant. Flesh *mottled*, thick at the umbo, thin elsewhere. Spores pink, elliptical, 6 × 3·5μ. Amongst grass in plantations. Rare.

III. P. thin, hygrophanous, rather silky when dry, often irregular and repand.

744. **E. clypeatum** (Linn.) Fr. Cke. Illus. no. 337, t. 319, as *Entoloma clypeum* Fr. *Clipeatum*, furnished with a shield.

P. 5–8 cm., *lurid* when moist, *grey* when dry, *variegated, or streaked with darker spots, or lines*, fleshy, *campanulate, then flattened*, umbonate, fragile. St. 4–8 × ·5–1·5 cm., *whitish, becoming cinereous*, *wholly fibrous*, equal, fragile, *longitudinally fibrillose*, apex pulverulent. Gills *whitish*, or *dingy*, becoming red-pulverulent with the spores, *rounded-adnexed*, separating free, 4–8 mm. broad, ventricose, subdistant, edge serrulate. Flesh *dark, becoming white*, thin. Spores pink, angular, globose, 7–9 or 8–9 × 6–7μ, 1-guttulate. Taste somewhat acid. Edible. Woods, gardens, pastures, and waste places. Caespitose, or solitary. April—Oct. Common. (*v.v.*)

745. **E. nigrocinnamomeum** Kalchbr. (= *Pluteus umbrosus* (Pers.) Fr. sec. Quél.) Kalchbr. Icon. t. 11, fig. 1.

Niger, black; *cinnamomeum*, cinnamon.

P. 5–7 cm., *umber brown, becoming blackish*, thin, tough, convex, then flattened and *depressed round the somewhat prominent umbo*, rather silky and shining; margin incurved, often splitting. St. 3–4 cm. × 5–8 mm., *grey, becoming tawny*, fibrillose, often twisted. Gills *reddish cinnamon*, adnexed, rounded behind, soon seceding from the stem, 6–8 mm. broad, rather distant. Flesh *darkish, becoming yellowish*. Spores pink, angular, oblong, 11–13 × 7–8μ, 1-guttulate. Smell pleasant, of new meal. Pastures, and heaths. Aug.—Oct. Uncommon. (*v.v.*)

746. **E. rhodopolium** Fr. Cke. Illus. no. 338, t. 342.

ῥόδον, rose; *πολιόν*, grey.

P. 3–12 cm., *fuscous, or livid*, becoming pale, *isabelline-livid, silky-shining when dry*, slightly fleshy, campanulate, then expanded and subumbonate, or gibbous, at length somewhat plane, and sometimes depressed, *fibrillose* when young, then smooth; margin bent inwards, and when larger undulated. St. 5–10 × ·5–1·5 cm., *white*, equal, or attenuated upwards, slightly striate, apex *white pruinose*. Gills *white, then rose colour*, adnate, then separating, somewhat sinuate, *flexuose*. Flesh *white, darkish under the cuticle of the p.* Spores pink, angular, elliptical, 8–9 × 7–8μ, 1-guttulate. Smell like new meal, or burnt sugar, or none. Woods. May—Oct. Common. (*v.v.*)

747. **E. pluteoides** Fr. Fr. Icon. t. 91, fig. 2.

Pluteus, the genus *Pluteus*; *εἶδος*, like.

P. 2·5–8 cm., *whitish grey, becoming dirty yellowish when dry*, slightly fleshy, *scissile*, convex, then expanded, obtuse, *slightly fibrillose at first*, then smooth. St. 5–8 cm. × 4–8 mm., *white, becoming yellow when touched*, rigid, equal, straight, or curved and ascending, *covered with a white, fibrillose, subtomentose cuticle*, base swollen, villose, commonly obliquely and shortly rooting. Gills *white*, then flesh colour, emarginato-adnexed, crowded, attenuated in front. Flesh dark. Spores pink. Taste mild. On fir stumps and rotten wood. Rare.

748. **E. majale** Fr. Fr. Icon. t. 94, fig. 2. *Majus*, the month of May.

P. 4–6 cm., *somewhat cinnamon, ochraceous pale yellow when dry*, fleshy-membranaceous, *scissile, campanulate*, then convex, subumbonate, somewhat fragile; margin repand, easily rimoso-incised. St. 7–10 cm. × 4–6 mm., *whitish*, somewhat compressed, *twisted, striate*, somewhat fibrillose, *often connate* at the thickened, white tomentose base. Gills *pallid*, then flesh-coloured with the rosy spores, free, *ventricose*, crowded, crenate. Flesh very thin. Spores pink, "subspheric, 5–6-angular, 7–10 × 7–8μ" Lange. Pastures, open woods, and mossy places in fir woods. April—Sept. Rare.

749. **E. costatum** Fr. Cke. Illus. no. 340, t. 320, upper figs. *Costatum*, ribbed.

P. 3–8 cm., *fuscous-livid, becoming grey when dry*, or *shining black when scorched by the sun*, fleshy-membranaceous, *convexo-bullate*, undulated, irregularly shaped, then rather plane, more or less *umbilicate*. St. 4–6 cm. × 4–8 mm., *grey*, often compressed, somewhat striate, apex *white mealy*. Gills *pallid, or greyish*, then flesh colour, emarginate, 5–7 mm. broad, *transversely veined with raised ribs, undulate*. Flesh *grey, becoming white*. Spores pink, angular, globose, 6–7μ, or oblong, 9–10 × 7–8μ. Pastures, and heaths, occasionally in woods. May—Nov. Common. (*v.v.*)

750. **E. sericeum** (Bull.) Fr. Cke. Illus. no. 340, t. 320, lower figs. *Sericeum*, silky.

P. 2–4 cm., *umber*, becoming pale *with a silky appearance* when dry, fleshy-membranaceous, convex, then plane, obtuse, often umbonate, somewhat repand; margin involute, striate at first. St. 2·5–5 cm. × 3–6 mm., *grey*, fibroso-fissile, equal, or thickened upwards, *fibrillose*, shining. Gills *grey*, then rufescent, emarginate, slightly adnexed, *equally attenuated from the stem to the margin*. Flesh *umber, becoming whitish*. Spores pink, angular, elliptical, 8–9 × 6μ, 1-guttulate. Smell pleasant, of new meal, or bitter almonds. Pastures, and woods. May—Oct. Common. (*v.v.*)

751. **E. venosum** Gillet. *Venosum*, full of veins.

P. 3–4 cm., *brown, or blackish, disc darker when moist*, thin, convex, slightly umbonate, silky and shining when dry. St. 4–6 cm. × 3–4 mm., *grey, very fragile*, easily splitting, fibrillosely striate, apex *slightly squamulose*. Gills *dirty reddish grey*, free, broad, *covered with prominent, transverse veins*. Spores pink. *Smell strong, of new meal*. Woods, and heaths. Sept.—Oct. Uncommon. (*v.v.*)

752. **E. nidorosum** Fr. Cke. Illus. no. 341, t. 321. *Nidorosum*, reeking.

P. 3–7 cm., *fawn cinereous, livid, silky shining when dry*, submembranaceous, convex, then expanded, at length often concave, and irregularly shaped, rimose. St. 5–13 cm. × 3–15 mm., *pale white*, equal, or attenuated upwards, apex *white pruinose*. Gills *pallid*, then flesh colour, emarginato-free, 6–12 mm. broad, at length distant, sometimes undulato-flexuose. Flesh *white*. Spores pink, angular, subglobose, 8–10 × 7–9μ, 1–3-guttulate. Smell strong, alkaline, or none. Woods, heaths, and lawns. Aug.—Dec. Common. (*v.v.*)

753. **E. speculum** Fr. Fr. Icon. t. 95, fig. 2. *Speculum*, a looking glass.

P. 2–5 cm., *watery, or straw white, silvery when dry*, submembranaceous, convex, soon flattened and depressed, obtusely, and obsoletely umbonate; margin thin, bent inwards, flexuose, pellucid-striate. St. 5–8 cm. × 4–8 mm., yellowish, round, often compressed, shining, very fragile. Gills *white*, then flesh colour, slightly adnexed, broadly emarginate, 6–8 mm. broad, ventricose, the shorter ones narrower, *edge becoming fuscous*. Flesh *brownish*. Spores pink, angular, oblong, 12–14 × 7–8μ, 1–2-guttulate. Woods, and pastures. Sept.—Oct. Not uncommon. (*v.v.*)

754. **E. tortipes** Massee. *Tortus*, twisted; *pes*, foot.

P. 6–7 cm., *dark brown*, and shining as if oiled when moist, then *cinnamon and silky shining when dry*, convex and broadly umbonate, then expanded and depressed round the umbo; margin arched, flexuose, often splitting. St. 5–6 × ·5 cm., *whitish, tinged with cinnamon, flexuose, or angularly bent*, silky-fibrillose. Gills *pale dingy pink*, broadly adnate with a minute sinus, then free, 8–10 mm. broad behind, tapering towards the margin, crowded. Flesh *dark when moist*, paler when dry. Spores pink, *elliptical*, 5 × 3μ. Amongst grass.

Spores ochraceous, or ferruginous.

Hebeloma Fr.

(ἥβη, youth; λῶμα, a fringe.)

Pileus fleshy, regular, margin incurved. Stem central, fibrous, or fleshy. Gills sinuate, sinuato-adnate, or adnexed. Spores ochraceous, ferruginous, or fuscous, elliptical, pip-shaped, pruniform, almond-shaped, elliptic oblong, or fusiform, smooth, continuous. Cystidia present, or absent. Growing on the ground, solitary, caespitose, or subcaespitose.

I. Furnished with a cortina from the manifest veil, by which the p. is often superficially silky round the margin.

755. **H. mussivum** Fr. (= *Cortinarius percomis* Fr. sec. Quél.) Fr. Icon. t. 111, fig. 1. *Musso*, I mutter.

P. 5–10 cm., *yellow, sometimes darker, and brownish at the disc*, fleshy, convex, then plane, unequal, very obtuse, viscid, *smooth at first*, then generally repand and *broken up into squamules*. St. 10–13 × 2·5 cm., *light yellow*, equal, or ventricose, *wholly fibrillose*, *apex pruinose*. Veil fibrillose, very fugacious. Gills *light yellow, then somewhat ferruginous*, emarginate, 6 mm. broad, arid, somewhat crowded. Flesh *becoming yellow*, *compact*, firm, very thick in the st. Spores ferruginous, "elliptical, 12 × 6μ" Massee. Smell weak, not unpleasant. Coniferous woods. Oct. Uncommon. (*v.v.*)

756. **H. sinuosum** Fr. (= *Hebeloma senescens* Batsch sec. Quél.) *Sinuosum*, full of curves.

P. 7·5–15 cm., never equalling the length of the st., *pale yellow, or brick-red becoming pale, rarely clay colour*, fleshy, irregular, convex, then plane, obtuse, smooth, viscid, soon dry, *very sinuosely repand*; margin whitish, membranaceous, inflexed, *exceeding the gills, crenulate, striate*. St. 5–15 × 2·5–3·5 cm., *white*, equal, generally oblique at the very base, fibrillosely striate, *apex at first floccosely squamulose*. Gills *dirty white, then becoming ferruginous*, slightly adnexed, *broadly emarginate and appearing free* and distant, but connected with the st. by a slender tooth, 6–12 mm. *broad*, dry, crowded, often undulate. Flesh *white*, thick, soft, somewhat fragile. Spores ferruginous, "almond-shaped-oval, 10–12 × 7–9μ, slightly rough. Cystidia on edge of gill clavate, 50–60 × 7–9μ" Rick. Smell fruity. Coniferous woods. Aug. Rare.

757. **H. fastibile** Fr. Fr. Icon. t. 111, fig. 2. *Fastibile*, disagreeable.

P. 4–8 cm., *pale yellowish tan, or becoming pale*, fleshy, convexo-plane, obtuse, somewhat repand, smooth; margin involute, *pubescent*. St. 4–8 cm. × 6–12 mm., *white*, or *pallid*, fleshy fibrous, somewhat bulbous, often twisted, *white silky and fibrillose*, white scaly upwards. Cortina *white*, silky, often in the form of a ring. Gills *pale-white, then dingy clay colour, very emarginate*, rather broad, *subdistant, edge whitish, distilling drops in rainy weather*. Flesh *white*, compact. Spores earth colour, pale under the microscope, pip-shaped, 9–11 × 5–6μ, "punctate. Cystidia on edge of gill filamentous-clavate, 50–75 × 6–9μ" Rick. Smell unpleasant, taste of radish, bitterish. Poisonous. Woods. July—Nov. Common. (*v.v.*)

var. **album** Fr. *Album*, white.

Differs from the type in the *long, equal stem, fibrillosely squamose at the apex, and in the distant gills*.

var. **sulcatum** Lindgr. *Sulcatum*, furrowed.

Differs from the type in the *sulcate, or rugose plicate margin of the p*.

var. **elegans** Massee. *Elegans*, nice.

Differs from the type in the *purple-brown p*.

758. **H. senescens** (Batsch) B. & Br. (= *Hebeloma sinuosum* Fr. sec. Quél.) *Senescens*, becoming old.

P. 5–10 cm., *ochraceous tawny*, convex, then flattened, slightly glutinous, delicately tomentose, *margin white*. St. 7–12·5 × 1–2·5 cm., *fuscous downwards, apex shining white, tomentose*, at first bulbous, attenuated upwards, *covered with paler, transversely arranged squamules*. Gills *pallid, then cinnamon*, adnexed, 4–5 mm. broad, crowded. Flesh *white*, thick at the disc. Spores pale ferruginous, pip-shaped, 8–9 × 5μ, with a large central gutta. Smell strong, acrid. Coniferous woods, and under conifers. Sept.—Oct. Uncommon. (*v.v.*)

759. **H. glutinosum** (Lindgr.) Fr. (= *Flammula lenta* (Pers.) Fr. sec. Quél.) Fr. Icon. t. 112, fig. 1. *Glutinosum*, sticky.

P. 5–9 cm., *yellow white, disc darker*, fleshy, convex, then plane, *regular*, obtuse, *glutinous*, slimy in wet weather, *sprinkled with white, superficial, fugacious squamules*. St. 7–9 × 1–1·5 cm., *whitish, becoming ferruginous downwards*, firm, *subbulbous, white squamulose*, fibrillose, apex white mealy, base strigose. Partial veil manifest, fugacious. Gills *pallid, then light yellowish, at length clay cinnamon*, sinuato-adnate, subdecurrent, broad, crowded. Flesh *whitish, becoming light yellow, ferruginous in the st. especially towards the base*, thick at the disc. Spores pale cinnamon, elliptical, 8 × 3–4μ, 1-guttulate, "punctate. Cystidia filamentous, 30–40 × 2–3μ" Rick. Smell not unpleasant, taste mild. Woods, especially oak and beech. Sept.—Dec. Common. (*v.v.*)

760. **H. testaceum** (Batsch) Fr. Cke. Illus. no. 449, t. 408. *Testaceum*, brick red.

P. 3–5 cm., *brick pale, often ochraceous tan, or tan*, somewhat opaque, *whitish at the margin*, fleshy, *campanulato-convex*, then flattened, regular, obtuse, smooth, obsoletely viscid, or dry. St. 3–7·5 cm. × 6–8 mm., *whitish, becoming somewhat tawny at the somewhat thickened, fibrillose base*, somewhat fragile, apex white mealy. Cortina thin, fugacious. Gills *pallid, then clay colour, subferruginous, attenuato-free, lanceolate, very thin*, at first ascending, very crowded. Flesh *whitish, then brownish*, thin at the margin. Spores pale ferruginous, pip-shaped, 8–9 × 4·5–5μ, multi-guttulate, "rough. Cystidia on edge filamentous-clavate, 40–60 × 6–10μ" Rick. Smell faint, of radish. Woods, and heaths. Sept.—Nov. Not uncommon. (*v.v.*)

761. **H. firmum** Fr. Fr. Icon. t. 112, fig. 3. *Firmum*, hard.

P. 5–7 cm., *brick-red*, fleshy, convex, then plane, at length depressed, smooth, *viscid*. St. 5–8 cm. × 4–12 mm., *whitish, brown and fibrillose at the base*, firm, somewhat attenuated downwards, *covered with white, floccose squamules*. Cortina *white*, fugacious. Gills *whitish, then clay colour and ferruginous*, sinuate, thin, *arid, crowded*. Flesh *whitish, then tawny*. Spores "pale dirty colour under the microscope, subfusiform, 9–11 × 4–5μ, minutely punctate. Cystidia only on the edge of the gill, filamentous-clavate, 36–40 × 3–5μ" Rick. Smell faint. Coniferous woods. Feb.—Oct. Rare.

762. **H. claviceps** Fr. Cke. Illus. no. 451, t. 410.
Clavus, a nail; *caput*, head.

P. 2–4 cm., *yellowish white, umbo darker*, fleshy, convex, then expanded, umbonate, or gibbous, viscid, *naked* ("more or less scaly, and fibrillose" Gillet). St. 5–7 cm. × 4–6 mm., *white, brownish at the base*, equal, *white mealy*. Gills *pallid, then ochraceous brownish*, emarginate, arid, crowded. Flesh *pale, yellowish under the cuticle of the p.*, thick at the disc. Spores "reddish under the microscope, almond-shaped, 10–12 × 6–7μ, roughish. Cystidia on edge of gill remarkably long, filamentous-clavate, 75–90 × 6–12μ" Rick. Woods, especially beech. Sept.—Oct. Uncommon.

763. **H. punctatum** Fr. Fr. Icon. t. 113, fig. 1. *Punctatum*, dotted.

P. 2–5 cm., *tan colour, disc umber*, becoming pale when dry, fleshy, convex, soon plane, *disc* obtuse, or gibbous, *viscoso-papillose*, at length depressed at the centre; *margin white, superficially silky with the fibrillose veil*. St. 5–10 cm. × 4–8 mm., *pallid*, equal, flexuose, dry, *silky-fibrillose with the adpressed veil*, apex white pruinose. Cortina *white*, fugacious. Gills *pallid, then pale ferruginous, or brownish*, arcuato-adnate, narrowed behind, slightly ventricose, *narrow*, 4–6 mm. broad, plane, *crowded*. Flesh *white*, thick at the disc. Spores "almond-shaped, 10–12 × 5–6μ, rough, with a thick membrane. Cystidia on edge of gill short, filamentous, 30–36 × 3–4μ, filled with yellow juice" Rick. Smell faint, not unpleasant. Gregarious. Pine woods. Sept. Uncommon.

764. **H. versipelle** Fr. *Versipelle*, changeable in appearance.

P. 2·5–8 cm., *reddish tan, becoming pale at the disc, then dingy tan and opaque when old*, fleshy, convexo-plane, obtuse, *at first covered with a tenacious, glutinous pellicle*, then becoming dry; *margin at first covered with glued white-silky, villose down*, then smooth. St. 5–8 cm. × 4–6 mm., *white, becoming fuscous at the base*, tough, equal, *at first remarkably white-silky with the evident cortina*, then longitudinally fibrillosely striate, and easily splitting up into fibres, *white mealy above the ring formed by the cortina*. Gills *whitish, then clay cinnamon*, rounded, arid, 6–10 mm. broad, crowded. Flesh *whitish, becoming fuscous* in the st. Spores pale ferruginous, elliptical, 8–12 × 6–7μ, 1–2-guttulate, rough. Smell faint, not unpleasant. Subcaespitose. Grassy places, and among fir needles. Aug.—Oct. Uncommon. (*v.v.*)

765. **H. strophosum** Fr. στρόφος, a belt.

P. 2–4 cm., *bay, white and silky from the veil at the margin*, fleshy, fragile, convex, then plane, subumbonate, viscid when wet, then dry. St. 2·5–5 cm. × 6 mm., *whitish, at length becoming fuscous downwards*, equal, fragile, often curved at the base, *clothed with the white silky veil which forms an apical ring. Ring white, floccose, reflexed.* Gills *flesh colour, then clay colour*, slightly adnexed, leaving a wide, bare space at the apex of the st., ventricose, 4–8 mm. broad, plane. Flesh *watery white*, thin at the margin. Spores pale cinnamon, elliptical, 8–9 × 5μ, 1-guttulate. Cystidia "only on the edge of the gill, clavate-filamentous, 40–50 × 5–7μ, thin walled" Rick. Bare soil, and grassy places. Oct. Rare. (*v.v.*)

766. **H. mesophaeum** Fr. Cke. Illus. no. 452, t. 411.
μέσος, middle; φαιόν, dusky.

P. 2·5–4 cm., *pale yellowish, or becoming pale, disc date brown*, fleshy, *conical, then convex*, then plane, or depressed and *darker at the disc*, viscid, smooth. St. 5–7·5 cm. × 4 mm., *whitish, then ferruginous, tough*, equal, fibrillose, *base becoming fuscous*, apex pruinose. Cortina *manifest*, thin, fugacious. Gills *clay ferruginous*, rounded, or emarginate, thin, plane, arid, crowded. Flesh *greyish, becoming white*, thin at the margin. Spores pale ferruginous, elliptical, 9–10 × 5–6μ, with a large central gutta, roughish. Cystidia on edge of gill, "clavate-filamentous, rarely fusiform-filamentous, 50–70 × 5–9μ, thin walled" Rick. Smell faint, sometimes of radish. Taste acrid. Woods, especially coniferous woods, and charcoal heaps. Sept.—Oct. Common. (*v.v.*)

var. **holophaeum** Fr. Fr. Icon. t. 113, fig. 3.
ὅλος, entirely; φαιόν, dusky.

Differs from the type in the *umbonate p. being entirely dark fuscous, in the st. becoming fuscous and subannulate with the cortinate veil, and in the gills being slightly sinuate.*

var. **minus** Cke. Cke. Illus. no. 453, t. 412. *Minus*, smaller.

Differs from the type in its *smaller size*. Common along with the type. (*v.v.*)

767. **H. subcollariatum** B. & Br. Cke. Illus. no. 454, t. 506, as a var.
Sub, somewhat; *collariatum*, having a collar.

P. 2·5 cm., *pallid, disc subfuscous*, fleshy, convex, then more or less expanded, *slightly glutinous*, the floccose veil soon vanishing. St. 5 cm. × 4 mm., *pale, brown at the base*, often flexuose, *pulverulent*. Gills *clay colour, edge shining white*, very slightly rounded behind, broadly adnate, soon separating from the st. and *forming a short interrupted collar*, ventricose, 4–6 mm. broad. Spores pale, ferruginous, elliptical, 12–13 × 6μ, 1-guttulate. On naked soil. May—Oct. Uncommon.

II. P. smooth, at the first with no cortina.

768. **H. sinapizans** (Paul.) Fr. Cke. Illus. no. 455, t. 413.
σίναπι, mustard.

P. 7·5–20 cm., *clay colour, disc sometimes pale yellowish,* fleshy, convex, then plane, very obtuse, for the most part *repand,* often excentric, smooth, slightly viscid when fresh. St. 7·5–12·5 × 2·5 cm., *white, rigid,* equal, or fusiform rooted, *fibrilloso-striate,* apex white squamulose. Gills *clay cinnamon,* opaque, *deeply emarginate,* sometimes arcuato-decurrent, 6–10 mm. broad, crowded, fragile, arid. Flesh *white, compact.* Spores ferruginous, almond-shaped, 10–11 × 6μ. Smell strong, of radish. Woods, and under trees. Sept.—Oct. Uncommon. (*v.v.*)

769. **H. crustuliniforme** (Bull.) Fr. Cke. Illus. no. 456, t. 507.
Crustulum, a small cake; *forma,* shape.

P. 5–10 cm., *pale whitish tan, pale yellowish, or brick colour, disc deeper coloured,* fleshy, convexo-plane, obtuse, or slightly gibbous, rarely repand, smooth, at first viscid. St. 4–7 × 1–2·5 cm., *whitish,* equal, or subbulbous, firm, apex white squamulose. Gills *whitish, then clay colour, at length date brown,* rounded-adnexed, almost adnate, *narrow,* linear, 2–4 mm. broad, crowded; *edge unequal, guttate, distilling watery drops in wet weather, spotted when dry.* Flesh *whitish, watery,* thick. Spores ferruginous, elliptical, 10–12 × 5–6μ, rough. Cystidia "only on the edge of the gill, filamentous-capitate, sometimes also subventricose, 50–60 × 7–10μ, thin walled" Rick. Smell strong of radish, or like the flowers of the common laurel. Taste acrid. Poisonous. Woods, heaths, and pastures. Aug.—Nov. Common. (*v.v.*)

var. **minus** Cke. (= *Hebeloma hiemale* Bres. sec. Big. & Guill.) Cke. Illus. no. 457, t. 414. *Minus,* smaller.

Differs from the type in its *smaller size, in the floccose edge of the gill and in its faint smell.* Woods, and heaths. Oct.—Nov. Not uncommon. (*v.v.*)

770. **H. subsaponaceum** Karst. Karst. Icon. t. 44.
Sub, somewhat; *sapo,* soap.

P. 3 cm., *gilvous, then pallid, darker when dry,* expanded, obtuse, naked, *dry.* St. 2–3 cm. × 3–4 mm., *pale, becoming umber below when touched,* equal, rather wavy, adpressedly fibrillose, apex rather mealy. Gills pale alutaceous, then ferruginous, adnate, 2 mm. broad, crowded, dry. Spores oval oblong, 6–10 × 4–6μ. Smell strong, of soap. Fir, and mixed woods. Aug. Uncommon.

771. **H. elatum** (Batsch) Fr. Cke. Illus. no. 1165, t. 962. *Elatum,* tall.

P. 2·5–8 cm., *tan colour, becoming pale tan when dry,* fleshy, convex, then flattened, obtuse, smooth, slightly viscid, opaque, margin very thin. St. 8–10 cm. × 8–12 mm., *whitish,* equal, *cylindrical,* tense and straight, *twisted with spiral fibres,* base with an ovately, villose bulb when growing amongst fir leaves, *adpressedly fibrillose,* apex white mealy. Gills *pale cinnamon,* rounded, with a small decurrent tooth, 6 mm. broad, crowded. Flesh *whitish,* thick at the disc. Spores brownish ferruginous, elliptical, 8–10 × 5μ. Smell very strong of radish. Woods, especially coniferous woods, and heaths. Sept.—Nov. Uncommon. (*v.v.*)

772. **H. longicaudum** (Pers.) Fr. Cke. Illus. no. 458, t. 415.

Longus, long; *cauda,* a tail.

P. 4–12 cm., *clay colour, becoming whitish, disc sometimes becoming fuscous,* fleshy, convex, then expanded, *umbonate,* at length repand, smooth, viscid; margin *whitish, pruinose.* St. 8–11 × 1–2 cm., *white, at length becoming tawny at the base, fragile,* equal, or thickened at the base, *obsoletely fibrillose,* apex mealy. Gills *white clay, then cinnamon,* arcuato-adnate, 6 mm. broad, crowded, *edge serrulate, somewhat dotted.* Flesh *white,* soft, watery, thin at the margin. Spores tawny ochraceous, oblong elliptical, 11–12 × 6μ, minutely verrucose. Cystidia "on edge of gill long, filamentous-clavate, 60–90 × 6–8μ" Rick. Smell faint, not unpleasant. Taste mild. Edible. Woods, especially coniferous woods. Sept.—Nov. Not uncommon. (*v.v.*)

773. **H. radicatum** (Cke.) Maire. (= *Hebeloma elatum* Quél. non Batsch sec. Maire.) Cke. Illus. no. 459, t. 416, as *Hebeloma longicaudum* Fr. var. *radicatum* Cke. *Radicatum,* rooting.

P. 3–10 cm., *ochraceous flesh colour, whitish at the margin,* convex, gibbous, glutinous. St. 7–11 × ·5–1 cm., *white, becoming brownish, fusiform, rooting,* soft, *twisted,* fibrillose, villose, mealy at the apex. Gills *pale flesh colour, then brownish,* adnate by a tooth, undulate. Flesh *white,* thick, soft. Spores fuscous ferruginous, elliptical, 10 × 5μ. *Smell of radish, or of honey, like that of Pholiota radicosa. Taste bitter.* Caespitose. Coniferous woods. Sept. Rare. (*v.v.*)

774. **H. lugens** (Jungh.) Fr. *Lugens,* mourning.

P. 4–7·5 cm., *brown, becoming somewhat yellow,* fleshy, convex, then plane, gibbous, often repand and irregular, smooth, subviscid. St. 4–7·5 cm. × 6–8 mm., *white, shining,* subbulbous, *fibrillosely striate,* apex white mealy. Gills *pallid, then ferruginous,* somewhat free, fragile, 4 mm. broad, *edge crenulate, darker.* Flesh *white.* Spores "10 × 6μ" Massee. Smell strong. In troops. Woods, especially beech. Sept.—Oct. Rare.

H. truncatum (Schaeff.) Fr. = **Tricholoma truncatum** (Schaeff.) Quél.

775. **H. nudipes** Fr. Kalchbr. Icon. t. 14, fig. 3.

Nudus, naked; *pes,* foot.

P. 4–7 cm., *pale tan, or clay colour,* fleshy, convex, then flattened, obtuse, unequal, *smooth,* slightly viscid, scarcely perceptibly streaked;

margin *membranaceous, exceeding the gills.* St. 5–8 cm. × 8–16 mm., *white,* equal, fibrillose at the base, *smooth above,* straight, or curved and ascending, *pellicle separable.* Gills *tan colour,* broadly emarginate, crowded, dry. Flesh *watery, white when dry,* compact at the disc, *very thin at the margin.* Spores fuscous, elliptical, 12 × 6μ. Smell weak, not unpleasant. Taste mild. Edible. Woods. Oct. Uncommon. (*v.v.*)

776. **H. sacchariolens** Quél. Quél. Soc. sc. n. de Rouen, 1879, t. 1, fig. 2. *Saccharum,* sugar; *olens,* smelling.

P. 2–6 cm., *whitish, disc buff colour, but the whole surface deepens in colour with age,* fleshy, campanulate, then convex, smooth, viscid. St. 4–5 × 1 cm., *white, streaked with fawn fibrils below,* attenuated above and at the base, *striate, silky,* apex pruinose. Gills *whitish, then buff colour, and finally ferruginous,* sinuato-adnate, 6–10 mm. *broad,* crenate, *edge whitish.* Flesh *yellowish, or somewhat buff,* thin at the margin. Spores deep ferruginous, almond-shaped, 10–11 × 7–8μ, with a hyaline basal apiculus. *Smell very peculiar and strong, like that of Entoloma ameides or "of burnt sugar, or orange flowers"* Quél. Woods, heaths, and pastures. Oct. Uncommon. (*v.v.*)

777. **H. nauseosum** Cke. Cke. Illus. no. 1166, t. 963. *ναυσία,* sea-sickness.

P. 2·5–3·5 cm., *ochrey-white,* fleshy, convex, gibbous, more or less expanded, smooth, viscid. St. 3–5 cm. × 6–10 mm., *concolorous, in decay turning black at the base,* equal, or attenuated at the base, faintly striate downwards, mealy above. Gills *pallid, then clay colour, and at length ferruginous,* sinuate, ventricose, 6–10 mm. broad, subdistant. Flesh *white,* thick at the disc. Spores pale ferruginous, elliptical-fusiform (20 × 10μ Cke.), 12 × 6μ, 1–2-guttulate. *Smell very unpleasant.* Woods, and heaths. Sept.—Oct. Uncommon. (*v.v.*)

778. **H. ischnostylum** Cke. Cke. Illus. no. 463, t. 420. *ἰσχνός,* thin; *στῦλος,* a pillar.

P. 2·5–5 cm., *white, or a little pallid at the disc,* fleshy, convex, then expanded, broadly umbonate, slightly viscid. St. 5 cm. × 3–4 mm., *whitish,* equal, or a little thickened at the base, *smooth.* Gills *whitish, then clay colour,* rounded behind, adnexed, *edge slightly serrate.* Flesh *white,* thin at the margin. Spores fuscous, elliptical, 12 × 7·5μ. Smell none, or with a faint odour of *Spiraea.* Amongst grass under alders. Sept. Uncommon.

779. **H. capniocephalum** (Bull.) Fr. Cke. Illus. no. 462, t. 419. *καπνός,* smoke; *κεφαλή,* head.

P. 4–7·5 cm., *pale yellowish, or reddish, disc darker, margin at length becoming black,* fleshy, convex, then plane, obtuse, *smooth.* St. 5–6 cm. × 4–6 mm., *whitish, attenuated downwards, striate with rufescent fibrils*, becoming pale. Gills *ferruginous*, emarginate, broad, scarcely crowded. Flesh *whitish*, thin at the margin. Spores fuscous, elliptical, 9 × 5μ. "Smell of musk" Secrétan. Mixed woods, and under firs, and pines. Oct.—Nov. Rare.

780. **H. diffractum** Fr. Fr. Icon. t. 114, fig. 1.

Diffractum, broken in pieces.

P. 3–5 cm., *tan colour*, fleshy, thin, convex, then plane, obtuse, smooth, somewhat dry, *at length squamosely broken up*. St. 2·5–4 cm. × 6–8 mm., *white*, subfusiform, *attenuated downwards*, *hollow*, squamulose with white flocci above. Gills *pallid*, *then ferruginous*, emarginate, ventricose, broad, crowded, dry. Spores ferruginous, "10–12 × 4–5μ" Massee. Smell weak, unpleasant, somewhat of radish. Pine needles in woods. Sept. Rare.

III. P. scarcely 2·5 cm. broad. Stature that of the Naucoriae.

781. **H. magnimamma** Fr. Fr. Icon. t. 114, fig. 2.

Magnus, large; *mamma*, breast.

P. 1–2·5 cm., *brick colour*, *at length becoming pale yellowish*, fleshy, convex, then plane, *with a prominent*, *breast-shaped umbo*, smooth, but becoming somewhat streaky towards the margin. St. 2·5–7 cm. × 2–3 mm., *pale yellowish*, *becoming pale*, equal, or flexuose, often substriate, fibrous, smooth. Gills *pallid*, *then ferruginous*, sinuato-adnate, 2–3 mm. broad, crowded. Flesh *yellowish*, *tawny under the cuticle of the p.*, very thin at the margin. Spores pale ferruginous, broadly elliptical, 9–12 × 5–6μ, with a hyaline basal, or subbasal, apiculus, multi-guttulate. Pastures, under oaks, and apple trees. Sept.—Oct. Uncommon. (*v.v.*)

H. petiginosum Fr. = **Astrosporina petiginosa** (Fr.) Rea.

Spores purple, or fuscous.

Hypholoma Fr.

(ὑφή, a web; λῶμα, a fringe.)

Pileus fleshy, regular, firm, or fragile. Stem central, fibrous, or fleshy. Gills sinuate, sinuato-adnate, or adnexed. Spores purple, fuscous, or cinereous purple; elliptical, subglobose, pip-shaped, or reniform, smooth, rarely verrucose, with an apical germ-pore. Cystidia present, or absent. Growing on wood, more rarely on the ground, often densely caespitose, or fasciculate.

*Colour of the tough, smooth, dry (except *Hypholoma silaceum*) p. bright, not hygrophanous.

782. **H. silaceum** (Pers.) Fr. (= *Flammuloides sublateritia* Schaeff. sec. Quél.)

Silaceum, ochraceous.

P. 6–8 cm., *orange rufous*, fleshy, convex, *viscid*; margin *whitish*, silky. St. 7–10 cm. × 6–8 mm., *concolorous*, *bulbous*, shining, fibrilloso-striate. Gills *grey*, *then olivaceous*, adnate, crowded. Spores pale purplish brown. Smell pleasant, of meal. Solitary, "caespitose from a common tuber" Secrétan. Old pastures, and under firs. Aug.—Nov. Rare.

783. **H. sublateritium** (Schaeff.) Fr. Rolland, Champ. t. 73, no. 162. *Sub*, somewhat; *lateritium*, brick colour.

P. 3–10 cm., *tawny-brick-red*, *paler round the margin*, fleshy, convexo-plane, obtuse, *discoid*, *dry*, covered with a superficial, somewhat silky, whitish cloudiness (arising from the veil), *becoming smooth*; margin often appendiculate with the veil. St. 5–10 cm. × 8–15 mm., *yellow*, *ferruginous downwards*, firm, attenuated downwards, rarely equal, *scaly-fibrillose*, fibrils pallid. Cortina *white*, *at length becoming black*, superior. Gills *dingy yellowish*, *and darker at the base*, *then fuliginous*, *at length inclining to olivaceous*, adnate, sinuate, somewhat crowded. Flesh *yellowish*, *ferruginous at the base of the st.*, *compact*. Spores fuscous purple, elliptical, 6–7 × 3–4μ, 1–2-guttulate. Cystidia cylindric-clavate, or flask-shaped, apex obtuse, 6–7μ in diam., 35–50 × 10–15μ, on the edge of the gill, fusiform-capitate, 30–36 × 7–9μ, contents yellow. Taste bitter. Poisonous. Subcaespitose. Woods, hedgerows, and old posts. Jan.—Dec. Common. (*v.v.*)

var. **Schaefferi** B. & Br. Schaeff. Icon. t. 49, figs. 4–5. J. C. Schaeffer.

Differs from the type in its *smaller size*, *in the light yellowish*, *conical*, *at length depressed*, *wrinkled p.*, *and the narrow decurrent gills*. Stumps. July—Oct. Not uncommon. (*v.v.*)

var. **pomposum** Fr. *Pomposum*, stately.

Differs from the type in the *thicker*, *almost entirely tawny p.*, *the thick stem up to 2·5 cm. broad*, *becoming pale above*, *the membranaceous ring*, *and the gills at length becoming a beautiful olive*. Stumps. July—Oct. Not uncommon. (*v.v.*)

var. **squamosum** Cke. Cke. Illus. no. 573, t. 558. *Squamosum*, scaly.

Differs from the type in the *p. being spotted with scales*, *especially towards the margin*. Trunks. Oct. Uncommon. (*v.v.*)

784. **H. capnoides** Fr. Fr. Icon. t. 133, fig. 1. *καπνώδης*, smoky.

P. 2·5–8 cm., *ochraceous-yellowish*, fleshy, convex, then flattened, obtuse, dry, *smooth*, often wrinkled, margin appendiculate with the veil. St. 5–7 cm. × 4–8 mm., *pallid*, *becoming ferruginous under the surface covering when old*, apex whitish, equal, often curved and flexuose, *becoming silky-even*, here and there striate. Cortina *white*, *then becoming fuscous purple.* Gills *whitish, or bluish-grey, then fuscous purple,* adnate, easily separating, rather broad, somewhat crowded. Flesh *whitish, often somewhat ferruginous towards the base of the st.* Spores pale fuscous, elliptical, 7–8 × 3–4μ, 1-guttulate. Cystidia "clavate, with a prominent point, 36–50 × 10–15μ" Rick. Smell and taste mild. Caespitose, or fasciculate. Coniferous stumps. April—Dec. Common. (*v.v.*)

785. **H. epixanthum** Fr. Fr. Icon. t. 133, fig. 2. *ἐπίξανθον*, tawny.

P. 5–7·5 cm., *light yellow, or becoming pale, disc generally darker,* fleshy, convexo-plane, obtuse, or gibbous, sometimes depressed at the disc, *slightly silky, then becoming smooth*; margin appendiculate with the veil. St. 7–14 cm. × 6–10 mm., *yellow, pale ferruginous, or becoming fuscous below,* equal, or attenuated from the thickened base, *floccoso-fibrillose,* apex pruinose. Cortina *white,* silky. Gills *light yellowish white, or citron yellow, then cinereous,* adnate, crowded. Flesh *yellow, often ferruginous* at the base of the st., thin at the margin. Spores cinereous purple, broadly elliptical, 6–7 × 4μ, 1-guttulate. Cystidia "clavate, with a prominent point, 36–40 × 9–12μ" Rick. Smell strong. Caespitose. Stumps in frondose, and coniferous woods, hedgerows, and parks. June—Dec. Common. (*v.v.*)

786. **H. elaeodes** Fr. (= *Flammuloides fascicularis* Huds. sec. Quél.) *ἐλαία*, the olive-tree; *εἶδος*, like.

P. 4–8 cm., *brick-red, or tan tinged reddish at the disc,* fleshy, convex, then plane, or slightly depressed, obtuse, or subumbonate, *dry, smooth,* opaque; margin for a long time incurved, *undulate,* folded. St. 5–9 cm. × 5–10 mm., *dirty yellow, more or less ferruginous,* equal, or attenuated at the base, incurved, or flexuose, fibrillose. Cortina *white,* apical. Gills *green, or greenish yellow, then olivaceous, and finally brownish purple,* adnate, crowded, thin. Flesh *yellow, ferruginous in the st.,* thin at the margin. Spores brownish purple, broadly elliptical, 6–8 × 4–5μ, 1–2-guttulate. Cystidia "nearly flask-shaped, 30 × 8–10μ, obtuse, filled with yellow juice" Rick. Smell and taste bitter. Caespitose. On stumps, and on the ground. Woods, and pastures. Aug.—Oct. Uncommon. (*v.v.*)

787. **H. fasciculare** (Huds.) Fr. Cke. Illus. no. 576, t. 561. *Fasciculare*, in little bundles.

P. 2–5 cm., *light yellow, disc often darker,* fleshy, convex, then flattened, obtuse, or subumbonate, smooth, dry; margin often appendiculate with the veil. St. 5–22 cm. × 4–10 mm., *concolorous,* equal, base attenuated, or thickened, incurved, or flexuose, fibrillose. Cortina *yellowish-white,* fibrillose, torn. Gills *sulphur yellow, then green,* adnate, linear, 3–4 mm. broad, very crowded, *subdeliquescent*

Flesh *yellow*, thin. Spores purple, elliptical, 6–7 × 4·5μ, 1-guttulate. Cystidia clavate, 28–35 × 7–10μ, contents yellowish. Smell and taste very bitter. On stumps, and on the ground. Woods, pastures, and hedgerows. Jan—Dec. Common. (*v.v.*)

788. **H. instratum** Britz. Cke. Illus. no. 1181, t. 1157.
Instratum, bare.

P. 2–3·5 cm., *dark brown*, fleshy, hemispherical, convex, broadly umbonate, *radiately rugose*, margin appendiculate with the white veil. St. 5–7·5 cm. × 4 mm., *white above, brownish at the base*, equal, apex smooth, *fibrillose, or squamulose below*. Gills *brown, then purple brown*, adnate, subventricose, 6 mm. broad. Flesh *brownish*, thin. Spores purple brown, elliptical, 8 × 4μ. Caespitose. On stumps. Sept.—Oct. Rare.

789. **H. aellopum** Fr. ἀελλόπουν, storm-footed.

P. 2–5 cm., *rufescent*, fleshy, convex, then plane, obtuse, smooth. St. 5–6 cm. × 4–8 mm., *variegated with minute, red squamules*, fusiform, rooting, *with a separable tube inside*. Gills *yellowish*, then *fuscous-olivaceous*, adnate. Subcaespitose. Stumps, especially fir, and larch. Sept. Rare.

790. **H. dispersum** Fr. Fr. Icon. t. 133, fig. 3. *Dispersum*, scattered.

P. 2–4 cm., *tawny honey colour*, not hygrophanous, fleshy, campanulate, then convex, at length expanded, smooth, *superficially white silky with the veil round the margin*. St. 5–7 cm. × 4–6 mm., *somewhat ferruginous, becoming fuscous at the base, apex pale*, equal, *tense and straight*, tough, fibrillosely silky, *besprinkled with white, zone-like markings from the remains of the veil*. Gills *pallid straw colour, at length clouded, obsoletely green*, adnate, ventricose, 4–8 mm. broad, thin, crowded, *edge often white* Flesh *yellowish, ferruginous under the cuticle of the p. and in the st.*, thin. Spores purple, pip-shaped, 8–9 × 4–5μ. Cystidia "subclavate, 30–45 × 7–10μ, often drawn out into a point, filled with a yellow juice" Rick. Solitary, rarely caespitose. Amongst coniferous needles and twigs, rarely on stumps, and sawdust. April—Nov. Uncommon. (*v.v.*)

791. **H. irroratum** Karst. *Irroratum*, bedewed.

P. 4 cm., *tawny honey colour*, convex, then plane, gibbous, even; margin silky, pruinose. St. 13 cm. × 5 mm., *pallid*, equal, rooting, tough, fibrillosely silky, wavy below, and *with dense rusty down*. Gills *straw colour, then darker, and tinged green*, adnate, crowded; margin *dentate*. Spores hyaline under the microscope, elliptical, 6–9 × 4–5μ. Smell and taste very sour. Rare.

**P. naked, viscid.

792. **H. incomptum** Massee. *Incomptum*, unadorned.

P. 7·5–10 cm., *deep bay brown, orange-tawny when dry*, campanulate, then expanded, broadly gibbous, *viscid*, silky when dry; margin usually flexuose. St. 6–7·5 cm. × 16 mm., *pale above, dark ferruginous below*, equal, *covered with minute, spreading, ferruginous, fibrillose squamules, mixed with primrose yellow tomentum.* Gills *pallid, then deep olive, finally clouded with purple from the spores*, adnate, slightly rounded behind, 3–4 mm. broad, crowded, thin. Flesh *tawny*, 2 mm. thick. Spores purplish, obliquely elliptical, 8 × 3·5μ. Stumps. Sept.—Oct. Rare.

793. **H. oedipus** Cke. Cke. Illus. no. 579, t. 587, fig. A. *οἰδίπους*, swollen footed.

P. 1–2·5 cm., *umber, or brownish olivaceous, paler, and subochraceous at the striate margin, which exceeds the gills*, fleshy, turbinate, or hemispherical, then convex, glutinous, smooth, dull; margin at first appendiculate with the veil. St. 3–5 cm. × 4–6 mm., *pallid, tawny at the base*, attenuated upwards *from the bulbous base*, apex pruinose, fibrillose below the ring. Ring *whitish*, median, torn, fugacious. Gills *whitish, then umber*, broadly adnate, sometimes with a minute decurrent tooth, plane, edge somewhat granular, or crenulate, *often whitish.* Flesh *brownish, becoming white*, thick at the disc. Spores dark brown, elliptical oblong, 9–10 × 5–6μ. Sticks, and dead leaves. Solitary, or caespitose. Feb. Rare.

***P. silky with innate fibrils, or streaked.

794. **H. lacrymabundum** Fr. non Quél. (= *Stropharia cotonea* Quél., *Hypholoma storea* Fr. var. *caespitosum* Cke., *Hypholoma hypoxanthum* Phill. & Plowr., *Hypholoma pseudostorea* W. G. Sm.) Fr. Icon. t. 134, fig. 1. *Lacrymabundum*, weeping.

P. 4–7·5 cm., *whitish, becoming fuscous and pale round the margin*, fleshy, *convex*, obtuse, *piloso-scaly, the innate scales darker*; margin appendiculate with the veil. St. 5–11 cm. × 6–12 mm., *whitish, then fuscous whitish*, slightly attenuated upwards from the somewhat thickened *base, which is often yellowish*, curved, *fibrillosely scaly*, apex smooth. Cortina *white*, separate, fibrillose. Gills *whitish, then fuscous purple*, adnate, 6 mm. broad, crowded, *edge whitish*, often distilling drops in wet weather. Flesh *white, greyish when moist*, soft. Spores fuscous purple, elliptical, 7–9 × 4–4·5μ. Cystidia cylindrical, or flask-shaped, base ventricose, apex obtuse, 5–6μ in diam., 28–40 × 8–18μ. Taste pleasant. Densely caespitose. On stumps, and on the ground. Beech, birch, and coniferous woods. Sept.—Dec. Not uncommon. (*v.v.*)

795. **H. pyrotrichum** (Holmsk.) Fr. (= *Stropharia pyrotricha* (Holmsk.) Quél.) Cke. Illus. no. 583, t. 564. *πῦρ*, fire; *θρίξ*, hair.

P. 5–10 cm., *fiery tawny*, fleshy, hemispherical, obtuse, then expanded, *densely clothed with somewhat adpressed, tawny fibrils*, which are here and there fasciculate in the form of scales. St. 5–7·5 cm. × 4–15 mm., *pallid, becoming tawny*, equal, *fibrillose, commonly squarrose with small, fiery tawny scales*. Cortina *tawny*. Gills *pallid, then becoming brown*, adnate, at length free, 10–12 mm. broad, somewhat crowded, *edge white, flocculose*. Flesh *tawny, deeper coloured in the st.*, thin at the margin. Spores fuscous purple, pip-shaped, 10–11 × 6·5–7μ. Cystidia clavate, 12–13μ in diam. at apex, 44–70 × 7–9μ, contents often yellowish. Caespitose. Woods, and about roots of trees. Aug.—Oct. Not uncommon. (*v.v.*)

var. **egregium** Massee. *Egregium*, distinguished.

Differs from the type in the *st. below the ring being covered with spreading, squarrose, whitish scales, and in the purple brown, broadly elliptical, or subglobose, apiculate spores*, 6 × 4–5μ. Fasciculate, near to stumps. Fir woods. Oct. Rare.

796. **H. velutinum** (Pers.) Fr. (= *Stropharia lacrimabunda* (Bull.) Quél.) Cke. Illus. no. 582, t. 563. *Velutinum*, velvety.

P. 5–15 cm., *lurid, becoming tawny, then clay colour isabelline, hygrophanous*, fleshy, campanulate, then expanded, at length obtusely umbonate, *adpressedly, tomentosely fibrillose*, becoming smooth; margin appendiculate with the white veil. St. 5–12·5 cm. × 4–15 mm., *dingy clay colour*, equal, fragile, *fibrillosely silky*, apex tomentose above the veil. Cortina *white, then black*, woolly. Gill *inclining to fuscous, then date brown fuscous, dotted black*, adnexed, easily separating, 8–10 mm. broad, subdistant, *edge white, floccose*, distilling watery, hyaline drops. Flesh *pallid*, very thin, fragile. Spores brownish purple, elliptical, 8–10 × 6–7μ, verrucose, with an apical germ pore. Cystidia capitate-clavate, 50–60 × 12–15μ. Taste mild. Poisonous. Solitary, or in groups of two or three. Woods, pastures, roadsides, rarely on stumps. May—Nov. Common. (*v.v.*)

var. **leiocephalum** B. & Br. λεῖος, smooth; κεφαλή, head.

Differs from the type in its *smaller size, in the very rugose disc, and pallid p., smooth except at the fibrillose margin, and the pallid st., farinose at the apex*. Densely caespitose. Old stumps, and in woods. Sept.—Oct. Uncommon. (*v.v.*)

797. **H. melantinum** Fr. Fr. Icon. t. 134, fig. 2.
μέλας, black; ἴς, a fibre.

P. 2–6 cm., *umber, then pale*, fleshy, campanulato-convex, then plane, obtuse, *covered with innate, adpressed, hairy, black squamules*. St. 4–6 cm. × 4–12 mm., *pallid*, equal, or slightly attenuated upwards, *fibrillosely hispid with whitish, then fuscous fibrils*. Cortina *white*, tender, fugacious. Gills *pallid, then umber*, adnexed, almost free, ventricose, crowded. Flesh *white*, very thin. Spores fuscous purple, reniform, 7–8 × 3–4μ, 1–2-guttulate. Cystidia "on surface of gill sparse, flask-shaped, on edge of gill vesiculose, 45–55 × 12–15μ" Rick. Solitary. Base of birch, and elm trees. Parks. Sept. Rare. (*v.v.*)

****P. covered with floccose, superficial, fugacious scales.

798. **H. cascum** Fr. Cke. Illus. no. 584, t. 544. *Cascum*, old.

P. 4–8 cm., *livid grey, tan whitish when dry*, fleshy, oval, then expanded, obtuse, *covered when young with superficial, white, fugacious scales*, then smooth, slightly wrinkled when dry, disc persistently even; margin appendiculate with the white, squamulose veil. St. 7·5–10 cm. × 4–6 mm., *white*, equal, fragile, *fibrillose*, apex white-mealy. Gills *grey, then black fuscous*, rounded-adnexed, ventricose, 4–8 mm. broad, arid, fragile. Flesh *white*, thin. Spores purple, elliptical, 7–8 × 4–5μ. Cystidia "fusiform-pointed, 40–50 × 9–13μ, somewhat thick walled" Rick. Taste bitter. Coniferous woods, and pastures. July—Oct. Uncommon. (*v.v.*)

H. punctulatum (Kalchbr.) Cke. = **Stropharia punctulata** (Kalchbr.) Fr.

*****P. smooth, hygrophanous, margin appendiculate with the veil.

799. **H. lanaripes** Cke. (= *Hypholoma appendiculatum* Bull. sec. Quél.) Cke. Illus. no. 585, t. 545. *Lana*, wool; *pes*, foot.

P. 3–6 cm., pallid, fleshy, campanulate, then expanded, hygrophanous, *squamose with superficial scales arising from the breaking up of the cuticle*; margin appendiculate with the fugacious veil. St. 5–7·5 cm. × 4 mm., *white*, equal, fragile, subfibrillose, base tomentose. *Gills whitish, then purplish brown*, reaching the st., crowded. Flesh *pallid*, thin. Soil in conservatories. Subcaespitose. June—July. Uncommon.

800. **H. Candolleanum** Fr. Cke. Illus. no. 586, t. 546. Alphonse de Candolle.

P. 5–10 cm., *date brown, then white, disc ochraceous*, fleshy acorn-shaped, then campanulate, soon convex, at length flattened, obtuse, unequal, smooth; margin appendiculate with the *white, at length fuscous veil*. St. 4–7·5 cm. × 4–8 mm., *white*, somewhat thickened at the base, fragile, fibrillose, *apex striate*. Gills *violaceous, then fuscous cinnamon, edge at first whitish*, rounded-adnexed, then separating, crowded. Flesh *white*, thin. Spores brownish violet, "elliptical, 8 × 4μ" Karst. Cystidia "only on edge of gill, subcylindrical or subventricose, 30–45 × 9–10μ" Rick. Taste mild. Edible. In troops, or subcaespitose. Woods and stumps. April—Nov. Common.

801. **H. appendiculatum** (Bull.) Fr. Cke. Illus. no. 587, t. 547.
Appendiculatum, having a small appendage.

P. 5–8 cm., *date brown, then tawny, ochrey pale when dry*, fleshy-membranaceous, ovate, then expanded, at length flattened, obtuse, pruinose, sprinkled with a few, fugacious flecks, then smooth, slightly wrinkled when dry; margin appendiculate with the white, fugacious veil. St. 4–7·5 cm. × 4–6 mm., *white*, equal, fragile, fibrillose, *apex pruinose*. Gills *white, then flesh colour, at length fuscous*, subadnate, crowded. Flesh *white*, very thin. Spores fuscous purple, elliptical, 7–8 × 4–4·5μ. Cystidia cylindrical, often slightly constricted below the apex, base subventricose, 35–45 × 10–13μ. Taste mild. Edible. In troops, or caespitose. Woods, hedgerows, and wood heaps. June—Dec. Common. (*v.v.*)

var. **lanatum** B. & Br. *Lanatum*, woolly.

Differs from the type in being *densely woolly when young, traces of the woolly coat remaining at the apex when the p. is expanded.*

var. **flocculosum** Boud. Boud. Icon. t. 137. *Flocculosum*, flocculose.

Differs from the type in the *white squamulose, longitudinally striate, or sulcate, somewhat lobed p., and the striate st. often flocculose.* Woods. Sept.—Oct. Uncommon. (*v.v.*)

802. **H. catarium** Fr. Cke. Illus. no. 1180, t. 1176.
Catarium, belonging to a cat.

P. 1·5–2·5 cm., *ochraceous, then paler*, fleshy membranaceous, hemispherical, then expanded, smooth, hygrophanous; margin appendiculate with the white veil. St. 2·5–4 cm. × 2–3 mm., *white, rather shining, base incrassated* and white floccose, apex striate. Gills *white, then fuscous*, adnate, narrow, rather crowded. Flesh *white*, thin. Spores fuscous purple, elliptic-oblong, 6 × 3μ. Gregarious, or subcaespitose. Amongst grass in parks, and roadsides. Sept. Rare.

803. **H. leucotephrum** B. & Br. Cke. Illus. no. 588, t. 548.
λευκός, white; *τεφρόν*, ash-coloured.

P. 3–7 cm., *dark grey, whitish when dry*, fleshy, somewhat campanulate, then convexo-expanded, wrinkled; margin appendiculate with the white veil. St. 7–10 cm. × 6–10 mm., *white*, equal, attenuated at the base, *silky-fibrillose downwards*, apex striate, or sulcato-striate. Gills *cinereous whitish, then grey, turning black*, slightly adnate, 3–6 mm. broad. Flesh *pallid, becoming white*, thick at the disc. Spores cinereous purple, elliptical, 8–9 × 5μ. Cystidia "on edge of gill subcylindrical, constricted-capitate, 30–40 × 6–8μ" Rick. Caespitose. Base of ash trees, and amongst beech leaves, and pine needles. Sept.—Nov. Uncommon. (*v.v.*)

804. **H. egenulum** B. & Br. (= *Hypholoma appendiculatum* Bull. sec. Quél.) Cke. Illus. no. 589, t. 605, fig. A. *Egenulum*, poor.

P. 3–4 cm., *watery white, snow white when dry*, fleshy, hemispherical, then expanded, *umbonate*, smooth; margin finely striate, appendiculate with the white veil. St. 4–5 cm. × 3 mm., *white*, attenuated upwards, or nearly equal, *minutely adpressedly scaly*. Gills *purplish umber, edge white*, adnate, with a tooth, slightly ventricose, subdistant. Flesh *white*, thick at the disc. Spores brown purple. Solitary. Amongst grass. May. Rare.

805. **H. pilulaeforme** (Bull.) Fr. (= *Hypholoma hydrophilum* Bull. sec. Quél.) Bull. Hist. Champ. Fr. t. 112.
Pilula, a little ball; *forme*, shaped.

P. 1–2 cm., *fuscous, dingy ochraceous when dry*, submembranaceous, *globose*, then convexo-plane, obtuse, smooth; *margin pale, striate*, slightly appendiculate with the white veil. St. 2·5–5 cm. × 2 mm., *white*, equal, flexuose, often slightly thickened at the base. Cortina white, membranaceous, often forming a ring, fugacious. Gills *white, then cinereous, at length fuscous*, adnexed, easily separating, narrow, 2–3 mm. broad, thin, crowded. Flesh *yellowish*, thin. Spores cinereous purple, elliptical, 6–8 × 4μ. Densely caespitose. On stumps, and buried wood. Woods, and pastures. Sept.—Oct. Not uncommon. (*v.v.*)

806. **H. hydrophilum** (Bull.) Fr. (= *Bolbitius hydrophilus* (Bull.) Fr. Hym. Eur.) Cke. Illus. no. 606, t. 610, as *Psilocybe spadicea* Fr.
ὕδωρ, water; *φίλον*, loving.

P. 3–6 cm., *date brown when moist, tawny, or tan colour when dry*, fleshy, globose, then convex and expanded, pruinose, striate near the margin when moist, wrinkled when dry; margin often undulate, appendiculate with the white, fugacious veil. St. 5–10 cm. × 4–8 mm., *white, becoming somewhat ferruginous downwards*, rigid, but fragile, equal, or attenuated slightly upwards, curved, fibrillose at the base. Gills *whitish, then date brown fuscous*, adnate, ventricose, crowded, often distilling hyaline drops. Flesh *pallid, white when dry*, very thin at the margin. Spores ferruginous purple, elliptical, 5–6 × 3–3·5μ, 1–2-guttulate. Cystidia cylindrical, ventricose, often constricted below the apex, apex obtuse, 8–9μ in diam., 25–28 × 12–14μ. Densely caespitose. Stumps in woods, and on sawdust. Aug.—Dec. Common. (*v.v.*)

H. fibrillosum (Pers.) Quél. = **Psathyra fibrillosa** (Pers.) Fr.

H. nolitangere (Fr.) Quél. = **Psathyra nolitangere** Fr.

H. fatuum (Fr.) Quél. = **Psathyra fatua** Fr.

H. ammophilum (Mont.) Quél. = **Psilocybe ammophila** (Mont.) Fr.

H. gossypinum (Bull.) Quél. = **Psathyra gossypina** (Bull.) Fr.

H. pennatum (Fr.) Quél. = **Psathyra pennata** Fr.
H. semivestitum (B. & Br.) Quél. = **Psathyra semivestita** B. & Br.
H. bifrons (Berk.) Big. & Guillem. = **Psathyra bifrons** Berk.
H. Gordonii (B. & Br.) Big. & Guillem. = **Psathyra Gordonii** B. & Br.

****Gills decurrent, or adnato-decurrent by a tooth.

Spores white; hymenium not waxy, nor pulverulent.

Clitocybe Fr.

(κλίτος, a slope; κύβη, head.)

Pileus fleshy, regular, or irregular; margin incurved. Stem central, externally fibrous. Gills decurrent, rarely adnate, with an acute edge. Spores white, rarely yellowish, or greenish, elliptical, pip-shaped, globose, subglobose, or oblong; smooth, punctate, verrucose, or echinulate, continuous. Growing on the ground, rarely on wood, solitary, caespitose, or forming rings.

A. P. fleshy, often pale and silky when dry, not hygrophanous.

a. P. convex, then plane, or depressed, regular, obtuse.

*P. cinereous, or fuscous.

807. **C. nebularis** (Batsch) Fr. Grev. Scot. Crypt. Fl. t. 9, as *Agaricus turgidus*. *Nebularis*, clouded.

P. 7·5–20 cm., *fuliginous, or fuscous, then grey*, fleshy, somewhat compact, convex, then plane, very obtuse, at length depressed at the disc, dry, *at first pruinosely grey*, becoming smooth, more rarely innately streaked, or shining when scorched by the sun. St. 7–12 × 2–3 cm., *whitish*, thickened at the base, attenuated upwards, spongy, elastic, *fibrillosely striate*. Gills *whitish, sometimes becoming yellow*, shortly and equally *decurrent, arcuate, very crowded*, thin. Flesh white, thick. Spores white, elliptical, 7–8 × 3–4μ, 1–2-guttulate. Smell and taste pleasant. Edible. Woods, heaths, and pastures. Aug.—Dec. Common. (*v.v.*)

808. **C. clavipes** (Pers.) Fr. Fr. Icon. t. 47, upper figs. *Clavus*, a nail; *pes*, foot.

P. 4–6 cm., *fuscous, fuliginous, cinereous-livid, sometimes whitish round the margin*, very rarely wholly white, fleshy, slightly convex, soon plane, at length *almost obconical*, very obtuse, sometimes umbonate, smooth. St. 4–6 × 1 cm., *concolorous, conical, base bulbous*, elastic, *somewhat fibrillose*. Gills *white, sometimes yellowish, deeply decurrent, subdistant*, flaccid, broad. Flesh *cinereous, then white*, lax, thin at the margin. Spores white, elliptical, 6–7 × 4–5μ, 1-guttulate. Smell faint, pleasant. Taste mild. Beech, and coniferous woods. Aug.—Nov. Common. (*v.v.*)

809. **C. comitialis** (Pers.) Fr. Fr. Icon. t. 47, lower figs.
Comitialis, belonging to the comitia.

P. 2–5 cm., *umber fuliginous, almost becoming black*, fleshy, *convex, then plane*, obtuse, smooth, somewhat moist. St. 4–7 cm. × 6–15 mm., *concolorous*, equally attenuated upwards, *elastic*, smooth. Gills *white, adnato-decurrent, horizontal, plane, crowded*, thin. Flesh *white*, firm, compact. Spores white, elliptical, 6–7 × 4μ, 1-guttulate. Pine woods. Oct. Uncommon. (*v.v.*)

810. **C. obscurata** Cke. Trans. Brit. Myc. Soc. III, t. 5, fig. C.
Obscurata, darkened.

P. 5 cm., *greyish umber*, plane, then *infundibuliform*, moist, smooth. St. 3–4 cm. × 6–8 mm., *concolorous*, slightly attenuated downwards, sometimes substriate. Gills *white*, decurrent, rather distant. Spores white, subelliptical, 3 × 1·5–2μ. Smell of meal. Amongst grass and dead leaves. Sept. Rare.

811. **C. gangraenosa** Fr. *γάγγραινα*, a gangrene.

P. 4–8 cm., *whitish, tinged with bistre, or livid, sometimes green*, fleshy, convexo-plane, obtuse, *white, pulverulent, then covered with brownish hairs especially at the margin*, then naked, *variegated, or streaked*. St. 4–5 × 1 cm., *white, subbulbous*, soft, striate, or slightly sulcate, curved, sometimes excentric. Gills *dingy white*, subdecurrent, arcuate, *very crowded*. Flesh *white, becoming blackish or spotted with black*. Spores white. Smell stinking, foetid. Woods, and larch plantations. Oct.—Nov. Rare.

var. **nigrescens** (Lasch) Cke. *Nigrescens*, becoming black.

Differs from the type in its *whitish colour, and rather sweet smell*. Larch plantations. Rare.

812. **C. polia** Fr. (= *Paxillus inornatus* (Sow.) Quél.) Fr. Icon. t. 48, fig. 1. *πολιά*, grey.

P. 2–5 cm., *grey*, fleshy, convexo-plane, regular, smooth; margin *white*, incurved. St. 3–9 × ·5–1 cm., *greyish white, equal*, base subbulbous, *smooth*. Gills *white*, decurrent, *very crowded, very narrow*, 1 mm. broad. Flesh *greyish, becoming white*. Spores white, "6–8 × 3–4μ" Sacc. Caespitose. Woods. Sept.—Oct. Rare.

813. **C. inornata** (Sow.) Fr. (= *Paxillus inornatus* (Sow.) Quél.) Bres. Fung. Trid. t. 155. *Inornatus*, unadorned.

P. 4–8 cm., *cinereous grey, then somewhat tan colour*, fleshy, convex, then plane and depressed, sometimes slightly gibbous; margin at first involute, *pubescent, striate with evanescent veins*. St. 4–6 cm. × 8–12 mm., *concolorous*, subequal, *fibrillosely tomentose*, base white, floccose. Gills *concolorous*, rounded behind, adnate, or adnato-decurrent, crowded, easily separable from the hymenophore. Flesh *whitish grey*, thick at the disc. Spores white, oblong, apiculate at the base, 8–10 × 3μ. Smell rancid, taste insipid. Woods, and pastures. Oct.—Nov. Rare.

C. nimbata (Batsch) Quél. = **Tricholoma panaeolum** Fr. sec. Quél.

814. **C. luscina** Fr. *Luscus*, one-eyed.

P. 2·5–6 cm., *brown*, then grey, *fleshy*, convex, then plane, obtuse, sometimes excentric, smooth, moist. St. 5 cm. × 4–6 mm., *white*, *slightly attenuated downwards, entirely white pulverulent, or only so at the apex*. Gills *white-hyaline, decurrent, horizontal, straight*, crowded, thin, 2–4 mm. broad. Flesh *pallid*, thin. Spores white. Pastures. Sept.—Oct. Rare.

815. **C. curtipes** Fr. Fr. Icon. t. 48, fig. 5. *Curtus*, short; *pes*, foot.

P. 2–7·5 cm., *fuscous, then pale and becoming whitish, fleshy*, convex, then plane, obtuse, *oblique*, silky. St. 2·5 cm. × 4–6 mm., *reddish fuscous, rigid, cartilaginous, attenuated downwards*, somewhat fibrillose, *apex white pruinose*. Gills *shining white, adnate*, scarcely decurrent, very crowded, 2 mm. broad. Flesh *white*, firm. Spores white. Amongst grass. Sept.—Oct. Rare.

816. **C. hirneola** Fr. *Hirneola*, a small jug.

P. 1–2 cm., *cinereous, or grey, becoming pale and hoary*, slightly fleshy, *plano-convex, then depressed in the centre and umbilicate*, very smooth, *shining*, semi-viscid when fresh, the cuticle as if with a glued silkiness; margin involute, very thin. St. 2·5–5 cm. × 2–3 mm., *concolorous, elastic*, equal, flexuose, smooth, *apex white pruinose*. Gills *whitish-grey*, subdecurrent, crowded, thin, rather broad. Flesh *white, often bistre when moist, thin*. Spores *dirty white in the mass*, subglobose, 4–5 × 4μ, multi-guttulate. Edible. Amongst grass, and leaves. Sept.—Oct. Uncommon. (*v.v.*)

var. **undulata** (Bull.) Fr. (= var. *major* Fr. Mon.) *Undulata*, waved.

Differs from the type in its *larger size, the p. being 3–4 cm., flattened, flexuose, subzonate, grey, becoming whitish.*

817. **C. zygophylla** Cke. & Massee. Cke. Illus. no. 1137, t. 948. *ζυγόν*, a yoke; *φύλλον*, leaf.

P. 5–10 cm., *greyish when moist, pale ochraceous white when dry*, fleshy, convex, then expanded, disc often slightly depressed, tough, flaccid, hygrophanous; margin thin, involute at first, *rugose, or plicate, as if pinched up at regular intervals*. St. 5–6 × 1 cm., *white, then pallid*, equal, *expanding into the p.*, smooth, base with a thin white tomentum. Gills *cinereous, deeply decurrent*, rather distant, 4 mm. broad, *distinctly connected by veins*. Flesh *white, greyish under the cuticle of the p.*, thick at the disc, thin at the margin. Spores white, elliptical, 8–9 × 4μ, with a large central gutta. Amongst leaves. Oct.—Nov. Uncommon. (*v.v.*)

**P. violet, or rufescent.

818. **C. cyanophaea** Fr. Gonn. & Rabenh. Heft. 8–9, t. 17, fig. 3, as *Tricholoma nudum*. *κύανος*, dark blue; *φαιά*, dusky.

P. 7–10 cm., *fuscous, becoming azure-blue,* rather fleshy, convex, then plane, obtuse, smooth. St. 7–8 × 1–2 cm., *bluish when young,* attenuated upwards from the thickened base, *apex abruptly white,* smooth. Gills *violet, then pale, deeply decurrent,* crowded. Woods. Rare.

var. **Pengellei** B. & Br. Cke. Illus. no. 131, t. 264.
T. W. Pengelly, the geologist.

Differs from the type in having the st. *attenuated at the base.*

C. opipara Fr. = **Tricholoma opiparum** (Fr.) Quél.

C. amara (A. & S.) Fr. = **Tricholoma amarum** (A. & S.) Quél.

819. **C. socialis** Fr. Fr. Icon. t. 49, lower figs. *Socialis*, sociable.

P. 2–3 cm., *reddish yellow,* fleshy, convex, then expanded, *acutely umbonate when young,* smooth, dry. St. 2–3 cm. × 4–6 mm., *reddish,* ascending, fibrous, *rooting base peronately hairy.* Gills *becoming yellow,* plano-decurrent, scarcely crowded. Flesh *white,* thin. Spores white. Densely gregarious. Amongst pine needles. Sept.—Oct. Rare.

***P. becoming yellow.

820. **C. amarella** (Pers.) Fr. (= *Paxillus amarellus* (Pers.) Quél., *Clitopilus popinalis* Fr. sec. Quél.) *Amarella*, bitterish.

P. 4–5 cm., *pale yellowish, or pallid fawn colour,* fleshy, plane, firm, *subumbonate,* smooth. St. 5 cm. × 4–6 mm., *concolorous,* equal, tough, *white villous at base.* Gills *pallid, somewhat shining,* subdecurrent, crowded, here and there dichotomous. Spores white, "5–6 × 2–3μ" Rick. Smell strong, like prussic acid, taste very bitter. Woods. Oct. Rare.

821. **C. vernicosa** Fr. Fr. Icon. t. 50, upper figs. *Vernicosa*, varnished.

P. 5–6 cm., *pale yellowish, shining,* fleshy, convex, then flattened or plano-depressed, obtuse, sometimes infundibuliform, smooth, margin involute. St. 2–3 cm. × 6–8 mm., *yellow,* firm, tough, equal, smooth. Gills *light yellow, adnato-decurrent, or deeply decurrent,* equally attenuated behind, *subdistant.* Flesh *whitish,* firm. Spores white, "subglobose, 3–4μ, nearly spinulose" Rick. Fir and larch woods. Sept.—Oct. Rare.

822. **C. venustissima** Fr. Fr. Icon. t. 50, lower figs. *Venustissima*, most beautiful.

P. 2–5 cm., *rich orange-reddish, becoming pale*, slightly fleshy, convex, then expanded, obtuse, or somewhat umbilicate, becoming slightly silky-even; margin even, in large specimens striate, and crenate, or toothed in a crisped manner. St. 3–4 cm. × 4–8 mm., *reddish*, equal, smooth, *base often white villous*. Gills *concolorous*, decurrent, subdistant, 3 mm. broad, subarcuate. Flesh *reddish*, thin. Spores white, broadly elliptical, 11–16 × 8–10μ, multi-guttulate. Amongst fir and larch leaves. Sept.—Oct. Uncommon. (*v.v.*)

823. **C. alutacea** Cke. & Massee. *Aluta*, soft leather.

P. 1–1·5 cm., *tan colour*, convex, then *umbilicate*, smooth; margin incurved. St. 3–4 cm. × 1–2 mm., *pale*, smooth. Gills *paler than the p.*, decurrent, arcuate, *narrow*, crowded. Spores white, elliptical, 6 × 4μ. Amongst grass and moss. Sept. Uncommon.

824. **C. subalutacea** (Batsch) Fr. Barla, Champ. Alp. Marit. t. 50, figs. 10–15. *Sub*, somewhat; *aluta*, soft leather.

P. 2·5–5 cm., *pale tan, becoming whitish*, fleshy, soft, tough, convex, then flattened, obsoletely umbonate, or depressed, smooth. St. 5–8 cm. × 6–8 mm., *reddish white, deeper coloured at the base*, cylindrical, flexuose, firm, *elastic*, fibrillose, smooth. Gills *pallid*, adnate, then subdecurrent, distant, *broad*. Flesh *whitish*, soft, tough. Spores white, "subglobose, 3–4 × 3μ" Rick. Smell pleasant of anise, sometimes disagreeable like rancid meal, sometimes obsolete. Woods, and under trees. Nov. Uncommon.

825. **C. aurantiaca** (Wulf.) Studer. (= *Cantharellus aurantiacus* (Wulf.) Fr.) Rolland, Champ. t. 53, no. 117, as *Cantharellus aurantiacus*. *Aurantiaca*, orange coloured.

P. 2–8 cm., *orange-ochraceous*, fleshy, soft, depressed, often excentric and undulated, *subtomentose*; margin involute. St. 5 cm. × 6–8 mm., *ochraceous, or bistre*, somewhat incurved and unequal. Gills *deep orange*, decurrent, tense, straight, *repeatedly dichotomous, crowded*, often crisped at the base. Flesh *yellowish*, soft, thick at the disc. Spores elliptical, 6–7 × 4–5μ, with a large central gutta. Taste unpleasant. Heaths, and woods, especially coniferous woods. Aug.—Nov. Common. (*v.v.*)

var. **albida** (Gillet) Rea. Cke. Illus. no. 1057, t. 1104, fig. B, as *Cantharellus aurantiacus* var. *pallidus*. *Albida*, whitish.

Differs from the type in the *white gills*. Heaths, and woods. Sept.—Oct. Uncommon. (*v.v.*)

var. **lactea** (Quél.) Rea. *Lactea*, milk white.

Differs from the type in being *entirely white*. Heaths, and coniferous woods. Aug.—Oct. Not uncommon. (*v.v.*)

var. **nigripes** (Pers.) Rea. *Niger*, black; *pes*, foot.

Differs from the type in the *st. becoming black towards the base.* Boggy places in woods, and on heaths. Sept.—Nov. Common. (*v.v.*)

826. **C. hypnorum** (Brond.) Rea. *Hypnum*, moss.

P. 3–4 cm., *pale primrose yellow, sometimes verging on pale ochre,* campanulato-convex, then expanded and slightly depressed; margin incurved, *minutely downy,* the down sometimes collected in little fascicles. St. 2–3 cm. × 4–6 mm., *yellow, sometimes darker than the p. at the base,* often slightly flexuose, almost glàbrous. Gills *yellow,* decurrent, branched, thin, somewhat crowded, edge *acute.* Flesh *whitish,* thin. Spores white, oblong, 7 × 4μ, with a minute apiculus. Sept. Uncommon.

****P. greenish.

827. **C. odora** (Bull.) Fr. (= *Clitocybe viridis* (Scop.) sec. Quél.) Barla, Champ. Alp. Marit. t. 51, figs. 10–15. *Odora*, fragrant.

P. 5–9 cm., *greenish,* fleshy, convex, then flattened, obtuse, or obsoletely umbonate, then depressed; margin incurved, *pubescent.* St. 3–5 cm. × 6–8 mm., *concolorous,* somewhat cylindrical, flexuose, flocculoso-fibrillose, then smooth, *white villous at the thickened base.* Gills *paler than the p.,* adnato-decurrent, *subdistant,* broader than the flesh of the p. Flesh *dirty white,* tough. Spores white, elliptical, 8 × 4–4·5μ, 1–2-guttulate. Smell pleasant, of anise. Taste pleasant. Edible. Deciduous woods. Aug.—Nov. Common. (*v.v.*)

828. **C. viridis** (With.) Fr. (= *Agaricus virens* Scop. sec. Fr.) Barla, Champ. Alp. Marit. t. 51, figs. 1-9. *Viridis*, green.

P. 3–6 cm., *pale greenish blue,* fleshy, convex, then expanded, obtuse, smooth; margin *naked.* St. 3–8 cm. × 3–7 mm., *whitish,* firm, cylindrical, attenuated at the base, smooth. Gills *white, with a tinge of greenish,* adnato-decurrent, thin, crowded. Flesh *white,* thick at the disc, firm. Spores white, elliptical, 7–8 × 4–5μ, with a large central gutta. Smell and taste pleasant, of anise. Edible. Deciduous woods. Sept.—Oct. Uncommon. (*v.v.*)

829. **C. Trogii** Fr. (= *Clitocybe subalutacea* (Fr. non Batsch) Quél.) Cke. Illus. no. 135, t. 102.

Jacob Gabriel Trog, an eminent Swiss botanist.

P. 6–8 cm., cinereous, then white, fleshy, compact, convex, then expanded, obtuse, smooth, opaque. St. 3–5 × 1 cm., *white,* thickened and villous at the base. Gills *white,* subdecurrent, crowded. Flesh *whitish,* thick at the disc, compact. Spores white. Smell very fragrant, spicy. Woods. Sept.—Nov. Uncommon.

*****P. whitish, white, or becoming white.

830. **C. rivulosa** (Pers.) Fr. Barla, Champ. Alp. Marit. t. 51, figs. 16–23. *Rivulosa*, rivulose.

P. 2·5–5 cm., *flesh colour, becoming whitish, or rarely fuscous*, slightly fleshy, *convexo-plane, then depressed*, undulato-lobed, repand, minutely *tomentose, at length rivulosely-streaked*; margin at first incurved, *villous*. St. 3–5 cm. × 4–8 mm., *concolorous*, tough, *elastic*, equal, somewhat fibrillose, *minutely tomentose*. Gills *flesh colour, becoming whitish*, adnate, then subdecurrent, obliquely acute behind, broad, somewhat crowded. Flesh *white*, soft, spongy. Spores white, elliptical, 4–6 × 2–3μ, 1-guttulate. Smell and taste pleasant. Poisonous. Heaths, and pastures. Aug.—Nov. Very common. (*v.v.*)

var. **Neptunea** (Batsch) Fr. *Neptune*, god of the sea.

Differs from the type in its *smaller size*.

831. **C. cerussata** Fr. (= *Clitocybe tornata* Fr. sec. Quél.) Barla, Champ. Alp. Marit. t. 51, figs. 24–28. *Cerussata*, painted with white-lead.

Entirely white. P. 5–8 cm., *with a white-lead appearance*, fleshy, convex, then expanded, obtuse, or gibbous, smooth, *at first floccoso-fibrillose*; margin involute, *villous*. St. 5–8 × ·5–1 cm., *fleshy-fibrous*, elastic; base *thickened*, more or less white tomentose. Gills adnate, then slightly decurrent, *very crowded*, thin. Flesh *white*, soft, compact. Spores white, elliptical, 5–6 × 3–4μ. Taste and smell pleasant. Poisonous. Deciduous, and coniferous woods. April—Oct. Not uncommon. (*v.v.*)

var. **difformis** (Schum.) Fr. *Difformis*, deformed.

Caespitose, often *gigantic*. P. 5–18 cm., *undulato-lobed*, often at first sprinkled with flocci. St. 2·5 × 2·5 cm., *sulcate, or longitudinally* wrinkled. Gills *at length pallid*. Spores white, 4 × 3μ. Rich ground, and gardens. Oct. Uncommon.

var. **obtexta** (Lasch) Fr. *Obtexta*, covered.

Differs from the type in its *snow-white colour, the fibrillose texture of the subviscid p., the tomentose st., and the narrow gills*. Heaps of leaves. Rare.

832. **C. phyllophila** Fr. Barla, Champ. Alp. Marit. t. 52, figs. 1–6. φύλλον, a leaf; φίλος, loving.

P. 5–10 cm., *tan, or cream colour, becoming pale white*, fleshy, convex, then plano-depressed, obtuse, often excentric and repand, *hoary with superficial villose down*; margin silky. St. 5–8 cm. × 5–8 mm., *concolorous*, elastic, fibrous, equal, villous at the base, incurved. Gills *white, then becoming pale and yellowish, adnate,* subdecurrent, 4–6 mm. broad, *subdistant.* Flesh *white,* thin. Spores white, elliptical, 6 × 4μ. Taste mild. Poisonous. Woods, especially beech. Sept.—Dec. Common. (*v.v.*)

833. **C. pithyophila** (Secr.) Fr. (= *Clitocybe cerussata* Fr. sec. Quél.) Barla, Champ. Alp. Marit. t. 52, figs. 7–10.

πίτυς, pine; φίλος, loving.

P. 5–7·5 cm., *dead white when moist, shining whitish when dry,* fleshy, thin, rather plane, *umbilicate,* at length irregularly shaped, repand, undulato-lobed, *smooth, flaccid*; margin slightly striate when old. St. 4–5 cm. × 5–8 mm., *white, often compressed,* equal, smooth, apex obsoletely, or scarcely pruinose, base white tomentose. Gills *always white,* adnate, subdecurrent, *very crowded,* plane, 4–6 mm. broad. Flesh *whitish,* thin. Spores white, elliptical, 6–7 × 4μ. Smell and taste pleasant. Poisonous. Pine woods. Sept.—Nov. Common. (*v.v.*)

834. **C. tornata** Fr. (= *Clitocybe cerussata* Fr. sec. Quél.) Fr. Icon. t. 51, fig. 1. *Tornata,* turned in a lathe.

P. 2·5–5 cm., *pure white,* fleshy, *convex, then becoming plane, at length depressed round the gibbous disc,* dry, somewhat silky with a glued film, at length *delicately rimoso-rivulose.* St. 4 cm. × 4–6 mm., *concolorous,* tough, equal, or attenuated downwards, round, smooth, base pubescent. Gills *white,* adnate with a small decurrent tooth, horizontal, *plane,* very crowded, 2 mm. broad. Flesh *white,* thick at the disc. Spores white, elliptical, 4–6 × 3–4μ. Taste mild. Poisonous. Woods, and amongst grass. Sept.—Oct. Uncommon. (*v.v.*)

var. **opala** Fr. *Opala,* opal.

Differs from the type in the *viscid p.* Fir woods. Rare.

835. **C. candicans** (Pers.) Fr. Fr. Icon. t. 51, fig. 3.

Candicans, shining white.

Entirely white. P. 2·5 cm., slightly fleshy, convex, then plane, depressed, *umbilicate,* regular, or a little excentric, *pruinose, silky, shining.* St. 2·5–5 cm. × 2–5 mm., *cartilaginous, polished, like an Omphalia,* equal, base incurved, villous, rooting. Gills adnate, then decurrent, very thin, crowded, *narrow,* straight, almost linear. Flesh *whitish,* thin. Spores white, "elliptical, 4–5 × 2–3μ, smooth" Rick., "ovoid, 6–7μ, minutely rough" Quél. Taste mild. Deciduous woods. July—Nov. Common. (*v.v.*)

836. **C. dealbata** (Sow.) Fr. Sow. Eng. Fung. t. 123.

Dealbata, white-washed.

P. 2–3 cm., *whitish, with greyish, or flesh coloured zones towards the margin, shining white when dry, slightly fleshy,* tough, convex, then plane, at length revolute, undulated, *dry*, smooth, but innately pruinose under a lens. St. 2·5–3·5 cm. × 8–10 mm., *whitish, becoming flesh colour, equal*, often ascending, *apex white-mealy*. Gills *greyish, then whitish, adnate*, scarcely decurrent, thin, crowded. Flesh *white*, thin, arid. Spores white, elliptical, 5–6 × 3μ. Smell and taste pleasant, of new meal. Edible. Woods, and pastures. July—Nov. Common. (*v.v.*)

var. **minor** Cke. Cke. Illus. no. 143, t. 173. *Minor*, smaller.

Differs from the type in its *smaller size*. Woods, and heaths. Sept. —Oct. Not uncommon. (*v.v.*)

837. **C. gallinacea** (Scop.) Fr. Hussey, Illus. Brit. Myc. I, t. 39. *Gallinacea*, pertaining to domestic fowls.

P. 1–2·5 cm., *dingy white, becoming whitish when dry, slightly fleshy*, convex, then plane, *obtuse*, unequal, dry, *opaque*, pruinose. St. 5–6 cm. × 6–10 mm., *white*, equal, ascending, or flexuose, excentric, *incurved, striate, mealy*. Gills *white, adnato-decurrent*, thin, plane, crowded. Flesh *white*, thin, compact. Spores white, oblong elliptical, 9 × 4μ, 1–2-guttulate. Smell strong, taste bitter. Woods, and pastures. Sept.—Nov. Not uncommon. (*v.v.*)

β. P. fleshy at the disc, margin thin, at first umbonate, then expanded, depressed, and irregular; generally caespitose.

C. ampla (Pers.) Fr. = **Tricholoma amplum** (Pers.) Rea.

C. molybdina (Bull.) Fr. = **Tricholoma amplum** (Pers.) Rea.

838. **C. decastes** Fr. (= *Tricholoma decastes* (Fr.) Quél.) Fr. Icon. t. 52. δεκάς, a company of ten men.

Caespitose. P. 10–20 cm., *mouse grey, or livid, becoming whitish tan colour when dry*, fleshy, *fragile*, convex, then plane, gibbous, or obtuse, *smooth*; margin membranaceous, at first incurved, then expanded, very undulate and lobed. St. 7·5–10 × 2·5–4 cm., *white*, fibrous, *connate at the base*, attenuated, or *curved-ascending*, often compressed, *smooth*, rarely pruinose at the apex. Gills *white, adnato-decurrent, or sinuate*, 6–8 mm. broad, crowded, or subdistant, attenuated towards the margin, often undulated and crenulate at the edge. Flesh *white, fragile*, thick at the disc, very thin at the margin, scissile. Spores white, globose, 6–8μ. Taste pleasant. Edible. Woods, pastures, and gardens. Oct.—Nov. Not uncommon. (*v.v.*)

839. **C. subdecastes** Cke. & Massee. Cke. Illus. no. 1131, t. 958. *Sub*, near to; *decastes*, the species *C. decastes*.

Caespitose. P. 3–6 cm., *pale ochraceous, becoming paler, and whitish* towards the margin, fleshy, campanulate, or convex, very obtuse, smooth; margin more or less lobed. St. 8–12 cm. × 12–16 mm., *whitish*, equal, fibrillose, connate at the base. Gills *white*, rounded behind, *adnate*, 4–6 mm. broad, narrower in front, rather crowded. Flesh *white*, thin. Spores white, globose, 4–5μ. On the ground. Sept. Uncommon.

840. **C. cartilaginea** (Bull. non Fr.) Bres. (= *Tricholoma loricatum* Fr. sec. Bres.) Bres. Fung. Trid. t. 110, 111.
Cartilaginea, cartilaginous.

Caespitose or in troops. P. 4–12 cm., *fuliginous black, or chestnut fuliginous, becoming paler*, fleshy, convex, then expanded, depressed, or gibbosely-umbonate, *cuticle cartilaginous*. St. 4–8 × 1–1·5 cm., *white, becoming greyish, or horn colour*, connate at the ventricose base and *somewhat rooting*, apex white and villosely-furfuraceous, *cuticle cartilaginous*. Gills *white, then straw-, or horn-colour*, adnate, or sinuato-adnate, rarely rounded behind, tough, *subcartilaginous*, crowded, often undulate. Flesh *white*, firm. Spores white, globose, 6–8μ, 1-guttulate. Smell of mice, or like fresh nuts. Taste sweet, then bitter. Edible. Mixed woods. Sept.—Oct. Not uncommon. (*v.v.*)

841. **C. aggregata** (Schaeff.) Fr. (= *Tricholoma aggregatum* (Schaeff.) Quél.) Schaeff. Icon. t. 305, 306. *Aggregata*, heaped together.

Caespitose. P. 7–12 cm., *livid-grey, then rufescent*, fleshy, convex, then expanded, at first umbonate, then depressed, often excentric, *flaccid, somewhat silky-streaked*. St. 7–10 × 1·5 cm., *white, often rufescent*, attenuated downwards, connate at the base, often branched, curved, compressed, *subfibrillose*. Gills *ashy-white, then flesh colour, and becoming light yellow, unequally decurrent*, 6–8 mm. broad, thin, crowded. Flesh *whitish*, thin. Spores white, subglobose, 6–7 × 5–6μ. Smell strong. Oak woods, sawdust heaps, and garden soil. July—Oct. Not uncommon. (*v.v.*)

C. tabescens (Scop.) Bres. = **Armillaria mellea** (Vahl.) Fr. var. **tabescens** (Scop.) Rea.

842. **C. elixa** (Sow.) Berk. Sow. Eng. Fung. t. 172. *Elixa*, soaked.

Not caespitose. P. 5–8 cm., *fuliginous, becoming pale and somewhat silky when dry*, fleshy, convex, then flattened, or depressed, *umbonate*, undulato-repand, *delicately virgate*. St. 3–5 cm. × 10–12 mm., *fuliginous whitish*, firm, subequal, *apex velvety*. Gills *white, unequally decurrent, distant, connected by veins*, 4 mm. broad. Flesh *dingy white*, thick at the disc, soft. Spores "white, elliptical, 7 × 4μ" Massee. Woods. Oct.—Nov. Not uncommon.

Ag. fumosus Pers. = **Collybia fumosa** (Pers.) Quél.

C. fumosa Fr. = **Tricholoma cinerascens** (Bull.) Quél.

843. **C. conglobata** (Vitt.) Bres. (= *Agaricus pes caprae* Fr., *Agaricus humosus* Fr., *Agaricus tumulosus* Kalchbr. sec. Bres.) Bres. Fung. Trid. t. 32. *Conglobata*, crowded together.

Caespitose. P. 5–10 cm., *umber, becoming blackish, or spotted with cinereous, or grey, or becoming pale and livid cinnamon*, fleshy, fragile at first, then tough, convex, then expanded, plane, or depressed, sometimes umbilicate, often irregular and lobed, dry, margin at first involute, *white pruinose.* St. 4–9 × 1–2 cm., *white, or greyish*, equal, or attenuated downwards, connate at the tuberous base, sometimes branched, *floccosely pulverulent*, then smooth. Gills *whitish cinereous, or cream colour*, crowded, sinuate, adnate, or decurrent. Flesh *white, greyish at the circumference*, fragile, then tough. Spores white, globose, 5–6μ (somewhat angular with age according to Bresadola). Smell faint, of new meal. Taste pleasant, Edible. Woods. Sept.—Nov. Not uncommon. (*v.v.*)

844. **C. tumulosa** (Kalchbr.) Fr. (= *Clitocybe conglobata* (Vitt.) Bres., *Tricholoma humosum* Fr. sec. Quél.) Kalchbr. Icon. t. 5. *Tumulosa*, like a mound.

Caespitose. P. 2·5–8 cm., *dark umber, becoming pale lurid fuscous*, conical, then expanded and *umbonate*, or depressed round the umbo, smooth; margin sometimes repand and torn. St. 2–8 × ·5–2 cm., *pallid*, fleshy, subequal, either attenuated, or *ventricosely thickened at the connate, half-buried base*, curved, ascending, floccosely pruinose, then smooth. Gills *white, then pale cinereous*, adnate, or decurrent. Flesh *white, livid, or cinereous at the circumference when moist*, thick at the disc. Spores white, oval, 6–7 × 4μ. Smell faint, of new meal. Taste pleasant. Edible. Woods. Aug.—Oct. Not uncommon. (*v.v.*)

845. **C. connata** (Schum.) Fr. Bres. Fung. Mang. t. 37. *Connata*, joined together.

Caespitose. P. 5–10 cm., *white, slightly tinged with ochre, or bistre*, convex, *pruinose.* St. 3–6 × ·5–1 cm., *white*, connate at the *swollen base, mealy.* Gills *glaucous white, then cream colour*, adnate, or decurrent, arcuate, distant. Flesh *white*, firm, elastic. Spores white, globose, 6 × 5μ, minutely punctate. Smell pleasant. Parks, and woods. Sept.—Oct. Uncommon. (*v.v.*)

846. **C. pergamena** Cke. Cke. Illus. no. 1132, t. 643. *Pergamena*, parchment.

Caespitose. P. 3–8 cm., *ochraceous, whitish at the margin*, subcartilaginous, convex, then plane, obtusely umbonate, smooth. St. 5–12 × 1–2 cm., *concolorous*, equal, ascending, *apex punctate squamose, cuticle cartilaginous.* Gills *white*, broadly adnate with a decurrent tooth, 4 mm. broad, rather crowded. Flesh *yellowish*, thick at the disc. Spores white, elliptical, 6 × 3–4μ, 1-guttulate. On the ground, and on stumps. Sept.—Oct. Uncommon. (*v.v.*)

847. **C. cryptarum** (Letell.) B. & Br. *Crypta*, a cellar.

Caespitose. P. 3–4 cm., *brown*, somewhat conical, then depressed, spotted, *floccose*. St. 6–9 × 2 cm., *white*, attenuated upwards, more or less compressed, *somewhat striate, virgate*. Gills *white*, subdecurrent, arcuate, narrow. Flesh of stem *mottled*. Spores white. Taste insipid. Sawdust. Oct. Rare.

848. **C. monstrosa** (Sow.) Gillet. Sow. Eng. Fung. t. 283.
Monstrosa, strange.

Often caespitose. P. 3–10 cm., *white, opaque as if whitewashed, often tinged with ochre*, fleshy, convex, umbonate, then waved and lobed; margin incurved. St. 2–6 × 2–2·5 cm., *concolorous, compressed, streaked, downy-squamulose* above, slightly rooting. Gills *white, or cream colour*, scarcely rounded behind, not truly decurrent, rather distant, *broad*, margin waved. Flesh *white*, thick at the disc. Spores white, elliptical, 7–9 × 5μ. On the ground. Sept.—Oct. Uncommon. (*v.v.*)

849. **C. opaca** (With.) Fr. Sow. Eng. Fung. t. 142. *Opaca*, shady.

Caespitose, rarely solitary, *entirely white*. P. 3–7 cm., fleshy, convex, then expanded, umbonate, repand, *covered over with a floccose lustre*. St. 3–8 cm. × 5–8 mm., unequal, *flexuose*, connate at the base, *subfibrillose*. Gills adnato-decurrent, *very crowded*, 3–4 mm. wide. Flesh *white*, thick at the disc. Spores white, elliptical, 6 × 4μ, 1-guttulate. Woods, and pastures. Sept.—Nov. Uncommon. (*v.v.*)

850. **C. occulta** Cke. Cke. Illus. no. 1133, t. 1184. *Occulta*, hidden.

P. 5–7 cm., *whitish, disc smoky*, fleshy, convex, then plane and depressed, smooth, *but innately streaked, or virgate, viscid*; margin *whitish*. St. 4–6 × 1 cm., *white*, equal, or slightly expanded into the p., often curved, *fibrillosely striate*. Gills *white*, adnate, very slightly decurrent, scarcely emarginate, 5 mm. broad, *subdistant*. Flesh *white*, thick at the disc, *cartilaginous*. Spores white. Gregarious. On charred ground. Nov. Rare.

γ. P. attenuated from a fleshy disc towards the margin, at length infundibuliform, or deeply umbilicate.

*P. coloured, or becoming pallid, innately floccose, or silky, bibulous, not moist.

C. gigantea (Sow.) Quél. = **Paxillus giganteus** (Sow.) Fr.

851. **C. maxima** (Fl. Wett.) Fr. (= *Clitocybe geotropa* (Bull.) Quél.)
Maxima, greatest.

P. 12–30 cm., *tan colour, becoming paler, or whitish,* fleshy, *somewhat flaccid,* broadly infundibuliform, gibbous, *umbo central, very dry, becoming silky, or squamulose*; margin involute, *pubescent.* St. 7–10 × 1 cm., *whitish,* attenuated upwards, *fibrillosely-striate, elastic.* Gills *whitish, deeply decurrent,* pointed at both ends, somewhat crowded, soft. Flesh *white,* thick at the disc, soft. Spores white, elliptical, 4–6 × 3–4μ. Smell and taste pleasant. Edible. Woods, and pastures. July—Nov. Not uncommon. (*v.v.*)

852. **C. infundibuliformis** (Schaeff.) Fr. Rolland, Champ. t. 26, no. 51. *Infundibuliformis,* funnel-shaped.

P. 3–6 cm., *flesh colour, then pale tan,* fleshy, moderately firm, convexo-depressed, *gibbous with an umbo,* at length infundibuliform, *silky,* bibulous; margin at first involute. St. 3–8 cm. × 4–8 mm., *concolorous, conico-attenuated,* rarely equal, firm, *elastic*; base swollen, and white tomentose. Gills *shining white,* very decurrent, *somewhat crowded,* very pointed at each end, soft. Flesh *white,* thick at the disc, soft. Spores white, ovoid, 6–7 × 6μ, pointed at the base, punctate, 1-guttulate. Smell and taste pleasant. Edible. Woods, heaths, and pastures. June—Dec. Very common. (*v.v.*)

var. **membranacea** (Fl. Dan.) Fr. Cke. Illus. no. 1135, t. 646. *Membranacea,* skinny.

Differs from the type in being *thinner in all its parts, in the equal st., and the brighter coloured, and not umbonate p.* Pine woods, and pastures. June—Dec. Not uncommon. (*v.v.*)

853. **C. trullaeformis** (Fr.) B. & Br. *Trullaeformis,* ladle-shaped.

P. 3–5 cm., *fuscous cinereous, fleshy,* infundibuliform, flattened at the margin, always obtuse, *flocculosely villous,* dry. St. 5 cm. × 8–10 mm., *cinereous,* attenuated upwards, firm, elastic, *fibrillosely-striate,* base villous. Gills *shining white,* decurrent, *distant,* 4–6 mm. broad, *connected by veins.* Flesh *snow white,* equal. Spores white, elliptical, 6 × 3–4μ, minutely punctate. Borders of fir wood, hedgerows, and thickets. Oct. Uncommon.

854. **C. incilis** Fr. *Incilis,* incised.

P. 2·5–5 cm., *brick-red,* fleshy, *plano-umbilicate,* then infundibuliform, *silky-flocculose,* obtuse; margin involute, *crenate.* St. 1–2 cm. × 4–6 mm., *concolorous, attenuated downwards,* often compressed, tough, at first *covered with an evanescent, flocculose pruina.* Gills *white, becoming pale,* decurrent, arcuate, 4–6 mm. broad, *distant, often reticulated with veins.* Flesh *whitish,* thin at the margin. Spores white, elliptical, 8–9 × 5–6μ, 1-guttulate. Smell of new meal, sometimes absent. Coniferous woods. March—Nov. Uncommon. (*v.v.*)

855. **C. sinopica** Fr. *Sinopica*, of Sinope, where red lead is found.

P. 3–5 cm., *brick-red, becoming paler*, fleshy, *plano-depressed*, slightly or deeply umbilicate, then repand, flocculose and bibulous, then *rimosely rivulose* and broken up into squamules; margin undulate, silky. St. 3–5 × ·5–1 cm., *concolorous*, equal, firm, *fibrillosely striate.* Gills *white, becoming yellow*, decurrent, arcuate, *very crowded*, 3–4 mm. broad. Flesh *white, reddish under the cuticle*, thick at the disc. Spores white, subglobose, 8–9 × 6–7μ. Smell strong, of new meal. Heaths, woods, and burnt ground. May—Oct. Uncommon. (*v.v.*)

856. **C. parilis** Fr. Fr. Icon. t. 48, fig. 6. *Parilis*, equal.

P. 1·5–3 cm., *fuscous, then greyish white*, slightly fleshy, convex, then plane and depressed at the disc, or umbilicate, disc *atomate, or flocculose*; margin involute, *deflexed.* St. 2·5 cm. × 3–6 mm., *fuliginous, becoming fuscous grey*, tough, equal, smooth. Gills *grey, becoming whitish*, deeply decurrent, *very crowded, narrow.* Flesh *greyish, becoming white.* Spores white, oblong, 9 × 3μ, 1–2-guttulate; "greyish in the mass, subglobose, 6 × 5μ" Rick. Woods, and hilly pastures. Sept. —Oct. Uncommon. (*v.v.*)

**P. coloured, or pallid, glabrous, moist in wet weather.

857. **C. gilva** Fr. (= *Clitocybe subinvoluta* Batsch sec. Quél.) *Gilva*, pale yellow.

P. 4–10 cm., *pale yellowish, fleshy, compact, convex, then depressed, very obtuse*, smooth, *dull*, moist, polished and *shining when dry*, often spotted as with drops; margin very involute, *swollen, villose.* St. 2·5–5 × 1–2·5 cm., *paler than the p., fleshy*, subequal, smooth, base villous. Gills *pallid, then ochraceous especially at the edge*, decurrent, thin, narrow, arcuate, *often branched and anastomosing.* Flesh *concolorous*, compact, at length fragile. Spores white, globose, 4–6μ, punctate. Pine woods. Oct. Uncommon. (*v.v.*)

858. **C. subinvoluta** W. G. Sm. non Batsch. Saund. & Sm. t. 36. *Subinvoluta*, somewhat rolled in.

P. 5–10 cm., *creamy flesh colour*, fleshy, convex, then plane, or depressed, gibbous, or umbonate, leathery, margin incurved. St. 5–10 × 1·5–2 cm., *pinkish flesh colour, zoned with spots*, attenuated upwards from the subbulbous base, fibrillose. Gills *white, then yellowish*, deeply decurrent, broad, rather crowded. Flesh *buff-white, darker below*, thick, firm. Spores white, pip-shaped, 6–7 × 4–5μ, 1-guttulate. Smell and taste pleasant. Edible. Pastures, and under firs. Oct.—Nov. Not uncommon. (*v.v.*)

Ag. subinvolutus Batsch = **Paxillus involutus** (Batsch) Fr. var. **subinvolutus** (Batsch) W. G. Sm.

C. spinulosa Stev. & Sm. = **Clitocybe subinvoluta** W. G. Sm. "Saunders wrongly described the spores as echinulate" W. G. Sm. in litt.

859. **C. geotropa** (Bull.) Fr. Grev. Scot. Crypt. Fl. t. 41, as *Agaricus gilvus*. γῆ, earth; τρόπος, turned.

P. 3–20 cm., *tan flesh colour, very fleshy*, convex, then plano-depressed, generally *gibbous, very smooth*, moist in wet weather, when young spotted as with drops, the spots vanishing with age; margin involute, thin, *pubescent*. St. 5–12 × 2–3 cm., *white, becoming yellow*, fleshy, slightly attenuated upwards, *subfibrillose*. Gills *white, becoming pale*, deeply decurrent, 4–6 mm. broad, somewhat crowded. Flesh *white*, thick, firm. Spores white, subglobose, 5–7μ. Smell and taste pleasant. Edible. Woods, and pastures, often forming large rings. Sept.—Dec. Common. (*v.v.*)

860. **C. splendens** (Pers.) Fr. Fr. Icon. t. 55, upper figs. *Splendens*, shining.

P. 5–8 cm., *pale yellowish, becoming yellow*, somewhat fleshy, convex, then plano-depressed, at length infundibuliform, smooth, shining; margin reflexed, *white, mealy*. St. 4–5 × 1–2 cm., *white, becoming light yellow, or ochraceous*, equal, or attenuated upwards, elastic, smooth. Gills *white, becoming light yellow*, deeply decurrent, thin, crowded, often forked at the base. Flesh *white, becoming concolorous*, thin at the margin. Spores white, subglobose, 4–6 × 4–5μ, 1-guttulate, punctate. Smell and taste pleasant. Woods. Sept.—Nov. Uncommon. (*v.v.*)

861. **C. inversa** (Scop.) Fr. Barla, Champ. Alp. Marit. t. 60, figs. 6–8. *Inversa*, inverted.

P. 5–8 cm., *brick colour, or liver-rufescent, fleshy, somewhat fragile*, convexo-plane, obtuse, then infundibuliform and undulated, *very smooth, moist when fresh*, sloping towards the margin, sometimes excentric. St. 4–6 × 1–1·5 cm., *whitish*, compressed, *cuticle rigid*, somewhat rooted and white villous at base. Gills *whitish, becoming reddish at the edge*, decurrent, 3–4 mm. broad, *crowded*. Flesh *of the same colour as the p. but paler*, thin, rigid, *fragile*. Spores white, globose, 4μ, minutely echinulate, 1-guttulate. Smell and taste acid. Coniferous woods. Aug.—Dec. Common. (*v.v.*)

862. **C. flaccida** (Sow.) Fr. Sow. Eng. Fung. t. 185. *Flaccida*, flabby.

P. 5–8 cm., *tawny ferruginous, shining*, not becoming pale, slightly fleshy, *tough*, orbicular, *flaccid* especially when dry, *umbilicate*, then infundibuliform, *smooth*, rarely rimuloso-squamulose; margin spreading, slightly convex. St. 2·5–5 × ·5–1 cm., *rubiginous ferruginous*, *elastic, tough*, subequal, polished, base thickened and villous. Gills *whitish, becoming yellow especially at the edge*, deeply almost *obconico-decurrent, very arcuate, very crowded, narrow*, 1–2 mm. broad. Flesh *pallid*, thin, fragile when fresh, flaccid when dry. Spores white, globose, 3–4μ, minutely warted, 1-guttulate. Woods, and heaths, often forming rings. Sept.—Dec. Common. (*v.v.*)

var. **lobata** (Sow.) Cke. Sow. Eng. Fung. t. 186. *Lobata*, lobed.

Differs from the type in its *more caespitose habit, in the darker colour, the lobed, or contorted margin of the p., and the st. thickened upwards*. Woods, and heaths. Sept.—Nov. Common. (*v.v.*)

863. **C. vermicularis** Fr. Bres. Fung. Trid. t. 49.
Vermicularis, belonging to a little worm.

P. 2–4 cm., *deep flesh colour, then tan flesh colour*, slightly fleshy, umbilicato-convex, then expanded and infundibuliform, undulato-lobed, smooth, moist, slightly hygrophanous; margin involute, *pruinosely tomentose*. St. 3–5 cm. × 3–6 mm., *whitish*, equal, often *compressed* and curved, fibrillosely striate, *apex mealy* and often surrounded by a floccose zone, *base white tomentose and arising from stout, palmately branched, strigose mycelia*. Gills *white*, then *cream colour, edge ochraceous, slightly decurrent*, attenuated, or obtuse at the base, very crowded, easily separating from the hymenophore, thin. Flesh *concolorous*, thin. Spores white, elliptical, 5 × 3μ. Smell slight, of new meal. Taste somewhat acid. Edible. Coniferous woods. Sept.—Oct. Uncommon.

864. **C. senilis** Fr. Fr. Icon. t. 56, fig. 1. *Senilis*, aged.

P. 4–8 cm., *dingy fuscous tan, or brown becoming paler*, fleshy-membranaceous, flaccid, disc depressed, soon infundibuliform, smooth, *concentrically cracked*; margin spreading. St. 4–5 × ·5–1 cm., *whitish*, equal, often ascending, smooth. Gills *whitish, then concolorous with the p.*, deeply decurrent, linear, *narrow*, very crowded. Flesh *white*, thin, flaccid. Spores white, pip-shaped, 6 × 3–4μ, 1-guttulate. Woods, and lawns. Sept.—Oct. Uncommon. (*v.v.*)

***P. shining white.

865. **C. catinus** Fr. Fr. Icon. t. 51, fig. 4. *Catinus*, a bowl.

P. 5–8 cm., *white*, becoming discoloured with age, fleshy, plane, then infundibuliform, always obtuse, *smooth*. St. 5–8 × 5–1 cm., *white, elastic*, tough; base thickened and tomentose. Gills *white*, decurrent, *straight, descending*, broad, not much crowded. Flesh *white*, thin, *flaccid*. Spores white, pip-shaped, 4–5 × 3μ, minutely punctate. Smell and taste pleasant. Edible. Woods, and among dead leaves. Aug.—Oct. Not uncommon. (*v.v.*)

866. **C. tuba** Fr. Cke. Illus. no. 164, t. 112. *Tuba*, a trumpet.

Entirely white. P. 5–8 cm., fleshy, *thin*, convexo-plane, *umbilicate, dead white when moist, shining whitish when dry*, smooth, slightly silky when young. St. 2·5–5 cm. × 5–6 mm., *very tough, equal, at length compressed, smooth.* Gills *becoming pale, deeply decurrent, horizontal, very crowded*, 6 mm. broad. Spores white, elliptical, 4–5 × 2–3μ, "punctate" Quél. Smell none, or of new meal. Coniferous woods. Oct.—Nov. Uncommon.

867. **C. ericetorum** (Bull.) Fr. Bres. Fung. Trid. t. 113. *Ericetorum*, of heaths.

P. 2–5 cm., *shining white, becoming slightly yellowish with age*, somewhat fleshy, convex, then *umbilicate* and cup-shaped, smooth, at length striate; margin *undulate, or lobed.* St. 2–3 cm. × 4–5 mm., *white*, attenuated downwards, sometimes compressed, *pubescent.* Gills *white*, more or less decurrent, subdistant, *often connected by veins.* Flesh *white*, thin, firm. Spores white, oval, 4–5 × 2·5–3μ, sparsely and minutely rough. Smell pleasant, like *Anthoxanthum odoratum.* Taste somewhat acrid. Edible. Heaths, and lawns. Sept.—Nov. Rather uncommon. (*v.v.*)

B. P. fleshy-membranaceous, truly hygrophanous.

δ. P. thin, depressed, then cup-shaped. Colour dingy when moist.

868. **C. cyathiformis** (Bull.) Fr. κύαθος, a cup; *forma*, shape.

P. 2–7 cm., *fuscous cinereous, or dark bistre, becoming paler when dry*, slightly fleshy, plano-depressed, then cup-shaped, often undulated, somewhat shining when moist, opaque when dry, very hygrophanous; margin *persistently incurved.* St. 5–10 cm. × 6–9 mm., *concolorous, or paler*, elastic, attenuated upwards, *fibrillosely-reticulated*, base white villous. Gills cinereous fuscous, adnate, or decurrent, *connate at the base, distant*, sometimes branched. Flesh *concolorous*, watery, thin. Spores white, elliptical, 10–11 × 5–6μ, punctate. Smell pleasant, or none. Edible. Woods, pastures, rarely on rotten wood. Aug.—Feb. Common. (*v.v.*)

var. **cinerascens** (Batsch) Fr. (= *Clitocybe cinerascens* (Batsch) W. G. Sm.) *Cinerascens*, becoming ash-coloured.

Differs from the type in its *smaller size, in the plane then depressed p., and yellowish gills.* Autumn. Uncommon.

869. **C. expallens** (Pers.) Fr. (= *Clitocybe vibecina* Fr. sec. Quél.) *Expallens*, becoming pale.

P. 2–5 cm., *cinereous fuscous, becoming whitish*, at first sprinkled with white-silky dew, slightly fleshy, convexo-plane, obtuse, then plano-infundibuliform, somewhat zoned when dry; margin *mem-*

branaceous, striate, soon expanded. St. 4–7 cm. × 4–6 mm., *whitish,* tough, equal, smooth, *apex white-silky.* Gills *greyish, decurrent,* acute at both ends, thin, subdistant, soft. Flesh *greyish,* thin. Spores white, "broadly elliptical, 7–9 × 6–7μ, smooth" Rick. Taste mild. Edible. Woods, and pastures. Aug.—Nov. Not uncommon. (*v.v.*)

870. **C. albo-cinerea** Rea. Trans. Brit. Myc. Soc. IV, t. 8.
Albus, white; *cinerea,* ash-coloured.

P. 2–3 cm., *cinereous fuscous, becoming pale,* fleshy, convexo-umbilicate, then expanded and cup-shaped, silky; margin involute. St. 5–6 cm. × 3–5 mm., *white,* equal, smooth, base white-tomentose. Gills *white,* decurrent, *narrow,* 1–2 mm. broad, crowded. Flesh *white,* thin, firm. Spores white, elliptical, 5–6 × 4μ, minutely punctate. Smell and taste pleasant. Edible. Woods, and pastures. Sept.—Oct. Not uncommon. (*v.v.*)

871. **C. obbata** Fr. Fr. Icon. t. 57, fig. 1. *Obbata,* a kind of cup.

P. 2–3 cm., *fuscous blackish, or cinereous, becoming very pale when dry, submembranaceous,* convexo-plane, with a *broadly umbilicate disc,* smooth, *striate to the middle.* St. 4–5 cm. × 2–4 mm., *fuscous cinereous,* equal, often compressed, ascending, or flexuose, *striate with white,* base often subbulbous. Gills *dark cinereous, slightly decurrent, distant,* broad. Flesh *ochraceous,* thin. Spores white, broadly elliptical, 9–10 × 6–7μ. Taste mild. Edible. Woods, and pastures. Sept.—Jan. Uncommon. (*v.v.*)

872. **C. pruinosa** (Lasch) Fr. (= *Omphalia litua* Fr. sec. Quél.)
Pruinosa, rimy.

P. 2·5–5 cm., *brown, becoming cinereous,* hygrophanous, fleshy-membranaceous, umbilicate, and *covered with a lead-grey pruina,* then broadly infundibuliform and smooth, sometimes squamulose. St. 2·5–5 cm. × 2–4 mm., *concolorous, or paler,* equal, often ascending, or curved, *fibrillose.* Gills *white, then dingy, or bistre,* decurrent, *crowded, narrow,* arcuate, then scythe-shaped. Flesh *becoming cinereous,* thin. Spores "6–8 × 2–4μ" ex Britz. in Sacc. (1915). Taste mild. Edible. Pine woods, and on rotten wood. Nov.—Dec. Rare.

873. **C. concava** (Scop.) Fr. Fr. Icon. t. 57, fig. 2.
Concava, hollowed out.

P. 3–5 cm., *fuliginous, then cinereous, or hoary-clay,* hygrophanous, slightly fleshy, very thin, *flaccid,* plano-convex, *widely and deeply umbilicate,* then *wholly concave, the convexo-plane border undulated,* smooth; margin even. St. 3–6 cm. × 3–8 mm., *cinereous,* tough, equal, smooth, *base attenuated.* Gills *dark fuliginous, then greyish,* decurrent, *arcuate,* very crowded, 2–4 mm. broad. Flesh *pallid,* tough, very thin. Spores "greenish, ovoid, 8–12 × 6–8μ" Sacc. Coniferous woods. Sept.—Oct. Rare. (*v.v.*)

874. **C. suaveolens** (Schum.) Fr. (= *Clitocybe fragrans* Sow. sec. Quél.)
Suaveolens, sweet smelling.

P. 2–3 cm., *white when moist with the disc darker, becoming pure white when dry*, fleshy, thin, convexo-plane, then depressed, *often umbilicate* and somewhat infundibuliform, *discoid*, smooth; margin *pellucidly striate*, at length reflexed. St. 4–5 cm. × 2–4 mm., *white, becoming pinkish, elastic, base swollen and villous*. Gills *whitish, then discoloured, adnato-decurrent, crowded*, thin. Flesh *white*, thin. Spores white, "elliptical, 6–7 × 3–4μ" Rick. Smell very pleasant, of aniseed. Taste pleasant. Edible. Woods, especially coniferous. Sept.—Dec. Uncommon.

875. **C. brumalis** Fr. Cke. Illus. no. 170, t. 114.
Brumalis, pertaining to winter.

P. 3–5 cm., *livid when moist, becoming whitish, and at length yellowish when dry, disc generally darker*, fleshy-membranaceous, convex, umbilicate, reflexed at the circumference, then infundibuliform, often irregular and undulated, smooth. St. 3–6 cm. × 3–8 mm., *greyish, then whitish*, equal, or slightly thickened at the apex, at length compressed, somewhat incurved, smooth, elastic, base white-villous. Gills *livid, becoming yellowish, or whitish*, decurrent, arcuate, then descending, 2 mm. broad, crowded. Flesh *whitish*, thin. Spores white, elliptical, 5–6 × 3μ, 1-guttulate. Smell faint, pleasant. Taste pleasant. Edible. Woods, and pastures. Aug.—Jan. Common. (*v.v.*)

ε. P. rather fleshy, convex then flattened, or depressed, polished. Colour dingy, or becoming pale.

*Gills becoming cinereous.

876. **C. orbiformis** Fr. *Orbiformis*, round-shaped.

P. 4–5 cm., *greyish fuliginous*, slightly fleshy, convexo-plane, *very obtuse, orbicular*, scarcely depressed, smooth, hygrophanous; margin *spreading, finally striate*. St. 6–8 cm. × 8–10 mm., *grey*, attenuated upwards from the villous, thickened base, elastic, *fibrillosely* striate. Gills *whitish, then greyish, adnate*, subdecurrent, plane, horizontal, little crowded. Flesh thin. Spores white, ovoid, "6–7 × 3–4μ" Sacc. Pine woods. Sept.—Nov. Rare.

877. **C. metachroa** (Fr.) Berk.
μετὰ, change; χρώς, colour of the skin.

P. 2–8 cm., *fuscous cinereous, then livid, whitish when dry*, slightly fleshy, convex, and subumbonate, soon plane, or depressed; margin finally *slightly striate*. St. 3–4 cm. × 4–6 mm., *grey*, equal, often compressed, tough, *cuticle horny*, fibrous, *apex white-mealy*. Gills *whitish cinereous, adnate*, scarcely decurrent, crowded, linear, plane, thin.

Flesh *greyish, becoming whitish,* thin. Spores white, elliptical, or pip-shaped, 6 × 3μ, 1-guttulate. Pine woods. Aug.—Nov. Common. (*v.v.*)

878. **C. incana** Quél. Trans. Brit. Myc. Soc. IV, t. 5. *Incana*, hoary.

P. 3–5 cm., *mouse grey, margin white,* convexo-plane, then depressed, hygrophanous, *pruinose.* St. 4–6 cm. × 4–6 mm., *pearl grey,* straight, or slightly curved, *base white floccose.* Gills *greyish, becoming somewhat ochraceous,* decurrent with a tooth, 2–4 mm. broad. Flesh *greyish,* soft, thin. Spores white, globose, 3μ. Amongst fir needles. Oct.—Nov. Uncommon. (*v.v.*)

879. **C. pausiaca** Fr. Fr. Icon. t. 58, fig. 2. *Pausiaca*, olive colour.

P. 2·5–4 cm., *cinereous, then olivaceous, becoming somewhat ochraceous* when dry, fleshy, thin, convex, sometimes umbonate, then plane and depressed, hygrophanous, silky hoary when young, then smooth. St. 5–7 cm. × 3–6 mm., *concolorous,* tough, equal, striate, often undulated, *apex white pruinose.* Gills *olivaceous, or brownish,* obtusely adnate, very broad behind, very crowded, *semicircular.* Flesh *ochraceous,* thin. Spores white, globose, 3–4μ, 1-guttulate. Smell weak, frumentaceous. Pine woods. Oct. Uncommon. (*v.v.*)

880. **C. ditopus** Fr. διττός, double; πούς, foot.

P. 5–6 cm., *cinereous, drying deep ochre from the centre outwards,* somewhat fleshy, tough, convexo-plane, obtuse, then inverted, infundibuliform, and often undulato-lobed. St. 3–5 cm. × 5–15 mm., *pale cinereous, compressed,* equal, *naked, often very white floccose at the basal half of the st.* Gills *dark cinereous, adnate, crowded,* thin, at length turned upwards, and divergent in the lobes, often undulate. Flesh *greyish, then white,* thin. Spores white, globose, 3–4μ. Smell strong, of new meal. Woods, and amongst dead leaves. Sept.—Nov. Common. (*v.v.*)

**Gills whitish.

881. **C. diatreta** Fr. Cke. Illus. no. 173, t. 232.
διατρητός, pierced through.

P. 2–3 cm., *flesh colour, then tan colour,* slightly fleshy, tough, convex, regular, obtuse, then plano-depressed, often flexuose, becoming flaccid, smooth, hygrophanous; margin *white,* incurved, *pruinose.* St. 3–5 cm. × 4–8 mm., *pallid,* elastic, flexile, equal, round, smooth, villose at the base. Gills *whitish flesh colour, then whitish, adnate,* sharp pointed behind, decurrent with a tooth, 2 mm. broad, crowded. Flesh *concolorous, becoming whitish,* thin. Spores white, elliptical, 7–8 × 6μ, often pointed at one end, 1-guttulate. Coniferous woods. Oct.—Nov. Uncommon. (*v.v.*)

882. **C. fragrans** (Sow.) Fr. (= *Clitocybe suaveolens* (Schum.) Fr. sec. Quél.) Sow. Eng. Fung. t. 10. *Fragrans*, scented.

P. 2–5 cm., *watery pallid when moist, whitish when dry, of one colour, not darker at the disc,* slightly fleshy, convex, then plane, or subdepressed smooth; margin *slightly striate when moist.* St. 5–7 cm. × 4–6 mm., *concolorous, or yellowish,* equal, elastic, smooth, *apex obsoletely pruinose,* base very often villous. Gills *whitish,* adnate, sharp-pointed behind, subdecurrent, *rather crowded,* broader than the flesh of the p. Flesh *white,* watery, thin. Spores white, elliptical, 6–7 × 4μ. Smell and taste very pleasant, of aniseed. Edible. Woods, and pastures. July—Jan. Common. (*v.v.*)

883. **C. angustissima** (Lasch) Fr. Fr. Icon. t. 59. *Angustissima,* very narrow.

P. 3–5 cm., *cream, or flesh coloured, shining whitish when dry,* fleshy, thin, plano-depressed, smooth; margin spreading, slightly striate when old. St. 5–7 cm. × 2–3 mm., *concolorous,* often curved, or *flexuose, internally fibrous,* base sometimes pubescent. Gills *white, subdecurrent, very crowded, narrow,* thin. Flesh *whitish,* very thin, moderately firm. Spores white, elliptical, 3–4 × 2–3μ. Woods, and amongst leaves. Sept.—Oct. Uncommon. (*v.v.*)

884. **C. obsoleta** (Batsch) Fr. *Obsoleta,* worn out.

P. 2–3 cm., *grey,* or cream colour, *soon turning whitish, clay white when dry, sometimes inclining to flesh colour,* somewhat fleshy, *soft,* convex, or gibbous, then plane, or depressed, hygrophanous, smooth. St. 5 cm. × 6–8 mm., *whitish, elastic,* tough, round, often compressed, equal, smooth, *apex pruinose.* Gills *greyish, then whitish, obtusely adnate, almost rounded behind,* then adnato-decurrent, *broad,* crowded. Flesh *whitish,* thin, soft. Spores white, elliptical, 7 × 4–5μ. Smell faint, pleasant. Coniferous woods, and amongst grass and leaves. Oct.—Nov. Uncommon.

ζ. P. deformed, more or less squamulose.

885. **C. ectypa** Fr. (= *Collybia ectypa* (Fr.) Quél.) Fr. Icon. t. 59, fig. 1. ἔκτυπος, wrought in relief.

P. 4–7·5 cm., *dingy, or light yellow honey colour, then rufescent, or brownish, fleshy,* somewhat thin, convex, then rather plane, or depressed, *disc streaked with innate fibrils radiating from the centre,* as if sprinkled with soot, or squamulose; margin very thin, *striate.* St. 5–10 × ·5–1 cm., *dingy light yellow, then olivaceous, becoming black at the often bulbous base, elastic,* equal. Gills *white, soon pale, then spotted rufous, somewhat mealy,* adnate, or decurrent with a rather delicate tooth, *distant,* often connected by veins. Flesh *pallid straw colour, thin at the margin.* Spores white, elliptical, 9 × 6–7μ, 1-guttulate. Smell pleasant, of aniseed, at length foetid. Meadows, damp places, and peat bogs. June—Dec. Rare.

C. Sadleri B. & Br. = **Hypholoma fasciculare** (Huds.) Fr.

Spores white; hymenium pulverulent.

Laccaria B. & Br.

(*Lac*, a resinous excretion left by the lac insect.)

Pileus fleshy, regular, or irregular. Stem central, externally fibrous. Gills adnate with a decurrent tooth, pulverulent. Spores white, globose, or elliptical, echinulate, or verrucose. Growing on the ground, and on wood.

886. **L. laccata** (Scop.) B. & Br. (= *Clitocybe laccata* (Scop.) Fr.; *Collybia laccata* (Scop.) Quél.) Cke. Illus. no. 179, t. 139, figs. coloured red, as *Clitocybe laccata* Scop.
Lac, the exudation from the lac insect.

P. 3–5 cm., *rufous flesh colour when moist, ochraceous when dry*, fleshy, convex, then rather plane, more or less umbilicato-depressed, dry, *very hygrophanous*, becoming pale in drying, *the cuticle often breaking up into mealy squamules*, or somewhat silky, sometimes undulato-crisped and irregularly shaped. St. 7–10 cm. × 6–10 mm., *concolorous, tough*, fibrous, equal, often flexuose, or twisted, *fibrillose, base white villous*. Gills *flesh colour, then white mealy*, adnate with a decurrent tooth, very broad, distant, plane, thick. Flesh *concolorous*, somewhat thin, firm in the st. Spores white, globose, 8–9μ, echinulate. Cystidia "on edge of gill clavate-vermiform, 50–60 × 9–12μ" Rick. Taste mild. Edible. Heaths, and woods. June—Dec. Very common. (*v.v.*)

var. **proxima** (Boud.) Maire. Boud. Icon. t. 60, as *Laccaria proxima* Boud. *Proxima*, very near.

Differs from the type in its *brighter colour, and the larger elliptical spores*, 10–15 × 6–7μ, minutely echinulate[1]. Heaths, and woods. Sept.—Nov. Not uncommon. (*v.v.*)

var. **amethystina** (Vaill.) B. & Br. *Amethystina*, amethyst colour.

Differs from the type in the *whole of the plant, including the flesh, being of a beautiful deep violet colour, becoming paler when dry*. Woods, and pastures. June—Dec. Common. (*v.v.*)

887. **L. tortilis** (Bolt.) Boud. Boud. Icon. t. 59. *Tortilis*, twisted.

P. 1–2·5 cm., *pale rose, or slightly yellowish, striate to the deeper coloured* disc, membranaceous, thin, convex, then plane and depressed;

[1] Both elliptical and globose spores have been found on the same plant.

margin *often undulate*. St. 1–2·5 cm. × 2–3 mm., *reddish yellow, or pale*, equal, or attenuated downwards, *slightly fibrillose*. Gills *concolorous, becoming white mealy*, adnate, *with a very slight tooth, broad, often connected by veins*. Flesh *pinkish*, very thin. Spores white, globose, 8–10μ, echinulate. Charcoal heaps, roadsides, and bare soil in woods. Aug.—Sept. Not uncommon. (*v.v.*)

888. **L. bella** (Pers.) B. & Br. (= *Clitocybe bella* (Pers.) Fr.; *Collybia bella* (Pers.) Quél.) *Bella*, lovely.

P. 3–5 cm., *dark yellow, or golden, sometimes rufescent, sprinkled with darker, or orange coloured squamules, becoming pale*, somewhat fleshy, pliant, convex, then expanded, depressed at the disc, then undulato-repand. St. 5 cm. × 4–6 mm., *bright yellow, or becoming yellow, tough*, equal, fibrous, *rivulose with the fibrils*. Gills *yellow, then rufescent and white mealy, adnate*, then decurrent with a tooth, very broad, *distant, connected by veins*, sometimes branched. Spores white, subglobose, 7 × 5–7μ, minutely warted. Smell *foetid*. On decaying coniferous stumps. Sept. Rare.

889. **L. nana** Massee. Massee, Kew Bull. (1913), t. to face p. 195, figs. 17–20. νάννος, dwarf.

P. 1 cm., *livid cinnamon, becoming paler*, somewhat fleshy, hemispherical, then plane and concave, smooth; margin at first *covered with white meal*. St. 1 cm., *white*, fibrillose. Gills *pale, at length white mealy*, adnate, attenuated at the base, rather distant. Spores white, globose, 15–16μ, echinulate. Naked soil under trees.

Spores white; hymenium waxy.

Hygrophorus Fr.

(ὑγρός, moist; φέρω, I bear.)

Pileus fleshy, regular, viscid, or dry. Stem central, fleshy. Gills decurrent, or adnato-decurrent. Spores white, very rarely slightly coloured, elliptical, oval, globose, clavate, pip-shaped, or oblong-elliptical, smooth, continuous. Cystidia present, or absent. Growing on the ground, very rarely on wood.

I. Universal veil *viscid*, with occasionally a floccose partial one, which is annular, or marginal. St. *clothed with scales*, or more frequently *rough with dots above*. Gills adnato-decurrent.

*White, or yellowish white.

890. **H. chrysodon** Fr. Cke. Illus. no. 872, t. 885. χρυσός, gold; ὀδούς, a tooth.

P. 5–7 cm., *white, or yellowish, covered with evanescent, yellow, floccose squamules*, which are more permanent at the involute margin, convex, then plane, viscid. St. 5–7·5 × 1–1·5 cm., *white, covered with minute, light yellow squamules,* which form a zone at the apex. Gills *white, somewhat yellowish at the edge,* adnate, or decurrent, broad, distant, sometimes crisped. Flesh *white, sometimes reddish.* Spores white, elliptical, 6–7 × 3μ. Smell pleasant, taste mild. Edible. Oak, and beech woods. Aug.—Nov. Not uncommon. (*v.v.*)

var. **leucodon** (A. & S.) Fr. λευκός, white; ὀδούς, a tooth.

Differs from the type in *having white squamules.*

891. **H. eburneus** (Bull.) Fr. Cke. Illus. no. 873, t. 886.
Eburneus, ivory white.

Entirely shining white, becoming yellowish with age. P. 3–10 cm., convexo-plane, somewhat repand, *very glutinous,* margin involute, *at first pubescent.* St. 3–8 × 1–1·5 cm., *glutinous, rough at the apex with dots in the form of squamules,* unequal. Gills decurrent, distant, veined at the base. Spores white, elliptical, 8 × 4μ. Smell not unpleasant, taste mild. Woods, and pastures. Aug.—Nov. Common. (*v.v.*)

892. **H. cossus** (Sow.) Fr. Boud. Icon. t. 30.
Cossus ligniperda, the Goat moth.

P. 4–8 cm., *white, disc ochraceous,* convexo-plane, then expanded and depressed, *umbonate, very viscid.* St. 5–10 cm. × 5–12 mm., *white, or becoming tinged with yellow, viscid,* equal, or slightly attenuated at the base, *furfuraceous and granular* at the apex. Gills *white,* decurrent, distant, thick, connected by veins. Spores white, oval, 8–9 × 5–6μ, 1-guttulate. Smell strong, like that of the larva of *Cossus ligniperda.* Woods, and under conifers. Aug.—Nov. Common. (*v.v.*)

893. **H. melizeus** Fr. Fr. Icon. t. 165, fig. 3. μέλι, honey.

Internally and externally becoming yellowish tan. P. 2·5–4 cm., disc fleshy, convex, then plane, obtuse, often repand, viscid; margin thin, *at first pubescent.* St. 7–8 cm. × 6–10 mm., *attenuated downwards,* subfusiform, apex *rough with innate, floccose, white granules.* Gills deeply decurrent, distant, connected by veins. "Spores elliptic-oblong, apiculate, creamy-white, 10 × 5μ" Mass. & Crossl. Smell pleasant. Woods. Nov. Uncommon.

894. **H. discoxanthus** (Fr.) Rea. Trans. Brit. Myc. Soc. III, t. 3.
δίσκος, disc; ξανθός, yellow.

P. 4–6 cm., *white, then yellowish, deeper coloured at the centre, the extreme margin becoming brownish with age,* viscid, convex, then expanded and revolute, disc depressed. St. 3–4 cm. × 6–12 mm., *soon becoming reddish brown,* apex white farinaceous, viscid, gradually attenuated downwards, often curved. Gills *white, then yellowish, edge turning reddish when bruised,* and *then finally reddish brown, especially towards the margin of the pileus.* Flesh *white, becoming reddish in the stem.* Spores white, pruniform, apiculate, 6–7 × 4μ, 1-guttulate. Smell *pleasant, like aniseed.* Parks, and pastures. Sept.—Oct. Uncommon. (*v.v.*)

895. **H. penarius** Fr. Sverig. ätl. Svamp. t. 48.
Penarius, for provisions.

P. 7–10 cm., *white, then tan colour, opaque,* umbonate, then obtuse, hemispherical, then flattened, *generally dry,* hard; margin at first involute, *exceeding the gills,* undulate when flattened. St. 4 cm. × 12 mm. at apex, *pale white, often yellowish at the base,* compact, *hard, attenuated at the base into a fusiform root,* ventricose to the neck, then attenuated upwards, *or wholly fusiform-attenuated,* smeared with tenacious easily dried slime, *scabrous.* Gills *white, or tan,* adnato-decurrent, *distant,* thick, 6–8 mm. broad, *rigid,* veined. Flesh *white, compact,* thick. Spores white, "ovate-spherical or ovate-oblong, 7–8 × 3–4μ" Sacc. Smell pleasant, taste sweet. Edible. Oak woods. Oct. Uncommon. (*v.v.*)

896. **H. pulverulentus** B. & Br. Quél. Soc. sc. n. de Rouen (1879), t. 3, fig. 9. *Pulverulentus,* dusted.

P. 8–18 mm., *shining white,* pulvinate, *viscous*; margin involute, tomentose. St. 18 × 2–4 mm., *white, wholly powdered with rose-coloured meal,* nearly equal, attenuated at the extreme base. Gills *whitish,* decurrent, thick, obtuse at the edge. Spores white, globose, 7μ. Amongst pine leaves. Nov.—Dec. Rare.

**Reddish.

897. **H. russula** (Schaeff.) Quél. (= *Tricholoma russula* Fr.) Cke. Illus. no. 1116, t. 926, as *Tricholoma russula* Schaeff.
Russula, reddish.

P. 10–20 cm., *flesh colour, or purplish with deeper coloured streaks, paler and whitish at the tomentose margin,* viscid, gibbous, convexo-plane, then depressed. St. 6–12 × 1–2 cm., *white, stained reddish, apex white, farinaceous.* Gills *whitish,* then *spotted with bright red, sinuate, or emarginate,* thin, *rather crowded.* Flesh white. Spores white, elliptical, 7–8 × 4–5μ or 6–7 × 4–5μ, slightly depressed on one side, with a large central gutta. Taste sweet, or slightly bitter. In deciduous woods. Sept.—Oct. Uncommon. (*v.v.*)

898. **H. erubescens** Fr. (= *Limacium rubescens* (Pers.).) Cke. Illus. no. 876, t. 888. *Erubescens,* becoming red.

P. 5–10 cm., *whitish, spotted with rose, slightly viscid,* gibbous, then convexo-plane. St. 5–8 × 2 cm., *whitish stained reddish, tinged yellowish when bruised, or rubbed,* equal, or attenuated at the base. Gills *whitish, washed with flesh colour, decurrent,* somewhat distant. Flesh *yellowish.* Spores white, elliptical, 8–11 × 6μ, with a large central gutta. Taste bitter, then sometimes slightly acrid. Coniferous woods. Sept.—Oct. Uncommon. (*v.v.*)

899. **H. pudorinus** Fr. Cke. Illus. no. 877, t. 911.
Pudorinus, modest.

P. 5–9 cm., *bright reddish flesh colour, disc deeper coloured,* convex, then depressed, viscid; margin *white, pubescent.* St. 5–8 × 1·5–3 cm., *white,* or *flesh colour,* firm, *viscid; apex contracted, rough with floccose granules.* Gills *white, flesh colour near the edge,* adnate, *wide,* thick, often crisped. Flesh *white, rose colour under the cuticle of the pileus.* Spores white, elliptical, 6–7 × 4–5μ. Smell very pleasant, taste sweet. Edible. Coniferous woods. Sept.—Oct. Uncommon. (*v.v.*)

900. **H. glutinifer** Fr. Cke. Illus. no. 878, t. 889.
Gluten, glue; *fero,* I bear.

P. 5–9 cm., *rufescent, whitish round the margin,* convexo-expanded, thin with the exception of the gibbous disc, *pellicle glutinous, disc wrinkled dotted.* St. 7–10 × 1–1·5 cm., *concolorous, apex white-squamulose,* somewhat elastic, *ventricose downwards, with a viscid veil.* Gills *shining white, or pale grey,* arcuato-decurrent, rather thick. Spores white, clavate, 9–10 × 7μ, 3-guttulate. Taste mild. Woods. Sept.—Oct. Uncommon. (*v.v.*)

901. **H. persicinus** Beck. *Persicinus,* pertaining to a peach.

P. 5 cm., *peach colour,* or *somewhat orange,* conical, then hemispherical, shining, even; margin incurved. St. 10 × 2 cm., *pale lilac-peach colour,* base *yellowish,* constricted below the gills. Gills *fuscescent,* adnato-decurrent, edge very obtuse. Spores 15–20 × 5–6μ. Amongst grass in woods. Rare.

***Tawny, or light yellow.

902. **H. arbustivus** Fr. Cke. Illus. no. 879, t. 896, fig. A.
Arbustivus, belonging to plantations.

P. 3–10 cm., *slightly tawny-brick colour, paler round the pubescent margin,* convexo-plane, *obtuse, umbonate,* somewhat repand, viscid, *disc streaked with innate fibrils.* St. 4–9 cm. × 6–15 mm., *pale white,* elastic, cylindrical, viscid, *apex covered with white, free, mealy granules.* Gills *white,* adnate, scarcely decurrent, distant, thick. Spores white, elliptical, 8 × 4–5μ, 2-guttulate. Smell and taste pleasant. Edible. Woods. Sept.—Dec. Not uncommon. (*v.v.*)

903. **H. discoideus** (Pers.) Fr. Gonn. & Rabenh. VIII–IX, t. 10, fig. 4. δίσκος, disc; εἶδος, like.

P. 2·5–6 cm., *pale yellowish inclining to pale, disc darker, somewhat ferruginous*, campanulate, then plane, *obtusely umbonate, very glutinous*. St. 4–6 cm. × 6–10 mm., *pale white, viscid, flocculose, apex with white dots*. Gills *pale yellowish white, or flesh colour*, adnate, decurrent, distant, soft. Flesh *under the cuticle of the umbo ferruginous, yellowish white, or flesh colour elsewhere*. Spores white, elliptical, 6–9 × 5μ, 1-guttulate. Woods. Oct.—Nov. Not uncommon. (*v.v.*)

904. **H. aureus** (Arrh.) Fr. Fr. Icon. t. 166, fig. 2. *Aureus*, golden.

P. 2–4 cm., *bright golden yellow, becoming reddish*, convex, then plane, *glutinous*. St. 4–6 cm. × 6 mm., *becoming tawny, apex white pruinose* above the *glutinous, ring-like, fugacious, tawny-reddish veil*. Gills *white, or yellowish*, adnato-decurrent, distant. Flesh *white*, or *pale ochraceous*. Spores white, elliptical, 8–10 × 5–6μ. Woods. Nov. Rare.

905. **H. aromaticus** (Sow.) Berk. Sow. Eng. Fung. t. 144. ἀρωματικός, fragrant.

P. 5–8 cm., *cinnamon*, convex, then expanded and plane, very fragile, *glutinous*, the gluten in drying sometimes contracting and *forming raised, anastomosing ribs*. St. 3–5 cm. × 4–8 mm., *concolorous*, sub-equal. Gills *white, with a pink tinge*, slightly decurrent, 2–3 mm. broad. Flesh *bruising blackish*. Smell spicy, taste like peppermint, acrid. Amongst grass. Rare.

****Olivaceous umber.

H. latitabundus Britz. = **Hygrophorus Clarkii** (B. & Br.) W. G. Sm.

906. **H. limacinus** Fr. Saund. & Sm. t. 28. *Limacinus*, slimy.

P. 4–6 cm., *disc umber, then fuliginous, margin paler*, convex, then plane, obtuse, *viscid*. St. 5–8 × 1–1·5 cm., *white, greyish, or bistre*, firm, *ventricose, viscid, flocculose, fibrilloso-striate*, apex *squamulose*. Gills *white inclining to cinereous, or yellowish*, adnate, then decurrent, subdistant. Flesh firm, *white*. Spores white, elliptical, 12 × 8μ. Amongst leaves in woods. Oct.—Nov. Uncommon. (*v.v.*)

907. **H. squamulosus** Rea. Trans. Brit. Myc. Soc. IV, t. 6. *Squamulosus*, covered with little scales.

P. 5–7 cm., *yellow olivaceous, disc fuscous*, convex, then expanded, subumbonate, glutinous, *floccosely squamulose beneath the gluten, tomentose* at the incurved margin *over the base of the gills*. St. 6–8 × 1·5–2 cm., *concolorous*, apex *white, mealy*, equal, or enlarged downwards, glutinous. Gills *white*, 5–10 mm. wide, sinuato-adnate, margin irregular, somewhat crowded. Flesh *whitish, becoming yellowish towards the lower half of the stem.* Spores white, globose, 3·5–4 × 3·5μ. Smell and taste pleasant. Amongst short grass. Oct. Uncommon. (*v.v.*)

908. **H. olivaceo-albus** Fr. Boud. Icon. t. 31.
Olivaceus, olivaceous; *albus*, white.

P. 3–10 cm., *olivaceous-fuscous, becoming pale especially towards the margin, at first acorn-shaped*, then expanded, *umbonate*, at length depressed round the umbo, glutinous and often forming tear-like drops at the margin, which becomes striate when old. St. 5–13 cm. × 6–15 mm., *white, sheathed with the squamulose, spotted, fuscous, viscid veil* which terminates at the apex in the form of a ring, equal, or attenuated at the base, *apex shining white.* Gills *white, or olivaceous from the gluten*, decurrent, *distant*, broad, connected by veins at the base. Flesh *white.* Spores white, elliptical, 7–8 × 4–5μ, 2-guttulate. Woods, especially of conifers. Aug.—Nov. Common. (*v.v.*)

var. **obesus** Bres. Bres. Fung. Trid. t. 92. *Obesus*, stout.

Differs from the type in having a *thick, squat stem.* Pine woods. Sept.—Oct. Not uncommon. (*v.v.*)

909. **H. hypothejus** Fr. Boud. Icon. t. 32.
ὑπό, under; *θεῖον*, brimstone.

P. 3–6 cm., *at first covered with olivaceous gluten, cinereous when the gluten disappears, becoming pale and yellowish, orange, or rarely* (*when rotting*) *rufescent*, convex, then *depressed*, obtuse, *somewhat streaked.* St. 5–10 cm. × 4–10 mm., *whitish, becoming yellowish*, equal, *viscid*, rarely spotted with the veil; partial veil floccose, at the first *cortinate and annular, soon fugacious.* Gills *pallid, soon yellow, sometimes flesh colour*, decurrent, distant. Flesh *white, then light yellow.* Spores white, elliptic-oblong, 10–11 × 4–5μ, 1-many-guttulate. Woods, and heaths, under conifers. Sept.—Jan. Common. (*v.v.*)

var. **expallens** Boud. Boud. Icon. t. 33. *Expallens*, becoming pale.

Differs from the type in its *smaller size, paler colour, and decreased viscidity.* Under pines. Nov.—Jan. Uncommon. (*v.v.*)

910. **H. cerasinus** Berk. (= *Hygrophorus agathosmus* (Fr.) Quél.) Cke. Illus. no. 884, t. 898. *Cerasus Laurocerasus*, the cherry laurel.

P. 4–6 cm., *pale umber, then grey*, convex, broadly umbonate, often more or less undulate, sometimes depressed, viscid, shining when dry; *margin minutely tomentose.* St. 2·5–8 × 1 cm., *white*, attenuated below, sometimes ventricose, punctato-squamulose above. Gills *white, tinged with pink,* decurrent, broad, sometimes forked, very distant. Spores white, elliptical, 8 × 4μ. Smell like that of cherry laurel leaves. Fir woods. Sept.—Oct. Uncommon. (*v.v.*)

*****Fuscous cinereous, or livid.

911. **H. fusco-albus** Fr. *Fuscus,* dark; *albus,* white.

P. 4–8 cm., *fuscous, then cinereous,* convexo-plane, then depressed, moderately firm, viscid; margin *white-floccose.* St. 5–12 cm. × 8–15 mm., *white,* equal, *when dry white-floccose at the apex.* Gills *snow-white,* decurrent, broad, rather thick. Flesh *greyish white.* Spores white, pip-shaped, 7–8 × 5μ, 1-guttulate. Woods, and amongst grass under conifers. Sept. Uncommon. (*v.v.*)

912. **H. agathosmus** Fr. (= *Hygrophorus cerasinus* Berk. sec. Quél.) Gonn. & Rabenh. VIII–IX, t. 11, fig. 4.
ἀγαθός, good; ὀσμή, scent.

P. 4–7 cm., *livid grey, unicolorous, dotted with minute, raised, crowded, viscid, pellucid, whitish papillae,* convex, then plane, gibbous, viscid; margin at first involute, *villose,* at length reflexed, and undulated. St. 5–12 cm. × 6–15 mm., *white,* equal, or slightly thickened downwards, *somewhat fibrillosely striate, granularly farinose at the apex, the squamules at length becoming cinereous.* Gills *shining white,* decurrent, distant, 6–8 mm. wide, soft, somewhat veined at the base. Flesh *watery whitish,* soft. Spores white, pip-shaped, 8–9 × 4–5μ. Smell very pleasant. Coniferous woods. Sept.—Nov. Not uncommon. (*v.v.*)

913. **H. pustulatus** (Pers.) Fr. Trans. Brit. Myc. Soc. III, t. 13.
Pustulatus, blistered.

P. 2–5 cm., *livid grey, disc fuscous, broken up into papillae,* convex, then expanded, *umbonate,* viscid. St. 3·5–4·5 cm. × 5–13 mm., *white, rough with black points,* equal, or fusiform. Gills *white, sometimes glaucous,* adnato-decurrent, 5–6 mm. wide, distant, soft. Spores white, ovoid pruniform, 8–9 × 5μ. Fir woods. Sept.—Oct. Uncommon. (*v.v.*)

914. **H. mesotephrus** B. & Br.
μέσος, middle; τεφρός, ash-coloured.

P. 2–3 cm., *white, disc brown,* convex, somewhat hemispherical, viscid, *striate,* the extreme margin often remaining quite even. St. 5 cm. × 4–6 mm., *white, often stained yellowish,* flexuose, attenuated at the base, viscid, *floccoso-granulated at the apex.* Gills *pure white,* shortly decurrent, moderately broad, ventricose, rather distant. Flesh *white, hygrophanous.* Spores white, elliptical, somewhat pointed at the one end, 9 × 6μ. Woods. Oct. Uncommon. (*v.v.*)

915. **H. livido-albus** Fr. Cke. Illus. no. 888, t. 915.
Lividus, livid; *albus*, white.

P. 4–7 cm., *livid, umbo bistre, with darker streaks*, convex, viscid. St. 6–10 cm. × 6–10 mm., *grey, becoming yellowish downwards*, equal, or attenuated at the base, *fibrillosely striate*. Gills *shining white*, decurrent, distant. Flesh *greyish when moist, white when dry, often tinged with yellow in the stem*. Spores white, elliptical, 10–11 × 5–6μ, multi-guttulate. Woods. Oct.—Nov. Uncommon. (*v.v.*)

II. Veil none. St. even, smooth, or fibrillose, *not rough with dots*. P. *firm*, opaque, *moist* in rainy weather, not viscous. Gills distant, arcuate.

*Gills deeply, and at length obconically decurrent.

916. **H. camarophyllus** (A. & S.) Fr. (= *Hygrophorus caprinus* (Scop.) Fr.) Cke. Illus. no. 889, t. 916, as *Hygrophorus caprinus* Scop.
καμάρα, a vault; φύλλον, a leaf.

P. 3–10 cm., *blackish-fuliginous, or blackish, at length cinereous-fuliginous, but varying azure-blue*, convex, then plane and depressed, sometimes more or less umbonate, especially when young, firm, moist, or dry, *more or less radiately streaked with innate fibrils*, pellicle hardly separable; margin *at first white*, pruinose, incurved, then expanded, revolute, concolorous and undulating. St. 4–8 × 1–1·5 cm., *fuliginous*, equal, or attenuated downwards, *longitudinally fibrillose*, apex finally whitish, base white, pubescent. Gills *white, becoming glaucous, or grey, deeply decurrent, distant*, 5–8 mm. *wide, thick*, connected by veins. Flesh *white*. Spores elliptical, slightly apiculate at the one end, 6–9 × 4–5μ, multi-guttulate. Smell *strong*, like that of *Cortinarius purpurascens*, taste mild, *slightly sweet*. Heaths and pastures under firs. Oct. Uncommon. (*v.v.*)

917. **H. leporinus** Fr. Cke. Illus. no. 891, t. 931, as *Hygrophorus nemoreus* Fr. *Leporinus*, belonging to a hare.

P. 3–6 cm., *yellow-rufescent, varying fuscous*, equally fleshy, convex, gibbous, *fibrilloso-floccose*, firm, opaque, margin scalloped, then split, silky. St. 4–5 × 1 cm., *whitish flesh colour*, firm, rigid, attenuated commonly downwards, more rarely upwards, fibrillose, base white. Gills *pale yellowish*, decurrent, branched. Flesh *reddish flesh colour*. Spores "pale umber, subglobose, 6–8 × 4–4·5μ" Sacc. Woods, and downs. Sept.—Oct. Rare.

H. bicolor Karst. = **Hygrophorus Karstenii** Sacc. & Cub.

918. **H. nemoreus** (Lasch) Fr. *Nemoreus*, belonging to a wood.

P. 3–8 cm., *somewhat orange, equally fleshy*, convex, then expanded, gibbous, at length depressed. St. 4–6 × 1–1·5 cm., *pale*, firm, *squamu-*

lose, fibroso-striate, attenuated at the base. Gills *somewhat concolorous,* decurrent, thick, distant, wide. Flesh *yellowish, deeper coloured under the cuticle of the pileus.* Spores white, globose, 5–6 × 5μ. Taste pleasant. Edible. Woods, and pastures. Oct.—Dec. Uncommon. (*v.v.*)

919. **H. pratensis** (Pers.) Fr. (= *Hygrophorus ficoides* (Bull.) Schroet.) Cke. Illus. no. 892, t. 917. *Pratensis,* growing in meadows.

P. 2·5–8 cm., *tawny yellow, or buff,* compactly fleshy especially at the disc, thin towards the margin, convex, then flattened, gibbous, *almost turbinate* from the stem being thickened upwards, moist, rimosely incised when dry. St. 4–6 × 1–1·5 cm., *concolorous but paler, attenuated downwards,* dilated into the pileus. Gills *concolorous, or white, deeply decurrent,* at first arcuate, then extended in the form of an inverted cone, very distant, thick, firm, brittle, connected by veins. Flesh *ochraceous.* Spores white, elliptical, 7–8 × 5μ. Smell and taste pleasant. Edible. Pastures, and heaths. Aug.—Dec. Common. (*v.v.*)

var. **cinereus** Fr. Cke. Illus. no. 893, t. 932, fig. B.
Cinereus, ash colour.

Differs from the type in the *thinner, cinereous p., at length striate at the margin, the white st. sometimes yellowish at the base, and the cinereous gills.* Heaths and downs. Sept.—Oct. Not uncommon. (*v.v.*)

var. **pallidus** B. & Br. Cke. Illus. no. 893, t. 932, fig. A.
Pallidus, pallid.

Differs from the type in the *pallid, infundibuliform p., with undulate, deflexed margin, the dilated, fibrillosely striate st., and the pallid, decurrent, branched, distant gills.* On downs and hillsides. Sept.—Oct. Uncommon. (*v.v.*)

var. **umbrinus** W. G. Sm. *Umbrinus,* umber.

Differs from the type in the *umber p., substriate at the margin, the white, equal st., and the flesh being coloured umber under the cuticle of the p.* Hilly pastures. Sept.—Oct. Uncommon. (*v.v.*)

var. **meisneriensis** Fr. *Meisneriensis,* belonging to Mount Meisner.

Differs from the type in the *much smaller p., at length striate at the margin, the white st., and the easily separable pellicle of the p.*

920. **H. Karstenii** Sacc. & Cub. (= *Hygrophorus bicolor* Karst. sec. W. G. Sm.) Karst. Icon. t. XIII.
P. A. Karsten, the eminent Finnish mycologist.

P. 5–8 cm., *white, or whitish,* convexo-plane, at length often depressed, disc compact. St. 8–14 × 1·5–2 cm., *white, or whitish,* attenuated downwards, wavy. Gills *yellow*, deeply decurrent, thick, distant. Spores white, elliptical, 10 × 6μ. Woods. Nov. Uncommon.

921. **H. foetens** Phill. Cke. Illus. no. 904, t. 903, fig. B.
Foetens, stinking.

P. 1·5–2·5 cm., *dark brown*, convex, then plane, *at length broken up into squamules*. St. 2·5–7·5 cm. × 4 mm., *paler than the pileus*, attenuated downwards, shining, *clothed with transversely arranged fibrous scales*. Gills *somewhat of the same colour as the pileus, or paler, somewhat glaucous-pruinose*, decurrent, distant, rather thick. Flesh *dark coloured*. Spores white, globose, 4–5μ. Smell very foetid, nauseous. Grassy places. Sept.—Nov. Uncommon. (*v.v.*)

922. **H. virgineus** (Wulf.) Fr. (= *Hygrophorus ericeus* (Bull.) Schroet.) Boud. Icon. t. 37. *Virgineus*, maidenly.

Entirely white. P. 3–7 cm., convex, then plane, *obtuse, subumbonate*, moist, *at length depressed*, cracked into patches, floccose and becoming yellowish when dry. St. 5–11 cm. × 5–10 mm., *firm*, attenuated at the base, pruinose, striate. Gills decurrent, distant, *rather thick*, veined at the base. Spores white, oblong elliptic, 9–12 × 5–6μ. Smell and taste pleasant. Edible. Pastures, heaths, and woods. Aug.—Dec. Common. (*v.v.*)

var. **roseipes** Massee. Cke. Illus. no. 895, t. 893.
Roseus, rose-coloured; *pes*, foot.

Differs from the type in the *stem being rose-coloured externally, and internally towards the base*. Woods, and pastures. Sept.—Nov. Not uncommon. (*v.v.*)

923. **H. ventricosus** B. & Br. Cke. Illus. no. 897, t. 901.
Ventricosus, ventricose.

Entirely white. P. 5–7·5 cm., convex, unequal. St. 6 × 1–1·5 cm., *ventricose, attenuated at both ends*. Gills deeply decurrent, *narrow, sometimes forked*. Spores white, elliptical, 7 × 4μ. Pastures, and amongst grass. Sept.—Oct. Uncommon.

924. **H. niveus** (Scop.) Fr. Cke. Illus. no. 896, t. 900, fig. A.
Niveus, snow-white.

Entirely white. P. 1·5–3 cm., *submembranaceous*, campanulate, then convex, *umbilicate, striate and viscid when moist*. St. 4–6 cm. × 2–4 mm., *equal*. Gills decurrent, distant, *thin*. Spores white, elliptical, 7–8 × 4–5μ. Smell and taste mild. Edible. Heaths, and pastures. Sept.—Dec. Common. (*v.v.*)

925. **H. russocoriaceus** Berk. & Miller. Cke. Illus. no. 896, t. 900, fig. B. *Russus*, Russian; *coriaceus*, leathery.

P. 1–2 cm., *ivory-white*, convex, slightly viscid. St. 1·5–4 cm. × 2–6 mm., *pure white, thickened upwards*. Gills *concolorous*, decurrent, arched, thick, very few, distant. Spores white, elliptical, 7–8 × 4–5μ, multi-guttulate. Smell very pleasant, like Russian leather. Edible. Pastures, and heaths. Sept.—Dec. Not uncommon. (*v.v.*)

**Gills ventricose, sinuato-arcuate, or plano-adnate.

926. **H. fornicatus** Fr. *Fornicatus*, arched.

P. 2·5–5 cm., *white, or pallid livid, campanulate, then expanded*, obsoletely umbonate, subrepand, viscid. St. 5–8 cm. × 5–13 mm., *shining white*, firm, tough, equal, subundulate. Gills *white, almost free*, or *sinuato-adnexed, thick, ventricose, distant*, exceeding the margin. Spores white, elliptical, 6 × 3–4μ. Pastures. Oct.—Nov. Uncommon. (*v.v.*)

927. **H. clivalis** Fr. Cke. Illus. no. 898, t. 933, as *Hygrophorus fornicatus* Fr. *Clivalis*, belonging to hills.

P. 3–4 cm., *whitish, disc yellowish*, campanulate, *silky*, umbonate, *becoming split*, scalloped, thin, *striate*. St. 4–6 × 1 cm., *white, fragile*, attenuated at the base. Gills *pale ochraceous*, sinuate, almost free, ventricose, distant, wide. Spores white, elliptical, 6 × 3μ. Parks, and pastures. Sept.—Oct. Uncommon. (*v.v.*)

928. **H. distans** Berk. Berk. Outl. t. 13, fig. 1.
Distans, being distant.

P. 3–5 cm., *white with a silky lustre, here and there stained with brown*, plane, or depressed, often umbilicate, viscid. St. 3–4 cm. × 5–6 mm., *white above, cinereous below* and attenuated. Gills *pure white, then tinged with cinereous*, decurrent, *few, very distant, subventricose*, interstices obscurely rugose. Spores white, broadly elliptical, 10 × 8μ. Smell sometimes pleasant, like the essence of almonds. Woods. Oct. Rare.

929. **H. Clarkii** (B. & Br.) W. G. Sm. (= *Hygrophorus latitabundus* (Britz.) sec. W. G. Sm.) J. Aubrey Clark.

P. 9–10·5 cm., *opaque umber, or livid cinereous*, margin *white*, obtuse, convexo-plane, viscid. St. 4–7 × 2–3 cm., *white-squamulose* above, *pale umber-scaly* below, equal, or attenuated downwards, viscid. Gills *ivory-white*, adnate, distant, broad, thick, veined. Spores white, elliptical, 10 × 6μ. Woods, and pastures. Sept.—Oct. Rare.

930. **H. metapodius** Fr. Cke. Illus. no. 901, t. 918.
μετά, reversed; πούς, foot.

P. 4–8 cm., *cinereous-fuscous*, convex, then plane, obtuse, at first viscid and slightly shining, *then silky and squamulose*, irregular. St. 4–10 × 1–2 cm., *concolorous*, attenuated downwards, sometimes swollen at the base, fibrillosely striate. Gills *dark grey, becoming stained with red*, adnate, or arcuato-decurrent, distant, thick, veined, broad. Flesh *pallid grey, reddish when broken and at length becoming black*. Spores white, broadly elliptical, 7–8 × 6μ, with a large central gutta. Smell of new meal, taste mild. Old mossy pastures. Oct.—Jan. Not uncommon. (*v.v.*)

931. **H. ovinus** (Bull.) Fr. Hussey, Illus. Brit. Myc. II, t. 50.
Ovinus, belonging to sheep.

P. 4–5 cm., *fuscous, campanulate, then expanded*, somewhat umbonate, at first slightly viscid, then dry and *squamulose*, at length revolute, undulated, rimosely incised. St. 3–4 × ·5–1 cm., *pallid, or greyish*, subequal, or slightly thickened at both ends, curved, or twisted, fibrillosely striate. Gills *grey, then rufescent*, arcuato-adnate, decurrent with a tooth, thick, broad, connected by veins. Flesh fragile, *grey, then tinged reddish, and finally black*. Spores white, subglobose, 6 × 4–5μ. Smell strong, of new meal, taste mild. Pastures, and hillsides. Sept.—Dec. Not uncommon. (*v.v.*)

932. **H. connatus** Karst. *Connatus*, born together.

P. 3–4 cm., *grey*, margin submembranaceous, convex, then plane, unequal, *dry*, silky, then smooth. St. 6–7 cm. × 4–5 mm., *concolorous*, equal, silky fibrillose, apex farinose. Gills *dark grey*, decurrent, somewhat thick, distant, branched. Spores white, pip-shaped, 7–8 × 4–5μ. Caespitose. Woods. Oct. Uncommon. (*v.v.*)

933. **H. subradiatus** (Schum.) Fr. Cke. Illus. no. 902, t. 935, fig. A.
Sub, somewhat; *radiatus*, rayed.

P. 3–4 cm., *white, livid, grey, or flesh colour, disc fuscous, submembranaceous*, convex, then expanded, *slightly umbonate, radiato-striate* from the translucent gills. St. 4–5 cm. × 4–5 mm., *pale*, equal, twisted, attenuated at the white base. Gills *white*, deeply decurrent with a tooth, plane, ventricose, *somewhat thin*, distant, connected by veins. Spores white, subglobose, 7–8 × 6μ, with a large central gutta. Heaths, and pastures. Sept.—Oct. Not uncommon. (*v.v.*)

var. **lacmus** Fr. Kalchbr. Icon. t. 25, fig. 3, as *Hygrophorus lacmus*.
Lac, varnish.

P. 2–5 cm., *lilac, then becoming pale*, convex, umbonate, unequal at the circumference, *radiato-striate when moist*, even when dry, shining, disc often fibrillose, or squamulose. St. 3–5 cm. × 7–15 mm., *white, or greyish, often tinged with yellow at the base*, narrowed at the base and apex. Gills *cinereous*, plano-decurrent, thin, connected by veins, subdistant. Flesh *tinged with grey, yellowish at the base of the stem*. Spores white, subglobose, 7–8 × 6μ, with a large central gutta. Heaths, and pastures. Sept.—Oct. More common than the type. (*v.v.*)

934. **H. irrigatus** (Pers.) Fr. Fr. Icon. t. 168, fig. 3.

Irrigatus, bedewed.

P. 2·5–5 cm., *livid, becoming dingy white as the gluten separates*, fragile, campanulate, expanded, obtuse, *viscid*, margin striate. St. 5–8 cm. × 2–4 mm., *livid, very viscid*, equal, tough. Gills *white, or grey*, adnate with a decurrent tooth, subdistant, *wide*, fragile, rather thick, connected by veins. Spores white, "elliptical, 6–7 × 4μ" Massee. Fir woods, grassy places, and pastures. Oct.—Nov. Uncommon.

III. Veil none. Whole fungus thin, watery, succulent, *fragile*. P. *viscid* when moist, *shining when dry*, rarely floccosely squamose. St. hollow, soft, not punctate. Gills soft. *Most of the species brightly coloured*, and shining.

*Gills decurrent.

935. **H. Colemannianus** Blox. Cke. Illus. no. 904, t. 903, fig. A.

W. H. Coleman.

P. 2·5–5 cm., *reddish umber, paler when dry except in the centre*, subcampanulate, then expanded, *strongly umbonate, striate* when moist, and slightly viscid. St. 2·5–4 cm. × 2–4 mm., *white, very slightly tinged with umber*, nearly equal, brittle, fibrous, *somewhat silky*. Gills *umber, paler than the pileus*, deeply decurrent, rather broad, distant, interstices veined and rugose. Flesh *white, tinged with umber under the cuticle of the pileus*. Spores white, broadly elliptical, 6–8 × 6μ. Pastures, and hillsides. Oct.—Nov. Not uncommon. (*v.v.*)

936. **H. sciophanus** Fr. Fr. Icon. t. 167, fig. 1.

σκιά, shade; *φαίνω*, I appear.

P. 1–4 cm., *deep tawny, or brick colour, becoming paler when dry, opaque, hemispherical*, then expanded, *obtuse*, slightly viscid, *somewhat fleshy; margin paler, striate*. St. 3–7·5 cm. × 2–6 mm., *tawny yellowish, or paler*, equal, flexuose, *viscid*. Gills *of the same colour as the pileus, or yellowish*, attenuato-adnate, distant. Flesh *yellowish, reddish near the cuticle*. Spores white, elliptical, 8–9 × 4–5μ. Amongst grass. Sept.—Oct. Uncommon. (*v.v.*)

937. **H. sciophanoides** Rea. Cke. Illus. no. 905, t. 937, fig. A, as *Hygrophorus sciophanus* Fr.

Sciophanus, the species *H. sciophanus*; *εἶδος*, like.

P. 1–3 cm., *rosy pink, campanulate*, then expanded, *striate to the disc*, interstices paler, *subumbonate*, or *papillate, membranaceous, fragile*. St. 2–5 cm. × 2–3 mm., *concolorous*, equal, *base white*. Gills *pale pink*, uncinato-adnate, *broadest in front*. Flesh *pale yellow, becoming white*. Spores white, elliptical, 6–7 × 4μ. Heaths, and hilly pastures. Sept.—Oct. Uncommon. (*v.v.*)

938. **H. laetus** (Pers.) Fr. (= *Hygrophorus Houghtoni* Berk. sec. Quél.) Fr. Icon. t. 167, fig. 2. *Laetus*, cheerful.

P. 2–3 cm., *tawny*, *shining*, convex, then flattened, obtuse, viscid; margin almost membranaceous and slightly pellucid-striate. St. 3·5–7·5 cm. × 4 mm., *concolorous*, *apex externally and internally bluish green*, equal, *tough*, *viscid*. Gills *flesh colour*, *whitish*, *or fuliginous*, subdecurrent, distant, *thin*, somewhat connected by veins. Flesh *of the same colour as the pileus but paler*. Spores white, elliptical, 7–9 × 4–5μ, 1–2-guttulate. Heaths, and hilly pastures. Sept.—Dec. Common. (*v.v.*)

939. **H. vitellinus** Fr. Fr. Icon. t. 167, fig. 3. *Vitellinus*, of yolk of egg.

P. 1–2·5 cm., *citron-egg-yellow*, *becoming white when dry*, very thin, convex, *umbilicate*, viscid; margin *plicato-striate*. St. 5 cm. × 2 mm., *pallid light yellow*, *becoming white when dry*, equal, flexuose, fragile. Gills *yellow*, *then egg-yellow*, *deeply decurrent*, subdistant. Spores white, broadly elliptical, with an apiculus at one end, 8–9 × 6μ, with a large central gutta. Woods, and pastures. Sept.—Dec. Uncommon. (*v.v.*)

940. **H. citrinus** Rea. Trans. Brit. Myc. Soc. III, t. 11. *Citrinus*, lemon yellow.

P. 1–2 cm., *citron yellow*, convex, then plane, viscid; margin *striate*. St. 1–2 cm. × 2–3 mm., *citron yellow*, apex *white*, base attenuated, *viscid*. Gills *whitish citron yellow*, adnato-decurrent, somewhat crowded, 2–3 mm. broad. Flesh *concolorous*. Spores white, elliptical, apiculate at the one end, 7–7·5 × 5μ, 1-guttulate. Roadsides, and hilly pastures. Sept.—Oct. Uncommon. (*v.v.*)

941. **H. ceraceus** (Wulf.) Fr. Boud. Icon. t. 39. *Ceraceus*, waxy.

P. 2–4 cm., *waxy-yellow*, shining, viscid, convexo-plane, obtuse; margin slightly striate, pellucid. St. 3–5 cm. × 4 mm., *concolorous*, often *unequal*, flexuose, at length *compressed*, base attenuated, white. Gills *yellow*, *adnato-decurrent*, *broad*, connected by veins, almost triangular. Flesh *concolorous*. Spores white, elliptical, 6–8 × 4μ. Woods, and pastures. Sept.—Dec. Common. (*v.v.*)

942. **H. coccineus** (Schaeff.) Fr. (= *Hygrophorus miniatus* (Scop.) Schroet.) Boud. Icon. t. 38. *Coccineus*, scarlet colour.

P. 2–6 cm., *bright scarlet*, *soon changing colour and becoming yellowish*, convex, then plane, often unequal, *obtuse*, at first viscid. St. 5–7 cm. × 6–8 mm., *concolorous*, *becoming yellowish*, *compressed*, *base always yellow*. Gills *purplish at the base*, *light yellow in the middle*, *glaucous at the edge*, adnate, *decurrent by a tooth*, distant, broad, connected by veins, trama *red*. Flesh *concolorous*. Spores white, elliptical, 8–11 × 5–6μ. Woods, and pastures. June—Dec. Common. (*v.v.*)

943. **H. miniatus** Fr. (= *Hygrophorus flammans* (Scop.) Schroet.) Cke. Illus. no. 910, t. 921, fig. A.

Miniatus, coloured with red-lead.

P. ·5–2 cm., *vermilion, then becoming pale, and opaque,* convex, often umbonate, *then umbilicate,* glabrous, or *squamulose.* St. 3–5 cm. × 2–4 mm., *vermilion, shining, equal, round.* Gills *yellow, or yellow-vermilion, adnate.* Flesh *reddish, then yellow.* Spores white, elliptical, 6–7 × 4–5μ, 1-guttulate. Heaths, pastures, and peat bogs. June—Oct. Common. (*v.v.*)

944. **H. Reai** Maire. Trans. Brit. Myc. Soc. III, t. 11.

Carleton Rea.

P. 1·5–2·5 cm., *scarlet,* fleshy, thin, convex-campanulate, then plane; margin *orange yellow, or yellow,* slightly striate when moist; no separable pellicle. St. 3–6 cm. × 2–3 mm., *orange scarlet to yellow,* base *whitish, viscid,* shining, somewhat tough. Gills *flesh colour, then orange, edge whitish, then yellow, broadly adnate* with a decurrent tooth, broad, thin, unequal. Flesh *orange.* Spores white, elliptical, apiculate, 7–8 × 3·5–4·5μ, 2-guttulate. Cystidia none. Taste *bitter.* In woods, and pastures. Aug.—Nov. Not uncommon. (*v.v.*)

945. **H. turundus** Fr. *Turunda*, a kind of sacrificial cake.

P. 1–2·5 cm., *yellow, or tawny, variegated with cinereous-fuscous squamules,* slightly fleshy, sometimes viscid at first, convex, then expanded, *umbilicate,* or depressed; *margin often elegantly crenate.* St. 3–5 cm. × 2–4 mm., *tawny-reddish,* rigid-fragile, attenuated at the base. Gills *white, then cream colour, decurrent,* narrow, 1–2 mm. wide, distant. Flesh *yellowish, or reddish.* Spores white, elliptical, 8–11 × 5–6μ, 1–2-guttulate. Heaths, pastures, and peat-bogs. July—Oct. Uncommon. (*v.v.*)

var. **mollis** B. & Br. Cke. Illus. no. 910, t. 921, fig. B. *Mollis*, soft.

Golden. P. 12–18 mm., nearly plane, at length depressed, *clad with soft, short, radiating hairs of the same colour.* St. 2·5–3 cm. × 2–4 mm., equal. Gills *whitish,* decurrent, narrow, distant. Flesh *yellow.* Spores white, elliptical, 8 × 4μ. Woods, and pastures. July—Oct. More common than the type. (*v.v.*)

var. **sphaerosporus** Rea. *σφαῖρα*, a ball; *σπορά*, seed.

P. 18 mm., *reddish, covered with golden yellow fibrils,* plane, then depressed. St. 3–4 cm. × 3–5 mm., *reddish,* base *white,* slightly incrassated upwards. Gills *white,* decurrent, distant. Flesh *bright yellow.* Spores white, *subglobose,* 6–7 × 5–6μ, *verrucose.* Amongst short grass, and moss. Sept.—Oct. Uncommon. (*v.v.*)

var. **lepidus** Boud. Bull. Soc. Myc. Fr. XIII, t. 1, fig. 2, as *Hygrophorus lepidus* Boud. *Lepidus,* charming.

P. 1·5–4 cm., *brilliant golden orange, becoming paler,* disc *deeper colour,* convex, umbilicate, *minutely squamulose, or hirsuto-tomentose.* St. 5–7 cm. × 2–3 mm., *concolorous,* slightly incrassated upwards. Gills *pallid, slightly tinged with the colour of the pileus at the base, deeply and abruptly decurrent,* thick, distant. Flesh *of stem orange, paler elsewhere.* Spores white, oblong-elliptic, 9–10 × 5–6μ. Woods, and bogs. Aug.—Oct. Uncommon. (*v.v.*)

946. **H. mucronellus** Fr. Cke. Illus. no. 905, t. 937, fig. B.
Mucronellus, having a little sharp point.

P. 2–8 mm., *scarlet, or yellow, then pale, becoming hoary, acutely conical* when small, when larger *campanulate,* obtuse, then expanded, pellucidly striate when moist, somewhat silky when dry. St. 2–4 cm. × 1–2 mm., *concolorous, white at the base,* flexuose, subattenuated downwards, somewhat silky. Gills *yellow,* decurrent, thick, triangular. Flesh *concolorous.* Spores white, globose, 3 × 2–3μ. Amongst moss and short grass in pastures, and heaths. Aug.—Dec. Uncommon. (*v.v.*)

947. **H. micaceus** B. & Br. *Micaceus,* like mica.

P. 8–12 mm., *light yellow, becoming cinereous,* hemispherical, *like a small Leotia lubrica, glittering with micaceous granules,* wrinkled. St. 18–20 × 2–3 mm., *light yellow, then brown towards the base,* granulated. Gills *pallid umber,* decurrent. Flesh *yellowish, somewhat brownish under the cuticle of the pileus.* Spores white, subglobose, 4 × 3μ, 1-guttulate. On bare earth, and clayey soil. Oct. Rare. (*v.v.*)

948. **H. Wynniae** B. & Br. (= *Omphalia bibula* Quél.) Cke. Illus. no. 911, t. 905, fig. A. Mrs Lloyd Wynne, of Coed Coch.

Entirely lemon-yellow, hygrophanous. P. 1–2·5 cm., thin, hemispherical, *umbilicate,* or somewhat infundibuliform, pellucidly striate. St. 3–4·5 cm. × 1–3 mm., attenuated upwards from the *white, swollen, strigose base.* Gills decurrent, narrow, thin. Spores white, broadly elliptical, 7–8 × 6μ. Smell foetid when decayed. On fir needles, twigs, chips, and stumps. Sept.—Nov. Uncommon. (*v.v.*)

**Gills adnexed, somewhat separating.

949. **H. puniceus** Fr. Cke. Illus. no. 912, t. 922.
Puniceus, blood-red.

P. 5–11 cm., *blood-red scarlet, becoming pale with age especially at the fleshy disc, campanulate, obtuse,* generally repand, or lobed, very irregular, viscid. St. 7–11 × 1–2·5 cm., *concolorous, or light yellowish, base always white,* attenuated at both ends, often incurved, *striate,* apex often squamulose. Gills *white-light-yellow, or yellow, often reddish at the base,* ascending, appearing free, ventricose, broad, thick, distant. Flesh *concolorous.* Spores white, broadly elliptical, 6–7 × 5μ, with a large central gutta. Taste mild. Edible. Woods, heaths, and pastures. July—Dec. Common. (*v.v.*)

950. **H. nigrescens** Quél. (= *Hygrophorus puniceus* Fr. var. *nigrescens* (Quél.) Massee.) *Nigrescens,* becoming black.

P. 5–10 cm., *white, then citron or jonquil yellow, streaked with pinkish, or orange fibrils, becoming grey and silky and finally black,* campanulate, scalloped, lobed. St. 5–7 × 1–2 cm., *citron yellow, streaked with orange,* base *white,* substriate, wrinkled, splitting, tough. Gills *cream, or citron yellow,* orange at the base, *then grey, becoming black.* Flesh *orange, white in the stem, becoming black*; *juice becoming lilac colour* on exposure to the air. Spores white, elliptical, 11–13 × 6–7μ. Heaths, and pastures. Sept. Rare. (*v.v.*)

951. **H. obrusseus** Fr. ὄβρυζον, pure gold.

P. 5–12 cm., *golden-sulphur-yellow with a tinge of green,* campanulate, then expanded and somewhat revolute, lobed and often splitting at the margin, obtuse, fragile, shining. St. 5–11 × 1–3·5 cm., *sulphur yellow, becoming tawny at the base,* often compressed, unequal. Gills *white, tinged with yellowish green towards the base,* adnexed, at length separating, free, very broad, 10–12 mm., distant, thick. Flesh *concolorous.* Spores white, elliptical, 8–9 × 5–6μ, 1–2-guttulate. Woods, and pastures. Aug.—Oct. Not uncommon. (*v.v.*)

952. **H. intermedius** Pass. *Intermedius,* intermediate.

P. 3·5–5 cm., *golden yellow, becoming greyish, or bright orange,* very thin, campanulate, then expanded, obtuse, or subumbonate, *fibrillosely-silky*; margin often wavy. St. 4–7·5 cm. × 6–9 mm., *yellow,* equal, *fibrillosely-striate.* Gills *whitish, then yellow,* adnate, ventricose, distant. Spores white, elliptical, 8–9 × 6μ. Smell of meal. Roadsides, grassy places, and damp ground. Sept.—Oct. Uncommon.

953. **H. conicus** (Scop.) Fr. Boud. Icon. t. 40, as var. *nigrescens* Boud. *Conious,* conical.

P. 3–6 cm., *scarlet, yellow, tawny, sulphur-greenish, livid, or fuliginous-light yellow, becoming black,* submembranaceous, campanulate, *conical, acute,* often lobed, then expanded and cracked, viscid when moist, shining when dry. St. 6–9 cm. × 4–9 mm., *concolorous, or yellow, becoming black,* cylindrical, tense and straight, *fibrillosely-striate.* Gills *white, or yellow, sometimes reddish at the base, becoming black when bruised,* attenuato-free, ventricose, thin, somewhat crowded. Flesh *concolorous, becoming black.* Spores white, broadly elliptical, 10–11 × 7–8μ, with a large central gutta. Pastures, heaths, roadsides, and woods. July—Nov. Common. (*v.v.*)

954. **H. calyptraeformis** Berk. (= *Hygrophorus amoenus* (Lasch) Quél.) Cke. Illus. no. 916, t. 894.

καλύπτρα, a woman's veil; *forma*, shape.

P. 3–10 cm., *pink, becoming pallid,* thin, campanulate, *acutely conical,* lobed below, then expanded and revolute, *minutely innato-fibrillose,* moist. St. 6–12 × 1 cm., *white, often with a rosy tinge within the p., striate,* brittle, often splitting longitudinally, easily separating from the pileus. Gills *rose coloured, at length pallid,* acutely attenuated behind, distant. Flesh *of pileus pink, of stem white.* Spores white, elliptical, 7–8 × 5μ, 1-guttulate. Pastures, heaths, and woods. Aug.—Oct. Not uncommon. (*v.v.*)

var. **niveus** Cooke. Cke. Illus. no. 917, t. 923. *Niveus*, snow-white.

Differs from the type in being *entirely snow-white.* Pastures, and lawns. Aug.—Oct. Uncommon. (*v.v.*)

955. **H. chlorophanus** Fr. Boud. Icon. t. 41.

χλωρός, pale green; φαίνω, I appear.

Entirely rich yellow, becoming pale, rarely scarlet, fragile. P. 3–5 cm., submembranaceous, *convex,* then plane and depressed, *obtuse,* orbicular, lobed, at length cracked, *viscid,* often striate. St. 3–8 cm. × 4–8 mm., equal, often compressed, *viscid,* sometimes sulcate in the middle. Gills *white, then sulphur yellow, emarginato-adnexed, very ventricose,* thin, distant. Flesh *yellow, deeper coloured under the cuticle.* Spores white, broadly elliptical, 7–8 × 5–6μ. Taste mild. Edible. Pastures, heaths, and woods. July—Nov. Common. (*v.v.*)

956. **H. psittacinus** (Schaeff.) Fr. Boud. Icon. t. 42.

ψίττακος, a parrot.

P. 2–5 cm., *green at first from the gluten, then yellowish, whitish, or brick colour, and finally purplish,* campanulate, then expanded, *umbonate, striate.* St. 4–7 cm. × 4–7 mm., *green at first from the gluten,* which is *persistent at the apex, then yellowish,* equal, often bent, toughish. Gills *yellow, greenish at the base,* adnate, ventricose, thick, broad, subdistant. Flesh *white, tinged with green and yellow.* Spores white, elliptical, 8–9 × 4–5μ. Taste mild. Edible. July—Nov. Common. (*v.v.*)

957. **H. spadiceus** (Scop.) Fr. Fr. Icon. t. 168, fig. 1.

σπάδιξ, date-brown.

P. 1–6 cm., *olivaceous date-brown, black and shining when dry,* fragile, campanulate, then expanded, obtuse, or acute, *very glutinous, distinctly virgate with black fibrils.* St. 4–7 cm. × 6–10 mm., *yellowish, striato-virgate with fuscous fibrils,* equal. Gills *citron yellow,* sinuate, broad, ventricose, rather thick, distant. Flesh *citron yellow.* Spores white, elliptical, "10–12 × 6–7μ" Sacc. Mossy meadows. July. Uncommon.

958. **H. unguinosus** Fr. Fr. Icon. t. 168, fig. 2. *Unguinosus,* oily.

P. 3–6 cm., *smeared with dense fuliginous gluten,* slightly fleshy, campanulate, then convex, obtuse, *very fragile,* even, or at length rimosely incised. St. 5–9 cm. × 5–10 mm., *concolorous, glutinous,* attenuated at the base and apex, unequal, *somewhat compressed.* Gills *shining white becoming glaucous,* adnate, very ventricose, distant, thick, broad, connected by veins. Flesh *greyish.* Spores white, broadly elliptical, 9 × 7μ, with a large central gutta. Woods, and pastures. Aug.—Oct. Common. (*v.v.*)

959. **H. obscuratus** Karst. *Obscuratus,* darkened.

P. 3–4 cm., *sooty,* or *livid blackish, mouse colour when dry,* fragile, convex, obtuse, *dry, squamulose.* St. 4–6 cm. × 4–8 mm., *pallid, centre often tinged smoky,* unequal, usually inflated below, wavy, glabrous. Gills *whitish, or glaucous,* sinuato-adnate. Spores white, elliptical, 7–10 × 3–5μ. Pastures. Sept. Rare.

960. **H. nitratus** (Pers.) Fr. *Nitratus,* nitrous.

P. 1–6 cm., *fuscous-cinereous, becoming pale,* scarcely fleshy, very fragile, convex, obtuse, or depressed in the centre, at first slightly viscid, soon flocculose, then *squamulose, or fibrillosely striate,* rimosely incised, irregularly shaped, somewhat repand. St. 2·5–10 cm. × 2–12 mm., *whitish, grey, or yellowish,* equal, often twisted, fragile, sometimes compressed, base attenuated. Gills *whitish, then becoming glaucous,* broadly emarginate, broad, distant, thick, mucid-soft, connected by veins. Flesh *grey, darker under the cuticle of the pileus.* Spores white, pip-shaped, 6–7 × 4–5μ, 1-guttulate. Smell *strong, nitrous.* Pastures, and heaths. Aug.—Nov. Not uncommon. (*v.v.*)

var. **glauco-nitens** Fr. γλαυκός, pale green; *nitens,* shining.

Stiff. P. *olivaceous black, or fuliginous, becoming pale, streaked with fibrils.* St. equal, shining. Gills becoming glaucous. Mixed woods. Aug. Rare.

Spores pink.

Clitopilus Fr.

(κλίτος, a slope; πῖλος, cap.)

Pileus fleshy, regular, or irregular. Stem central fleshy. Gills decurrent. Spores pink, elliptical, fusiform, globose, oblong, angular, smooth, or verrucose, continuous. Growing on the ground.

I. P. irregular; margin at first flocculose. Gills deeply decurrent.

961. **C. prunulus** (Scop.) Fr. (= *Clitopilus orcella* (Bull.) Fr.) Cke. Illus. no. 343, t. 322. *Prunulus*, a little plum.

P. 3–11 cm., *white, or yellowish,* or more rarely becoming cinereous, fleshy, compact, convex, then flattened, at length depressed and repand or unequal, viscid when moist, delicately pruinose, often spotted, or zoned; margin involute, thin, mealy. St. 2–6 × 1–1·5 cm., *white,* ventricose, or thickened upwards, pruinose, or villose, often striate, cottony at the base. Gills *white,* then flesh colour, deeply decurrent, attenuated at both ends. Flesh *white.* Spores pink, fusiform, 11–13 × 5μ, 1-2-guttulate. Cystidia none. Smell and taste of new meal. Edible. Woods, and pastures. June—Nov. Common. (*v.v.*)

962. **C. mundulus** (Lasch) Fr. (= *Clitopilus pseudo-orcella* Fr. sec. Quél.) Cke. Illus. no. 345, t. 375, fig. A. *Mundus*, neat.

P. 3–5 cm., becoming *pale white, then spotted cinereous, at length becoming black, fleshy, thin,* convex, gibbous, soon flattened and depressed, unequal repand, often excentric, rivulose, or even, *floccoso-soft*; margin involute. St. 2–3 cm. × 4 mm., *white,* subequal, *floccoso-villose,* base white-villose. Gills *pallid, deeply decurrent, very crowded,* narrow, thin, with many shorter ones intermixed. Flesh *white,* soft, *becoming black in the stem.* Spores elliptical, 8–11 × 4–5μ. Taste bitter. Woods, amongst leaves. Aug.—Oct. Uncommon. (*v.v.*)

var. **nigrescens** (Lasch) Fr. *Nigrescens*, becoming black.

Differs from the type in the *whole of the flesh becoming black.*

963. **C. popinalis** Fr. (= *Paxillus amarellus* (Pers.) Quél.) Fr. Icon. t. 96, fig. 1. *Popinalis*, belonging to a cook-shop.

P. 2–5 cm., *cinereous, here and there mottled with guttate spots,* slightly fleshy, flaccid, convex, then depressed, somewhat repand; margin thin, inrolled, pruinose, *grey.* St. 2–5 cm. × 5–12 mm., *paler than the p.,* subequal, often flexuose, attenuated, or somewhat bulbous at the white, cottony base. Gills *ochraceous, then grey,* deeply decurrent, thin, narrow. Flesh *grey, becoming white.* Spores pink, globose, 4–6μ, warted. Cystidia none. Smell of new meal, or rancid. Downs, fields, and sandy sea-shores. Aug.—Oct. Uncommon. (*v.v.*)

964. **C. undatus** Fr. (= *Eccilia undata* (Fr.) Quél.; *Clitopilus vilis* Fr. sec. Quél.) Fr. Icon. t. 96, fig. 4. *Undatus*, waved.

P. 2–4 cm., *fuliginous cinereous,* becoming pale, membranaceous, convex, then depressed, *umbilicate,* sometimes infundibuliform, unequal, *undulated,* often somewhat zoned, pruinose, silky. St. 2–3 cm. × 2–4 mm., *concolorous,* entirely fibrous, unequal, compressed, mealy; base cottony, white, and attenuated. Gills *dark cinereous*, deeply decurrent, 4 mm. broad, thin, entire, or undulate. Spores "distinctly rusty-brown, pure yellow under the microscope, elliptical, 7-8 × 4–5μ" Rick. Downs, and hilly pastures. Sept.—Oct. Uncommon. (*v.v.*)

var. **viarum** Fr. *Via*, a way.

Differs from the type in the *greyish hoary, smooth, shining, zoned p., and glabrous stem.*

965. **C. cancrinus** Fr. (=*Eccilia cancrina* (Fr.) Quél.) Fr. Icon. t. 95, fig. 4. *Cancer*, a crab.

P. 2–3 cm., *whitish tan, or wholly white*, becoming pale, slightly fleshy, *submembranaceous*, convex, then plane, very irregularly shaped, at length broken into cracks, becoming *flocculoso-even*. St. 2–2·5 cm. × 2–4 mm., *white*, round, or compressed, equal, or enlarged upwards, base white-villose, the mycelium often gathering the soil into a ball. Gills *white*, then flesh colour, *truly decurrent, distant*, 3 mm. broad, rather thick, arcuate when young then straight. Flesh *white*, hyaline near the gills. Spores pink, angular, oblong, 9 × 5–6μ, multi-guttulate. Pastures, and roadsides. July—Oct. Not uncommon. (*v.v.*)

966. **C. cretatus** B. & Br. Cke. Illus. no. 345, t. 375, fig. B. *Cretatus*, marked with chalk.

P. 6–18 mm., *dead white*, but *shining*, membranaceous, convex, then umbilicate, margin involute. St. 4–6 × 1–2 mm., *white*, often curved at the base, sometimes thickened, *tomentose, especially below.* Gills *rose colour*, very decurrent, narrow. Flesh *white.* Spores pink, elliptical, 7–8 × 3–4μ. Woods, and pastures. Aug.—Oct. Uncommon. (*v.v.*)

II. P. regular, silky, or hygrophanous-silky; margin naked. Gills adnate, slightly decurrent.

967. **C. carneo-albus** (With.) Fr. (= *Eccilia carneo-alba* (With.) Quél.) Cke. Illus. no. 349, t. 324, upper figs. *Caro*, flesh; *albus*, white.

P. 1·5–3 cm., *white, disc often becoming reddish, or yellowish*, convex, then expanded and depressed, slightly silky. St. 2–3 cm. × 4–6 mm., *white*, unequal, *fibroso-striate*, silky. Gills *white*, then flesh colour, adnato-decurrent, narrow. Flesh *white*, thin. Spores pink, angular, "10 × 6μ" Sacc. Woods, and heaths. Sept.—Oct. Uncommon. (*v.v.*)

968. **C. angustus** (Pers.) Fr. Fr. Icon. t. 96, fig. 3. *Angustus*, narrow.

P 1–2 cm., *bluish-grey-cinereous*, somewhat fleshy, convexo-plane, subumbonate, hygrophanous, silky-shining when dry. St. 7–8 × 1–

1·5 cm., *white*, curved, *strigosely rooting at the hairy base*. Gills *somewhat flesh colour*, adnato-decurrent, crowded, narrow. Flesh *brownish*. Spores pink, 7–8 × 5μ. Woods. Sept. Rare.

969. **C. Sarnicus** Massee. *Sarnicus*, belonging to Guernsey.

P. 2–3 cm., *mouse colour*, paler with a ruddy tinge when dry, campanulate, then quite plane, subumbonate, often more or less depressed round the umbo, slightly striate when moist, minutely silky flocculose. St. 2–3 cm. × 2 mm., *white*, equal, slightly flexuose. Gills *pinkish salmon colour*, *plane nearly up to the stem then suddenly decurrent*, 3–4 mm. broad, rather crowded. Flesh very thin. Spores pink, nodulose, with an apiculus, 7–8 × 6μ. Amongst grass. Rare.

970. **C. vilis** Fr. (= *Eccilia undata* (Fr.) Quél.) Cke. Illus. no. 351, t. 487. *Vilis*, of small value.

P. 2–3 cm., *grey*, *submembranaceous*, convex, *umbilicate*, silky-fibrillose when dry. St. 5–7·5 cm. × 2–4 mm., *concolorous*, equal, tough, but fibroso-fissile, fibrilloso-striate, base *white-villose*. Gills *whitish*, plano-decurrent, or adnate with a decurrent tooth, nearly triangular, crowded, *almost extending beyond the margin of the p*. Flesh *white*. Spores pink, "quadrangular, almost quadrilateral, 8–9μ" Rick. Downs, and open spaces. July—Sept. Not uncommon. (*v.v.*)

971. **C. stilbocephalus** B. & Br. Cke. Illus. no. 349, t. 324, lower figs. στίλβω, I shine; κεφαλή, the head.

P. 2–6 cm., *yellowish white*, or *greyish*, *sparkling with atoms*, whitish and rather silky when dry, *campanulate*, obtuse, sometimes umbonate, hygrophanous; margin *straight*. St. 5–8 cm. × 3–8 mm., *white*, or *greyish*, somewhat equal, undulato-fibrous. Gills *salmon colour*, adnate, sometimes emarginate behind, 3–5 mm. broad, veined. Flesh *whitish*. Spores pink, angular, oblong, 9–12 × 6–9μ, 1-guttulate. Smell pleasant of new meal. Pastures. Aug.—Sept. Uncommon. (*v.v.*)

972. **C. Smithii** Massee. Cke. Illus. no. 350, t. 599, as *Clitopilus stilbocephalus* Berk. var.
Worthington G. Smith, the eminent mycologist.

P. 2–4 cm., *whitish*, or *with a dingy yellow tinge*, *soon becoming plane* and orbicular, sometimes undulated, *atomate*. St. 5–7·5 cm. × 3–4 mm., *pallid*, *with a reddish tinge below*, tapering very slightly upwards, undulated, base white, downy. Gills *salmon colour*, broadly adnate with a slight decurrent tooth, 4 mm. broad. Flesh *white*, rather thick except at the margin. Spores pink, globose, 4μ. Oct. Rare.

973. **C. straminipes** Massee. Cke. Illus. no. 1159, t. 960.
Stramen, straw; *pes*, foot.

P. 3–5 cm., *whitish*, submembranaceous, fragile, at length expanded and depressed in the centre, *shining*. St. 5 cm. × 3 mm., *straw colour* below, *sprinkled with white meal* above, equal, often compressed. Gills *whitish* then rosy, shortly decurrent, scarcely crowded. Flesh *white*. Spores pink, angular, globose, 11–12μ. Amongst grass. Sept. Uncommon.

Spores ochraceous, or ferruginous.

Flammula Fr.

(**Gymnophilus** (Karst.) Murr. sec. Maire.)

(*Flammula*, a little flame.)

Pileus fleshy, regular, viscid, or dry. Stem central, fleshy, or fibrous. Gills decurrent, or adnate with a decurrent tooth. Spores ochraceous, ferruginous, or fuscous, elliptical, oblong elliptical, globose, or navicular; smooth, punctate, or verrucose; continuous, or with a germ-pore. Cystidia present, rarely none. Growing on the ground, or on wood; solitary, gregarious, fasciculate, or caespitose.

I. Veil *none*; p. dry, most frequently squamulose. Spores ferruginous, in *Flammula decipiens* fuscous ferruginous.

974. **F. gymnopodia** (Bull.) Fr. (= *Armillaria mellea* (Vahl.) Fr. var. *tabescens* (Scop.) Rea sec. Quél.) γυμνός, naked; πούς, foot.

Entirely dark ferruginous. P. 5–7·5 cm., fleshy, campanulato-convex, *squamulose*. St. 5–6 × 1 cm., becoming smooth, ascending equal. Gills *deeply decurrent*, arcuate, crowded. Caespitose. Pine sawdust, and on the ground. Sept.—Oct. Rare.

975. **F. Aldridgei** Massee. (= *Flammula veluticeps* Cke. & Massee.)
Miss Emily Aldridge.

P. 2–5 cm., *brick red with a tinge of orange, or tawny orange*, fleshy, convex, *then infundibuliform*, with a subinvolute margin, *minutely velvety*. St. 7–10 cm. × 6–8 mm., *concolorous*, equal, flexuose, smooth, base with a white floccose mycelium. Gills *golden yellow, then ferruginous-orange*, deeply decurrent, lanceolate, 3 mm. broad, rather crowded. Spores ferruginous orange, elliptical, slightly apiculate at the base, 16 × 5μ. Gregarious. Amongst moss on the ground in woods. Sept. Rare.

976. **F. vinosa** (Bull.) Fr. Cke. Illus. no. 466, t. 437.
Vinosa, full of wine.

P. 2–4 cm., *ferruginous fawn*, fleshy, expanded, at length depressed, dry, *delicately flocculose*. St. 2–3 cm. × 6 mm., *pale*, firm, somewhat thickened at the base, *delicately flocculose*. Gills *ferruginous*, decurrent, simple, narrow, crowded. Spores "pale brown, 5μ long, ovate" Sacc. On the ground. Rare.

F. paradoxa Kalchbr. = **Paxillus paradoxus** (Kalchbr.) Quél.

F. Tammii Fr. = **Paxillus paradoxus** (Kalchbr.) Quél.

977. **F. clitopila** Cke. & Sm. Cke. Illus. no. 468, t. 500.

κλίτος, a slope; *πῖλος*, cap.

P. 2·5–5 cm., *purplish brown, or madder brown*, fleshy, convex, then expanded, *disc depressed and umbilicate*, smooth, dry. St. 5–7·5 × 1–1·5 cm., *fuliginous, ventricose*, erect, with a few scattered fibrils towards the base. Gills *pallid, or yellowish, slightly adnexed, ventricose*, scarcely crowded. Flesh *white, brown in the st.*, fairly thick. Spores brown, elliptical, 10 × 4μ. Amongst firs. Nov. Rare.

978. **F. purpurata** Cke. & Massee. Cke. Illus. no. 1167, t. 964.

Purpurata, clad in purple.

P 2·5–5 cm., *purple, or purple brown*, fleshy, convex, then expanded, obtusely umbonate, *clad with minute, floccose, concolorous scales*, dry. St. 2·5–5 cm. × 4–6 mm., *pallid above, purple below*, equal, curved, ascending, apex smooth, *granular downwards*. Ring imperfect, fibrillose. Gills *lemon yellow, at length bright* ferruginous, adnate, somewhat rounded behind, subdistant, narrow. Flesh *purplish, yellow at the apex of the st.*, thick at the disc. Spores ferruginous, elliptical, 8 × 5μ. Taste very bitter. Tree-fern stems. May. Rare.

979. **F. floccifera** B. & Br. Cke Illus. no. 467, t. 438, upper figs.

Floccus, a flock of wool; *fero*, I bear.

P. 4–5 cm., *tawny*, fleshy, convex, then expanded, *sprinkled with snow-white fibrils*, becoming somewhat zoned in drying. St. 3–4 cm. × 6 mm., *white*, attenuated downwards, silky scaly, apex furfuraceous. Gills *ferruginous, edge white*, adnate, rounded behind, scarcely ventricose, moderately broad, *wrinkled transversely*. Flesh *white, tawny at the edge, and beneath the cuticle of the p., umber in the st.*, fleshy at the disc. Spores ferruginous. Caespitose. On lime stumps. Oct. Rare.

980. **F. decipiens** W. G. Sm. Cke. Illus. no. 467, t. 438, lower figs.

Decipiens, deceiving.

P. 2·5–3 cm., *rich brown, becoming pale, and almost white at the disc*, fleshy, convex, very obtuse, or umbonate, at length sometimes depressed round the umbo, dry, *minutely squamulose*. St. 3–6 cm. × 4–6 mm., *rich tawny*, attenuated downwards, often twisted, striate. Gills *orange brown*, decurrent, 4 mm. broad, crowded. Flesh *golden yellow, bright brown at base of st.*, thick at the disc. Spores orange brown, elliptical, apiculate at the base, 6–7 × 4μ. Inclined to be fasciculate. Charcoal heaps, and burnt earth. June—Oct. Uncommon. (*v.v.*)

981. **F. nitens** Cke. & Massee. Cke. Illus. no. 1168, t. 1154. *Nitens*, shining.

P. 2–5 cm., *dark purple brown*, fleshy, hemispherical, convex, then expanded, obtuse, shining, dry, somewhat silky. St. 4–7·5 × 1 cm., *flesh colour*, or *pale pinkish brown*, equal, fibrillose, incurved. Gills *pallid, then umber*, adnate, 4–6 mm. broad, crowded. Flesh *white*, thin at the margin. Spores pale brown, almond-shaped, 10 × 5–7μ. Caespitose. On the ground. Aug.—Sept. Rare.

II. P. *covered with a continuous, somewhat separable, smooth, viscid pellicle; cortina manifest fibrillose.* Spores ferruginous, not tawny; fuscous ferruginous in *Flammula carbonaria. Gregarious, growing on the ground, rarely on wood.*

F. lenta (Pers.) Fr. = **Hebeloma glutinosum** (Lindgr.) Fr.

982. **F. lubrica** (Pers.) Fr. Fr. Icon. t. 116, fig. 1. *Lubrica*, slimy.

P. 5–10 cm., *brick-red tawny, or bright cinnamon*, sometimes pallid with the disc tawny, fleshy, convex, then flattened, obtuse, or slightly umbonate, sometimes depressed and repand, *viscid*, smooth, sometimes spotted with glued down scales; margin sometimes striate. St. 5–10 cm. × 6–10 mm., *whitish, at length becoming fuscous*, equal, or slightly attenuated upwards, dry, *laxly fibrillose*, base pubescent. Gills *pallid, then clay colour*, adnate, subdecurrent, 6 mm. broad, crowded. Flesh *white*, thick at the disc, tough. Spores pale rusty brown, "cylindrical-elliptical, nearly reniform, 5–6 × 3–3·5μ, smooth. Cystidia lanceolate-fusiform, 50–65 × 12–18μ, contents at first yellowish" Rick. Smell scarcely strong. On and near trunks, and in pastures. Sept.—Oct. Uncommon.

983. **F. lupina** Fr. *Lupina*, pertaining to a wolf.

P. 7–10 cm., *brown, tan fuscous, or tawny*, fleshy, convex, obtuse, then plano-depressed, *smooth*, covered with a viscid, easily separable pellicle. St. 2·5 cm. × 12 mm., *whitish at the apex, elsewhere ferruginous with dense adpressed fibrils, sometimes light yellowish*, firm, thickened either upwards or downwards. Gills *clay colour, or light yellowish*, adnato-decurrent, broad, moderately crowded. Flesh *white, becoming ferruginous in the st.*, soft. Spores "nearly elliptical-oval, 9–10 × 5–6μ, smooth, almost colourless under the microscope. Cystidia on edge of the gill ventricose-fusiform, with a long pointed clavate apex, 50–60 × 9–12μ" Rick. Smell very strong or mild. Taste very bitter. Pastures. Aug.—Oct. Uncommon.

984. **F. mixta** Fr. Cke. Illus. no. 474, t. 476. *Mixta*, mixed.

P. 2·5–5 cm., *dingy tan*, fleshy, convexo-plane, obtuse, disc unequal, *darker, rugulose*, smooth; margin sloping, *paler*. St. 2·5–7·5 cm. × 6–

8 mm., *whitish*, equal, either short, ascending, curved, or elongated, flexuose, *with lax, fuscous fibrils, clothed below with reflexed, rufous fuscous scales*, base somewhat thickened. Cortina manifest, fibrillose. Gills *white, then clay colour*, subdecurrent, 6–8 mm. broad, somewhat crowded, unequal at the edge. Flesh *watery*, rather firm. Spores yellow brown, "almost almond-shaped, 12–15 × 6–7μ, smooth. Cystidia flask-shaped-lanceolate, 50–60 × 13–15μ" Rick. Subcaespitose. Pine and mixed woods. Aug.—Nov. Rare.

985. **F. juncina** W. G. Sm. Cke. Illus. no. 472, t. 475.

Juncina, pertaining to a rush.

P. 3–4 cm., *sulphury yellow, disc rich brown*, fleshy, hemispherical, convex. St. 8–10 cm. × 4–6 mm., *sulphur yellow, base tawny*, attenuated downwards, clothed with a few fibres. Gills *red brown*, decurrent, 4–6 mm. broad, very thin. Flesh *sulphur whitish, brownish towards the base of the st.*, thin at the margin. Taste nauseous and disagreeable, somewhat bitter. Dead bulrushes in an old clay pit. Nov. Rare.

986. **F. gummosa** (Lasch) Fr. Fr. Icon. t. 116, fig. 2.

Gummosa, sticky.

P. 3–6 cm., *pallid light yellow, or becoming green, at length ferruginous with the spores, paler at the circumference*, fleshy, regular, campanulate, then soon flattened, obtuse, or depressed, covered with a separable, viscid pellicle, *sprinkled with superficial floccose scales*, then smooth. St. 4–7·5 cm. × 4–6 mm., *ferruginous, rubiginous at the base, paler upwards*, equal, tense, straight, rigid, silky fibrillose. Gills *pale yellowish white, then cinnamon*, adnate, narrow, crowded. Flesh *becoming yellow*, thin. Spores yellow, elliptical, 5–7 × 3–4μ, smooth. Cystidia on the surface of the gill sparse, subulate-fusiform, 30–40 × 7–8μ, on edge of the gill cylindrical, capitate, flexuose, apex 6–7μ in diam., 40–45 × 4–5μ. On and about old stumps, and in grassy places. Oct.—Dec. Uncommon. (*v.v.*)

987. **F. decussata** Fr. (=*Flammula carbonaria* Fr. var. *decussata* Fr. sec. Quél.) Kalchbr. Icon. t. 15, fig. 1.

Decussata, divided crosswise.

P. 3–4 cm., *crust colour*, fleshy, convex, then plane, viscid, *virgate with innate, radiating, darker fibrils*; disc gibbous, *darker*. St. 3–5 cm. × 4 mm., *pallid above, elsewhere becoming fulvous*, equal, *adpressedly fibrillose*. Cortina manifest. Gills *yellowish, then clay colour*, adnate, *narrow*, crowded. Flesh *white, becoming yellowish under the separable pellicle*. Spores "7–8 × 3μ" Sacc. Beech woods. Rare.

988. **F. spumosa** Fr. Fr. Icon. t. 116, fig. 3. *Spumosa*, full of foam.

P. 3–5 cm., *pallid light yellow, disc often darker*, fleshy, convex, then plane, subumbonate, *very viscid*, pellicle separable, *naked*. St. 5–

10 cm. × 4 mm., *light yellow, or concolorous, sometimes olivaceous fuscous*, attenuated downwards, more or less fibrillose, remarkably cortinate. Gills *light yellow, then ferruginous*, adnate, crowded. Flesh *light yellow, becoming green*, watery, thin. Spores pale ferruginous, bluntly elliptical, 8–9 × 4–5μ, 2-guttulate. Cystidia "flask-shaped, 50–60 × 10–15μ, long-necked" Rick. Gregarious, or subcaespitose. Woods, especially fir, sawdust, pastures, and rarely on trunks. Sept.—Dec. Rare. (*v.v.*)

989. **F. carbonaria** Fr. Cke. Illus. no. 475, t. 442
Carbonaria, pertaining to charcoal.

P. 3–9 cm., *tawny*, fleshy, convex, then soon plane, and often depressed at the disc, smooth, *viscid*; margin incurved, often floccosely fimbriate. St. 2·5–11 cm. × 2–14 mm., *pallid, often blackish at the base, rigid*, equal, or slightly thickened upwards, *fibrillosely-squamulose*, the mycelium at the base often forming a pseudo-bulb with the soil. Cortina fibrillose, fugacious. Gills *clay, then fuscous clay colour*, adnate, rather broad, crowded. Flesh *yellowish*, firm, thin at the margin. Spores fuscous ferruginous, subelliptical, 6–7 × 4μ, 1-guttulate. Cystidia flask shaped, apex obtuse, 4–6μ in diam., 35–50 × 10–16μ. Densely gregarious. Charcoal heaps, and burnt earth. Aug.—Dec. Common. (*v.v.*)

III. *Cuticle of the p. continuous*, not distinct, nor separable, *smooth* (here and there with a superficial covering), *moist, or a little viscid in wet weather. Cortina manifest, appendiculate.* Spores not tawny, nor ochraceous. Caespitose, growing on wood.

990. **F. filia** Fr. Fr. Icon. t. 117, fig. 1. *Filia*, a daughter.

P. 5–7 cm., *pale yellow, disc rufescent*, fleshy, convex, soon plane, *moist*, smooth. St. 7·5–15 cm. × 12 mm., *pallid, base reddish*, equal, or attenuated at the base, smooth. Veil terminated by an incomplete ring, fugacious. Gills *white*, then pallid, adnate, somewhat crowded. Flesh *whitish, reddish in the st.*, thin. Spores "tawny orange, elliptic-fusiform, 10 × 5μ" Massee. Woods, and on logs. Oct. Rare.

991. **F. fusus** (Batsch) Fr. *Fusus*, a spindle.

P. 5–9 cm., *somewhat brick colour*, fleshy, convex, then plane, obtuse, smooth, slightly viscid. St 4–6 × 1–1·5 cm., *pallid*, firm, attenuated downwards in a fusiform manner, rooting, *fibrillosely striate*. Cortina manifest, appendiculate. Gills *pallid, or light yellow, then ferruginous, sometimes becoming green grey*, subdecurrent, not very crowded. Flesh *pallid, becoming yellowish, compact*, firm. Spores dingy ferruginous, subelliptical, 8–9 × 4–5μ. Cystidia "flask-shaped or clavate with a prominent point, 30–36 × 10–15μ, filled with olive yellow juice" Rick. Taste mild. Gregarious. On the ground, and on stumps. Sept.—Nov. Uncommon. (*v.v.*)

var. **superba** Massee. Cke. Illus. no. 478, t. 434. *Superba*, splendid.

Differs from the type in the *bright deep orange p. with darker disc, the pale orange st., the bright yellow gills, and the reddish tinge of the flesh.* On the ground. Nov. Rare.

992. **F. astragalina** Fr. Fr. Icon. t. 117, fig. 2.

ἀστραγαλῖνος, a goldfinch.

P. 3–8 cm., *blood saffron, or golden flesh colour, darker at the disc, pale at the circumference,* fleshy, convex, or lens-shaped, then flattened, obtuse, somewhat moist in rainy weather, *smooth,* at first *superficially-silky round the margin with the very thin, adpressed, whitish veil.* St. 5–10 cm. × 4–6 mm., *concolorous, or paler,* equal, or attenuated downwards, flexuose, *floccosely fibrillose.* Cortina *white,* manifest, appendiculate. Gills *pallid light yellow,* concolorous with the p. at the base, adnate, broad, crowded, edge obtuse, flocculose when young. Flesh *concolorous, becoming black when wounded, or bruised,* firm. Spores pale ferruginous, broadly elliptical, 6 × 3–4μ, 1-guttulate. Cystidia "clavate-lanceolate, 50–75 × 12–15μ, filled with olive brown juice" Rick. Taste bitter. Subcaespitose. Pine and fir stumps, and dead branches. Aug.—Oct. Uncommon.

993. **F. rubicundula** Rea. Grevillea, XXII (1894), t. 185, fig. 2.

Rubicundula, somewhat ruddy.

P. 4–6 cm., *yellow, then tinged with red, at length tawny orange,* fleshy, convex, then plane, often splitting at the margin, *viscid at first and innately fibrillose,* soon becoming smooth; margin at first veiled. St. 5–6 × 1·5–2·5 cm., *whitish, then tinged with red and becoming red at the base,* equal, or attenuated downwards, fibrillose below the veil, *apex white mealy.* Veil *white, then yellowish and at length reddening.* Gills *light ochre, then ferruginous,* adnate with a sinus, or adnato-decurrent, often forming a ring-like zone at the apex of the st., often separating, 3–4 mm. broad, crowded; *edge unequal, tinged red with age or when bruised.* Flesh *bright yellow, then lighter.* Spores ferruginous, elliptical, 9–10 × 4–5μ, 1–2-guttulate. Taste acrid. The whole plant becoming reddish with age, or when touched. Woods, under scrub oak. July—Sept. Uncommon. (*v.v.*)

994. **F. alnicola** Fr. Cke. Illus. no. 480, t. 443.

Alnus, alder; *colo,* I inhabit.

P. 3–8 cm., *yellow, at length becoming ferruginous, and sometimes green,* fleshy, convex, then flattened, obtuse, slimy when moist, at first superficially fibrillose towards the margin. St. 4–9 cm. × 6–

12 mm., *yellow, becoming ferruginous, attenuato-rooting,* sometimes subbulbous at the base, commonly curved, flexuose, *fibrillose.* Cortina *concolorous,* either fibrillose, or woven into an arachnoid veil. Gills *dingy pallid, then ferruginous,* somewhat adnate, broad, plane. Flesh *concolorous,* thick at the disc, not very compact. Spores ferruginous, elliptical, 9 × 4–5μ, 1–2-guttulate. Cystidia flask-shaped, 40–50 × 7–15μ. Taste bitter. Often fasciculate. On stumps, and trunks. Sept.—Oct. Uncommon. (*v.v.*)

var. **salicicola** Fr. *Salix,* willow; *colo,* I inhabit.

Differs from the type in the *glabrous, rarely at the first floccosely squamulose, gibbous p., and the gills being at first yellowish pallid.* On willow. Sept. Rare.

995. **F. flavida** (Schaeff.) Fr. Cke. Illus. no. 481, t. 444.
Flavida, light yellowish.

P. 2·5–12·5 cm., *bright light yellow,* fleshy, convex, then expanded, obtuse, *smooth,* moist, generally regular. St. 4–9 cm. × 6–10 mm., *light yellow, becoming ferruginous towards the base,* either attenuated, or thickened downwards, subflexuose, fibrillose. Cortina *white,* manifest, woven, adhering to the margin of the p., rarely almost forming a ring. Gills *whitish, then light yellow,* at length tawny ferruginous, adnate, not much crowded. Flesh *white, becoming light yellow,* thin at the margin. Spores ferruginous, broadly elliptical, 5–8 × 4μ, 1–2-guttulate. Cystidia "clavate, 36–40 × 8–9μ, filled with golden yellow juice" Rick. Caespitose. On trunks, stumps, and buried wood. Aug.—Nov. Common. (*v.v.*)

996. **F. inaurata** W. G. Sm. Cke. Illus. no. 482, t. 477.
Inaurata, gilded.

Entirely sulphur yellow. P. 2–3 cm., fleshy, convex, then expanded, moist, smooth, furnished with a distinct veil. St. 2·5–3·5 cm. × 4 mm., incurved, *clothed with innate scales.* Veil slight, fibrillose, fugitive. Gills *pale yellowish clay colour,* adnate with a decurrent tooth, broad. Flesh *yellowish, ferruginous at base of the st.* Taste mild. Single, or caespitose. Willows. Nov. Rare.

997. **F. conissans** Fr. Cke. Illus. no. 483, t. 445. *κόνις,* dust.

P. 1–7·5 cm., *light yellowish tan,* fleshy, hemispherico-expanded, obtuse, or umbilicate, moist, smooth. St. 5–8 cm. × 4–10 mm., *becoming light yellow white,* equal, or attenuated downwards, often compressed, irregular, twisted, *silky,* base white-villose. Cortina *white,* silky-fibrillose, appendiculate. Gills *whitish, then fuscous clay colour,* adnate with a decurrent tooth, *linear,* 3–4 mm. broad, very crowded. Flesh *white, or pale yellow,* equal, 2 mm. thick. Spores dark ferruginous, elliptical, 8 × 4μ. Cystidia "on edge of gill filamentous-clavate, subcapitate, or undulating, 36–45 × 5–7μ" Rick. Smell acid. Densely caespitose. Woods, dead stumps, and on willows. Sept.—Nov. Uncommon. (*v.v.*)

998. **F. inopus** Fr. (= *Flammula fusus* (Batsch) Fr. sec. Quél.) Fr. Icon. t. 118, fig. 1. ἴς, a fibre; πούς, foot.

P. 3–10 cm., *honey tan, or reddish tan, paler round the margin,* fleshy, convex, then expanded, obtuse, slippery (almost viscid) when moist, and *smooth* when dry. St. 7–25 cm. × 2–10 mm., *pallid, brick colour downwards,* equal, or slightly enlarged before continuing *into the long, tapering, rooting base,* tough, flexuose, *adpressedly fibrillose.* Cortina fugacious. Gills *pale yellowish white, sometimes green, then becoming purplish,* adnate, emarginate, 4–6 mm. broad, thin, crowded. Flesh *concolorous, becoming whitish, ferruginous in the st.,* thin at the margin. Spores purple, broadly elliptical, 8 × 5μ, 1-guttulate. Solitary, or caespitose. Pine trunks, and stumps. May—Dec. Not uncommon. (*v.v.*)

999. **F. apicrea** Fr. (=*Flammula alnicola* Fr. var. *salicicola* Fr. sec. Quél.) Cke. Illus. no. 485, t. 436. ἄπικρος, not bitter.

P. 3–7·5 cm., *dingy orange, or deep tawny, disc darker,* fleshy, convex, then expanded and almost plane, gibbous, or obtusely umbonate, smooth, moist; margin often splitting. St. 5–10 cm. × 4–10 mm., *pallid, ferruginous downwards,* equal, or attenuated downwards, *covered with ferruginous fibrils,* somewhat striate. Gills *ferruginous, shining,* adnate, or sinuate, 4–5 mm. broad, thin, crowded, edge often uneven. Flesh *yellow, tawny under the cuticle of the p., and ferruginous in the base of the st.,* thin at the margin. Spores ferruginous, elliptical, 8 × 5μ, 2–3-guttulate. Taste mild. Subcaespitose. Stumps, base of trees, and deal boards. Sept.—Dec. Uncommon. (*v.v.*)

1000. **F. carnosa** Massee. *Carnosa,* fleshy.

P. 2–3·5 cm., *dull tawny orange,* very fleshy, soon expanded, broadly gibbous, edge remaining more or less incurved for some time, even, smooth. St. 5–7·5 cm., *concolorous,* subequal, fibrous. Gills *rust coloured,* powdered with the spores, slightly decurrent, thin, somewhat crowded. Flesh *yellowish,* compact, 1 cm. or more thick at the disc, thin at the extreme edge. Spores brown, elliptical, 7 × 5μ. Tufted in small clusters. On wood. Sept. Rare.

1001. **F. azyma** Fr. ἄζυμος, unleavened.

P. 2–3 cm., *ferruginous, tan colour when dry,* fleshy, convex, then plane, obtuse, *smooth when in full vigour, becoming silky and rimosely squamulose when dry.* St. 2·5–4 cm. × 4–6 mm., *concolorous, or paler,* firm, somewhat equal, often curved, or flexuose, *slightly fibrillose,* base white woolly. Cortina fugacious, sometimes forming a ring-like zone on the st. Gills *yellowish, then ferruginous, broadly adnate,* connected behind, 4 mm. broad, subdistant, *edge whitish.* Flesh *yellowish, ferruginous under the cuticle of the p., and in the st.,* thin at the margin. Spores ferruginous, navicular, 8–9 × 5μ, 1–multi-guttulate, "subverrucose. Cystidia on edge of gill ventricose-subulate, 36–45 × 8–9μ" Rick. Taste mild. Gregarious, or subcaespitose. On *Tilia cordata, Pyrus Malus,* and rotten wood. Oct. Uncommon. (*v.v.*)

IV. P. scarcely pelliculose, flesh scissile, or torn above into scales, not viscid, at first somewhat hoary. Veil fibrillosely adpressed to the st., *not furnished with an appendiculate cortina, almost none, or forming an annular zone on the st.* Gills *light yellow, or yellow, then tawny. Spores ochraceous, or tawny.* Subcaespitose, *always on conifers,* or on the ground amongst conifer branches.

1002. **F. penetrans** Fr. Fr. Icon. t. 118, fig. 2. *Penetrans,* penetrating.

P. 5–8 cm., *yellowish tawny, or golden, becoming pale and yellowish,* fleshy, convex, then plane, obtuse, often irregular, dry, *smooth, hoary under a lens when young.* St. 5 cm. × 6–10 mm., *pallid, or yellowish becoming pale,* firm, somewhat equal, base white villous and often *rooting,* sometimes fusiform when on the ground, silky, *striate with tawny fibrils.* Cortina *white, flocculose,* submembranaceous, *very fugacious.* Gills *whitish, then pale yellow, spotted tawny when old,* adnate, emarginate, 4–6 mm. broad, crowded. Flesh *whitish* ("pale sulphur yellow" Quél.), thick at the disc. Spores ochraceous, elliptical, "8–9 × 4–5μ" Sacc. Taste bitter. Gregarious. Coniferous stumps, and humus. Oct. Uncommon.

1003. **F. hybrida** Fr. *Hybrida,* a mongrel.

P. 4–5 cm., *tawny cinnamon, then tawny orange,* fleshy, hemispherical, with the margin involute, then expanded, obtuse, regular, well formed, *smooth, moist.* St. 5–7·5 cm. × 5–10 mm., *becoming tawny,* equal, or *attenuated upwards, somewhat striate,* apex often somewhat mealy, base white villous. Cortina *white, at length coloured with the ferruginous spores, manifest, forming a ring at the apex of the st.* Gills *light yellow, then tawny,* adnate, somewhat crowded. Flesh *pallid, or yellow,* moderately compact. Spores ferruginous, oblong-elliptical, 9 × 4μ, "roughish. Cystidia on edge of gill filamentous, subcapitate, 45–50 × 4–6μ" Rick. Taste bitter. Growing in troops. On fir stumps, and fallen branches. Aug.—Dec. Uncommon. (*v.v.*)

1004. **F. sapinea** Fr. Fr. Icon. t. 118, fig. 3.
Sapinea, pertaining to a fir tree.

P. 2·5–10 cm., *golden tawny, opaque at the disc, paler and shining towards the margin,* fleshy, hemispherical, then convexo-plane, very obtuse, dry, *covered with thin, squamulose, adpressed floccules,* often rimosely scaly, with a few remnants of the yellowish cortina at the margin. St. 4–6 cm. × 4–12 mm., *becoming yellow pallid, turning fuscous when bruised,* irregularly shaped, often compressed, very fleshy, fibrous, sulcate, or lacunose, naked, often rooting at the base. Gills *golden, at length tawny-cinnamon,* adnate, plane, 8 mm. *broad,* crowded. Flesh *becoming yellow,* thick, firm, but at length soft, not scissile. Spores deep ochraceous, elliptical, 7–8 × 4–5μ, "roughish. Cystidia ventricose-fusiform, 36 × 9μ" Rick. Smell strong, taste often bitter. Subcaespitose. Coniferous stumps, branches, and sawdust. Aug.—Nov. Common. (*v.v.*)

var. **terrestris** Fr. *Terrestris,* pertaining to the earth.

Differs from the type in the *long, fusiform st.* Growing on coniferous humus.

1005. **F. liquiritiae** (Pers.) Fr. Fr. Icon. t. 119, fig. 1.
Liquiritia, liquorice.

P. 2·5–7·5 cm., *bay brown, or orange tawny, becoming pale,* fleshy, convex, then flattened, subumbonate, *very smooth, moist*; margin at length flaccid, *slightly striate.* St. 4–5 cm. × 4–6 mm., *tawny, then ferruginous,* attenuated upwards, often unequal, curved, *striate, somewhat naked,* or obsoletely pruinose at the apex, base thickened and villose. Cortina *none.* Gills *golden, then tawny,* obtusely adnate, sometimes rounded, separating, 6 mm. *broad,* plane, crowded. Flesh *yellow, yellow tawny in the st.,* thin, scissile. Spores ochraceous, "subelliptical, 8–9 × 5–6μ. Cystidia on edge of gill subcylindrical, slightly ventricose-capitate, 30–40 × 6–9μ" Rick. Taste slightly bitter, then sweet like liquorice. Subcaespitose. Fir stumps, rarely pine. Oct. Uncommon.

1006. **F. picrea** Fr. Fr. Icon. t. 119, fig. 2. πικρός, bitter.

P. 2–3 cm., *rufous, or bay brown cinnamon, becoming pale and tawny,* fleshy, *campanulate, then convex,* obtuse, regular, *smooth,* rarely rimuloso-papillate, moist in rainy weather. St. 5–7·5 cm. × 2–4 mm., *umber,* slightly attenuated upwards, tense and straight, *white-pulverulent when young.* Cortina *none.* Gills *yellow, then ferruginous,* adnate, or decurrent and separating, ascending, *narrow,* 1–2 mm. broad. Flesh *concolorous,* very thin, not easily scissile. Spores ferruginous, "elliptical, 8–10 × 5–6μ" Schroet. Taste acid. Caespitose. Pine stumps, and old deal boards. Sept.—Dec. Uncommon. (*v.v.*)

V. Furnished with a cortina. Cuticle of the p. slightly silky, dry, or at the first viscid.

1007. **F. tricholoma** (A. & S.) Fr. (= *Paxillus tricholoma* (A. & S.) Quél.) Cke. Illus. no. 444, t. 404, fig. B, as *Inocybe tricholoma* A. & S. θρίξ, hair; λῶμα, fringe.

P. 1–4 cm., *whitish,* fleshy, orbicular, rather plane, depressed in the centre, *fibrillose with white, adpressed, fugacious hairs,* viscid when moist, shining when dry; *margin fringed with* strigose hairs. St. 2·5–7·5 cm. × 4–5 mm., *whitish,* slightly attenuated upwards, *fibrillosely scaly at the apex, often becoming reddish in places.* Gills *whitish, becoming pallid fuscous, then clay fuscous, decurrent,* 1–2 mm. broad, thin, crowded. Flesh *white,* thin at the margin. Spores pale ochraceous, globose, 3–5μ, minutely verrucose. Cystidia none. Woods. Sept.—Nov. Not uncommon. (*v.v.*)

1008. **F. strigiceps** Fr. (= *Paxillus tricholoma* (A. & S.) Quél.)
Strix, a furrow; *caput,* head.

P. 1–2 cm., *obsoletely rufescent,* slightly fleshy, convex, obtuse, then plane, *dry, silky with long, strigose hairs*; margin at first involute, *fringed with long, deflexed ciliate hairs.* St. 3–5 cm. × 4 mm., *white,* equal, firm, *densely villose,* especially when young. Gills *whitish, becoming fuscous,* adnato-decurrent, arcuate, crowded. Beech woods. Rare.

1009. **F. helomorpha** Fr. (= *Paxillus helomorphus* (Fr.) Quél.) Fr. Icon. t. 120, fig. 4.
ἧλος, a nail; μορφή, form.

P. 1–3 cm., *white,* fleshy, convexo-plane, gibbous, or with a broad, obtuse, prominent umbo, often angular, viscid, *becoming adpressedly fibrilloso-even when dry*; margin thin, unequal, inflexed, naked. St. 2–3 cm. × 4–6 mm., *whitish,* equal, or not perceptibly attenuated from the base, sometimes enlarged upwards, ascending from the incurved base, adpressedly silky, or pruinose upwards under a lens, *smooth.* Gills *whitish, scarcely clay colour,* plano-decurrent, 1–2 mm. broad, very crowded. Flesh *whitish,* thick at the disc. Spores pale ochraceous, globose, 4–5μ, minutely verrucose. Fir woods. Oct.—Nov. Uncommon. (*v.v.*)

1010. **F. scamba** Fr. (= *Paxillus scambus* (Fr.) Quél.) Fr. Icon. t. 120, fig. 3.
σκαμβός, crooked.

P. 1–4 cm., *whitish, then clay white,* fleshy, convex, then plane and depressed, sometimes umbonate, slightly silky, *viscid in wet weather when young,* soon becoming dry and opaque. St. 1–3 cm. × 2–3 mm., whitish, equal, curved ascending, *flocculose, or sprinkled with white mealy squamules,* base pubescent, sometimes attenuated and becoming ferruginous downwards. Gills *light yellow clay colour,* adnate, or sub-decurrent, somewhat repand, crowded. Flesh *yellowish,* thin. Spores pale ferruginous, elliptical, 9 × 5μ. Pine woods, and on larch branches. Sept.—Oct. Uncommon. (*v.v.*)

1011. **F. ochrochlora** Fr. Fr. Icon. t. 120, fig. 2.
ὦχρος, pale; χλωρός, green.

P. 2·5–5 cm., *straw colour, becoming greenish,* fleshy, convex, then plane, obtusely umbonate, *dry, silky,* squamulose. St. 5–6 cm. × 4–10 mm., *yellowish, becoming ferruginous towards the base,* attenuated upwards, often curved, or flexuose, *squamulose and white floccose.* Cortina *white,* manifest, *Hypholoma*-like. Gills *whitish, then becoming greenish, and at length olivaceous,* adnate, or somewhat sinuate, 2–4 mm. broad, crowded. Flesh *whitish, becoming greenish, and ferruginous at the base of the st.* Spores pale ferruginous, elliptical, 6–7 × 4μ, 1–2-guttulate. Caespitose. On old trunks, and buried wood. Aug.—Nov. Common. (*v.v.*)

1012. **F. filicea** Cke. Cke. Illus. no. 491, t. 450. *Filix,* a fern.

P. 2–4 cm., *deep yellow, disc tawny orange,* fleshy, convex, then plane, or slightly depressed, minutely squamuloso-fibrillose. St. 3–5 cm. × 3 mm., *sulphur yellow, base often tawny,* equal, almost smooth. Veil *reddish,* adhering to the st. and the margin of the p. in fugacious fragments. Gills *sulphur yellow,* then *tawny cinnamon,* adnate, 3 mm. broad, crowded. Flesh *sulphur yellow,* thin. Old tree fern stems. Spring and summer. Rare.

F. chrysophylla (Fr.) Quél. = **Omphalia chrysophylla** Fr.

Spores greenish fuscous, or blackish; gills mucilaginous.

Gomphidius Fr.

(γόμφος, a large wedge-shaped nail.)

Pileus fleshy, regular, viscid. Stem central, fleshy. Gills decurrent, mucilaginous. Spores fuscous, olivaceous, or blackish, fusiform, or oblong, smooth, continuous. Cystidia cylindrical, projecting. Growing on the ground.

1013. **G. glutinosus** (Schaeff.) Fr. Rolland, Champ. t. 74, no. 165. *Glutinosus,* glutinous.

P. 5–12·5 cm., *purple fuscous, or fuscous, often mottled with black spots,* fleshy, convex, obtuse, at length plane, smooth, *very glutinous.* St. 5–10 × 1–2 cm., *whitish, yellow at the base,* equal, thickened, or attenuated at the base, glutinous, fibrillose, sometimes with black scales. Cortina annular, fugacious. Gills *whitish, then cinereous,* deeply decurrent, forked, distant, mucilaginous, 6–8 mm. broad. Flesh *white, yellow towards the base of the stem,* thick. Spores deep olivaceous, spindle-shaped, 18–24 × 5–6μ, 4–5-guttulate. Cystidia "cylindrical, 130–160 × 12–16μ" Rick. Taste bitter. Coniferous woods. July—Nov. Common. (*v.v.*)

1014. **G. roseus** (Fr.) Quél. Cke. Illus. no. 857, t. 880. *Roseus,* rose-coloured.

P. 2–5 cm., *rose, or rose-red colour,* convexo-plane, obconical, obtuse, at length sometimes depressed, slightly glutinous. St. 3–5 × 1–

1·5 cm., *white, often tinged with rose at the base*, attenuated downwards. Cortina thin, slightly glutinous. Gills *whitish cinereous, then olivaceous*, decurrent, 4–5 mm. broad, distant, forked. Flesh *white, rosy under the cuticle and at the base of the st.* Spores pale greyish olivaceous, fusiform, 15–17 × 4–5μ, 1–3-guttulate. Cystidia "cylindrical, 90–160 × 12–15μ" Rick. Taste pleasant. Coniferous woods, and under conifers. Aug.—Oct. Not uncommon. (*v.v.*)

1015. **G. viscidus** (Linn.) Fr. Cke. Illus. no. 858, t. 881.
Viscidus, viscid.

P. 5–15 cm., *fuscous rufous*, fleshy, campanulate, or obconical, then expanded, umbonate, slightly viscid, paler and shining when dry. St. 7–12 × 2–3 cm., *concolorous, paler, yellowish at the base*, equal, or attenuated downwards, fibrillosely scaly, slightly viscid. Cortina floccose, forming a fugacious ring. Gills *olivaceous, then fuscous purple*, deeply decurrent, distant, often branched, edge often paler. Flesh *reddish, deep yellow in the lower two-thirds of the st.* Spores brownish olivaceous, subfusiform, 18–22 × 6–7μ, 3-guttulate. Cystidia obtusely cylindrical, apex often subcapitate, 135–150 × 15–17μ. Taste mild, often slightly astringent. Edible, indigestible to some people. Coniferous woods, and under conifers. July—Dec. Common. (*v.v.*)

var. **testaceus** Fr. *Testaceus*, brick-red.

Differs from the type in the *brick-red colour of the flatter p., and base of st. both externally and internally.* Coniferous woods, and under conifers. July—Nov. Common. (*v.v.*)

1016. **G. maculatus** (Scop.) Fr. (= *Gomphidius gracilis* B. & Br. sec. Quél.)
Maculatus, spotted.

P. 3–6 cm., *reddish brown, disc paler, often spotted with black, and becoming black at the edge*, campanulate, or obconic, then plane, glutinous. St. 6–8 × 1–1·5 cm., *white, yellow at the base*, becoming *blackish when touched*, ventricose downwards, expanding into the p. at the apex, floccose, slightly viscid. Gills white cinereous, then olivaceous, *deeply decurrent*, often forked, somewhat crowded. Flesh *whitish, stained bistre in the p. and yellow towards the base of the st.*, thick at the disc, thin at the margin. Spores olivaceous, fusiform, somewhat blunt at the end, 17–20 × 6μ, 1–3-guttulate. Cystidia obtusely cylindrical, or fusiform, 120–140 × 18–20μ. Coniferous woods, and under conifers. Sept.—Oct. Uncommon. (*v.v.*)

var. **Cookei** Massee. Cke. Illus. no. 859, t. 882, as *Gomphidius maculatus* Scop. var. M. C. Cooke, the eminent mycologist.

P. 2·5–5 cm., *whitish, with black stains especially near the margin*, convex, then subdepressed, or gibbous, viscid. St. 6–8 × 1 cm., *pale above, becoming blackish towards the base*, attenuated upwards. Gills *whitish, then brownish,* decurrent, *distant.* Flesh *pallid, blackish at the base of the st.,* thick at the disc, very thin at the margin. Spores brownish, fusiform, 20 × 5–6μ, 1-guttulate. Woods. Sept. Rare.

1017. **G. gracilis** B. & Br. Cke. Illus. no. 860, t. 883.

Gracilis, slender.

P. 2·5–5 cm., *pale vinous brown, or dingy tan colour,* conico-hemispherical, clothed with dingy gluten, at length spotted with black, especially near the margin, the spots often forming an irregular black border. St. 4–5 cm. × 3–6 mm., *pale above, yellow at the base, and often becoming blackish,* slightly attenuated downwards, flexuose, *apex white squamulose,* virgate below with the remains of the gluten. Gills whitish cinereous, decurrent, arched, forked, thick, obtuse, *clothed (under a lens) with short, washy bistre hairs.* Flesh *white, yellow, or reddish at the base of the st.,* thin at the margin. Spores dingy olive, or brown, fusiform, 18–19 × 5–7μ, 2–3-guttulate. Fir woods, and heaths. July—Nov. Common. (*v.v.*)

C. Pileus confluent with, but heterogeneous from, the cartilaginous stem.

*Gills adnate, or sinuato-adnate.

†Margin of pileus at first incurved, or exceeding the gills.

Spores white.

Collybia Fr.

(κόλλυβος, a small coin.)

Pileus fleshy, membranaceous, regular; margin incurved. Stem central, cartilaginous. Gills adnate, adnexed, or free. Spores white, rarely yellowish, greenish, or brownish red; elliptical, globose, oblong, pip-shaped; smooth, verrucose, punctate, or echinulate; continuous. Cystidia present, or absent. Growing on the ground, or on wood; solitary, or caespitose.

a. St. stout, *sulcate, or fibrillosely striate.*

A. Gills white, or brightly coloured, not cinereous. Flesh often white.

*Gills broad, subdistant.

1018. **C. radicata** (Relh.) Berk. Rolland, Champ. t. 45, no. 98.

Radicata, rooted.

P. 3–10 cm., *fuscous-olivaceous, bistre, or whitish,* fleshy, thin, convex, then flattened, gibbous, often irregular, *glutinous, radiato-rugose.* St. 10–20 × ·5 cm., *white, or paler than the p., attenuated upwards, and downwards from the level of the soil,* and *forming a long tail-like fusiform root,* smooth, at length *striato-sulcate,* cuticle cartilaginous, often twisted. Gills *shining white, sometimes bistre at the edge*, attenuated behind, and adfixed, often with a decurrent tooth, at length somewhat separating, ventricose, *distant, rather thick.* Flesh *white*, thin, soft, elastic. Spores white, broadly elliptical, 14–15 × 8–9μ, 1-guttulate; "cystidia inflated, cylindric-sack-shaped, 20μ broad" Lange. Woods, and pastures. June—Nov. Common.

1019. **C. retigera** Bres. (= *Collybia plexipes* (Fr.) Quél.) Bres. Fung. Trid. t. 4. *Rete*, a net; *gero*, I bear.

P. 3–6 cm., *fuscous cinereous, becoming pale, disc somewhat tawny*, fleshy, thin, campanulate, then expanded, umbonate, *reticulated with swollen, pale, anastomosing veins, especially when old*, dry, smooth; margin striate. St. 4–6 cm. × 5–7 mm., *livid-pallid*, equal, *somewhat rooting*, often compressed, *white-fibrillose.* Gills *cinereous, edge paler, fimbriate*, rounded behind, almost free, broad, ventricose, somewhat crowded. Flesh *concolorous*, thin, soft, somewhat watery. Spores white, elliptical, 7 × 5–6μ. Beech stumps. Dec. Uncommon.

1020. **C. Henriettae** W. G. Sm. Henrietta Smith.

P. 10 cm., *somewhat yellowish umber*, convex, then expanded, dry, even, *somewhat downy.* St. 18–19 cm. × 6–7 mm., *pale pallid yellowish brown, darker below*, attenuated upwards, even, *slightly rooting, subpruinose.* Gills broadly adnate, slightly rounded behind, broad, *distant.* Flesh very thin, *pale pallid yellowish brown in the st.* Spores white, 18 × 12μ. On and about trees, stumps, etc. Sept. Uncommon.

1021. **C. longipes** (Bull.) Berk. (= *Marasmius longipes* (Bull.) Quél.) *Longus*, long; *pes*, foot.

P. 5–10 cm., *pale brown*, fleshy, thin, conico-expanded, then plane, umbonate, *dry, somewhat velvety-villous.* St. 8–12 cm. × 6–8 mm., *dark brown*, attenuated upwards, *with a long fusiform root, velvety*, at length sulcate. Gills *milk white*, free, rounded behind, *very distant*, ventricose. Flesh *white, yellowish in the stem*, firm. Spores white, globose, 12–15μ. Cystidia "very sparse, cylindrical-subulate, 50–60 × 8–10μ" Rick. Taste nutty. Edible. Heaths, and pastures. Aug.—Oct. Not uncommon. (*v.v.*)

var. **badia** Lucand. *Badia*, bay brown.

Differs from the type in being *thinner and smaller, and in the deep chestnut brown p. and st. covered with long, bay brown, shining hairs.* Spores globose, 10–11μ. Hedgerows. Sept.—Oct. Not uncommon. (*v.v.*)

1022. **C. eriocephala** Rea. ἔριον, wool; κεφαλή, head.

P. 3–6 cm., *fulvous tawny*, convex, then expanded, *velvety*; margin involute. St. 4–7 × ·5–1·5 cm., *concolorous, paler above, fusiform, extending into the long, branched, rhizomorphoid, brown mycelium, striate*, only slightly velvety at the thickest part. Gills *deep ochre*, sinuato-adnate, 5–8 mm. broad, distant. Flesh *pale, then yellowish, somewhat rufous at the base of the st.*, thick at the disc, firm. Spores white, oblong, 7–8 × 3–4μ, 1–2-guttulate. Caespitose. Interior of a rotten elm stump, and in timber yards. Sept.—Jan. Uncommon. (*v.v.*)

1023. **C. platyphylla** (Pers.) Fr. (= *Collybia grammocephala* (Bull.) Quél.; *Collybia platyphylla* var. *repens* Fr.) Rolland, Champ. t. 47, no. 101, as *Collybia grammocephala*.

πλατύς, broad; *φύλλον*, leaf.

P. 5–20 cm., *fuscous, or cinereous, becoming whitish*, fleshy membranaceous, thin, *fragile*, convex, soon flattened, obtuse, watery when moist, *streaked with bistre fibrils*. St. 7–12 × 1–2 cm., *whitish*, equal, *fibrillosely striate*, apex sometimes pruinose, *arising from a network of white, creeping, string-like mycelium*. Gills *white*, obliquely truncate behind, *slightly adnexed, very broad*, 10–15 mm., *distant*, soft. Flesh *white*, thin at the margin. Spores white, broadly elliptical, 8–10 × 6–8μ, 1-guttulate; "cystidia sack-shaped-club-shaped, 14μ broad" Lange. Woods. May—Nov. Common. (*v.v.*)

1024. **C. fumosa** (Pers.) Quél. (= *Collybia semitalis* Fr. sec. Quél.; *Tricholoma immundum* Berk. sec. Bres.) Bres. Fung. Trid. t. 156. *Fumosa*, smoky.

P. 3–9 cm., *pitch black, lurid grey, or smoky greyish, becoming paler and spotted fuscous*, fleshy, convexo-campanulate, then expanded and depressed, silky, then smooth; margin *undulate, finally splitting*. St. 4–8 × ·5–1·5 cm., *concolorous, or paler*, subequal, *subcartilaginous, somewhat fibrillosely striate*, base sometimes bulbous. Gills *greyish-cinereous*, rounded behind, or truncate and free, *veined at the sides, spotted with black when touched*. Flesh *cinereous, becoming whitish*, thick at the disc. Spores white, globose, 6–7μ, 1-guttulate. Smell rancid, taste bitterish. Caespitose. Woods, and pastures. Sept.—Oct. Not uncommon. (*v.v.*)

1025. **C. crassifolia** (Berk.) Bres. (= *Tricholoma crassifolium* Berk.) Bres. Fung. Trid. t. 157. *Crassus*, thick; *folium*, leaf.

P. 4–7 cm., *lurid ochraceous, disc fuscous, becoming concolorous*, fleshy, convex, or campanulate and *umbonate*, then expanded and depressed silky, becoming smooth; margin *undulate, or lobed*. St. 2·5–5 × ·5–1·5 cm., *white, becoming fuscous*, often attenuated at the base, *pruinose*, becoming smooth, round, or compressed, *subcartilaginous*. Gills *whitish-grey, becoming bluish and finally blackish when touched*, rounded behind, adnexed, sometimes forked, *distant, broad, thick*, fleshy. Flesh *white, spotted black when broken*, thick at the disc. Spores white, globose, 5–7μ, 1-guttulate. Smell strong, rancid, taste mild. Coniferous woods. Oct. Uncommon. (*v.v.*)

1026. **C. semitalis** Fr. (= *Collybia fumosa* (Pers.) Quél.) Bres. Fung. Trid. t. 158. *Semitalis*, pertaining to footpaths.

P. 3–7 cm., *whitish fuliginous, or fuscous, becoming pale cinereous yellow, or isabelline when dry*, fleshy-membranaceous, convex, or convexo-campanulate, then expanded and *umbonate*, or depressed, smooth, moist, sometimes innately fibrillose; margin striate. St. 3–8 cm. × 6–8 mm., *white, becoming fuscous*, subequal, *fibrillose, base white-strigose*. Gills *white, becoming yellowish, and finally spotted black when touched*, adnate, or sinuato-adnate, somewhat crowded. Flesh *white, becoming black when broken*, thin. Spores white, elliptical, 7–8 × 3–4μ, pointed at one end, 1-guttulate. Smell rancid, taste bitterish. Coniferous woods. Sept.—Nov. Uncommon. (*v.v.*)

1027. **C. fusipes** (Bull.) Berk. Cke. Illus. no. 185, t. 141. *Fusus*, a spindle; *pes*, foot.

P. 4–10 cm., *rufescent reddish brown, or liver colour, becoming pale, or dingy tan*, fleshy, convex, then flattened, umbonate, *the umbo evanescent*, smooth, dry, often splitting. St. 7–15 × 1 cm., *concolorous*, very cartilaginous, *swollen, ventricose in the middle, attenuated at both ends*, often twisted, *longitudinally striato-sulcate, fusiformly attenuated at the base and blackish*, often arising from the remains of underground stems of a previous year's growth, the so-called sclerotium of Léveille. Gills *whitish, becoming concolorous and often spotted, annulato-adnexed*, soon separating, free, broad, distant, firm, connected by veins, crisped. Flesh *concolorous*, becoming whitish, firm. Spores white, elliptical, 5–6 × 3–4μ. Cystidia filiform, flexuose, clavate, 10–44 × 1–2μ. Taste mild. Edible. Caespitose, at the base of oaks and on old stumps. May—Dec. Common. (*v.v.*)

var. **oedematopus** (Schaeff.) Fr. Bulliard, t. 76, as *Agaricus fusiformis*. *οἴδημα*, a swelling; *πούς*, foot.

Differs from the type in the *rufous date brown, conical, then plane, pulverulent p., the pulverulent, very ventricose stem, and the pallid gills*. Stumps. Sept.—Oct. Not uncommon. (*v.v.*)

var. **contorta** (Bull.) Gill. & Lucand. Bulliard, t. 36. *Contorta*, twisted together.

Differs from the type in the *equal, contorted stems, connate at the base, the white, crowded gills, and the deeper coloured, thinner p.* Stumps.

1028. **C. lancipes** Fr. *Lancea*, a spear; *pes*, foot.

P. 4–7 cm., *pale reddish brown, or flesh colour, becoming paler, often white at the striate margin*, fleshy, convex, then plane, often umbonate, *radiately rugose*, smooth. St. 4–10 cm. × 5–12 mm., *concolorous, or paler*, equal, attenuated at the base, *striate, twisted*. Gills *pale flesh colour, or yellowish*, emarginate, adnexed, becoming free, broad, thick, distant, often connected by veins. Flesh *whitish, reddish under the cuticle*. Spores white, pip-shaped, 6 × 4μ, 1–2-guttulate. Taste mild. Edible. On the ground, and near stumps. Sept.—Oct. Uncommon. (*v.v.*)

**Gills narrow, crowded.

1029. **C. maculata** (A. & S.) Fr. Cke. Illus. no. 186, t. 142.
Maculata, spotted.

P. 7–12 cm., *white, then spotted rufescent*, rarely becoming wholly rufescent, fleshy, *very compact*, convexo-plane, obtuse, repand, smooth; margin thin, involute at first. St. 7–12 × 1–2 cm., *white, spotted rufescent, somewhat ventricose*, attenuated downwards to the *praemorse base, hard*, externally cartilaginous, *striate*. Gills *cream colour, often spotted rufescent*, emarginato-free, linear, 2–4 mm. broad, *very crowded, denticulate*. Flesh *white*, thick, firm. Spores white, subglobose, 5–6μ, punctate. Cystidia none. Smell pleasant, or none. Taste unpleasant, bitter. Beech, and pine woods. May—Nov. Common. (*v.v.*)

var. **immaculata** Cke. Cke. Illus. no. 187, t. 221.
Immaculata, unspotted.

Differs from the type in *not being spotted, and in the broader gills*. Pine woods. Aug.—Oct. Not uncommon. (*v.v.*)

var. **scorzonerea** (Batsch) Fr. *Scorzon*, a serpent.

Differs from the type in its *smaller size, and in becoming yellowish, in the long, rooting often flexuose st., and the yellowish gills*. Beech woods. Sept.—Oct. Uncommon. (*v.v.*)

1030. **C. fodiens** Kalchbr. Kalchbr. Icon. t. 36, fig. 2.
Fodiens, digging,

P. 5–8, *flesh colour, becoming yellowish, disc darker yellow*, fleshy, firm, convex, obtuse, smooth; margin involute. St. 10–12 cm. × 10–12 mm., *white*, firm, subventricose, *often longitudinally ribbed*, smooth, *attenuated downwards in a long root deeply sunk in the ground*. Gills *yellowish white*, emarginate, rounded behind, narrow, crowded. Flesh *yellowish*, thick at the disc, firm. Spores white, elliptical, 6–8 × 4–5μ, 1-guttulate. Smell and taste pleasant. Grassy places. Oct. Uncommon. (*v.v.*)

1031. **C. prolixa** (Fl. Dan.) Fr. *Prolixa*, stretched out.

P. 5–12 cm., *brick-red ferruginous, becoming paler*, fleshy, *fragile*, convex, then plane, gibbous, *lax*, smooth, margin often irregular. St. 10 × 1–3 cm., *brick-red*, firm, *subequal, sulcate, often scrobiculate*, minutely pubescent, fibrillose, *base praemorse*. Gills *white*, free, crowded. Flesh *white*, rather thick. Spores white, "subglobose, 3–4μ, smooth" Rick. In dense clusters on leaf heaps. Aug.—Sept. Uncommon. (*v.v.*)

1032. **C. distorta** Fr. *Distorta*, twisted.

P. 5–9 cm., *bay brown, becoming pale*, fleshy, thin, convex, then expanded, umbonate, *very lax*, smooth. St. 5–8 × 1 cm., *pallid, fragile, externally cartilaginous*, attenuated upwards from the tomentose base, *contorted, sulcate*. Gills *white, then spotted rubiginous*, slightly adnexed, *crowded, somewhat linear*, toothed. Flesh *white, reddish under the cuticle of the p. and in the centre of the stem*, thin. Spores white, broadly elliptical, 5–6 × 4–5μ, 3–4-guttulate. Gregarious, or growing in rings. Pine woods. July—Nov. Not uncommon. (*v.v.*)

1033. **C. butyracea** (Bull.) Fr. Cke. Illus. no. 189, t. 143. *Butyracea*, buttery.

P. 5–8 cm., *rufous brown, fuscous livid, bistre, or bay, becoming pale and almost white when dry*, fleshy, convex, then expanded, more or less *umbonate*, smooth, *greasy*. St. 5–8 × ·5–1 cm., *rufous*, or *bistre, conico-attenuated upwards from the swollen, white-tomentose base, cuticle rigid, cartilaginous, striate*, smooth, rarely villous. Gills *white*, slightly adnexed, *somewhat free*, broad, thin, *crowded, crenulate*. Flesh *pinkish, or pale brown, becoming whitish*, soft, watery, with a horn-like line at the base of the gills. Spores white, elliptical, 9 × 4–5μ, 1-guttulate. Cystidia none. Woods, heaths, and hilly pastures. Jan.—Dec. Common. (*v.v.*)

var. **bibulosa** Massee. *Bibulosa*, sodden.

Differs from the type in the *dingy olive p.*

var. **aurorea** (Larb.) Fr. *Aurorea*, like the dawn.

Differs from the type in the *thinner p., and striate margin.*

C. phaeopodia (Bull.) Fr. = **Tricholoma phaeopodium** (Bull.) Quél.

1034. **C. stridula** Fr. Fr. Icon. t. 62, lower figs. *Stridula*, creaking.

P. 3–6 cm., *blackish, or fuliginous, becoming pale*, fleshy, soft, convex, then plane, slightly umbonate, *smooth, moist, or slightly viscid, hygrophanous*. St. 5–7 cm. × 4–6 mm., *concolorous*, cylindrical, rigid, but fragile, subcartilaginous, *fibrillosely striate*, base thickened, praemorse. Gills *white*, arcuato-adnexed, *crowded, broad*. Flesh *brown, then whitish*, soft. Spores white, "8–10 × 4μ" Sacc. On the ground. Oct. Rare.

1035. **C. pulla** (Schaeff.) Fr. Trans. Brit. Myc. Soc. II, t. 1. *Pulla*, dusky.

P. 3–6 cm., *purplish bay, nearly black, becoming paler when dry*, fleshy, thin, fragile, campanulate, then expanded, obtuse, smooth, hygrophanous. St. 6–8 cm. × 4–8 mm., *whitish*, equal, *twisted*, somewhat striate, apex mealy, attenuated at the praemorse base. Gills *whitish*, adnexed, rather broad, *transversely pellucid-striate and veined*, crowded. Flesh *white*, thick at the disc. Spores white, elliptical, 10 × 6–7μ. Smell none, or strong of garlic. Caespitose, or solitary. Birch stumps. Aug.—Oct. Uncommon. (*v.v.*)

1036. **C. xylophila** (Weinm.) Fr. (= *Mycena rugosa* Fr. sec. Quél.) Fr. Icon. t. 63, lower figs. *ξύλον*, wood; *φίλος*, loving.

P. 6–10 cm., *whitish, or becoming fuscous tan at the disc*, slightly fleshy, *campanulate, lax*, obtuse, or with a minute umbo, then expanded, broadly gibbous, smooth, moist; margin often rimosely split. St. 4–8 cm. × 5–8 mm., *whitish*, equal, often flexuose, *fibrillosely* striate. Gills *white*, adnate, often decurrent with a small tooth, *very narrow*, 2 mm. broad, *very crowded*. Flesh *becoming watery fuscous*, thin, fragile. Spores white, "elliptical, 4 × 2·5μ" Massee. Caespitose. Old stumps. Sept.—Oct. Not uncommon. (*v.v.*)

β. St. thin, *velvety, floccose, or pruinose*.

*Gills broad, subdistant.

1037. **C. velutipes** (Curt.) Fr. (= *Pleurotus velutipes* (Curt.) Quél.) Cke. Illus. no. 191, t. 184, fig. A. *Vellus*, a fleece; *pes*, foot.

P. 2–10 cm., *fulvous, or tawny, sometimes paler at the margin*, fleshy, convex, soon becoming plane, often excentric, irregular and repand, smooth, *viscid*; margin spreading, at length slightly striate. St. 5–10 cm. × 4–8 mm., *lemon yellow, then umber and blackish*, equal, often ascending, or twisted, tough, cartilaginous, *densely velvety*. Gills *pallid yellow, becoming tawny*, broader and rounded behind, slightly adnexed, *subdistant, very unequal*. Flesh *yellowish*, thin at the margin, watery, soft. Spores white, elliptical, 7–8 × 5μ, 1–2-guttulate. "Cystidia conic, rather acute, almost subulate, 8–12μ broad, protruding part 18–30μ long" Lange. Taste and smell very pleasant. Edible. Caespitose. On old stumps, fallen trunks, and pales. Aug.—April. Common. (*v.v.*)

var. **lactea** Quél. *Lactea*, milk-white.

Differs from the type in being *creamy white*. Stumps. Oct. (*v.v.*)

var. **rubescens** Cke. Cke. Illus. no. 1141, t. 650.
Rubescens, becoming reddish.

Differs from the type in the *bright ferruginous brown p., the darker blackish cinnamon st., and the gills becoming spotted with brown*. Amongst fir leaves.

1038. **C. laxipes** (Batt.) Fr. (= *Marasmius laxipes* (Batt.) Quél.) Quél. Jur. et Vosg. II, t. 2, fig. 2. *Laxus*, loose; *pes*, foot.

P. 1–3 cm., *whitish, often yellowish at the disc,* slightly fleshy, convexo-plane, obtuse, smooth, *moist,* sometimes striate. St. 6–12 cm. × 2–4 mm., *rufous,* lax, stiff, *velvety,* apex white pruinose, channelled, and twisted when dry. Gills *milk white,* separating free, broad, ventricose, distant. Flesh *white, rufous in the st.,* thin. Spores white, elliptical, 5–6 × 3μ, 1-guttulate. Amongst pine chips, twigs, and on wood. Feb.—Sept. Not common. (*v.v.*)

1039. **C. mimica** W. G. Sm. Cke. Illus. no. 192, t. 129. *Mimica*, mimic.

P. 2–4 cm., *pale yellow-buff, disc brownish buff,* slightly fleshy, plane, smooth, with a thin separable cuticle. St. 5 cm. × 3–4 mm., *deep brown, apex yellow buff and smooth, or slightly pruinose,* base fibrillose, fibrillosely striate in the middle. Gills *dingy ochraceous, very broad,* subdistant. Flesh *rufous,* very thin. Spores white, elliptical, 8 × 4–5μ. Smell and taste strong, *like fish.* Amongst deal shavings. Nov. Uncommon.

1040. **C. floccipes** Fr. Cke. Illus. no. 1142, t. 1168.
Floccus, a flock of wool; *pes*, foot.

P. 1–2 cm., *fuliginous black, becoming livid,* rather fleshy, campanulate, then convex, *umbonate,* smooth. St. 3–5 cm. × 2–3 mm., *pallid,* equal, straight, rooting, *rough with black, punctiform, floccose scales.* Gills *white,* adnexed, *ventricose,* thick, *subdistant.* Flesh *white, greyish under the cuticle of the p.,* thin. Spores "white, subglobose, prominently apiculate, 5–6 × 4–5μ. Cystidia abundant, narrowly lanceolate, 60–90 × 7–11μ, apex subobtuse" Kauffm. On the ground, and about trunks, in beech woods. Sept. Rare.

C. undata Berk. = **Marasmius undatus** (Berk.) Quél.
C. vertirugis Cke. = **Marasmius undatus** (Berk.) Quél.
C. stipitaria Fr. = **Crinipellis stipitarius** (Fr.) Pat.

1041. **C. leucomyosotis** Cke. & Smith. Cke. Illus. no. 1144, t. 651.
λευκός, white; *Myosotis*, the Forget-me-not.

P. 2·5–3 cm., *pale mouse-grey, disc darker, paler at the margin, the whole plant becoming pallid, almost white when dry,* fleshy, convex, then expanded, sometimes obtusely umbonate; margin faintly striate. St. 10–13 cm. × 4–6 mm., pallid, equal, *very brittle, apex slightly pruinose,* base white, obtuse. Gills *white,* adnate, sinuate behind, thick, subdistant. Flesh *dingy,* rather thick. Spores white, elliptical, 6–7 × 4μ, 1-guttulate. Cystidia flask-shaped, apex obtuse, 3–3·5μ in. diam., 25–30 × 7μ. Smell strong, rather fragrant. On *Sphagnum* in bogs. May—Sept. Not uncommon. (*v.v.*)

**Gills very narrow, crowded.

C. hariolorum (DC.) Fr. = **Marasmius hariolorum** (DC.) Quél.

C. confluens (Pers.) Fr. = **Marasmius hariolorum** (DC.) Quél.

C. ingrata (Schum.) Fr. = **Marasmius ingratus** (Schum.) Quél.

C. esculenta (Wulf.) Fr. = **Marasmius esculentus** (Wulf.) Karst.

C. conigena (Pers.) Bres. = **Marasmius conigenus** (Pers.) Karst.

1042. **C. cirrhata** (Schum.) Fr. Fr. Icon. t. 68, fig. 1. *Cirrata*, curled.

P. ·5–1 cm., *white, disc rufescent, or ochraceous*, slightly fleshy, conico-convex, then plane, *umbilicato-depressed*, and *often with a small central protuberance, slightly silky*, at length very delicately, and often concentrically rivulose. St. 2·5–5 cm. × 1 mm., pallid, *filiform*, flexuose, *white-pulverulent*, rooted with a *fibrillose twisted tail*. Gills *white*, adnate, at length occasionally separating, linear, *very narrow, very unequal, crowded*. Flesh *whitish*, very thin. Spores white, elliptical, 4–5 × 2–3μ. Amongst leaves and on bare ground. Aug.—Nov. Common. (*v.v.*)

1043. **C. tuberosa** (Bull.) Fr. Grev. Scot. Crypt. Fl. t. 23, as *Agaricus tuberosus*. *Tuberosa*, having a swelling.

P. 4–12 mm., *white, disc ochraceous*, slightly fleshy, convex, then plane, *umbonate*, slightly silky, becoming smooth, opaque. St. 1·5–3 cm. × 1 mm., *white, or rufescent*, equal, commonly ascending, *pruinose, arising from a purple brownish, or ochraceous, pear-shaped or roundish lobed sclerotium*. Flesh *whitish, or reddish becoming whitish*, very thin. Spores white, elliptical, 4–6 × 2·5–3μ, punctate. Cystidia "on edge of gill scattered, filamentous" Rick. On dead Agarics, chiefly *Russula adusta, Russula nigricans, Lactarius vellereus, Polyporus squamosus* and Hydnei. Aug.—Nov. Common. (*v.v.*)

1044. **C. racemosa** (Pers.) Fr. Sow. Eng. Fung. t. 287. *Racemosa*, clustered.

P. 5–8 mm., *grey*, submembranaceous, convex, then plane, *often imperfectly formed, papillate, tomentose*, striate. St. 3–5 cm. × 1 mm., *grey, springing from a swollen black sclerotium, racemose with simple, small, capitate hairs*, which are globose at the apex, hyaline, glutinous, and are really oblong, 12–15μ long, guttulate, greenish conidia. Gills *concolorous*, adnate, *very narrow*, crowded. Spores "oval, incurved, 5μ, minutely echinulate, greyish" Quél. On the ground, and rotten fungi. Sept.—Oct. Rare. (*v.v.*)

γ. St. thin, *glabrous*.

*Gills broad, rather distant.

1045. **C. collina** (Scop.) Fr. Cke. Illus. no. 198, t. 205. *Collina*, belonging to hills.

P. 2·5–5 cm., *pale fuscous, or pale tan*, fleshy-membranaceous, *campanulate, then expanded* and *often umbonate*, smooth, *subviscid*, striate when moist, shining when dry. St. 7–10 cm. × 4–6 mm., *pallid whitish, or cream colour*, subequal, or slightly attenuated upwards, *somewhat fragile*, smooth, apex mealy, base pubescent, *praemorse*. Gills *whitish*, adnexed, then *free, broad*, lax, *often veined at the base, subdistant*. Flesh *white*, thin. Spores white, "pruniform, 10μ, 1-guttulate" Quél. Smell like burnt meat. Edible. Beech stumps, and forming rings in pastures and on grassy slopes. May—Oct. Uncommon.

1046. **C. thelephora** Cke. & Massee. Cke. Illus. no. 1143, t. 1167.
θηλή, a nipple; φέρω, I bear.

P. 2–3 cm., *pale dingy ochraceous, disc darker*, slightly fleshy, *campanulate*, lax, *with a small, acute, papillate umbo*, then expanded and wavy, often depressed round the umbo, smooth, slightly striate; margin at first incurved. St. 6–10 cm. × 2–3 mm., *purplish brown at the base, paler upwards*, equal, smooth. Gills *whitish*, adnate, *narrower in front*, 3 mm. broad, thin, rather crowded. Flesh *pinkish*, thin. Spores white, elliptical, 9 × 7μ. Gregarious. Peat bogs, and partly dried up *Sphagnum* swamps. Sept. Uncommon.

1047. **C. ventricosa** (Bull.) Fr. Bulliard, t. 411, fig. 1.
Ventricosa, ventricose.

P. 1–4 cm., *tan, or isabelline*, slightly fleshy, campanulato-convex, *umbonate*, smooth; margin slightly striate. St. 6–10 cm. × 2–3 mm., *concolorous, or rufescent, base ventricose and attenuated into a long, slender, tapering root*, smooth. Gills *rufescent*, arcuato-adfixed, ventricose, subdistant, *undulate*. Flesh white, thin. Spores white. Solitary, or gregarious. Woods. Oct. Uncommon.

1048. **C. Stevensonii** B. & Br. Cke. Illus. no. 199, t. 145, fig. B.
Rev. John Stevenson, the eminent Scotch mycologist.

P. 1–1·5 cm., *pallid yellow*, slightly fleshy, *semi-ovate*, obtuse, *viscid*, here and there spotted by the viscous matter. St. 3–5 cm. × 2 mm., *slightly rufous, attenuated at the base into a somewhat long, thread-like root deeply immersed in the soil*, fibrillose, *pulverulent upwards*. Gills *white*, adnate with a decurrent tooth, *subventricose, very broad*, distant. Flesh *white, reddish in the st.*, thin. Spores white, elliptical, 10–11 × 7–8μ. Old pastures. Aug. Rare.

1049. **C. psathyroides** Cke. Cke. Illus. no. 200, t. 266.
Psathyra, the genus *Psathrya*; εἶδος, like.

Entirely ivory white. P. 2·5 cm. high, 18 mm. broad, slightly fleshy, *campanulate*, obtuse, *rather viscid*; margin regular, even. St. 7–10 cm. × 3–4 mm., equal, straight, rather tough. Gills adnate with a decurrent tooth, *very broad*, 6–8 mm., *triangular*, subdistant. Flesh whitish, thin. Spores white, elliptical, 15 × 7μ. On the ground in woods. Oct. Rare.

1050. **C. xanthopus** Fr. ξανθός, yellow; πούς, foot.

P. 2·5–5 cm., *tan, becoming pale,* slightly fleshy, *campanulato-convex,* then expanded, lax, *umbonate,* smooth, dry; margin at length spreading, slightly striate. St. 6–10 cm. × 4–6 mm., *tawny yellow,* equal, tough, smooth, *strigosely rooting at the base.* Gills *whitish, truncate behind,* adnexed, then free, *very broad, crowded,* lax. Flesh *white, yellowish under the cuticle of the p., rufous in the st.,* thin. Spores white, elliptical, 5 × 3μ, 1-guttulate. Cystidia "flask-shaped, 45–50 × 10–15μ" Rick. On stumps, and amongst leaves, chiefly in pine woods. July—Nov. Uncommon. (*v.v.*)

1051. **C. nitellina** Fr. Fr. Icon. t. 65, figs. 1, 2.
Nitellina, belonging to a dormouse.

P. 1·5–4 cm., *tawny, or brick tawny,* becoming tan colour when dry, submembranaceous, convexo-plane, obtuse, *often umbonate,* elastic, *flaccid,* hygrophanous, smooth, *polished, somewhat rugulose under a lens,* pellucido-striate when moist. St. 2·5–7·5 cm. × 3–5 mm., *ferruginous tawny, becoming yellow when dry,* equal, flexuose, cartilaginous, *shining,* polished, apex often pruinose, base white villous. Gills *whitish, or citron yellow,* then *flesh colour,* adnate, *very obtuse behind, broad,* attenuated in front, somewhat crowded, often undulate. Flesh *concolorous,* thin. Spores "*bright brownish red in the mass, yellowish under the microscope,* elliptical, with a basal apiculus, 7–8 × 4–5μ, or sometimes 10 × 5μ, 1-many-guttulate, warted" René Maire. Smell strong, rancid, or "like melon" Quél. Taste mild. Edible. Forming rings on the ground in coniferous woods. May—Oct. Uncommon.

1052. **C. succinea** Fr. Cke. Illus. no. 203, t. 151, upper figs.
Succinea, of amber.

P. 2·5–5 cm., *rufous, or brown fuscous, becoming pale,* fleshy, thin, *convex, then flattened,* obtuse, at length depressed and unequal, rimosely split when dry, smooth. St. 2·5–5 cm. × 2–5 mm., *rufescent, apex paler,* equal, attenuated at the base, tough, smooth, *shining,* sometimes arising from nodules of compact mycelium. Gills *cream colour, adnate,* obtuse behind, *very broad, rather thick,* not much crowded, *delicately toothed.* Flesh *reddish,* thin. Spores white, pip-shaped, 7–8 × 4μ, depressed on one side, multi-guttulate. Cystidia none. Solitary or gregarious in coniferous woods and under conifers. May—Aug. Not uncommon. (*v.v.*)

1053. **C. nummularia** (Lam.) Fr. Cke. Illus. no. 203, t. 151, lower figs.
Nummularia, like a coin.

P. 1–4 cm., *whitish, or very pale ochre, becoming white, tinged yellow at the umbilicate disc,* slightly fleshy, convex, then plane, *orbicular,* then *depressed round the obsolete umbo,* hygrophanous, smooth. St. 3–5 cm. × 3–4 mm., *whitish,* attenuated downwards to the bulbous, tomentose base, curved, tough, smooth. Gills *white*, free, subdistant, narrow, *minutely toothed.* Flesh *pallid*, thin. Spores white, elliptical, 7 × 4–5μ. In troops, in mixed woods. July—Nov. Not uncommon. (*v.v.*)

C. tenacella (Pers.) Fr. = **Marasmius conigenus** (Pers.) Karst.

C. tenacella (Pers.) Fr. var. *stolonifera* (Jungh.) = **Marasmius conigenus** (Pers.) Karst.

1054. **C. planipes** (Brig.) Fr. *Planus*, flat; *pes*, foot.

P. 2–3 cm., *bay*, slightly fleshy, convexo-plane, *orbicular*, *somewhat viscid*, smooth; margin *paler*, *crenate*. St. 3 cm. × 2–3 mm., *concolorous*, equal, *compressed*, *rooting*. Gills *whitish*, *free*, ventricose. Spores white. Caespitose. Woods. Sept. Rare.

**Gills narrow, crowded.

1055. **C. acervata** Fr. (= *Collybia erythropus* (Pers.) Quél.) Fr. Icon. t. 64, lower figs. *Acervata*, heaped up.

P. 3–7 cm., *reddish flesh colour*, *whitish when dry*, *slightly fleshy*, convex, then flattened, obtuse, or at length gibbous; margin at first involute, at length flattened and slightly striate. St. 5–10 cm. × 2–5 mm., *rufous*, *sometimes brown*, *rigid-fragile*, slightly attenuated upwards, rarely compressed, *very smooth*, base white-tomentose. Gills *flesh colour*, *then whitish*, adnexed, soon free, *linear*, *narrow*, plane, very crowded. Flesh *pallid*, *reddish in the st.*, thin. Spores white, elliptical, 6–8 × 3–4μ. *Caespitoso-fasciculate.* Pine stumps. Aug.—Oct. Not uncommon. (*v.v.*)

C. dryophila (Bull.) Fr. = **Marasmius dryophilus** (Bull.) Karst.

C. dryophila (Bull.) Fr. var. *funicularis* Fr. = **Marasmius dryophilus** (Bull.) Karst. var. **funicularis** (Fr.) Rea.

C. dryophila (Bull.) Fr. var. *aurata* Quél. = **Marasmius dryophilus** (Bull.) Karst. var. **auratus** (Quél.) Rea.

C. dryophila (Bull.) Fr. var. *oedipus* Quél. = **Marasmius dryophilus** (Bull.) Karst. var. **oedipus** (Quél.) Rea.

C. dryophila (Bull.) Fr. var. *alvearis* Cke. = **Marasmius dryophilus** (Bull.) Karst. var. **alvearis** (Cke.) Rea.

C. dryophila (Bull.) Fr. var. *aquosa* (Bull.) Quél. = **Marasmius dryophilus** (Bull.) Karst. var. **aquosus** (Bull.) Rea.

1056. **C. extuberans** (Batt.) Fr. Cke. Illus. no. 202, t. 146, as *Collybia nitellina* Fr. *Extuberans*, swelling out.

P. 2–5 cm., *rufous fuscous*, *bay brown*, *occasionally becoming pale*, slightly fleshy, convex, then flattened, orbicular, at length depressed round the *prominent umbo*, smooth, *slightly viscid when moist.* St. 4–5 cm. × 3–5 mm., *concolorous*, *or paler*, equal, tense and straight, smooth, *shining*, base attenuated and *rooting*. Gills *white, then cream colour*, somewhat free, reaching the st. with a small tooth, *crowded, narrow*, plane. Flesh *reddish, becoming white*, thin. Spores white, pip-shaped, 6 × 3μ ("yellowish" Quél). Smell pleasant, or none. Edible. On the ground, and on trunks, in pastures, and coniferous woods. May—Sept. Uncommon. (*v.v.*)

C. exsculpta Fr. = **Marasmius exsculptus** (Fr.) Rea.

1057. **C. luteifolia** Gillet. *Luteus*, yellow; *folia*, leaves.

P. 3–5 cm., *reddish, or cinnamon, becoming paler and white or whitish*, slightly fleshy, convex, soon plane, smooth, *glabrous*; margin often lobed, irregular. St. 3–4 cm. × 2–3 mm., *concolorous*, equal, smooth. Gills *sulphur yellow*, free, rounded at the base, pointed at the margin, *very crowded*. Flesh *white, reddish in the st.* Taste pleasant. Woods, and parks. Rare.

1058. **C. macilenta** Fr. Fr. Icon. t. 66, fig. 1. *Macilenta*, lean.

P. 1–2·5 cm., *dark yellow, bright yellow at the margin*, slightly fleshy, convex, then plane, *obtuse*, orbicular, smooth, dry. St. 4 cm. × 2–3 mm., *concolorous*, or *bright light yellow, becoming brownish at the fibrillose, rooting base*, tough, cartilaginous, flexuose, smooth. Gills *pure yellow*, separating-free, *narrow, linear, very crowded*, very unequal. Flesh *yellow*, thin. Spores white, elliptical, 5–6 × 4μ. "Cystidia hair-shaped, subnodulose or wavy" Lange. Amongst pine needles. Autumn. Rare.

C. clavus (Linn.) Fr. = **Mycena clavus** (Linn.) Rea.

1059. **C. ocellata** Fr. (= *Collybia cirrhata* (Schum.) Quél. var. *ocellata* (Fr.) René Maire.) Cke. Illus. no. 209, t. 147, middle figs. *Ocellata*, having little eyes.

P. 1–2 cm., *whitish, fuscous, rufous, or yellowish at the depressed, eye-like, umbonate disc*, slightly fleshy, conico-convex, then plane, smooth; *margin crenulate*. St. 3–5 cm. × 1–2 mm., *white, becoming yellowish, or fuscous*, equal, *filiform*, tough, smooth, *often pruinose, base fibrillose and rooting*. Gills *white*, adnate, at length separating, *crowded*, the alternate ones shorter. Flesh *white*, thin. Spores white, elliptical, 5 × 3μ. Amongst mosses, and *Jungermannia* in woods. Sept.—Oct. Uncommon. (*v.v.*)

1060. **C. muscigena** (Schum.) Fr. (= *Mycena muscigena* (Schum.) Quél.) Cke. Illus. no. 209, t. 147, lowest figs. *Muscus*, moss; *γίγνομαι*, to be born.

Entirely white. P. 3–6 mm., *submembranaceous, pellucid*, globoso-hemispherical, then flattened, obtuse, smooth, *withering up*. St. 3–4 cm. × 1 mm., *capillary, flexuose, flaccid*, smooth, *base attenuated*, *rooting*. Gills *adnate*, linear, *somewhat crowded*, the alternate ones shorter. Flesh *very thin*. Spores white, elliptical, 10 × 6–7 μ, minutely echinulate. Amongst moss, and short grass. Aug.—Oct. Uncommon. (*v.v.*)

C. ludia Fr. = **Mycena lactea** (Pers.) Fr. var. **pithya** (Pers.) Fr.

B. Gills becoming cinereous. Hygrophanous.

δ. P. fuscous, or becoming cinereous.

*Gills crowded, rather narrow.

1061. **C. rancida** Fr. Cke. Illus. no. 210, t. 153, upper figs.

Rancida, stinking.

P. 2–5 cm., *lead colour, or fuliginous, disc blackish, or fuscous, at first covered with a delicate, silky, white pruina*, then becoming paler, slightly *fleshy-cartilaginous, tough*, convex, then plane, *broadly and obtusely umbonate*, smooth, viscid when very wet. St. 7–15 cm. × 4–6 mm., *livid, becoming greyish, rigid*, equal, tense and straight, *smooth*, attenuated at the base into a *long, fusiform, villous root*. Gills *dark cinereous*, somewhat pruinose, *free, crowded, narrow*, but ventricose. Flesh *greyish*, thin at the margin. Spores white, oblong elliptical, 9–10 × 4–5 μ, 1–2-guttulate, minutely punctate. Smell like rancid meal. Woods, and hedgerows. Aug.—Dec. Common. (*v.v.*)

1062. **C. eustygia** Cke. Cke. Illus. no. 1146, t. 1185.

εὖ, truly; *στύγια*, belonging to the nether world.

P. 3–5 cm., *dingy white, disc a little darker, shining when dry*, fleshy, convex, then plane, sometimes depressed, sometimes wavy, smooth. St. 5–8 cm., *white above, sprinkled with small punctate scales, darker below and often becoming sooty*, attenuated downwards into a rooting base, often curved, *somewhat longitudinally striate, or fibrous*. Gills *dark grey*, rounded behind, adnexed, or almost free, *rather broad*, not crowded. Flesh *white*, thick at the disc. Spores white, globose, 4–5 μ. Smell like rancid meal. The whole plant becoming black in drying. On the ground. Oct. Uncommon.

1063. **C. coracina** Fr. *κόραξ*, a raven.

P. 2–4 cm., *fuscous and shining, becoming grey and opaque when dry*, somewhat fleshy-cartilaginous, convexo-expanded, umbonate, or depressed, often irregular and undulate, *smooth, hygrophanous*; margin sometimes wrinkled. St. 2–4 cm. × 4–8 mm., *becoming fuscous, very cartilaginous*, tough, rigid, at length fragile, *often compressed, or twisted, apex mealy with white squamules*, attenuated downwards to the *swollen base*. Gills *whitish grey, obtusely adnate*, separating-free, *broad chiefly behind*, scarcely crowded, distinct, then connected by veins. Flesh *white*, scissile, thin. Spores white, elliptical, 6–7 × 3–4 μ, punctate, 1-guttulate ("greenish" Quél.). Smell strong of new meal. Grassy places, and fir plantations. Oct.—Nov. Uncommon. (*v.v.*)

1064. **C. ozes** Fr. *ὄζω*, I have a smell.

P. 2–3 cm., *grey-fuscous, becoming clay-fuscous, pallid when dry,* slightly fleshy, convex, then plane, *umbonate, hygrophanous, smooth;* margin *striate when moist.* St. 6–11 cm. × 2 mm., *fuliginous grey,* equal, or scarcely attenuated from the base, lax, *flexuose, fragile, slightly striate, containing a pith when young, apex white mealy.* Gills *fuliginous-olivaceous, adnate,* subventricose, 4–6 mm. broad, *crowded* often veined. Spores white, "elliptical, 6–7 × 3–4μ, smooth. Cystidia none" Rick. Smell strong of new meal. On the ground and on pine needles. Feb. Rare.

1065. **C. mephitica** Fr. Trans. Brit. Myc. Soc. II, t. 14.
Mephitis, a noxious exhalation from the ground.

P. 1·5–4 cm., *greyish ochre, becoming whitish,* fleshy, convex, subumbonate, hygrophanous, smooth, dry, *adpressedly and innately silky.* St. 5–7 cm. × 2–5 mm., *grey, filiform,* equal, *rigid, pruinosely velvety with white flocci,* base enlarged, covered with the white mycelium. Gills *grey,* obtusely adnate, separating from the st., attenuated in front, 4–7 mm. broad, crowded. Flesh *yellowish,* thin. Spores white, elliptical, 7–8 × 4μ. Smell strong of new meal. Amongst pine needles in woods. Sept.—Oct. Uncommon. (*v.v.*)

1066. **C. inolens** Fr. *Inolescens,* growing in.

P. 2–5 cm., *livid, becoming pale tan and slightly silky when dry, but opaque, slightly fleshy,* campanulato-convex, then plane, *obtusely and broadly umbonate,* hygrophanous, very smooth; margin inflexed, then expanded, striate, undulate. St. 5–8 cm. × 2–8 mm., *concolorous, becoming pale when dry, rigid,* equal, often compressed, *undulated, apex white-squamulose, base white-strigose.* Gills *grey, adfixed, separating,* somewhat free, linear, or slightly ventricose, 2–4 mm. broad. Flesh *greyish,* thin. Spores white, elliptical, 7 × 4–5μ, 1-guttulate. Smell of new meal. Pine woods, and under conifers. Sept.—Nov. Uncommon. (*v.v.*)

1067. **C. plexipes** Fr. (= *Collybia retigera* Bres. sec. Quél.) Cke. Illus. no. 211, t. 154, lower figs. *Plexus,* twisted; *pes,* foot.

P. 3–5 cm., *blackish, whitish at the margin, becoming fuliginous livid,* fleshy-membranaceous, *campanulate, umbonate, somewhat wrinkled,* slightly striate. St. 7–10 cm. × 2–4 mm., *livid, cartilaginous,* equal, *covered with a network of silky-fibrils,* slightly striate, *base shortly, and bluntly rooted.* Gills *white, then glaucous, free,* very much attenuated behind, ventricose, *somewhat crowded.* Spores white, elliptical, 8–9 × 5μ. Woods, especially beech. Sept.—Nov. Uncommon.

1068. **C. atrata** Fr. Cke. Illus. no. 212, t. 155, upper figs.
Atrata, clothed in black.

P. 2–4 cm., *pitch-black and shining, becoming fuscous when dry, slightly fleshy, firm, plano-depressed at the disc*, convex at the margin, orbicular, smooth, viscid when very wet. St. 2·5–5 cm. × 3–7 mm., *fuscous*, very cartilaginous, *tough*, equal, or thickened upwards, round, smooth. Gills *whitish, then grey, becoming fuscous*, adnate, scarcely decurrent, arcuate, then plane, rather broad, *subdistant*. Flesh *fuscous, especially in the st.*, thin. Spores white, globose, 5μ, with a large central gutta. Smell none, or strong, and unpleasant. Charcoal heaps, and burnt soil. July—Dec. Not uncommon. (*v.v.*)

1069. **C. ambusta** Fr. Cke. Illus. no. 212, t. 155, lower figs.
Ambusta, scorched.

P. 1–2·5 cm., *fuscous, becoming greyish, submembranaceous*, convex, then plane, at length depressed, *umbonate with a minute papilla*, smooth, becoming slightly striate. St. 2–3 cm., *concolorous*, cartilaginous, *tense, straight, pruinose when young*. Gills *pallid, becoming fuscous, adnate*, with a decurrent tooth, lanceolate, plane, *crowded*. Flesh *becoming whitish*, thin. Spores white, globose, 5μ, very minutely warted, "4–5-angled" Rick. Burnt soil, and charcoal heaps. July—Nov. Common. (*v.v.*)

**Gills broad, rather distant.

1070. **C. lacerata** (Lasch) Berk. Bres. Fung. Trid. t. 19.
Lacerata, torn to pieces.

P. 2–5 cm., *fuscous fuliginous, becoming pale, fleshy-membranaceous*, campanulate, then convex and *umbilicate*, somewhat moist, *streaked with fuscous lines*; margin fimbriately torn, splitting with age. St. 4–7 cm. × 4–6 mm., pallid, equal, at length compressed, fibrillosely striate, firm, often twisted, apex floccoso-pruinose, base white-tomentose, somewhat rooting. Gills *white-grey*, rounded behind, adnate, broad, thick, somewhat crowded, or distant. Flesh *greyish white*, thin, firm. Spores white, subglobose, 6–7μ, 1-guttulate, "sub-granular" Rick. Caespitose. Fir woods, often on stumps. Autumn. Rare.

1071. **C. murina** (Batsch) Fr. *Murina*, of mice.

P. 3–4 cm., *fuscous brown, becoming pale, slightly fleshy*, campanulato-convex, then expanded, obtuse, or umbilicate, *slightly wrinkled, or very thinly squamulose*, tough; margin at first involute. St. 5–8 cm. × 3–4 mm., *white, becoming cinereous*, equal, *tense, straight, delicately fibrillose*, apex flocculose when young, base pubescent. Gills *white, becoming cinereous, attenuato-adnexed*, very broad, almost obovate, rather thick, *distant*. Flesh greyish white, thin, tough. Spores white, "subfusiform-elliptical, 8–9 × 3–4μ" Rick. Woods, and under oaks. Oct. Uncommon.

1072. **C. protracta** Fr. Fr. Icon. t. 67, fig. 2. *Protracta*, drawn out.

P. 2 cm., *grey-fuscous, shining*, submembranaceous, convexo-plane, *disc depressed, often with a little central umbo*; *margin paler*, striate. St. 15–16 cm. × 2–3 mm., *livid grey*, very cartilaginous, *tense, straight*, smooth, *attenuated at the base into a tapering, subterranean, strigosely fibrous root*. Gills *grey, delicately white-pruinose*, adfixed, very ventricose,—as if truncate behind,—*very broad*, 6 mm., *subdistant*. Spores white, "elliptical, 7–9 × 5–6μ" Bres. Mossy ground near stumps in fir woods. Aug.—Nov. Rare.

1073. **C. tesquorum** Fr. Fr. Icon. t. 70, fig. 3. *Tesqua*, deserts.

P. 6–10 mm., *fuscous black, becoming pale*, fleshy-membranaceous, slightly firm, convex, very obtuse, smooth. St. 2–4 cm. × 2 mm., *fuscous*, somewhat filiform, equal, flexuose, smooth, *apex mealy*. Gills *cinereous fuscous, free*, very ventricose, 4 mm. broad, *subdistant*. Flesh *concolorous*, thin. Spores white, broadly elliptical, 7–8 × 5–6μ, echinulate. Waste ground, and open pastures. Uncommon. (*v.v.*)

1074. **C. clusilis** Fr. Cke. Illus. no. 215, t. 247, lower figs.
Clusilis, easily closing.

P. 1–3·5 cm., *livid, becoming pale, grey clay colour when dry, submembranaceous*, rather plane, *disc depressed, or broadly umbilicate*, very much sloped downwards towards the margin, smooth, soft, fragile; margin at first *incurved*, slightly striate when moist. St. 4 cm. × 2 mm., *livid*, cartilaginous, *soft, flexile*, equal, smooth, polished, *stuffed with a white floccose pith*. Gills *white, becoming cream colour, adnate*, plane, with a decurrent tooth, 4–8 mm. *broad, in the form of a segment*, somewhat crowded. Flesh *white*, thin at the margin. Spores white, elliptical, 8–9 × 4–5μ, with a large central gutta. Amongst moss and grass on heaths, and hillsides. Sept.—Oct. Uncommon. (*v.v.*)

1075. **C. tylicolor** Fr. Cke. Illus. no. 215, t. 247, upper figs.
Tylus, a crustacean allied to the woodlouse; *color*, colour.

P. 1–3 cm., *grey cinereous, slightly fleshy*, convex, then flattened, *subumbonate*, smooth, opaque. St. 2–5 cm. × 2 mm., *grey*, somewhat fragile, equal, *whitish-pulverulent*. Gills *grey, paler than the p., free, broad*, plane, *distant*, rather thick. Spores white, "oval, 5·5–6 × 3–3·5μ, minutely echinulate" Sacc. Deciduous woods amongst grass. Oct. Rare.

Introduced species.

1076. **C. Dorotheae** Berk. Lady Dorothy Neville.

P. 2–3 cm., *dark brown, becoming paler*, globose, then flatly hemispherical, at length expanded, slightly umbonate, finally depressed, *radiately sulcate almost to the disc, granulated, covered with short, white bristles pointing in every direction when young*; margin crenate. St. 5–6 cm. × 2 mm., *brownish, white below, becoming white above and yellowish or rufous below, granulated, covered with white bristles*, base with a minute disc-like swelling. Gills *white*, adnexed, slightly ventricose, *connected behind*, distant. Dead fern stems in a hot-house. Rare.

1077. **C. caldarii** Berk. *Caldarium*, a hot bath.

P. 12–15 mm., *brown*, hemispherical, *umbonate, rugose*. St. 4–5 cm. × 2 mm., *paler*, cartilaginous, smooth. Gills *somewhat ash-coloured, adnato-decurrent, interstices veined near the margin*. On *Sphagnum* in an orchid pot. Rare.

Spores pink.

Leptonia Fr.

(λεπτός, thin.)

Pileus slightly fleshy, regular; margin incurved. Stem central, cartilaginous. Gills adnate, sinuato-adnate, or adnexed. Spores pink, angular, elliptical, subglobose, or oblong; continuous. Cystidia rarely present. Growing on the ground, or on wood.

*Gills whitish. P. slightly fleshy.

1078. **L. placida** Fr. Fr. Icon. t. 97, fig. 1. *Placida*, gentle.

P. 2–3 cm., *grey, becoming bluish, disc densely villose, blackish*, fleshy membranaceous, campanulate, then convex, obtuse, *squamulose with dark concentric scales and dark fuliginous black fibrils*. St. 5–7·5 cm. × 2–6 mm., *dark azure-blue*, or *black-blue*, equal, very rigid; apex thickened, *white pruinose and black-dotted*. Gills *whitish*, then *purplish*, adnexed, very broad behind, plane crowded. Flesh *brownish in the pileus, bluish in the stem*. Spores pink, "angular, 7–12 × 6–7μ" Herpell. On and near beech, and fir stumps. Sept.—Oct. Rare.

1079. **L. anatina** (Lasch) Fr. *Anatina*, belonging to a duck.

P. 3–4 cm., *greyish fuscous*, somewhat fleshy, conico-campanulate, broadly umbonate, *longitudinally fibrillose and squamulose*, often rimose. St. 3–4 cm. × 4–6 mm., *blue*, equal, or attenuated downwards, at first *pruinose, then squamoso-fibrillose*, apex smooth, base *white-villose*. Gills *whitish*, then flesh colour, adnexed, then separating, broad, ventricose. Flesh *bluish, becoming whitish*. Spores pink, angular, broadly elliptical, 10–11 × 9–10μ, 1–2-guttulate. Heaths, and pastures. July—Oct. Uncommon. (*v v.*)

1080. **L. lappula** Fr. Fr. Icon. t. 97, fig. 2. *Lappa*, a bur.

P. 2·5–4 cm., *grey*, somewhat fleshy, hemispherical, convexo-plane, *umbilicate*, flocculoso-soft, then *roughish with short erect fibrils, which become black and crowded on the disc.* St. 4–5 cm. × 2–4 mm., *fuscous lilac*, or *dark purple*, equal, moderately tough, *striate and black-dotted upwards under a lens*, base white-villose. Gills *whitish-grey, then purplish*, adnate with a small tooth, then separating, plane, *very broad*, ovate, crowded. Flesh *white*. Spores pink, "angular, elliptical, 12μ" Quél. Amongst beech leaves. July—Oct. Uncommon.

1081. **L. Reaae** Maire. Trans. Brit. Myc. Soc. III, t. 11.
Mrs E. A. Rea, the artist who has made many original paintings of fungi.

P. ·5–1 cm., *dark blackish blue, convex, then expanded*, submembranaceous disc fleshy, *not, or only slightly hygrophanous*, rarely umbonate or papillate at maturity; margin slightly incurved, then expanded and sometimes substriate. St. 2–3 cm. × 1–5 mm., *deep blue, or blue black, then often vinous*, equal, flexuose, *wavy*, shining, obsoletely whitish mealy at the apex. Gills *whitish then greyish-pink, broadly and deeply sinuate, narrowly adnate*, then free, somewhat crowded, *short, broad*. Flesh *vinous*. Spores pink, obsoletely polygonal, subglobose, 8–10 × 7–8μ, including the apiculus, containing many oil drops. Pastures. Sept. Uncommon. (*v.v.*)

1082. **L. lampropus** Fr. Cke. Illus. no. 353, t. 331.
λαμπρός, bright; πούς, foot.

P. 1–3 cm., *mouse colour, or steel-blue, then fuliginous-grey*, somewhat fleshy, convex, then expanded and depressed, *becoming more or less squamulose*. St. 2·5–4 cm. × 2–4 mm., *becoming azure-blue, commonly steel-blue-violaceous*, cartilaginous. Gills *whitish*, then slightly rose colour, adnate, readily separating, then free, ventricose. Flesh *bluish*. Spores pink, angular, broadly elliptical, 9 × 7μ, 1-guttulate, with somewhat rounded angles. Heaths, and pastures. July—Nov. Common. (*v.v.*)

var. **cyanulus** (Lasch) Fr. κύανος, dark blue.

Differs from the type in the *more slender, membranaceous, blackish-blue, subumbilicate, floccosely-villose p., the capillary, pruinose st., and the glaucous, then flesh colour, adnate*, distant gills. On the ground near alders.

1083. **L. aethiops** Fr. Fr. Icon. t. 97, fig. 3. αἰθίοψ, an Ethiop.

P. 1–3 cm., *black, then fuliginous*, slightly fleshy, *plano-depressed, streaked with fibrils*, shining when dry. St. 4–5 cm. × 2 mm., *fuscous blackish, black dotted upwards*. Gills *whitish, then purplish*, adnexed, or adnate, linear, or ventricose. Flesh *whitish*. Spores pink, angular, 9–10 × 6μ, 1-guttulate. Woods, and heaths. Sept.—Oct. Not uncommon. (*v.v.*)

1084. **L. solstitialis** Fr. *Solstitialis,* belonging to midsummer.

P. 1–3 cm., *becoming fuscous,* slightly fleshy, at length depressed, *papillate in the centre, slightly wrinkled, obsoletely innato-fibrillose.* St. 2–3 cm. × 2–4 mm., *smoke colour.* Gills *whitish,* then flesh colour, *emarginate,* broad. Spores pink, angular, oblong, 10 × 7μ, 1-guttulate. Pastures, and amongst stones. Aug.—Sept. Uncommon. (*v.v.*)

**Gills at the first azure-blue, or slightly dark-blue.

1085. **L. serrulata** (Pers.) Fr. *Serrula,* a small saw.

P. 1–3 cm., *blackish-blue* (shining when dry), *fuliginous when old* or in wet weather, and then slightly striate, slightly fleshy, convex, *umbilicato-depressed,* squamulose, or fibrillose. St. 2–3 cm. × 2–4 mm., *paler than the p.,* cartilaginous, equal, *apex black dotted,* base white-woolly. Gills *bluish-grey-whitish,* then grey flesh colour, adnate, in the form of a segment, broad in the middle; *edge black, serrulate.* Flesh *whitish.* Spores pink, angular, 8–11 × 7μ, 1-guttulate. Cystidia "pale grey, fasciculate, clavate, 11–12μ broad" Lange. Woods, and pastures. June—Oct. Common. (*v.v.*)

var. **Berkeleyi** Maire. Cke. Illus. no. 355, t. 333, as *Leptonia serrulata* Fr. Rev. Miles Joseph Berkeley, the father of British mycology.

P. 2·5–4 cm., *whitish with a lilac tinge,* umbilicate, slightly sprinkled with fibrils; margin vaulted. St. 7–10 cm. × 3 mm., *whitish with a lilac tinge,* flexuose, *smooth.* Gills *salmon colour,* broad, adnate, *little or not serrulate.* Flesh *whitish.* Spores pink, angular, oblong, 11–12 × 6–7μ, 1–2-guttulate. Pastures. July—Sept. Not uncommon (*v.v.*)

var. **laevipes** Maire. *Laevis,* smooth; *pes,* foot.

Differs from the type in the *smooth* (*not black dotted*) *apex of the stem.* Woods, and pastures. July—Sept. Not uncommon. (*v.v.*)

1086. **L. euchroa** (Pers.) Fr. Boud. Icon. t. 98.

εὔχρως, well coloured.

P. 1–4·5 cm., *violaceous, then purple-fuliginous,* slightly fleshy, convex, obtuse, *squamuloso-fibrillose.* St. 2–6 cm. × 2–4 mm., *concolorous,* equal, fibrillosely mealy especially at the apex, tough; base white, hairy. Gills *dark violaceous, becoming pale, the edge retaining the darker colour,* adnate, ventricose. Flesh *bluish.* Spores pink, angular, 10–15 × 7–9μ. On stumps, and branches of alder, hazel, and birch. Aug.—Oct. Not uncommon. (*v.v.*)

1087. **L. chalybaea** (Pers.) Fr. χάλυψ, steel.

P. 2–3 cm., *dark violaceous, or blackish blue,* slightly fleshy, convex, *subumbonate, flocculose,* then squamulose. St. 4–5 cm. × 2 mm., *dark blue,* cartilaginous, slightly firm. Gills *bluish-grey-whitish, edge paler,* adnate, crowded, broad, ventricose. Spores pink, "longish, 5–6-angled, 9–10 × 7–8μ, with prominent angles" Rick. Pastures. July—Oct. Not uncommon. (*v.v.*)

1088. **L. lazulina** Fr. *Lapis lazuli*, ultramarine.

P. 1·5–2 cm., *becoming black fuliginous, at first black blue, or date-brown-mouse colour, with the disc darker, submembranaceous*, campanulate, then expanded and obtuse, *striate*, obsoletely umbilicate, rimoso-squamulose. St. 4–5 cm. × 2–3 mm., *dark blue*, cartilaginous, base white-woolly. Gills *pallid deep blue*, adnate, separating, equally attenuated from the stem to the margin of the pileus. Flesh *dark blue*. Spores pink, angular, oblong, 11–12 × 7–8μ, 1-guttulate. Heaths, and pastures. Sept. Uncommon. (*v.v.*)

***Gills pallid. Becoming pale, yellow or green.

1089. **L. incana** Fr. (= *Leptonia chloropolia* (Fr.) Quél.) Cke. Illus. no. 359, t. 336 *Incana*, hoary.

P. 2–3 cm., *variegated fuscous and green, becoming cinereous when dry*, submembranaceous, fragile, convex, then expanded, *umbilicate, striate*, slightly silky when dry. St. 2·5–5 cm. × 2–4 mm., *green, or fuscous green*, cartilaginous, base white-floccose. Gills *whitish green*, then flesh colour, adnate, decurrent with a tooth, at length separating, 4–6 mm. broad at the middle, *distant*. Flesh *green*, thin. Spores pink, angular, 8–12 × 7-8μ, 1-guttulate. Smell like that *of mice*. Woods, heaths, and pastures. July—Oct. Common. (*v.v.*)

1090. **L. euchlora** (Lasch) Fr. Boud. Icon. t. 99. *εὖ*, well; *χλωρά*, pale green.

P. 1·5–3·5 cm., *olivaceous, becoming paler*, submembranaceous, campanulato-convex, then plane, *fuscous fibrillose, subsquamulose*, especially at the darker, finally depressed disc. St. 3–6 cm. × 3–5 mm., *greenish, apex yellowish, becoming deep blue or verdigris when bruised or handled*, equal, slightly thickened at the white, tomentose base, hollow, fragile, smooth. Gills *whitish, or very pale yellowish, then pink*, 5–6 mm. wide, broadly adnate, subdistant. Flesh *greenish, becoming deep blue or verdigris when bruised or pressed*, thin. Taste and smell none. Spores pink, oblong, angular, 11–15 × 8–10μ, multi-guttulate. Amongst short grass in woods and open downs. Sept.—Oct. Uncommon. (*v.v.*)

1091. **L. sericella** (Fr.) Quél. (= *Entoloma sericellum* Fr.) Cke. Illus. no. 335, t. 307, as *Entoloma sericellum* Fr. *Sericus*, silken.

P. 1·5–3 cm., *white, or becoming yellow white*, somewhat fleshy, *convex, then plane*, obtuse, at length depressed, often unequal, *silky*, often squamulose; margin inflexed, floccose. St. 2·5–5 cm. × 2–3 mm., *white, then becoming pale, waxy*, equal, fibrillose, at length somewhat polished, pellucid. Gills *white*, then flesh colour, *at first adnate*, even, decurrent with a tooth, then separating and somewhat emarginate, very broad, subdistant. Flesh *white*, thin. Spores pink, angular, oblong, 9–11 × 6–7μ, 1-guttulate. Woods, and pastures. July—Oct. Common. (*v.v.*)

var. **decurrens** (Boud.) Rea. Boud. Icon. t. 94.
Decurrens, running down.

Differs from the type in the *distinctly decurrent gills*. Woods, pastures, and roadsides. July—Sept. Not uncommon. (*v.v.*)

var. **lutescens** Fr. *Lutescens*, becoming yellow.

Differs from the type *in the yellowish, more regular, convex, even p., and almost adnate gills.*

var. **sublutescens** Henn. *Sub*, somewhat; *lutescens*, becoming yellow.

Differs from the type in the *white, silky-floccose p. becoming smooth and dingy yellow, and the white st. becoming yellowish.*

1092. **L. formosa** Fr. Fr. Icon. t. 98, fig. 1. *Formosa*, beautiful.

P. 2–3 cm., *yellow wax colour, sprinkled over with minute fuscous squamules or fibrils*, submembranaceous, slightly tough, convex, then plane, *slightly umbilicate*, striate. St. 4–5 cm. × 1–2 mm., *yellow*, cartilaginous, equal, shining. Gills *light-yellow-pallid*, then flesh colour, adnate, decurrent with a tooth, subdistant. Spores pink, angular, oblong, 10 × 8μ, 1-guttulate. Coniferous woods, and heaths. Sept.—Oct. Uncommon. (*v.v.*)

var. **suavis** (Lasch) Fr. Cke. Illus. no. 360, t. 488. *Suavis*, pleasant.

Differs from the type in the *stem becoming blue*. Amongst *Equisetum*. Sept. Rare.

1093. **L. chloropolia** Fr. (= *Leptonia incana* Fr. sec. Quél.) Fr. Icon. t. 98, fig. 2. χλωρός, pale green; πολιός, grey.

P. 2–3 cm., *livid, disc black squamulose*, membranaceous, convex, then flattened, striate; margin at first inflexed. St. 5–7·5 cm. × 2–4 mm., *bluish-grey-green*, slightly firm, rigid. Gills *whitish*, then flesh colour, adnate. Spores pink, angular, globose, 9–10 × 8μ, 1-guttulate. Heaths, and pastures. Sept.—Oct. Not uncommon. (*v.v.*)

****Gills grey, or glaucous. Hygrophanous, p. somewhat striate.

1094. **L. asprella** Fr. Quél. Jur. et Vosg. t. 6, fig. 4.
Asprella, somewhat rough.

P. 2–4 cm., *fuliginous, or mouse colour, then livid-grey*, submembranaceous, convex, then flattened, *darker umbilicus villose, at length squamulose*, marked with spots, *striate*, often fibrillose. St. 2·5–5 cm. × 1–2 mm., *fuscous, green, or azure-blue*, cartilaginous, equal, tense and straight, base white-villose. Gills *whitish grey*, then flesh colour, adnate, separating free, subdistant, *equally attenuated from the stem towards the margin, edge often black.* Spores pink, angular, oblong, 10 × 6–7 μ. Pastures, and heaths. Aug.—Oct. Uncommon. (*v.v.*)

1095. **L. nefrens** Fr. *Nefrens*, having no teeth.

P. 2·5–5 cm., *fuliginous, then livid-grey*, membranaceous, campanulate, then flattened, *with a deep darker umbilicus, at length infundibuliform*, striate, *obsoletely fibrillose.* St. 2–3 cm. × 2–4 mm., *fuscous-livid*, fragile, equal. Gills *pallid grey*, adnexed, separating, broad, *edge slightly black.* Spores pink, "elliptical, 4–5 × 3 μ" Massee. Grassy places, and marshy pastures. July—Sept. Uncommon.

Spores ochraceous, or ferruginous.

Naucoria Fr.

(*Naucum*, a flock of wool.)

Pileus fleshy, regular; margin at first incurved. Stem central, cartilaginous. Gills adnate, sinuato-adnate, or adnexed. Spores ochraceous, ferruginous or fuscous; elliptical, pip-shaped, almond-shaped, or oblong elliptical, smooth, punctate, or verrucose; continuous, or with a germ-pore. Cystidia present. Growing on the ground, more rarely on wood; solitary, gregarious, or caespitose.

I. P. smooth. Veil none. Spores ferruginous, not becoming fuscous ferruginous.

*Gills free, or slightly adnexed.

1096. **N. lugubris** Fr. Fr. Icon. t. 121, fig. 1. *Lugubris*, mournful.

P. 5–8 cm., *pallid, then ferruginous, at length almost date brown*, fleshy, *campanulate*, then expanded, gibbous, smooth, rarely bullate, undulated, and tenaciously viscid. St. 5–10 cm. × 6–10 mm., *pallid, becoming ferruginous downwards when old, with a long, attenuated, fusiform root*, externally very cartilaginous, rigid, smooth. Gills *pallid, then ferruginous, quite free*, ventricose, very broad behind, 12 mm. and more broad, crowded, edge for the most part serrated. Flesh *white.* Spores pallid, then ferruginous, "nearly almond-shaped, 7–8 × 4–5 μ. Cystidia only on edge of gill, clavate filamentous" Rick. Often caespitose. Mountainous fir woods. Sept.—Oct. Rare.

1097. **N. festiva** Fr. Bres. Fung. Trid. t. 22. *Festiva*, handsome.

P. 2–5 cm., *olivaceous fuscous, becoming olivaceous straw colour, isabelline, bay, or rufous when dry*, fleshy, *convex*, slightly gibbous, smooth, *glutinous when fresh.* St. 5–9 cm. × 4–8 mm., rufous, violaceous, *olivaceous, or pallid, with reticulately adpressed black fibrils*, equal, or attenuated downwards, somewhat rooting, sometimes ventricose and compressed, very cartilaginous, either fuscous squamulose, or smooth. Gills *whitish, then olivaceous, rufous, or ferruginous blood red, free,* attenuated behind, generally ventricose, 6 mm. broad, crowded, edge often white, pubescent. Flesh *pallid, becoming reddish in the st.*, thin at the margin. Spores ferruginous, almond-shaped, 8 × 4μ, 1-guttulate, "roughish. Cystidia only on edge of gill, filamentous" Rick. Smell weak, of radish. Coniferous woods, and amongst grass. Sept. Rare.

1098. **N. obtusa** Cke. & Massee. Cke. Illus. no. 1171, t. 1155.

Obtusa, blunt.

P. 2–3 cm., *rufous, or orange tawny, becoming pale*, campanulate, obtuse, margin faintly striate. St. 5 cm. × 4–6 mm., *flesh colour, or pale cinnamon*, equal, smooth. Gills *cinnamon*, broadly adnate, or with a tooth, 4–6 mm. broad, ventricose, edge serrate. Flesh *concolorous, becoming pale, darker at the base of the st.* Spores ferruginous, elliptical, 7–8 × 4μ. On the ground. Sept. Rare.

1099. **N. subglobosa** (A. & S.) Fr.

Sub, somewhat; *globosa*, spherical.

P. 2–3 cm., *light yellow, darker at the disc*, fleshy, hemispherical, *smooth, moist*. St. 2·5 cm. × 2–4 mm., *concolorous, brownish at the base*, equal, rigid, *striate at the apex*. Gills *concolorous*, somewhat free, *very broad*, convex, rhomboidal. Flesh *lemon yellow*, thin. Spores ochraceous, "unequal-elliptical, 6–7 × 3–4μ, smooth. Cystidia on edge of gill large, lanceolate-fusiform, 50–60 × 12–15μ, without a globose head" Rick. Pine woods, and on the ground. Sept.—Oct. Uncommon.

1100. **N. hamadryas** Fr. Fr. Icon. t. 121, fig. 3.

ἁμαδρυάς, a wood nymph.

P. 4–5 cm., *bay brown ferruginous, pale yellowish when old, and becoming pale*, fleshy, convex, then expanded, gibbous, smooth. St. 5–7·5 cm. × 6 mm., *pallid, somewhat fragile*, equal, smooth. Gills *ferruginous*, opaque, *attenuato-adnexed*, somewhat free, slightly ventricose, 3–4 mm. broad, crowded. Flesh *paler*, thin, that of the p. easily separating from the st. Spores "ferruginous, elliptical, 13–14 × 7μ" Massee. Woods, and under trees. Sept.—Nov. Uncommon.

1101. **N. cidaris** Fr. Fr. Icon. t. 123, fig. 2. κίδαρις, a tiara.

P. 2–5 cm., *clay cinnamon, or dark tawny cinnamon, tan colour when dry*, fleshy, *conical, then campanulate*, slightly striate when moist at the undulate, membranaceous margin, pruinose, or smooth. St. 4–5 cm. × 4 mm., *fuscous black, apex concolorous* and pruinose, *attenuated from the apex to the base*, or fusiform, subcompressed, flexuose, or straight, tough, smooth. Gills *honey colour, or cinnamon clay, adfixed*, separating free, ascending, ventricose, 4–6 mm. broad, crowded. Flesh *whitish*, scissile, thin at the disc. Spores ferruginous, "elliptical,

5–6 × 3–4μ, smooth. Cystidia only on edge of gill, filamentous" Rick. Pine woods, and on the ground round trunks. Oct.—Nov. Uncommon.

1102. **N. Cucumis** (Pers.) Fr. (= *Nolanea nigripes* (Trog) Fr.; *Nolanea pisciodora* (Ces.) Fr.; *Nolanea picea* Kalchbr. sec. Quél.) Cke. Illus. no. 364, t. 378, upper figs., as *Nolanea pisciodora* Ces. *Cucumis*, cucumber.

P. 1–4 cm., *tawny cinnamon, pitch black, bay-brown-fuscous, becoming paler towards the margin, umber, fawn, or tan colour when dry*, fleshy, campanulate, then convex, umbonate, or obtuse, *pruinose*; margin incurved, often striate when moist. St. 3–6 cm. × 3–6 mm., *date brown, chestnut brown, or fuscous blackish*, tough, equal, *pruinose, velvety*, apex often paler, white floccose at the base. Gills *pale, yellowish flesh colour, then saffron yellow, or tawny*, emarginate, *ventricose*, crowded. Flesh *concolorous*, thin at the margin. Spores pale, ferruginous, oblong, elliptical, 8–10 × 3–4μ, 1–3-guttulate. Cystidia "broadly lanceolate, 60–75 × 18–23μ" Rick. Smell unpleasant, *of fish, or cucumber*. Coniferous, and damp woods, amongst dead leaves, and bare soil in gardens. Aug.—Nov. Common. (*v.v.*)

1103. **N. echinospora** W. G. Sm. ἐχῖνος, hedgehog; σπορά, seed.

P. 12 mm., *buff, then pale*, flat, subumbonate, moist, hygrophanous, *slightly furfuraceous*; margin substriate. St. 2 cm. × 4–5 mm., *brownish salmon, paler above, rufescent below, white flocculose*. Gills *ochre, olive-shaded*, sinuate, subdistant. Spores 7 × 6μ, rough. Greenhouses. Aug. Rare.

1104. **N. anguinea** Fr. Fr. Icon. t. 122, fig. 1. *Anguinea*, snaky.

P. 3–6 cm., *rufous, or pale yellowish, somewhat tan colour when dry*, fleshy, campanulate, then convex, gibbous, smooth, *covered near the margin when young with a superficial silky zone from the fibrils of the veil*. St. 5–8 cm. × 4–6 mm., *bay brown*, equal, base thickened, often flexuose, *densely white-fibrillose, and forming numerous zone-like marks*; often with silky spots when dry. Gills *pallid isabelline, or yellow, then ferruginous*, somewhat free, ascending into the top of the cone, *somewhat linear*, 3–4 mm. broad, crowded. Flesh *concolorous*, thin except at the disc. Spores ferruginous. Damp places in woods, and heaths. Sept.—Nov. Uncommon. (*v.v.*)

1105. **N. centunculus** Fr. Cke. Illus. no. 495, t. 601, fig. A. *Centunculus*, patch-work.

P. 8–20 mm., *lurid, or olivaceous fuscous, becoming light yellow green, finally becoming pale, but not hygrophanous*, fleshy, *convex, then plane*, obtuse, often excentric, dry, slightly silky under a lens; margin incurved, often striate, occasionally yellow-pulverulent. St. 2·5–3 cm. × 2–4 mm., *cinereous light yellow*, somewhat equal, often curved, *pulverulent with white mealy squamules at the apex*, base white-villous. Gills *light yellow cinereous*, adnate, separating, *broad*, *rather thick*, convex, undulated when old, *edge slightly toothed with greenish yellow floccules* ("pulverulent with crystalline particles under a lens" Quél.). Flesh *pallid*, thin. Spores ochraceous, elliptical, 8–10 × 6μ. Cystidia "undulating-clavate, or fusiform-capitate, 30–36 × 4–6μ" Rick. Gregarious, or caespitose. Rotten wood, especially beech. Oct. Rare.

1106. **N. horizontalis** (Bull.) Fr. (= *Galera horizontalis* (Bull.) Quél.) Cke. Illus. no. 495, t. 601, fig. B. *Horizontalis*, horizontal.

Entirely watery cinnamon. P. 5–15 mm., fleshy, convexo-plane, obtuse, smooth. St. 6–12 × 2 mm., *incurved*, smooth. Gills rounded, free, plane, broad, subdistant. Flesh *concolorous*, *paler*, thin. Spores ferruginous, punctate, broadly elliptical, 6–8 × 5–6μ. Cystidia clavate, or conical, flexuose, 20–35 × 2–4μ. On bark of elm, and pear trees. Nov.—Dec. Rare. (*v.v.*)

1107. **N. rimulincola** (Lasch) Rabenh. (= *Galera horizontalis* (Bull.) Quél.) Cke. Illus. no. 496, t. 509, fig. B.
Rimula, a small crack; *colo*, I inhabit.

Entirely cinnamon. P. 10–12 mm., hemispherical, *umbilicate*, *plicate*, *slightly wrinkled*, *tomentose*. St. 10–12 × 2 mm., somewhat excentric, incurved, somewhat thickened at the base. Gills adnexed, thick, *very broad*, subdistant, *edge whitish crenulate*. Flesh *white*, thin. Spores "cinnamon, elliptical, 10 × 5μ" Massee. On elm, and pear twigs. Oct.—Dec. Rare.

1108. **N. semiflexa** B. & Br. Cke. Illus. no. 496, t. 509, fig. A.
Semi, half; *flexa*, bent.

P. 12 mm., *chestnut*, hygrophanous, fleshy, subcampanulate, then hemispherical, or flattened; *margin adorned with the delicate white veil*. St. 6–18 × 1–2 mm., *pale*, semi-horizontal. Gills *tawny*, adnexed, rather broad, distant. Flesh *white*, thin. Spores "amber, elliptical, 8 × 5μ, verrucose" Massee. On wood, and on the ground. Oct. Rare.

N. rubricata B. & Br. = **Marasmius rubricatus** (B. & Br.) Massee.

**Gills adnate, p. convexo-plane.

1109. **N. abstrusa** Fr. Fr. Icon. t. 122, fig. 2. *Abstrusa*, hidden.

P. 2–4 cm., *ferruginous clay*, fleshy, convex, then plane, orbicular, smooth, *viscid*. St. 2·5–3·5 cm. × 2–4 mm., *pallid ferruginous*, *base darker*, very cartilaginous, tough, rigid, equal, round, tense, straight, *polished*, naked. Gills *watery ferruginous*, *or cinnamon*, adnate, plane, crowded. Flesh *concolorous*, *becoming pale*, thin at the margin. Spores ferruginous, "elliptical, 10 × 5μ" Sacc. Woods, and on sawdust. Oct. Uncommon.

1110. **N. innocua** (Lasch) Fr. Cke. Illus. no. 498, t. 489, fig. A.
Innocua, harmless.

P. 3–4 cm., *rufous, becoming pale when dry*, fleshy, convex, obtuse, rather smooth, fibrillosely-smooth under a lens, *striate*. St. 4–5 cm. × 3–4 mm., *white fibrillose*, base woolly. Gills *light yellow ochraceous*, adnate, somewhat crowded. Spores ferruginous, elliptical, 10 × 4–6μ. Damp places. Sept.—Oct. Uncommon.

1111. **N. cerodes** Fr. Cke. Illus. no. 498, t. 489, fig. B.
κηρώδης, wax-like.

P. 1–3 cm., *watery cinnamon, tan colour when dry*, submembranaceous, *campanulato-convex*, then flattened, at length depressed, *obtuse*, smooth, pellucidly striate at the margin when moist, *slightly silky-atomate* when dry. St. 5–8 cm. × 2–4 mm., *pallid, or pale yellowish, becoming bay-brown-fuscous, sometimes only at the base*, slightly firm, equal, somewhat flexuose, fibrillosely striate under a lens, apex mealy. Gills *pallid, then cinnamon*, adnate, separating, *very broad behind*, hence almost triangular, *subdistant*, broad, plane, edge minutely fimbriate under a lens. Flesh *pallid*, thin. Spores brownish ferruginous, broadly elliptical, 9–12 × 6–7μ, 1–2-guttulate. Cystidia "on surface of gill fusiform, 35–40 × 10–12μ, sparse, on edge of gill the majority filamentous-clavate" Rick. Woods, heaths, burnt ground, and on stumps. May—Oct. Uncommon. (*v.v.*)

1112. **N. melinoides** Fr. Cke. Illus. no. 499, t. 457, upper figs.
μέλι, honey; εἶδος, like.

P. 1–2 cm., *tawny, ochraceous when dry*, fleshy, *convex, then plane*, sometimes globose then hemispherical, obtuse, or gibbous, *striate at the margin when old*. St. 4–7·5 cm. × 2 mm., *concolorous, or ochraceous, base paler, white*, equal, or slightly attenuated, sometimes attenuated at both ends, slightly firm, smooth, *apex white pruinose*. Gills *somewhat tawny, or light yellowish ochraceous*, adnate, ventricose, crowded, *edge often denticulate*. Flesh *yellowish*, thin at the margin. Spores pale ferruginous, elliptical, 9–13 × 4–8μ, with a large central gutta. Cystidia flask-shaped, capitate, apex 5–8μ in diam., base ventricose, 50–55 × 8–19μ. Heaths, pastures, lawns, and roadsides. June—Nov. Common. (*v.v.*)

1113. **N. pusiola** Fr. Fr. Icon. t. 124, fig. 4. *Pusio*, a little boy.

P. 6–12 mm., *yellow, or tawny yellow*, submembranaceous or slightly fleshy, hemispherical, or campanulato-hemispherical, then expanded, obtuse, smooth, *slightly viscid*. St. 2·5–4 cm. × 1–2 mm., *shining light yellow, or lemon yellow*, tough, equal, or attenuated upwards, flexuose, smooth, *slightly viscid*, apex often pruinose, base often becoming brownish. Gills *yellow white, or watery cinnamon, then brown*, adnate, broad, plane. Flesh *concolorous*, thin. Spores "brown, pruniform, 8 × 4μ" Sacc. Pastures, and mossy hillsides. Sept.—Oct. Uncommon. (*v.v.*)

***Gills adnate. P. campanulate, then expanded.

1114. **N. nucea** (Bolt.) Fr. Bolt. Hist. Fung. t. 70.
Nucea, belonging to a nut.

P. 1–2·5 cm., *pale chestnut*, submembranaceous, *globoso-campanulate*, never flattened, *umbilicate*, smooth, slightly dotted; margin *incurved*, somewhat lobed. St. 7–8 cm. × 2 mm., *white*, tough, equal, *silky fibrillose, becoming even*, base with a small bulb. Gills *cinnamon*, adnate, *semicircular*, 6–8 mm. broad, plane, often undulate. Spores ferruginous, "elliptical, base apiculate, 10–11 × 6μ" Massee. Pine, and fir woods. Oct. Rare.

1115. **N. glandiformis** W. G. Sm. Cke. Illus. no. 500, t. 490, fig. B.
Glans, acorn; *forma*, shape.

P. 2–5 cm., *nut brown, disc darker*, obtusely campanulate, becoming somewhat hemispherical, *or filbert-shaped*, smooth. St. 7·5–10 cm. × 4 mm., *pallid*, equal, sometimes twisted, splitting. Gills *umber*, adnate, *very broad*, 12 mm. and more, rounded behind, serrate. Flesh *yellowish*, thick at the disc. Spores ferruginous, broadly almond-shaped, 10–12 × 6–8μ. On the ground. Oct. Rare.

1116. **N. scolecina** Fr. Fr. Icon. t. 124, fig. 1. σκώληξ, a worm.

P. 1–2 cm., *bay-brown-ferruginous, becoming pale*, opaque, fleshy, campanulate, then convex, *often umbonate*, obtuse, smooth, fragile; *margin paler, at length striate.* St. 3–7·5 cm. × 1–3 mm., *rufous ferruginous, base becoming fuscous*, equal, often flexuose and curved, *white mealy everywhere, then only at the paler apex.* Gills *whitish flesh colour, then ferruginous*, adnate, broader behind, 2–4 mm. broad, *edge flocculose, ciliate.* Flesh *pallid*, thin at the margin. Spores ferruginous, subglobose, 4 × 3μ, 1-guttulate. Alder swamps, and damp places. Sept.—Oct. Not uncommon. (*v.v.*)

1117. **N. striaepes** Cke. Cke. Illus. no. 502, t. 478.
Stria, a furrow; *pes*, foot.

P. 2·5–6 cm., *ochraceous*, fleshy, campanulate, obtuse, then expanded, smooth. St. 4–8 cm. × 4–6 mm., *white*, equal, straight, or flexuose, *longitudinally striate.* Gills *tawny ferruginous*, slightly adnate, subdistant, 4–6 mm. broad. Flesh *concolorous, then white*, thin at the margin. Spores pale ferruginous, elliptical, 7–9 × 4–5μ. Caespitose, or gregarious. Woods, lawns, pastures, and roadsides. Sept.—Nov. Not uncommon. (*v.v.*)

1118. **N. amarescens** Quél. Boud. Icon. t. 127.
Amarescens, becoming bitter.

P. 1·5–4 cm., *reddish brown, or cinnamon, becoming pale when dry*, campanulate, then expanded, often umbonate, minutely rugose, then torn and cracked, moist. St. 3–7 cm. × 4–6 mm., *concolorous, or ochraceous, then blackish bistre,* apex often slightly velvety, equal, base white tomentose. Gills *ochraceous, then tawny,* emarginate, adnate, ventricose, very broad. Flesh *concolorous,* scissile. Spores brownish ferruginous, oblong elliptical, 10–12 × 5–6μ. Taste insipid, then *very bitter.* In troops. Old charcoal heaps in woods. May—Sept. Uncommon.

1119. **N. sideroides** (Bull.) Fr. Cke. Illus. no. 503, t. 458, upper figs. σίδηρος, iron; εἶδος, like.

P. 1–2·5 cm., *pale yellowish, honey colour, or cinnamon, tan ochraceous when dry, somewhat shining,* fleshy, campanulate, then expanded, *umbonate,* smooth, very slightly viscid when moist; margin incurved at first, *then slightly striate.* St. 5–8 cm. × 2–4 mm., *pallid, then becoming yellow and ferruginous downwards, base at length becoming fuscous,* slightly firm, equal, sometimes undulated, or slightly thickened at the apex, smooth, apex often white pruinose. Gills *becoming watery-yellow ochraceous, at length somewhat cinnamon,* adnate, *with a small decurrent tooth,* sometimes sinuate and *uncinato-adfixed,* ascending, *linear, crowded.* Flesh *pallid,* thin. Spores pale yellow, "cylindrical-elliptical, 7–8 × 3–4μ. Cystidia on edge of the gill very delicate, filamentous-fusiform, 30–36μ, apex with a globose head" Rick. Stumps, ash and pine trunks, twigs, and chips. Oct.—Nov. Uncommon.

1120. **N. badipes** Fr. (= *Galera badipes* (Fr.) Rick.) Fr. Icon. t. 123, fig. 3. *Badius,* bay brown; *pes,* foot.

P. 8–20 mm., *yellowish ferruginous, tan when dry,* submembranaceous, campanulate, then convex, umbonate, *pellucidly striate to the disc when moist,* smooth; *margin almost straight* and adpressed to the st. St. 5–7·5 cm. × 2 mm., *ferruginous, darker and becoming fuscous towards the base,* equal, firm, rigid, often flexuose, *covered up to the middle with white floccose scales, apex paler, naked.* Gills *pale yellowish ferruginous,* adnate, *very ventricose,* thin, *subdistant.* Flesh *concolorous,* thin. Spores ferruginous, elliptical, 10–12 × 5μ. Cystidia "thin, fusiform, with long, blunt point, 50–60 × 10–15μ" Rick. Damp places in coniferous woods, and on heaths. Sept.—Nov. Not uncommon. (*v.v.*)

1121. **N. camerina** Fr. (= *Galera camerina* (Fr.) Rick.) Fr. Icon. t. 124, fig. 2. καμάρα, an arched roof.

P. 1–2 cm., *honey colour, tan colour when dry, obtuse umbo often darker,* fleshy, campanulato-convex, obtuse, umbonate, smooth, hygrophanous; *margin somewhat striate when moist.* St. 3–4 cm. × 1–2 mm., *umber, apex pallid,* equal, or attenuated downwards, somewhat curved, or flexuose, *tough, adpressedly fibrillose.* Gills *yellowish cinnamon, then ferruginous,* adnate, plane, ascending, attenuated behind, *very crowded*, often crenulate. Flesh *concolorous*, thin at the margin. Spores ferruginous, "elliptical, 6–7 × 3–4μ, smooth. Cystidia on the edge of the gill subulate-capitate, 36–40 × 5–6μ" Rick. Coniferous stumps. May—Sept. Uncommon. (*v.v.*)

1122. **N. hydrophila** Mass. (= *Naucoria nasuta* Kalchbr. sec. Cke.) Cke. Illus. no. 1173, t. 1172, fig. A, as *Naucoria nasuta* Kalchbr.
ὕδωρ, water; φίλος, loving.

P. 1–3·5 cm., *pale ochraceous tan with a distinct tinge of green here and there*, fleshy, campanulate, then slightly expanded, *acutely umbonate*, smooth; margin striate when moist. St. 3–5 cm. × 2 mm., *rather ferruginous, or with red and green tints*, equal, flexuose, smooth. Gills *pallid, then brownish*, adnexed, rather crowded, 3 mm. broad, *edge pale*. Flesh *greenish*, very thin. Spores brown, elliptical, 13–14 × 6–7μ. Gregarious. Swampy places under trees. Sept.—Oct. Rare.

1123. **N. triscopa** Fr. (= *Galera triscopa* (Fr.) Quél.) Fr. Icon. t. 124, fig. 3.
τρι-, three; σκοπός, faced.

P. 4–10 mm., *deep bay, tawny or ochraceous when dry*, always opaque, fleshy, hemispherical, obtuse, then convexo-plane, *with a prominent umbo*, smooth. St. 1–3 cm. × 1–2 mm., *ferruginous*, opaque, *base umber*, often velvety, equal, curved, or flexuose, *smooth*. Gills *yellowish, then dark ferruginous*, adnate, ventricose, plane, thin, *somewhat crowded*. Flesh *yellowish, ferruginous in the st.*, very thin at the margin. Spores "ferruginous, pruniform, 10μ" Quél. Stumps, and rotten wood of frondose trees, rarely on humus. May—Sept. Uncommon.

II. P. naked. Gills and spores fuscous ferruginous.
Veil potential, rarely manifest.

*Growing in fields, and plains.

1124. **N. vervacti** Fr. Cke. Illus. no. 504, t. 617, fig. A.
Vervactum, fallow ground.

P. 2–3 cm., *light yellow, or pallid yellow*, fleshy, convex, then plane, obtuse, or umbonate, soft, smooth, *slightly viscid*, shining when dry. St. 2·5–4 cm. × 4–6 mm., *whitish*, cartilaginous, *rigid*, attenuated either upwards or downwards, smooth, often striate. Gills *pallid, then ferruginous fuscous*, adnate with a decurrent tooth, 6 mm. *broad*, plane, at length ventricose, crowded. Flesh *white*, thick at the disc. Spores "olive brown in the mass, elliptic-oval, 12–17 × 8–12μ. Cystidia on edge of gill fusiform, 40–45 × 9–10μ, with or without a head" Rick. Gardens, and pastures. May—Nov. Uncommon.

1125. **N. pediades** Fr. (= *Cantharellus Brownii* B. & Br. sec. Pat.; *Naucoria semi-orbicularis* (Bull.) Fr. sec. Quél.) Cke. Illus. no. 505, t. 492. *πεδίον*, a plain, or field.

P. 2–5 cm., *yellow, or pale yellowish ochraceous, then becoming pale*, fleshy, convex, then plane, obtuse, dry, smooth, at length rimoso-rivulose. St. 5–8 cm. × 2–4 mm., *yellowish, stuffed with a pith, subflexuose*, tough, equal, base bulbous from the mycelium being rolled together, *slightly silky, becoming even.* Gills *somewhat fuscous, then dingy cinnamon*, adnexed, 4–10 mm. broad, crowded, then subdistant. Flesh *white*, thin at the margin. Spores fuscous ferruginous, broadly elliptical, 8–9 × 5–6μ, with a large central gutta. Cystidia "ventricose-fusiform, or with an enlarged head, 45–50 × 8–10μ" Rick. Pastures, and roadsides. July—Nov. Not uncommon. (*v.v.*)

1126. **N. semi-orbicularis** (Bull.) Fr. (= *Naucoria pediades* Fr. sec. Quél.) Cke. Illus. no. 507, t. 493, fig. A.

Semi-, half; *orbicularis*, round.

P. 2–5 cm., *tawny ferruginous, then ochraceous*, fleshy, convexo-expanded, obtuse, slightly viscid when fresh and moist, then dry, smooth, corrugated when dry. St. 7–10 cm. × 2–3 mm., *ochraceous, becoming pallid ferruginous, shining, often darker at the base*, cartilaginous, tough, equal, tense, straight, smooth, *internally with a separate fistulose tube* which is easily broken up into fibrils. Gills *pallid, then ferruginous*, adnate, rarely sinuate, 4–6 mm. broad, crowded. Flesh white, thin at the margin. Spores brownish, elliptical, 9–12 × 5–7μ. Cystidia flask-shaped, or fusiform, 45–50 × 13–15μ, apex obtuse, 6–8μ in diam. Pastures, heaths, roadsides. June—Oct. Common. (*v.v.*)

1127. **N. arvalis** Fr. *Arvum*, an arable field.

P. 1–2 cm., *yellow fuscous, pallid ochraceous when dry, disc yellow, or concolorous*, fleshy, firm, convex, then expanded, orbicular, obtuse, smooth, slightly viscid. St. 3–4 cm. × 1–2 mm., *becoming yellowish*, equal, smooth, *often pulverulent, attenuated into a long, cottony root, internally with a not easily separable fistulose tube.* Gills *pallid grey, then umber fuscous, or becoming ferruginous, adnexed*, plane, *very broad*, quaternate, *distant, edge often lemon yellow.* Flesh *lemon yellow.* Spores "light yellow, almond-shaped, 10–12μ, oblong" Quél. Arable fields, gardens, and sea-sands. July—Oct. Uncommon.

1128. **N. tabacina** (DC.) Fr. Cke. Illus. no. 507, t. 493, fig. B.

Tabacum, tobacco.

P. 6–20 mm., *umber, then bay-brown-cinnamon*, very moist, *tan colour when dry*, fleshy, convex, then plane, disc very obtuse, smooth; *margin involute*, often covered with a silky veil. St. 2·5–5 cm. × 2 mm., *brown, darker and fuscous at the base*, equal, attenuated downwards when shorter, somewhat flexuose, *smooth.* Gills *bay-brown-cinnamon, then ferruginous*, adnate, plane, linear, or ovate, *crowded.* Flesh *pallid reddish, deeper in the st.*, thin. Spores tobacco coloured in mass, "elliptical, 8–9 × 4–5μ" Massee. Cystidia "on edge of gill fusiform-subulate, 30–40 × 5–6μ" Rick. Lawns, waysides, and heaths. Sept.—Oct. Not uncommon. (*v.v.*)

**Growing in moist, uncultivated, wooded places.

1129. **N. tenax** Fr. *Tenax*, firm.

P. 2–5 cm., *varying between cinnamon and an olivaceous, or somewhat fuscous yellow*, dirty, *becoming pale when dry*, fleshy, hemispherical, then expanded, obtuse, rarely umbonate, sometimes becoming depressed at the disc, smooth, or slightly wrinkled, slightly viscid when young. St. 5–10 cm. × 4–10 mm., *concolorous, or dingy pallid, becoming fuscous, or olive, apex paler*, equal, or thickened upwards, *striate with adpressed fibrils*, and sprinkled with white fibrils, the remains of the fugacious veil. Gills *whitish fuscous, then ferruginous with the edge whitish*, adnate, becoming somewhat rounded and separating, 6–10 mm. *broad*, plane, triangular, or oblong. Flesh *yellowish, becoming whitish*, thick at the disc. Spores ferruginous, pip-shaped, 13–16 × 7–8μ, sometimes depressed on one side, 1–2-guttulate. Bogs on dead stems of *Potentilla Comarum*, and in ditches amongst sticks. May—Oct. Not uncommon. (*v.v.*)

1130. **N. Myosotis** Fr. Fr. Icon. t. 125, fig. 1.
Myosotis, the Forget-me-not.

P. 2–4 cm., *olivaceous, or fuscous green, becoming pale, or light* yellow, disc darker, fleshy, convex, then plane, subumbonate when flattened, smooth, *with a viscid pellicle.* St. 7–15 cm. × 2–6 mm., *pallid, then fuscous, apex white pruinose*, slightly firm, equal, often flexuose, *either fibrillose (the fibrils here and there blackish) or scaly.* Gills *pallid umber olivaceous, then brown ferruginous, with the edge whitish and serrate*, adnate, decurrent with a tooth, 5–6 mm. broad, subdistant. Flesh *pallid*, thin. Spores fuscous ferruginous, pip-shaped, 16–18 × 8–10μ, 1-guttulate. Cystidia on gill edge subcylindrical, flexuose, 36–40 × 7–9μ, apex obtuse, 5–6μ in diam. Bogs amongst *Sphagnum*, and *Potentilla Comarum.* July—Sept. Uncommon. (*v.v.*)

1131. **N. temulenta** Fr. Fr. Icon. t. 125, fig. 2. *Temulenta*, sodden.

P. 1–3 cm., *ferruginous, ochraceous, or whitish when dry*, hygrophanous, *submembranaceous*, campanulate, then convex, sometimes subumbonate, smooth; *margin striate when moist.* St. 4–8 cm. × 2 mm., *yellow, or ferruginous, tough*, equal, *flexuose, polished, fistulose with a pith, apex pruinose*, base white-villous. Gills *yellow, or lurid ferruginous, then umber, or ferruginous*, adnate, *attenuated in front*, subdistant. Flesh *pallid*, thin at the disc. Spores ochraceous, elliptical, 9 × 4–5μ, 1–2-guttulate. Cystidia "on edge of gill flask-shaped, on surface of gill quite differently shaped, ventricose with three points at the apex, 60 × 20μ" Rick. Woods and heaths. Aug.—Oct. Not uncommon. (*v.v.*)

1132. **N. subtemulenta** Lamb. *Sub*, somewhat; *temulenta*, sodden.

P. *brown ochre, tan when dry*, campanulate, then convex, slightly umbonate, striate, hygrophanous. St. *concolorous, dark rusty at the thickened base*. Gills *concolorous, almost free, narrow*, crowded.

1133. **N. latissima** Cke. Cke. Illus. no. 510, t. 482. *Latissima*, very broad.

P. 1–3·5 cm., *deep chestnut brown*, fleshy, subglobose, then hemispherical; margin at first incurved. St. 2–5 cm. × 4–8 mm., *dark brown below, paler above*, attenuated downwards, *rooting*, smooth. Gills *tawny umber*, rounded behind, slightly adnate, *very broad*, 12 mm. Flesh *white, brownish at the base of the stem*, thick at the disc. Amongst grass. Sept. Rare.

1134. **N. reducta** Fr. Fr. Icon. t. 125, fig. 3. *Reducta*, reduced.

P. 8–20 mm., *olivaceous, or fuscous honey colour, dirty tan colour when dry, membranaceous*, convex, then plane, sometimes umbonate, hygrophanous, smooth, then pruinose under a lens, *striate to the disc when moist*. St. 5–8 cm. × 2–4 mm., *amber fuscous, apex paler*, pruinose, slightly attenuated upwards, tough, flexuose, soft and splitting, smooth. Gills *dirty yellow, or pale ochraceous, then ferruginous*, adnate, or rounded and separating, broad, ventricose, somewhat crowded. Flesh *subconcolorous*, slightly thick at the disc. Spores "ochraceous, pruniform, oblong, 10μ, punctate" Quél. Boggy woods. Sept.—Oct. Rare.

III. P. flocculose, or squamulose. Veil manifest.
Spores ferruginous.

*Squamules of p. superficial, separating.

1135. **N. porriginosa** Fr. Cke. Illus. no. 511, t. 510. *Porriginosa*, full of scurf.

P. 2·5–5 cm., *tawny, tan colour when dry*, fleshy, hemispherical, then convex, hardly expanded, very obtuse, *viscid when moist, covered with superficial, fugacious, saffron coloured squamules*; margin striate when old. St. 5–7·5 cm. × 2–3 mm., *pallid*, equal, more or less fibrillosely silky, base white villose. Gills *yellow, then tawny cinnamon, adnate*, often with a small decurrent tooth, crowded. Flesh *lemon yellow*, thick at the disc. Amongst twigs, and rubbish. Oct. Rare.

1136. **N. sobria** Fr. Cke. Illus. no. 512, t. 511, fig. A.

Sobria, sober, not bibulous.

P. 6–20 mm., *honey colour, disc darker, becoming pale*, fleshy, convex, obtuse, or umbonate, slightly viscid, *margin appendiculate with the silky, fugacious veil.* St. 3–4 cm. × 2 mm., *pallid upwards, ferruginous fuscous downwards*, slightly firm, somewhat tough, straight, or slightly bent, equal, often sprinkled with whitish spots, the remains of the veil, apex often mealy. Gills *paler than the p., then saffron, with the edge whitish or yellowish floccose, obtusely adnate*, broader behind, plane, 3 mm. broad, subdistant. Flesh *pallid, somewhat ferruginous towards the base of the st.*, thin. Spores ochraceous, elliptical, or pip-shaped, 6–7 × 4μ. Charcoal heaps, and on the ground in woods. July—Nov. Uncommon. (*v.v.*)

var. **dispersa** B. & Br. *Dispersa*, scattered.

Differs from the type in its *smaller size, in the punctulate p. and the appendiculate ring on the st.* Lawns. July. Rare.

**P. with innate squamules.

N. erinacea Fr. = **Pholiota erinacea** (Fr.) Quél.

1137. **N. siparia** Fr. Fr. Icon. t. 126, fig. 2.

Siparium, a little curtain.

Entirely rufous ferruginous. P. 6–20 mm., fleshy, convex, then plane, obtuse, *densely villoso-squamulose, moist.* St. 1–2·5 cm. × 2 mm., fragile, equal, tense and straight, *densely sheathed with the scaly, villose, downy veil*; apex ochraceous, pruinose. Gills *ochraceous, then ferruginous*, adnate, quaternate, subdistant, *edge flocculose.* Flesh *concolorous, becoming yellowish*, thick, soft. Spores pale ferruginous, pip-shaped, 8–9 × 6–7μ, 1-guttulate. On wood, dead branches, earth, dead fern stems, and caddis worm cases. July—Oct. Rare. (*v.v.*)

1138. **N. conspersa** (Pers.) Fr. Cke. Illus. no. 514, t. 512, fig. A.

Conspersa, besprinkled.

P. 1–2·5 cm., *bay brown, or rufous cinnamon, ochraceous when dry, fragile, very hygrophanous*, fleshy, campanulato-convex, then flattened, obtuse, even, *soon furfuraceous and broken up into small scales.* St. 2·5–5 cm. × 2 mm., *cinnamon, ochraceous when dry*, equal, *fibrillose, apex squamuloso-furfuraceous*, base white tomentose. Gills *dark cinnamon*, adnate, then emarginato-*separating*, linear, or ventricose, crowded. Flesh *whitish*, thin. Spores "ferruginous, elliptical, 9–11 × 5–6μ" Karst. Gregarious. Woods, heaths, and pastures. Aug.—Oct. Not uncommon. (*v.v.*)

var. **uliginosa** Fr. *Uligo*, marshy ground.

Differs from the type in being *twice or thrice as large in all its parts, in the fuscous rufescent, umbonate pileus, the long, twisted, umber st.*

and the broader gills. Bogs, and amongst *Sphagnum.* Aug. Uncommon. (*v.v.*)

1139. **N. escharoides** Fr. ἐσχαρώδης, scab-like.

P. 1–2 cm., *tan, then whitish, disc at length becoming fuscous,* fleshy, soft, conico-convex, soon flattened, obtuse, *flocculoso-furfuraceous.* St. 2·5–5 cm. × 2–3 mm., *pallid, at length becoming fuscous, fragile,* equal, flexuose, *adpressedly fibrillose, or floccose,* becoming smooth, apex pruinose. Gills *pallid tan, then somewhat cinnamon, edge often pale and floccose,* adfixed, or decurrent with a tooth, at length emarginate, somewhat free, *ventricose,* 3–4 mm. broad, *lax.* Flesh *whitish,* thin at the margin. Spores pale ferruginous, elliptical, 11–12 × 6μ, 1-guttulate. Alder swamps, and bare damp ground. Aug.—Oct. Not uncommon. (*v.v.*)

1140. **N. Wieslandri** Fr. Fr. Icon. t. 126, fig. 3. J. Wieslander.

P. 2–4 mm., *tawny,* fleshy, convex, then plane, obtuse, *smooth,* but the cuticle is soon areolately rivulose, hence *spotted with darker, crowded, wart-like papillae.* St. 5 cm. × 1 mm., *becoming black,* flaccid, *almost naked.* Gills *at length dark ferruginous,* adnate, broad. Flesh *ferruginous,* thin. Amongst short grass in woods. Sept. Rare.

***P. destitute of scales, silky, or sprinkled with atoms.

1141. **N. carpophila** Fr. (= *Galera carpophila* (Fr.) Quél.) Fr. Icon. t. 126, fig. 4. καρπός, fruit; φίλος, loving.

P. 2–10 mm., *tan, pallid, or whitish when dry,* hygrophanous, submembranaceous, convex, obtuse, *furfurate with shining atoms, sometimes also floccoso-squamulose,* striate, diaphanous; margin crenulate with furfuraceous, fugacious flocci. St. 2–2·5 cm. × 1 mm., *pallid,* firm, *furfuraceous,* then naked. Gills *pallid, then ochraceous,* rounded, adnexed, somewhat free, *ventricose,* broad, subdistant, often crenulate. Flesh *yellowish,* thin. Spores "rusty-yellowish in the mass, subpyriform, 7–8 × 4–5μ, smooth. Cystidia on the edge of the gill filamentous-subulate" Rick. Beech leaves and mast. May—Oct. Rare. (*v.v.*)

1142. **N. graminicola** (Nees) Fr. Cke. Illus. no. 515, t. 513, fig. B. *Gramen,* grass; *colo,* I inhabit.

P. 5–8 mm., *fuscous, then fawn ochraceous,* submembranaceous, convex, *papillate, shaggy tomentose.* St. 1·5–2·5 cm. × 1 mm., *becoming fuscous, apex pale, tough,* equal, *hairy.* Gills *pale ochraceous,* then *pale cinnamon,* slightly adnexed, subdistant. Flesh *whitish,* thin. Spores pale ferruginous, broadly elliptical, 6–7 × 4–5μ, 1–3-guttulate. On grass stalks, and dead *Pteris* stalks. Aug.—Oct. Uncommon. (*v.v.*)

1143. **N. effugiens** Quél. (= *Crepidotus Rubi* Berk. sec. Quél.) Quél. Jur. et Vosg. II, t. 2, fig. 3. *Effugiens*, escaping notice.

P. 5–8 mm., *ochraceous, then pale olivaceous greyish*, convex, then plane, globose, diaphanous, *covered with shining crystalline grains*. St. 5 × 1 mm., *cream olivaceous*, curved, *mealy*, base villose. Gills *cream colour, then brown, or olive*, sinuato-free, or decurrent by a tooth. Flesh *olivaceous*, thin. Spores brown, broadly elliptical, 9–10 × 5–6μ, with a large central gutta. Dead twigs and branches of pear. Sept.—Oct. Uncommon. (*v.v.*)

Spores purple, or fuscous.

Psilocybe Fr.

(ψιλός, naked; κύβη, head.)

Pileus fleshy, regular; margin at first incurved. Stem central, cartilaginous. Gills adnate, sinuato-adnate, or adnexed. Spores purple, fuscous, rarely pinkish fuscous; elliptical, pip-shaped, almond-shaped or oblong elliptical; smooth, or verrucose, with an apical germ-pore. Cystidia present. Growing on the ground, or on wood, solitary, gregarious, caespitose, or subcaespitose.

I. Veil accidental, rarely conspicuous. St. thick-skinned, flexile, most frequently coloured. P. pelliculose, most frequently slightly viscid in wet weather, becoming somewhat pale. Colour of p. bright.

*Gills ventricose, not decurrent.

1144. **P. sarcocephala** Fr. (= *Psathyra sarcocephala* (Fr.) Quél.) Fr. Icon. t. 135, fig. 1. σάρξ, flesh; κεφαλή, head.

P. 3–12 cm., *ferruginous, becoming pale*, fleshy, convex, then expanded, obtuse, smooth, *dry*. St. 5–12 × ·5–2 cm., *whitish, sometimes becoming slightly ferruginous*, equal, or slightly attenuated downwards, firm, smooth, *apex white mealy, and somewhat squamulose*. Gills *whitish, then flesh colour and at length fuscous*, adnate, ventricose, 8–13 mm. broad, thick, fragile, not crowded. Flesh *white*, thick, firm. Spores pinkish fuscous, oblong elliptical, 9–10 × 4–5μ, 1–2-guttulate. Cystidia *broadly* lanceolate, acute, 50–60 × 12–18μ. Taste pleasant. Edible. Solitary, or caespitose. Often at the base of trees. Woods, and pastures. Sept.—Nov. Uncommon. (*v.v.*)

var. **Cookei** Sacc. Cke. Illus. no. 591, t. 620.
M. C. Cooke, the eminent English mycologist.

Differs from the type in the *larger, ochraceous, radiately rivulose, at length umbilicate p*. Base of trees. Sept.—Nov. Uncommon. (*v.v.*)

1145. **P. atrobrunnea** (Lasch) Fr. *Ater*, black; *brunnea*, brown.

P. 2–5 cm., *brownish*, fleshy, thin, campanulate, then convex, umbonate. St. 6–12·5 cm. × 4–5 mm., *paler than the p.*, *fibrillose*, apex white-mealy. Gills *becoming brownish*, adnexed, then separating, subdistant, somewhat thick, wide. Spores dark brown, "9–12 × 5–6μ" Sacc. Smell and taste of radish. Marshes amongst *Sphagna*. Rare.

1146. **P. nemophila** Fr. νέμος, wood; φίλος, loving.

P. 4–10 cm., *brick red*, *paler at the margin*, fleshy, convex, obtuse, *smooth*. St. 7–10 × 1 cm., *concolorous but paler*, *fusiformly attenuated downwards*, naked. Gills *pallid*, *deeply decurrent*, very narrow, edge crisped and dentate. On the ground, under old trees. Rare.

1147. **P. helvola** (Schaeff.) Massee. Schaeff. Icon. t. 210.
Helvola, pale yellow.

P. 2–4 cm., *tawny ochraceous*, *disc darker*, fleshy, campanulate, then expanded, obtuse, smooth, *the cuticle often cracked into patches near the disc*. St. 4–6 cm. × 3–4 mm., *white*, *or tinged ochraceous*, equal, fibrillose, often twisted; base white, thickened, downy. Gills *purplish umber*, clouded, *edge white*, adnate, then separating from the st., broad, rather distant. Flesh *pallid*, thin at the margin. Spores "purplish brown, obliquely elliptical, ends pointed, 9–10 × 4–5μ, warted" Massee. Amongst grass. Sept. Rare.

1148. **P. ericaea** (Pers.) Fr. Fr. Icon. t. 136, fig. 1. ἐρείκη, heath.

P. 2–4 cm., *tawny ferruginous*, *tawny yellow*, *or date brown*, fleshy, convex, then expanded, obtuse, slightly viscid when wet, dry and shining when dry. St. 7–10 cm. × 2–4 mm., *paler than the p.*, *becoming yellowish*, equal, tough, somewhat smooth, or silky. Gills *pallid*, *becoming blackish*, adnate, narrowed behind, 6–8 mm. broad, plane, *edge whitish*. Flesh *yellow*, thin. Spores fuliginous, broadly elliptical, 12–14 × 7–8μ. Cystidia "on surface of gill sparse, clavate with a prominent point, 30–36 × 9–10μ, contents yellowish, on edge of gill filamentous" Rick. Heaths, and damp pastures. May—Oct. Common. (*v.v.*)

1149. **P. subericaea** Fr. (= *Psilocybe ericaea* (Pers.) Fr. sec. Quél.) Fr. Icon. t. 136, fig. 2.
Sub, somewhat; *ericaea*, the species *P. ericaea*

P. 3–5 cm., tawny, fleshy, convex, then plane, smooth. St. 2·5–6 cm. × 2–5 mm., *becoming yellow*, equal, tough, smooth. Gills *pallid*, *then fuscous purple*, sinuato-adnate, 5–10 mm. broad. Flesh *yellow*, thin. Spores fuliginous purple, elliptical, or pip-shaped, 7–8 × 4μ. Heaths, and grassy places. June—Sept. Common. (*v.v.*)

1150. **P. uda** (Pers.) Fr. (= *Flammuloides uda* (Pers.) Quél.) Cke. Illus. no. 594, t. 569. *Uda*, moist.

P. 1–2·5 cm., *brick tawny, becoming pale,* fleshy, convex, then flattened, often more or less umbonate, smooth, slightly wrinkled when old; margin often striate. St. 4–7·5 cm. × 2–4 mm., *tawny ferruginous, apex paler,* equal, flexuose, *fibrillose.* Gills *pallid, then becoming fuscous purple,* adfixed, *ventricose,* 4–6 mm. broad, *lax,* plane, or convex. Flesh *yellow, ferruginous towards the base of the st.,* thin. Spores purple, broadly elliptical, 8–11 × 5–6μ, 1-2-guttulate. Cystidia "on the edge of the gill clavate-filamentous, 45–50 × 3–4μ" Rick. Swampy places and amongst *Sphagna.* Woods, and heaths. Sept.—Nov. Not uncommon. (*v.v.*)

var. **Polytrichi** Fr. *Polytrichum*, a genus of mosses.

Differs from the type in the p. *being pallid yellow, becoming whitish especially at the margin, and at length plane and depressed, in the glabrous, pallid tawny st., and the whitish clouded gills at length becoming decurrent, never becoming purple but often greenish.* Amongst *Polytricha.* Sept.—Oct. Uncommon. (*v.v.*)

var. **elongata** (Pers.) Fr. *Elongata*, tall.

Differs from the type in the *submembranaceous p. being livid, or greenish yellow, and striate when moist, becoming yellowish when dry.* Amongst *Sphagna* in mountain pine woods.

1151. **P. canofaciens** Cke. Cke. Illus. no. 595, t. 621.
Canus, hoary; *faciens*, making.

P. 2–4 cm., *dark bay brown, disc ferruginous,* fleshy, campanulate, then expanded, scarcely umbonate, *clad everywhere as well as the st. with delicate, scattered, white hairs,* which are soon evanescent at the apex; margin appendiculate with the white, fibrillose veil. St. 5–7·5 cm. × 4–6 mm., *concolorous, base very dark,* subequal. Gills *dark umber,* adnate, 6 mm. broad, ventricose. Flesh *of the p. pallid, of the st. rufescent, gradually darker downwards,* thick at the disc. Spores umber brown, elliptic oblong, 17 × 8, 10 × 4μ. Caespitose. Rotten straw, and rubbish. May. Rare.

1152. **P. areolata** (Klotzsch) Berk. Cke. Illus. no. 596, t. 570.
Areolata, divided into small patches.

P. 3–7·5 cm., *ochraceous, or fuscous,* fleshy, convex, minutely fibrillose, *cuticle cracking into nearly equal, square patches,* interstices and margin pale. St. 5–7·5 cm. × 4–6 mm., *dirty white,* equal, often thickened at the base, fibrillose. Gills *umber, at length blackish, edge white and beaded with drops of moisture,* adnate, 4–6 mm. broad. Flesh *tinged brown,* thick at the disc. Spores "blackish umber with a purple tinge, broadly almond-shaped, 12–13 × 8μ" Massee. Densely caespitose. On wood. Gardens. May—Oct. Rare.

1153. **P. virescens** (Cke. & Massee) Massee. Cke. Illus. no. 1182, t. 1177, as *Psilocybe areolata* Klot. var.

Virescens, growing green.

P. 3–5 cm., *bright dark brown*, fleshy, convex, obtuse, then expanded, minutely silky when young, *the brown cuticle breaking up into persistent, angular patches, the interstices being clear pale green*, becoming yellowish with age. St. 4–5 cm. × 6 mm., *pale green, and strongly striate at the apex, ferruginous below, base snow-white*, downy, equal, smooth, firm, straight, or slightly incurved. Gills *pallid, then smoky purple, edge pale*, adnexed, 6 mm. broad, rather crowded, soft. Flesh *whitish, ferruginous towards the base of the st.*, thick at the disc. Spores sooty purple, elliptical, ends obtuse, 9 × 5μ. Solitary. On rotten chips, and stumps. Sept.—Oct. Rare.

1154. **P. agraria** Fr. Fr. Icon. t. 137, fig. 1.

Agraria, pertaining to land.

P. 2·5–5 cm., *white, becoming cinereous when dry*, fleshy, convex, at length flattened and then often umbonate, smooth. St. 4–5 cm. × 2–4 mm., *white*, equal, smooth. Gills *white, at length fuscous*, obtusely adnate, linear. Flesh *white*, firm, not 2 mm. thick. Clay fields, and about the roots of decayed trees. Sept.—Oct. Rare.

1155. **P. chondroderma** B. & Br. Cke. Illus. no. 599, t. 606, fig. A.

χόνδρος, grain; *δέρμα*, skin.

P. 2–3 cm., *dark date brown*, fleshy, campanulate, very smooth, *cracked here and there in different directions*; margin appendiculate with the woven, jagged veil. St. 4–5 cm. × 5–6 mm., *paler than the p.*, subequal, fibrillose, *base squamulose.* Gills *dark brown, edge white*, adfixed, separating, ventricose. Flesh *yellowish, ferruginous towards the base of the st.*, thick at the disc. Spores "purple black, elliptical, 6–7 × 3–3·5μ" Massee. Fir woods. Sept. Rare.

1156. **P. scobicola** B. & Br. Cke. Illus. no. 598, t. 607.

Scobis, saw-dust; *colo*, I inhabit.

P. 2·5–4 cm., *white*, fleshy, convex, *umbilicate*, smooth. St. 2·5–4 cm. × 4 mm., *whitish*, subequal, or dilated at the apex, fibrillose. Gills *brown with a red tinge*, adnexed, broad. Flesh *white*, thin. Spores pale, elliptical, 8 × 5μ. On pine sawdust, and branches. Nov. Rare.

**Gills plane, very broad behind, somewhat decurrent.

(Deconica W. G. Sm.)

1157. **P. ammophila** (Dur. & Mont.) Fr. (= *Hypholoma ammophilum* (Mont.) Quél.) Cke. Illus. no. 599, t. 606, fig. B.

ἄμμος, sand; *φίλος*, loving.

P. 2–4 cm., *reddish tan, becoming paler*, fleshy, campanulate, then convex, and at length flattened, *fibrillose.* St. 4–5 cm. × 2–3 mm., *white, becoming yellowish except at the apex*, equal, *basal half clavate, densely covered with matted mycelium and sunk in the sand*, sometimes fusiform and rooting, striate. Gills *smoky, then bistre purple*, subdecurrent with a tooth, 4 mm. broad. Flesh *white, yellowish at the base of the st.*, very thin at the margin. Spores purple, broadly elliptical, 11–12 × 7–8μ. Sands on the sea shore. Aug.—Oct. Not uncommon. (*v.v.*)

1158. **P. coprophila** (Bull.) Fr. Cke. Illus. no. 600, t. 608, fig. A.
κόπρος, dung; φίλος, loving.

P. 2–3 cm., *rufescent tan*, fleshy, hemispherical, then expanded, umbonate, slightly viscid, smooth. St. 2·5–4 cm. × 2–3 mm., *rufescent, becoming pale*, attenuated upwards, *at first containing a pith, shaggy-flocculose*, the smooth and *shining* apex pruinose. Veil *reddish*, floccose, very fugacious Gills *livid, then fuscous, somewhat arcuato-*decurrent, *very broad*, 4–6 mm., crowded. Flesh *white*, thin. Spores purple, "nearly lemon-shaped, 11–12 × 7–8μ, smooth, with a very thick membrane. Cystidia cylindric-fusiform, 40–50 × 8–10μ" Rick. On dung, especially cow and rabbit. Pastures, and manure heaps. Sept.—Jan. Uncommon. (*v.v.*)

1159. **P. bullacea** (Bull.) Fr. Cke. Illus. no. 600, t. 608, fig. B.
Bullacea, having a knob.

P. 1–2 cm., *dark bay brown tawny, then ochraceous brick and tan when dry*, fleshy, hemispherical, obtuse, then expanded, umbonate, *covered with a viscid, separable pellicle, striate*; margin at first appendiculate with the white, floccose, fugacious veil. St. 2·5–3 cm. × 2 mm., *slightly tawny, becoming yellow, fuscous ferruginous at the base*, equal, or attenuated at the base, *slightly fibrillose*, apex pruinose. Gills *livid-whitish, then fuscous purple, adnate, somewhat triangular*, plane, crowded. Flesh *brown*, very thin at the margin. Spores purple, pip-shaped, 7–8 × 4–5μ. Cystidia "on the edge of the gill, filamentous" Rick. Gregarious. Horse dung, pastures, and amongst *Polytricha*. April—Oct. Common. (*v.v.*)

1160. **P. physaloides** (Bull.) Fr. Cke. Illus. no. 601, t. 609, fig. A.
φυσαλίς, a bladder; εἶδος, like.

P. 1–1·5 cm., *fuscous purple, then tan, or flesh colour, paler round the margin*, fleshy, campanulate, then expanded, at length flattened, *with a prominent umbo*, finally depressed round the umbo, *often striate, pelliculoso-viscid, smooth*, shining. St. 2–3 cm. × 2 mm., *becoming pale, base date brown*, equal, *filiform*, flexile, *adpressedly fibrillose*. Gills *pallid, or flesh colour, then ferruginous-fuscous*, adnate, *subdecurrent*, equally attenuated from the st. Flesh *whitish, subferruginous at the base of the st.*, thin. Spores "yellowish under the microscope, oval, 8–9 × 5–6μ. Cystidia on edge of gill filamentous" Rick. Heaths, pastures, rarely on dung. Feb.—Oct. Not uncommon. (*v.v.*)

1161. **P. atrorufa** (Schaeff.) Fr. Schaeff. Icon. t. 234.
Ater, black; *rufa*, red.

P. 1–1·5 cm., *black-rufous, or purple-fuscous, becoming very pale when dry*, fleshy, hemispherico-convex, obtuse, or with a knob, smooth, *striate at the margin when in full vigour*, without striae when dry. St. 2·5–5 cm. × 2 mm., *pallid date brown*, equal, fibrillose, or smooth, fragile, *apex pruinose*. Gills *greyish, then umber, or purple umber, adnate*, subdecurrent, triangular, broad, *edge white*. Flesh thin. Spores "yellowish under the microscope, oval-elliptical, 7–8 × 4–5μ. Cystidia on edge of gill fusiform-subulate, 30–36 × 4–6μ" Rick. Mixed woods. Nov. Rare.

1162. **P. nuciseda** Fr. Cke. Illus. no. 601, t. 609, fig. B.
Nux, a nut; *sedeo*, I sit.

P. 1–2 cm., *light yellowish*, fleshy, convex, subumbonate, *slightly silky when dry*. St. 2–3 cm. × 2 mm., *pallid, becoming fuscous, attenuated downwards*, tough, base white villose. Gills *fuscous, then umber black*, adnate, scarcely decurrent, broad, plane. Flesh yellowish, thin. Spores "brown, elliptical, 8 × 4μ" Massee. Beech mast, hazel nuts, and among chips. Rare.

***Gills somewhat linear, ascending.

1163. **P. tegularis** (Schum.) Fr. *Tegula*, a tile.

P. *tan colour*, fleshy, acorn-shaped, then campanulate, smooth, *rimosely areolate*. St. *pallid, attenuated upwards from the thickened base*, firm. Gills *becoming fuscous*, attenuato-adnexed, ascending, crowded. Grassy places. Sept.—Oct. Rare.

1164. **P. compta** Fr. (= *Agaricus comptulus* B. & Br. non Fr.) Cke. Illus. no. 603, t. 589, fig. A. *Compta*, adorned.

P. 2·5–4 cm., *pallid, then ochraceous*, submembranaceous, conical, then expanded, *striate, sprinkled with shining spots*; margin subcrenulate. St. 5 cm. × 2–3 mm., *pallid, pale rufous downwards*, flexuose, silky-shining. Gills *rosy umber*, adnate, ascending, distant. Flesh *white*, thin. Spores purple brown. Amongst grass. Woods. Sept.—Oct. Rare.

1165. **P. semilanceata** Fr. Cke. Illus. no. 604, t. 572.
Semi-, half; *lanceata*, spear-shaped.

P. 1–2·5 cm., *yellow, green, or fuscous*, submembranaceous, *acutely conical, almost cuspidate*, 10–15 mm. high, never expanded, *covered with a viscid pellicle, separable in wet weather*; margin incurved when young, slightly striate. St. 4–7·5 cm. × 2 mm., *pallid*, equal, often flexuose, *containing a pith, capable of being twisted round the finger*, smooth, *cortinate when young*. Gills *cream colour, then purple black, ascending into the apex of the cone*, adnexed, *almost linear*, crowded. Flesh *white*, thin. Spores purple, 11–13 × 6–7·5μ. Cystidia on edge of gill flask-shaped, or fusiform-subulate, 18–22 × 5–7μ. Woods, heaths, pastures, and roadsides. Aug.—Dec. Common. (*v.v.*)

var. **caerulescens** Cke. Cke. Illus. no. 605, t. 573.
Caerulescens, becoming blue.

Differs from the type in the *base of the st. turning indigo-blue*. Heaths, and pastures. July—Oct. Not uncommon. (*v.v.*)

1166. **P. callosa** Fr. Pers. Myc. Eur. t. 27, fig. 3.
Callosa, thick-skinned.

P. 1–2 cm., *livid, becoming fuscous, yellow, or whitish*, fleshy, *conical, then campanulato-convex*, obtuse, or broadly gibbous, *smooth*, dry. St. 5–7·5 cm. × 1–2 mm., *yellowish, becoming pallid, equal*, often flexuose, *smooth, tough*. Gills *cream colour, then fuliginous black, adnate*, ascending, ventricose, crowded. Flesh thin. Spores "dark to dark brown, elliptic-oval, triangular-globose, of various sizes and shapes, 5–11 × 4–6μ" Herpell. Pastures, lawns, and roadsides. Aug.—Oct. Uncommon.

II. No veil. St. rigid. P. scarcely with a pellicle, but the flesh most frequently scissile, hygrophanous. Gills adnexed, very rarely adnate.

1167. **P. canobrunnea** (Batsch) Fr. (= *Psathyra canobrunnea* (Batsch) Quél.) *Canus*, hoary; *brunnea*, brown.

P. 5–8 cm., *watery pallid, or fuscous flesh colour, becoming pallid tan, dry*, fleshy, convex, then plane, obtuse, smooth, *sometimes cracked into small squares, somewhat viscid when moist*. St. 5–6 cm. × 6–10 mm., *whitish, rigid*, equal, *rooting at the base, squamulose*. Gills *pallid, then fuscous purple, somewhat free*, ventricose, 6 mm. broad, somewhat crowded. Flesh *white*, thick, firm. Spores "very dark in the mass, narrowly elliptical, 8–9 × 4–5μ, brown, almost opaque. Cystidia on edge of gill clavate-filamentous, 45–50 × 5–7μ" Rick. Solitary, or laxly gregarious. Grassy places in woods, and burnt ground in beech woods. Sept.—Oct. Uncommon.

1168. **P. spadicea** Fr. (= *Psathyra spadicea* (Fr.) Quél.)
Spadicea, date brown.

P. 3–12 cm., *date-brown-umber, becoming pale when dry*, fleshy, convex, then plane, obtuse, *smooth, moist* in rainy weather, often broken up in cracks when dry, *hygrophanous*; margin inflexed when young. St. 4–10 cm. × 3–10 mm., *white*, firm, subcartilaginous, equal, often curved, smooth, silky. Gills *whitish, then flesh colour, at length umber, rounded adnexed, crowded*, arid. Flesh *whitish*, soft, somewhat thick at the disc. Spores brownish purple, elliptical, 9 × 5–6μ, 1-guttulate. Solitary, or subcaespitose. On stumps, and on the ground. Woods, and pastures. April—Nov. Uncommon. (*v.v.*)

var. **hygrophila** Fr. ὑγρός, moisture; φίλος, loving.

Differs from the type in its *larger size, the bullate p., the long*, 10–15 *cm., subfusiform rooting st., and the emarginate gills deeply decurrent in the form of lines.* Base of ash trees, and stumps. Oct. Rare.

var. **polycephala** Fr. πολύς, many; κεφαλή, head.

Differs from the type in the *more slender, densely crowded pilei, and the connately branched, thinner, flexuose stems.* Prostrate trunks. Rare.

1169. **P. cernua** (Fl. Dan.) Fr. (= *Psathyra cernua* (Fl. Dan.) Quél.) Cke. Illus. no. 607, t. 574. *Cernua*, nodding.

P. 2·5–6 cm., *livid, inclining to pale when moist, white when dry*, hygrophanous, fleshy, fragile, campanulate, then flattened, obtuse, smooth, or atomate under a lens, slightly wrinkled when dry, often slightly pellucid-striate. St. 5–6 cm. × 3–4 mm., *shining white*, equal, round, rigid-fragile, sometimes curved, smooth, *apex mealy*. Gills *white, then cinereous-blackish, adnate*, at first linear, then ventricose, 2–4 mm. broad. Flesh *white*, thin at the margin. Spores "dark brown in the mass, brown under the microscope, subcylindrical, 7–8 × 3–4μ, smooth. Cystidia on edge of gill clavate-bottle-shaped, 36–40 × 12–20μ" Rick. Solitary, or gregarious. On dead wood, chips, leaves. Aug.—Dec. Uncommon.

1170. **P. squalens** Fr. Fr. Icon. t. 137, fig. 2. *Squalens*, dirty.

P. 2·5–5 cm., *ferruginous, with a lurid tinge, becoming pale and dirty tan when dry*, fleshy, convex, then plane, obtuse, or depressed, smooth. St. 2·5–5 cm. × 2–4 mm., *concolorous*, equal, slightly rigid, obsoletely fibrillose, apex somewhat pruinose, striate. Gills *clay colour, then umber cinnamon, adnato-decurrent*, ventricose, 4 mm. broad, crowded. Flesh *becoming whitish*, thin. Spores fuscous ferruginous. Solitary, or subcaespitose. On rotten trunks. Rare.

1171. **P. hebes** Fr. Fr. Icon. t. 137, fig. 3, minor. *Hebes*, blunt.

P. 2–3 cm., *lurid, pale when dry*, hygrophanous, fleshy, convex, then expanded, obtuse, smooth, somewhat *viscid*, slightly striate at the margin when moist, even when dry. St. 4–5 cm. × 4 mm., *becoming pale white*, cartilaginous, *rigid*, equal, smooth. Gills *white, becoming fus-*

cous, wholly adnate, very broad behind, *triangular, rigid*, arid, crowded. Flesh *cinereous*, thin at the margin. Spores black purple, "elliptical, 14–16 × 7μ" Massee. Commonly solitary. On trunks in beech woods, and amongst grass and leaves near chestnut trees. Nov. Rare.

1172. **P. foenisecii** (Pers.) Fr. (= *Psathyra foenisecii* (Pers.) Quél.) Cke. Illus. no. 608, t. 590. *Foenisecia*, hay-harvest.

P. 1·5–2·5 cm., *pale fuliginous fuscous, or brown, becoming pale from the disc outwards in drying*, fleshy, campanulato-convex, obtuse, smooth, slightly wrinkled in very dry weather. St. 5–7·5 cm. × 2–4 mm., *rufescent, at first paler and white pulverulent*, somewhat pubescent, then naked, equal, rigid fragile, tense and straight. Gills *inclining to fuscous, livid fuscous at the sides, then umber*, adnate, ventricose in front, *hence appearing broadly emarginate*, subdistant, *edge white*. Flesh *whitish*, thin at the margin. Spores cinereous purple, almond-shaped, 12–15 × 7–9μ. Cystidia on edge of gill cylindrical, flexuose, apex obtuse, 4–7μ in diam., base ventricose, 29–42×9–11μ. Taste mild. Edible. Pastures, lawns, heaths, and roadsides. Feb.—Dec. Common. (*v.v.*)

1173. **P. clivensis** B. & Br. Cke. Illus. no. 1183, t. 969. *Clivensis*, belonging to a hill.

P. 2–3 cm., *pallid brown, then pallid ochre inclining to white*, subhemispherical, smooth, *sprinkled with shining particles*; margin striate. St. 2·5–4 cm. × 2–3 mm., *whitish*, nearly equal, base slightly clavate, somewhat silky. Gills *umber, edge white*, adnate, widely emarginate, 4–5 mm. broad, ventricose in front. Flesh *brownish, becoming whitish*, thin. Spores cinereous purple, broadly elliptical, 9–10 × 6–7μ, 1–2-guttulate. Heaths, and pastures. June—Nov. Uncommon. (*v.v.*)

1174. **P. catervata** Massee. *Catervata*, crowded.

P. 1–1·5 cm., *snow white*, campanulate, obtuse, smooth, *satiny*. St. 4–5 cm. × 2 mm., *white*, equal, usually rather wavy, *shining, brittle*. Gills *grey, then brown with a tinge of purple, edge white*, slightly adnexed, rather broad, crowded. Flesh *white*, rather thick. Spores brown with a purple tinge, elliptic-oblong, 12 × 4μ. *Densely fasciculate*. On the ground. Oct. Rare.

P. spadiceo-grisea (Schaeff.) Boud. = **Psathyra spadiceo-grisea** (Schaeff.) Fr.

Spores black, or blackish.

Panaeolus Fr.

(παναίολος, all variegated.)

Pileus slightly fleshy, regular, viscid, or dry, margin *exceeding the gills*. Gills adnate, or adnexed, variegated with the dark spores. Spores black, or fuscous black, elliptical, oblong ovate, boat-shaped, or almond-shaped; smooth, with an apical germ-pore. Cystidia present. Growing on dung and rich soil, solitary, or caespitose.

*P. viscid, shining when dry.

1175. **P. leucophanes** B. & Br. Cke. Illus. no. 625, t. 927, fig. A.

λευκός, white; φαίνω, I appear.

P. 2–3 cm., *white, here and there somewhat ochraceous*, campanulate, obtuse, *viscid*, shining when dry, *innately silky*; *margin appendiculate with the veil*. St. 5–8 cm. × 2–4 mm., *white*, attenuated upwards, fibrillose, *sprinkled with mealy particles, somewhat transversely undulated*. Gills *pallid grey flesh colour, then black*, adnate, *edge white*. Flesh *white*, thin at the margin. Spores black, "red-brown under the microscope, elliptical, with a flattened germ-pore, 10–12 × 6–7μ. Cystidia on edge of gill clavate-vesiculose, 30–36 × 9–10μ" Rick. Pastures. Aug. Rare.

1176. **P. egregius** Massee. Cke. Illus. no. 624, t. 624.

Egregius, distinguished.

P. 4–5 cm., *bright orange brown, disc darker*, fleshy, ovate-campanulate, 6 cm. high, smooth, *viscid when moist*, with a trace of agglutinated down, slightly wrinkled when dry; margin exceeding the gills. St. 12·5 cm. × 12 mm., *pale brown, duller than the p.*, slightly thickened at the base, fibrillose, readily splitting longitudinally, base white and cottony, apex smooth. Gills *brownish black, with a tinge of purple at maturity*, adnexed, ventricose, 12 mm. broad, thin, crowded, dry, not deliquescent; *edge paler*. Flesh *ochraceous*, thick at the disc. Spores brown, then blackish, oblong ovate, with a minute apiculus, 15–17 × 7–8μ. Solitary. On the ground. Jan. Rare.

1177. **P. phalaenarum** Fr. *φάλαινα*, a moth.

P. 1–4 cm., *clay white*, fleshy, campanulato-convex, obtuse, *viscid*, smooth; *margin appendiculate with the fugacious veil*. St. 6–10 cm. × 3–4 mm., *pallid rufescent*, equal, slightly firm, *pruinose*. Gills *grey, then cinereous black*, adnexed, broad. Flesh *pallid*, thin. Spores black, "elliptical, 10 × 6μ" Massee. On dung, especially cow. Rich pastures. July—Oct. Not uncommon. (*v.v.*)

**P. moist, opaque, bibulous, subflocculose when dry.

1178. **P. retirugis** Fr. Cke. Illus. no. 628, t. 627.

Rete, a net; *ruga*, a wrinkle.

P. 1–3 cm., *flesh tan colour, or pale grey*, fleshy, globose, then hemispherical, often subumbonate, *reticulate with raised ribs, atomate, opaque*; *margin appendiculate with the dentate, torn, fugacious veil*. St. 5–9 cm. × 2–4 mm., *flesh colour, becoming purple*, equal, pruinose. Gills *cinereous blackish*, adfixed, ascending. Flesh thin. Spores black, "olive black under the microscope, nearly lemon-shaped, 12–14 × 8–

9μ, smooth. Cystidia on edge of gill filamentous-clavate, 30–36 × 5–6μ" Rick. On dung. Pastures, and parks. April—Nov. Not uncommon. (*v.v.*)

1179. **P. sphinctrinus** Fr. Cke. Illus. no. 629, t. 628.

σφιγκτήρ, a band.

P. 1–3 cm., *fuliginous, or fuliginous grey, livid when dry,* hygrophanous, fleshy, oval, then campanulate, 2–2·5 cm. high, *never expanded,* obtuse, *always opaque,* moist in rainy weather, *somewhat silky when dry; margin crenate with the appendiculate, fugacious, white veil.* St. 2·5–7·5 cm. × 2–4 mm., *fuliginous grey,* tense and straight, equal, fragile, pruinose, *apex smooth.* Gills *cinereous blackish,* adnate, ascending, crowded, *edge often white.* Flesh *reddish,* fairly thick. Spores black, globose-elliptical, 13–14 × 9–10μ, with a hyaline germ-pore at each end. Cystidia on the edge of the gill cylindrical, flexuose, apex obtuse, 4–5μ in diam., base often subventricose, 30–38 × 6–7 μ. Rich pastures, and garden soil. April—Nov. Common. (*v.v.*)

***P. dry, smooth, slightly shining, not zoned.

1180. **P. campanulatus** (Linn.) Fr. Cke. Illus. no. 630, t. 629.

Campanulatus, bell-shaped.

P. 1–3 cm., *fuscous fuliginous, rufescent when dry,* fleshy, campanulate, at length convex, often umbonate, slightly viscid when moist, then *somewhat shining,* often excoriated when dry. St. 6–9 cm. × 2–4 mm., *rufescent, whitish pruinose at first,* equal, tense and straight, apex *striate.* Gills *varying grey and black,* adnate, ascending, crowded, *edge often white, and distilling watery drops.* Flesh *reddish,* thin. Spores black, almond-shaped, 10–12 × 7–8μ. Cystidia cylindrical, flexuose, apex obtuse, 4–6μ in diam., 35–40 × 7–9μ. On dung, especially horse. Pastures, and gardens. June—Nov. Common. (*v.v.*)

1181. **P. papilionaceus** (Bull.) Fr. Cke. Illus. no. 631, t. 630.

Papilionaceus, like a butterfly.

P. 2–4 cm., *pallid, or pale grey, disc reddish, hemispherical,* pruinose, *rimosely cracked when dry.* St. 2·5–7 cm. × 3–6 mm., *whitish,* equal, or attenuated downwards, *apex white pulverulent.* Gills *blackish, broadly adnate,* 6–15 *mm. broad,* at length plane, *edge often white.* Flesh *pallid, then whitish,* thick at the disc. Spores black, almond-shaped, 14–15 × 7–8μ, with a large central gutta. Cystidia "on edge of gill clavate-capitate, 30–36 × 6–7μ" Rick. On dung, and on the ground. Pastures, and woods. June—Nov. Not uncommon. (*v.v.*)

1182. **P. caliginosus** (Jungh.) Fr. Cke. Illus. no. 632, t. 631, fig. A.

Caliginosus, dark.

P. 1–2 cm., *brown,* fleshy, campanulate, obtuse, *smooth.* St. 4–7 cm. × 2 mm., *concolorous,* equal, *even, naked.* Veil very thin, or obsolete.

Gills *fuliginous black*, slightly adnexed, ascending, lanceolate. Flesh thin. Spores "blackish, elliptical, 10 × 6–7μ" Massee. Rich pastures, and lawns. Oct.—Nov. Rare.

****P. dry, smooth, zoned round the margin.

1183. **P. subbalteatus** B. & Br. Cke. Illus. no. 632, t. 631, fig. B.
Sub, somewhat; *balteatus*, belted.

P. 2–5 cm., *dull deep fawn colour, pallid when dry*, hygrophanous, fleshy, convex, margin slightly incurved, then expanded, obtuse, or slightly umbonate, irregular, slightly wrinkled, *marked near the margin with a dark, narrow zone.* St. 5–8 cm. × 2 mm., *red brown*, equal, *brittle*, stringy, splitting longitudinally, *marked with short white fibrils.* Gills *brownish*, adnate, slightly ventricose, *edge white, slightly toothed.* Flesh *brownish*, thick at the disc. Spores "black, opaque, almost lemon-shaped, 13–14 × 8–9μ. Cystidia on edge of gill filamentous, with a brown apex, 36–45 × 6–7μ" Rick. Caespitose. In a tare-field. Sept.—Oct. Rare.

1184. **P. acuminatus** Fr. Cke. Illus. no. 633, t. 632, fig. A.
Acuminatus, pointed.

P. 1·5–2·5 cm., *flesh tan colour*, fleshy, *conical, acuminate*, smooth, shining, *zoned with a blackish line round the margin which is at first crenulate.* St. 2·5–7·5 cm. × 2–3 mm., *pallid above, fuscous below*, equal, *pruinose*, base thickened. Gills *whitish, then blackish*, adnexed, ventricose, broad. Flesh *pallid, fuscous towards the base of the st.*, thin. Spores black, "opaque when mature, lemon-shaped, 12–15 × 8–10μ. Cystidia on edge of gill filamentous, 50–70 × 4–6μ" Rick. On dung. Pastures, and roadsides. Oct.—Nov. Rare.

1185. **P. fimicola** Fr. Cke. Illus. no. 633, t. 632, fig. B.
Fimus, dung; *colo*, I inhabit.

P. 1–2 cm., *fuliginous grey, clay hoary when dry*, fleshy, campanulate, then convex, *obtuse*, smooth, opaque, *marked round the margin with a narrow fuscous zone, and inside this with a white one.* St. 5–10 cm. × 2 mm., *becoming dingy pale*, equal, *soft-fragile*, obsoletely *slightly silky striate*, apex white pruinose. Gills *variegated grey and* fuliginous, adnate, slightly rounded, somewhat ventricose, broad, almost semi-ovate, *edge often white.* Flesh *grey white*, thin. Spores brownish black, "nearly transparent, subelliptical, 11–12 × 7–8μ, smooth. Cystidia on edge of gill, cylindrical-filamentous, rarely flask-shaped, 40–50 × 6–12μ" Rick. On dung. Rich pastures. April—Oct. Uncommon.

var. **cinctulus** (Bolt.) Cke. Bolt. Hist. Fung. t. 152.
Cinctulus, a little girdle.

Differs from the type in the *reddish cinnamon p. with broad brown marginal zone, and the dingy brown st.* On dunghills after rain. June—July. Not found since the time of Bolton.

P. caudatus (Fr.) Quél. = **Psathyrella caudata** Fr.

P. atomatus (Fr.) Quél. = **Psathyrella atomata** Fr.

††Margin of pileus straight, at first adpressed to the stem. Spores white.

Mycena Fr.

(μύκης, a fungus.)

Pileus fleshy, or submembranaceous, regular; margin straight, never incurved. Stem central, cartilaginous. Gills adnate, or sinuato-adnate with a decurrent tooth. Spores white, elliptical, oval, globose, or oblong elliptical; smooth, punctate, or verrucose; continuous. Cystidia present, very rarely absent. Growing on the ground, or on wood; solitary, or caespitose.

I. St. juiceless, base not dilated into a disc. *Edge of gills darker, denticulate.*

1186. **M. pelianthina** Fr. (= *Mycena denticulata* (Bolt.) Quél.) Cke. Illus. no. 216, t. 156, fig. 1. πελιαίνομαι, to be livid.

P. 2–4 cm., *pale purple livid, becoming whitish when dry, diaphanous,* fleshy, *convex,* obtuse, or obsoletely umbonate; margin striate. St. 5–8 cm. × 2–5 mm., *concolorous,* equal, sometimes incurved at the base, *firm,* apex fibrillosely striate. Gills *dark violaceous, edge blackish violet, denticulate,* truncato-adnexed, *very sinuate,* distant, *very elegantly connected by a network of veins.* Flesh *whitish, somewhat yellowish under the pellicle of the p.*, thick at the disc. Spores white, elliptical, 6–7 × 3μ, 1–2-guttulate. Basidia clavate, with 4-sterigmata. Cystidia filled with a dark purplish juice or colourless, cylindric-fusiform, 60–100 × 10–12μ. Smell of radish. Amongst dead leaves in woods, especially beech. Aug.—Nov. Common. (*v.v.*)

1187. **M. carneosanguinea** Rea. *Caro,* flesh; *sanguinea,* blood-red.

P. 2·5–3 cm., *livid grey, disc tinging rufous, becoming paler,* fleshy, convex, *subumbonate,* smooth. St. 4 cm. × 5–9 mm., *grey,* equal, base yellowish, incrassated, clad with short mycelial strands. Gills *dull purplish brown,* adnate, broad in front; *edge denticulate, blood-red.* Flesh of p. and apex of st. *white, changing to blood-red.* Spores white, elliptical, 4–5 × 2–3μ, 1-guttulate. Woods. Aug.—Sept. Uncommon. (*v.v.*)

1188. **M. marginella** Fr. (= *Mycena mirabilis* Cke. & Quél.) Cke. Illus. no. 1148, t. 951, fig. A, as *Mycena mirabilis* Cke. & Quél. *Margo,* a border.

P. 6–8 mm., *somewhat fuscous when young,* fleshy, campanulate, *umbo darker,* smooth, slightly striate elsewhere and *somewhat azure-blue-floccose.* St. 6–7 cm. × 1–2 mm., *azure-blue-floccose,* equal, tomentosely rooting. Gills *white, or cinereous,* slightly adnexed, distant; *edge fringed with azure-blue, or red floccules.* Flesh *tinged with blue,* thin at the margin. Spores white. Fir trunks amongst *Hypnum cupressiforme.* Aug.—Sept. Uncommon.

1189. **M. avenacea** (Fr.) Schroet. *Avenacea,* oaten.

P. 1–2·5 cm., *dirty yellowish brown, disc often darker and obtusely umbonate,* submembranaceous, campanulate, 1 cm. high, wrinkled when dry; margin striate. St. 5–6 cm. × 1–2 mm., *yellowish brown, apex paler,* equal, tough, *shining,* base white floccose. Gills *dirty white,* or *greyish,* slightly adnexed, fairly distant, 1·5–2 mm. broad; *edge floccose, brown.* Spores white, ovate ellipsoid, 9–11 × 5–6μ. Cystidia filled with a brownish juice, flask-shaped, pointed above, sometimes branched, attenuated at base, 45–70 × 6–12 × 2·5–3μ at apex. Woods, hedgerows and wood stacks. Sept.—Nov. Not uncommon. (*v.v.*)

var. **olivaceo-marginata** (Massee) Rea. Cke. Illus. no. 1153, t. 959, fig. A, as *Mycena olivaceo-marginata* Massee.
Olivaceus, olive-coloured; *marginata,* bordered.

Differs from the type in the *smaller spores,* 6–7 × 4–5μ. Amongst short grass in pastures, hill sides, and hedgerows. Sept.—Nov. Not uncommon. (*v.v.*)

1190. **M. aurantio-marginata** Fr. Fl. Dan. t. 1292.
Aurantius, orange; *marginata,* bordered.

P. 1–2 cm., *olivaceous-fuscous, or olivaceous tan, becoming paler,* fleshy, campanulate, then *convex,* obtuse, or obsoletely umbonate, smooth; *margin striate when moist.* St. 2·5–5 cm. × 2–4 mm., *yellowish, or greyish,* firm, equal, smooth, *base inflated ventricose, clothed with strigose yellow hairs.* Gills *grey, then greenish livid, very attenuato-adnexed,* very ventricose, crowded, connected by veins; edge *orange, floccose.* Flesh *fuscous, becoming whitish,* thick at the disc. Spores white, elliptical, or pip-shaped with a lateral apiculus, 6–8 × 4–5μ. Cystidia filled with a yellowish juice, broadly clavate, or bludgeon-shaped, *coarsely verrucose,* 30–34 × 12μ. Smell none, or strong. Grassy places in woods and amongst conifer needles. Aug.—Nov. Uncommon. (*v.v.*)

1191. **M. elegans** (Pers.) Fr. *Elegans,* graceful.

P. 10–12 mm., *yellow fuscous, or light yellow livid,* opaque, *membranaceous, campanulate,* more or less umbonate; *margin saffron coloured, slightly sulcate.* St. 5–6 cm. × 2 mm., *deep, or light yellow, apex paler,* rigid, tense and straight, equal, base attached to its support *by yellow bristling filaments.* Gills *greyish,* adnate, decurrent with a tooth, linear, 2 mm. broad; *edge darker, saffron yellow.* Flesh *white*, thin. Spores white, elliptical, with a basal, or subbasal apiculus, 8–9 × 4–5μ. Cystidia "content dark yellow, obovate or bludgeon-shaped, prickly, warted, about 9–11μ broad" Lange. Coniferous woods. July—Nov. Common. (*v.v.*)

1192. **M. atro-marginata** Fr. (= *Mycena balanina* Berk. sec. Quél.)
Fr. Icon. t. 78, fig. 3. *Ater*, black; *marginata*, bordered.

P. 1·5–3 cm., *cinereous-fuscous, or purplish fuscous, paler at the margin*, very membranaceous, conical, then conico-campanulate, 1·5–3 cm. high, lax, *deeply sulcate*, smooth, slightly viscid when moist. St. 4–10 cm. × 2–4 mm., *concolorous, apex paler*, strict, *very fragile*, equal, or slightly attenuated upwards, sometimes twisted, *sulcate*; base rooting, white tomentose. Gills *whitish grey, then flesh colour, slightly adnexed*, attenuated behind, almost linear, narrow, 2–3 mm. broad, subdistant, often slightly connected by veins; *edge black, very thin.* Flesh *white, purplish in the stem*, thin. Spores white, oblong-elliptic, 10–12 × 7μ, with a large central gutta. Pine woods. Sept.—Oct. Uncommon. (*v.v.*)

1193. **M. balanina** Berk. (= *Mycena atro-marginata* Fr. sec. Quél.)
Cke. Illus. no. 216, t. 156, fig. 2. *βάλανος*, an acorn.

P. 2–4 cm., *ochraceous with a slight tinge of umber*, fleshy membranaceous, convex, somewhat campanulate, obtusely umbonate, at length more or less expanded, *minutely pulverulent*, slightly rugulose, striate when moist. St. 6–7 cm. × 2–4 mm., *white and pruinose above, deep sienna-brown below*, attenuated downwards, flexuose, rigid, shining, quite smooth; *base dark brown, villose.* Gills *pale, or whitish with a pinkish tinge*, quite free, rounded, 3 mm. broad, *connected by veins, edge fringed with dull purple spiculae, which are also sprinkled over the surface.* On beech mast, and amongst oak leaves.

1194. **M. rubro-marginata** Fr. Fr. Icon. t. 78, fig. 4.
Ruber, red; *marginata*, bordered.

P. 1–2 cm., *red-livid, or purple fuscous, becoming pale, sometimes almost white, hygrophanous, membranaceous*, campanulate, obtuse, *striate*, smooth. St. 2·5–5 cm. × 2–3 mm., *pallid livid*, rigid-fragile, equal, often curved, smooth, base slightly thickened. Gills *whitish, then grey*, adnate, with a small decurrent tooth, *distant*; *edge fuscous purple, then brown.* Flesh *whitish*, thin. Spores white, boat-shaped, 12–13 × 3–4·5μ, 1-guttulate. Cystidia "claviform, pointed at the free end, 50–60 × 10–16μ, filled with brownish juice" Barbier; "with one or more sterigma-like projections on the top" Rick. Pine, and larch woods. July—Nov. Common. (*v.v.*)

var. **fusco-purpurea** (Lasch) Cke. *Fuscus*, brown; *purpurea*, purple.

Differs from the type in the *purple brown p., the finely striate st. villosely rooting at the base, and the eroded, brown edge of the gills.* Willow trunks, and amongst dead leaves. Sept. Uncommon.

1195. **M. strobilina** Fr. *στροβίλινος*, belonging to a pine cone.

P. 6–12 mm., *persistently scarlet, often paler at the circumference,* membranaceous, conical, then campanulate, *acutely umbonate,* slightly striate, smooth, dry. St. 3–5 cm. × 1–2 mm., *concolorous, slightly rigid,* equal, smooth, base white strigose. Gills *rosy red, adnate,* decurrent with a tooth, distant, alternate; *edge darker, deep blood colour.* Flesh *reddish,* very thin at the margin. Spores white, elliptical, 7–9 × 4–4·5μ, 1-guttulate. Cystidia "on edge of gill elongate-vesiculose, 45–50 × 15–18μ" Rick. Gregarious. On pine needles, occasionally in beech woods. Sept.—Oct. Not uncommon. (*v.v.*)

1196. **M. rosella** Fr. (= *Mycena rosea* (Pers.) Sacc.)
Rosella, somewhat rose-coloured.

Rose colour, becoming pale. P. 7–10 mm., membranaceous, campanulate, then hemispherical, *obtusely umbonate,* slightly hygrophanous, striate. St. 2–3·5 cm. × 1 mm., equal, soft, base white tomentose. Gills *rose colour, adnate, with a tooth,* subdistant, *edge blackish purple.* Flesh *white, reddish* in the st., very thin at the margin. Spores white, elliptical, 7–8 × 4μ. Cystidia "obovate or bludgeon-shaped, 42 × 7–15μ, occasionally the apex drawn out, somewhat bottle-neck-like, the free portion more or less warted, filled with a reddish or pinkish juice" Lange and Schroeter. On coniferous needles. June—Dec. Not uncommon. (*v.v.*)

1197. **M. atrovirens** Rea. Trans. Brit. Myc. Soc. VI, t. 7.
Ater, black; *virens*, green.

P. 8 mm., *blackish green, paler and whitish at the striate margin, bright green at the circumference,* slightly fleshy, hemispherical, smooth, somewhat viscid on the disc at first. St. 3 cm. × 1 mm., *cinereous, or greyish fuliginous,* equal, smooth. *Gills whitish, green and minutely toothed on the edge* especially towards the margin of the p., adnate, 2 mm. broad, subdistant, attenuated in front. Flesh *fuscous,* thin. Smell and taste none. Spores white, elliptical, or pip-shaped, often with an oblique apiculus, 5–6 × 3μ, minutely punctate; basidia clavate, 23–25 × 6–7μ, with 4-sterigmata. Cystidia on edge of gill abundant, often fasciculate, slightly clavate, or cylindrical, 35–40 × 3–4μ, flexuose, *filled with a greenish juice,* thin walled. Beech stumps. Oct. Rare. (*v.v.*)

II. St. juiceless, base not dilated into a disc. Gills *unicolorous*, not changing colour. P. *pure coloured, bright, not becoming fuscous, nor cinereous.*

1198. **M. pura** (Pers.) Fr. (= *Mycena pseudopura* Cke.) *Pura*, clean.

P. 2–8 cm., *rose, purple, lilac, bluish-grey, or white*, fleshy, campanulate, then expanded, at length rather plane, sometimes umbonate; margin striate. St. 3–10 cm. × 2–6 mm., *concolorous, or whitish, tough, polished*, equal, or attenuated upwards when larger, smooth, base white villose. Gills *pallid, or whitish*, adnate, *broad*, ventricose, *connected by veins*. Flesh *white*, thin at the margin. Spores white, elliptical, 6–9 × 3·5–4μ, 1–2-guttulate. Cystidia on gill edge only, hyaline, cylindrical, broadly fusiform, or bladder-like, obtuse, more rarely somewhat pointed, becoming larger with age, 45–60 × 11–20 × 5–10μ at apex. Smell and taste of radish. Poisonous. Woods and pastures. June—Dec. Common. (*v.v.*)

var. **carnea** Rea. *Carnea*, fleshy.

Differs from the type in the *fleshy, fibrillose st.* Amongst oak and beech leaves in deciduous woods. Aug.—Nov. Not uncommon. (*v.v.*)

var. **multicolor** Bres. Bres. Fung. Trid. t. 114. *Multus*, many; *color*, colour.

P. 3·5–5 cm., *bright greyish blue, umbo fulvous.* St. 5–7 cm. × 4–8 mm., *rosy purple, base becoming yellowish*, white tomentose. Gills *grey.* Spores white, elliptical, 5 × 3μ (7–9 × 4–5μ Bres.). Cystidia cylindrical, or ventricosely fusiform, rarely subclavate. Amongst leaves. Oct. Uncommon. (*v.v.*)

1199. **M. zephirus** Fr. (= *Mycena spiripes* (Schwartz) Sacc.) Fr. Icon. t. 78, fig. 6. Ζέφυρος, the west wind.

P. 2–3 cm., *livid reddish, whitish flesh colour, or greyish, disc occasionally becoming fuscous, diaphanous*, submembranaceous, campanulate, then convex, *striate to the middle.* St. 3–8 cm. × 2–4 mm., *lilac, becoming rufescent*, fragile, equal, or slightly attenuated upwards, *slightly striate, at first clad with deciduous, white scales*; base incurved, woolly. Gills *white, or flesh coloured, adnate*, subdecurrent with a small tooth, at length separating, broad, *slightly connected by veins.* Flesh *pallid*, thin at the margin. Spores white, pip-shaped, 7–9 × 4μ, 1–4-guttulate. Cystidia "obovate, globose, or clavate, 40–60 × 16–25μ, crowned with 1–4-finger-like, often branched protuberances, sometimes only toothed" v. Hoehnel. Gregarious. Amongst fir needles, or decayed fir wood, and amongst dead leaves. Sept.—Dec. Uncommon. (*v.v.*)

1200. **M. Seynii** Quél. Quél. Soc. bot. Fr. XXIII, t. 2, fig. 9. M. de Seynes.

P. 1–2 cm., *rosy vinous, disc greyish*, pellucid, very thin, campanulate, then convex, *silky*. St. 3–5 cm. × 2 mm., *hyaline white, becoming purplish,* often flattened, bristling with white hairs at the base. Gills *rose, or lilac,* adnate by a tooth, distant, reticulately connected by veins, firm. Spores white, punctate, elliptical, 7–13μ, barrel-shaped. Smell none, taste like turnips. Fasciculate. Pine cones. Sept.—Oct. Rare.

1201. **M. flavipes** Quél. (= *Mycena Renati* Quél.) Quél. Jur. et Vosg. II, t. 1, fig. 4. *Flavus*, light yellow; *pes*, foot.

P. ·5–2 cm., *rosy pink, or violaceous, disc brownish,* membranaceous, campanulate, *striate, pellucid,* smooth, wrinkled when dry. St. 2–5 cm. × 1–2 mm., *yellow amber, pellucid,* tough, curved, *shining, villose base swollen*. Gills *white, then flesh colour,* adnate with a tooth, distant, connected by veins. Flesh *yellow in the st.*, very thin. Spores white, elliptical, 11–12 × 6–7μ, punctate. Cystidia "only on the edge of the gill, very numerous, threadlike above, long and ventricose below, 26–35 × 10–16μ" v. Hoehnel. Smell faint, of radish. On fir branches, and stumps. June—Sept. Uncommon. (*v.v.*)

1202. **M. clavus** (Linn.) Rea. (= *Mycena rubella* Quél.) Boud. Icon. t. 68, as *Mycena rubella* Quél. *Clavus*, a nail.

P. 5–12 mm., *orange scarlet, disc often darker,* fleshy, conico-convex, then plane, *umbo acute, diaphanous,* smooth; *margin pale,* striate. St. 2–4 cm. × 2 mm., *white, hyaline, with a fugacious, rosy tint at the apex,* equal, smooth, base woolly. Gills *white, then rose colour, edge paler,* adnexed. Flesh *of p. red, of st. white,* very thin. Spores white, elliptical, or pip-shaped, 6–7 × 3–4μ, multigranular. Cystidia "only on the edge of the gill, sparse, rigid and threadlike above, ventricose below, 28–32 × 10μ" v. Hoehnel. On twigs, leaves, and amongst mosses. Aug.—Dec. Not uncommon. (*v.v.*)

1203. **M. coccinea** (Sow.) Quél. *Coccinea*, scarlet colour.

P. 5–12 mm., *rosy-red, somewhat orange,* membranaceous, campanulate, *striate*; margin incurved, wrinkled. St. 2–3 cm. × 1 mm., *rose colour,* equal, smooth; *base subbulbose,* white strigose. Gills *rosy,* emarginate. Flesh of *p. red, whitish or yellowish in the st.*, very thin. Spores white, pip-shaped, or elliptical, with a basal apiculus, 8–10 × 5–6μ. Cystidia hyaline, flask-shaped, ventricose, apex pointed, or obtuse, often constricted at about one-third of its length from the apex, 25–35 × 7–10 × 3–5μ at apex. On larch cones, twigs and needles. Sept.—Oct. Uncommon. (*v.v.*)

1204. **M. Adonis** (Bull.) Fr. *Adonis*, son of Cinyras, king of Cyprus.

P. 6–10 mm., *rose-red, margin whitish,* membranaceous, campanulate, 6–8 mm. high, minutely and almost obsoletely papillate, smooth, *pellucidly striate*. St. 3–5 cm. × 1 mm., *shining white, pellucid, flexuose,* equal, smooth; base swollen, strigose. Gills *white*, or *flesh coloured*, *uncinato-adnexed*, ascending, very thin, narrow, somewhat distant. Flesh of *pileus red*, *of st. white*, very thin. Spores white, elliptical, often with a basal apiculus, 7–8 × 4μ, 1–3-guttulate. Basidia "2-spored. Cystidia awl-shaped-conical, long, pointed, up to 60μ" Lange. Woods and pastures. Sept.—Nov. Uncommon. (*v.v.*)

1205. **M. chlorantha** Fr. (= *Mycena virens* (Bull.) Quél.) Trans. Brit. Myc. Soc. III, t. 7, as *Mycena virens* (Bull.) Quél.

χλωρός, grass green; ἄνθος, flower.

P. 10–15 mm., *olive green, becoming paler, disc yellow*, membranaceous, *conico-campanulate*, 4–8 mm. high, then expanded, *obtuse*, smooth, striate when moist. St. 2·5–10 cm. × 2–3 mm., *bluish*, transparent, firm, *straight*, smooth, base white villose. Gills *white with a slight tinge of green*, adnate, narrow, 2 mm. broad, subventricose, thin, crowded. Flesh *bluish in the st.* Spores white, elliptical, 8–9 × 5–6μ, 1-guttulate, white in the mass, greenish by transmitted light. Cystidia "only on the edge of the gill, rarely obtuse, obovate oblong, generally fusiform, conical or bluntly pointed at the apex, 27–35 × 9–15μ" v. Hoehnel. Woods, and hedgerows. Aug.—Oct. Uncommon. (*v.v.*)

1206. **M. lineata** (Bull.) Fr. *Lineata*, striate.

P. 6–15 mm., *yellow, olivaceous, rarely whitish, generally becoming light yellow*, very membranaceous, hood-shaped, then campanulate, obtuse, *lineato-sulcate to the disc*, smooth. St. 4–7 cm. × 2–3 mm., *concolorous*, equal, smooth, base white villose. Gills *white, or pale cream, adnate*, linear, 2–3 mm. broad, subdistant. Flesh *yellowish at the disc*, very thin at the margin. Spores white, pip-shaped, 7–8 × 4μ or elliptical, often slightly depressed on one side, 9–12 × 5–6μ, often 1-guttulate, punctate. Cystidia pyriform, setulose, 20–25μ in diam. Amongst moss. Woods, and pastures. Aug.—Oct. Not uncommon. (*v.v.*)

var. **expallens** Fr. Fr. Icon. t. 78, fig. 5. *Expallens*, becoming pale.

P. 1–2 cm., *pale yellow*, campanulate, *striate up to the disc*. St. 5 cm. × 1 mm., pallid; base white, downy. Gills *white*, linear, *very narrow*, rather distant. Amongst moss.

1207. **M. farrea** (Lasch) Fr. Fr. Icon. t. 80, fig. 4, wrongly cited as t. 79, fig. 4, in the text[1]. Trans. Brit. Myc. Soc. I, t. 2, fig. 1, surface cells of the pileus magnified. *Farrea*, mealy.

P. 1–2 cm., *whitish, or yellowish, often with a tinge of flesh colour*, membranaceous, campanulate, then expanded, *subumbonate, sulcate, covered with a shining pruina* ("composed of globose, hollow bodies,

[1] Plates 79 and 80 are interchanged in some copies of Fries' Icones.

40–50μ" Plowright); *margin light yellow, or flesh colour, becoming whitish when dry, at first floccose, crenulate.* St. 5–7·5 cm. × 2–3 mm., *white,* somewhat fragile, equal, *silky-striate,* often somewhat rooting. Gills *white,* adnate, thin, somewhat crowded, *connected by veins, often fimbriate.* Flesh *white,* thick at the disc. Amongst moss and heather on heaths. Sept. Rare.

1208. **M. luteo-alba** (Bolt.) Fr. *Luteus,* yellow; *alba,* white.

P. 6–10 mm., *yellow, somewhat shining, not becoming pale,* membranaceous, acutely campanulate, *umbonate* when expanded, *slightly pellucidly striate,* smooth. St. 2·5–5 cm. × 2 mm., *canary yellow, or yellowish,* equal, *subflexuose, tough,* smooth, base subfibrillose. Gills *shining white, adnate,* somewhat uncinate, *at first joined behind,* broad, alternate. Spores white, "elliptical, 6–8 × 3·5–4·5μ" Schroeter. Cystidia "scattered, conical, often crowned with threadlike protuberances, 22–36 × 9–12μ, or slightly ventricose, conical, numerous, obtuse, 16–21 × 7–9μ" v. Hoehnel. Amongst moss. Pine woods. Aug.—Oct. Not uncommon. (*v.v.*)

1209. **M. flavo-alba** Fr. Fr. Icon. t. 80, fig. 5, wrongly cited as t. 79, fig. 5, in text. *Flavus,* light yellow; *alba,* white.

P. 1–2 cm., *ochraceous, light yellow-white, or wholly white,* submembranaceous, campanulate, then convex, at length flattened, *umbonate,* smooth, often striate, *scalloped when dry* and rimosely split. St. 2–3 cm. × 2 mm., *pellucidly white, or with a yellowish tinge, slightly rigid,* fragile, *tense and straight,* equal, *apex pruinose.* Gills *white, adnato-decurrent, soon separating-free, at length plane,* ventricose, *distant.* Flesh *white,* thick at the disc. Spores white, elliptical, 6–8 × 3–4μ, 1-guttulate. Cystidia flask-shaped, apex subulate, 3–4μ in diam., ventricose at the base, 35–55 × 12–15μ. Pastures, heaths, and woods. July—Dec. Common. (*v.v.*)

1210. **M. chelidonia** Fr.[1] (= *Mycena pumila* (Sow.) Quél.; *Mycena raeborhiza* (Lasch) Gill.) Fr. Icon. t. 83, fig. 4, as *Mycena raeborhiza* Lasch. *χελιδών,* a swallow.

P. 1–2·5 cm., *yellow flesh colour, or somewhat tawny, becoming paler flesh colour, or whitish at the margin,* submembranaceous, campanulate, conical, fragile, *pruinose, pellucidly striate when moist,* even when dry. St. 3–5 cm. × 2 mm., *pallid, or becoming yellow,* equal, tough, rooting, smooth, *apex mealy,* base villose. Gills *whitish, becoming yellowish,* or flesh colour, adnate, or somewhat free, subdistant, at length connected by veins. Flesh *yellowish,* thin at the margin. Spores white, elliptical, 3 × 1·5–2μ, 1-guttulate ("subglobose, 6–8μ, or 6–7 × 5–6μ" Gill.). Beech, and alder stumps. Oct. Rare. (*v.v.*)

[1] Fries placed this species in the section having milk, or a coloured juice in the st., but this factor is so slight that the fungus is far better ranged in this position.

1211. **M. lactea** (Pers.) Fr. Cke. Illus. no. 222, t. 159, bottom figs.
Lactea, milk-white.

Entirely white. P. 1–2 cm., membranaceous, campanulate, *disc sometimes becoming light yellowish, subumbonate, striate when moist,* even when dry, *margin scalloped.* St. 3–7 cm. × 1–2 mm., equal, *flexile,* smooth, *base fibrillosely rooting.* Gills *adnate,* ascending, *narrow, crowded.* Flesh *white,* very thin at the margin. Spores white, elliptical, 8–9 × 3–3·5μ, 1-guttulate. Basidia with 2-sterigmata. Cystidia sparse, subulate, apex 3μ in diam., base subventricose, 35–40 × 9μ. Gregarious. On needles, and twigs. Coniferous woods. July—Dec. Common. (*v.v.*)

var. **pithya** (Pers.) Fr. (= *Collybia ludia* Fr. sec. Quél.) Fr. Icon. t. 68, fig. 4, as *Collybia ludia* Fr. πίτυς, a pine, or fir tree.

Snow white, drying up and becoming yellowish. P. 6–10 mm., becoming almost plane, margin *incurved.* St. 2–2·5 cm. × 1–2 mm., *pulverulent,* base villose, subbulbous, *villosely strigose.* Gills *somewhat wide, distant.* On coniferous needles and chips. Sept. Rare.

var. **pulchella** Fr. Fr. Icon. t. 80, fig. 3, wrongly cited as t. 79, fig. 3, in text as *Mycena lactea* Pers. *Pulchella*, beautiful little.

Differs from the type in the *much thinner p., prominent umbo, and the gills broader at the base and subdistant.* Caespitose. On wood.

M. muscigena (Schum.) Quél. = **Collybia muscigena** (Schum.) Fr.

1212. **M. gypsea** Fr. γύψος, chalk.

P. 1–2 cm., *white, or yellowish, disc yellow,* membranaceous, *somewhat fragile,* conico-campanulate, umbonate, umbo not prominent, *striate to the disc.* St. 6–8 cm. × 1–2 mm., *white,* equal, or attenuated upwards, *rigid, fragile, smooth,* or pruinose, base villose. Gills *shining white,* adnate, *broadest behind,* equally attenuated to the margin. Spores white, elliptical, 8–10 × 4–5μ. Cystidia "obtusely conical, 30 × 11μ, or obtuse and broadly thread-shaped above, slightly conical below, 30–62 × 8–16μ" v. Hoehnel. Gregarious and caespitose. Amongst fragments of wood, and on trunks. June—Nov. Uncommon.

1213. **M. nivea** Quél. Quél. Soc. bot. Fr. XXIII, t. 2, fig. 1.
Nivea, snow-white.

Shining white, transparent. P. 15 mm., *very thin,* campanulate, never expanding, *sulcate from the apex to the base,* pruinose. St. 2–3 cm. × 1 mm., *rigid,* smooth, *apex pruinose,* base recurved, *fibrillose.* Gills adnate, uncinate, *narrow,* distant. Spores pruniform, 10–12μ, granular. On twigs. Deciduous woods.

M. galeropsis Fr. = **Marasmius dryophilus** (Bull.) Karst. var. **oedipus** (Quél.) Rea.

III. St. firm, rigid, somewhat tough, juiceless, somewhat strigose and rooted at the base. Gills changing colour, white, then grey, or reddish, commonly at length connected by veins. P. not hygrophanous. Generally lignicolous and caespitose.

M. cohaerens (A. & S.) Fr. = **Marasmius cohaerens** (A. & S.) Cke.

1214. **M. prolifera** (Sow.) Fr. Sow. Eng. Fung. t. 169.

Proles, offspring; *fero*, I bear.

P. 1–2 cm., *pallid, disc fuscous*, fleshy, campanulato-expanded, *broadly umbonate*, dry, slightly striate; *margin at length sulcate*, or rimosely split, pale yellowish, or becoming fuscous tan. St. 5–8 cm. × 4–5 mm., *pallid, apex greyish, base tawny bay*, firm, rigid, *smooth, shining, slightly striate, rooting, often proliferous*. Gills *white, then ochraceous*, adnexed. Flesh pallid, thin. Spores white. Densely caespitose, glued together by villose down. Woods, and old logs in gardens. July—Oct. Uncommon. (*v.v.*)

1215. **M. excisa** (Lasch) Gillet. *Excisa*, cut out.

P. 2–3 cm., *brownish, rugulose, disc pallid fuscous, or greyish fuscous*, membranaceous, campanulate, then convex, subumbonate, *tough*; margin paler, striate. St. 3–4 cm. × 3–4 mm., *greyish brown, becoming fuscous, apex paler*, firm, tough, smooth, rooting. Gills *paler than the p., hoary*, somewhat free, *very attenuated at the base*, ventricose, *connected by veins, very distant*, thick. Flesh *pallid*, thick at the disc. Spores white, broadly elliptical, 9–11 × 7–8μ. Basidia with 4-sterigmata. Cystidia subulate, filiform, or bottle-shaped, rarely furcate, base swollen, 20–30 × 10–15μ. Caespitose, rarely solitary. On trunks and stumps of pine. Aug.—Oct. Uncommon. (*v.v.*)

1216. **M. fagetorum** (Fr.) Gillet. *Fagetum*, a beech wood.

P. 1–2 cm., *yellowish, pale livid, or fuliginous*, membranaceous, campanulate, then convex, *striate half way to the disc*, smooth. St. 3–6 cm. × 3–4 mm., *pallid, incurved, base villose, attached at right angles to the beech leaves*. Gills *white, or glaucous*, attenuated at the base and *attached to a collar*. Spores white, "oblong elliptical, 9·5–11 × 4–4·5μ. Cystidia few, small, insignificant, club, or pear-shaped, their not much protruding free portion set with short setae" Lange. Gregarious. On dead beech leaves. Sept.—Nov. Not uncommon. (*v.v.*)

1217. **M. Berkleyi** Massee. Cke. Illus. no. 224, t. 148, as *Mycena excisa* Lasch.

The Rev. Miles Joseph Berkeley, the founder of British mycology.

P. 3–7 cm., *dingy brown, umbo darker, paler when dry*, fleshy, campanulate, then more or less expanded, slightly umbonate, *slightly and distantly striate to the umbo*, hygrophanous. St. 8–13 cm. × 6–8 mm., *dingy brown with a purple tinge*, almost equal, or slightly thickened below, more or less striate; *base long, tapering, rooting.* Gills *tinged purplish, or flesh colour,* broadly sinuate behind, and adnate with a decurrent tooth, ventricose, 4–6 mm. broad, rather distant, thin. Flesh *of st. purplish.* Spores white, elliptical, 5 × 3·5μ, with an oblique basal apiculus. Cystidia none. Solitary or subcaespitose. On trunks. Mixed woods. Sept.—Oct. Rare.

1218. **M. psammicola** B. & Br. Cke. Illus. no. 225, t. 186, upper figs.
ψάμμος, sand; *colo,* I inhabit.

P. 4–9 mm., *brown, becoming paler towards the margin,* hygrophanous, somewhat hemispherical, *sprinkled with very minute particles;* margin striate. St. 1–2 cm. × 1 mm., *white upwards, umber downwards,* firm, *rooting, wholly white pulverulent.* Gills *white,* shortly adnate, sinuate behind, in the form of a segment. Smell strong, but not nitrous. Amongst moss on a sandbank. Sept. Rare.

1219. **M. rugosa** Fr. Cke. Illus. no. 225, t. 186, lower figs.
Rugosa, wrinkled.

P. 2–6 cm., *cinereous, becoming pale, very tough,* membranaceous, campanulate, then expanded, at length rather plane, somewhat obtuse, more or less corrugated, *rugosely wrinkled, dry,* striate at the circumference. St. 3–8 cm. × 3–4 mm., *pallid, very cartilaginous,* rigid, tough, straight, *at length compressed,* smooth, *with a short, oblique, strigose root.* Gills *white, then grey, arcuato-adnate,* with a decurrent tooth, united behind in a collar, broad, ventricose, connected by veins, edge sometimes serrulate. Flesh *whitish,* thick at the disc. Spores white, oblong elliptical, 9–11 × 6–7μ, with a large central gutta. Basidia with 2–4-sterigmata. Cystidia hyaline, broadly clavate, or bludgeon-shaped, apex covered with short, simple, more rarely branched setae, 24–50 × 10–12μ. On stumps and old posts. July—Dec. Common. (*v.v.*)

1220. **M. sudora** Fr. *Sudor,* sweat.

P. 2–5 cm., *whitish, or yellowish, diaphanous,* submembranaceous, convex, umbonate, often irregular, striate, *viscid.* St. 6–11 cm. × 3–6 mm., *concolorous,* equal, firm, dry, smooth, rooting. Gills *white, then flesh colour, obtusely adnate,* broad, subdistant. Flesh *white,* thick at the disc. Spores white, broadly elliptical, 9–11 × 7–8μ. Cystidia "small, little protruding, 24 × 8μ, elliptical-vesiculose, mostly bluntly conical, somewhat ventricose below and often oblique" v. Hoehnel. On and near beech, and beech stumps. Aug.—Nov. Uncommon. (*v.v.*)

1221. **M. galericulata** (Scop.) Fr. (= *Mycena simillima* Karst.)
Galericulum, a cap.

P. 2–5 cm., *fuscous-livid, or changeable in colour, often becoming yellow, or rubiginous, sometimes white,* submembranaceous, conicocampanulate, then expanded, *striate to the umbo,* dry, smooth. St.

5–12 cm. × 3–5 mm., *concolorous*, often becoming yellowish, or rubiginous, *somewhat fragile, polished*, often curved, *smooth*; base strigose, fusiform-rooted. Gills *whitish, then flesh colour, adnate, decurrent with a tooth*, sometimes connected by veins. Flesh *greyish*, very thin at the margin. Spores white, broadly elliptical, 10–11 × 6–8μ, 1-guttulate. Basidia generally with 2-sterigmata only. Cystidia hyaline, broadly clavate, or bludgeon-shaped covered with short setae, 15–40 × 9–12μ. Caespitose, the numerous stems often glued together with villose down at the base. On stumps, trunks, and pollards. Jan.—Dec. Common. (*v.v.*)

var. **calopus** Fr. (= *Mycena inclinata* Fr. sec. Lange.) Fr. Icon. t. 80, fig. 2. καλός, beautiful; πούς, foot.

Differs from the type in the *fasciculate, fusiform chestnut coloured stem.* "P. viscid, gills cinereous becoming whitish, spores ovate globose, 8 × 6μ" Sacc. On stumps. Sept.—Nov. Uncommon. (*v.v.*)

1222. **M. simillima** Karst. *Simillima*, very like.

P. *livid*, or *dingy pallid*, conico-campanulate, even, dry, glabrous. St. fragile, polished, even, glabrous, base curved, rooting. Gills *white, very slightly tinged with rose*, emarginato-decurrent, crowded. Tree stumps. Sept. Rare.

1223. **M. polygramma** (Bull.) Fr. Cke. Illus. no. 228, t. 223. πολύς, many; γραμμή, a line.

P. 2–5 cm., *fuscous, cinereous, livid, or becoming light yellow, margin white when young*, submembranaceous, conical, then campanulate, subumbonate, striate, smooth, rarely pruinose; margin often toothed. St. 6–10 cm. × 2–4 mm., *silvery, livid, blue grey, or becoming azure blue, rigid*, tense and straight, equal, *longitudinally striato-sulcate*, base strigose-rooted. Gills *white, or pinkish*, attenuated behind, uncinate, subdistant, sometimes serrulate. Flesh *greyish*, very thin at the margin. Spores white, broadly elliptical, 9–12 × 6–8μ, 1–2–multi-guttulate. Basidia with 4-sterigmata. Cystidia hyaline, flask-shaped, base ventricose, apex prolonged and attenuated upwards, flexuose, simple or branched, 20–60 × 4–10 × 1·5–3μ at apex. Gregarious or solitary. On stumps and twigs. Aug.—Feb. Common. (*v.v.*)

1224. **M. inclinata** Fr. Cke. Illus. no. 234, t. 225, upper figs., as *Mycena alcalina* Fr. *Inclinata*, bent in.

P. 2–3 cm., *fuscous, livid fuscous, or bistre*, submembranaceous, *globose*, then campanulate, obtuse, rarely gibbous, at length expanded and depressed at the apex, smooth, *striate to the disc*, shining when dry; *margin at first white, exceeding the gills, delicately crenulate.* St. 6–10 cm. × 2–4 mm., *whitish*, or *brownish, becoming fulvous from the base upwards*, slightly attenuated upwards, twisted, flexuose, apex incurved at first, somewhat tough when young, then fragile, *pruinosely fibrillose, apex interruptedly striate*; base rooting, villose. Gills *whitish, greyish at the base*, sometimes pinkish, adnate, crowded, at length soft. Flesh *whitish, tawny in the st.*, thick at the disc. Spores white, subglobose, 8–10 × 6–8μ, 1-guttulate; basidia with 4-sterigmata. Cystidia clavate, apex covered with short setae, 30–40 × 9–16μ. Smell somewhat alkaline. Densely caespitose on oak stumps, and at the base of posts. Aug.—Dec. Not uncommon. (*v.v.*)

1225. **M. parabolica** Fr. Fr. Icon. t. 79, fig. 3, wrongly cited as t. 80, fig. 3, in the text. *Parabolica*, like a parabola.

P. 2–5 cm., *somewhat violaceous, disc black, margin whitish, or lilac*, submembranaceous, at first erect and oval, then *conical*, never expanded, moist, somewhat shining when dry, smooth, *striate to the disc.* St. 5–8 cm. × 2–3 mm., *whitish, or lilac, apex dark violaceous*, tense and straight, not very rigid, *white mealy when young*, smooth, dry; base thickened, bearded-rooting. Gills *white, greyish at the base*, adnate, ascending, subdistant, rarely connected by veins. Spores white, elliptical, 7–8 × 5–6μ. "Basidia 2-spored. Cystidia obovate, crowned with minute wart-like setae" Lange; "conical (not, or slightly ventricose), mostly sharp pointed, 40–60 × 10–20μ" v. Hoehnel. Gregarious, or caespitose. On needles and rotten wood, in coniferous woods. Sept.—Dec. Uncommon. (*v.v.*)

1226. **M. tintinnabulum** Fr. Fr. Icon. t 79, fig. 4, wrongly cited as t. 80, fig. 4, in the text. *Tintinnabulum*, a door bell.

P. 2–3 cm., *date brown, becoming pale, yellowish fuscous, azure blue, or whitish*, submembranaceous, *very tough, campanulato-convex*, then plane, smooth, *subviscid when moist.* St. 2–3 cm. × 1–2 mm., *pallid, very tough*, smooth; base shortly white-strigose. Gills whitish, *then cream, or flesh colour*, adnate, decurrent with a tooth, *horizontal*, narrow, very thin and crowded. Spores white, "elliptical, 9–10 × 5–7μ; basidia with 4-sterigmata. Cystidia on edge of gill subulate" Rick. For *Mycena tintinnabulum* Fr. sensu Schroet., v. Hoehnel and Lange give the following dimensions. Spores long-cylindrical, or ovate, 5–7 × 2·5–3μ. Cystidia vesiculose, obovate, subglobose, or cylindrical, often set with wart-like setae, 9–15μ across. Caespitose, or solitary. On fallen beech trunks. Oct.—April. Uncommon.

IV. St. *fragile*, dry, juiceless, fibrillose at the base, scarcely rooting, but not dilated nor inserted. P. hygrophanous. Gills *changing colour*, at length somewhat connected by veins. Usually strong scented, solitary and terrestrial, a few caespitose and lignicolous.

1227. **M. atroalba** (Bolt.) Fr. Bolt. Hist. Fung. t. 137. *Ater*, black; *alba*, white.

P. 2–3 cm., *bistre blackish, whitish at the margin*, submembranaceous, *conico-campanulate, obtuse*, smooth; *margin pellucidly striate.*

St. 7–10 cm. × 3–4 mm., *pallid, apex dark and occasionally pruinose, tense and straight, shining*, smooth, *base with a hairy, bulbous, swollen root*. Gills *white, becoming glaucous*, free, ventricose, crowded. Spores white, "oval, 12–14 × 7–8μ" Sacc. Cystidia "lanceolate-subulate, 75–105 × 15–20μ" Rick. Solitary, or gregarious. Amongst moss. Mixed woods. July—Nov. Uncommon. (*v.v.*)

1228. **M. dissiliens** Fr. Fr. Icon. t. 81, fig. 2. *Dissiliens*, flying apart.

P. 2–5 cm., *cinereous-fuscous, margin whitish*, submembranaceous, *very fragile*, acorn-shaped, *then conico-campanulate, sulcate to the disc*, pruinose; margin revolute. St. 4–5 cm. × 2–5 mm., *cinereous*, attenuated upwards from the *strigose base, somewhat incurved*, smooth, or pruinate, slightly striate under a lens, *split and breaking up into revolute flaps* when compressed or bent, often twisted. Gills *whitish, or cinereous at the base, rounded behind*, separating free, broader in front, soft, watery. Flesh *white, greyish in the st.*, thin at the margin. Spores white, elliptical, 7–8 × 4–5μ, multi-guttulate. Cystidia "globose-ovate, 20μ, crowned with a few, short, finger-like protuberances" v. Hoehnel. Smell weak. Amongst grass in woods and heaths, and on trunks. July—Nov. Uncommon. (*v.v.*)

1229. **M. atrocyanea** (Batsch) Fr. (= *Mycena nigricans* Bres. sec. Quél.) Cke. Illus. no. 231, t. 236, lower figs.
Ater, black; *cyanea*, dark blue.

P. 5–13 mm., *fuscous, then azure-blue-grey*, membranaceous, campanulato-convex, at length flattened, gibbous, *with an irregularly shaped, somewhat angular, wrinkled, obtuse, fuscous blackish umbo, deeply sulcate to the umbo, sprinkled with a white, evanescent pruina*. St. 3–5 cm. × 1–2 mm., *dark blue-black*, slightly attenuated from *the subbulbous base*, almost equal, rigid, fragile, smooth. Gills *white, grey at the base, attenuato-adnate, joined in a collar*, ventricose, distant. Flesh *greyish*, thin. Spores white, oblong, often apiculate at one end, 10–12 × 6–7μ. Cystidia subulate-fusiform, or cylindrical, apex acute, 2–3μ in diam., 80–100 × 10–12μ. Amongst pine needles and on stumps. Sept.—Nov. Uncommon. (*v.v.*)

1230. **M. pullata** Berk. & Cke. Cke. Illus. no. 232, t. 237.
Pullus, dark coloured.

P. 18 mm., *dark brown with a tinge of purple, disc almost black, becoming paler, sometimes with a glaucous bloom*, membranaceous, campanulate, obtusely umbonate, *sulcato-striate to the middle*. St. 7·5 cm. × 2 mm., *concolorous*; *base thickened*, whitish floccose, sometimes rooting. Gills *white*, adnexed, rather broad, scarcely crowded. Spores white, elliptical, 6 × 3μ. Smell slightly nitrous. Amongst dead leaves. Oct.—Nov. Uncommon. (*v.v.*)

1231. **M. cinerella** Karst. Cke. Illus. no. 264, t. 210, upper figs., as *Omphalia grisea* Fr. sec. Lange. *Cinerella*, somewhat ash colour.

P. 5–15 mm., *grey, or pallid greyish*, submembranaceous, campanulate, *entirely striate*. St. 2–4 cm. × 1–2 mm., *greyish white*, equal, base fibrillose. Gills *greyish white, broadly adnato-decurrent*. Spores elliptical, 7–10 × 4–6μ. Cystidia "ovate oblong, generally conical, ventricose below, apex obtusely conical, often rough, 50–60 × 12–16μ" v. Hoehnel; "globose, finely warted, not protruding" Pearson "in litt." Smell very strong, of meal. Woods, and pastures. Sept.—Nov. Not uncommon. (*v.v.*)

1232. **M. paupercula** Berk. (= *Mycena metata* Fr. sec. Quél.) Cke. Illus. no. 231, t. 236, upper figs. *Paupercula*, poor.

P. 2–5 mm., *pale ochraceous white*, becoming almost tawny with age, submembranaceous, obtusely conical, or hemispherical, *minutely innato-fibrillose*, sometimes translucidly striate. St. 1–2·5 cm. × 1 mm., *white*, curved, rooting, smooth; base thicker, villose. Gills *white*, at first free, then adnexed. Smell of new meal. Inside decayed stumps. July—Sept. Uncommon.

1233. **M. leptocephala** (Pers.) Fr. Pers. Ic. et Desc. t. 12, fig. 4. λεπτός, thin; κεφαλή, head.

P. 1–2·5 cm., *cinereous*, submembranaceous, campanulato-expanded, repand, umbonate, fragile, *sulcate, pruinose, opaque*. St. 4–6 cm. × 2–3 mm., *concolorous*, equal, *slightly striate*, opaque, dry. Gills *white cinereous, becoming white at the edge, emarginate*, connected by veins. Flesh *grey*, thin at the margin. Spores white, elliptical, 6–9 × 3–4μ, with a large central gutta. Cystidia "acute awl-shaped, somewhat fusiform, 60–70 × 10–14μ" Lange; "lanceolate, 60–100 × 10–18μ" Rick. Smell nitrous. Solitary. On trunks, and on the ground, especially in coniferous woods. Sept.—Oct. Uncommon. (*v.v.*)

1234. **M. alcalina** Fr. Fr. Icon. t. 81, fig. 3. *Alcalina*, alkaline.

P. 2–5 cm., *cinereous, fuscous, date brown, inclining to olivaceous, often tinged with yellow or pink*, submembranaceous, campanulate, obtusely umbonate, *deeply striate when moist, shining when dry*. St. 5–8 cm. × 2–4 mm., *normally yellow, often cinereous*, slightly firm, *rigid*, slippery when moist, shining when dry, smooth, base villose. Gills *glaucous white, or dark cinereous, becoming white and sometimes brown at the edge*, adnate, slightly ventricose, sometimes connected by veins, subdistant. Flesh whitish, thick at the disc. Spores white, broadly elliptical, 8–10 × 6–7μ. Basidia with 4-sterigmata. Cystidia hyaline, flask-shaped, or fusiform, base often ventricose; apex prolonged, obtuse or pointed, 35–45 × 10–18 × 4–5μ at apex. Smell strong, alkaline. Caespitose, rarely solitary. Coniferous stumps, trunks, and needles. Jan.—Dec. Common. (*v.v.*)

1235. **M. ammoniaca** Fr. Cke. Illus. no. 235, t. 238, upper figs. *Ammoniaca*, ammoniacal.

P. 1·5–2 cm., *fuscous, becoming blackish, varying cinereous, disc fuscous blackish, paler round the striate margin*, submembranaceous, acutely conical, papillate, then campanulate, *naked, discoid, opaque*. St. 2·5–5 cm. × 1–2 mm., *whitish*, slightly firm, equal, polished, dry, smooth; *base rooting*, strigose. Gills *whitish, or grey, edge whitish*, adnate, linear, distant. Flesh *greyish in the p., becoming whitish*, thin at the margin. Spores white, elliptical, 6–11 × 4–7μ, often 1–2-guttulate. Cystidia hyaline, flask-shaped, ventricose at the base; apex acute, obtuse or subglobose, 40–55 × 15–18 × 3–7μ at apex. Smell strong, alkaline. In troops on coniferous needles, and amongst short grass. July—Nov. Common. (*v.v.*)

1236. **M. metata** Fr. (= *Mycena paupercula* Berk. sec. Quél.) *Metata*, conical.

P. 1–2 cm., *cinereous and slightly striate when moist, opaque, whitish, and somewhat silky in appearance when dry*, submembranaceous, hemispherico-campanulate, obtuse, then plane, disc *papillate* or *somewhat umbilicate*, very hygrophanous. St. 5–7·5 cm. × 1–2 mm., *white, becoming cinereous, rarely yellowish, or flesh colour, soft-flaccid*, equal, smooth, base white fibrillose. Gills *whitish, or yellowish grey*, adnate, linear, subdistant. Spores white, elliptical, 6–8 × 4–5μ, with a large central gutta. Cystidia "obovate or pyriform, 12–19μ across, set with setulose warts" Lange; "spinulose, 30 × 12–15μ" Rick.; "conical, ventricose, 20–50 × 12–16μ" v. Hoehnel. Smell faintly alkaline. In pastures, and amongst short grass. Sept.—Nov. Common. (*v.v.*)

1237. **M. plicosa** Fr. Fr. Icon. t. 81, fig. 4. *Plicosa*, folded.

P. 1·5–3 cm., *fuscous cinereous, opaque when dry*, membranaceous, *fragile*, campanulate, then expanded, broadly and obtusely umbonate, *deeply lineato-sulcate, plicate with the distant furrows*, often split. St. 2·5 cm. × 2–3 mm., *grey, then fuscous, rigid but fragile*, equal, smooth; *base abrupt*, white villose. Gills *grey, at length whitish pruinose*, adnate, *thick, distant*, connected by veins Spores oblong-elliptical, 9–11 × 4–5μ, "minutely punctate" Quél. Cystidia "on edge of gill clavate, 40–45 × 12–18μ, with finger-like appendages" Rick. On bare soil in woods. Sept.—Oct. Uncommon. (*v.v.*)

1238. **M. cinerea** Massee & Crossl. *Cinerea*, ash colour.

Entirely grey. P. 1·5–2 cm., submembranaceous, subgibbous, or obtuse, soon expanded, paler and silky when dry; margin striate. St. 5–7 cm. × 2 mm., cylindrical, smooth; base white, downy. Gills adnate, subdistant, *edge pale*, mealy with the spores. Flesh *greyish*, *thin.* Spores white, elliptical, 8 × 5μ, 1–2-guttulate. Cystidia fusiform. Smell of radishes. Pastures, and amongst short grass. Sept.—Oct. Uncommon. (*v.v.*)

1239. **M. peltata** Fr. *Peltata*, having a shield.

P. 1·5–3 cm., *black fuscous when moist, grey when dry,* membranaceous, convex, *soon exactly plane, disc orbicular, even, flat,* rather umbilicate than umbonate; *margin up-turned and becoming black when dry, very closely striate.* St. 4–5 cm. × 2 mm., *livid,* equal, rigid, somewhat fragile, often flexuose, smooth. Gills *grey, paler at the edge,* adnate, with a small decurrent tooth, ventricose. Flesh *greyish,* thick at the disc. Spores white, elliptical, 8–10 × 4–6μ. Cystidia "broadly lanceolate, 40 × 12–15μ, sometimes slightly capitate" Rick. Smell none, or alkaline. Woods, and heaths. Aug.—Oct. Uncommon. (*v.v.*)

1240. **M. consimilis** Cke. Cke. Illus. no. 1150, t. 1186.
Consimilis, entirely similar.

P. 2·5–3 cm., *cinereous, umbo darker,* membranaceous, conically campanulate, umbonate, striate to the middle, smooth, opaque; *margin soon upturned,* at length splitting. St. 2–3 cm. × 2 mm., *paler than the p.,* attenuated upwards, often compressed below, rather rigid, dry, smooth. Gills *cinereous,* adnexed, or nearly free, *linear,* 1–2 mm. broad, scarcely crowded. Flesh *white,* thick at the disc. Spores white. Amongst grass. Sept. Rare.

1241. **M. aetites** Fr. (= *Mycena umbellifera* (Schaeff.) Quél.) Fr. Icon. t. 81, fig. 5. ἀετίτης, the eagle-stone.

P. 1–2 cm., *fuscous grey, becoming pale,* membranaceous, campanulate, then convex, *sulcate to the broad, obtuse, prominent umbo,* hygrophanous, smooth; *extreme margin becoming black.* St. 4–5 cm. × 2 mm., *whitish, becoming fuscous downwards, shining,* often compressed, unequal, *fragile,* smooth. Gills *white,* grey at the sides, adnate, *subuncinate,* thin, *at first cohering in the form of a collar, beautifully reticulated* by veins, linear, subdistant. Spores white, elliptical, often with an oblique apiculus, 8–10 × 5–6μ. Cystidia hyaline, flask-shaped, ventricose at the base; apex prolonged, acute, or obtuse, 25–50 × 6–8 × 1μ at apex. Smell alkaline, or none. Taste bitterish, or obsolete. Amongst moss, and short grass in woods and upland pastures. June—Nov. Uncommon. (*v.v.*)

1242. **M. stannea** Fr. Fr. Icon. t. 82, fig. 2. *Stannea*, tin-colour.

P. 3–4 cm., *grey when moist, tin colour and silky shining when dry, hygrophanous,* membranaceous, campanulate, then flattened, obsoletely umbonate, fragile, often rimose, smooth, pellucidly striate when moist. St. 5–8 cm. × 2–4 mm., *grey, becoming pale, slightly rigid,* not very fragile, sometimes compressed, smooth, *shining*. Gills *whitish grey*, adnate, *with a small decurrent tooth*, *connected by veins*, scarcely crowded. Flesh *pallid*, thin at the margin. Spores white, elliptical, 8–10 × 4–5μ. Cystidia "flask-shaped-fusiform, 45–50 × 10–18μ, blunt, sometimes capitate" Rick. Smell, like fresh trout, or none. Amongst grass in woods. June—Oct. Not uncommon. (*v.v.*)

1243. **M. vitrea** Fr. Fr. Icon. t. 82, fig. 1. *Vitrea*, glassy.

Very fragile. P. 1–2·5 cm., *fuscous, then livid or bluish grey, membranaceous*, campanulate, obtuse, *entirely lineato-striate*, *opaque*, smooth, dry. St. 5–10 cm. × 1 mm., *whitish, hyaline*, equal, smooth, *glistening*, *striate under a lens*, base fibrillose. Gills *whitish*, adnate, linear, subdistant. Flesh *fuscous in the p.*, very thin. Spores white, "oblong oval, 10 × 4·5μ. Cystidia nearly globose, with short spines in circles, 45μ broad" v. Hoehnel. In woods amongst *Sphagnum*. Sept.—Oct. Uncommon.

1244. **M. tenuis** (Bolt.) Fr. Cke. Illus. no. 237, t. 160, lower figs. *Tenuis*, thin.

Very fragile, caespitose, white. P. 1–2 cm., *hyaline, or becoming fuscous white*, very membranaceous, campanulato-convex, obtuse, *lineato-striate*, smooth; *margin slight, beautifully fringed in a crenate manner*, as if appendiculate with the fragments of the veil. St. 7–8 cm. × 1-2 mm., *hyaline, base becoming yellowish, membranaceous, pellucid*, tense and straight, smooth. Gills adnate, with a small decurrent tooth which is often obsolete, linear, rather thick, comparatively *distant*, soft. Spores white, subglobose, 4 × 3μ, with a large central gutta. Woods, and pastures. Sept.—Oct. Uncommon. (*v.v.*)

V. St. filiform, scarcely a line thick (and not more), *flaccid*, somewhat tough, rooting, dry, juiceless, commonly very long in proportion to the p. Gills paler at the edge and *changing colour*. *Very slender, tense and straight, terrestrial, and amongst moss, inodorous, solitary. P. fuscous, becoming somewhat pale, not hygrophanous*, in the last species *orange*.

1245. **M. filopes** (Bull.) Fr. Cke. Illus. no. 238, t. 161, upper figs. *Filum*, a thread; *pes*, foot.

P. 1–1·5 cm., *livid fuscous, or livid grey, rarely whitish*, very membranaceous, conical then campanulate, obtuse, striate, dry, smooth. St. 7–9 cm. × 1–2 mm., *livid, or becoming fuscous, filiform*, tense and straight, *flaccid, not very tough*, equal, *rooting with a long pilose tail*, filled with a watery juice when in full vigour. Gills *white, at length grey at the base*, free, or only reaching the st., ventricose, *or lanceolate*, crowded. Flesh whitish, thin at the margin. Spores white, elliptical,

8–10 × 4–5μ. Basidia with 2–4-sterigmata. Cystidia on gill edge, hyaline, crowded, forming a compact layer, obovate, or pyriform, minutely setulose towards the apex, 20–48 × 16–32μ. Amongst dead leaves in deciduous woods, hedgerows and plantations. Aug.—Jan. Common. (*v.v.*)

1246. **M. amicta** Fr. (= *Mycena Iris* Berk. sec. Quél.) Fr. Icon. t. 82, fig. 3. *Amicta*, clothed.

P. 6–12 mm., *green, bluish grey, or livid,* membranaceous, conico-campanulate, *slightly pellucidly striate to the disc, covered with fugacious pruina.* St. 6–8 cm. × 1–2 mm., *livid, equal, flexile, covered with a delicate white pruina*; base straight, or with a long tortuose root, smooth. Gills *grey, edge paler, free,* or only reaching the st., *linear, narrow, crowded.* Flesh *fuscous,* thin at the disc. Spores white, elliptical, 7–8 × 4–5μ, 1–2-guttulate. Basidia with 2-sterigmata. Cystidia hyaline, flask-shaped, apex acutely conical, 20 × 6μ. Amongst mosses in woods, and pastures. Sept.—Nov. Not uncommon. (*v.v.*)

1247. **M. Iris** Berk. (= *Mycena amicta* Fr. sec. Quél.) Cke. Illus. no. 238, t. 161, lower figs. *ἶρις*, the rainbow.

P. 10–15 mm., *grey, becoming yellowish,* membranaceous, hemispherical, obtuse, *covered with blue, evanescent fibrils, viscid*; margin denticulate. St. 4–9 cm. × 2–3 mm., *grey,* equal, *covered with evanescent blue fibrils,* which are often only apparent at the base. Gills *grey, edge becoming pale,* free, or slightly adnexed, linear, edge sometimes denticulate. Flesh *greyish,* thick at the disc. Spores white, elliptical, 6–8 × 3–4μ. Cystidia "crowded, conical, threadlike, obtuse, 20 × 4–4·5μ" v. Hoehnel. On fir stumps, and sticks. Sept.—Nov. Not uncommon. (*v.v.*)

var. **caerulea** Rea. *Caerulea*, azure blue.

Differs from the type in the *pure blue colour of the p., in the absence of blue fibrillae on the p. and st., and in the white pulverulent apex of the st.* Spores white, elliptical, with an oblique apiculus, 8–9 × 5μ. Inside a hollow tree. May. Rare. (*v.v.*)

1248. **M. urania** Fr. *Οὐρανία*, the Heavenly One.

P. 6–10 mm., *dark blue, then becoming violaceous, and at length pallid lilac, rarely becoming fuscous,* membranaceous, campanulate, then hemispherical, obtuse, *striate,* dry, smooth. St. 5–8 cm. × 1–2 mm., *dark blue, then becoming somewhat azure blue,* equal, *flexile,* flaccid, smooth, slightly rooting; base white floccose. Gills *white, uncinato-adnate,* thin. Amongst alder leaves, *Jungermanniae,* and twigs in damp woods. July—Sept. Rare.

1249. **M. plumbea** Fr. (= *Omphalia plumbea* (Fr.) Rick.)
Plumbea, lead colour.

P. 1–3 cm., *cinereous lead colour, covered with a white pruina, sometimes bluish ash colour*, membranaceous, convex, *then plane*, obtuse, *sulcate*. St. 7–10 cm. × 2 mm., *becoming cinereous*, equal, fragile, *pulverulent*, apex hyaline, base white strigose. Gills *concolorous*, adnate, *horizontal*. Flesh *whitish*, very thin at the margin. Spores white, elliptical, 10–11 × 5–6μ, 2–multi-guttulate. Mossy pastures, and amongst leaves. Sept.—Nov. Uncommon. (*v.v.*)

1250. **M. debilis** Fr. Quél. Jur. et Vosg. I, t. 14, fig. 6.
Debilis, weak.

P. 4–6 mm., *whitish livid, or somewhat flesh colour, becoming fuscous, withered and corrugated when dry*, membranaceous, *very thin*, campanulate, then convex, obtuse, *striate when moist*, even when dry, smooth. St. 5–10 cm. × 1 mm., *concolorous, capillary-filiform, weak, lax*, base fibrillose. Gills *whitish, or concolorous, broadly adnate*, rather broad, subdistant. Spores white, elliptical, 10–12 × 5μ. Cystidia "thin, lanceolate, 60–75 × 9–12μ, much projecting and making the gills appear rough" Rick. Amongst dead leaves in woods, and hedgerows. Sept.—Nov. Not uncommon. (*v.v.*)

1251. **M. vitilis** Fr. Cke. Illus. no. 240, t. 189, fig. 2. *Vitilis*, plaited.

P. 6–10 mm., *fuscous, or livid, becoming pale, or whitish*, membranaceous, *conical*, then campanulate, *papillate*, striate to the middle, dry, smooth. St. 7–15 cm. × 1–2 mm., *livid, filiform, rooted, tense and straight*, rigid, *tough*, easily flexile, smooth, *shining*. Gills *whitish, or grey, edge becoming whitish, attenuato-adnate*, ascending, *linear*, thin. Flesh *white*, very thin at the margin. Spores white, broadly elliptical, apiculate at base, 9–12 × 5–7μ. Cystidia "very like those of *Mycena polygramma*, free portion hair-shaped, c. 10μ long, 2μ broad, basal part slightly thickened" Lange. Amongst dead leaves in deciduous woods. Sept.—Feb. Common. (*v.v.*)

var. **amsegetes** Fr. *Amsegetes*, field by the roadside.

Differs from the type in the *obsoletely umbonate p., and the shorter, and thicker st.* Meadows, and roadsides.

1252. **M. collariata** Fr. *Collariata*, possessing a collar.

P. 1–2 cm., *fuscous, becoming pale, often greyish white, becoming fuscous only at the disc*, membranaceous, campanulate, *then convex*, subumbonate, striate, rigid when dry, smooth. St. 5 cm. × 1–2 mm., *grey, becoming pale, filiform, tough*, dry, smooth, slightly striate under a lens. Gills *hoary-whitish, or obsoletely flesh colour*, adnate, *joined in a collar behind*, thin, crowded. Spores white, "elliptical, 8–10 ×

4–6μ" Berk. Cystidia on gill edge "lanceolate subulate, 50–60 × 10–13μ" Rick. In woods, amongst grass, and on oak bark. Oct.—Nov. Uncommon.

1253. **M. speirea** Fr. (= *Omphalia speirea* (Fr.) Quél.) Fr. Icon. t. 78, fig. 2. *σπεῖρα*, a coil.

P. 4–10 mm., *pallid cinereous, or whitish variegated with fuscous striae, umbo fuscous*, membranaceous, conico-convex, *then plane, at length depressed at the disc*, smooth, sometimes pruinose. St. 5 cm. × 1 mm., *white, base becoming fuscous and ending in a tail-like fibrillose root, tough, filiform*, equal, smooth, *shining*. Gills *shining white*, adnate, then deeply decurrent, *distant*, the alternate ones shorter. Flesh *white*, thin at the margin. Spores white, "globose, 6μ, or broadly elliptical, 6–9 × 4–6μ. Cystidia numerous, cylindrical, conical, generally sharp pointed, full of small oil globules, on the edge of the gill, 60 × 20μ, on the surface, 85 × 20μ, or cylindrical, conical, with protruding points, 40 × 20μ" v. Hoehnel. Woods, and mossy trunks. Aug.—Oct. Uncommon.

1254. **M. tenella** Fr. Cke. Illus. no. 241, t. 190, middle figs. *Tenella*, rather tender.

Entirely white, or livid flesh colour, caespitose. P. 5–12 mm., membranaceous, very tender, campanulato-convex, obtuse, *pellucid*; margin slightly striate. St. 2 cm. × 1 mm., filiform, soft, smooth, base villose. Gills *white, then flesh colour*, uncinate, very thin, crowded. Flesh *white*, very thin. Spores white, elliptical, 8–9 × 4–5μ, minutely punctate. Cystidia "on the edge, in several rows, globose, 16–23μ, crowned with numerous, short spines" v. Hoehnel. On felled trunks, and twigs in wood heaps. Aug.—Dec. Common. (*v.v.*)

1255. **M. acicula** (Schaeff.) Fr. (= *Mycena coccinea* (Scop.) Sacc.) *Acicula*, a small pin.

P. 2–10 mm., *vermilion-orange*, membranaceous, campanulate, then convex, *with a very small slightly fleshy umbo*, striate, smooth, shining. St. 2–5 cm. × 1 mm., *bright yellow, becoming pale*, filiform, rooting, smooth, shining, apex somewhat pruinose, base villose. Gills *yellow, becoming whitish at the edge, or wholly white, rounded-adnexed*, almost free, comparatively broad, *ventricose*, somewhat ovate, distant, the alternate ones shorter. Flesh *reddish in the p.*, very thin. Spores white, oblong-fusiform, 9–12 × 2–4μ, attenuated at the base, 1-guttulate. Cystidia hyaline, flask-shaped, base ventricose or fusiform; apex acute, obtuse, or subglobose; 25–30 × 8–12 × 2–4μ at apex. On dead leaves and twigs in woods and hedgerows. May—Dec. Common. (*v.v.*)

VI. St. and gills exuding a *milky, usually coloured juice* when broken. St. dry, rooting.

1256. **M. hematopus** (Pers.) Fr. Fr. Icon. t. 83, fig. 1.
αἷμα, blood; *πούς*, foot.

P. 2–4 cm., *greyish, or white flesh colour with a purplish tinge, disc bistre,* fleshy-membranaceous, conical, then campanulate, *obtuse,* smooth; *margin denticulate,* slightly striate. St. 5–10 cm. × 2–5 mm., *white, greyish, flesh colour, or violaceous, becoming cinereous,* rigid, fragile, recurved, *white pruinose,* becoming smooth, *containing a blood-like juice,* base strigose. Gills *white, then flesh colour, or violaceous,* adnate, often with a small decurrent tooth, the alternate ones shorter. Flesh *turning blood red,* thick at the disc. Spores white, broadly elliptical, 10 × 6μ, rounded at both ends, with a large central gutta. Cystidia "conical, sharp pointed, often ventricose below, 45 × 15μ, or rigid, long, threadlike and pointed above, ventricose below, 40–45 × 12μ, contents colourless, seldom reddish" v. Hoehnel. Caespitose. On trunks, and stumps, especially birch. Aug.—Dec. Common. (*v.v.*)

var. **marginata** Lange. *Marginata,* bordered.

Differs from the type in the *dark edge of the gills.* Cystidia "drawn out to a sharp point, below the middle fusiformly inflated, 10–17μ broad, their free portion 45μ long, contents pale brownish red" Lange. On stacked birch logs. Oct. Probably not uncommon. (*v.v.*)

1257. **M. cruenta** Fr. Fr. Icon. t. 83, fig. 2. *Cruenta,* bloody.

P. 6–20 mm., *bay brown, or fuscous, then red, becoming pale,* submembranaceous, conical, then campanulate, obtuse, striate, smooth. St. 5–8 cm. × 2 mm., *paler than the p.,* slightly firm, tense and straight, *smooth, containing a dark red juice,* base villose-rooted. Gills *whitish, or pinkish,* adnate, linear, crowded. Flesh *dark red,* thin. Spores white, broadly elliptical, 9–10 × 6μ, with a large central gutta. Cystidia "on gill edge only, contents granular, 35 × 9 × 2μ at apex" v. Hoehnel. Generally solitary. Pine woods. Sept.—Oct. Uncommon. (*v.v.*)

1258. **M. sanguinolenta** (A. & S.) Fr. Fr. Icon. t. 83, fig. 3.
Sanguinolenta, bloody.

P. 4–20 mm., *pallid reddish, becoming fuscous, umbo and striae commonly darker, very membranaceous,* campanulato-convex, or hemispherical, papillate, smooth. St. 5–12·5 × 1–2 mm., *pallid, flaccid,* weak, almost capillary, moderately tough, smooth, *containing a pale reddish juice,* base subfibrillose. Gills *whitish, or paler than the p., edge black purple,* adnate, linear, subdistant, the alternate ones shorter. Flesh *reddish, becoming whitish*, very thin. Spores white, elliptical, pip-shaped, or pyriform, 8–9 × 4–6μ, 1-2-guttulate. Cystidia filled with a reddish juice, or colourless, flask-shaped, often prolonged at the base; apex long, conical, pointed, 35–50 × 6–8 × 1–2·5μ at apex. Gregarious or solitary. Amongst dead leaves in woods, especially coniferous woods, and in hedgerows. May—Dec. Common. (*v.v.*)

1259. **M. crocata** (Schrad.) Fr. *Crocata*, saffron yellow.

P. 1–2·5 cm., *olivaceous, cinereous, or shining white, umbonate disc reddish*, submembranaceous, conical, then campanulate, smooth; margin striate. St. 7–12·5 × 2–3 mm., *saffron-blood-colour*, especially towards the rooting, creeping, fibrillose base, slightly attenuated upwards, apex whitish, *containing a saffron-blood juice* that readily stains the rest of the fungus. Gills *white*, attenuato-adnexed, broader in front, subventricose, subdistant. Flesh *saffron-blood-colour, yellowish in the centre of the st.*, thick at the disc. Spores white, broadly elliptical, or pip-shaped, 9–11 × 6–7μ, punctate. Basidia with 4-sterigmata. Cystidia "club-shaped, or somewhat pyriform set with minute wart-like setae, apex occasionally with a hair-shaped appendix" Lange. On dead leaves, and twigs in woods, especially beech. Sept. Uncommon. (*v.v.*)

1260. **M. galopus** (Pers.) Fr. Cke. Illus. no. 244, t. 207.
γάλα, milk; πούς, foot.

P. 6–15 mm., *fuscous, or greyish, the indistinct umbo darker*, membranaceous, conical, then campanulate, striate, smooth, or pruinose. St. 5–11 cm. × 1–2 mm., *fuscous, or grey, apex white*, firm, somewhat fragile, smooth or pruinose; base thickened, white villose and rooting, *containing a milk white juice*. Gills *white, sometimes becoming glaucous*, attenuated behind, slightly adnexed, broader towards the margin of the p. Flesh *white*, very thin. Spores white, oblong elliptic, 12–14 × 6–7μ. Cystidia hyaline, subulate, fusiform; apex acute, simple or forked, 30–90 × 10–12 × 2·5–3μ at apex. Amongst dead leaves, on twigs and stumps. Woods, hedgerows, and wood piles. July—Jan. Common. (*v.v.*)

var. **alba** Fl. Dan. *Alba*, white.

Differs from the type in being *entirely white*. In woods amongst leaves. Sept.—Nov. Not uncommon. (*v.v.*)

var. **nigra** Fl. Dan. (= *Mycena leucogala* Cke.) Cke. Illus. no. 1151, t. 653, as *Mycena leucogala* Cke. *Nigra*, black.

Differs from the type in the *dark colour of all its parts*. In woods, on stumps, twigs, and leaves. Aug.—Nov. Common. (*v.v.*)

VII. St. juiceless, *glutinous, or viscid.* Gills at length decurrent with a tooth.

1261. **M. epipterygia** (Scop.) Fr. Cke. Illus. no. 245, t. 208, upper figs. ἐπί, upon; πτερύγιον, a little wing.

P. 1–3 cm., *cinereous, grey, or yellow, becoming whitish,* membranaceous, campanulate, then more or less expanded, striate, *covered with a viscid, separable pellicle; margin often denticulate.* St. 5–8 cm. × 1–2 mm., *yellow, sometimes cinereous, pallid, or whitish, covered with a viscid separable pellicle,* equal, tough, often flexuose; base rooted, white fibrillose. Gills *white,* adnate with a decurrent tooth, straight, or slightly arcuate, little crowded. Flesh *white,* very thin. Spores white, oblong elliptic, 8–11 × 4–5μ, 1-guttulate. Cystidia only on gill edge, subglobose, 10–13μ, setulose, soon fugacious. "None" Lange. Smell none, or of rancid fat. Woods, pastures, and on leaves, and twigs. Aug.—Dec. Common. (*v.v.*)

1262. **M. viscosa** (Secr.) R. Maire. *Viscosa,* viscid.

P. 2–3 cm., *whitish, pearl grey, then greyish brown, and finally reddish brown,* hygrophanous, membranaceous, campanulate, then expanded, striate, *covered with a viscid separable pellicle.* St. 5–8 cm. × 1–2 mm., *citron yellow, or golden,* equal, *viscid, apex whitish;* base white fibrillose. Gills *whitish, then greyish or flesh colour,* adnate with a more or less decurrent tooth, slightly arcuate, narrow, little crowded. Flesh *whitish, becoming reddish brown with age,* thin. Spores white, shortly elliptic, 8–12 × 6–8μ, 1-multi-guttulate. Cystidia none. Smell of rancid fat. On needles, and rotten stumps in coniferous woods. Sept.—Oct. Uncommon. (*v.v.*)

1263. **M. epipterygioides** Pearson.
Epipterygia εἶδος, like the species *M. epipterygia.*

P. 1–2 cm., *greenish yellow, disc darker,* membranaceous, persistently hemispherical, depressed at the centre, viscid pellicle separable; margin striate, or sulcate, often crenate. St. 5–8 cm. × 1–2 mm., *greenish, usually with reddish stains at the base,* cylindrical, or compressed, hollow, *viscid.* Gills *white, then delicate greenish yellow,* adnate, with a decurrent tooth, subdistant, with intermediate shorter ones. Spores white, broadly elliptical, 9–10 × 7·5–8μ, contents granular. Basidia 30 × 6·5μ, with two prominent sterigmata. Cystidia on edge of gill brush-like; on gill face none. Damp places in pine woods. Oct.—Nov. Common. (*v.v.*)

1264. **M. plicato-crenata** Fr. (= *Mycena plicata* (Schaeff.) Quél.) Fr. Icon. t. 84, fig. 2. *Plicata,* folded; *crenata,* crenate.

P. 6–10 mm., *white, becoming yellow,* membranaceous, conical, subumbonate, *very sulcato-plicate,* somewhat viscid; *margin crenate.* St. 4 cm. × 1–2 mm., *pallid reddish,* filiform, smooth, viscid. Gills *white,* adnate, with a small decurrent tooth, narrow, attenuated behind, distant. Flesh *of st. light yellow*. Spores white, elliptical, "9–12 × 6μ" Sacc. Amongst moss on heaths, and in coniferous woods. Sept.—Oct. Uncommon. (*v.v.*)

1265. **M. clavicularis** Fr. Fr. Icon. t. 84, fig. 1.
Clavicula, a small key.

P. 1–3 cm., *whitish, or light yellowish, becoming fuscous*, membranaceous, campanulate, *then convex and umbonate*, at length depressed, striate, *dry*. St. 5–8 cm. × 1–2 mm., *whitish, or yellowish*, equal, tough, smooth, *slightly viscid*, base fibrillose. Gills *whitish, adnate, subdecurrent*, often connected by veins. Flesh *fuscous*, very thin. Spores white, elliptical, 6 × 4μ, "cylindric-lanceolate, 10–12 × 3–4μ. Cystidia on gill edge vesiculose-bottle-shaped, 30–36 × 9–11μ, rarely lanceolate without a head" Rick. Woods, and damp places. Sept.—Oct. Uncommon. (*v.v.*)

1266. **M. pelliculosa** Fr. Cke. Illus. no. 246, t. 191, upper figs.
Pelliculosa, having a thin skin.

P. 1–2·5 cm., *fuscous, then grey*, membranaceous, campanulate, then convex, *obtuse, lineato-striate to the middle, covered with a viscid, separable pellicle*. St. 2–6 cm. × 2–3 mm., *white-livid, becoming fuscous, viscid*, rigid, tense and straight, smooth, apex somewhat thickened. Gills *glaucous white, adnate*, alternate, very distant, *fold-like, joined in a collar behind*, decurrent. Spores white, boat-shaped, 8–9 × 5–6μ, 2-guttulate. Cystidia "none" Rick. On heaths, and in heathy woods. Sept.—Oct. Not uncommon. (*v.v.*)

1267. **M. vulgaris** (Pers.) Fr. Cke. Illus. no. 246, t. 191, lower figs.
Vulgaris, common.

P. 6–10 mm., *fuscous, or cinereous, sometimes whitish with the papilla fuscous, often rufescent when old*, submembranaceous, campanulate, then convex, *disc depressed, papillate*, slightly striate, *viscid pellicle separable*. St. 2·5–5 cm. × 2 mm., *cinereous, very viscid*; base rooting and white strigose. Gills *white, or grey*, uncinato-adfixed, then decurrent, thin. Flesh *whitish*, thick at the disc. Spores white, elliptical, 6–9 × 3–4μ, 1-guttulate. Cystidia "globose, 10–12μ, provided with numerous short, spiny, protuberances" v. Hoehnel. Gregarious. Coniferous woods. July—Dec. Common. (*v.v.*)

1268. **M. citrinella** (Pers.) Fr. (= *Mycena tenella* (Batsch) Sacc.) Pers. Icon. et Desc. t. 11, fig. 3. *Citrinella*, lemon-coloured.

P. 4–10 mm., *lemon yellow, disc often darker*, membranaceous, campanulate, then hemispherical and flattened, *striate, slightly viscid*. St. 2–3 cm. × 1 mm., *lemon yellow, filiform, viscid when moist*, base villose. Gills *shining white*, uncinate, moderately broad, distant.

Flesh *white*, thin at the margin. Spores white, "broadly elliptical, 8–8·5 × 7μ, or 6–8 × 4μ (Britz.)" Sacc. Cystidia "filiform-clavate, or fusiform, 30–40 × 5–6μ" Rick. Gregarious. On pine needles, and wood. Oct. Rare.

var. **candida** Fr. *Candida*, shining white.

Differs from the type in being *shining white, and becoming yellow when dry.*

1269. **M. rorida** Fr. Quél. Jur. et Vosg. t. 4, fig. 4. *Rorida*, bedewed.

Entirely white, or with a greyish tinge, but varying with the p. becoming yellow. P. 3–8 mm., membranaceous, conico-campanulate, then convex, *dry*, sulcate when moist, even when dry; *margin crenate.* St. 1–3 cm. × 1 mm., *filiform, covered over with a thick, fluid, hyaline gluten, base inserted.* Gills arcuate, decurrent, distant, the alternate ones shorter. Flesh *whitish*, very thin. Spores white, oblong-elliptic, 8–12 × 4–5μ. Cystidia "in dense, large groups, slightly conical, often somewhat ventricose, 18–25 × 6–9μ, or threadlike, obtuse, slightly ventricose, 22–25 × 6–7μ" v. Hoehnel. On twigs, in woods, hedgerows, and wood heaps. May—Dec. Common. (*v.v.*)

VIII. St. dry, rootless, *the base naked, and dilated into a disc*, or strigose and *swollen into a little bulb. Tender, solitary, becoming flaccid.*

1270. **M. stylobates** (Pers.) Fr. Cke. Illus. no. 248, t. 249, fig. A. *στῦλος*, a pillar; *βάσις*, a pedestal.

Entirely white, sometimes grey. P. 4–10 mm., membranaceous, campanulate, then convex, pellucidly striate, striae often dichotomous, *generally sprinkled with spreading hairs.* St. 2·5–5 cm. × 1 mm., *filiform*, equal, smooth, dry, *arising from a round, striate, white-villose disc.* Gills *free, wholly separate behind, ventricose*, broader in front, distant, alternate. Flesh *white*, very thin. Spores white, elliptical, 4 × 2μ ("7–9 × 3·5–4·5μ" Sacc.). Cystidia "on the edge of the gills hair-shaped" Schroet. On twigs, and leaves. June—Nov. Common. (*v.v.*)

1271. **M. dilatata** Fr. Trans. Brit. Myc. Soc. vi, t. 7. *Dilatata*, spread out.

Wholly white. P. 5–10 mm., membranaceous, convexo-plane, obtuse, smooth; margin striate. St. 10–15 × 1 mm., filiform, straight, *arising from a convex, smooth, glabrous, orbicular disc.* Gills ·5–1 mm. wide, sublinear, *attached to a free collar behind.* Flesh white, thin. Spores white, oblong, obtuse at both ends, 7–8 × 3·5μ. Cystidia hyaline, clavate, obtuse, or produced into an acute point, 70–80 × 5–7μ. Dead twigs and leaves in woods and hedgerows. Sept.—Dec. Not uncommon. (*v.v.*)

1272. **M. tenerrima** Berk. (= *Mycena setosa* (Sow.) Quél.) Berk. Outl. t. 6, fig. 6. *Tenerrima*, very delicate.

Pure white. P. 2–3 mm., very tender, convex, *frosted with minute granules*. St. 1–2·5 cm. × 1 mm., flexuose, *pilose, arising from a minute, pubescent disc*. Gills *free, ventricose*, distant, unequal. Flesh *white*, very thin. Spores white, subglobose, 4–5 × 3–4μ. Cystidia "50–55 × 10μ" Sacc. Dead twigs, fallen branches, and felled trunks. Aug.—April. Common. (*v.v.*)

1273. **M. discopus** Lév. (= *Mycena setosa* (Sow.) Quél.) Cke. Illus. no. 249, t. 192, middle figs. *δίσκος*, a disc; *πούς*, foot.

Entirely shining white. P. 2–4 mm., membranaceous, conical, obtuse, *mealy-pulverulent*. St. 1–2 cm. × 1 mm., very tender, *mealy-pulverulent, inserted with a small hairy bulb*. Gills *adnate*, few, *fold-like*, very distant. Flesh *white*, very thin. Spores white, globose, 3μ. On twigs, and dead herbaceous stems. Sept.—Dec. Not uncommon. (*v.v.*)

1274. **M. saccharifera** B. & Br. Cke. Illus. no. 249, t. 192, top figs. *Saccharon*, sugar; *fero*, I bear.

Whitish, everywhere beset with shining granules. P. 4 mm., hemispherical. St. 4 × 1 mm., filiform, *fixed at the base by a few flocci*. Gills *arcuato-decurrent*, 8–9, very distant, rather thick, *margin and surface granulated*. Spores white, globose, 3μ. On bramble, rose, furze, and nettle stalks. Nov.—March. Uncommon.

var. **electica** Bucknall. Cke. Illus. no. 248, t. 249, fig. C, as *Mycena electica* Bucknall. *Electica*, choice.

Differs from the type in the *sulcate p., and adnate gills*. On dead furze, and sticks.

1275. **M. pterigena** Fr. Fr. Icon. t. 85, fig. 4. *πτέρις*, a fern; *γίγνομαι*, I am born.

Entirely rose colour. P. 2–6 mm., globose, then campanulate, 4 mm. high, very tender, pellucidly striate, obtuse, sometimes at length umbilicate. St. 1–7·5 cm. × 1 mm., tense and straight, or flexuose, smooth; *base bulbous, white strigose*. Gills adnate, broad, distant, *edge darker*. Flesh *pinkish*, very thin. Spores white, pip-shaped, 9–12 × 4–6μ. Cystidia "ovate, or subglobular with numerous, minute, erect setae, contents pinkish" Lange. On dead leaves, and dead fern stems. Sept.—Nov. Not uncommon. (*v.v.*)

IX. St. very thin, *inserted* (*i.e.* growing on other plants without a root, or tubercle, or flocci at the base), dry. Gills *adnate*, uncinate with a small decurrent tooth. *Very tender*, becoming flaccid as soon as the sun touches them.

1276. **M. corticola** (Schum.) Fr. Fr. Icon. t. 85, fig. 2.

Corticola, growing on bark.

P. 4–10 mm., *blackish, becoming azure blue, fuscous, or cinereous*, thin, hemispherical, obtuse, *at length slightly umbilicate*, pellucid, *sulcate*, sometimes flocculoso-pruinate. St. 1–3 cm. × 1 mm., *paler than the p.*, sometimes furfuraceous and incurved. Gills *paler than the p.*, adnate, with a small decurrent tooth, *broad, somewhat ovate*, distant. Flesh *concolorous*, very thin. Spores white, globose, 9–10μ. Cystidia "club-shaped, set with short warts and occasionally some few hair-shaped appendices" Lange; "on edge of gill clavate, 30–40 × 9–10μ, without brush-like head" Rick. On living trunks of deciduous trees. June—Jan. Common. (*v.v.*)

1277. **M. hiemalis** (Osbeck) Fr. Fr. Icon. t. 85, fig. 1.

Hiemalis, pertaining to winter.

P. 3–7 mm., *whitish, flesh colour, rufescent, rarely azure blue, or fuscous*, membranaceous, campanulate, *disc darker, slightly umbonate*; *margin* striate. St. 2–3 cm. × 1 mm., *white*, ascending, incurved, *pubescent downwards*. Gills *whitish, or flesh colour*, adnate, and uncinate, *narrow*, linear. Flesh *whitish*, thin. Spores white, globose, 8–9μ, or broadly elliptical, 10–12 × 8–10μ. Basidia with two long, curved sterigmata. Cystidia on gill edge only, hyaline, cylindrical; apex obtuse, or acute, 20–34 × 3–8μ. On trunks in woods. Sept.—March. Common. (*v.v.*)

1278. **M. codoniceps** Cke. Cke. Illus. no. 1149, t. 952, fig. B.

κώδων, a bell; *caput*, head.

P. 1–2 mm., *umber*, submembranaceous, campanulate, scarcely expanding, 2–3 mm. high, *sulcate, sprinkled with short, erect hairs*. St. 4–10 × 1 mm., *umber, apex whitish*, attenuated downwards. Gills *white*, adnate, *linear*, not crowded. Spores white, elliptical, 5 × 2·5–3μ. Gregarious. On tree-fern stems. June. Rare.

1279. **M. setosa** (Sow.) Fr. (= *Mycena tenerrima* Berk.; *Mycena discopus* Lév. sec. Quél.) Cke. Illus. no. 251, t. 193, fig. 1.

Setosa, bristly.

Entirely white. P. 1–2 mm., very tender, *often becoming fuscous*, hemispherical, obtuse, smooth. St. 1·5–3 cm. × ·5 mm., filiform, *covered with distant spreading hairs*. Gills distant. Spores white, pip-shaped, 7–8 × 3–4μ. Amongst dead leaves, especially beech. Sept.—Nov. Not uncommon. (*v.v.*)

1280. **M. capillaris** (Schum.) Fr. Fr. Icon. t. 84, fig. 6.

Capillaris, hair-like.

Entirely white. P. ·5–2 mm., very tender, but *tough, like a small pin's head*, then campanulate, rarely at length umbilicate, slightly striate when moist. St. 2–7 cm. × ·5 mm., filiform, flexuose, flaccid, *base inserted*, rarely girt with radiating fibrils, apex becoming fuscous. Gills adnate, *few*, broad, equal in length. Spores white, "obovate-lanceolate, 7·5–9 × 3–3·7μ, or 9–11 × 3–3·7μ. Cystidia crowded, obovate globular, set with wart-like setae" Lange. On dead leaves, especially beech. Sept.—Dec. Common. (*v.v.*)

1281. **M. juncicola** Fr. Fr. Icon. t. 85, fig. 6.
Juncus, a rush; *colo*, I inhabit.

P. 2–3 mm., *rufescent, or blood red*, convex, striate, smooth. St. 12 × ·5 mm., *fuscous*, filiform, inserted, *smooth*. Gills *white, or yellowish white*, adnate, distant. On dead rushes in bogs, and twigs. June—July. Rare.

Spores pink.

Nolanea Fr.

(*Nola*, a little bell.)

Pileus fleshy, or submembranaceous, regular; margin straight, at first adpressed to the stem. Stem central, cartilaginous. Gills adnate, adnexed, or sinuato-adnate. Spores pink, angular, elliptical, or globose, smooth, or rough, continuous. Growing on the ground, rarely on wood; solitary, or gregarious.

*Gills grey or fuscous. P. dark coloured, hygrophanous.

1282. **N. pascua** (Pers.) Fr. Boud. Icon. t. 96. *Pascua*, of pasture.

P. 2–8 cm., *fuliginous when moist, hoary, or becoming pale fawn when dry*, membranaceous, conical, then campanulate and more or less expanded, *striate* when moist, *silky shining when dry*. St. 3–8 cm. × 2–6 mm., *pallid fuliginous*, or *silvery tinged with smoke colour*, equal, or compressed, *soft, silky-fibrous, striate*. Gills *grey*, or *whitish-fuliginous*, sprinkled with the rosy spores, very much attenuated behind, almost free, *crowded*, thin, ventricose, or rather broader and obtuse towards the margin. Flesh *whitish, fuliginous when moist*. Spores pink, angular, oblong, 10–13 × 7–8μ. Woods, pastures, and roadsides. May—Nov. Common. (*v.v.*)

var. **umbonata** Quél. Quél. Jur. et Vosg. t. 6, fig. 5, as *Nolanea mammosa* Fr. *Umbonata*, having an umbo.

Differs from the type in the *umbonate, bay p., and fibrillosely striate, silvery st.* Heaths, woods, and pastures. May—Oct. Not uncommon. (*v.v.*)

1283. **N. proletaria** Fr. (= *Nolanea staurospora* Bres. sec. Quél.; *Nolanea cetrata* Schroet.) Boud. Icon. t. 95. *Proletaria*, poor.

P. 1·5–4 cm., *grey, disc umber and villose*, submembranaceous, scissile, campanulate, then expanded, *very obtuse*, striate at the margin when moist. St. 4–10 cm. × 2–5 mm., *fuliginous-grey, dirty white, or pale yellow*, equal, slightly thickened at the base, *very fragile, fibrillose*, striate. Gills *fuliginous, then greyish*, becoming rosy, separating, free, *subdistant*, watery. Flesh *pale, deeper coloured at the periphery*, very thin. Spores pink, *very angular, generally subquadrangular* or stellate, 10–11 × 8–10μ. Basidia generally with 2-sterigmata only. Woods, and pastures. June—Oct. Not uncommon. (*v.v.*)

1284. **N. versatilis** Fr. Trans. Brit. Myc. Soc. III, t. 12.
Versatilis, variable.

P. 2·5–5 cm., *livid aeruginous, fuscous* when dry, submembranaceous, convex, then expanded, obtuse, or obtusely umbonate, *shining*. St. 3–5 cm. × 2–4 mm., *greyish white, with a silvery sheen when dry*, rigid. Gills *grey*, then sprinkled with the rosy spores, adnate, ventricose, 3–4 mm. broad, widest in front. Flesh dark, fuscous. Spores pink, angular, oblong, 9–10 × 7μ, 1-guttulate. Cystidia "abundant, clavate, 45–70 × 9–12μ, with darkish olive, granular contents" Rick. Heaths, pastures, and lawns. Sept.—Oct. Uncommon. (*v.v.*)

1285. **N. Babingtonii** Blox. Cke. Illus. no. 363, t. 377, upper figs.
Professor C. C. Babington, the eminent botanist.

P. 10–15 mm., *cinereous*, shining like silk, *adorned with dark brown, fasciculate fibrils* which are free at one end, conico-campanulate, disc rather squamulose. St. 2·5 cm. × 2 mm., *clothed with dark brown down*, equal, somewhat strigose, slightly wavy. Gills *cinereous, darker at the base*, adnate, ventricose, distant, *glittering with little points*. Spores pink, angular, elliptical, 7–9μ. Woods. Oct.—Nov. Rare.

1286. **N. araneosa** Quél. Trans. Brit. Myc. Soc. III, t. 12.
Araneosa, full of spiders' webs.

P. 1–2 cm., *dark grey*, membranaceous, campanulate, *fibrillosely silky*. St. 3·5–4 cm. × 2 mm., *grey, with a greyish fugacious cortina*, fragile, fibrillose. Gills *greyish-bistre*, then dusted with the rosy spores, adnate, 2–3 mm. wide. Flesh *dark, then yellowish*. Spores pink, angular, often pentagonal, oblong, 13–16 × 8–9μ, 1-guttulate. Coniferous woods and under conifers. July—Oct. Uncommon. (*v.v.*)

1287. **N. strigosissima** Rea. Trans. Brit. Myc. Soc. VI, t. 7.
Strigosissima, very rough haired.

P. 4–8 mm. broad, 3–5 mm. high, *reddish brown, or ferruginous*, somewhat fleshy, conical, *densely clothed with erect, reddish brown strigose hairs*; hairs elongate, apex blunt, septate, 450–600 × 15–20μ; margin incurved. St. 1·5–2·5 cm. × 1 mm., *concolorous*, equal, slightly thickened at the base, *densely clothed with similar hairs*. Gills *brown, becoming cinereous*, pruinose with the spores, adnate, 1 mm. broad.

Flesh *concolorous, becoming cinereous,* thin, firm. Spores pink, oblong, angular, 15–17 × 7–8μ, often apiculate, 2-guttulate. Basidia pyriform, or broadly clavate, 36–40 × 15–18μ, with 4-sterigmata. Cystidia on edge of gill sparse, fusiform, or lanceolate, 60–70 × 10–12μ; apex acute, thin walled. Cells of the cuticle of the p. pyriform, 25μ in diam. Old pine logs. Oct. Uncommon. (*v.v.*)

1288. **N. mammosa** (Linn.) Fr. Bres. Fung. Trid. t. 81.
Mammosa, having large breasts.

P. 2–8 cm., *umber, or fuliginous when moist, isabelline-silky when dry,* submembranaceous, conico-campanulate, *papillate,* striate. St. 5–15 cm. × 1·5–3 mm., *fuliginous-livid,* or *yellowish grey,* rigid, very cartilaginous, equal, sometimes compressed; *apex thickened, white-mealy; base enlarged, white tomentose.* Gills *grey,* then hoary-rose-colour, adnexed, separating-free, ventricose, subdistant. Flesh *concolorous.* Spores pink, angular, oblong, 9–11 × 6–7μ, 1-guttulate. Smell none, or like rancid meal. Woods, pastures, and lawns. Feb.—Oct. Not uncommon. (*v.v.*)

1289. **N. papillata** Bres. Bres. Fung. Trid. t. 82, fig. 1.
Papillata, having a nipple.

P. 2–3 cm., *fuscous bay, somewhat cinnamon when dry,* submembranaceous, convexo-subcampanulate, then expanded, *papillate,* striate. St. 3–5 cm. × 2 mm., *concolorous,* shining, apex obsoletely white-mealy, base white-tomentose. Gills *livid white, then fuscous flesh colour,* sinuato-adnate, somewhat crowded. Flesh *concolorous.* Spores pink, angular, oblong, 8–11 × 6–7μ, 1-guttulate. Smell none, or pleasant. Pastures, and lawns. Sept.—Oct. Not uncommon. (*v.v.*)

1290. **N. juncea** Fr. *Juncea*, like a rush.

P. 1·5–2 cm., *umber-fuliginous,* then *livid* when dry, hygrophanous, submembranaceous, conical, then expanded, disc *somewhat umbilicate and somewhat squamulose, radiately striate.* St. 7–8 cm. × 1–2 mm., *fuscous,* then *livid fuscous,* cartilaginous, equal, round, or compressed. Gills *grey,* ascending, adnexed, separating, subdistant. Spores pink, "angular, globose, 11–13μ" Quél. In *Sphagnum* swamps, and in woods. Oct. Rare.

var. **cuspidata** Fr. Fr. Icon. t. 99, fig. 2. *Cuspidata*, having a point.

Differs from the type in *the papillato-cuspidate pileus.*

1291. **N. fulvo-strigosa** B. &. Br. *Fulvus*, tawny; *strigosa*, strigose.

P. 18 mm., *grey,* conical, 12 mm. high, *slightly wrinkled.* St. 5 cm. × 2 mm., *reddish,* furfuraceo-squamulose, *clothed at the base with rigid red hairs.* Gills *grey,* adnate. Spores pink, 13 × 9μ. Woods. Sept. Rare.

1300. **N. rubida** Berk. Cke. Illus. no. 367, t. 340, lower figs.
Rubida, reddish.

P. 8 mm., *white or greyish, at length with a pale ruddy tinge*, membranaceous, convex, at length umbilicate, *finely silky*. St. 3–4 × 1 mm., *white, or greyish*, thickest above, *minutely silky*. Gills *whitish*, then rose colour, adnate, broad, ventricose, attenuated behind, with frequently a more or less distinct tooth, sometimes subdecurrent. Spores pink, "elliptical, 4–5 × 3μ, smooth. Cystidia none" Massee. Smell of new meal. Among grass in a conservatory. March—Nov. Rare.

1301. **N. rhodospora** Br. & W. G. Sm. ῥόδον, rose; σπορά, seed.

P. 2·5 cm., *sooty-fibrillose, or rufescent pilose*. St. 3 cm. × 2 mm., *white, subbulbous*. Gills *salmon or rose*, sinuate or free. Spores pink. On earth, and wooden borders in stoves. May—Sept. Rare.

1302. **N. minuta** Karst. *Minuta*, small.

P. 1–1·5 cm., *pallid fuscous*, paler when dry and shining, convex, sometimes umbilicate, striate up to the umbilicus. St. 3–5 cm. × 1·5 mm., *pallid fuscous*. Gills pallid, adnate. Spores pink, globose, angular, 7–9μ. On peaty soil in woods. Sept. Uncommon.

Spores ochraceous, or ferruginous.

Galera Fr.

(*Galerus*, a cap.)

Pileus fleshy, or submembranaceous; margin straight, at first adpressed to the stem. Stem central, cartilaginous. Gills adnate, or adnexed. Spores ochraceous, cinnamon, or ferruginous, elliptical, pruniform, or almond-shaped; smooth; with a germ-pore, rarely continuous. Cystidia generally present. Growing on the ground.

*P. conico-campanulate, hygrophanous, rather even, when dry dotted with soft particles; st. tense and straight; gills ascending, inserted at the top of the cone, somewhat crowded. Veil none.

1303. **G. hapala** Fr. (= *Bolbitius apalus* (Fr.) Quél.) Fr. Icon. t. 127, fig. 1, as *Galera apala* Fr. ἁπαλός, tender.

P. 1–6 cm., *livid becoming pale, quite white and shining when dry*, submembranaceous, conico-campanulate, then campanulate, obtuse, regular, smooth, hygrophanous. St. 10–15 cm. × 2–4 mm., *shining white, rather fragile*, slightly and equally attenuated upwards, very straight, *clothed with dense, erect, white flocci*, base sometimes subbulbous. Gills *whitish, then bright ochraceous*, adnexed, then free, *very narrowly lanceolate*, thin, crowded. Flesh *concolorous*, very thin. Spores tawny, pruniform, "12–14 × 7–8μ" Sacc. Rich grassy places. Sept. Rare.

var. **sphaerobasis** v. Post. *σφαῖρα*, a globe; *βάσις*, base.

Differs from the type in the *smooth stem, and bulbous base*. Grassy places.

1304. **G. lateritia** Fr. Fr. Icon. t. 127, fig. 2. *Lateritia*, brick-red.

P. 1–2·5 cm., *ferruginous*, or *pale yellowish, ochraceous when dry*, hygrophanous, membranaceous, *acorn-shaped, then campanulate*, obtuse, smooth; margin slightly and densely striate when moist. St. 7–11 cm. × 2 mm., *whitish*, attenuated upwards, tense and straight, *very fragile*, even, *white pruinose*. Gills *cinnamon, or tawny ferruginous, adnexed*, then free, ascending, very narrow, almost adpressed to the st. Flesh *white*, thin. Spores ochraceous, "elliptical, with a flattened germ-pore, 12–15 × 8–10μ, smooth. Cystidia on edge of gill basidia-like; apex prominent, small, stalked, capitate" Rick. Rich pastures, and grassy places. June—Oct. Rare.

1305. **G. tenera** (Schaeff.) Fr. Cke. Illus. no. 518, t. 461, upper figs. *Tenera*, tender.

P. 1–2 cm., *pallid ferruginous, becoming pale when dry*, hygrophanous, submembranaceous, *conico-campanulate*, smooth, slightly striate when moist, opaque, somewhat atomate, or pulverulent. St. 7·5–10 cm. × 2 mm., *concolorous*, fragile, equal, or when larger thickened downwards, *tense and straight, somewhat shining, striate upwards, pulverulent*. Gills *cinnamon, adnate, then free*, ascending, *linear*, somewhat crowded. Flesh *yellowish, slightly reddish in the st.*, thin. Spores pale ferruginous, elliptical, with a flattened germ-pore, 14–15 × 8–9μ. Cystidia on gill edge flask-shaped, apex subglobose, or obtuse; 18–20 × 9–10 × 4–6μ at apex. Woods, pastures, roadsides and gardens. April—Dec. Common. (*v.v.*)

1306. **G. pilosella** (Pers.) Rea. Cke. Illus. no. 518, t. 461, lower figs., as *Galera tenera* Schaeff. var. *pilosella*. *Pilosella*, hairy.

P. 1·5–2 cm., *ferruginous, becoming paler when dry*, hygrophanous, submembranaceous, *hemispherical, densely covered with short, erect hairs*. St. 4–5 cm. × 2 mm., *concolorous*, equal, *densely covered with short, erect hairs*. Gills *ferruginous, margin paler*, adnexed, then free, *ventricose*, 4–5 mm. wide, subdistant. Flesh *of p. whitish, concolorous in the stem*. Spores pale ferruginous, elliptic oblong, 13–15 × 8μ. Basidia broadly clavate, 20–25 × 12–14μ, with 2–4-sterigmata. Cystidia on gill edge only, sparse, fusiform, apex globose; 20–22 × 9–10 × 4–5μ at apex. Amongst grass in pastures and on rotten wood. March—Oct. Uncommon. (*v.v.*)

1307. **G. flexipes** Karst. *Flexus*, bent; *pes*, foot.

P. 1–1·5 cm., *ferruginous, ochraceous when dry*, fleshy membranaceous, campanulate, obtuse, pellucidly striate when moist. St. 2–

3 cm. × 1·5 mm., *pallid, becoming ferruginous*, equal, flexuose, *white fibrillose*, apex white pruinose. Gills *pallid, becoming ferruginous*, adnate, crowded, oblong. Spores 10–12 × 5–6μ. Amongst grass and rotten wood. May—Sept. Rare.

1308. **G. siliginea** Fr. *Siligo*, a kind of very white wheat.

P. 1–2 cm., *pallid grey*, membranaceous, *globoso-campanulate*, then convex and expanded, unequal, smooth; margin often flexuose. St. 5–7 cm. × 1–2 mm., *whitish, or pallid*, equal, *often flexuose, sprinkled with white pruina*. Gills *pallid ochraceous, broadly adnate*, broadly linear, somewhat crowded. Flesh *pallid*, thin. Spores ochraceous, broadly elliptical, 10–12 × 6–7μ. Cystidia "stalked, capitate" Rick. Pastures and roadsides. Sept.—Oct. Uncommon. (*v.v.*)

1309. **G. campanulata** Massee. Cke. Illus. no. 1174, t. 1156, as *Galera siliginea* Fr. *Campanula*, a little bell.

P. 1–2 cm., *deep cinnamon, almost white and atomate when dry*, persistently campanulate, subacute, smooth, hygrophanous, slightly rugulose. St. 5 cm. × 1–2 mm., *pallid, base darker, whitish when dry*, equal, or slightly incrassated at the base, flexuose, almost glabrous. Gills *tawny cinnamon*, adnate, 2 mm. broad, rather crowded. Flesh *white when dry*, thin. Spores ochraceous, elliptical, ends rather acute, 12 × 7μ. Smell *strong*. Gregarious. Road scrapings, and dry places by roadsides. Sept. Uncommon. (*v.v.*)

1310. **G. ovalis** Fr. Cke. Illus. no. 519, t. 462. *Ovalis*, oval.

P. 2–3 cm., *ferruginous, becoming yellow when dry*, submembranaceous, *ovato-campanulate*, obtuse, smooth; margin straight and adpressed to the st. St. 7–10 cm. × 2–3 mm., *concolorous*, equal, *tense and straight, slightly striate, very fragile*. Partial veil here and there in the form of a ring, fugacious. Gills *ferruginous, somewhat free, very ventricose and broad, crowded*, subdeliquescent. Flesh *reddish*, thin. Spores ferruginous, elliptical, 10 × 6μ. Pastures, and on dung. Sept.—Oct. Not uncommon. (*v.v.*)

1311. **G. antipus** (Lasch) Fr. Fr. Icon. t. 128, fig. 2.
ἀντί, opposite; πούς, foot.

P. 1–3 cm., *deep ochraceous, pale almost white when dry*, hygrophanous, fleshy, campanulate, then expanded, disc prominent, smooth. St. 2·5–8 cm., *paler than the p.*, tense and straight, equal, or bulbous at the ground level, *then continued into a long, tortuose, smooth, tail-like root, apex white-mealy*. Gills *light yellowish ochraceous, then cinnamon*, almost free, attenuated behind, semi-lanceolate, crowded. Flesh *white when dry*, thick at the disc. Spores cinnamon, "nearly angular-lemon-shaped, 8–10 × 6–7μ. Cystidia on edge of gill, basidia-like-pyriform, apex prominent, small, stalked, capitate, stalk very short, head 4–5μ, base 12 × 9μ" Rick. Pastures, bare soil in gardens, and on dung. March—Sept. Uncommon. (*v.v.*)

1312. **G. conferta** (Bolt.) Fr. Bolt. Hist. Fung. t. 18.
Conferta, crowded.

P. 2–3 cm., *fuscous, fuscous ochraceous when dry*, hygrophanous, submembranaceous, acutely conico-campanulate, *fragile*, striate, smooth, *often glittering with micaceous particles*. St. 2–5 cm. × 1–2 mm., *whitish, or cream colour*, very fragile, silky, shining, *naked, attenuated at the base into a long root, striate*, apex mealy. Gills *white, then fuscous ochraceous*, slightly adnexed, then free, subdistant. Flesh *whitish*, thin. Spores fuscous ferruginous, "ochraceous, pruniform, 10μ" Quél. Very crowded, subcaespitose. Stoves, and on tan. Nov.

1313. **G. spicula** (Lasch) Fr. Quél. Jur. et Vosg. I, t. 7, fig. 5, as *Naucoria furfuraceus* Pers. *Spiculum*, a little sharp point.

P. 5–15 mm., *brown ochre*, membranaceous, conico-campanulate, then expanded, hygrophanous, smooth, striate when moist, flocculose when dry and atomate. St. 2–3 cm. × 2–3 mm., *white*, equal, thickened at the base, firm, *densely covered with white flocci*. Gills *ochraceous, then cinnamon*, adnate, ventricose, 1·5–2 mm. broad. Flesh *concolorous, whitish in the st.*, very thin. Spores cinnamon, elliptical, 6–8 × 4μ, "with an apical germ pore. Cystidia stalked-capitate; head 8–9μ, stalk 3–4 × 3–4μ, base 18–20 × 15–18μ" Rick. Coconut fibre trunks, and fallen leaves. Nov. Uncommon. (*v.v.*)

1314. **G. spartea** Fr. σπάρτος, esparto grass.

P. 5–12 mm., *watery ferruginous, or cinnamon, tan when dry*, hygrophanous, membranaceous, campanulato-convex, then expanded, obtuse, *pellucidly striate when moist*, smooth. St. *pale tawny, date brown at the base*, tense and straight, equal, *smooth, polished*, flexile, diaphanous. Gills *darker than the p., wholly adnate*, somewhat linear, then plane, crowded. Flesh *concolorous, becoming paler*, thin, very fragile. Spores ferruginous, "subelliptical, 6–8 × 3–4μ, smooth. Cystidia stalked-capitate, base subglobose, 15 × 12–15μ, head 5–6μ" Rick. Amongst moss on heaths, pastures, and on burnt soil. Sept.—Oct. Not uncommon. (*v.v.*)

1315. **G. pygmaeoaffinis** Fr. Fr. Icon. t. 128, fig. 1.
Affinis, allied to *Naucoria pygmaea*.

P. 2–4 cm., *subfuscous, or honey colour, then tan*, fleshy membranaceous, campanulate, then flattened, *dry, delicately and—under a lens—conspicuously reticulato-wrinkled*, almost rugged or minutely granular. St. 5–7·5 × 2 mm., *shining white, fragile*, equal, *often striate and pruinose at the apex*. Veil scarcely any. Gills *clay-ochraceous, then ferruginous ochraceous*, just reaching the st., *almost free*, thin, crowded. Flesh *concolorous*, very thin at the margin. Spores pale ochraceous, "elliptical, with a flattened germ-pore, 15–18 × 8–12μ, smooth. Basidia 2-spored. Cystidia on edge of gill pyriform, apex prominent, stalked, small, capitate, stalk 3–4 × 1μ, head round, 3–4μ, base 9μ broad" Rick. Grassy places at the base of trees, heaths, thickets, and cucumber house. July—Nov. Uncommon.

**P. membranaceous, campanulate, striate, smooth, hygrophanous, even when dry, opaque, slightly silky; st. thin, lax, flexile; gills broadly and planely adnate, broad, somewhat denticulate; cortina very fugacious. Slender, growing amongst moss.

1316. **G. vittaeformis** Fr. Cke. Illus. no. 522, t. 464, upper figs.

Vitta, a fillet; *forma*, shape.

P. 1–2·5 cm., *date brown, tawny, or reddish*, membranaceous, conical, then hemispherical, obtuse, rarely papillate, pellucid, disc even, smooth; margin *striate*, often delicately villose. St. 4–7·5 cm. × 1–2 mm., *rubiginous, opaque*, equal, *somewhat straight*, smooth, *or sometimes pubescent or pruinose, slightly striate under a lens*. Veil scarcely conspicuous. Gills *watery cinnamon, then ferruginous*, adnate, *ventricose*, subdistant. Flesh *concolorous*, very thin. Spores pale ferruginous, "almond-shaped, 11–15 × 7–9μ, rough. Cystidia lanceolate, 50–60 × 10–12μ, with a long, blunt point" Rick. Amongst moss, and on burnt ground in pastures. May—Nov. Uncommon.

1317. **G. rubiginosa** (Pers.) Fr. Fr. Icon. t. 128, fig. 3, as var. *major*.

Rubiginosa, rusty.

P. 6–30 mm., *cinnamon, or honey colour, tan colour when dry*, hygrophanous, membranaceous, campanulate, obtuse, *striate throughout*, smooth. St. 5 cm. × 1–2 mm., *bay brown, or dark ferruginous*, equal, tough, flaccid, *shining*, smooth or pubescent under a lens. Gills *ochraceous*, adnate, ascending, rather broad, but almost linear. Flesh *concolorous, becoming pale*, thin. Spores ferruginous, elliptical, 10 × 5μ. Woods, heaths, and pastures. Sept.—Oct. Not uncommon. (*v.v.*)

1318. **G. hypnorum** (Schrank) Fr. Cke. Illus. no. 523, t. 465.

Hypnum, a moss.

P. 6–12 mm., *ochraceous pale yellowish, or watery cinnamon, tan when dry*, hygrophanous, membranaceous, campanulato-convex, *often papillate at the umbo*, or obtuse, *lineato-striate* except at the disc, smooth. St. 5 cm. × 1–2 mm., *slightly tawny, lemon yellow, or ochraceous*, equal, *flexuose, lax*, smooth, *apex pruinose*. Gills *cinnamon tawny*, adnate, *broad, ventricose, distant*, often connected by veins, *edge flocculose*. Flesh *yellowish*, thin. Spores ferruginous, almond-shaped, 11–15 × 6–8μ. Cystidia fusiform, ventricose, 50–65 × 15–17 × 5–7μ at apex. Woods, heaths, and pastures. May—Nov. Common. (*v.v.*)

var. **bryorum** (Pers.) Fr. *Bryum*, a moss.

Differs from the type in its *larger size, and rather horny papilla.* Woods, heaths, and hedgerows. Sept.—Oct. Not uncommon. (*v.v.*)

var. **sphagnorum** (Pers.) Fr. *Sphagnum*, a moss.

Differs from the type in being *twice or thrice as large, and in the long, subfibrillose tawny st.* Bogs, and amongst *Sphagna* in woods. June—Oct. Uncommon. (*v.v.*)

1319. **G. mniophila** (Lasch) Fr. Cke. Illus. no. 524, t. 466, upper figs. *μνίον*, moss; *φίλος*, loving.

P. 1–1·5 cm., *fuscous light yellowish, almost clay colour when dry*, membranaceous, campanulate, *almost papillate, striate*, disc even. St. 4–7·5 cm. × 2 mm., *yellow*, equal, flexile, *fibrillose, apex mealy*, base floccose. Gills *light yellow ochraceous, then often fuscous clay colour*, obtusely adnate, plano-ascending, broad, subdistant. Flesh *whitish*, thick at the disc. Spores ochraceous, oblong elliptical, 10–12 × 6μ. Cystidia "on edge of gill cylindrical-filiform, 30–36 × 3–4μ" Rick. Amongst mosses especially *Mnium*. Sept.—Nov. Uncommon. (*v.v.*)

1320. **G. minuta** Quél. Quél. Jur. et Vosg. III, t. 1, fig. 5. *Minuta*, little.

P. 2–3 mm., *ochraceous flesh colour, or chamois-bistre*, membranaceous, campanulate, glabrous, striate. St. 1 cm. × 1 mm., *tawny, shining*, smooth, *arising from an arachnoid white pellicle.* Gills *cream bistre*, adnate, triangular, *edge minutely fringed under a lens.* Spores ochraceous, pruniform, 6μ. In troops. Amongst moss, and on the ground. Sept.—Oct. Rare.

***P. submembranaceous, veil manifest, superficial, separating, at the first (chiefly round the margin) silky, and squamulose.

1321. **G. pityria** Fr. *πίτυρον*, bran.

P. 2·5 cm., *lurid, or becoming ferruginous, pallid tan when dry, fleshy-membranaceous*, campanulate, then expanded, obtuse, smooth, *viscid*; margin appendiculate with the fugacious, partial veil, at length striate. St. 5–6 cm. × 4–6 mm., *silvery-shining, firm*, cartilaginous, but at length splitting into fibrils, tough, equal, *smooth*, rarely fibrillose, *apex white pulverulent.* Gills *watery cinnamon, then ferruginous, slightly adnexed*, ascending, *crowded.* Spores ferruginous, "almond-shaped, 12–13 × 8–9μ, verrucose. Cystidia on edge of gill filiform-clavate, 36–45 × 4–7μ" Rick. Damp, frondose woods. Oct.—Nov. Rare.

1322. **G. ravida** Fr. Cke. Illus. no. 525, t. 467, fig. A. *Ravida*, greyish.

P. 1–4 cm., *of a peculiar greyish colour, dirty ochraceous when dry*, fleshy membranaceous, campanulate, then hemispherical, moist, *somewhat slightly viscid*, very hygrophanous, *somewhat silky when dry*, margin appendiculato-toothed with the white veil when young. St. 4–7·5 cm. × 2 mm., *pallid, becoming somewhat yellow, but silvery shining, very fragile*, ascending, *or twisted*, equal, fibrillosely striate, apex somewhat pruinose. Gills *ochraceous saffron, or pale yellowish, somewhat free, broad, ventricose, distant*. Flesh *white*, thick at the disc. Spores ochraceous, "subfusiform-elliptical, 8–9 × 4–5μ, smooth" Rick. Gregarious. Amongst chips, or rotten wood. Sept.—Oct. Uncommon.

1323. **G. mycenopsis** Fr. Fr. Icon. t. 129, fig. 1.

Mycena ὄψις, like a Mycena.

P. 6–20 mm., *pallid honey colour*, slightly fleshy membranaceous, subglobose, then campanulate, *at length convexo-plane, obtuse*, or gibbous with a broadly elevated disc, naked at the disc, striate and *silky to the middle with superficial, white, villose down*; margin often clothed with little white scales the remains of the veil. St. 5–7·5 cm. × 2–4 mm., *yellowish, white silky with adpressed, villose down*, attenuated upwards, straight or undulated, *soft*, apex obsoletely pruinose, or slightly furfuraceous, base white villose. Gills *pallid*, adnexed, then free, so ventricose at the middle as almost to be triangular, distant. Flesh *greyish in the p., whitish in the st.*, thick at the disc. Spores deep ochraceous, elliptical, 9–13 × 5–8μ. Cystidia bottle-shaped, apex often globose, base ventricose, 46–52 × 15–18 × 6–10μ at apex. *Sphagnum* swamps and in woods. Aug.—Oct. Uncommon. (*v.v.*)

1324. **G. Sahleri** Quél. Trans. Brit. Myc. Soc. III, t. 13. Sahler.

P. 4–8 mm., *tawny chestnut, honey colour when dry, disc brighter coloured*, membranaceous, campanulate, *often acutely conical*, smooth, hygrophanous, striate; margin at first covered with silky, fugacious fibrils. St. 1–3 cm. × 1–2 mm., *amber coloured*, shining, filiform, fragile, fibrillose. Gills *cream colour, then tawny ochraceous*, adnate, 1 mm. broad, crowded. Flesh *yellowish*, very thin. Spores tawny ochre, oval, 9–11 × 6–7μ, with an apical germ-pore. On mossy stumps, especially fir. May—Sept. Uncommon. (*v.v.*)

Spores purple, or fuscous.

Psathyra Fr.

(ψαθυρός, fragile.)

Pileus fleshy, or submembranaceous, regular; margin straight, at first adpressed to the stem. Stem central, cartilaginous. Gills adnate, adnexed, or free. Spores purple, fuscous, or cinereous purple; elliptical, oval, or oblong elliptical; smooth; with an apical germ-pore. Cystidia present. Growing on the ground, or on wood; solitary, or caespitose.

P. sarcocephala (Fr.) Quél. = **Psilocybe sarcocephala** Fr.
P. canobrunnea (Batsch) Quél. = **Psilocybe canobrunnea** (Batsch) Fr.
P. spadicea (Fr.) Quél. = **Psilocybe spadicea** Fr.
P. cernua (Fl. Dan.) Quél. = **Psilocybe cernua** (Fl. Dan.) Fr.
P. foenisecii (Pers.) Quél. = **Psilocybe foenisecii** (Pers.) Fr.

I. P. conico-campanulate, gills ascending, adnexed, often free. St. tense and straight. Veil none.

1325. **P. elata** Massee. (=*Psathyra conopilea* Fr. var. *superba* (Jungh.) Cke.) Cke. Illus. no. 1185, t. 1158, as *Psathyra conopilea* Fr. var. *superba* Jung. *Elata*, tall.

P. 2·5–5 cm., *dark clear brown, pale ochraceous and minutely atomate when dry,* submembranaceous, *obtusely campanulate,* very symmetrical, smooth. St. 10–17·5 cm. × 5–6 mm., *snow white, silky shining,* slightly and uniformly attenuated upwards, straight, *rigid,* smooth. Gills *whitish, then purplish brown, broadly adnate,* 3–4 mm. broad, soft, crowded. Flesh *brownish,* becoming *whitish,* thin at the margin. Spores brown with a purple tinge, elliptical, 18 × 8–9μ. Amongst grass in hedge banks. Aug. Rare.

1326. **P. conopilea** Fr. (= *Psathyra superba* Jungh. sec. Quél.) Cke. Illus. no. 609, t. 575. *κῶνος*, a cone; *pileus*, cap.

P. 2–5 cm., *bay brown, then pale ochraceous when dry,* submembranaceous, *conico-campanulate,* scarcely expanded, obtuse, smooth, *fragile.* St. 10–15 cm. × 2–4 mm., *silvery-shining, becoming yellowish, slightly attenuated upwards, tense and straight, polished,* smooth. Gills *white, then flesh colour and finally fuscous purple, adnexed in the top of the cone,* 4–5 mm. broad, only slightly ventricose, crowded. Flesh *yellowish, then whitish,* thin. Spores fuscous purple, broadly elliptical, 12–15 × 7–8μ. Pastures, roadsides, ditches. Sept.—Nov. Uncommon. (*v.v.*)

1327. **P. mastigera** B. & Br. Cke. Illus. no. 610, t. 591, fig. A. *μαστός*, a breast; *gero*, I bear.

P. 2–3 cm., *dark rich brown, umber tan when dry,* fleshy, nearly cylindrical, obtuse, conico-campanulate, *with a strong mammiform umbo,* repand; margin straight. St. 6–8 cm. × 3–4 mm., *white,* attenuated upwards, smooth, or fibrillose and furfuraceous. Gills *umber, edge paler,* affixed, ascending, rather narrow. Flesh *pale umber,* thick at the disc. Spores fuscous, "elliptical, 15–16 × 7–8μ" Massee. Roadsides amongst grass. July—Nov. Rare.

1328. **P. Loscosii** Rabenh. Francisco Loscos.

P. 5 cm., *greyish fuscous,* membranaceous, campanulate, then expanded, *radiately sulcate, folds at length granularly crenate*; margin *involute.* St. 7·5–12·5 cm. × 4–5 mm., *pallid, becoming fuscous,* equal, tough, striate. Gills *fuscous, becoming black,* adnate, somewhat crowded. Flesh *sienna,* thin. Smell and taste slight, fungoid. Caespitose. Gardens, on mushroom beds. Nov. Rare.

1329. **P. corrugis** (Pers.) Fr. Cke. Illus. no. 611, t. 576.
Corrugis, full of wrinkles.

P. 1–4 cm., *rose colour, or pallid flesh colour, becoming pale when dry,* submembranaceous, fragile, *campanulate, often subumbonate,* smooth, slightly striate when moist, *wrinkled when dry,* sprinkled with shining atoms. St. 4–10 cm. × 2–5 mm., *whitish, or rufescent,* equal, *tense and straight,* slightly firm, smooth. Gills *white, then violaceous, at length blackish, edge white, adnate,* or sinuato-adnate, ventricose. Flesh *whitish,* thin. Spores brownish purple, elliptical, 12–14 × 6–7μ. Cystidia "ventricose-fusiform, 60–75 × 10–12μ, often with a clavate, swollen apex" Rick. Woods, pastures, hedgerows, and gardens. April—Jan. Common. (*v.v.*)

var. **vinosa** (Cda.) B. & Br. Cke. Illus. no. 612, t. 592.
Vinosa, wine colour.

Differs from the type in the *somewhat roseate p.* Gardens, and pastures. Sept.—Oct. Not uncommon. (*v.v.*)

var. **gracilis** B. & Br. *Gracilis,* thin.

Differs from the type in being *more slender.* Gardens, and roadsides. Aug.—Oct. Not uncommon. (*v.v.*)

These forms are not really worthy of varietal names.

1330. **P. pellosperma** (Bull.) B. & Br. Cke. Illus. no. 613, t. 577.
πελλός, dark coloured; σπέρμα, seed.

P. 1·5–3 cm., *white, or ochrey white, becoming fuliginous with age,* subcampanulate, or subovate, smooth, then striate, sometimes rugose. St. 6–12 cm. × 2–3 mm., *white, or concolorous,* nearly equal, naked. Gills *cinereous, then fuliginous, at length black,* free, broad, much narrowed at the tips. Flesh *white,* thin. Spores cinereous fuscous, elliptical, 8 × 4–5μ. Woods, and gardens. Sept.—Oct. Uncommon.

1331. **P. gyroflexa** Fr. Cke. Illus. no. 1184, t. 970.
γυρός, round; *flexa,* bent.

P. 1–1·5 cm., *white, then pallid, or greyish, disc rufescent,* submembranaceous, conical, then campanulate, obtuse, smooth, atomate; *margin striate.* St. 4–5 cm. × 1–2 mm., *white, shining, fragile, flexuose, twisted,* smooth. Gills *greyish, then purple, adnate,* ascending, broad. Flesh *white,* thin. Spores "brown, elliptical, 9–10 × 5–6μ, smooth. Cystidia on edge of gill ventricose-flask-shaped, 36–40 × 10–15μ, blunt" Rick. Scattered, or subcaespitose. Pastures, and at the roots of trees. Aug. Uncommon.

1332. **P. tenuicula** Karst. *Tenuicula*, slight.

P. *whitish, then livid, or smoky, pale when dry*, campanulate, then somewhat expanded, *everywhere striate*. St. *hyaline, pellucid*, usually wavy. Gills *pallid, then grey*, adnate. Flesh, very thin. Spores, elliptical, 5–6 × 3μ.

II. P. campanulato-convex, flattened, smooth, or atomate; gills plano- or arcuato-adfixed. Veil none.

1333. **P. spadiceo-grisea** (Schaeff.) Fr. Boud. Icon. t. 135, as *Psilocybe spadiceo-grisea* (Schaeff.) Fr. *Spadicea*, date brown; *grisea*, grey.

P. 3–6 cm., *date brown, whitish grey when dry*, very hygrophanous, submembranaceous, very fragile, campanulate, then convex, at length flattened, obtuse, or with a darker umbo, smooth; *margin striate*. St. 4–7·5 cm. × 4–6 mm., *whitish, shining*, equal, *apex striate, sometimes pulverulent*, base slightly swollen and white hairy. Gills *umber fuscous*, adnexed, *attenuated behind*, at first ascending, *narrow, crowded*. Flesh *more or less fuliginous, becoming whitish, rather thick*. Spores brownish purple, oblong-elliptic, 8–11 × 4–6μ, 1-multi-guttulate. Cystidia "on surface of gill ventricose-cylindrical, 40–50 × 9–12μ, on edge of gill vesiculose-clavate, 30–40 × 15–20μ" Rick. Taste mild. Edible. Solitary, or gregarious. On stumps, or at the base of trees. Woods, and plantations. March—Nov. Uncommon. (*v.v.*)

1334. **P. obtusata** Fr. Cke. Illus. no. 615, t. 593. *Obtusata*, blunted.

P. 1–3 cm., *date brown fuscous, or umber fuscous, paler at the margin, somewhat shining, submembranaceous*, conical, then convex, *at length flattened*, obtuse, *wrinkled*, disc even, *hygrophanous*; margin striate. St. 5–7·5 cm. × 2–4 mm., *whitish*, equal, round, fragile, fibrilloso-silky. Gills *cinereous fuscous, then umber, adnate*, broad, distinct, *subdistant*. Flesh *concolorous*, very thin. Spores "reddish brown under the microscope, elliptical, 9–10 × 5μ, smooth. Cystidia lanceolate-flask-shaped, 45–60 × 12–15μ" Rick. Solitary, or caespitose. On oak trunks, and on the ground. Woods, and hedgerows. Aug.—Nov. Uncommon.

var. **minor** (Vaill.) Fr. *Minor*, smaller.

Differs from the type in its *smaller size*.

1335. **P. neglecta** Massee. *Neglecta*, overlooked.

P. 6–8 mm., *pale ochraceous, white when dry except the disc*, convex, then almost plane, smooth, *atomate when dry*. St. 2–3 cm. × 1–2 mm., *white, tinged with rufous below, pellucid*, rather wavy, smooth. Gills *purple brown at maturity*, slightly attached, rather broad, ventricose, crowded. Spores purple brown, elliptical, 12 × 6μ. On the ground. Gardens. Oct.—Nov. Rare.

III. P. and st. at the first floccose or fibrillose from the universal veil.

1336. **P. frustulenta** Fr. *Frustulenta*, full of small pieces.

P. 2–3 cm., *watery ferruginous, but somewhat pallid*, becoming pale when dry, submembranaceous, *very fragile*, campanulate, then hemispherical, obtuse, *somewhat striate when moist, whitish floccose at or about the margin*. St. 5–7·5 cm. × 2 mm., *whitish*, equal, *somewhat undulate, fibrillose, or sprinkled with white flocci*. Gills *watery cinnamon, then fuscous, adnate*, ascending, crowded. Flesh thin at the disc. Spores "brown, short, elliptical, almost round, 6–7 × 4–5μ, smooth. Cystidia fusiform, 45–50 × 10–12μ" Rick. Amongst damp gravel. Woods. Sept. Rare.

1337. **P. bifrons** Berk. Cke. Illus. no. 616, t. 594, fig. A. *Bifrons*, with two faces.

P. 5–20 mm., *ochraceous brown, tinged with red, pale tan when dry*, submembranaceous, campanulate, obtuse, slightly wrinkled, covered with a delicate evanescent veil when young; margin thin, transparent. St. 4–6 cm. × 2–3 mm., *white*, thickest at the base, straight, *very brittle, minutely satiny*, naked. Gills *pinkish cinereous*, adnate, moderately broad; *edge white, composed of minute wavy teeth* Flesh *yellowish*, thin. Spores cinereous purple, elliptical, obtuse at the one end, subapiculate at the other, 9–10 × 4–5μ. Cystidia "on edge of gill subulate, 36–40 × 6–8μ, blunt" Rick. Woods, hedgerows, and wood heaps. Aug.—Oct. Not uncommon. (*v.v.*)

var. **semitincta** Phill. Cke. Illus. no. 616, t. 594, fig. B. *Semi-*, half; *tincta*, dyed.

Differs from the type in the *pinkish p., with ochraceous disc*. Woods, and hedgerows. Sept.—Oct. Not uncommon. (*v.v.*)

1338. **P. fatua** Fr. (= *Hypholoma fatuum* (Fr.) Quél.) Cke. Illus. no. 618, t. 595, fig. A. *Fatua*, foolish.

P. 3–8 cm., *tan fuscous, ochraceous clay when dry*, submembranaceous, oval, then campanulate, at length expanded, obtuse, *everywhere adpressedly fibrillose when young* (*the fibrils soon fugacious*), then *smooth, rugulose and whitish clay colour when full grown*; margin somewhat undulate, sometimes appendiculate with the veil. St. 5–7·5 cm. × 4–6 mm., *shining white*, somewhat firm, *soon smooth, apex striate and white mealy*, base white villose. Gills *white, then fuscous, adnate*, linear, 3–4 mm. broad, crowded, *edge often white*. Flesh *concolorous*, thin. Spores brownish purple, elliptical, "12–13 × 6–7μ" Sacc. Caespitose, rarely solitary. Thickets, gardens, and rich pastures. Sept.—Oct. Uncommon.

1339. **P. semivestita** B. & Br. (= *Hypholoma semivestitum* (B. & Br.) Quél.) Cke. Illus. no. 617, t. 578. *Semi-*, half; *vestita*, clothed.

P. 1–2 cm., *dark brown, becoming pale*, ovate, obtuse, *sprinkled with little snow-white fibrils more than half way up*. St. 5–7 cm. × 3 mm., *snow-white, with a pale under tinge of brown*, nearly straight, *fibrilloso-silky*, the walls within white with down. Gills *umber brown*, tinged with the dark spores, adnate, ascending, broad behind. Flesh *white*, thick at the disc. Spores brownish purple, elliptical, 10–12 × 5μ. Cystidia "fusiform, 45–60 × 10–13μ" Rick. Caespitose, or solitary. Amongst grass. Rich pastures, and woods. Aug.—Oct. Uncommon. (*v.v.*)

1340. **P. fibrillosa** (Pers.) Fr. (= *Hypholoma fibrillosum* (Pers.) Quél.) Cke. Illus. no. 618, t. 595, fig. B.
Fibrillosa, full of fibrils.

P. 2–3 cm., *livid, or becoming white*, submembranaceous, *fragile*, campanulate, then convex, at length flattened, obtuse, *striate, covered with long, white, fugacious fibrils*, soon smooth. St. 6–10 cm. × 4–6 mm., *white*, equal, fragile, *clothed throughout with fibrilloso-fasciculate, spreading, fugacious, white squamules*, then smooth. Gills *cinereous, then becoming black purple*, adnate, broader behind, 6–10 mm. broad, at length plane, *edge often white*. Flesh *greyish, becoming white*, thin at the margin. Spores black purple, pip-shaped, 6 × 3μ. Cystidia "on edge of gill, vesiculose-clavate" Rick. Solitary. Woods. Sept.—Nov. Not uncommon. (*v.v.*)

1341. **P. Gordonii** B. & Br. (= *Hypholoma Gordonii* (B. & Br.) Big. & Guillem.) Cke. Illus. no. 620, t. 580, fig. A.
Marchioness of Huntly.

P. 2–4 cm., *pale cinereous, then white*, membranaceous, campanulate, *sulcato-striate, sprinkled with white floccose scales*. St. 4–5 cm. × 3 mm., *white*, equal, brittle, *transversely undulated*, white pruinose above, floccose below, becoming at length smooth and shining. Gills *cinereous*, narrowly adnate, ascending, moderately broad, distant. Flesh *yellowish*, somewhat thick at the disc. Spores "broad, elliptical, 11–13 × 7–8μ, smooth, subopaque" Rick. Smell faint, nauseous. Densely caespitose. Stumps, and amongst chips. Oct. Rare.

1342. **P. glareosa** B. & Br. Cke. Illus. no. 610, t. 591, fig. B.
Glareosa, belonging to gravel.

P. 12–15 mm., *grey, disc pale chestnut*, campanulate, obtuse, or umbonate, striate, *with flocci like little crumbs*. St. 2·5–5 cm. × 2 mm., *brown, clothed with white fibrils*. Gills *umber*, adnate, broad behind. Flesh *brown*, especially close to the gills. Spores black. On gravelly soil after wet weather. June. Rare.

1343. **P. helobia** Kalchbr. (= *Psathyra corrugis* (Pers.) Fr. sec. Rick.) Kalchbr. Icon. t. 17, fig. 4. ἕλος, a marsh; βίος, life.

P. 4–6 cm., *fuliginous umber, becoming pallid clay colour, or somewhat rufescent when dry*, scarcely fleshy, *hygrophanous*, campanulate, soon plane or depressed, slightly umbonate, *radiately rugose, with concentric, elevated ridges towards the spreading, striate margin.* St. 10–20 cm. × 2–3 mm., *paler umber than the p., rufescent, becoming pallid when dry*, equal, undulate, flexuose, *covered with lax, whitish, fugacious flocci*, fragile. Gills *fuliginous*, adnate, rounded behind, ventricose, somewhat crowded. Flesh *watery reddish.* Spores black, "elliptical, ends rather acute, 12 × 6μ" Massee. Gregarious. Moist places in pine woods. Sept. Rare.

1344. **P. pennata** Fr. (= *Hypholoma pennatum* (Fr.) Quél.) Cke. Illus. no. 620, t. 580, fig. B. *Pennata*, feathered.

P. 2–3 cm., *inclining to livid, then white, or becoming fuscous-brick when young*, submembranaceous, ovate, then campanulate, 12 mm. high, *for a long time densely clothed with white, fugacious, plumose scales towards the margin*, at length naked. St. 2·5–4 cm. × 2–4 mm., *white, then silvery*, fragile, equal, *villose*, apex white pulverulent. Gills *livid, then fuscous blackish*, adnexed, ventricose, 4–5 mm. broad, crowded, *edge often white.* Flesh *pallid*, thin at the margin. Spores blackish purple, pip-shaped, 8–10 × 4–5μ, 1-guttulate. Cystidia "on surface of gill lanceolate-pointed, 50–70 × 10–20μ, on edge of gill vesiculose-clavate, 40–50 × 8–10μ, sometimes with reddish contents" Rick. Gregarious. Burnt soil, and sawdust. Woods and gardens. June—Oct. Uncommon. (*v.v.*)

1345. **P. gossypina** (Bull.) Fr. (= *Hypholoma gossypinum* (Bull.) Quél.) Cke. Illus. no. 621, t. 612, fig. A. *Gossypina*, cottony.

P. 1·5–3 cm., *ochraceous clay, disc darker*, submembranaceous, campanulate, then expanded, *tomentose with white, fugacious flocci, soon becoming smooth*; *margin striate.* St. 4–5 cm. × 3–4 mm., *whitish, densely tomentose with white, erect flocci*, equal, or slightly attenuated at the base, fragile. Gills *white, then fuscous-black*, adnate, 3–4 mm. broad, ventricose, crowded. Flesh *yellowish*, thick at the disc. Spores purple, elliptical, 8–9 × 4μ. Subcaespitose. On the ground, and on twigs. Woods and heaths. May—Oct. Not uncommon. (*v.v.*)

1346. **P. noli-tangere** Fr. (= *Hypholoma noli-tangere* (Fr.) Quél.) Fr. Icon. t. 138, fig. 3. *Noli-tangere*, touch not.

P. 1–2·5 cm., *pallid umber, or dark fuscous, becoming pale when dry*, very hygrophanous, *fragile*, membranaceous, campanulate, then expanded, obtuse, smooth, *striate throughout*, becoming even when dry, *covered with white, fugacious, thin flocci round the margin.* St. 2–4 cm. × 1–2 mm., *pallid fuscous, base darker*, equal, very fragile, often curved, *smooth*. Gills *pallid, then dark fuscous*, adnate, broad, plane. Flesh *grey*, very thin at the margin. Spores lilac, "subcylindrical, 7–9 × 4–5μ, smooth, transparent brown. Cystidia on edge of gill fusiform, 40–45 × 10–13μ" Rick. Gregarious. Oak chips, and damp shady ground. Sept.—Dec. Rare.

1347. **P. microrhiza** (Lasch) Fr. Cke. Illus. no. 622, t. 596, fig. A.
μικρός, small; ῥίζα, root.

P. ·5–3 cm., *ochraceous, or rufous brown, becoming pale*, membranaceous, campanulate, dry, *shining with atoms, at first yellow pilose*. St. 4–10 cm. × 2–3 mm., *whitish*, fragile, *rooting, silky*. Gills *pallid, then black brown*, adnexed, narrow, crowded. Spores *fuscous*, "broadly elliptical, 10–12 × 6–7μ, smooth, transparent brown. Cystidia lanceolate, 45–50 × 10–12μ, blunt" Rick. Gregarious. Bare soil in gardens. Sept. Rare.

P. urticaecola B. & Br. = **Coprinus urticaecola** (B. & Br.) Buller.

Spores black, or blackish.

Psathyrella Fr.

(Diminutive of *Psathyra*.)

Pileus fleshy, or submembranaceous, regular; margin straight, at first adpressed to the stem. Stem central, confluent with the pileus. Gills adnate, or free. Spores black, or fuscous black, elliptical, or oval; smooth; with an apical germ-pore. Cystidia present. Growing on the ground, or on wood; solitary, or caespitose.

*St. tense and straight, smooth.

1348. **P. subatrata** (Batsch) Fr. Fr. Icon. t. 139, fig. 1.
Sub, somewhat; *atrata*, clothed in black.

P. 2·5–5 cm., *umber-rufescent, fuliginous, or somewhat olivaceous, pallid rufescent when dry*, membranaceous, campanulate, 2·5 cm. high, then expanded, obtuse, or somewhat umbonate, smooth, slightly striate round the margin. St. 2·5–12·5 cm. × 2–4 mm., *becoming pale white*, tense and straight, equal, *smooth*. Gills *fuliginous blackish, almost umber*, adnexed in the top of the cone, adnate when the p. is more expanded, linear, usually 2 mm. broad, sometimes ventricose, 4 mm. broad. Flesh *yellowish white, somewhat fuliginous under the cuticle of the p.*, thin. Spores fuliginous black, elliptical, 14–17 × 7–9μ. Cystidia "on edge of gill bluntly fusiform, 45–55 × 8–15μ" Rick. Taste bitter. Gregarious. Rich pastures, woods, and hedgerows. Sept.—Oct. Uncommon. (*v.v.*)

1349. **P. gracilis** Fr. Cke. Illus. no. 635, t. 634. *Gracilis*, slender.

P. 1–4 cm., *fuliginous, livid, or pale grey, tan, rosy, or whitish when dry*, hygrophanous, membranaceous, *campanulate*, obtuse, smooth, slightly and pellucidly striate only round the margin. St. 6–8 cm. × 2–3 mm., *whitish, remarkably tense and straight, fragile*, equal, smooth, *naked, base white villose.* Gills *whitish, then cinereous-blackish*, wholly adnate, *commonly broader behind*, rarely linear, *subdistant, edge rose-coloured.* Flesh *white*, thin. Spores black, oblong elliptical, 11–14 × 5–6·5μ. Cystidia on edge of gill abundant, fusiform, or cylindrical, apex obtuse, 5–9μ in diam., base subventricose, 36–50 × 8–16μ. Gregarious. Woods, hedgerows, waysides, and wood heaps. May—Dec. Common. (*v.v.*)

1350. **P. hiascens** Fr. (= *Coprinus hiascens* (Fr.) Quél.) Cke. Illus. no. 636, t. 635. *Hiascens*, splitting.

P. 2–3 cm., *livid, then becoming yellow*, membranaceous, *conico-campanulate*, 2·5 cm. high, obtuse, smooth, *soon split and opening in furrows often to the middle*, the divided margin at length revolute. St. 4–7·5 cm. × 2–3 mm., *whitish*, tense and straight, *rigid-fragile*, naked, smooth. Gills *whitish, then shining black, at length very dead black*, adnate, narrow, *linear*, or somewhat attenuated in front, *distant*. Flesh *white*, very thin at the margin. Spores black, "wedge-shaped-rounded, 10–12 × 7–11μ" Karst. Grassy places, hedgerows, damp woods, and rubbish heaps. April—Nov. Rare.

1351. **P. arata** Berk. Cke. Illus. no. 637, t. 636. *ἀρόω*, I plough.

P. 2 cm., *bright brown*, membranaceous, campanulato-conic, 2·5 cm. high, rather acute, *deeply sulcate.* St. 12·5 cm. × 2–3 mm., *white*, thickened at the base, smooth. Gills *purplish black, quite free*, lanceolate. Flesh *concolorous at the disc*, thin at the margin. Under hedges. Sept.—Oct. Rare.

1352. **P. trepida** Fr. Fr. Icon. t. 139, fig. 2. *Trepida*, trembling.

P. 2–3 cm., *fuliginous, disc date brown*, membranaceous, *very fragile*, campanulate, obtuse, smooth, *slightly but densely striate up to the even disc.* St. 6–7·5 cm. × 1–2 mm., *whitish, diaphanous*, equal, *tense and straight*, rarely flexuose, quite smooth, naked. Gills *greyish, then fuliginous shining black*, adnate, *crowded, ventricose*, very thin. Flesh *brownish in the p.*, very thin. Spores dead black, "elliptical, 12–14 × 6–7μ, smooth, opaque. Cystidia on edge of gill fusiform, 40–50 × 9–10μ" Rick. Muddy marshes, and on twigs in woods. July—Oct. Uncommon. (*v.v.*)

1353. **P. hydrophora** (Bull.) Fr. (= *Coprinus hydrophorus* (Bull.) Quél.) Bull. Hist. Champ. Fr. t. 358.

ὕδωρ, water; *φέρω*, I bear.

P. 2–3 cm., *rufescent, becoming greyish towards the margin, submembranaceous, conico-campanulate*, disc broad, obtuse, smooth, *at length expanded and revolute*; *margin striate, at first appendiculate with the fugacious veil.* St. 6–7·5 cm. × 2–3 mm., *white*, fragile, straight, equal, smooth, *beaded with dew-like drops in wet weather*. Gills *pale grey, then livid black*, adnate, ascending, narrow, *linear*, 2 mm. broad, *crowded*. Spores bay purple, "elliptical, 9–10 × 5–6μ, smooth, transparent brown. Cystidia on edge of gill subulate, 50–60 × 8–10μ" Rick. Caespitose. Gardens, and woods. Sept.—Oct. Rare.

**St. flexuose, pruinate at the apex.

1354. **P. caudata** Fr. (= *Panaeolus caudatus* (Fr.) Quél.) Cke. Illus. no. 639, t. 637. *Caudata*, having a tail.

P. 2–5 cm., *date brown, tan colour obsoletely turning to flesh colour when dry*, membranaceous, very tender, *conical, then campanulate*, at length flattened, smooth, *disc subgibbous, even*, otherwise *pellucidly striate, dry*, often splitting and subdeliquescent in wet weather. St 7–11 cm. × 3–4 mm., *whitish*, attenuated upwards from the *thickened, rooting, fibrillose base*, very fragile, curved, *at length twisted*, undulate, *apex white pruinose*. Gills *grey, then cinereous black*, adnate, 8 mm. broad. Spores fuscous black, "elliptical, 13–17 × 8–9μ, smooth, opaque. Cystidia on edge of gill, ventricose-fusiform, 30–40 × 9–10μ" Rick. In troops, or caespitose. Gardens, charcoal heaps, and stumps of a wooden pavement. May—Dec. Uncommon. (*v.v.*)

1355. **P. prona** Fr. (= *Psathyrella prona* Fr. var. *Smithii* Massee.) Fr. Icon. t. 139, fig. 3. *Prona*, bending downwards.

P. 5–12 mm., *fuliginous, hoary when dry*, hygrophanous, membranaceous, campanulate, then *hemispherical*, very obtuse, smooth, *pellucidly striate, obsoletely silky-atomate and opaque when dry*. St. 4 cm. × 1 mm., *white, hyaline, becoming pale, equal, flexuose, lax*, very smooth, *apex pruinose*. Gills *greyish, then livid fuliginous*, adnate, plane, *subtriangular*, 4 mm. broad, *distant, edge often rose-coloured*. Flesh *yellowish*, very thin. Spores very dead black, "elliptical, 12–16 × 7–8μ, smooth, opaque. Cystidia on edge ventricose-fusiform, 40–60 × 8–10μ" Rick. Rich pastures, and in ruts of roads in woods. May—Oct. Rare.

1356. **P. empyreumatica** B. & Br. Cke. Illus. no. 641, t. 657, fig. A. *ἔμπυρος*, burnt.

P. 4 cm., *rufous, then becoming pale, hygrophanous*, membranaceous, expanded, *atomate*; *margin crenate*. St. 6 cm. × 3 mm., *pallid, silky furfuraceous*. Gills *rufous, then brown purple*, adnate, with a decurrent tooth, 4 mm. broad, thick, distant, *connected by veins, edge pallid*. Flesh *concolorous*, thin. Spores black. Smell strong. Wooden pavement. Oct. Rare.

1357. **P. atomata** Fr. (= *Panaeolus atomatus* (Fr.) Quél.) *ἄτομος*, an atom.

P. 1–3 cm., *livid, or reddish, becoming pale tan or pale flesh colour* when dry, hygrophanous, membranaceous, campanulate, obtuse, *slightly striate*, slightly wrinkled and without striae when dry, *sprinkled with shining atoms*. St. 4–7 cm. × 3–4 mm., *white*, equal, *lax*, fragile, slightly bent, *apex white pulverulent*. Gills *whitish, then cinereous-blackish*, adnate, broad, *ventricose, slightly distant*. Flesh *pallid*, thin. Spores black, elliptical, 11–15 × 6–8μ. Cystidia "fusiform, 40–50 × 8–10μ" Rick. Solitary, or gregarious. Woods, pastures, roadsides, and hedgerows. May—Dec. Common. (*v.v.*)

var. **expolita** Fr. *Expolita*, polished.

Differs from the type in its *smaller size, conical p., and undulate, smooth st.* Woods, pastures and hedgerows. Sept.—Oct. Not uncommon. (*v.v.*)

1358. **P. crenata** (Lasch) Fr. (= *Coprinus crenatus* (Lasch) Rick.) Cke. Illus. no. 643, t. 847. *Crenata*, notched.

P. 1–3 cm., *ochraceous, or rufescent, then pale*, hygrophanous, membranaceous, hemispherical, *sulcate, atomate, crenate at the margin*. St. 4–10 cm. × 2–4 mm., *whitish, or brownish*, fragile, equal, base thickened and villose, *striate and mealy above*. Gills *yellowish fuscous, then blackish*, adnate, subventricose. Flesh *yellowish in the p.*, thin. Spores brownish black, elliptical, 9–12 × 6μ. Cystidia "bottle-shaped, 50–150 × 22–33μ" Rick. Woods, pastures, roadsides, and amongst beech leaves. Sept.—Nov. Not uncommon. (*v.v.*)

1359. **P. disseminata** (Pers.) Fr. (= *Coprinus disseminatus* (Pers.) Quél.) Boud. Icon. t. 140. *Disseminata*, spread abroad.

P. 1–2 cm., *whitish, or yellowish, then becoming cinereous, commonly livid, disc becoming yellow*, membranaceous, oval, then campanulate or convex, *scurfy*, then becoming smooth, deeply striate, *sulcate*. St. 2·5–6 cm. × 2 mm., *white*, fragile, often curved, lax, somewhat flexuose, *slightly scurfy, then smooth*, arising from a byssoid, white mycelium. Gills *whitish, then blackish*, adnate, linear, 2 mm. broad. Flesh *white, yellowish at the disc*, very thin. Spores black, pip-shaped, 9–10 × 5–6μ. Cystidia "cylindrical-vesiculose, 60–75 × 8–12μ" Rick. Densely crowded, or caespitose. Old stumps, and bare ground. April—Nov. Common. (*v.v.*)

**Gills decurrent.

Spores white.

Omphalia (Pers.) Fr.

(ὀμφαλός, the navel.)

Pileus fleshy, or submembranaceous, often umbilicate. Stem central, cartilaginous. Gills decurrent. Spores white, rarely yellowish, elliptical, reniform, pip-shaped, boat-shaped, subglobose, or oblong elliptical; smooth, punctate, verrucose, or echinulate; continuous. Cystidia present, or absent. Growing on the ground, or on wood; solitary, caespitose, subcaespitose, or fasciculate.

I. P. at the first spread out, margin incurved.

A. Generally comparatively large; gills narrow, very crowded.

1360. **O. hydrogramma** (Bull.) Fr. Fr. Icon. t. 71.
ὕδωρ, water; γραμμή, a line.

Livid, or whitish livid when moist, whitish when dry. P. 5–7 cm., *submembranaceous,* flaccid, *deeply umbilicate,* very hygrophanous; *margin spreading,* undulate, *striate.* St. 6–8 cm. × 6 mm., very cartilaginous, smooth, generally compressed, undulated, base rooted and white tomentose. Gills *livid-whitish,* deeply decurrent, *very crowded,* narrow, arcuate, very unequal. Flesh *white,* thick at the disc. Spores white, elliptical, 5 × 3μ. Subcaespitose. Amongst dead leaves, especially beech. Oct. Uncommon. (*v.v.*)

1361. **O. detrusa** Fr. Fr. Icon. t. 73, fig. 1. *Detrusa,* thrust down.

P. 2–5 cm., *dark cinereous, subzonate,* somewhat fleshy, convex, then umbilicate. St. 2·5–3·5 cm. × 4 mm., *concolorous,* firm, attenuated upwards, smooth, *whitish at the base.* Gills *whitish, decurrent by a tooth,* thin, crowded. Flesh *whitish,* thin at the margin. Spores white, "7–8 × 4μ" Sacc. Woods. Sept.—Oct. Rare.

1362. **O. umbilicata** (Schaeff.) Fr. Schaeff. t. 207.
Umbilicata, having a navel.

P. 2–3 cm., *livid when moist, the disc becoming somewhat fuscous, whitish or yellowish* when dry, hygrophanous, submembranaceous, convexo-plane, *deeply umbilicate* at first, then infundibuliform, smooth. St. 2–5 cm. × 2–4 mm., *white,* here and there flexuose, twisted, or incurved, *apex silky-striate with white fibrils,* base somewhat rooting, or cohering with villose down. Gills *whitish,* at first *shortly,* then *deeply decurrent, crowded,* thin, unequal. Spores white, "kidney-shaped, 6–8 × 2·5–4μ" Sacc. Caespitose. In woods amongst moss. Sept. Uncommon.

1363. **O. maura** Fr. Fr. Icon. t. 73, fig. 2. *Maura,* Moorish.

P. 2–4 cm., *fuliginous and striate when moist, livid and silky shining when dry,* hygrophanous, submembranaceous, *convex, deeply umbilicate,* smooth. St. 2·5–5 cm. × 2 mm., *fuliginous-blackish,* very cartilaginous, *somewhat horny, rigid,* smooth. Gills *shining white,* very acutely and deeply decurrent, *arcuate,* attenuated at both ends, *very crowded.* Flesh *fuliginous,* thin at the margin. Spores white, subglobose, 5–6 × 5μ, punctate. Smell none, or of new meal. Pastures, heaths, and lawns. Sept.—Nov. Not uncommon. (*v.v.*)

1364. **O. offuciata** Fr. Fr. Icon. t. 72, fig. 3. *Offuciata*, painted.

P. 2–3 cm., *dark, then pale flesh colour, becoming pale and almost whitish when old and dry*, hygrophanous, slightly fleshy, *convex, then plano-depressed*, smooth. St. 2–5 cm. × 2–4 mm., *reddish, or concolorous*, very cartilaginous, round, then compressed, equal, apex obsoletely pruinose, *smooth*. Gills *of the same colour as the pileus*, moderately decurrent, narrow, *straight, crowded*. Under beech. Oct.—Nov. Uncommon.

1365. **O. scyphoides** Fr. Fr. Icon. t. 75, fig. 3, as *Omphalia scyphiformis*. σκύφος, a cup; εἶδος, like.

Shining white, becoming yellowish when dry. P. 8–50 mm., membranaceous, umbilicate, then *infundibuliform, undulate, silky*. St. 2–5 cm. × 1–3 mm., *flexuose, villose*; base white, tomentose. Gills decurrent, *narrow, crowded, linear*. Flesh *white*, thin at the margin. Spores white, boat-shaped, 8–9 × 5μ, and 6 × 2–3μ, 1-guttulate. On bare soil and amongst leaves in woods. Aug.—Oct. Uncommon. (*v.v.*)

B. Medium size; gills rather distant, narrow, attenuated at both ends.

1366. **O. chrysophylla** Fr. (= *Flammula chrysophylla* (Fr.) Quél.) Fr. Icon. t. 74, fig. 1. χρυσός, gold; φύλλον, leaf.

P. 2–5 cm., *yellow-fuscous when moist, tan-hoary* or *hoary whitish when dry*, submembranaceous, at the first *deeply umbilicate, flocculose*, subsquamulose, the spreading border somewhat reflexed. St. 2·5–5 cm. × 4 mm., *golden egg-yellow, tough*, equal, somewhat incurved; base villose, rooting. Gills *golden egg-yellow*, truly decurrent, *distant, broad*. Spores white, "pale yellow" Quél., "elliptical, 11–12 × 5μ, smooth. Cystidia none" Rick. On pine sawdust, and stumps. Aug.—Oct. Uncommon.

1367. **O. Allenii** René Maire. Trans. Brit. Myc. Soc. III, t. 11. W. B. Allen, the mycologist of Benthall, Broseley, Shropshire.

P. 1–2 cm., *olive-greenish, whitish when dry*, hygrophanous, convex, then plane, *somewhat umbilicate*, thin. St. 2–4 cm. × 2–5 mm., *lemon-yellow*, cylindrical, subcartilaginous; base white, strigose. Gills *lemon-yellow*, decurrent, *very narrow*, somewhat thick, subdistant, unequal, *more or less undulating*, united by veins. Flesh *yellow in the stem, greenish yellow in the pileus*. Spores white, elliptical, 6·5–7·5 × 3·5–4μ. Cystidia none. Taste mild. On a stump of a deciduous tree. Sept. Rare. (*v.v.*)

1368. **O. Postii** Fr. Fr. Icon. t. 74, fig. 2. H. von Post, the Swedish mycologist.

P. 2–6 cm., *bright orange*, membranaceous, at first umbilicate, then depressed, convex, *smooth, striate* towards the margin. St. 5–8 cm. ×

2–4 mm., *light yellow*, becoming pale, equal, tense and straight, smooth. Gills *whitish*, deeply decurrent, 2 mm. broad, linear, arcuate, subdistant. Flesh *concolorous*, thin. Spores white, elliptical, 6–8 × 4–5μ, 1-guttulate. Charcoal heaps, and boggy places. July—Oct. Not uncommon. (*v.v.*)

var. **aurea** Massee. Cke. Illus. no. 1151, t. 1152, fig. B, as *Omphalia Postii* Fr. *Aurea*, golden.

P. 2·5–5 cm., *golden-yellow*, very regular, *infundibuliform*, margin drooping. St. 5 cm. × 3–4 mm., *concolorous*; base white, tapering. Gills *white*, slightly decurrent, *crowded*. Spores white, elliptical, 7 × 3·5μ. On *Sphagnum* in swamps. Rare.

1369. **O. pyxidata** (Bull.) Fr. (= *Omphalia hepatica* (Batsch) Quél.) Cke. Illus. no. 254, t. 194, lower figs. *Pyxidata*, box-shaped.

P. 2–3 cm., *brick-rufescent, or rufous fuscous and radiato-striate when moist, becoming pale*, opaque, flocculose or slightly silky when dry, membranaceous, pellucid, umbilicate, then infundibuliform. St. 2·5 cm. × 2 mm., *pallid, then rufescent, tough*, sometimes pruinose. Gills *flesh colour, then pale yellowish*, decurrent, *subdistant*, narrow. Flesh *pallid*, thin. Spores white, elliptical, 6–7 × 4–5μ, 1-guttulate. Amongst grass on lawns, and in woods. July—Nov. Common. (*v.v.*)

1370. **O. leucophylla** Fr. Fr. Icon. t. 73, fig. 4.
λευκός, white; φύλλον, leaf.

P. 2–3 cm., *dark cinereous*, submembranaceous, infundibuliform; *margin reflexed, involute*. St. 4 cm. × 2 mm., *cinereous*, slightly rigid. Gills *shining white*, decurrent, arcuate, subdistant. Spores white, elliptical, 6–7 × 3–4μ. Woods, and pastures. April—Oct. Uncommon. (*v.v.*)

1371. **O. telmatiaea** Berk. & Cke. Cke. Illus. no. 256, t. 240.
τελματιαῖος, marshy.

P. 2–6 cm., *brown, then mouse-coloured*, rather membranaceous, soon *infundibuliform, silky*, margin *reflexed*. St. 3–4 cm. × 3–6 mm., *cinereous*, compressed; base white, tomentose. Gills *pallid*, decurrent, distant. Flesh *brownish*, thick at the disc. Spores white, "elliptical with an oblique apiculus, 7 × 4μ" Massee. On *Sphagnum*. Aug. Rare.

1372. **O. striaepilea** Fr. Fr. Icon. t. 73, fig. 3.
Strix, a furrow; *pileus*, a cap.

P. 2–3 cm., *livid fuscous, becoming pale-white when dry*, submembranaceous, convex, then *flattened*, umbilicate, *the whole elegantly striate*, smooth. St. 5 cm. × 2 mm., *becoming fuscous*, slightly tough, often flexuose. Gills *whitish*, slightly decurrent, somewhat crowded, 2–3 mm. broad. Flesh *whitish*, thin at the margin. Spores white, "globose, 7–8μ, echinulate. Basidia with 2-sterigmata" Rick. Amongst moss, and leaves in woods. Oct.—Nov. Not uncommon. (*v.v.*)

1373. **O. epichysium** (Pers.) Fr. Pers. Icon. pict. t. 13, fig. 1.
ἐπίχυσις, a vessel for pouring out.

P. 1–3 cm., *cinereous-fuliginous and striate when moist, becoming pallid, silky, or flocculosely-squamulose when dry*, membranaceous, somewhat plane, umbilicate; margin somewhat reflexed. St. 2·5–3 cm. × 2 mm., *cinereous*, tough, base white tomentose. Gills *whitish cinereous, shortly plano-decurrent.* Spores white, elliptical, 8–10 × 4–5μ. On rotten stumps, and logs. Sept.—Oct. Uncommon. (*v.v.*)

1374. **O. sphagnicola** Berk. (= *Omphalia philonotis* (Lasch) Quél.) Cke. Illus. no. 257, t. 289, upper figs.
Sphagnum, Sphagnum; *colo*, I inhabit.

P. 2·5–4 cm., *dirty pale-ochre, becoming darker*, somewhat fleshy, moist, *tough*, infundibuliform, *obscurely striate, minutely squamulose.* St. 2·5–5 cm. × 2 mm., *concolorous*, somewhat crooked, apex minutely squamulose at first. Gills *dirty ochraceous*, decurrent, narrow, subdistant, *thick, edge flattish.* Spores white, "elliptical, 6–9 × 3–5μ" Karst. On *Sphagnum* in bogs, and woods. May—Sept. Uncommon. (*v.v.*)

1375. **O. philonotis** (Lasch) Fr. (= *Omphalia sphagnicola* Berk. sec. Quél.) Fr. Icon. t. 76, fig. 1. φίλος, loving; νοτίς, wet.

Cinereous-fuliginous, fragile. P. 1–3 cm., membranaceous, *the whole deeply infundibuliform*, hygrophanous, *floccose when dry*; margin *erect.* St. 4 cm. × 2 mm., sometimes attenuated upwards; base white, floccose. Gills *deeply decurrent*, subdistant, narrow, lanceolate. Flesh *greyish*, thin at the margin. Spores white, elliptical or pip-shaped, 7–8 × 4–5μ. On *Sphagnum* in bogs and amongst short grass. May—Sept. Uncommon. (*v.v.*)

1376. **O. oniscus** Fr. (= *Omphalia caespitosa* Bolt. sec. Quél.) Fr. Icon. t. 76, fig. 3. ὀνίσκος, a wood-louse.

P. 2–3 cm., *dark cinereous, becoming pale, grey-hoary when dry*, submembranaceous, *flaccid, fragile* when old, *convexo-umbilicate*, or infundibuliform, often irregular, *undulato-flexuose or lobed, smooth*; margin striate. St. 2·5–3 cm. × 2 mm., *grey, somewhat firm*, tough, sometimes compressed, curved. Gills *cinereous, shortly decurrent*, subdistant. Flesh *grey*, thick at the disc. Spores white, elliptical, 7–8 × 4–5μ, 1-guttulate. Woods, and boggy places. Oct.—Dec. Uncommon. (*v.v.*)

1377. **O. Luffii** Massee. John Luff.

P. 2–3 cm., *pallid, then white*, convex, then *depressed*; margin *upturned* at extreme edge. St. 2–3 cm. × 2 mm., *pallid*, thickened above, *polished*, often wavy. Gills *pallid*, decurrent, *crowded*. Flesh *white*, very thin. Spores white, elliptical, 5 × 3μ. Smell fragrant, of anise. Amongst grass. Rare.

1378. **O. caespitosa** (Bolt.) Cke. (= *Omphalia oniscus* Fr. sec. Quél.) Cke. Illus. no. 258, t. 209, lower figs. *Caespitosa*, tufted.

P. 1–2·5 cm., *yellowish-white, opaque white when dry*, submembranaceous, *sulcate nearly to the disc*, convex, *subhemispherical*, umbilicate; margin *crenate*. St. 1–2 cm. × 3 mm., *concolorous*, generally curved, base *subbulbose*. Gills *whitish*, shortly decurrent, *very broad, very distant*, triangular. Spores white, "subglobose, 6 × 5μ" W. G. Sm. Moors, and on peat in sandy heaths. May—Oct. Uncommon. (*v.v.*)

1379. **O. glaucophylla** (Lasch) Fr. Cke. Illus. no. 1153, t. 959, fig. B. γλαυκός, pale green; φύλλον, a leaf.

P. 1 cm., *mouse colour, becoming pale when dry*, membranaceous, infundibuliform, *plicato-striate*, hygrophanous, slightly smooth. St. 10–15 × 2–3 mm., *concolorous*, firm. Gills *olivaceous*, decurrent, lanceolate, subdistant. Spores white, "nearly comma-shaped, 4–5 × 2μ" Rick. On the ground in woods. Sept.—Oct. Uncommon. (*v.v.*)

1380. **O. rustica** Fr. Cke. Illus. no. 1153, t. 959, fig. C. *Rustica*, belonging to the country.

P. 1 cm., *fuscous, then grey and striate when moist, becoming either fuscous, or silky and hoary when dry*, membranaceous, umbilicate at the disc, otherwise *convex*. St. 10–15 × 1 mm., *fuscous, then grey, polished*, equal, often thickened upwards, base white, villose. Gills *grey*, decurrent, thick, subdistant, *edge arcuate*. Flesh *white*, thin at the margin. Spores white, elliptical, 8–10 × 4–5μ, often curved, 2–many-guttulate. Woods, and pastures. Aug.—Oct. Common. (*v.v.*)

1381. **O. scyphiformis** Fr. σκύφος, a cup; *forma*, shape.

Entirely snow-white. P. 5–20 mm., membranaceous, convex, then *infundibuliform*, pellucid; margin *striate, crenulate*. St. 3–4 cm. × 2 mm., flexuose, apex thickened. Gills very decurrent, distant, *thin*. Spores white, elliptical, 8 × 4μ, 1-guttulate. Flesh *white*, very thin at the margin. On bare ground, and amongst moss in deciduous woods. Sept.—Oct. Uncommon. (*v.v.*)

1382. **O. alutacea** Cke. & Massee. (= *Clitocybe alutacea* Cke. & Massee.)[1] *Alutacea*, like tanned leather.

P. 10–15 mm., *tan-coloured*, membranaceous, convex, then umbilicate; margin incurved. St. 3–4 cm. × 1–2 mm., *rather paler than the pileus*. Gills *paler than the pileus*, decurrent, narrow, *crowded*, arcuate. Spores white, elliptical, 6 × 4μ. Amongst grass, and moss in woods. Sept. Uncommon. (*v.v.*)

[1] By an oversight this species was described under Clitocybe (no. 823), but its correct position is here.

C. Gills very distant, broad, generally thick.

1383. **O. atropuncta** (Pers.) Quél. (= *Eccilia atropuncta* (Pers.) Fr.) Boud. Icon. t. 70. *Ater*, black; *puncta*, spotted.

P. 1–1·5 cm., *cinereous, or grey fuliginous*, campanulate, then depressed and cup-shaped, often slightly squamulose. St. 2–4 cm. × 2–3 mm., *blackish grey*, apex *paler, covered with black punctiform squamules*, thickened upwards; base *pulverulent*, white. Gills *greyish flesh colour*, decurrent, *thick, narrow*, distant. Flesh of pileus *pale fuliginous, blackish towards the basal portion of the stem*. Spores white or yellowish, subglobose, or angularly-globose, 4·5–5·5 × 4–5μ, 1–many-guttulate. Smell unpleasant. Woods, and heaths. Sept.—Oct. Uncommon. (*v.v.*)

1384. **O. demissa** Fr. (= *Ag. rufulus* B. & Br.) Bres. Fung. Trid. t. 35, fig. 1. *Demissa*, let down.

P. 8–15 mm., *fuscous-rufescent, submembranaceous*, convex, then expanded, obtuse, *at length umbilicate, striate when moist, subflocculose, obsoletely pruinose when dry*; margin *crenate*. St. 1·5–3 cm. × 2–3 mm., *liver-rufescent, becoming pale, shining*, flexuose, base white tomentose. Gills *becoming purple, often forked, thick*, subdecurrent, becoming very broad behind, *distant, interstices veiny*. Flesh *purple-vinous, becoming pale*. Spores white, "ovoid, 10–12 × 6–8μ, granular" Bres. In woods and waste places. Aug.—Oct. Uncommon.

1385. **O. hepatica** (Batsch) Fr. (= *Omphalia pyxidata* (Bull.) Fr. sec. Quél.; *Omphalia subhepatica* (Batsch) Sacc.) Cke. Illus. no. 259, t. 250, fig. B. *Hepatica*, like liver.

P. 1–4 cm., *rufous-flesh-colour when moist, slightly tawny, or tan and somewhat shining when dry, coriaceo-membranaceous, tough*, umbilicato-convex, then infundibuliform, *often undulato-lobed*, smooth. St. 2·5 cm. × 2 mm., *fuscous-flesh-colour, very tough*, becoming compressed, broader and dilated at the apex, rarely white-pruinose. Gills *whitish, becoming pale*, deeply decurrent, *distant*, prominently *connected by veins*, narrow, linear, sometimes crisped. Spores white, elliptical, 5–8 × 4–5μ. On lawns, and amongst short grass in woods. Sept.—Dec. Uncommon. (*v.v.*)

1386. **O. muralis** (Sow.) Fr. Cke. Illus. no. 259, t. 250, fig. C. *Muralis*, belonging to a wall.

P. 8–20 mm., *rufous brown*, submembranaceous, tough, convex, umbilicate, then infundibuliform, *radiato-striate*, smooth; *margin crenulate*. St. 6–12 × 1–2 mm., *concolorous*, equal, smooth; base white, floccose. Gills *pallid*, or *flesh colour*, decurrent, distant. Flesh *concolorous*, thin. Spores white, elliptical, with an oblique basal apiculus, 9–10 × 4·5–5μ. On old walls, sandy banks, and bare soil in woods. Jan.—Dec. Common. (*v.v.*)

1387. **O. umbellifera** (Linn.) Fr. (= *Omphalia pseudoandrosacea* Bull. sec. Quél.) Boud. Icon. t. 69. *Umbellifera*, umbel-bearing.

P. 1–2 cm., *grey, bistre, straw colour, whitish or ochraceous, becoming whitish, slightly fleshy-membranaceous*, convex, then *plane, broadly obconic, faintly umbilicate, rayed with darker striae*; when dry silky, flocculose, rarely squamulose; *margin inflexed* at first, *crenate*. St. 2·5–3 cm. × 2 mm., *concolorous, dilated towards the apex into the pileus*, sometimes pubescent, base white villose. Gills *white, then cream, or yellowish*, decurrent, *very broad behind, triangular, very distant*, sometimes dichotomous, connected by veins. Flesh *pallid*, thick at the disc. Spores white, elliptical, 7–9 × 4–5μ, 2–3-guttulate. Boggy ground in woods and on mountains, also on rotten wood. April—Dec. Common. (*v.v.*)

var. **nivea** Fl. Dan. Fl. Dan. t. 1015, fig. A. *Nivea*, snow white.

Differs from the type in being *entirely snow white*. In bogs. Not uncommon. (*v.v.*)

var. **citrina** Quél. *Citrina*, citron yellow.

Differs from the type in being *glabrous, citron yellow, and pellucid*.

var. **viridis** Fl. Dan. Fl. Dan. t. 1672, fig. 1. *Viridis*, green.

Differs from the type in being *pubescent, and bluish, then greenish*. Boggy ground, in woods and on hills. Sept. Uncommon. (*v.v.*)

var. **chrysoleuca** (Pers.) Fr. (= var. *abiegna* B. & Br.) χρυσός, gold; λευκός, white.

Differs from the type in being *bright yellow then whitish*. Growing on fir stumps.

var. **pallida** Cke. Cke. Illus. no. 260, t. 271, top figs. *Pallida*, pallid.

Differs from the type in being *entirely pale grey*. Bogs in woods. Not uncommon. (*v.v.*)

var. **flava** Cke. Cke. Illus. no. 260, t. 271, lowest figs. *Flava*, yellow.

Differs from the type in the *golden yellow p. and st*. On mountains. Uncommon. (*v.v.*)

var. **pyriformis** (Pers.) Fr. *Pyriformis*, pear-shaped.

Differs from the type in being *entirely dark umber*. Rotten beech trunks and shady places.

1388. **O. myochroa** (Fr.) Rea. μῦς, mouse; χρώς, colour.

P. 5–15 mm., *reddish brown, or rufescent, becoming whitish with age, somewhat fleshy*, convex, then plane, *umbilicate, striate to the middle*, margin crenulate. St. 1–2 cm. × 2–4 mm., *whitish, apex rufescent*; base white, strigose. Gills *yellowish, narrow, furcate at the margin*, arcuato-decurrent, distant. Flesh *of pileus rufescent, yellowish in the stem.* Spores white, broadly elliptical, 7–8 × 5–6μ, 1–2-guttulate. On rotten beech stumps. April—Oct. Uncommon. (*v.v.*)

1389. **O. velutina** Quél. Trans. Brit. Myc. Soc. III, t. 3.
Velutina, velvety.

P. 10–12 mm., *greyish, or yellowish grey*, convex, umbilicate, striate. St. 10–15 × 1–2 mm., *concolorous, finely tomentose*; base covered with the white mycelium, often subbulbose. Gills *yellowish grey*, narrow, 1–2 mm., arcuate, distant. Flesh *dark grey*. Spores white, ovoid pruniform, 10 × 6μ, 1–2-guttulate. Parks, heaths, and woods. Sept.—Oct. Uncommon. (*v.v.*)

1390. **O. infumata** B. & Br. *Infumata*, smoked.

P. 4 mm., *greenish, then smoky*, obtuse. St. 2·5 cm. × 1–2 mm., *yellow*, base *dilated, tomentose* especially below. Gills *yellow*, decurrent, few, broad, distant. On bark amongst moss. Sept. Rare.

1391. **O. retosta** Fr. Fr. Icon. t. 76, fig. 2. *Retosta*, scorched.

Entirely umber. P. 1–3 cm., slightly fleshy, plano-depressed, *polished when dry*, smooth; margin convex, *involute*. St. 2–3 cm. × 2–4 mm., *paler*, tough, equal. Gills *pallid umber*, slightly decurrent, distant, attenuated at both ends and *resembling a segment of a circle*. Flesh *concolorous*. Spores white, globose, "5–6μ" Sacc. Amongst dead leaves, and on lawns. Sept.—Oct. Uncommon. (*v.v.*)

1392. **O. buccinalis** (Sow.) Cke. Sow. Brit. Fung. t. 107.
Buccinalis, trumpet-like.

Entirely white. P. 5–10 mm., *trumpet-shaped*, plane, or depressed. St. 5–15 × 1–2 mm., expanding into the pileus. Gills deeply decurrent, *triquetrous*, distant. Spores white. On twigs, etc. Common.

1393. **O. abhorrens** B. & Br. Cke. Illus. no. 261, t. 272, fig. C.
Abhorrens, disgusting.

P. 1–1·5 cm., *fuscous, then pale*, umbilicate. St. 2 cm. × 1–2 mm., *concolorous*, apex thickened, sometimes pruinose when young, base white-tomentose. Gills *pale*, decurrent, distant, *thick, narrow*. Spores white. Smell very foetid, stercoraceous. Caespitose. On lawns under yews. Oct. Rare.

1394. **O. pseudoandrosacea** (Bull.) Fr. (= *Omphalia umbellifera* (Linn.) Fr. sec. Quél.)
ψευδής, false; *androsacea*, *Androsaceus androsaceus*.

Entirely whitish, or grey. P. 8–15 mm., fleshy-membranaceous, convex, deeply umbilicate, *at length infundibuliform*, smooth, *striato-plicate*; margin crenulate. St. 2–3 cm. × 1–2 mm. Gills deeply decurrent, *segment-like*, distant. Spores white, elliptical, "6–7 × 3–4μ, or 8–10 × 4–5μ" Sacc. Amongst moss on lawns, and in short pastures. July—Nov. Not uncommon. (*v.v.*)

1395. **O. griseo-pallida** (Desm.) Fr. (= *Omphalia griseola* (Pers.) Quél.) *Griseo*, grey; *pallida*, pallid.

P. 5–10 mm., *fuscous-grey, then becoming hoary, slightly fleshy*, convex, then plane, *umbilicate, smooth*, slightly shining; *margin deflexed*. St. 8–12 × 2 mm., *fuscous*, firm, equal, or thickened upwards, smooth. Gills *concolorous when moist, darker when dry*, decurrent, *broader behind*, distant, rather thick. Flesh *fuscous*, thin at the margin. Spores white, pip-shaped, or elliptical with an oblique basal apiculus, 9–11 × 6–7μ, 1-guttulate. On the ground, rubbish heaps, and mossy wall tops. Sept.—Dec. Not uncommon. (*v.v.*)

1396. **O. albidopallens** Karst.
Albido, whitish; *pallens*, becoming pallid.

P. 1 cm., *hyaline white, or pallid*, convex, orbicular, slightly umbilicate, *pellucidly striate*. St. 3–4 cm. × 1–2 mm., *pallid*. Gills *pallid*, adnate, decurrent, crowded. Spores white, elliptical, 4–5 × 3μ. Amongst moss. Aug.—Sept. Rare.

O. bibula Quél. = **Hygrophorus Wynniae** B. & Br.

1397. **O. stellata** Fr. Cke. Illus. no. 262, t. 241, bottom figs.
Stellata, set with stars.

Entirely white and diaphanous. P. 6–10 mm., membranaceous, convex, *umbilicate, striate*, smooth. St. 2–2·5 cm. × 1 mm., *filiform*, equal, fragile, *the dilated base strigoso-radiate*. Gills decurrent, *distant, broad, thin*. Spores white, "subglobose, irregular, 4–6μ, or 6–8 × 3–5μ" Sacc. On twigs, dead wood, and herbaceous stems. Feb.—Nov. Not uncommon. (*v.v.*)

II. P. at the first campanulate, margin straight, pressed to the stem.

A. Gills broad, perfect, unequal.

1398. **O. campanella** (Batsch) Fr. Cke. Illus. no. 263, t. 273, top figs.
Campanella, a little bell.

P. 1–2 cm., *yellow-ferruginous*, hygrophanous, membranaceous, tough, campanulate, then soon *convex, umbilicate*, striate. St. 2·5–3 cm. × 2 mm., *date brown, horny, rigid, polished*, attenuated, rooted, apex paler; *base tawny, strigose*. Gills *yellow*, deeply decurrent, somewhat crowded, *prominently connected by veins*. Spores white, elliptical, "8–9 × 3–4μ" Maire. Caespitose. On stumps of firs. Aug.—Sept. Not uncommon. (*v.v.*)

var. **badipus** Fr. *Badius*, bay; πούς, foot.

Differs from the type in the *thickened base of the st. being clothed with ferruginous down*.

var. **papillata** Fr. *Papillata*, having a nipple

Differs from the type in the *acutely conical p.*, *and papillate umbilicus.*

var. **myriadea** Kalchbr. μυριάς, ten thousand.

Differs from the type in being *half the size, densely caespitose, and pale tawny in colour with gills pale brick-red with a fleshy tinge.*

1399. **O. Kewensis** Massee. *Kewensis*, belonging to Kew.

P. 3–5 mm. high, *ochraceous, becoming whitish*, somewhat fleshy, *cylindrically-campanulate*, very smooth, *deeply sulcate*; margin crenate. St. 2–3 cm. × 1–2 mm., *pale*, round, more or less flexuose. Gills *pale*, subdecurrent, distant, membranaceous, edge entire. Spores white, elliptical, 7 × 5μ. Basidia subclavate, 28–32 × 6–7μ. Gregarious on dead rhizomes in Filmy Fern House. Rare.

1400. **O. picta** Fr. Fr. Icon. t. 77, fig. 4. *Picta*, painted.

P. 4–8 mm., *fuscous*, the *umbilicate disc generally light yellow, deeply campanulate*, 5–9 mm. high, striate, membranaceous; margin paler. St. 5–8 cm. × 1–2 mm., *date brown, horny, rigid*, smooth; apex thickened, paler; *inserted at the base and arising from a little, fuscous tawny, radiating membrane.* Gills *whitish, turning light yellow*, adnate, subdecurrent, *very broad* (much broader than long), distant. Spores white, elliptical, "7–10 × 4μ, minutely echinulate" Sacc. On twigs, and rotten wood in mixed woods. Aug.—Oct. Uncommon. (*v.v.*)

1401. **O. camptophylla** Berk. (= *Omphalia speirea* (Fr.) Quél.; *Mycena speirea* Fr. sec. Quél.) Cke. Illus. no. 264, t. 210, upper figs. καμπτός, bent; φύλλον, a leaf.

P. 12 mm., *brown, with a grey margin*, convexo-plane, *deeply striate, smooth.* St. 5–6 cm. × 1–2 mm., at first *yellow, then pale above, rufescent below*, subflexuose, somewhat rigid, *minutely pubescent*, base radiato-strigose. Gills *white*, at first adnate, nearly plane, then ascending and suddenly decurrent, subdistant. Spores white, elliptical, 8–10 × 6–8μ. On twigs, and sticks. Aug.—Oct. Uncommon. (*v.v.*)

1402. **O. umbratilis** Fr. Fr. Icon. t. 77, fig. 3, as *Omphalia umbratilis* Fr. var. *minor*. *Umbratilis*, remaining in the shade.

P. 1–3 cm., *black-fuscous*, or *umber-fuscous, hoary* when dry, submembranaceous, *obtusely campanulate*, then convex and *umbilicate*, smooth; margin substriate. St. 2·5–5 cm. × 2 mm., *fuscous-black, becoming greyish, tough, smooth.* Gills *becoming fuscous-white*, adnato-decurrent, *crowded, arcuate, broad*, acute at both ends. Flesh *concolorous.* Spores white, pip-shaped, 8–9 × 5μ, 1-guttulate. In pastures, and roadsides. Sept.—Nov. Uncommon. (*v.v.*)

1403. **O. grisea** Fr. Fr. Icon. t. 78, fig. 1. *Grisea*, grey.

P. 1–2 cm., *livid grey, then hoary,* submembranaceous, campanulate, then convex, *subpapillate* and *at length slightly umbilicate,* smooth, striate. St. 5–7 cm. × 2 mm., *whitish-cinereous, slightly firm,* smooth, shining, *longitudinally brittle, apex slightly thickened,* base white-floccose. Gills *whitish-grey, shortly decurrent, distant,* broad, rather thick. Spores white, elliptical, 6–9 × 4μ, 1-guttulate. Woods, and hedgerows. Aug.—Nov. Not uncommon. (*v.v.*)

1404. **O. fibula** (Bull.) Fr. Quél. Jur. et Vosg. I, t. 4, fig. 3. *Fibula*, a pin.

P. 4–20 mm., generally *orange-yellow, becoming pale when dry,* membranaceous, campanulate, then umbilicate, and finally infundibuliform, smooth, *striate* when moist. St. 3–4 cm. × 1–2 mm., *concolorous, bristle-like, weak,* often pubescent under a lens. Gills *whitish, or yellowish,* deeply decurrent, broad, distant. Spores white, elliptical, 3–4 × 2μ. Cystidia "on edge of gill sparse, subulate" Rick. Woods, pastures, heaths, and charcoal heaps. Jan.—Dec. Common. (*v.v.*)

var. **nivalis** Fl. Dan. (= var. *candida* Sacc.) Fl. Dan. t. 1072, fig. 2. *Nivalis*, snowy.

Differs from the type in the *whitish, or yellowish* p. and *white, or tinged with orange yellow* st. Amongst moss, and on charcoal heaps. July—Nov. Not uncommon. (*v.v.*)

var. **Swartzii** Fr. Fr. Icon. t. 75, fig. 4, as *Omphalia setipes* var. Fr. O. Swartz.

Differs from the type in the *firmer, at length plane p. with umbilicate, fuscous disc, and in the whitish st. externally and internally violaceous at the apex.* Spores white, elliptical, 4–5 × 2·5–3μ. Cystidia fusiform, apex subcapitate, 8–9μ in diam., base ventricose, 50–55 × 12–14μ. Amongst moss, short grass, and on charcoal heaps. Aug.—Dec. Not uncommon. (*v.v.*)

1405. **O. directa** B. & Br. *Directa*, straight.

White, very slender. P. 1–3 mm., *nail-shaped,* apex plane. St. 2·5 cm. × 1 mm., *slightly rufous,* filiform, ascending, *clothed with long hairs towards the base.* Gills deeply decurrent. Spores white. On dead leaves. May—Nov. Rare.

1406. **O. pseudo-directa** W. G. Sm. Cke. Illus. no. 266, t. 251, upper figs., as *Omphalia directa* B. & Br.

ψευδής, false; *directa*, *Omphalia directa*.

P. 2 mm., *white-pruinose.* St. 12 mm. × 2μ, *white, mealy-granular below, springing from a white, floccose, evanescent disc or volva.* Gills *white,* then *saffron,* few, adnate, pruinoso-sparkling. On *Encephalartos* cone. May. Rare.

1407. **O. Belliae** Johnst. Cke. Illus. no. 266, t. 251, lower figs.
The Misses Bell, of Coldstream.

P. 12 mm., *of a pale wood-brown hue,* membranaceous, inverted, *deeply cyathiform*; margin *waved, furrowed.* St. 4 cm. × 2 mm., *white, or very pale wood-brown above, dark brown towards the base, becoming paler when dry,* then apparently mealy, erect, stiff, elastic; root slightly incrassated, bent, *fixed by a dense cottony web.* Gills *dull chalky white,* decurrent, 2 mm. wide, rather distant, thick, more or less undulated, wrinkled on the sides and in the interstices with flexuose veins, once or twice divided near the edge. Spores white, oblong. On dead stems of reed. Oct. Rare.

1408. **O. gracilis** Quél. Trans. Brit. Myc. Soc. III, t. 2.
Gracilis, thin.

Entirely snow-white. P. 3–8 mm., membranaceous, campanulate, *papillate, striate,* transparent. St. 20–30 × 1 mm., filiform, transparent, *pruinose, base fibrillose.* Gills 1 mm. wide, very decurrent, distant, thin. Spores white, oblong, or pip-shaped, 8 × 3–3·5μ, 1–2-guttulate. On dead grass leaves, and twigs. Sept.—Nov. Uncommon. (*v.v.*)

1409. **O. gracillima** (Weinm.) Fr. Cke. Illus. no. 267, t. 252, top figs.
Gracillima, very thin.

Entirely snow-white. P. 4–6 mm., membranaceous, hemispherical, either minutely papillate, or umbilicate, *flocculose,* striate; *margin sulcate.* St. 6–12 × 1 mm., *bristle-like, inserted by a floccose base.* Gills *subdecurrent, broad,* distant, thin, the alternate ones dimidiate. Spores white, oblong-elliptical, 11–12 × 4–5μ. On twigs, dead herbaceous stems, and dead bramble stalks. Aug.—Oct. Uncommon. (*v.v.*)

1410. **O. bullula** (Brig.) Cke. Cke. Illus. no. 267, t. 252, middle figs.
Bullula, a watery vesicle.

Entirely shining white. P. 3–4 mm., membranaceous, hemispherical, *diaphanous.* St. 2 cm. × 1 mm., filiform. Gills arcuato-decurrent, *very distant.* Spores white. On twigs, and dead sticks. Sept.—Oct. Rare.

B. Gills fold-like, narrow.

1411. **O. integrella** (Pers.) Fr. Fr. Icon. t. 75, fig. 6.
Integrella, entire.

Entirely white. P. 3–6 mm., membranaceous, conical, then hemispherical, most frequently irregularly shaped, when flattened 12 mm. broad, umbilicate, *pruinose, diaphanous*; *margin sulcate.* St. 1·5–2·5 cm. × 1–2 mm., *pruinose, pellucid, slightly firm,* with a *small villose bulb at the base.* Gills decurrent, *narrow, fold-like, distant,* often branched, commonly disappearing short of the margin of the pileus, edge *acute*. Spores white, pip-shaped, 6–7 × 4–5μ, with a large central gutta. On twigs, and amongst leaves in woods, and hedgerows. May—Nov. Common. (*v.v.*)

1412. **O. polyadelpha** (Lasch) Fr. Cke. Illus. no. 1088, t. 1137, fig. B, as *Marasmius polyadelphus* Lasch.

πολυάδελφος, with many brothers.

Entirely snow-white. P. 2–3 mm., very tender, *hemispherical, umbilicate, sulcate*, pruinose, *tomentose under a lens*. St. 10–15 × 1 mm., filiform, curved, flaccid, *pruinose*, thickened and floccose at the base. Gills decurrent, *very narrow, wrinkle-like, distant*. Spores white, "fusiform-lanceolate, 7–9 × 3–4μ" Rick. Fasciculate, and in troops. On dead oak, and beech leaves. Oct.—Dec. Not uncommon. (*v.v.*)

1413. **O. Nevillae** Berk. Lady Dorothy Neville.

P. 1–1·5 cm., *brown*, hemispherical, disc *depressed, rugose, minutely granulated, striate, margin becoming pale*. St. 2–3 cm. × 1 mm., *brownish, rough with black granules*, base rather dilated and clothed with villose hairs. Gills *white*, arcuato-decurrent, *interstices and sides venoso-rugose*. Flesh *of stem white*. Spores white. On *Sphagnum*, in an orchid pot.

Spores pink.

Eccilia Fr.

(ἔγκοιλος, hollowed out.)

Pileus fleshy, or submembranaceous, umbilicate; margin incurved. Stem central, cartilaginous. Gills decurrent. Spores pink, angular, continuous. Cystidia rarely present. Growing on the ground, rarely on wood.

1414. **E. parkensis** Fr. Fr. Icon. t. 100, fig. 5.

Parkensis, belonging to a park.

P. 2–3 cm., *fuscous when moist, blackish when dry*, membranaceous, plano-convex, *deeply umbilicate*, slightly striate to the middle. St. 2·5 cm. × 1–2 mm., *fuscous*, attenuated downwards. Gills *whitish*, then *becoming dingy, flesh colour*, decurrent, *crowded*, linear, 1–2 mm. wide. Spores pink, subspheroid-angled, irregular, 6–9μ, 1-guttulate. Grassy places, pastures, and roadsides. July—Aug. Uncommon.

1415. **E. carneogrisea** B. & Br. Cke. Illus. no. 368, t. 380, lower figs.

Carneus, fleshy; *grisea*, grey.

P. 2–3 cm., *grey-flesh-colour*, umbilicate, striate, delicately dotted; *margin slightly glittering with dark particles*. St. 4 cm. × 2–3 mm., *concolorous, shining*, base white-tomentose. Gills *rosy*, adnato-decurrent, *somewhat undulated, the irregular margin darker*, distant. Spores pink, angular, elliptical, 8–9 × 6μ, 1–2-guttulate. Amongst fir leaves, and grass. July—Oct. Not uncommon. (*v.v.*)

1416. **E. griseorubella** (Lasch) Fr. Fr. Icon. t. 100, fig. 4.
Griseus, grey; *rubella*, reddish.

P. 2–3 cm., *umber, or fuscous when moist, grey when dry, hygrophanous*, membranaceous, deeply umbilicate, at first convex at the circumference, then plane, striate. St. 3–5 cm. × 2–3 mm., *concolorous, or a little paler*, equal. Gills *grey*, then flesh colour, slightly decurrent, subdistant. Spores pink, angular, elliptical, or subglobose, 7–9 × 7μ, 1-guttulate. Woods, and amongst pine leaves. Sept.—Oct. Not uncommon. (*v.v.*)

E. Smithii (Massee) W. G. Sm. = **Clitopilus Smithii** Massee.

1417. **E. atrides** (Lasch) Fr. *Ater*, black.

P. 15–25 mm., *black, fuscous*, becoming pale, *black-streaked*, submembranaceous, plane, deeply umbilicate, striate. St. 3 cm. × 3 mm., *pallid, black dotted upwards*, base white-cottony. Gills *pallid, deeply and truly decurrent*, attenuated behind, somewhat crowded, *the black edge slightly toothed*. Spores pink, "tuberculate-angular, elongated, 11–13 × 6–7μ (incl. apiculus), bright flesh colour in mass" Kauffm. Woods, and sandy heath. Sept.–Oct. Rare.

E. atropuncta (Pers.) Fr. = **Omphalia atropuncta** (Pers.) Quél.

1418. **E. nigrella** (Pers.) Gillet. *Nigrella*, blackish.

P. 10–15 mm., *reddish black, blackish with a tinge of lilac*, becoming blackish, submembranaceous, convex, then plane, deeply umbilicate, slightly striate, covered with an adnate, fibrillose silk, or whitish striate with blackish; margin incurved, *violet*, finally scaly. St. 2–3 cm. × 2 mm., *pale, grey horn colour, or pale lilac grey, punctate with black above*, base white-cottony. Gills *lilac or flesh colour*, decurrent, fairly thick, *edge toothed black, or blackish*. Spores pink. Pastures. Sept.—Oct. Rare.

1419. **E. rhodocylix** (Lasch) Fr. Fr. Icon. t. 100, fig. 6, as *Eccilia rhodocalix* Lasch. *ῥόδον*, rose; *κύλιξ*, cup.

P. 12–15 mm., *somewhat fuscous when moist, grey when dry*, hygrophanous, membranaceous, deeply umbilicate, *or rather infundibuliform* with the margin reflexed, *remotely striate when moist, flocculose when dry*. St. 2–3 cm. × 1–2 mm., *cinereous*, tough, *thickened upwards*. Gills *whitish*, then flesh colour, deeply decurrent, *very distant*, broad, few, the alternate ones shorter. Spores pink, "subspheric, pentagonal, 8–10μ. Cystidia coarsely hair-shaped" Lange. On rotten wood, alder stumps, and wall tops. Sept. Rare.

1420. **E. flosculus** W. G. Sm. Cke. Illus. no. 369, t. 613, fig. B.
Flosculus, a little flower.

P. 1·5–2 cm., *black-brown, becoming white with age*, submembranaceous, deeply umbilicate, somewhat irregular, *pruinoso-crystalline*.

St. 2 cm. × 1 mm., *reddish*, cartilaginous, attenuated downwards, pruinose, or innato-fibrillose. Gills *pink*, decurrent, somewhat waved, thick. Trama *dark brown*. Spores pink, nodulose. On the ground, at the foot of and upon the stems of tree ferns in conservatories. June. Rare.

1421. **E. acus** W. G. Sm. Cke. Illus. no. 369, t. 613, fig. C. *Acus*, a needle.

P. 5–15 mm., *snow-white*, submembranaceous, deeply umbilicate, *densely pruinose*; margin incurved, *striate*. St. 2–3 cm. × 1 mm., *white*, cartilaginous. Gills *pink*, deeply decurrent, thick, distant. Spores pink, nodulose. Amongst germinating coffee-seeds in coconut fibre in conservatories. Aug. Rare.

Spores ochraceous, or ferruginous.

Tubaria W. G. Sm.

(*Tuba*, a trumpet.)

Pileus fleshy, or submembranaceous; margin incurved. Stem central, cartilaginous. Gills decurrent, or broadly adnate. Spores ochraceous, ferruginous, or rarely fuscous; elliptical, pip-shaped, or almond-shaped; smooth, continuous. Cystidia present. Growing on the ground, or on wood.

1422. **T. cupularis** (Bull.) Fr. (= *Lactarius cupularis* (Bull.) Quél.) Cke. Illus. no. 526, t. 602, as var. *Cupularis*, cup-shaped.

P. 1–2 cm., *rufescent, tawny, or reddish yellow, becoming light yellowish*, slightly fleshy, convex, then plane, obtuse, *disc sometimes depressed, smooth*. St. 3–6 cm. × 3–4 mm., *tawny, or reddish tawny*, rarely whitish, attenuated upwards, naked. Gills *tawny, or a little deeper coloured than the p., decurrent*, crowded, thin, edge often serrulate. Spores "rusty, elliptical, 6 × 3μ" Massee. Mountainous heaths, amongst grass, and under firs. Aug.—Oct. Rare.

1423. **T. furfuracea** (Pers.) W. G. Sm. (= *Naucoria pellucida* (Bull.) Quél.) Boud. Icon. t. 129. *Furfuracea*, scurfy.

P. 1–4 cm., *pale cinnamon, or tawny, becoming pale, hoary tan when dry*, fleshy, convex and obtuse, then flattened and sometimes umbilicate, slightly and somewhat pellucidly striate when moist, even and slightly silky when dry, *covered round the margin with the hoary, silky squamulose, fugacious veil*, very hygrophanous. St. 2–5 cm. × 2–4 mm., *concolorous*, but deeper in colour as the pileus becomes pale, equal, floccosely furfuraceous when young, base villose with the effused white mycelium. Gills *concolorous, or bright cinnamon*, subdecurrent, broad near the st., more or less distant. Flesh *concolorous*, *becoming paler when dry*, thin. Spores ochraceous, elliptical, 6–9 × 5–6μ, 1-multi-guttulate. Cystidia on edge of gill flask-shaped, or cylindrical, apex obtuse, 4–8μ in diam., base ventricose, 40–60 × 15–20μ. Taste mild. Edible. Gregarious. Woods, fields, heaths, hedgerows, and roadsides. Jan.—Dec. Common. (*v.v.*)

var. **heterosticha** Fr. *ἕτερος*, different; *στίχος*, rank.

Differs from the type in the *umbonate and depressed pileus, and the somewhat naked st.* Woods, and heaths. July—Nov. Not uncommon. (*v.v.*)

var. **trigonophylla** (Lasch) Fr. Cke. Illus. no. 528, t. 483.
τρίγωνος, triangular; *φύλλον*, a leaf.

Differs from the type in its *smaller size, in becoming pale, and in the very broad, triangular, more distant, somewhat tawny ochraceous gills.* Waysides, charcoal heaps, and old brick pits. July—Nov. Not uncommon. (*v.v.*)

1424. **T. anthracophila** Karst. *ἄνθραξ*, charcoal; *φίλος*, loving.

P. 1–4 cm., *yellowish or ferruginous cinnamon, becoming paler when dry*, fleshy, convex, then expanded, often irregular, and repand, flexuose, dry; margin pellucidly striate when moist, *covered with concentric, white, fugacious squamules.* St. 2–4 cm. × 3–5 mm., *ferruginous, becoming paler*, equal, or enlarged upwards, flexuose, curved, sometimes twisted, *at length often compressed*, white fibrillose, apex somewhat naked and striate, base white villose. Gills *pallid, then concolorous*, adnate, broadest behind or at the middle, somewhat crowded, *edge unequal*, often dentate and floccosely crenate. Flesh *concolorous, becoming paler when dry*, thin. Spores pale ferruginous, broadly elliptical, or pip-shaped, 6–8 × 4–5μ, with a large central gutta. Charcoal heaps, footpaths, and burnt places. Aug.—Oct. Not uncommon. (*v.v.*)

1425. **T. paludosa** Fr. (= *Galera paludosa* (Fr.) Quél.) Fr. Icon. t. 129, fig. 3. *Paludosa*, marshy.

P. 5–15 mm., *pale yellowish fuscous, or honey colour*, submembranaceous, conical, then convex, *umbonate with a very prominent papilla, everywhere silky with superficial, fugacious, pallid flocci.* St. 4–8 cm. × 1–2 mm., *ochraceous*, attenuated at the apex, flexuose, *paler, white villose below*, flocculose above with the remains of the veil, which often forms a ring-like zone. Gills *watery ochraceous, decurrent*, very broad behind, triangular and with a decurrent tooth, thin, *crowded.* Flesh *concolorous*, very thin at the margin. Spores pale ferruginous, elliptical, or almond-shaped, 9–10 × 4–5μ. Cystidia "on edge of gill filiform-subulate" Rick. *Sphagnum* swamps, and boggy ground. May—Oct. Not uncommon. (*v.v.*)

1426. **T. stagnina** Fr. (= *Galera stagnina* (Fr.) Quél.) Fr. Icon. t. 129, fig. 2. *Stagnina*, belonging to swamps.

P. 6–20 mm., *bay-brown-ferruginous, or brown, somewhat ochraceous when dry*, submembranaceous, conical, then hemispherical, *obtuse*, sometimes rather depressed at the centre, *somewhat viscid*, slightly striate when moist; *margin elegantly clothed and appendiculate with floccose, superficial, fugacious, concentric, white scales*. St. 9–17·5 cm. × 2–3 mm., *rubiginous, then date brown*, slightly tough, equal, apex somewhat pruinose; base attenuated, white villose. ("Veil forming a membranaceous, fugacious, white ring" Quél.) Gills *ferruginous, decurrent, very broad*, triangular. Flesh *concolorous*, very thin at the margin. Spores dingy ferruginous, almond-shaped, 10–15×5–6μ. Bogs, and amongst *Sphagnum* in woods. July—Sept. Uncommon. (*v.v.*)

1427. **T. pellucida** (Bull.) Fr. (= *Naucoria pellucida* (Bull.) Quél.; *Tubaria furfuracea* (Pers.) W. G. Sm. sec. Quél.; *Naucoria conspersa* (Pers.) Fr. sec. Rick.) *Pellucida*, transparent.

P. 1–2 cm., *cinnamon*, submembranaceous, conico-campanulate, *umbonate*, hygrophanous; *margin striate, silky and squamulose*. St. 3–4 cm. × 2 mm., *pale*, attenuated upwards, *shining apex pruinose*. Gills *paler*, subdecurrent, broadest behind, triangular. Spores pale ochraceous, elliptical, 7–8 × 4–5μ. Cystidia on edge of gill cylindrical, flexuose, often capitate, apex 6–9μ in diam., base ventricose, 30–50 × 6–8μ. Roadsides, amongst leaves, especially beech. Sept.—Oct. Uncommon. (*v.v.*)

1428. **T. muscorum** (Hoffm.) Fr. (= *Galera muscorum* (Hoffm.) Quél.; *Tubaria pellucida* (Bull.) Fr. sec. Rick.) *Muscus*, moss.

P. 2–3 cm., *tawny brown, then honey yellow, or wax colour*, membranaceous, campanulate, then convex, hygrophanous, striate. St. 5 cm. × 1·5 mm., *cream colour, then ochraceous, base tawny brown*, soft, *fibrillosely silky*. Gills *yellow, then rust colour, uncinato-adnate*, broad, ventricose, *thick, distant*. Spores "ferruginous, elliptical, 8–9μ" Quél.; "6–8 × 4μ" Sacc. Amongst mosses in damp places, and on trunks of trees. July—Sept. Rare.

1429. **T. embola** Fr. Cke. Illus. no. 531, t. 514, fig. A. ἔμβολος, a wedge.

Entirely pale yellowish tawny when mature. P. 12 mm., *ochraceous tan when dry*, membranaceous, hygrophanous, campanulate, then hemispherical, obtuse, *smooth, lineato-striate*. St. 5 cm. × 2–3 mm., *shining yellow when dry, base becoming ferruginous, thickened upwards, smooth*, naked. Gills *tawny cinnamon when dry, adnate, very broad behind, triangular*, thick, very distant. Spores cinnamon, elliptical,

"10 × 4–6μ" Massee. On heathy ground, and marshy thickets. June—Nov. Uncommon.

1430. **T. autochthona** (B. & Br.) W. G. Sm. (= *Naucoria autochthona* (B. & Br.) Quél.) Cke. Illus. no. 531, t. 514, fig. B.
αὐτός, self; *χθών*, earth.

P. 6–12 mm., *ochrey white*, hemispherical, obtuse, then plane, silky; margin striate, *flocculose*. St. 1·5–2·5 cm. × 1 mm., *white*, equal, or thickened upwards, flexuose, pruinose, villose above; base thickened, white woolly. Gills *honey colour, then tawny*, adnate with a tooth, horizontal, *edge often white*. Flesh *white*, very thin. Spores ochraceous, elliptical, or pip-shaped, 6–7 × 3–4μ, 1-guttulate, minutely punctate. Naked soil, and open downs. June—Oct. Not uncommon. (*v.v.*)

1431. **T. crobulus** Fr. (= *Naucoria crobulus* (Fr.) Quél.) Cke. Illus. no. 532, t. 496.
κρωβύλος, a braid of hair gathered to a knot on the crown of the head.

P. 1–2 cm., *yellowish tawny, becoming hoary tan*, fleshy, convex, then flattened, obtuse, slightly viscid, *covered with floccose, somewhat squarrose, separating, fugacious, white scales*, then naked, shining. St. 2–3 cm. × 1–2 mm., *fuscous*, tough, equal, incurved, flexuose, *densely besprinkled with white floccose scales, apex paler*. Gills *ochraceous, then fuscous ferruginous*, adnate, *subdecurrent*, 2 mm. broad, *crowded*, edge unequal under a lens. Flesh *concolorous*, very thin. Spores brown, elliptical, 6–8 × 4μ. Cystidia filiform, flexuose, often capitate, apex 4–5μ in diam., 35–40 × 2–3μ. Chips, twigs, and on wood heaps. Sept.—Oct. Not uncommon. (*v.v.*)

1432. **T. inquilina** (Fr.) W. G. Sm. (= *Naucoria inquilina* (Fr.) Quél.) Cke. Illus. no. 533, t. 497. *Inquilina*, a lodger.

P. 1–2 cm., *livid fuscous, somewhat brick colour, becoming hoary, or tan colour when dry*, hygrophanous, membranaceous, convex, then plane, at length often umbonate, *slightly viscid*, smooth, *striate when moist*, pellicle separable. St. 2–3 cm. × 1–2 mm., *date brown*, tough, *attenuated downwards*, flexuose, *white fibrillose*, or slightly silky, apex at first flocculose, thickened, base white floccose. Gills *clay fuscous, then umber, broadly adnate, subdecurrent*, broad behind, *triangular*, 2–3 mm. broad, subdistant. Flesh *yellowish*, thick at the disc. Spores ferruginous, pip-shaped, 5–6 × 3μ. Cystidia "on edge of gill fusiform-filiform, 30–40 × 5–7μ" Rick. On twigs, and sticks. Jan.—Dec. Common. (*v.v.*)

var. **ecbola** Fr. *ἔκβολος*, thrown out.

Differs from the type in the *clay coloured p., the rooting, equal st.*, and *the crowded, ferruginous gills*. On grass roots. Sept.—Oct. Uncommon.

D. Pileus confluent with the excentric, or lateral stem, dimidiate, sessile, or resupinate.

Spores white, gill edge entire.

Pleurotus Fr.

(πλευρόν, side; οὖς, ear.)

Pileus fleshy, or submembranaceous, excentric, dimidiate, or resupinate. Stem excentric, lateral, or wanting; with or without a ring. Gills sinuate, adnate, decurrent, or radiating from a central point. Spores white, rarely pink, yellowish, lilac, or dingy; elliptical, globose, subglobose, pip-shaped, oblong elliptical, cylindrical, or reniform, smooth, granular, verrucose, or echinulate; continuous. Cystidia present, or absent. Growing on wood, more rarely on the ground, or on dung.

I. P. entire, laterally extended, excentric, not truly lateral. Lignicolous.

A. Veil forming a ring.

1433. **P. corticatus** Fr. (= *Pleurotus dryinus* (Pers.) Fr. sec. Quél.) Boud. Icon. t. 76. *Corticatus*, possessed of a bark.

P. 5–20 cm., *whitish grey, sometimes becoming yellowish, covered with dense grey down which separates into floccose scales*, very compact, convex, then flattened, somewhat disc-shaped, horizontal, *always entire* although excentric, rarely infundibuliform; margin involute, often denticulate with the remains of the ring. St. 2·5–9 × 2·5–3 cm., *whitish*, hard, rooted, *more or less excentric, curved-ascending*, squamuloso-fibrillose. Ring *white*, silky-floccose, moderately thick, ruptured in a torn manner, adhering to the st. and the margin of the p., at length vanishing. Gills *white, becoming yellow when old*, deeply decurrent, *dichotomosely branched, anastomosing at the base*, subdistant. Flesh *white*, hard. Spores white, oblong, cylindrical, often slightly curved and apiculate at the base, 13–15 × 4–5μ, or 9–10 × 3–4μ, often with a large central gutta. Smell and taste pleasant, rather strong. Edible. Caespitose. On trunks of ash, elm, lime, and apple. Sept.—Oct. Not uncommon. (*v.v.*)

var. **Albertinii** (Fr.) Quél. (= *Pleurotus corticatus* Fr. var. *tephrotrichus* Fr. sec. Quél.) Bres. Fung. Trid. t. 80, as *Pleurotus corticatus* Fr. var. *tephrotrichus* Fr.

J. Albertini, an early mycologist.

Differs from the type only in its *smaller size, p.* 7–10 *cm., in the densely villose p., soon covered with subfuscous squamules, the hairy stem, and villose edge of the gills*. Solitary. At the base of fir trunks, and on oak piles. July—Sept. Uncommon.

1434. **P. dryinus** (Pers.) Fr. (= *Pleurotus dimidiatus* (Schaeff.) Sacc.; *Pleurotus corticatus* Fr. sec. Quél.) Cke. Illus. no. 269, t. 226. δρυϊνος, oaken.

P. 5–10 cm., *whitish, variegated with spot-like scales, which become fuscous*, lateral, oblique, rather plane. St. 2·5–4 × 1–3 cm., *white*, sublateral, somewhat woody, *squamulose*, with a short, blunt root. Ring scarcely apparent on the st., but appendiculate round the margin of the p. when young. Gills *white, becoming yellow when old, not very decurrent, simple, narrow*. Flesh *white*, thick, firm. Spores white, oblong, cylindrical, 12–13 × 3–4μ, 1–3-guttulate. Taste pleasant, like mushrooms. Edible. On oaks, ash, willow, and walnut. Sept.—Feb. Not uncommon. (*v.v.*)

1435. **P. spongiosus** Fr. *Spongiosus*, spongy.

P. 5–18 cm., *at first whitish, tomentose with persistent, cinereous down, then becoming brownish and fibrillose with age*, excentric, somewhat lateral, pulvinate; *margin paler*. St. 1–3 × 1–4 cm., *white, becoming greyish*, very excentric, incurved, *tomentose, base abrupt*. Ring *white*, soon torn, appendiculate at the margin of the p., fugacious. Gills *whitish*, becoming beautifully yellow when old and dried, 3 mm. broad on one side of the st., 20 mm. wide on the other, *sinuato-adnexed*, crowded. Flesh *spongy, greyish marbled in the p., tinged yellowish in the st.* Spores white, oblong, cylindrical, rounded, or sometimes pointed at one end, 12–14 × 4μ, 1–2-guttulate. On rotten beech, and mossy trunks. Oct.—Nov. Rare. (*v.v.*)

B. Veil none; gills sinuate, or obtusely adnate.

1436. **P. ulmarius** (Bull.) Fr. Cke. Illus. no. 271, t. 227. *Ulmarius*, belonging to elm.

P. 6–20 cm., *ochraceous becoming pale-livid*, often marbled with round spots, convex, then plane, disc-shaped, *compact*, horizontal, often cracked in a tesselated manner, smooth. St. 5–11 × 1·5–4 cm., *white, becoming tinged with yellow*, firm, *elastic*, subexcentric, curved, ascending, base *somewhat fusiform*, or *thickened* and *tomentose*, often villose throughout. Gills *pale ochraceous, or whitish emarginate, broad*, somewhat crowded. Flesh *white*, tough. Spores white, globose, 5–6μ. Smell pleasant, or somewhat acid, taste pleasant. Edible. On trunks, especially elm. June—Dec. Common. (*v.v.*)

1437. **P. tessulatus** (Bull.) Fr. Bull. Hist. Champ. Fr. t. 513, fig. 1. *Tesselatus*, checkered.

P. 5–10 cm., *grey, becoming pale tawny*, convex, then plane, and in a form somewhat lateral, depressed behind, irregular, horizontal, *variegated with round and hexagonal spots*. St. 2–3 × 1·5 cm., *white*, compact, *equal*, or *attenuated at the base*, very excentric, curved-ascending, *smooth*. Gills *white, or becoming yellow, sinuate behind,* thin, *crowded*. Flesh *white*, thick. Spores white, "obovate-globose, 5·5 × 4·5μ" Sacc. Smell of new meal, taste pleasant. Edible. On trunks. Oct.—Nov. Rare.

P. decorus Fr. = **Tricholoma decorum** (Fr.) Quél.

1438. **P. palmatus** (Bull.) Fr. (= *Pleurotus subpalmatus* Fr.; *Pluteus reticulatus* Cke.; *Entoloma Cookei* Rich.) Cke. Illus. no. 273, t. 255, as *Pleurotus subpalmatus* Fr.

Palmatus, having the shape of a hand.

Entirely more or less rufescent. P. 5–12 cm., of *a beautiful orange-buff or nankeen colour*, convex, then flattened, obtuse, imbricated and glued together, horizontal, more or less excentric, pruinose; margin involute, *reticulato-corrugated.* Cuticle *gelatinous, thick, tough, diaphanous*, distilling limpid rufescent drops with an astringent taste. St. 3–7 × 1–1·5 cm., *whitish*, becoming *rufescent, fibrilloso-striate*, pruinose, *equal*, curved-ascending. Gills *paler than the pileus, joined in a collar behind*, sinuate, *connected by veins*, broad, crowded. Flesh *white, then tinged with red.* Spores *pink, or pale yellowish*, globose, 4–6μ, *verrucose*, 1-guttulate. Smell pleasant, taste bitter and acrid. Caespitose. On elm trunks, old posts and beams. Sept.—Jan. Not uncommon. (*v.v.*)

1439. **P. craspedius** Fr. Fr. Icon. t. 86, fig. 2.

κράσπεδον, the margin of a thing.

P. 7–13 cm., *brick colour, becoming pale tan, sometimes cinereous*, more or less excentric, sometimes sublateral, but marginate behind, *thin*, at length almost membranaceous towards the margin, flaccid, plane, depressed behind when very excentric, smooth, somewhat moist; margin at first involute, then evolute, *elegantly crenato-lobed, fimbriate.* St. 2·5–7·5 × 1–2·5 cm., *pallid*, firm, *elastic*, very unequal, either thickened at the base, or equal, sometimes villose at the base. Gills *shining white, wholly adnate*, very thin, crowded, narrow, at length lacerated. Flesh watery, *white when dry.* Spores white, "5 × 4–5μ" Sacc. Smell "strong, of cucumber" W. G. Sm. Caespitose. On rotten wood, and trunks, especially old poplars. Sept.—Oct. Rare.

1440. **P. fimbriatus** (Bolt.) Fr. (= *Clitocybe fimbriata* (Bolt.) Quél.) Cke. Illus. no. 275, t. 178, fig. 1. *Fimbriatus*, fringed.

P. 5–8 cm., *whitish, hyaline, hygrophanous, slightly fleshy*, convexo-plane, then *infundibuliform*, more or less excentric, occasionally lateral, pruinose; margin *sinuato-lobed, incised.* St. 1–4 cm. × 4–10 mm., *concolorous*, tough, round, or compressed, base pubescent. Gills *white, wholly adnate, very narrow, very thin, very crowded.* Flesh thin, tough, *watery-pallid.* Spores white, "oval, 3·5–5 × 2·5–3μ, minutely rough" Sacc. Smell of new meal. Edible. On dead trunks, especially beech. Aug.—Jan. Rare.

1441. **P. Ruthae** B. & Br. Cke. Illus. no. 275, t. 178, fig. 2. Miss Ruth Berkeley.

P. 4 cm., *whitish, or yellowish buff, fan-shaped*, slightly *hispid above the gelatinous stratum*; margin very thin, striate. St. 1–3 cm. × 5–6 mm., *reddish*, lateral, *hispid*, arising from a fibrous mycelium. Gills *white with a reddish tinge*, rather broad, acute behind, *anastomosing, interstices veined*. On sawdust. Oct. Rare.

1442. **P. lignatilis** Fr. (= *Clitocybe lignatilis* (Pers.) Quél.) Saund. & Sm. Myc. Illus. t. 6, figs. 4–6. *Lignatilis*, woody.

Dingy whitish. P. 3–10 cm., rarely central, generally more or less excentric, occasionally wholly lateral, often reniform, thin but compact, tough, convex, then plane, obtuse, often umbilicate, *flocculoso-pruinose*; margin involute, then expanded and undulato-lobed. St. 5–7·5 × 1–3 cm., 6–8 × 3–4 mm., sometimes absent, unequal, curved, or flexuose, tough, *pruinosely villose*; base rooting, somewhat tomentose. Gills *shining white, often with a tinge of yellow, adnate, very crowded, narrow*, divergent in the lobes, undulate. Flesh *white*, firm. Spores white, subglobose, 4 × 3μ. Smell strong of new meal. On trunks especially beech, and on rotten wood. Sept.—Oct. Uncommon. (*v.v.*)

var. **tephrocephala** Fr. τεφρός, ash-coloured; κεφαλή, head.

Differs from the type in the *more compact p., the black disc becoming cinereous, and the white margin*.

1443. **P. circinatus** Fr. (= *Clitocybe circinata* (Fr.) Quél.) Fr. Icon. t. 88, fig. 1. *Circinatus*, rounded.

Entirely white. P. 6–9 cm., *orbicular*, horizontal, tough, convex, then plano-disc-shaped, obtuse, *covered with a shining, whitish, slightly silky lustre*. St. 3–8 cm. × 6–10 mm., *elastic, central*, or slightly excentric, *generally straight, sometimes curved*, bluntly or attenuato-rooted at the base. Gills adnate, slightly decurrent, crowded, *broad*. Flesh *white*, firm. Spores white, globose, 3–4μ. Smell pleasant. On beech, and brick stumps. Sept.—Oct. Uncommon.

C. Veil none; gills deeply decurrent; stem distinct, somewhat vertical.

1444. **P. sapidus** Schulz. (= *Pleurotus cornucopiae* (Paul.) Quél.) Kalchbr. Icon. t. 8, fig. 1. *Sapidus*, pleasant.

P. 5–12 cm., *white, or light yellow, becoming fuscous, or umber*, excentric, lateral, subsessile, deformed, convex, disc depressed. St. 2·5–5 cm. × 6–16 mm., *white, or pallid yellow, incrassated upwards and dilated in the pileus*, often branched, curved, ascending, *arising from a fleshy, bulbous, white base*. Gills *pallid*, decurrent, rather distant, *broad.* Flesh *white.* Spores white, oblong ovate, 7–8 × 4μ or 10–12 × 4–5μ, lilac colour in the mass. Smell and taste pleasant. Edible. On elm trunks. June—Sept. Not uncommon. (*v.v.*)

1445. **P. pantoleucus** Fr. Fr. Icon. t. 88, fig. 2.
πᾶς, all; λευκός, white.

Entirely white. P. 4–8 cm., excentric, dimidiate, *spathulate,* slightly convex, subdepressed and marginate behind. St. 2–3 × 1–1·5 cm., *ascending,* very excentric, equal, or attenuated downwards. Flesh *white,* compact. Gills decurrent, somewhat crowded, *broad.* On trunks, willow. Oct. Uncommon.

1446. **P. mutilus** Fr. (= *Omphalia mutila* (Fr.) Quél.) Fr. Icon. t. 88, fig. 4. *Mutilus,* maimed.

Entirely white. P. 1–3 cm., very excentric, or wholly lateral, soft, tough, *reniform, spathulate,* and depressed behind, *subumbilicate,* otherwise ascending, *silky when dry.* St. 6–8 × 2–4 mm. *erect,* or ascending, excentric, or somewhat lateral, *round,* tough, base villose. Gills decurrent, *somewhat crowded, narrow,* thick. Flesh *white,* thin. Spores white, pip-shaped, 6–7 × 4μ, 1-guttulate. Pastures, and hill-sides. July—Dec. Not uncommon. (*v.v.*)

D. Veil none; gills deeply decurrent; p. lateral, sessile, or extended behind into a short, oblique stem-like base.

1447. **P. ostreatus** (Jacq.) Fr. *Ostreatus,* rough.

P. 7–13 cm., *when young almost black, soon becoming pale, fuscous-cinereous, passing into yellow when old, soft, conchate,* somewhat dimidiate, *ascending,* moist, cuticle sometimes torn into squamules. St. 2–4 × 2 cm., *often wanting, white,* firm, elastic, ascending obliquely, *dilated upwards into the pileus,* base strigosely villose. Gills *white, becoming yellowish,* margin sometimes umber, *decurrent, anastomosing at the base, subdistant,* broad. Flesh *white.* Spores lilac in the mass, elliptical, 9–11 × 4·5–6μ, 1-many-guttulate. Taste and smell pleasant. Edible. On stumps, trunks, and logs. Jan.—Dec. Common. (*v.v.*)

var. **glandulosus** (Bull.) Fr. *Glandulosus,* having glands.

Differs from the type in the *glandular gills.* On stumps. July—Nov. Uncommon. (*v.v.*)

var. **euosmus** (Berk.) Cke. (= *Pleurotus columbinus* Quél.) Hussey, Ill. Brit. Myc. I, t. 75. εὔοσμος, sweet smelling.

Differs from the type in its strong *smell like that of tarragon.* Poisonous. On elm stumps. April—Oct. Not uncommon. (*v.v.*)

var. **columbinus** (Quél.) Cke. (= *Pleurotus columbinus* Quél.) Bres. Fung. Trid. t. 6, as *Pleurotus columbinus* Bres.

Columbinus, pertaining to a pigeon.

Differs from the type in the *dark bluish grey pileus, and glaucous gills*. Edible. On stumps. Feb.—Sept. Not uncommon. (*v.v.*)

1448. **P. revolutus** Kickx. *Revolutus*, rolled back.

P. 9–15 cm., at first *smoky, then lead and mouse colours, disc darker*, firm, elastic, convexo-plane, depressed behind, shining; margin incurved. St. 2–5 × 2–3 cm., *whitish*, sometimes pubescent. Gills *white*, decurrent, *serrulated*. On old trunks, poplar, beech. Sept.—Oct. Rare.

var. **anglicus** Massee. Cke. Illus. no. 281, t. 180, as *Pleurotus revolutus* Kickx. *Anglicus*, English.

Differs from the type in the *margin of the pileus being only very slightly, or not at all incurved, and in its pallid ochraceous gills*. On trunks. Rare.

1449. **P. salignus** (Pers.) Fr. Cke. Illus. no. 282, t. 228.

Salignus, belonging to willow.

P. 5–8 cm., *fuliginous-cinereous, or ochraceous, compact, spongy*, subdimidiate, *horizontal, at first pulvinate, at length depressed behind*, here and there strigose. St. 1–1·5 × 3–4 cm., *tan*, firm, more or less tomentose. Gills *dingy-fuliginous*, horizontal, *branched in the middle*, crowded, *edge often eroded*. Spores dingy, "oblong cylindrical, often curved, 8–14 × 3–4·5μ, 1-guttulate" Karst. On willow, and alder trunks. Sept.—Jan. Uncommon.

1450. **P. acerinus** Fr. *Acerinus*, belonging to maple.

Entirely shining white, tough. P. 2·5–10 cm., thin, unequal, *silky-villose*. St. 1 × 1 cm., often obsolete, somewhat lateral, whitish, *villose*. Gills *white, becoming yellow*, decurrent, crowded, thin. Spores white, elliptical, 6 × 3μ, or 6–7 × 4–5μ, 1–2-guttulate. On trunks, and logs of maple, ash, and hornbeam. Sept.—Oct. Uncommon. (*v.v.*)

II. Pileus definitely lateral, immarginate behind, not resupinate at first.

1451. **P. petaloides** (Bull.) Fr. Bull. Hist. Champ. Fr. t. 226.

πέταλον, a leaf; *εἶδος*, like.

P. 2·5–5 cm., *fuscous, becoming pale*, dimidiate, *somewhat spathulate, continuous with the stem*; margin at first involute, then expanded. St. 1·5–2·5 × 1·5–2·5 cm., sometimes very short, *whitish*, firm, *compressed, channelled* when larger, villose. Gills *white, or yellowish, then cinereous, decurrent, very crowded, very narrow*, very unequal. Flesh of the pileus *with a gelatinous layer under the cuticle*. Spores white, elliptical, 6–8 × 4–5μ, granular. Cystidia fusiform, 50–60 × 12–15μ, apex pointed, very thick walled. On stumps, and on the ground. Aug.—Jan. Rare. (*v.v.*)

1452. **P. pulmonarius** Fr. *Pulmonarius*, belonging to the lungs.

P. 4–8 cm., *cinereous, then tan colour*, continuous with the stem, soft but tough, flaccid, *obovate, or reniform*; margin plane, or reflexo-conchate. St. 1·5 cm. × 12 mm., or wanting, exactly lateral, *horizontal*, or ascending, expanded into the pileus, *round*, villose. Gills *whitish, then livid, or cinereous*, decurrent, but *ending determinately, moderately broad*. Spores white, "8–10 × 2–3μ, or 10–12 × 3–4μ" Sacc. On beech, and birch stumps. Sept. Rare.

var. **juglandis** Fr. Fr. Icon. t. 87, fig. 2. *Juglans*, a walnut.

Sessile, smaller than the type, caespitose. P. *greyish-brown*, obovate, attenuated into a very short stem-like base. Gills *concolorous, or paler*. On walnut trunks. Rare.

1453. **P. serotinus** (Schrad.) Fr. Pat. tab. anal. t. 629. *Serotinus*, late.

P. 3–7 cm., *yellow-green, fuliginous olive, then olive, thick*, gibbous-convex, then plane and ascending, reniform, or obovate, *pellicle viscid in wet weather*; margin involute, then expanded and revolute. St. 1–2·5 × 1 cm., or wanting, lateral, *yellow, dotted with fuliginous or brownish squamules*, forming a fuliginous zone near the gills. Gills *bright yellow*, adnate, *narrow, crowded, often branched*. Flesh *white, with a gelatinous layer under the cuticle of the p.* Spores white, sausage-shaped, curved, 5–6 × 1·5–2μ, becoming 2-septate. Cystidia abundant on gill edge, scattered elsewhere, with yellowish contents in the upper part, cylindrical, or subfusiform, obtuse, or slightly clavate at the apex, base ventricose, or attenuated; 40–53 × 8–10 × 5–8μ at apex. On trunks and fallen logs. Sept.—Dec. Uncommon. (*v.v.*)

var. **Alméni** (Fr.) Big. & Guill. Fr. Icon. t. 87, fig. 3, as *Agaricus (Pleurotus) Almeni* Fr. Professor A. Almén.

Differs from the type in its *larger size, tawny fuscous* p. and paler st. and gills. Fallen logs. Nov. Rare. (*v.v.*)

1454. **P. mitis** (Pers.) Berk. Outl. t. 6, fig. 9. *Mitis*, mild.

Entirely white, or becoming rufescent. P. 1–2 cm., thin, continuous with the st. in a straight line, horizontal, *reniform*. St. 6–12 × 6–12 mm., definitely lateral, *compressed and dilated upwards, sprinkled with white, mealy squamules*. Gills adnate, linear-lanceolate, very crowded. Flesh *white, with a gelatinous layer under the cuticle of the p.* Spores white, reniform, 4 × 2μ. On coniferous twigs and stumps. Sept.—Dec. Common. (*v.v.*)

1455. **P. rufipes** Massee & W. G. Sm. *Rufus*, red; *pes*, foot.

P. 3–4 mm., *white, disc salmon*, membranaceous, *dimidiate, or reniform*, convex, *very glutinous when moist*; margin incurved. St. 4–6 × ·5 mm., *reddish, viscous*, becoming recurved, base white downy. Gills *white, interstices pale salmon*, adnate, broad, distant. Flesh *of stem salmon-red, centre white*. Spores white, oblong, 2–2·5 × ·75μ. On wood. Sept.—Oct. Rare.

1456. **P. gadinoides** W. G. Sm. Cke. Illus. no. 286, t. 276, top figs. *Gadinia*, a species of bivalve; εἶδος, like.

Entirely white. P. 1–1·5 cm., *dimidiate, shell-shaped*, hygrophanous, *smooth*, or *clothed with fine adpressed flocci*. St. minute, lateral, or none. Gills adnate, somewhat crowded, slightly branched. Spores white, elliptical, 7 × 3μ. On tree-fern stems. May. Rare.

1457. **P. limpidus** Fr. Fr. Icon. t. 88, fig. 3. *Limpidus*, clear.

Entirely hyaline white, shining white when dry. P. 2–3 cm., *obovate*, or *reniform*, horizontal, *narrowed behind into a stem-like base*, pruinose; margin shortly inflexed, very thin. Gills *decurrent at the base, crowded*, thin, linear. Spores white, "globose, 6μ, with a large central gutta" Quél.; "subcylindrical, obtusely rounded at both ends, obliquely apiculate at the base, 7–8 × 3–4μ" Sacc. On ash, beech, and willow stumps. Nov.—Jan. Uncommon. (*v.v.*)

1458. **P. reniformis** Fr. Fr. Icon. t. 89, fig. 3. *Reniformis*, kidney-shaped.

P. 6–10 mm., *cinereous, horizontal, reniform*, plane, emarginate behind, *villose*; margin spreading. St. *rudimentary*, lateral, *villose*. Gills *grey*, linear, *running out from the stem-like tubercle*, thin. Flesh thin, *somewhat gelatinous*, diaphanous. Spores white, globose, 3–4μ, warted, 1-guttulate. On buried twigs, and branches of silver-fir. Aug.—Oct. Uncommon. (*v.v.*)

1459. **P. Laurocerasi** B. & Br. Cke. Illus. no. 287, t. 242, top figs. *Cerasus Laurocerasus*, the cherry laurel.

P. 2–3 cm., *brown, oyster-shaped, sulcate*, cuticle very thin, cracking at the furrows. St. obsolete. Gills *pinkish*, adnate, *connected by veins, broad*. Spores white, ovate, 8 × 5μ. On a trunk of cherry laurel. Oct. Rare.

1460. **P. tremulus** (Schaeff.) Fr. Sow. Eng. Fung. t. 242. *Tremulus*, shaking.

P. 1–4 cm., *fuscous-grey, becoming pale*, submembranaceous, somewhat horizontal, *reniform, plane*, sometimes infundibuliform and lobed, hygrophanous, diaphanous, *tomentose* under a lens. St. 8–12 × 4–6 mm., *grey*, exactly lateral, *ascending-vertical*, round, *dilated upwards.* Gills *grey, adnate* or decurrent, very unequal, linear, narrow, *somewhat distant.* Flesh *pallid,* thin. Spores pip-shaped, 7–8 × 3–4μ, 1–2-guttulate. On the ground, moss, and fungi. Aug.—Dec. Uncommon. (*v.v.*)

1461. **P. acerosus** Fr. Fr. Icon. t. 89, fig. 2.

Acerosus, acerose, coniferous.

P. 2–3 cm., *grey, or brown, silky white when dry,* membranaceous, reniform, *somewhat lobed, striate,* flaccid. St. 2–6 × 2 mm., often wanting, lateral, *whitish,* base *strigose-rooting.* Gills *grey,* adnate, linear, *crowded.* Flesh *pallid,* thin. Spores white, globose, 4–6μ, 1-guttulate. On twigs, needles, and stumps in coniferous woods, and on *Sphagnum* in bogs. Aug.—Dec. Not uncommon. (*v.v.*)

1462. **P. dictyorhizus** (DC.) Fr. (= *Calathinus dictyorhizus* (DC.) Quél.) Bolt. Hist. Fung. t. 72, fig. 2.

δίκτυον, net-work; *ῥίζα*, root.

Entirely shining white. P. ·5–1 cm., *orbicular, or dimidiate, lobed,* membranaceous, very delicate, *villosely silky, reticulately fibrillose* at the base. St. 1–3 × ·5 mm., generally wanting, *villose.* Gills extending to the base, distant, linear, lanceolate. Flesh *white,* thin. Spores white, pip-shaped, 6–7 × 4μ, 1-guttulate. On twigs, and dead wood. Oct. Uncommon. (*v.v.*)

III. P. at first resupinate, then reflexed, sessile; gills meeting at an excentric point.

A. P. fleshy, uniform in texture.

1463. **P. porrigens** (Pers.) Fr. (= *Calathinus porrigens* (Pers.) Quél.) Cke. Illus. no. 288, t. 259, fig. A. *Porrigens*, stretching out.

Entirely shining white. P. 2·5–10 cm., *at first resupinate,* sessile, adnate behind, *forming excentric orbicular shields,* with the gills concurrent in an umbilicus, *soon extended laterally,* ascending, *ear-shaped,* narrow at the base, dilated above, at length undulato-lobed, tough, flaccid, pruinose, tomentose towards the base, *diaphanous.* Gills at first concurrent, then decurrent, often branched, somewhat veined, *very narrow,* crowded, linear. Flesh *white,* thin, compact. Spores white, subglobose, 7–8 × 6μ. On coniferous stumps. June—Nov. Common in Scotland, uncommon elsewhere. (*v.v.*)

1464. **P. septicus** Fr. (= *Calathinus pubescens* (Sow.) Quél.) Cke. Illus. no. 288, t. 259, fig. B. *σηπτικός*, putrefying.

Entirely shining white. P. 2–10 mm., *at first resupinate,* attached to the wood, then reflexed and appearing sessile with flaxy rootlets, *villose.* St. 2–4 × ·5 mm., filiform, incurved, *villose,* becoming erect and at length vanishing. Gills converging round the rudiment of a stem, *comparatively broad*, somewhat distant. Flesh *white*, thin at the margin. Spores elliptical, 9–10 × 5μ, often depressed on the one side. On dead twigs, decayed wood, rabbit dung and fungi. March—Nov. Not uncommon. (*v.v.*)

P. nidulans (Pers.) Fr. = **Crepidotus nidulans** (Pers.) Quél. Ench.

B. P. fleshy, striate, with an upper gelatinous layer, or viscous pellicle.

1465. **P. mastrucatus** Fr. (= *Calathinus mastrucatus* (Fr.) Quél.) Cke. Illus. no. 289, t. 243, upper figs.

Mastrucatus, clothed in a sheep-skin.

P. 3–12 cm., *mouse grey, as if prickly with floccose, squarrose, concolorous scales*, obovate, or tongue-shaped, soft, flaccid; margin involute, *lobed*. Gills *whitish-grey*, concurrent in an excentric umbilicus, then converging to the base of the pileus, broad, *somewhat distant*. Flesh, upper layer gelatinous, *mouse-fuscous*, the lower a little thicker, *pallid*. Spores white, "pruniform, 7–9μ" Quél. On old beech stumps. Rare.

1466. **P. atrocaeruleus** Fr. (= *Calathinus atrocaeruleus* (Fr.) Quél.) Cke. Illus. no. 289, t. 243, lower figs.

Ater, black; *caeruleus*, azure-blue.

P. 2·5–5 cm., *dark azure-blue, rarely fuscous*, at first resupinate, soon reflexed, horizontal, obovate, or reniform, *villose*, slightly wrinkled when dry. Gills *whitish, becoming light yellow*, at first concurrent, then reaching the base, *broad*. Flesh, upper layer toughly gelatinous, as much as 4 mm. thick, *fuscous-blackish*, the lower thinner, *whitish*. Spores white, elliptical, 7–8 × 4–5μ, often depressed on one side, with a large central gutta. Cystidia abundant, fusiform, 46–60 × 8–11μ, very thick walled, sometimes septate at the base, encrusted in the upper portion. Smell sometimes pleasant. On beech, birch, and poplar stumps. Oct.—Dec. Uncommon. (*v.v.*)

1467. **P. Leightonii** Berk. Cke. Illus. no. 290, t. 260, upper figs.

The Rev. W. A. Leighton, the eminent Shropshire lichenologist.

P. 10 mm., *umber, then lead-coloured*, at first cyphellaeform, obliquely conical, *furfuraceous, with short, scattered, black bristles intermixed*. Gills *pallid tan colour*, rather thick, distant, somewhat forked at the base, slightly undulated, obscurely wrinkled at the base, the interstices scarcely reticulated. Flesh, upper layer gelatinous, *of the colour of the pileus*, the lower *white*. Spores white, "somewhat sausage-shaped, slightly curved, 10 × 3μ" Sacc. On wood, rotten rails. Dec. Rare.

1468. **P. algidus** Fr. (= *Calathinus algidus* (Fr.) Quél.) Cke. Illus. no. 290, t. 260, lower figs. *Algidus*, cold.

P. 1–5 cm., *umber, or rufous brown*, at first resupinate, then expanded, subreniform and reflexed, fleshy, *velvety*, then smooth, *viscid*. Gills *pallid yellow*, concurrent, then appearing adnate, crowded, rather broad, sometimes crisped at the base. Flesh, upper layer gelatinous, *brownish*, lower *whitish*. Spores white, elliptical, 7–9 × 4–5μ, 1–2-guttulate. On trunks of willow, mountain ash. Aug.—Oct. Uncommon. (*v.v.*)

1469. **P. fluxilis** Fr. (= *Calathinus fluxilis* (Fr.) Quél.) Cke. Illus. no. 291, t. 244, top figs. *Fluxilis*, fluid.

P. 2–3 cm., *somewhat umber, pale grey, or olivaceous*, thin, dimidiate, sessile, reniform, *covered with a fluid, gelatinous stratum*. Gills *whitish, linear*, 2 mm. broad, rounded behind, *distant*. Flesh thin, soft, *yellowish*. Spores white, "elliptical-oblong, 10μ, guttate" Quél. On mossy beech trunks, and on wood, and sawdust. Oct. Rare.

1470. **P. cyphellaeformis** Berk. (= *Dictyolus cyphellaeformis* (Berk.) Cost. & Duf.) Cke. Illus. no. 291, t. 244, middle figs. *κύφελλα*, the hollow of the ear; *forma*, shape.

P. 4–10 mm., *cinereous*, cup-shaped, then dependent, very minutely strigose, especially at the base; margin *paler*, sprinkled with a few meal-like scales. Gills *pure white*, rather distant, the alternate ones shorter, narrow, linear. Flesh, upper layer gelatinous, *cinereous*, the lower *white* and very thin. Spores white, sausage-shaped, curved, 7–8 × 4μ, 2-guttulate. On dead herbaceous stems, and sticks. Feb.—Oct. Uncommon. (*v.v.*)

1471. **P. applicatus** (Batsch) Berk. (= *Calathinus applicatus* (Batsch) Quél.) Cke. Illus. no. 291, t. 244, bottom figs. *Applicatus*, attached to.

Entirely dark cinereous. P. 4–7 mm., *cup-shaped*, orbicular, *adnate behind*, villose at the base, then *reflexed*, slightly *villose*, or subpruinose, substriate when moist. Gills *whitish at the edge*, few, radiating from a white umbilicus, rather *thick, broad*. Flesh *grey*, upper layer gelatinous. Spores white, elliptical, 7–9 × 4–5μ, often depressed on one side, 1–3-guttulate. On dead branches, and twigs. Jan.—Dec. Common. (*v.v.*)

C. P. membranaceous, not viscid.

1472. **P. Hobsonii** Berk. Cke. Illus. no. 292, t. 212, fig. A. Lieut. Julian C. Hobson.

P. 2–8 mm., *pale grey*, membranaceous, *reniform*, or dimidiate, *sessile, minutely downy*; margin involute. Gills pallid, rather distant. On larch stumps. Sept. Rare.

1473. **P. striatulus** Fr. (= *Calathinus striatulus* (Fr.) Quél.) Fr. Icon t. 89, fig. 5. *Striatulus*, somewhat striate

Entirely pale cinereous. P. 4–7 mm., sessile, cup-shaped, *very tender, pellucid, striate, wrinkled when dry.* Gills *sometimes whitish,* few, *distant.* Spores white, "oval, 5μ" Quél. On twigs, branches, and stumps of fir, hazel, elm. May—Dec. Uncommon.

1474. **P. hypnophilus** Berk. (= *Calathinus hypnophilus* (Berk.) Quél.) Cke. Illus. no. 292, t. 212, fig. C.
Hypnum, the name of a moss genus; φίλος, loving.

Entirely white. P. 5–10 mm., sessile, resupinate, somewhat reniform, *rugose,* slightly striate. Gills thin, crowded, radiating from a central point. Spores white, elliptical, 5 × 3μ. On the larger mosses, and fallen leaves. Sept.—Dec. Uncommon.

1475. **P. chioneus** (Pers.) Fr. (= *Calathinus chioneus* (Pers.) Quél.) Cke. Illus. no. 292, t. 212, fig. D. χιών, snow.

Entirely snow-white. P. 4–5 mm., *very tender,* lateral, then resupinate, orbicular, becoming reniform, *villose;* margin involute. St. 1–2 × ·5 mm., *villose, vanishing.* Gills radiating, crowded, rather broad, sometimes with intermediate shorter ones. Spores white, pip-shaped, depressed on one side, 5–8 × 3μ, minutely verrucose. On twigs, dead leaves, and dung. Sept.—Dec. Uncommon. (*v.v.*)

Spores white, gill edge longitudinally split.

Schizophyllum Fr.

(σχίζω, I split; φύλλον, a leaf.)

Pileus coriaceous, resupinate. Stem lateral or none. Gills radiating from a central point, becoming longitudinally split and revolute at the edge. Spores white, cylindrical, smooth, continuous. Cystidia none. Growing on wood.

1476. **S. commune** Fr. (= *Schizophyllum commune* Fr. var. *multifidum* Massee.) Grev. Scot. Crypt. Fl. t. 61.
Commune, common.

P. 1–3 cm., *greyish, or flesh colour, becoming white,* more or less fan-shaped, or reniform, often much lobed, very arid, pendulous, commonly extended behind into a stem-like base, *covered with white-grey down, then strigose.* Gills *fuscous-grey, then purplish, or whitish, splitting and revolute at the edge,* radiating, narrow. Flesh *brownish, becoming whitish.* Spores white, cylindrical, straight, or curved, 6 × 3μ. On fallen trunks, and dead branches in woods, also in timber yards on imported timber. May—Jan. Not uncommon. (*v.v.*)

Spores pink.

Claudopus W. G. Sm.

(*Claudus*, lame; πούς, foot.)

Pileus fleshy, excentric, lateral, or resupinate. Stem lateral, or none. Gills radiating from a central point, or decurrent. Spores pink, elliptical, globose, oblong, smooth, angular or verrucose, continuous. Cystidia present. Growing on wood, or on the ground.

1477. **C. variabilis** (Pers.) W. G. Sm. (= *Crepidotus variabilis* (Pers.) Quél.) Cke. Illus. no. 371, t. 344, top figs.

Variabilis, variable.

P. 1–2 cm., *white*, slightly fleshy, resupinate, then reflexed, *tomentose*, putting forth from the centre a short, incurved, villose stem which is obliterated when the pileus is reflexed. Gills *whitish, then rubiginous*, at first concurrent in an excentric point, then reaching the base, broad, distant. Flesh *white*. Spores pink, elliptical, 5–6 × 3μ, "warted" Maire. Cystidia "on edge of gill clavate-bottle-shaped, 36–45 × 7–9μ" Rick. On dead sticks, fallen branches, and leaves, in woods, hedgerows, and wood-yards. Jan.—Dec. Common. (*v.v.*)

var. **sphaerosporus** Pat. σφαῖρα, a ball; σπορά, seed.

Differs from the type in the *subglobose spores*, 7–8 × 6–7μ. Woods, and hedgerows. Sept.—Dec. Not uncommon. (*v.v.*)

1478. **C. depluens** (Batsch) W. G. Sm. Cke. Illus. no. 371, t. 344, middle figs. *Depluens*, raining down.

P. 2–3 cm., *rufescent-hoary*, submembranaceous, resupinate, then reflexed, changeable in form, *delicately silky*; at first with a *villose stem then stemless*. Gills *grey, then rufescent*, scarcely decurrent, diverging, *broad*, ventricose, somewhat crowded. Flesh *thin, watery, fragile*. Spores pink, "subelliptical-reniform, 10–12 × 5–6μ. Cystidia on edge of gill undulating-filiform, 45–50 × 4–6μ" Rick. On the ground, amongst moss, sawdust, wood-ashes, sometimes in stoves. Oct. Uncommon.

1479. **C. byssisedus** (Pers.) Fr. Pers. Icon. et Descr. t. 14, fig. 4.

βύσσος, fine flax; *sedeo*, I sit.

P. 1–4 cm., *grey, becoming pale when dry*, slightly fleshy, at length *horizontal, reniform*, plane, villose. St. 12 mm. long, *incurved, villose, attenuated upwards, zoned at the base with white cottony fibrils*. Gills *whitish-cinereous*, then *rubiginous with the spores, adnato-decurrent*, ventricose, rather broad. Flesh *concolorous*, thin. Spores pink, angular, broadly elliptical, 10–11 × 7μ, 1–2-guttulate. On dead beech, and wood of hornbeam, also on the ground. Sept.—Oct. Uncommon. (*v.v.*)

Spores ochraceous.

Crepidotus Fr.

(κρηπίς, a man's boot; οὖς, ear.)

Pileus fleshy, excentric, lateral, or resupinate. Stem lateral, or none. Gills more or less decurrent, or radiating from a central point. Spores ochraceous, ferruginous, or fuscous; elliptical, subglobose, oval, or fusiform; smooth, granular, verrucose, or echinulate; continuous. Cystidia present. Growing on wood, rarely on the ground.

C. palmatus (Bull.) Fr. = **Pleurotus palmatus** (Bull.) Quél.

1480. **C. nidulans** (Pers.) Quél. Ench. (= *Pleurotus nidulans* (Pers.) Fr.; *Crepidotus jonquilla* (Paul.) Quél.) Pers. Icon. et Descr. t. 6, fig. 4. *Nidulans*, nestling.

P. 1–8 cm., *yellow, or yellow orange, becoming pale,* fleshy, resupinate, cup-shaped, then expanded and reflexed, dimidiate, kidney-shaped, sessile, *tomentose,* the tomentum concolorous, or becoming whitish; margin inrolled, *often lobed and orange-coloured.* Gills *orange tawny,* at first concurrent, then adnate, 2–4 mm. broad, subdistant, often veined on the sides. Flesh *yellowish, becoming whitish when dry,* staining paper a yellow colour, thick, soft. Spores bright ochraceous, broadly elliptical, 5–6 × 4μ, 1–2-guttulate. Smell pleasant, "of melon" Quél. Gregarious, sometimes imbricate. On rotten pine, and beech wood. Sept.—Oct. Uncommon. (*v.v.*)

1481. **C. alveolus** (Lasch) Fr. Cke. Illus. no. 534, t. 499, upper figs. *Alveolus*, a little trough.

P. 2–6 cm., *ochraceous fuscous, occasionally becoming olive at the margin, becoming pale when dry,* fleshy, obovate, somewhat cuneiform, sometimes repand, rather plane, *moist,* smooth, *dimidiate,* laterally somewhat sessile, *or extended behind with a short, stem-like tomentoso-villous base* and horizontal. Gills *clay-fuscous, determinate,* 4 mm. broad, crowded. Flesh *whitish,* thick, *soft.* Spores brownish, elliptical, 8–10 × 6μ. Stumps and logs, especially oak. Aug.—Nov. Not uncommon. (*v.v.*)

1482. **C. mollis** (Schaeff.) Fr. Cke. Illus. no. 535, t. 498. *Mollis*, soft.

P. 3–7 cm., *pallid, then becoming hoary,* fleshy, convexo-plane, obovate, or reniform, undulate and lobed when larger, flaccid, *smooth,* dimidiate, *subsessile,* or extended behind into a short, 12 mm., strigose st., often imbricated. Gills *whitish grey, then watery cinnamon, commonly decurrent to the base, linear,* 2–4 mm. broad, often branched. Flesh *watery whitish, subgelatinous* especially under the cuticle, thick. Spores ochraceous, elliptical, 8–9 × 5μ. Cystidia "on edge of gill cylindrical-filiform, 45–54 × 5–6μ" Rick. Taste mild. Edible. Stumps, twigs, fallen branches, and sawdust. May—Dec. Common. (*v.v.*)

1483. **C. applanatus** (Pers.) Fr. *Applanatus*, flattened.

P. 1–8 cm., *watery cinnamon, or fuliginous, whitish when dry*, very hygrophanous, fleshy, *fragile, wholly plane* and horizontal, *extended behind* in a straight line *into a very short, white, tomentose st., reniform*, or cuneiform, at length depressed behind, subsessile, minutely tomentose; margin slightly striate when moist. Gills *whitish, then watery cinnamon, ending determinately behind*, linear, crowded, thin. Flesh *whitish*, watery, soft. Spores brownish, elliptical, 7–8 × 5μ, with a large central gutta. Cystidia "on edge of gill cylindrical-filiform, 36–40 × 5–7μ" Rick. On twigs and dead wood. Oct. Uncommon. (*v.v.*)

1484. **C. calolepis** Fr. Fr. Icon. t. 129, fig. 4.
καλός, beautiful; λεπίς, scale.

P. 1–7 cm., *cream, or pale yellow, beautifully variegated with minute, crowded, rufescent scales*, fleshy, reniform, convex, almost shell-shaped, dimidiate, sessile on a small villose knot, margined with white behind. Gills *pallid fuscous, then fuscous ferruginous, concurrent at the base*, rounded behind, comparatively broad. Flesh *white, firm*, fairly thick. Spores brownish, elliptical, 7–9 × 5–6μ. Stumps, and fallen branches, especially poplar, and ash. Common. (*v.v.*)

1485. **C. putrigenus** Berk. & Curt. *Puter*, rotten; *genus*, birth.

P. 12–19 mm., *whitish*, subreniform, imbricate, tomentose, beset at the base with a delicate white tomentum. Gills *whitish, becoming ferruginous brown*, broad. Spores ferruginous, subglobose, 7μ. Damp wood. Sept. Rare.

1486. **C. versutus** Peck. *Versutus*, deceitful.

P. 9–20 mm., *white*, resupinate, then reflexed, sessile, thin, *covered with a soft villose tomentum*; margin incurved. Gills *pallid, then ferruginous, concurrent in an excentric point*, rounded behind, rather broad, subdistant. Flesh *white*, thin. Spores ferruginous brown, subelliptical, 9–10 × 4–5μ. On dead wood, and rotten branches. June—Nov. Not uncommon. (*v.v.*)

1487. **C. epigaeus** (Pers.) B. & Br. Cke. Illus. no. 537, t. 516, fig. A.
ἐπίγαιος, upon the earth.

P. 1–2 cm., *reddish grey*, fragile, reniform, or flabellate; base *whitish*, downy. Gills *watery rufescent, divergent*, narrow. Flesh *pallid*, very thin. Spores pale cinnamon, broadly elliptical, 10 × 7μ. On marlstone clay. Nov. Rare.

1488. **C. haustellaris** Fr. (= *C. flurstedtiensis* (Batsch) Sacc.) Cke. Illus. no. 536, t. 515, fig. A. *Haurio*, I draw water.

P. 2–3 cm., *pale yellowish tan, becoming pale, but often cinnamon when old*, fleshy, almost pellucid, flaccid, *exactly lateral, reniform*, plane, *delicately villose.* St. 4–8 × 2–4 mm., *white, attenuated upwards,* almost conical, round, *villose,* somewhat ascending when young, then straight and horizontal. Gills *pallid, then fuscous cinnamon,* determinate, *rounded,* somewhat crowded. Flesh *pallid,* watery, thin. Spores "ellipsoid, ochraceous, 6–7 × 4·5–5·5μ." Sacc. On dead trunks, and fallen branches of poplar. June—Oct. Rare.

C. Rubi Berk. = **Naucoria effugiens** Quél.

C. variabilis (Pers.) Quél. = **Claudopus variabilis** (Pers.) W. G. Sm.

1489. **C. chimonophilus** B. & Br. Cke. Illus. no. 536, t. 515, fig. D.
χειμών, winter; φίλος, loving.

Entirely pure white. P. 4–6 mm., convex, fleshy, *clothed with villose down*; margin inflexed. St. extremely short, or obsolete, excentric. Gills *attenuated behind,* narrow, *distant, few.* Flesh *white, rather thick.* Spores pale cinnamon, elliptical, 5 × 3μ. Dead branches of *Pyrus torminalis.* Dec. Rare.

1490. **C. epibryus** Fr. Cke. Illus. no. 537, t. 516, fig. C.
ἐπί, upon; βρύον, moss.

P. 4–10 mm., *shining white,* membranaceous, *cup-shaped,* resupinate, sessile, pellucid, adnate at the vertex, becoming *silky-even.* Gills *whitish, then pale yellowish, concurrent in the centre,* thin, *crowded.* Flesh *white,* very thin. Spores "ochraceous, elliptical, fusiform, 10–12μ, minutely echinulate" Quél. On mosses, grass, holly-leaves, *Vaccinium,* twigs, and herbaceous stems. Oct.—Jan. Uncommon.

1491. **C. pezizoides** (Nees) Fr. Cke. Illus. no. 537, t. 516, fig. D.
πέζις, Peziza; εἶδος, like.

P. 4–6 mm., *whitish, or reddish,* sessile, thin, *cup-shaped, then reflexed, mealy* subtomentose. Gills *olivaceous fuscous, then tawny,* concurrent at the centre, subdistant. Gregarious. On rotten branches, and old wood. Rare.

1492. **C. Phillipsii** B. & Br. Cke. Illus. no. 536, t. 515, fig. C.
W. Phillips of Shrewsbury, the eminent mycologist.

Slightly umber. P. 4–6 mm., oblique, *striate,* smooth. St. 2–3 × ·5 mm., incurved at the base, sometimes obsolete. Gills shortly adnate, narrow, ventricose. Flesh *concolorous,* very thin. Spores pale ochraceous, elliptical, 5–6 × 2·5–3μ. Dead grass leaves, and stems. May—Oct. Uncommon. (*v.v.*)

1493. **C. Ralfsii** B. & Br. Cke. Illus. no. 537, t. 516, fig. B.
J. Ralfs, an eminent botanist.

P. 5–15 mm., *yellow, or fuscous,* semi-reflexed, *delicately furfuraceous, slightly hispid,* the involute margin spreading, adfixed by cottony flocci. Gills *clay colour*, ventricose, *edge whitish*. Flesh *concolorous*, thin. Spores brown, broadly elliptical, 7–8 × 4–5μ, with a large central gutta. Dead branches, and decaying wood. March. Rare. (*v.v.*)

1494. **C. luteolus** Lamb. *Luteolus*, yellowish.

P. *clear yellow, then pale*, thin, stipitate at first, then resupinate and st. disappearing, *tomentose*. Gills *orange yellow, then cinnamon*, crowded. Spores rusty.

1495. **C. Parisotii** Pat. Parisot.

P. 5–6 mm., *bright lemon yellow*, sessile, *velvety tomentose with simple, or branched thin hairs*; margin incurved, attached to the support above by some white fibrils, with an extremely short stem below which terminates on the under side of the p. in a yellowish white, projecting, velvety heap. Gills *reddish, unequal*, the longer ones reaching the central mass, *very distant*, thick. Spores ochraceous, hyaline, or granular, oval, apiculate, 4–5 × 2μ. On twigs. Rare.

1496. **C. proboscideus** Fr. προβοσκίς, an elephant's trunk.

P. 2–3 cm., *ochraceous, sometimes white*, fleshy, elongate shell-shaped, or trumpet-shaped, more prolonged on the one side, *cottony*. St. conical, very short, really only a continuation of the p., *base cottony*. Gills *whitish, then concolorous, or watery cinnamon*, thin, fairly distant. Rotten wood, fallen branches, and wooden ceiling. Dec. Rare.

II. Receptacle fleshy, trama vesiculose, and traversed by lacticiferous vessels. Spores white, or yellow.

Latex watery, uncoloured.

Russula Fr.

(*Russulus*, reddish.)

Pileus fleshy, regular. Stem central, fleshy. Gills adnate, sinuato-adnate, adnexed, free, or decurrent, rigid, fragile, edge acute. Spores white, or yellow, rarely greenish; globose, subglobose, or elliptical, echinulate, verrucose, subreticulate, or with anastomosing ridges and spines, continuous. Cystidia present, or absent. Growing on the ground, rarely on wood.

I. P. fleshy throughout, margin *more or less involute, pellicle slightly developed, dry, adnate*. Flesh compact, firm. Gills unequal, *alternate*. Spores white in the mass.

*Flesh not changing colour, gills narrow, decurrent.

1497. **R. delica** Fr. Bres. Fung. Trid. t. 201. *Delica*, weaned.

P. 5–8 cm., *white, becoming spotted with light brown*, convex, umbilicate, then somewhat infundibuliform, at first delicately tomentose, then only in little patches. St. 2–3·5 × ·5–2 cm., *white, becoming somewhat light brown*, attenuated at the base, finely tomentose under a lens. Gills *white*, exuding watery drops when young, decurrent, or adnate, crowded, *narrow*, sometimes branched near the margin, rarely bifid at the base. Flesh *white*. Spores hyaline, subglobose, 8–9 × 7–8μ, verrucose, with a large central gutta. "Hyphae containing oil globules traverse the tissue and terminate in cystidia-like bodies, 60–70 × 6–8μ." Bres. Smell pleasant, taste slowly acrid. Deciduous, and pine woods. Aug.—Oct. Not uncommon. (*v.v.*)

1498. **R. chloroides** (Krombh.) Bres. (= *Lactarius exsuccus* (Otto) Fr.; *Russula delica* Aut. plur. pr. p. Bres.) Bres. Fung. Trid. t. 202. χλωρός, pale green; εἶδος, like.

P. 6–15 cm., *pallid then ochraceous*, convexo-plane, then somewhat infundibuliform, pubescent, soon smooth, areolately cracked in dry weather. St. 3–6 × 2–3·5 cm., *white then concolorous, the extreme apex incircled by a greenish zone*, equal, or attenuated at the base, rugulose. Gills *whitish*, or *greenish, becoming pallid and often spotted fuscous in old age*, subdecurrent, or sinuato-adnate, broad, somewhat crowded, *connected by veins*. Flesh *white*, cheesy. Spores hyaline, subglobose, 8–11 × 8–10μ, echinulate. Cystidia fusiform, 70–90 × 8–12μ. Smell at length unpleasant, taste acrid. Woods. Aug.—Dec. Common, especially in some years. (*v.v.*)

1499. **R. elephantina** Fr. (= *Russula chloroides* Krombh. sec. Bataille.) ἐλεφάντινος, of ivory.

P. 7·5 cm., *fuscous-tan, paler at the margin*, convexo-umbilicate; margin undulated, exceeding the gills. St. 5–7·5 × 2·5 cm., *shining white*, obese, very hard. Gills *white, spotted pale yellowish when touched, obtusely or sinuato-adnate, arcuate, somewhat crowded, thin*, divided behind. Flesh *of stem shining white*. Spores "14 × 10μ" Sacc. Woods. Sept. Rare.

**Flesh becoming black.

1500. **R. nigricans** (Bull.) Fr. Cke. Illus. no. 970, t. 1015. *Nigricans*, becoming black.

P. 5–20 cm., *olivaceous-fuliginous, at length black*, convex, then flattened and umbilicato-depressed, slightly viscid when moist, at length rimoso-squamulose. St. 3–7 × 2·5 cm., *pallid, at length black*, equal. Gills *ochraceous, reddening when touched, rounded behind*, slightly adnexed, *thick, distant, wide*. Flesh firm, *white, becoming red on exposure to the air, and finally black*. Spores white, globose, 8–9μ, verrucose, with a large central gutta. Cystidia "only on edge of gill, vesiculose, then ventricose, pointed, 45–60 × 15–30μ" Rick. Taste mild, then acrid. Woods. June—Dec. Common. (*v.v.*)

1501. **R. adusta** (Pers.) Fr. (= *Russula albo-nigra* Krombh. sec. Quél.) Cke. Illus. no. 972, t. 1051. *Adusta*, scorched.

P. 8–15 cm., *pallid, or whitish, becoming cinereous-fuliginous*, convex, then depressed, and somewhat infundibuliform. St. 3–5 × 2–3 cm., *concolorous*, obese. Gills *white, then dingy*, adnate, then decurrent, *thin, crowded, narrow*. Flesh *white, then brownish, and finally black*. Spores white, globose, 8μ, verrucose. Cystidia "sparse, subulate, 45–50 × 7μ" Rick. Taste mild. Woods. Aug.—Nov. Common. (*v.v.*)

var. **caerulescens** Fr. *Caerulescens*, becoming blue.

Differs from the type in the flesh *becoming dark blue* when cut or broken. Deciduous woods. Rare. (*v.v.*)

var. **albo-nigra** (Krombh.) Fr. Cke. Illus. no. 971, t. 1016. *Albus*, white; *nigra*, black.

Differs from the type in the *white pileus becoming smoky near the margin, the stem fuscous from the first, and the flesh immediately becoming black when broken*. Cystidia "only on the edge of the gill, subulate-pointed, 75–90 × 9–10μ, filled with dark juice" Rick. Woods. Aug.—Oct. Uncommon. (*v.v.*)

1502. **R. densifolia** (Secr.) Gill. Cke. Illus. no. 973, t. 1017. *Densus*, crowded; *folium*, leaf.

P. 7–10 cm., *whitish, then dingy brown, and finally black*, convex, then depressed, slightly viscid at first; margin *elastic, villose, white*. St. 3–5 × 1–2 cm., *white, then concolorous*, equal, pruinose. Gills *white, becoming grey when touched, then dingy, and finally black*, adnate, decurrent by a tooth, *narrow, crowded, thin*. Flesh *white, becoming red when broken, and finally black*. Spores white, globose, 7–8μ, echinulate, 1-guttulate. Smell pleasant, taste slowly acrid. Woods. Aug.—Oct. Not uncommon. (*v.v.*)

1503. **R. semicrema** Fr. Fr. Icon. t. 172, fig. 1. *Semi*, half; *crema*, burnt.

P. 6–11 cm., *persistently white*, convex then plane, disc umbilicate. St. 5–8 × 5–6 cm., *white, becoming black*, firm. Gills *persistently white*, decurrent, crowded, thin. Flesh of *pileus persistently white, becoming black in the stem*. Spores white, globose, 8–9μ, verruculose. Taste mild. Woods. Aug.—Sept. Rare.

II. Pellicle of the pileus *dry, adnate*, rarely possessing cystidia, *usually breaking up into flocci, granules, or areolae*. Margin rounded, *never striate* (except 1511), *or involute*. Gills *with a very broad, rounded apex*. Spores whitish cream-colour in mass.

1504. **R. lactea** (Pers.) Fr. *Lactea*, milk-white.

P. 5–12 cm., *milk-white, then tan-white*, convex, then plane, often excentric, *pruinose, appearing as if stippled under a lens*, then minutely cracked. St. 3–5 × 2–4 cm., *white*, equal, or ventricose, pruinose. Gills *whitish cream colour, free, very broad, thick, distant*, forked. Flesh *white*, compact. Spores *very pale ochraceous*, globose, 7–8μ, echinulate, 1-guttulate. Taste mild. Edible. Woods. Aug.—Oct. Uncommon. (*v.v.*)

1505. **R. incarnata** Quél. Cke. Illus. no. 990, t. 1071, as *Russula lactea* Pers. var. *incarnata* Quél. *Incarnata*, flesh-colour.

P. 6–10 cm., *white, tinged with rose, at length tan colour*, convex, then depressed, *minutely mealy, then cracked into areolae*. St. 4–6 × 1·5–2 cm., *white*, firm, pruinose. Gills *whitish cream colour, adnate, broad*, forked, rigid. Flesh *white*. Spores *very pale ochraceous*, globose, 8–10μ, echinulate, 1-guttulate. Taste mild. Edible. Woods. Aug.—Oct. Uncommon. (*v.v.*)

1506. **R. virescens** (Schaeff.) Fr. Cke. Illus. no. 991, t. 1039. *Virescens*, green.

P. 6–12 cm., *deep or pallid green, globose*, then expanded, at length depressed, often unequal, *the flocculose cuticle broken up into patches, or warts*. St. 5–10 × 2–3 cm., *white, or whitish cream colour*, firm, pruinose, *subrivulose*. Gills *white, then whitish cream colour*, free, or adnate, thick, somewhat crowded, sometimes forked. Flesh *white*, not very compact. Spores *very pale ochraceous*, globose, 6–8μ, verrucose, 1-guttulate. Cystidia narrowly fusiform, apex obtuse, 2–3μ in diam., 55–65 × 8–10μ. Taste mild. Edible. Woods. July—Oct. Common. (*v.v.*)

1507. **R. lepida** Fr. (= *Russula lepida* Fr. var. *pulcherrima* Gillet.) Bres. Fung. Trid. t. 204. *Lepida*, charming.

P. 5–10 cm., *blood-red-rose, becoming pale, and somewhat tan-leather colour at the disc*, convex, then expanded, rarely depressed, *pruinose, appearing under a lens as if stippled, at length often rimoso-squamulose*. St. 3·5–7 × 1·5–2 cm., *white, often tinged with rose colour, especially on one side, or at the base*, equal, or attenuated at the base, *very firm*. Gills *whitish cream colour, the edge often minutely dentate and red, especially towards the margin of the pileus*, rounded behind, or attenuate, rather thick, somewhat crowded, often forked, connected by veins. Flesh *white*, firm, cheesy. Spores *very pale ochre in the mass*, hyaline under the microscope, globose, 8–10μ, echinulate. Cystidia "cylindrical, rounded, 60–90 × 15μ" Rick. Smell pleasant. Taste pleasant, then very slowly acrid. Edible. Woods. Aug.—Oct. Common. (*v.v.*)

var. **alba** Quél. *Alba*, white.

Differs from the type in the *pruinose, milk-white p. sometimes tinged with rose colour, and the white, mealy st.* Woods. Aug.—Sept. Uncommon. (*v.v.*)

1508. **R. Linnaei** Fr. Fr. Icon. t. 172, fig. 3.
Carlos Linnaeus, the eminent Swedish botanist.

P. 7–12 cm., *unicolorous, dark purple, blood-red, or bright rose,* opaque, *not becoming pale,* convex, then plane and depressed, sometimes repand, *dry, pruinose.* St. 4–6 × 2–3 cm., *blood-red, rarely white,* somewhat ventricose, firm, spongy within, *obsoletely fibrilloso-reticulate.* Gills *white, then ochraceous, adnate, subdecurrent,* rather thick, *broad,* fragile, slightly connected by veins, not crowded, somewhat anastomosing behind. Flesh *white,* compact, firm. Spores *pale ochraceous,* elliptically globose, echinulate, 8–11 × 8μ. Cystidia "on surface of gill sparse, cylindrical, pointed, 50–60 × 8–12μ" Rick. Taste mild. Woods. Oct. Rare.

1509. **R. azurea** Bres. Bres. Fung. Trid. t. 24. *Azurea,* sky-blue.

P. 3–6 cm., *bright blue, margin sometimes lilac, becoming pale,* convex, then plane or depressed, fleshy, soon dry, *constantly minutely granular;* margin scarcely striate in old age. St. 3–5 × 1–1·5 cm., *white,* ventricose, or attenuated at the base, somewhat rugulose, firm. Gills *white, attenuato-adnate,* crowded, equal, forked. Flesh *white.* Spores white, subglobose, 8–9 × 8μ, verrucose. Cystidia fusiform, 60–70 × 12–13μ. Taste mild. Edible. Coniferous woods, and under conifers. Aug.—Oct. Uncommon. (*v.v.*)

1510. **R. olivacea** (Schaeff.) Fr. Cke. Illus. no. 1001, t. 1041.
Olivacea, olivaceous.

P. 6–12 cm., *dingy purple, then olivaceous, or wholly fuscous-olivaceous,* convex, then plane and depressed, fleshy, *slightly silky and squamulose.* St. 5–8 × 1·5–2 cm., *rose colour, or pallid,* firm, ventricose. Gills *bright yellow,* adnexed, wide, with shorter and forked ones intermixed, crowded. Flesh *white, becoming yellowish.* Spores *pale ochraceous,* globose, 10μ, punctate. Cystidia "subulate, 50–75 × 8–12μ" Rick. Taste mild. Edible. Fir woods. Aug. Uncommon.

1511. **R. elegans** Bres. Bres. Fung. Trid. t. 25. *Elegans,* neat.

P. 3–5 cm., *bright rosy flesh colour, soon becoming ochraceous at the circumference,* convex, then somewhat depressed, fleshy, thin, viscid, *everywhere densely granulate;* margin tuberculosely striate when old. St. 3–5 × 1 cm., *white, becoming ochraceous at the somewhat thickened base,* rather rugulose. Gills *whitish, becoming* either wholly, or partially *orange ochre,* attenuated behind, adnexed, or slightly rounded, very crowded, equal, rarely furcate. Flesh *white, becoming ochraceous with age. Spores whitish in the mass, pale greenish hyaline or yellowish under the microscope,* globose, 8–10μ, strongly echinulate, 1-guttulate. Cystidia "sparse, subulate, 50 × 8–9μ" Rick. Taste acrid when old. Coniferous woods. Sept. Uncommon.

1512. **R. serotina** Quél. Cke. Illus. no. 1003, t. 1042, lower figs.
Serotina, late.

P. 2–3 cm., *violet, lilac, bistre, or olivaceous, margin lilac with the extreme edge white*, globose, then plane, *white pruinose at first*. St. 2–3 cm. × 3–4 mm., *white, minutely pubescent*, equal. Gills *white, then tinged yellowish*, adnate, crowded. Flesh *white*. Spores *pale ochraceous*, globose, 7μ, echinulate. Taste acrid. Beech woods, and on old willow, and poplar stumps. Aug.—Sept. Uncommon. (*v.v.*)

III. Pellicle of the pileus viscid, separable at the margin and possessing cystidia. Margin subacute, rarely striate in old age. Flesh firm. Taste *mild*. Gills *attenuated in front, often forked and unequal*. Spores verrucose, small, *white in the mass*.

1513. **R. cyanoxantha** (Schaeff.) Fr. Cke. Illus. no. 1007, t. 1076.
κύανος, blue; *ξανθός*, yellow.

P. 5–15 cm., *lilac, or purplish, then olivaceous green, disc commonly becoming pale, often yellowish, margin commonly becoming azure-blue, or livid-purple*, convex, then plane and depressed, or infundibuliform, viscid, sometimes wrinkled, or streaked. St. 5–9 × 2–3 cm., *white, rarely tinged with lilac, elastic*, equal. Gills *shining white*, rounded behind, connected by veins, *forked*, broad, not much crowded. Flesh *white, purple or reddish under the pellicle*. Spores white, globose, 7–10μ, verrucose. Cystidia abundant, conical, 70–80 × 7–8μ. Taste pleasant. Edible. June—Nov. Common. (*v.v.*)

1514. **R. lilacea** Quél. Cke. Illus. no. 1004, t. 1054.
Lilacea, lilac-coloured.

P. 4–8 cm., *violet, or lilac, often brownish, margin becoming whitish*, convex, then depressed, thin, viscid; margin *striate*, thin. St. 4–6 × 1·5–2 cm., *white, often rosy at the base*, corticate, *fragile*, apex pruinose, *wrinkled-striate*. Gills *white*, free, ventricose, connected by veins, often forked. Flesh *white, violet under the pellicle*. Spores white, subglobose, 8–9μ, verrucose. Smell pleasant, of apple. Taste mild. Edible. Aug.—Sept. Uncommon. (*v.v.*)

1515. **R. citrina** Gillet. Cke. Illus. no. 1031, t. 1078.
Citrina, lemon yellow.

P. 5–10 cm., *bright citron yellow, colour usually uniform, sometimes paler at the margin, occasionally with a greenish tint, disc at length becoming pale ochraceous*, convex, then more or less depressed in the centre, slightly viscid, pellicle separable; margin thin, becoming tuberculosely striate with age. St. 5–8 × 1–1·5 cm., *white*, equal, or slightly attenuated at the base, striate. Gills *white*, slightly decurrent, forked at the base, and sometimes also near the middle, attenuated at both ends. Flesh *white*. Spores white, globose, 7–8μ, verrucose.

Cystidia abundant, conical, 50–60 × 7–8μ. Taste mild, becoming acrid. Woods, and heaths. Aug.—Oct. Uncommon. (*v.v.*)

1516. **R. fingibilis** Britz. Cke. Illus. no. 1030, t. 1048.
Fingibilis, imaginary.

P. 5 cm., *yellow*, *disc darker*, convex, then plane or depressed, fleshy at the disc, viscid. St. 2·5–4 × 1 cm., *white*, equal, soft. Gills *white*, narrowed behind, almost free, rather crowded, unequal, thin. Flesh *white*. Spores white, broadly elliptical, 9 × 7μ, minutely echinulate. Taste mild. Amongst grass under trees. July. Uncommon.

1517. **R. furcata** (Pers.) Fr. (= *Russula bifida* (Bull.) Schroet.) Barla, Champ. Nice, t. 16, figs. 1–9. *Furcata*, forked.

P. 6–12 cm., *green, becoming somewhat ochraceous at the disc with age*, convex, then plane or depressed, sometimes infundibuliform, fleshy, viscid in wet weather, polished in dry weather; margin thin. St. 4–6 × 1·5–2 cm., *white*, equal, or attenuated downwards, firm. Gills *shining white, sometimes becoming spotted with brown when old, attenuated at both ends, adnato-decurrent, forked from the base*, more rarely higher up, somewhat distant, rather thick. Flesh *white, brownish under the separable pellicle*. Spores white, globose, 6–8μ, minutely verrucose, 1-guttulate. Cystidia clavate, 45–60 × 8–11μ. Taste mild, becoming slightly bitter when old especially in the gills. Woods, and lawns. Aug.—Oct. Common. (*v.v.*)

var. **pictipes** Cke. Cke. Illus. no. 979, t. 1086.
Pictus, painted; *pes*, foot.

Differs from the type in the *slightly striate margin of the pileus, in the stem being rosy at the apex and tinted with green at the base, and in the rosy flesh beneath the cuticle of the pileus*. Woods and under trees. Aug.—Sept. Uncommon. (*v.v.*)

var. **ochroviridis** Cke. Cke. Illus. no. 980, t. 1100.
ὠχρός, pale yellow; *viridis*, green.

Differs from the type in the *paler greenish ochre pileus, the narrower gills, rugose stem, and fuliginous flesh when cut*. Woods. Aug.

1518. **R. mitis** Rea. *Mitis*, mild.

P. 6–8 cm., *disc yellowish, surrounded by purplish mouse colour, or pale rose and purple, becoming yellowish towards the circumference*, convex, then plano-expanded and depressed, spongy but firm, viscid, pellicle easily separable; margin thin, *pellucidly striate*, tuberculate when old. St. 4–5 × 1·5 cm., *white*, attenuated downwards, rugulose. Gills *white, then tinged faint straw colour*, adnate, attenuated at both ends, branched from the base or higher up, scarcely any intermediate ones, veined at the base. Flesh *white, somewhat rust colour at the base of the stem, ochraceous under the pellicle.* Spores white, globose, 6μ, verrucose. Taste mild. Woods. Aug.—Oct. Uncommon. (*v.v.*)

1519. **R. heterophylla** Fr. (= *Russula livida* (Pers.) Schroet.) Cke. Illus. no. 1010, t. 1045. ἕτερος, different; φύλλον, leaf.

P. 5–8 cm., *greenish or yellowish brown, disc becoming ochraceous,* very variable in colour but *never becoming reddish or purple,* fleshy, firm, convex, then plane and depressed; margin thin, sometimes densely but slightly striate. St. 2–5 × 1·5–2·5 cm., *shining white,* equal, or attenuated at the base, firm, delicately striate. Gills *shining white, decurrent, very narrow, very crowded,* thin, often forked. Flesh *white.* Spores white, globose, 6–7μ, verrucose. Cystidia "on edge of gill filiform-clavate, often constricted and capitate, 50–60 × 7–9μ" Rick. Smell and taste pleasant. Edible. Woods. July—Oct. Uncommon. (*v.v.*)

1520. **R. galochroa** Fr. Cke. Illus. no. 1011, t. 1089.
γάλα, milk; χρώς, colour.

P. 4–6 cm., *milk white, then greenish,* convex, then plane, viscid in wet weather, sometimes sprinkled with white floccose spots; margin sometimes striate. St. 2·5–5 × 1–1·5 cm., *white,* firm. Gills *white,* adnate, crowded, *narrow,* more or less forked. Flesh *white.* Spores white, globose, 6–7μ, verrucose. Taste mild. Edible. Woods. Aug.—Oct. Not uncommon. (*v.v.*)

1521. **R. virginea** Cke. & Massee. Cke. Illus. no. 1197, t. 1197.
Virginea, maidenly.

Entirely pure white. P. 5 cm., convex then depressed, fleshy, firm, viscid when moist, polished when dry. St. 5 × 2 cm., attenuated upwards, firm, finely rugulose. Gills subdecurrent, *very narrow,* 1–2 mm. wide, repeatedly forked, connected by veins, brittle, crowded. Spores white, globose, 4μ, minutely apiculate at the base, *almost smooth.* Cystidia none. Taste mild. Woods. Oct. Rare. (*v.v.*)

IV. Pellicle of the pileus viscid, more or less separable, possessing cystidia, *yellowish, ochraceous, or brownish,* sometimes olivaceous brown, never red or violet. Margin straight, more or less striate, subacute. Taste *acrid.* Spores pure white, or cream.

1522. **R. foetens** (Pers.) Fr. Cke. Illus. no. 1015, t. 1046.
Foetens, stinking.

P. 8–15 cm., *dingy yellow, often becoming pale,* thinly fleshy, globose, then expanded and depressed, *rigid,* viscid in wet weather; margin broadly membranaceous, at first incurved, at length *tuberculately-striate.* St. 5–9 × 1–3 cm., *whitish,* ventricose. Gills *whitish, or straw colour, often dingy when bruised,* at the first *exuding watery drops,* adnexed, crowded, connected by veins, often forked. Flesh *white, then ochraceous.* Spores *pale ochraceous,* subglobose, 8–11 × 8–9μ, echinulate. Cystidia clavate, often slightly constricted below the apex, 45–50 × 8–10μ, contents yellowish. Smell very strong. Taste acrid. Woods. July—Nov. Common. (*v.v.*)

1523. **R. consobrina** Fr. (= *Russula livescens* (Batsch) Quél.) Cke. Illus. no. 1012, t. 1055. *Consobrina,* cousin.

P. 7–12 cm., *dark cinereous, or fuscous olivaceous,* fleshy, *fragile,* campanulate, then expanded, at length depressed, viscid; margin membranaceous. St. 5–8 × 2–2·5 cm., *white, at length becoming cinereous,* equal. Gills *white, then greyish,* free, *forked,* broad, *crowded,* thick. Flesh *white, cinereous under the pellicle.* Spores *pale ochraceous,* globose, 10μ, verrucose. Smell faint, taste very acrid. Coniferous woods. July—Oct. Not uncommon. (*v.v.*)

var. **sororia** (Larb.) Fr. (= *Russula consobrina* Fr. var. *intermedia* Cke.) Fr. Icon. t. 173, fig. 1. *Sororia,* sisterly.

Differs from the type in the *striate margin of the pileus, and in the subdistant gills, connected by veins.* Woods, and pastures. July—Oct. Not uncommon. (*v.v.*)

1524. **R. pectinata** (Bull.) Fr. (= *Russula consobrina* Fr. var. *sororia* (Larb.) Cke.) Cke. Illus. no. 1024, t. 1101.
Pectinata, like the teeth of a comb.

P. 4–8 cm., *toast brown, becoming pale tan, disc always darker,* fleshy, viscid, *rigid,* convex, then flattened and depressed, or concavo-infundibuliform; margin thin, *tuberculately-sulcate.* St. 3–5 × 1–2·5 cm., *shining white,* equal, or attenuated at the base, *rigid, substriate longitudinally.* Gills *whitish,* attenuato-free, broader towards the margin, *equal,* somewhat crowded. Flesh *white, light yellowish under the pellicle.* Spores pale ochraceous, subglobose, 6–8 × 6–7μ, verrucose. Cystidia sparse, conico-cylindrical, 40–50 × 8–13μ, obtuse. Smell unpleasant. Taste very acrid. Woods, and pastures. Aug.—Oct. Not uncommon. (*v.v.*)

1525. **R. ochroleuca** (Pers.) Fr. Cke. Illus. no. 1025, t. 1049.
ὠχρός, sallow; λευκός, white.

P. 3–9 cm., *yellow, becoming pale,* convex, then flattened or depressed, polished, viscid. St. 4–7 × 1·5–2 cm., *white, becoming cinereous,* firm, equal, sometimes enlarged at the base, *slightly reticulately rugose.* Gills *white, becoming pale,* rounded behind, free, broader in front, *somewhat equal,* fragile. Flesh *white, yellowish under the pellicle.* Spores white, globose, 8–11μ, echinulate. Smell pleasant, taste acrid. Cystidia conical, 55–70 × 8–10μ. Beech, and coniferous woods. July—Dec. Common. (*v.v.*)

var. **claro-flava** (Grove) Cke. Cke. Illus. no. 1198, t. 1196.
Clarus, bright; *flavus*, yellow.

Differs from the type in the *bright chrome-yellow pileus, and in the gills becoming pale lemon-yellow*. Amongst grass in damp places.

var. **granulosa** (Cke.) Rea. Cke. Illus. no. 1026, t. 1038.
Granulosa, mealy.

Differs from the type in the *cuticle of the pileus and stem breaking up into minute granules, which are snow-white at the apex of the stem, fuscous below*. Woods, and pastures. Aug.—Sept. Not uncommon. (*v.v.*)

1526. **R. fellea** Fr. Fr. Icon. t. 173, fig. 2. *Fellea*, full of gall.

Entirely straw-colour. P. 3–9 cm., *often with a deeper yellowish tinge, disc darker*, thinly fleshy, convex, then plane, viscid; margin *striate when old*. St. 5–6 × 1–2·5 cm., equal. Gills *exuding watery drops, then spotted with yellow*, adnate, crowded, thin, narrow, forked, obsoletely connected by veins. Flesh *whitish, then concolorous with the gills*. Spores *very pale ochraceous*, globose, 8μ, echinulate, 1-guttulate. Cystidia conical, 55–65 × 7–9μ. Taste very acrid and bitter. Beech woods. Aug.—Dec. Common. (*v.v.*)

1527. **R. subfoetens** W. G. Smith. Cke. Illus. no. 1016, t. 1047.
Sub, somewhat; *foetens*, stinking.

P. 4–8 cm., *yellowish white, then ochraceous, especially on the disc, firm, rigid*, convex, then plane or depressed, viscid; margin thin, translucid, *tuberculately sulcate*. St. 5–6 × 1–2·5 cm., *white, becoming tinged with yellow*, subequal, or attenuated at the base, firm. Gills *white, becoming yellow*, adnate, thick, distant, narrow, branched. Flesh *white*. Spores *white*, subglobose, 7–8 × 6–7μ, echinulate. Smell somewhat disagreeable, taste slightly acrid. Grassy places, and on lawns under beeches. Aug.—Oct. Uncommon. (*v.v.*)

V. Pellicle of the pileus viscid, more or less separable, possessing numerous cystidia, *purplish*. Margin *straight, acute*, somewhat striate. Gills *more or less unequal and forked*, generally narrow and acutely attenuated in front, often adnate. Flesh *firm*. Taste *acrid*. Spores white cream, or ochraceous yellow in mass.

1528. **R. sanguinea** (Bull.) Fr. Cke. Illus. no. 981, t. 1019.
Sanguinea, bloody.

P. 5–9 cm., *blood-red*, or *becoming pale* round the spreading *acute margin*, fleshy, firm, convex, obtuse, then depressed and infundibuliform, disc generally gibbous, polished, moist in damp weather. St. 4–10 × 1–2 cm., *reddish, rarely white*, at first contracted at the apex, then equal, firm, *wrinkled striate*, pruinose. Gills *white, then cream colour, decurrent*, rarely forked, *crowded*, narrow, connected by veins, fragile. Flesh *white, reddish under the cuticle,* cheesy. Spores *pale ochraceous,* subglobose, 6–7 × 6μ, echinulate, 1-guttulate. Cystidia conical, 55–65 × 10–12μ. Taste acrid. Woods especially pine. Aug.—Sept. Uncommon. (*v.v.*)

1529. **R. rosacea** (Pers.) Fr. *Rosacea,* rosy.

P. 4–10 cm., *rosy flesh colour, varying in intensity, becoming whitish, variegated with darker spots when dry,* convex, then plane and umbilicate, or flexuose and incised, often irregular, compactly fleshy, firm, *viscid; margin acute.* St. 4–5 × 2 cm., *white, or reddish,* equal, or attenuated at the base, occasionally ventricose, pruinose. Gills *white, often coloured reddish on the edge near the margin of the pileus, adnate,* fairly broad, *forked,* edge unequal. Flesh *white, reddish under the pellicle,* cheesy. Spores white, globose, 8μ, echinulate, 1-guttulate. Taste acrid. Woods. Aug.—Oct. Not uncommon. (*v.v.*)

1530. **R. drimeia** Cke. (= *Russula expallens* Gill.) Cke. Illus. no. 988, t. 1023. δριμύς, pungent.

P. 5–11 cm., *bright purple to dark rose colour, becoming decoloured with age,* convex, then expanded and more or less depressed, scarcely viscid when moist, opaque when dry, compact, firm; margin incurved, slightly striate when old. St. 5–10 × 1–3 cm., *tinged with purple,* equal, firm, sometimes rather mealy. Gills *pale sulphur yellow, then deeper yellow,* adnexed, scarcely crowded, narrow, furcate at the base. Flesh *yellowish, then white, reddish under the cuticle of the p. and st.* Spores *pale ochraceous,* subglobose, 8–9 × 8μ, verrucose, or slightly echinulate. Cystidia fusiform, 50–65 × 8–10μ. Taste very acrid. Coniferous woods. Sept.—Dec. Common. (*v.v.*)

var. **Queletii** (Fr.) Bataille. Cke. Illus. no. 1019, t. 1028.
Lucien Quélet, the eminent French mycologist.

Differs from the type in the *white or wax coloured gills which exud drops that on drying leave azure-blue-cinereous, or pallid olivaceous spots.* Woods. Aug.—Nov. Not uncommon. (*v.v.*)

1531. **R. rubra** (Krombh.) Bres. (non Lam. et DC.). Fung. Trid. t. 203. *Rubra,* red.

P. 4–10 cm., *red, rosy or whitish at the margin,* fleshy, convex then plane and depressed, *dry,* sometimes somewhat pruinose St. 3·5–7 × 1–3 cm., *white, often becoming somewhat cinereous with age,* equal, often attenuated or incrassated at the base, rugulose. Gills *white, then ochraceous,* sinuato-adnexed, or rounded behind, crowded or somewhat crowded, forked, connected by veins. Flesh *white, rosy under the cuticle.* Spores *ochraceous,* subglobose, 8–9 × 7–8μ, verrucose or slightly echinulate. Cystidia clavate, 60–70 × 6–10μ. Smell pleasant, taste very acrid. Deciduous woods. Aug.—Oct. Not uncommon. (*v.v.*)

VI. Pellicle of the pileus viscid, generally separable, possessing numerous cystidia, *red or purple*. Margin rounded, generally striate. Flesh *fragile*. Taste *acrid*. Gills *generally equal*, fragile, rounded in front, free. Spores pure white, rarely cream-white in mass.

1532. **R. fragilis** (Pers.) Fr. Cke. Illus. no. 1028, t. 1091.
Fragilis, brittle.

P. 3–5 cm., *flesh colour, or red*, changing colour, convex, *often umbonate*, then plane and depressed, *very thin*, fleshy only at the disc, slightly viscid; margin very thin, *tuberculoso-striate*. St. 4–5 × 1 cm., *white, very fragile*, pruinose, often slightly striate. Gills *shining white*, slightly adnexed, *very thin, crowded, ventricose, all equal*. Flesh *white*. Spores white, subglobose, 7–9 × 7–8μ, echinulate, 1-guttulate. Cystidia "sparse, with a short lanceolate point, 60–70 × 10–12μ" Rick. Taste very acrid. Woods, and pastures. Aug.—Dec. Common. (*v.v.*)

var. **nivea** (Pers.) Cke. Cke. Illus. no. 1029, t. 1060, fig. B.
Nivea, snow-white.

Differs from the type in *the white pileus*. Woods. Aug.—Nov. Not uncommon. (*v.v.*)

var. **fallax** (Schaeff.) Massee. Cke. Illus. no. 1023, t. 1059, as *Russula fallax* Schaeff. *Fallax*, deceptive.

Differs from the type in the *olivaceous disc of the pileus*. Woods. Aug.—Nov. Not uncommon. (*v.v.*)

1533. **R. violacea** Quél. (= *Agaricus fragilis violascens* Secr.) Cke. Illus. no. 1029, t. 1060, fig. A, as *Russula fragilis* Fr. var. *violacea* Quél. *Violacea*, violet.

P. 3–5 cm., *bright violet, with a narrow whitish margin, often spotted with yellow, green, or olive*, convex, then plane and depressed, *thin*, viscid, *striate*. St. 3–4 × ·5–1 cm., *white*, fragile, striate, pruinose. Gills *white, adnate, crowded*, thin. Flesh *white*. Spores *pale straw in the mass*, globose, 8–9μ, verrucose. Smell "of laudanum" Quél. Taste very acrid. Woods. Aug.—Dec. Common. (*v.v.*)

1534. **R. emetica** (Schaeff.) Fr. Cke. Illus. no. 1021, t. 1030.
ἐμετική, provoking sickness.

P. 4–10 cm., *rosy, then blood colour, tawny when old, sometimes becoming yellow, and at length white*, campanulate, then flattened, or depressed, polished, sometimes rugulose; *margin at length tubercularly sulcate*. St. 3–7 × 1–1·5 cm., *white, or reddish*, rigid. Gills *shining white*, free, or adnate, broad, subdistant. Flesh *white, reddish under the separable pellicle*. Spores white, globose, 8μ, echinulate, 1-guttulate. Cystidia "lanceolate, 60–75 × 12–18μ, not very abundant" Rick. Taste very acrid. Beech woods, and under beeches. July—Dec. Common. (*v.v.*)

var. **Clusii** Fr. Vitt. t. 38, fig. 1, as *Agaricus emeticus*.

Clusius, one of the earliest illustrators of fungi.

Differs from the type in the *gills and flesh becoming yellow*. Woods. Sept.—Oct. Uncommon. (*v.v.*)

1535. **R. luteo-tacta** Rea. (= *Russula sardonia* Bres. non Fr.) Bres. Fung. Trid. t. 94, as *Russula sardonia* Fr.

Luteus, yellow; *tacta*, touched.

P. 4–7 cm., *rosy, or blood-red, soon becoming whitish in places and spotted with yellow*, convex, then plane and depressed, fleshy. St. 4–5 × 1–1·5 cm., *white, or rosy, spotted with yellow*, equal, rugose. Gills *white, exuding watery drops in wet weather, then spotted with yellow, becoming yellowish when cut or bruised*, adnate, crowded, somewhat forked. Flesh *white, tinged yellowish when cut or bruised, reddish under the cuticle*. Spores white, globose, 6–8μ, echinulate, 1-guttulate. Taste very acrid. Woods, and parks. Aug.—Oct. Not uncommon. (*v.v.*)

1536. **R. atropurpurea** (Krombh.) Maire. (= *Russula rubra* Cke. non Fr.; *Russula depallens* Cke. an Fr.?; *Russula purpurea* Gill.; *Russula Clusii* Bataille, an Fr.? Maire.) Cke. Illus. no. 997, t. 1087. *Ater*, black; *purpurea*, purple.

P. 5–9 cm., *deep blood-red, almost black at the disc, and often yellowish at first at the margin*, hemispherical, then convexo-plane, and finally depressed or infundibuliform, *fleshy*, firm, viscid, slightly rugosely wrinkled; margin thin, hardly striate in old age, often exceeding the gills. St. 4–7 × 1–3 cm., *white, unchangeable, or sometimes becoming slightly stained with ochraceous brown, sometimes rosy in the middle, base ochraceous*, firm, somewhat equal, slightly rugoso-striate, apex pruinose. Gills *white, then yellowish, sinuato-free*, attenuated behind, broader in front, *equal*, rather crowded. Flesh *whitish, either unchangeable, or becoming slightly stained with ochraceous brown, reddish purple under the cuticle*. Spores pure white, subglobose, 9 × 8μ, verrucose, 1-guttulate. Smell slight, pleasant. *Taste either mild, or acrid*. Woods, and under conifers. Aug.—Oct. Common. (*v.v.*)

var. **depallens** (Cke.) Maire. Cke. Illus. no. 985, t. 1021.

Depallens, becoming pale.

Differs from the type in the *pileus soon loosing its colour, and in the stem and flesh becoming grey with age*. Woods. Aug.—Oct. Common. (*v.v.*)

VII. Pellicle of the pileus viscid, separable, possessing cystidia, *variously coloured*. Margin rounded, generally striate. Flesh *fragile*. Gills *equal*, fragile, rounded in front, free or somewhat free. Spores cream ochraceous, or yellow ochraceous in the mass (rarely whitish yellow in the mass but then the taste is mild, or only slightly acrid when young and the pileus is never red).

*Flesh becoming black, taste mild or slightly acrid when young.

1537. **R. decolorans** Fr. Cke. Illus. no. 1039, t. 1079.
Decolorans, discolouring.

P. 5–10 cm., *orange-red, then light yellow, and becoming pale*, spherical, then expanded and depressed, remarkably regular, fleshy, viscid; margin thin, at length striate. St. 6–10 × 1–2 cm., *white, becoming cinereous*, cylindrical, often *rugoso-striate*. Gills *white, then yellowish*, adnexed, often in pairs, thin, crowded, fragile. Flesh *white, becoming cinereous when broken, especially in the stem, and more or less variegated with black spots when old*. Spores *ochraceous*, elliptical, "11–13 × 8–9μ" Maire, verrucose, 1-guttulate. Cystidia "sparse, subulate, 50–60 × 6–8μ" Rick. Taste mild, then slightly acrid. Edible. Coniferous woods, and peat bogs. Aug.—Sept. Uncommon.

**Flesh not becoming black, taste *mild*, or somewhat acrid when young.

1538. **R. integra** (Linn.) Bataille. *Integra*, entire.

P. 8–12 cm., *bay, brown, or olivaceous, becoming pale*, convex, then plane, fleshy, *firm*, viscid; margin thin, *becoming tuberculately striate*. St. 9–10 × 2–3 cm., *white*, clavate, or ventricose, fragile, *wrinkled-striate*. Gills *white, then mealy and ochraceous cream, free, very broad*, connected by veins. Flesh *white*. Spores *ochraceous cream in mass*, subglobose, 8–10 × 7–9μ, echinulate, 1-guttulate. Cystidia "clavate apex obtuse, 50–60 × 10–15μ" Rick. Smell pleasant, taste mild, then slightly acrid. Edible. Coniferous woods. Uncommon. (*v.v.*)

1539. **R. Romellii** Maire. (= *Russula olivascens* Quél. sec. Maire.) Cke. Illus. no. 1036, t. 1034, no. 1037, t. 1093, as *Russula integra* Linn.; no. 1038, t. 1094, as *Russula integra* Linn. var. *alba* Cke.
Lars Romell, the eminent Swedish mycologist.

P. 8–15 cm., *reddish, purple, violet, becoming olivaceous, isabelline, whitish*, convex, then plane and more or less depressed, *soft, fragile*, viscid, disc often streaked with innate fibrils; margin rounded, *often striate*, rarely tuberculoso-striate. St. 6–9 × 1·5–2 cm., *white, pruinose*, becoming glabrous, somewhat cylindrical, *wrinkled-striate*, sometimes distinctly corticate. Gills *white, at length light yellow, somewhat ochraceous-pulverulent with the spores*, free or somewhat adnate, *very broad, equal*, somewhat distant, rarely forked, or unequal, more or less connected by veins. Flesh *white, rarely slightly violaceous under the cuticle*. Spores *deep yellow ochre in the mass*, yellow under the microscope, elliptical, 7–9 × 6–7μ, *marked with anastomosing ridges and spines*. Cystidia often with an appendage, 60–90 × 8–10μ. Smell weak, taste pleasant. Deciduous woods. June—Nov. Common. (*v.v.*)

1540. **R. erythropus** (Fr.) Peltereau. ἐρυθρός, red; πούς, foot.

P. 8–16 cm., *dark blood-red, disc darker, decolouring very slightly with age*, never tinted ochraceous or olivaceous, convex, then depressed, *firm, dull, unpolished*; margin slightly striate when old. St. 6–10 × 2–3 cm., *rose-red, rarely rose-red on one side only, firm*, equal. Gills *white, then deep ochraceous*, very broad, *attenuated near the stem*, rounded near the margin of the pileus. Spores *ochraceous*, globose, 8–10 × 8–9μ, echinulate. Taste pleasant. Edible. Aug.—Oct. Not uncommon. (*v.v.*)

1541. **R. xerampelina** (Schaeff.) Fr. (= *Russula Barlae* Cke. sec. Maire.) Cke. Illus. no. 1000, t. 1074, upper figs.
ξηραμπέλινος, of the colour of withered vine leaves.

P. 6–12 cm., *rosy purple, disc becoming pale, yellowish white, sometimes inclining to olivaceous*, convex, then flattened, at length depressed, fleshy, compact, *without a distinct pellicle*, slightly viscid at first, then very slightly *rimulose, so that the cuticle under a lens is very thinly granular or punctate*; margin spreading. St. 4–8 × 1·5–2·5 cm., *white*, or *reddish*, equal, or thickened at the base, firm. Gills *whitish, then yellowish*, adnexed, broader in front, forked behind, somewhat crowded. Flesh *white, becoming brownish with age*. Spores *pale ochraceous*, globose, 7–9μ, echinulate. Cystidia obtusely conical, 68–78 × 10–12μ. Smell *strong, when old like crab*. Taste mild. Edible. Woods. July—Oct. Not uncommon. (*v.v.*)

1542. **R. cutifracta** Cke. Cke. Illus. no. 992, t. 1024.
Cutis, skin; *fracta*, broken.

P. 7–12 cm., *purple, green, or dull red*, convex, then a little depressed in the centre, fleshy, firm, *dry, pulverulent, dull*, viscid in wet weather, opaque, *cuticle sometimes cracking from the margin inwards into minute firmly adnate areolae*. St. 6–8 × 2·5 cm., *white, often slightly tinged with purple or rose on one side*, nearly equal, or a little attenuated above, firm. Gills *white, then cream*, adnexed, or nearly free, narrowed behind, furcate, somewhat crowded, often tinted yellow on the cracks at the edge. Flesh *white, tinged with purple under the cuticle*. Spores *ochraceous*, globose, 10μ, *marked with anastomosing ridges and spines*. Taste mild. Woods. Sept.—Oct. Uncommon. (*v.v.*)

1543. **R. grisea** (Pers.) Bres. (= *Russula palumbina* Quél.) Trans. Brit. Myc. Soc. III, t. 13. *Grisea*, grey.

P. 6–9 cm., *greyish lilac, or bluish grey, mixed with rose, yellow, or olive, then becoming greenish*, convex, then expanded and depressed, fleshy, *fragile*, slightly viscid, shining when dry. St. 8–10 × 2–3 cm., *white*, fragile, *rugoso-striate*. Gills *cream colour, with a tint of apricot flesh colour*, adnate, sometimes forked, broadest towards the margin.

Flesh *white, lilac beneath the thin, separable pellicle.* Spores *ochraceous*, elliptical, 8 × 7μ, echinulate. Taste mild, slightly acrid in the gills of young specimens. Cystidia "lanceolate, 60–90 × 10–15μ" Rick. Coniferous woods, and under conifers. Sept.—Oct. Not uncommon. (*v.v.*)

1544. **R. graminicolor** (Secr.) Quél. (= *Russula aeruginea* (Lindb.) Fr.) Fr. Icon. t. 173, fig. 3, as *Russula aeruginea* Lindb.

Gramen, grass; *color*, colour.

P. 5–14 cm., *aeruginous-green, disc darker, slightly brownish bistre*, convex, then plane and depressed, fleshy, *fragile, pellicle separable*; margin *paler, striate.* St. 5–12 × 2·5–4 cm., *white, firm*, equal, or attenuated at the base, *rugose.* Gills *white, then cream, sometimes spotted with brown when old*, slightly adnexed, attenuated behind, broad in front, often connate two by two at the base. Flesh *white*, fragile. Spores *cream colour in the mass*, elliptical, 5–8 × 6–7μ, echinulate. Cystidia "abundant, lanceolate, 69–95 × 8–12μ, granular in the upper part" Rick. Taste acrid when young, then only in the gills when old. Under birches, and in pine woods. May—Oct. Not uncommon. (*v.v.*)

1545. **R. chamaeleontina** Fr. Cke. Illus. no. 1054, t. 1098.

χαμαιλέων, the chameleon.

P. 2·5–4 cm., *flesh colour, rosy blood-red, purplish lilac, then soon changing colour, becoming yellow at the disc, and at length wholly yellow*, soon plane, thinly fleshy, viscid, pellicle separable; margin slightly striate when old. St. 2–6 cm. × 6–8 mm., *white*, equal, pruinose, slightly striate, *fragile.* Gills *light yellow, then darker yellow*, more or less adnexed, *thin, crowded, equal.* Flesh *white*, fragile. Spores *ochraceous*, globose, 6–7μ, verrucose, 1-guttulate. Cystidia "sparse, subulate, 50–60 × 8–10μ" Rick. Taste mild. Edible. Woods, and downs. Sept.—Nov. Uncommon. (*v.v.*)

1546. **R. roseipes** (Secr.) Bres. Bres. Fung. Trid. t. 40.

Roseus, rosy; *pes*, foot.

P. 4–7 cm., *rosy flesh colour, rosy orange, or rosy with a tinge of ochre*, at first with whitish spots, at length blanched, convex, then plane and depressed, fleshy, viscid, soon dry; margin thin, becoming somewhat *tuberculosely striate.* St. 3–6 cm. × 8–15 mm., *white*, either *entirely or here and there sprinkled with rosy meal*, equal. Gills *whitish, then ochraceous egg-yellow, edge often rosy*, free, rounded and furcate behind, equal, or a few dimidiate, rather crowded, sometimes with an adnate tooth, ventricose, connected by veins. Flesh *whitish*, then *becoming yellowish.* Spores *ochraceous*, globose, 8–10μ, echinulate. Cystidia fusiform, 60 × 8μ. Smell and taste pleasant. Edible Beech, and pine woods. Aug.—Oct. Uncommon.

***Flesh not becoming black, taste *distinctly acrid.*

1547. **R. veternosa** Fr. Bres. Fung. manger. t. 75. *Veternosa*, languid.

P. 5–8 cm., *rose, or flesh colour, soon becoming pale, commonly whitish or yellowish at the disc*, convex, then plane and depressed in the middle, slightly fleshy, viscid, pellicle adnate. St. 5–8 cm. × 12 mm., *white, often tinged with rose towards the base, fragile*, equal. Gills *white, then bright yellow*, adnate, narrowed behind, broader in front. Flesh *white*, soft. Spores *ochraceous*, subglobose, 7–8 × 6–7μ, echinulate. Cystidia "lanceolate, 45–70 × 9–15μ, shorter on the edge of the gill" Rick. Taste acrid. Woods, and heaths. Aug.—Oct. Uncommon. (*v.v.*)

1548. **R. nauseosa** (Pers.) Fr. Bres. Fung. Trid. t. 129. *Nauseosa*, nauseous.

P. 3–5 cm., *variable in colour*, typically *purplish at the disc, then livid, but becoming pale and often whitish*, plano-gibbous, then depressed, viscid, laxly fleshy; margin *submembranaceous, tuberculoso-sulcate*. St. 2–3 cm. × 8 mm., *white, becoming grey with age, fragile*, slightly striate. Gills *light yellow*, then *dingy ochraceous*, adnexed, ventricose, somewhat distant. Flesh *white*. Spores *yellow*, subglobose, 7–10μ, verrucosely echinulate. Cystidia "fusiform, 50–55 × 10μ" Bres. Smell often unpleasant, taste mild, then acrid. Coniferous woods. Sept.—Oct. Uncommon.

var. *flavida* Cke. = **Russula lutea** (Huds.) Fr.

1549. **R. puellaris** Fr. Bres. Fung. Trid. t. 64. *Puellaris*, girlish.

P. 2·5–6 cm., *livid purplish, becoming yellowish, disc brown*, conically convex, then flattened or depressed, membranaceous except at the disc, viscid, *margin tuberculosely-striate*. St. 4–5 cm. × 7–10 mm., *white, becoming yellowish*, and *stained brownish when touched*, attenuated upwards, rugulose. Gills *white, then pallid yellow*, adnate, attenuated behind, thin, crowded. Flesh *white, ochraceous at the base of the stem*. Spores *ochraceous*, subglobose, 8–10 × 7–8μ, echinulate, 1-guttulate. Cystidia abundant, conical, 50–65 × 9–11μ. Taste mild, then slightly acrid. Woods, and damp places. Aug.—Oct. Common. (*v.v.*)

var. **intensior** Cke. Cke. Illus. no. 1047, t. 1066. *Intensior*, deeper.

Differs from the type in the *darker, deep purple p., nearly black at the disc.*

1550. **R. nitida** (Pers.) Fr. (= *Russula nitida* Fr. var. *cuprea* Cke.) Krombh. t. 66, figs. 1–3. *Nitida*, shining.

P. 3–6 cm., *bay-brown-purplish, disc darker*, convex, then plane or slightly depressed, thin, viscid, shining when dry; margin *striate*, *somewhat tubercular.* St. 5–7·5 × 1 cm., *white,* equal, or attenuated downwards, *rigid,* minutely wrinkled, pruinose. Gills *pallid, then bright sulphur-yellow,* adnexed, thin, crowded, *equal.* Flesh *white.* Spores *ochraceous,* elliptical or globose, 7–9 × 7μ, echinulate, 1-guttulate. Cystidia conical, 60–70 × 10–12μ. Smell none, or somewhat unpleasant. Taste mild, then acrid. Woods. July—Oct. Not uncommon. (*v.v.*)

var. **pulchralis** (Britz.) Cke. Cke. Illus. no. 1044, t. 1095, fig. A, as *Russula pulchralis* Britz. *Pulchralis,* beautiful.

P. 4–8 cm., *ochraceous, centre spotted with red or purple,* convex, then flattened and depressed, thin, viscid; margin thin, *deeply striate* and often split. St. 5–6 × 1–2 cm., *white,* fragile, equal, ventricose, or thickened at the base. Gills *whitish, then ochraceous yellow,* broad, distant, rather thick. Flesh *white.* Spores *ochraceous,* subglobose, 9 × 8μ, echinulate. Taste mild. Woods. July—Oct. Not uncommon. (*v.v.*)

1551. **R. maculata** Quél. Quél. Soc. bot. Fr. (1877), t. 5, fig. 8. *Maculata,* spotted.

P. 5–9 cm., *pale reddish flesh colour, then decoloured yellow or ivory white, spotted with purple or brown,* convex, then plane, thick, *firm,* viscid; margin undulate, *generally remaining red.* St. 3–4 × 1–1·5 cm., *white, rarely tinged with rose, at last spotted with red or bistre, firm,* polished, *reticulately striate.* Gills *pale sulphur, then yellow apricot or pink,* attenuato-adnate, forked. Flesh *white,* fragile. Spores *citron-yellow,* subglobose, 10μ, echinulate. Smell pleasant, like apple, or sweet-briar. Taste mild, then acrid. Edible. Woods. Sept.—Oct. Uncommon. (*v.v.*)

1552. **R. ochracea** (A. & S.) Fr. Richon et Roze, t. 43, figs. 17–20. ὠχρός, pale yellow.

P. 6–7 cm., *ochraceous, with a tinge of yellow, disc usually becoming darker,* convex, then plane or depressed, *soft,* viscid; margin *thin, sulcate.* St. 3–4 × 1–1·5 cm., *ochraceous, rarely white,* equal, or thickened at the base, striate. Gills *concolorous,* slightly adnexed, broad, scarcely crowded. Flesh *ochraceous.* Spores *ochraceous,* globose, 9–10μ, verrucose, 1-guttulate. Cystidia "cylindrical, 50–60 × 7–8μ, rounded above" Rick. Taste mild, or slightly acrid. Pine, and mixed woods. Aug.—Sept. Uncommon. (*v.v.*)

VIII. Pellicle of the pileus viscid, separable, destitute of cystidia. Margin rounded, generally striate. Flesh *fragile.* Taste *mild,* rarely acrid. Gills *equal, or subequal,* rounded in front, somewhat free. Spores whitish cream to yellow ochre, rarely pure white.

1553. **R. alutacea** (Pers.) Fr. Cke. Illus. no. 1048, t. 1096. *Alutacea,* like tanned leather.

P. 5–18 cm., *purple, or blood-red, tinted with olive, green, or bistre, sometimes entirely olivaceous*, convex, then plane or depressed, fleshy, rigid, slightly viscid, pellicle separable, soon dry; margin thin, at length *tuberculosely striate*. St. 5–12 × 2–5 cm., *white, generally reddish at the apex or on one side, sometimes yellowish at the base*, firm, equal. Gills *pallid light yellow, soon becoming ochraceous egg-yellow*, at first *free, thick, very broad, equal*, somewhat distant. Flesh *white*, firm, becoming soft with age. Spores *deep ochre yellow in the mass*, yellow under the microscope, subglobose, 10 × 9μ, verrucose, 1-guttulate. Cystidia "sparse, cylindrical-fusiform, 60–75 × 8–10μ" Rick. Taste pleasant, nutty. Edible. Woods, especially beech. July—Oct. Common. (*v.v.*)

var. **purpurata** Bres. Bres. Fung. Trid. t. 96.
Purpurata, clad in purple.

Differs from the type in the *light purple p. and st.* Coniferous woods. Aug.—Sept. Uncommon. (*v.v.*)

var. **olivascens** (Fr.) Rea. Fr. Icon. t. 172, fig. 2, as *Russula olivascens* Pers. *Olivascens*, becoming olive coloured.

P. 6–10 cm., *olivaceous, becoming yellowish at the disc*, convex, then expanded and umbilicate, fleshy, rigid. St. 3–7 × 1·5–2 cm., *white*, firm, equal. Gills *cream colour, then yellowish*, slightly adnexed, narrowed behind, broader in front, *nearly equal*, rarely forked. Flesh *white*. Spores *deep ochraceous*, globose, 9–10μ, echinulate. Taste mild, then slightly acrid. Woods. Aug.—Oct. Uncommon. (*v.v.*)

1554. **R. aurata** (With.) Fr. Cke. Illus. no. 1041, t. 1080.
Aurata, golden.

P. 5–9 cm., varying *lemon-yellow, orange or red, disc darker*, convex, then plane, or depressed, fleshy, *rigid*, viscid in wet weather; margin thin, slightly striate when old. St. 5–9 × 1·5 cm., *white, tinged with lemon yellow especially towards the base*, equal, or attenuated downwards, *firm*, obsoletely striate. Gills *whitish, inclining to light yellow, vivid lemon-yellow at the edge*, rounded, free, equal, connected by veins, sometimes forked behind. Flesh *lemon-yellow under the separable pellicle, yellowish, then white below*. Spores *ochraceous*, globose, 8–9μ, *marked with ridges and spines*, 1-guttulate. Cystidia on edge of gill abundant, cylindrical-fusiform, 55–70 × 8–12μ, apex obtuse, contents yellowish. Smell pleasant. Taste mild, or very slowly acrid. Edible. Woods. June—Oct. Not uncommon. (*v.v.*)

1555. **R. fusca** Quél. Cke. Illus. no. 1000, t. 1074, bottom fig., as *Russula xerampelina* Schaeff. *Fusca*, dark.

P. 6–8 cm., *ochraceous brown, speckled, darker at the disc*, convex, then infundibuliform, fleshy, viscid. St. 4–6 × 2 cm., *milk-white*, rigid, at length slightly wrinkled. Gills *milk-white, then ochraceous cream, or yellow wax colour*, sinuate, adnate by a tooth, forked, connected by veins. Flesh *white-cream*, firm. Spores *white-cream colour in the mass*, elliptical, 9 μ. Smell and taste pleasant. Edible. Coniferous woods, and under conifers. Sept. Rare.

1556. **R. vesca** Fr. (= *Russula rosea* (Schaeff.) Quél.) Sverig. ätl. Svamp. t. 63. *Vescor*, I feed.

P. 2–11 cm., *red flesh colour, disc darker*, fleshy, firm, convex, then plano-depressed, *slightly wrinkled with veins*, viscid; margin at length spreading. St. 2–8 × 1–3 cm., *shining white, often foxed with age*, equal, often attenuated downwards, *rugosely striate*. Gills *whitish*, adnate, *rather narrow*, thin, *crowded*, connected by veins. Flesh *white, foxing when cut or bruised*. Spores *white*, globose, 8–9 μ, echinulate. Cystidia flask-shaped, apex elongate, obtuse, 2·5–3 μ in diam., 35–60 × 8–10 μ. Smell none, or unpleasant like crab. Taste mild. Edible. Aug.—Nov. Common. (*v.v.*)

var. **Duportii** (Phill.) Massee. Cke. Illus. no. 1003, t. 1042, fig. A, as *Russula Duportii* Phill.

Rev. Canon J. M. Du Port, an enthusiastic mycologist.

P. 4–6·5 cm., *disc rufous or flesh red, obtuse margin bluish*, compact, fleshy, firm, convexo-plane, then depressed, *dry*. St. 2·5–5 cm. × 10–16 mm., *white*, minutely striate. Gills *white*, rounded behind, *broad, distant*. Flesh *turning reddish brown when cut*. Spores white, globose, 9 μ, echinulate. Smell of crab. Woods. Sept.

1557. **R. Barlae** Quél.[1] Quél. As. Fr. (1883), t. 6, fig. 12.

J. B. Barla, the eminent mycologist of Nice.

P. 6–9 cm., *yellow apricot, or bright nankeen yellow, tinged with orange passing into rosy flesh colour*, convex, then plane, or infundibuliform, compact, slightly viscid, cuticle separable, often cracked. St. 4–5 × 1–1·5 cm., *cream colour, then streaked with bistre*, firm, wrinkled striate, silky pruinose. Gills *cream colour, then saffron yellow with a tinge of rosy flesh colour*, sinuate, free. Flesh *white*. Spores *cream colour*, globose, 9 μ, verrucose. Smell pleasant, like melilot ("mousse de Corse" Barla). Mountainous woods. Summer.

1558. **R. punctata** (Gill.) Maire. (= *Russula amoena* Quél. sec. Maire; *Russula punctata* Gill. var. *leucopus* Cke.) Gillot et Lucand, Catal. Champ. Autan, t. 2, fig. 3, as *Russula amoena* Quél.

Punctata, dotted.

P. 3–8 cm., *rose, purple, purple-violaceous, lilac, often more or less mixed with olive green or yellowish green, sometimes entirely greenish olive, yellowish green, or yellow citron colour*, fleshy, convex, then plane,

[1] This is recorded as British by M. C. Cooke in Handbook of British Fungi, Ed. II, p. 335, but his diagnosis is referred by René Maire in Bull. Soc. Myc. Fr. XXVII (1910), 172 to *Russula xerampelina* Fr.

sometimes slightly depressed, either *viscid, or dry, granularly mealy, often punctate with deeper coloured granules*; margin sometimes somewhat sulcate when old. St. 3–5 × ·5–1·5 cm., *rose, purple, purple-violaceous, often partially or quite white*, equal, or obconic, *pruinosely mealy*, sometimes rugosely striate. Gills *whitish cream, then cream colour*, adnate, or subadnate, *edge sometimes purple, or purplish-violaceous and floccose*, either throughout its length or near the margin of the pileus only, *thin, crowded*, often forked at the base. Flesh *white, sometimes reddish near the cuticle of the pileus*. Spores *whitish cream in the mass*, somewhat hyaline under the microscope, elliptical, 7·5–9 × 7–8μ, verrucose, subreticulate. Cystidia rather rare, fusiform, or subclavate, sometimes with a short and broad appendage at the apex, 90–130 × 13–15μ. Smell slight, or very pleasant. Taste mild. Edible. Coniferous woods. July—Oct. Uncommon. (*v.v.*)

var. **violeipes** (Quél.) Maire. Quél. Ass. Fr. (1897), 450, pro forma *R. citrinae*. *Violeus*, violet; *pes*, foot.

Differs from the type in the *citron yellow p. sometimes tinted lilac, and the lilac, or white tinged with lilac stem*. Coniferous woods. Sept.—Oct. Uncommon. (*v.v.*)

1559. **R. carnicolor** Bres. Bres. Fung. Trid. t. 128, as *Russula lilacea* Quél. var. *carnicolor* Bres. *Caro*, flesh; *color*, colour.

P. 3–7 cm., *flesh colour, disc fuscous livid, then concolorous*, fleshy, convex, then plane and depressed, viscid; margin at length slightly tuberculately striate. St. 4–5 cm. × 6–12 mm., *white, base sometimes rosy*, equal, subpruinose, somewhat rugulose. Gills *shining white*, rounded behind, adnexed, forked, somewhat distant. Flesh *white*. Spores white, subglobose, 6–8μ, echinulate. Smell and taste pleasant. Edible. Sept.—Nov. Not uncommon. (*v.v.*)

1560. **R. mustelina** Fr. Cke. Illus. no. 976, t. 1018. *Mustelina*, pertaining to a weasel.

P. 5–10 cm., *bright brown, or dingy yellowish*, convex, then plane and depressed, fleshy, firm, dry; margin at first incurved, *minutely tomentose*, then straight. St. 4–6 × 1·5–2·5 cm., *white*, equal, somewhat rugose. Gills *white, then cream colour*, rounded behind, adnexed, broad in front, connected by veins. Flesh *white, ochraceous at the margin*. Spores *ochraceous cream in the mass*, hyaline under the microscope, subglobose, 7–8μ, verrucose, 1-guttulate. Taste pleasant. Edible. Woods. Sept.—Oct. Uncommon. (*v.v.*)

1561. **R. caerulea** Cke. Cke. Illus. no. 987, t. 1052. *Caerulea*, azure-blue.

P. 5–8 cm., *bright purple, or bluish purple, darker or sometimes brownish at the umbonate disc*, convex, then expanded, or somewhat depressed, *umbonate*, polished; margin thin, at length slightly striate with age. St. 5–11 × 1–2·5 cm., *white*, equal, firm. Gills *yellowish*, adnate, equal, rounded at the apex. Flesh *white, brownish, or purplish under the cuticle.* Spores *pale ochraceous*, globose, 9–10μ, echinulate. Taste mild. Coniferous woods, and under conifers. Aug.—Oct. Not uncommon. (*v.v.*)

1562. **R. lutea** (Huds.) Fr. Cke. Illus. no. 1051, t. 1082.

Lutea, golden yellow.

P. 2–5 cm., *yellow, at length becoming pale, and occasionally wholly white*, convex, then plane, or plano-depressed, thin, viscid; margin sometimes obsoletely striate when old. St. 2–4 cm. × 6–8 mm., *white*, equal, *fragile*. Gills *ochraceous egg-yellow*, somewhat free, connected by veins, *crowded, equal*, thin. Flesh *white*. Spores *ochraceous*, globose, 8–9μ, echinulate. Cystidia "clavate, 45–50 × 10–12μ, with a blunt apex" Rick. Smell pleasant, like apricots. Taste mild. Edible. Woods, and lawns. July—Nov. Common. (*v.v.*)

var. **armeniaca** (Cke.) Rea. Cke. Illus. no. 1045, t. 1064, as *Russula armeniaca* Cke.

Armeniaca, of Armenia, the native country of the apricot.

Differs from the type only in the *rich apricot colour of the pileus*. Woods, and lawns. July—Oct. Common. (*v.v.*)

var. **vitellina** (Pers.) Bataille. Cke. Illus. no. 1052, t. 1102, fig. B, as *Russula vitellina* (Pers.) Fr. *Vitellina*, egg-yellow.

Differs from the type in the *egg-yellow colour, and tuberculately striate margin of the pileus, the distant, saffron yellow gills, and the strong unpleasant smell*. Coniferous woods, and under conifers. Aug.—Oct. Uncommon. (*v.v.*)

R. Turci Bres.[1]

Latex milk-white, or coloured, rarely like serum.

Lactarius Fr.

(*Lac*, milk.)

Pileus fleshy, regular. Stem central, rarely excentric, fleshy. Gills adnate, or decurrent, somewhat rigid, milky, acute at the edge. Spores white, or yellowish, rarely pinkish in the mass; globose, subglobose, or elliptical, echinulate, verrucose, punctate, or reticulate; continuous. Cystidia present, or absent. Growing on the ground, more rarely on wood; solitary, or caespitose.

[1] Bres. refers Cke.'s Illus. no. 1199, t. 1147, *Russula nauseosa* Fr. to this species. There is no other British record of its occurrence.

I. St. central. Gills unchangeable, naked, not changing colour and not pruinose. Milk at the first white, (commonly) acrid.

*P. viscid when moist, margin at first involute, tomentose.

1563. **L. scrobiculatus** (Scop.) Fr. Cke. Illus. no. 922, t. 971.
Scrobiculatus, pitted.

P. 10–30 cm., *yellow, becoming pale*, zoned, or zoneless, convex, umbilicate, at length infundibuliform, very viscid when moist, covered with agglutinated down; *margin bearded when young*. St. 4–8 × 2·5–5 cm., *light yellow, pitted with darker yellow broad roundish spots*, incrassated upwards, *somewhat viscid*, base pubescent. Gills *whitish, or flesh colour*, decurrent, crowded, thin. Flesh *whitish, becoming yellow when broken*. Milk *white, soon sulphur-yellow* when exposed to the air. Spores *light yellow*, subglobose, 9 × 7–8μ, echinulate, 1-guttulate. Taste very acrid. Under birches, and in coniferous woods. Sept.—Oct. Not uncommon. (*v.v.*)

1564. **L. torminosus** (Schaeff.) Fr. Cke. Illus. no. 923, t. 972.
Torminosus, griping.

P. 4–12 cm., *pallid flesh colour, or strawberry colour, sometimes pale ochraceous, or white*, convex, then depressed, at length infundibuliform, viscid when moist, *zoned*; *margin white fibrillosely bearded*. St. 6–9 × 1·5–2·5 cm., *pale flesh colour*, equal or attenuated downwards, sometimes pitted, delicately tomentose, then smooth. Gills *paler than the pileus*, adnato-decurrent, thin. Flesh *pallid*. Milk white, acrid. Spores white, subglobose, 8–9 × 7μ, echinulate, 1-guttulate. Cystidia "sparse, subulate, 50–60 × 8μ" Rick. Mixed woods, and heaths. Aug.—Nov. Common. (*v.v.*)

1565. **L. cilicioides** Fr. Cke. Illus. no. 924, t. 973.
κιλίκιον, goat's-hair cloth; *εἶδος*, like.

P. 5–15 cm., *flesh colour inclining to fuscous*, convex, then flattened and depressed in the centre, viscid, tomentose, *margin white fibrillosely woolly*. St. 5–7·5 × 2·5 cm., *pale flesh colour, becoming yellowish*, equal, pruinato-silky under a lens. Gills *white, becoming yellowish*, decurrent, crowded, *branched*. Flesh *white, then yellow*. Milk *white, or light yellow*, acrid. Spores elliptical, 8–10 × 6–7μ, minutely echinulate. Cystidia "subulate-lanceolate, 30–40 × 7–9μ" Rick. Pine woods. Aug.—Nov. Not uncommon. (*v.v.*)

var. **intermedius** (Krombh.) B. & Br. Krombh. t. 58, figs. 11–13.
Intermedius, intermediate.

P. 10–14 cm., *ochraceous yellow*, infundibuliform, viscid, margin tomentose. St. 3–5 × 2 cm., *yellowish, becoming tinged with rufous, covered with spot-like depressions*. Gills *lurid whitish*, subdecurrent, broad. Flesh *white, then yellowish*. Milk *white, then yellowish*, acrid.

Spores elliptical, 8–10 × 6–8μ, minutely echinulate, 1-guttulate. Woods. Sept.—Oct. Uncommon. (*v.v.*)

1566. **L. lateritioroseus** Karst. *Lateritius*, brick-red; *roseus*, rosy.

P. 6–9 cm., *flesh colour, or brick-red with a rosy tinge*, becoming pale, *convexo-umbilicate*, then depressed and somewhat infundibuliform, wavy, often unequal, *disc broken up into minute granule-like squamules, scales larger towards the margin* and eventually disappearing. St. 6–8 × 1·5 cm., *concolorous, or paler*, unequal, incrassated at the base, curved, or flexuose, *very slightly flocculose.* Gills *pinkish, becoming yellowish*, decurrent, rather distant, often furcate and connected by veins. Milk white, acrid. Spores white, subglobose, 8–9 × 6–8μ, echinulate, 1-guttulate. Woods. Sept.—Oct. Uncommon.

1567. **L. turpis** (Weinm.) Fr. (= *Lactarius plumbeus* (Bull.) Quél.; *Lactarius necator* (Pers.) Schroet.) Cke. Illus. no. 925, t. 987. *Turpis*, ugly.

P. 6–30 cm., *olivaceous inclining to umber, sometimes tawny towards the margin, at length entirely inclining to umber*, convex, then plane, disc-shaped, or umbilicate, at length depressed, sometimes somewhat zoned, tomentose, viscid; margin *at first villose, olivaceous light yellow*, at length densely rivuloso-sulcate. St. 4–8 × 1–2·5 cm., *pallid, or dark olivaceous, apex ochraceous whitish*, equal, or attenuated downwards, often viscid and pitted. Gills *white straw colour*, spotted fuscous when broken or bruised, adnato-decurrent, much crowded, forked. Flesh white. Milk white, acrid. Spores white, globose, 6–7μ, echinulate, 1-guttulate. Cystidia "subulate, 60–75 × 6–8μ" Rick. Edible. Woods, heaths, and roadsides, especially under birches. Aug.—Dec. Common. (*v.v.*)

1568. **L. controversus** (Pers.) Fr. Cke. Illus. no. 926, t. 1003. *Contra*, over against; *versus*, turned.

P. 6–30 cm., *whitish, becoming reddish with blood-coloured spots and zones* especially towards the margin, convex, broadly umbilicate, then *somewhat infundibuliform, oblique*, viscid in wet weather; margin acute, involute, more or less villose. St. 2–6 × 2–4 cm., *white becoming concolorous*, attenuated downwards, apex pruinose. Gills *pallid-white-flesh-colour*, decurrent, thin, very crowded. Flesh *pallid, reddish under the cuticle.* Milk white, acrid. Spores white, *or tinged rosy*, subglobose, 8 × 6–7μ, verrucose, 1-guttulate. Smell pleasant, taste acrid. Woods, and pastures, especially under poplars. Aug.—Nov. Not uncommon. (*v.v.*)

1569. **L. pubescens** Fr. Cke. Illus. no. 927, t. 974. *Pubescens*, becoming pubescent.

P. 4–6 cm., *whitish, passing into flesh colour, rather plane*, depressed in the centre, *then broadly infundibuliform*, shining; margin *fibrilloso-*

pubescent. St. 2–4 × 1–2 cm., *flesh colour, then white,* attenuated downwards, often compressed, pruinato-pubescent when young. Gills *pallid, slightly flesh-coloured,* adnate, or slightly decurrent, crowded, narrower than the flesh of the pileus. Flesh white, *pinkish under the cuticle.* Milk white, acrid. Spores white, globose, 7–8μ, echinulate, 1-guttulate. Taste very acrid. Woods, heaths, and pastures, especially under birches. Sept.—Nov. Common. (*v.v.*)

1570. **L. aspideus** Fr. (= *Lactarius uvidus* Fr. sec. Quél.) ἀσπίς, a round shield; εἶδος, like.

Entirely straw-colour, sometimes tinged with lilac. P. 5–10 cm., convex, then slightly depressed, viscid; *margin incurved, tomentose and white,* becoming smooth. St. 5–8 × 1 cm., equal, viscid. Gills adnate. Flesh white, *then lilac.* Milk white, *then lilac,* acrid ("sweet" W. G. Sm. and Massee). Spores white, subglobose, 10 × 9μ, verrucose. Taste acrid. Damp meadows, and moist places. Sept. Uncommon.

**P. viscid when moist, pelliculose, margin naked.

1571. **L. insulsus** Fr. (= *Lactarius zonarius* (Bull.) Quél.) Cke. Illus. no. 929, t. 975. *Insulsus,* insipid.

P. 5–15 cm., *yellowish-brick-colour, zoned,* deeply umbilicate, at length infundibuliform, pellicle somewhat separable, viscid. St. 4 × 2·5, rarely 7–8 × 1·5 cm., *white, becoming pallid, often pitted-spotted.* Gills *whitish, becoming pale and tinged with flesh colour,* decurrent, very crowded, *forked,* often crisped and anastomosing. Flesh *pallid, somewhat zoned under the pellicle.* Milk white, acrid. Spores *yellow,* subglobose, 10 × 8μ, echinulate. Cystidia none. Smell pleasant, taste acrid. Mixed woods, and pastures. Aug.—Oct. Common. (*v.v.*)

1572. **L. zonarius** (Bull.) Fr. (= *Lactarius insulsus* Fr. sec. Quél.) *Zonarius,* zoned.

P. 5–10 cm., *pallid orange, or pale yellowish to deeper yellow,* convex, becoming plane then depressed, somewhat umbilicate, pellicle adnate, viscid, beautifully *zoned* most frequently towards the margin, *at length minutely ruguloso-flocculose*—at first only at the circumference—margin *thin, long involute,* naked. St. 5–8 × 1–1·5 cm., equal, or 2·5 × 1 cm. attenuated downwards, *white, then yellowish,* pale upwards, *firm,* elastic. Gills *whitish, at length dingy yellowish, becoming dingy or even somewhat aeruginous when bruised,* rounded-adnate, or adnato-decurrent, arcuate, thin, narrow, somewhat crowded. Flesh *white,* compact. Milk white, acrid. Spores whitish, globose, 9 × 8μ, echinulate. Cystidia "sparse, subulate" Rick. Smell strong, taste very acrid. Woods, and grassy places. Aug.—Oct. Uncommon. (*v.v.*)

1573. **L. utilis** (Weinm.) Fr. (= *Lactarius pallidus* (Pers.) Fr. sec. Quél.) Cke. Illus. no. 930, t. 1084. *Utilis*, useful.

P. 12–20 cm., *tan colour, pale dull ochre, or livid*, convex, then plane, obtuse, at length infundibuliform, humid, often cracked at maturity. St. 5–8 × 2·5 cm., *concolorous or darker*, fragile, longitudinally striate. Gills *pallid*, adnate, crowded. Milk white, somewhat acrid. Woods. Sept.—Oct. Rare.

1574. **L. blennius** Fr. Cke. Illus. no. 931, t. 988.
βλέννος, mucous matter.

P. 4–11 cm., *pallid olivaceous* or *aeruginous-grey*, plano-depressed, *glutinous, often concentrically guttate*, or somewhat zoned; *margin at first incurved and slightly downy*, then naked. St. 4–5 × 1–1·5 cm., *pallid olive, or pallid grey, viscid*, equal, or attenuated downwards. Gills *white, becoming cinereous when wounded*, subdecurrent. Flesh white, *becoming grey*. Milk white, *then grey*, acrid. Spores white, elliptical, 8 × 6–7μ, verrucose. Cystidia "sparse, fusiform-subulate, 60–75 × 8–10μ" Rick. Taste slowly acrid. Woods, especially beech. Aug.—Nov. Common. (*v.v.*)

var. **viridis** (Schrad.) Quél. *Viridis*, green.

Differs from the type in *its bright green slightly olivaceous pileus*. Beech woods. Aug.—Oct. Not uncommon. (*v.v.*)

1575. **L. lividus** Lamb. *Lividus*, livid.

P. *pale livid, disc fuscescent*, convex, then plane or depressed, not distinctly zoned, viscid (?). St. *livid*, curved. Gills *pale livid*, subdecurrent, crowded. Milk *white, acrid*. Woods. Oct. Rare.

1576. **L. fluens** Boud. Trans. Brit. Myc. Soc. III, t. 12. *Fluens*, lax.

P. 5–10 cm., *blackish olive*, either somewhat zoned, or zoneless and unicolorous, but always *paler ochraceous towards the margin*, convex, scarcely flattened with age, *rough, granularly punctate on the epidermis, viscid*, not glutinous. St. 5–8 × 1–2 cm., *greyish ochre, becoming brown when bruised*, somewhat viscid, unequal, attenuated at the base. Gills *ochraceous, then cinereous ochraceous*, adnate, or subdecurrent. Flesh white, *brown when bruised*. Milk plentiful when wounded like *Lactarius volemus*, white, *then brownish, at first mild, then acrid and bitter*. Spores white, round or oval, 7–8 × 6μ, "10–11 × 7–8μ" Boud., echinulate, netted. Sometimes caespitose. Amongst grass under beeches. Sept.—Oct. Uncommon. (*v.v.*)

1577. **L. hysginus** Fr. Cke. Illus. no. 932, t. 989.
ὕσγινον, a vegetable dye of scarlet colour.

P. 6–10 cm., *reddish flesh colour, or reddish brown*, umbilicate, becoming plane, viscid, often zoned and spotted; *margin thin, inflexed*.

St. 3–10 × 1·5–2·5 cm., *ochraceous cream, or flesh colour,* here and there *pitted or somewhat spotted with rose,* apex constricted, attenuated at the base, pruinose. Gills *white, then light yellow-ochraceous,* adnato-decurrent, thin, crowded, branched. Flesh *white, then yellowish, reddish under the cuticle of the p. and st.* Milk white, slowly acrid. Spores *pale ochre,* globose, 7μ, echinulate, 1-guttulate. Cystidia "subulate, 60–75 × 8–9μ" Rick. Taste very acrid. Woods. Aug.—Nov. Uncommon. (*v.v.*)

1578. **L. trivialis** Fr. Cke. Illus. no. 933, t. 976. *Trivialis,* common.

P. 6–17 cm., *at first dark lurid, becoming pale when full grown, pallid yellowish, tan-flesh-colour,* convex, then soon *depressed,* at length infundibuliform, viscid; margin involute, *at length only the pellicle inflexed.* St. 2·5–15 × 2·5 cm., *paler than the pileus,* slippery. Gills *whitish, becoming pale,* subdecurrent, *rather broad,* somewhat thin, crowded. Flesh white. Milk white, *sometimes becoming yellow,* acrid. Spores *ochraceous,* globose, 6–8μ, echinulate, 1-guttulate. Cystidia "subulate, 60–75 × 9–11μ" Rick. Taste acrid. Coniferous woods, and heaths. Aug.—Oct. Uncommon. (*v.v.*)

1579. **L. circellatus** Fr. Cke. Illus. no. 934, t. 990. *Circellatus,* ringed.

P. 5–10 cm., *rufous inclining to fuscous* in wet weather, *becoming pale, variegated with darker zones, umbilicato-convex, then becoming plane, depressed in the centre,* often repand when older, very viscid in wet weather. St. 4–5 × 1–1·5 cm., *pale,* equal, or attenuated at the base, *tough.* Gills *whitish then becoming yellow,* subdecurrent with a tooth, *horizontal, very thin and crowded, narrow,* often forked. Flesh white, cheesy. Milk white, acrid. Spores *ochraceous,* globose, 7–8μ, echinulate. Taste very acrid. Woods. Sept.—Oct. Uncommon. (*v.v.*)

1580. **L. uvidus** Fr. (= *Lactarius aspideus* Fr. sec. Quél.; *Lactarius flavidus* Boud. sec. Quél.; *Lactarius violascens* (Otto) Fr. sec. Bataille.) Cke. Illus. no. 935, t. 991. *Uvidus,* moist.

P. 3–10 cm., *hoary whitish, grey flesh-colour-livid, becoming fuscous, somewhat thin,* convex, plane, then depressed, viscid; margin at first involute, very soon almost straight. St. 4–9 cm. × 12–18 mm., *whitish, becoming light yellow, sometimes with yellow ferruginous spots,* equal, *viscid,* sometimes pitted. Gills *shining white to dead white, spotted with lilac when wounded,* sometimes yellowish, or ochraceous ferruginous or tinged with pallid brick colour, adnate and subdecurrent, arcuate, thin, crowded, *very unequal,* here and there branched and anastomosing by veins. Flesh *white, lilac when broken.* Milk *white, then lilac,* acrid. Spores *pale ochraceous,* broadly elliptical, 10–12 × 8–9μ, echinulate. Cystidia "subulate, pointed, 50–75 × 6–10μ" Rick. Smell nauseous, or aromatic, taste slowly acrid. Woods, and heaths. Aug.—Nov. Not uncommon. (*v.v.*)

1581. **L. flavidus** Boud. (= *Lactarius uvidus* Fr. sec. Quél.) Trans. Brit. Myc. Soc. IV, t. 9. *Flavidus*, yellowish.

P. 5–10 cm., *pale citron, or sulphur yellow, becoming stained with violet* on injury or rubbing, convex, then expanded and slightly depressed at the centre, which often remains umbonate. St. 3–8 × 1–2 cm., *white, or yellowish, soon stained with violet on handling or other injury,* more or less attenuated at the base. Gills *yellowish, bruising violet on injury,* adnato-decurrent, narrow, crowded. Flesh *white, becoming quickly violet* on exposure to the air. Milk *white, then violet,* acrid. Spores white, ovoid, 9–10 × 8–9 μ, verrucose, reticulate, 1-guttulate. Taste mild, then acrid. Woods. Sept.—Oct. Uncommon. (*v.v.*)

***P. without a pellicle, hence absolutely dry, most frequently unpolished.

1582. **L. flexuosus** Fr. *Flexuosus*, full of turns.

P. 5–10 cm., *lead-grey, or violet-grey, becoming pale,* zoned, or zoneless, convex, becoming plane, depressed, somewhat repand, *dry,* somewhat shining, then *rivuloso-scaly,* or *floccose*; margin incurved, *velvety, and whitish.* St. 5–9 × 1·5–2·5 cm., *pallid grey, apex whitish, base somewhat yellowish,* obese, or equally attenuated downwards, often lacunose or pitted, *delicately pubescent.* Gills *light yellowish, at length becoming whitish-flesh-colour,* adnate, *thick, distant,* branched. Flesh white, cheesy, hard. Milk white, very acrid. Spores *pale ochraceous,* globose, 6–7 μ, verrucose. Cystidia "thin, clavate, 50–80 × 7–8 μ" Rick. Taste very acrid. Pine, and beech woods. July—Oct. Uncommon. (*v.v.*)

var. **roseozonatus** Fr. Fr. Icon. t. 169, fig. 3.
Roseus, rose-colour; *zonatus*, zoned.

Differs from the type in the *rose colour, or rosy violet pileus marked with darker zones.* Pine woods. Sept. Rare.

1583. **L. pyrogalus** (Bull.) Fr. Cke. Illus. no. 937, t. 993.
πῦρ, fire; *γάλα*, milk.

P. 5–10 cm., *cinereous-grey,* at length *becoming dingy yellow,* firm, convex, becoming plane, depressed, *somewhat zoned, delicately grumose* under a lens, *moist* in wet weather; margin soon spreading. St. 4–6 cm. × 6–12 mm., *pallid white, sometimes dingy,* often attenuated downwards, grumoso-tubercular under a lens. Gills *light yellow-wax-colour, the colour rich inclining to ochraceous or flesh colour,* adnato-decurrent, thin, *somewhat distant.* Flesh white, *greyish near the pileus.* Milk white, very acrid. Spores *pale ochraceous,* subglobose, 7–8 μ, echinulate. Cystidia "only on the edge of the gill, 45–70 × 7–8 μ, obtuse" Rick. Taste very acrid. Woods, and pastures. Aug.—Dec. Common. (*v.v.*)

1584. **L. squalidus** (Krombh.) Fr. Krombh. t. 40, figs. 23–25.
Squalidus, dirty.

P. 2·5–9 cm., *pale greyish olive, or lurid, margin sprinkled with saffron-yellow dots*, convexo-plane, umbilicate. St. 5–8 cm. × 5–10 mm., *white, or concolorous*, equal, firm. Gills *becoming yellow, narrow*, adnate. Flesh white. Milk whitish, *sweet*. Spores white, globose, 6–10μ, echinulate. Amongst moss in damp woods. Oct. Rare.

1585. **L. capsicum** Schulz. Kalchbr. Icon. t. 26, fig. 1.
Capsicum, red pepper.

P. 5–10 cm., *chestnut colour, darker at the closely involute margin*, compact, pulvinate, dry. St. 3–9 × 2·5 cm., *whitish, rufous striate, apex fulvous*, subequal, or incrassated upwards. Gills *fulvous, somewhat orange colour*, adnato-decurrent, crowded, anastomosing at the base, 3–4 mm. wide. Flesh *yellow, becoming fuscous* on exposure to the air. Milk white, very acrid. Spores globose, 6μ, rough. Under birches. Sept. Rare.

1586. **L. chrysorheus** Fr. (= *Lactarius theiogalus* (Bull.) Quél.) Cke. Illus. no. 940, t. 984. χρυσός, gold; ῥέω, I flow.

P. 5–8 cm., *pale yellowish flesh colour, with darker zones or spots*, convex, umbilicate, then infundibuliform. St. 5–7·5 × 1–2·5 cm., *white*, equal, delicately pruinose under a lens. Gills *pallid yellowish*, decurrent, very thin and crowded. Flesh white, *bright sulphur-yellow when broken*. Milk white, *then bright sulphur-yellow* (*golden*), very acrid. Spores white, subglobose, 6–7 × 6μ, echinulate, 1-guttulate. Cystidia "lanceolate, 50–60 × 8–12μ" Rick. Taste acrid. Woods, especially oak. Aug.—Nov. Common. (*v.v.*)

L. glaucescens Crossland = **Lactarius piperatus** (Scop.) Fr.

1587. **L. acris** (Bolt.) Fr. Cke. Illus. no. 941, t. 1005. *Acris*, sharp.

P. 5–9 cm., *cinereous fuliginous, sometimes darker, sometimes paler*, convex, then plane, at length obliquely infundibuliform, irregular, often excentric, or emarginate on one side, firm, rigid, moist, here and there spotted. St. 4–5 × 1–1·5 cm., *pallid, apex white, attenuated downwards, often oblique*, ascending, or curved. Gills *pallid, then yellow flesh colour*, subdecurrent, thin, somewhat crowded, forked. Flesh *white becoming reddish on exposure to air*. Milk white, *soon reddish*, acrid. Spores *ochraceous*, subglobose, 8–11μ, echinulate. Smell strong, stinking, taste acrid. Woods. Aug.—Nov. Rare.

1588. **L. violascens** (Otto) Fr. (= *Lactarius uvidus* Fr. sec. Bataille; *Lactarius uvidus* Fr. var. *violascens* (Otto) Quél.; *Lactarius luridus* (Pers.) Fr. sec. Rick.) *Violascens*, becoming violet.

P. 6–8 cm., *grey, or pale brown, with darker zones*, convex, then expanded, or more or less depressed at the disc, dry. St. 5–6 cm. ×

12–15 mm., *greyish white*, equal. Gills *white, then lilac*, slightly decurrent, crowded. Flesh *white*, thick, firm. Milk *white, becoming violet* on exposure to the air, mild (becoming acrid?). Spores "white, globose, 8–9μ, echinulate" Sacc. Cystidia "ventricose-subulate, 50–70 × 8–15μ" Rick. Woods. Oct. Rare.

1589. **L. umbrinus** (Pers.) Fr. Cke. Illus. no. 942, t. 1006.
Umbrinus, umber brown.

P. 5–7·5 cm., more or less *olivaceous umber, paler and yellowish when old*, convex, then plane, disc slightly depressed, often wavy and excentric, dry, *flocculoso-rivulose*. St. 2·5–3·5 × 1–2 cm., *concolorous, or paler, apex white*, attenuated upwards. Gills *pallid, dingy yellowish*, slightly decurrent, thin, crowded, forked behind, 2 mm. wide. Flesh white, *becoming tinged with brown on exposure to the air*, firm. Milk *white, forming grey spots* when it has escaped, acrid. Spores white, globose, 8μ, rough. Cystidia none. Taste acrid. Pine woods. Sept.—Oct. Rare.

L. plumbeus (Bull.) Fr. = **Lactarius turpis** (Weinm.) Fr.

1590. **L. piperatus** (Scop.) Fr. (= *Lactarius glaucescens* Crossland.) Cke. Illus. no. 944, t. 979. *Piperatus*, peppery.

Entirely white. P. 5–22·5 cm., *becoming yellowish with age, umbilicate*, then reflexed and *infundibuliform*, rigid; margin involute at first. St. 3–7 × 2–5 cm., equal, or obconical, obsoletely pruinose. Gills *becoming pale ochraceous, decurrent, crowded, narrow, dichotomous*, edge *obtuse*. Flesh *white, then yellowish, becoming sometimes greenish grey*. Milk *white, often becoming greenish when dry*, very acrid. Spores white, globose elliptical, 6–9 × 6–8μ, minutely punctate, with a large central gutta. Cystidia "clavate, or fusiform-filiform, 60–70 × 8–9μ, obtuse, very sparse" Rick. Taste very acrid. Said to be edible. Woods. Aug.—Oct. Common some years. (*v.v.*)

var. **pergamenus** (Swartz) Quél. Cke. Illus. no. 943, t. 978.
Pergamena, parchment.

Differs from the type in the *rugose wrinkled pileus, the longer thinner stem*, and the *adnate, horizontal gills*. Woods. Aug.—Oct. Uncommon. (*v.v.*)

1591. **L. vellereus** Fr. Cke. Illus. no. 945, t. 980. *Vellus*, a fleece.

Entirely white. P. 10–30 cm., *becoming stained or spotted with yellow, convexo-saucer-shaped, innato-pubescent*, compact; margin strongly incurved. St. 5–8 × 2·5–6 cm., becoming yellowish, equal, *hard, finely pubescent*. Gills *watery white, then pale ochraceous*, adnato-decurrent, *arcuate*, rather thick, *somewhat distant, rather broad, branched, edge acute*. Flesh *white, yellowish on exposure to the air*. Milk white, *scanty, turning litmus paper red, very acrid*. Spores white, subglobose, 8 × 7–8μ, minutely echinulate, 1-guttulate. Cystidia "cylindrical, with a short point, 70–105 × 6–8μ" Rick. Taste very peppery. Said to be edible. Woods. Aug.—Dec. Common. (*v.v.*)

var. **velutinus** Bertillon. *Velutinus*, velvety.

Differs from the type in the *more crowded gills, in the flesh turning reddish tawny on exposure to the air, and in the sweet milk* only slowly becoming somewhat acrid, and *not turning litmus paper red.* Woods. Uncommon. (*v.v.*)

L. exsuccus (Otto) Fr. = **Russula chloroides** (Krombh.) Bres.

1592. **L. scoticus** B. & Br. Cke. Illus. no. 938, t. 1004, fig. B. *Scoticus*, Scotch.

P. 2·5–5 cm., *whitish*, convex, then depressed, *tomentose*, then smooth; margin involute, tomentose. St. 2·5–3 cm. × 6 mm., *somewhat flesh colour*, somewhat unequal, curved. Gills *whitish*, very slightly decurrent, thin, scarcely branched, about 2 mm. broad. Flesh whitish. Milk white, acrid. Spores white, subglobose, 7–8μ, apiculate, minutely echinulate. Smell *pungent*, taste very acrid. Amongst moss. Sept. Rare.

1593. **L. involutus** Soppitt. Cke. Illus. no. 1195, t. 1194. *Involutus*, incurved.

Entirely white, or with a pale ochraceous tinge. P. 2·5–5 cm., convex, then plane, or slightly depressed; margin *strongly and persistently involute*, minutely silky. St. 2–3 cm. × 4–6 mm., equal, or slightly thickened at the base, very firm. Gills very slightly decurrent, *very crowded*, 1 mm. broad, sometimes forked. Milk white, very acrid. Spores white, pip-shaped, 5 × 3μ, *smooth.* Taste acrid. Woods. Sept. Rare.

II. St. central. Gills naked. Milk always deeply coloured.

1594. **L. deliciosus** (Linn.) Fr. Cke. Illus. no. 947, t. 982. *Deliciosus*, delicious.

P. 5–15 cm., *orange-brick-colour, becoming pale and stained with verdigris, concentrically zoned with darker markings*, convex, then plano-depressed, or broadly infundibuliform, slightly viscid. St. 2–8 × 2–2·5 cm., *concolorous, or paler, often stained with verdigris*, equal, or attenuated at the base, fragile, often spotted in a pitted manner. Gills *saffron yellow, becoming stained with verdigris*, subdecurrent, crowded, narrow, arcuate, often branched. Flesh *white, then reddish and becoming verdigris.* Milk *red-brick-saffron, aromatic*, mild, then slightly acrid. Spores white, pinkish in the mass, subglobose, 8–9 × 7–8μ, echinulate, 1-guttulate. Cystidia "sparse, fusiform-subulate, 30–40 × 4–6μ" Rick. Smell pleasant, taste slightly acrid. Edible. Coniferous woods, and elsewhere under conifers. July—Dec. Common. (*v.v.*)

1595. **L. sanguifluus** (Paul.) Fr. Boud. Icon. t. 50.

Sanguis, blood; *fluus*, flowing.

P. 3–8 cm., *reddish tawny, spotted with darker markings*, rarely zoned, *becoming stained with verdigris*, plane, then depressed, firm, slightly viscid; margin at first involute and white pruinose. St. 3–6 × 1–2·5 cm., *rosy flesh, or blood colour, at length concolorous and stained verdigris*, at first equal, then dilated at the apex, and attenuated at the base, *pruinose*, often pitted. Gills *pale ochraceous, then concolorous, and finally stained verdigris*, adnate, then decurrent, very crowded, rather narrow, often branched. Flesh white, *tinged with blood-red especially near the cuticle of the stem*, firm, cheesy. Milk *blood-red*, slightly acrid. Spores pale ochraceous, globose, 8–9 × 8μ, echinulate, 1-guttulate. Cystidia "sparse, subulate, 45 × 5–8μ" Rick. Smell pleasant, *often like Mentha piperita*, taste slightly acrid. Edible. Pine woods. Sept.—Oct. Uncommon. (*v.v.*)

III. St. central. Gills pallid, then changing colour, afterwards darker, glancing when turned to the light, at length white pruinose. Milk at the first white, mild, or mild becoming acrid.

*P. at the first viscid.

1596. **L. pallidus** (Pers.) Fr. Cke. Illus. no. 948, t. 1007.

Pallidus, pale.

P. 6–15 cm., *flesh colour, or clay colour to pallid, somewhat tan, umbilicato-convex*, depressed, obtuse, viscid; margin broadly and for a long time involute. St. 5–6 × 1·5 cm., *concolorous*, somewhat equal. Gills *whitish, then concolorous*, pruinose, subdecurrent, rather broad, somewhat thin, crowded, somewhat branched. Flesh *pallid*. Milk white, acrid. Spores white, globose, 9–10μ, echinulate, 1-guttulate. Cystidia "abundant, subulate-fusiform, 70–75 × 7–9μ" Rick. Taste mild, then acrid. Woods. Sept.—Oct. Common. (*v.v.*)

1597. **L. quietus** Fr. Boud. Icon. t. 51. *Quietus*, calm.

P. 3–9 cm., *somewhat cinnamon, flesh colour, disc darker, somewhat zoned*, convex, then depressed, obtuse; margin deflexed, at first very slightly viscid, then *somewhat silky*, opaque, *becoming pale*. St. 4–9 × 1–1·5 cm., *rufescent, at length beautifully rubiginous*, spongy, equal, sometimes attenuated at the base which is covered with concolorous hairs. Gills *white, then soon brick-rufescent*, adnato-decurrent, somewhat forked at the base. Flesh *white, then rufescent*. Milk white, sweet. Spores white, globose, 8–10μ, verrucose. Cystidia "lanceolate-subulate, 50–60 × 6–7μ" Rick. Smell oily, taste pleasant. Woods. Aug.—Dec. Common. (*v.v.*)

1598. **L. aurantiacus** (Fl. Dan.) Fr. *Aurantiacus*, orange.

P. 3–6 cm., *bright golden orange colour*, convex, then plane, or depressed, sometimes umbonate, slightly viscid. St. 6–8 × 1 cm., *concolorous*, equal, pruinose. Gills *yellowish flesh colour, decurrent*, thin, crowded, narrow. Flesh *pallid*. Milk white, slowly acrid. Spores *ochraceous*, globose, 8–9μ, echinulate. Cystidia "abundant, lanceolate-subulate, 75–90 × 10–11μ" Rick. Smell pleasant, taste mild. Woods. Sept.—Oct. Not uncommon. (*v.v.*)

1599. **L. theiogalus** (Fr.) Plowr. (= *Lactarius chrysorheus* Fr. sec. Quél.; *Lactarius hepaticus* (Plowr.) Boud.) Trans. Brit. Myc. Soc. I, t. v. *θεῖον*, brimstone; *γάλα*, milk.

P. 2·5–7 cm., *liver coloured when moist, drying to rufous tawny and lighter at the margin*, convex, then expanded, and finally depressed, umbonate, umbo sometimes wanting, viscid at first; *margin crenulate*, thin. St. 3–4 × ·5–1 cm., *slightly pinker than the colour of the pileus*, equal, attenuated at the base. Gills *pale, then rufescent*, adnato-decurrent, each terminating at the crenulations of the margin of the pileus, 3–6 mm. broad, thin, rather distant. Flesh of the pileus *pallid, then ochraceous*, of the stem *becoming rufous* especially downwards. Milk white, *very slowly changing to sulphur yellow*, mild, then slightly acrid. Spores white, globose, 6–7μ, echinulate, 1-guttulate. Cystidia "sparse, subulate, 45–75 × 6–8μ" Rick. Fir woods. Aug.—Nov. Not uncommon. (*v.v.*)

1600. **L. cremor** Fr. *Cremor*, thick juice.

P. 3–6 cm., *tawny orange*, convex, then plane and obtuse, sometimes umbonate when young, often unequal and excentric, viscid, *minutely punctate*; *margin striate*. St. 3–5 cm. × 6–8 mm., *concolorous*, equal, fragile, silky upward under a lens. Gills *white, then flesh colour*, adnate, somewhat distant, fragile, pruinose. Flesh *concolorous, or paler*, thin. Milk *whitish, often watery*, somewhat mild. Spores white, globose, 9–10μ, echinulate. Cystidia "subulate, 45–60 × 7–8μ" Rick. Taste mild, or slowly acrid. Beech woods. Sept. Rare.

var. **pauper** Karst. Cke. Illus. no. 951, t. 1008. *Pauper*, poor.

Differs from the type in being *entirely yellowish flesh colour*, ochraceous when dry, *the margin of the pileus at length sulcate*, and *the juiceless, white flesh*. Under larches. Sept. Rare.

1601. **L. vietus** Fr. Cke. Illus. no. 952, t. 1009, fig. A. *Vietus*, shrunken.

P. 3–6 cm., *flesh colour, or livid grey, becoming pale*, subpapillate, becoming plane, then *umbilicate*, at length *somewhat infundibuliform*, viscid, opaque, *slightly silky when dry*; margin somewhat deflexed. St. 5–7 cm. × 4–10 mm., *concolorous*, attenuated upwards, or somewhat equal. Gills *whitish, then yellowish*, adnato-decurrent, thin, somewhat crowded, somewhat flaccid. Flesh *whitish, then grey*. Milk *white, then grey*, slowly acrid. Spores white, globose, 7–8μ, echinulate. Cystidia "sparse, subulate, 60–70 × 10–12μ" Rick. Smell somewhat pungent, taste slightly acrid. Woods. Sept.—Oct. Not uncommon. (*v.v.*)

1602. **L. cyathula** Fr. (= *Lactarius cupularis* (Bull.) Quél.) Cke. Illus. no. 952, t. 1009, fig. B, and no. 953, t. 1085.

κύαθος, a cup.

P. 1–6 cm., *rufescent brick, or flesh colour, somewhat zoned, when dry becoming pale, livid or flesh colour, hoary tan, rimoso-rivulose*, convexo-plane, *umbonate*, at length *plano-depressed*, umbo often vanishing, slightly viscid, sometimes striate. St. 5 cm. × 2–10 mm., *pale, at length whitish*, equal, pruinose. Gills *white flesh colour, then yellowish*, decurrent, *very crowded*, thin, linear. Flesh *white flesh colour*. Milk white, acrid. Spores pale ochraceous, globose, 6–10μ, echinulate. Cystidia "very sparse, subulate" Rick. Smell strong of bugs when drying. Woods. Aug.—Nov. Rare.

**Pileus unpolished, squamulose, villose, or pruinose.

1603. **L. rufus** (Scop.) Fr. Cke. Illus. no. 954, t. 985. *Rufus*, red.

P. 5–10 cm., *bay-brown-rufous, umbonate* when young, soon depressed with an umbo, and *at length infundibuliform*, dry, *at first flocculoso-silky*, but soon polished; margin *involute when young, somewhat whitish-tomentose*. St. 5–8 × 1 cm., *rufescent, paler than the pileus, white pubescent at base*, obsoletely pruinate. Gills *ochraceous, or pallid, then rufescent*, adnato-decurrent, crowded, scarcely branched. Flesh *pallid*, not compact. Milk white, very acrid. Spores white, broadly elliptical, 9–10 × 7–8μ, verrucose. Cystidia abundant, "lanceolate-fusiform, 60–70 × 7–10μ" Rick. Taste very acrid. Coniferous woods. June—Dec. Common. (*v.v.*)

var. **exumbonatus** Boud. Boud. Icon. t. 52.

Exumbonatus, without an umbo.

Differs from the type in *the absence of an umbo*. Coniferous woods. Sept—Oct. Not uncommon. (*v.v.*)

1604. **L. helvus** Fr. Cke. Illus. no. 955, t. 994. *Helvus*, light bay.

P. 5–10 cm., *pale yellowish-brick-colour, becoming pale, fragile*, convex, then *flattened, somewhat umbonate, the surface wholly broken up into granuloso-squamulose flocci*. St. 5–8 × 1–1·5 cm., *brick colour inclining to pale*, equal, *pruinose*, base white, tomentose. Gills *whitish, then somewhat flesh colour, at length yellowish*, decurrent, thin, crowded. Flesh *concolorous but paler*. Milk white, mild, then slightly acrid, somewhat watery. Spores *pale ochraceous*, globose, 7–8μ, echinulate. Cystidia "cylindrical-rounded, 50–70 × 9–12μ" Rick. Smell strong, resinous, somewhat like Foenugreek, taste mild. Coniferous woods. Aug.—Oct. Not uncommon. (*v.v.*)

1605. **L. tomentosus** (Otto) Cke. Cke. Illus. no. 956, t. 1010.

Tomentosus, woolly.

P. 7–9 cm., *dingy flesh colour, rufescent, or brownish*, umbonate, then depressed, or infundibuliform, *finely tomentose*. St. 5 × 1–1·5 cm., *pallid*, equal. Gills *yellow flesh colour*, rather decurrent. Flesh *white, then tinged brown*. Milk whitish, mild. Spores white, globose, 8–9μ, verrucose. Smell pleasant, taste mild, then slightly acrid. Damp places. Sept. Rare.

1606. **L. mammosus** Fr. Fr. Icon. t. 170, fig. 2.

Mammosus, having large breasts.

P. 3–6 cm., *greyish fuscous*, convex, *acutely umbonate*, umbo at length vanishing, then depressed, *clothed with appressed down*; margin involute, *white-pubescent*. St. 5–8 × 1–1·5 cm., *white, inclining to pale, pubescent*, firm. Gills *whitish, then pallid ferruginous*, adnate, crowded. Flesh *reddish white*. Milk white, mild, then acrid. Spores white, "subglobose, 6–7 × 5–6μ, echinulate. Cystidia subulate, 60 × 8–9μ" Rick. Birch, and pine woods. Oct. Rare.

var. **monstrosus** Cke. Cke. Illus. no. 957, t. 995.

Monstrosus, strange.

Differs from the type in its *larger size*. Woods. Oct. Rare.

var. **minor** Boud. Boud. Icon. t. 58. *Minor*, smaller.

Differs from the type in its *smaller size, and the pileus clothed with shorter hairs*. Under birches. Nov. Rare. (*v.v.*)

1607. **L. glyciosmus** Fr. Fr. Icon. t. 170, fig. 3.

γλυκύς, sweet; ὀσμή, scent.

P. 2–7 cm., *grey, brick colour, fuscous, most frequently passing into violet*, convex, becoming somewhat plane, *acutely papillate*, then depressed, the papilla vanishing, *delicately innato-squamulose, or unpolished*, opaque. St. 2·5–5 cm. × 4–8 mm., *light yellowish, becoming tawny when bruised, sometimes silvery-whitish*, especially at the apex, somewhat equal, *pubescent*. Gills *straw colour, then tawny-flesh-colour, or somewhat ochraceous*, adnato-decurrent, arcuate becoming plane, thin, often branched, *crowded*. Flesh *white, then slightly tinged with the external colour*. Milk *white, rarely greenish*, mild, then slightly acrid. Spores *pale ochraceous*, globose, 7–9μ, echinulate, 1-guttulate. Cystidia "subulate, 69–70 × 7–9μ" Rick. Smell very pleasant, aromatic, taste mild. Edible. Woods. Aug.—Nov. Common. (*v.v.*)

var. **flexuosus** Massee. Cke. Illus. no. 958, t. 1011, as type.
Flexuosus, full of turns.

Differs from the type in the *silky, umbilicate, flexuose, more or less zoned pileus*. Woods. Sept.—Oct. Not uncommon. (*v.v.*)

1608. **L. lignyotus** Fr. Fr. Icon. t. 171, fig. 1. λιγνύς, smoke.

P. 4–8 cm., *fuliginous umber, elastic-fragile*, convexo-plane, somewhat depressed; margin at first incurved, then depressed, *acutely-umbonate, plicately-rugulose*, or *wrinkled sulcate, pruinosely velvety*. St. 7–12 × 1·5–2 cm., *concolorous*, base paler, becoming whitish, *fragile*, corticate, *constricted and plicate at the apex, pruinosely velvety*. Gills *snow-white, then whitish ochre, reddish when wounded*, rounded behind and adnate, then subdecurrent, thin, rather crowded. Flesh *white, then slowly becoming reddish, ochraceous, or ferruginous*. Milk *watery white, becoming reddish, or saffron colour*, sparse, *sweet*. Spores *ochraceous*, globose, 9μ, strongly echinulate, 1-guttulate. Taste pleasant. Under fir, and beeches. Sept.—Oct. Uncommon. (*v.v.*)

1609. **L. fuliginosus** Fr. (= *Lactarius azonites* (Bull.) Quél.) Cke. Illus. no. 959, t. 996. *Fuliginosus*, sooty.

P. 3–10 cm., *tan whitish, sprinkled with innate, fuliginous pruina, coffee and milk colour*, finely velvety, then somewhat rugulose, *soapy livid, fawn brick colour*, the disc sometimes at last brown, convex, then *somewhat repand* and depressed; margin at first inflexed, then soon spreading. St. 4–8 cm. × 6–10 mm., *dead white to shining white, then dingy, tan, somewhat rufescent-brick-colour, fuliginous*, somewhat equal, sometimes rugulose. Gills *white, at length light yellow ochraceous*, rounded adnexed, then decurrent, somewhat thin, somewhat distant, branched, connected by veins, the intermediate ones at length crisped. *Flesh and milk white, then rose colour, and at length saffron yellow*. Spores ochraceous, globose, 9–10μ, echinulate. Cystidia "sparse, subulate" Rick. Taste mild, then slightly acrid. Woods, and pastures. Aug.—Nov. Common. (*v.v.*)

1610. **L. picinus** Fr. Cke. Illus. no. 960, t. 997. *Picinus*, pitch-black.

P. 4–8 cm., *umber, or blackish umber, convex becoming plane, umbonate*, orbicular, at first everywhere *villose, somewhat velvety*, then becoming smooth. St. 5–8 × 1–1·5 cm., *paler than the pileus*, equal, *pruinose*. Gills *ochraceous*, adnate, thin, very crowded, straight. Flesh *pallid, becoming reddish on exposure to the air*. Milk *white, acrid*. Spores ochraceous, globose, 7–10μ, echinulate and ribbed. Taste acrid. Coniferous woods. Sept.—Oct. Uncommon. (*v.v.*)

1611. **L. retisporus** Massee. *Rete*, a net; σπορά, seed.

P. 5–9 cm., *dark smoky-brown*, convex, then plane, disc depressed, minutely *velvety, radially rugose* from disc to margin. St. 3–5 ×

1·5 cm., *paler than the pileus*, equal. Gills *pale ochraceous with darker spots, deeply sinuate* and slightly adnexed, subdistant. Flesh *dingy yellow, becoming reddish brown when cut*. Milk *white, then brown, sweet*, becoming very thick and tenacious. Spores *colourless*, globose, 10μ, *with raised bands forming a network*. Under beeches. Oct. Rare.

1612. **L. lilacinus** (Lasch) Fr. Boud. Icon. t. 53.
Lilacinus, lilac-coloured.

P. 3–8 cm., *rosy lilac, covered with a concolorous tomentum*, often spinulose when young, plane, then depressed, papillate, sometimes very obsoletely zoned. St. 3–8 cm. × 5–12 mm., *paler and more ochraceous than the pileus*, equal, somewhat rugulose, apex *white-mealy*. Gills *ochraceous, rarely tinted rosy lilac*, adnato-decurrent, narrow. Flesh *rosy white*, acrid. Milk white, acrid, plentiful. Spores white, globose, or oval, 7–10μ, verrucose, reticulated. Cystidia "fusiform, 60–75 × 7–9μ" Rick. Taste acrid. Woods, and damp places. Aug.—Oct. Not uncommon. (*v.v.*)

1613. **L. spinosulus** Quél. Trans. Brit. Myc. Soc. II, t. 11.
Spinosulus, full of little spines.

P. 2–4 cm., *brick red, or rosy lilac*, convex, *acutely umbonate*, then depressed, *covered with minute erect spines*, especially towards the margin, *zoned* and *spotted*. St. 3–4 cm. × 4–8 mm., *concolorous*, rugulose, granular. Gills *yellowish flesh colour, then yellowish*, decurrent, narrow. Flesh *paler*. Milk white, slowly acrid. Spores *pale ochraceous*, globose, 7–8μ, echinulate. Taste slightly acrid. Woods. Sept.—Oct. Not uncommon. (*v.v.*)

var. **violaceus** Cke. Cke. Illus. no. 961, t. 998, fig. B. *Violaceus*, violet.

Differs from the type in the *rosy-violet p., the incurved margin, and the pale st.* On the ground. Sept.

***P. polished, smooth.

1614. **L. volemus** Fr. (= *Lactarius lactifluus* (Schaeff.) Quél.) Cke. Illus. no. 962, t. 999. *Volema pira*, a species of large pear.

P. 5–12 cm., *rufous tawny, golden, becoming pale, compact, rigid, obtuse*, paler at the margin, plano-convex, at length depressed and *rimoso-rivulose*; margin at first incurved. St. 6–10 × 1·5–3 cm., *concolorous*, somewhat equal, or attenuated upwards, *obese, hard, pruinose*. Gills *white to yellowish*, adnato-decurrent, thin, crowded, *becoming dingy when wounded*. Flesh *white, becoming brownish on exposure to the air*. Milk *white, sweet, plentiful, rarely becoming yellow*. Spores *very pale ochraceous*, globose, 5–6μ, verrucose. Cystidia "very abundant, subulate-fusiform, 60–100 × 8–9μ, very undulating and thick walled" Rick. Smell and taste very pleasant. Edible. Woods. Aug.—Oct. Not uncommon. (*v.v.*)

1615. **L. ichoratus** (Batsch) Fr. Cke. Illus. no. 963, t. 1000.

ἰχώρ, serum.

P. 5–10 cm., *tawny-brick-colour, disc often brown, brick colour and zoned*, plano-depressed, often unequal, excentric, occasionally repand, opaque, *thin*, obtuse. St. 4–7·5 cm. × 6–10 mm., *tawny, then rufescent, equal, or fusiform below.* Gills *white, then ochraceous*, adnate, decurrent with a tooth, scarcely crowded. Flesh *pallid, becoming brownish on exposure to the air.* Milk white, sweet. Spores *ochraceous*, elliptical, 8–10 × 6–7μ, echinulate. Cystidia "sparse, subulate, 50 × 5–6μ" Rick. Smell strong, taste pleasant. Woods. Oct.

1616. **L. serifluus** (DC.) Fr. Cke. Illus. no. 964, t. 1012.

Serum, lymph; *fluus*, flowing.

P. 2·5–6 cm., *brown tawny*, plane, then depressed, sometimes slightly umbonate, somewhat flexuose; margin inflexed. St. 3–4 cm. × 4–10 mm., *concolorous, or paler*, somewhat incurved, *base often strigose with tawny hairs.* Gills *yellowish flesh colour, then reddish*, adnate, decurrent with a tooth. Flesh *reddish tawny.* Milk *watery, insipid, scanty.* Spores *pale ochraceous*, globose, 6–7μ, echinulate, 1-guttulate. Cystidia "vesiculose-pyriform, then vesiculose-flask-shaped, 30 × 14–20μ" Rick. Smell unpleasant, like bugs. Woods, and boggy places. July—Dec. Common. (*v.v.*)

1617. **L. mitissimus** Fr. Cke. Illus. no. 965, t. 1001.

Mitissimus, very mild.

P. 3–8 cm., *golden tawny*, convex, *papillate*, depressed, papilla often vanishing, *somewhat slippery* when moist. St. 2·5–8 cm. × 8–12 mm., *concolorous.* Gills *a little paler than the pileus*, often stained with minute rufous spots, adnato-decurrent, somewhat arcuate, thin, crowded. Flesh *pallid.* Milk *white, plentiful, mild, then somewhat bitterish.* Spores *pale ochraceous*, elliptical, 8–9 × 6–7μ, verrucose, 1-guttulate. Cystidia "sparse, subulate, 45–50 × 5–6μ" Rick. Taste slightly acrid. Edible. Woods. Aug.—Dec. Common. (*v.v.*)

1618. **L. subdulcis** (Pers.) Fr. *Sub*, somewhat; *dulcis*, sweet.

P. 3–8 cm., *rufescent*, not becoming pale, *papillate*, at length depressed. St. 2·5–5 cm. × 4–10 mm., *concolorous*, equal, *somewhat pruinose.* Gills *paler*, adnate, crowded, *fragile.* Flesh rufescent. Milk white, *somewhat mild.* Spores *pale ochraceous*, globose, 9–10μ, echinulate. Cystidia "sparse, subulate, 50–60 × 8–10μ" Rick. Taste somewhat bitterish. Edible. Woods. July—Dec. Common. (*v.v.*)

var. **concavus** Fr. *Concavus*, hollowed out.

Differs from the type in the *rufous bay colour of all its parts* recalling *Lactarius rufus, the inflexed margin of the pileus, the smooth stem*, the *very pruinose gills,* and the *constant mild* taste. Damp places in woods. Aug.—Oct. Not uncommon. (*v.v.*)

var. **sphagneti** Fr. *Sphagnetum,* a *Sphagnum* swamp.

Differs from the type in the *obtuse, red bay pileus shining as if varnished, and the crenate, inflexed margin.*

1619. **L. camphoratus** (Bull.) Fr. Cke. Illus. no. 967, t. 1013, fig A. *Camphoratus,* strong scented.

P. 2·5–6 cm., *brown-brick-red,* convex, then depressed, sometimes somewhat zoned. St. 2–3 cm. × 4–6 mm., *concolorous,* somewhat undulated. Gills *yellowish-brick-colour,* adnate, crowded. Flesh *reddish.* Milk *white, mild, watery.* Spores *ochraceous,* globose, 8–9μ, echinulate. Smell strong, like Melilot when dried. Taste pleasant. Woods, especially conifers. Aug.—Nov. Common. (*v.v.*)

var. **Terrei** (B. & Br.) Cke. Michael Terrey.

Differs from the type in the *corrugated pileus, and the swollen base of the stem clad with orange down.*

1620. **L. subumbonatus** Lindgr. (= *Lactarius cimicarius* (Batsch) Quél.; *Lactarius rubescens* (Bres.) Bataille.) Cke. Illus. no. 968, t. 986, fig. A. *Sub,* rather; *umbonatus,* umbonate.

P. 2–3 cm., *dark cinnamon colour,* convex, then depressed, rather umbonate, *rugose,* punctate. St. 2·5–3 cm. × 3 mm., *concolorous,* attenuated at the base. Gills *rufous flesh colour,* adnate. Flesh *grey, then yellowish.* Milk *white, watery, mild.* Spores white, subglobose, 5–6μ, rough. Smell foetid when old, taste mild. Woods. Sept.

1621. **L. cimicarius** (Batsch) Cke. (= *Lactarius subumbonatus* (Lindgr.) Quél.) Cke. Illus. no. 967, t. 1013, fig. B. *Cimex,* a bug.

P. 2–6 cm., *dark bay brown,* convex, then depressed, or infundibuliform; margin often waved and lobed. St. 4–6 cm. × 4–14 mm., *paler than the pileus,* equal. Gills *dingy ochraceous with a red tinge,* slightly decurrent. Flesh *tinged brownish.* Milk *white, acrid.* Spores white, globose, 9 × 7–8μ, verrucose, 1-guttulate. Smell strong, like bugs. Taste acrid. Woods. Sept.—Nov. Not uncommon. (*v.v.*)

1622. **L. obnubilus** (Lasch) Fr. Boud. Icon. t. 55. *Obnubilus,* overclouded, dark.

P. 2–6 cm., *brown fuliginous, darker at the disc, paler at the striate margin,* convex, at first papillate, then umbilicate, slightly viscid. St. 3–7 cm. × 6–9 mm., *paler than the pileus,* equal, slightly rugulose, covered with concolorous hairs at the base. Gills *yellowish,* adnato-decurrent, crowded. Flesh *reddish.* Milk *white, rather acrid.* Spores *pale ochraceous,* subglobose, 7–9 × 7–8μ, verrucose, 1-guttulate.

Taste somewhat bitter. Pine woods, and under alders. Sept.—Nov. Uncommon. (*v.v.*)

var. **crenatus** Massee. *Crenatus*, notched.

Differs from the type in the *coarsely sulcate margin of the pileus.* Fir woods.

1623. **L. tabidus** Fr. Boud. Icon. t. 57. *Tabidus*, wasting away.

P. 1–5 cm., *pale brick colour, then tan, becoming paler, submembranaceous*, somewhat plane, *acutely papillately umbonate*, at length expanded and somewhat depressed, *somewhat rugulose*; margin *pellucidly striate.* St. 2–4 cm. × 4 mm., *concolorous*, equal, or attenuated upwards, white tomentose at base. Gills *concolorous but paler*, adnato-decurrent, *flaccid*, narrow, somewhat distant. Flesh *concolorous.* Milk *white, sweet, then slightly acrid, scanty.* Spores white, globose, or oval, 10–12 × 7–10μ, verrucose, 1-guttulate. Taste pleasant, then slightly acrid. Woods, and under willows and alders. Sept. Uncommon.

1624. **L. minimus** W. G. Sm. Cke. Illus. no. 968, t. 986, fig. B. *Minimus*, smallest.

P. 6–15 mm., *pallid clay colour*, pulvinate, rounded, or slightly umbonate; margin incurved. St. 6 × 2 mm., *concolorous, generally excentric.* Gills *pallid*, slightly decurrent, arcuate, distant. Milk *white, mild, abundant.* Spores white, globose, 3–4μ, echinulate. Woods, and pastures. Oct. Uncommon.

IV. St. excentric, or lateral. Growing on trunks.

1625. **L. obliquus** Fr. *Obliquus*, slanting.

Entirely white, becoming yellowish. P. 5 cm., *thin*, plane, then depressed, oblique, *zoned with grey*, lobed, *silky.* St. 2·5 cm. × 4 mm., *rather excentric*, curved. Gills very slightly decurrent, crowded, narrow. Flesh *whitish.* Milk *white, slightly acrid.* Spores white, globose, 6μ, echinulate. Smell pleasant. Caespitose. On trunks of beech, and on banks. Oct. Uncommon.

III. Receptacle membranaceous, or fleshy membranaceous, fragile, rapidly putrescent, or shrivelling up.

Spores ochraceous, or ferruginous.

Bolbitius Fr.

(*βόλβιτον*, dung, especially cow dung.)

Pileus membranaceous, regular. Stem central, not confluent with the pileus. Gills free, or slightly adnate, acute at the edge. Spores ochraceous, ferruginous, fuscous, or salmon colour; elliptical, elliptic-oblong, or almond-shaped; smooth; with an apical germ-pore. Cystidia present, or absent. Growing on dung, and on rich soil.

B. hydrophilus (Bull.) Fr. = **Hypholoma hydrophilum** (Bull.) Fr.

1626. **B. vitellinus** (Pers.) Fr. Kalchbr. Icon. t. 19, fig. 2.
Vitellus, the yolk of an egg.

P. 2–4 cm., *egg yellow*, submembranaceous, *deeply campanulate*, then expanding and convex, viscid, smooth, *then furrowed and splitting at the margin*. St. 6–11 cm. × 2–4 mm., *cream colour*, attenuated upwards from the subbulbous base, *covered with white, fugacious, mealy flocci*. Gills *ochraceous, then somewhat ferruginous*, free, attenuated at both ends, thin, crowded. Flesh *yellowish*, thick at the disc. Spores ferruginous, yellow under the microscope, broadly elliptical, 12–14 × 7–8μ, often truncate at the one end. On horse dung, dung heaps, and amongst grass. May—Oct. Not uncommon. (*v.v.*)

var. **olivaceus** Gillet. *Olivaceus*, olivaceous.

Differs from the type in its *olivaceous colour*. Horse dung. Sept. Uncommon. (*v.v.*)

1627. **B. Boltonii** (Pers.) Fr.
James Bolton of Halifax, the author of "An History of Fungusses."

P. 2–4 cm., *yellow, becoming pale, disc darker*, fleshy, conical, or convex, then plane and *depressed at the disc*, viscid, smooth, becoming sulcate and splitting at the membranaceous margin, finally withering and becoming like paper. St. 5–8 cm. × 4–6 mm., *yellowish*, equal, *attenuated at the base*, often twisted, *flocculose with the fugacious cortina*, which sometimes forms a ring-like zone. Gills *light yellow, then livid fuscous*, slightly adnate. Flesh *yellowish, especially in the st.*, very thin. Spores fuscous ferruginous, broadly elliptical, or pip-shaped, 12–13 × 7–8μ. On dung, and manured soil. May—Oct. Uncommon. (*v.v.*)

1628. **B. flavidus** (Bolt.) Massee. Cke. Illus. no. 677, t. 689, as *Bolbitius Boltoni* Fr. *Flavidus*, yellowish.

P. 2·5–5 cm., *pale yellow, glutinous*, conical, then expanded, *disc usually slightly elevated*; margin striate, and usually splitting. St. 5–6 cm. × 4–6 mm., yellow, slightly thinner upwards. Veil very fugacious, *white*, leaving no mark on the st. Gills *white, then yellow, at length dusky brown, almost or quite free*, 4 mm. broad. Flesh *yellow in the st.* Spores *brown*, elliptical, 10 × 6μ. The whole plant deliquescing. Dung-hills, and rotten cloth in woods. June—Aug. Uncommon. (*v.v.*)

1629. **B. grandiusculus** Cke. & Massee. Cke. Illus. no. 1187, t. 1159.
Grandiusculus, pretty well grown up.

P. 3–5 cm., *pale yellow, disc rufous*, submembranaceous, campanulate, then expanded, smooth; margin slightly striate. St. 7·5–10 cm. × 6 mm., *white*, gradually attenuated upwards, straight, *smooth.*

Gills *pale, then rusty ochraceous, quite free*, attenuated behind, 2 mm. broad, linear, crowded. Flesh *white*, very thin. Spores rusty, elliptic-oblong, 15 × 5–6μ. Amongst grass. Sept. Rare.

1630. **B. fragilis** (Linn.) Fr. Cke. Illus. no. 679, t. 720, fig. A. *Fragilis*, fragile.

P. 1·5–2·5 cm., *light yellow, then becoming pale*, submembranaceous, pellucid, conical, then expanded, *subumbonate*, smooth, viscid; *margin striate, often crenulated.* St. 5–7·5 cm. × 2–4 mm., *yellow, very fragile*, attenuated upwards, *naked, smooth.* Gills *yellow, then pale cinnamon*, attenuato-adnexed, almost free, ventricose, 2–4 mm. broad. Flesh *yellowish*, thin at the disc. Spores ferruginous, almond-shaped, 10–11 × 6–7μ. Cystidia "vesiculose, 30–36 × 12–20μ" Rick. The whole plant rapidly withering. Roadsides, road-scrapings, horse dung, and pastures. April—Nov. Common. (*v.v.*)

1631. **B. affinis** Massee. *Affinis*, related.

P. 1–2 cm., *yellowish tawny, whitish when dry*, campanulate, then expanded, *umbonate*, glabrous, *dry*; *margin striate.* St. 4–7 cm., *white*, attenuated upwards from a *marginate bulb, shining.* Gills *yellowish tawny*, adnexed, narrow. Spores 8 × 6μ. Rare.

1632. **B. titubans** (Bull.) Fr. Cke. Illus. no. 680, t. 690. *Titubans*, tottering.

P. 2–4 cm., *light yellow at the disc, paler and becoming greyish at the margin, membranaceous*, very tender, ovato-campanulate, then *flattened and split*, diaphanous, slightly viscid, *striate*; *margin plicate.* St. 5–7·5 cm. × 1–2 mm., *white*, equal, very fragile, *shining*, sometimes mealy. Gills *ochraceous, or purplish, then fuscous flesh colour, or ferruginous, adnexed*, or free, 2–3 mm. broad. Flesh *whitish, scarcely any except at the disc.* Spores deep ochraceous, broadly elliptical, 11–15 × 8–9μ, with a hyaline apical germ-pore. Rich pastures, gardens, and roadsides. May—Oct. Common. (*v.v.*)

1633. **B. apicalis** W. G. Sm. Cke. Illus. no. 679, t. 720, fig. B. *Apex*, the top of a thing.

P. 8–15 cm., *brown, disc ochraceous bounded by a darker line*, membranaceous, *conical*, 18 mm. high, obsoletely umbonate, striate, then plicate to the apical disc, splitting at the margin. St. 5–6 cm. × 2 mm., *white*, attenuated upwards from the thickened base, *striate*, minutely pruinose under a lens. Gills *brown*, free, at first pressed to the stem, ventricose, 3 mm. broad. Flesh *white*, thick at the disc. Spores brown, elliptical, 9 × 6–7μ. Pastures. June. Rare.

1634. **B. rivulosus** B. & Br. Cke. Illus. no. 678, t. 928, fig. B. *Rivulosus*, rivulose.

P. 2–3 cm., *tan colour*, campanulate, striate, *rivulose.* St. 7 cm. × 4–6 mm., *white*, attenuated upwards, smooth. Gills *cinnamon*, slightly adnexed, narrowed behind, 3 mm. broad, rather distant. Flesh *white*, rather thick at the disc. Spores ferruginous, elliptic-oblong, 10–12 × 6–7μ. Earth in an orchid house. July. Rare.

1635. **B. niveus** Massee. Cke. Illus. no. 1186, t. 1160, as *Bolbitius conocephalus* Bull. *Niveus*, snow white.

P. 2–3 cm., *pure white*, fleshy, campanulate, obtusely umbonate, smooth, slightly viscid; margin striate. St. 7–9 cm. × 3–4 mm., *pure white*, gradually attenuated upwards from the *clavato-bulbous base*, rather tough, *shining*. Gills *pallid, then salmon colour, free*, narrowed behind, 2 mm. broad, broadest in front, subdistant. Flesh *white*, thick at the disc. Spores salmon colour, elliptical, 18 × 9–10μ. Earth in palm house. Rare.

1636. **B. tener** Berk. Cke. Illus. no. 681, t. 691. *Tener*, tender.

P. 7–15 mm., *white, yellowish white, or flesh colour, then whitish*, very delicate, *conical, elongated*, 8–25 mm. high, moist, *smooth*. St. 4–7 cm. × 2 mm., *white*, base bulbous or slightly thickened. Gills *salmon colour*, nearly free, attenuated behind, narrow, not crowded. Flesh *white*, very thin. Spores "salmon colour, elliptical, 15–16 × 8–10μ" Massee. Lawns, and rich pastures. May—Sept. Uncommon. (*v.v.*)

Spores black, or blackish fuscous. Gills auto-digested from below upwards.

Coprinus (Pers.) Fr.

(κόπρος, dung.)

Pileus fleshy, or membranaceous, regular. Stem central, confluent, or distinct from the pileus, with or without a ring or volva. Gills free, adnate, or attached to a collar, very thin, parallel-sided, or subparallel-sided, and auto-digested from below upwards. Spores black, violet black, chocolate, or fuscous; oval, elliptical, subglobose, angularly subglobose, pip-shaped, almond-shaped, or cordiform, smooth, very rarely echinulate; with an apical germ-pore. Cystidia usually large, rarely absent. Growing on the ground, on dung, and on wood. Solitary, or caespitose.

I. Gills covered above with a fleshy or membranaceous cuticle, hence the p. does not open into furrows along the gills but becomes torn and revolute.

*Furnished with a ring arising from the volva; the cuticle torn into scales.

1637. **C. comatus** (Fl. Dan.) Fr. Cke. Illus. no. 644, t. 658. *Comatus*, hairy.

P. 4–6 cm., *white, becoming pinkish at the margin and finally black*, fleshy, cylindrical, 5–15 cm. high, then campanulate, the *continuous cuticle soon separating into adpressed, shaggy scales, the scales becoming ochraceous at their apices, disc deep ochraceous, remaining persistently entire,* at length striate; margin often torn and unequal. St. 12–25 × 1–2 cm., *white, then dingy or lilac white,* attenuated upwards from the *bulbous, rooting base,* fibrillose, silky, shining. Ring *white,* membranaceous, thin, torn, *movable,* fugacious. Gills *white, then pink, at length black and deliquescent,* free, separate from the st., linear, 6–10 mm. broad. Flesh *white,* thick at the disc. Spores black, elliptical, often with a lateral apiculus, 11–13 × 6–7μ. Cystidia vesiculose, 50–65 × 20–30μ. Taste mild. Edible. Woods, pastures, roadsides, especially ground made up with night soil. April—Dec. Common. (*v.v.*)

var. **ovatus** (Schaeff.) Quél. Schaeff. Icon. t. 7. *Ovatus,* egg-shaped.

Differs from the type only in its *smaller size and the ovate p.*

var. **clavatus** (Batt.) Quél. Schaeff. Icon. t. 8. *Clavatus,* club-shaped.

Differs from the type in the *elongate elliptical p., the rootless st., the shaggy volva-like ring, and the gills never becoming pink.*

1638. **C. umbrinus** Cke. & Massee. *Umbrinus,* umber.

P. 2·5–4·5 cm., *dark umber,* fleshy, conico-hemispherical, then almost plane, finally splitting at the margin and revolute, *coarsely sulcate up to the disc,* universal veil generally remaining as a *large white patch at the apex, elsewhere covered with scattered, snow white, floccose scales.* St. 10–15 cm. × 6–8 mm., *dark umber,* slightly and gradually attenuated upwards from the bulbous, slightly rooting base, polished, shining. Volva *persistently white, sheathing* the base of the st., and *free and reflexed* about 2 cm. from the base. Gills *becoming black, edge persistently white,* free, distant from the st., 3 mm. broad, thin, crowded. Flesh *umber,* exceedingly thin. Spores sooty-black, elliptic-oblong, obliquely apiculate, 17–18 × 9μ. Cystidia absent. Manured ground. Aug. Rare.

1639. **C. sterquilinus** Fr. Cke. Illus. no. 646, t. 660.
Sterquilinum, a dung-pit.

P. 2–5 cm., *white, then silvery grey, disc tinged fuscous,* submembranaceous, ovato-conical, then campanulate, scarcely expanded, obtuse, fragile, *deeply sulcate, the furrows forked, silky-villose, disc squarrose with divergent, imbricated scales*; margin at length split. St. 8–15 cm. × 6–8 mm., *white,* attenuated upwards, fragile, fibrillose, base bulbous, *becoming blackish when touched.* Ring *white,* membranaceous, *narrow, near the base and volva-like.* Gills *white, then pink, and at length purplish umber,* free, slightly ventricose, 5–6 mm. broad. Flesh *greyish, then whitish,* very thin. Spores black, broadly elliptical,

14–23 × 9–14μ. Cystidia "on edge of gill vesiculose, filled with a reddish juice" Rick. Dung-heaps, and gardens. July—Sept. Uncommon. (*v.v.*)

1640. **C. oblectus** (Bolt.) Fr. Bolt. Hist. Fung. t. 142.
Oblectus, delightful.

P. 3–5 cm., *whitish, then bright flesh colour, at length black*, membranaceous, cylindrical, then conico-campanulate, *covered with white silky scales*, sulcate nearly up to the disc. St. 8–12 cm. × 5–6 mm., *white*, slightly attenuated upwards, soft, silky. Volva *brownish* on the outside, *white* on the inside, large, free, spreading, up to 6 mm. broad. Gills *becoming blackish*, with a tinge of flesh colour, free, linear. Flesh *white*, thin. Spores black, elliptic-oblong, with an oblique basal point, 16 × 8–9μ. Newly made dung-hills. Aug.—Nov. Rare. (*v.v.*)

1641. **C. squamosus** Morg. Journ. Cincinnati Soc. Nat. Hist. VI, t. 8.
Squamosus, scaly.

P. 2–8 cm., *cinereous, covered with reddish brown scales*, submembranaceous, ovoid, 2–3 cm. high, then expanded, at length splitting and revolute. St. 6–15 cm. × 6–12 mm., *white*, equal, often attenuated at the base, *covered with reddish brown scales below the ring*, smooth above. Ring distant, narrow, fugacious. Gills *white, then reddish brown, and finally black*, free, ventricose. Flesh *brownish under the cuticle, white elsewhere*, thin. Spores black, boat-shaped, 9–10 × 5μ. Caespitose. Base of trunks, elms, and old palings. Sept.—Oct. Uncommon. (*v.v.*)

**Somewhat ringed, but not with a volva. P. dotted, or spotted with minute, innate squamules.

1642. **C. atramentarius** (Bull.) Fr. (= *Coprinus fuscescens* (Schaeff.) Quél.) Cke. Illus. no. 648, t. 662. *Atramentarius*, inky.

P. 5–8 cm., *lurid fuliginous, becoming hoary with adpressed, silky lustre*, fleshy, *ovate*, obtuse, wholly longitudinally and *deeply sulcate* and ribbed, soft to the touch, minutely mealy, repand-unequal at the margin, *disc often squamulose*, slightly adpressedly silky at the sides. St. 7–20 cm. × 8–18 mm., *white, at first ventricose*, fusiform, attenuated shortly downwards, and for a greater distance upwards, *furrowed, then elongato-attenuated upwards*, firm, longitudinally fibrillose, apex smooth, base often with a few tawny squamules. Ring *white*, basal, very fugacious. Gills *whitish, then brown-black*, quite free, *ventricose*, 10–15 mm. broad, edge flocculose. Flesh *fuliginous*, scissile. Spores black, elliptical, 9–10 × 5μ, 1-multi-guttulate. Cystidia cylindric-saccate, 50–120 × 25–37μ. Taste mild. Edible. Caespitose rarely solitary. Woods, pastures, and gardens, almost always connected with buried wood. May—Dec. Common. (*v.v.*)

var. **soboliferus** (Fr.) Rea. Cke. Illus. no. 649, t. 848.
Soboles, offshoot; *fero*, I bear.

Differs from the type in the *truncate, more squamulose p., and in the stems arising from a common tuberous base.* Base of stumps, rotten palings, and in hot-houses. May—Sept. Uncommon. (*v.v.*)

1643. **C. fuscescens** (Schaeff.) Fr. (= *Coprinus atramentarius* (Bull.) Fr. sec. Quél.) *Fuscescens*, becoming fuscous.

P. 5–7·5 cm., *whitish, or greyish, disc becoming fuscous, or rufescent,* submembranaceous, globose, then ovate and expanded, at length revolute when deliquescent, *at first covered with a somewhat mealy pruina, then smooth, or rimosely squamulose on the disc,* obtuse. St. 4–7·5 cm. × 4–6 mm., *white*, equal, fragile, slightly silky under a lens, *at first obsoletely ringed towards the base.* Gills *white, then umber,* free, very broad, semi-ovate. Flesh *fuscous in the p. and base of st.*, thick at the disc. Spores fuscous black, "elliptical, apiculate, 10 × 6μ" Massee. Caespitose. Elm stumps, rotten wood, wood in cellars. May—Dec. Uncommon. (*v.v.*)

var. **rimoso-squamosus** Cke. Cke. Illus. no. 651, t. 664.
Rimosus, cracked; *squamosus*, scaly.

Differs from the type only in the *p. becoming cracked into angular patches.* Hardly worthy of a varietal name. About stumps. Oct. Uncommon.

***Universal veil floccose, at first continuous, then broken up into superficial scales which form patches on the p.

1644. **C. picaceus** (Bull.) Fr. Cke. Illus. no. 652, t. 665.
Picaceus, appertaining to a magpie.

P. 5–10 cm., *fuliginous black, variegated with broad, unequal, superficial, separating, white scales, from the breaking up of the universal woven veil, striate.* St. 10–25 cm. × 6–12 mm., *white*, attenuated upwards from the *bulbous base*, fragile, smooth. Gills *white, then pinkish, at length black,* free, ventricose, 8–12 mm. broad. Flesh *brownish under the cuticle of the p., pallid elsewhere,* thick only at the disc. Spores black, broadly elliptical, with a basal apiculus, 14–18 × 8–12μ; "basidia pyriform, 30–45 × 15–17μ. Cystidia conical-cylindrical, 100–150 × 30–50μ" Rick. Said to be poisonous. Smell none, or foetid. Frondose woods, especially beech, and roadsides. Sept.—Dec. Rather uncommon. (*v.v.*)

1645. **C. aphthosus** Fr. *ἄφθαι*, the thrush.

P. 2–3 cm., *livid*, submembranaceous, *ovate*, 2–5 cm. high, *then campanulate*, expanded, deliquescing slowly, *covered with the universal veil, soon separating into floccose, white, fugacious scales.* St. 5 cm. × 4 mm., *white, soft,* somewhat fragile, equal, *often twisted,* fibrillose. Gills *white, then blackish,* adnate, linear. Spores "black, lemon-shaped, 8·5–10 × 5·5–6·5μ. Cystidia vesiculose, cylindric-oval, 50–75 × 20–27μ" Lange. Subcaespitose. Hollow trees, especially willow. Sept.—Oct. Rare.

var. **Boltonii** Massee. Bolt. Hist. Fung. t. 26. James Bolton.

Differs from the type in the *repand, olivaceous p.* Decaying pieces of moist wood in cellars, cold kitchens, etc. Common.

1646. **C. flocculosus** (DC.) Fr. Cke. Illus. no. 654, t. 667. *Flocculosus,* flocculose.

P. 4–10 cm., *dingy white,* membranaceous, *ovate,* then expanded and splitting in the direction of the gills, *striate, covered over with floccose scales.* St. 6–10 × 1 cm., *white,* attenuated upwards from the swollen base, finely silky under a lens. Gills *violaceous, then fuscous black,* free, ventricose. Spores fuscous black, elliptical, 10 × 7–8μ. Cystidia cylindrical, 30×12μ. Solitary, rarely caespitose. Pastures, and gardens. June—Sept. Rare. (*v.v.*)

1647. **C. similis** B. & Br. *Similis,* like.

P. 2·5–4 cm., *pallid, disc darker,* hygrophanous, ovato-campanulate, lineato-striate, *clothed with acute separating warts which are fuscous at the apex.* St. 6–7 cm. × 4–5 mm., *white,* broader at the base. Gills *brownish near the margin,* adnate, attenuated behind, somewhat linear. Spores "brown, elliptic-oval with a germ-pore, 8–9 × 5–6μ. Cystidia vesiculose, 30–36 × 15–25μ" Rick. Trunks of dead trees. Sept. Rare.

****P. at first clothed with distinct flocci, or lax villous down, which fall off and disappear. Ring none.

1648. **C. exstinctorius** (Bull.) Fr. Bull. Hist. Champ. Fr. t. 437, fig. 1. *Exstinctorius,* like an extinguisher.

P. 2–7·5 cm., *pale, disc darker, sublivid,* submembranaceous, *clavate, then campanulate, at length rimosely split,* expanded, scarcely revolute, *firm, floccoso-scaly with the universal veil, becoming bare from the margin upwards;* margin *striate.* St. 8–12 cm. × 5–6 mm., *white,* attenuated upwards from the *rooting base, smooth,* base cottony. Gills *white, then fuscous blackish,* reaching the st., narrow, *lanceolate.* Spores fuscous purple, almond-shaped, 8–11 × 6–7μ, "mitriform" Rick. Generally solitary. On the ground, at the base of ash trees, or in woods. May—Sept. Rare.

1649. **C. macrorhizus** (Pers.) Rea. Cke. Illus. no. 657, t. 670, as *Coprinus fimetarius* Fr. var. *macrorhizus* Pers.

μακρός, long; ῥίζα, root.

P. 2–5 cm., *cinereous, or livid, then tinged fuscous,* submembranaceous, *oval*, then campanulate, 1·5–5 cm. high, *at length revolute, ribbed and furrowed from the apex to the margin, at first floccosely mealy with the fugacious veil*, becoming naked from the apex downwards; margin thin, deliquescing. St. 2–6 cm. × 4–15 mm., *white*, equal, or slightly attenuated upwards, *continued into a long, thin, tapering root from the base*, 1·5–10 cm. long, somewhat silky. Gills *white, then black, edge white*, free, ventricose. Flesh *greyish, becoming white*, very thin. Spores black, almond-shaped, 11–15 × 8–9μ. Cystidia vesiculose. Smell often strong. Solitary, or caespitose. On decaying vegetable matter, more rarely where dung-heaps have been. Woods, and homesteads. July—Oct. Not uncommon. (*v.v.*)

1650. **C. cinereus** (Schaeff.) Cke. Cke. Illus. no. 658, t. 671.
Cinereus, ash colour.

P. 1–4 cm., *ashy grey, disc often fuscous*, membranaceous, *cylindrical*, 1·5–4 cm. high, then campanulate, and at length revolute, *densely covered with white, fugacious flocci*, then naked and striate. St. 4–11 cm. × 3–6 mm., *white*, equal, or slightly attenuated upwards from the thickened base, *densely covered with white, fugacious, downward pointing flocci*. Gills *white*, then *black*, free, lanceolate. Flesh *of p. fuscous, white in the st.*, very thin at the margin. Spores black, pip-shaped, 9–11 × 6–7μ. Cystidia vesiculose, 60–70 × 30–40μ. Woods, heaths, pastures, and manure beds. Feb.—Nov. Common. (*v.v.*)

1651. **C. echinosporus** Buller. ἐχῖνος, hedgehog; σπορά, seed.

P. 3 cm. broad, 18 mm. high, *white, then grey, and finally dirty yellowish brown*, oval, then conico-campanulate, becoming flattened, and finally revolute and radially splitting along the lines of the longest gills, at first clothed with short, dense down, then breaking up into small, delicate, thin, fugacious tufts or scales; hairs *sometimes branched*, consisting of slender cells, 80–150 × 5–10μ. St. 9 cm. × 3 mm. at base, *white*, slightly attenuated upwards, straight, or flexuose, firm, adpressedly hairy. Gills blackish at maturity, adnexed, very thin, very slightly wedge-shaped, auto-digesting on the edge. Flesh *brownish yellow, brownish at the apex of the p.*, becoming finally *dirty ochraceous*. Spores black in the mass, very dark and opaque under the microscope, *finely warted*, or *echinulate*, oval, more or less pip-shaped, *apex truncate*, 9–11 × 5–7μ, with an apical germ-pore *through which a transparent membrane often protrudes*; basidia of three lengths, surrounded by 3–4 paraphyses. Cystidia abundant, conical, rounded at both ends, generally parallel-sided, rarely globose, 70–95 × 23–30μ, varying up to 105μ in length and 45–57μ in diam. Sticks dredged from a pool. Oct. Rare.

C. fimetarius (Linn.) Fr. = **Coprinus macrorhizus** (Pers.) Rea, and **Coprinus cinereus** (Schaeff.) Cke.

1652. **C. tomentosus** (Bull.) Fr. Bolt. Hist. Fung. t. 156.
Tomentosus, downy.

P. 1·5–4 cm., *grey white*, submembranaceous, *cylindrical*, 3–4 cm. high, *then conical, or narrowly pyramidal*, not expanded, *striate*, at length longitudinally cracked, *entirely covered with a greyish felty veil, which becomes torn into scales during expansion*. St. 5–7·5 cm. × 4–6 mm., *greyish*, subequal, *velvety*, base thickened, rooting. Gills *whitish, then fuscous blackish, free*, linear, *edge at first white-micaceous*. Flesh *white*, very thin at the margin. Spores "reddish brown, transparent, 25 × 7–8μ. Cystidia vesiculose-bottle-shaped, 30–40 × 20–30μ" Rick. On dung. Rich pastures, roadsides, and woods. Sept. Uncommon.

1653. **C. niveus** (Pers.) Fr. Cke. Illus. no. 659, t. 672, fig. B.
Niveus, snow white.

Entirely snow white. P. 1·5–5 cm., submembranaceous, *ovate, soon campanulato-expanded*, at length revolute, split and torn, *mealy floccose*, often squamulose. St. 2·5–7·5 cm. × 3–6 mm., attenuated upwards, very fragile, *densely covered with fugacious, upward pointing flocci*, becoming smooth. Gills *white, then flesh colour, and finally blackish*, adnexed, narrow. Flesh *white*, very thin at the margin. Spores black, broadly elliptical, 15 × 10–12μ, often apiculate at one end. Cystidia vesiculose. On dung, especially horse. Woods, and pastures. May—Dec. Common. (*v.v.*)

var. **astroideus** Fr. ἀστήρ, star; εἶδος, shape.

Differs from the type in the *squamose, grey p. becoming inverted and smooth, and in the elongate, thin, smooth st., stellate at the base*. Woody places.

1654. **C. roseotinctus** Rea. *Roseus*, rose; *tinctus*, coloured.

P. 5–8 mm., *fuscous, densely powdered with rose coloured meal*, membranaceous, *cylindrical*, 7–11 mm. high, then campanulate, at length revolute, umbonate, silky, striate, becoming sulcate along the back of the gills; margin torn. St. 2–5 cm. × 2–3 mm., *white, at first densely powdered with deep rose coloured meal*, then only sparingly powdered near the apex at maturity, equal, becoming elongate and flexuose; base bulbous, white floccose. Gills *white, then black*, adnexed, 1–1·5 mm. broad, deliquescent. Flesh *greyish*, thin. Spores black, pip-shaped, slightly apiculate, 9–11 × 5–6μ. Ash plantations, and under trees. Aug.—Oct. Uncommon. (*v.v.*)

1655. **C. cothurnatus** Godey. Gillet, Champ. Fr. Hym. t. 175.
κόθορνος, a high hunting boot.

P. 2–3 cm., membranaceous, *dingy white, reddish, flesh coloured* (or *yellowish* sec. Massee), *conico-campanulate*, then expanded, umbonate, and irregularly split, *densely furfuraceous*. St. 3–4 cm. × 2–4 mm., *white*, attenuated upwards, *squamulose, base sheathed with white, fibrillose squamules*. Gills *white, then flesh coloured, at length blackish*, free, sublanceolate, crowded. Flesh very thin. Spores black, elliptical. Cow dung. Pastures. Sept.—Oct. Uncommon.

*****P. covered with small micaceous scales, or granules which fall off and disappear. Ring none.

1656. **C. micaceus** (Bull.) Fr. Cke. Illus. no. 660, t. 673.

Micaceus, glittering.

P. 3–6 cm., *yellow ferruginous, yellowish livid, disc darker, at length date-brown-fuscous, ferruginous ochraceous when dry*, submembranaceous, *oval*, then campanulate, undulato-lobed, rimosely split, *striate, at first covered with glistening micaceous particles, soon naked and becoming sulcate*, disc even; margin *plicate and irregular*. St. 5–20 cm. × 4–8 mm., *white, or whitish*, equal, *silky*, often curved, fibrillose, becoming smooth. Gills *white, or isabelline, then brown, or livid at the edge, and finally fuscous blackish*, adnexed, lanceolate. Flesh pallid, thin at the disc. Spores fuscous black, pip-shaped, or elliptical with a long apiculus, 9–10 × 5μ. Cystidia cylindric-oblong, apex rounded, 85–140 × 48–75μ. Edible. Densely caespitose. Stumps, old posts, and buried wood. Woods, pastures, and hedgerows. Jan.—Dec. Common. (*v.v.*)

1657. **C. truncorum** (Schaeff.) Fr. Schaeff. Icon. t. 6.

Truncorum, of tree trunks.

P. 2–4 cm., *globose*, then campanulate, *ferruginous ochraceous, densely covered with micaceous meal*, soon naked, striate, *not becoming sulcate*. St. 7–10 × 3–4 mm., *white, very fragile, somewhat striate*, smooth. Gills *white, then rosy, at length black, free*. Spores fuscous, elliptical, 12–14×6μ. Caespitose. Rotten willow trunks. Uncommon.

1658. **C. frustulosum** Sacc. Sacc. Myc. Ven. Spec. t. 6, figs. 10–14, from Atti della Soc. Ven.-Trent. II.

Frustulosum, consisting of small fragments.

P. 3–6 cm., *covered up to the yellow umbo with rosy red micaceous meal*, ovate, 1 cm. high, then campanulate, umbonate, even. St. 1·5–12·5 cm. × 4–6 mm., *white, glistening*, very brittle, conical, then cylindrical, smooth, apex mealy. Gills reddish white, then black, free. Spores 8 × 6μ. Caespitose. Amongst long grass near a post, and under *Rhododendra*. Sept. Rare.

1659. **C. aratus** B. & Br. Cke. Illus. no. 661, t. 674. *ἀρόω*, I plough.

P. 5–7·5 cm., *umber*, submembranaceous, campanulate, then expanded, *deeply sulcate up to the darker, usually wrinkled disc, sprinkled with large micaceous particles*, revolute in decay. St. 10–15 cm. × 4–6 mm., *snow white*, attenuated upwards from the slightly bulbous base, silky. Gills *deep rich brown, then black*, attached, then seceding and becoming free, narrow, attenuated at both ends. Flesh *brownish*, thin. Spores 15 × 10–11μ. Solitary, or clustered. Hollow trees, rich ground, and manure heaps. May—Aug. Uncommon.

1660. **C. radians** (Desm.) Fr. *Radians*, radiant.

P. 2–3 cm., *yellow tawny, becoming pale*, membranaceous, ovate, then campanulate and expanded, *micaceous, disc granular*, sulcate; margin striate. St. 2–4 cm. × 4–8 mm., *white*, slightly attenuated upwards from the swollen base, and *arising from a dense mass of tawny mycelium*, the *Ozonium* of old authors. Gills *white, then violaceous black*, adnate, linear, 3–4 mm. broad. Flesh *white, tawny under the cuticle of the p. and in the st.*, thin at the margin. Spores black, elliptic fusiform, 9–10 × 4–5μ, 1–2-guttulate, with a hyaline germ-pore at each end. Caespitose. On stumps, especially elm, old stacked logs, and plaster walls. April—Dec. Not uncommon. (*v.v.*)

1661. **C. papillatus** (Batsch) Fr. Cke. Illus. no. 663, t. 676, fig. B.
Papillatus, having a nipple.

P. 4–15 mm., *fuscous, disc darker*, membranaceous, ovate, then campanulate, at length flattened and revolute, torn, striate, *scurfy, beset with minute warts, which are more crowded on the disc*. St. 2–3 cm. × 2 mm., *hyaline-pellucid*, equal, smooth except at the base. Gills *blackish, then black*, free, *few*. Spores "brownish black, angularly oval, 7–8 × 6–7μ. Cystidia vesiculose" Rick. On the ground, and on dung. Pastures, and gardens. June—Oct. Uncommon. (*v.v.*)

var. **oxygenus** Fr. ὀξύς, sharp; γένος, race.

Differs from the type in the *whitish p., inclining to grey, slightly flocculose as well as the st.*

1662. **C. Patouillardii** Quél. Trans. Brit. Myc. Soc. III, t. 8.
N. Patouillard, the eminent French mycologist.

P. 5–20 mm., *ashy-grey, disc yellowish, rough with minute reddish granules*, membranaceous, conico-campanulate, then expanded, plicato-sulcate up to the disc at maturity. St. 1–4 cm. × 2 mm., *white*, fragile, slightly attenuated upwards, smooth. Gills *cream colour, then black*, free, *attached to a collar*, 2 mm. broad, distant. Flesh *white, reddish at the disc and base of the st.*, very thin. Spores black, with a hyaline apiculus, angularly globose, 6–7μ. Spent tea leaves. July. Uncommon. (*v.v.*)

******P. smooth, without floccose, or micaceous squamules. Veil none.

1663. **C. alternatus** (Schum.) Fr. *Alternatus*, alternate.

P. 3–4 cm., *chalky-pallid, disc pale umber*, fleshy, hemispherical, discoid, smooth, *striped with alternately broad and narrow striae*. St.

7–10 × 4–6 mm., whitish, attenuated upwards from the thickened base, *smooth.* Gills *cinereous, then black,* adnate, linear. Flesh thin. Spores "black, broadly pip-shaped, 10 × 6–7μ" Massee. Subcaespitose. On the ground. Rare.

1664. **C. erythrocephalus** (Lév.) Fr. (= *Coprinus oblectus* (Bolt.) Fr. sec. Quél.) Lév. Ann. sc. nat. (1841), t. 14, fig. 3.

ἐρυθρός, red; κεφαλή, head.

P. 1 cm., *reddish vermilion, at length grey,* membranaceous, conico-campanulate, 1 cm. high, smooth, striate. St. 2–3 cm., *paler than the p., naked.* Gills *fuscous, then black,* slightly adnexed, crowded. Gregarious. On limed soil. Spring. Rare.

1665. **C. deliquescens** (Bull.) Fr. *Deliquescens,* melting away.

P. 5–10 cm., *livid fuliginous,* membranaceous, ovato-campanulate, then expanded, smooth, *disc papillate with minute points,* never split, but revolute and striate, the striae broad but not deep. St. 7–10 cm. × 4–8 mm., *shining white,* equally attenuated upwards, *corticate, smooth.* Gills *clay colour, then lurid blackish,* free, *at length remote from the st.,* flexuose, *very narrow,* only 1 mm. broad, very crowded. Flesh thin. Spores black, "elliptical, 8 × 5μ, obliquely apiculate" Massee. Subcaespitose. On trunks, stumps, and heaps of leaves. Aug.—Oct. Not uncommon. (*v.v.*)

1666. **C. digitalis** (Batsch) Fr. *Digitalis,* belonging to the finger.

P. 2–3 cm., *whitish, or straw coloured, disc often darker, becoming olivaceous livid or yellowish cinereous,* submembranaceous, fragile, ovate, then campanulate, 2–3 cm. high, *quite smooth and naked,* moist, *striate,* except at the even disc. St. 2·5–12·5 cm. × 2–3 mm., *shining white, equal, somewhat flexuose,* smooth, *corticate,* fragile, base villose. Gills *whitish brown, then black,* somewhat free, *reaching the st.,* ventricose; *edge white, micaceous.* Spores "brown, elliptical, 8–9 × 5μ" Rick. Caespitose. Damp places in woods, and pastures. Sept.—Oct. Uncommon.

1667. **C. congregatus** (Bull.) Fr. Cke. Illus. no. 667, t. 679.

Congregatus, collected into a flock.

P. 1–2 cm., *pale ochraceous,* membranaceous, cylindrical, then campanulate, 1·5–2 cm. high, *viscid,* smooth; margin striate, split when expanded. St. 2–3 cm. × 2 mm., *white,* equal, smooth. Gills *white, then black,* reaching the st., *linear.* Flesh *white, yellowish at the disc,* very thin. Densely caespitose. Woods, roadsides, and gardens. Sept.—Nov. Uncommon. (*v.v.*)

1668. **C. tardus** Karst. Cke. Illus. no. 666, t. 719. *Tardus,* late.

P. 2·5–5 cm., *bright brown, becoming pale ochraceous tan,* fleshy, fragile, ovate, then campanulate, 2·5–5 cm. high, *sulcate, or deeply striate*, smooth, rather dry. St. 10–15 cm. × 4 mm., *whitish*, equal, somewhat flexuose, *slightly downy*, *apex minutely striate*. Gills *whitish*, *then tinged with brown*, *at length black*. Flesh *whitish*, thin at the margin. Spores blackish brown, elliptical, or sometimes subangular, 12–15 × 7–9μ. Cystidia "vesiculose, very large, conically flask-shaped, up to 24μ broad" Lange. Caespitose. Naked soil. Oct. Rare.

C. hydrophorus (Bull.) Quél. = **Psathyrella hydrophora** (Bull.) Fr.

II. P. very thin, without a pellicle, at length opening into furrows along the back of the gills, and becoming plicato-sulcate. St. thin, fistulose. Gills melting away into very thin lines.

*St. annulate, or volvate.

1669. **C. dilectus** Fr. (= *Coprinus oblectus* (Bolt.) Fr. sec. Quél.) Fr. Icon. t. 140, fig. 2. *Dilectus*, beloved.

P. 1–2·5 cm., *whitish, then rosy, at length reddish*, submembranaceous, *campanulate*, obtuse, *floccosely mealy*, at length split, revolute, and naked. St. 2·5–5 cm. × 2–4 mm., *whitish*, attenuated downwards, *sprinkled with red fibrils*. Volva reduced to whitish, spreading, fugacious squamules at the base. Gills *white, then reddish fuscous, at length black*, free, somewhat lanceolate, crowded. Flesh *white, reddish at the disc*, very thin at the margin. Spores black, elliptical, 10 × 6μ. Cystidia "ovate, vesiculose, average breadth 23μ" Lange. On the ground. Burnt ground in beech woods, and bare soil. Nov. Rare. (*v.v.*)

1670. **C. ephemeroides** (Bull.) Fr. Bull. Hist. Champ. Fr. t. 582, fig. 1. ἐφήμερος, lasting but a day; εἶδος, like.

P. 4–6 mm., *whitish, or livid*, membranaceous, pellucid, *cylindrical*, 5–20 mm. high, then campanulate, *sprinkled with superficial flocci*, *plicato-sulcate*, at length split and revolute. St. 2–3 cm. × 1–2 mm., *white*, smooth, *base bulbous and hairy*. Ring *white*, *very tender*, *movable*, narrow. Gills *white*, *then black*, free, distant from the st., very tender. Spores black, subglobose, often somewhat angular, 7 × 5–6μ. "Cystidia globose, 23–30μ" Lange. On horse and cow dung. July—Oct. Not uncommon. (*v.v.*)

1671. **C. bulbillosus** Pat. Trans. Brit. Myc. Soc. III, t. 7. *Bulbillosus*, with a little bulb.

P. 5–10 cm., *grey*, *disc yellow*, convex, then expanded, *at first covered with white meal*; margin striate, incurved. St. 10–20 × 1 mm., *white*, smooth, *base bulbous*. Ring *white*, *movable*, median. Gills *grey*, *then black*, narrow, 1 mm. broad. Flesh *white*, *yellowish at the disc*, very thin. Spores black, subglobose, angular, 8–9 × 8μ. Horse dung. Aug.—Oct. Uncommon. (*v.v.*)

1672. **C. Hendersonii** Berk. Cke. Illus. no. 668, t. 680, fig. A.
J. L. Henderson.

P. 2–12 mm., *cinereous, disc brownish*, membranaceous, cylindrical, then ovali-campanulate, at length plane, *minutely granular under a lens*, striate half way up; margin folded. St. 2·5–4 cm. × 1–2 mm., *white*, attenuated upwards, nearly or quite smooth. Ring *white, cup-shaped*, more or less distant, permanent, *fixed*. Gills *white, then black*, free, narrow, rather distant, *edge white*. Flesh *white*, very thin. Spores black, "spherical, apiculate, 8–9 μ" Quél. Hotbeds, and horse dung. Feb.—Sept. Uncommon. (*v.v.*)

1673. **C. Bresadolae** Schulz.
L'Abbé J. Bresadola, the eminent mycologist of Trieste.

P. 8 mm., *greyish white, disc tinged brown*, membranaceous, *sub-cylindrical*, 17 mm. high, at first covered with a very thin universal veil, which does not break up into squamules, but splits from apex to base, and becomes obliterated. St. 12 cm. × 4 mm., *white*, tapering upwards, smooth. Ring loose, deciduous. Gills *black*, edge white. Spores black, cylindrical, ends rounded, 12–17 × 6 μ. Gregarious. Worked wood, and on the ground amongst rotten branches. Sept. Rare.

1674. **C. volvaceo-minimus** Crossland.
Volvaceus, having a volva; *minimus*, least.

P. 4–6 mm., *grey, inclining to cinereous, disc darker*, membranaceous, *ovate*, then campanulate, *sprinkled with white squamules, striate*; margin at length split and reflexed. St. 2–2·5 cm. × 1 mm., *white, hyaline*, apex slightly swollen, smooth; *bulbous base furnished with a distinct, sheathing, persistent ring, or collar, half the width of the p., and finally reflexed*. Gills *becoming blackish purple*, almost free, narrow, attenuated at each end. Spores blackish purple, subglobose, 6–7 μ, minutely apiculate. Manure heaps. Rare.

**P. clothed with superficial, separating floccules.
Gills free. Ringless.

1675. **C. lagopus** Fr. Saund. & Sm. Myc. Ill. t. 19.
λαγώς, a hare; πούς, foot.

P. 2–5 cm., *whitish, or greyish, disc livid*, very tender, pellucid, cylindrical, then campanulate, *covered with white, fugacious fibrils*, then naked, flattened and split, radiato-sulcate. St. 5–12·5 cm. × 2–6 mm., *white*, attenuated upwards rarely at the base, very fragile, *everywhere white woolly*. Gills *white, then black*, free, *at length remote*, linear. Flesh *white*, very thin. Spores black, elliptical, 10–12 × 6–7 μ. Cystidia "large, vesiculose, ovate, or oblong, about 12–25 μ broad" Lange. Rich soil, rotten wood, sawdust, and on dung. Woods, and pastures. July—Oct. Uncommon. (*v.v.*)

1676. **C. lagopides** Karst. *Lagopus*, the species *C. lagopus*, εἶδος, like.

P. 4–7 cm., *whitish, disc cinereous*, membranaceous, campanulate, sulcate, then splitting, disc naked, *clothed elsewhere with free, white, pointed, feathery squamules*; margin subrevolute. St. 17 cm. × 3–5 mm., slightly attenuated above, hollow, *densely floccose*. Gills black, free, crowded, linear. Spores black, ovoid, 6–8×5–6 μ, basidia 4-spored. On the ground. Dec. Rare.

1677. **C. narcoticus** (Batsch) Fr. Cke. Illus. no. 668, t. 680, fig. B. ναρκωτικός, making numb.

P. 2–3 cm., *white*, or *greyish*, very tender, pellucid, conico-cylindrical, *villose with white floccose, fugacious, recurved scales*, then flattened, naked, striate. St. 4–5 cm. × 2 mm., *white, pellucid*, equal, or attenuated upwards, *villosely fibrillose*, becoming smooth. Gills *white, then blackish*, free, *reaching the st., narrow*. Flesh *white*, very thin at the margin. Spores blackish brown, elliptical, 11 × 5μ, "with a hyaline epispore. Cystidia subglobose, 20–40μ" Lange. Smell *very strong, foetid, narcotic-alkaline*. Manure heaps. Oct.

1678. **C. macrocephalus** Berk. Cke. Illus. no. 670, t. 682, fig. A. μακρός, long; κεφαλή, head.

P. 12–15 mm., *ashy grey, or slate colour, disc brown*, submembranaceous, cylindrical, 18 mm. high, then cylindrico-campanulate, *sprinkled with adpressed, or patent, pointed scales*; margin slightly striate. St. 2·5–5 cm. × 4 mm., *dirty white*, attenuated upwards, *clothed with* short cottony down, and with *longer, sometimes deflexed, loose fibres*, base strigose. Gills *at length black, quite free*, linear, 1–2 mm. broad. Flesh *whitish*, thin at the margin. Spores black, "broadly elliptical, or obliquely pip-shaped, 11–15 × 7–9μ" Massee. Subcaespitose. Putrid dung. March--Sept. Rare.

1679. **C. nycthemerus** Fr. Bull. Hist. Champ. Fr. t. 542, fig. D, as *Agaricus ephemerus*. νυχθήμερον, a day and night.

P. 8–15 mm., *grey, disc fuscous, or tawny*, very tender, cylindrico-conical, 12–15 mm. high, soon opening into furrows and flattened, *furfuraceo-flocculose*, at length naked and *forked-striate*. St. 4 cm. × 1–2 mm., *whitish cream colour, becoming pale white*, equal, flaccid, *smooth*, base slightly bulbous. Gills *ochraceous cream colour, then brownish black*, free, *at length remote*, narrow, linear. Spores "brownish black, pruniform ovoid, 10μ" Quél. Subcaespitose. On dung. Pastures, and gardens. July—Oct. Uncommon.

1680. **C. cordisporus** Gibbs. Lange, Dansk. Bot. Ark. bind 2, no. 3, t. 1, fig. g. *Cor*, the heart; σπορά, seed.

P. 3–9 mm., *whitish, or pallid ochraceous, disc sprinkled with tawny, furfuraceous papillae*, membranaceous, cylindric-ovate, 3–8 mm. high, then expanded, at length upturned, plicato-sulcate, splitting along the backs of the gills; margin crenate. St. 2–3 cm. × 1–2 mm., *whitish-hyaline*, glabrous, equal, *base* slightly thickened and *densely strigose-squamulose*. Gills *pale, then blackish*, free, ending close to the st., rather narrow, 25–30 in number in the larger pilei; intermediate shorter ones few or none. Flesh *whitish*, very thin. Spores dark brown-purple, laterally compressed, front view *obtusely cordate*, 9–10μ diameter, side view elliptical, 9–11 × 5–6μ. Cystidia cylindric-fusiform, 50 × 10μ. On cow, horse, sheep, and rabbit dung. April—Oct. Not uncommon. (*v.v.*)

1681. **C. radiatus** (Bolt.) Fr. Cke. Illus. no. 671, t. 683, fig. A.
Radiatus, beaming.

P. 2–6 mm., *dingy yellowish, or greyish, disc darker, often rufescent*, very tender, membranaceous, *clavate*, then campanulate, soon opening into furrows, flattened, *radiato-plicate*, pellucid, *covered with cinereous down*, soon naked. St. 5–25 × 1 mm., *hyaline*, equal, often pruinose when young, *becoming smooth*, base silky and villose. Gills *whitish, then pallid blackish*, free, few, distant. Flesh *white*, very thin. Spores black, elliptical, 7–8 × 4–5μ. In troops. Cow, and horse dung. Woods, and pastures. May—Nov. Common. (*v.v.*)

1682. **C. Spraguei** Berk. & Curt. Cke. Illus. no. 671, t. 683, fig. B.
Charles J. Sprague.

P. 1–2 cm., *greyish, disc tawny*, very tender, membranaceous, conical, then campanulate, at length expanded and revolute, tomentose, plicate; margin coarsely striate. St. 3–4 cm. × 1–2 mm., *pale cinnamon*, equal, smooth. Gills *white, then* blackish, free, few, distant, narrow. Flesh *whitish*, tawny at the disc, very thin. Spores black, "elliptical, slightly curved, 10 × 5μ" Massee. Gardens. July. Rare.

1683. **C. urticaecola** (B. & Br.) Buller. Cke. Illus. no. 622, t. 596, fig. B, as *Psathyra urticaecola* B. & Br.
Urtica, nettle; *colo*, I inhabit.

P. 4–6 mm., *chalky white*, spherical, then hemispherical, becoming revolute and radially split, *beset with numerous, small, white scales*, composed of matted hyphae. St. 1·5–2 cm. × 2 mm., *white*, attenuated upwards, smooth, hollow, base floccose. Gills *white, then chocolate*, ventricose, 1·5 mm. wide, very thin. Spores chocolate, oval, 7 × 4μ; basidia dimorphic. Cystidia cylindrical-oval, firmly fixed by both ends. Nettle roots, sticks, dead leaves and grass haulms. June—Aug. Uncommon.

1684. **C. platypus** Berk. Cke. Illus. no. 675, t. 687, fig. B.
πλατύς, broad; πούς, foot.

P. 4–6 mm., *white, then ochraceous flocculose*, campanulate, convex, then expanded. St. 6-8 × 1 mm., *whitish, flattened and discoid at the base*, smooth. Gills *becoming black*, free, narrow, distant. Flesh *white*, thin. Spores "blackish, 8 × 6μ" Massee. On dead stems of *Phalaris arundinacea*, and on Palm stems in conservatories. Oct. Rare. (*v.v.*)

1685. **C. Spegazzinii** Karst.

C. Spegazzini, the South American mycologist.

P. 3 cm., *greyish, cylindrical, or oval*, 2 cm. high, then expanding and splitting up to the disc, at first with a cobweb-like covering, and even, soon naked and *grooved*. St. 6–7 cm., *white, thickened below and rooting*, adpressedly silky. Gills free. Flesh very thin. Spores elliptical, 9–14 × 5–6μ. On soil in a plant pot. Rare.

***P. micaceous or furfuraceous. Gills commonly adnate to the apex of the st., which (in some species) is dilated into a ring or collar. Ringless.

1686. **C. domesticus** (Pers.) Fr. Fr. Icon. t. 140, fig. 3.

Domesticus, pertaining to the house.

P. 3–6 cm., *fuliginous, disc date brown*, submembranaceous, ovate, then campanulate, *furfuraceo-squamulose*, then opening into furrows, and flattened, *undulato-sulcate*, disc obtuse, even. St. 5–7·5 cm. × 2–6 mm., *white*, slightly firm, attenuated upwards, *adpressedly silky*, then polished. Gills *white, then flesh colour, at length fuscous blackish, adnexed*, linear, 2 mm. broad. Flesh *white, fuliginous in the p.*, thin. Spores black, elliptical, 9–10 × 5–6μ. Cystidia "only on edge of gill, globular, about 5μ broad, with or without a 5–16μ long, 5–6μ broad, appendix" Lange. Often caespitose. On damp carpets, in cellars, on old walls, and in gardens. April—Dec. Not uncommon. (*v.v.*)

1687. **C. stercorarius** Fr. *Stercorarius*, belonging to dung.

P. 2–3 cm., *whitish*, very tender, membranaceous, ovate, then campanulate, at length expanded, pellucid, *covered with dense, micaceous, somewhat persistent, shining white meal*; margin striate. St. 4–5 cm. × 2 mm., *shining white*, fragile, attenuated upwards from the thickened base, *pruinose*, becoming smooth, often springing from a darkish sclerotium. Gills *white, soon black*, adnexed, attenuated behind, ventricose, 1–2 mm. broad. Flesh *white, greyish at the disc*, very thin. Spores black, broadly elliptical, or subglobose, 7–10 × 7–8μ. Cystidia "vesiculose-clavate, 50–75 × 30–40μ" Rick. Rich soil, dung, roadsides. May—Oct. Not uncommon. (*v.v.*)

1688. **C. tuberosus** Quél. (= *Coprinus stercorarius* Fr. sec. Lange.) Quél. Soc. bot. xxv, t. 3, fig. 2. *Tuberosus*, having a swelling.

P. 3–5 mm., *white, then greyish*, membranaceous, elliptical, then campanulate, finely striate, *covered with hyaline, warted, granular vesicles, the remains of the universal veil.* St. 2–4 cm. × 1 mm., *white*, *hyaline*, equal, flexuose, silky, villose, *springing from a small, blackish brown sclerotium.* Gills *white, then purplish black*, free, narrow, *edge micaceous.* Flesh very thin. Spores black, elliptical, 12μ. On dung, and decaying vegetable matter. Pastures. May—Dec. Uncommon.

1689. **C. stellaris** Quél. Quél. Soc. bot. xxiv, t. 5, fig. 6.

Stellaris, starry.

P. 1–2 mm., *snow white, then greyish*, ovate, then campanulate, *striate*, at length split in a star-like manner, *crowned with minute, pellucid vesicles, the remains of the universal veil.* St. 1–2 cm. × 1 mm., *hyaline, velvety with long, silky white hairs.* Gills *greyish, then brown*, adnate, narrow. Spores for a long time hyaline, then brownish bistre, elliptical, 8μ. Fox, cow, and human dung. Caves, and pastures. April—July. Rare.

1690. **C. Friesii** Quél. Quél. Jur. et Vosg. I, t. 23, fig. 5.

Elias Fries, the eminent Swedish mycologist.

P. 1–1·5 cm., *snow white, then striate and violaceous at the margin, at length grey*, ovate elliptical, then revolute, *floccosely mealy.* St. 5–20 × 1 mm., *white*, equal, *pulverulent*, base surrounded by a floccose collar. Gills *white, then violaceous, at length brownish black*, free, narrow, crowded. Flesh *white*, very thin. Spores brownish black, angularly globose, 8–10·5 × 6–7·5μ. Dead grass stems. Aug.—Oct. Uncommon. (*v.v.*)

1691. **C. tigrinellus** Boud. Boud. Icon. t. 139.

Tigrinellus, spotted like a tiger.

P. 1 cm., *snow white, covered with small, scattered, blackish flocci, especially at the disc, becoming rosy towards the striate margin and at length greyish*, oblong, then slightly campanulate, at length revolute at the margin, *pulverulent.* St. 2–3 cm. × 1–2 mm., *white*, equal, smooth; base somewhat marginately bulbous, velvety, and *often with blackish flocci like the p.* Gills *white, then brownish*, free, narrow. Flesh *white*, thin. Spores brownish black, fuliginous under the microscope, broadly elliptical, 11 × 7μ. Dead leaves of *Carex riparia, Carex paludosa*, more rarely of *Iris Pseudacorus.* July—Aug. Uncommon.

1692. **C. Gibbsii** Massee & Crossland.

Thomas Gibbs, a Yorkshire mycologist.

P. ·5 mm., *pale ochraceous, disc darker*, hemispherical, then expanded, striate, smooth, *minutely atomate.* St. 4–7 mm., *white, pellucid*, smooth, attached by a few white strands of mycelium. Gills adnate, 5–7 in number. Spores purplish brown, subcircular, compressed, 8–9μ in diameter, 5μ thick. Cystidia pyriform. Horse, and sheep dung. Nov. Uncommon.

1693. **C. ephemerus** (Bull.) Fr. Cke. Illus. no. 673, t. 685, fig. B.
ἐφήμερος, lasting but a day.

P. 6–18 mm., *greyish, disc rufescent*, very tender, membranaceous, ovali-clavate, then soon campanulate, and on opening into furrows flattened, *radiato-sulcate, disc elevated, often umbonate*, even, *at first slightly scurfy*, then naked (but minutely mealy under a lens). St. 2·5–5 cm. × 1–2 mm., *whitish, pellucid*, equal, smooth. Gills *whitish, then fuscous, at length black, reaching the st.*, remote, linear. Flesh *white*, very thin. Spores black, ovate, or cylindrical-elliptical, 8–10 × 5·5–8μ. Cystidia vesiculose, globular or subconical, sometimes with a bottle-neck apex, 20–50 × 16–30μ. On horse, and rabbit dung. Dung-hills, and pastures. May—Oct. Common. (*v.v.*)

1694. **C. bisporus** Lange. (= *Coprinus bisporiger* Buller.)
Bis, twice; σπορά, seed.

P. 5–12 mm. high and broad, *pallid, or ochraceous, then greyish-hyaline*, ovate-conical, then revolute and radially sulcate up to the prominent disc, covered with erect, minute hairs, 45–120 × 12–24μ. St. 3–8 cm. × 1–3 mm., *white*, equal, base strigose. Gills *white, then blackish*, adnexed, narrow, 2 mm. wide. Flesh *white, ochraceous under the pellicle of the pileus*, thin except at the disc. Spores purplish-brown in the mass, dark brown under the microscope, oval, or oblong elliptical, 12–14 × 6–7μ; basidia broadly ovate, 8–10μ in diam., *with* 2-*sterigmata*. Cystidia inflated, ovate, 80–90 × 45–55μ. Wood and dung. Aug.—Oct. Uncommon. (*v.v.*)

1695. **C. velox** Godey. Gillet, Champ. Fr. Hym. t. 175
Velox, fleeting.

P. 3–4 mm., *greyish*, membranaceous, cylindrical then obovate, soon expanded, *striate, furrowed, disc and ribs scurfy*. St. 1·5–3 cm. × 1 mm., *white*, somewhat pellucid, *covered with short, whitish flocci, especially towards the radiating fibrillose base*. Gills *grey, then black*, reaching the stem, thin. Flesh very thin. Spores "dark brown, elliptical, 7·5–9 × 4·5μ" Lange. Cystidia "none" Massee and Salmon. Cow, and horse dung. Oct.—Jan. Uncommon.

1696. **C. aquatilis** Peck. Peck, 27th Rep. New York State Mus. t. 1, figs. 26–28. *Aquatilis*, found in or near water.

P. 1·5–2 cm., *yellowish brown*, campanulate, sulcate-plicate almost to the apex, *scurfy*. St. 5–7 cm. × 2 mm., *whitish*, equal, *scurfy*. Gills *brownish, then black*, reaching the st. Flesh of p. *pale orange*, thin. On decaying sticks, or twigs partly submerged, or lying in wet mossy places, also on *Luzula sylvatica*. June. Rare.

1697. **C. sociatus** Fr. (= *Psathyrella crenata* (Lasch) Fr. sec. Rick.) *Sociatus*, gregarious.

P. 3–6 cm., *fuscous, then becoming pale, disc date brown, even, and umbilicate,* very tender, ovali-cylindrical, then expanded, *densely split into furrows,* the raised ribs slightly scurfy. St. 4–6 cm. × 1–2 mm., *white, delicately attenuated from the base, smooth.* Gills *grey, then cinereous black, adnexed in the form of a ring,* attenuated behind, slightly ventricose. Flesh very thin. Spores "brownish black, ovate-oval, 12 × 7μ. Cystidia somewhat bottle-shaped with a broad neck, 20–25μ broad" Lange. Damp ground in gardens, old walls, and woods. July—Oct. Uncommon.

1698. **C. plicatilis** (Curt.) Fr. Cke. Illus. no. 674, t. 686, fig. A. *Plicatilis*, folded.

P. 1–3 cm., *fuscous, then bluish-grey-cinereous, disc darker, fuscous or rufescent,* ovali-cylindrical, then campanulate, soon expanded, opening into furrows, *sulcato-plicate, for the most part smooth,* disc broad, even, *at length depressed.* St. 2·5–7·5 cm. × 1–2 mm., *pallid, somewhat pellucid,* equal, smooth. Gills *cream colour, then grey,* at length *grey-blackish, remote from the st., and adnate to a collar formed by the dilated apex of the st.,* distant. Flesh *whitish,* very thin. Spores black, broadly elliptical, 10–12 × 8–9μ. Cystidia vesiculose, 60–85 × 25–35μ. Woods, pastures, roadsides, and gardens. April—Dec. Common. (*v.v.*)

1699. **C. curtus** Kalchbr. (= *Coprinus plicatiloides* Buller.) Lange, Dansk. Bot. Ark. bind 2, no. 3, t. 1, fig. h. *Curtus*, short.

P. ·5–1·5 cm. broad, 3–8 mm. high, *foxy-red, or rufescent to tan colour at first, becoming grey to dark grey, disc tan coloured,* oval, to cylindrical, or elliptical, then expanded and flattened with a strongly depressed disc, splitting along the lines of the gills and becoming plicate, *bearing a certain number of minute, scattered, flaky, separable, rufescent or whitish scales,* consisting of globose, angular, or elliptical cells, often in chains, 12–30μ in diam., some brown and some colourless, not ornamented with crystals of calcium oxalate, the pileus also villose or downy with many colourless hairs, 70–100 × 5μ, enlarged at the apex where minute drops of a clear fluid are exuded under moist conditions. St. 2–8 cm. × 1–2 mm., *white, becoming stained with dull yellow,* equal, smooth, hollow. Gills *grey, then black,* at first attached to the stem by the margin for its entire length, then adnexed, and finally free, linear, narrow; margin, before autodigestion begins, slightly divided, fimbriate. Flesh white, thin. Spores black in the mass, dark brownish to black under the microscope, elliptical, 9–15 × 6–9μ. Cystidia on gill surface none. Horse dung. Sept.—Oct. Uncommon. (*v.v.*)

1700. **C. filiformis** B. & Br. Cke. Illus. no. 674, t. 686, fig. B.
Filum, a thread; *forma*, shape.

P. 1–2 mm., *grey, shining with white mealy particles*, very tender, cylindrical, 1–2 mm. high, striate. St. 10–15 × ·5 mm., *white, hyaline, sprinkled with a few, short, delicate* hairs. Gills "*blackish*, adnate, linear" W. G. Sm. Spores "subglobose, 5 × 4μ" Massee. On the ground. Woods, and pastures. April—Sept. Uncommon (*v.v.*)

****P. always smooth. Ringless.

1701. **C. hemerobius** Fr. Cke. Illus. no. 675, t. 687, fig. A.
ἡμέρα, a day; *βίος*, life.

P. 1·5–3 cm., *greyish, disc and surface of ribs date brown*, ovate, then campanulate, *radiately sulcate up to the even, not depressed disc, smooth*. St. 5–9 cm. × 3–4 mm., *pallid*, fragile, attenuated upwards, *smooth*. Gills *pallid, then blackish, adnate to the dilated apex of the st.*, attenuated at the margin, linear, 2–4 mm. broad. Flesh *white*, very thin at the margin. Spores black, elliptical, with a hyaline apical or oblique germ-pore, 10–11 × 5μ. Cystidia "bottle-shaped, or cylindrical-conical, 60–70 × 15–20μ" Rick. Pastures, and woods, especially beech. Sept.—Nov. Not uncommon. (*v.v.*)

C. hiascens (Fr.) Quél. = **Psathyrella hiascens** Fr.

C. disseminatus (Pers.) Quél. = **Psathyrella disseminata** (Pers.) Fr.

1702. **C. eburneus** Quél. Quél. As. fr. (1883), t. 6, fig. 9.
Eburneus, white as ivory.

Entirely white and shining. P. 3–4 cm., elliptic campanulate, *firm, striate*, polished, besprinkled with small, recurved, fugacious flocci. St. 5 cm. × 6–8 mm., *firm*, smooth. Gills *white, then deep bay*, free, lanceolate, tardily deliquescing. Spores violet, almond-shaped, 14μ. Mountainous pastures. July—Sept. Rare.

1703. **C. Schroeteri** Karst. J. Schröter, the Silesian mycologist.

P. 1 cm., *dingy ochraceous, becoming pale, at length sooty grey*, elliptical, then expanded, *sulcate*, smooth. St. 1–2 cm., *minutely pulverulent at first, slightly striate upwards*. Gills *brown*, free. Spores black, angularly subglobose, 13–15 × 8–12μ. On cow dung. Rare.

var. **proximellus** (Karst.) Massee.
Proximellus, somewhat nearly related.

Differs from the type in the elliptical spores, 10–13 × 5–7μ. Manured ground. Rare.

IV. Receptacle membranaceous, tough, reviving with moisture, not putrescent. Spores white.

*Pileus with a thin, unspecialized cellular pellicle.

Marasmius Fr.

(*μαραίνω*, I die away.)

Pileus membranaceous, or coriaceous, regular, or resupinate. Stem central, or wanting, cartilaginous, or horny. Gills adnate, adnexed, decurrent, or free, pliant, rather tough. Spores white; elliptical, pip-shaped, oblong elliptical, almond-shaped, tear-drop-shaped, globose, or subglobose; smooth, punctate, or echinulate; continuous. Cystidia present, or absent. Growing on the ground, or on wood; solitary, gregarious, caespitose, or fasciculate.

A. Margin of p. incurved at first. St. cartilaginous. Mycelium floccose.

a. St. externally villose, or pruinose. Gills separating, free.

*St. woolly, or strigose, at the base.

1704. **M. urens** (Bull.) Fr. (= *Marasmius peronatus* Bolt. sec. Quél.) Gonnerm. & Rabenh. Heft. 8–9, t. 8, fig. 1. *Urens*, burning.

P. 3–7·5 cm., *deep yellow, or pinkish buff*, becoming paler, *disc darker* and often slightly depressed, slightly fleshy, convex, then plane, here and there squamulose, or rimoso-squamulose when dry; margin thin, involute. St. 4–5 cm. × 2–6 mm., *concolorous*, equal, or slightly bulbous at the base, covered with white farinose down. Gills *pale wood colour, then brown*, free, united behind, *at length remote, distant*, tough. Flesh *yellowish*. Spores white, elliptical, or pip-shaped, 8 × 4μ, 1–2-guttulate. Taste very acrid. Woods. May—Oct. Uncommon. (*v.v.*)

1705. **M. peronatus** (Bolt.) Fr. (= *Marasmius urens* Bull. sec. Quél.) Cke. Illus. no. 1070, t. 1117. *Peronatus*, wearing boots of untanned leather.

P. 3–6 cm., *light yellowish, or pallid brick rufescent*, then *becoming pale wood colour*, or *tan*, at first fleshy-pliant, then *coriaceo-membranaceous*, convex, then plane, obtuse, flaccid, slightly wrinkled, *at length lacunose*; *margin striate*. St. 5–9 cm. × 2–6 mm., *white, clothed with dense white, or yellowish villose, strigose hairs in the basal third*, attenuated upwards, incurved at the base. Gills *cream colour*, then *pallid wood colour and rufescent, adnexed, then separating*, free, moderately *thin, crowded*. Flesh *yellowish*. Spores white, pip-shaped, or tear-drop-shaped 7–10 × 4–5μ. Taste acrid. Woods. July—Dec. Common. (*v.v.*)

1706. **M. porreus** (Pers.) Fr. *Porrum*, a leek.

P. 2–5 cm., *dingy yellowish*, pallid when dry, *coriaceo-membranaceous*, flaccid, convex, then flattened, obtuse; margin striate. St. 7–11 cm. × 3–5 mm., *red-fuscous*, paler at the apex, tough, *pubescent*, somewhat thickened at both ends, villose at the base, *containing a red juice*. Gills *light yellowish, becoming pale*, separating free, *distant*, rather thick, tough, broadly linear, at length coriaceous. Flesh *reddish*. Spores white, pip-shaped, 8–9 × 5μ, 2–many-guttulate. Smell *strong, of garlic*. Deciduous woods. Sept.—Dec. Uncommon. (*v.v.*)

**St. naked at base.

1707. **M. oreades** (Bolt.) Fr. Cke. Illus. no. 1072, t. 1118. 'Ορειάς, belonging to mountains.

P. 2–6 cm., *rufescent, then becoming pale, hygrophanous, whitish when dry*, fleshy, pliant, convex, then plane, subumbonate; margin *striate* when moist. St. 4–10 cm. × 2–4 mm., *pallid*, very tough, equal, *everywhere clothed with a villose-woven cuticle*. Gills *pallid-white*, free, *broad, distant*, the alternate ones shorter, *at first soft*, then firmer. Flesh *pale-ochraceous*. Spores white, elliptical, with an oblique basal apiculus, 7–8 × 5μ. Smell and taste pleasant. Edible. In rings in pastures, on lawns, and roadsides. May—Oct. Common. (*v.v.*)

1708. **M. globularis** Fr. (= *Marasmius Wynnei* B. & Br.) Trans. Brit Myc. Soc. III, t. 13. *Globularis*, globular.

P. 1·5–3 cm., *milk white*, then shining, *often tinted with rose or greyish violet, and finally fuscous violaceous, globose, then campanulate*, hygrophanous, pellucidly striate. St. 2–5 cm. × 3–4 mm., *white*, then *brownish at the base*, flexuose, *pulverulent*. Gills *white, then dingy*, free, distant, ventricose. Spores white, ovoid pruniform, 9 × 7μ, 1-guttulate. Smell pleasant like that of *Marasmius oreades*. Taste mild. Edible. Beech woods, and under beeches. Aug.—Dec. Not uncommon. (*v.v.*)

var. **carpathicus** (Kalchbr.) Cost. & Dufour. Kalchbr. Icon. t. 26, fig. 4. *Carpathicus*, Carpathian.

Differs from the type in the stem becoming blackish at the base. On stumps and amongst dead leaves. Oct. Uncommon. (*v.v.*)

1709. **M. plancus** Fr. Cke. Illus. no. 1073, t. 1119, fig. A. *Plancus*, flat footed.

P. 2–3 cm., *rufescent, then becoming pale*, thin, flexuose, plane, or depressed, somewhat repand, obtuse. St. 3–5 cm. × 4–6 mm., *pale yellow*, soon *compressed*, unequal, *twisted*, covered with a white villose cortex; base *naked, somewhat attenuated*. Gills *yellow*, then *bright bay, or rust colour, narrow*, distant, becoming free. Taste mild. Deciduous woods. Sept.—Oct. Uncommon.

1710. **M. scorteus** Fr. Cke. Illus. no. 1073, t. 1119, fig. B.
Scorteus, made of leather.

P. 6–10 mm., *pallid, often whitish*, slightly fleshy, convex, then plane, *obtuse*, at length rugulose. St. 2·5 cm. × 1–2 mm., *white, becoming fuscous and contorted* when dry, *equal, delicately pruinose at the apex*, tough. Gills *white, quite free*, remarkably *broad*, rounded behind, *ventricose*. Spores white, elliptical, 8 × 6μ. Taste mild. Deciduous woods, moist places, and amongst grass under trees. Aug.—Jan. Uncommon.

b. St. rooting, distinctly cartilaginous. Gills separating-free.

*St. woolly downwards, smooth upwards.

1711. **M. prasiosmus** Fr. (= *Marasmius archyropus* Pers. sec. Quél.)
πράσον, a leek; *ὀσμή*, smell.

P. 1–3 cm., *pale dingy yellow, or whitish, disc darker*, submembranaceous, campanulate, then convex and plane, obtuse, rugulose; margin striate when moist. St. 5–8 cm. × 2 mm., pallid, becoming reddish at the base, equal, tough, *tomentose*, thickened and often incurved at the base, and adherent to the leaves. Gills *white, then pallid*, slightly adnexed, somewhat crowded. Spores white, pip-shaped, 9–10 × 4–5μ, 1–many-guttulate. Cystidia none. Smell and taste strong, *of garlic*, persistent. Edible. Beech woods. July—Oct. Uncommon. (*v.v.*)

1712. **M. varicosus** Fr. (= *Marasmius fuscopurpureus* Pers. sec. Quél.) Boud. Icon. t. 72. *Varicosus*, full of dilated veins.

P. 1–3 cm., *fuscous purple, becoming darker when dry*, submembranaceous, pliant, campanulate, then convex and plane, *umbonate*, even. St. 3–6 cm. × 2–4 mm., *rubiginous, blackish when dry, filled with dark blood-coloured juice, which forms drops when it is broken*, very minutely tomentose, often longitudinally grooved; base reddish, strigose. Gills *white, then becoming concolorous with the pileus*, adnate, or almost free, *very crowded, linear, very narrow*. Flesh *concolorous*. Spores white, elliptical, 4–5 × 2·5–3μ. Damp mossy places. Sept.—Oct. Uncommon. (*v.v.*)

1713. **M. fuscopurpureus** (Pers.) Fr. (= *Marasmius varicosus* Fr. sec. Quél.) Cke. Illus. no. 1075, t. 1121, fig. B.
Fuscus, dark; *purpureus*, purple.

P. 1–3 cm., *dark purple, becoming pale when dry*, slightly fleshy, at first hemispherical, then plane, *obsoletely umbilicate*, slightly wrinkled. St. 2·5–8 cm. × 2–4 mm., *pallid*, then *rufous, or dark purple, sheathed towards the base with strigose, rubiginous down*. Gills *rufescent*, adnexed in the form of a ring, then free, *distant, narrow*. Flesh *dark coloured*. Spores white, elliptical, 4 × 3μ, 1-guttulate. Cystidia none. Oak, and beech woods. Aug.—Oct. Uncommon. (*v.v.*)

1714. **M. terginus** Fr. (= *Marasmius Stephensii* Berk. sec. Fr.) Fr. Icon. t. 174, fig. 4. *Tergum*, hide.

P. 1–3 cm., *flesh colour* when moist, *whitish when dry*, *shining*, tough, slightly fleshy, convex, then plane, obtuse, at length somewhat depressed; margin *striate when moist*. St. 5–8 cm. × 2–4 mm., *pallid upwards*, *reddish downwards*, slightly attenuated upwards, longitudinally fissile, *white villose at the rooting base*. Gills *pallid*, separating free, *somewhat crowded*, *narrow*. Flesh *white*. Spores white, elliptical, 6–7 × 4μ, punctate. Taste mild. Edible. Amongst leaves, and on twigs in deciduous woods. Oct.—Nov. Uncommon.

1715. **M. esculentus** (Wulf.) Karst. (= *Collybia esculenta* (Wulf.) Fr.; *Collybia conigena* Fr. (non Pers.) sec. Bres.; *Collybia clavus* Schaeff. sec. Quél.) Bres. Fung. Trid. t. 198, fig. 1. *Esculentus*, edible.

P. 1–2·5 cm., *lurid ochraceous*, *or pale yellowish*, somewhat fleshy, convex, then plane, moist, *sometimes subumbonate*; margin *at first pellucidly striate*. St. 2–7 cm. × 2–4 mm., *white*, then *concolorous*, equal, apex pruinose; base long, rooting, tomentosely fibrillose. Gills *white*, *often becoming yellowish at the edge*, rounded behind, adnexed, crowded. Spores white, oblong, 5–6 × 2–3μ. Cystidia *capitate at the apex*, ventricosely fusiform, 40–66 × 10–18μ, very thick walled. Taste slightly bitter. Edible. On buried cones of conifers in woods, and plantations. Sept.—May. Common. (*v.v.*)

1716. **M. conigenus** (Pers.) Karst. (= *Collybia conigena* (Pers.) Bres.; *Collybia tenacella* Fr. and *Collybia stolonifera* Jungh. sec. Bres.) Bres. Fung. Trid. t. 198, fig. 2. *κῶνος*, a cone; *γίγνομαι*, to be born.

P. 1–3 cm., *fuscous*, then paler, very rarely white, somewhat fleshy, convex, or conico-campanulate, then expanded and *subumbonate* or plane, moist. St. 2–6 cm. × 2–3 mm., *white*, *soon ochraceous*, equal, apex *white mealy*; base long, rooting, tomentosely fibrillose. Gills *cinereous*, then white, *often becoming yellowish*, sinuato-adnexed, crowded. Spores white, oblong, depressed on one side, 4–5 × 2·5–3μ. Cystidia ventricosely fusiform, 48–72 × 10–14μ, thick walled. Taste slightly bitter. Edible. On buried cones of conifers in woods, and plantations. Sept.—May. Common. (*v.v.*)

**St. (at least when dry) everywhere pruinosely velvety.

M. Wynnei B. & Br. = **Marasmius globularis** Fr.

1717. **M. erythropus** (Pers.) Fr. (= *Collybia erythropus* (Pers.) Quél.) Cke. Illus. no. 1077, t. 1123, fig. B. *ἐρυθρός*, red; *πούς*, foot.

P. 2–3 cm., *pallid*, *becoming whitish when dry*, hygrophanous, slightly fleshy, convex, then plane, obtuse, slightly wrinkled when dry, striate. St. 5–7·5 cm. × 4 mm., *dark red*, apex paler, firm, tough, often compressed, shining, pruinose when dry, base white strigose. Gills *whitish, becoming flesh colour*, separating free, *broad, lax*, connected by veins, *subdistant*. Flesh *concolorous*. Spores white, pip-shaped, 6 × 3μ. Taste mild. Edible. Deciduous woods, and heaths. Sept.—Nov. Common. (*v.v.*)

1718. **M. undatus** (Berk.) Quél. (= *Collybia undata* Berk.; *Collybia vertirugis* Cke.) Cke. Illus. no. 193, t. 149, upper figs., as *Collybia vertirugis* Cke. *Undatus*, waved.

P. 1–2 cm., *dull brown, or cinereous*, campanulate, then convexo-plane, tough, submembranaceous, *minutely pulverulent, radiato-wrinkled*. St. 5–6 cm. × 2–3 mm., *rufous*, apex whitish, sometimes compressed, *minutely velvety*, base strigose. Gills *white with a yellowish tinge*, adnate, connected by veins, subdistant, broad. Spores white, subglobose, 6–7 × 4–5μ, apiculate at the one end. On dead, and cut off bracken stems. Sept.—Oct. Common. (*v.v.*)

1719. **M. hariolorum** (DC.) Quél. (= *Collybia hariolorum* (DC.) Fr.; *Collybia confluens* (Pers.) Fr. sec. Quél.) Cke. Illus. no. 194, t. 150, lower figs., as *Collybia confluens* Pers. *Hariolus*, a soothsayer.

P. 2–3 cm., *rufescent, then flesh coloured, and finally whitish when dry*, submembranaceous, *tough, pliant*, campanulato-convex, then flattened, obtuse, rather depressed. St. 7–12 cm. × 2–4 mm., *flesh colour*, or *purplish*, cartilaginous, equal, *often compressed, pulverulent with white villose down, confluent at the base;* apex pallid, naked. Gills *flesh colour, then whitish*, linear, narrow, slightly adnexed, then free, somewhat crowded. Spores white, elliptical, 5–7 × 3–4μ, 1-guttulate. Cystidia "on edge of gill moniliform, 50–65 × 10–12μ" Rick. Fasciculate. Amongst dead leaves in deciduous woods. June—Dec. Common. (*v.v.*)

1720. **M. ingratus** (Schum.) Quél. (= *Collybia ingrata* (Schum.) Fr.) Fr. Icon. t. 64, fig. 1, as *Collybia ingrata* (Schum.) Fr. *Ingratus*, unpleasant.

P. 2–6 cm., *dingy fuscous-tan*, or *reddish, becoming whitish flesh colour*, slightly fleshy, pliant, globoso-campanulate, then expanded, *umbonate*. St. 4–10 cm. × 2–8 mm., *fuscous*, or *purplish*, cartilaginous, *twisted, covered with a white pruina*, often compressed, *villose internally*, apex *white mealy*. Gills *reddish brick colour*, becoming pallid, *free, very crowded, narrow*. Flesh *reddish*, or *the same colour as the pileus*. Spores white, "lanceolate-fusiform, 7–8 × 3–4μ. Cystidia none" Rick. Smell mouldy, taste bitter. Caespitose. On dead twigs in woods, and in damp places. Aug.—Oct. Uncommon.

1721. **M. pruinatus** Rea. Trans. Brit. Myc. Soc. v, t. 8.

Pruinatus, covered with hoar frost.

P. 5–10 mm., *white, becoming tinged with yellow*, fleshy-horny, convex, obtuse, or obsoletely papillate, *pruinose*; margin thin, incurved. St. 1·5–3 cm. × 1–2 mm., *white*, equal, rigid, *pruinose*, base white villose. Gills *shining white*, decurrent, very narrow, 1 mm. wide, subdistant. Flesh *greyish, very tough, elastic*. Spores *white*, pip-shaped, elongated into a long acute point at one end, blunt and rounded at the other, 9–12 × 4μ, with a large central gutta. Oak woods. Sept.—Oct. Uncommon. (*v.v.*)

1722. **M. archyropus** (Pers.) Fr. (= *Marasmius prasiosmus* Fr. sec. Quél.) Gonn. & Rabenh. t. 8, fig. 6.

ἄργυρος, silver; *πούς*, foot.

P. 2·5 cm., *tan colour, becoming pale*, slightly fleshy, convex, then plane, or depressed. St. 6–10 cm. × 2 mm., *pallid rufescent under the white tomentose pruina, which forms an outer covering, similar at the base*, rigid, tense, straight. Gills *pallid*, adnexed, separating, crowded, *linear*. Flesh *yellowish, deeper coloured in the stem*. Spores white, pip-shaped, 8–10 × 3–4μ, 1-3-guttulate. Deciduous woods. Sept.—Oct. Uncommon. (*v.v.*)

1723. **M. suaveolens** Rea. (=*Marasmius ingratus* (Weinm.) Quél. var. *suaveolens* Rea sec. Bataille.) Trans. Brit. Myc. Soc. II, t. 12, as *Marasmius archyropus* (Pers.) Fr. var. *suaveolens* Rea.

Suaveolens, sweet smelling.

P. 4·5–6 cm., *flesh colour, becoming pale*, convex, then plane, or depressed, tough; *margin striate*. St. 6–7 cm. × 2–3 mm., *reddish, apex paler*, everywhere covered with a white tomentose pruina, becoming twisted when dried. Gills *pallid, then fuscous*, adnexed, separating, crowded, 6–8 mm. *wide, ventricose*. Flesh *whitish in the pileus, reddish in the stem*. Spores white, globose, 3–4μ. *Smell very pleasant, like Marasmius oreades*. Beech woods, and under beeches. Sept.—Nov. Not uncommon. (*v.v.*)

1724. **M. torquescens** Quél. Quél. Jur. et Vosg. I, t. 22, fig. 3.

Torquescens, becoming twisted.

P. 1–2 cm., *pallid, disc tawny*, membranaceous, thin, convex, then plane, striate, then *sulcate*. St. 4–6 cm. × 1–2 mm., *brown*, apex smooth and *whitish, delicately* velvety, twisted and sulcate when dry, white floccose at the base. Gills *white, then reddish*, free, *thin*, ventricose, distant, minutely serrulate. Flesh *whitish, reddish in the stem*. Spores white, almond-shaped, 9–10 × 6–7μ. On twigs in woods, and hedgerows. Sept.—Nov. Uncommon. (*v.v.*)

1725. **M. obtusifolius** Rea. Trans. Brit. Myc. Soc. VI, t. 7.
Obtus, blunt; *folius*, leaf, gill.

P. 1 cm., *whitish, disc tawny*, membranaceous, convexo-plane, with a prominent umbo, smooth, *sulcate*; margin at first incurved. St. 3–4 cm. × 1 mm., *tawny, apex white*, equal, minutely velvety. Gills *pallid, adnate to a collar behind*, 2 mm. wide, distant, simple; edge *very obtuse*, blunt, ciliate with the prominent cystidia. Flesh *white*, tough, thin. Spores white, broadly oval, or subglobose, 14–15 × 10–12μ, with a large central gutta, thick walled; basidia clavate, 40–60 × 18–23μ, with 2–4-sterigmata. Cystidia abundant, fusiform, ventricose, 95–145 × 17–25μ, apex capitate, 14–18μ, thin walled. Sept.—Oct. Uncommon. (*v.v.*)

1726. **M. impudicus** Fr. Cke. Illus. no. 1078, t. 1124, fig. B.
Impudicus, disgusting.

P. 1–3 cm., *bay-brown-rufous, tinted with purple or lilac, becoming pale when dry*, slightly fleshy at the disc, convex, then plane, often depressed in the centre, membranaceous, *striato-plicate*. St. 5 cm. × 2 mm., *rufous, or rufous-fuscous* when moist, sometimes violaceous-purple, *wholly covered over with white villose down when dry*, equal, often compressed, tough, flexile, rooting at the attenuated base. Gills *flesh colour, or greyish, then whitish*, at first reaching the stem, then free, truncate behind, connected by veins, ventricose, at first crowded, then distant. Spores white, elliptical, 8 × 4–5μ, "minutely echinulate" Quél. Smell *strong, disgusting*, resinous. On pine trunks, and needles in coniferous woods. Sept.—Oct. Uncommon.

***St. smooth.

1727. **M. dryophilus** (Bull.) Karst. (= *Collybia dryophila* (Bull.) Fr.) Cke. Illus. no. 206, t. 204, as *Collybia dryophila* Fr.
δρῦς, oak; φίλος, loving.

P. 2–4 cm., *bay-brown-rufous, yellow, ochraceous, or white, becoming pale*, slightly fleshy, tough, convexo-plane, obtuse, *disc generally depressed*. St. 4–7 cm. × 2–4 mm., *whitish, becoming yellow, or rufescent*, cartilaginous, somewhat rooting. Gills *white*, or *becoming pale*, somewhat free, *crowded, narrow*. Flesh *white*, thin. Spores white, elliptical, 5–6 × 3–4μ. Woods, pastures, and hedgerows. May—Nov. Common. (*v.v.*)

var. **funicularis** (Fr.) Rea. (= *Collybia dryophila* (Bull.) Fr. var. *funicularis* Fr.) *Funiculus*, a slender rope.

Larger than the type, *caespitose*. St. lax, decumbent, villose at the base. Gills *sulphur-yellow*. Spores white, elliptical, 6 × 3μ. Woods, and pastures. May—Nov. Uncommon. (*v.v.*)

var. **auratus** (Quél.) Rea. (= *Collybia dryophila* (Bull.) Fr. var. *aurata* Quél.) *Auratus*, golden.

P. *golden yellow*. St. *bright yellow*. Gills *cream colour*. Woods. Sept.—Oct. Uncommon. (*v.v.*)

var. **oedipus** (Quél.) Rea. (= *Collybia dryophila* (Bull.) Fr. var. *oedipus* Quél.) Fr. Icon. t. 80, fig. 1, as *Mycena galeropsis* Fr. *οἰδίπους*, swollen footed.

Pale *yellow amber*. St. *bulbous, vesiculosely swollen at the base*. Gills *cream colour*. In *Sphagnum* bogs. May—Sept. Not uncommon. (*v.v.*)

var. **alvearis** (Cke.) Rea. (= *Collybia dryophila* (Bull.) Fr. var. *alvearis* Cke.) Trans. Brit. Myc. Soc. III, t. 5, fig. B, as *Collybia dryophila* Bull. var. *alvearis* Cke. *Alveare*, a bee-hive.

P. 7·5 cm., *dome-shaped*, resembling the old straw bee-hive. St. 1 cm. at the base, *rufescent*. Amongst dead leaves. Sept.—Nov. Not uncommon. (*v.v.*)

var. **aquosus** (Bull.) Rea. (= *Collybia aquosa* (Bull.) Fr.) Fr. Icon. t. 66, fig. 2, as *Collybia aquosa* (Bull.) Fr. *Aquosus*, watery.

Pale tan, becoming white. P. *hygrophanous*; *margin striate*. Spores elliptical, 5–6 × 3–4μ, 1-guttulate. Amongst moss in woods, and on heaths. May—Oct. Common. (*v.v.*)

1728. **M. exsculptus** (Fr.) Rea. (= *Collybia exsculpta* Fr.; *Collybia dryophila* (Bull.) Fr. sec. Quél.) Fr. Icon. t. 66, fig. 3, as *Collybia exsculpta* Fr. *Exsculptus*, cut out.

P. 2·5–5 cm., *tawny brown, or sulphur yellow*, slightly fleshy, tough, convex, then expanded and *umbilicate, unchangeable*. St. 2–3 cm. × 3–4 mm., *bright sulphur yellow*, cartilaginous, incurved, base sometimes enlarged. Gills *bright sulphur yellow*, somewhat free (decurrent with a small tooth), *arcuate, linear*, very crowded, narrow. Flesh *rufous*. Spores white, 6·5–7 × 3–3·75μ. Taste strong, mushroom-like. Edible. Gregarious. On old pine stumps, and in coniferous woods. May—Oct. Uncommon.

c. St. abrupt, often furnished with a floccose tubercle at the base. Gills adnate, or subdecurrent.

*Stem very smooth, and shining upwards, base simple.

1729. **M. Vaillantii** (Pers.) Fr. Cke. Illus. no. 1080, t. 1126, fig. A. Sebastian Vaillant.

P. 10–15 mm., *whitish*, somewhat membranaceous, pliant, convex, soon flattened and depressed at the disc, *plicato-rugose*. St. 2–3 cm. × 1–2 mm., *date-brown*, bright, shining, *thickened and paler upwards*, here and there obsoletely pruinate; *base inserted*, naked, *blackish*. Gills *white*, adnate, *somewhat decurrent on account of their triangular form*, broad, *distant*, distinct, simple. Spores white, oblong, 10–13 × 3–4μ. On twigs, leaves, and grass stems in woods. Sept.—Oct. Uncommon. (*v.v.*)

1730. **M. angulatus** (Batsch) B. & Br. (= *Marasmius graminum* Lib. sec. Quél.) Pers. Myc. Eur. III, t. 26, figs. 3, 4. *Angulatus*, angled.

P. 1–1·5 cm., *fuscous whitish*, somewhat membranaceous, hemispherical, then flattened, *at length plicato-angular*. St. 2–3 cm. × 1 mm., *rufescent grey, thickened at both ends*, base hairy. Gills whitish, distant. Spores white, elliptical, 7 × 4μ. On grass, submerged rushes, roots, on sandhills. Aug. Uncommon.

1731. **M. languidus** (Lasch) Fr. *Languidus*, weak.

P. 10–12 mm., *white, inclining to flesh colour, or light yellow*, slightly fleshy, convex, then expanded and *umbilicate, flocculose, rugoso-sulcate*; margin involute. St. 2–3 cm. × 1–2 mm., *pallid, becoming fuscous*, thickened upwards, base generally white villose. Gills *white*, adnate, then decurrent, distant, connected by veins, *narrow*. Spores white, elliptical, 8μ, minutely punctate. On dead leaves of grass, twigs, and dead herbaceous stems in woods. Aug.—Oct. Uncommon.

1732. **M. rubricatus** (B. & Br.) Massee. Cke. Illus. no. 496, t. 509, fig. C, as *Naucoria rubricata* B. & Br. *Rubricatus*, coloured red.

P. 6–10 mm., *whitish, then tinged with red or buff*, convex, then plane. St. 6–12 × 1 mm., *tan* colour, incurved, and sprinkled with delicate mealy granules at the base and about half way up. Gills *white, then brownish*, adnexed, narrowed in front. Spores "colourless, pyriform, 6 × 3μ" Massee. Dead twigs, and brambles. Sept.—Oct. Rare.

**St. velvety, or pruinate, base subtuberculose.

1733. **M. foetidus** (Sow.) Fr. Boud. Icon. t. 73. *Foetidus*, stinking.

P. 1·5–4 cm., *bay-brown-tawny, membranaceous, somewhat pellucid*, pliant, convex, then expanded, at length *umbilicate, striato-plicate*. St. 2–4 cm. × 1–2 mm., *date brown, becoming blackish, everywhere velvety*, horny, attenuated downwards, abrupt, or sometimes inserted with a very small floccose tubercle at the base. Gills *yellowish rufescent*, adnate, or subdecurrent, distant, connected by veins. Flesh *yellowish, becoming blackish in the stem*. Spores white, elliptical-oblong, or tear-drop-shaped, 9–12 × 4–6μ, 1-guttulate. Smell strong, *very foetid*. Dead twigs, and rotten branches in woods, and hedgerows. Aug.—Jan. Uncommon. (*v.v.*)

1734. **M. inodorus** Pat. *Inodorus*, without smell.

P. 1–2 cm., *reddish brown*, membranaceous, convex, then expanded, *covered with adpressed silky down*. St. 1–2 cm. × 1–2 mm., *blackish red, apex pale*, rigid, *entirely covered with a white pruinosity* consisting of hyaline, short, flexuose hairs. Gills *white*, adnate, unequal, somewhat crowded. Flesh *white, blackish in the stem.* Spores white, pip-shaped, guttulate. On tree trunks. Sept.—Oct. Rare.

1735. **M. xerotoides** von Post. Fr. Icon. t. 174, fig. 3.
Xerotus, the genus *Xerotus*, εἶδος, like.

P. 4–8 mm., *fuscous-umber and striate when moist, becoming pale fuscous grey* and somewhat undulate when dry, submembranaceous, convex, obtuse, soon expanded and *umbilicate*; margin involute. St. 10–14 × 1 mm., *grey* or *cinereous-fuscous, velvety*; base swollen, and strigose. Gills *whitish, becoming cinereous*, broadly adnate, subdecurrent, distant, intermixed with shorter ones. Spores white, "elliptical, 5 × 3μ" Massee. On sandy soil in woods. Rare.

1736. **M. lagopinus** von Post. λαγώς, a hare.

P. 2 cm., *pallid*, slightly fleshy, convex, then plane. St. 2·5 cm. × 1–2 mm., *pallid, squamulose with white flocci for the lower half.* Gills *pallid*, adnate, somewhat ventricose. Spores "pale straw colour, subglobose, 3 × 2μ" Massee & Crossl. Dead branches, and on fir trunks. Oct. Rare.

1737. **M. amadelphus** (Bull.) Fr. Bres. Fung. Trid. t. 130, fig. 2.
ἀ, with; ἀδελφός, brother.

P. 6–12 mm., *pale yellowish, becoming pale, margin whitish, fleshy-membranaceous*, convex, hemispherical, then expanded and depressed, or umbilicate, *sulcately striate, pruinose* under a lens. St. 1–1·5 cm. × ·5–1 mm., *concolorous, somewhat darker at the base, somewhat mealy* but becoming smooth, equal, undulate. Gills *white, adnate*, or *subdecurrent*, broad, *distant, margin fimbriate.* Spores white, oblong, 10–12 × 2·5μ. On trunks, and branches in coniferous woods. Sept.—Dec. Uncommon. (*v.v.*)

var. **insignis** Fr. *Insignis*, striking.

Differs from the type in the *whitish, thinner pileus, and the more distant, separating, pallid umber gills with tumid veins.* On fir twigs.

1738. **M. ramealis** (Bull.) Fr. Cke. Illus. no. 1082, t. 1127, fig. B.
Ramus, a branch.

P. 6–15 mm., *white, disc rufescent, somewhat fleshy*, convex, then plane, obtuse, or depressed, *slightly wrinkled*, minutely silky under a lens. St. 6–10 × 1–2 mm., *whitish, base rufescent*, tubercular when young, often incurved, *mealy*, squamulosely hairy under a lens. Gills *white*, adnate, connected behind, *slightly distant, narrow.* Spores white, elliptical, 8–10 × 3–4μ, minutely punctate. On dead twigs, branches, and bramble stems in woods, hedgerows, and wood stacks. Jan.—Dec. Common. (*v.v.*)

1739. **M. candidus** (Bolt.) Fr. Boud. Icon. t. 75.
Candidus, shining white.

P. 3–15 mm., *white*, hemispherical, then plane, or slightly depressed, pellucid, submembranaceous, *at length wrinkled, sulcate*. St. 5–15 × 1–2 mm., *white*, incurved, *delicately pruinose*; base floccose, *at length becoming fuscous*. Gills *white*, adnexed, ventricose, distant. Flesh *white*. Spores white, pip-shaped, 9 × 3μ, multi-guttulate. On pine branches, and twigs. Sept.—Nov. Uncommon. (*v.v.*)

1740. **M. sclerotipes** Bres. (= *Collybia cirrhata* recent. auct. non Fr. sec. Bres.) Bres. Fung. Trid. t. 11, fig. 1.
σκληρότης, hardness; *pes*, foot.

P. 6–8 mm., *shining white, umbilicate disc yellow*, membranaceous, convex, then plane, *rugosely striate*, somewhat flocculose. St. 12–18 × ·5–1 mm., *pallid rufous*, apex whitish, equal, *springing from a rufescent sclerotium, pruinose under a lens*. Gills *white*, adnate, distant, edge *fimbriate*. Spores white. On the ground in swampy places. Aug.—Oct. Uncommon.

B. Margin of p. straight, and adpressed to the stem at first. St. cartilaginous. Mycelium rhizomorphoid, corticate.

1741. **M. alliaceus** (Jacq.) Fr. Cke. Illus. no. 1083, t. 1128, fig. A.
Allium, garlic.

P. 1–4 cm., *whitish inclining to fuscous, often milk-white when young*, submembranaceous, campanulate, then expanded, subumbonate, at length striate and sulcate. St. 4–20 cm. × 2–4 mm., *blackish, horny*, rigid, attenuated upwards, *pruinato-velvety*; base rooting, incurved, naked. Gills *whitish*, adnexed in the form of a ring, then *free*, subventricose, slightly distant, crisped when dry. Flesh *white, blackish in the stem*. Spores white, pip-shaped, 8–9 × 3μ, 2–3-guttulate, "globose-elliptical, 7–9 × 6–7μ" Rick. Cystidia "bluntly fusiform, 45–60 × 12–15μ" Rick. Smell *very unpleasant, of garlic, persistent*. Amongst leaves, and on stumps in woods. Aug.—Nov. Common. (*v.v.*)

1742. **M. molyoides** Fr. μῶλυ, wild garlic; εἶδος, like.

P. 2–3 cm., *brownish, then pale*, submembranaceous, convex, then plane and depressed, sometimes slightly striate. St. 5–6 cm. × 2–3 mm., *blackish fuscous*, paler when young, *white at the apex*, equal, *base clavate*, becoming twisted when old. Gills *white, or yellowish, free*, crowded, ventricose; margin *ciliate and darker coloured*. Flesh *pale ochraceous*. Spores white. Smell *faint, of garlic* when young, then inodorous. Amongst leaves in woods. Aug.—Oct. Uncommon.

1743. **M. cohaerens** (A. & S.) Cke. (= *Marasmius ceratopus* Pers. sec. Quél.) Fr. Icon. t. 79, fig. 1, as *Mycena cohaerens* Fr.
Cohaerens, sticking together.

P. 2–3 cm., *cinnamon*, or *umber-tawny*, becoming pale, slightly fleshy, campanulate, obtuse, pruinose, *velvety under a lens*. St. 10–12 cm. × 4 mm., *bay brown*, *horny*, *very rigid*, shining, apex *whitish*, *pruinose*, *caespitoso-fasciculate*, base date-brown, *glued together with white villose down*. Gills *white*, becoming pale, *sprinkled with fulvous bristles* under a lens, rounded behind, *somewhat free*, distant, generally connected by veins, *very broad*. Spores white, "lanceolate, 8–9 × 4μ. Cystidia brown, fusiform-subulate, very pointed, 60–100 × 7–14μ, thick walled" Rick. On trunks, and amongst leaves in deciduous woods. Oct.—Feb. Uncommon.

1744. **M. cauticinalis** (With.) Fr. Bres. Fung. Trid. t. 41, fig. 2.
Cautes, a rough pointed rock.

P. 1–2 cm., *dingy yellow*, *becoming ferruginous*, then ochraceous, membranaceous, thin, pliant, campanulate, then convex, obtuse, at length plane, umbilicate, and striato-sulcate. St. 3–5 cm. × 2–3 mm., *bay brown*, *paler and mealy upwards*, rigid, tough, *floccoso-villose* at the tubercular base, and arising from a filiform, *dark bay mycelium*. Gills *pallid light yellow*, *adnato-decurrent*, subdistant, connected by a network of veins. Flesh *yellow*, or *rufescent*. Spores white, pip-shaped, 6–7 × 3–4μ, 1-guttulate. Amongst needles in coniferous woods. Sept.—Nov. Uncommon. (*v.v.*)

1745. **M. torquatus** Fr. *Torquatus*, adorned with a collar.

P. 1–2 cm., *whitish*, or *greyish white*, *truly campanulate*, 1–2 cm. high, membranaceous, tough, obtuse; margin *plicato-sulcate*. St. 3–4 cm. × 2 mm., *whitish*, equal, shining, base *generally fuscous*, arising from *a minute*, *round tubercle*. Gills *white*, *adnate to a free collar*, connected by veins, distant, ascending, unequal. Flesh of pileus *pale yellow*. Spores white. On stems, and leaves. Sept.—Oct. Uncommon.

1746. **M. scorodonius** Fr. (= *Marasmius alliatus* Schaeff. sec. Quél.) Cke. Illus. no. 1079, t. 1125, fig. A. σκόροδον, garlic.

P. 1–2 cm., *rufous*, *soon becoming pale whitish*, membranaceous, pliant, convex, soon plane, obtuse, at length wrinkled and crisped. St. 2·5 cm. × 1–2 mm., *rufous*, *shining*, *horny*, tough, equal, *inserted and naked at the base*. Gills *whitish*, *adnate*, often separating, connected by veins, crowded, narrow. Spores white, "lanceolate, 5–7 × 3μ" Rick. Smell *strong*, *of garlic*. Edible. On twigs, and needles, on heaths, and in pine woods. Sept.—Oct. Rare.

C. P. sessile, resupinate.

1747. **M. spodoleucus** B. & Br. (= *Marasmius Broomei* Berk. sec. Cke.) Cke. Illus. no. 1088, t. 1137, fig. C.

σποδός, ashes; λευκός, white.

P. 4–5 mm., *cinereous, conchate, resupinate*, pulverulent, or slightly furfuraceous; margin free, arched. Gills *white*, few, narrow, entire, so short as to leave a naked space at the base. Flesh *umber*. Spores white, globose, 3–4μ. Dead twigs and stumps of elm. Nov.—March. Uncommon. (*v.v.*)

**Pileus with a thick, cellular pellicle.

†Cells of the pellicle upright, echinulate or verrucose.

Androsaceus (Pers.) Pat. (= **Marasmius** Fr. p.p.)

(ἀνδρόσακες, an unidentified sea plant.)

Pileus membranaceous, thin, regular, cells of pellicle echinulate, or tuberculose. Stem central, horny. Gills adnate, emarginate, decurrent, or attached to a collar. Spores white, elliptical, pip-shaped, oblong elliptical, subglobose, or club-shaped; smooth; continuous. Cystidia present, or absent. Growing on wood, fallen leaves, etc.

*Fleshy-membranaceous.

1748. **A. calopus** (Pers.) Pat. (= *Marasmius calopus* (Pers.) Fr.) Cke. Illus. no. 1079, t. 1125, fig. B, as *Marasmius calopus* Pers.

καλός, beautiful; πούς, foot.

P. 1–1·5 cm., *whitish*, slightly fleshy, tough, convex, then flattened, obtuse, rarely depressed, slightly wrinkled when dry. St. 2–3 cm. × 2 mm., *bay-brown-rufous*, tough, *shining*, attenuated upwards. Gills *white*, slightly emarginate, thin, subdistant. Spores white, elliptical, 7 × 4μ. On twigs, grass roots, etc. Sept.—Nov. Common. (*v.v.*)

**Tender, pellucid.

1749. **A. polyadelphus** (Lasch) Pat.[1] (= *Marasmius polyadelphus* (Lasch) Cke.) Cke. Illus. no. 1088, t. 1137, fig. B, as *Marasmius polyadelphus* Lasch. πολυάδελφος, with many brothers.

Entirely snow white. P. 2–3 mm., very tender, *hemispherical, umbilicate, sulcate*, pruinose, *tomentose under a lens*. St. 10–15 × 1 mm., filiform, curved, flaccid, pruinose, thickened and floccose at the base. Gills decurrent, *very narrow, wrinkle-like, distant*. Spores white, "fusiform-lanceolate, 7–9 × 3–4μ" Rick. Fasciculate, and in troops. On dead oak, and beech leaves. Oct.—Dec. Not uncommon. (*v.v.*)

1750. **A. flosculinus** (Bataille) Rea. (= *Marasmius flosculus* Quél.) Quél. Soc. bot. 1878, t. 3, fig. 4. *Flosculus*, a little flower.

P. 4–5 mm., *white, shining*, very thin, diaphanous, campanulate,

[1] By an oversight this species was described under Omphalia (no. 1412), but its correct position is here.

then convex and umbilicate, *ribbed, sulcate.* St. 2–3 × ·5 mm., *bay brown,* shining, *apex white,* thickened, incurved, horny, base downy. Gills *white,* adnate, distant, wide, *thick.* Spores "ovoid lanceolate, 10μ" Quél. On leaves of grasses. June—Aug. Rare.

***Membranaceous.

1751. **A. rotula** (Scop.) Pat. (= *Marasmius rotula* (Scop.) Fr.) Cke. Illus. no. 1084, t. 1129, fig. A, as *Marasmius rotula* Scop.
Rotula, a little wheel.

P. 5–15 mm., *whitish,* unicolorous, or with the *umbilicus becoming fuscous,* membranaceous, convex, *umbilicate,* plicate; margin undulato-crenulate. St. 2–5 cm. × 1 mm., *blackish, horny,* equal, shining, striate when dry. Gills *white, adnate to a collar free from the stem,* broad, few (often equal), very distant. Spores white, pip-shaped, acutely attenuated at the one end, 7–9 × 3·5–4·5μ. Cystidia vesiculose, apex echinulate, 14–16 × 7–8μ. On dead twigs and roots in woods and hedgerows. May—Jan. Common. (*v.v.*)

1752. **A. graminum** (Lib.) Pat. (= *Marasmius graminum* (Lib.) Berk.) Cke. Illus. no. 1084, t. 1129, fig. B, as *Marasmius graminum* Lib. *Graminum,* of grasses.

P. 5–8 mm., *rufous, or very pale rufous,* the furrows paler, *umbo brown,* nearly plane, *umbonate, sulcate.* St. 2·5–4 cm. × 1 mm., *bay or brownish tawny, white above,* shining. Gills *cream-coloured, adnate to a collar free from the stem,* few, subventricose, *interstices veined.* Spores white, pip-shaped, 8–10 × 4μ. On leaves, and stems of grasses. July—Feb. Not uncommon. (*v.v.*)

1753. **A. androsaceus** (Linn.) Pat. (= *Marasmius androsaceus* (Linn.) Fr.) Cke. Illus. no. 1084, t. 1129, fig. C, as *Marasmius androsaceus* Bull. ἀνδρόσακες, an unidentified sea plant.

P. 4–10 mm., *whitish, or somewhat fuscous,* membranaceous, subumbilicate, *wrinkled striate.* St. 3–6 cm. × 1 mm., *black,* horny, very tough, equal, contorted and striate when dry. Gills *whitish or greyish flesh colour,* adnate, crowded, narrow. Spores white, pip-shaped, 7 × 3–4μ. On leaves, and twigs, etc. April—Dec. Common. (*v.v.*)

1754. **A. splachnoides** (Hornem.) Rea. (= *Marasmius splachnoides* (Hornem.) Fr.) Cke. Illus. no. 1085, t. 1130, fig. A, as *Marasmius splachnoides* Fr. σπλάγχνον, intestines; εἶδος, like.

P. 5–10 cm., *white, disc yellowish flesh colour,* submembranaceous, convex, then expanded and slightly umbilicate, *sulcate.* St. 2–4 cm. × 1 mm., *red* (becoming fuscous), apex *whitish flesh colour,* shining. Gills white, *subdecurrent,* narrow, crowded, simple and *anastomosing.* Spores white, elliptical, "8 × 5μ" Cke. On pine, oak, and beech leaves in woods. Oct. Uncommon.

1755. **A. Curreyi** (B. & Br.) Rea. (= *Marasmius Curreyi* B. & Br.) Cke. Illus. no. 1085, t. 1130, fig. B, as *Marasmius Curreyi* B. & Br. Frederick Currey, the eminent mycologist.

P. 4–10 mm., *pallid rufous, furrows paler, umbo fuscous*, somewhat plane, *sulcate*, somewhat radiate. St. 2–3 cm. × 1 mm., *black, apex white*, shining. Gills *cream-coloured, attached to a collar*, few, subventricose, interstices veined. Spores white, elliptical, 9 × 5–6μ. On leaves of grasses. Aug. Uncommon.

1756. **A. perforans** (Fr.) Pat. (= *Marasmius perforans* Fr.; *Marasmius abietis* Batsch sec. Quél.) Cke. Illus. no. 1085, t. 1130, fig. C, as *Marasmius perforans* Fr. *Perforans*, boring through.

P. 8–12 mm., *whitish, becoming pale rufescent*, submembranaceous, convexo-plane, minutely umbonate, then flattened, at length *slightly wrinkled*. St. 2–3 cm. × 1 mm., *bay brown, then black, apex flesh colour*, tough, equal, *velvety*. Gills *whitish, adnate*, narrow, numerous, *simple, unequal* (the alternate ones shorter), not very distant. Spores white, "lanceolate, 6 × 3μ" Rick. Smell *very foetid*. On fir leaves. Aug.—Oct. Rare.

1757. **A. insititius** (Fr.) Rea. (= *Marasmius insititius* Fr.) Cke. Illus. no. 1086, t. 1135, fig. A, as *Marasmius insititius* Fr. *Insititius*, ingrafted.

P. 10–20 mm., white, membranaceous, pliant, convexo-plane, somewhat umbilicate, then plicato-sulcate. St. 2–3 cm. × 1 mm., *rufous, inclining to fuscous*, horny, attenuated downwards into an inserted base, *floccoso-furfuraceous*. Gills *pale, to cream*, broadly adnate, attenuated at the margin, *distant, simple, unequal*. Spores white, pip-shaped, 6–8 × 3·5–4μ. Cystidia hyaline, cylindrical, clavate or lanceolate, 40–50 × 5–6μ. On leaves, decayed grass, etc. Aug.—Oct. Uncommon. (*v.v.*)

var. **albipes** (Fr.) Rea. (= *Marasmius insititius* Fr. var. *albipes* Fr.) Berk. Outl. t. 14, fig. 6, as *Marasmius insititius* Fr. *Albus*, white; *pes*, foot.

Differs from the type in *the white stem*. Spores white, pip-shaped, 8–10 × 4–5μ, 1-multi-guttulate. On dead oak leaves and grasses. Aug.—Oct. Uncommon. (*v.v.*)

1758. **A. Hudsonii** (Pers.) Pat. (= *Marasmius Hudsoni* (Pers.) Fr.; *Marasmius pilosus* (Huds.) Quél.) Cke. Illus. no. 1086, t. 1135, fig. B, as *Marasmius Hudsoni* Pers. William Hudson, author of "Flora Anglica."

P. 2–6 mm., *pale fuscous-rufescent*, membranaceous, hemispherical, wrinkled, *beset with scattered, long, purplish, or brownish hairs*. St.

1–2·5 cm. × ·5 mm., *dark purple, or reddish,* horny, *beset with the same scattered hairs as the pileus, apex pale.* Gills *white,* adnexed, narrow, single, the alternate ones dimidiate, distant. Spores white, oblong elliptical, 9–12 × 4–6μ, 1–2-guttulate. On fallen, dead holly-leaves in woods, and under hollies elsewhere. March—Dec. Not uncommon. (*v.v.*)

1759. **A. epichloe** (Fr.) Rea. (= *Marasmius epichloe* Fr.; *Marasmius scabellus* (A. & S.) Quél. sec. Bataille.) Cke. Illus. no. 1087, t. 1136, fig. A, as *Marasmius epichloe* Fr.

ἐπί, upon; χλόη, grass.

P. 4–5 mm., *whitish, disc bay-brown-fuscous,* thin, plano-convex, *subpapillate.* St. 2–3 cm. × ·5 mm., *bay brown, paler at the base, coarsely striate,* striae *setulose.* Gills *whitish,* rounded, broader behind, somewhat crowded. Spores white, "elliptical, 3 × 2μ" Cke. On dead grass stems, and spines of *Robinia.* Aug.—Oct. Rare.

1760. **A. actinophorus** Rea. (= *Marasmius actinophorus* (B. & Br.) Massee, nec B. & Br. sec. Petch.) Cke. Illus. no. 1087, t. 1136, fig. B, as *Marasmius actinophorus* B. & Br.

ἀκτίς, a ray; φέρω, I bear.

P. 2–4 mm., *pale bay brown, with distant darker radiating lines,* very thin, convex, then plane, *umbilicate,* wrinkled when dry. St. 1–2 cm. × ·5 mm., *paler than the pileus,* equal. Gills *whitish,* adnexed, narrow, alternate ones shorter. Spores white, "subglobose, 3μ" Massee. On fallen twigs. Aug. Rare.

1761. **A. saccharinus** (Batsch) Rea. (= *Marasmius saccharinus* (Batsch) Fr.) Cke. Illus. no. 1087, t. 1136, fig. C, as *Marasmius saccharinus* Batsch. *Saccharum,* sugar.

P. 2–5 mm., *snow white,* membranaceous, *convex, subpapillate, sulcate,* and plicate. St. 1·5–2·5 cm. × ·5 mm., *white, reddish at the bulbose base,* pruinose, villose under a lens. Gills *whitish,* broadly adnate, narrow, thick, very distant, reticulato-united. Spores white, "ovoid lanceolate, 12μ, guttulate" Quél., "elliptical, 5 × 3μ" Massee. On dead twigs, and leaves. Aug. Rare.

1762. **A. epiphyllus** (Fr.) Pat. (= *Marasmius epiphyllus* Fr.)

ἐπί, upon; φύλλον, a leaf.

P. 4–10 mm., *milk white,* membranaceous, very thin, convex, then plane, *at length umbilicate,* smooth, *at length plicato-rugose.* St. 1–2·5 cm. × 1 mm., *date brown,* apex *whitish,* somewhat horny, equal, *velvety* under a lens. Gills *white,* adnate, *few, very distant, entire,* veined, branched. Spores white, "oval-oblong, 5–6μ, minutely aculeolate" Quél., "3 × 2μ" Cke. On dead leaves, twigs, etc. Sept.—Nov. Common. (*v.v.*)

1763. **A. epiphylloides** Rea. (= *Marasmius epiphyllus* Fr. sec. Lange. Trans. Brit. Myc. Soc. III, t. 14.
Epiphyllus, the species *A. epiphyllus*; εἶδος, like.

P. 2–5 mm., *white*, membranaceous, subspherical, then convex and expanded. St. 3–8 × ·5 mm., *chestnut brown*, apex *white*, equal. Gills *white*, adnate, ·5–1 mm. wide, sometimes connected by veins, few, distant. Spores hyaline, *club-shaped*, 13–15 × 3·5–4 μ, multi-guttulate. Cystidia 43–45 × 9–10 μ, attenuated at the apex, ventricose at the base. On dead leaves, and twigs. Sept.—Dec. Common. (*v.v.*)

††Cells of the pellicle decumbent, very long, fibrillose.

Crinipellis Pat.

(*Crinis*, hair; *pellis*, skin.)

Pileus membranaceous, regular, cells of pellicle long, thick, tough. Stem central, firm. Gills adnate, or free. Spores white, pip-shaped, smooth, or punctate; continuous. Cystidia present. Growing on wood, twigs, etc.

1764. **C. stipitarius** (Fr.) Pat. (= *Marasmius scabellus* (A. & S.) Quél.; *Collybia stipitaria* Fr.) Cke. Illus. no. 193, t. 149, lower figs., as *Collybia stipitaria* Fr. *Stipitarius*, possessing a stem.

P. 8–10 mm., *ochraceous, disc becoming fuscous*, membranaceous, convexo-plane, *umbonate*, then *umbilicate, zoned, velvety squamulose*, or *fuscous fibrillose*; margin *white*, scalloped. St. 2·5–5 cm. × 2–3 mm., *dark brown, shaggy-fibrillose*, tough, *channelled*. Gills *white, or with a yellowish tinge*, separating free, ventricose, somewhat distant. Flesh *brownish*. Spores white, pip-shaped, 10–12 × 6–7 μ, multi-guttulate. Cystidia "subulate, 30–40 × 6–8 μ" Rick. On dead grass stems, thatch, and twigs. July—Nov. Common. (*v.v.*)

1765. **C. caulicinalis** (Bull.) Rea. (= *Marasmius caulicinalis* (Bull.) Quél.) καυλός, a stalk.

P. 10–15 mm., *tan, or fawn colour*, campanulate, then convex, thin, *delicately tomentose* and *pubescent*. St. 2–3 cm. × 2–3 mm., *concolorous*, brown at the attenuated base, fibrous, *pubescent*. Gills *whitish yellow, then tan*, free, ventricose, thick. Flesh *tan*. Spores white, pip-shaped, 7–8 × 5 μ, punctate, 1-guttulate. On dead grass stems. Aug.—Oct. Not uncommon. (*v.v.*)

V. Receptacle coriaceous, fleshy-coriaceous, or woody. Spores white.

Pileus fleshy-coriaceous, gills somewhat soft.

Panus Fr.

(πᾶν, all; οὖς, ear.)

Pileus fleshy-coriaceous, excentric, dimidiate, or resupinate, sessile or stipitate. Stem when present lateral, confluent with the pileus.

Gills soft, then coriaceous, decurrent, or arising from a central point. Spores white, cylindrical, or elliptical; smooth, continuous. Cystidia present, or absent. Growing on wood, often caespitose.

*P. irregular. St. excentric.

1766. **P. conchatus** (Bull.) Fr. (= *Panus flabelliformis* (Schaeff.) Quél.) Krombh. t. 42, figs. 1, 2. *Conchatus*, shell-shaped.

P. 5–10 cm., *cinnamon, then becoming pale*, fleshy-pliant, thin, unequal, *excentric*, or *dimidiate, flaccid, squamulose when old*. St. 12 × 8 mm., *pale*, unequal, often compressed, base pubescent. Gills *whitish, or pale flesh colour*, at length *ochraceous wood-colour*, deeply decurrent in parallel lines, here and there *branched*, crisped when dry. Flesh *white*. Spores white, cylindrical, 6 × 3μ, 1–2-guttulate. On beech, and poplar stumps and willows. June—Oct. Rare. (*v.v.*)

1767. **P. torulosus** (Pers.) Fr. (= *Panus flabelliformis* (Schaeff.) Quél.) Cke. Illus. no. 1096, t. 1149, fig. B. *Torulus*, a tuft of hair.

P. 5–8 cm., *somewhat flesh colour*, varying *rufescent-livid*, and *becoming violet, entire*, but very excentric, fleshy-pliant, then coriaceous, *plano-infundibuliform*. St. 2–5 × 2–2·5 cm., *pale, covered with grey often violaceous down*, oblique, tough, firm. Gills *reddish, then tan colour*, decurrent, subdistant, simple, separate behind. Flesh *pallid*. Spores white, cylindrical, 6 × 3μ, 1–2-guttulate. On old stumps of birch and pollard willows. May—Oct. Common. (*v.v.*)

1768. **P. rudis** Fr. (= *Panus hirtus* (Secr.) Quél.) Quél. Jur. et Vosg. I, t. 14, fig. 1. *Rudis*, rough.

P. 5–10 cm., *ochraceous fawn, or reddish*, fleshy, coriaceous, then corky, thin, unequal, excentric, or dimidiate; margin *incurved, lilac, and bristling with hairs*. St. 1–2 × 2–3 cm., *ochraceous fawn*, unequal, *shaggy with a rough, hairy, lilac velvet*. Gills *whitish pink, then pale ochraceous fawn*, very decurrent, narrow. Flesh *white*. Spores white, cylindrical, 5–6 × 3μ, 1–3-guttulate. Cystidia "on edge of gill cylindrical-clavate, 45–50 × 12μ, very thick walled" Rick. Beech stumps. May. Uncommon. (*v.v.*)

**St. definitely lateral.

1769. **P. stipticus** (Bull.) Fr. (= *Panus farinaceus* Schum. sec. Quél.) *στυπτικός*, astringent.

P. 1–4 cm., *cinnamon, becoming pale*, thin, elastic, *reniform*, sometimes infundibuliform and lobed, pruinose, *the cuticle breaking up into furfuraceous scales*. St. 5–20 × 2–3 mm., *pale*, coriaceous, *dilated at the apex*, ascending, pruinose. Gills *ochraceous, or cinnamon*, ending determinately, thin, *very narrow*, crowded, connected by veins. Flesh *concolorous*. Spores white, elliptical, 4–5 × 2–2·5μ. Cystidia "on edge of gill lanceolate, clavate at first, 30 × 15μ, then 40–70 × 7–8μ" Rick. Taste very astringent. Poisonous. On dead stumps, and fallen branches. Jan.—Dec. Common. (*v.v.*)

var. **farinaceus** (Schum.) Rea. *Farinaceus*, mealy.

Differs from the type in the *cuticle of the pileus breaking up into whitish-bluish-grey scurf*. Trunks, and fir branches. Oct.—Dec. Uncommon. (*v.v.*)

var. **albido-tomentosus** (Cke. & Massee) Rea. Cke. Illus. no. 1097, t. 1144, fig. B, as *Panus farinaceus* Fr.
Albidus, whitish; *tomentosus*, hairy.

Differs from the type in the *pileus being densely clothed with a short, whitish, velvety tomentum*. Trunks. Uncommon.

***P. resupinate, sessile, or extended behind.

1770. **P. patellaris** Fr. Fr. Icon. t. 176, fig. 3.
Patellaris, like a little dish.

P. 10–15 mm., *pallid externally, viscid, furfuraceo-villose, resupinate*, coriaceous, orbicular, *plano-cup-shaped*, adnate by the sessile vertex; the free margin involute, villose, white. Gills *dark ochraceous*, concurrent in a central point, crowded. Flesh *ochraceous*. Spores white, "elliptical, 8μ" Quél. On beech, and cherry branches. Oct.—March. Uncommon.

1771. **P. Stevensonii** B. & Br.
Rev. John Stevenson, the eminent Scotch mycologist.

P. *olivaceous-light-yellow*, spathulate. St. *golden*, dilated upwards, convex, slightly hispid. Gills narrow, entire. Flesh *greenish-yellow*. On oak. Sept.—Oct. Rare.

Pileus membranaceous-coriaceous, gills coriaceous, branched, obtuse.

Xerotus Fr.

(ξηρός, dry; οὖς, an ear.)

Pileus membranaceous-coriaceous, regular. Stem central, confluent with the pileus. Gills coriaceous, broadly plicaeform, dichotomous, edge entire, obtuse. Spores white, elliptical, irregular. Growing on the ground.

1772. **X. degener** Fr. (= *Cantharellus carbonarius* (A. & S.) Fr. sec. Quél.) *Degener*, degenerate.

P. 1·5–4 cm., *date-brown-grey* when moist, *grey* when dry, *somewhat zoned*, coriaceo-membranaceous, very thin, but very tough, plano-infundibuliform, *striate when moist*, flocculose when dry. St. 4–20 × 2 mm., *fuscous*, somewhat white-velvety, very tough, equal. Gills *whitish-grey*, decurrent, few, very distant, when properly developed thin, rather broad, edge acute, flaccid. Spores white, "elliptical, irregular, 8–12 × 4–6μ" Berk. On bare gravelly soil, and in peat mosses. Jan. Very rare.

Pileus coriaceous, or woody, pliant; gills firm, often toothed.

Lentinus Fr.

(*Lentus*, pliant, or tough.)

Pileus coriaceous, pliant, more or less irregular, stipitate, or sessile. Stem when present, central, excentric, or lateral, confluent with the pileus. Gills tough, adnate, or decurrent, often toothed at the edge. Spores white, elliptical, pip-shaped, oblong cylindrical, or globose; smooth, or echinulate, continuous. Cystidia present, or absent. Growing on wood, rarely on the ground; solitary, or caespitose.

I. P. nearly entire. St. distinct.

*P. scaly, more or less manifestly veiled.

1773. **L. tigrinus** (Bull.) Fr. (= *Lentinus Dunalii* DC. sec. Quél.) Cke. Illus. no. 1089, t. 1138. *Tigrinus*, spotted like a tiger.

P. 3–8 cm., *white, or cream colour, variegated with somewhat adpressed, brownish, or blackish, fibrillose squamules*, fleshy-coriaceous, *thin*, commonly orbicular and central, convex, then infundibuliform; margin often split when dry. St. 3–5 × ·5–1·5 cm., *whitish, becoming fuscous at the base*, very hard, often attenuated downwards and rooting, *minutely squamulose, furnished at the apex with an entire, reflexed, fugacious ring*. Gills *white, then yellowish, decurrent*, narrow, crowded, *serrate*. Flesh *white, fuscous at base of stem*. Spores white, pip-shaped, 7–9 × 3μ, 1–3-guttulate. Smell strong, acid. On oak, ash, willow, and poplar stumps, and on railway sleepers. April—Oct. Not uncommon. (*v.v.*)

var. **Dunalii** (DC.) Fr. Berk. Outl. t. 15, fig. 2. Dunal.

Differs from the type in the *evanescent, adpressed spot-like scales of the pileus, the subsilky stem*, and the *scarcely manifest veil*. On willows, and poplars. Rare.

1774. **L. lepideus** Fr. (= *Lentinus squamosus* (Schaeff.) Quél.) λεπίς, a scale.

P. 5–10 cm., *pallid ochraceous, variegated with darker, adpressed, spot-like scales sometimes becoming rufescent, fleshy, very compact*, firm, *irregular*, commonly excentric, convex, then plane, or depressed, sometimes broken up into cracks. St. 2–8 × 1–3 cm., *whitish*, covered with *tomentose scales that become rufescent*, apex smooth, base woody, sometimes rooting, *at the first furnished with a cortina towards the apex*.

Gills *whitish, or yellowish, sinuate,* decurrent by a tooth, *broad, transversely striate,* serrate. Flesh *white,* pliant. Spores white, elliptical, 10–11 × 5μ, 1–3-guttulate. Cystidia none. Smell pleasant. On pine stumps, railway sleepers and paving blocks. March—Oct. Not uncommon. (*v.v.*)

var. **contiguus** Fr. *Contiguus,* neighbouring.

Differs from the type in being *entirely white* and *destitute of scales both on the pileus and stem.*

**P. villose, or pulverulent.

1775. **L. leontopodius** Schulz. Kalchbr. Icon. t. 28.
λέων, a lion; πούς, foot.

P. 7–20 cm., *tan-clay-colour,* rather darker towards the margin, fleshy, compact, tough, irregular, very excentric, or almost lateral, *broadly* umbilicate; *margin arched and bent downwards, more or less lobed, distinctly tomentose.* St. 7–10 × 2·5–3 cm., *pale chestnut, blackish downwards,* curved and ascending, hard, tough, subequal, *pulverulently furfuraceous, or tomentose.* Gills *pale reddish ochraceous,* deeply decurrent, especially on the lower side of the stem, 6–8 mm. wide, connected by veins, wrinkled at the sides; edge serrated, *darker.* Flesh *white.* Spores white, "cylindrical, obliquely apiculate, 12–15μ long" Massee. Smell very pleasant. Taste sweet, pleasant. On decayed willow. Sept.—Oct. Rare.

1776. **L. pulverulentus** (Scop.) Fr. *Pulverulentus,* dusty.

P. *yellow, mealy with white dust,* fleshy-pliant, convex. St. *mealy with white dust,* elongated, stout, equal, rigid. Gills *white,* slightly toothed. Trunks. Caespitose. Oct. Rare[1].

1777. **L. adhaerens** (A. & S.) Fr. (= *Lentinus resinaceus* Trog sec. Quél.) Bres. Fung. Trid. t. 131. *Adhaerens,* sticking to.

P. 8–13 cm., *lurid whitish, then hazel, becoming fuscous,* fleshy pliant, somewhat irregular, convexo-subcampanulate, then depressed and infundibuliform, *pulverulently villose, covered with a resinous, amber coloured gluten.* St. 2–5 cm. × 5–12 mm., *concolorous, covered with a resinous gluten, rooting,* subequal, *pulverulently tomentose.* Gills *white, then yellowish,* sinuato-decurrent, somewhat crowded; edge serrate, *glutinous.* Flesh *white.* Spores white, "oblong cylindrical, 7–10 × 2·5–3μ" Bres. Smell pleasant, fragrant. Taste somewhat bitter and astringent. Coniferous woods. Autumn—Spring. Rare.

[1] Berkeley and Broome describe P. 5 cm., *fuliginous,* at first infundibuliform then lateral, flabelliform, floccoso-pulverulent with little umber particles. St. 7–5 cm. at length smooth. Gills pallid, deeply decurrent, thick, edge crenulate but not torn.

1778. **L. suffrutescens** (Brot.) Fr. Bull. Soc. Myc. Fr. XVI (1900), t. 3 and 4. *Suffrutescens*, arborescent.

P. 4–8 cm., *whitish cream colour*, fleshy, compact, conical, *convex, umbonate, disc covered with reddish brown squamules*, then *depressed and infundibuliform*; margin often revolute. St. 7–11 cm. × 7–9 mm., *concolorous, covered with reddish brown scales* (which sometimes project) *in the lower half or up to the sulcate apex*, equal, or more or less bulbose at the base, slightly incurved, strongly flexuose, or twisted in abnormal specimens. Gills *white, then yellowish*, deeply decurrent, crowded, serrate. Flesh *white*. Spores white, elliptic cylindrical, 7·5 × 2·5μ. Squared wood in cellars. Nov.—Feb. Rare. (*v.v.*)

***P. smooth.

1779. **L. umbellatus** Fr. *Umbellatus*, like a sunshade.

Very much branched, fleshy coriaceous, tough. Pileoli very numerous, 1·5–2·5 cm., *becoming yellowish cinereous*, entire, *umbilicate*. St. 5–9 cm. × 3–8 mm., *white, caespitosely connate at the base, branched, each branch giving rise to a separate p., sulcate*. Gills *white*, deeply decurrent, *very narrow*, 1 mm., minutely serrated, crowded. Flesh *white*. Spores white, globose, 4–5μ, with a large central gutta. Smell and taste pleasant. Old stumps. Aug.—Oct. Uncommon. (*v.v.*)

1780. **L. cochleatus** (Pers.) Fr. Cke. Illus. no. 1093, t. 1142, fig. A. κοχλίας, a snail with a spiral shell.

P. 2·5–9 cm., *flesh colour, becoming pale, somewhat tan, fleshy pliant*, thin, commonly excentric, imbricated, very unequal, somewhat lobed or contorted, sometimes plane, sometimes infundibuliform-umbilicate. St. 3–9 × ·5–1·5 cm., *flesh colour, rufous fuscous downwards*, firm, sometimes central, sometimes wholly lateral, *sulcate*, often connate at the base. Gills *white flesh colour*, decurrent, crowded, serrated. Flesh *pinkish*. Spores white, globose, 5–6μ, with a large central gutta. Cystidia none. Smell very pleasant, of anise, or tonquin bean. Taste mild. Edible. Stumps. July—Nov. Common. (*v.v.*)

II. Dimidiate, sessile, or furnished with a sublateral stem.

1781. **L. scoticus** B. & Br. Cke. Illus. no. 1094, t. 1143. *Scoticus*, Scottish.

P. 1–4 cm., *pallid, then brownish*, hygrophanous, *umbilicate*, sometimes infundibuliform, at length flattened; extremely variable in form, either quite stemless and *reniform*, or variously stipitate, lobed at the margin and sinuate, or plicate. St. ·5–5 cm. × 2–3 mm., *darker, cylindrical, pulverulent, springing from a brown, fibrillose mycelium*. Gills *pallid*, decurrent when the stem is developed, rather distant, strongly toothed, and irregularly torn. Spores white, elliptical, 5–6 × 4μ. On decayed *Ulex*, birch, and spruce. Nov.—Jan. Rare. (*v.v.*)

1782. **L. fimbriatus** Currey. Cke. Illus. no. 1095, t. 1148, fig. A. *Fimbriatus*, fringed.

P. 1–2·5 cm., *fawn colour, covered with darker floccose scales*, somewhat dimidiate, subcoriaceous, thin, depressed, sometimes very much so and almost cyathiform; margin slightly involute, *almost strigose*. St. 4–6 × 2–3 mm., *concolorous with the gills, or paler*, lateral, *rough with somewhat reflexed scales*; apex with a delicate, white, fimbriate collar or fringe when young. Gills *pale brown*, descending, irregularly serrated and torn at the margin. On a stump standing in a pond. Sept. Rare.

1783. **L. vulpinus** (Sow.) Fr. (= *Lentinus castoreus* Fr. sec. Quél.) Fr. Icon. t. 176, fig. 1. *Vulpinus*, foxy.

Sessile, *many times imbricated*. Pilei *tan*, fleshy, very pliant, *reniform-conchate*, very convex; margin deflexed and almost perpendicular, hence concave beneath; *surface* wholly peculiar, *with raised longitudinal ribs, which are broken up into scales or fibrous teeth towards the thin, incurved, fuscescent margin*, hence *entirely rough and corrugated, velvety tomentose*, and white-warty behind. Gills *whitish, then reddish*, extended to the base, *broad*, crowded, torn into teeth. Flesh *white*. Spores white, "somewhat needle-shaped, 9–10 × 1·5μ" Rick. Smell none or strong of field mint. Trunks, and stumps. March—Oct. Rare.

1784. **L. auricula** Fr. Fr. Icon. t. 175, fig. 2. *Auricula*, the ear.

White, becoming tinged with yellow when old, caespitosely imbricate. P. 2–4 cm., fleshy-coriaceous, *ear-shaped, dimidiate, oblong, ascending, sessile*, cucullately-revolute beneath. Gills *very narrow*, 1 mm. wide, very closely crenulately-serrated, decurrent to the base, and separate. Flesh *white*. Spores white, globose, 3–4μ, with a large central gutta. Lime stumps. Aug.—Oct. Uncommon. (*v.v.*)

1785. **L. flabelliformis** (Bolt.) Fr. Cke. Illus. no. 1095, t. 1148, fig. B. *Flabelliformis*, fan-shaped.

Subsessile. P. 2–3 cm., *pallid fawn colour, membranaceous*, pliant, *reniform*, plane, even; margin *fimbriato-toothed*. St. commonly *rudimentary*, 4–6 mm. long. Gills *whitish, or pallid*, broad, somewhat distant, rather thick, torn into teeth at the edge. Spores white, "cylindrical, 8-9 × 2–2·5μ" Rick, "minutely echinulate" Quél. On stumps. Feb. Rare.

CANTHARELLINEAE.

Hymenium inseparable from the pileus, spread over the surface of narrow, obtuse veins, gills, or folds, or quite smooth.

CANTHARELLACEAE.

Same characters as the suborder.

*Spores white.

Receptacle fleshy, stipitate; gills simple. Parasitic on other Agarics.

Nyctalis Fr.

(*νύξ*, night.)

Pileus fleshy, regular. Stem central, fleshy, confluent with the pileus. Gills adnate, or decurrent, thick, soft, edge obtuse. Spores white, elliptical, smooth. Chlamydospores often present. Growing on other Agarics, more rarely on the ground.

1786. **N. parasitica** (Bull.) Fr. Cke. Illus. no. 1068, t. 1113.
Parasitica, parasitic.

P. 1·5–3 cm., *whitish fuscous*, then *becoming pale whitish*, somewhat fleshy, conico-campanulate, then convex, plane, and obtuse, or obsoletely umbonate, unequal, *pellicle persistent, pruinose, grey*. St. 2·5–6 cm. × 2–4 mm., *white*, straight, or curved, equal, or slightly attenuated upwards, pubescent, bristling with hairs at the base. Gills *white, becoming fuscous*, adnate, thick, distant, with alternate shorter ones intermixed, at length contorted and anastomosing. Flesh *dark grey*. Spores white, elliptical, 5–7 × 3–4μ. Smell like *Polyporus squamosus*. On dead *Russula adusta, Russula foetens, Russula chloroides and Lactarius vellereus*. Aug.—Dec. Common. (*v.v.*)

1787. **N. asterophora** Fr. ἀστήρ, a star; φέρω, I bear.

P. 1–1·5 cm., *white*, then *fawn colour from the large stellate chlamydospores*, 15–20μ, fleshy, conical, then hemispherical, *floccoso-pulverulent*. St. 1–2 cm. × 2–4 mm., *white, then fuscous*, equal, often twisted, pruinose. Gills *white, then dingy*, adnate, distant, thick, tense and straight, somewhat forked, *often wanting*. Flesh *dark grey*. Spores white, "elliptical, 6 × 4μ" Rick. On dead *Russula nigricans*. July—Nov. Common. (*v.v.*)

N. caliginosa W. G. Sm. =? a diseased state of some Clitocybe.

Trogia Fr. = **Plicatura** Peck.

Receptacle fleshy, stipitate; gills forked.

Cantharellus Adans.

(*κάνθαρος*, a drinking cup.)

Pileus fleshy, regular, excentric, or lobed. Stem central, confluent with the pileus. Gills decurrent, thick, branched. Spores white, rarely pale ochraceous in the mass, elliptical, oval, or pip-shaped, smooth; basidia with 4–8-sterigmata. Cystidia present, or absent. Growing on the ground.

*P. and solid st. fleshy.

1788. **C. cibarius** Fr. Cke. Illus. no. 1055, t. 1103.
Cibarius, pertaining to food.

Entirely egg-yellow. P. 5–10 cm., convex, *turbinate*, then plane or somewhat depressed, repand; margin often lobed. St. 4–7 × 2–4 cm., attenuated downwards. Gills decurrent, *fold-like, thick, distant*, branched, often anastomosing. Flesh *yellowish, drying whitish*, firm. Spores white, elliptical, 10 × 8μ, multi-guttulate; basidia with 5–6-sterigmata. Smell pleasant, like that of apricots. Taste mild. Edible. Woods. July—Dec. Common. (*v.v.*)

var. **albus** Fr. *Albus*, white.

Differs from the type in being *entirely white, or here and there tinged with pink*. Woods. Aug.—Oct. Uncommon. (*v.v.*)

var. **rufipes** Gillet. Cke. Illus. no. 1056, t. 1131.
Rufus, red; *pes*, foot.

Differs from the type in the *rufous base of the stem*. Woods. Sept.—Oct. Not uncommon. (*v.v.*)

var. **ramosus** Schulz. Kalchbr. Icon. t. 27, fig. 4. *Ramosus*, branched.

Differs from the type in the *branched stem, and in the pileoli becoming finally infundibuliform*. Woods. Sept. Rare. (*v.v.*)

1789. **C. amethysteus** Quél. Trans. Brit. Myc. Soc. III, t. 12.
ἀμέθυστος, amethyst.

P. 5–10 cm., *egg-yellow, covered with a lilac down* either in zones, or more especially at the margin, fleshy, firm, *turbinate*, then plane and somewhat depressed; margin often scalloped. St. 3–4 × 2·5–3 cm., *egg-yellow*, obconic, attenuated downwards. Gills *egg-yellow, vein-like*, branched, thick. Flesh *white, then yellowish*. Spores white, oval, 10 × 5–6μ, filled with granular protoplasm. Smell and taste pleasant. Beech woods. Sept.—Oct. Not uncommon. (*v.v.*)

1790. **C. Friesii** Quél. Quél. Jur. et Vosg. I, t. 23, fig. 2.
Elias Fries, the eminent Swedish mycologist.

P. 2–4 cm., *orange, soon becoming ochraceous*, convex, then depressed, thin, *villose*; margin scalloped. St. 2–4 cm. × 2–4 mm., *yellow*, pruinose, base white villose. Gills *yellow, flesh colour, or orange*, decurrent, *fold-like*, narrow, branched. Flesh *white, yellowish under the cuticle*. Spores pale ochre in the mass, hyaline under the microscope, elliptical, 6–7 × 3–4μ, 1-guttulate. Taste somewhat sour. Edible. Beech woods. Aug.—Nov. Uncommon. (*v.v.*)

C. aurantiacus (Wulf.) Fr. = **Clitocybe aurantiaca** (Wulf.) Studer.

C. hypnorum Brond. = **Clitocybe hypnorum** (Brond.) Rea.

1791. **C. Brownii** B. & Br. (= *Naucoria pediades* Fr. sec. Pat., a monstrous form described as a *Ptychella.*) Cke. Illus. no. 1058, t. 1106, fig. A. J. Brown.

Entirely ochraceous white, or cream coloured. P. 10–15 mm., thin, convex, subumbonate, obscurely silky. St. 4–5 cm. × 1–2 mm., tough, nearly equal, *somewhat furfuraceous*, furnished with a little white, fibrillose mycelium at the base, which sometimes forms a small earthy ball. Gills *fold-like*, obtusely decurrent, rather distant, *linear, very narrow*, sometimes forked. Hymenium nearly white. Spores white, broadly elliptical, 7 × 5–6μ. Amongst grass. Autumn. Rare.

1792. **C. carbonarius** (A. & S.) Fr. (= *Xerotus degener* Fr. sec. Quél.; *Cantharellus radicosus* (B. & Br.) Fr.) Cke. Illus. no. 1059, t. 1105. *Carbonarius*, pertaining to charcoal

P. 1–6 cm., *date brown, then black, umbilicate*, or *infundibuliform, coriaceous, minutely squamulose*; margin lobed. St. 3–6 cm. × 3–10 mm., *paler than the pileus, rooting, striate*, sometimes branched. Gills *white, then glaucous, or grey*, decurrent, straight, narrow. Flesh *whitish.* Spores white, elliptical, 9–10 × 5–6μ, 2–3-guttulate. Cystidia fusiform, apex acute, 3–4μ in diam., 95–120 × 13–14μ, very thick walled, upper portion incrusted. Charcoal heaps, and burnt ground. July—Dec. Common. (*v.v.*)

1793. **C. umbonatus** (Gmel.) Fr. *Umbonatus*, umbonate.

P. 1·5–4 cm., *cinereous blackish*, convex, *umbonate, at length depressed, flocculosely-silky*; margin incurved, *white.* St. 5–8 cm. × 7–10 mm., *concolorous*, base white floccose, elastic, equal. Gills *shining white*, decurrent, thin, straight, *crowded, repeatedly dichotomous.* Flesh *white, often becoming red when wounded.* Spores white, pip-shaped, 8–9 × 3–4μ, 1-guttulate. Cystidia none. Woods, and heaths. April—Oct. Uncommon. (*v.v.*)

1794. **C. albidus** Fr. *Albidus*, whitish.

P. 1–2·5 cm., *whitish, inclining to yellowish, or rufescent*, convex, *umbilicate*, thin, lobed, slightly villose. St. 2–4 cm. × 4–8 mm., *white, rarely yellowish*, tough, flexuose. Gills *white, then yellowish*, decurrent, crowded, repeatedly dichotomous. Flesh *white, yellowish under the cuticle.* Spores white, elliptical, 6–7 × 4–5μ, 1-guttulate. Cystidia none. Taste mild. Edible. Woods, and pastures. Sept.—Oct. Uncommon. (*v.v.*)

**P. submembranaceous; st. tubular, polished.

1795. **C. tubaeformis** Fr. Cke. Illus. no. 1061, t. 1108.
Tuba, a trumpet; *forma*, shape.

P. 2–6 cm., *fuscous when moist, becoming pale when dry*, fleshy-membranaceous, *infundibuliform, deeply umbilicate*, repand and lobed, *flocculose.* St. 3–7 cm. × 3–8 mm., *orange-tawny*, at length compressed and *lacunose.* Gills *yellow,* then pruinose and *greyish,* fold-like, decurrent, thick, distant, branched. Spores white, elliptical, 8–10 × 6μ. Woods. Aug.—Nov. Common. (*v.v.*)

var. **lutescens** (Bull.) Fr. *Lutescens*, becoming yellowish.

Differs from the type in the *convexo-umbilicate, somewhat regular and rather even pileus, in the more equal stem, attenuated upwards and in the dichotomous gills being less divided.* Woods. Sept.—Oct. Uncommon. (*v.v.*)

1796. **C. infundibuliformis** (Scop.) Fr. Cke. Illus. no. 1062, t. 1109.
Infundibulum, a funnel; *forma*, shape.

P. 2–5 cm., *yellowish-cinereous,* or *fuliginous when moist, becoming pale when dry,* submembranaceous, umbilicate, then infundibuliform, here and there pervious to the base, *rugose, fibrillose.* St. 4–8 cm. × 4–6 mm., *light yellow,* base somewhat thickened. Gills *light yellowish,* then pruinose and *grey,* fold-like, decurrent, thick, distant, branched, anastomosing. Spores white, broadly elliptical, or subglobose, 7–9 × 7μ. Woods. July—Jan. Common. (*v.v.*)

var. **subramosus** Bres. Bres. Fung. Trid. t. 97.
Sub, somewhat; *ramosus*, branched.

Differs from the type in the *somewhat branched stems.* Woods. Sept. —Oct. Uncommon. (*v.v.*)

1797. **C. lutescens** (Pers.) Fr. (= *Craterellus lutescens* (Pers.) Fr. Hym. Eur.) Pers. Myc. Eur. II, t. 13, fig. 1, as *Merulius xanthopus.*
Lutescens, becoming yellowish.

P. 2–10 cm., *fuscous,* submembranaceous, *tubaeform,* soon pervious, undulated, flocculose. St. 5–10 × 1 cm., *golden yellow,* attenuated at the base, undulate. Gills *yellow, inclining to reddish, or orange,* fold-like, thin, flexuose, anastomosing. Flesh *yellowish, deeper at the periphery.* Spores white, elliptical, 10–12 × 6–7·5μ. Smell *strong, spirituous.* Mountainous fir woods. Sept.—Oct. Uncommon. (*v.v.*)

1798. **C. cinereus** (Pers.) Fr. Cke. Illus. no. 1063, t. 1110, fig. A.
Cinereus, colour of ashes.

P. 2–5 cm., *hoary fuliginous, becoming whitish,* submembranaceous, infundibuliform, *often pervious to the base of the stem, villoso-squamulose.* St. 3–8 cm. × 4–8 mm., *concolorous, then blackish,* attenuated downwards, curved, fibrillosely-striate. Gills *cinereous, then whitish pruinose,* fold-like, decurrent, thick, *distant,* connected by veins, slightly branched. Flesh *blackish* Spores white, elliptical, 7 × 5μ. Smell pleasant, "like the Mirabelle plum" Quél. Woods. Sept.—Nov Uncommon. (*v.v.*)

1799. **C. leucophaeus** Nouel. Cke. Illus. no. 1064, t. 1111, fig. A.
λευκός, white; φαιός, dusky.

P. 2–3 cm., *dusky brown*, submembranaceous, tough, plane, then depressed or infundibuliform; margin incurved, then reflexed. St. 3–4 cm. × 2–4 mm., *concolorous*, or *paler*, slightly thickened at the base. Gills *white*, decurrent, distant, simple, or forked, with intermediate shorter ones. Spores white, elliptical, 9 × 5μ. Woods, and heaths. Sept.—Oct. Uncommon. (*v.v.*)

1800. **C. Houghtonii** Phill. Cke. Illus. no. 1060, t. 1107, fig. B.
Rev. William Houghton.

P. 2–3 cm., *dirty white with a tinge of flesh colour*, thin, convex, umbilicate. St. 5 cm. × 2 mm., *whitish*, thickened at the apex, at first delicately fibrillose, base rooting, *cottony*. Gills *pallid flesh colour*, decurrent, narrow, scarcely forked. Spores white, elliptical, 7 × 4μ. On the ground. Oct.

1801. **C. cupulatus** Fr. (= *Cantharellus helvelloides* (Bull.) Quél.) Boud. Icon. t. 71, as *Cantharellus helvelloides* (Bull.) Quél.
Cupulatus, pertaining to a little tub or cask.

P. 10–15 mm., *pallid fuscous when moist, becoming pale, somewhat rufescent when dry*, membranaceous, *plano-infundibuliform* (exactly *cup-shaped*), repand, lobed, flocculose when dry; *margin striate*. St. 2–3 cm. × 3–5 mm., *paler than the pileus, expanding into the pileus*, tough, pruinose. Gills *grey*, decurrent, very distant, narrow, branched, *with intermediate simple ones*. Flesh *greyish, becoming white*. Spores white, pip-shaped, or elliptical with a lateral basal apiculus, 8–10 × 5–6μ. Open heathy ground and old walls. Oct. Uncommon. (*v.v.*)

1802 **C. Stevensonii** B. & Br. Cke. Illus. no. 1064, t. 1111, fig. B.
Rev. John Stevenson, the eminent Scotch mycologist.

P. 4 mm., *pallid*, orbicular, *umbilicate*; margin inflexed. St. 6 × 1 mm., *white, then darker*, cylindrical, *delicately pulverulent*. Gills *pallid, becoming fuscous in front*, decurrent. On rotten wood amongst moss. March—April. Rare.

1803. **C. replexus** Fr. *Replexus*, bent back.

P. 1·5–2·5 cm., *fuscous, then cinereous*, membranaceous, *campanulate, convex*, expanded and inverted, striate. St. 3–5 cm. × 2 mm., *grey*, thickened above. Gills *white, then glaucous*, adnato-decurrent, connected by veins, distant, branched, and dimidiate. On the ground amongst grass, moss, and leaves. Oct.—Nov.

var. **devexus** Fr. Cke. Illus. no. 1098, t. 1150, fig. A.
Devexus, inclining downwards.

Differs from the type in the *cucullate pileus, and the simple, cinereous* gills. Spores white, elliptical, 9–10 × 6μ, 1-guttulate. Burnt ground, and amongst moss on heaths. Nov. Rare. (*v.v.*)

C. muscigenus (Bull.) Fr. = **Dictyolus muscigenus** (Bull.) Quél.
C. glaucus (Batsch) Fr. = **Dictyolus glaucus** (Batsch) Quél.
C. retirugus (Bull.) Fr. = **Dictyolus retirugus** (Bull.) Quél.
C. lobatus (Pers.) Fr. = **Dictyolus lobatus** (Pers.) Quél.

Receptacle fleshy, membranaceous, funnel-shaped, or umbilicate. Hymenium veined, or smooth.

Craterellus Fr.

(κρατήρ, a large bowl.)

Pileus fleshy, or membranaceous, funnel-shaped, or umbilicate. Stem central, confluent with the pileus. Hymenium smooth, becoming wrinkled. Spores white, rarely pale ochraceous in the mass, elliptical, ovoid, or oblong elliptical, smooth, or punctate; basidia with 2–4-sterigmata. Cystidia none. Growing on the ground.

*Tubaeform, pervious to the base of the stem.

C. lutescens (Pers.) Fr. = **Cantharellus lutescens** (Pers.) Fr.

1804. **C. cornucopioides** (Linn.) Fr. Berk. Outl. t. 19, fig. 6.
Cornu copiae, horn of plenty; εἶδος, like.

P. 3–5 cm., *blackish fuliginous when moist, brownish when dry*, submembranaceous, *tubaeform*, pervious, *squamulose*. St. 5–10 × 1 cm., *black*, dilated upwards into the pileus. Hymenium *cinereous*, even, at length wrinkled. Spores white, oblong elliptical, 11–15 × 6–8μ. Taste mild. Edible. Woods. Aug.—Nov. Common. (*v.v.*)

**Infundibuliform, st. stuffed.

1805. **C. sinuosus** Fr. Fr. Icon. t. 196, fig. 2. *Sinuosus*, undulated.

P. 2–3 cm., *fuscous grey*, slightly fleshy, infundibuliform, undulated, floccose. St. 2·5–4 cm. × 3–8 mm., *grey, apex ochraceous*. Hymenium *grey, then ochraceous*, at length with interwoven wrinkles. Spores *pale ochraceous* in the mass, elliptical, 7–9 × 5–6μ, punctate. Smell none, or of musk. Woods. Aug.—Nov. Uncommon. (*v.v.*)

1806. **C. crispus** (Sow.) Fr. *Crispus*, curled.

P. 2–4 cm., *fuliginous becoming fuscous*, fleshy-membranaceous, infundibuliform, somewhat pervious, *lobed*, pruinose. St. 2–3 cm. × 4–8 mm., *yellow, becoming greyish*, pruinose. Hymenium *pallid, even*. Spores white, elliptical, 9–10 × 6–7μ. Woods. Aug.—Nov. Uncommon. (*v.v.*)

1807. **C. pusillus** Fr. Trans. Brit. Myc. Soc. III, t. 2.
Pusillus, very little.

P. 10–12 mm., *cinereous*, convex, umbilicate, thin, rugose, *villose*. St. 5–20 × 2–4 mm., *grey*, somewhat compressed. Hymenium *bluish-*

grey, smooth, or slightly wrinkled, pruinose. Spores white ovoid, or elliptical, 8–10 × 6–7μ, finely punctate. Beech woods, and under beeches. Aug.—Oct. Not uncommon. (*v.v.*)

Receptacle membranaceous, spathulate, or cup-shaped pendant. Hymenium veined, or smooth.

Dictyolus Quél. (= **Cantharellus** p.p.).

(δίκτυον, network.)

Pileus membranaceous, spathulate, or cup-shaped and pendant. Hymenium consisting of vein-like gills, anastomosing in a reticulate manner, or almost smooth. Spores white, elliptical, or pip-shaped, smooth. Cystidia none. Growing on wood, or on mosses.

*P. attached by the apex, resupinate, then reflexed.

1808. **D. retirugus** (Bull.) Quél. (= *Cantharellus retirugus* (Bull.) Fr.) Cke. Illus. no. 1066, t. 1112, fig. A, as *Cantharellus retirugus* Fr. *Rete*, a net; *ruga*, a wrinkle.

P. 1–2 cm., *cinereous, becoming whitish*, pellucid, membranaceous, cup-shaped, then expanded, repando-lobed, very tender. Gills *grey, or whitish*, fold-like, radiating from the centre, *connected by veins and reticulated*, very tender. Spores white, elliptical, 7–8 × 5–6μ. On twigs, and mosses. April—June. Uncommon. (*v.v.*)

1809. **D. lobatus** (Pers.) Quél. (= *Cantharellus lobatus* (Pers.) Fr.) *Lobatus*, lobed.

P. 1·5–3 cm., *dark fuscous when moist, becoming pale when dry*, pellucid, membranaceous, somewhat round, or reniform, then somewhat lateral and lobed, *white cottony at the base*. Gills *concolorous*, fold-like, subdistant, *distinct, branched*. Spores white, broadly elliptical, 8–10 × 6–7μ. On mosses, and *Carices* in bogs. April—Oct. Uncommon. (*v.v.*)

D. cyphellaeformis (Berk.) Cost. & Duf. = **Pleurotus cyphellaeformis** Berk.

**P. lateral, substipitate, or sessile.

1810. **D. muscigenus** (Bull.) Quél. (= *Cantharellus muscigenus* (Bull.) Fr.) *Muscus*, moss; *genus*, born.

P. 1–2·5 cm., *fuscous when moist, cinereous whitish and zoned when dry*, membranaceous, tough, *spathulate*, slightly undulate when full grown. St. 2–4 × 2–4 mm., lateral, *villose at the base*, horizontal and continuous with the pileus. Gills *concolorous*, slightly swollen, diverging from the apex of the stem, *distant, branched*. Spores white, pip-shaped, 7–9 × 4–6μ, 1–2-guttulate. On mosses. June—Nov. Uncommon. (*v.v.*)

1811. **D. glaucus** (Batsch) Quél. (= *Cantharellus glaucus* (Batsch) Fr.) Cke. Illus. no. 1065, t. 1115, fig. B, as *Cantharellus glaucus* Batsch. *γλαυκός*, pale green.

P. 1 cm., *grey*, pellucid, membranaceous, *ligulate*, ascending, *silky*. St. 2 × 1 mm., *white*, lateral, *pruinose*. Gills *glaucous, then grey*, fold-like, tumid, distant, forked. Spores white, "elliptical, 5–6 × 4μ" Karst. On sandy slopes, and on mosses. Sept.—Nov. Uncommon.

**Spores ochraceous.

Receptacle fleshy coriaceous, stipitate. Hymenium fold-like.

Neurophyllum Pat.

(*νεῦρον*, a sinew; *φύλλον*, a leaf.)

Pileus fleshy coriaceous, irregular. Stem central, confluent with the pileus. Hymenium fold-like, thick, decurrent. Spores ochraceous, elliptical, or fusiform, smooth. Cystidia none. Growing on the ground.

1812. **N. clavatum** (Pers.) Pat. (= *Craterellus clavatus* (Pers.) Fr.) Krombh. t. 45, figs. 13–17, as *Cantharellus clavatus*. *Clavatum*, club-shaped.

P. 3–15 cm., *lilac, or rose colour, then flesh colour, and finally ochraceous*, fleshy, *turbinate, truncate, or depressed*, flexuose, *attenuated into the solid stem*. St. 1–5 × 1–3 cm., *whitish lilac, or amethyst, then pale*, occasionally branched, *obconic*. Hymenium *purplish, then concolorous*, fold-like, thick, reticulated, reticulations anastomosing, decurrent. Spores ochraceous, elliptical, or fusiform, 10–12 × 4–5μ, 1–2-guttulate. Smell and taste pleasant. Edible. Mountainous fir woods; the British record is beech woods. June—Oct. Rare. (*v.v.*)

BOLETINEAE.

Hymenium soft, separable from the pileus, and lining the inside of pores, or pore-like gills.

BOLETACEAE.

Same characters as the suborder.

1. Hymenium spread over gills, which anastomose by veins, and form irregular pores, especially at the apex of the stem. Spores white, ochraceous, or ferruginous.

Paxillus Fr.

(*Paxillus*, a small stake.)

Pileus fleshy, regular, excentric, dimidiate, or resupinate. Stem central, excentric, lateral, or none, confluent with the pileus. Gills decurrent, soft, almost mucilaginous, separable, often anastomosing. Spores white, ochraceous, reddish, or ferruginous; elliptical, pip-shaped, or globose, smooth. Cystidia present, or absent. Growing on the ground, or on wood.

I. P. entire, central. Spores dirty white, only in *P. panaeolus* with a tendency to ferruginous.

1813. **P. giganteus** (Sow.) Fr. (= *Clitocybe gigantea* (Sow.) Quél.) Cke. Illus. no. 150, t. 106, as *Clitocybe gigantea* Sow.

γίγας, a giant.

Entirely tan white. P. 3–30 cm., fleshy, convex, then plane, or depressed, then plano-infundibuliform, soft, minutely adpresso-squamulose, often guttate, the whole surface under a lens clothed with a fine matted silkiness; *margin strongly involute* and pubescent at first, then spreading, and becoming smooth, at length revolute and *sulcate with small shallow channels*, often splitting. St. 3–7·5 × 2–5 cm., equal, or attenuated upwards, base subbulbous, smooth, or minutely pubescent. Gills *whitish, then tan colour*, subdecurrent, *often branched and anastomosing*, narrow, or broad, very crowded. Flesh *white, very firm*, thin at the margin. Spores whitish, broadly elliptical, 7–8 × 5–6μ, 1–3-guttulate. Smell pleasant, taste mild. Edible. Often forming large rings. Pastures, heaths, rarely in woods. Aug.—Nov. Not uncommon. (*v.v.*)

1814. **P. Alexandri** Fr. (= *Clitocybe gilva* Fr. sec. Quél.)

P. Alexandre.

P. 5–7·5 cm., *fawn colour*, fleshy, *compact*, plane, then depressed, dry, unpolished; margin closely involute, somewhat striate when expanded. St. 12 mm. × 2·5–3 cm., ventricose. Gills *boxwood colour*, subdecurrent, crowded. Flesh *white, becoming yellow*. Spores whitish. Amongst moss in woods. Sept.—Oct. Rare.

1815. **P. lepista** Fr. Fr. Icon. t. 164, fig. 1.

λεπαστή, a drinking vessel.

P. 4–10 cm., *dingy whitish*, fleshy, convex, then plane and depressed, obtuse, rimuloso-squamulose towards the circumference, dry; margin involute, often undulato-flexuose, smooth. St. 2·5–10 cm. × 12–15 mm., *whitish, sometimes rufescent, or inclining to fuscous*, equal, or attenuated downwards, always blunt at the white villose base, *cuticle somewhat horny and continuous with the hymenophore*. Gills *dingy white, at length darker, deeply decurrent, somewhat branched*, simple at the base, 3–6 mm. broad, very crowded. Flesh *whitish*, compact, or thin, spongy-elastic in the st. Spores reddish, becoming fuscous pallid, elliptical, 7–8 × 5μ. Cystidia none. Smell mealy, often rather rancid. Woods, and pastures. Oct.—Nov. Uncommon. (*v.v.*)

1816. **P. extenuatus** Fr. Fr. Icon. t. 164, fig. 2. *Extenuatus*, thinned.

P. 3–7·5 cm., *clay, or becoming fuscous tan*, fleshy, *convex, gibbous*, then expanded, obtuse, tough, smooth, moist; margin involute, pubescent, at length expanded, rigid-fragile. St. 3–5 cm. × 6–12 mm., *pallid*, somewhat horny, elastic, fibrillose, firm, *conico-elongate*, or clavate, becoming more equal; *base incrassated, rooting*, and condensing the earth into a large ball. Gills *whitish, then mouse colour, arcuate*, very narrow, 3–4 mm. broad, linear, very crowded. Flesh *watery whitish, rigid*, compact at the disc, very thin at the margin, spongy in the st., scissile. Spores white, "elliptical, 6–7 × 3–4μ, smooth. Cystidia none" Rick. Grassy places in pine woods. Oct. Rare.

1817. **P. panaeolus** Fr. Cke. Illus. no. 863, t. 874, fig. A.
πavaίoλoς, all-variegated.

Entirely whitish, the st. becoming rufescent. P. 2–5 cm., fleshy, convexo-plane, then somewhat depressed, smooth; margin thin, involute. St. 3–5 cm. × 3–6 mm., attenuated upwards or downwards, striato-fibrillose. Gills at length *watery-ferruginous*, slightly decurrent, narrow, crowded, separated from the hymenophore by a horny line, and readily separating from the p. Flesh *becoming black*, thin. Spores very pale ferruginous, globose, 5μ, 1-guttulate. Coniferous woods. Oct.—Nov. Rare. (*v.v.*)

var. **spilomaeolus** Fr. σπίλωμα, a spot; αἰόλος, variegated.

Differs from the type in the *yellowish white p. spotted as with drops, in the slender, yellowish white st. and in the gills being horny grey at the base.* Amongst fir leaves. Oct. Rare.

1818. **P. orcelloides** Cke. & Massee. Cke. Illus. no. 863, t. 874, fig. B
Orcella, the species *Cliptopilus orcella*; εἶδος, like.

P. 2–5 cm., *snow white, becoming stained with livid or greyish blotches*, fleshy, convex, then flattened, minutely silky, shining; *margin thin, persistently incurved.* St. 2·5–6 cm. × 3–6 mm., *ochraceous, gradually tapering to the base*, elastic, silky fibrillose. Gills *whitish, then livid, at length dingy yellowish brown*, adnato-decurrent, crowded, 3 mm. broad, separated from the flesh of the p. by a horny line. Flesh *white*, thin at the margin. Spores pale dingy ochraceous, elliptical, 5–8 × 3–4μ, 1-guttulate. Woods, and amongst grass. Sept.—Nov. Rare. (*v.v.*)

1819. **P. lividus** Cke. Cke. Illus. no. 864, t. 861.
Lividus, of a leaden colour.

P. 2·5–5 cm., *dingy white, or livid ochraceous*, opaque, convex, at length slightly depressed at the disc. St. 7·5–10 × 1 cm., *white*, attenuated downwards, fibrillose. Gills *white*, decurrent, arcuate, 3 mm. broad, almost crowded. Flesh *whitish*, thick at the disc. Spores nearly white, globose, 3–3·5μ. Smell pleasant. Woods. Oct. Uncommon. (*v.v.*)

1820. **P. revolutus** Cke. Cke. Illus. no. 865, t. 862.

Revolutus, rolled back.

P. 2·5–4 cm., *pale ochraceous, disc slightly darker*, fleshy, convex, obtuse; margin thin, even, sometimes at first tinged with violet, a little revolute. St. 3–5 × 1 cm., *paler than the p., often tinted violet at the base*, attenuated downwards. Gills *pallid, then clay coloured, very decurrent*, scarcely crowded. Flesh *white*, thin at the margin. Spores pale, globose, 3·5–4μ. Smell mealy. Fields. Oct. Uncommon.

II. P. commonly excentric, or resupinate. Spores ferruginous.

1821. **P. paradoxus** (Kalchbr.) Quél. (= *Flammula paradoxa* Kalchbr.; *Flammula Tammii* Fr. sec. Bres.; *Phylloporus Pelletieri* (Lév.) Quél.) Bres. Fung. Trid. t. 207, as *Phylloporus rhodoxanthus* (Schw.) Bres. παράδοξος, strange.

P. 5–8 cm., *deep rufous umber, or yellowish brown*, fleshy, convex, or pulvinate, then plane, *adpressedly tomentose*, becoming smooth, dry; margin often lobed and sinuate. St. 3–5 cm. × 6–15 mm., *yellow, often with a dingy purplish tinge*, bulbous, or attenuated at the base, equal, or fusiform, somewhat rooted, adpressedly fibrillose, the cuticle often breaking up into squarrosely revolute flaps. Gills *yellow, then golden, becoming reddish when bruised*, decurrent, distant, the alternate ones broader, connected by veins at the base, and in some cases *anastomosing to form pores towards the margin*. Flesh *reddish, becoming yellow*, soft, juicy. Spores lurid ochraceous, oblong, 12–13 × 4–5μ, often 2-guttulate. Cystidia "almost flask-shaped-lanceolate, 60–70 × 9–15μ" Rick. Woods. Aug.—Nov. Rather uncommon. (*v.v.*)

1822. **P. involutus** (Batsch) Fr. *Involutus*, rolled in.

P. 7–20 cm., *ochrey ferruginous*, fleshy, convexo-plane, then depressed, for the most part central, pubescent, soon becoming smooth, somewhat viscid when moist, shining when dry; *margin obtuse, villous, closely involute*, then extenuated and acute. St. 5–8 × 1–4 cm., *dingy yellowish, generally spotted*, thickened upwards, more rarely bulbous at the base, naked. Gills *pallid, then ferruginous, at once dingy-spotted when touched*, decurrent, rather broad, branched behind, and *often anastomosing and forming pores near the st.* Flesh *pallid, or yellowish*, compact, soft. Spores deep ochraceous, elliptical, 8–10 × 5–7μ. Cystidia "lanceolate, 60–75 × 8–15μ" Rick. Taste mild. Edible. Woods, heaths, and pastures. June—Dec. Common. (*v.v.*)

var. **excentricus** Fr. ἔκκεντρος, out of the centre.

Differs from the type in the *excentric p. and short st.* On trunks, and stumps. Aug.—Oct. Not uncommon. (*v.v.*)

var. **subinvolutus** (Batsch) W. G. Sm.

Sub, somewhat; *involutus*, rolled in.

Differs from the type in the *less involute margin of the p.*

1823. **P. porosus** Berk. πόρος, a pore.

P. 10–11 cm., *reddish claret, or olive brown*, fleshy, viscid when moist; *margin thin, even, not involute*. St. 8–9 × 1·5–2 cm., *claret brown, darker below*, excentric, tough, equal, or attenuated downwards, somewhat reticulate above with the pores. Gills *yellow to sulphur green, changing to pale blue, and then brownish when bruised, dull green when old*, decurrent, *shallow*, *poriform*, pores round to elongate, irregular large to small. Flesh *dull pale vinous brown*, mottled and streaked, darker and changing colour in the st. Smell very strong, unpleasant. Moist woods under firs. Sept.—Oct. Rare.

1824. **P. leptopus** Fr. Fr. Icon. t. 164, fig. 3.
λεπτός, thin; πούς, foot.

P. 4–8 cm., *fuscous yellowish*, always excentric, or lateral, at length depressed, but gibbous at the disc, fleshy, dry, *covered with dense down, soon torn up into dense, villose, fuscous, or yellowish scales*. St. 1–2·5 × 1 cm., *lemon-yellow-olivaceous, short*, attenuated downwards, somewhat incurved. Gills *yellowish, then darker*, not spotted when touched, decurrent, *simple*, not anastomosing, *tense and straight*, very narrow, crowded. Flesh *yellow*, thin at the margin. Spores "pale dingy yellow, pip-shaped, 8–9 × 5μ" Massee. Woods, bogs, and on stumps. Aug.—Oct. Not uncommon. (*v.v.*)

1825. **P. atrotomentosus** (Batsch) Fr. Cke. Illus. no. 869, t. 876.
Ater, black; *tomentosus*, woolly.

P. 5–30 cm., *ferruginous*, fleshy, excentric, convex, then plano-infundibuliform, sometimes wholly lateral and ascending, dry, *rivuloso-granular*, sometimes also slightly tomentose; margin thin, involute. St. 5–8 × 1–2·5 cm., *covered over with dense, soft, umber blackish, or inclining to violaceous, velvety down*, elastic, somewhat equal, curved, ascending, *rooting*. Gills *yellowish*, adnate, scarcely decurrent, branched at the base, somewhat anastomosing, 6 mm. broad, crowded, easily separating from the sulcate hymenophore. Flesh *white*, compact, firm. Spores pale ochraceous, broadly elliptical, 4–6 × 3–4μ, 1-guttulate. Cystidia none. Taste mild. Edible. Pine woods and on pine stumps. Aug.—Nov. Not uncommon. (*v.v.*)

1826. **P. crassus** Fr. Cke. Illus. no. 870, t. 877. *Crassus*, thick.

P. 5–7·5 cm., *becoming ferruginous*, fleshy, oblique, almost plane, becoming smooth. St. 1–2 × 1 cm., *concolorous*, tapering downwards, excentric, ascending. Gills *cinnamon*, decurrent, straight, 4 mm. broad, *subdistant*. Flesh *somewhat concolorous*, thick, soft, spongy. Spores "ferruginous, elliptical, 15–18 × 7–8μ" Cke. On trunks, worked wood, ground of rifle butts, and in woods. Nov. Rare.

1827. **P. panuoides** Fr. (= *Paxillus lamellirugus* (DC.) Quél.) Cke. Illus. no. 871, t. 878. *Panus*, the genus Panus; εἶδος, like.

Entirely dingy yellow. P. 2–5 cm., fleshy, *sessile, or extended, and at the first resupinate*, soon conchate, dimidiate and obovate, at length broadly expanded, undulato-lobed, and often imbricated, pubescent, becoming smooth, somewhat rivulose. Gills *yellow*, decurrent to the base, anastomosing behind, *branched, crisped*, crowded. Flesh *cream colour, becoming whitish*, soft, equal, thin. Spores ochraceous, elliptical, 5 × 3μ. Cystidia none. Smell often very fragrant. On sawdust, rotten pine stumps, and wood. June—Nov. Not uncommon. (*v.v.*)

var. **fagi** (B. & Br.) Cke. *Fagus*, beech.

Differs from the type in being *crisped, pallid upwards, orange beneath and having orange gills*. Gregarious. Beech stumps. Sept.—Oct. Rare.

2. Hymenium lining the inside of fleshy tubes.

Spores white, or pale yellowish.

Gyroporus (Quél.) Pat. (=**Boletus** p.p.).

(γυρός, round; πόρος, pore.)

Pileus fleshy, tomentose, or smooth. Stem central, velvety, or glabrous, externally firm, fragile, internally spongy, often cavernous, base immersed in the soil. Pores white, then often yellowish, entire, round; tubes concolorous, free. Flesh white, firm, sometimes becoming blue on exposure to the air. Spores white, or pale yellowish, oval, elliptical, pip-shaped, or elliptic-oblong, smooth. Cystidia clavate. Growing on the ground.

1828. **G. cyanescens** (Bull.) (= *Boletus cyanescens* (Bull.) Fr.) Quél. Rostk. Bol. t. 44, as *Boletus cyanescens* Bull. κύανος, dark blue.

P. 5–13 cm., *subfuscous, or tan*, convexo-expanded, adpressedly tomentose, floccosely-scaly, opaque. St. 5–9 × 2–3 cm., *concolorous, ventricose*, villoso-pruinose, firm, fragile, stuffed with a spongy pith, at length cavernous. Tubes *white, at length light yellow, becoming bluish when touched*, free; orifice of pores minute, round. Flesh *white, instantly becoming deep indigo blue on exposure to the air, pouring out an azure blue juice when compressed, firm*. Spores whitish, elliptical, often with a basal apiculus, 8–9 × 4–5μ, multi-guttulate. Woods. July—Oct. Uncommon. (*v.v.*)

1829. **G. lacteus** (Lév.) Quél. (= *Boletus cyanescens* (Bull.) Fr.) *Lacteus*, milk-white.

P. 10–15 cm., *pure white*, convex, gibbous, minutely tomentose. St. 9–12 × 4–6 cm., *pure white, incrassated at the base*, firm, *velvety*, stuffed with a spongy pith and cavernous, at length hollowed out. Tubes *white*, free, short; orifice of pores *white*, minute, round, or angular. Flesh *white, becoming deep indigo blue on exposure to the air,* spongy, thick at the disc, thin at the margin of the p. Spores white, pip-shaped, 8–9 × 4–5μ, 3–5-guttulate. Woods. Aug.—Oct. Uncommon. (*v.v.*)

1830. **G. castaneus** (Bull.) Quél. (= *Boletus castaneus* (Bull.) Fr.; *Gyroporus fulvidus* (Fr.) Pat. sec. Quél.) Rolland, Champ. t. 79, no. 177. *Castanea,* chestnut.

P. 5–10 cm., *cinnamon, or chestnut coloured,* convex, then expanded, or depressed, firm, *minutely velvety.* St. 5–7 × 2–3 cm., *concolorous, attenuated upwards from the somewhat bulbous base, minutely velvety,* stuffed, then hollow. Tubes *white, becoming yellowish,* free, short; orifice of pores *white, then yellowish,* minute, round. Flesh *white, very firm.* Spores white, or pale yellow, elliptical, 8–9 × 5–6μ, often 1-guttulate. Taste pleasant, nutty. Edible. Woods, rarely in pastures under trees. Aug.—Oct. Not uncommon. (*v.v.*)

1831. **G. fulvidus** (Fr.) Pat. (= *Boletus fulvidus* Fr.; *Gyroporus castaneus* (Bull.) Quél.) Rostk. Bol. t. 45, as *Boletus fulvidus* Fr. *Fulvidus,* somewhat tawny.

P. 5–9 cm., *becoming tawny,* convex, then plane, rigid, firm, *smooth, shining.* St. 4–7 × 1·5–2·5 cm., *concolorous, equal,* firm, *smooth, shining,* stuffed, then hollow. Tubes *white, then citron yellow,* free, elongated; orifice of pores *white, then lemon yellow,* angular. Flesh *white, becoming yellowish,* firm. Spores "yellowish, elliptic-oblong, 10–11 × 5μ" Massee. Taste pleasant. Edible. Heaths, and under trees. Aug.—Sept. Uncommon.

Spores pink.

Tylopilus Karst (= **Boletus** p.p.).

(τύλος, a knot; πῖλος, cap.)

Pileus villose, or glabrescent. Stem central, reticulate, apex granular, or smooth. Tubes white, then pinkish, adnate, or sinuate, long, or short; orifice of pores concolorous, angular, or round. Flesh unchangeable, or slightly pinkish when exposed to the air. Spores pink, fusiform, or oblong, smooth. Growing on the ground.

1832. **T. felleus** (Bull.) Karst. (= *Boletus felleus* (Bull.) Fr.) Rostk. Bol. t. 43, as *Boletus felleus* Bull. *Fel,* gall.

P. 6–10 cm., *pale yellowish, chestnut, or tawny,* pulvinate, or hemispherical, then expanded, fleshy, *smooth.* St. 7–8 × 2–3 cm., *concolorous,* attenuated upwards from the thickened base, *tomentosely reticulated.* Tubes *white, then flesh colour,* adnate, convex, long; orifice of pores *white, then pinkish,* fairly broad, *angular.* Flesh *white, becoming flesh colour,* thick, soft. Spores pink, fusiform, or oblong, 12–14 × 3–4·5μ, 2–3-guttulate. Taste *very bitter.* Poisonous. Woods, especially on calcareous soil. July—Dec. Not uncommon. (*v.v.*)

1833. **T. alutarius** (Fr.) Rea. (= *Boletus alutarius* Fr.)
Aluta, tanned leather.

P. 7–10 cm., *fuscous tan*, pulvinate, or convex, then expanded, soft, *velvety*, becoming smooth. St. 6–8 × 2–3 cm., *concolorous*, bulbous, somewhat smooth, *apex rugose*. Tubes *white, becoming pinkish*, depressed round the st., short, plane; orifice of pores *white, becoming fuscous when bruised, round*. Flesh *white, unchangeable*, soft. Spores pink. Taste *mild*. Woodland pastures. Oct. Rare.

Spores purple.

Phaeoporus Bataille (=**Boletus** p.p.).

(φαιός, dusky; πόρος, pore.)

Pileus tomentose, or velvety-silky. Stem central, blackish bistre, velvety, or glabrous. Tubes grey, or pinkish grey, sinuate, or free, fairly long; orifice of pores concolorous, becoming greenish blue when touched. Flesh compact, becoming blue, or grey when exposed to the air. Spores fuscous purple, elliptic-fusiform, smooth. Growing on the ground.

1834. **P. porphyrosporus** (Fr.) Bat. (= *Boletus porphyrosporus* Fr.) Boud. Icon. t. 149, as *Boletus porphyrosporus* Fr.
πορφύρα, purple; σπορά, seed.

P. 5–15 cm., *olivaceous, or brownish fuliginous*, becoming blackish when bruised, hemispherical, then convex, fleshy, *minutely velvety*. St. 7–12 × 1·5–3 cm., *concolorous*, attenuated upwards from the thickened, paler base, *velvety*. Tubes *pale grey or olivaceous*, adnate, fairly long; orifice of pores *yellowish, becoming bluish green when bruised and staining white paper an emerald green colour*, broad, angular. Flesh *dirty white, becoming bluish, or fuliginous*, compact. Spores dull, or brownish purple, fusiform, 14–16 × 3–4μ, 2–4-guttulate. Smell strong. Woods and pastures under trees. July—Oct. Uncommon. (*v.v.*)

var. **fuligineus** (Fr.) Bat. (= *Boletus fuligineus* Fr.) *Fuligineus*, sooty.

Differs from the type in the *villosely silky p. and glabrous st.* Under conifers. Sept. Rare. (*v.v.*)

Spores blackish, or fuscous. Pileus covered with imbricate scales.

Strobilomyces Berk.

(στρόβιλος, a fir cone; μύκης, fungus.)

Pileus fleshy, firm, floccose, clothed with large, imbricate scales. Stem firm, rigid, woolly, or scaly, annulate. Tubes white, then greyish bistre, adnate, long, orifice of pores concolorous, angular. Flesh floccose, not putrescent, firm, light, becoming reddish, or bluish grey, and finally blackish on exposure to the air. Spores blackish purple, subglobose, verrucose. Cystidia present. Growing on the ground.

1835. **S. strobilaceus** (Scop.) Berk. (= *Boletus strobilaceus* (Scop.) Fr.) Rostk. Bol. t. 38, as *Boletus strobilaceus* Scop.

στρόβιλος, a fir cone.

P. 5–10 cm., *white, becoming brownish or blackish umber*, pulvinate, then convex, *broken up into large, thick, floccose scales*; margin appendiculate with the white floccose veil. St. 7–15 × 1–2 cm., *concolorous*, equal, apex *white*, sulcately reticulated, floccosely scaly below the ring. Ring *white*, floccose, thick. Tubes *white, becoming brownish*, adnate, or with a decurrent tooth, long; orifice of pores *white, becoming reddish when touched or bruised*, broad, angular. Flesh *white, becoming reddish and finally blackish bistre*, thick, floccose. Spores blackish purple, subglobose, verrucose, 9–11 × 8–9μ. Smell pleasant. Deciduous and coniferous woods. Aug.—Oct. Uncommon. (*v.v.*)

Spores ochraceous, ferruginous, or olivaceous.

Tubes short, alveolar, decurrent.

Boletinus Kalchbr.

(*Boletinus*, diminutive of *Boletus*.)

Pileus dry, fibrillosely scaly. Stem central, hollow, bulbous, woolly. Ring white, floccose, thick. Pores large, alveolar, compound, reticulately decurrent on the stem. Flesh yellow, unchangeable. Spores yellow, elliptic-fusiform, smooth. Cystidia present. Growing on the ground and on mossy trunks.

1836. **B. cavipes** (Opatowski) Klotzsch. Kalchbr. Icon. t. 31.

Cavus, hollow; *pes*, foot.

P. 3–8 cm., *tawny*, or *brownish tawny*, convex, *subumbonate, fibrillosely scaly*, fleshy. St. 5–8 × ·5–1 cm., *lemon yellow above the ring, concolorous below*, subequal, or attenuated upwards, thickened at the base and rooting, incurved, tough, stuffed, then hollow especially at the base, apex reticulate, rough or fibrillosely scaly. Ring *white, floccose, thick*, evanescent. Tubes *yellow*, or *sulphur coloured, becoming greenish or olivaceous, compound, broad, honey-comb-like*, decurrent. Flesh *becoming yellow in the p., white in the st.*, firm. Spores yellow, elliptic-fusiform, 10μ. Taste pleasant. Edible. Under larches and on mossy beech trunks. Sept.—Oct. Rare.

Tubes very short, gyroso-plicate.

Gyrodon Opatowski.

(*γυρός*, round; *ὀδών*, a tooth.)

Pileus fleshy, viscid, or villose. Stem central, smooth, or punctate. Tubes very short, 1–2 mm. long; orifice of pores sinuous, torn, or gyroso-plicate. Spores ochraceous, or olivaceous, elliptical, elliptic cylindrical, or fusiform, smooth. Growing on the ground, often fasciculate.

1837. **G. caespitosus** Massee. *Caespes*, a clump.

P. 2·5–5 cm., *olivaceous umber, becoming paler towards the margin*, hemispherical, dry, *tomentose*; margin *pale pink*, wavy, acute. St. 4–8 × 2–3 cm., *yellow, dingy red at the base, connate*, ventricose, glabrous. Tubes *yellow*, subadnate, 1 mm. long; orifice of pores *yellow, very irregular*, elongate, sinuous. Flesh *yellow, instantly changing to intense blue, then fading to a dirty white, and finally rufous, dingy red at the base of the st.*, thick, firm. Spores pale olive, narrowly elliptical, 12 × 4μ. Densely fasciculate. Under trees amongst grass. Aug. Rare.

1838. **G. sistotrema** Fr. *σειστός*, shaking; *τρῆμα*, a hole.

P. 6–8 cm., *reddish, or brownish olive*, convex, *then flattened*, glabrous, or minutely pubescent, thin, *dry*. St. 5–7 × 1 cm., *pale lemon yellow, becoming reddish*, equal, or enlarged below, smooth. Tubes *yellow tawny, or pale sulphur*, adnate, very short; orifice of pores *yellow, becoming tawny, sinuous, round, becoming gyroso-plicate*. Flesh *cream colour, tawny under the cuticle of the p.*, firm. Spores "cream-olive, elliptic cylindrical, 10–14μ long, guttulate" Quél. Taste somewhat sharp. Coniferous woods. Aug.—Oct. Rare.

var. **brachyporus** (W. G. Sm.) Rea. *βραχύς*, small; *πόρος*, pore.

Differs from the type in the *white p., the white st. sulphur coloured below, and in the flesh changing to pale green.*

1839. **G. rubellus** McWeeney. *Rubellus*, reddish.

P. 1–1·5 cm., *red, with a tinge of purple at the disc, becoming yellowish towards the margin*, convex, even, smooth, dry. St. 1 cm. × 3 mm., *bright yellow*, equal, smooth, even. Tubes *pale yellow*, 1 mm. long; orifice of pores *bright yellow, linear*, elongate, sinuous, dissepiments thick. Flesh *yellow*, unchangeable, firm. Spores greenish olive, cylindric fusiform, with a minute basal apiculus, 10 × 4μ. Amongst moss. Oct. Rare.

Tubes long.

Boletus (Dill.) Pat.

(*βῶλος*, a clod.)

Pileus fleshy, dry, glabrous, tomentose, silky, viscid, or glutinous. Stem central, equal, ventricose, or bulbous; dry, glabrous, tomentose, or viscid, sometimes reticulate; with or without a ring. Tubes long, adnate, sinuato-adnate, or decurrent, rarely free; orifice of pores round, angular, unequal, or toothed, often compound. Flesh thick, soft, putrescent. Spores ochraceous, ferruginous, olivaceous, or fuscous, rarely colourless, fusiform, oblong-elliptic, elliptical, or pip-shaped, smooth. Cystidia present. Growing on the ground, solitary, gregarious, caespitose, or subcaespitose.

I. Pores *angular*, or *large*, rarely round, often unequal, or toothed; tubes often *compound, sometimes connate*, long or rather short, *more or less adnate*, sometimes decurrent, rarely slightly sinuate. P. *dry*, glabrous, tomentose, silky or *glutinous*. St. often slightly thick, sometimes furnished with a ring, rarely reticulate at the apex. Plants of medium, or small size, rarely large.

A. P. *glutinous*, or *viscid*, at least in wet weather, never pruinosely granular. Pores, tubes, and flesh of various colours.

1. St. annulate.

*Pores yellow; spores yellow olivaceous.

1840. **B. sphaerocephalus** Barla. Barla, Champ. Nice, t. 36.
σφαῖρα, a ball; *κεφαλή*, head.

P. 10–20 cm., *ochraceous yellow, deeper coloured at the centre, and sometimes tinged with brown*, globose, viscid; margin *light yellow*, appendiculate with fragments of the ring. St. 4–6 × 3–4 cm., *tawny yellow, becoming darker*, furrowed. Ring *yellowish*, membranaceous, shaggy, usually fugacious. Tubes *bright yellow, becoming tawny or brownish with age*, short, decurrent; orifice of pores round, or angular. Flesh *light yellow, bluish under the cuticle*, very thick, soft, watery. Spores olivaceous, fusiform, 8–10 × 3–4μ, 2-guttulate. On the ground, and on rotten sawdust. Sept.—Oct. Uncommon. (*v.v.*)

1841. **B. luteus** (Linn.) Fr. Rolland, Champ. t. 78, no. 174.
Luteus, yellow.

P. 4–14 cm., *fuscous with the dark separating gluten, becoming paler*, convex, gibbous, then pulvinate. St. 5–10 × 2–3 cm., *whitish, becoming fuscous below the ring, apex light yellow and granular*, equal, firm. Ring *cream colour, becoming fuscous*, membranaceous, large. Tubes *yellow*, adnate; orifice of pores *yellow*, round. Flesh *whitish*, or *yellowish*, thick, soft. Spores yellowish brown, elliptic-fusiform, 8–10 × 3–3·5μ, 1–3-guttulate. Taste pleasant. Edible. Coniferous woods, and under conifers. June—Dec. Common. (*v.v.*)

1842. **B. elegans** (Schum.) Fr. Grev. Scot. Crypt. Fl. t. 183, as *Boletus luteus*.
Elegans, nice.

P. 4–10 cm., *golden, or inclining to ferruginous*, convexo-plane, viscid. St. 5–7 × 1·5–2 cm., *golden, then rufescent*, firm, unequal, *apex reticulate with granules*. Ring *cream colour*, often torn and fugacious. Tubes *golden-sulphur-yellow, decurrent*; orifice of pores angular. Flesh *light yellowish*, soft. Spores yellowish, oblong elliptic, 8–9 × 3–3·5μ, 2–4-guttulate. Taste pleasant. Edible. Coniferous woods, and under conifers. May—Dec. Common. (*v.v.*)

var. **flavus** (With.) Rea. Bres. Fung. Trid. t. 132, as *Boletus flavus* With. *Flavus*, light yellow.

Differs from the type in the *adnate tubes, the larger orifice of the greyish yellow pores, and the paler flesh which is rosy when broken.* Coniferous woods. Aug.—Nov. Uncommon. (*v.v.*)

var. **pulchellus** (Fr.) Rea. Fr. Icon. t. 178, fig. 1, as *Boletus pulchellus* Fr. *Pulchellus*, beautiful little.

Differs from the type in its *smaller size,* 3–4 cm. *broad, its short, smooth, yellow st., its less viscid, greenish yellow p., its narrow, linear ring and its rosy flesh.* Coniferous woods. July—Sept. Rare. (*v.v.*)

1843. **B. flavidus** Fr. Krombh. t. 4, figs. 35–37. *Flavidus*, light yellow.

P. 2–5 cm., *livid light yellowish,* campanulate, *umbonate,* then plane and gibbous, viscid, *radiately wrinkled.* St. 5–7·5 × 4–6 mm., *whitish tinged with yellow,* subequal, tough, mealy, sometimes striate, *apex sprinkled with fugacious glandules*; base white, cottony. Ring *gelatinous, greenish white,* thin, narrow; margin floccose, viscid. Tubes *dirty light yellow,* decurrent; orifice of pores *large,* angular, *compound.* Flesh *yellowish, reddish on exposure to the air, thin,* firm. Spores "subhyaline, elongato-ellipsoid, straight, 8–10 × 3·5–4μ" Karst. Taste pleasant. Pine woods. July—Oct. Uncommon.

**Pores white, grey, or green. Spores brownish.

1844. **B. viscidus** (Linn.) Fr. (= *Boletus laricinus* Berk. sec. Quél.) Fr. Icon. t. 178, fig. 3. *Viscidus*, viscid.

P. 5–10 cm., *dirty white with livid stains, or dingy yellowish,* campanulato-convex, pulvinate, viscid, *floccose,* rugose. St. 5–9 × 1–1·5 cm., *white, becoming yellow or greyish,* equal, or thickened at the base, *viscid, floccose,* apex reticulate. Ring *white,* membranaceous, large, thin, often torn. Tubes *white, then greyish or tinged greenish,* adnate, subdecurrent; orifice of pores large, unequal, compound, often toothed. Flesh *white,* watery, soft. Spores brownish, oblong elliptical, 11–13 × 4–5μ, multi-guttulate. Taste pleasant. Edible. Woods, especially under larches. May—Dec. Common. (*v.v.*)

2. St. exannulate.

*Pores yellowish, yellow, golden, or orange; spores yellow. P. never white.

1845. **B. collinitus** Fr. Lucand, Champ. t. 240. *Collinitus*, besmeared.

P. 5–6 cm., *chestnut, becoming pale when the fuscous gluten separates, and veined with brown,* convex, pulvinate. St. 5–7·5 × 1·5–2·5 cm., *white, becoming fuscous,* firm, attenuated downwards, *somewhat reticulated with adpressed squamules.* Tubes *pallid, then yellow,* adnate, elongated; orifice of pores *divided into two,* rather large. Flesh *white, brownish under the adnate cuticle of the p.,* firm. Spores "8–10 × 4μ" Guill. Taste pleasant. Edible. Pine woods. Nov. Rare.

1846. **B. granulatus** (Linn.) Fr. Rolland, Champ. t. 78, no. 175. *Granulatus,* granulated.

P. 4–8 cm., *fuscous ferruginous, becoming yellowish when the gluten disappears,* convexo-expanded, *smooth.* St. 5–8 × 1–2 cm., *light yellowish,* subequal, often attenuated upwards, *dotted with granules upwards;* base white, cottony. Tubes *sulphur yellow,* adnate, short; orifice of pores *at first dripping with white milk, granulated when the milk dries, simple,* subcircular. Flesh *light yellowish under the separable pellicle of the p.,* soft. Spores yellow ferruginous, oblong elliptical, 8–10 × 3–4μ, 2-guttulate. Taste pleasant. Edible. Gregarious. Coniferous woods, and under pines. May—Dec. Common. (*v.v.*)

1847. **B. tenuipes** (Cke.) Massee. *Tenuis,* thin; *pes,* foot.

P. 2·5–5 cm., *yellowish brown, or gilvous,* convex, then almost plane, viscid, *streaked with minute fibrils when dry.* St. 5–8 cm. × 6–8 mm., *yellow,* attenuated at the base, *smooth.* Tubes *yellowish,* adnate, shortened round the st., about 6 mm. long; orifice of pores rather large, angular. Flesh *white, rosy under the cuticle of the p.,* thick. Spores pale yellow, fusiform, 10 × 3μ. Taste mild. Edible. Woods, and heaths. July—Oct. Uncommon. (*v.v.*)

1848. **B. badius** Fr. Rostk. Bol. t. 5. *Badius,* bay-brown.

P. 3–15 cm., *bay-brown-tawny,* hemispherical, pulvinate, viscid, then pubescent, often shining when dry. St. 6–10 × 2–3 cm., *paler,* subequal, sometimes attenuated upwards, sometimes downwards, *brown pruinate.* Tubes *cream,* or *citron yellow, turning immediately bluish green when touched,* adnate, or sinuato-depressed, *long;* orifice of pores angular, rather large. Flesh *yellowish, becoming azure blue, then somewhat brownish,* thick, soft. Spores light yellow, oblong fusiform, 13–15 × 4·5–6μ, 1-multi-guttulate. Taste pleasant. Edible. Woods, especially coniferous woods. Aug.—Nov. Common. (*v.v.*)

1849. **B. paludosus** Massee. *Paludosus,* marshy.

P. 7·5–10 cm., *bright rufous brown, paler when dry,* slightly convex, then quite plane, slightly viscid, smooth. St. 7·5–12·5 × 1–1·5 cm., *rather paler than the p.,* equal, smooth, base attenuated. Tubes *yellow, then olive green,* adnate, or subdecurrent, *short,* about 4 mm. long; orifice of pores large, angular, compound. Flesh *with a very pale tinge of brown,* firm, *thin.* Spores olive, elongato-fusiform, 16 × 4μ. Gregarious. Bogs, amongst *Sphagnum.* Sept. Rare.

1850. **B. rutilus** Fr. *Rutilus*, red, inclining to golden yellow.

P. 5–7·5 cm., *rufescent brick colour*, pulvinate, viscid, smooth. St. 4 cm. × 6–8 mm., *sulphur above, red below*, equal, smooth, firm. Tubes *sulphur yellow*, opaque, adnate, *sinuato-depressed*; orifice of pores minute, regular. Flesh *of st. somewhat ochraceous, almost becoming black when broken*. Smell strong. Oak woods. Oct. Rare.

1851. **B. aurantiporus** Howse. *Aurantius*, orange; πόρος, pore.

P. 5–6 cm., *fulvous ferruginous, then pale gilvous*, convex, then expanded, *adpressedly squamulose*, viscid. St. 4–7·5 × 1–1·5 cm., *yellow*, equal, *beautifully reticulated with yellow and red*. Tubes *golden yellow, then orange, turning red when bruised*, deeply decurrent; orifice of pores broad, angular. Flesh *yellowish, often tinged reddish*, thin at the margin, firm. Spores yellow, oblong elliptical, or subfusiform, 9–10 × 4–5μ, 1–4-guttulate. Smell and taste pleasant. Coniferous woods, and under yew trees. Aug.—Oct. Uncommon. (*v.v.*)

**Pores becoming olive, bistre olive, brown olive, or olive rust or rust colour; tubes short; spores concolorous. P. sometimes white.

1852. **B. bovinus** (Linn.) Fr. Krombh. t. 75, figs. 1–6. *Bovinus*, pertaining to oxen.

P. 5–10 cm., *pale reddish yellow, dull orange yellow, or deep buff*, hemispherical, then convex, smooth, viscid; *margin white*, often tomentose. St. 5–10 × 1 cm., *concolorous, or paler*, equal, smooth, base whitish. Tubes *grey light yellow, becoming ferruginous*, subdecurrent, at first very shallow like those of *Merulius lacrymans*, then elongating with age; orifice of pores compound, toothed. Flesh *yellowish-flesh colour, reddish in the st.*, soft. Spores yellow, or olivaceous, oblong fusiform, 8–10 × 3–3·5μ, 1-multi-guttulate. Taste pleasant. Edible. Gregarious, or subcaespitose. Woods and heaths, especially near pines. Aug.—Nov. Common. (*v.v.*)

1853. **B. piperatus** (Bull.) Fr. Rostk. Bol. t. 6. *Piperatus*, peppery.

P. 2–10 cm., *cinnamon, or yellow to pale yellowish*, convexo-plane, smooth, slightly viscid. St. 4–12 × ·5–1·5 cm., *concolorous*, fragile, equal, or attenuated at the base, *containing yellow milk at the base*, and *springing from a yellow mycelium*. Tubes *ferruginous*, decurrent; orifice of pores *large*, angular, often toothed. Flesh *sulphur yellow, tinged reddish in the p.*, soft. Spores brownish, oblong elliptic, 8–10 × 3–4μ, 1–3-guttulate. Taste very acrid, or peppery. Woods, and heaths. Aug.—Nov. Common. (*v.v.*)

B. P. *dry*, rarely moist when wet, *tomentose, silky, pruinose, powdery, or granular*, sometimes squamulose. Pores *yellowish, or yellow*, rarely slightly olivaceous or red. Flesh yellow, rarely cream, or white. Spores yellow, rarely tawny bistre.

1854. **B. variegatus** (Swartz) Fr. Rostk. Bol. t. 16.

Variegatus, variegated.

P. 6–12·5 cm., *dark yellow, or ochraceous, sprinkled with fasciculate-hairy, superficial, brown squamules*, convex, then plane, obtuse, slightly moist; margin acute, at first flocculose. St. 5–8 × 1–2·5 cm., *dark yellow, or straw colour, sometimes reddish*, firm, equal, base white. Tubes *brown, or yellow olive, then cinnamon*, adnate; orifice of pores round. Flesh *yellow becoming here and there azure blue*. Spores greenish ochre, oblong elliptic, 9–10 × 3–4μ. Smell unpleasant, "of chlorine" Quél. Coniferous woods, heaths, and moorlands. July—Nov. Common. (*v.v.*)

1855. **B. sulphureus** Fr. Quél. As. fr. (1887), t. 9.

Sulphureus, sulphur-yellow.

P. 5–10 cm., *sulphur yellow*, compact, convex, then plane, *silky-tomentose with innate flocci*. St. 4–10 × 1–5 cm., *sulphur yellow, at length becoming dingy ferruginous*, firm, ventricose, smooth, *springing from a golden, woolly mycelium*. Tubes *sulphur yellow, becoming spotted ferruginous, and at length becoming green*, adnate, short, 2–4 mm. long; orifice of pores minute, compound. Flesh *yellow*, becoming *greenish, or azure blue when broken, but golden when exposed to the air, here and there reddish under the tubes*, firm. Spores light yellow, elliptical, 6–7 × 3μ, 1-guttulate. Caespitose. Pine sawdust, and twigs. Sept.—Nov. Uncommon. (*v.v.*)

1856. **B. chrysenteron** (Bull.) Fr. Rolland, Champ. t. 80, no. 180.

χρυσός, gold; *ἔντερον*, intestine.

P. 3–10 cm., *fuscous, or somewhat brick colour*, convexo-plane, soft, *minutely tomentose, often cracked into patches with the interstices red*. St. 5–8 cm. × 6–12 mm., *scarlet, or light yellow*, subequal, or attenuated at the base, rigid, fibroso-striate. Tubes *sulphur yellow, then greenish yellow*, subadnate, or depressed round the st.; orifice of pores rather large, *angular*, compound. Flesh *yellow, scarcely turning blue, red beneath the cuticle of the p.*, soft. Spores deep ochraceous, fusiform, or oblong elliptic, 13–14 × 4–5μ, 1–4-guttulate. Taste mild. Edible. Woods, heaths, and pastures. May—Dec. Common. (*v.v.*)

var. **nanus** Massee. *νάννος*, a dwarf.

Differs from the type in its *smaller size, in the flesh becoming red when cut and in the elongated, narrow, sinuous, or gyrose orifice of the pores*. Parks, gardens, and pastures. Aug.—Sept. Not uncommon. (*v.v.*)

1857. **B. sanguineus** (With.) Quél. non Fr. *Sanguineus*, blood red.

P. 2–7 cm., *blood red, or brightish crimson, disc becoming brownish*, hemispherical, convex, then plane, firm, opaque, *dry*, rough under a lens; margin almost even. St. 2·5–6·5 × 1–2 cm., *yellowish, streaked or blotched with dilute crimson, or brownish,* slightly bulbous at the base, and with traces of long reticulations at the apex. Tubes *golden yellow, or dullish yellow,* becoming bluish or greenish when bruised, and finally orange, adnate, or slightly depressed round the st., 4–7 mm. long; orifice of pores angular, medium in size. Flesh *yellow, then rosy,* "*becoming bluish when cut or broken*" Perceval, soft. Spores "15–18μ, guttulate" Quél. Taste somewhat acid. Woods, and pastures. July—Sept. Rare.

1858. **B. subtomentosus** (Linn.) Fr. Rolland, Champ. t. 80, no. 181.
Sub, somewhat; *tomentosus,* downy.

P. 3–10 cm., *more or less deep brownish olivaceous,* convex, then pulvinato-expanded, soft, dry, *villoso-tomentose,* sometimes cracked into patches with the interstices yellow. St. 5–12 × 1–2 cm., *yellowish, usually streaked with red,* attenuated downwards, *sulcately ribbed, the tawny ribs sometimes anastomosing,* rough with dots under a lens. Tubes *golden sulphur yellow,* adnate; orifice of pores *large, angular.* Flesh *white, or yellowish, rust colour under the cuticle of the p.,* soft. Spores pale yellow, oblong elliptical, 12–14 × 5μ. Taste mild. Edible. Woods, heaths, and pastures. July—Dec. Common. (*v.v.*)

var. **radicans** (Krombh.) Massee. Krombh. t. 48, figs. 1–6.
Radicans, rooting.

Differs from the type in the *usually undulated, bright yellowish olive green p., the st. whitish below, yellow and strongly grooved above and the tubes greenish-olive at maturity.* Woods.

var. **striaepes** (Secr.) Quél. *Stria,* a line; *pes,* foot.

Differs from the type in the *bistre olive, silky p., and the yellow stem with thin bistre ribs, brownish red at the base.* Woods. Oct. Rare.

var. **marginalis** Boud. Boud. Icon. t. 142 *Marginalis,* bordered.

Differs from the type in being *more slender, in the fuliginous p. being bordered with a pale, tomentose zone at the margin, in the longer almost smooth st., and in the narrower pores.* Woods. Uncommon. (*v.v.*)

1859. **B. cruentus** Vent. Venturi, t. 43, figs. 3 and 4.
Cruentus, bloody.

P. 7–10 cm., *olivaceous with a reddish tinge, becoming instantly red where bruised,* convex, then plane, soft, *minutely tomentose.* St. 5–8 × 2·5-4 cm., *yellow with reddish markings,* gradually attenuated upwards from the incrassated, *rooting base, minutely flocculose.* Tubes *pale yellowish olive,* sinuato-free, long; orifice of pores *yellowish,* minute, subangular. Flesh *yellow, becoming red when cut,* thick, firm. Spores "pale olive, elliptic-fusiform, 14–16 × 5μ" Massee, 1-guttulate. Smell strong. Under beeches. Aug.

1860. **B. spadiceus** (Schaeff.) Fr. Krombh. t. 36, figs. 19, 20, as *Boletus tomentosus* Krombh. *Spadiceus*, date brown.

P. 5–10 cm., *date brown*, opaque, convex, then pulvinato-expanded, dry, tomentose, then widely cracked. St. 5–8 × 1·5–2·5 cm., *yellow, becoming fuscous, flocculoso-furfuraceous*, firm, *with anastomosing, thin, tawny ribs*. Tubes *golden sulphur yellow*, adnate; orifice of pores *wide*, round, toothed. Flesh *white, yellowish in the st., and fuscous reddish under the cuticle of the p.* Spores yellow, "10–12μ, 2–4-guttulate" Quél. Woods. July—Nov. Uncommon.

1861. **B. Rostkovii** Fr. Rostk. Bol. t. 18, as *Boletus lividus* Bull. Dr Friedrich Wilhelm Theophilus Rostkovius.

P. 7·5–10 cm., *dingy olive brown, or rufous*, convex, or almost plane, very minutely tomentose, often areolately cracked, interstices pale. St. 2·5–5 × 2·5 cm., *pale reddish yellow, obconic*, tapering almost to a point at the base. Tubes pale yellow green, adnate, about 12 mm. long; orifice of pores irregularly angular, compound. Flesh *white, becoming tinged red when cut*, with here and there a shade of blue. Spores "pale olive, elongato-fusoid, 20 × 5μ" Massee, 2-guttulate. Under beeches. Aug. Rare.

1862. **B. radicans** (Pers.) Fr. *Radicans*, rooting.

P. 6–8 cm., *olivaceous cinereous, then pale yellowish*, convex, pulvinate, dry, subtomentose; margin thin, *incurved*. St. 7–10 × 1·5–2·5 cm., *light yellow, attenuato-rooted, flocculose with reddish pruina*, naked and dark when touched, firm. Tubes *lemon yellow, becoming greenish or bluish when touched*, adnate; orifice of tubes unequal, large, angular. Flesh *pale lemon yellow, becoming immediately deep blue on exposure to the air*, soft, watery. Spores yellow, oblong fusiform, 13–14 × 4–5μ, 1–2-guttulate. Taste bitter. Woods, and hedgerows. July—Sept. Rare. (*v.v.*)

1863. **B. rubinus** W. G. Sm. W. G. Sm. in Journ. Bot. (1868), t. 75, figs. 1–4. *Rubinus*, ruby coloured.

P. 4–8 cm., *yellow fuscous*, pulvinato-gibbous, then plane, dry, subtomentose, slightly cracked. St. 5–7·5 × 2 cm., *yellow, smeared with crimson*, equal, or attenuated downwards. Tubes *wholly carmine*, subdecurrent, short at first; orifice of pores medium size, compound. Flesh *yellow, becoming whitish when dry in the p.* Spores pale, oval, 6 × 4–5μ. Woods, roadsides, and under oaks. Aug.—Sept. Uncommon. (*v.v.*)

1864. **B. versicolor** Rostk. Boud. Icon. t. 143. *Versicolor*, of various colours.

P. 2–7 cm., *blood red, or deep rose-pink, becoming tawny at the disc, or all over with age*, convex, then plane, minutely tomentose, rarely cracked. St. 4–8 cm. × 6–12 mm., *yellow, rose-red in the middle and at the base, becoming blue when rubbed*, equal, often attenuated at the base, slightly viscid in wet weather, *minutely granular, or pruinose*. Tubes *yellow, becoming blue when touched*, adnate, decurrent by a tooth; orifice of pores fairly large, angular. Flesh *yellow, reddish at the base of the st., turning blue when cut especially near the tubes*, soft. Spores yellow, or olivaceous, oblong fusiform, 9–10 × 4–5μ, 1–2-guttulate. Woods, pastures, and roadsides. Aug.—Oct. Not uncommon. (*v.v.*)

1865. **B. parasiticus** (Bull.) Fr. Boud. Icon. t. 145.
Parasiticus, parasitic.

P. 2–8 cm., *dingy yellow, or tawny, and more or less brown*, convex, then plane, minutely tomentose, dry, often cracked in a tessellated manner. St. 3–7 × 1–2 cm., *yellow, or paler than the p.*, equal, attenuated at the base, rigid, incurved, slightly fibrillose, apex mealy. Tubes *yellow, then vinous*, decurrent, short; orifice of pores compound, of medium size, round or angular. Flesh *yellow, often becoming reddish in the st.*, firm. Spores deep olivaceous, fusiform, 12–15 × 4–5μ, 1-guttulate. Parasitic on *Scleroderma aurantium* and *S. verrucosum*. Woods, and heaths. Sept.—Oct. Not uncommon. (*v.v.*)

1866. **B. pruinatus** Fr. *Pruinatus*, covered with hoar-frost.

P. 5–6 cm., *purplish bay brown, covered with a whitish, or greyish bloom*, convex, then plane, rigid, *dry*. St. 5–10 × ·5–1·5 cm., *variegated yellow and reddish*, equal, or ventricose, firm, smooth. Tubes *light yellow*, adnate; orifice of pores small, slightly angular. Flesh *yellow, red under the cuticle of the p., turning bluish and reddish*, firm. Spores olivaceous, pip-shaped, 9–10 × 4μ, 1-guttulate. Amongst grass in woods, and parks. Sept.—Oct. Uncommon. (*v.v.*)

1867. **B. pusio** Howse, ex B. & Br. *Pusio*, a little boy.

P. *reddish buff*, hemispherical, pulvinate, *pulverulent*. St. *dull yellowish white, white above, becoming blackish*, slightly thickened downwards, *plicate, pulverulent*. Tubes *dull sulphur white*.

1868. **B. purpurascens** Rostk. Rostk. Bol. t. 8.
Purpurascens, becoming purple.

P. 6–16 cm., *bay purple*, convex, then plane, *smooth*, dry. St. 3–5 × 1·5–2 cm., *deep purple red*, firm, *attenuated downwards to the rooting base*, smooth. Tubes *dingy yellow*, adnate, 12–15 mm. long, *becoming greenish when touched*; orifice of pores small. Flesh *dirty whitish with darker streaks, reddish under the cuticle of the p.*, thick, compact. Spores brownish grey, subfusiform, 10 × 5μ. Woods, especially pine. Sept.—Oct. Uncommon.

II. Pores *small, round, equal*; tubes *free, or sinuate*, sometimes adnate, *long*, connate. P. *dry*, smooth, or tomentose, rarely viscid or moist in wet weather. St. thick, smooth, tomentose, or floccose, often ribbed or reticulate. Spores *yellow, ochraceous, or olivaceous*, rarely brown, or white, elliptic fusiform. Generally large in size and thick.

A. St. *fibrillosely fleshy*, generally *firm, thick, ovoid at first*; either covered *with a network of* white, straw-coloured, yellow, or red *veins*, or *minutely punctate, or granular*, rarely smooth. Pores small, round, white, or coloured; tubes concolorous, free, or sinuate, rarely adnate. Flesh generally *firm*, white, or coloured, changing colour or not, but *never becoming black*. Generally large in size.

†Flesh white, or yellow, *unchangeable*, sometimes reddish or vinous rosy under the cuticle, tasty, often fragrant. Pores white, cream, or yellow, sometimes becoming greenish with age, *never becoming blue or green when touched*. St. with a white, straw-coloured, or brownish cream network, sometimes smooth, rarely floccose, or reddish.

1869. **B. regius** Krombh. Krombh. t. 7. *Regius*, royal.

P. 7–12·5 cm., *bright rose-pink, reddish purple, or olivaceous*, convex, pulvinate, dry, smooth, or minutely tomentose. St. 5–9 × 3–5 cm., *pale yellow, becoming purplish at the base, reticulate*. Tubes *golden yellow*, almost free, short; orifice of pores small, subangular. Flesh *pale yellow*, very thick. Spores "pale yellow, elongate fusiform, 16 × 5μ" Massee. Taste pleasant. Edible. Gregarious. Woods, and open places. Aug.—Oct. Uncommon.

1870. **B. edulis** (Bull.) Fr. Rolland, Champ. t. 81, no. 182. *Edulis*, eatable.

P. 10–20 cm., *bay, brown, fuliginous, or bistre, rarely white, the margin often white*, convex, pulvinate, *smooth*, often rugose, *somewhat viscid in wet weather*. St. 10–15 × 3–6 cm., *pallid fuscous*, delicately reticulated, equal, or attenuated upwards from the bulbous base. Tubes *white, then yellow, and finally greenish*, somewhat free, long; orifice of pores small, round. Flesh *white, often faintly tinged reddish under the cuticle of the p.*, compact, then softer, thick. Spores yellow, fusiform, 13–16 × 4–4·5μ, 1–3-guttulate. Smell and taste pleasant. Edible. Woods, especially beech. June—Nov. Common. (*v.v.*)

var. **laevipes** Massee. *Laevis*, smooth; *pes*, foot.

Differs from the type in *the absence of reticulations on the perfectly even, white, or faint buff st.* Woods. Aug.—Oct. Common. (*v.v.*)

var. **bulbosus** (Bull.) Big. & Guill. (= *Boletus crassus* Massee.) *Bulbosus*, bulbous.

Differs from the type in the *bulbous st. often exceeding, or equalling in width the diameter of the p., in the flesh becoming pale primrose yellow, in the smaller spores, and in the strong acid smell.* This variety is a condition of the type brought about by the attack of a parasitic *Hypomyces*. Woods. July—Nov. Common. (*v.v.*)

1871. **B. pinicola** (Vitt.) Rea. (=? *Boletus fusco-ruber* Quél.) Trans. Brit. Myc. Soc. IV, t. 6. *Pinus*, pine; *colo*, I inhabit.

P. 9–20 cm., *rich chestnut colour, bordered by a narrow white line at the margin*, convex, slightly viscid when moist, *then dry and floccose.* St. 9–15 × 4–5 cm., *concolorous*, subbulbous, rugose, slightly reticulate. Tubes *greenish*, adnate, 15–20 mm. long, ventricose; orifice of pores round, or angular, 1 mm. broad. Flesh *white, reddish under the cuticle of the p.*, thick. Spores olivaceous, fusiform, 15–18 × 4–5μ, 1–3-guttulate. Smell and taste pleasant. Edible. Coniferous woods. Sept.—Oct. Not uncommon. (*v.v.*)

1872. **B. reticulatus** (Schaeff.) Boud. Trans. Brit. Myc. Soc. IV, t. 5. *Reticulatus*, netted.

P. 8–15 cm., *ochraceous yellow, or greyish fawn*, convex, *finely tomentose*, often cracked in dry weather. St. 6–9 × 4–6 cm., *concolorous, or paler*, slightly constricted at the base, *reticulated to the base*. Tubes *greenish yellow*, free, or almost free, fairly long; orifice of pores round, small, 1 mm. across. Flesh *white, slightly coloured under the cuticle of the p. and at the base of the tubes*, firm, thick. Spores olivaceous, oblong fusiform, 13–18 × 4–5μ, 1–3-guttulate. Smell and taste pleasant. Edible. Deciduous woods. May—Oct. Not uncommon. (*v.v.*)

1873. **B. aestivalis** (Paul.) Fr. Hussey, Illus. Brit. Myc. II, t. 25. *Aestivalis*, pertaining to summer.

P. 10–20 cm., *whitish, bistre cream, or reddish*, convex, pulvinate, somewhat repand, smooth, *then granular in dry weather*. St. 8–11 × 5–6 cm., *light yellow*, ovoid, bulbous, *smooth, or minutely reticulate*. Tubes *yellow, or greyish*, somewhat free, long; orifice of pores small, round, equal. Flesh *yellow, white above, reddish at the base of the st.*, thick. Spores yellow, oblong fusiform, or oblong elliptical, 12–14 × 4–5μ, 1–2-guttulate. Smell and taste pleasant. Edible. Woods, and heaths. June—Oct. Uncommon. (*v.v.*)

1874. **B. aereus** (Bull.) Fr. Krombh. t. 36, figs. 1–7. *Aereus*, made of copper.

P. 6–9 cm., *olivaceous fuscous, somewhat blackish*, hemispherical, then convex, pulvinate, minutely pubescent, or villose. St. 7–9 × 2·5–3·5 cm., *yellowish, becoming fuscous downwards*, beautifully reticulate. Tubes *white, then sulphur yellow*, somewhat free; orifice of pores minute, round, or angular. Flesh *white, reddish under the cuticle of the p. and st., and reddish purple when the surface of the p. has been eaten by slugs*, firm, compact. Spores yellow, oblong elliptical, or oblong fusiform, 12–15 × 4μ, 2–3-guttulate. Smell and taste pleasant. Edible. Woods. Aug.—Oct. Uncommon. (*v.v.*)

1875. **B. carnosus** Rostk. Rostk. Bol. t. 14. *Carnosus*, fleshy.

P. 10–12 cm., *fuscous, or bay brown*, pulvinate, convex, glabrous. St. 6–7 × 2–3 cm., *yellow, streaked with reddish brown*, subequal, firm, *substriate*. Tubes *dark yellow*, sinuato-adnate, long; orifice of pores *dark yellow, large*, angular. Flesh *pallid, dirty yellowish*. Woods. Sept.—Oct. Rare.

1876. **B. vaccinus** Fr. Fr. Sverig. ätl. Svamp. t. 51. *Vaccinus*, pertaining to a cow.

P. 5–10 cm., *chestnut*, convex, then expanded, minutely tomentose, margin obtuse. St. 5–8 × 1–2 cm., *concolorous, or paler than the p.*, bulbous, or attenuated downwards, smooth; *base becoming tawny*, lacunose. Tubes *white, then light yellow*, free; orifice of pores round or subangular, small. Flesh *whitish, reddish under the cuticle of the p.*, firm. Spores *yellow*, oblong fusiform, 12–14 × 4–5μ. Taste mild. Edible. Subcaespitose. Woods. Sept.—Oct. Uncommon. (*v.v.*)

1877. **B. impolitus** Fr. Fr. Sverig. ätl. Svamp. t. 42. *Impolitus*, unpolished.

P. 8–20 cm., *pale yellow brown, or tawny brown*, convex, then more or less expanded, flocculose, at length granuloso-rivulose, sometimes cracking into areolae. St. 6-9 × 3–5 cm., *yellow, often tinged with brownish red when full grown and forming a ring-like zone at the apex*, equal, or subbulbous, *pubescent*. Tubes *pale lemon yellow, then tinged with olivaceous green*, free, or slightly adnate, depressed near the st.; orifice of pores small, round. Flesh *whitish yellow, pale yellow under the cuticle of the p.*, thick. Spores olivaceous, oblong-fusiform, 16–17 × 5–6μ. Smell and taste pleasant. Edible. Woods. June—Oct. Uncommon. (*v.v.*)

1878. **B. candicans** Fr. *Candicans*, shining-white.

P. 8–12 cm., *whitish*, or *pale tan colour*, convex, then expanded, *subtomentose*, becoming smooth. St. 8–10 × 3–4·5 cm., *concolorous, delicately reticulated*, ventricose, or bulbous. Tubes *lemon yellow*, adnate, or sinuate, long; orifice of pores *white*, then *lemon yellow*, roundish. Flesh *white, becoming blue when broken, and then whitish*, thick. Spores dark olivaceous, oblong-fusiform, blunt at the one end, 13–14 × 4μ, 1-guttulate. Taste mild. Woods, and under trees. June—Oct. Uncommon. (*v.v.*)

1879. **B. fragrans** Vitt. *Fragrans*, scented.

P. 6–9 cm., *fuscous umber*, pulvinate, repand, *subtomentose*; margin incurved. St. 7–9 × 3–4 cm., *variegated yellowish and red*, stout, ventricose, often fusiform at the base and ovato-bulbous, *even*. Tubes *yellow*, sinuato-free, fairly long; orifice of pores *yellow*, round. Flesh *yellow*, either unchangeable, or becoming greenish, *at length becoming reddish*, especially at the base of the st. and under the cuticle. Spores dark olivaceous, oblong-fusiform, 10–11 × 4–5μ, 2–3-guttulate. Taste mild, smell pleasant. Edible. Woods, and under oaks. Sept.—Oct. Uncommon. (*v.v.*)

1880. **B. rubiginosus** Fr. *Rubiginosus*, rusty.

P. 5–12·5 cm., *reddish brown*, pulvinate, or convex, soft, pubescent, soon becoming very glabrous, dry; margin acutely incurved, then patent. St. 5–8 × 2·5–3 cm., *whitish, then yellowish*, becoming slightly greyish or yellowish olive when bruised, attenuated upwards, glabrous, *very distinctly reticulated*. Tubes *white*, adnate, short; orifice of pores *white*, angular, unequal. Flesh *white*, *unchangeable*, thick, spongy. Spores "colourless, elongato-fusiform, 12 × 4μ" Massee. Beech woods. Sept.—Oct. Rare.

††Flesh generally yellow, *changing colour on exposure to the air, often immediately turning bluish, or greenish*, sometimes poisonous, or bitter. Pores more or less yellowish, yellow, orange, or red, *immediately tinged blue, green or black when touched*. St. reticulately veined, or punctate, rarely smooth, generally *red*, at least in part.

*Pores at first *cream, lemon yellow, or yellow*, rarely finally becoming reddish. St. *reticulate with white, or yellow veins*, sometimes flesh colour or punctate on the veins. Flesh often bitter.

1881. **B. appendiculatus** (Schaeff.) Fr. Rostk. Bol. t. 26, as *Boletus radicans* Pers. *Appendiculatus*, with a small appendage.

P. 5–15 cm., *brown, bright bay, or fuscous brick colour*, convex, pulvinate, then expanded, *subtomentose*. St. 6–9 × 2·5 cm., *sulphur yellow*, ventricose, rooting, *apex minutely reticulate with white veins*, often tinged rosy towards the base. Tubes *sulphur yellow, becoming greenish when touched*, adnate; orifice of pores round, small. Flesh *yellow, becoming blue on exposure to the air, tinged with rose at the base of the st.*, compact, firm. Spores light yellow, oblong elliptic, 9–11 × 4μ, 1–3-guttulate. Smell and taste pleasant. Edible. Woods. Aug.—Oct. Uncommon. (*v.v.*)

1882. **B. variecolor** B. & Br. B. & Br. no. 1020, t. 13, fig. 3. *Variecolor*, with diverse colours.

P. 4–6 cm., *olivaceous*, convex, subtomentose; margin *involute*. St. 5–8 × 2 cm., *yellowish downwards, rufescent and delicately pubescent upwards*, bulbous, attenuated upwards, apex reticulated. Tubes *yellow*, free; orifice of pores minute. Flesh *pale, here and there inclining to yellow and partially marbled, dark purple under the cuticle of the p.*

Spores pale olive, elliptic fusiform, slightly oblique, 10 × 4μ. Woods. Aug.—Sept. Uncommon.

1883. **B. calopus** Fr. Bull. Soc. Myc. Fr. XXXIV (1918), t. 2.

καλός, beautiful; *πούς*, foot

P. 6–15 cm., *olivaceous*, globose, then convex, pulvinate, subtomentose. St. 7–10 × 2–3 cm., *scarlet throughout, or at the apex, apex often yellow*, conical, then subequal, *reticulated with white, or flesh coloured veins*. Tubes *yellow, becoming bright green, adnate*; orifice of pores minute, *angular*, becoming spotted with greenish blue. Flesh *yellow, becoming blue on exposure to the air*, compact. Spores olivaceous, 10–14 × 4μ, 1–3-guttulate. Said to be poisonous. Woods, especially coniferous woods. July—Nov. Not uncommon. (*v.v.*)

1884. **B. olivaceus** (Schaeff.) Fr. Bull. Soc. Myc. Fr. XXXIV (1918), t. 1.

Olivaceus, olivaceous in colour.

P. 3–8 cm., *olivaceous fuscous*, convex, *pruinose*, becoming smooth; *margin at first inflexed*. St. 5–8 × 2–3 cm., *red, or pale citron yellow, yellowish at the apex and reticulate with white veins, punctate with rose red at the base*, firm, clavate, bulbous, or fusiform. Tubes *light yellow, then olivaceous*, adnate; orifice of pores minute, round, unequal. Flesh *cream colour, becoming blue on exposure to the air*, firm. Spores elliptical, "17μ long, 2–5-guttulate" Quél. Woods, especially beech. Aug.—Sept. Uncommon.

1885. **B. pachypus** Fr.

παχύς, thick; *πούς*, foot.

P. 10–20 cm., *fuscous, then pallid tan*, convex, pulvinate, subtomentose; *margin at first incurved, exceeding the tubes*. St. 5–12·5 × 3–5 cm., *variegated light yellow and red, often wholly intensely blood red, often with a rosy purple zone at the apex of the st.*, firm, ovato-bulbous, then elongated, equal, *reticulated with white veins*. Tubes *light yellow, then somewhat green, free*, somewhat elongated, *shortened round the st.*; orifice of pores *round*, becoming spotted with green, or blue. Flesh *yellow, then azure blue when exposed to the air, becoming reddish at the base of the st.*, compact, thick. Spores olivaceous, oblong elliptic, 9–12 × 4μ. Said to be poisonous. Woods, especially pine. Sept.—Oct. Uncommon. (*v.v.*)

1886. **B. albidus** (Roques) Quél. Roques, Champ. com. et vén. t. 8, fig. 2.

Albidus, whitish.

P. 6–9 cm., *whitish, with a slight greenish tinge*, convex, subtomentose. St. 6–7 × 3–4 cm., *pale citron yellow*, ventricose, *finely reticulated with veins*, which become brownish on handling. Tubes *pale citron yellow*, adnate; orifice of pores round, small. Flesh *cream colour, or pale citron, becoming blue when cut or broken*. Spores deep ochre, boat-shaped, 10–11 × 4–5μ, 1–2-guttulate. Woods, and pastures. July—Oct. Uncommon. (*v.v.*)

**Pores *at first red, or orange, rarely yellow.* St. *reticulated with red veins, often punctate with red,* rarely white, or yellow.

1887. **B. satanas** Lenz. (= *Boletus tuberosus* (Bull.) Quél.) Krombh. Icon. t. 38, figs. 1–6, as *Boletus sanguineus* Pers.

Σατανᾶς, the Devil.

P. 9–20 cm., *at first more or less tinged with red, then becoming brownish, and finally whitish,* globose, then convex, pulvinate, *somewhat viscid, smooth.* St. 7–10 × 5–6 cm., *dingy yellow, reticulated with blood red veins, ovato-ventricose.* Tubes *yellow,* free; orifice of pores *yellow, then rubiginous, and finally orange,* round, minute. Flesh *white, then cream colour, becoming bluish or greenish on exposure to the air, reddish in the st.,* thick, firm. Spores olivaceous, oblong elliptic, 11–13 × 4–5μ, 2–3-guttulate. Taste mild. Woods, and heaths. July—Oct. Uncommon. (*v.v.*)

1888. **B. luridus** (Schaeff.) Fr. Rolland, Champ. t. 85, no. 189.

Luridus, lurid in colour.

P. 5–20 cm., *umber olivaceous, or fuliginous,* hemispherical, convex, then plane, *tomentose.* St. 5–15 × 3–6 cm., *yellow, reticulated with blood red veins,* equal, or incrassated at the base. Tubes *yellow, at length becoming green,* free; orifice of pores *at first vermilion, then orange,* round, small. Flesh *yellow, becoming immediately deep indigo on exposure to the air, and then again yellow,* reddish at the *base of the tubes and at the base of the st.,* thick, compact. Spores yellowish, oblong fusiform, or oblong elliptical, 12–13 × 4μ, 2–3-guttulate. Taste pleasant. Edible. Woods, especially deciduous woods, heaths, and pastures. May—Dec. Common. (*v.v.*)

1889. **B. erythropus** (Pers.) Quél. ἐρυθρός, red; πούς, foot.

P. 10–15 cm., *brown, or bay, often tawny rufescent,* convex, *minutely* pubescent. St. 5–12 × 2–4 cm., *yellow, punctate with red,* ventricose, *minutely tomentose.* Tubes *yellow,* free; orifice of pores *dark blood red,* round, or subangular. Flesh *yellow, becoming deep indigo in the p. and upper part of the st. on exposure to the air, blood red in the rest of the st., yellow at the base of the tubes,* thick, compact. Spores yellow, fusiform, 12–13 × 4μ, 2–3-guttulate. Taste pleasant. Woods, especially coniferous woods. Aug.—Nov. Not uncommon. (*v.v.*)

1890. **B. purpureus** Fr. Fr. Sverig. ätl. Svamp. t. 41.

Purpureus, purple.

P. 7–12 cm., *purplish red, or violet, rarely brownish,* hemispherical, pulvinate, *somewhat velvety,* opaque, dry. St. 6–11 × 2–3 cm., *yellow, reticulate with purple veins and dots, and often dotted on the veins,* equal, attenuated at the base. Tubes *light yellow, becoming greenish,* somewhat free; orifice of pores *purple orange,* round, minute. Flesh *yellow, becoming bluish on exposure to the air when young and reddish at the base of the st.*, compact. Spores olivaceous, oblong-fusiform, 11–12 × 5–6μ, 2–4-guttulate. Woods, and pastures. June—Oct. Uncommon. (*v.v.*)

1891. **B. Queletii** Schulzer, var. **rubicundus** René Maire. (= *Boletus purpureus* Fr. (forma) Massee, Brit. Fung. Fl. I, 290.) Bull. Soc. Myc. Fr. XXVI, 195, t. V, figs. 5–6. *Rubicundus*, ruddy.

P. 5–16 cm., *reddish purple*, or *reddish brown*, hemispherical, then convex, *pruinose, then subtomentose*; margin at first involute, pruinose and *flesh colour*. St. 7–10 × 2–3 cm., *yellow straw colour*, then *pale ochraceous, densely punctate with red and dark purple at the base*, more or less bulbous and fusiform, slightly rooting. Tubes *yellowish*, then *pure yellow, becoming blue and finally black* when touched, free, sinuate; orifice of pores *greyish orange*, then *purple orange* or *saffron colour*, small, round, or slightly irregular. Flesh *yellow, becoming blue and finally blackish, reddish purple at the base of the st.* Spores olivaceous, elliptical-oblong, 8–10 × 4–5μ, 1–3-guttulate. Taste pleasant. Calcareous woods. Uncommon.

B. St. *fibrillosely fleshy*, generally *floccosely squamulose and mucronate, rarely rugosely or reticulately ribbed.* Pores *white, or whitish*, rarely yellow. Tubes concolorous, free, or sinuate, rarely adnate. Flesh generally *white, and soft*, often *becoming bistre colour on exposure to the air, then becoming black.* Large or medium in size.

1892. **B. duriusculus** Schulz. Boud. Icon. t. 150. *Duriusculus*, somewhat hard.

P. 5–15 cm., *grey fuliginous, or grey bistre*, convex, *minutely tomentose*, viscid in wet weather, often areolately cracked when dry. St. 10–20 × 1·5–4 cm., *white, densely striate and black punctate*, the striae often anastomosing in a reticulate manner, *often spotted greenish* at the attenuated, or incrassated base, *very firm*, equal, or subventricose. Tubes *dirty white*, somewhat free, fairly long; orifice of pores minute, round, or subangular. Flesh *white, becoming reddish on exposure to the air especially in the p. and apex of the st., then becoming blackish, very firm.* Spores fuliginous in the mass, pale under the microscope, oblong fusiform, 13–15 × 4·5–6μ, 2–multi-guttulate. Taste pleasant. Edible. Woods, heaths, and pastures, especially under poplars. Aug.—Oct. Uncommon. (*v.v.*)

1893. **B. versipellis** Fr. (= *Boletus rufus* (Schaeff.) Quél.; *Boletus aurantiacus* Bull. sec. Quél.) Rolland, Champ. t. 87, no. 193, as *Boletus aurantiacus*. *Versipellis*, changeable in appearance.

P. 5–15 cm., *rufous*, hemispherical, pulvinate, *dry, tomentose*, then scaly, and becoming even; *margin often appendiculate with the remains of the membranaceous, fugacious veil.* St. 8–12 × 3–5 cm., *whitish*, *covered with rufous or greyish, mucronate flocci,* attenuated upwards, often tinged greenish when eaten by slugs or snails. Tubes *dingy white,* free, long; *orifice of pores often grey or blackish at first,* minute, round. Flesh *white, often greenish near the cuticle of the stem,* thick, compact. Spores ochraceous, oblong fusiform, 16–18 × 5–7μ, 1–4-guttulate. Smell and taste pleasant. Edible. Woods, heaths, and pastures. July—Nov. Common. (*v.v.*)

1894. **B. scaber** (Bull.) Fr. (= *Boletus nigrescens* Roze & Rich. sec. Quél.) Rolland, Champ. t. 87, no. 192. *Scaber,* rough.

P. 5–20 cm., *ochraceous fuliginous, greyish bistre, or brownish bistre,* hemispherical, pulvinate, *smooth, viscid when moist,* at length rugulose, or rivulose; margin at first furnished with a cortina. St. 7–20 × 2–4 cm., *whitish, or greyish, rough with fibrous scales that become blackish with age, often greenish or bluish especially towards the base when eaten by slugs or snails,* attenuated upwards. Tubes *white, then dingy,* free, long; orifice of pores minute, round. Flesh *white,* watery, soft. Spores ochraceous, oblong fusiform, 16–18 × 5–6·5μ, multi-guttulate. Taste pleasant. Edible. May—Dec. Common. (*v.v.*)

var. **niveus** Fr. Rostk. Bol. t. 48, as *Boletus holopus* Rostk. *Niveus,* snow-white.

Differs from the type in the *white p. becoming greenish grey at the disc, and in the white granularly punctate st. which becomes greenish grey at the base.* Spores pale ochraceous, oblong fusiform, 14–16 × 5μ, multi-guttulate. Deciduous woods. Sept.—Oct. Not uncommon. (*v.v.*)

1895. **B. nigrescens** Roze & Rich. (= *Boletus scaber* (Bull.) Fr. sec. Quél.) Trans. Brit. Myc. Soc. III, t. 20. *Nigrescens,* becoming black.

P. 4–12 cm., *yellowish,* convex, *tomentose,* cracking with age. St. 6–11 × 2–4 cm., *yellowish, dotted with grey scales,* ventricose, attenuated at both ends, striate. Tubes *white, soon becoming bright yellow,* free; orifice of pores small, ·5 mm. across, round, or oblong, unequal. Flesh *yellowish white, becoming red on exposure to the air, and finally dark brown.* Spores olivaceous, fusiform, 12–16 × 5–6μ, 1–3-guttulate. Taste pleasant. Edible. Deciduous woods. July—Sept. Uncommon. (*v.v.*)

1896. **B. rugosus** Fr. Rostk. Bol. t. 41. *Rugosus,* wrinkled.

P. 5–6 cm., *bay, or brown,* convex, pulvinate, *dry,* smooth. St. 7–12 × 2–3 cm., *whitish, or ochraceous,* attenuated upwards from the subbulbous base, *longitudinally ribbed; ribs dark, anastomosing, or reticulate.* Tubes *whitish, then ochraceous,* free; orifice of pores small, round. Flesh *white, reddish under the cuticle of the p.,* compact. Spores olivaceous, fusiform, 9–10 × 3–4μ. Taste mild. Edible. Woods. Aug.—Oct. Uncommon. (*v.v.*)

APHYLLOPHORALES.

Hymenium indefinite, increasing by centrifugal growth, fully exposed from the first, amphigenous, or unilateral, lining the interior of pores, covering the surface of teeth, tubercles or anastomosing gills, or forming a smooth surface.

I. POROHYDNINEAE.

Receptacle pileate, stipitate, sessile, or resupinate; hymenium inferior.

1. Polyporaceae.

Hymenium lining tubes coherent throughout their length, forming a layer distinct from the substance of the pileus, sometimes becoming torn into teeth, or gill-like plates, and separated by dissepiments sterile on the edge.

Polyporus (Micheli) Fr.

(πολύς, many; πόρος, a pore.)

Pileus fleshy, cheesy, coriaceous or corky, often at length becoming hard with age; entire, lobed, excentric, or dimidiate, simple, or branched. Stem central, lateral, or none, simple or branched. Tubes homogeneous, or heterogeneous, long, or short; orifice of pores round, angular, entire, torn, or toothed. Flesh white, or coloured. Spores white, or coloured, elliptical, pip-shaped, globose, subglobose, pruniform, oblong, or elliptic fusiform; smooth, punctate, or verrucose. Cystidia present, or absent, hyaline, or coloured. Annual, or perennial. Growing on wood, or on the ground; solitary, caespitose, imbricate, or connate at the base.

A. Stipitate, or caespitose.

I. P. *thin*, hemispherical, generally *depressed, cup-shaped*. St. *thin, fibrillosely corky*. Flesh *coloured*. Tubes *homogeneous*, short; pores polygonal, or rounded, *tawny* or *brown*. Spores hyaline, or yellowish. Cystidia coloured, or wanting. Terrestrial. Perennial.

*P. tomentose, or velvety. Cystidia coloured.

1897. **P. tomentosus** Fr. Kalchbr. Icon. t. 38, fig. 1, as *Polyporus Kalchbrenneri* Fr. *Tomentosus*, woolly.

P. 5–10 cm., *tawny ferruginous, nankeen yellow*, or *yellowish cinnamon*, convex, then plane and cyathiform, leathery, *zoned, radiately rugose, clothed with tawny ferruginous hairs*; margin thin. St. 2·5 cm. × 4–8 mm., *fuscous umber, becoming blackish*, subequal, firm, *pulverulent*, or *tomentose*, sometimes a mere base or central point. Tubes *greyish, becoming fuscous*, adnate, 1 mm. long; orifice of pores *grey*, glistening, minute, entire. Flesh *fuscous umber*, 2 mm. thick, firm. Spores hyaline, "oblong pruniform, 9–10 × 2·5–3μ" Sacc. Cystidia coloured, abundant. Coniferous woods. Sept.—Oct. Rare.

**P. silky, or glossy. Cystidia wanting.

1898. **P. Montagnei** Fr. (= *Polystictus cinnamomeus* (Jacq.) Sacc. sec. Lloyd.) Trans. Brit. Myc. Soc. II, t. 11, as *Polystictus Montagnei* Fr.

J. F. C. Montagne, the eminent French mycologist.

P. 3–8 cm., *ferruginous*, cyathiform, irregular, uneven, *zoneless, tomentose, becoming smooth*; margin thin. St. 2–3 cm. × 5–9 mm., *concolorous*, unequal, pubescent. Tubes *yellowish white, then ferruginous*, short, decurrent; orifice of pores *large*, round, obtuse, *entire*. Flesh *ferruginous*, somewhat corky, thick. Spores hyaline, ovoid pruniform, 5–6 × 4–5μ, punctate. Woods, and heaths. Sept.—Oct. Uncommon. (*v.v.*)

1899. **P. cinnamomeus** (Jacq.) Sacc. Lloyd, Myc. Notes, Polyporoid Issue, no. 1, fig. 200. *Cinnamomeus*, cinnamon.

P. 2–3 cm., *bright cinnamon*, plano-depressed, then subinfundibuliform, *somewhat corky, flaccid*, velvety becoming glabrous, silky, shining, at first with darker fuscous zones, then becoming tawny, and zoneless. St. 3–4 cm. × 4–5 mm., *concolorous*, attenuated at the base, or somewhat bulbous, *velvety*. Tubes *fuscous cinnamon, becoming tawny when dry*, adnate, 1–2·5 mm. long; orifice of pores *greyish*, somewhat large, pentagonal, or hexagonal. Flesh *concolorous*, spongy. Spores *yellowish*, broadly elliptical, with an oblique apiculus, 6–7 × 4–5μ. Smell unpleasant. Under trees, and burnt places. Oct. Rare. (*v.v.*)

1900. **P. perennis** (Linn.) Fr. Rolland, Champ. t. 89, no. 197.

Perennis, perennial.

P. 3–8 cm., *cinnamon, then date brown, often becoming whitish with age*, hemispherical, plane, or cyathiform, coriaceous, tough, thin, *zoned*, velvety, becoming smooth; margin fimbriate, then entire. St. 2·5–5 cm. × 3–8 mm., *tawny*, or *concolorous*, attenuated upwards, often bulbous at the base, velvety, firm. Tubes *tawny*, decurrent, 2–3 mm. long; orifice of pores *silvery, becoming tawny*, minute, angular, *acute, then torn*. Flesh *tawny*, fibrillosely leathery, tough. Spores hyaline, elliptical, 8–9 × 4–5μ, 1-guttulate. Charcoal heaps, and about stumps. Jan.—Dec. Common. (*v.v.*)

II. P. *thin*, hemispherical, or excentric, firm, squamulose, villose, or smooth. St. generally *thin*, corky, rarely branched, sometimes black at the base. Flesh *leathery, white*. Tubes *heterogeneous*; pores round, or polygonal, small, *white*. Spores hyaline. Perennial. Growing on wood.

*P. at first villose, or squamulose.

1901. **P. brumalis** (Pers.) Fr. *Brumalis*, pertaining to winter.

P. 2–10 cm., *fuliginous, becoming pale and tan colour with age*, convex, then plane, more or less umbilicate, fleshy pliant, then coriaceous, *villose*, or squamulose, becoming smooth; margin fimbriato-ciliate, or velvety. St. 1·5–5 cm. × 4–8 mm., *concolorous, velvety*, or squamulose. Tubes white, decurrent, 1 mm. long; orifice of pores *white, becoming yellowish*, round, angular, or oblong, small, toothed. Flesh *whitish*, very firm, thin at the margin. Spores white, oblong, often curved, 6–8 × 2–3μ, 1-2-guttulate. Dead branches, and twigs in woods, and wood heaps. Sept.—May. Not uncommon. (*v.v.*)

1902. **P. arcularius** (Batsch) Fr. Trans. Brit. Myc. Soc. II, t. 16. *Arcula*, a casket.

P. 1–3 cm., *fuscous, becoming yellow with age*, convex, subumbilicate, pliant, then coriaceous, *at first covered with fuscous squamules*, then becoming smooth and ochraceous; *margin strigose*. St. 1–2·5 cm. × 1–2 mm., *greyish fuscous, or bistre*, subsquamulose, becoming smooth. Tubes *whitish*, adnato-decurrent, 1–2 mm. long; orifice of pores *white, becoming tawny, oblong, rhomboidal*, thin, rather large, entire. Flesh *white*, thin, leathery. Spores white, globose, 3μ. Dead twigs, and branches in woods, and shrubberies. Sept.—May. Uncommon. (*v.v.*)

1903. **P. lentus** Berk. Berk. Outl. t. 16, fig. 1. (? = *Polyporus tubarius* Quél.) *Lentus*, pliant.

P. 2·5–5 cm., *reddish brown, becoming ochraceous*, convex, umbilicate, thin, tough, furfuraceo-squamulose, becoming smooth. St. 1–2·5 cm. × 4–9 mm., *concolorous*, central, or excentric, straight, or curved, *hispid*, or *furfuraceous*, often covered with pores to the base. Tubes *white*, decurrent, 2–3 mm. long; orifice of tubes *white, large*, angular, irregular. Flesh *white*, leathery, tough. Spores "white, elliptic fusiform, 12 × 4–5μ" Massee. Dead roots, fallen branches, and gorse stems. April—Sept. Not uncommon. (*v.v.*)

1904. **P. melanopus** (Swartz) Fr. *μέλας*, black; *πούς*, foot.

P. 3–10 cm., *white*, then *yellowish fuscous* or *greyish bistre*, convexo-plane and umbilicate, then infundibuliform, fleshy pliant, *at first minutely flocculose, or pruinose*. St. 2–4 cm. × 6–10 mm., *dark brown, bistre*, or *black*, gradually incrassated upwards, or thickened downwards, excentric, *minutely velvety*. Tubes *white*, decurrent, ·5–1 mm. long; orifice of pores *white*, minute, round, fimbriate under a lens. Flesh *white*, thick, soft. Spores white, oblong, or pip-shaped, 7–8 × 3μ, 1-guttulate. Smell often pleasant. Dead pine roots, and branches in woods. Aug.—Oct. Uncommon. (*v.v.*)

**P. and st. glabrous. St. often black at the base.

1905. **P. fuscidulus** (Schrad.) Fr. *Fuscidulus*, somewhat dark.

P. 2·5–6 cm., *fuscous yellowish*, convexo-plane, fleshy, pliant, subcoriaceous, smooth. St. 2·5–5 cm. × 4 mm., *fuscous*, then *yellow*, equal, or thickened at both ends, *smooth*. Tubes *yellowish*, adnate, ·7 mm. long; orifice of pores *yellowish*, subangular, *quite entire*, minute. Flesh *yellowish white*, thin, tough. Spores "hyaline, elliptic-oblong, 5–6 × 2μ" Massee. On twigs, and chips of wood. Sept.—Feb. Rare.

1906. **P. leptocephalus** (Jacq.) Fr. λεπτός, thin; κεφαλή, head.

P. 2–3 cm., *pale*, then *fawn colour*, convexo-plane, pliant, then coriaceous, *thin*, smooth; margin rather wavy. St. 1–2·5 cm. × 3–4 mm., *pallid*, smooth. Tubes *whitish*, adnate; orifice of pores *whitish*, round, minute, obtuse. Flesh *white*, leathery. Spores "hyaline, oblong pruniform, 8μ, guttulate" Quél. On stumps, and dead birch trunks in woods. Oct.—March. Rare.

1907. **P. nummularius** (Bull.) Quél. Rostk. Polyp. t. 12. *Nummularius*, like money.

P. 1–2·5 cm., *whitish cream colour*, then *ochraceous and becoming whitish*, convexo-plane, hemispherical, rarely umbonate, thin, smooth. St. 1–2 cm. × 3–5 mm., *blackish bistre*, *whitish cream at the apex*, equal, or attenuated downwards, *firm*, pruinose. Tubes *white, then yellowish*, decurrent, 1–2 mm. long; orifice of pores *white*, then *straw colour*, minute, round, ciliate under a lens. Flesh *white*, *hard*, woody. Spores white, oblong, 7–9 × 2·5–3μ, 1–2-guttulate, sometimes curved. Dead twigs, and branches in woods, and wood heaps. Aug.—Dec. Common. (*v.v.*)

1908. **P. picipes** Fr. Pers. Icon. pictae rar. fung. t. IV, fig. 1, as *Boletus infundibulis*. *Pix*, pitch; *pes*, foot.

P. 5–10 cm., *pallid, then chestnut*, or *pale yellowish livid* and *chestnut at the disc*, cyathiform, or depressed at the disc or behind, fleshy coriaceous, then *rigid*, smooth; margin *scalloped*. St. 2–7 × ·5–2 cm., *brownish bistre, or olivaceous*, punctate with black up to the pores, excentric, or lateral, *pruinosely velvety*, becoming smooth, firm. Tubes *white, then yellowish*, decurrent, 1·5 mm. long; orifice of pores *white, then yellowish, or pinkish*, very small, round. Flesh *white*, tough. Spores hyaline, pruniform, 7–8 × 4μ, 1-guttulate. Smell pleasant. On pollarded willows, and stumps. July—Dec. Common (*v.v.*)

1909. **P. varius** Fr. (= *Polyporus calceolus* (Bull.) Quél.; *Polyporus elegans* (Bull.) Fr.) Grev. Scot. Crypt. Fl. t. 202. *Varius*, variable.

P. 5–12 cm., *pallid ochraceous, then tan colour, or cinnamon streaked with brown*, very coriaceous, cyathiform, or plane and depressed at the disc or beyond, smooth. St. 1·5–5 cm. × 4–12 mm., *pale ochraceous at the apex, black lower down*, central, excentric, or lateral, *smooth*, tough, firm. Tubes *whitish*, then *ochraceous or cinnamon*, decurrent, 1–3 mm. long; orifice of pores *white, becoming yellowish*, small, round. Flesh *white, then yellowish*, very tough, woody, thin. Spores white, oblong or elliptical, 7 × 2·5–3μ. Smell slight. Taste bitter. On stumps, trunks, and fallen branches. July—Nov. Common. (*v.v.*)

1910. **P. petaloides** Fr. πέταλον, a leaf; εἶδος, like.

P. 5–6 cm., *chestnut fuscous*, spathulate, *submembranaceous, rugose, smooth*, flaccid when moist. St. 2 cm. × 8–10 mm., *whitish*, lateral, ascending, compressed, expanding into the p., smooth, *dilated at the base into a shield-like organ of attachment.* Tubes *shining white*, decurrent, very short; orifice of pores *white*, very small. Spores "almost colourless, elliptical, 6 × 2·5–3μ" Massee. Old stumps. Jan. Rare.

1911. **P. osseus** Kalchbr. (= *Polyporus albidus* (Schaeff.) Quél.) Kalchbr. Icon. t. 34, fig. 2. *Osseus*, like bone.

P. 2·5–8 cm., *white, becoming yellowish*, convex, often imbricate, lobed, smooth. St. *white*, simple, branched, or obsolete, smooth, hard, firm. Tubes *white*, decurrent, 1 mm. long; orifice of pores *white*, round, at length torn, or denticulate, minute. Flesh *white*, compact, *becoming firm*. Spores white, subglobose, 4–5μ. Smell somewhat acid. Taste becoming bitter. Larch stumps. Oct. Rare.

III. P. simple, rarely compound. *Stipitate, or caespitose.* Flesh *soft*, or slightly leathery, *fragile*, or firm, *white*. Tubes *heterogeneous*, pores *round*, or *polygonal, of medium size, white*. Spores white, or coloured. Annual. Growing on the ground, rarely on wood.

1912. **P. leucomelas** (Pers.) Fr. Trans. Brit. Myc. Soc. II, t. 15. λευκός, white; μέλας, black.

P. 4–12 cm., *black fuliginous*, convex, then expanded, often irregular and lobed, fleshy, fibrillose, disc often squamulose. St. 2·5–10 × 2–3 cm., *concolorous, pale at the apex*, equal, or tuberous, *subtomentose*, sometimes squamulose from the breaking up of the cuticle, firm. Tubes *white*, decurrent, 1–2 mm. long; orifice of pores *white, becoming grey*, rather large, entire, then torn. Flesh *white, reddish when broken*, and *often blackish in the* st., thick, soft. Spores white, subglobose, 5–6μ, *warted*. Taste slightly bitter. Edible. Coniferous woods. Sept.—Oct. Rare. (*v.v.*)

1913. **P. flavo-virens** Berk. & Rav. *Flavus*, yellow; *virens*, green.

P. 8–10 cm., *dirty yellowish green*, pulvinate, or depressed, irregularly lobed, fleshy, subtomentose. St. 5 × 2–3 cm., *pallid, subcon-*

colorous, incrassated upwards. Tubes *white, then yellowish green*, very decurrent, ·5 mm. long; orifice of pores *yellowish*, angular, *very irregular*, finally torn. Flesh *white*, thick, soft. Spores dirty green, broadly fusiform, 15–18 × 7–8μ. Under pines. Sept. Rare. (*v.v.*)

IV. P. simple, rarely compound. St. generally *thick*. Flesh soft, or leathery, *white*. Tubes *heterogeneous*; pores *honey-comb-like, broad*. Spores *white* or *pale coloured*. Annual. Growing on the ground, or on wood.

*P. velvety, hispid or squamulose.

1914. **P. squamosus** (Huds.) Fr. Grev. Scot. Crypt. Fl. t. 207; and forma *erecta* Bres. Fung. Trid. t. 133. *Squamosus*, scaly.

P. 10–60 cm., *ochraceous, variegated with broad, adpressed, centrifugal, darker, fuscous scales*, fan-shaped, or hemispherical, convex, then plane and at length concave, often umbilicate when young, fleshy pliant, often imbricate and flattened. St. 1–5 × 1–5 cm., *ochraceous, base blackish*, excentric, lateral, or wanting, apex *reticulate*. Tubes *white, then yellowish*, adnato-decurrent, 5–10 mm. long; orifice of pores *pallid*, at first minute, then large, angular, and torn. Flesh *white*, soft, becoming leathery, thick. Spores white, oblong, 10–12 × 4–5μ, 1–2-guttulate. Smell strong. Said to be edible. On trunks of ash, apple, walnut, maple, elm, yew, oak, birch, lime, etc. April—Dec. Common. (*v.v.*)

1915. **P. Boucheanus** (Klotzsch) Fr. (= *Polyporus Forquignoni* Quél. sec. Lloyd.) Augustus Bouché.

P. 3–7 cm., *bright tawny*, or *yellowish*, plane, or cyathiform, fleshy, *smooth, then breaking up into floccose, erect squamules*. St. 3–4·5 × 1–2 cm., *concolorous, becoming fuscous at the base*, excentric, or lateral, rarely central, *tomentose*. Tubes *whitish becoming yellow*, adnato-decurrent, 2–5 mm. long; orifice of pores *yellowish, large*, angular, toothed. Flesh *yellowish*, soft, becoming firm. Spores *white*, oblong, or oblong fusiform, 14–16 × 6μ, 1-guttulate. On dead oak branches, twigs, and burnt gorse stems. July—Oct. Uncommon. (*v.v.*)

1916. **P. Michelii** Fr. Rostk. Polyp. t. 1.
Pier Antonio Micheli, an early mycologist.

P. 5–10 cm., *yellowish white*, depressed, repand, fleshy pliant, *minutely silky, subsquamulose*. St. 2–5 × 1 cm., *white, becoming fuscous at the base*, somewhat lateral, bulbous, rough. Tubes *white*, adnato-decurrent, 1–2 mm. long; orifice of pores *white, large*, 2 mm. across, *round, or oblong*. Flesh *white*, firm. Spores "almost colourless, elongato-elliptical, 16–17 × 7μ" Massee. Trunks, and stumps, especially willow. Sept. Rare.

**P. smooth.

1917. **P. Rostkovii** Fr. Rostk. Polyp. t. 17, as *Polyporus infundibuliformis* Rostk.
Dr Friedrich Wilhelm Theophilus Rostkovius, one of the editors and illustrators of Sturm's Deutschlands Flora.

P. 3–15 cm., *smoke colour*, or *sometimes yellowish*, dimidiato-infundibuliform, fleshy pliant, thin, *smooth, even.* St. 7–15 cm., *black*, excentric, often connate and caespitose at the thickened base, *reticulated.* Tubes *white*, then *dingy yellowish*, very decurrent, 4–8 mm. long; orifice of pores *white*, then *ochraceous, large, pentagonal*, or oblong, acute, toothed. Flesh *white*, soft. Spores "almost colourless, elongato-elliptical, 14–16 × 5–6μ" Massee, "guttulate" Quél. Smell pleasant. Stumps, and trunks, especially ash. July—Jan. Uncommon.

V. P. *compound*, many pileoli arising from a common trunk, or tubercle, or caespitose and imbricate. Spores white, or pale yellowish. Of large size, growing on wood or the roots of trees.

A. P. fleshy, *firm*, many pileoli *arising from a common stalk.*

1918. **P. umbellatus** Fr. Rolland, Champ. t. 91, no. 200.
Umbella, a parasol.

P. 1–4 cm., *fuliginous, rufous*, or *pallid light yellow, rarely white*, convex, *umbilicate, entire*, pruinose, or villose. St. 2–3 × ·5–1 cm., *white, branched*, arising from a common stalk which is *often developed from a sclerotium, each branch giving rise to a separate p.*, pruinose. Tubes *white*, decurrent, ·5 mm. long; orifice of pores *white*, minute, round, or angular. Flesh *white*, soft, elastic. Spores white, oblong, or pip-shaped, 7–9 × 3μ, 1–2-guttulate. Smell pleasant. Edible. Forming large tufts on and around old stumps. July—Oct. Not uncommon. (*v.v.*)

1919. **P. frondosus** (Fl. Dan.) Fr. Rolland, Champ. t. 91, no. 201.
Frondosus, leafy.

P. 2–6 cm., *fuliginous grey*, or *greyish tan colour*; *margin often white*, dimidiate, spathulate, lobed, intricately recurved, *rugose*, pruinose, or villose. St. 10–30 × 5–10 cm., *white, becoming discoloured, sparingly branched*, smooth. Tubes *white*, decurrent, ·5 mm. long; orifice of pores *white, very small*, round, or polygonal, then denticulate. Flesh *white*, becoming discoloured, fibrous, firm. Spores white, subglobose, 6 × 5μ, punctate. Smell of new meal. Edible, but rather tough. Forming large tufts on oaks, hornbeam, and old stumps. Sept.—Oct. Not uncommon. (*v.v.*)

1920. **P. intybaceus** Fr. Rolland, Champ. t. 92, no. 202.
ἔντυβον, chicory.

P. 3–6 cm., *pale yellowish inclining to fuscous, nut colour becoming brownish, much branched, and divided up into numerous spathulate lobes,* undulate, sinuous, often conchate. St. 2–3 × 1·5–4 cm., *white, gradually spreading and dividing into the lobes of the p.* Tubes *white,* decurrent, ·5 mm. long; orifice of pores *white, becoming fuscous,* minute, round. Flesh *white, often becoming reddish,* soft, elastic. Spores white, elliptical, 6–7 × 3μ. Smell often like that of mice. Edible but rather tough. Forming large tufts on beech, oak trees and on stumps. Sept—Dec. Not uncommon. (*v.v.*)

1921. **P. cristatus** (Pers.) Fr. Krombh. t. 48, figs. 15 and 16.
Cristatus, crested.

P. 5–10 cm., *rufous greenish, or greenish yellow,* entire, or dimidiate, or spathulate, scalloped, or lobed, *depressed, subpulverulently villose, then rimosely squamulose.* St. 2·5–6 × 1–2 cm., *white,* or *lemon yellow, becoming discoloured,* irregularly shaped, connate, rarely simple, glabrous. Tubes *whitish,* decurrent, 1–2 mm. long; orifice of pores whitish, becoming discoloured, minute, angular, toothed. Flesh *white, becoming discoloured,* soft, fragile. Spores white, subglobose, 5–6 × 5μ, with a large central gutta. Beech woods. Sept.—Oct. Rare. (*v.v.*)

B. P. *at first soft and succulent,* then dry and fragile, *arising from a tubercle, or caespitose.* Pores yellow, or flesh colour.

*Flesh white, or yellowish.

1922. **P. sulphureus** (Bull.) Fr. Grev. Scot. Crypt. Fl. t. 113.
Sulphureus, like sulphur.

P. 10–40 cm., *reddish yellow, or orange, becoming paler with age,* imbricated, undulated, pruinose, sessile, rarely stalked. Tubes *sulphur yellow, becoming paler,* ·5 mm. long; orifice of pores *bright sulphur yellow,* becoming pale, minute, round. Flesh *light yellowish, then white,* soft, *cheesy,* often exuding a sulphur yellow milk when broken and quite fresh. Spores white, elliptical, 7–8 × 5μ, minutely papillose. Taste acid. Said to be edible. On stumps, and trunks of willows, oaks, alders, walnuts, poplars, apples, ashes, yews, pines, etc. May—Nov. Common. (*v.v.*)

var. **ramosus** (Bull.) Quél. *Ramosus,* branching.

Differs from the type in *dividing up into several digitate-like, cylindrical branches covered with the pores.* On oaks, and willows. Aug.—Sept. Not uncommon. (*v.v.*)

var. **albolabyrinthiporus** Rea.
Albus, white; λαβύρινθος, intricate; πόρος, a pore.

Differs from the type in the *white, labyrinthiform, torn pores,* and in *the flesh being white from the first.* On an oak. Nov. Uncommon. (*v.v.*)

1923. **P. imbricatus** (Bull.) Fr. (= *Polyporus sulphureus* (Bull.) Fr. sec. Lloyd.) Rostk. Polyp. t. 21.

Imbricatus, covered with tiles.

P. 50–100 cm., *yellowish tawny*, or *buff*, *becoming pale*; *margin pale*, somewhat zoned, imbricated, very broad, lobed, and undulated, glabrous, sessile, rarely stalked. Tubes *pale*, *becoming dirty yellowish*, long, thin; orifice of pores *pale*, *becoming concolorous with the p.*, minute, round. Flesh *becoming fuscous when moist*, *white when dry*, somewhat firm, *fibrillosely cheesy*. Taste bitter. Smell "like Gentian root" Bulliard, or "seed-cake" W. G. Sm. Oak trunks. June—Nov. Rare.

1924. **P. Herbergii** (Rostk.) B. & Br. Rostk. Polyp. t. 18.

Herbergius.

P. 10–20 cm., *bright rusty bay*, *becoming sulphur yellow towards the margin*, imbricated, *minutely velvety*, becoming almost glabrous, sessile. Tubes *pale grey*, 4–8 mm. long; orifice of pores pale grey, *labyrinthiform*, unequal, torn, and toothed. Caespitose. On trunks. Oct. Rare.

**Flesh deeply coloured.

1925. **P. spongia** Fr. (= *Polyporus Schweinitzii* Fr. sec. Lloyd.) Fr. Icon. t. 180, fig. 2. *σπογγιά*, a sponge.

P. 5–30 cm., *brownish ferruginous*, *becoming tawny ferruginous when dry*, *dimidiate*, connate in broad, dense tufts, or imbricate, then flattened, *wrinkled*, *rugulose*, strigosely tomentose, sessile, or stalked. Tubes *light yellow*, *becoming brownish*, 2 mm. long; orifice of pores *light yellow*, *soon becoming brownish*, small, round, or angular, entire. Flesh *rhubarb colour*, *becoming paler*, *spongy*, *soft*, finally fragile. Spores very pale yellow, elliptical, $7 \times 4\mu$. Caespitose. Coniferous stumps. Oct.—Nov. Uncommon. (*v.v.*)

1926. **P. Schweinitzii** Fr. Fr. Icon. t. 79, fig. 3.

Ludwig David von Schweinitz, the American mycologist.

P. 10–40 cm., *bright tawny*, *disc becoming date brown with the exception of the yellowish tawny margin*, and finally *becoming entirely fuscous*, regular and plano-cup-shaped, or irregular, dimidiate, and imbricate, *rugose*, *strigosely tomentose*, fibrillose. St. $3\text{–}12 \times 5\text{–}6$ cm., *ferruginous*, sometimes wanting, strigose. Tubes *greenish yellow*, decurrent, 3–5 mm. long; orifice of pores *greenish yellow*, broad, angular, often irregular. Flesh *rhubarb colour*, *becoming fuscous*, spongy, then fibrillose, and finally fragile. Spores white, elliptical, $7\text{–}8 \times 4\mu$, with a large central gutta. On stumps, and roots of conifers. July—Nov. Not uncommon. (*v.v.*)

P. rufescens Fr. = **Daedalea biennis** (Bull.) Quél.

C. P. *firm, tough, subcoriaceous,* corky, or leathery. *Caespitose.*

1927. **P. giganteus** (Pers.) Fr. (= *Polyporus acanthoides* (Bull.) Quél.) Boud. Icon. t. 153. γίγας, a giant.

P. 10–80 cm., *date brown, at first pale, then brownish yellow, disc at length black,* densely imbricated, dimidiate, very broad, flaccid, subzoned, rivulose, depressed behind, cuticle breaking up into granules or fibrillose squamules. St. 3–10 × 2–5 cm., *whitish,* connato-branched from a common tubercle, sometimes wanting. Tubes *whitish,* decurrent, 1–2 mm. long; orifice of pores *whitish, becoming fuliginous and black when touched or rubbed,* round, or angular, minute. Flesh *white, becoming black,* tough, subcoriaceous. Spores white, globose, 4–5μ, with a large central gutta. Smell sour. Taste unpleasant. Forming dense masses at the base of beeches, oaks, elms, chestnuts, and robinias. July—Jan. Common. (*v.v.*)

1928. **P. acanthoides** (Bull.) Fr. ἄκανθος, acanthus; εἶδος, like.

P. 5–90 cm., *ferruginous,* or *pale chestnut,* densely imbricated, infundibuliform, inciso-dimidiate, subzoned, longitudinally rugose, thin. St. *white, then rufescent,* connato-branched. Tubes *white, then rufescent,* short; orifice of pores *white, then rufescent, lamelloso-sinuate, thin,* toothed. Flesh *faintly rufous,* thin, 4–6 mm. thick, pliant, then coriaceous. Spores white, "subglobose, 4 × 3μ" Massee. In dense clusters on trunks, roots, and buried wood. Sept. Rare.

1929. **P. alligatus** Fr. (= *Polyporus imberbis* (Bull.) Quél.) *Alligatus,* bound up.

P. 2·5–8 cm., *tan isabelline,* imbricated, unequal, very variable, irregularly club-shaped, or variously expanded, dilated, often circular in outline, undulate, *villose,* sessile. Tubes *white,* short; orifice of pores *white,* minute, readily stopped up with flocci. Flesh *paler,* rigid, fibrous. Spores "pale, elliptical, 6 × 7μ" Massee. On roots, often wrapping round stipules and grasses. Sept. Rare.

1930. **P. heteroclitus** (Bolt.) Fr. Bolt. Hist. Fung. t. 164, as *Boletus heteroclitus* Bolt. ἑτερόκλιτος, leaning to one side.

P. 6 cm., *orange, sessile,* flat, *expanded on all sides from a radical tubercle,* lobed, *villose.* Tubes *golden yellow,* short; orifice of pores *yellow, becoming brownish,* irregular, and elongate. On the ground under oaks. Jan. Rare.

P. salignus Fr. = **Daedalea saligna** Fr.

B. Sessile.

VI. P. with a *rigid crust,* often resinous. Tubes *heterogeneous,* separable; pores round, rarely polygonal. Spores white, or slightly coloured. Cystidia coloured, or none. Annual. Growing on wood.

1931. **P. betulinus** (Bull.) Fr. Grev. Scot. Crypt. Fl. t. 229.
Betula, birch.

P. 7–30 cm., *pale, becoming brownish with age and often mottled*, roundish, or reniform, attached by a narrow, bossy base which sometimes forms a short stalk; pellicle smooth, thin, separating; margin very obtuse, sterile. Tubes *white*, 2–8 mm. long, often separating; orifice of pores *white, becoming darker*, minute, round. Flesh *white*, soft, then corky. Spores white, oblong, 5–7 × 2μ, often curved. Birch, rarely beech. Common. (*v.v.*)

1932. **P. quercinus** (Schrad.) Fr. Boud. Icon. t. 154. *Quercus*, oak.

P. 7–15 × 5–12 cm., *pale tan*, or *tinged with red*, and *becoming reddish when bruised or with age*, tongue-shaped, convexo-plane, *narrowed behind into a thick horizontal stem*, at first floccoso-granular, or minutely squamulose, becoming smooth; margin obtuse, sometimes lobed. Tubes *whitish*, 4 mm. long; orifice of pores *whitish*, or *yellowish, becoming reddish when bruised*, and finally *concolorous*, minute, round. Flesh *whitish, lemon yellow under the cuticle and at the base of the tubes, often pinkish elsewhere when young, especially in the stem*, 1·5–2·5 cm. thick, floccose, soft, then hardened. Spores *white, often yellowish*, oblong, 10–12 × 4–5μ, slightly granular inside. Taste very bitter. Oaks, and oak logs. May—Dec. Rare. (*v.v.*)

1933. **P. dryadeus** (Pers.) Fr. Bull. Hist. Champ. Fr. t. 458. δρῦς, oak.

P. 7–30 cm., *yellowish, then ferruginous and becoming brown*, dimidiate, horizontal, pulvinate, imbricate, cuticle thin, soft, pruinose, *rugged, becoming even*, smooth; margin often exuding watery drops. Tubes ferruginous, 10–30 mm. long; orifice of pores *whitish*, round, small. Flesh *ferruginous*, subzoned, soft, becoming corky, thick. Spores pale ferruginous in the mass, yellowish under the microscope, globose, 6–8 × 6–7μ, 1-guttulate. Cystidia "sparse, straight, 40 × 8μ" Lloyd. Taste acid, the drops very astringent from the tannic acid they contain. At the base of oaks. May—Dec. Common. (*v.v.*)

VII. P. *villose, velvety, or strigose*, without a cuticle. Flesh *coloured*, moist, then firm, and fragile. Tubes *heterogeneous*, separable, *coloured*. Spores white, or coloured. Cystidia coloured, or wanting. Annual. Growing on wood.

1934. **P. hispidus** (Bull.) Fr. Boud. Icon. t. 158. *Hispidus*, shaggy.

P. 10–30 cm., *yellowish, then ferruginous and finally blackish*, dimidiate, pulvinate, *thick, very hispid, shaggy*. Tubes *ferruginous*, 2–3 cm. long; orifice of pores *yellowish, becoming concolorous*, small, round, becoming torn, often exuding watery drops. Flesh *ferruginous*, 2·5–10 cm. thick, spongy, *fibrous*, becoming dry and fragile. Spores brown, subglobose, 9–10 × 7–8μ, often apiculate, 1–multi-guttulate.

Cystidia sparse, or absent. Ashes, apples, and walnuts. May—Feb. Common. (*v.v.*)

1935. **P. cuticularis** (Bull.) Fr. *Cuticula*, a thin external skin.

P. 7–30 cm., *tawny, then ferruginous fuscous and finally blackish*, applanate, dimidiate, rather triquetrous, imbricate, becoming plane, *thin*, hairy, tomentose, obsoletely zoned, becoming smooth; margin *incurved, fimbriate*. Tubes *dark brown*, 3–10 mm. long; orifice of pores *whitish*, glistening, *then concolorous*, small, round, or angular, often torn. Flesh *dark brown, thin*, 3–10 mm. thick, fibrillose, hard. Spores ferruginous, elliptical, 6–7 × 4–5μ. Cystidia sparse, or absent. Beeches, birches, and hornbeams. Aug.—Feb. Not uncommon. (*v.v.*)

1936. **P. benzoinus** (Wahlenb.) Fr. (= *Polyporus fuliginosus* (Scop.) Quél.) Trans. Brit. Myc. Soc II, t. 12.

Benzoin, a fragrant, resinous juice.

P. 7–12 cm., *fuscous rubiginous*, shell-shaped, dimidiate, often constricted at the base, subimbricate, *often marked with metallic, bluish zones*, tomentose, becoming rugose when old, and darker. Tubes *whitish*, or *yellowish*, 6–10 mm. long; orifice of pores *whitish, becoming ferruginous*, minute, round, or deltoid. Flesh *fuscous, then pale wood colour*, firm. Spores *white*, oblong, 4–5 × 2–2·5μ, curved. Smell very pleasant, aromatic. Cedars and *Abies*. Oct.—March. Uncommon. (*v.v.*)

1937. **P. cryptarum** (Bull.) Fr. κρύπτη, a vault.

P. 10–20 cm., *tawny, or brown, becoming paler*, effuso-reflexed, imbricate, wrinkled, *silky*, distilling drops. Tubes *cinnamon*, very long; orifice of pores *ochraceous*, minute, round. Flesh *cinnamon*, thick, or thin, spongy, then corky. *Abies*, and rotting coniferous wood. Not uncommon.

1938. **P. rutilans** (Pers.) Fr. (= *Polyporus nidulans* Fr. sec. Quél.) Pers. Icon. et Desc. fung. min. cogn. t. 6, fig. 3.

Rutilans, becoming reddish.

P. 2–6 cm., *tawny-cinnamon* or *reddish grey, becoming pale*, convex, effused behind, *imbricate*, rarely solitary, villose, becoming smooth; margin *inflexed*, obtuse, unequal. Tubes *white, becoming concolorous*, 1–3 mm. long; orifice of tubes *white*, glistening, *soon becoming concolorous*, minute, *round*. Flesh *concolorous*, firm, pliant, giving a blue juice when extracted with alcohol. Spores white, *globose*, 4μ. Smell pleasant. Fallen branches, and dead aspens. Jan.—Dec. Not uncommon. (*v.v.*)

1939. **P. nidulans** Fr. (= *Polyporus rutilans* (Pers.) Fr. sec. Quél.) Saund. & Sm. t. 45, as *Polyporus rutilans*. *Nidus*, a nest.

P. 2·5–5 cm., *pale yellowish, or flesh colour, pulvinate, solitary, convex above and below*, villose, becoming smooth; margin *spreading*, obtuse.

Tubes *cinnamon*, 2–5 mm. long; orifice of pores *concolorous, becoming purplish when bruised*, rather large, *angular*. Flesh *pale cinnamon, very soft*, easily compressed, giving a blue juice when extracted with alcohol. Spores white, *pip-shaped*, 4–5 × 1·5–2·5μ. Smell pleasant when dried. Twigs, and fallen branches. Jan.—Dec. Not uncommon. (*v.v.*)

1940. **P. gilvus** Schwein. *Gilvus*, pale yellow.

P. 5–10 cm., *brown, becoming pale yellowish*, applanate, often imbricate, even, often rugulose. Tubes *brown*, 3–10 mm. long; orifice of pores *brown*, small, round. Flesh *bright yellow, becoming cinnamon brown when old*, hard, firm. Spores white, subglobose, 4–5 × 3·5–4μ, 1-guttulate. "Cystidia abundant, slender, sharp, projecting, 12–16μ" Lloyd. Deciduous trees, especially beech. Jan. Rare. (*v.v.*)

1941. **P. radiatus** (Sow.) Fr. *Radiatus*, radiate.

P. 2–6 cm., *tawny, margin yellow, becoming ferruginous fuscous, dimidiate*, very imbricate, *radiately rugose, minutely velvety, becoming smooth*; margin spreading, repand. Tubes *ferruginous*, 4–5 mm. long; orifice of pores *silvery, glistening*, minute, round, or angular. Flesh *pale cinnamon*, fibrous, hard, rigid. Spores white, subglobose, 5 × 4μ. Cystidia coloured, fusiform, 20–30 × 5–8μ, sparse. Beeches, birches, and especially on alders. Sept.—April. Common. (*v.v.*)

1942. **P. nodulosus** Fr. (= *Polyporus polymorphus* Rostk.) Trans. Brit. Myc. Soc. II, t. 16. *Nodulosus*, full of little knobs.

P. 1–3 cm., *fulvous, then rust coloured, triquetrous, nodular, connate, villose*, rugose, rough. Tubes *light cinnamon*, 1–5 mm. long; orifice of pores *silvery, glistening*, minute, round, unequal, acutely torn. Flesh *paler*, very hard. Spores white, elliptical, 4–5 × 3μ. Beeches. Sept.—Nov. Not uncommon. (*v.v.*)

P. polymorphus Rostk. = **Polyporus nodulosus** Fr.

VIII. P. *villose, floccose, or fibrillose*, rarely smooth, without a cuticle. Flesh *white*, fibrous, soft, zoned, *putrescent*. Tubes *heterogeneous*, often separable. Spores white, pale blue in 1954. Cystidia present, or none. Annual. Growing on wood.

†Pores *coloured*.

1943. **P. amorphus** Fr. *ἄμορφος*, misshapen.

P. 3–4 cm., *white*, effuso-reflexed, or dimidiate, imbricate, sometimes resupinate, silky, or tomentose. Tubes *white, becoming golden, or pinkish*, short; orifice of pores *concolorous*, round, or irregular and torn. Flesh *white*, soft, pliant, "subgelatinous" Lloyd. Spores white, subglobose, 4–5μ, 1-guttulate. Trunks, stumps, and needles of various conifers. Sept.—March. Not uncommon. (*v.v.*)

1944. **P. armeniacus** Berk. (= *Polyporus amorphus* Fr. sec. Cke.) *Armeniaca*, apricot.

P. 8 cm., *white*, broadly effused, suborbicular, confluent; margin minutely downy. Tubes *white, then bright buff, changing to deep cinnamon during drying*, short; orifice of pores *concolorous*, minute, round, rather irregular, often confined to the centre. Flesh *whitish*, very thin. Spores "white, elliptical, 7 × 4·5μ" Massee. Fir, and pine bark. Sept.—March. Uncommon. (*v.v.*)

1945. **P. adiposus** B. & Br. (= *Polyporus undatus* Pers. sec. Bres.) Lloyd, Synop. Sec. Apus. Gen. Polyp. figs. 662 and 663, as *Polyporus undatus*. *Adiposus*, fat.

P. 1–1·5 cm., *white, here and there acquiring a foxy tinge*, effuso-reflexed, often entirely resupinate, obscurely tomentose. Tubes *whitish, tinged in places with brown*, short, or long; orifice of pores *whitish, becoming brownish*, either small and round, or angular and torn. Flesh *white, waxy*, soft. Spores white, globose, 4–5μ, 1-guttulate. The whole fungus turns brown in drying. Ditch sides, ground beside stumps, and mosses. Sept.—Dec. Uncommon. (*v.v.*)

1946. **P. albus** (Huds.) Fr. *Albus*, white.

P. 3–9 cm., *white, becoming greyish*, dimidiate, shell-shaped, *smooth*. Tubes *white, becoming reddish*, short; orifice of pores *white, then reddish*, small, round, becoming irregular. Flesh *white*, soft, zoned. Spores "white, oval, 6μ, punctate" Quél. Willows, and beeches. Nov.—March. Rare.

1947. **P. fumosus** (Pers.) Fr. *Fumosus*, smoky.

P. 4–12 cm., *pale ochraceous, then fuliginous, and becoming black at the margin*, dimidiate, adnate and dilated behind, imbricate, minutely tomentose, becoming smooth. Tubes *whitish cream*, then *smoky*, short; orifice of pores *whitish, becoming fuliginous*, minute, round. Flesh *whitish*, firm, *fibrous, somewhat zoned*. Spores white, elliptical, 6–7 × 3–4μ, often with a basal apiculus. Smell strong, or none. Willows, beeches, birches, and poplars. July—March. Common. (*v.v.*)

var. **fragrans** (Peck) Rea. *Fragrans*, scented.

Differs from the type in the *sweet smell, and concolorous margin of the p*. Willow, and elm stumps. Sept.—Dec. Not uncommon. (*v.v.*)

1948. **P. adustus** (Willd.) Fr. Quél. Jur. et Vosg. t. 18, fig. 2. *Adustus*, swarthy.

P. 3–7 cm., *cinereous pallid, becoming black at the margin, effuso-reflexed*, dimidiate, orbicular, imbricate, sometimes entirely resupinate, villose, obsoletely zoned, slightly wrinkled. Tubes *cinereous*, short; orifice of pores *at first whitish pruinose, soon cinereous fuscous*, *becoming blackish when dry,* minute, round, marginal ones obsolete and leaving a whitish margin on the underside. Flesh *white, then grey or black,* soft, floccose, pliant, 3–5 mm. thick. Spores white, elliptical, 4–5 × 2·5–3μ. Smell none, rarely fragrant. Trunks, stumps, and fallen branches. Jan.—Dec. Common. (*v.v.*)

var. **crispus** (Pers.) Quél. Rostk. Polyp. t. 37. *Crispus,* crisped.

Differs from the type in the *lobed, crisped margin of the p., and the larger, unequal pores becoming labyrinthiform.* Stumps, and fallen branches. Jan.—Dec. Not uncommon. (*v.v.*)

††Pores *white.*

*Orifice dentate.

1949. **P. lacteus** Fr. Fr. Icon. t. 182, fig. 1. *Lacteus,* milk white.

P. 3–7 cm., *shining white,* triangular, transversely elongated, sloping downwards, gibbous behind, *pubescent,* at length smooth and uneven; margin inflexed, acute. Tubes *white,* 2–6 mm. long; orifice of pores *concolorous,* medium sized, *toothed, at length labyrinthiform and torn into Sistotrema-like teeth.* Flesh *white,* soft, fragile, generally thinner than the length of the tubes. Spores white, elliptical, or pip-shaped, 4–5 × 2–2·5μ, 1-guttulate. Taste astringent. Stumps, and fallen branches. Sept.—Dec. Uncommon. (*v.v.*)

1950. **P. fragilis** Fr. Fr. Icon. t. 182, fig. 2. *Fragilis,* brittle.

P. 3–6 cm., *whitish, becoming spotted with fuscous when touched,* plano-depressed, reniform, dimidiate, sometimes attenuated behind into a stem-like base and pendulous, convex beneath, *villose, rugose.* Tubes *whitish,* 2–6 mm. long; orifice of pores *whitish, becoming fuscous when bruised,* round, or angular, becoming sinuous and labyrinthiform, pubescent. Flesh *white, becoming discoloured, fragile,* fibrous. Spores *white,* elliptical, 5–6 × 2·5–3μ, 1–2-guttulate. Taste somewhat bitter. Stumps, and branches of conifers. Aug.—Nov. Uncommon. (*v.v.*)

1951. **P. Wynnei** B. & Br. Mrs Lloyd Wynne.

P. 1–6 cm., *tan colour, sometimes whitish at first,* effuso-reflexed, adnate behind, confluent, incrusting, marked with silky raised lines. Tubes *white,* 1–2 mm. long; orifice of pores *white, becoming tan colour when dried,* angular, fimbriate. Flesh *white,* soft, becoming hard and fragile. Spores *white,* elliptical, or pip-shaped, 3–4 × 2–3μ, 1-guttulate. Incrusting twigs, leaves, and branches. Sept.—Dec. Uncommon. (*v.v.*)

1952. **P. mollis** (Pers.) Fr. *Mollis,* soft.

P. 2·5–8 cm., *white, becoming reddish when bruised or with age,* dimidiate, imbricate, rugose, *silky,* fibrillose; margin acute. Tubes *white,* 3–10 mm. long; orifice of pores *white, spotted with red when touched,* large, elongate, unequal, flexuose. Flesh *white, becoming reddish when cut, soft,* firm when dry, fragile, *thick.* Spores white, elliptical, 5–6 × 2–3μ. Pine stumps. Sept.—Nov. Uncommon. (*v.v.*)

1953. **P. Keithii** B. & Br. Rev. Dr James Keith.

P. 12 mm., *bright red brown,* shell-shaped, effuso-reflexed, narrowed behind, *rough with rigid, tooth-shaped processes.* Tubes and pores *pallid,* large, angular, lacerated. Spores "white, elliptical, 6 × 3μ" Massee. Fallen sticks. Rare.

1954. **P. caesius** (Schrad.) Fr. *Caesius,* bluish grey.

P. 1–8 cm., *white, then tinged with bluish grey,* dimidiate, often imbricate, rarely stipitate, sometimes resupinate, *villose,* or *silky.* Tubes *white,* 3–9 mm. long; orifice of pores *white, becoming bluish grey when touched,* small, unequal, flexuose, toothed. Flesh *white, bluish when broken,* soft, watery, then firm. Spores *pale blue,* oblong, 4–5 × 1–1·5μ, 1-guttulate, often curved. Stumps, and dead branches of conifers, more rarely on deciduous trees. March—Dec. Common. (*v.v.*)

1955. **P. trabeus** Fr. Rostk. Polyp. t. 28. τράφηξ, a beam.

P. 5–10 cm., *white, becoming pallid, often tinged with ochre or bistre,* effuso-reflexed, transversely elongated, minutely pubescent, or smooth. Tubes *white,* 2–6 mm. long; orifice of pores *white,* somewhat round, or toothed and labyrinthiform. Flesh *white, obsoletely zoned,* floccose, then firm. Spores white, elliptical, 5–6 × 3–4μ. Conifers, and yews. Oct.—Feb. Rare. (*v.v.*)

1956. **P. destructor** (Schrad.) Fr. Krombh. t. 5, fig. 8.
Destructor, destroyer.

P. 5–15 cm., *fuscous whitish,* effuso-reflexed, sometimes resupinate, rugose, subundulate, pubescent. Tubes *white,* 3–8 mm. long; orifice of pores *white, becoming discoloured with age,* somewhat round, toothed, or torn. Flesh *whitish, watery, fleshy,* thick, zoned. Spores white, subglobose, 3–4 × 3μ. Smell sometimes strong. Conifers, and worked wood. May—Dec. Uncommon. (*v.v.*)

var. **undulatus** (Fr.) Sacc. *Undulatus,* wavy.

Differs from the type in the *broadly expanded, marginate, whitish bay brown p.*

**Orifice *entire.*

1957. **P. epileucus** Fr. ἐπίλευκος, whitish.

P. 7–12 cm., *whitish, or yellowish, becoming ochraceous when dried,* dimidiate, pulvinate, concave below, *villous-rugged.* Tubes *whitish,* 4–18 mm. long; orifice of pores *whitish,* becoming *yellowish,* minute, round. Flesh *whitish, becoming yellowish,* cheesy-soft, 2·5–5 cm. thick, scarcely zoned. Spores white, oval, 4μ. Beech, birch, elm, poplar, willow, and fir stumps. Sept.—Nov. Rare. (*v.v.*)

1958. **P. spumeus** (Sow.) Fr. Sow. Eng. fung. t. 211.
Spumeus, frothy.

P. 7–16 cm., *whitish*, dimidiate, pulvinate, gibbous, *rugosely hispid, or floccose*, becoming smooth; margin incurved. Tubes *whitish*, 2–8 mm. long; orifice of pores *whitish, becoming discoloured*, minute, round, or linear, separable. Flesh *whitish*, soft, becoming hard, and discoloured, zoned towards the margin. Spores white, globose, 7–9μ, multi-guttulate. Beech, pear, apple, ash, hornbeam, elm, oak, willow, and birch trunks and stumps. April—Oct. Uncommon. (*v.v.*)

P. borealis Fr. = **Daedalea borealis** (Wahlenb.) Quél.

1959. **P. tephroleucus** Fr. Rostk. Polyp. t. 26.
τεφρός, ash-coloured; λευκός, white.

P. 5–10 cm., *grey*, triquetrous, applanate, often imbricate, plane beneath, *unequal, villose*, becoming smooth; margin obtuse, flexuose, often white, and becoming blackish when touched. Tubes *white*, 10–15 mm. long; orifice of pores *white*, round, small, becoming toothed and fimbriate. Flesh *white*, zoned with grey or bistre, subgelatinous, soft, becoming firm and fragile. Spores white, allantoid, 4–5 × 1–1·5μ, slightly curved. Beech, and pine stumps, and logs. Sept.—Oct. Not uncommon. (*v.v.*)

1960. **P. alutaceus** Fr. Rostk. Polyp. t. 30, as *Polyporus epixanthus* Rostk.
Aluta, tanned leather.

P. 2·5–5 cm., *tan*, reniform, convex, or flattened, often connate, subimbricate, plane beneath, *somewhat velvety* and rugose; margin *acute, even*. Tubes *whitish tan colour*, 3–18 mm. long; orifice of pores *yellowish*, minute, round. Flesh *white*, or *yellowish*, soft, then tough and fragile, obsoletely zoned. Spores "with a slight ochraceous tinge, subglobose, 4μ" Massee. Beech, and pine stumps and trunks. Sept.—Oct. Uncommon.

1961. **P. stipticus** (Pers.) Fr. Fr. Icon. t. 181, fig. 2.
στυπτικός, astringent.

P. 3–6 cm., *white*, dimidiate, pulvinate, often imbricate, minutely pubescent, becoming smooth; margin obtuse, *becoming reddish*. Tubes *white*, 6–8 mm. long; orifice of pores *at first with white milk-like drops, becoming slightly rufescent when dry*, small, round, or irregular. Flesh *white*, soft, then hard, 1·5 cm. thick. Spores white, elliptical, 3–4 × 1·5μ, slightly curved, guttulate. Taste astringent. Pine trunks and stumps. Sept.—Oct. Uncommon. (*v.v.*)

1962. **P. chioneus** Fr. Pers. Myc. Eur. II, t. 15, figs. 4, 5. *χιών*, snow.

P. 2–5 cm., *white*, dimidiate, sometimes constricted behind into a stem-like base, *becoming even, smooth*; margin inflexed, thin, acute. Tubes *white*, short; orifice of pores *white*, minute, round, becoming toothed with age. Flesh *white*, soft, watery, then rigid. Spores white, "elliptical oblong, 5μ, incurved" Quél. Taste astringent. Birch stumps, and fallen branches. June—Dec. Uncommon. (*v.v.*)

1963. **P. pallescens** Fr. *Pallescens*, growing pale.

P. 4–7·5 cm., *yellowish*, dimidiate, subcaespitose, *even, smooth*; margin acute. Tubes *white*, short; orifice of pores *white, becoming yellowish*, minute, round. Flesh *yellowish*, soft, then corky. Spores "ellipsoid, 6–8 × 4μ" Sacc. Old stumps. Rare.

1964. **P. albidus** Trog. Schaeff. Icon. t. 124. *Albidus*, whitish.

P. 6–10 cm., *white*, globose, shell-shaped, triquetrous, or subapplanate, sometimes slightly stalked, dry, dull smooth. Tubes *white*, short; orifice of pores *white*, round, at length sinuate and splitting. Flesh *white*, hard, corky-woody. Spores white, elliptical, 5–6 × 3–4μ. *Abies*. Sept.—Oct. Rare. (*v.v.*)

1965. **P. cerebrinus** B. & Br. *Cerebrinus*, brain-like.

P. 2·5 cm., *snow white*, pulvinate, resupinate, *delicately tomentose*, becoming smooth; margin crenate. Tubes and orifice of pores *white*, rather large, round, entire, smooth. Flesh *white*, 6 mm. thick. Spores "white, subfusiform, 5 × 2·5μ" Massee. Fir. Aug. Rare.

Sistotrema (Pers.) Fr.

(*σειστός*, shaking; *τρῆμα*, a hole.)

Pileus fleshy, hemispherical, spathulate, effuso-reflexed, or resupinate. Stem central, lateral, or none. Tubes becoming broken up into teeth, or plates, and anastomosing at the base. Flesh pale, or coloured. Spores white, subglobose, oboval, or oblong; smooth, or echinulate; basidia with 4–8-sterigmata. Cystidia none. Growing on the ground, or on wood.

1966. **S. confluens** (Pers.) Fr. Boud. Icon. t. 169. *Confluens*, confluent.

P. 1–2·5 cm., *white, then yellowish*, hemispherical, spathulate, or irregular, horizontal, villose; margin *often yellowish*, flexuose. St. ·5–2·5 cm. × 2–3 mm., *white, or ochraceous*, central, or lateral, often connate, attenuated downwards, pruinose. Tubes *concolorous, becoming broken up into teeth, or plates*, flexuose, entire, or toothed, pruinose. Flesh *white, often yellowish at the base of the teeth*, thin, firm. Spores white, oval, subglobose, 4–5 × 3–4μ, 1–multi-guttulate. Coniferous woods. Sept.—Nov. Not uncommon. (*v.v.*)

1967. **S. sulphureum** (Quél.) Bourd. & Galz. Quél. Ass. fr. (1893), t. III, fig. 10, as *Daedalea sulphurea* Quél.

Sulphureum, sulphur colour.

P. 1–2 cm., *whitish sulphur, or citron yellow*, effused, little adnate; margin *concolorous*, similar, or fibrillosely fringed. Spines *sulphur, then ochraceous orange, or tawny, apex white*, pubescent, obtuse, scattered, forming flexuose plates. Flesh *concolorous*, floccose, spider-web-like, fibrillose, membranaceous, thin. Spores "*light yellow*, subhyaline, at first smooth, then *rough*, spines hyaline, fugacious, oboval, oblong, apiculate at the base" Bourd. & Galz. Bare earth, stones, herbaceous roots and buried twigs. Jan.—Dec. (The type has not yet been recorded for Britain.)

var. **variecolor** (Fr.) Bourd. & Galz. (= *Hydnum variecolor* Fr.)

Variecolor, of different colours.

Differs from the type in the *white subiculum, the variable, scattered, yellow, then tawny spines, and the oboval, echinulate spores*, 7–8 × 4–6μ. Dead oak stumps. Oct. Rare.

Fomes Fr.

(*Fomes*, tinder.)

Pileus hard, woody, or corky, dimidiate, hoof-shaped, or resupinate, sessile, often concentrically zoned, and covered with a rigid crust. Tubes homogeneous, or heterogeneous, often stratose. Flesh white, or coloured. Spores white, or coloured, globose, subglobose, elliptical, or elliptic-oblong, smooth. Cystidia present, or absent, coloured or hyaline. Perennial. Growing on wood.

*Flesh deeply coloured.

1968. **F. fomentarius** (Linn.) Fr. Fr. Sverig. ätl. Svamp. t. 62.

Fomentum, touch-wood.

P. 10–60 cm., *greyish, becoming hoary, hoof-shaped*, or dimidiate, attached by a broad base, 7–20 cm. thick, remotely and concentrically sulcate, opaque, pruinose, *cuticle thick and very hard*. Tubes *ferruginous*, 1–3 cm. long, stratose; orifice of pores *glaucous pruinose, then ferruginous*, minute, round. Flesh *dark brown, soft, floccose, very thick*. Spores hyaline, elliptic oblong, 16–18 × 5μ, 1–3-guttulate. Beeches, oaks, limes, hornbeams, and birches. Jan.—Dec. Common. (*v.v.*)

var. **nigrescens** (Klotzsch) Lloyd. Lloyd, Polyp. Issue, fig. 210.

Nigrescens, becoming black.

Differs from the type in its *black, shining, strongly concentrically sulcate crust*. Beeches. Jan.—Dec. Not uncommon. (*v.v.*)

1969. **F. igniarius** (Linn.) Fr. Sow. Eng. Fung. t. 132.
Igniarius, belonging to fire.

P. 10–30 cm., *floccosely hoary, then ferruginous*, and *at length blackish* especially at the base, tuberculoso-globose, immarginate, hoof-shaped, or flattened, rarely resupinate, covered with a very hard, rough, uneven, often rimose cuticle; margin *at first whitish, rounded, obtuse*. Tubes *cinnamon, becoming whitish with deposits of lime with age*, 2–8 mm. long, stratose; orifice of pores *hoary, then cinnamon*, minute, round. Flesh *dark brown, very hard*, zoned. Spores hyaline, globose, 5–7μ, 1-multi-guttulate. Cystidia dark coloured, sparse, subulate, base ventricose, 25–30 × 7–8μ. Willows, and occasionally on ashes. Jan.—Dec. Common. (*v.v.*)

var. **nigricans** (Fr.) Lloyd. Fr. Icon. t. 184, fig. 2.
Nigricans, blackish.

Differs from the type in the *smooth, shining, black crust*. Birches, rarely on willows. Jan.—Dec. Common. (*v.v.*)

var. **roburneus** (Fr.) Lloyd. *Robur*, oak.

Differs from the type in the *slight, resinous exudation on the crust* and the *strongly silvery, glancing orifice of the pores*. Willows, and oaks. Jan.—Dec. Rare.

1970. **F. robustus** Karst. (= *Polyporus Hartigii* Allesch.; *Fomes Hartigii* (Allesch.) Sacc. & Trav.; *Polyporus igniarius* Linn. var. *Pinuum* Bres.) *Robustus*, firm.

P. 10–40 cm., *fuscous, becoming black with age*, hoof-shaped, or sub-hemispherical, sessile, remotely and concentrically sulcate; cuticle concrete, rough, uneven, rigid, very hard, pilose at first, then glabrous, cracked; orifices of pores *silvery white, then concolorous*, minute, round. Tubes *rhubarb root colour*, or *pale fulvous, stratose*, cylindrical. Flesh *concolorous with the tubes*, very firm, zoned. Spores white, globose, 7–8μ, with a large central gutta. Setae none. *Abies pectinata* and oak. Jan.—Dec. Uncommon. (*v.v.*)

1971. **F. fulvus** Fr. *Fulvus*, tawny.

P. 8–9 cm., *tawny, at length becoming hoary*, convex above and below, adnate by a broad base, *triangular in section*, even, not concentrically sulcate, at first hairy, or villose. Tubes *cinnamon*, short, *not distinctly stratose*; orifice of pores *at first covered with cinereous yellow pruina*, minute, round. Flesh very hard, woody-corky. On decaying trunks, especially poplar. Oct.—March. Rare.

1972. **F. salicinus** (Pers.) Fr. Fr. Icon. t. 185, fig. 1.
Salicinus, belonging to willows.

P. 5–30 cm. and more, *cinnamon, then hoary, blackish bay at the base*, undulated, smooth, pubescent, for the most part resupinate, or in vertical positions incircled above with a narrow, undulated, short, obtuse, spreading margin. Tubes *ferruginous cinnamon*, short; orifice of pores *silvery grey, glistening*, minute, round. Flesh *ferruginous cinnamon*, woody, thin. Spores *yellowish tawny*, "elliptical, 5 × 3μ. Cystidia abundant, slightly thickened at the base, 12–35 × 6μ" Massee. Willow trunks. Sept.—Oct. Common.

1973. **F. conchatus** (Pers.) Fr. κόγχη, a mussel-shell.

P. 5–8 cm., *date brown, effuso-reflexed*, the reflexed portion *somewhat shell-shaped*, concentrically sulcate, often imbricate, sometimes wholly resupinate, *tomentose*; margin acute. Tubes *cinnamon*, short; orifice of pores *ashy pruinose, glistening*, minute, round. Flesh *light brown*, hard, corky. Spores *ferruginous*, subglobose, 5–6 × 4μ, 1-guttulate. Cystidia abundant, slightly thickened at the base, 15–30 × 7–9μ. Willows, and beeches. Feb.—Nov. Not uncommon. (*v.v.*)

1974. **F. Ribis** (Schum.) Fr. (= *Polyporus pectinatus* (Klotzsch) Quél.) *Ribes*, currant.

P. 5–25 cm., *ferruginous fuscous, bright yellow at the margin, becoming dark at the base*, horizontal, imbricate, flattened, concentrically zoned, velvety, strigose. Tubes *cinnamon*, 2–4 mm. long, often stratose; orifice of pores *yellow, then cinnamon*, minute, round. Flesh *cinnamon*, floccose, soft. Spores *fuscous*, globose, 3–4 × 3μ, 1-guttulate. Currants, gooseberries, hawthorns, and spindle. Jan.—Dec Not uncommon. (*v.v.*)

F. Euonymi (Kalchbr.) Cke. = **Fomes Ribis** (Schum.) Fr.

F. pectinatus (Klotzsch) Fr. = **Fomes Ribis** (Schum.) Fr.

1975. **F. pomaceus** (Pers.) Big. & Guill. *Pomum*, fruit.

P. 3–5 cm., *fuscous, becoming cinereous*, dimidiate, triquetrous, or somewhat hoof-shaped, often subresupinate, delicately silky, or almost smooth. Tubes *cinnamon*, 4–6 mm. long, stratose; orifice of pores *whitish, then cinnamon*, minute, round, pruinose. Flesh *light brown*, firm, *woody*. Spores white, globose, 6 × 5–6μ. Cystidia abundant, deep mahogany brown, apex hyaline, flask-shaped, 15–20 × 7–8μ. Plums, rarely cherries. Jan.—Dec. Common. (*v.v.*)

var. **fulvus** (Quél.) Rea. *Fulvus*, tawny.

Differs from the type in its *brighter tawny colour, and more tomentose p.* Plums. Jan.—Dec. Common. (*v.v.*)

1976. **F. ferruginosus** (Schrad.) Massee. (= *Poria ferruginosa* (Schrad.) Fr.) Grev. Scot. Crypt. Fl. t. 155, as *Polyporus ferruginosus*. *Ferruginosus*, iron rust colour.

P. 1–3 cm., *bright ferruginous brown, becoming dusky ferruginous*, effuso-reflexed, imbricate, often entirely resupinate, subtomentose, rough; margin sterile. Tubes *cinnamon*, 2–6 mm. long, stratose; orifice of pores *ferruginous*, round, torn. Flesh *pale ferruginous*, fibrous, firm. Spores white, subglobose, 3–5μ. Cystidia deep mahogany brown, apex paler, acutely conical, 30–52 × 6–9μ. Logs, fallen branches, and posts. Sept.—May. Common. (*v.v.*)

1977. **F. resupinatus** (Bolt.) Massee. Bolt. Hist. Fung. t. 165, as *Boletus resupinatus* Bolt. *Resupinatus*, supine.

P. 1–2 cm., *ferruginous*, nodulose, often imperfect, commonly entirely resupinate. Tubes *ferruginous*, 2–5 mm. long; orifice of pores *concolorous*, minute, round. Flesh *concolorous*, thin, fibrous. Spores "colourless, elliptical, 4 × 1·5μ. Cystidia none" Massee. Trunks, and fallen branches. Feb. Uncommon.

F. obliquus (Pers.) Fr. = **Poria obliqua** (Pers.) Quél.

**Flesh white, pallid, rosy, or violet.

1978. **F. ulmarius** (Sow.) Fr. (= *Polyporus incanus* Quél.) Hussey, Illus. Brit. Myc. t. 64. *Ulmus*, elm.

P. 7–30 cm., *white, becoming yellowish and discoloured with age*, effused, flattened, incrusted, tubercular, becoming smooth; margin *obtuse*, sometimes free. Tubes *cinnamon*, 5–20 mm. long, stratose; orifice of pores *whitish, becoming yellowish*, minute, round. Flesh *white, becoming yellowish with age*, corky-woody, very hard. Spores white, globose, 6–7μ, 1-guttulate. At the base of old elms, and elm stumps, rarely higher up. Jan.—Dec. Common. (*v.v.*)

1979. **F. fraxineus** (Bull.) Fr. (= *Polyporus incanus* Quél.; *Polyporus cytisinus* Berk.) *Fraxinus*, an ash tree.

P. 7–25 cm., *whitish, becoming rubiginous and fuscous*, applanate, dimidiate, sometimes imbricate, glabrous, often coarsely tuberculated; margin incurved, pubescent at first. Tubes *pale cinnamon*, 5–25 mm. long, stratose; orifice of pores *white, becoming greyish*, minute, round, or oblong. Flesh *yellowish, somewhat zoned*, soft, becoming hard and woody. Spores white, subglobose, 6–7 × 6μ. Smell often strong. Ashes, and laburnums. Jan.—Dec. Uncommon. (*v.v.*)

F. cytisinus (Berk.) Massee = **Fomes fraxineus** (Bull.) Fr.

1980. **F. annosus** Fr. Fr. Icon. t. 186, fig. 2. *Annosus*, full of years.

P. 7–45 cm., *bay brown, becoming blackish*, convex, then becoming plane, imbricate, sometimes resupinate, *rugoso-tubercular*, sulcately zoned, silky, then with a rigid, smooth crust; margin *at first white*, thin. Tubes *yellowish*, 4–8 mm. long, stratose; orifice of pores *whitish*, round, or polygonal, obtuse. Flesh *white*, or *yellowish*, soft, becoming hard. Spores white, subglobose, 4–5 × 4μ, 1-guttulate. At the base of conifer trunks, and on felled and worked conifer wood, rarely on frondose trees, and wood. Jan.—Dec. Common. (*v.v.*)

1981. **F. castaneus** Fr. *Castaneus*, chestnut.

P. 4–5 cm., *chestnut*, reniform, or applanate, imbricate, connate at the base, 4 mm. thick, *smooth*, *glabrous*. Tubes *yellowish*, short; orifice of pores *yellowish*, *becoming fuscous*, minute, round. Flesh *white*, coriaceous woody. Taste pleasant, bitter. Black poplar. Rare.

1982. **F. carneus** Nees. (= *Fomes roseus* (A. & S.) Fr. sec. Bres.) *Carneus*, flesh colour.

P. 5–15 cm., *flesh colour*, effuso-reflexed, or dimidiate, imbricate, rarely solitary, rugose, *smooth*. Tubes *pale flesh colour*, short, stratose; orifice of pores *concolorous*, minute, round. Flesh *concolorous*, thin, woody, hard. Trunks, stumps, junipers and *Picea*. Nov. Rare.

1983. **F. roseus** (A. & S.) Fr. Fr. Icon. t. 186, fig. 1, as *Polyporus rufopallidus* Trog. *Roseus*, rose-coloured.

P. 5–10 cm., *rose colour, then reddish and finally blackish brown*, hoof-shaped, triangular, sulcately zoned when old, crust thin, pruinose. Tubes *rose colour*, short, stratose; orifice of pores *concolorous*, minute, round. Flesh *rose colour*, corky-woody, hard. Spores white, "oblong, $10 \times 4\mu$" Lloyd. Caespitose. On worked wood. Sept. Rare.

1984. **F. connatus** Fr. Boud. Icon. t. 157. *Connatus*, connate.

P. 2–10 cm., *white, becoming cinereous or blackish with age*, dimidiate, *densely imbricated*, *villose*. Tubes *white, then ochraceous*, 1–3 mm. long, stratose; orifice of pores *white*, glistening, minute, round. Flesh *white, becoming ochraceous*, hard, woody, sometimes slightly zoned. Spores *white*, globose, 5–6μ, with a large central gutta. Cystidia hyaline, capitate, 10μ wide. Poplars, elms, limes, and apple trees. Jan.—Dec. Not uncommon. (*v.v.*)

1985. **F. populinus** Fr. *Populus*, poplar.

P. 1·5–5 cm., *white*, dimidiate, imbricate, connate at the base, *villose*; margin *obtuse*. Tubes *white*, short; orifice of pores *white*, minute, round. Flesh *white*, corky-woody, rigid. Spores "globose, hyaline, 3–4μ diam." Sacc. Black and white poplars and *Robinia*. July—Jan. Rare. (*v.v.*)

1986. **F. variegatus** (Secr.) Fr. (= *Ganoderma resinaceum* Boud. sec. Lloyd.) Sow. Eng. Fung. t. 368, as *Polyporus variegatus*. *Variegatus*, variegated.

P. 7–12 cm., *orange, variegated with bay*, dimidiate, somewhat flattened, imbricate, even, smooth, *shining*; margin wavy. Tubes *yellowish*, short; orifice of pores *yellowish*, minute, round, unequal, torn. Flesh *pallid*, corky-woody. Trunks. Rare.

Ganoderma (Karst.) Pat.

(γάνος, shining; δέρμα, skin.)

Pileus corky, stipitate, or sessile, covered with a resinous, laccate crust. Stem lateral, rarely central, or none. Tubes heterogeneous, often stratose. Flesh coloured. Spores coloured, elliptical, ovate oblong, or obovate, *truncate at the base*, smooth, punctate, verrucose, or echinulate, thick-walled. Cystidia none, or very rare. Annual or perennial. Growing on wood, rarely on the ground.

*Spores verrucose.

1987. **G. lucidum** (Leyss.) Karst. Grev. Scot. Crypt. Fl. t. 245, as *Polyporus lucidus*. *Lucidus*, shining.

P. 5–28 cm., *light yellow, becoming blood-red-chestnut*, more or less reniform, sometimes flabelliform, rarely orbicular, or imbricate and sessile, *polished, shining*, sulcato-rugose. St. 5–18 × 1–5 cm., *concolorous and shining like the p.*, lateral, rarely central, sometimes wanting, rugose. Tubes *white*, then *cinnamon*, adnate, 4–12 mm. long; orifice of pores *white, becoming discoloured*, minute, round. Flesh *whitish, at length reddish*, spongy, becoming corky and woody, zoned. Spores brown, minutely verrucose, elliptical, truncate at the base, 10–12 × 6–8μ, with a large central gutta. At the base and roots of elms, oaks, hornbeams, and also in peat beds. July—April. Common. (*v.v.*)

1988. **G. applanatum** (Pers.) Pat. (= *Polyporus applanatus* (Pers.) Fr.) *Applanatum*, flattened.

P. 10–40 cm., *cinnamon, margin white, becoming hoary*, and often dusted with the spores, dimidiate, or orbicular, often imbricate, attached by a broad base, *flattened*, tubercular, obsoletely zoned, pulverulent, or smooth, *covered with a laccate crust*. Tubes ferruginous, 1–4 cm. long, stratose; orifice of pores *white, becoming fuscous when bruised*, minute, round, or angular. Flesh *cinnamon, becoming paler, very firm, thick*. Spores ferruginous, minutely echinulate, broadly elliptical, truncate at the base, 9–13 × 6–8μ, 1-guttulate. Beeches, oaks, and ashes. July—March. Common. (*v.v.*)

var. **vegetum** (Fr.) Romell. *Vegetum*, vigorous.

Differs from the type in the *white mycelial layer interposed between each stratum of the tubes*. Limes, and elms. Jan.—Dec. Uncommon. (*v.v.*)

var. **laccatum** (Kalchbr.) Rea. (= *Polyporus resinosus* (Schrad.) Quél.) *Lac*, a resinous excretion left by the lac insect.

Differs from the type in the yellow orifice of the pores. Beeches. July—Oct. Not uncommon (*v.v.*)

**Spores smooth.

1989. **G. australe** (Fr.) Pat. (= *Polyporus australis* Fr.; *Polyporus vegetus* Fr. sec. Pat.) *Australe*, southern.

P. 15–30 cm. and more, *deep umber chestnut*, or paler, convexo-plane, dimidiate, sessile, incrusted on the surface with a sticky resinous coating, which dries up into tubercular ridges, and becomes *laccate* and shining; margin sterile. Tubes *reddish umber*, 2–3 cm. or more long, stratose; orifice of pores *white, then fuscous*, minute, round. Flesh *dark umber chestnut, soft, very thin*, 2–4 mm. thick. Smell aromatic. Spores ochraceous, broadly elliptical, truncate at the base, 10–12 × 7–8μ, 1-guttulate. Elms. Aug.—May. Uncommon. (*v.v.*)

1990. **G. resinaceum** Boud. *Resina*, resin.

P. 15–30 cm., *yellow, white at the margin, becoming blood-red-umber-chestnut*, and *finally concolorous*, semicircular, somewhat flattened, sessile, rarely stalked, or imbricate, concentrically sulcate, the primary furrows wide, becoming shallower and more crowded with age, *viscid, then varnished*, very shining, becoming duller and dusted with the spores; margin at first delicately pruinose and rounded, becoming glabrous and more acute. Tubes *fuscous cinnamon*, ·5–3 cm. long, stratose; orifice of pores *white, then fuscous cinnamon*, minute, round Flesh *pale cinnamon, becoming paler, thick*, soft. Spores fuscous, ovate oblong, or obovate, truncate at the base, 10–12 × 6–8μ, eguttulate, or 1-guttulate, epispore thick. Oaks, beeches, and pines. Sept.—Oct. Uncommon. (*v.v.*)

Poria (Pers.) Fr.

(πόρος, a pore.)

Pileus membranaceous, coriaceous, or corky, entirely resupinate. Tubes round, or angular, often directly inserted on the mycelium. Spores white, or coloured, elliptical, pruniform, globose, subglobose, obovate, elliptic oblong, or cylindrical; smooth, or punctate. Cystidia present, or absent, hyaline, rarely coloured. Growing on wood, rarely on the ground.

I. Fleshy, soft; pores minute, equal, round.

*Pores persistently white, or at length becoming yellowish or reddish.

1991. **P. vulgaris** Fr. Rostk. Polyp. t. 60, as *Polyporus vulgaris* Fr. *Vulgaris*, common.

P. 1–30 cm., *white, sometimes yellowish*, broadly effused, consisting almost entirely of the pores, thin, 1 mm. thick, arid, *closely adnate*, inseparable, minutely tomentose; margin *smooth*. Pores *white*, glistening, 1–2 mm. long, *very small*, round, subequal, sometimes oblique and gaping open. Spores white, "ovoid pruniform, 6μ" Quél., "allantoid, 4–6 × 1·25–1·5μ, hymenial hyphae firm, undulate, not fibulate, 2–4μ broad" Romell. Dead wood, and branches of frondose trees. Jan.—Dec. Common. (*v.v.*)

1992. **P. mollusca** (Pers.) Fr. *Mollusca*, soft.

P. 1–11 cm., *white, sometimes yellowish*, effused, *thin*, soft; margin *white, byssoid, radiately fibrillose*. Pores *white, then pale ochraceous*, ·5–1 mm. long, often confined to the centre, or here and there in patches, very small, thin, round, unequal, torn. Spores subglobose, 4 × 3·5μ. Stumps, dead wood, and branches of conifers, and on dead leaves. Jan.—Dec. Not uncommon. (*v.v.*)

1993. **P. hybrida** (B. & Br.) Massee. Sow. Eng. Fung. t. 289, and t. 387, fig. 6, as *Boletus hybridus* Sow. *Hybrida*, a mongrel.

P. 1–18 cm., *white*; mycelium thick, forming a dense membrane or creeping branched strings. Pores *white*, 2–4 mm. long, *in scattered patches*, slender, minute. Spores "colourless, elliptic-oblong, 4 × 2μ" Massee. Oak wood, causing the dry rot of ships. Jan.—Dec. Rare.

1994. **P. medulla-panis** (Pers.) Fr. Fr. Icon. t. 190, fig. 2, as *Polyporus medulla panis* (Jacq.). *Medulla*, pith, crumb; *panis*, bread.

P. 5–10 cm., *white, becoming yellowish*, effused, consisting almost entirely of the pores, *determinate, subundulate, firm*, separable, smooth; margin *naked, distinct*. Pores *white*, 2–4 mm. long, straight, or oblique, medium sized, entire, pruinose. Spores white, elliptical, 3–4 × 1·5–2μ. Rotten wood, and dead branches of frondose trees, rarely on the ground. Sept.—Feb. Uncommon. (*v.v.*)

1995. **P. mucida** (Pers.) Fr. (= *Irpex obliquus* (Schrad.) Fr. & sec. Bourd. & Maire.) *Mucida*, mucid.

P. 2–15 cm., *white, becoming pale*, effused, *rather thick*, up to 12 mm. in depth, subimmersed, soft; margin *white, indeterminate, byssoid*. Pores *white, then pale ochraceous*, 1–3 mm. long, medium sized (seated on the crust formed of the mycelium), round, unequal, torn. Spores white, "5–6 × 3–4μ" Karst. Stumps, and dead branches of conifers. Oct.—Jan. Uncommon. (*v.v.*)

1996. **P. vitrea** (Pers.) Fr. *Vitrea*, glassy.

P. 1–10 cm., *whitish, subhyaline*, waxy, broadly and unequally effused, 2–4 mm. or more thick, subundulate, indeterminate; margin *shining white*, thin, villose. Mycelium *forming a woody stroma, tough and separable*. Pores *whitish, hyaline*, ·5–2 mm. long, straight, or oblique, very small, round, obtuse, entire. Spores "globose, ocellate, hyaline, 4μ; or ovoid, 4 × 2·5μ" Bres. Rotten beech trunks, rarely on dead fir wood. Aug.—March. Uncommon. (*v.v.*)

1997. **P. gilvescens** Bres. Trans. Brit. Myc. Soc. VI, text figs. p. 321.
Gilvescens, becoming pale yellow.

P. 6–7 cm., *white, then flesh coloured, at length brownish*, effused, bleeding, waxy, fleshy, then slightly coriaceous, contracted and inrolled when dry (often tubercularly nodular and *Ptychogaster*-like); margin persistently *white*, tomentose. Pores *white, becoming yellowish flesh coloured or reddish*, 2–8 mm. long, angular, orifice pulverulent, often oblique, soft, subfleshy. Spores hyaline, cylindric-curved, 4·5–5 × 1·5–2μ. Basidia clavate, 12–16 × 4μ. Subhymenial hyphae hyaline, 2·5–3·5μ in diam., thick or thin walled, gelatinous. Hyphae of pore walls yellowish, 3μ in diam., often incrusted with mineral matter. Beech stumps. Oct.—Dec. Rare. (*v.v.*)

**Pores yellowish.

1998. **P. nitida** (Pers.) Fr. Boud. Icon. t. 160. *Nitida*, shining.

P. 3–10 cm., *whitish, then yellowish orange*, effused, thin, subadnate, subundulate, determinate; margin *villose*. Pores *white, then yellowish orange*, shining, curt, sometimes stratose, minute, *round*, equal. Spores *whitish, tinged yellowish*, oblong, 8–9 × 3·5–4μ. Dead wood, and branches of willow, aspen, and chestnut. Nov.—May. Rare.

1999. **P. Laestadii** Fr. & Berk. C. P. Laestadius.

P. 5–20 cm., *bright yellow*, effused, forming confluent patches, 2–4 mm. thick, separable, *tubercular*. Pores *bright lemon yellow*, very short, sometimes stratose, round, or elongate and curved. Flesh *white*, very brittle, compact. Spores "5 × 2·5μ" Massee. Underside of deal boards in a hot-house. Aug. Rare.

2000. **P. eupora** Karst. (= *Poria nitida* (Pers.) Fr. sec. Quél.)
εὖ, typical; πόρος, a pore.

P. 2–15 cm., *bright buff, or yellow*, effused, adherent, or adnate, thin, somewhat shining; margin *white*, byssoid, *at length free, glabrous*. Pores *bright buff or yellow*, short, minute, round or subangular. Spores hyaline, elliptical, 3–4 × 1·5–2μ. "Cystidia hyaline, clavate, 15–105 × 6–15μ, upper part incrustate and verruculose" Romell. Decorticated logs, and branches. June—Dec. Not uncommon. (*v.v.*)

2001. **P. cincta** Berk. *Cincta*, girded.

P. 3–9 cm., *white, turning pallid*, or *pale ochraceous* and *more or less tawny when dry*, forming *small, erect, scattered tufts*, each *surrounded by radiating, strigose fibres*, at length confluent, up to 3 mm. thick in the centre of the tufts. Pores *pallid ochraceous, darker when dry*, 2–4 mm. long, *extremely minute*, scarcely visible to the naked eye, angular, dissepiments extremely thin, edge ragged. Spores "colourless, subglobose, 4–5μ" Massee. Old deal boards. Rare.

2002. **P. callosa** Fr. *Callum*, hardened skin.

P. 2–9 cm., *white*, broadly effused, 2–4 mm. thick, even, tough, *entire*, *separable*, like soft leather. Pores *white*, *hyaline*, *firm*, round, equal, quite entire, obtuse. Spores hyaline, "obliquely elliptical, 6 × 3·5μ" Massee. Rotten wood, and branches. April—May. Uncommon. (*v.v.*)

2003. **P. obducens** (Pers.) Fr. *Obducens*, covering over.

P. 2–8 cm., *white*, effused, incrusting, innate, inseparable, firm. Pores *pallid tan*, 2 mm. long, *distinctly stratose*, very small, round, equal. Spores hyaline, "elliptical, 4 × 2μ" Massee, "subglobose, 4μ, 1-guttulate. Cystidia with incrusted top, 12–15 × 9–15μ, or when the crust is removed, 4·5–7μ broad. Hyphae 3μ broad, not fibulate" Romell. Old stumps, and rotten branches of oak, elm, ash, pear, and maple. July—Feb. Uncommon. (*v.v.*)

***Pores flesh colour.

2004. **P. placenta** Fr. Fr. Icon. t. 188, fig. 3, as *Polyporus placenta* Fr. *Placenta*, a flat-cake, placenta.

P. 5–20 cm., *rosy flesh colour*, widely effused, rather thick, *soft*, *separable*; margin *white*, byssoid, sterile. Pores *rosy flesh colour*, *fuscous when dry*, 2–6 mm. long, angular, unequal, *irregularly stratose*. Smell very pleasant. Spores white, elliptical, 5 × 3μ, minutely punctate. Larch stumps. Aug.—Oct. Rare. (*v.v.*)

2005. **P. rhodella** Fr. *ῥόδον*, a rose

P. 5–11 cm., *white flesh colour*, effused, thin, adnate, *soft*; margin determinate, *naked*. Pores *white flesh colour*, short, minute, somewhat round, pruinose, continuous, or in patches. Spores hyaline, "ovoid spherical, 6μ" Quél. Beech, and conifer trunks. Aug. Rare.

****Pores red.

2006. **P. rufa** (Schrad.) Fr. Rostk. Polyp. t. 62, as *Polyporus haematodus* Rostk. *Rufa*, red.

P. 2-8 cm., *blood-red-rufous*, effused, coriaceous, thin, *adnate*, even, smooth, *determinate*; margin byssoid when young. Pores *concolorous*, very sm ll, thin, acute. Dead branches, and prostrate trunks of beech, broom, and bird cherry. Rare.

2007. **P. fusco-carnea** (Pers.) Fr. *Fuscus*, dusky; *carnea*, flesh colour.

P. 5–15 cm., *fuscous purple*, effused, thin; margin definite, here and there inflexed, thick, tomentose. Pores *fawn colour*, or *vinous brown*, 1·5–4 mm. long, somewhat round. Rotten wood, and putrid trunks. Rare

*****Pores ferruginous.

2008. **P. umbrina** Fr. Rostk. Polyp. t. 27, fig. 6, as *Polyporus ferruginosus* Fr. *Umbrina*, umber-coloured.

P. 5–8 cm., *rufous umber*, effused, up to 12 mm. thick, determinate, *undulato-tubercular*; margin *paler*, distinct, smooth. Pores *concolorous*, 2–4 mm. long, minute, somewhat round, unequal. Spores hyaline, "ovoid, 6–7μ" Quél. Trunks, and dead wood. Sept.—Oct. Uncommon. (*v.v.*)

II. Flesh thin; pores unequal, angular, or rather large.

*Pores white, or yellowish, and becoming yellowish red or greenish.

2009. **P. radula** (Pers.) Fr. *Radula*, a rasp.

P. 2–8 cm., *white*, effused, thin, *made up of the naked, tomentose* mycelium, closely compacted, soft, separable, villose beneath. Pores *white, then yellowish*, ·5–2 mm. long, sometimes oblique, medium sized, angular, *toothed*, pubescent when young. Spores white, elliptical, 5–6 × 3–4μ, 1-guttulate. Dead branches, and bark of willow, oak, aspen, hornbeam, beech, and fir. Aug.—March. Uncommon. (*v.v.*)

2010. **P. vaporaria** (Pers.) Fr. *Vaporarium*, a steam pipe.

P. 5–10 cm., *white*, effused, innate, inseparable, the white floccose mycelium creeping into the wood. Pores *white, then cream colour*, ·5–1 mm. long, *large, angular*, toothed, *forming a continuous, firm, persistent stratum*. Spores white, "allantoid, 6 × 1·5–2μ" Romell. Dead branches, and worked wood of conifers. Jan.—Dec. Uncommon. (*v.v.*)

var. **secernibilis** B. & Br. *Secernibilis*, separable.

Differs from the type in being *separable. Shining white, becoming honey colour when dry*. Fir leaves under moss. Sept.—Oct. Uncommon.

2011. **P. Eyrei** Bres. Trans. Brit. Myc. Soc. III, t. 14.
Rev. W. L. W. Eyre, a former president of the British Mycological Society.

P. 3–10 cm., *yellowish*, effused; margin subtomentose, soon similar; subiculum very thin, scarcely visible. Pores *concolorous*, 1 mm. long, straight, or oblique, oblong, or sinuate, entire. Spores hyaline, *obovate*, 4–5 × 3–3·5μ, 1-guttulate; basidia clavate, 12–15 × 4μ. Cystidia *clavate*, or *fusoid-ventricose*, 15–18 × 4–5μ; hyphae septate, often nodular on one side, 2–3μ thick. Oak wood. May. Uncommon. (*v.v.*)

2012. **P. sericeo-mollis** Romell. Romell, in Arkiv f. Bot. XI, t. 2, fig. 7, and Svensk Bot. Tidsk. 1912, Bd. 6, H. 3, fig. 4. *Sericeus*, silky; *mollis*, soft.

P. 1–4 cm., *white*, effused, very soft, 1–3 mm. thick, loosely adherent; margin often separating and reflexed, incurved when dry.

Pores *white, at length somewhat cream, or pallid,* usually angular, variable in diameter. Spores white, 4–6 × 2–3μ; basidia 4-spored, 20 × 5μ; hyphae septate, with clamp connections. Rotten coniferous wood. Nov.—Jan. Not uncommon. (*v.v.*) "Some specimens which seem to belong to this species are partly or totally reduced into a floccose-pulveraceous state of sulphurous or pallid colour, which contains abundant subglobose or ellipsoidal, apparently asperulate, 1-guttulate, chlamydospores 5–7·5 × 4–5μ, not unlike those of *Ptychogaster albus*, though more hyaline[1]." "The conidia are smooth, not asperulate, but have granular contents, which give a rough appearance" Wakefield & Pearson.

2013. **P. rancida** Bres. Bres. Fung. Trid. t. 208, fig. 1.
Rancida, stinking.

P. 2–8 cm., *white, then pale tan colour,* effused, *coriaceous*; margin subfimbriate, at length separating; subiculum thin, submembranaceous. Pores *concolorous,* 2–4 mm. long, round, oblong, or subangular, entire, or at length torn. Spores hyaline, cylindrical, somewhat curved, 5–7 × 2·5–·75μ; basidia clavate, 15–18 × 4–6μ; hyphae thin, 2·5–4μ thick. Smell *strong, of rancid meal.* Larch, and pine bark and needles. Sept. Rare.

2014. **P. hibernica** B. & Br. *Hibernica*, Irish.

P. 2–8 cm., *white,* broadly effused, orbicular, then confluent, *adnate, inseparable*; margin *narrow,* thin, *tomentose.* Pores *white,* very short, *small, angular, dissepiments somewhat rigid, almost entire.* Spores white, "elliptical, 5 × 3μ" Massee. Decorticated branches of pine. Sept. Rare. (*v.v.*)

2015. **P. Gordoniensis** B. & Br. Marchioness of Huntly.

P. 2–5 cm., *persistently shining white,* effused, superficial, *membranaceous,* very thin, *separable*; margin *shortly fringed.* Pores *shining white, becoming faintly yellowish white,* minute, unequal, angular, *dissepiments very thin, fimbriato-toothed.* Pine poles. Feb. Rare.

2016. **P. Vaillantii** (DC.) Fr. (= *Porothelium Vaillantii* (Fr.) Quél.) Sow. Eng. Fung. t. 326.
Sebastian Vaillant, a French mycologist.

P. 2–15 cm., *white, or slightly rufescent,* broadly effused, thin, translucid, *the free mycelium resulting in root-like ribs which are somewhat united by a membrane.* Pores *white,* here and there crowded together, curt, rather large, thin, unequal. Spores "hyaline, elliptical, 4–6 × 2–3μ" Karst. Dead wood, and on the ground. April—Oct. Rare.

[1] "I am still not fully satisfied that the chlamydosporic specimens really belong here" Romell, S.B.T. p. 643.

2017. **P. sanguinolenta** (A. & S.) Fr. *Sanguinolenta*, bloody.

P. 2–10 cm., *whitish, bleeding when touched*, nodulose, soon confluent, effused, soft; margin flaxy, soon vanishing. Pores *white, blood red when touched*, 1–3 mm. long, somewhat round, unequal; orifice *pubescent, pruinose*, at length torn. Spores white, oblong, 4–6 × 1·5–2μ, 2-guttulate. Smell strong. Dead wood, branches, and rails. Aug.—Nov. Not uncommon. (*v.v.*)

2018. **P. bombycina** Fr. (= *Trametes bombycina* (Fr.) Quél.) Sow. Eng. Fung. t. 387, fig. 5, as *Boletus terrestris*. *Bombycina*, silky.

P. 2–7 cm., *dingy yellowish*, effused, *silky-membranaceous*, adhering laxly; margin *spider-web-velvety*. Pores *whitish cream*, then *pale ochraceous*, somewhat round, becoming angular and flexuose, *large*. Spores straw colour, "elliptic oblong, 6–7 × 4μ" Massee. Dead wood. Sept.—Nov. Uncommon. (*v.v.*)

2019. **P. hymenocystis** B. & Br. *ὑμήν*, a membrane; *κύστις*, bladder.

P. 1–10 cm., *snow-white*, effused, very thin, *arachnoid*; margin minutely byssoid, almost indeterminate. Pores *white*, then *pallid*, large, *scarious dissepiments collapsing*. Spores white, *rough*, subglobose, 3 × 2μ, 1-guttulate. "Hyphae soft, fibulate, 2–3μ broad" Romell. Dead wood. Sept.—Oct. Not uncommon. (*v.v.*)

2020. **P. aneirina** (Sommerf.) Fr. (= *Trametes aneirina* (Sommerf.) Quél.) *ἀ*, not; *εἰρίνεος*, woolly.

P. 1–10 cm., *white*, effused, orbicular, then confluent, thin, subinnate; margin *byssoid*. Pores *white, then tawny or fulvous, large, cell-like, waxy, angular, often exactly hexagonal*, acute, smooth. Spores "obovate, 5–6 × 3·5–4·5μ" Bres. Dead branches of poplar, and willow. Oct. Rare.

2021. **P. ramentacea** B. & Br. *Ramentum*, chips.

P. 2–3 cm., *white*, effused, suborbicular; margin obsolete; subiculum *white*, tomentose, *cartilaginous and horny when dry*. Pores *honey colour*, large, ·5–·75 mm. across, subhexagonal, dissepiments thin, slightly rigid, acute. Spores white, "6 × 3μ" Massee. Dead pine branches. Sept. Rare.

2022. **P. viridans** Berk. *Viridans*, becoming green.

P. 2–6 cm., *white, becoming pallid green when dry*, effused, crustaceo-adnate, thin; margin *pulverulento-tomentose*. Pores *white, becoming pallid green*, minute, angular, dissepiments very thin. Spores white, elliptical, 4–5 × 2·5μ. Rotten wood, and sticks. Sept.—Dec. Rare. (*v.v.*)

2023. **P. Rennyi** B. & Br.
James Renny, an eminent English mycologist.

P. 2–6 cm., *white, becoming lemon yellow when dry,* at first forming a thick, somewhat frothy, then pulverulent mass. Pores *white, then yellowish,* 2–3 mm. long, sparingly produced, dissepiments thin Spores "colourless, elliptical, 3 × 1·5μ" Massee. Pine stumps, and on the ground. Oct.—Nov. Rare.

**Pores flesh colour.

2024. **P. incarnata** (A. & S.) Fr. Fr. Icon. t. 189, fig. 1, as *Polyporus incarnatus* Fr. *Incarnata,* flesh colour.

P. 2·5–10 cm., *flesh colour,* effused, *corky-coriaceous,* persistent, firm, smooth; margin *white,* silky, often shortly reflexed. Pores *flesh colour,* long, unequal, round, or angular, generally oblique. Spores "elongate, hyaline,. 7 × 2μ" Bres. in Sacc. "Basidia ovoid, subglobose. Cystidia colourless, terminating in a small point" Pat. Rotten conifer trunks. June—Nov. Rare.

2025. **P. micans** (Ehrenb.) Fr. *Micans,* sparkling.

P. 2–8 cm., *whitish flesh colour,* effused, suborbicular, becoming confluent, thin, adnate, soft, fugacious; margin *white, byssoid.* Pores *whitish flesh colour,* very shallow, very thin, *resembling honeycomb,* angular, *subcrenate.* Spores white, sausage-shaped, 7–8 × 3μ. Dead wood, and rotten trunks. Oct.—Nov. Uncommon. (*v.v.*)

***Pores violaceous, or purple.

2026. **P. violacea** (A. & S.) Fr. Rostk. Polyp. t. 27, fig. 3, as *Boletus purpureus* Fr. *Violacea,* violet colour.

P. 2–10 cm., *violaceous,* effused, determinate, *waxy gelatinous,* thin, closely adnate, even, *smooth,* destitute of a distinct subiculum. Pores *violaceous,* translucid, very shallow, *cellular, or veined,* quite entire. Spores tinged yellowish, punctate, elliptical, 7 × 4–4·5μ. Fir stumps, trunks and poles. Aug.—May. Not uncommon. (*v.v.*)

2027. **P. purpurea** Fr. Fr. Icon. t. 189, fig. 2, as *Polyporus rhodellus* Fr. *Purpurea,* purple.

P. 10–30 cm., *purple lilac,* very broadly and widely effused, *the mucedinous, flocculose, white mycelium creeping over the surface* of rotten wood; margin *white, silky.* Pores *purple lilac,* 1–2 mm. long, minute, unequal, round, or angular, *interruptedly* scattered, or conglomerate. Spores "ellipsoid-oblong, cylindric, curved, 6–7 × 2μ" Sacc. Decayed trunks and stumps of beech, willow, oak, and alder. Oct—Feb. Uncommon. (*v.v.*)

III. Effused, dry, tough; pores rather large, rigid, roundish, angular.

*Pores whitish, or greyish brown.

2028. **P. corticola** Fr. *Cortex*, bark; *colo*, I inhabit.

P. 2–8 cm., *white, becoming pale*, widely effused, equal, firm, smooth; *mycelium forming a bare, xylostramatoid layer*. Pores *whitish*, very minute, *superficial*, often obsolete, punctiform. Bark of poplar, beech, willow, oak, birch, and fir. Rare.

P. sinuosa Fr. = **Trametes sinuosa** (Fr.) Quél.

2029. **P. subfusco-flavida** (Rostk.) Massee. Rostk. Polyp. t. 27, fig. 11, as *Polyporus subfusco-flavidus* Rostk.

Subfuscus, somewhat dusky; *flavida*, yellowish.

P. 6–30 cm., *white, then light yellow fuscous*, broadly effused, becoming confluent, thin, *coriaceous*, arid, adnate; margin *white*, byssoid, determinate, thin. Pores *greyish brown*, or *whitish*, minute, irregular. Dead oak wood, and planks. Rare.

**Pores brown, or cinnamon.

2030. **P. obliqua** (Pers.) Quél. (= *Fomes obliquus* (Pers.) Fr.) Fr. Icon. t. 188, fig. 1, as *Polyporus obliquus* Fr. *Obliqua*, slanting.

P. 5–10 cm., *pallid, then date brown, becoming blackish*, widely spreading, *throwing off the bark*, very thin, coriaceous corky; margin *often reflexed, wrinkled, and laciniate*. Pores *brown*, 2–5 mm. long, extending to the wood, pervious to the base, often oblique; orifice of the pores *grey, glistening*, very small, obtuse, subpentagonal, sometimes obscurely stratose. Spores white, globose, 4–5μ. Trunks, and dead branches, under the bark, especially beech. Sept.—Oct. Uncommon. (*v.v.*)

P. ferruginosa (Schrad.) Fr. = **Fomes ferruginosus** (Schrad.) Massee.

P. resupinata (Bolt.) W. G. Sm. = **Fomes resupinatus** (Bolt.) Massee.

2031. **P. contigua** (Pers.) Fr. *Contigua*, touching together.

P. 6–8 cm., *cinnamon, becoming dingy*, effused, firm, 12 mm. *thick, smooth, submarginate*; margin at first *villose*; mycelium *ochraceous*. Pores *cinnamon, rather large*, round, equal, obtuse, entire. Spores "cylindrical, hyaline, often 1-guttulate, 5–7 × 3–3·5μ" Sacc. Rotten wood, and sticks. Sept.—Oct. Uncommon.

2032. **P. laevigata** Fr. *Laevigata*, made smooth.

P. 2–6 cm., *cinnamon*, broadly effused, *coriaceous rigid*, determinate, not marginate, *separating when old, smooth*, very glabrous, *with a rigid cuticle underneath*, 1–2 mm. thick. Pores *cinnamon*, very minute, round, entire. Spores white, "3–5 × 3–4μ. Cystidia like those in *Fomes igniarius*" Romell. Fallen birch branches. Nov. Uncommon. (*v.v.*)

IV. Unequally effused, membranaceous, thin, mostly incrusting; pores rather large, very short, often vein-like.

2033. **P. reticulata** (Pers.) Fr. Fr. Icon. t. 190, fig. 3, as *Polyporus reticulatus* Fr. *Reticulata*, netted.

P. 2–10 cm., *snow white, becoming pallid,* orbicular, thin, *fugacious*; margin byssoid, radiating. Pores *white, then yellowish, distant, cup-shaped.* Spores "allantoid, 7–9 × 2–3μ. Hyphae fragile, not fibulate, 4–5μ broad" Romell. Rotten wood. Jan.—Dec. Uncommon.

2034. **P. farinella** Fr. Trans. Brit. Myc. Soc. VI, text figs. p. 321. *Farinella*, mealy.

P. 1–11 cm., *snow white,* widely effused, very thin, fugacious when touched; mycelium naked, *flocculoso-pulverulent, not interwoven.* Pores *white,* thin, shallow, continuous, unequal, *hexagonal,* sub-flexuose, intricate. Spores white, oblong elliptical, 6–7 × 3–3·5μ, 1–2-guttulate, "cylindrical, curved, 8–9 × 2–2·5μ. Hyphae rather straight, 4–5μ in diam., no clamp connections" Wakef. & Pears. Dead wood, and logs of beech, lime, and fir. July—April. Not uncommon. (*v.v.*)

2035. **P. collabefacta** B. & Br. *Collabefacta*, brought to ruin.

P. 2–8 cm., *white,* forming *Corticium*-like patches, quite smooth; margin obtuse. Pores *white, arising from the mere collapsing of the substance,* shallow, obtuse. Spores "colourless, elliptic-oblong, 4 × 1·5μ" Massee. Dead wood. Oct. Rare.

2036. **P. blepharistoma** B. & Br. βλεφαρίς, eyelash; στόμα, mouth.

P. 1–5 cm., *snow white,* very thin; mycelium arachnoid, somewhat mealy. Pores *white,* small, dissepiments thin; orifice of pores *ciliato-dentate.* Spores white, elliptical, 4–5 × 3μ. Dead wood, and fallen branches. April—Nov. Not uncommon. (*v.v.*)

2037. **P. subgelatinosa** B. & Br. *Subgelatinosa*, somewhat gelatinous.

P. 4 cm., *pallid, becoming black, subgelatinous,* orbicular, forming little pulvinate patches, tomentose at first; margin raised, obtuse. Pores *delicate grey,* very shallow, angular, acute, entire. Spores "colourless, broadly elliptical, 4 × 2·5–3μ" Massee. Parasitic on *Polyporus amorphus.* Rare.

2038. **P. terrestris** (DC.) Fr. *Terrestris*, pertaining to the land.

P. 2–10 cm., *white,* effused, very thin, *spider-web-flaxy,* rather tender, *fugacious.* Pores *white, then rufescent,* central, extremely small, very shallow, round, or angular, becoming torn. Spores white, subglobose, 4–5 × 4μ, 1-guttulate. Naked soil, and rotten wood. May—Nov. Uncommon. (*v.v.*)

2039. **P. bathypora** (Rostk.) Massee. Rostk. Polyp. IV, t. 59, as *Polyporus bathyporus* Rostk. βαθύς, deep; πόρος, a pore.

P. 7–8 cm., *white*, effused; margin thin, byssoid. Pores *white, becoming brownish*, rather large, cup-shaped, toothed, sometimes stratose. Dead oak, and beech branches. Rare.

2. Polystictaceae.

Hymenium lining tubes, or covering gills, or teeth, homogeneous with the substance of the pileus, not forming a distinct layer, sterile on the edge.

Polystictus Fr.

(πολύστικτος, with many punctures.)

Pileus coriaceous, membranaceous, or somewhat spongy, dimidiate, sessile, surface often zoned. Tubes *homogeneous, developing from the centre outwards*. Spores white, elliptical, pruniform, oblong, or oblong-elliptical; smooth, or punctate. Cystidia sparse, or none. Annual. Growing on wood, often imbricate.

2040. **P. hirsutus** (Wulf.) Fr. *Hirsutus*, hairy.

P. 3–8 cm., *whitish, often brownish or tawny at the margin, sometimes becoming blackish with age*, dimidiate, convexo-plane, often imbricate, *shaggy with rigid hairs*, furrowed with concentric and *concolorous zones*. Tubes *whitish*, short; orifice of pores *whitish, becoming brownish or yellow*, round, or angular, obtuse. Flesh *whitish*, thin, very coriaceous, soft. Spores white, oblong elliptical, 6–7 × 2·5–3μ. Trunks, and stumps. June—March. Not uncommon. (*v.v.*)

2041. **P. velutinus** Fr. *Velutinus*, velvety.

P. 2–5 cm., *white, becoming yellowish or greyish*, dimidiate, plane on both sides, *minutely velvety, or pubescent, with obscure, slightly darker zones*; margin *thin*, acute. Tubes *white, or yellowish*, very short; orifice of pores *concolorous*, round, minute. Flesh *whitish*, thin, corky coriaceous, then rigid. Spores white, oblong elliptical, 6–8 × 2–2·5μ. Birch, beech, and willow trunks, and stumps. Sept.—Dec. Uncommon. (*v.v.*)

2042. **P. zonatus** Fr. ζώνη, a belt.

P. 3–8 cm., *pale tan colour, margin becoming whitish*, dimidiate, *convex, tuberculose and gibbous behind, villose*, or pruinose, *opaque, somewhat zoned with ochraceous and grey bands*. Tubes *whitish*, short; orifice of pores *whitish, becoming ochraceous bistre*, small, round, or angular, obtuse. Flesh *whitish*, corky, rather thick. Spores white, elliptical, 5–6 × 3–4μ. Elms, birches, and poplar trunks, and stumps. June—Nov. Uncommon. (*v.v.*)

2043. **P. versicolor** (Linn.) Fr. Hussey, Illus. Brit. Myc. I, t. 24.
Versicolor, of various colours.

P. 2–8 cm., *variously coloured*, dimidiate, orbicular, often imbricate, depressed behind, becoming plane, *velvety*, or pubescent, *marked with concentric, smooth, shining, satiny zones of various colours*. Tubes *white*, very short; orifice of pores *whitish, becoming yellowish*, small, round, becoming torn and irregular. Flesh *whitish*, thin, coriaceous. Spores *white*, oblong, 6–9 × 3μ. Trunks, stumps, twigs, pales, and branches. Jan.—Dec. Common. (*v.v.*)

var. **fuscatus** Fr. *Fuscatus*, dusky.

Differs from the type in the *fuscous, zoneless or obscurely zoned p.* and *the torn yellow pores*. Twigs, and fallen branches. Jan.—Dec. Not uncommon. (*v.v.*)

var. **nigricans** Lasch. *Nigricans*, becoming black.

Differs from the type in the *greyish black p. with black zones, and the smoke grey pores*. Stumps, and branches. Jan.—Dec. Not uncommon. (*v.v.*)

2044. **P. stereoides** Fr. (= *Trametes mollis* (Sommerf.) Fr. sec. Bres.) Fr. Icon. t. 187, fig. 3, as *Polyporus stereoides* Fr.
Stereum, the genus *Stereum*; εἶδος, like.

P. 2–3 cm., *greyish fuscous, becoming black*, effuso-reflexed, reniform, imbricate, pubescent, then glabrous, *with concolorous, depressed, narrow zones*, scarcely 1 mm. thick. Tubes *white*, short; orifice of pores *white, medium sized, obtuse, deformed, and daedalioid*. Flesh *yellowish*, thin, rigid, coriaceous. Spores white, "elliptical, 9μ" Quél., "oblong, 9–12 × 3·5–4μ" Romell. Trunks, and stumps of *Abies* and deciduous trees. Aug.—Oct. Rare.

2045. **P. fibula** Fr. Sow. Eng. Fung. t. 387, fig. 8. *Fibula*, a buckle.

P. 1–3 cm., *whitish, or greyish, becoming yellowish*, reniform, or orbicular, adnate behind, or affixed by the centre, *velvety hairy*, often radiato-rugose; margin entire, acute. Tubes *white*, very short; orifice of pores *whitish, becoming yellowish*, small, round, absent at the margin. Flesh *whitish, soft*, tough, coriaceous. Spores "ovate, internally granular, externally punctato-roughened, hyaline, 8–10 × 5–6μ or 7–9 × 2–3μ, cylindrical, curved, rarely straight" Bres. Elm stumps, oak branches, and worked wood. Oct.—Jan. Uncommon. (*v.v.*)

2046. **P. gossypinus** (Lév.) Massee. (= *Daedalea gossypina* (Lév.) Quél.) *Gossypium*, the cotton plant.

P. 3–10 cm., *white*, effuso-reflexed, becoming plane, *tomentose*. Tubes *white*, 2–4 mm. long; orifice of pores *greyish, labyrinthiform*, then angular, rather large, *denticulate*. Flesh *white*, thin, coriaceous. Spores white, "pruniform, 6μ, punctate" Quél. Trunks, fallen branches, and furze stems. Jan. Rare.

2047. **P. ravidus** Fr. Sow. Eng. Fung. t. 367, as *Boletus heteroclitus*. *Ravidus*, greyish.

P. 10–13 cm., *becoming dirty yellow*, applanate, effused at the base, imbricate, *with rugoso-villose, zone-like markings near the margin*. Tubes *whitish*, short; orifice of pores *becoming yellowish*, unequal, torn. Flesh *white*, corky coriaceous, tough. "Spores hyaline, yellow in the mass, cylindrical, slightly curved, 6–8 × 3μ" Sacc. Old willow stumps. Rare.

2048. **P. abietinus** (Dicks.) Fr. (= *Irpex violaceus* (Pers.) Quél.) Grev. Scot. Crypt. Fl. t. 226, as *Polyporus abietinus*. *Abies*, fir.

P. 2·5–8 cm., *cinereous white*, effuso-reflexed, dimidiate, sometimes resupinate, imbricate, villose, obsoletely zoned. Tubes *violaceous, becoming pale*, ·5–1 mm. long; orifice of pores *concolorous*, unequal, *torn*. Flesh *tinged brownish or purplish*, thin, coriaceous. Spores *white*, oblong, 3–4 × 2–2·5μ, curved, 2–3-guttulate. Trunks, and fallen branches of conifers, rarely on beeches. Jan.—Dec. Common. (*v.v.*)

Irpex Fr.

(*Irpex*, a harrow.)

Pileus corky coriaceous, or membranaceous, dimidiate, or resupinate, sessile. Tubes homogeneous, alveolar at first, then *becoming torn into teeth, or plates*. Flesh white, or coloured. Spores white, elliptical, oval, globose, cylindrical, or elliptic-oblong; smooth, or punctate. Cystidia present, or absent. Growing on wood, rarely on the ground.

I. Pendulous with the p. extended behind.

2049. **I. pendulus** (A. & S.) Fr. *Pendulus*, hanging down.

P. 2·5–4 cm., *pale yellow*, margin *white*, effused, more or less circular, *extended behind, pendulous*, free above, plicate, adpressedly squamuloso-pilose, or slightly rugulose. Teeth *shining white*, 2 mm. long, in irregular rows, large, incised. Flesh *concolorous*, membranaceous, elastic, very thin. Spores white, "3–5 × 1·5–2μ" Karst. Pine, and larch sticks. Rare.

II. Sessile, or effuso-reflexed, marginate.

2050. **I. fusco-violaceus** Fr. (= *Irpex violaceus* (Pers.) Quél.; *Polystictus abietinus* (Dicks.) Fr. sec. Quél.) *Fuscus*, dark; *violaceus*, violet.

P. 5–8 cm., *white, becoming greyish, or hoary*, dimidiate, effuso-reflexed, often imbricate and confluent, *zoned*, silky. Teeth *fuscous violaceous*, in rows in the form of plates, incised at the apex. Flesh *white*, corky coriaceous, firm. Spores *white*, "elliptical, cylindrical, curved, 9–10μ" Quél., "3–5 × 1μ" Karst. Coniferous trunks, and branches, rarely beech. Sept.—Oct. Uncommon. (*v.v.*)

2051. **I. lacteus** Fr. *Lacteus*, milk white.

P. 3–5 cm., *white*, effused, shortly reflexed, or dimidiate, sometimes imbricate, *villose, concentrically* sulcate; margin byssoid. Teeth *milk white*, subulate, or compressed, toothed, thin. Flesh *white*, coriaceous, thin. Spores white, "ovoid, globose, punctate, 5μ" Quél., "4–5 × 2–3μ" Karst. Birch, fir, pine, beech, and mountain ash. Oct.—Dec. Uncommon. (*v.v.*)

III. Resupinate.

2052. **I. hypogaeus** Fuck. ὑπόγαιος, under the earth.

R. 10–11 cm., *white, then pale yellowish, or dark brown*, widely effused; margin determinate. Teeth *concolorous*, 2–7 mm. long, irregular, sublabyrinthiform, lax, variable in size, straight, incised, base usually compressed, thin. Flesh *white, or yellowish*, byssoid. Incrusting pine leaves, twigs, grass, earth, pebbles. Oct. Rare.

2053. **I. Johnstonii** Berk. Dr George Johnston.

R. 2·5–5 cm., *white*, resupinate, effused, separable; margin reflexed, naked. Teeth *white, arranged in rows*, 2–3 mm. long, compressed, unequal, crowded. Flesh *white*, coriaceo-membranaceous. Dead beech. Rare.

2054. **I. candidus** (Ehrenb.) Fr. *Candidus*, shining white.

R. 3–5 cm., snow white, broadly effused, separable, thin, *arachnoid*; margin *byssoid*. Teeth *snow white*, subulate, or compressed, toothed, thin. Flesh *white*, membranaceous, floccose. Dead pine wood. Feb. Rare. (*v.v.*)

2055. **I. spathulatus** (Schrad.) Fr. Fr. Icon. t. 194, fig. 3. σπάθη, a broad blade.

R. 5–10 cm., *shining white, becoming yellowish when dry*, effused, adnate, inseparable; margin byssoid. Teeth *white, becoming yellowish*, 3–6 mm. long, *spathulate*, compressed, equal, entire, reticulato-connected with obsolete veins. Flesh *white*, membranaceous, thin. Spores white, elliptical, 4–5 × 2–3μ. Dead coniferous branches. Oct. —Dec. Uncommon. (*v.v.*)

2056. **I. obliquus** (Schrad.) Fr. *Obliquus*, slanting.

R. 5–20 cm., *white, then yellowish, or wood colour*, broadly effused, adnate; margin byssoid. Teeth *concolorous*, 2–6 mm. long, *at first very pore-like*, then becoming *compressed, incised*, or torn, oblique, lamellar at the base. Flesh *whitish, crustaceous*, thin. Spores white, elliptical,

4–5 × 3–3·5μ, 1-guttulate. Stumps, dead branches, and leaves. Jan.—Dec. Common. (*v.v.*)

2057. **I. deformis** Fr. *Deformis*, misshapen.

R. 5–15 cm., *whitish*, effused, adnate; margin byssoid, pubescent. Teeth *concolorous*, 2–4 mm. long, *subulate, arising from a minutely porous base*, somewhat digitato-incised, and often torn into shreds almost to the base. Flesh *concolorous*, crustaceous, thin. Spores white, "ovoid, punctate, 10μ" Quél. Oak branches, and cherry. Sept.—Feb. Uncommon. (*v.v.*)

2058. **I. carneus** Fr. (= *Phlebia merismoides* Fr. sec. Quél.) *Carneus*, flesh colour.

R. 2·5–7·5 cm., *reddish*, effused, adnate. Teeth *concolorous*, subulate, obtuse, entire, united at the base. Flesh *cartilaginous, gelatinous*, thin. Wood, and bark. Sept.—Oct. Rare.

Lenzites Fr.

(Harold Othmar Lenz, a German botanist.)

Pileus corky, or coriaceous, dimidiate, or resupinate, sessile. Gills *coriaceous, often anastomosing at the base*, homogeneous with the substance of the pileus, and not forming a distinct layer. Flesh white, or coloured. Spores white, elliptical, subglobose, cylindrical, or oblong-elliptical, smooth. Cystidia sparse, or none. Growing on wood; often imbricate.

*Growing on wood of deciduous trees.

2059. **L. betulina** (Linn.) Fr. Cke. Illus. no. 1100, t. 1145, fig. A. *Betulina*, of the birch.

P. 2·5–10 cm., *whitish grey, becoming pale, corky coriaceous*, firm, *rigid*, dimidiate, sessile, becoming plane, sometimes resupinate, *tomentose*, commonly *obsoletely zoned*, zones sometimes darker. Gills *dingy white*, reaching the base, straight, *simple*, or branched, *often anastomosing*, edge acute. Flesh white, floccose. Spores white, "globose or elliptic-spheric, 5–6μ" Karst. On stumps, trunks, posts, and rails, especially birch. Jan.—Dec. Common. (*v.v.*)

2060. **L. flaccida** (Bull.) Fr. *Flaccida*, flabby.

P. 10–30 cm., *whitish, then dingy, with quite concolorous zones*, coriaceous, *thin*, scarcely 2 mm. thick, unequal, dimidiate, sessile, *easily bent, strigosely hairy*. Gills *shining white, becoming pale*, thick, firm, straight, very broad, simple, or branched at the base, *with shorter ones intermixed*. Spores "white, 12μ" Quél. On beech stumps. Jan.—Dec. Common. (*v.v.*)

var. **variegata** (Fr.) Cost. & Dufour. *Variegata*, with diverse colours.

Differs from the type in the *silky, velvety zones and white flesh.* On fallen logs of beech and birch. Sept.—March. Not uncommon. (*v.v.*)

L. cinerea (Fr.) Quél. = **Daedalea cinerea** Fr.

L. quercina (Linn.) Quél. = **Daedalea quercina** (Linn.) Fr.

**Growing on coniferous wood.

2061. **L. saepiaria** (Wulf.) Fr. Cke. Illus. no. 1101, t. 1146, fig. A. *Saepes*, a fence.

P. 3–8 cm., *yellow tawny*, then *date brown with a yellow tawny margin*, becoming *black* when old, dimidiate, lateral, corky coriaceous, hard, convex, becoming plane, sometimes orbicular, more frequently extended longitudinally, sometimes resupinate, zoned, *strigosely tomentose*, at length squamulose and pitted. Gills *yellowish, becoming umber*, extended to the base, very rigid, firm, branched, more or less anastomosing, 2–4 mm. broad, edge entire, or slightly toothed. Flesh *tawny*. Spores white, cylindrical, curved, $10 \times 3\text{–}4\mu$. Coniferous stumps, branches, and worked wood. Jan.—Dec. Not uncommon. (*v.v.*)

2062. **L. abietina** (Bull.) Fr. Cke. Illus. no. 1101, t. 1146, fig. B. *Abies*, a fir tree.

P. *umber-tomentose*, then becoming smooth, *effuso-reflexed*, often lengthened out to 30 × 1 cm., sometimes resupinate, hoary, coriaceous, thin, and comparatively *soft*. Gills *yellowish red, becoming glaucous with dense pruina*, decurrent in the effused base, distant, simple, unequal, here and there torn into teeth. Flesh *concolorous*, very thin. Spores white, oblong elliptical, $10 \times 4\mu$, apiculate at the one end. Dressed fir wood. Oct. Rare.

2063. **L. heteromorpha** Fr. Fr. Icon. t. 177, fig. 3. ἑτερόμορφος, of different shape.

P. 2–3 cm., *whitish, becoming pale, and finally yellowish when old*, effuso-reflexed, imbricate, connate, corky soft, then hard, *nodular*, often pectinately incised at the margin, always gibbose, almost glabrous with adpressed tufts of hairs, coarsely rugose. Gills *white*, very firm, thick, very broad, triquetrous, somewhat crowded, somewhat branched, incised, or forming pores, sometimes falling short of the margin. Spores white, "subglobose, $3\text{–}5\mu$" Karst. Flesh white. On fir stumps. Oct. Rare. (*v.v.*)

Trametes Fr.

(*Trama*, the woof.)

Pileus woody, or corky, dimidiate, or resupinate, sessile. Tubes homogeneous with the substance of the pileus, and not forming a distinct layer, regular, round, or oblong. Flesh white, or coloured. Spores white, rarely yellowish, elliptical, ovoid, globose, subglobose, cylindrical, or oblong, smooth. Cystidia present, or absent, hyaline, or coloured. Annual, or perennial. Growing on wood, very rarely on the ground; sometimes imbricate.

I. Dimidiate, sessile.

*Flesh whitish.

2064. **T. Trogii** Berk. (= *Trametes hispida* (Bagl.) Quél.)
J. G. Trog, the Swiss mycologist.

P. 5–10 cm., *fuscous, somewhat olivaceous,* dimidiate, convex, solitary, or imbricate, somewhat zoned, zones at first very indistinct, then becoming evident, *concolorous, clothed with rigid, fasciculate,* 6 mm. *long, hairs*; margin acute. Pores *cream colour, then coffee and milk colour,* unequal, subangular, toothed. Flesh pale tan or wood colour. Spores "white, elliptic-cylindric, 12–13μ, guttulate" Quél. Dead poplar trunks. Oct. Rare.

2065. **T. gibbosa** (Pers.) Fr. Boud. Icon. t. 162. *Gibbosa,* humped.

P. 10–20 cm., *whitish, becoming greyish,* dimidiate, flattened, extended behind, *gibbose,* villose, obsoletely zoned; margin *often brownish,* obtuse. Pores *whitish,* 2–8 mm. long, *linear,* straight, equal. Flesh *whitish,* corky, compact, thick, very firm. Spores white, oblong, sometimes curved, 5–7 × 2·5–3μ. On stumps, and posts of beech, and poplar, more rarely on oak, and willow. Aug.—March. Not uncommon. (*v.v.*)

2066. **T. rubescens** (A. & S.) Fr. Trans. Brit. Myc. Soc. II, t. 16.
Rubescens, turning red.

P. 5–12 cm., *whitish, becoming red,* dimidiate, flattened, at first *white pruinose,* at length zoned; margin thin. Pores *white, becoming crimson lake, or blood red when touched,* pruinose, 1–3 mm. long, roundish, then elongate and daedaliform, narrow, obtuse. Flesh *white, becoming crimson lake, or blood red when broken,* corky, soft, zoned. Spores *white,* oblong, curved, 10 × 2μ, 3-guttulate. On willow, and alder trunks. Aug.—Nov. Not uncommon. (*v.v.*)

2067. **T. Bulliardii** Fr. (= *Trametes rubescens* (A. & S.) Fr. sec. Quél.) Bull. Hist. Champ. Fr. t. 310, as *Boletus suaveolens.*
Pierre Bulliard, the eminent French mycologist.

P. 5–14 cm., *whitish, becoming fuscous,* dimidiate, flattened, often gibbose at the base, *at length zoned,* even, *smooth*; margin thin, subacute. Pores *pallid, then rufescent,* 3–10 mm. long, somewhat round, or linear, unequal. Flesh *yellowish, at length becoming fuscous,* corky, thick Spores white, elliptical, 4–5 × 3μ, 1-guttulate. Smell *pleasant.* On willow, and alder trunks. Aug.—Oct. Uncommon. (*v.v.*)

2068. **T. suaveolens** (Linn.) Fr. Boud. Icon. t. 163.
Suaveolens, sweet smelling.

P. 4–15 cm., *whitish*, dimidiate, *pulvinate*, triquetrous, *villose*; margin *becoming yellowish*, thin. Pores *white, becoming yellowish, or fuscous*, 3–12 mm. long, round, rather large, obtuse. Flesh *white, soft*, corky, thick. Spores white, oblong, often incurved, 10–12 × 3–4μ. Smell *strong, of anise*. Trunks of willow, rarely lime. Sept.—Feb. Common. (*v.v.*)

2069. **T. odora** (Sommerf.) Fr. Bolt. Hist. Fung. t. 162, as *Polyporus odorus*. *Odora*, fragrant.

P. 5–10 cm., *pallid*, dimidiate, gibbose, uneven, villose, becoming smooth; margin *yellow*, thin. Pores *whitish, then ochraceous*, 4–8 mm. long, *minute*, round, often becoming toothed, equal. Flesh *white*, corky, elastic, thick. Spores "yellowish, ovoid, 7–8μ, 1-guttulate" Quél., "hyaline, oval, 5–6 × 3μ" Karst. Smell *strong, of anise*. Willow, and ash trunks. Jan. Rare.

2070. **T. inodora** Fr. Fr. Icon. t. 191, fig. 1. *Inodora*, without smell.

P. 3–8 cm., *white, or yellowish*, dimidiate, triquetrous, sometimes imbricate, minutely tomentose, often obsoletely zoned, becoming smooth; margin thin, acute. Pores *white, unchangeable*, 2–6 mm. long, small, *round, or oblong*, pubescent. Flesh *white*, corky, *firm*. Spores white, globose, 5–6μ, 1-guttulate. Beech, and oak stumps. Sept.—Nov. Uncommon. (*v.v.*)

**Flesh ferruginous.

2071. **T. Pini** (Brot.) Fr. Boud. Icon. t. 161. *Pinus*, pine.

P. 5–10 cm., *ferruginous fuscous, then blackish*, dimidiate, *pulvinate*, concentrically sulcate, *rimoso-rugged*, rough, becoming incrusted with age; margin *bright yellowish at first*, and tomentose. Pores *yellow-brick-red*, 6–15 mm. long, large, somewhat round, or oblong; orifice at first pubescent. Flesh *tawny ferruginous*, corky woody, *very hard*. Spores *pale yellowish*, oval, or subglobose, 4–6 × 4–5μ, with a large central gutta. Cystidia *dark brown*, conical, pointed, 30–40 × 8–9μ. Smell slightly pleasant. Pine trunks. Sept.—Nov. Uncommon. (*v.v.*)

2072. **T. odorata** (Wulf.) Fr. *Odorata*, scented.

P. 7·5–13 cm. long, 5–8 cm. broad, *blackish umber, edge tawny cinnamon*, dimidiate, *downy*, then vaguely concentrically zoned, rugulose, tomentose, sometimes attenuated behind. Pores *tawny cinnamon*, subrotund, oblong, more or less decurrent, uneven. Flesh *fulvous*, somewhat corky. Spores "tawny, elliptical" Quél. Smell strong, pleasant, like hay, or spicy. On decaying coniferous wood. Jan. Rare.

***Flesh red.

2073. **T. cinnabarina** (Jacq.) Fr. (= *Phellinus cinnabarinus* (Jacq.) Quél.) Trans. Brit. Myc. Soc. IV, t. 9. *κιννάβαρι*, dragon's blood.

P. 5–9 cm., *bright reddish orange*, becoming darker, dimidiate, slightly pubescent, then glabrous, rugulose, indistinctly zoned towards the margin. Pores *deep blood red*, 1–3 mm. long; orifice *vermilion*, minute, round, pubescent. Flesh *red*, corky, pliant, thick. Spores white, oblong, curved, 6 × 2μ. Birch, and beech trunks. July—Oct. Uncommon. (*v.v.*)

II. Resupinate.

2074. **T. sinuosa** (Fr.) Quél. (= *Polyporus sinuosus* Fr.) Trans. Brit. Myc. Soc. IV, t. 10. *Sinuosa*, full of curves.

R. 3–6 cm. and more, *pure white, becoming yellowish*, resupinate, broadly effused, *furnished with long, white, string-like, mycelial rhizoids on the underside.* Pores *white, then yellowish*, 2–3 mm. long; orifice large, *flexuose, irregularly torn, often daedaliform* or *sistotremiform, pruinose.* Spores white, elliptical, 5–6 × 3–4μ, with a large central gutta. Smell very pleasant, "like liquorice" Fries, "of balsam" Quél. Ivy trailing on the ground, and conifer stumps, and branches. Oct.—Nov. Uncommon. (*v.v.*)

2075. **T. mollis** (Sommerf.) Fr. (= *Polystictus stereoides* Fr. sec. Bres.) *Mollis*, soft.

R. 2–10 cm., *pallid wood colour, at length becoming fuscous, or black*, resupinate, broadly effused, determinate, adnate behind in the centre, separable, submembranaceous; margin *umber, at length revolute, pubescent beneath.* Pores *whitish cream, becoming greyish*, large, shallow, angular, or round, often irregular, unequal, torn. Flesh *white*, soft, then coriaceous. Spores white, cylindrical, slightly curved, 8–11 × 3–4μ. Dead beech wood, and branches. Jan.—Dec. Common. (*v.v.*)

2076. **T. serpens** Fr. Fr. Icon. t. 192, fig. 3. *Serpens*, creeping.

R. 10–30 cm., *white, then pale ochraceous*, resupinate, closely adnate, inseparable, arid, pruinose, at first *erumpent in the form of a tubercle*, orbicular, then confluent; margin determinate, pubescent. Pores *white*, then *cream fuliginous*, rather large, very shallow, round, or angular, then labyrinthiform, unequal, obtuse. Flesh *white*, corky, coriaceous. Spores "white, elliptical, 14 × 6μ" Rabenh. Oak, beech, hornbeam, and privet logs. Aug.—Jan. Not uncommon. (*v.v.*)

2077. **T. Terryi** B. & Br. Michael Terrey.

R. 7–8 cm., *whitish*, resupinate, broad, suborbicular, pulvinate; margin determinate, undulate. Pores *pallid*, angular, here and there sinuate, rather large, shallow. Flesh *white*, corky, firm. Beech. Rare.

2078. **T. purpurascens** B. & Br. *Purpurascens*, becoming purple.

R. 2 cm., *chestnut*, resupinate, subcoriaceous, subtomentose. Pores *becoming purple*, rigid, small. Dead willow. Rare.

T. bombycina (Fr.) Quél. = **Poria bombycina** Fr.

T. aneirina (Sommerf.) Quél. = **Poria aneirina** (Sommerf.) Fr.

Daedalea (Pers.) Fr.

(δαίδαλος, curiously wrought.)

Pileus spongy, cork, coriaceous, or woody, dimidiate, or resupinate, stipitate, or sessile. Stem central, lateral, or none. Tubes homogeneous with the substance of the pileus, and not forming a distinct layer, *irregularly sinuous, and more or less labyrinthiform, often becoming torn, or toothed.* Flesh white, or coloured. Spores white, oval, pip-shaped, subglobose, elliptic-oblong, or sausage-shaped, smooth, or punctate. Cystidia present, or absent. Annual, or perennial. Growing on wood, very rarely on the ground; sometimes imbricate.

I. Dimidiate, sessile, or substipitate.

2079. **D. biennis** (Bull.) Quél. (= *Polyporus rufescens* Fr.) Sow. Eng. Fung. t. 191, as *Boletus biennis*. *Biennis*, two years.

P. 5–12 cm., *flesh colour, whitish towards the margin*, convex, then plane or depressed, sometimes dimidiate, *strigose, or hairy*. St. 1–5 × 1·5–2 cm., *ferruginous*, irregularly shaped, subcentral, or lateral, or wanting, subtomentose. Pores *white*, then *flesh colour*, 2–4 mm. long, *labyrinthiform*, or *sinuate*, at length torn, pruinose. Flesh *reddish, becoming whitish*, consisting of a firm, coriaceous lower layer, with a soft spongy upper layer. Spores white, broadly oval, or subglobose, 6–7 × 4–5μ, with a large central gutta. Smell pleasant. Stumps, roots, and buried wood. Sept.—Jan. Not uncommon. (*v.v.*)

2080. **D. quercina** (Linn.) Fr. (= *Lenzites quercina* (Linn.) Quél.) Grev. Scot. Crypt. Fl. t. 238. *Quercina*, pertaining to oak.

P. 9–50 cm., *pale wood colour, or brownish becoming paler*, dimidiate, sessile, rarely substipitate, or resupinate, *smooth*, rugulose, uneven, *marked with concentric, raised, or depressed zones*. Pores *greyish, fuliginous*, or *paler than the p.*, 6–50 mm. long, sinuate, or lamellose, branched, and anastomosing, thick, woody. Flesh *pale reddish brown, or concolorous*, corky, woody, thick, firm. Spores white, pip-shaped, 6 × 2–3μ. Oaks, oak stumps, and posts. Jan.—Dec. Common. (*v.v.*)

2081. **D. borealis** (Wahlenb.) Quél. (= *Polyporus borealis* Fr.) Kalchbr. Icon. t. 35, fig. 2, as *Polyporus borealis* Fr. *Borealis*, northern.

P. 5–15 cm., *white, then yellowish*, dimidiate, reniform, or subpulvinate, sessile, or attenuated behind into a short more or less distinct stem, *velvety, or strigose,* becoming matted when old; margin acute, spreading. Pores *white, then yellowish,* 4–6 mm. long, unequal, round, or angular, becoming *sinuate and daedaliform.* Flesh white, spongy, then *corky, composed of parallel fibres,* compact, thick, fissile. Spores white, subglobose, 5–6 × 4–5μ, *minutely punctate.* Smell slightly pleasant when dried. *Abies* trunks, and stumps. Sept.—Oct. Uncommon. (*v.v.*)

2082. **D. unicolor** (Bull.) Fr. Bolt. Hist. Fung. t. 163, as *Boletus unicolor* Bolt. *Unicolor,* of one colour.

P. 5–15 cm., *cinereous, fuliginous when moist, whitish grey when dry, with zones of the same colour,* shell-shaped, dimidiate, usually imbricate, *villoso-strigose*; margin *sometimes whitish.* Pores *whitish cinereous, sometimes fuscous,* very short, labyrinthiform, flexuose, intricate, narrow, acute, at length torn into teeth. Flesh *white,* coriaceous, thin. Spores white, "6–9 × 3–5μ" Karst. Stumps, trunks, and rails of birch, beech, maple, oak, willow, *Robinia,* chestnut, and hornbeam. Jan.—Dec. Not uncommon. (*v.v.*)

2083. **D. polyzona** (Pers.) Fr. πολύς, many; ζώνη, belt.

P. 13–18 cm., *yellowish brown, darker or reddish at the base,* dimidiate, sessile, imbricate, tomentose, *with many dark brown zones.* Pores *pallid wood colour,* or *buff white, very short,* equal, thin, sublabyrinthiform, sometimes 2-stratose. Flesh *buff white, coriaceous,* somewhat thin. Jan.—Dec. Rare.

D. gossypina (Lév.) Quél. = **Polystictus gossypinus** (Lév.) Massee.

2084. **D. saligna** Fr. (= *Polyporus salignus* Fr. Hym. Eur.; *Polyporus fumosus* (Pers.) Fr. sec. Lloyd.) Fr. Icon. t. 181, fig. 1, as *Polyporus salignus* **Holmiensis* Fr. *Saligna,* of willows.

P. 5–15 cm., *whitish,* dimidiate, imbricate, dilated reniform, sessile, adpressedly villose, *depresso-sulcate round the margin*; margin *swollen,* lobed. Pores *white,* long, round, or *intricately flexuose,* labyrinthiform, pruinose, thin. Flesh *whitish,* coriaceous, soft, elastic. Spores white, "elliptic-oblong, 7–8μ" Quél. Caespitose. Willows. Oct.—Feb. Uncommon.

2085. **D. confragosa** (Bolt.) Fr. Bolt. Hist. Fung. t. 160, as *Boletus confragosus* Bolt. *Confragosa,* rough.

P. 5–13 cm., *unicolorous, brick red fuscous, becoming ferruginous,* dimidiate, reniform, constricted at the base, gibbose, convex, sessile, rough, zoned. Pores *cinereous pruinose, then rufous fuscous,* 5–20 mm. long, sinuous, narrow, then labyrinthiform, torn, toothed. Flesh *wood colour, or reddish, then brown,* fibrous, corky, thin. Spores white, sausage-shaped, curved, 7–8 × 2μ. Beech, oak, willow, and service trunks. Oct.—Feb. Uncommon. (*v.v.*)

var. **angustata** (Sow.) Fr. Sow. Eng. Fung t. 193, as *Boletus angustatus* Sow. *Angustata*, narrowed.

Differs from the type in *the brownish rufescent, repand p., the subtomentose paler margin of the p., and the subolivaceous pores.* Poplars. Rare.

2086. **D. aurea** (Batt.) Fr. *Aurea*, golden.

P. 2–5 cm., *unicolorous, golden,* triangular, gibbose, imbricate, sessile, *velvety, subzoned*; margin swollen. Pores *light yellow,* somewhat long, round, then narrowly sinuato-labyrinthiform. Flesh *light yellow,* corky coriaceous, thin. Dead oak. Rare.

2087. **D. cinerea** Fr. (= *Lenzites cinerea* (Fr.) Quél.) Fr. Icon. t. 192, fig. 2. *Cinerea*, ash colour.

P. 2·5–13 cm., *cinereous,* dimidiate, sessile, sometimes imbricate, subundulate, *zoned, tomentose*; margin *paler,* thin. Pores *white, or cinereous,* 5–10 mm. long, round, or very long, labyrinthiform, flexuose, intricate, obtuse, *entire,* sometimes stratose. Flesh *pale buff, or ochraceous,* corky woody, thick. Spores white, globose, 10μ. Beech, and oak trunks, and stumps. Jan.—Dec. Not uncommon. (*v.v.*)

2088. **D. ferruginea** (Schum.) Fr. Fl. Dan. t. 2029. *Ferruginea*, iron rust colour.

P. 4–8 cm., *whitish flesh colour at first, then yellow ferruginous,* effuso-reflexed, horizontal, imbricate, zoned; *white villose when young*; margin *white villose,* swollen, flexuose. Pores *tawny,* 4 mm. long, deformed, narrowly labyrinthiform, flexuose, anastomosing. Flesh *yellowish sienna,* 3–4 mm. thick, coriaceous. Beech trunks, and dead wood. Rare.

II. Resupinate.

2089. **D. latissima** Fr. (= *Trametes latissima* (Fr.) Quél.) *Latissima*, very broad.

P. 12·5–60 cm., *pale wood colour,* broadly effused, undulated. Pores *pale wood colour,* very long, somewhat round, flexuose, *sinuose, narrow, distant.* Flesh *wood colour,* corky or woody, thick, *zoned,* with parallel filaments. Beech trunks. Rare.

2090. **D. vermicularis** (Pers.) Fr. Sow. Eng. Fung. t. 424, as *Boletus resupinatus* Sow. *Vermicularis*, pertaining to worms.

P. 10–15 cm., *flesh-colour-rufescent,* broadly effused, adnate, becoming even. Pores *concolorous,* short, attenuated at both ends, flexuose. Flesh *thin.* Adhering to the soil by root-like fibres. Aug. Rare.

3. Meruliaceae.

Hymenium spread over veins, anastomosing pores, or quite smooth; *edge of veins or pores fertile.*

Merulius Fr.

(*Merus*, pure.)

Receptacle gelatinous, coriaceous gelatinous, waxy, membranaceous, or floccose, resupinate, or effuso-reflexed. Hymenium at first smooth, *becoming reticulated with irregular, obtuse folds or pores,* at length gyrose or obsoletely toothed, and fertile on the edge. Spores white, or coloured, elliptical, ovoid, pip-shaped, globose, subglobose, elliptic-oblong, cylindrical, or sausage-shaped, smooth. Cystidia present, or absent. Growing on wood, rarely on the ground.

*Spores white.

†P. effuso-reflexed, margin determinate.

2091. **M. confluens** Schwein. *Confluens,* becoming confluent.

R. 2·5–10 cm., *vinous biscuit colour,* resupinate, longitudinally effused, becoming confluent, somewhat fleshy; margin *biscuit colour,* free, inflexed, subtomentose. Folds becoming *pinkish cinnamon to pecan brown* when dry, very small, uneven, reticulate. Flesh coriaceous, thin. Spores "hyaline, even, cylindric, flattened on one side, 4·5–5 × 2·5μ. Subhymenial hyphae incrusted" Burt. Alder branches. Aug. Rare.

2092. **M. tremellosus** (Schrad.) Fr. Hussey, Illus. Brit. Myc. I, t. 10. *Tremellosus,* trembling.

R. 2·5–15 cm., *white,* translucent, resupinate, then free and reflexed, often connate and imbricate, *tomentose*; margin *often pinkish,* dentato-radiate. Folds *ruddy, pinkish, or pale,* porous, twisted, toothed. Flesh *gelatinous,* cartilaginous when dry. Spores white, sausage-shaped, curved, 4–5 × 1μ. Cystidia "even or incrusted, sparse, 3·5–4·5μ in diam., emerging 15–25μ above the basidia" Burt. Stumps, and dead branches of birch, beech, and oak. Aug.—Feb. Not uncommon. (*v.v.*)

2093. **M. aurantiacus** Klotzsch. *Aurantiacus,* golden.

R. 2·5–4 cm., *between yellow and dirty white, here and there cinereous,* effuso-reflexed, *tomentose, obsoletely zoned.* Folds *dull orange,* minute, subporiform. Flesh *coriaceous,* tough. Dead beech trunks. May—Sept. Rare.

2094. **M. corium** (Pers.) Fr. (= *Merulius papyrinus* (Bull.) Quél.) Grev. Scot. Crypt. Fl. t. 147, as *Thelephora corium* Pers. *χόριον,* leather.

R. 5–20 cm., *whitish,* resupinato-effused, often imbricate; margin at length free, reflexed, villose beneath. Folds *flesh colour, or pale tan, reticulato-porous*, thin. Flesh soft, *leathery*, flexible, tough. Spores white, oblong elliptical, 6–9 × 3·5–4μ. Cystidia "none. Hyphae loosely interwoven, hyaline, septate, 3–4μ in diam." Burt. Dead wood, and branches. Jan.—Dec. Common. (*v.v.*)

2095. **M. niveus** Fr. *Niveus*, snow white.

R. 1–5 cm., *snow white*, resupinate, effuso-reflexed, *adnate at the centre*, free elsewhere, smooth. Folds *snow white*, rugose, subreticulate. Flesh very soft, spongy, thin, becoming membranaceous and papery when dry. Spores white, broadly elliptical, 8 × 5–6μ, "slightly curved, 4·5 × ·5–1μ" Burt. Dead alder branches. Nov. Rare. (*v.v.*)

††Resupinato-effused, flaxy membranaceous, *separable*, margin and underside byssoid.

2096. **M. laeticolor** B. & Br. (= *Merulius fugax* Fr. sec. Romell.) *Laetus*, bright; *color*, colour.

R. 7–8 cm., *bright orange*, resupinate, effused, adnate; margin *white*, byssoid. Folds *concolorous*, at first even, then plicato-rugose, *distant*. Flesh thin. Spores "white, subglobose, 6–7μ" Massee. Sawdust, and leaves and branches of oak, pine, and mountain ash. Oct.—Nov. Rare. (*v.v.*)

†††Crustoso-adnate, margin somewhat byssoid.

2097. **M. porinoides** Fr. πόρος, a pore; εἶδος, like.

R. 2–11 cm., *light dingy yellow*, resupinate, crustaceo-adnate; margin *white*, byssoid. Folds *concolorous*, poriform, large, round, distant. Flesh very thin. Spores white, globose, 3μ. Dead wood, chips, bark, and leaves. Sept.—Dec. Uncommon. (*v.v.*)

2098. **M. rufus** (Pers.) Fr. Pers. Myc. Eur. II, t. 16, figs. 1, 2, as *Xylomyzon isoporum* Pers. *Rufus*, red.

R. 2–8 cm., *red flesh colour*, resupinate, effused, crustaceo-adnate, often immersed in the wood, smooth; margin *somewhat naked*. Folds *concolorous, porose*, equal, angular. Flesh *waxy soft*. Spores white, pip-shaped, slightly curved, 5–6 × 2–3μ. Cystidia none. "Hyphae loosely interwoven, hyaline, 3–3·5μ in diam." Burt. Rotten oak, and hornbeam wood and posts. May—Feb. Uncommon. (*v.v.*)

2099. **M. serpens** (Tode) Fr. Fr. Icon. t. 193, fig. 3. *Serpens*, creeping.

R. 3–15 cm., *pallid, becoming red*, resupinate, *crustaceo-adnate*, becoming smooth; margin *white*, byssoid. Folds *concolorous*, at first in the form of wrinkles, then porous, angular, entire. Flesh thin. Spores white, cylindrical, curved, 4 × 2μ, 2-guttulate. Dead branches of conifers, lime and ash, and on cones. Oct.—March. Rare. (*v.v.*)

2100. **M. pallens** Berk. *Pallens*, pale.

R. 4–8 cm., *pale reddish*, resupinate, adnate, *inseparable*; margin indeterminate. Folds *concolorous*, poriform, minute. Flesh *subgelatinous*, thin. Spores white, globose, 4μ. Fir, and oak branches. Sept.—Dec. Rare. (*v.v.*)

2101. **M. Carmichaelianus** (Grev.) Berk. Grev. Scot. Crypt. Fl. t. 224, as *Polyporus Carmichaelianus* Grev.
Captain Dugald Carmichael, a friend of Greville.

R. 2·5–10 cm., *white, becoming pinkish brown when dry*, resupinate, irregularly effused; margin byssoid, laciniate. Folds *concolorous, forming regular, hexagonal reticulations or pores*, very shallow. Flesh membranaceous, very thin. Spores white, "globose, very minute" Grev. Bark. Rare.

M. crispus (Pers.) Quél. = **Plicatura crispa** (Pers.) Rea.

**Spores coloured.

†P. effuso-reflexed.

2102. **M. lacrymans** (Wulf.) Fr. Rolland, Champ. t. 98, no. 216.
Lacrymans, weeping.

R. 5–50 cm., *yellow ferruginous*, effuso-reflexed, more rarely arising from a stalk-like central tubercle; margin *white*, tomentose, swollen. Folds *concolorous*, porous, gyroso-toothed, large. Flesh *greyish white, spongy-fleshy, slightly moist*, exuding drops of water when growing, 2–12 mm. thick. Spores reddish rust colour in the mass, yellow under the microscope, elliptical, often subapiculate at the base, 8–10 × 5–6μ, 1-guttulate. Cystidia none. Hyphae either yellowish, thick walled, 5–6μ in diam., or hyaline, 3·5–4·5μ in diam., septate, with clamp connections. Smell often strong. Worked wood in buildings, logs in timber yards, rarely on stumps, and on the ground. Jan.—Dec. Common. (*v.v.*)

var. **minor** Falck. (= *Merulius lacrymans* (Wulf.) Quél.)
Minor, smaller.

Differs from the type in the *smaller, elliptical spores*, 5–6 × 4–4·5μ. Dead wood, and on the ground. July—Oct. Not uncommon. (*v.v.*)

var. **Guillemotii** Boud. Boud. Icon. t. 165. Jules Guillemot.

Differs from the type in the *well developed, dimidiate, imbricate pilei*. Worked wood. June—Oct. Not uncommon. (*v.v.*)

var. **pulverulentus** (Fr.) Quél. (= *Merulius pulverulentus* Fr.)
Pulverulentus, full of dust.

Differs from the type in being *membranaceous, becoming even, zoned, arid, gradually decaying from the centre to the margin, and in the folds being reticulated to the margin.* Rare.

2103. **M. aureus** Fr. Fl. Dan. t. 2027, fig. 2. *Aureus*, golden.

R. 2·5–5 cm., *golden yellow*, resupinate, effused, or effuso-reflexed, easily separable; margin *concolorous, thin, villose.* Folds *golden yellow, ochraceous orange to russet when dried,* plicato-porous, gyroso-crisped. Flesh membranaceous, soft, thin. Spores "yellowish in mass, cylindric, 3–4·5 × 1·5–2μ. Cystidia none. Hyphae loosely interwoven, nodose-septate, 2·5–4μ in diam." Burt. Pine wood, leaves, and cones. Oct.—Nov. Uncommon. (*v.v.*)

††Resupinate.

2104. **M. terrestris** (Peck) Burt. (= *Merulius lacrymans* var. *terrestris* Peck, non Ferry.) *Terrestris*, pertaining to the earth.

R. 3–10 cm., *bright ferruginous, drying amber brown,* resupinate, widely effused, membranaceous; margin *whitish.* Folds *concolorous,* gyrose, with intermediate, shallow, labyrinthiform depressions. Flesh *yellowish,* membranaceous, thin. Spores brownish in the mass, yellow under the microscope, broadly elliptical, with often a basal apiculus, 7–9 × 4·5–6μ. "Basal hyphae loosely interwoven, thick walled, rigid, 4·5–6μ in diam., nodose-septate, aniline-yellow under the microscope; subhymenial hyphae thin walled, often collapsed, 3μ in diam. Cystidia none" Burt. On bare soil. July—Oct. Rare. (*v.v.*)

2105. **M. papyraceus** Fr. *πάπυρος*, the paper reed.

R. 5–20 cm., *umber ferruginous,* resupinate, widely effused, *dry, glabrous*; margin *paler.* Folds *concolorous,* reticulato-porous; pores equal, dilated. Flesh *of the consistence of paper,* tough. Spores ferruginous, elliptical, 8–10 × 6–7μ, 1–2-guttulate. Old beams amongst grass. Sept.—Oct. Uncommon. (*v.v.*)

2106. **M. squalidus** Fr. *Squalidus*, dirty.

R. 7–30 cm., *hyaline flesh colour,* resupinate, effused, loosely adnate, adpressedly fibrillose and *becoming cinereous on the underside, smooth*; margin *white,* membranaceous. Folds *flesh colour, becoming subolivaceous,* sinuoso-porous. Flesh membranaceous, soft, loose, watery. Spores ferruginous. Hornbeam posts, and worked wood. Feb. Rare.

2107. **M. himantioides** Fr. Fr. Icon. t. 193, fig. 1. *ἱμάς*, leather thong; *εἶδος*, like.

R. 2–5 cm., *lilac, becoming raw umber when dried,* resupinate, effused, separable, *fibrillosely silky beneath*; margin *whitish,* byssoid. Folds *dingy yellow, then subolivaceous,* porous, then gyrose. Flesh very soft, silky, thin. Spores "honey yellow under the microscope, elliptical, 9–10 × 6μ. Basal hyphae narrow, few, honey yellow, up to 6–7μ in diam., not incrusted; subhymenial hyphae, loosely interwoven, hyaline, 4μ in diam." Burt. Dead conifer and cherry wood. Sept.—Dec. Rare.

2108. **M. fugax** Fr. (= *Merulius molluscus* Fr. sec. Burt; *Merulius laeticolor* Berk. sec. Romell.) Fr. Icon. t. 193, fig. 2, as *Merulius molluscus* Fr. *Fugax*, fleeting.

R. 3–10 cm., *white, or yellowish*, resupinate, effused; margin *whitish, byssoid*. Folds *flesh colour, drying cream colour, pinkish buff, with or without a tinge of orange, or dark brown*, gyroso-plicate. Flesh membranaceous, very soft, thin. Spores yellowish, or hyaline, broadly elliptical, 4–5 × 3–4μ. Cystidia none. "Hyphae loosely interwoven, long-celled, nodose-septate, 3–4μ in diam., sparingly and coarsely granule-incrusted towards the substratum" Burt. Coniferous wood, and branches. Sept.—Feb. Uncommon. (*v.v.*)

M. molluscus Fr. = **Merulius fugax** Fr.

2109. **M. pinastri** (Fr.) Burt. (= *Hydnum pinastri* Fr.; *Hydnum sordidum* Weinm. sec. Burt.) *Pinastri*, of pines.

R. 2–20 cm., *pinnard yellow, then olive ochre, becoming darker, and finally Dresden brown to raw-umber*, resupinate, effused, loosely attached to the substratum, *whitish* and tomentose beneath; mycelium often *reddish, or deep brownish vinaceous*; margin *whitish*, or *flesh pink*. Hymenium for a long time smooth, at length raised in shallow folds forming irregular, angular pores, or reticulations, ·5–1·5 mm. in diam., *or prolonged into subulate, or Irpex-like teeth*. Spores pale ochraceous in the mass, yellow brown, broadly ovoid to subglobose, 5–7 × 4–5μ. Cystidia none. Hyphae hyaline, 2–5μ in diam., loosely interwoven, nodose-septate. Pine wood and leaves, and cedar chips. Sept.—Nov. Rare.

Phlebia Fr.

(φλέψ, a vein.)

Receptacle waxy, or subgelatinous, becoming cartilaginous when dry, erect, or resupinate and effused. Hymenium from the first covering *radiating, obtuse wrinkles or veins*, continuous or broken up into tubercles, rarely smooth, fertile on the edge. Spores white, elliptical, reniform, oblong, or cylindrical, smooth. Cystidia none. Growing on wood, rarely on the ground.

2110. **P. merismoides** Fr. (= *Phlebia aurantiaca* (Sow.) Karst. sec. Pat.; *Phlebia radiata* Fr.; *Phlebia contorta* Fr.) Grev. Scot. Crypt. Fl. t. 280. *Merisma*, an old genus of *Thelephora*; εἶδος, like.

R. 2·5–9 cm., *flesh colour, then livid*, widely effused, smooth, or uneven, *villose and white beneath*; margin *orange, strigose*. Wrinkles *purplish flesh colour*, simple, straight, or tubercular, crowded. Flesh *concolorous, subgelatinous*, then membranaceous. Spores white, cylindrical, somewhat curved, 4–5 × 1·5–2μ. Stumps, branches, and logs. Sept.—Feb. Common. (*v.v.*)

2111. **P. radiata** Fr. (= *Phlebia aurantiaca* (Sow.) Karst. sec. Pat.) *Radiata*, rayed.

R. 2·5–20 cm., *red flesh colour, or almost orange*, somewhat round, effused, often confluent, equal, *smooth on both sides*; margin *radiately toothed*. Wrinkles *purplish flesh colour*, straight, *radiating in rows*. Flesh *paler*, membranaceous, tough, thin. Spores white, cylindrical, slightly curved, 4–6 × 1–2μ. Dead wood, branches, and logs, especially alder. July—April. Common. (*v.v.*)

2112. **P. erecta** Rea. *Erecta*, upright.

R. 1–3 cm., *bright flesh colour, becoming blackish, clavate, erect*, effused; clubs cylindrical, 2–3 mm. thick, apex obtuse, or acute, *white floccose at base, smooth*. Flesh *concolorous, or paler*, waxy. Spores white, reniform, 4–5 × 2–3μ, 2-guttulate. Basidia with 4-sterigmata. Burnt ground amongst mosses. Oct.—Nov. Uncommon. (*v.v.*)

2113. **P. contorta** Fr. (= *Phlebia aurantiaca* (Sow.) Karst. sec. Pat.) *Contorta*, twisted.

R. 2·5–10 cm., *rufous, then fuscous*, widely effused, *smooth on both sides*; margin indeterminate. Wrinkles *concolorous, here and there conglomerated, branched*, subflexuose, irregularly arranged. Flesh *membranaceous*, firm. Spores white, oblong, 5–6 × 3μ, bi-guttulate. Dead wood, and fallen branches. Sept.—Dec. Uncommon.

P. vaga Fr. = **Hypochnus fumosus** Fr.

2114. **P. albida** Fr. *Albida*, whitish

R. 2–4 cm., *white*, orbicular, effused, becoming confluent, adnate; margin determinate, becoming slightly free. Wrinkles *white, simple*, elevated, irregularly dispersed. Flesh *white, waxy coriaceous, then cartilaginous*. Spores white, elliptical, obtuse at both ends, 4–5 × 2·5–3μ, 1–2-guttulate. Fallen trunks. Oct. Rare. (*v.v.*)

2115. **P. lirellosa** (Pers.) B. & Br. Pers. Myc. Eur. II, t. 17, fig. 2, as *Daedalea lirellosa* Pers. *Lirellosa*, with little ridges.

R. 2·5–5 cm., *umber grey*, resupinate; margin free. Wrinkles *very small*, linear, thin, straight, branched, or anastomosing to form pores. Flesh thin. Wood, and branches, especially black poplar. Rare.

Plicatura Peck (= **Trogia** Fr. p.p.).

(*Plicatus*, folded; *οὐρά*, tail.)

Pileus *spongy coriaceous*, soft, flaccid, dimidiate, sessile, or substipitate. Hymenium covering *obtuse veins, gill-like in front*, crisped and branched behind, fertile on the edge. Spores white, oblong, or cylindrical, smooth. Cystidia none. Growing on wood.

2116. **P. crispa** (Pers.) Rea. (= *Plicatura faginea* (Schrad.) Karst.; *Trogia crispa* (Pers.) Fr.; *Merulius crispus* (Pers.) Quél.) Cke. Illus. no. 1099, t. 1114 A, as *Trogia crispa* Fr. *Crispa*, curled.

R. ·5–3 cm., *light yellow rufescent, whitish at the margin*, cup-shaped, sessile, or substipitate, then reflexed, dimidiate, often lobed, *villose, slightly zoned.* Gills *white, or bluish grey*, vein-like, *dichotomous*, narrow, *crisped*, very much swollen, edge obtuse. Flesh *white*, fibrillose, firm, tough. Spores white, oblong, depressed on one side, 5–6 × 3μ, 1-guttulate. Beech, and birch logs. Jan.—Dec. Uncommon. (*v.v.*)

Coniophora (DC.) Pers.

(κόνις, dust; φέρω, I bear.)

Receptacle fleshy, waxy, subcoriaceous, or membranaceous, resupinate, effused. Hymenium *smooth, subundulate tubercular, or granular.* Spores coloured, elliptical, navicular, or subfusiform, smooth. Cystidia none. Growing on wood, or on the ground.

2117. **C. puteana** (Schum.) Karst. (= *Corticium* (*Coniophora*) *puteanum* (Schum.) Fr.; *Coniophora cerebella* Pers.) *Puteana*, pertaining to a well.

R. 4–20 cm., *light yellowish pallid, at length fuscous olivaceous*, broadly effused, roundish, *separable*; margin *white, mucedinous.* Hymenium *fuscous olivaceous*, even, subundulate, or gyrose, often subtubercular, pulverulent. Flesh *whitish*, membranaceous, rather thick. Spores fuscous olivaceous, or ferruginous, broadly elliptical, obtuse, or pointed at one end, 11–13 × 7–8μ. Hyphae hyaline, 4–7μ in diam., densely interwoven. Stumps, felled trees, logs and worked wood. Jan.—Dec. Common. (*v.v.*)

2118. **C. arida** Fr. (= *Corticium* (*Coniophora*) *aridum* Fr.; *Coniophora Cookei* Massee.) Fr. Icon. t 199, fig. 1. *Arida*, dry.

R. 3–20 cm., *sulphur yellow, then umber inclining to ferruginous*, effused, *adnate*, continuous; margin *paler, or whitish*, byssoid. Hymenium *concolorous*, even. Flesh *whitish*, membranaceous, *thin.* Spores olivaceous, elliptical, obtuse, or pointed at one end, 9–12 × 6–7μ. Hyphae hyaline, 2–3μ in diam., loosely interwoven, thin walled. Stumps, felled trees, logs, and worked wood, especially of conifers. Jan.—Dec. Not uncommon. (*v.v.*)

2119. **C. laxa** (Fr.) Quél. (= *Corticium* (*Coniophora*) *laxum* Fr.) *Laxa*, loose.

R. 5–10 cm., *white, then ferruginous*, effused, adhering laxly, arachnoid beneath; margin *white*, byssoid. Hymenium *pallid, then ochra-*

ceous ferruginous, papillose, pulverulent. Flesh *whitish*, membranaceous, floccose, loose. Spores deep rusty purple, or ferruginous in the mass, elliptical, 10–11 × 6–7μ; basidia with 2–4-sterigmata. Bark, twigs, lichens and moss. Nov.—Jan. Uncommon. (*v.v.*)

2120. **C. Bourdotii** Bres.
L'abbé H. Bourdot, the eminent French mycologist.

R. 2–10 cm., *whitish, then umber and fuscous*, broadly effused, adnate; margin *white*, delicately fimbriate. Hymenium *umber*, even. Flesh *pale*, soft, loose. Spores brown, navicular, or subfusiform, 17–23 × 6–9μ. Bark, and fallen branches. Sept.—Oct. Rare. (*v.v.*)

2121. **C. pulverulenta** (Lév.) Massee. *Pulverulenta*, dusty.

R. 4–18 cm., *rusty brown*, broadly effused, dry; margin *whitish* byssoid, thin. Hymenium *concolorous, or dark brown*, pulverulent, even entire. Flesh thin. Spores "yellow brown, elliptical, 14–15 × 9–10μ" Massee. Wood. Rare.

2122. **C. membranacea** (DC.) Massee. Sow. Eng. Fung. t. 214, as *Auricularia pulverulenta* Sow. *Membranacea*, having a skin.

R. 5–30 cm., *pallid, then dirty pale ferruginous*, broadly effused, subrotund, fragile, separable; margin *yellowish*, minutely fibrillose. Hymenium *concolorous*, minutely pulverulent. Flesh thin. Spores "yellow brown, elliptical, 10–15 × 5–6μ" Massee. Wood, walls, paper, etc. Rare.

2123. **C. incrustans** Massee. *Incrustans*, covering with a coat.

R. 3–15 cm., *pale ochraceous, or dirty white*, effused, indeterminate, *inseparable*. Hymenium concolorous, compact, waxy, pulverulent. Flesh thin. Spores "very pale ochraceous, elliptical, 15–17 × 8–10μ" Massee. Running over leaves, twigs, etc. Rare.

2124. **C. stabularis** Fr. *Stabularis*, pertaining to a stall.

R. 3–10 cm., *white, then vinous fuscous*, effused, flaxy; margin *white*, byssoid. Hymenium *concolorous, white pruinose*, the tubercles collapsing. Flesh soft, floccose, thin, smell foetid. Fir wood. Rare.

2125. **C. subdealbata** (B. & Br.) Massee. (= *Corticium subdealbatum* B. & Br.) *Sub*, somewhat; *dealbata*, whitewashed.

R. 4–12 cm., *ochraceous olive*, effused, determinate. Hymenium *concolorous*, often with paler barren patches, pulverulent. Flesh thin. Spores "ochraceous, elliptical, with a minute basal apiculus, 11–12 × 7–8μ" Massee. Pine bark, and wood. Dec. Rare.

2126. **C. Berkeleyi** Massee.
Rev. Miles Joseph Berkeley, the father of British mycology.

R. 3–10 cm., *yellow brown, becoming purplish with age*, effused, determinate; margin sometimes minutely byssoid. Hymenium *concolorous*, becoming much cracked. Flesh thick, silky, compact. Spores "yellow brown, elliptical, with a minute basal apiculus, 11–12 × 6–7μ" Massee. Decorticated wood. Rare.

2127. **C. ochracea** Massee. Linn. Soc. Bot. Jour. xxv, t. 47, fig. 13.
ὠχρός, pale.

R. 4–12 cm., *whitish, then ochraceous*, very broadly effused, inseparable, usually indeterminate. Hymenium concolorous, pulverulent. Flesh *yellowish*, membranaceous, thin. Spores "yellowish, subglobose, 8 × 6–7μ; subhymenial hyphae pale yellow, very thick, up to 18μ in diam." Massee. Inside elm bark. July. Rare.

2128. **C. sulphurea** (Fr.) Massee. (= *Corticium sulphureum* Fr.)
Sulphurea, brimstone colour.

R. 3–13 cm., *bright sulphur yellow*, broadly effused, adnate; margin *bright sulphur yellow*, fibrillosely byssoid, and *running out in cord-like radiating strands*. Hymenium *concolorous, brownish yellow on the fertile portions*, often imperfect, waxy, cracking when dry. Flesh spongy, fibrillose, thick. Spores "brownish yellow, broadly elliptical, 11–12 × 8–10μ" Massee. Wood, bark, and leaves. Oct.—Jan. Rare.

var. **ochroidea** (Berk.) Massee. ὠχρός, pale; εἶδος, form.

Differs from the type in the *pale ochraceous hymenium, and the larger olive spores, elliptical, with a minute apiculus at the base*, 16–18 × 9–10μ. Wood, and bark. Rare.

C. sulfurea (Pers.) Quél. = **Hypochnus fumosus** Fr.

Coniophorella Karsten.

(*Coniophorella*, diminutive of *Coniophora*.)

Like *Coniophora*, but with long, cylindrical, cystidia.

2129. **C. umbrina** (A. & S.) Bres. (= *Corticium* (*Coniophora*) *umbrinum* (A. & S.) Fr.) *Umbrina*, umber.

R. 3–8 cm., *umber*, effused, not easily separable, *villose beneath*; margin *concolorous*, narrow, radiating. Hymenium *concolorous, or ferruginous*, even, sometimes granular, tomentose, setulose. Flesh *brownish*, soft, loose, fairly thick. Spores umber, elliptical, or pip-shaped, 9–13 × 5–8μ. Cystidia *concolorous*, cylindrical, 90–170 × 9–12μ, sometimes incrusted, obtuse, septate. Hyphae brownish, 3–6μ in diam., loosely interwoven. Wood, branches, and twigs. Sept.—May. Not uncommon. (*v.v.*)

2130. **C. olivacea** (Fr.) Karst. (= *Corticium* (*Hypochnus*) *olivaceum* Fr.) Trans. Brit. Myc. Soc. vi, figs. in text, p. 73.
Olivacea, olive coloured.

R. 4–30 cm., *distinctly olive when fresh, drying to a colour varying between Saccardo's olive, buffy citrine and Isabella colour*, thinly effused, following the inequalities of the matrix, adnate; margin *whitish*, very thin. Hymenium *concolorous*, tomentose, setulose. Flesh *brownish*, loose, thin. Spores olivaceous, or yellow brown, elliptical, flattened on one side, obtuse, or pointed at the one end, 9–13 × 4–8μ; basidia elongated, tapering gradually downwards, 40–80 × 7–8μ, with 4 curved sterigmata, 5–8μ long. Cystidia brownish, paler and blunt at the apex, 160–290 × 12–18μ, projecting about 75–130μ, many-septate, slightly constricted at the septa. Basal hyphae clear dark brown, 4–7μ in diam., branched, frequently septate with clamp connections. Nov.—Jan. Uncommon. (*v.v.*)

C. byssoidea (Pers.) Bres. = **Peniophora byssoidea** (Pers.) v. Hoehn. & Litsch.

4. Fistulinaceae.

Hymenium inferior, lining free and *separate* tubes.

Fistulina (Bull.) Fr.

(*Fistulina*, a little pipe.)

Pileus fleshy, subgelatinous in the upper layer, stipitate, or sessile. Stem lateral, or none. Tubes at first papillose, then cylindrical, distinct and free from each other. Spores coloured, elliptical, smooth. Conidia present in the tissues. Growing on wood.

2131. **F. hepatica** (Huds.) Fr. Berk. Outl. t. 17, fig. 1.

ἡπατικός, *belonging to the liver.*

P. 5–30 cm., *blood red, pale purplish red, liver colour*, or *chocolate, becoming blackish*, roundish, dimidiate, or subspathulate, sessile, or stipitate, rough, thick, fleshy, viscid. St. when present, 3–7 × 2–4 cm., *concolorous*, punctate. Tubes *pallid, becoming reddish, separate*; orifice of tubes *pale, round*. Flesh *reddish, marbled like beet root*, fibrous, distilling a red pellucid juice, 2–3 cm. thick. Spores pink, subglobose, 4·5–5 × 4μ, with a large central gutta. Taste somewhat acrid, especially when young. Edible. On trunks of trees, oak, ash, walnut, willow, beech, sweet chestnut, hornbeam, elm. Aug.—Nov. Common. (*v.v.*)

5. Hydnaceae.

Hymenium spread over the surface of spines, granules, warts, or other protuberances, or quite a smooth surface, with the intervening spaces fertile. Receptacle fleshy, coriaceous, waxy, crustaceous, or floccose, rarely none.

Mucronella Fr.

(*Mucronella*, a little sharp point.)

Receptacle *none*, consisting of a floccose, fugacious mycelium. Spines simple, cylindrical, subulate, acute, scattered, or fasciculate, and then more or less connate at the base. Spores white, oblong, or subglobose, smooth, or punctate; basidia with 1–4-sterigmata. Cystidia present. Growing on wood.

2132. **M. calva** (A. & S.) Fr. *Calva*, bald.

Spines *white, then pale*, 1–3 mm. long, scattered, rigid, thin. Spores white, "oblong, hardly depressed, 4–6 × 3μ. Hyphae thin walled, 3–6μ, emerging in a sterile bundle" Bourd. & Galz. Rotten pine stumps. Sept.—Oct. Rare.

2133. **M. aggregata** Fr. *Aggregata*, clustered.

R. ·5–2 cm., *white, then pale*, subiculum absent, or occasional. Teeth subulate, short, free, *arranged in groups*. Spores hyaline, elliptical, 4–6 × 2·5–4μ; basidia cylindrical or clavate, 10–20 × 3·5–5μ. Hyphae 2–4μ in diam., thin walled, clamp connections sparse. Very old rotten logs. Oct. Rare.

Hydnum (Linn.) Fr.

(*ὕδνον*, the old name for truffles.)

Receptacle fleshy, coriaceous, or corky, simple, or branched, pileate, or coralloid, stipitate, or sessile. Stem central, lateral, or none. Spines subulate, acute, distinct at the base. Flesh white, or coloured. Spores white, or coloured, elliptical, oval, globose, subglobose, or angularly globose, smooth, granular, verrucose, or echinulate; basidia with 2–5-sterigmata. Cystidia present, or absent. Micro- and macro-conidia present in some species. Growing on the ground, or on wood.

I. St. central.

A. P. fleshy.

*Spores white.

2134. **H. repandum** (Linn.) Fr. Grev. Scot. Crypt. Fl. t. 44. *Repandum*, bent backwards.

P. 5–15 cm., *pale buff flesh colour, or subrufescent*, convex, *somewhat repand*, often irregular and excentric, smooth, or minutely floccose and pruinose, firm; margin often lobed. St. 3–12 × 1·5–4 cm., *white, or pallid, ochraceous at the base*, irregularly shaped. Spines *white, then flesh colour*, 4–8 mm. long, decurrent, unequal, conical, entire, rarely bifid, or tubular, brittle. Flesh *white*, firm, fragile, thick. Spores white, subglobose, apiculate, 6–7 × 5–6μ. Smell pleasant. Taste bitter. Edible. Woods. Aug.—Nov. Common. (*v.v.*)

var. **album** Quél. *Album*, white.

Differs from the type in the *milk white pileus*. Woods. Sept.—Oct. Uncommon. (*v.v.*)

var. **rufescens** (Pers.) Fr. Bolt. Hist. Fung. t. 88, as *Hydnum repandum.* *Rufescens*, becoming reddish.

Differs from the type in being *rufescent, in the smaller size of all its parts, the non-decurrent spines* and *the slightly larger spores*, 8–10μ. Woods. Aug.—Nov. Common. (*v.v.*)

2135. **H. fuligineo-album** Schmidt. Boud. Icon. t. 168. *Fuligineus*, sooty; *album*, white.

P. 5–15 cm., *whitish, tinged rosy, or somewhat fuscous towards the margin*, convex, then expanded and depressed, often finally infundibuliform, repand, wavy, rather silky, smooth, *disc often with darker scales*, or spots; margin at first involute. St. 4–5 × 1–1·5 cm., *whitish rosy*, or *concolorous and slightly fuscous at the base*, central, or excentric, subequal, or attenuated at the base, *subsquamulose*, or glabrous. Spines *white, becoming rosy reddish, slightly fuscous at the base*, 6–8 mm. long, decurrent, subulate, or compressed, crowded. Flesh *white, rosy when broken*, thick, firm. Spores "white, verrucose, subreticulate, globose, 3–5μ" Boud. Smell strong, unpleasant, "of liquorice" Quél. Taste somewhat pleasant. Coniferous woods, and adjoining pastures. Sept.—Oct. Rare.

2136. **H. fragile** Fr. *Fragile*, brittle.

P. 4–30 cm., *pallid, soon cinereous, or brick rufescent*, convex, then plane and depressed, *pubescent at first, becoming smooth, often zoned* towards the margin, and minutely squamulose, or *wrinkled*; margin undulate, lobed. St. 4–8 × 1·5–10 cm., *pallid, becoming rufescent or cinereous*, often incrassated at the base, unequal, smooth. Spines *whitish, then grey*, 4–8 mm. long, scarcely decurrent, subulate, slender, fragile. Flesh *grey, or reddish*, soft, thick, firm, sometimes zoned. Spores white, subangularly globose, 3–4μ, 1-guttulate. Pine woods, and moors. Sept.—Nov. Rare, but occasionally abundant. (*v.v.*)

2137. **H. molle** Fr. Fr. Icon. t. 2, upper figs. *Molle*, soft.

P. 6–11 cm., *white, becoming greyish or tinged with chocolate*, convex, then umbilicate, or irregularly depressed, often wavy, *covered with a dense tomentose coat*. St. 4–5 × 2 cm., *white, then grey*, equal, conical, glabrous. Spines *white, then grey*, 6–8 mm. long, decurrent, acuminate, thin, crowded. Flesh *white, becoming yellowish when broken*, soft, thick. Spores *white*, "globose, with a basal apiculus, 7μ" Massee. Taste pleasant. Edible. Coniferous woods. Sept.—Oct. Uncommon.

**Spores coloured.

2138. **H. imbricatum** (Linn.) Fr. Grev. Scot. Crypt. Fl. t. 71. *Imbricatum*, covered with tiles.

P. 7–50 cm., *umber*, convex, then plane, often subumbilicate, and finally infundibuliform, *floccose, tessulato-scaly*. St. 2·5–7·5 × 2·5–

5 cm., *whitish, or concolorous*, firm, smooth. Spines *cinereous white*, 10–12 mm. long, decurrent, subulate, thin. Flesh *pale, then buffish or reddish*, thick, firm, sometimes zoned. Spores reddish brown, verrucose, oval or globose, 5–6 × 5μ, 1-guttulate. Taste bitter. Edible. Coniferous woods. Sept.—Nov. Not uncommon. (*v.v.*)

2139. **H. squamosum** (Schaeff.) Fr. Schaeff. Icon. t. 273.
Squamosum, scaly.

P. 3–8 cm., *rufous fuscous*, convex, gibbous, then irregular and depressed, *smooth, breaking up into irregular, fibrillose, chestnut coloured scales*. St. 3–4 × 1–1·5μ, *white*, attenuated downwards, smooth. Spines *greyish fuscous, apex whitish*, thin. Flesh *whitish*, thick, firm. Spores "yellowish tawny, subglobose, 7μ, granular" Quél. Taste pleasant. Edible. Coniferous woods. Sept.—Oct. Rare.

2140. **H. scabrosum** Fr. *Scabrosum*, rough.

P. 3–4 cm., *umber ferruginous, turbinate*, then plane, very convex beneath, tomentose, then rough with fasciculate flocci, which form minute crowded scales. St. 2·5 × 2·5 cm., *cinereous, blackish at the base*, attenuated downwards, round, or compressed, dotted with the rudimentary decurrent spines. Spines *fuscous ferruginous, apex whitish*, 8 mm. long, decurrent, subulate, equal, crowded. Flesh *white*, becoming blackish at the base of the stem, very thick, firm. Spores reddish brown, verrucose, globose, 4–5μ. Pine woods. Sept. Rare. (*v.v.*)

2141. **H. laevigatum** (Swartz) Fr. Bres. Fung. Trid. t. 138.
Laevigatum, made smooth.

P. 5–13 cm., *greyish umber, at first often tinged with fuscous purple*, somewhat irregular, convex, then plane, or depressed, smooth, then breaking up into minute squamules; margin incurved, pubescent. St. 4–6 × 2–3 cm., *greyish, or lilac colour*, often excentric, subequal, somewhat glabrous. Spines *umber fuscous, apex whitish*, 1–2·5 cm. long, decurrent, large. Flesh *whitish grey, pale lilac purple when young and broken*, thick, compact. Spores "*somewhat fuscous*, angularly globose, tuberculose, 6–7 × 4–5μ" Bres. Taste somewhat bitter. Smell strong, unpleasant, "d'immortelle sauvage when dried" Quél. Edible. Pine woods. Aug.—Nov. Rare.

2142. **H. acre** Quél. Quél. Soc. bot. (1877), no. 36, t. 6, fig. 1.
Acre, sharp.

P. 10–12 cm., *light yellow, then olivaceous, or bistre*, plane, *shaggy, velvety*. St. 3 × 2 cm., *cream olivaceous*, oval, often branched, villose. Spines *white, then brown, apex light yellow*, decurrent, thin. Flesh *light yellow*, watery. Spores light yellow, spinulose, 6μ. Taste very pungent, bitter and peppery. Poisonous. Pine, and chestnut woods, and sandy places. Rare.

2143. **H. infundibulum** (Swartz) Fr. (= *Hydnum fusipes* Pers. sec. Quél.) Pers. Myc. Eur. II, t. 20, figs. 4–6, as *Hydnum fusipes* Pers. *Infundibulum*, a funnel.

P. 5–20 cm., *brown*, infundibuliform, lobed, unequal, *smooth*. St. 5–7·5 × 2·5 cm., *white, then reddish or brownish*, constricted at the base, unequal, smooth. Spines *white, then bay or brown*, decurrent. Flesh *white*, fleshy, fibrous, firm. Pine woods. Sept. Rare.

H. fusipes Pers. = **Hydnum infundibulum** (Swartz) Fr.

B. P. corky, or coriaceous, tough.

*Spores white.

2144. **H. cinereum** (Bull.) Fr. *Cinereum*, ash colour.

P. 5–9 cm., *white, becoming greyish and tinged with lilac or chocolate*, convex, umbilicate, repand, at length often infundibuliform, minutely tomentose, or pubescent; margin thin. St. 2–3 × ·5 cm., *white, then grey*, attenuated at the base, often branched, firm, glabrous. Spines *white, then grey*, 1–2 mm. long, decurrent, thin. Flesh *whitish, becoming ferruginous*, corky, thin. Spores white, "6μ, granular" Quél. Coniferous woods. Sept.—Oct. Rare. (*v.v.*)

2145. **H. nigrum** Fr. Fr. Icon. t. 5, lower figs. *Nigrum*, black.

P. 2–10 cm., *whitish, soon becoming azure-blue-black with the margin whitish*, club-shaped, then turbinate, at length flattened, plano-depressed, tubercular, tomentose, sometimes zoned. St. 2–3 × ·5–1·5 cm., *black*, equal, often rooting at the base, unequal, thickened at the tomentose base. Spines *white, then grey*, 2 mm. long, subulate, thin. Flesh *black, corky rigid*. Spores white, minutely verrucose, subglobose, 4μ. Often connate. Coniferous woods. Sept.—Nov. Not uncommon. (*v.v.*)

2146. **H. graveolens** (Delast.) Fr. Fr. Icon. t. 6, upper figs. *Graveolens*, strong smelling.

P. 2–5 cm., *bistre becoming black, then cinereous with the margin whitish*, hemispherical, sinuate, often depressed at the centre, thin, *soft*, rugose, silky. St. 1·5–3 cm. × 1–3 mm., *fuscous black*, thickened at the apex, tough, equal, smooth. Spines *white, then grey*, ·5–1 mm. long, decurrent, thin. Flesh *concolorous*, or *paler*, coriaceous, rigid, thin. Spores white, echinulate, globose, 3–4μ, 1-guttulate. Smell strong of Fenugreek, or tincture of Belladonna. Often connate. Coniferous woods. Sept.—Nov. Uncommon. (*v.v.*)

2147. **H. melaleucum** Fr. Schaeff. Icon. t. 272, as *Hydnum pullum* Schaeff. μέλας, black; λευκός, white.

P. 1·5–3 cm., *greyish violet, or bistre, then black with the margin white,* plane, irregular, *striate*, rigid, *with little elevations at the disc*, silky. St. 1·5–3 cm. × 2–3 mm., *black*, enlarged at the apex, *smooth*. Spines *white, then flesh colour*, ·5–1 mm. long, slender, thin. Flesh *violaceous, or black*, coriaceous, thin, firm, sometimes zoned. Spores *white*, globose, 2·5–3μ. Caespitose. Pine woods. Sept.—Oct. Rare. (*v.v.*)

2148. **H. cyathiforme** (Schaeff.) Fr. Schaeff. Icon. t. 139.
κύαθος, a cup; *forma*, shape.

P. 2·5–6 cm., *pale cinereous, or lilac, with the margin white*, plane, then cup-shaped, or infundibuliform, sometimes zoned, silky, disc subtomentose. St. 2–3 cm. × 5–8 mm., *pale cinereous, often violet*, equal, smooth. Spines white, ·5–1 mm. long, very slender. Flesh *white, becoming greyish or ferruginous*, coriaceous, *thin*. Spores white, echinulate, globose, 3μ, 1-guttulate. Often connate. Woods. Aug.—Nov. Uncommon. (*v.v.*)

**Spores coloured.

2149. **H. compactum** (Pers.) Fr. *Compactum*, strongly built.

P. 2·5–15 cm., *white, soon becoming olivaceous cinereous or fuscous*, plane, flat, irregularly shaped, sometimes almost sessile, undulated, tuberculose, *densely covered with whitish down when young*; margin *becoming bluish*, lobed, waved, thick. St. 2–3 × 2–3·5 cm., *tawny inclined to fuscous*, irregularly shaped. Spines *fuscous*, apex pallid, 2–7 mm. long, decurrent, subulate. Flesh *zoned with azure blue, intervals between the zones whitish, somewhat blood red in the stem*, corky, compact, firm, very thick. Spores light reddish brown, minutely verrucose, angularly globose, 4–5μ, 1-guttulate. Often connate. Coniferous woods, and moors. Aug.—Nov. Uncommon. (*v.v.*)

2150. **H. aurantiacum** (A. & S.) Fr. Bres. Fung. Trid. t. 142.
Aurantiacum, golden.

P. 2·5–15 cm., *whitish, soon orange yellow with a white margin*, turbinato-dilated, *with small elevations, at first covered with white down*. St. 2–5 × 1–3 cm., *orange*, obconic, or equal. Spines *whitish, becoming orange, and at length fuscous*, 2–4 mm. long, decurrent, subulate. Flesh *tawny orange, zoned*, corky, compact, thick. Spores yellowish, verrucose, angularly globose, 4–6 × 4–5μ, multi-guttulate. Taste pleasant. Woods. Sept.—Nov. Uncommon. (*v.v.*)

2151. **H. ferrugineum** Fr. (= *Hydnum floriforme* (Schaeff.) Quél.) Bres. Fung. Trid. t. 143.
Ferrugineum, of the colour of iron rust.

P. 2·5–10 cm., *white, and exuding blood red drops, then variegated with blood red zones, and finally entirely reddish brown or ferruginous, with the margin white*, obconic, hemispherical, then expanded and depressed or cyathiform, *at first covered with white down*, rugose. St.

4–8 × 1·5–2·5 cm., *fuscous ferruginous*, often *becoming blackish*, unequal, compressed, or sulcate, attenuated, or incrassated at the base, slightly tomentose, becoming smooth. Spines *white, soon fuscous ferruginous*, 3–5 mm. long, decurrent, subulate, equal. Flesh *ferruginous, becoming blackish with age especially in the stem*, zoned, spongy corky, thick. Spores fuscous, minutely verrucose, angularly globose, 3–4μ. Smell slightly pleasant. Coniferous woods, and moors. Aug.—Oct. Uncommon. (*v.v.*)

2152. **H. scrobiculatum** Fr. Fr. Icon. t. 5, upper figs.
Scrobiculatum, pitted.

P. 2·5–10 cm., *ferruginous, becoming paler*, plane, depressed and cyathiform, or rarely infundibuliform, *pubescent, disc slightly pitted and scaly*. St. 1–4 × 1–2·5 cm., *concolorous, or darker*, equal, often rooting, smooth. Spines *concolorous, apex flesh colour, becoming fuscous*, 2–5 mm. long, subdecurrent, thin, fragile. Flesh *ferruginous, becoming paler, zoned*, corky coriaceous, thick. Spores pale reddish brown, angularly globose, 4μ. Often confluent. Coniferous woods, and moors. Aug.—Nov. Uncommon. (*v.v.*)

2153. **H. zonatum** (Batsch) Fr. Rolland, Champ. t. 99, no. 218.
Zonatum, zoned.

P. 2·5–5 cm., *ferruginous, becoming paler when dry*, plane, then depressed and cyathiform, *zoned, radiato-rugose*, silky, *becoming smooth*; margin thin. St. 2–5 cm. × 4–6 mm., *concolorous, or paler*, thickened at the base, equal, floccose. Spines *pallid, apex grey and glistening, then ferruginous*, 1–3 mm. long, thin. Flesh *concolorous, coriaceous*, fibrous, *thin*. Spores pale reddish brown, echinulate, globose, 3–4μ, 1-guttulate. Coniferous, and frondose woods. Aug.—Nov. Not uncommon. (*v.v.*)

2154. **H. Queletii** Fr. Quél. Jur. et Vosg. I, t. 20, fig. 2.
Lucien Quélet, the eminent mycologist.

P. 2–3 cm., *bright chestnut, becoming dark fuscous*, plane, then umbilicate, silky, *disc with thin, crowded, radiating tufts and wrinkles*; margin *white*. St. 1–2 cm. × 2–3 mm., *concolorous*, equal, silky, base floccose. Spines *grey, then bay brown*, 1–2 mm. long, decurrent, thin. Flesh *concolorous, or reddish, paler at the apex of the p.*, corky membranaceous, thin. Spores yellowish, minutely verrucose, globose, 3–4μ. Often confluent. Frondose woods. Sept.—Oct. Uncommon. (*v.v.*)

II. St. lateral.

2155. **H. auriscalpium** (Linn.) Fr. Grev. Scot. Crypt. Fl. t. 196.
Auriscalpium, ear-pick.

P. 1–2 cm., *pallid, or flesh colour, then date brown and blackish, dimidiate, reniform*, rarely hemispherical, *hairy*; margin sometimes lobed. St. 3–8 cm. × 1–4 mm., *ochraceous, then concolorous*, vertical, lateral, rarely central, *hairy*, rooted. Spines *yellowish, or flesh colour, then cinereous and brown*, 1–3 mm. long, coriaceous, thin, crowded. Spores white, minutely echinulate, globose, 4–5 μ. Cones of conifers. May—Feb. Common. (*v.v.*)

III. Very much branched, or tuberculiform, and immarginate.

2156. **H. coralloides** (Scop.) Fr. Rolland, Champ. t. 100, no. 221.
κοράλλιον, coral; *εἶδος*, like.

P. 10–40 cm., *shining white, at length yellowish*, very much branched, *entirely broken up into attenuated, intricate branches*, arising from a thick trunk; primary branches, 12–20 mm. thick, ultimate ones, 1–2 mm. thick, pruinose. Spines *white, becoming yellowish*, 6–10 mm. long, *fasciculate*, subulate, entire, *unilateral*. Flesh *white*, fleshy fibrous. Spores white, globose, 4 μ, with a large central gutta. Edible. Decayed fir, beech, ash, birch, and oak trunks. Oct.—Nov. Rare. (*v.v.*)

2157. **H. erinaceus** (Bull.) Fr. Rolland, Champ. t. 100, no. 220.
Erinaceus, a hedgehog.

P. 5–30 cm., *white, then yellowish*, spathulate, or epaulet-shape, pendulous, tubercular, immarginate, *torn into fibrils above*. St. sometimes rudimentary. Spines *white, 3–6 cm. long, pendulous*, straight, equal, simple, crowded, pruinose. Flesh *white*, unchangeable, thick, lacunose, tough, very soft, elastic. Spores white, subglobose, 6–7 μ, with a large central gutta. Taste acid, then sweet. Edible. Beech, oak, hornbeam, and alder trunks. Sept.—Dec. Rare. (*v.v.*)

2158. **H. caput-Medusae** (Bull.) Fr. Bull. Hist. Champ. Fr. t. 412.
Caput, head; *Medusae*, of a Medusa.

P. 7–10 cm., *white, then fuliginous cinereous*, globose, tuberculiform, substipitate, *covered all over with spines*. Spines *on upper surface distorted*, the lower ones, 10–20 mm. long, straight. Flesh *white*, fibrillose, soft. Trunks. Rare.

2159. **H. setosum** (Pers.) Bres. (= *Hydnum Schiedermayeri* Heufl.; *Dryodon luteocarneum* (Secr.) Quél.) Kalchbr. Icon. t. 38, fig. 4, as *Hydnum Schiedermayeri* Heufl. *Setosum*, bristly.

P. 15–30 cm., *sulphur, then flesh colour, becoming rufescent on exposure to the sun*, broadly effused, immarginate, *tubercular, stalactite-like*, pruinose. Spines *sulphur flesh colour, apex white fimbriate, incised*, 3–5 mm. long, subulate, often fasciculate, compressed, channelled, intermixed with shorter conical spines, crowded. Flesh *white, sulphur near the exterior*, cheesy, juicy, firm, *lacunose*, sometimes bearing spines in the inside. Spores white, pale ochraceous, subglobose, 4–5 × 3·5–4μ, 1-guttulate. Mycelium *citron yellow*. Taste unpleasant, acid. Old apple trunks. Aug.—Oct. Rare. (*v.v.*)

2160. **H. squalinum** Fr. Ray, Syn. t. 1, fig. 5.
Squalinum, pertaining to a shark.

P. 2–8 cm., *pale wood colour*, suborbicular, then confluent, adnate, waxy, pruinose, villose, becoming smooth; margin *white*, villose, thin. Spines *yellowish amber*, *becoming fuscous*, base *brownish*, 2–3 mm. long, acute, *subdivided*, or entire, stout, compressed, connate, translucid. Flesh *white*, *coriaceous*, *firm*, thick. Spores "*yellowish*, echinulate, oval, 4μ" Quél. Dead beech trunks. Sept.—Oct. Rare.

IV. P. sessile, dimidiate, marginate, often effuso-reflexed.

2161. **H. cirrhatum** (Pers.) Fr. Fr. Sverig. ätl. Svamp. t. 71, fig. 1.
Cirrhatum, curled.

P. 5–10 cm., *pallid*, *varying white*, *light yellowish*, *or rufescent*, effused, then reflexed, dimidiate, shell-shaped, often imbricate, *bristling above with sterile spines or scattered*, *flexuose fibres*; margin *pink*, incurved, *fimbriate*. Spines *cream colour*, 10–15 mm. long, subulate, tough, elastic, equal, thin. Flesh *white*, *then pinkish cream colour*, *corky soft*, thick. Spores white, "subelliptical, 3·5–4 × 2·75–3μ, often 1-guttulate, becoming blue with iodine" Bourd. & Galz. Taste and smell pleasant. Edible. Oak, beech, birch, and fir trunks. Aug.—Sept. Uncommon.

2162. **H. diversidens** Fr. Fr. Sverig. ätl. Svamp. t. 71, fig. 2.
Diversus, different; *dens*, tooth.

P. 5–13 cm., *white*, *then yellowish*, *or flesh colour*, dimidiate, shell-shaped, often very irregularly shaped, here and there lobed, sessile, or substipitate, often imbricate, *densely beset above with erect*, *variously shaped*, *incised teeth*; margin *membranaceous*, lobed, *clothed with club-shaped spines*. Spines white, 6–12 mm. long, subulate, entire, regular, pubescent. Flesh *whitish*, soft, moist. Spores white, "oval globose, 3–4μ, with a large central gutta" Quél. Taste pleasant. Edible. Beech, birch, hornbeam, and oak stumps. Oct. Rare.

2163. **H. pulcherrimum** Berk. & Curt. *Pulcherrimum*, very beautiful.

P. 3 cm., *white*, *shaded pale tawny*, pulvinate, dimidiate, expanded, subimbricate, *stiffly downy*; margin lobed, thin. Spines *tawny*, 3–5 mm. long, variable in size. Flesh fibrous. Oct. Rare.

2164. **H. multiplex** Fr. Fr. Icon. t. 6, lower figs.
Multiplex, with many folds.

P. 8–10 cm., *date brown*, *becoming fuscous*, reniform, spathulate, wedge-shaped, densely imbricate, connate in very numerous flabelliform, connate pilei, *radiately striate*, *velvety*, produced behind into a common, fusiform stem, 12 mm. and more long; margin *at first becoming white, then concolorous*, acute. Spines *whitish cinereous, or lead colour, then date brown*, short, slender, very crowded. Flesh *becoming fuscous*, coriaceous, thin, pliant. Rare.

H. fusco-atrum Fr. = **Acia fusco-atra** (Fr.) Pat.

H. membranaceum Fr. = **Acia membranacea** (Fr. non Bull.) Bourd. & Galz.

H. membranaceum Bull. = **Radulum molare** Fr.

H. Weinmannii Fr. = **Acia fusco-atra** (Fr.) Pat. sec. Bres.

H. crinale Fr. = **Caldesiella crinalis** (Fr.) Bourd. & Galz.

H. ferruginosum Fr. = **Caldesiella crinalis** (Fr.) Bourd. & Galz.

H. variecolor Fr. = var. of **Sistotrema sulphureum** Quél. sec. Bourd. & Galz.

H. aureum Fr. = **Odontia aurea** (Fr.) Quél.

H. denticulatum (Pers.) Fr. = **Acia denticulata** (Pers.) Bourd. & Galz.

H. alutaceum Fr. = **Odontia arguta** (Fr.) Quél. var. **alutacea** (Fr.) Bourd. & Galz.

H. sulphureum Schwein. = **Odontia sulphurea** (Schwein.) Rea.

H. sordidum Weinm. = **Merulius pinastri** (Fr.) Burt.

H. limonicolor B. & Br. = **Odontia limonicolor** (B. & Br.) Quél.

H. pinastri Fr. = **Merulius pinastri** (Fr.) Burt.

H. spathulatum (Schwein.) Fr. = **Odontia spathulata** (Schwein.) Rea.

H. multiforme B. & Br. = **Odontia multiformis** (B. & Br.) Rea.

H. anomalum B. & Br. = **Odontia anomala** (B. & Br.) Rea.

H. melleum B. & Br. = **Odontia mellea** (B. & Br.) Rea.

H. viride (A. & S.) Fr. = **Caldesiella viridis** (A. & S.) Pat.

H. udum Fr. = **Acia uda** (Fr.) Bourd. & Galz.

H. Hollii (Schmidt) Fr. = **Odontia Hollii** (Schmidt) Rea.

H. bicolor (A. & S.) Fr. = **Odontia bicolor** (A. & S.) Bres.

H. nodulosum Fr. = **Acia stenodon** (Pers.) Bourd. & Galz. var. **nodulosa** (Fr.) Bourd. & Galz.

H. niveum (Pers.) Fr. = **Grandinia farinacea** (Pers.) Bourd. & Galz.

H. farinaceum (Pers.) Fr. = **Grandinia farinacea** (Pers.) Bourd. & Galz.

H. argutum Fr. = **Odontia arguta** (Fr.) Quél.

H. stipatum Fr. = **Odontia stipata** (Fr.) Quél.

H. subtile Fr. = **Odontia bicolor** (A. & S.) Bres.

H. Stevensonii B. & Br. = **Odontia Stevensonii** (B. & Br.) Rea.

H. plumosum Duby = **Odontia plumosa** (Duby) Rea.

Mycoleptodon Pat. (= **Hydnum** (Linn.) Fr. p.p.).

(*μύκης*, fungus; *λεπτός*, thin; *ὀδών*, tooth.)

Receptacle *membranaceous-coriaceous*, thin, firm, resupinate, or reflexed. Spines simple, firm, cylindrical, pointed, hispid at the apex; none, or reduced in size at the margin. Spores white, ovoid, oboval, subelliptical, or oblong, smooth; basidia with 2–4-sterigmata. Cystidia present, abundant at the apex of the spines. Growing on wood, more rarely on humus.

2165. **M. ochraceum** (Pers.) Pat. (= *Hydnum ochraceum* (Pers.) Fr.; *Hydnum pudorinum* Fr. sec. Bourd. & Galz.) Pers. Syn. t. v, fig. 5, as *Hydnum ochraceum*. *Ochraceum*, ochre-yellow.

R. 2·5–7·5 cm., *white, or pale ochraceous*, rounded, then confluent, effused, or effuso-reflexed and dimidiate, tomentose, sometimes narrowly grooved, *zoned*; margin *white*, membranaceous, subfimbriate, pubescent. Spines *ochraceous flesh colour*, subulate, *very small*, hispid at the apex, shorter at the margin. Flesh *whitish*, thin, *coriaceous*. Spores white, "oboval oblong, 3–4 × 2–2·5μ, often 1-guttulate. Cystidia claviform, or fusiform, 24–100 × 5–10μ, thick walled, or incrusted" Bourd. & Galz. Dead branches. Jan.—Dec. Common. (*v.v.*)

2166. **M. fimbriatum** (Pers.) Bourd. & Galz. (= *Odontia fimbriata* (Pers.) Fr.) Fr. Icon. t. 196, fig. 1, as *Odontia fimbriata* (Pers.) *Fimbriatum*, fringed.

R. 2–20 cm., *fawn colour, cinnamon, or pale buff, often tinged with lilac*, effused, separable, *veined, or traversed by root-like ribs*; margin fibrilloso-fringed. Spines rufescent, minute, blunt, *in the form of granules, crowned with hyaline hairs*. Flesh *membranaceous-coriaceous*, thin. Spores white, "ovoid, subelliptical, sometimes slightly depressed, 3·5–4·5 × 1·75–3μ. Cystidia claviform, or fusiform, 7–9μ in diam., thick walled, rugose, or incrusted, often obtuse and slightly bent" Bourd. & Galz. Dead wood, and humus. Jan.—Dec. Not uncommon. (*v.v.*)

Radulum Fr.

(*Radula*, a rasp.)

Receptacle resupinate, effused, *waxy*, or *membranaceous waxy*. Tubercles or spines, thick, deformed, obtuse, simple, or branched, irregularly scattered, or confluent and tooth-like. Spores white, or coloured, elliptical, subglobose, or cylindric oblong, smooth. Cystidia none, cystidioles (sterile basidia) sometimes present. Growing on wood.

R. pendulum Fr. = **Corticium subcostatum** Karst. sec. Bourd. & Galz.

2167. **R. orbiculare** Fr. Grev. Scot. Crypt. Fl. t. 278. *Orbiculare*, round.

R. 2·5–15 cm., *white, then yellowish,* orbicular, confluent; margin *white, byssoid, membranaceous.* Tubercles *concolorous, or dingy flesh colour,* 2–6 mm. long, cylindrical, scattered, or fasciculate. Flesh *whitish, or yellowish, waxy fleshy,* thin, 2–4 mm. thick. Spores white, cylindric oblong, slightly curved, 8–12 × 3·5μ. Dead bark of birch, cherry, willow, aspen, hornbeam, pine, and fir. Jan.—Dec. Common. (*v.v.*)

var. **junquillinum** Quél. *Junquillinum*, bright yellow.

Differs from the type in its *bright yellow colour.* Pine. March. Uncommon. (*v.v.*)

2168. **R. quercinum** Fr. (= *Radulum fagineum* (Pers.) Fr. sec. Bourd. & Galz.) *Quercinum*, pertaining to oak.

R. 5–30 cm., *white, then pallid or tan colour,* somewhat round, then broadly confluent, adnate, often throwing back the bark; margin *white,* villose, *floccose.* Tubercles *concolorous,* 4–6 mm. long, cylindrical, obtuse, pointed, or toothed, scattered, or fasciculate, *often villose at the apex.* Flesh *whitish, or yellowish, crustaceous waxy,* thin, 2–4 mm. thick. Spores white, oblong subcylindric, very slightly depressed on one side, 5–7–8·5 × 2·5–4μ, guttulate" Bourd. & Galz. Fallen branches, especially oak, also worked wood. Jan.—Dec. Common. (*v.v.*)

2169. **R. molare** Fr. (= *Radulum membranaceum* (Bull.) Bres.; ? *Corticium confluens* Fr. a form sec. Bourd. & Galz.) Pers. Myc. Eur. II, t. 22, fig. 1, as *Sistotrema molariforme* Pers. *Molare*, a molar tooth.

R. 5–10 cm., *pale, yellowish, or tan colour,* orbicular, confluent, widely effused, adnate, firm, cracked when dry; margin byssoid, or radiately fibrillose. Tubercles *concolorous,* 2–3 mm. long, *deformed,* cylindrical or conical, scattered, or confluent and connate, smooth, or fimbriate. Flesh *whitish,* waxy, thin. Spores white, "elliptical, subglobose, 7·5–9–13 × 5–7–8μ" Bourd. & Galz. Fallen oak, and birch branches. Jan.—Dec. Not uncommon. (*v.v.*)

2170. **R. mucidum** (Pers.) Bourd. & Galz. nec *Hydnum mucidum* Fr. *Mucidum*, mucid.

R. 5–10 cm., *yellow,* effused, separable, more or less nodular, glabrous, or pubescent; margin fibrillose. Tubercles *concolorous,* short, scattered, subulate, elongate when growing on an upright surface. Flesh *yellowish,* soft, thin. Spores ferruginous in the mass, very pale yellow under the microscope, elliptical to subglobose, with a lateral apiculus, 4–5 × 3·5–4μ, 1-guttulate; basidia clavate, 7μ in diam. with 4-sterigmata. Hyphae thin walled, frequently septate, with clamp connections, 4–7μ in diam. Inside a hollow stump, and growing over living stems of ivy. Nov. Rare. (*v.v.*)

2171. **R. tomentosum** Fr. (? = var. of *Odontia arguta* (Fr.) Quél. sec. Bourd. & Galz.) *Tomentosum*, downy.

R. 2·5–13 cm., *white, then yellowish, pallid wood colour when dry*, effused, irregular, innate; margin *whitish, sometimes becoming ferruginous, swollen, erect*, tomentose. Tubercles *white*, short, *angular*, obtuse, crowded, confluent, smooth. Flesh floccose, crustaceous. Spores white, "cylindric-oblong, slightly thinner, curved and apiculate at the base, 8 × 4μ" Massee. *Pyrus aucuparia*, willow, and pine sawdust. Nov.—Dec. Rare. (*v.v.*)

R. deglubens B. & Br. = **Eichleriella spinulosa** (Berk. & Curt.) Burt sec. Wakef.

2172. **R. corallinum** B. & Br. κοράλλιον, coral; εἶδος, like.

R. 5–15 cm., *white*, effused, shining. Tubercles *white*, 4–6 mm. long, *fasciculate* (fascicles 6 mm. across), very irregular, coralloid, divided downwards. Flesh very thin, pelliculose. Spores white, "subglobose, apiculate, 5μ" Massee. Lichen covered oak branches. Sept. Rare.

2173. **R. epileucum** B. & Br. ἐπίλευκος, whitish.

R. 5–20 cm., *ochrey white*, widely effused, adnate. Tubercles *pale ochraceous*, 2–4 mm. long, *scattered*, cylindrical, *fimbriate at the apex*, deciduous, brittle. Flesh *snow white*, very thin, upper portion waxy. Spores white, "cylindrical, slightly curved, 6–7 × 3–3·5μ" Massee. Decorticated wood. Aug. Rare.

R. fagineum (Pers.) Fr. = **Radulum quercinum** Fr. sec. Bourd. & Galz.

R. laetum Fr. = **Peniophora incarnata** (Pers.) Cke. var. **hydnoidea** (Pers.) Bourd. & Galz.

R. botrytes Fr. = **Corticium comedens** (Nees) Fr. sec. Quél.

R. aterrimum Fr. = **Corticium nigrescens** (Schrad.) Fr. sec. Quél.; *Eutypa hydnoidea* (Fr.) von Hoehn.

Acia Karst. (= **Hydnum** (Linn.) Fr. p.p.).

(ἀκή, a point.)

Receptacle resupinate, effused, *waxy, inseparable*. Spines subulate, generally entire, distinct, or connate at the base. Flesh dense. Spores white, elliptical, oblong elliptical, or oblong subelliptical, smooth; basidia with 2–4-sterigmata, with or without sterile basidia (cystidioles). Cystidia *none*. Growing on wood.

2174. **A. uda** (Fr.) Bourd. & Galz. (= *Hydnum udum* Fr.) *Uda*, moist.

R. 5–13 cm., *bright sulphur colour, lemon yellow, flesh colour*, or *olivaceous, becoming watery yellowish*, widely effused, adnate, smooth; margin *lemon yellow*, pruinose, or fibrillose. Spines *concolorous, or yellowish flesh colour, becoming tawny*, 1–2 mm. long, subulate, thin, entire, or toothed. Flesh *yellowish, white next the matrix*, waxy, soft, *subgelatinous*. Spores white, "elliptical, scarcely depressed on the side, 4–6·5 × 2–3·5μ; basidia 9–15–20 × 3–4·5μ. Hyphae thin walled, 1·5–3·5μ in diam., emerging as a sterile bundle at the apex of the spines, somewhat broader, 4·5–6μ in diam. and rough with prismatic crystals" Bourd. & Galz. Smell pleasant, often of anise. Dead branches. Sept.—April. Not uncommon. (*v.v.*)

2175. **A. denticulata** (Pers.) Bourd. & Galz. (= *Hydnum denticulatum* (Pers.) Fr.) *Denticulata*, toothed.

R. 3–6 cm., *light yellow ochraceous, then fawn colour*, longitudinally effused, pruinose; margin narrow, somewhat radiating. Spines *bright yellow, then tawny*, 2–3 mm. long, subulate, *toothed and ciliated in the upper half*, crowded. Flesh waxy, membranaceous, thin. Spores white, "oblong elliptical, slightly depressed on the side, 5–6 × 2μ; basidia 12–15 × 3–4μ, accompanied by fusiform, sterile basidia, often crowned by a resinous or oily globule. Hyphae thick walled, 2–3μ in diam., forming bundles which divide and give rise to sterile emergences along the spines and at their apex" Bourd. & Galz. Smell pleasant, of anise when fresh. Rotten wood, especially alder. June—April. Uncommon. (*v.v.*)

2176. **A. stenodon** (Pers.) Bourd. & Galz.
στενός, narrow; *ὀδών*, a tooth.

R. 7·5–10 cm., *yellowish*, effused, adnate; margin *white*, narrow, byssoid, radially fibrillose, or pubescent. Spines whitish hyaline, *then ochraceous, and finally tawny*, 1–3 mm. long, thin, crowded, or connate at the base, entire, or fimbriate and ciliate, sometimes branched. Flesh waxy, fleshy, thin. Spores white, "oblong elliptical, depressed on the side, 3–4·5–(6·5) × 1·5–2·75μ, often 2-guttulate; basidia 9–14–28 × 3–4(–7)μ. Hyphae thin walled, 2–3μ in diam., prolonged into a sterile point, and enlarged at the apex, 4–6μ in diam." Bourd. & Galz. Fallen branches. Jan.—Dec. Type not yet recorded for Britain.

var. **nodulosa** (Fr.) Bourd. & Galz. (= *Hydnum nodulosum* Fr.) *Nodulosa*, nodulose.

Differs from the type in its *nodulose, or tuberculose habit*, and its *pendant, connate, often compressed, and channelled, 2–5 mm. long spines*. Fir, oak, ash stumps, and fallen branches. Oct. Rare. (*v.v.*)

2177. **A. fusco-atra** (Fr.) Pat. (= *Hydnum fusco-atrum* Fr.; *Hydnum Weinmannii* Fr. sec. Bres.) *Fuscus*, dark; *atra*, black.

R. 5–15 cm., *glaucous, then ferruginous fuscous*, widely effused, very adnate, flocculoso-pruinose; margin *white, or greyish*, similar or fimbriate. Spines *greyish glaucous*, or *fawn colour, becoming brownish black, apex grey for a long time*, 1–2 mm. long, *conico-subulate, acute*, entire. Flesh crustaceous, waxy, thin. Spores "faintly coloured, subelliptical, scarcely depressed on the side, 4·5–6 × 2–3μ; basidia 12–24 × 3·5–4·5μ, accompanied at the apex of the spines, with fusiform sterile basidia, slightly projecting. Hyphae thin walled, or slightly thickened, 3–4μ in diam., with rare clamp connections" Bourd. & Galz. Fallen branches of ash. Nov.—March. Uncommon. (*v.v.*)

2178. **A. membranacea** (Fr.) Bourd. & Galz. (= *Hydnum membranaceum* Fr. non Bull.) *Membranacea*, membranaceous.

R. 3–7 cm., *tawny ferruginous, becoming livid and finally brown*, effused, very adnate, *smooth*; margin similar, attenuate. Spines *concolorous*, 1-2 mm. long, subulate, thin, crowded, acute. Flesh *waxy membranaceous*, thin. Spores white, "oblong subcylindrical, scarcely depressed on the side, 4·5–5 × 2–2·75μ; basidia 9–24 × 3·5–4·5μ, accompanied by subulate sterile basidia, slightly projecting. Hyphae thin walled, 2·5–4μ in diam." Bourd. & Galz. Fallen branches of elm, and oak. Sept.—Jan. Uncommon. (*v.v.*)

Grandinia (Fr.) Pat. (= **Hydnum** (Linn.) Fr. p.p.).

(*Grando*, hail.)

Receptacle resupinate, effused, *membranaceous*, or crustaceous. Tubercles or spines hemispherical, obtuse, or subulate and entire. Spores white, or yellowish, ovoid, elliptical, globose, subglobose, obovate, or oblong; smooth, punctate, verrucose, or echinulate; basidia with 2–8-sterigmata accompanied, or not, with sterile basidia (cystidioles). Cystidia *none*. Growing on wood.

*Spores subglobose.

†Spores echinulate.

2179. **G. farinacea** (Pers.) Bourd. & Galz. (= *Hydnum farinaceum* (Pers.) Fr.; *Hydnum niveum* (Pers.) Fr.) *Farinacea*, mealy.

R. 2–13 cm., *snow white, then cream, or tan*, widely effused, indeterminate, closely adnate, mealy; margin byssoid, minutely fibrillose, or pruinose. Spines *white*, 1–2 mm. long, generally crowded, subulate, rarely dentate, sometimes confluent and crested, or granular, very soft, fragile. Flesh *white*, thin, floccose, or membranaceous, containing oxalate crystals. Spores *white*, minutely echinulate, ovoid, or globose, 3–4μ; basidia 6–12–21 × 3–5μ, with 2–4-sterigmata, 3–4·5μ long. Hyphae very thin walled, with clamp connections and swollen at the septa. Dead wood, branches, sticks, and leaves. Jan.—Dec. Common. (*v.v.*)

††Spores smooth, rarely punctate, or rough.

2180. **G. helvetica** (Pers.) Fr. *Helvetica*, Swiss.

R. 2–5 cm., *pale yellowish to deep ochraceous when fresh*, drying *alutaceous, or sometimes with a faint greyish tinge*, effused, separable, margin reticulately fibrillose. Tubercles *yellowish*, irregular, subglobose, soon collapsing, small, pulverulent, crowded. Flesh *yellowish*, pelliculose, or membranaceous, waxy, thin. Spores *yellowish*, subglobose, or obovate, pointed at the base, 3·5–6 × 3–4μ, 1-guttulate; basidia cylindrical, or clavate, 20–35 × 6–9μ, with 4 slightly curved sterigmata, 3–7 × 1μ. Basal hyphae yellowish, 4–8μ in diam., with occasional clamp connections, often united to form long branching strands. Fallen branches. May—Jan. Not uncommon. (*v.v.*)

2181. **G. mutabilis** (Pers.) Bourd. & Galz. (= *Grandinia granulosa* Pers. sec. Bourd. & Maire.) Bres. Fung. Trid. t. 141, fig. 2, as *Odontia olivascens* Bres. *Mutabilis*, changeable.

R. 2–5 cm., *chalk white, or cream colour, then glaucous, becoming yellowish, tan, or apple green when dried*, effused, adnate, dry, friable; margin subsimilar, or pruinosely pubescent. Granules *concolorous*, hemispherical, rarely subcylindrical, scattered, or rather crowded. Flesh *yellowish*, somewhat waxy, then floccose. Spores white, "smooth, rarely rough with a few scattered warts, 3·5–5·5 × 3–5μ; basidia 9–12–21 × 4·5–6–8μ, with 2–4-sterigmata, 3–5μ long. Hyphae thin walled, 3–7μ in diam., with very rare clamp connections" Bourd. & Galz. Dead wood, and branches. Jan.—Dec. Uncommon. (*v.v.*)

**Spores oblong, or elliptical.

2182. **G. granulosa** Fr. *Granulosa*, granular.

R. 2–12 cm., *tan colour*, broadly effused, closely adnate; margin determinate, smooth. Granules *concolorous*, *hemispherical*, equal, crowded. Flesh *yellowish*, waxy, very thin. Spores white, "oblong, 6 × 4μ. Hyphae thick walled, 3–5μ in diam., dichotomously branched" Bourd. & Galz. Dead wood, and branches. Oct.—May. Uncommon. (*v.v.*)

2183. **G. Brinkmannii** (Bres.) Bourd. & Galz. Trans. Brit. Myc. Soc. VI, text figs. p. 74. W. Brinkmann.

R. 2–5 cm., *pure white, becoming yellowish with age*, effused, very adnate, indeterminate, pruinose, waxy, then dry and chalky; margin pruinose, or minutely fibrillose. Granules *concolorous*, minute at first, then wart-like, or forming short spines, crowded. Flesh *concolorous*, loose, sparse, containing numerous crystals of calcium oxalate. Spores white, elliptical, flattened on one side, 4 × 2μ; basidia clavate, 15 × 4μ, with 4–6–8 curved sterigmata, 2–3μ long. Hyphae indistinct, soon collapsing, 4μ in diam., septate, with clamp connections. Birch bark. Nov.—March. Not uncommon. (*v.v.*)

2184. **G. mucida** Fr. (Near *Corticium ochraceum* Fr. sec. Bres. ex Bourd. & Galz.) Fr. Icon. t. 195, fig. 3. *Mucida*, mucid.

R. 2–10 cm., *pale yellowish*, effused, subinnate, corrugated when dry; margin indeterminate, somewhat radiating. Granules *concolorous*, hemispherical, *large*, *unequal*, crowded, soft. Flesh *yellowish*, waxy, *subgelatinous*. Spores white, elliptical, 6–7 × 3μ, 1-guttulate. Rotten bark, and pine wood. May—Feb. Not uncommon. (*v.v.*)

G. ocellata Fr. = **Corticium lividum** (Pers.) Fr. sec. Bres.

G. papillosa Fr. = **Odontia papillosa** (Fr.) Bres.

G. crustosa (Pers.) Fr. = **Odontia crustosa** (Pers.) Quél.

Odontia (Pers.) Pat. (=**Hydnum** (Linn.) Fr. p.p.).

(ὀδούς, a tooth.)

Receptacle resupinate, effused, membranaceous, crustaceous, or pruinose, rarely waxy, gelatinous or subcartilaginous. Spines conical, multifid, *penicillate, or ciliate*. Spores white, elliptical, globose, subglobose, pip-shaped, oboval, or cylindrical; smooth, rough, muriculate, or echinulate; basidia with 2–4-sterigmata. Cystidia *present*. Growing on wood.

*Waxy membranaceous, gelatinous, or subcartilaginous.

2185. **O. sudans** (A. & S.) Bres. (= *Dacryobolus sudans* (A. & S.) Fr.; *Porothelium confusum* B. & Br.; *Porothelium Stevensonii* B. & Br. sec. Wakef.) *Sudans*, sweating.

R. 3–10 cm., *whitish cream, or pallid*, effused, scarcely separable, very smooth; margin similar, byssoid, or mealy. Spines *concolorous*, granular, cup-shaped, conical, or truncate, short, scattered, *bearing at the apex a viscid, resinous, diaphanous, amber yellow globule*, rarely terminated by a bundle of cystidia. Flesh *yellowish*, waxy membranaceous. Spores white, "cylindrical, slightly curved, 5–6–8 × 1–1·75μ; basidia 15–24 × 3–4μ, with 2–4 straight sterigmata, 2–3μ long. Cystidia tubular, 0–3-septate, 60–150 × 3·5–5μ, emerging in tufts. Hyphae either thick or thin walled, 1–3μ in diam., coherent" Bourd. & Galz. Dead conifer branches, and wood. Jan.—Dec. Uncommon. (*v.v.*)

2186. **O. Hollii** (Schmidt) Rea. (=*Hydnum Hollii* (Schmidt) Fr.) F. Holl.

R. 10–90 cm., *fuscous lilac*, orbicular, then confluent and very widely effused, adnate; margin *white*. Spines *concolorous*, 2–4 mm. long, deformed, fasciculate, incised. Flesh concolorous, *waxy* membranaceous, floccose, thin. Decorticated wood. Oct. Rare.

2187. **O. anomala** (B. & Br.) Rea. (= *Hydnum anomalum* B. & Br.) B. & Br. Ann. Nat. Hist. no. 1438, with fig. ἀνώμαλος, uneven.

R. 4–6 cm., *pallid light yellow*, effused. Spines *concolorous*, granular, then stipitate and obtusely divided upwards, tough. Flesh *concolorous, gelatinous*, thin. Spores "globose, shortly pedicellate" Massee. Inside of very rotten oak tree. March. Rare.

2188. **O. aurea** (Fr.) Quél. (= *Hydnum aureum* Fr.) *Aurea*, golden.

R. 5–13 cm., *golden*, at first nodular, then confluent and irregularly effused, adnate, mealy; margin *white*, becoming *violet when dry*, radiately strigose. Spines *concolorous, apex white*, 2 mm. long, *setaceous*, subulate, equal, entire, crowded. Flesh *concolorous, subcartilaginous*, thin. Spores white, "muriculate, subglobose, 4–5μ" Massee. Mycelium penetrating the wood, and forming a yellow flesh colour circumscribing zone. Dead branches. Dec. Rare.

**Membranaceous, floccose, or mealy.

2189. **O. stipata** (Fr.) Quél. (= *Hydnum stipatum* Fr.) Fr. Icon. t. 194, fig. 2, as *Hydnum stipatum* Fr. *Stipata*, crowded.

R. 3–8 cm., *white, then isabelline, or light yellowish*, very widely effused, flocculoso-furfuraceous; margin similar, sterile, sometimes swollen and tomentose, rarely membranaceous and silky. Spines *white, then concolorous*, granular, becoming subulate and pointed, *minutely toothed*, thin, crowded, soft. Flesh *concolorous, floccose*, thin. Spores white, "oblong, 3–4–6·5 × 2·5–3–4μ; basidia 9–18 × 3–4–6μ, with 2–4 straight sterigmata, 3μ long. Cystidia firm, 2–4μ in diam., emerging in tufts. Hyphae thin walled, 1·5–3·5μ in diam., with clamp connections, intermixed with yellowish, firm, thick walled hyphae which become coloured with eosin" Bourd. & Galz. Stumps and fallen branches of deciduous trees. Jan.—Dec. Not uncommon. (*v.v.*)

2190. **O. barba-Jovis** (With.) Fr. Sow. Eng. Fung. t. 328, as *Hydnum barba-Jovis*. *Barba*, beard; *Jovis*, of Jupiter.

R. 5–20 cm., *white, then yellowish*, effused, slightly adnate, floccose; margin narrow, byssoid, pubescent. Spines *concolorous*, 1–2 mm. long, subulate, with one or many very thin points, *more or less bristly on the sides*, sometimes with an orange fringe at the apex. Flesh *concolorous, floccose*, membranaceous, loose. Spores white, "oboval, or subglobose, obliquely attenuated, or apiculate at the base, 4–7 × 3·5–4·5μ, often 1-guttulate; basidia 15–24–30 × 4–6μ. Cystidia cylindrical, or narrowly clavate, 60–600 × 4·5–7μ, generally fasciculate (often poorly differentiated, with thin walls, 1–2-septate), thick walled at the base, with a narrow channel insensibly enlarged upwards where the walls become thinner. Hyphae thin walled, or scarcely thickened, 2·5–4μ in diam., with clamp connections, and coloured by eosin" Bourd. & Galz. Wood, and branches of conifers. July—Jan. Uncommon. (*v.v.*)

2191. **O. limonicolor** (B. & Br.) Quél. (= *Hydnum limonicolor* B. & Br.) Bres. Fung. Trid. t. 11, fig. 2, as *Hydnum Bresadolae* Quél. *Limonicolor*, lemon colour.

R. 3–7 cm., *bright lemon yellow*, widely and irregularly effused, adnate; margin *white*, floccose. Spines *concolorous, becoming golden when dry*, acute, rough on the sides, or somewhat incised, crowded, often oblique. Flesh *whitish*, floccose, thin. Spores white, "echinulate, globose, 3μ" Bres. Pine leaves, and larch trunks. Oct. Rare.

2192. **O. plumosa** (Duby) Rea. (= *Hydnum plumosum* Duby.) *Plumosa*, feathery.

R. 4–5 cm., *snow white*, resupinate, *tomentose*. Spines *white*, 2 mm. or more long, slender, *minutely feathered near the apex*, usually crowded. Flesh *white*, floccose, very delicate. Spores white, "globose, 4–5μ" Massee. Dead wood, and bark, etc. Rare.

2193. **O. Stevensonii** (B. & Br.) Rea. (= *Hydnum Stevensonii* B. & Br.) Rev. John Stevenson, the eminent Scotch mycologist.

R. 3–8 cm., *white*, effused, mealy beneath; margin byssoid, or pulverulent. Spines *white*, 2–3 mm. long, cylindrical, obtuse, or truncate, sometimes compressed, more or less confluent at the base, *pulverulent at the apex*, somewhat crowded. Flesh *concolorous*, floccose, very thin. Spores "white, subglobose, apiculate, 3–4μ" Massee. Dead wood, leaves, and mosses. March. Rare.

2194. **O. bicolor** (A. & S.) Bres. (= *Hydnum bicolor* (A. & S.) Fr.; *Grandinia mucida* Fr. of British authors sec. Wakef.; *Hydnum subtile* Fr.) *Bi-color*, two-coloured.

R. 5–20 cm., *white, or whitish, becoming glaucous, then tan colour*, widely effused, *subtomentose*, soft, pruinose, waxy, often cracked at the base of the spines; margin indeterminate, or whitish pruinose. Spines *concolorous, apex often brownish red*, small, granular, *minutely villose*, obtuse. Flesh *concolorous*, floccose, soft. Spores white, "oblong, scarcely depressed on the side, 4·5–7 × 2·75–4μ; basidia 10–24 × 3–5μ, with 2–4-sterigmata, 4–5μ long. Cystidia with a *globose head*, 8–15μ in diam., thin walled, contents becoming yellowish, and often crowned by radiate twin crystals. Hyphae of the subiculum 2–3μ in diam., in the axis of the spines amber coloured, fasciculate, agglutinated together by a resinous substance, and ending at the apex of the spines in an oil coloured tuft" Bourd. & Galz. Firs, and brambles. Oct.—March. Not uncommon. (*v.v.*)

2195. **O. papillosa** (Fr.) Bres. (= *Grandinia papillosa* Fr.) *Papillosa*, having nipples.

R. 2·5–5 cm., *milk white, or yellowish*, effused, *separating when entire, very much cracked*; margin *white*, very thin, pubescent, or pruinose. Spines *concolorous*, granular, subhemispherical, equal, becoming subulate, thin, small, very crowded. Flesh *concolorous*, membranaceous, floccose, firm. Spores white, "oblong, subcylindrical, depressed on the side, 4·5–6 × 2–2·75μ; basidia 10–20 × 3–4·5μ, with 2–4-sterigmata, 3–4·5μ long, accompanied by subulate, sterile basidia, sometimes capped with oxalate, and with numerous smooth, or rough paraphysis-like hyphae forming a tuft at the apex of the spines. Hyphae with walls slightly thickened, firm, with clamp connections; subhymenial hyphae denser, 3–4·5μ in diam." Bourd. & Galz. Fallen oak, beech, and fir branches. July—Nov. Rare.

2196. **O. arguta** (Fr.) Quél. (= *Hydnum argutum* Fr.; ? *Radulum tomentosum* Fr. sec. Bourd. & Galz.) *Arguta*, sharp.

R. 3–6 cm., *white, then yellowish*, effused, *tomentose, or minutely pubescent*, slightly adnate, finally minutely cracked; margin similar, or floccose. Spines *white*, then *ochraceous*, 1–2 mm. long, granular, *pubescent*, then cylindrical, or subulate, sometimes connate at the base, *apex penicillate*. Flesh *concolorous*, floccose, firm, thin. Spores white, "oboval, 4–6 × 3–5μ, often 1-guttulate; basidia 10–15–18 × 3–4–6μ. Cystidia fusiform, or capitate, crowned, or not, by an oil globule, 7–9μ in diam. Hyphae thin walled, or slightly thickened, 2–4μ in diam., with clamp connections" Bourd. & Galz. Trunks, stumps, and fallen branches. Sept.—Dec. Not uncommon. (*v.v.*)

var. **alutacea** (Fr.) Bourd. & Galz. (= *Hydnum alutaceum* Fr.) *Alutacea*, tanned leather.

Differs from the type in its *deeper ochraceous tan colour, and the stouter spines*. Pines. Aug.—Dec. Not uncommon. (*v.v.*)

2197. **O. spathulata** (Schwein.) Rea. (= *Hydnum spathulatum* (Schwein.) Fr.) *σπάθη*, a broad blade.

R. 2·5–5 cm., *whitish, becoming yellow*, effused, separable, adpressedly villose beneath; margin involute, *fimbriate*. Spines *brick red, or orange*, 2–4 mm. long, *spathulate*, oblique, sometimes flattened, acicular. Flesh *concolorous*, membranaceous, thin. Spores "white, broadly elliptical, apiculate, 8 × 5μ" Massee. Dead *Robinia* wood. Jan.—Dec. Rare.

2198. **O. multiformis** (B. & Br.) Rea. (= *Hydnum multiforme* B. & Br.) *Multiformis*, many shaped.

R. 5–10 cm., *ochrey white*, effused, inseparable, becoming cracked when dry; margin indeterminate, thin. Spines *pallid*, 2–4 mm. long, very acute, becoming fimbriate, crowded. Flesh *concolorous*, floccose, thin. Spores white, "subglobose, or very broadly pip-shaped, obliquely apiculate, 9 × 6–7μ" Massee. Dead wood. Rare.

2199. **O. mollusca** (Fr.) Rea. (= *Hydnum molluscum* Fr.)
Mollusca, soft.

R. 5–10 cm., *whitish*, effused, easily separable, dry. Spines *reddish*, short, slender. Flesh *white*, *membranaceous*, thin. Wood. Rare.

2200. **O. sulphurea** (Schwein.) Rea. (= *Hydnum sulphureum* Schwein.)
Sulphurea, sulphur yellow.

R. 10 cm., *sulphur yellow*, effused, adnate; margin *paler*, byssoid, sterile. Spines *concolorous*, subulate, minute, *few*. Flesh membranaceous, thin. Dead birch wood. Rare.

***Waxy crustaceous, very adnate.

2201. **O. crustosa** (Pers.) Quél. (= *Grandinia crustosa* (Pers.) Fr.)
Crustosa, having a crust.

R. 5–15 cm., *whitish cream colour, then yellowish, or tan colour*, effused, adnate, finally cracked and minutely areolate; margin *white*, distinct, *narrow*, pruinose, or *minutely pubescent*. Spines *concolorous*, granular, short, pointed, or obtuse, scattered, or somewhat crowded. Flesh *concolorous*, *crustaceous*, thin. Spores white, "oblong, subcylindrical, depressed on the side, 4·5–6–8 × 2–4μ; basidia 12–21–30 × 3–4–6μ, accompanied by numerous fusiform, or subulate cystidioles, 3–4–6μ in diam., sometimes branched, slightly projecting. Hyphae thin walled, 1·5–4μ in diam., with rather rare clamp connections" Bourd. & Galz. Fallen branches. Jan.—Dec. Uncommon. (*v.v.*)

2202. **O. cristulata** Fr. (= *Peniophora setigera* Bres. sec. Bourd. & Maire.)
Cristulata, crested.

R. 5–14 cm., *pale, or rosy flesh colour*, widely effused, adnate, mealy; margin *white*, narrowly byssoid, pubescent, or mealy. Spines *concolorous*, or *reddish brown*, short, crowded, or confluent and crested; apex penicillate, pointed. Flesh *white*, somewhat waxy, then *crustaceous*, thin. Spores white, "cylindrical, slightly depressed on the side, 8–10 × 3·5–4μ; basidia 25–32 × 4·5–7μ, contents granular. Cystidia fasciculate, cylindrical, 4–5μ in diam., 1–2-septate, with, or without clamp connections. Hyphae thin walled, 3–6μ in diam., with clamp connections" Bourd. & Galz. Birch branches. Feb. Rare.

2203. **O. mellea** (B. & Br.) Rea. (= *Hydnum melleum* B. & Br.)
Mellea, honey colour.

R. 5–10 cm., *honey colour*, effused, *pulverulent*; margin minutely byssoid. Spines *concolorous*, 1–2 mm. long, acute, sometimes divided at the apex, *pulverulent downwards*, naked at the middle. Flesh crustaceous, thin. Spores white, "cylindrical, 7–10 × 2·5μ" Massee. Fallen rails. Rare.

2204. **O. sepulta** (B. & Br.) Rea. (= *Hydnum sepultum* B. & Br.) *Sepulta*, buried.

R. 3–15 mm., *golden yellow*, resupinate, forming little, scattered patches; margin *white*. Spines *concolorous*, 1–2 mm. long, acute. Flesh very thin. Spores white, "globose, 5μ" Massee. Stones buried among pine leaves. Sept. Rare.

2205. **O. alliacea** Weinm. *Alliacea*, of garlic.

R. 3–10 cm., *white, translucid, becoming pale, or cinereous*, broadly effused, incrusting; margin silky. Spines *concolorous*, 1–2 mm. long, with some shorter ones, *incised*, villose. Flesh *concolorous, crustaceous*, membranaceous, thin. Spores white, elliptical, 3–4 × 2μ; basidia with 2-sterigmata. Smell faint, of garlic. Lichens on trees, and dead branches. Sept.—Oct. Uncommon. (*v.v.*)

2206. **O. Pruni** Lasch. *Pruni*, of plums.

R. *white, becoming pallid*, effused, adnate; margin byssoid. Spines *white*, granular, minute, rounded, apex penicillate. Flesh *concolorous, crustaceous*, thin. On *Prunus spinosa*. Rare.

Kneiffia Fr.

(Friederich Gotthard Kneiff, a German mycologist.)

Receptacle *subgelatinous*, effused. Spines or granules, scattered, *minute, sterile*. Spores white, elliptical, smooth. Growing on wood.

2207. **K. subgelatinosa** B. & Br. *Subgelatinosa*, somewhat gelatinous.

R. 10 cm., *yellowish, then cream colour*, broadly effused. Spines *concolorous*, granular, minute, scattered, *subgelatinous*, fringed at the apex. Flesh *concolorous*, subgelatinous, thin. Spores white, "broadly elliptical, apiculate, 4 × 2·5μ" Massee. Fir stumps. April. Rare.

K. setigera Fr. = **Peniophora setigera** (Fr.) Bres.

Hydnopsis (Schroet.) Rea.

(ὕδνον, the genus *Hydnum*; ὄψις, like.)

Receptacle floccose, resupinate, effused. Spines subulate, acute. Spores *coloured, elliptical, smooth*. Growing on dead leaves, and on the ground.

2208. **H. farinacea** Rea. Trans. Brit. Myc. Soc. III, t. 7. *Farinacea*, mealy.

R. 2·5–6 mm., *white*, effused, adnate. Spines *white, then wood colour, and finally umber*, 1 mm. long, subulate, acute, thin, subdistant. Flesh *white*, floccose, thin. Spores fuscous, elliptical, 6–7 × 3–4μ, 1–3-guttulate. Dead fallen beech leaves, and on the ground. May. Uncommon. (*v.v.*)

Caldesiella Sacc. (= **Hydnum** (Linn.) Fr.).

(L. Caldesi, an Italian botanist.)

Receptacle *floccose*, soft, resupinate. Spines conical, soft, villose, fimbriate at the apex. Spores *coloured, globose, verrucose, or echinulate*; basidia clavate, with 2–4-sterigmata. Growing on wood.

2209. **C. crinalis** (Fr.) Bourd. & Galz. (= *Hydnum crinale* Fr.; *Hydnum ferrugineum* Auct. pl. non Fr. nec Karst.; *Odontia barba-Jovis* Pat. Tab. Anal. f. 247; *Caldesiella ferruginosa* Sacc. sec. Bres., as *Odontia crinalis* (Fr.) Bres.) Pers. Myc. Eur. II, t. 17, fig. 3, as *Hydnum castaneum* Pers. var. *fuscum* Pers. *Crinalis*, hairy.

R. 5–10 cm., *tawny ferruginous*, effused, separable, *tomentose*. Spines *concolorous*, subulate, conical, acute, straight, or oblique, often somewhat compressed, *tomentose*, crowded. Flesh *concolorous*, floccose, lax. Spores deep brown, echinulate, globose, 8–9μ. Decayed wood especially under the bark. July—Dec. Not uncommon. (*v.v.*)

2210. **C. italica** Sacc. *Italica*, Italian.

R. 2–10 cm., *fuliginous*, widely effused, incrusting, resupinate. Spines *concolorous, becoming olivaceous with the snuff-coloured spores*, 1–1·5 mm. long, ·5–1 mm. thick, cylindrical, obtuse, often compressed, crowded, pruinose. Flesh *concolorous*, floccose, thick. Spores snuff-coloured in the mass, olivaceous-hyaline under the microscope, obtusely verrucose, angularly globose, 8–9 × 8μ; basidia clavate with 2–4-sterigmata. Basal hyphae concolorous, thick walled, 6–8μ in diam., septate, with clamp connections. Birch stumps. Oct. Uncommon. (*v.v.*)

2211. **C. viridis** (A. & S.) Pat. (= *Hydnum viride* (A. & S.) Fr.) Boud. Icon. t. 170, as *Odontia viridis* (A. & S.) Quél. *Viridis*, green.

R. 5–25 cm., *white, then indigo blue, soon greenish, and at length yellowish*, broadly effused, tomentose; margin *white*, membranaceous, thin. Spines *indigo blue, then greenish*, 1–2 mm. long, cylindrical, irregular, obtusely divided at the apex, often crowned with white hairs. Flesh *concolorous*, floccose, thin. Spores *indigo blue*, verrucose, globose, 4–5μ. Rotten wood. Sept.—Oct. Uncommon. (*v.v.*)

Phylacteria (Pers.) Pat. (= **Thelephora** (Ehrh.) Fr. p.p.).

(φυλακτήριον, an amulet.)

Receptacle *fibrous, or coriaceous*, pileate, stipitate, sessile, or resupinate, entire, or laciniate, destitute of a pellicle. Stem central, lateral, or none, confluent with the pileus. Flesh coloured. Hymenium inferior or amphigenous, smooth, faintly ribbed, or papillose. Spores coloured, elliptical, globose, subglobose, or angular, verrucose, or echinulate; basidia with 2–4-sterigmata. Cystidia none. Growing on the ground, or on wood.

I. Erect with usually a central st.

*P. divided into very narrow, branching, flattened, or cylindrical divisions.

2212. **P. palmata** (Scop.) Pat. (= *Thelephora palmata* (Scop.) Fr.) Grev. Scot. Crypt. Fl. t. 46, as *Merisma foetidum* Pers.

Palmata, having the shape of a hand.

R. 2–6 cm. high, 1–3 cm. broad, *fuscous purple, apex whitish*, fimbriate, very much divided into palmate, flattened, subfastigiate, even, flattened branches, dilated upwards. St. 1–1·5 cm. × 1–2 mm., *concolorous*, simple, or branched. Hymenium amphigenous. Flesh concolorous, coriaceous, soft. Spores fuscous purple in the mass, pale umber under the microscope, echinulate, globose, 8–9μ. Smell very foetid. Woods, especially of conifers. Aug.—Nov. Uncommon. (*v.v.*)

2213. **P. anthocephala** (Bull.) Pat. (= *Thelephora anthocephala* (Bull.) Fr.) Berk. Outl. Brit. Fung. t. 17, fig. 4, as *Thelephora anthocephala*. ἄνθος, a flower; κεφαλή, head.

R. 2–5 cm. high, 1–3 cm. broad, *somewhat ferruginous, becoming fuscous*, apex *whitish*, fimbriate, divided down to the stem *into flaps which are dilated upwards*, or *into irregular branched erect branches, pubescent*. St. ·5–1·5 cm. × 1–2 mm., concolorous, simple, equal, *villose*. Hymenium even. Flesh *concolorous*, coriaceous, soft. Spores purplish in the mass, pale umber under the microscope, echinulate, globose, or broadly elliptical, 6–8×6μ, 1-2-guttulate. Woods. Aug.—Nov. Not uncommon. (*v.v.*)

2214. **P. clavularis** (Fr.) Big. & Guill. (= *Thelephora clavularis* Fr.) Fr. Icon. t. 196, fig. 3, as *Thelephora clavularis* Fr.

Clavularis, a little nail.

R. 2·5–4 cm. high, 1-3 cm. broad, *rufous fuscous*, apex *whitish, acute*, divided down to the st. into *round, attenuate*, even, delicately pruinose branches. St. ·5–1·5 cm. × 1–3 mm., *concolorous*, base *somewhat tuberous*. Flesh *concolorous*, coriaceous, soft. Spores reddish purple in the mass, echinulate, subglobose, 6–7 × 6μ. Woods. Sept.—Oct. Uncommon. (*v.v.*)

**P. more or less infundibuliform, cup-shaped, or flabelliform, often splitting into lobes, or divisions.

2215. **P. caryophyllea** (Schaeff.) Pat. (= *Thelephora caryophyllea* (Schaeff.) Fr.) Schaeff. Icon. t. 325, as *Helvella caryophyllea* Schaeff. κάρυον, a nut; φύλλον, a leaf.

R. 1·5–4 cm. high, 1·5–5 cm. broad, *fuscous purple, becoming wood brown when dried*, infundibuliform, cup-shaped; margin lobed, or incised, often broken up into wedge-shaped, imbricate branches, or segments, *fibrillosely torn*, often radiately rugose, or striate, obsoletely zoned when moist. St. ·5–1 cm. × 2–3 mm., *concolorous*, simple, or branched, equal, *villose*. Hymenium inferior, even. Flesh *concolorous, or paler*, subcoriaceous. Spores purple in the mass, pale umber under the microscope, verrucose, globose, 6–7 μ. Woods, especially under conifers. Aug.—Nov. Uncommon. (*v.v.*)

II. Dimidiate, horizontal, subsessile, or effuso-reflexed.

2216. **P. intybacea** (Pers.) Pat. (= *Thelephora intybacea* (Pers.) Fr.) Bull. Champ. Fr. t. 278, as *Thelephora intybacea.*

ἔντυβον, chicory.

R. 2–4 cm., *whitish, then rufous ferruginous, at length fuliginous*; margin *whitish fimbriate at first*, then concolorous, dimidiate, confluent, imbricate, fibrous, *the fibrils often agglutinated into adpressed, adnate squamules*; margin dilated. St. short, sublateral, often confluent. Hymenium *concolorous*, inferior, papillose. Flesh firm, fibrillose, 1 mm. thick. Spores deep ochraceous, verrucosely echinulate, subglobose, or elliptical, 7–9 × 6–7 μ, with a large central gutta. Caespitose. Pine wood, and bare soil. Sept.—Nov. Uncommon. (*v.v.*)

2217. **P. terrestris** (Ehrh.) Big. & Guill. (= *Thelephora terrestris* (Ehrh.) Fr.; *Thelephora laciniata* (Pers.) Fr.) Rolland, Champ. t. 101, no. 224. *Terrestris*, pertaining to the earth.

R. 3–5 cm., *ferruginous fuscous, or fuscous, often becoming black* with age, dimidiate, sessile, or effuso-reflexed, laterally confluent, often imbricate, *fibrillosely scaly*, strigose; margin fimbriate, laciniate. Hymenium *fuscous, or pale fawn*, inferior, papillose. Flesh *concolorous*, coriaceous, fibrillose, soft, 1 mm. thick. Spores fuscous, verrucose, angularly globose, 8–9 × 6–8 μ. Woods, and heaths. July—Dec. Common. (*v.v.*)

P. biennis (Fr.) Big. & Guill. = **Hypochnus umbrinus** (Fr.) Quél.

2218. **P. atra** (Weinm.) Rea. (= *Thelephora atra* Weinm.)

Atra, black.

P. 5–8 cm., *black, becoming fuliginous*, imbricate, arising from a tuberous base, deformed, somewhat lobed, sessile, attenuated at the base; margin at first white, fimbriate. Hymenium *black*, white pruinose, *smooth, setulose*. Caespitose. Dead logs. Rare.

2219. **P. mollissima** (Pers.) Rea. (= *Thelephora mollissima* (Pers.) Fr.; *Phylacteria spiculosa* (Fr.) Bourd. & Maire.) Berk. Outl. Brit. Fung. t. 17, fig. 5, as *Thelephora mollissima.*

Mollissima, very soft.

R. 2–4 cm., *whitish, becoming brownish*, broadly effused, continuous, or effuso-reflexed, *forming flaps, subtomentose*. Hymenium *fuscous purple*, inferior, smooth, even. Flesh *concolorous*, soft, thin. Spores brownish purple, warted, subglobose, 7–8μ. Woods. Aug.—Oct. Uncommon. (*v.v.*)

III. Resupinate, and incrusting.

2220. **P. spiculosa** (Fr.) Bourd. & Maire. (=*Thelephora spiculosa* (Fr.) Burt.) Pers. Syn. Fung. t. 3, fig. 16, as *Merisma penicillatum*.
Spiculosa, having little sharp points.

R. 2–15 cm., *fuscous purple, whitish at the apex of the spicules*, effused, incrusting; margin *ramoso-spiculose, tips penicillate*. Hymenium concolorous, even or slightly rugose. Flesh floccose, 1 mm. thick. Spores umber, echinulate, irregularly globose, or elliptical, 7–9 × 6–7μ. Running over twigs, and dead leaves. Aug.—Nov. Not uncommon. (*v.v.*)

Hypochnus (Fr.) Karst. (= **Tomentella** (Pers.) Pat.).

(ὑπό, under; χνόος, fine down.)

Receptacle *floccose, or felt-like*, resupinate, effused. Hymenium smooth, or papillose. Flesh coloured, soft, loose. Spores coloured, rough, verrucose, or echinulate; globose, subglobose, elliptical, ovoid, or angular; basidia sometimes in scattered clusters, with 2–4-sterigmata. Growing on wood, mosses, or on the ground.

2221. **H. ferrugineus** (Pers.) Fr. *Ferrugineus*, iron rust colour.

R. 2–6 cm., *ferruginous*, effused, adnate, often suborbicular, dry, tomentose, hypochnoid. Hymenium *concolorous*, "Sudan-brown" when dry. Flesh *concolorous*, loose, thin. Spores concolorous, echinulate, subglobose, 7–10μ, with numerous hyaline spines. Basal hyphae, 5–8μ wide, septa with clamp connections. Decaying wood, and bark of deciduous trees. Sept.—Dec. Not uncommon. (*v.v.*)

2222. **H. umbrinus** (Fr.) Quél. (= *Thelephora biennis* Fr.)
Umbrinus, umber.

R. 3–10 cm., *brown, with more or less of a vinaceous tint* (varying from *drab to fuscous and* "*Chaetura-drab*" of Ridgway), effused, soft, separable; subiculum *warm sepia*, villose. Hymenium *concolorous*, membranaceous, compact. Flesh *concolorous*, dense. Spores fuscous, or dark brown, aculeate, or coarsely verrucose, globose, or subglobose, 6–8μ, or 6–8 × 5–7μ; basidia brownish, clavate, with 4-sterigmata. Hyphae brown, thick walled, 4–5μ in diam., septate, without clamp connections. Dead wood. Oct.—Feb. Uncommon. (*v.v.*)

2223. **H. fuscus** (Pers.) Fr. (= *Corticium fuscum* (Pers.) Fr.)
Fuscus, dark.

R. 2–10 cm., *subfuscous, cinnamon, or brown, somewhat vinaceous in colour*, effused, *separable*. Hymenium *concolorous*, loose. Flesh *concolorous*, membranaceous, loose, thin. Spores reddish brown, echinulate, subglobose, or broadly elliptical, 7–8 × 5–6μ. Hyphae with numerous clamp connections, 5–6μ in diam. Rotten wood, and fallen branches. Sept.—June. Not uncommon. (*v.v.*)

2224. **H. subfuscus** Karst. *Subfuscus*, somewhat dark.

R. 1–4 cm., *dark purplish brown, between* Ridgway's "*Natal Brown*" and "*Bone Brown*," effused; margin *concolorous*. Hymenium *concolorous, pulverulent*, mould-like. Spores sepia, aculeate, globose, 8–9μ without the hyaline spines, spines acute, 1–2μ long; basidia clavate, 10–12μ in diam., with 2–4 curved sterigmata. Subhymenial hyphae pale, 5–7μ in diam. Basal hyphae brown, 7–8μ in diam., thick walled, straight, with branches at right angles, sometimes slightly incrusted, with clamp connections. Bark of fallen logs. Sept.—Oct. Uncommon. (*v.v.*)

2225. **H. granulosus** (Peck) Burt. (= *Grandinia tabacina* Cke. & Ell.) Burt, Theleph. of North Am. VI, Hypochnus, text-fig. p. 219. *Granulosus*, granular.

R. 2–4 × 1–2 cm., *sepia*, effused, thin, membranaceous, *granular*, separable; margin somewhat radiate, *concolorous*. Spores *concolorous with the hyphae*, aculeate, angular-subglobose, the body about 6μ in diam. Flesh *concolorous or* paler, loose. Hyphae *yellowish under the microscope*, loosely interwoven, 2·5–4μ in diam., thin walled, occasionally with clamp connections, forming near the substratum some *rope-like mycelial strands* up to 15μ in diam. Pine sticks and rotten bark and wood of frondose trees. Aug.—Sept. Uncommon. (*v.v.*)

2226. **H. puniceus** (A. & S.) Sacc. (= *Corticium* (*Hypochnus*) *puniceum* (A. & S.) Fr.) *Puniceus*, reddish.

R. 1–3 cm., *dull red*, vaguely effused, mould-like; margin similar. Hymenium *concolorous, minutely granular*, loose and pulverulent under a lens. Flesh *pale*, fibrillose, loose, very thin. Spores *dull reddish*, with short spines, subglobose, 8–9·5μ, 7·5–8μ without the spines; basidia clavate, 40–50 × 8μ, with 2–4 stout curved sterigmata, 5–8 × 2·5–3μ. Subhymenial hyphae hyaline, or pale coloured, 3·5–5μ, wavy, much branched, with clamp connections. Decorticated pine-wood. July—Sept. Uncommon. (*v.v.*)

2227. **H. isabellinus** Fr. (= *Corticium isabellinum* Fr.) Fr. Obs. Myc. II, t. 6, fig. 3. *Isabellinus*, dirty yellowish.

R. 5–10 cm., *pale* "*Isabella colour*" of Ridgway, *or deep olive buff to dark olive buff*, effused, adnate, inseparable, *tomentose*; margin *concolorous*, thinner. Hymenium *concolorous*, loose, pulverulent. Flesh *concolorous*, loose, thin. Spores isabelline, echinulate, globose, 7–9μ

without the spines. Hyphae *concolorous*, thick walled, branched at right angles, 8–14μ in diam., without clamp connections. Rotten wood, and bark. Sept. Rare. (*v.v.*)

2228. **H. cyaneus** Wakef. *κύανος*, dark blue.

R. 1–3 cm., *deep dull violaceous blue, through "Eton blue" to glaucous* green, *becoming dull greyish green or yellow in parts when dried*, effused, easily separable, tomentose; margin *concolorous*, arachnoid. Hymenium *concolorous*, floccose. Flesh *concolorous, paler*, thin. Spores *dull bluish*, minutely and sparsely aculeate, elliptical, depressed on one side, 5–8 × 4μ; basidia hyaline, elongate-clavate, 30–40 × 7μ, with 2–4-sterigmata. Basal hyphae slightly tinged bluish, 1–3μ in diam., here and there incrusted, without clamp connections. Potassium hydrate solution takes away the colour of the spores but turns the other parts greenish when fresh; when dried, the hyphae and spores become a dull violet colour. Wet rotten coniferous logs. Oct. Rare.

2229. **H. cinerascens** Karst. *Cinerascens*, becoming ash colour.

R. 2–3 cm., *drab grey, to pale drab*, indefinitely effused, adnate, separable; margin *concolorous, or whitish*. Hymenium *concolorous*, loose. Flesh very thin, loose. Spores *grey-brown*, with minute spines, angularly subglobose, 6–7 × 5–6μ; basidia hyaline, cylindric-clavate, 40 × 6μ, with 2–4-sterigmata. Subhymenial and basal hyphae *hyaline*, 3–4μ in diam., much branched, frequently septate with clamp connections. Bark. Sept. Uncommon.

2230. **H. caesius** (Pers.) Wakef. (=*Thelephora caesia* (Pers.) Fr.) Pers. Obs. I, t. 3, fig. 6, as *Corticium caesium*. *Caesius*, bluish grey.

R. 3–10 cm., *cinereous-bluish-grey, or brownish bistre*, effused, suborbicular, *determinate*, tomentose. Hymenium *concolorous, becoming paler*, even, *minutely pubescent*. Flesh *brownish*, soft, thin. Spores hyaline, then lilac bistre, and finally brownish, spinulose, subglobose, 7–8μ. Wood, twigs, mosses, and on the ground. Aug.—Nov. Not uncommon. (*v.v.*)

2231. **H. botryoides** (Schwein.) Burt.
βότρυς, a bunch of grapes; *εἶδος*, like.

R. 1–5 cm., *yellow-brown* (*ochraceous-tawny to "Buckthorn-brown"* of Ridgway), effused, separable; margin *much paler, brownish*, floccose. Hymenium *fuscous, finely granular*, forming a delicate pellicle. Flesh *pale brown*, very soft, loose. Spores fuscous, aculeate, angularly subglobose, spore body 6 × 5·5μ; basidia clavate, 30–35 × 6μ, with 4-sterigmata. Basal hyphae yellow-brown, 3–4μ in diam., often united to form long slender strands, with clamp connections. "Potassium hydrate turns microscopic sections of the hymenium immediately *blue green*" Burt. Bark. Sept.—Oct. Uncommon.

2232. **H. crustaceus** (Schum.) Karst. (= *Thelephora crustacea* (Schum.) Fr.; *Phylacteria spiculosa* (Fr.) Bourd. & Maire sec. von Hoehn. & Litsch.) Fl. Dan. t. 1851, fig. 2. *Crustaceus*, having a bark.

R. 2–8 cm., *fuscous umber*, broadly effused, incrusting; margin *whitish, or black*, fibrillose. Hymenium *concolorous*, irregularly papillose. Flesh *brownish*, soft, floccose, thin. Spores brown, verrucose, globose, 8–10μ. Hyphae brown. Bare soil, and running over grass, leaves, and twigs. Feb.—Nov. Uncommon. (*v.v.*)

2233. **H. zygodesmoides** (Ellis) Burt.
ζυγόδεσμον, a yoke-band; εἶδος, like.

R. 2–3 cm., *pinkish*, to *vinaceous-buff, often with rusty stains*, broadly effused, easily separable, soft; margin "*Sayal*" *to* "*snuff-brown*," narrow, byssoid. Hymenium *concolorous, loose*. Flesh *pale brown*, arachnoid, membranaceous, floccose. Spores pale, or with a slight tinge of buff in the mass, very pale straw colour to almost hyaline under the microscope, with fairly long, blunt spines, elliptical to subglobose, spore body 5–7 × 4–6μ; basidia cylindric clavate, 40 × 6–8μ, with 4-sterigmata, 4–7μ long. Basal hyphae pale brown, little branched, and infrequently septate, 4–6μ in diam. Rotten bark. Sept. Uncommon.

2234. **H. echinosporus** (Ellis) Burt. (= *Corticium echinosporum* Ellis; *Hypochnus mollis* Fr. var. *pellicula* Fr.; *Hypochnus pellicula* Bres.) ἐχῖνος, hedge-hog; σπορά, seed.

R. 2–4 cm., *sulphur yellow, or rose pink*, indefinitely effused, very delicate, membranaceous, separable; margin *whitish, or concolorous*, very thin, indefinite, occasionally with very fine, white hyphal strands spreading over the subiculum. Hymenium *pale clear sulphur yellow*, becoming *spotted with brown when old*, or *dull rose pink, occasionally with a very faint lilac tinge, and with darker reddish, or brownish stains*, forming a fine pulverulent pellicle. Flesh *concolorous*, arachnoid, soft, thin, loose. Spores hyaline, or pale straw colour, contents sometimes golden yellow, or rosy, echinulate, subglobose, or in lateral view broadly elliptical, 5–7 × 4–6μ; basidia cylindrical to clavate, 20–30 × 6–8μ, with 2–4-sterigmata, slightly curved, 3–5·5 × 1–1·5μ. Basal hyphae hyaline, or very faintly coloured, 2·5–5μ in diam., branched, septate. Rotten wood. Oct. Rare.

2235. **H. roseo-griseus** Wakef. & Pearson. Trans. Brit. Myc. Soc. VI, text-figs. p. 141. *Roseus*, rosy; *griseus*, grey.

R. 3–8 cm., "*light vinaceous fawn*" and "*cinnamon drab*" of Ridgway, *with a paler* "*drab-grey*" *margin, becoming like H. fuscus, but paler with a greyish bloom, greyish white to dirty buff* when quite young, effused, pelliculose, or membranaceous, easily separable; margin somewhat radiating, *grey*. Hymenium *pale greyish vinaceous*, pulverulent. Flesh pale, thin, soft. Spores hyaline, or pale straw colour, coarsely verrucose, angularly-subglobose, 7–9 μ, often 1-guttulate; basidia subhyaline, clavate, 40–55 × 7–10 μ, 2–4-sterigmata, 7–9 μ long. Subhymenial hyphae subhyaline; basal hyphae greyish, scarcely branched, septate, 2·5–3 μ in diam., without clamp connections. Bark, wood, etc., especially pine. Oct.—Jan. Common.

var. **lavandulaceus** Pears. *Lavandulaceus*, lavender coloured.

Differs from the type only in the *greyish lavender colour* of the hymenium without a trace of pink. Ground in woods under *Castanea sativa*. Sept. Uncommon. (*v.v.*)

2236. **H. fumosus** Fr. (= *Corticium fumosum* Fr.; *Corticium sulphureum* (Pers.) Bres.; *Phlebia vaga* Fr.; *Coniophora sulfurea* (Pers.) Quél.) Fr. Icon. t. 198, fig. 3, as *Corticium fumosum* Fr. *Fumosus*, smoky.

R. 3–10 cm., *pale, yellow, tawny, cinnamon, grey, drab, brownish or fuscous*, effused, membranaceous, arachnoid, separable, more or less overrun with intricate, branching, anastomosing threads; margin *bright yellow, becoming whitish*, byssoid, fibrillose. Hymenium *concolorous*, granular, or reticulately veined. Flesh *pale, or slightly cinereous*, membranaceous, thin. Spores white, or brownish, minutely echinulate, ovoid, or globose, 3–7 × 3–5 μ. Hyphae longitudinally interwoven, occasionally with clamp connections, thin walled, hyaline, or slightly smoky, 2·5–3·5 μ in diam. Dead wood, branches, and twigs. Jan.—Dec. Common. (*v.v.*)

2237. **H. sphaerosporus** R. Maire. (= *Corticium sphaerosporum* (R. Maire) von Hoehn. & Litsch.) von Hoehn. & Litsch. Beit. zur Kennt. der Cort. in Sitzungsber. der k. Akad. d. Wissensch. Wien, Math.-Nat. Kl. Bd. CXVII (1908), 1106, and reprint 26, text-fig. 5. *σφαῖρα*, a ball; *σπορά*, seed.

R. 1–2 cm., *chalk white* or *snow white, becoming yellowish in the centre*, effused, arachnoid, and porous under a lens; margin similar or fibrillose. Hymenium *concolorous*, mealy, or granular. Flesh *concolorous*, thin, floccose. Spores hyaline, coarsely and minutely warted, globose, or angularly-globose, 3–6 × 2·5–4 μ (mostly 4·5 × 4 μ), 1-guttulate; basidia clavate or pyriform, 8–15 × 4–6 μ, with 2–4-sterigmata 2–5 μ long. Hyphae hyaline, 2–4 μ in diam., thin walled, with clamp connections, basal hyphae often forming rhizoidal strands, and inclosing acicular, or fusiform crystals of oxalate of lime, 10–25 μ long. Beech logs and on bare soil. Oct. Rare. (*v.v.*)

2238. **H. submutabilis** (von Hoehn. & Litsch.) Rea. (= *Corticium submutabile* von Hoehn. & Litsch.) *Submutabilis*, changeable.

R. 1–3 cm., *dirty whitish to yellowish*, effused, irregular; margin similar. Hymenium *concolorous*, very loose, pulverulent. Spores hyaline, rough with short conical warts, subglobose, broadly elliptical or oval and flattened on one side, attenuated at the base, 2–3·5μ in diam., or 2–3·5 × 2–2·5μ, usually 1-guttulate; basidia clavate, 8–16 × 4–6μ, with 2–4 thin, pointed sterigmata 1·5–3μ long. Hyphae hyaline, 1–3μ in diam., thin walled, septate-nodulose, rarely distinct (no clamp connections observed). Pine stick. Sept. Rare.

Hypochnella Schroet.

(*Hypochnella*, diminutive of *Hypochnus*.)

Same characters as *Hypochnus* but differing in the *smooth, elliptical, violet spores*. Growing on wood.

2239. **H. violacea** (Awd.) Schroet. *Violacea*, violet.

R. 2–10 cm., *rich lilac colour, becoming darker and duller when dry*, irregularly effused. Hymenium *concolorous*, smooth. Flesh *bluish*, very thin, floccose, loose. Spores deep violet, elliptical, with a lateral basal apiculus, 7–9 × 3–4μ; basidia hyaline, or faintly coloured, cylindric-clavate, 20–25 × 8–9μ; paraphyses obtuse, 10–12 × 6–7μ, often with a few crystals on the external walls. Subhymenial hyphae very faintly coloured, often slightly incrusted with small rounded crystals; basal hyphae pale lilac, 6–9μ in diam. Underside of fallen branches. Sept.—Oct. Uncommon. (*v.v.*)

Jaapia Bres.

(Otto Jaap.)

Resupinate, effused, immarginate, flocculose-pulverulent, with the habit of some Corticia or of a pale *Hypochnus*. Spores straw coloured, subelliptical, *hyaline-appendiculate*.

2240. **J. argillacea** Bres. Trans. Brit. Myc. Soc. VI, text-figs. p. 320. *Argillacea*, clay coloured.

R. 1 cm., *clay coloured*, irregularly effused, flocculose, sometimes with scattered granules. Hymenium *concolorous*, loose, then more continuous. Flesh *concolorous*, very thin, floccose. Spores straw coloured, fusiform, slightly curved, 22–25 × 7–8μ, consisting of a central oblong-elliptical portion, 14–18 × 7–8μ (mostly 15 × 7μ), containing faintly coloured, granular protoplasm, *divided off by a wall from a clear conical portion at either end*. Basidia clavate, up to 60μ long by 8–10μ wide, with 2–4 curved sterigmata, 8μ long. Cystidia hyaline, cylindrical, obtuse, 100–160 × 7–8μ, occasionally with a single septum. Basal hyphae flexuous, frequently septate, with clamp connections, 4–6μ in diam. Fallen sticks. Oct. Rare. (*v.v.*)

Aldridgea Massee.

(Miss Emily Aldridge.)

Receptacle *subgelatinous, becoming cartilaginous* when dry, resupinate, effused. Hymenium smooth, even. Spores coloured, elliptical, smooth; basidia with 4-sterigmata. Growing on wood.

2241. **A. gelatinosa** Massee. Massee, Brit. Fung. Fl. I, figs. 20 and 21, p. 97. *Gelatinosa*, jelly-like.

R. 5–13 cm., *pallid*, broadly effused; margin determinate. Hymenium *purple brown*, smooth, even. Flesh *subgelatinous*, then cartilaginous, or rigid and collapsed when dry. Spores "olive, broadly elliptical, obliquely apiculate, 10 × 6–7μ" Massee. Sawdust. Rare.

[**Ptychogaster** Corda.]

(πτύξ, a fold; γαστήρ, belly.)

Receptacle fleshy, or somewhat corky, round, or cushion-shaped, producing conidia and chlamydospores. Cystidia present, or absent. Growing on wood, or incrusting plants.

2242. **P. albus** Cda. *Albus*, white.

R. 2–15 cm., *white, becoming brownish*, globose, obconic, or pulvinate, soft, shaggy, or filamentous; internally *white, becoming brownish, formed of many concentric layers*. Conidia colourless, long, oval. Chlamydospores "brownish, elliptic, or oblong, 6μ long" Henn. Dead wood and branches. Aug.—Feb. Common. (*v.v.*)

6. Thelephoraceae.

Hymenium spread over a smooth, rugose, or ribbed surface, either resting upon an intermediate layer of hyphae running longitudinally between it and the mycelium, or seated directly upon the mycelium.

1. Hymenium separated from the mycelium by an intermediate layer of hyphae.

Sparassis Fr.

(σπαράσσω, I tear in pieces.)

Receptacle fleshy, erect, much branched; branches flattened in a lamellar, or plate-like manner, more or less confluent. Hymenium smooth, inferior. Spores white, or yellowish, ovoid, elliptical, globose, or subglobose; smooth; basidia with 2–4-sterigmata. Cystidia none. Mycelium cord-like, often attached to the roots of trees. Growing on the ground.

2243. **S. crispa** (Wulf.) Fr. Rolland, Champ. t. 102, no. 229. *Crispa*, curled.

R. 10–60 cm., *whitish, or pale ochraceous*, very much branched, resembling a cauliflower; branches 2·5–5 cm., broad, *intricate*, ribbon-like, apex tinged yellowish, crisped, and slightly zoned. St. *whitish, becoming blackish with age*, stout, rooting. Flesh *whitish*, or *yellowish*, fleshy, brittle. Spores pale ochraceous in the mass, hyaline under the microscope, subglobose, or elliptical, 6–7μ in diam., or $6 \times 4\mu$, 1–2-guttulate; basidia with 2–4-sterigmata. Smell very pleasant, of anise. Taste agreeable. Edible. Coniferous woods. Aug.—Nov. Not uncommon. (*v.v.*)

2244. **S. laminosa** Fr. Trans. Brit. Myc. Soc. II, t. 13.

Laminosa, having plates.

R. 10–60 cm., *yellowish straw colour*, very much branched; branches *laminar, patent*, more lax, and less dense than those of *S. crispa*. St. *whitish, becoming discoloured with age*, stout, rooting. Flesh *yellowish*, fleshy, brittle. Spores pale ochraceous in the mass, hyaline under the microscope, globose, 8μ, with granular contents. Smell pleasant. Taste agreeable. Edible. Mixed woods, especially near oaks. Sept.—Nov. Not uncommon. (*v.v.*)

Stereum (Pers.) Massee (= **Thelephora** (Ehrh.) Fr. pp.).

(*στερεόν*, firm.)

Receptacle coriaceous, pileate, stipitate, or sessile, infundibuliform, dimidiate, resupinate, or effuso-reflexed. Stem central, lateral, or none. Hymenium inferior, with an intermediate layer, smooth, rarely rugulose, or ribbed, sometimes setulose, pubescent, or velvety. Flesh pale. Spores white, oval, elliptical, globose, subglobose, cylindrical, oblong, or oblong elliptic; smooth, or granular; basidia with 2–4-sterigmata. Cystidia hyaline, rarely coloured in nos. 2261 and 2262, present, or absent. Annual, or perennial. Growing on wood, or on the ground.

I. R. infundibuliform. St. central.

2245. **S. Sowerbeii** (B. & Br.) Massee. (= *Thelephora Sowerbeii* B. & Br.; *Podoscypha Sowerbeji* (B. & Br.) Pat.; *Stereum pallidum* (Pers.) Lloyd sec. Burt.; *Thelephora vitellina* Plowr.) Rolland, Champ. t. 101, no. 225, as *Podoscypha Sowerbeji*.

James Sowerby, the well-known botanical illustrator.

R. 1–2·5 cm., *snow white, becoming yellow, or pale yellow, infundibuliform, flabelliform, or spathulate*, uneven; margin incised, often crenate. St. 1–2·5 cm. × 2–3 mm., *concolorous*, gradually expanding into the p., often confluent, smooth, or wrinkled. Hymenium *concolorous*, smooth, or rugulose. Flesh *yellowish*, fleshy coriaceous, thin. Spores white, oval, $3 \times 2\mu$. Under pine bark, and on the ground. Sept.—Nov. Uncommon. (*v.v.*)

2246. **S. pallidum** (Pers.) Cooke. (= *Thelephora pallida* (Pers.) Fr.) Pers. Icon. et Desc. I, t. 1, fig. 3, as *Craterella pallida*.

Pallidum, pale.

R. 1–5 cm., *pallid, then cream, or buff*, infundibuliform, *strigosely squamulose*. St. 2–6 × ·5–1 cm., *concolorous*, expanding upwards into the p., often confluent, smooth, base villose. Hymenium *pallid, rugulose*, with slight, very obtuse, radial folds, *more or less setulose* with hyaline hairs under a lens. Flesh *concolorous*, coriaceous-spongy, rather thick. Spores white, elliptical, often flattened on one side, 5–8 × 3–5μ. Cystidia hyaline, cylindrical, smooth, 6–8μ in diam., projecting 10–50μ above the hymenium. Often caespitose. Woods. July—Nov. Rare. (*v.v.*)

2247. **S. multizonatum** (B. & Br.) Massee. (= *Thelephora multizonata* B. & Br.) B. & Br. Ann. Nat. Hist. ser. 3, xv, t. XIII, fig. 4, as *Thelephora multizonata*.

Multus, many; ζώνη, a belt.

R. 9–20 cm., *bright rufous flesh colour, or rich brown, margin white at first*, deeply infundibuliform, variously cut and lobed, *zoned with darker bands*; margin lobed, crenulate. St. 5–9 × 1–3 cm., *concolorous*, gradually expanding into the p., often confluent, smooth. *Hymenium paler than the p., or somewhat cinereous*, slightly ribbed, smooth. Flesh *concolorous, or paler*, coriaceous, tough, thin. Spores white, broadly elliptical, 7 × 4–5μ, with a large central gutta. Caespitose. Woods. Sept.—Dec. Uncommon. (*v.v.*)

2248. **S. undulatum** (Fr.) Massee[1]. (= *Thelephora undulata* Fr.)

Undulatum, waved.

R. 1–3 cm., *whitish, then tan*, depressed, plano-infundibuliform, minutely fibrillose; margin entire, *undulate*. St. 1–3 × ·5–1 cm., *whitish*, equal, villose. Hymenium *pale tan, ribbed, setulose under a lens*. Flesh *concolorous*, subcoriaceous, firm. Spores white, "broadly pip-shaped, 10 × 6μ" Massee. On the ground. Oct. Rare.

2249. **S. tuberosum** (Grev.) Massee. (= *Thelephora tuberosa* (Grev.) Fr.) Grev. Scot. Crypt. Fl. t. 178, as *Merisma tuberosum* Grev.

Tuberosum, tuberous.

R. 1–2 cm., *grey, or with a slight brownish tinge, infundibuliform, broken up into narrow, compressed segments almost to the base of the st.*, segments acute, or obtuse at the apex. St. ·5–2·5 cm. × 2–4 mm., *concolorous*, subcylindrical, obscurely furrowed or lacunose, base *bulbous*. Hymenium *concolorous*, inferior, smooth. Flesh subcoriaceous, thin. Spores white, "elliptical, 7–8 × 5μ" Massee. Bare soil. Sept.—Nov. Rare.

[1] "The record in England is an error of determination" sec. Lloyd, Synopsis Stipitate Stereums, p. 20.

II. R. dimidiate, sessile, or resupinate and effuso-reflexed, marginate.

*Hymenium bleeding when touched.

2250. **S. spadiceum** Fr. (= *Stereum gausapatum* Fr.; *Stereum cristulatum* Quél.; *Stereum quercinum* Potter.) Rolland, Champ. t. 102, no. 227. *Spadiceum*, date brown.

R. 5–10 cm., *greyish, brownish, or subferruginous, margin often white at first*, effuso-reflexed, or subdimidiate, often imbricated, confluent, *villose or hirsute*; margin obtuse, often lobed. Hymenium *fuscous, or bistre, bleeding when fresh if cut or bruised, becoming snuff brown or more or less darker and discoloured with age*, smooth, or wrinkled. Flesh *whitish in the middle stratum*, coriaceous, soft. Spores white, elliptical, 7–8 × 4–5μ. Cystidia none. Lacticiferous hyphae, red, "coloured, 75–120 × 5μ, very numerous" Burt. Stumps, and fallen branches, especially oak, and ash. Jan.—Dec. Common. (*v.v.*)

2251. **S. rugosum** (Pers.) Fr. *Rugosum*, wrinkled.

R. 2–20 cm., *pinkish buff, base paler*, widely effused, or shortly reflexed, obtusely marginate, silky, then glabrous, and at length *concentrically furrowed*, radially pitted and weathering *grey*. Hymenium *pinkish buff to drab-grey, bleeding if bruised when fresh*, pruinose. Flesh *whitish*, becoming discoloured, coriaceous, rigid, "intermediate layer bordered on the upper side by a *dense golden zone* and on the lower side by a *two–many-zoned hymenial layer* 120–1200μ thick, hyphae of intermediate layer 2·5–3μ in diam." Burt. Spores white, oblong, incurved, 10–12 × 4–5μ. Cystidia none. Lacticiferous hyphae red, "dark coloured, very numerous, 3–6μ in diam." Burt. Stumps, trunks, logs, and fallen branches of frondose trees. Jan.—Dec. Common. (*v.v.*)

2252. **S. sanguinolentum** (A. & S.) Fr. Grev. Scot. Crypt. Fl. t. 225, as *Thelephora sanguinolenta* A. & S. *Sanguinolentum*, full of blood.

R. 1–8 cm., *pallid, white at the thin, acute margin*, effused, often circular, becoming confluent, then reflexed, *adpressedly villose, or silky*, substriate. Hymenium *cinereous, then fuscous, bleeding when wounded*, even, smooth, becoming cracked when dry. Flesh whitish, coriaceous, thin, "intermediate layer bordered on the upper side by a *narrow, dense golden zone*, and composed of densely arranged hyaline hyphae, 3μ in diam." Burt. Spores white, cylindrical, slightly curved, 8–9 × 3–4μ. Cystidia none. Lacticiferous hyphae red, "coloured, 3–4μ, usually numerous" Burt. Dead stumps, and branches, especially of conifers. Jan.—Dec. Common. (*v.v.*)

**Hymenium yellow, or grey.

2253. **S. hirsutum** (Willd.) Fr. Berk. Outl. Brit. Fung. t. 17, fig. 7. *Hirsutum*, hairy.

R. 2–10 cm., *pallid, yellowish, or greyish, margin yellow*, widely effused, then reflexed, sometimes entirely resupinate, *strigosely hairy*, subzoned; margin obtuse. Hymenium *bright ochraceous, pinkish*, or *tan colour, sometimes becoming grey*, even, smooth. Flesh *yellowish*, coriaceous, firm, tough, "intermediate layer bordered next to the hairy covering by a *very dense, narrow, golden zone*, composed of densely and longitudinally arranged hyaline hyphae, 3–4μ in diam., some of which in the subhymenium are thick walled, up to 5–6μ in diam., and very rarely have golden brown contents as seen between the basidia" Burt. Spores white, elliptical, incurved, 6–8 × 3–4μ. Cystidia none. Stumps, trunks, logs, posts, and fallen branches. Jan.—Dec. Common. (*v.v.*)

var. **subcostatum** (Karst.) Massee. (= *Corticium subcostatum* (Karst.) Bourd. & Galz.) *Subcostatum*, somewhat ribbed.

Differs from the type in its *vaguely costate, or rugose hymenium*. Fallen branches. Sept.—Nov. Uncommon. (*v.v.*)

var. **luteocitrinum** Sacc. *Luteus*, yellow; *citrinum*, lemon yellow.

Differs from the type in the *golden yellow margin, and dark coloured p*. Stumps. Sept. Uncommon. (*v.v.*)

2254. **S. ochroleucum** Fr. ὠχρός, pale; λευκός, white.

R. 3–5 cm., *whitish cream, then greyish tan colour*, orbicular, effuso-reflexed, confluent, then free, often entirely resupinate, *villose, or strigose*, zoned. Hymenium *pale ochre*, smooth, *cracked when dry*. Flesh *pale*, floccose, thin, "hyphae about 2·5μ in diam., *granule-incrusted* and interwoven throughout the thickness of the pileus" Burt. Spores white, elliptical or subglobose, 4·5–5 × 3μ. Dead wood and bark. Jan.—March. Uncommon. (*v.v.*)

***Hymenium purple, lilac, or brown.

2255. **S. purpureum** (Pers.) Fr. (= *Stereum vorticosum* Fr. sec. Burt.) Hussey, Illus. Brit. Myc. I, t. 20. *Purpureum*, purple.

R. 2–8 cm., *whitish, pallid, or greyish*, effuso-reflexed, more or less imbricate, sometimes entirely resupinate, zoned, *villosely tomentose*; margin entire, sometimes crisped or lobed. Hymenium *lilac, or purplish*, even, smooth. Flesh *whitish*, coriaceous-soft, somewhat thick. Spores white, oblong, or oboval, apiculate at one end, 6–8 × 3–4μ. Hymenial cystidia *none*, subhymenial cystidia vesiculose, 15–30 × 12–25μ. Dead branches, and felled trunks, especially birch, beech, elm, and poplar. Jan.—Dec. Common. (*v.v.*)

var. **atro-marginatum** W. G. Sm. Sow. Eng. Fung. t. 412, fig. 1, as *Auricularia elegans* Sow. *Ater*, black; *marginatum*, bordered.

Differs from the type in the *narrow, black zone near the white margin of the p.* Dead branches. Sept.—Nov. Uncommon. (*v.v.*)

2256. **S. rugosiusculum** Berk. & Curt. (= *Stereum purpureum* (Pers.) auct. pl.) *Rugosiusculum*, somewhat wrinkled.

R. 2–6 cm., *cartridge-buff to cinnamon buff when dry*, more or less broadly reflexed, rarely resupinate, *tomentose*, spongy, sometimes with projecting hairs collapsed together into a plane or wrinkled surface; margin entire. Hymenium *vinaceous-buff to fawn colour when dry*, even. Flesh *whitish*, coriaceous-soft, fairly thick. Spores white, elliptical, incurved, 5–7 × 3–4μ. Hymenial cystidia *cylindrical*, thin walled, 4μ in diam., subhymenial cystidia vesiculose, 15–30 × 10–20μ. Logs and stumps of frondose trees. Jan.—Dec. Common. (*v.v.*)

2257. **S. conchatum** Fr. κόγχη, a mussel shell.

R. 4–30 mm., *dirty yellowish*, effuso-reflexed, then shell-shaped, subimbricate, *rugose, glabrous*, obscurely zoned, somewhat crisped. Hymenium *fuscous*, smooth. Flesh coriaceous, thin. Fir. Rare.

2258. **S. bicolor** (Pers.) Fr. (= *Stereum fuscum* (Schrad.) Quél.) Fr. Icon. t. 197, fig. 2. *Bicolor*, two coloured.

R. 1–5 cm., *snuff-brown to bistre when dry*, sometimes resupinate, generally becoming conchate-reflexed, often imbricate, *villose*, becoming glabrous, *somewhat concentrically sulcate*. Hymenium *white, then cream colour to pallid mouse grey when dry*, even, smooth. Flesh submembranaceous, soft, spongy, "composed of longitudinally and loosely interwoven hyphae, 3μ in diam., coloured towards the upper surface, hyaline towards the hymenium. Spores hyaline, 3–4·5 × 2–3μ. *Gloeocystidia hyaline, flexuose*, 20–60 × 5–7μ" Burt. Rotting frondose limbs and sometimes on pine. April—Dec. Rare.

III. Effused, resupinate, margin scarcely, or not at all free.

*Hymenium stratose.

2259. **S. frustulosum** (Pers.) Fr. *Frustulosum*, full of pieces.

R. 2–5 mm., *date-brown-blackish*, resupinate, *tuberculose, crowded as if confluent*, and then broken up into frustules, sometimes growing outward from the place of attachment and narrowly reflexed, or with a free margin all round, *concentrically sulcate*, glabrous. Hymenium *pinkish buff to whitish, convex*, pruinose, *stratose*. Flesh *woody*, thick. Spores "hyaline, oboval, 5–6 × 3–3·5μ. Paraphyses *bottle-brush*, or aculeate, numerous" Burt. Oak logs and stumps. Rare.

2260. **S. stratosum** B. & Br. *Stratosum*, stratose.

R. 5–10 cm., *bright ochraceous white, becoming yellowish*, effused, smooth, here and there wrinkled. Flesh *pallid, stratose*, strata at length broken up. Rare.

**Hymenium pubescent, velvety.

2261. **S. Chailletii** (Pers.) Fr. D. Chaillet.

R. 2–15 cm., *somewhat fawn colour*, or *brownish*, broadly effused, resupinate the first year, then becoming *stratose*, and at length pileate, pilei sometimes well developed, tomentose, *more or less concentrically sulcate*; margin entire. Hymenium *pale ferruginous, or fawn colour*, pubescent, *velvety*. Flesh *pallid*, coriaceous, fairly thick, "composed of somewhat longitudinally and not densely interwoven hyphae, 3–4·5μ in diam., some of which are hyaline, thin walled, and with deeply staining protoplasm, and many thick walled, stiff, giving their colour to the fructification and curving into the hymenium where they terminate in cystidia" Burt. Spores "hyaline, elliptical, inequilateral, 6–7·5 × 3–4μ. Cystidia *yellowish*, rough, fusiform, cylindrical, 50–120 × 4–5μ, or in old stratose plants, 45–60 × 5–7μ" Bres. Felled fir trunk. Oct.—Dec. Rare. (*v.v.*)

S. disciforme (DC.) Fr. = **Aleurodiscus disciformis** (DC.) Pat.

***Hymenium pruinose.

2262. **S. abietinum** (Pers.) Fr. *Abietinum*, of firs.

R. 2–8 cm., *burnt umber*, resupinate, effused, rarely reflexed, tomentose, obscurely zonate, tuberculate or uneven. Hymenium *light drab to cinereous or glaucous, pruinose*. Flesh *coloured*, coriaceous-spongy, thick, "intermediate layer composed of longitudinally arranged, interwoven, coloured hyphae, 3–3·5μ in diam., bordered on its outer side by a darker, denser zone which connects with the tomentose covering; hymenial layer becoming zonate and containing numerous, *coloured*, cystidia. Spores hyaline, flattened on one side, 9–13 × 4–5μ. Cystidia *coloured*, cylindric, obtuse, even, rough walled or more or less incrusted, 90–150 × 6–8μ, protruding up to 60μ" Burt. Pine and Abies trunks and logs. Rare.

2263. **S. Pini** Fr. *Pini*, of pines.

R. 1–4 mm., *fuscous, then Benzo-brown*, resupinate, adnate, at first orbicular, then confluent, and *again broken up into bullate tubercles*, smooth beneath; margin fimbriate, lobed. Hymenium *purple flesh colour, becoming fuscous, pruinose*. Flesh coriaceous-cartilaginous, rigid, thin at the margin, "intermediate layer *bordered on each side by a narrow, coloured zone*, and composed of longitudinally arranged, densely interwoven, hyaline hyphae with walls gelatinously modified, the subhymenium olivaceous-coloured. Spores hyaline, curved, 5–6 × 2–2·5μ. Cystidia hyaline, *incrusted*, 24 × 8μ, sometimes very sparse. *Gloeocystidia* hyaline, fusoid, or irregular, 30–40 × 10–15μ, sparse" Burt. Pine bark. Nov. Rare.

2264. **S. rufum** Fr. (= *Stereum rufomarginatum* (Pers.) Quél.; British records of this plant = *Eichleriella spinulosa* (Berk. & Curt.) Burt, sec. Wakef.) Burt, The Thelephoraceae of North America, XII, Stereum, p. 121, text-fig. 11. *Rufum*, red.

R. 2–4 mm., *vinaceous-brown to hematite red, erumpent, tuberculiform*, then somewhat round, *marginate*, smooth beneath. Hymenium *vinaceous-brown, often greyish pruinose, becoming coarsely wrinkled.* Flesh coriaceous-fleshy, firm, fairly thick, "composed of ascending loosely interwoven, *incrusted*, hyaline hyphae, 4–4·5μ in diam. over the incrustation. Spores white, oblong, curved, 6–8 × 1·5–2μ. Cystidia none. *Gloeocystidia* hyaline, flexuose, 50–90 × 7–10μ, scattered, not protruding" Burt. Poplar. Sept.—March. Uncommon.

S. acerinum (Pers.) Fr. = **Aleurodiscus acerinus** (Pers.) von Hoehn. & Litsch.

Hymenochaete Lév. (= **Stereum** (Pers.) Massee p.p.).

(ὑμήν, a membrane; χαίτη, long flowing hair.)

Receptacle coriaceous, firm, sessile, effuso-reflexed, or resupinate. Hymenium inferior, with an intermediate layer, setulose, or velvety, even, rarely granular. Spores white, or coloured, elliptical, oval, subglobose, oblong, fusoid, or cylindrical ellipsoid; smooth. Cystidia or *setae present, coloured.* Perennial. Growing on wood.

I. Sessile, effused, free and reflexed.

2265. **H. rubiginosa** (Dicks.) Lév. (= *Stereum rubiginosum* (Schrad.) Fr.) Sow. Brit. Fung. t. 26, as *Auricularia ferruginea*. *Rubiginosa*, rusty.

R. 3–15 cm., *rubiginous*, or *brownish rust colour, margin ochraceous tawny*, effused, reflexed, sometimes entirely resupinate, separable, rigid, somewhat fasciate, *concentrically sulcate, velvety, becoming smooth and date brown.* Hymenium *ferruginous*, or bistre, setulose, subcolliculose. Flesh *tawny ferruginous*, coriaceous, *firm*, intermediate layer "composed of longitudinally arranged, coloured hyphae, 2·5μ in diam., and bordered above by a narrow, dense, dark zone" Burt. Spores white, elliptical, 4–6 × 2–3μ. Setae coloured, crowded, acutely conical, slightly curved, 50–70 × 5–7μ. Stumps, branches and logs of frondose trees. Jan.—Dec. Common. (*v.v.*)

2266. **H. tabacina** (Sow.) Lév. (= *Stereum tabacinum* (Sow.) Fr.; *Stereum avellanum* Fr. in part; *Hymenochaete avellana* (Fr.) Cke.) Sow. Brit. Fung. t. 25, as *Auricularia tabacina* Sow.
Tabacina, tobacco colour.

R. 3–30 cm., *subferruginous, becoming brown, margin golden*, effused, reflexed, often imbricate, sometimes entirely resupinate, *silky*, at length becoming smooth. Hymenium *paler, snuff brown*, or *sepia*, setulose, often deeply cracked into a series of radial anastomosing cracks when resupinate. Flesh *golden*, coriaceous, *flaccid*, thin, intermediate layer composed of "longitudinally arranged, orange-yellow hyphae, 2·5–3μ in diam., bordered on each side by a narrow, dark, dense zone" Burt. Spores white, oblong, often curved, 4–6 × 1·5–2μ. Setae coloured, conico-acuminate, 70–100 × 8–12μ. Stumps, trunks, and logs, especially of frondose trees. Sept.—March. Not uncommon. (*v.v.*)

H. avellana (Fr.) Cke. = **Hymenochaete tabacina** (Sow.) Lév.

2267. **H. Boltonii** (Fr.) Cke. (= *Corticium Boltonii* Fr.)
James Bolton, author of "An History of Fungusses growing about Halifax."

R. 3–4 cm., *white to ochre, or pale lavender, zoned brown, or black*, effused; margin shortly reflexed, villose. Hymenium *white to pale brown, becoming ferruginous fuscous*, or *dark red*, setulose. Setae "clavate, attenuated at the base, 70–80 × 10–11μ, smaller in resupinate forms" Cke. Bird cherry. Feb. Rare.

II. Resupinate.

A. Hymenium simple.

†Setae acuminate.

*Spores white.

2268. **H. nigrescens** Cke. *Nigrescens*, becoming black.

R. 2·5–5 cm., *fuscous, becoming black*, peltate, subcircular, solitary, or gregarious, sometimes confluent, adnate; margin sometimes free and slightly reflexed, smooth and *greyish beneath*. Hymenium *brown, then blackish*, or *blackish umber*, setulose, often cracked. Flesh rigid. Spores "white, elliptical, 10 × 5μ. Setae *blackish*, conical, 80–140 × 10–12μ" Massee. Dry wood. May.

2269. **H. Stevensonii** B. & Br. (= *Stereum rufo-hispidum* Stev.)
Rev. John Stevenson, the eminent Scotch mycologist.

R. 2–4 cm., *pale fawn colour*, effused, adnate; margin abrupt, sometimes a little thickened, and raised. Hymenium *livid*, or *greyish pink, with a tinge of lilac when dry*, setulose. Flesh rigid. Spores white, "elliptic fusoid, 6–7 × 3–4μ. Setae (*rufous*), rigid, 20–40 × 8–10μ" Massee. Bark of yew. Sept.—April. Rare. (*v.v.*)

2270. **H. leonina** Berk. & Curt. λέων, a lion.

R. 2–10 cm., *orange ferruginous, drying tawny olive to "Brussels-brown,"* entirely resupinate, widely effused, *separable*; margin tomentose. Hymenium *concolorous*, setulose, unequal. Flesh *concolorous, coriaceous*, loose, "composed of a compact setigerous layer 50–75μ thick, with setae starting at different levels within it, and of a broad supporting hyphal layer, 100–600μ thick, composed of loosely interwoven, rather longitudinally arranged hyphae, 3μ in diam., stiff, coloured like the fructification; in fully developed, thick fructifications the *hyphal layer is divided*, parallel with the substratum, *by a narrow, dark zone*" Burt. Spores white, elliptical, "5–6 × 3–3·5μ. Setae conical, tapering from the base to the apex, 60–80 × 7–9μ, emerging up to 50μ" Burt. Dead wood and holly. Rare.

2271. **H. fuliginosa** (Pers.) Lév. (? = *Hymenochaete fuliginosa* (Pers.) Bres.) *Fuliginosa*, sooty.

R. 4–5 cm., *obscure smoky brown*, effused, *closely adnate*; margin *yellowish rust*, often very much broken up into patches, and almost indeterminate. Hymenium *umber with rust, or purple tinge*, densely or sparsely setulose, appearing almost smooth under a lens, sometimes minutely cracked, and brighter in colour. Flesh coriaceous, compact. Spores white, "subglobose, 5 × 4μ. Setae *brown, often clear purple* by transmitted light, 30–50 × 6–8μ" Massee. Wood, and decorticated branches. Sept. Uncommon. (*v.v.*)

2272. **H. Mougeotii** (Fr.) Cke. (= *Corticium Mougeotii* Fr.) Trans. Brit. Myc. Soc. IV, t. 9.
J. B. Mougeot, part author of "Stirpes cryptogamicae Vogeso-rhenanae."

R. 5–20 mm., *rusty brown, reddish brown at the margin*, effuso-reflexed, *closely adnate*, minutely tomentose; margin silky. Hymenium *deep red*, tubercular, or granular, pruinosely pubescent, setulose. Flesh *concolorous, or paler*, waxy, then rigid. Spores white, cylindrically ellipsoid, or oblong, 5–7 × 2μ. Setae *red, apex hyaline and white*, gradually attenuated upwards, 30–60 × 5–8μ. Dead branches of *Picea excelsa*. Aug.—Sept. Rare. (*v.v.*)

2273. **H. corrugata** (Fr.) Lév. (= *Corticium corrugatum* Fr.) Grev. Scot. Crypt. Fl. t. 234, as *Thelephora Padi* Pers. *Corrugata*, wrinkled.

R. 5–20 cm., *pallid cinnamon*, widely effused, *closely adnate, when dry, cracked, into small polygonal areas, about* 1–3 *to a mm.*, sometimes grumous; margin thin, sometimes *paler*. Hymenium *cinnamon brown to bistre and "Rood's brown," sometimes weathering to mouse grey*, setulose. Flesh *concolorous*, or *paler*, firm, "composed of densely interwoven hyphae, 3μ in diam., coloured like the fructification" Burt. Spores white, allantoid, 4·5–7 × 1·5–2μ. Setae *brown*, cylindrical, acute, 55–75 × 7–12μ. Dead wood and branches. Jan.—Dec. Not uncommon. (*v.v.*)

**Spores coloured.

2274. **H. croceo-ferruginea** Massee. Linn. Soc. Bot. Jour. 27, t. v, figs. 9, a, b, c.

Croceus, saffron yellow; *ferruginea*, iron rust colour.

R. 5–8 cm., *orange ferruginous to brownish*, broadly effused, *closely adnate*; margin byssoid or indeterminate. Hymenium *concolorous*, very minutely setulose, cracked when dry. Flesh crustaceous, very thin. Spores "*olive*, subglobose, 7 × 6μ. Setae cylindrical, base very much swollen, 70–100 × 30–35μ" Massee. Dead stems of *Rosa canina*. Rare.

††Setae subclavate, sometimes rough.

2275. **H. crassa** (Lév.) Berk. Lév. Voy. Bonite, t. 139, fig. 1 B, as *Thelephora crassa* Lév. *Crassa*, thick.

R. 2·5–11 cm., *pale rufous*, resupinate, effused, *minutely velvety*; margin thickened, at length free. Hymenium *rufous*, unequal, setulose. Flesh coriaceous, soft, spongy. Spores white, "cylindric-ellipsoid, 7–8 × 4μ. Setae *subclavate, often rough at the apex*, 70–130 × 7–14μ" Massee. Trunks. Rare.

H. abietina (Pers.) Massee = **Stereum abietinum** (Pers.) Fr.

B. Hymenium stratose.

2276. **H. cinnamomea** (Pers.) Bres. (= *Corticium cinnamomeum* (Pers.) Fr.) *Cinnamomea*, cinnamon colour.

R. 3–7 cm., *cinnamon brown, or auburn*, resupinate, widely effused, adnate, velvety; margin *paler*, floccose. Hymenium *cinnamon-brown, drying antique brown to "Brussels-brown,"* setulose, *stratose*. Flesh *concolorous*, thick, loose, "stratose, ranging up to 6 strata, each composed of a setigerous layer 30–45μ broad, and of a hyphal layer of equal or greater breadth, with hyphae coloured like the fructification, loosely interwoven, 3μ in diam." Burt. Spores white, cylindric-ellipsoid, curved, 5–6 × 2–2·5μ. Setae *mahogany colour, apex paler*, tapering upwards into an acute point, 70–100 × 5–6μ. "Basidia clavate, 10–12 × 3·5μ, with 4-sterigmata, intermingled with long, cylindrical, blunt, paraphyses, brown below, more or less hyaline above, 4μ in diam." Wakef. Bark, decaying wood and fallen branches of frondose and coniferous trees. Sept.—Oct. Uncommon. (*v.v.*)

Cladoderris Pers.

(κλάδος, a branch; δέρρις, a leathern covering.)

Receptacle coriaceous, pileate, sessile, or produced behind into a stem-like base. Hymenium inferior, *with fan-like folds, or radiating, woody, branched ribs, or veins*. Spores white, elliptic oblong, smooth. Cystidia present. Growing on wood.

2277. **C. minima** B. & Br. Stevenson, Brit. Fung. II, p. 266, fig. 85. *Minima*, least.

R. 4–6 mm., *white*, flabelliform, resupinate, springing from a stem-like, or obsolete base, tomentose, somewhat zoned. Hymenium *white, becoming yellowish tan, radiated on branched ribs.* Flesh coriaceous, firm. Spores white, "elliptic-oblong, apiculate at the base, curved, 14–15 × 4–5μ" Massee. Birch. Dec. Rare.

2. Hymenium seated directly on the mycelium.

Epithele Pat.

(ἐπί, upon; θηλή, a nipple.)

Receptacle waxy, or floccose, resupinate, effused. Hymenium smooth, interspersed with scattered, *sterile protuberances, caused by the breaking through of fasciculate mycelial hyphae.* Spores white, fusiform, smooth; basidia with 2–4-sterigmata. Cystidia none. Growing on dead leaves, herbaceous stems, and wood.

2278. **E. Typhae** (Pers.) Pat. (= *Corticium Typhae* (Pers.) Fr.) Beit. zur Kennt. der Cort. in Sitzungsber. der k. Akad. d. Wissensch. Wien, Math.-Nat. Kl. Bd. cxv (1906), 1598, and reprint 50, text-fig. 3. *Typha*, the Mace-reed.

R. 1–4 cm., *white, becoming yellowish, or dull buff*, longitudinally effused, originating as byssoid spots, then confluent, minutely tomentose under a lens. Hymenium *concolorous*, smooth, then papillose, and often cracked. Flesh *whitish*, waxy, floccose, very thin. Spores white, fusiform, 20–25 × 7–8μ, 2–3-guttulate. Dead dry leaves of *Typha*, and *Carex*. Oct.—Nov. Uncommon. (*v.v.*)

Aleurodiscus Rabenh.

(ἄλευρον, flour, starch; δίσκος, a round plate.)

Receptacle waxy floccose, or crustaceous, becoming coriaceous; resupinate, saucer-shaped with a free margin, or effused and adnate. Hymenium smooth, pulverulent, *often containing much granular, or crystalline matter.* Spores white, *large*, ovoid, elliptical, or subglobose; smooth, or echinulate; basidia *large with* 4 *stout sterigmata*, intermixed with *torulose, moniliform, or racemose paraphyses*, or sterile basidia. Growing on wood.

I. Discoid, cup-shaped, pezizaeform.

*Spores smooth.

2279. **A. disciformis** (DC.) Pat. (= *Stereum disciforme* (DC.) Fr.) δίσκος, a quoit; *forma*, shape.

R. 1–2·5 cm., *white, or tan colour*, resupinate, disciform; margin *white*, free, narrow. Hymenium *white, becoming greyish*, rigid, uneven, pulverulent. Flesh *concolorous*, subcoriaceous, *hard*, firm. Spores white, "ovoid, or subglobose, 16–22 × 12–16μ, membrane coloured blue with iodine; basidia 60–90 × 10–14μ. Paraphyses or sterile basidia torulose, 5–9μ in diam. Hyphae rather thick walled, 3–5μ in diam." Bourd. & Galz. Trunks, and branches of oaks. Oct.—March. Uncommon.

**Spores rough.

2280. **A. amorphus** (Pers.) Rabenh. (= *Corticium amorphum* (Pers.) Fr.) *ἄμορφος*, misshapen.

R. 3–15 mm., *white, becoming pallid, cup-shaped,* then flattened, scattered, or confluent, externally *white tomentose,* and hairs incrusted with calcium oxalate; margin free, incurved. Hymenium *orange, or buff pink, becoming paler, especially at the margin and subolivaceous when dried, even, continuous,* pulverulent. Flesh *pale,* subcoriaceous, pliant, dense. Spores white, *minutely echinulate,* spines hyaline, subglobose, 20–30 × 17–25μ, the membrane colours blue with iodine; "basidia 100–150 × 15–24μ, with 2–4 subulate, curved sterigmata, 20–30 × 4–5μ. Paraphyses or sterile basidia torulose, 4–10μ in diam. Hyphae rather thick walled, 3–6μ in diam., basal hyphae slightly coloured, often incrusted with calcium oxalate" Bourd. & Galz. Silver fir, and larch. Aug.—Jan. Uncommon. (*v.v.*)

II. Resupinate, effused, margin never reflexed.

2281. **A. acerinus** (Pers.) von Hoehn. & Litsch. (= *Stereum acerinum* (Pers.) Fr.) *Acerinus*, pertaining to maples.

R. 3–10 mm., *white,* irregularly effused, scattered, resupinate, crustaceous, adnate; margin abrupt. Hymenium *white,* mealy, then smooth, and finally cracked. Flesh *white,* chalky, containing numerous crystals of calcium oxalate, compact, thin. Spores white, "ovoid, elliptical, 10–15 × 6–11μ, scarcely coloured by iodine; basidia 36–50–60 × 6–9–14μ, with 2–4-sterigmata, 6–7μ long. Paraphyses or sterile basidia branched, pointed, or capped by 1–2-globules. Hyphae very much branched, ·75–1·5μ in diam." Bourd. & Galz. Maple, and sycamore. Dec. Rare.

Corticium (Pers.).

(*Cortex*, bark.)

Receptacle waxy, crustaceous, or floccose, resupinate, effused. Hymenium smooth, or tubercular, waxy, *continuous,* often cracked. Spores white, very rarely coloured, ovate, elliptical, globose, oboval, pip-shaped, pyriform, boat-shaped, almond-shaped, subtriangular, cylindrical, cylindric ellipsoid, oblong, or sausage-shaped; smooth, rarely granular; basidia with 2–4–6–8-sterigmata, *forming a homogeneous hymenium,* sometimes accompanied with sterile basidia (cystidioles). Cystidia none. Growing on wood, more rarely on leaves, or on the ground.

I. Hymenium homogeneous, regular, consisting only of basidia. Hyphae distinct, or indistinct, with or without clamp connections at the septa, but never having abnormally large clamp connections, or becoming tuberosely swollen at the septa.

1. Receptacle membranaceous; trama fibrillose, or tomentose, hyphae always distinct; hymenium fairly thick.

2282. **C. caeruleum** (Schrad.) Fr. Sow. Eng. Fung. t. 350, as *Auricularia phosphorea* Sow. Trans. Brit. Myc. Soc. IV, t. 3, fig. 26. *Caeruleum*, dark blue.

R. 2–15 cm., *beautiful azure blue*, somewhat round, broadly effused, adnate, at first tomentose; margin *whitish, or azure blue*, byssoid. Hymenium *concolorous, paler when dry*, papillose. Flesh *bluish under the hymenium*, waxy, floccose, loose. Spores white, ovate-elliptical, 7–9 × 4–6μ; "basidia 30–48 × 6–7·5μ, with 2–4-sterigmata. Hyphae thin-, or slightly thick-walled, 3–4·5μ in diam., with clamp connections, loose, blue in the subhymenial layer" Bourd. & Galz. Dead wood, branches, sticks, and twigs, in woods, and hedgerows. Jan.—Dec. Common. (*v.v.*)

2283. **C. laeve** (Pers.) Quél. (= *Corticium evolvens* Fr.) Fr. Icon. t. 198, fig. 1, as *Corticium radiosum* Fr. Trans. Brit. Myc. Soc. IV, t. 3, figs. 23–24. *Laeve*, smooth.

R. 2–20 cm., *white, cream colour, flesh colour, or tan*, effused, entirely adnate, *or forming distinct, reflexed, strigose pileoli*; margin *white*, silky, radiating, or becoming obtuse, reflexed. Hymenium *cream colour, then pinkish ochre, or livid to brownish* when old; *pale buff with a pinkish tinge or lilac tinge when dry*, smooth, waxy, more or less undulate, sometimes coarsely tuberculate, and rarely *Radulum-like*, usually much cracked in an areolate manner when dry. Flesh *pale*, floccose, loose. Spores white, pyriform, or pip-shaped, usually slightly incurved at the base, 9–12 × 6–7·5μ, often slightly punctate; "basidia 25–40–90 × 5·5–9μ, with 2–4-sterigmata. Hyphae thin walled, hyaline, 2–3–6μ in diam., with clamp connections, parallel at the base, then ascending in a loose trama" Bourd. & Galz. Trunks, logs, and fallen branches. Jan.—Dec. Common. (*v.v.*)

2284. **C. roseum** (Pers.) Fr. (= *Corticium roseolum* Massee sec. Wakef. in litt.) *Roseum*, rose-coloured.

R. 2–12 cm., *rose colour*, effused, adnate; margin *white*, byssoid, fringed. Hymenium *rose pink, becoming pallid*, or *pale ochraceous with a pink tinge when dry, pruinose*, at length rimosely cracked. Flesh pale, floccose, loose. Spores "white, *sometimes tinged rosy*, oboval, 8–12–16·5 × 6–9–10μ; basidia at first bladder-shaped, sunk in the simple, or branched paraphysoid hyphae, then normal, 28–45 × 6–10μ, with 2–4 curved sterigmata, 6–8μ long. Hyphae with slightly thickened walls, 2–4·5μ in diam." Bourd. & Galz. Dead wood, and fallen branches. Oct.—April. Not uncommon. (*v.v.*)

2285. **C. bombycinum** (Sommerf.) Bres. Trans. Brit. Myc. Soc. VI, text-figs. p. 139. *Bombycinum*, silky.

R. 2–5 cm., *white, then cream colour, or pale alutaceous*, effused, separable, smooth, or slightly rough; margin pubescent, floccose, rarely fibrillose. Hymenium *concolorous*, smooth, or rough, pulverulent under a lens. Flesh membranaceous, fairly thick, floccose. Spores white, broadly elliptical, or ovate, somewhat irregular, 9–12 × 6–8μ, 1-guttulate; basidia cylindrical, 21–34–45 × 4–6–9μ, with 2–4-sterigmata, 6–8μ long. Hyphae rather thick walled, 4–6μ in diam., branched, frequently septate, with clamp connections. Trunks of living pollarded willows. Oct.—Jan. Uncommon. (*v.v.*)

2286. **C. vellereum** Ellis & Cragin. (= *Corticium chlamydosporium* Burt; *Corticium Bresadolae* Bourd.) *Vellereum*, woolly.

R. 2–10 cm., *white, cream, or tinged buff pink*, widely effused, adnate; margin *white*, silky, radiating. Hymenium *waxy-white, cream, or tinged buff pink*, not changing when dried, smooth, pulverulent. Flesh *concolorous, or pale*, waxy, floccose, loose, thick. Spores white, subglobose, or broadly elliptical, apiculate at the base, 5–6 × 5μ, or 5–9 × 5–7·5μ, 1-guttulate; basidia 18–30–54 × 5–7·5μ, with 2–4 curved sterigmata, 3–5μ long. Basal hyphae very loosely interwoven, sparingly branched, thin walled, 2–7μ in diam., with clamp connections. On bark, and felled elm trunks. Nov.—Feb. Not uncommon. (*v.v.*)

2. Receptacle fleshy-membranaceous, then rigid and fragile, thick, often reflexed, and *Stereum*-like. Hymenium tuberculose and radially crested.

2287. **C. subcostatum** (Karst.) Bourd. & Galz. (= *Stereum subcostatum* Karst.; *Radulum pendulum* Fr. sec. Bourd. & Galz.) Quél. Ass. Fr. (1882), t. XI, fig. 16, as *Stereum album* Quél. *Sub*, somewhat; *costatum*, ribbed.

R. 3–12 cm., *cream, or cream chamois colour*, resupinate, or reflexed, villose, or strigose on the outside; margin torn, fibrillose, or ciliate. Hymenium concolorous, becoming chamois, pinkish, or reddish when dried, tubercular in the centre, *radially rugose, and wrinkled towards the margin*, finally deeply cracked. Flesh *white*, fleshy membranaceous, then firm, fibrillose, brittle. Spores white, "oblong subcylindric, slightly depressed on the side, 5–6–8·5 × 2·75–4μ, contents homogeneous; basidia 12–25–45 × 3–4–7μ, with 2–4 straight sterigmata, 4–4·5μ long. Hyphae thin walled, 2–4μ in diam., with clamp connections" Bourd. & Galz. Dead branches of alder, birch, and pine. Oct.—Jan. Uncommon. (*v.v.*)

3. Receptacle waxy-membranaceous. Hyphae thin walled, distinct, soon agglutinated and collapsed.

2288. **C. lacteum** Fr. *Lacteum*, milk white.

R. 5–15 cm., *milk white*, effused, pruinose, *laxly fibrillose beneath*; margin fibrillose. Hymenium *deeper coloured*, waxy, often rugulose, or reticulately veined like a *Merulius* when moist. Flesh *whitish*, membranaceous, waxy, thin. Spores white, "oboval, 4·5–7 × 2·5–6μ, 1–2-guttulate, or contents granular; basidia 20–36 × 4–6μ, with 2–4 straight sterigmata, 5–6μ long. Hyphae thin walled, subhymenial, 2·5–3μ in diam., the basal 5–8μ in diam., with rather infrequent clamp connections" Bourd. & Galz. Trunks and fallen branches. Oct.—Feb. Uncommon. (*v.v.*)

2289. **C. Wakefieldiae** Bres.
Miss E. M. Wakefield, the well-known Kew mycologist.

R. 2–6 cm., *whitish*, then *isabelline*, broadly effused; margin *pallid*, pruinose, subfimbriate. Hymenium *concolorous*, smooth, at length widely cracked. Flesh membranaceous, soft. Mycelium *white*, pruinose. Spores hyaline, 6–8 × 5–6μ; basidia collapsed forming an indistinct layer. Hyphae very distinct, 4–7–9μ in diam., septate with clamp connections. Ground and wood. Rare.

2290. **C. fuciforme** (Berk.) Wakef. (= *Isaria fuciformis* Berk.; *Hypochnus fuciformis* McAlp.) φῦκος, sea-weed; *forma*, shape.

R. 1–5 cm., *pale, or bright rose colour*, effused, incrusting, forming small patches here and there. Hymenium *concolorous*, smooth. Flesh *concolorous, subgelatinous*, thin. Spores white, pip-shaped, depressed on one side, apiculate, 11–12·5 × 5–6μ; basidia slightly tinged pink, clavate, 5·5–7μ in diam., with 2–4 stout curved sterigmata. Hyphae *tinged pink*, with clamp connections, thin walled and rather vacuolate, 2–4μ in diam. Leaves and stalks of grasses. Aug. Uncommon. (*v.v.*)

4. Receptacle pelliculose, or arachnoid, slightly adnate to the substratum. Trama loose, consisting of thin walled, distinct hyphae, 2–6μ in diam. Basidia with 2-4-sterigmata.

2291. **C. Galzinii** Bourd. A. Galzin.

R. 3–10 cm., *whitish, with a more or less glaucous, or yellowish green tint*, effused, forming a delicate pellicle, loosely adherent to the substratum, smooth, or porous; margin byssoid. Hymenium *concolorous*, loose, and pulverulent under a lens. Flesh *whitish*, membranaceous, arachnoid. Spores white, obovate, or narrowly cylindrical, pointed at the base, 2–4 × 1–1·5μ; basidia borne in dense tufts, "candelabra" fashion, 7–9–14 × 3–4μ, with 2–4 straight sterigmata, 3–4μ long. Basal hyphae thin walled, 2–5μ in diam., loosely interwoven, septate, with clamp connections. Conifer wood, cones, and on birch. Sept.—March. Uncommon.

2292. **C. arachnoideum** Berk. (= *Corticium centrifugum* (Lév.) Bres.) *ἀράχνη*, a spider's web; *εἶδος*, like.

R. 2–18 cm., *white, greenish white, or greyish,* effused, subadnate, arachnoid, more rarely continuous; margin delicately byssoid or arachnoid. Hymenium *concolorous,* loose, rarely continuous. Flesh *concolorous,* floccose, very thin. Spores white, oblong, obliquely apiculate at the base, 5–7 × 3–4μ, "often cohering in 2–4; basidia clavate, 9–15–27 × 3–4–7μ, with 2–4-sterigmata, 4–6μ long. Basal hyphae regular, thin- or slightly thick-walled, clamp connections sparse, 3–8μ in diam.; subhymenial hyphae 2–3·5μ in diam." Bourd. & Galz. Stumps, logs, and fallen branches. Oct.—March. Not uncommon. (*v.v.*)

2293. **C. coprophilum** Wakef. *κόπρος*, dung; *φίλος*, loving.

R. 1–2 cm., *greyish white,* effused, arachnoid, easily separable. Hymenium *concolorous,* pulverulent. Flesh *whitish,* filamentous, very thin. Spores white, *subglobose,* apiculate at the base, 4μ, 1-guttulate; basidia with 3–6-sterigmata, 15–25 × 6μ, with 3–6 curved sterigmata, 2–5μ long. Basal hyphae 3·5–4μ in diam., scarcely nodose-septate. *Horse dung* and surrounding grass culms. July—Aug. Uncommon.

2294. **C. microsporum** (Karst.) Bourd. & Galz. *μικρός*, small; *σπορά*, seed.

R. 3–6 cm., *milk white, or cream colour,* irregularly effused, pelliculose; margin *white,* pruinose, or fibrillose. Hymenium *cream, with sometimes a faint pinkish tinge,* often imperfect, or cracked. Flesh *white,* delicate, very thin, fragile. Spores white, subglobose, 2μ, or 3 × 2μ, often with a small oil drop; basidia 12–15(–18) × 3–4(–5)μ, with 2–4 straight sterigmata, 3–4μ long. Basal hyphae 3–5μ, with clamp connections, sometimes incrusted with crystals. Stumps and branches. Sept.—Oct. Uncommon.

2295. **C. croceum** (Kunze) Bres. (= *Sporotrichum croceum* Kunze & Schmidt; *Corticium sulphureum* Fr.) *κρόκος*, saffron.

R. 1–4 cm., *white, becoming yellowish,* effused, arachnoid, then submembranaceous, separable when fresh, adnate when dry; margin *white or lemon yellow,* arachnoid, *running out into the bright yellow or saffron coloured, branched strands of the rhizomorphoid mycelium.* Hymenium *concolorous,* pruinose or mealy. Flesh *concolorous,* arachnoid, thin. Spores white, "subglobose or ovoid, 2·75–3·5 × 2·5–3μ; basidia 12–17 × 3–4·5μ. Hyphae thin walled, 2–3μ in diam., often verrucose or rough with small crystals" Bourd. & Galz. Fallen sticks, etc. Oct.—Jan. "Not uncommon" Berk.

2296. **C. atrovirens** Fr. Trans. Brit. Myc. Soc. III, t. 16.

Ater, black; *virens*, green.

R. 2–6 cm., *blue, greenish blue, or dark greenish*, irregularly effused, floccoso-fibrillose, or arachnoid, seated on a *concolorous*, profuse, mycelium. Hymenium *paler, or tinged with yellow*, submembranaceous. Flesh *concolorous*, arachnoid, thin. Spores *greenish, or bluish*, subglobose, 3·5–4μ in diam.; basidia clavate, 18–20 × 5–6μ, with 2–4-sterigmata, 3–4μ long. Hyphae *greenish blue*, 2–4μ in diam., thin walled, without clamp connections. Bark and fallen branches. Sept.—Dec. Uncommon. (*v.v.*)

5. Receptacle dry, subpelliculose, crustaceous, or pruinose, adnate. Basidia truncate at the apex, with 4–6, or 6–8-sterigmata.

2297. **C. niveo-cremeum** von Hoehn. & Litsch. Trans. Brit. Myc. Soc. VI, text-figs. p. 71.

Niveus, snow white; *cremeum*, cream colour.

R. 2–5 cm., *greyish white to cream colour*, effused, indeterminate, closely adnate, dry. Hymenium *concolorous*, waxy, slightly granular in places, *very minutely and abundantly cracked when dry*, giving a characteristic appearance under a lens, the cracks being bridged by numerous, fine, byssoid strands of the subiculum. Flesh *concolorous*, byssoid, very thin. Spores white, cylindric-ellipsoid, or slightly incurved and boat-shaped, 6–7 × 3–4μ, occasional spores up to 10 × 5μ; basidia 12–18(–30) × 4·5–7μ, truncate above, with 4–6–8 straight sterigmata, 4–5μ long. Basal hyphae thin walled, 4–5μ in diam., indistinct, branched, septate, with clamp connections. Rotten wood. Nov.—Jan. Uncommon. (*v.v.*)

6. Receptacle dry, chalky, or pubescent, adnate. Trama distinct. Basidia with 2–4-sterigmata.

2298. **C. Sambuci** (Pers.) Fr. (= *Corticium serum* (Pers.) Quél.; *Peniophora Chrysanthemi* Plowr. sec. Wakef. in litt.) Grev. Scot. Crypt. Fl. t. 242, as *Thelephora Sambuci* Pers.

Sambuci, of elder.

R. 2–18 cm., *pure snow white, or chalk white, becoming yellowish when dried*, effused, subinnate, *incrusting, chalky*, collapsing and more or less powdery when dry. Hymenium *concolorous*, granular, pruinose. Flesh *white*, crustaceous, very thin. Spores white, broadly elliptical, appearing almost globose under a low magnification, with a small lateral apiculus, 3–6 × 3–5μ, often 1-guttulate; basidia 15–22 × 3·5–5·5μ, accompanied by fusoid sterile basidia (cystidioles) often expanded into a knob at the apex, and incrusted with tiny crystals. Hyphae thin walled, 2–3·5(–4·5)μ in diam., loosely interwoven, with clamp connections, and sometimes with scattered minute crystals adhering to the outer walls. Stumps, rotten branches, logs and old herbaceous stems. Jan.—Dec. Common. (*v.v.*)

2299. **C. trigonospermum** Bres. Trans. Brit. Myc. Soc. IV, t. 3, figs. 3–5. τρίγωνος, triangular; σπέρμα, seed.

R. 2–5 cm., *chalk white, or becoming slightly tinged with cream colour*, irregularly effused, chalky, pulverulent; margin arachnoid, fugacious. Hymenium *concolorous*, granular, or mealy under a lens. Flesh *white*, crustaceous, thin, loose, fragile. Spores white, subtriangular, angles rounded viewed laterally, in profile more or less elliptical, flattened on the inner side, and swollen towards the base on the outer side, 4·5–6μ; basidia 16–25 × 4–6μ, with 2–4 straight sterigmata, 2–3·5μ long. Basal hyphae thin walled, 2·5–4μ in diam., with clamp connections, and sometimes slightly incrusted with minute crystals. Pine bark and on the ground. Sept.—Nov. Rare.

7. Hymenium consisting of more or less crowded, granular tufts, seated on an arachnoid subiculum. Mould-like rather than pelliculose, or submembranaceous. Hyphae yellowish, 6–15μ in diam., branching at right angles. Basidia large, in clusters.

*Hyphae without clamp connections.

2300. **C. vagum** Berk. & Curt. (= *Corticium vagum* Berk. & Curt. var. *Solani* Burt; *Hypochnus Solani* Prill. & Del.; *Corticium Solani* Prill. & Del.; *Corticium botryosum* Bres.; *Rhizoctonia Solani* Kühn.) *Vagum*, wandering.

R. 5–15 cm., *pale olive buff to cream colour*, effused, arachnoid, thin, perforate membrane more or less separable. Hymenium *concolorous*, smooth. Flesh *brownish, or hyaline*, arachnoid, filamentous, loose. Spores white, "elliptic oblong, or navicular, flattened on one side, 8–14 × 4–6μ; basidia not forming a compact hymenium, 10–20 × 7·5–11μ, with 4–6-sterigmata, 6–10μ long, more or less swollen towards the basidium. Basal hyphae slightly brownish, hyaline elsewhere, 6–10μ in diam., branches smaller, not incrusted, septate, without clamp connections" Burt. Bark, wood, herbaceous plants, and bare soil. Jan.—Dec. Not uncommon. (*v.v.*)

C. botryosum Bres. = **Corticium vagum** Berk. & Curt.

2301. **C. flavescens** (Bon.) Massee. (= *Hypochnus flavescens* Bon.) Trans. Brit. Myc. Soc. VI, text-figs. p. 318. *Flavescens*, becoming yellow.

R. 3–10 cm., *whitish to dirty buff*, irregularly effused, thin, pulverulent, with the habit of *C. vagum*. Hymenium *concolorous*, loose. Flesh *hyaline, or yellowish*, filamentous, loose. Spores yellowish, *somewhat lemon-shaped*, apiculate at either end, flattened on the inner side, 15–17 × 7–9μ (mostly 15 × 8μ). Basidia oblong, or clavate, 20–30 × 12–13μ, with 2–4 curved sterigmata, 8μ long. Basal hyphae hyaline, or yellowish, septate, *without clamp connections*, branched at right angles, loosely interwoven. Rotten wood. Feb. Uncommon. (*v.v.*)

**Hyphae with stout clamp connections at the septa.

2302. **C. subcoronatum** von Hoehn. & Litsch.

Sub, somewhat; *coronatum*, crowned.

R. 3–10 cm., *white, then cream colour, ochraceous, or pale greenish, tinged with brown when bruised,* effused, arachnoid, or slightly membranaceous, loosely adnate; margin similar, or minutely reticulated. Hymenium *concolorous,* loose. Flesh *pale, or yellowish,* arachnoid, filamentous, loose. Spores white, "almond-shaped, or subnavicular, rarely fusiform, 5–9 × 2·5–4·5μ; basidia 12–18–30 × 5–9μ, with 4–6-sterigmata, 3–5μ long. Hyphae thin walled, 4–14μ in diam., with numerous, stout clamp connections" Bourd. & Galz. Rotten wood, and fallen branches. Sept.—March. Not uncommon. (*v.v.*)

8. Receptacle waxy, dry. Trama indistinct. Spores clavate, fusiform, almond-shaped, or boat-shaped. Generally growing on dead, herbaceous plants.

2303. **C. aurora** Berk. *Aurora*, the dawn.

R. 3–5 cm., *rose colour, becoming pallid,* effused, adnate, indeterminate, waxy, then subpruinose. Hymenium *concolorous,* waxy, smooth. Flesh spongy, very thin. Spores white, "subclavate, attenuated at the base, generally slightly curved, 12–16 × 3–4·5μ, 2–3-guttulate; basidia 24–36 × 12–16μ, with 4 straight sterigmata, 4μ long" Bourd. & Galz. Dead leaves of *Carex*, and stems of *Juncus*. Feb. Rare.

2304. **C. Pearsonii** Bourd. Trans. Brit. Myc. Soc. VII, text-fig. I. p. 51. A. A. Pearson, the well-known British mycologist.

R. 2–10 cm., *greyish,* adnate, hiding in the crevices of rotten wood. Hymenium *concolorous,* soon furfuraceous and granular, *always beautifully reticulated with white crustaceous lines when dry,* consisting of basidia and equally long sterile hyphae. Flesh very thin, 20–50μ thick. Spores hyaline, narrowly clavate, laterally depressed, or subarcuate, 4·5–6 × 1·5–2(–2·5)μ; basidia obovate, 9–15 × 5–6μ, with 2–4-sterigmata, up to 6μ long and at length curved. Hyphae hyaline, closely interwoven, rarely distinct, 2–2·5μ in diam., thin walled, clamp connections sparse. Cracks of a rotten pine trunk. Sept.—Oct. Rare.

9. Receptacle waxy, delicate, closely adnate.

2305. **C. confluens** Fr. (? = *Radulum molare* Fr. sec. Bourd. & Galz.)

Confluens, running together.

R. 1–8 cm., *whitish,* effused, indeterminate, *agglutinated*; margin *white,* mealy. Hymenium *hyaline, white when dry,* smooth. Flesh *whitish,* submembranaceous, thin, loose. Spores white, broadly elliptical, or subglobose, 8–10 × 8–9μ, with a large central gutta, or contents granular or cloudy; "basidia 20–50–80 × 6–12μ, with 2–4-

sterigmata, 5–9μ long. Hyphae thin walled, 2–3·5μ in diam., with scattered clamp connections; superior hyphae dense, flexuose, coherent and collapsing" Bourd. & Galz. Stumps, and fallen branches. Jan.—Dec. Common. (*v.v.*)

2306. **C. lividum** (Pers.) Fr. (= *Grandinia ocellata* Fr. sec. Bres.) *Lividum*, black and blue.

R. 2–13 cm., *bluish grey, hyaline grey, then tinged reddish,* or bluish, widely effused, *agglutinated*; margin similar, or white fimbriate and fugacious. Hymenium *concolorous, subviscid when moist,* pruinose, smooth, tubercular, or radiately wrinkled. Flesh *paler, subgelatinous, then horny, dense.* Spores white, oblong elliptical, 3·5–5 × 2–3μ, or "elongate oblong, depressed on one side, 4–5 × 1·5–1·75μ; basidia 15–25–34 × 3–4·5μ, with 2–4 straight sterigmata, 3μ long. Basal hyphae thick walled, gelatinous, 3–5μ in diam., clamp connections rare; superior hyphae 2–3μ in diam., rarely distinct" Bourd. & Galz. Dead birch, and elm. Oct.—June. Uncommon. (*v.v.*)

2307. **C. seriale** Fr. *Seriale*, in series.

R. 5–10 cm., *pale tan, isabelline, or greenish, becoming ochraceous, tawny, brick red, chocolate, greenish cinereous, or bluish vinous, more rarely livid brown when dry,* longitudinally effused, agglutinated, often in series, waxy, rarely shining when dry; margin *white*, narrow, pubescent. Hymenium *concolorous,* smooth, or papillose, pruinose, *very much cracked when dry.* Flesh waxy, rigid, dense. Spores white, "narrowly oblong, depressed on the side, 4–7 × 2·5–3μ; basidia 15–21 × 3–4·5μ, without cystidioles, or 12–27–40 × 3–4·5μ, with 2–4 straight sterigmata, 5–7μ long, and accompanied with numerous fusiform, or subulate cystidioles, 3–4·5μ in diam., and projecting 10–35μ. Hyphae more or less agglutinated, with walls thin or slightly thickened, 2–5μ in diam., with clamp connections" Bourd. & Galz. Pine wood. Rare.

2308. **C. ochraceum** (Fr.) Bres. Bres. Fung. Trid. t. 170, fig. 1. ὠχρός, pale.

R. 3–10 cm., *pale, or cream colour, then ochraceous,* broadly effused, very adnate, waxy; margin *white, pruinose,* soon similar. Hymenium *concolorous,* papillose or tubercular, *very much cracked when dry.* Flesh *white,* waxy, then firm, agglutinated, thick. Spores white, elliptical, apiculate at the base, 5–6 × 3–4μ; "basidia 30–45 × 4–7μ, with 2–4 straight sterigmata, 3–4μ long. Hyphae with thin or slightly thickened walls, 3–3·5μ in diam., agglutinated, distinct only at the base" Bourd. & Galz. Conifer trunks, and logs. Sept.—Oct. Uncommon. (*v.v.*)

II. Hymenium homogeneous, regular, consisting of basidia only. Hyphae occasionally septate, with either normal clamp connections, or with clamp connections two to three times larger than the normal, and *tuberosely swollen at the septa*. Growing in humus and on very decayed wood and rubbish.

2309. **C. confine** Bourd. & Galz. Trans. Brit. Myc. Soc. IV, t. 3, figs. 12–14. *Confine*, nearly related.

R. 3–10 cm., *snow white, becoming yellowish, superficially like Grandinia farinacea*, widely effused, arachnoid; margin *white*, byssoid, somewhat radiating, gradually attenuated. Hymenium *white, becoming cream colour, or ochraceous, granular*, like a *Grandinia*, granules waxy, crowded when fresh, shrinking away from one another, and revealing the white subiculum when dry. Flesh *white*, fibrillose, loose. Spores white, subglobose, pointed at the base, 3–4 × 2–3μ, usually 1-guttulate; basidia 9–15 × 3–5μ, with 2–4 straight, or slightly curved sterigmata, 2–4μ long. Basal hyphae 2–4μ in diam., with clamp connections, and often swollen at the septa. Mycelium often forming fine branching cord-like strands beneath the bark. Rotten wood, bark, and twigs. Jan.—Dec. Not uncommon. (*v.v.*)

III. Hymenium heterogeneous, irregular, consisting of basidia originating at the base of the trama, and surrounded by sterile, undifferentiated, mycelial branches.

2310. **C. comedens** (Nees) Fr. (= *Vuilleminia comedens* (Nees) R. Maire; *Radulum botrytes* Fr. sec. Quél.; ? *Corticium Carlylei* Massee sec. Wakef. in litt.) Trans. Brit. Myc. Soc. IV, t. 3, fig. 25. *Comedens*, eating away.

R. 1–13 cm., *flesh colour, or dingy lilac, becoming pale, erumpent*, effused, innate, growing under the bark, inseparable, slightly viscid when moist. Hymenium *concolorous*, smooth, even. Flesh *paler, subgelatinous*, then rigid, firm. Spores white, sausage-shaped, curved, 15–22 × 6–7μ, 2–4-guttulate. "Basidia scattered, very long, 9–12μ in diam., with 2–4 curved sterigmata, 8–10 × 3μ" Bourd. & Galz. Dead branches, and felled trunks, especially oak. Jan.—Dec. Common. (*v.v.*)

IV. Doubtful British species insufficiently described.

2311. **C. nigrescens** (Schrad.) Fr. (? = *Radulum aterrimum* Fr. sec. Quél.; *Corticium comedens* (Nees) Fr. discoloured sec. Wakef. in litt.) *Nigrescens*, becoming black.

R. 2–7·5 cm., *yellowish*, erumpent, effused, interrupted, agglutinated, inseparable; margin indeterminate. Hymenium *yellowish, becoming blackish*, spuriously papillose, waxy, pruinose. Flesh waxy, very thin. Spores white, "cylindric-oblong, obtuse at both ends, curved, 18–20 × 5–6μ" Massee. Dead oak and beech branches, growing beneath the bark. Rare.

2312. **C. populinum** (Sommerf.) Fr.
Populinum, pertaining to poplars.

R. 1–3 cm., *cinereous ferruginous*, effused, tubercular, soon confluent, *at length involute*, marginate, white tomentose beneath. Hymenium *ferruginous*, uneven. Flesh soft, thin. Spores "white, subglobose, 7–8μ" Massee. Poplars. Rare.

2313. **C. foetidum** B. & Br. Massee, Linn. Soc. Bot. Jour. XXVII, t. 6, fig. 3. *Foetidum*, stinking.

R. 6–30 mm., *white, then ochraceous*, effused, crustaceous, arachnoid beneath. Hymenium *concolorous*, smooth. Flesh crustaceous, thin. Spores white, "elliptical, 7 × 4μ" Massee. Smell *very foetid* when fresh. Sawdust. Rare.

2314. **C. flaveolum** Massee. *Flaveolum*, yellowish.

R. 5–7·5 cm., *clear pale primrose yellow*, effused, loosely attached to the matrix; margin determinate. Hymenium *concolorous*, smooth. Flesh membranaceous, thin. Spores white, cylindric-ellipsoid, obtuse at both ends, 7 × 5μ. Trunk of tree fern in a conservatory. Rare.

2315. **C. anthochroum** (Pers.) Fr. (= *Hypochnus anthochrous* (Pers.) Quél.) *ἄνθος*, a flower; *χρώς*, colour.

R. 3–15 cm., *bright rose colour, or brick red, becoming pale*, broadly effused; margin *white*, byssoid, pruinose. Hymenium *concolorous*, waxy, sometimes cracked when dry, usually sterile and minutely velvety. Flesh membranaceous, very thin. Spores white, "elliptical, 11–13 × 8–9μ" Massee, "ovoid, globose, 5μ, with a large central gutta" Quél. Sycamore and birch sticks. Feb. Rare.

2316. **C. molle** Fr. *Molle*, soft.

R. 2–9 cm., *pale, or flesh colour, more or less spotted with red*, effused, subrotund, easily separable, villose underneath; margin naked. Hymenium *concolorous*, waxy, *papillose*, cracked when dry. Flesh membranaceous, floccose, loose, soft, thick. Spores white, "cylindric ellipsoid, obtuse at both ends, 7 × 5μ" Massee. Pine trunks and bark. Rare.

2317. **C. strigosum** (Pers.) W. G. Sm. var. *filamentosum* W. G. Sm. (= *Peniophora byssoidea* (Pers.) von Hoehn. & Litsch. sec. Wakef. in litt.) *Strigosum*, lean.

R. web-like, filamentous, string-like, *dull yellowish*, externally pulverulent. *Amaryllis.*

C. echinosporum Ellis = **Hypochnus echinosporus** (Ellis) Burt.
C. sulphureum (Pers.) Bres. = **Hypochnus fumosus** Fr.
C. amorphum (Pers.) Fr. = **Aleurodiscus amorphus** (Pers.) Rabenh.
C. evolvens Fr. = **Corticium laeve** (Pers.) Quél.
C. Typhae (Pers.) Fr. = **Epithele Typhae** (Pers.) Pat.

C. fastidiosum (Fr.) Bourd. & Galz. = **Cristella cristata** (Pers.) Pat.
C. salicinum Fr. = **Cytidia rutilans** (Pers.) Quél.
C. citrinum (Pers.) Fr. = **Corticium (Gloeocystidium) radiosum** (Fr.) Rea.
C. lacunosum B. & Br. = **Peniophora byssoidea** (Pers.) von Hoehn. & Litsch.
C. flocculentum Fr. = **Cytidia flocculenta** (Fr.) von Hoehn. & Litsch.
C. scutellare Berk. & Curt. "The British specimen so named by Berk. is different from the type" Wakef. in litt.
C. roseolum Massee = **Corticium roseum** (Pers.) Fr. sec. Wakef. in litt.
C. punctulatum Cke. = **Corticium (Gloeocystidium) albostramineum** (Bres.) Bourd. & Galz.
C. subalutaceum Karst. = **Peniophora subalutacea** (Karst.) von Hoehn. & Litsch.
C. (Coniophora) byssoideum (Pers.) Fr. = **Peniophora byssoidea** (Pers.) von Hoehn. & Litsch.
C. sanguineum Fr. = **Peniophora sanguinea** (Fr.) Bres.
C. velutinum (DC.) Fr. = **Peniophora velutina** (DC.) Cke.
C. puberum Fr. = **Peniophora pubera** (Fr.) Sacc.
C. Roumeguèrii Bres. = **Peniophora Molleriana** (Bres.) Sacc.
C. giganteum Fr. = **Peniophora gigantea** (Fr.) Massee.
C. incarnatum (Pers.) Fr. = **Peniophora incarnata** (Pers.) Cke.
C. nudum Fr. = **Peniophora nuda** (Fr.) Bres.
C. maculaeforme Fr. = **Peniophora nuda** (Fr.) Bres. var. **maculaeformis** (Fr.) von Hoehn. & Litsch.
C. violaceo-lividum (Sommerf.) Fr. = **Peniophora violaceo-livida** (Sommerf.) Bres. ex Bourd. & Galz.
C. Lycii (Pers.) Cke. = **Peniophora caesia** (Bres.) Bourd. & Galz.
C. cinereum Fr. = **Peniophora cinerea** (Fr.) Cke.
C. laevigatum Fr. = **Peniophora laevigata** (Fr.) Massee.
C. quercinum (Pers.) Fr. = **Peniophora quercina** (Pers.) Cke.
C. limitatum Fr. = **Peniophora limitata** (Fr.) Cke.
C. subdealbatum B. & Br. = **Coniophora subdealbata** (B. & Br.) Massee.
C. Carlylei Massee = ? **C. comedens** (Nees) Fr. sec. Wakef. in litt.
C. sphaerosporum (R. Maire) von Hoehn. & Litsch. = **Hypochnus sphaerosporus** R. Maire.
C. submutabile von Hoehn. & Litsch. = **Hypochnus submutabilis** (von Hoehn. & Litsch.) Rea.

Subgen. **Gloeocystidium** Karst.

(γλοιός, sticky; κύστις, bladder.)

Differs from *Corticium* in possessing gloeocystidia, generally immersed in the tissue, which resemble cystidia, but their walls are never thickened, nor incrusted with crystalline deposits.

*Spores turning blue with iodine.

2318. **C. (Gloeo.) porosum** Berk. & Curt. (= *Gloeocystidium stramineum* Bres.) πόρος, a pore.

R. 1–8 cm., *white, then cream colour, or straw colour*, effused, adnate; margin *white*, narrow, pruinose, or reticulately porous. Hymenium *concolorous*, smooth. Flesh *concolorous*, *subgelatinous*, firm, *dense*. Spores white, elliptical, 4–7 × 2–4μ, generally 2-guttulate, the membrane turning deep violet blue with iodine; basidia 12–18–28 × 3–6μ, with 2–4-sterigmata, 3–4μ long. Gloeocystidia abundant, tapering to an obtuse, narrow apex, 15–150 × 6–14μ, sometimes bifurcate, contents granular, yellowish, then resinous. Hyphae coherent, 1·5–3μ in diam. Fallen branches, and decorticated wood. Jan.—Dec. Not uncommon. (*v.v.*)

**Spores not turning blue with iodine.

2319. **C. (Gloeo.) polygonium** (Pers.) Fr. Trans. Brit. Myc. Soc. IV, t. 3, figs. 21–22. πολυγώνιον, with many angles.

R. 3–80 mm., *flesh colour, or lilac, erumpent* in small cushions, then confluent, and effused, very adnate, pruinose; margin *white, or flesh colour*, narrow, pruinose. Hymenium *concolorous, often reddish when dried, pruinose*, soft. Flesh *pale*, subgrumous, waxy, then hard, and firm. Spores white, cylindrical, slightly curved, with a lateral apiculus, 8–13 × 3–4μ; basidia 45–55 × 6–8μ. Gloeocystidia included, forming balloon-like, pear-shaped, or subglobose vesicular swellings, 20–30μ in diam. Basal hyphae hyaline, rather thick walled, 3–6μ in diam., with clamp connections. Dead branches, especially poplar. Jan.—Dec. Not uncommon. (*v.v.*)

2320. **C. (Gloeo.) roseo-cremeum** Bres.

Roseus, rose colour; *cremeum*, cream colour.

R. 3–5 cm., *pallid pink, or dull reddish when bruised*, effused, waxy; margin *white*, pruinose, or pubescent. Hymenium *concolorous*, smooth, or minutely porous, *minutely atomate under a lens*. Flesh *pale*, waxy membranaceous, soft. Spores white, cylindrical, straight, the inner side flattened, 8–11 × 3–4μ; basidia 22–45 × 4–7μ, with 2–4-sterigmata, 4–7μ long. Paraphyses long, slender, blunt at the apex, 2–3μ in diam. Gloeocystidia, when present, completely immersed in the tissue, cylindrical, wavy, 30–90 × 5–9μ, contents pale yellowish. Basal hyphae thin walled, 2·5–7μ in diam., with occasional clamp connections. Rotten wood. Oct.—Dec. Uncommon. (*v.v.*)

2321. **C. (Gloeo.) praetermissum** (Karst.) Bres. (= *Peniophora praetermissa* Karst.; *Corticium tenue* Pat.) *Praetermissum*, passed over.

R. 1–6 cm., *pure white, then yellowish, or greenish*, widely effused, adnate, smooth; margin very thin, indeterminate, somewhat porous under a lens. Hymenium *concolorous, becoming cream colour with age, or when dried.* Flesh *pale*, waxy, soft, loose. Spores white, *elliptical* to cylindric ellipsoid, slightly curved, or flattened on the one side, 7–12 × 3·5–6·5μ; basidia 18–38 × 6–11μ, with 2–4 rather straight sterigmata, 4·5 × 1μ. Gloeocystidia cylindrical, subfusiform, or ventricose, 21–150 × 4·5–21μ, contents hyaline, or pale yellowish. Basal hyphae loosely interwoven, much branched, 2·5–7μ in diam., with clamp connections. Bark, and fallen branches. Jan.—Dec. Common. (*v.v.*)

2322. **C. (Gloeo.) lactescens** Berk. Trans. Brit. Myc. Soc. IV, t. 3, figs. 6–8. *Lactescens*, turning to milk.

R. 1–20 cm., *whitish, or flesh colour*, widely effused, agglutinated, adnate; margin *white*, narrow, byssoid, pubescent. Hymenium *white, then cream, tan, flesh colour, or greenish, and finally* brownish pink, smooth, pruinose, cracked when dry. Flesh *pale*, waxy, fibrillose, rather thick, *giving out a watery, milk white juice when wounded.* Spores white, broadly elliptical, obtuse at both ends, with a lateral apiculus, 5–9 × 4–6μ, contents densely granular; basidia 20–40 × 5–8μ. Gloeocystidia cylindrical, sometimes swollen at the base, 80–600 × 4–9μ, contents oily, and granular. Basal hyphae coherent, 1μ in diam., other hyphae 1–3μ in diam. Smell like that of *Lactarius quietus*. Dead oak, ash, and willow trunks, and branches. Sept.—Feb. Not uncommon. (*v.v.*)

2323. **C. (Gloeo.) radiosum** (Fr.) Rea. (= *Gloeocystidium alutaceum* (Schrad.) Bourd. & Galz.; *Corticium citrinum* (Pers.) Fr. sec. Bres.) *Radiosum*, radiant.

R. 3–10 cm., *milk white, becoming yellowish*, widely effused, closely adnate, waxy; margin *white, broad*, fibrillose, *silky*, radiating. Hymenium *milk white, or bright yellow when fresh, becoming cream, tan, or dingy ochraceous, very smooth.* Flesh *white*, waxy, fibrillose, thin. Spores white, "subglobose, shortly apiculate at the base, 4–7 × 4–6μ, sometimes rough; basidia 35–60 × 5–9μ, with 2–4 straight sterigmata, 4–6μ long. Gloeocystidia very thin-walled and hyaline, oboval, fusiform, or prolonged into a neck, often constricted in the middle, 60–150 × 8–27μ, contents hyaline, not granular. Hyphae thin walled, 2–3μ in diam., soon collapsing" Bourd. & Galz. Rotten wood. Oct.—Feb. Uncommon.

2324. **C. (Gloeo.) albostramineum** (Bres.) Bourd. & Galz. (= *Hypochnus albostramineus* Bres.; *Corticium punctulatum* Cke. sec. Wakef. in litt.) Trans. Brit. Myc. Soc. IV, t. 3, figs. 9–11. *Albus*, white; *stramineum*, straw colour.

R. 3–6 cm., *whitish, then deep cream or pale straw colour*, widely effused, separable; margin similar, indeterminate, subreticulate, or fibrillose, thin. Hymenium *concolorous*, rather loose, pulverulent under a lens. Flesh *whitish*, floccose, loose, rather thick. Spores white, broadly elliptical, or subglobose, 7–9 × 6–8μ, contents granular, thick walled ("finely granular, or rough, becoming smooth" Bourd. & Galz.); basidia 25–35–60 × 5–9μ, with 2–4 slightly curved sterigmata, 6–12μ long. Gloeocystidia erect, cylindrical, elongate, 45–120 × 6–9μ, thin walled, contents staining rather deeply. Basal hyphae interwoven, 5–6μ in diam., much branched, with numerous clamp connections. Bark, fallen branches, especially pine. Sept.—April. Not uncommon. (*v.v.*)

2325. **C.** (**Gloeo.**) **coroniferum** von Hoehn. & Litsch. Trans. Brit. Myc. Soc. VI, text-figs. p. 140. *Corona*, a crown; *fero*, I bear.

R. 3–5 cm., *pure white, then cream*, effused, easily separable as a delicate pellicle; margin indeterminate, gradually thinning out to a cobweb-like film. Hymenium *concolorous*, pulverulent. Flesh very thin, fragile. Spores white, narrowly elliptical, with an oblique basal apiculus, 4·5–6–8 × 2–3–5μ; basidia cylindric-clavate, wavy, 3·5–4μ in diam., when mature elongated and projecting from the hymenium, apex truncate, sterigmata 4–8, in British specimens usually 4, rather long. Gloeocystidia rare, sometimes wanting, cylindrical, obtuse, very thin walled, 45–50 × 5–6μ, contents more or less yellowish. Basal hyphae frequently septate, with clamp connections, 4–5μ in diam. Bark, and rotten wood, often spreading on to the surrounding soil. Sept.—Jan. Uncommon.

Gloeocystidium croceo-tingens Wakef. sec. Bres. in Ann. Mycol. XVIII (1920), 48 = **Sebacina** (**Bourdotia**) **Eyrei** Wakef.

Cristella Pat. (=**Thelephora** (Ehrh.) Fr. p.p.)

(*Cristella*, a little crest.)

Receptacle waxy, firm, effused, incrusting. Hymenium smooth, or tubercular. Spores white, ovoid, or oboval, echinulate; basidia clavate, with 2–4-sterigmata. Cystidia none. Growing on the ground, on wood, mosses, or dead herbaceous stems.

2326. **C. cristata** (Pers.) Pat. (= *Thelephora fastidiosa* (Pers.) Fr.; *Corticium fastidiosum* (Fr.) Bourd. & Galz.) Pat. Essai tax. des Hymén. fig. 28. *Cristata*, crested.

R. 5–30 cm., *white, chalky in appearance, then becoming yellowish*, widely effused, incrusting, shapeless, or forming irregular, flattened, confluent, lobed, or subulate branches, fringed, or laciniate at the apex. Hymenium *concolorous*, papillose, granular, or reticulately veined. Flesh *white*, fibrillosely floccose, thin. Spores white, echinulate, ovoid, or obovate, 5–9 × 3–5μ; basidia clavate, 20–25 × 5–6μ, with 2–4 slightly bent sterigmata, 4–6μ long. Hyphae very thin walled, 1·5–

4·5μ in diam., sparingly septate, with clamp connections, and sometimes swollen up to 6–12μ in diam., often incrusted with crystals. Smell unpleasant, or slight, of garlic when quite fresh. On the ground, and running over sticks, dead leaves, twigs and herbaceous stems. Jan.—Dec. Uncommon. (*v.v.*)

Peniophora Cke.

(πηνίον, a shuttle; φέρω, I bear.)

Receptacle waxy, coriaceous, cartilaginous, membranaceous, submembranaceous, floccose, or filamentous; resupinate, effused. Hymenium waxy, floccose, or pulverulent; smooth, rarely tubercular. Spores white, rarely pink, or yellowish, elliptical, subelliptical, globose, subglobose, oboval, clavate, subcylindrical, fusiform, oblong, needle-shaped, or sausage-shaped; smooth; basidia with 2–4-sterigmata, sometimes accompanied by cystidioles. Cystidia hyaline, rarely coloured, fusiform, oboval, elliptical, subglobose, subulate, conical, acicular, filiform, cylindrical, clavate or capitate, sometimes septate, and with clamp connections, smooth, or incrusted with crystalline granules, generally thick walled, sometimes thin walled and then projecting, not immersed in the tissue. Growing on wood, more rarely on leaves, or on the ground.

1. Cystidia cylindrical, or conical, thick walled, not incrusted externally with crystalline deposits, often divided at the base into several roots. In *Peniophora Aegerita* and its allies the walls of the cystidia are more or less rugose, the central canal is narrow and not enlarged at the apex, and the trama is poor or indistinct. In *Peniophora glebulosa* and its allies the cystidia are very thick walled, vitreous, with a capillary canal always more or less abruptly dilated at the apex and with thinner walls.

2327. **P. Aegerita** von Hoehn. & Litsch. Beit. zur Kennt. der Cort. in Sitzungsber. der k. Akad. d. Wissensch. Wien, Math.-Nat. Kl. Bd. cxvi (1907), 813, and reprint 75, text-fig. 7.
Aegerita, a genus of fungi, with which this species is always associated.

R. 1–5 cm., *white to alutaceous*, effused, adnate. Hymenium *concolorous, finely bristling with the cystidia* under a lens, and porous. Flesh *white*, submembranaceous, very thin. Spores white, broadly elliptical, or subglobose, 6–9 × 5–6μ, 1-guttulate; basidia clavate, 24–30 × 7–8μ, with 2–4 straight sterigmata, 4–6μ long. Cystidia arising from the basal hyphae, cylindrical, or slightly swollen below, apex blunt, 42–100 × 6–12μ, thick walled, rugose throughout their length. Hyphae thin walled, 3–4·5μ in diam., soon collapsing. Rotten sticks, and fallen branches, generally in association with *Aegerita candida* Pers. Sept.—April. Uncommon. (*v.v.*)

2328. **P. glebulosa** (Fr.) Bres. (= *Thelephora calcea* Fr. var. *glebulosa* Fr.) Bres. Fung. Trid. II, t. 170, fig. 2.

Glebulosa, full of little clods.

R. 2–5 cm., *cream colour, dirty white, or greyish*, effused, closely adnate; margin mealy, or similar. Hymenium *concolorous*, pubescent with the cystidia, *cracked into small irregular areas* when dry. Flesh *pale*, membranaceous, floccose, rather thick, dense. Spores white, narrowly cylindrical, curved, 7–9·5 × 1·5–2μ; basidia 5–15 × 3–4μ, with 4-sterigmata, about 4μ long. Cystidia cylindrical to conical, 70–160 × 6–12μ, obtuse, or pointed, sometimes subventricose at the base, often forked below, some sunken, and some projecting 90μ above the hymenium, springing from the basal hyphae, thick walled, smooth, or slightly incrusted in the upper portion. Hyphae thin walled, 1–3μ in diam., very closely interwoven and scarcely distinct. Wood, and bark. Jan.—Dec. Not uncommon. (*v.v.*)

var. **subulata** Bourd. & Galz. Trans. Brit. Myc. Soc. VI, text-figs. p. 72. *Subulata*, awl-shaped.

Differs from the type in the *hymenium not cracking into small irregular areas when dry, and in the more acute, or subulate cystidia*. On wood. Nov.—Jan. Uncommon. (*v.v.*)

2329. **P. accedens** Bourd. & Galz. Trans. Brit. Myc. Soc. VI, text-figs. p. 140. *Accedens*, approaching.

R. 1–2 cm., *whitish, or greyish*, irregularly effused, spot-like, filmy. Hymenium *concolorous*, becoming cracked when dry, setulose, glistening with the cystidia under a lens. Flesh very thin, scarcely perceptible. Spores white, elliptical, with a lateral apiculus, 4–5 × 3–3·5μ, often 1-guttulate; basidia 9–15 × 4–4·5μ, with 2–4-sterigmata, 3–3·5μ long. Cystidia filiform, or linear, 50–60 × 4μ, *dilated at the apex into a globose head*, 10–11μ in diam., thick walled. Hyphae indistinct, 1·5–2μ in diam. Rotten wood. Nov.—Dec. Rare. (*v.v.*)

2330. **P. subalutacea** (Karst.) von Hoehn. & Litsch. (= *Corticium subalutaceum* Karst.) *Sub*, somewhat; *alutacea*, tanned leather.

R. 2–5 cm., *dirty white, or greyish, with a slight ochraceous tinge*, widely effused, adnate; margin very narrow, pruinose, or similar. Hymenium *concolorous*, loose, rather rough under a lens, with slight, irregularly scattered thickenings of the tissue (hardly granules). Spores white, narrowly cylindrical, slightly curved, 5–9 × 1·5–2·5μ; basidia 10–24 × 3–5μ, with 2–4 straight sterigmata, 4–5μ long. Cystidia cylindrical, slightly attenuated at the base, 95–150 × 5–7μ, projecting 60μ or more above the hymenium, smooth, thin walled and thinner at the rounded apex, often 1–2-septate. Basal hyphae much branched, wavy, rather rigid, thick walled, 2–3μ in diam., with clamp connections. Wood, and fallen branches of conifers. Sept.—Oct. Uncommon.

2. Trama always distinct. Cystidia long, arising from the basal hyphae and more or less similar, narrowly clavate, fusiform, or swollen into a ball at the apex or at the septa, often septate with, or without, clamp connections. The membranes are readily stained by a weak alkaline solution of eosin. Spores subglobose, oboval, or fusiform.

2331. **P. pallidula** Bres. ex Bourd. & Galz. (= *Gonatobotrys pallidula* Bres.) Beit. zur Kennt. der Cort. in Sitzungsber. der k. Akad. d. Wissensch. Wien, Math.-Nat. Kl. Bd. CXVI (1907), 827, and reprint 89, text-fig. 12, as *Gloeocystidium oleosum* von Hoehn. & Litsch. *Pallidula*, palish.

R. 1–6 cm., *pallid, yellowish cream colour, or clay*, regularly effused, or interrupted, *Hypochnus*-like; margin similar, rarely pruinose. Hymenium *concolorous*, pubescent, often granular, unequal. Flesh *pale*, filamentous, very thin. Spores white, oval, or subglobose, apiculate at the base, 4–6 × 3–4μ, often with a large central gutta; basidia 12–21 × 4μ, with 2–4-sterigmata, 3–4μ long. Cystidia cylindrical, 40–120 × 4–6μ, 1–4-septate, often constricted at the septa, or swollen, often incrusted. Hyphae thin walled, 2–4μ in diam., with scattered clamp connections. Rotten wood, dead branches, and fallen leaves. Oct.—March. Common. (*v.v.*)

2332. **P. detritica** Bourd. Trans. Brit. Myc. Soc. VI, text-figs. p. 319. *Detritica*, worn down.

R. 1–2 cm., *pure white*, effused, with scattered granules suggesting a *Grandinia*. Hymenium *concolorous*, not continuous, appearing farinaceous under a lens. Flesh very thin, floccose, membranaceous. Spores white, broadly elliptical, or obovate, 5–6 × 4μ, 1-guttulate; basidia 12–15–24 × 4–4·5μ, with 2–4-sterigmata, 3–4μ long. Cystidia cylindrical, or narrowly club-shaped, apex obtuse, 70–90 × 5–6μ, smooth, thin walled. Hyphae 2–4μ in diam., thin walled, septate, with clamp connections. Rotten wood. Feb. Rare.

2333. **P. sphaerospora** von Hoehn. & Litsch. Beit. zur Kennt. der Cort. in Sitzungsber. der k. Akad. d. Wissensch. Wien, Math.-Nat. Kl. Bd. CXV (1906), 1600, and reprint 52, text-fig. 5. σφαῖρα, a ball; σπορά, seed.

R. 1–5 cm., *chalk white*, broadly effused, firmly attached to the substratum, ·15–·30 mm. thick. Hymenium *concolorous*, smooth, or papillate, waxy when fresh, not cracked when dry; margin indeterminate. Flesh *concolorous*, membranaceous, thin. Spores hyaline, globose, apiculate, 4–7μ, smooth, 1-guttulate; basidia clavate, 25–35 × 6–8μ, with 4 long, subulate sterigmata. Cystidia abundant, cylindrical, apex usually narrowed, 35–85 × 5–8μ, thin walled, projecting 10–40μ above the hymenium. Hyphae 4–5μ in diam., smooth, thin walled, subnodulose, often anastomosing, with frequent clamp connections. Naked ground and fallen stick, probably alder. Nov. Rare.

2334. **P. byssoidea** (Pers.) von Hoehn. & Litsch. (= *Corticium* (*Coniophora*) *byssoideum* (Pers.) Fr.; *Coniophorella byssoidea* (Pers.) Bres.; *Corticium lacunosum* B. & Br.; *Peniophora tomentella* Bres.) βύσσος, fine flax; εἶδος, like.

R. 1–6 cm., *ochrey white, drying cream colour, to "Naples yellow,"* widely effused, dry, arachnoid, separable; margin *whitish*, byssoid. Hymenium *concolorous*, even, tomentose, or pulverulent. Flesh *yellowish*, floccose, loose. Spores *yellowish*, sometimes nearly hyaline under the microscope, broadly elliptical, or pip-shaped, 4–4·5 × 2·5–3·5μ; basidia 12–25 × 4·5–5μ, with 2–4-sterigmata, 2·5–3μ long. Cystidia *yellowish*, cylindrical, or narrowly fusiform, tapering, sharp pointed, 60–90 × 3–6μ, projecting 20–75μ above the hymenium, thin walled, or slightly thickened, 1–4-septate, generally with clamp connections. Hyphae yellowish, thin walled, 2·5–4μ in diam., very loosely interwoven, with clamp connections. Rotten wood, fallen twigs, and leaves, especially in conifer woods. Sept.—April. Not uncommon. (*v.v.*)

2335. **P. longispora** (Pat.) von Hoehn. & Litsch. (= *Hypochnus longisporus* Pat.) *Longus*, long; σπορά, seed.

R. 2–5 cm., *whitish yellow*, widely effused, thin, pubescent, then consisting of a membrane incompletely felted, scarcely adnate. Spores white, "fusiform, or acicular, straight, or slightly flexuose, 12–18 × 1–3μ, multi-guttulate; basidia 12–24 × 4–5μ. Cystidia needle-shaped, sometimes bulbous at the base, 60–75 × 2·5–6μ, fairly thick walled, rough with crystals, projecting 30–45μ. Hyphae rigid, 2·5–4μ in diam., walls slightly thickened, often verrucose, with clamp connections" Bourd. & Galz. Rotten, moist wood in cool places. Spring —Winter. Rare.

3. Receptacle pelliculose, or membranaceous; subiculum soft, more or less thick, fibrillose, and forming long, branched, rhizomorphoid strands.

2336. **P. sanguinea** (Fr.) Bres. (= *Corticium sanguineum* Fr.) Fr. Icon. t. 198, fig. 2, as *Corticium sanguineum* Fr. and Trans. Brit. Myc. Soc. IV, t. 3, figs. 18–20. *Sanguinea*, blood coloured.

R. 2–30 cm., *blood red*, effused, loosely adnate, arachnoid beneath; margin *blood red*, byssoid, or fibrillose, running out, and connected with the spreading strands of the blood red, rhizomorphoid mycelium. Hymenium *creamy white, or tinged with pink, rarely red*, smooth, becoming slightly cracked when dry. Flesh *concolorous*, membranaceous, floccose, loose, containing a red juice. Spores white, subelliptical, often with a curved apiculus, 5–6 × 2–4μ; basidia 16–40 × 4–7μ. Cystidia sparse, cylindrical-fusiform, pointed, 40–60 × 4–7μ, thin walled, smooth, rarely slightly incrusted. Basal hyphae with slightly thickened walls, 3–9μ in diam., with rather few clamp connections; subhymenial hyphae 3–4μ in diam. Dead wood, and fallen branches, especially of conifers. Jan.—Dec. Not uncommon. (*v.v.*)

2337. **P. leprosa** Bourd. & Galz. Trans. Brit. Myc. Soc. VI, text-figs. p. 318. *Leprosa*, rough.

R. 1–3 cm., *white, then ochraceous*, irregularly effused, crustaceous; margin *white*, indeterminate, occasionally prolonged into white rhizomorphic strands. Hymenium *pinkish ochraceous*, somewhat cracked when dry, rough with cystidia under a lens. Flesh somewhat thick, fragile. Spores white, elliptical, 4–6 × 2·5–3μ. Basidia inconspicuous, about 4μ in diam. Cystidia *very rough*, cylindrical to subfusiform, frequently occurring in clusters, so as to give an *Odontia*-like appearance, occasionally branched near the apex, 60–90 × 8–14μ. Basal hyphae 3–4(–7)μ, often strongly incrusted with crystals, clamp connections rare. Dead bark. April. Rare.

4. Receptacle membranaceous, fairly thick, easily separable when fresh. Cystidia often little differentiated from the cystidioles (sterile basidia) of *Corticium*, or scattered and unequally distributed.

2338. **P. cremea** Bres. Bres. Fung. Trid. II, t. 73, fig. 2, as *Corticium* (*Peniophora*) *cremeum* Bres. *Cremea*, cream colour.

R. 2–6 cm., *white, cream to ochraceous*, broadly effused, separable; margin *white*, arachnoid, then similar. Hymenium *concolorous*, smooth, or here and there slightly tubercular, velvety, widely cracked when dry. Flesh *white*, membranaceous, soft, thin. Spores white, oblong, or cylindric ellipsoid, slightly curved, 4–8 × 2–4μ; basidia 20–50 or more × 7μ, with 2–4-sterigmata, 3–4μ long. Cystidia cylindrical, or slightly elongated fusiform, tapering very gradually from the base to the blunt apex, 70–120 × 5–9μ, very thin walled, smooth, or slightly incrusted at the apex with easily detached crystals, usually projecting up to 60μ; sometimes thicker walled, embedded cystidia are present, shorter than the projecting ones, fusiform, much incrusted, 40–60 × 9–10μ. Subhymenial hyphae loosely interwoven, much branched, rather rigid, thick walled, constricted at the septa so as to appear somewhat jointed, with no clamp connections, 5–6μ in diam. Bark, and fallen branches. Jan.—Dec. Common. (*v.v.*)

var. **Allescheri** (Bres.) Wakef. Bres. Fung. Trid. II, t. 72, as *Corticium* (*Peniophora*) *Allescheri* Bres. A. Allescher.

Differs from the type in its *thicker subiculum, more sharply differentiated from the hymenium, and containing numerous, short, rough, thick walled cystidia*. Bark. Jan.—Dec. Not uncommon. (*v.v.*)

2339. **P. laevis** (Fr.) Burt. *Laevis*, smooth.

R. 1–3 cm., *white, then cream coloured*, broadly effused, not closely adnate; margin radiately fibrillose. Hymenium *concolorous*, more or less cracked when dry. Flesh *concolorous*, membranaceous, thin. Spores hyaline, elliptical oblong, 4·5–6 × 2·5–3·5μ, 1-guttulate; basidia very variable, 20–36 × 3–6μ (most frequently 35 × 4·5μ), with 2–4-sterigmata, 4–6μ long. Cystidia fusoid, 40–90 × 4–7μ, without incrustation, × 6–11μ with incrustation, walls thin or slightly thickened. Hyphae regular with few, or no, clamp connections, thin walled; subhymenial hyphae 3–4μ in diam.; basal hyphae up to 7–8μ in diam. Birch bark. Nov. Rare.

2340. **P. velutina** (DC.) Cke. (= *Corticium velutinum* (DC.) Fr.; *Peniophora scotica* Massee.) Grevillea, VIII, t. 125, no. 15. *Velutina*, velvety.

R. 3–15 cm., *white, or whitish*, broadly effused, adnate; margin *white, or flesh colour*, running out into long, branching strands. Hymenium *concolorous, becoming flesh colour, or reddish when dried, minutely velvety*. Flesh *concolorous*, soft, loose, fairly thick. Spores white, oblong, elliptical, apiculate at the base, 4–8 × 2·5–5μ; basidia "20–32–50 × 4–7μ. Cystidia fusiform, 30–140 × 6–9μ, generally thick walled, smooth, or incrusted with oxalate crystals (18μ in diam.), immersed in the tissue, or projecting. Basal hyphae more or less thick walled, 4–10μ in diam., with few clamp connections; subhymenial hyphae thin walled, 3–4μ in diam., soon collapsing" Bourd. & Galz. Wood, and fallen branches. Jan.—Dec. Common. (*v.v.*)

2341. **P. setigera** (Fr.) Bres. (= *Kneiffia setigera* Fr.) Pat. Essai tax. des Hymén. fig. 45, as *Corticium setigerum* (Fr.) Karst. *Setigera*, having bristles.

R. 2–10 cm., *white, yellowish when dry*, broadly effused, or indeterminate, closely adnate, incrusting. Hymenium *concolorous*, papillose, beset with scattered, or fasciculate hyaline bristles, often very much cracked. Flesh *concolorous*, floccose, loose, thin. Spores white, "subcylindrical, slightly curved, 8–11–16 × 3–4–6μ, contents granular, or 1-multi-guttulate; basidia 21–45 × 4–8μ, with 7–8-sterigmata. Cystidia cylindrical, 75–250 × 7–15μ, septate, with, or without, clamp connections, often incrusted with crystalline granules, immersed, or projecting. Basal hyphae distinct, thin walled, 2–4–8μ in diam., with clamp connections, medial and subhymenial hyphae soon collapsing, 2–3μ in diam." Bourd. & Galz. Dead wood, and fallen branches. Jan.—Dec. Common. (*v.v.*)

5. Receptacle waxy, very adnate, pubescent, hispid, or guttulate under a lens. Cystidia abundant, strongly incrusted with crystalline granules.

2342. **P. pubera** (Fr.) Sacc. (= *Corticium puberum* Fr.) Bres. Fung. Trid. II, t. 145, fig. 1, as *Corticium puberum* Fr. *Pubera*, grown up.

R. 2–8 cm., *white, becoming dirty yellowish,* broadly effused, closely adnate, indeterminate; margin mealy, soon similar. Hymenium *concolorous,* smooth, at first velvety, then setulose, finally widely cracked. Flesh *concolorous,* waxy, thin. Spores white, subcylindrical, depressed on one side, 7–9 × 4–5μ; basidia "18–25–60 × 4–6μ. Cystidia fusiform or elongate conical, pointed, 30–90–150 × 6–12–35μ, thick walled, with separable incrustations. Basal hyphae sparse, thick walled, 4–6μ in diam.; medial and upper hyphae vertical, thin walled, 2–4μ in diam., little distinct, with rare clamp connections" Bourd. & Galz. Dead wood. Uncommon. (*v.v.*)

2343. **P. Molleriana** (Bres.) Sacc. (= *Corticium Roumeguèrii* Bres.) Bres. Fung. Trid. II, t. 144, fig. 1, as *Corticium Roumeguèrii* Bres. A. F. Moller.

R. 1–6 cm., *cream to bright biscuit colour,* broadly effused, closely adnate; margin *white,* abrupt, or narrow, and pruinose. Hymenium *concolorous,* smooth, dry, opaque, almost farinaceous, cracked when dry. Flesh *whitish,* waxy, soft, then rigid, *porcelain-like,* brittle. Spores white, elliptical, 3–5·5 × 2–2·5μ; basidia clavate, 12–20–30 × 4–5μ, with 2–4 straight sterigmata, 2–4 × ·5μ. Cystidia very abundant, mostly immersed in the tissue, fusiform, or conical, apex acute, 60–70 × 10μ, thick walled, much incrusted in the upper portion. Hyphae closely agglutinated, scarcely distinct, 2–2·5μ in diam. Fallen logs. May. Rare. (*v.v.*)

2344. **P. hydnoides** Cke. & Massee. (= *Peniophora crystallina* von Hoehn. & Litsch.; *Odontia conspersa* Bres.; *Peniophora rimosa* Cke. and *Peniophora terrestris* Massee sec. Wakef. in litt.) Massee, Linn. Soc. Bot. Jour. XXV, t. 47, figs. 15–16. ὕδνον, the genus *Hydnum*; εἶδος, like.

R. 5–13 cm., *whitish, or greyish,* broadly effused, subinnate, indeterminate, closely adnate, waxy, hyaline. Hymenium *grey, often becoming somewhat cream coloured, setulose,* and finally cracked. Flesh thin, filamentous. Spores white, broadly elliptical, or subcylindrical, more or less depressed on one side, 4–5 × 1·5–2μ. Basidia clavate, 8–14 × 3–4μ, with 4 straight sterigmata, 4–5μ long. Cystidia hyaline, subconical, or fusiform, *aggregated in clusters,* 60–120 × 10–12μ, thick walled, strongly incrusted. Basal hyphae indistinct. Bark, and fallen branches. Sept.—June. Common. (*v.v.*)

2345. **P. gigantea** (Fr.) Massee. (= *Corticium giganteum* Fr.; *Peniophora Crosslandii* Massee sec. Wakef. in litt.) Fr. Icon. t. 197, fig. 3, as *Corticium giganteum* Fr. γίγας, a giant.

R. 3–30 cm., *hyaline white,* very broadly effused, swelling when moist, *cartilaginous when dry*; margin *white, fibrillose, radiating,* finally becoming free. Hymenium *concolorous, often tinged brownish or lilac when old*, smooth, minutely velvety. Flesh *whitish*, waxy, then horny and *parchment-like*, thick, tough. Spores white, oblong, subcylindrical, attenuated at the base, 5–8 × 2·5–4μ, "basidia 12–18–30 × 4–5μ. Cystidia fusiform, subulate, often contracted, 40–100 × 9–16μ, very thick walled, apex often incrusted. Hyphae very thick walled, 4–7μ in diam., with few clamp connections, subhymenial hyphae thin walled, 2·5–3μ in diam." Bourd. & Galz. Stumps, fallen branches, and needles of pines, rarely of other conifers. Jan.—Dec. Common. (*v.v.*)

6. Receptacle at first waxy, becoming hard and rigid, closely adnate, sometimes contracting when dry and becoming free at the margin, or splitting and becoming inrolled along the cracks; varying in colour from orange or brick red, to cinereous grey or brownish bistre, passing through rose colour, purple, violaceous livid, etc. The cystidia often commence as gloeocystidia with granular contents which concentrate in vitreous or amber coloured masses, more or less rugose and split up, along the inside of the walls and incrust either the whole of the cystidium, or else only the upper portion,—in the latter case the cystidium appears as if stipitate: the membrane of the cystidium is often torn and reabsorbed: other cystidia oboval in the basal hyphae, narrowly fusiform in the trama, have thick, smooth walls from the commencement.

2346. **P. aurantiaca** (Bres.) Bourd. & Galz. Bres. Fung. Trid. t. 144, fig. 2, as *Corticium aurantiacum* Bres. *Aurantiaca*, golden.

R. 1–4 cm., *orange, or vermilion, becoming paler or somewhat tan colour*, effused, forming small round patches, then confluent; margin *white*, broad, *radiating*. Hymenium *concolorous*, smooth, pruinose. Flesh *pale*, waxy, firm, dense. Spores white, or slightly tinged with pink in the mass, broadly elliptical, 14–18 × 9–11μ; basidia 55–90 × 12–15μ. Cystidia fusiform, 30–85 × 7–10μ, thick walled, incrusted. Hyphae irregular, dense, thin walled, 3–6μ. Fallen branches, especially alder. Sept.—Feb. Not uncommon. (*v.v.*)

2347. **P. incarnata** (Pers.) Cke. (= *Corticium incarnatum* (Pers.) Fr.) *Incarnata*, flesh colour.

R. 3–13 cm., *reddish, or orange*, effused, agglutinated, adnate; margin *white*, narrow, byssoid, radiating, fugacious, often wanting. Hymenium *concolorous*, pruinose, sometimes undulato-papillose, and becoming cracked. Flesh *slightly coloured*, waxy, then rigid, firm. Spores white, subcylindrical, laterally depressed, 7–12 × 4–5μ, 3–4-guttulate; basidia 20–40 × 5–7μ. Cystidia fusiform, or cylindrical, 25–60 × 6–15μ, thick walled, incrusted. Hyphae thin walled, 3–5μ in diam., basal hyphae coloured. Dead wood, and branches. Jan.—Dec. Common. (*v.v.*)

var. **hydnoidea** (Pers.) Bourd. & Galz. (= *Radulum laetum* Fr.) *ὕδνον*, the genus *Hydnum*; *εἶδος*, like.

Differs from the type in its *Radulum-like hymenium, and subcorticolous habit.* Dead wood, and branches of hornbeam, more rarely on alder. Oct.—March. Not uncommon. (*v.v.*)

2348. **P. nuda** (Fr.) Bres. (= *Corticium nudum* Fr.; ? *Peniophora ochracea* (Fr.) Mass. sec. Wakef. in litt.) *Nuda*, naked.

R. ·5-6 cm., *hyaline livid, then rose colour or pale lilac*, effused, confluent, adnate; margin similar, or narrow, pruinose. Hymenium *concolorous, becoming paler*, pruinose, cracked when dry. Flesh *brownish*, waxy, then rigid, firm. Spores white, "cylindrical, incurved, 7–12 × 3–5μ; basidia 15–27 × 4–7μ. Cystidia hyaline, the basal ones oboval, or elliptical, 15–45 × (6–)15–19μ, the others more elongate, 45–50 × 6–8μ, thin walled, contents granular, incrusting the walls, rugose, cracked. Hyphae little distinct, 3–5μ in diam." Bourd. & Galz. Dead wood, and branches. Nov.—April. Not uncommon. (*v.v.*)

var. **maculaeformis** (Fr.) von Hoehn. & Litsch. (= *Corticium maculaeforme* Fr.) *Macula*, a spot; *forma*, shape.

Differs from the type in *commencing as very small spots*, 1–2 *mm. broad, which become confluent and effused, and in the lilac pruina on the reddish violet hymenium.* Dead wood. Uncommon.

2349. **P. violaceo-livida** (Sommerf.) Bres. ex Bourd. & Galz. (= *Corticium violaceo-lividum* (Sommerf.) Fr.) *Violaceus*, violet; *lividum*, black and blue.

R. 2–12 cm., *violaceous livid, then cinereous lilac*, resupinate, round, tubercular, closely adnate. Hymenium *concolorous, becoming paler*, minutely pruinose, then cracked. Flesh *discoloured*, waxy, then rigid, *fairly thick.* Spores white, "cylindrical, slightly incurved, 9–12 × 3–4·5μ; basidia 20–26 × 6–8μ. Cystidia ovoid, or broadly fusiform, 24–45 × 12–21μ, thin walled, contents vitrified. Hyphae little distinct, 2–4μ in diam." Bourd. & Galz. Dead wood, and plum trees. Jan.—April. Uncommon. (*v.v.*)

2350. **P. caesia** (Bres.) Bourd. & Galz. (=*Corticium Lycii* (Pers.) Cke.) Bres. Fung. Trid. II, t. 145, fig. 2, as *Corticium caesium* Bres. *Caesia*, bluish grey.

R. 1–4 cm., *bluish grey, greyish lilac, bluish cinereous, becoming paler or hoary*, broadly effused, closely adnate; margin similar. Hymenium *concolorous*, delicately pruinose, at length cracked. Flesh *paler*, subgrumous, *thin.* Spores white, sausage-shaped, or cylindrical and incurved, 8–11 × 3–4·5μ; "basidia 25–32 × 4–6μ. Cystidia basal ones obovate, or globose, 5–18–32 × 3–14–24μ, hyaline, soon vitrified; the others cylindrical, or fusiform, and produced into a neck. Hyphae indistinct" Bourd. & Galz. Dead wood, branches, and on *Lycium* and *Syringa vulgaris*. Nov.—Feb. Uncommon. (*v.v.*)

2351. **P. cinerea** (Fr.) Cke. (= *Corticium cinereum* Fr.) Grevillea, VIII, t. 123, no. 8. *Cinerea*, colour of ashes.

R. 2–15 cm., *lurid, cinereous grey*, effused, confluent, agglutinated, closely adnate; margin similar. Hymenium *concolorous*, minutely pruinose. Flesh *brownish*, waxy, then rigid, firm, compact. Spores white, cylindrical, incurved, 6–10 × 3–4μ; "basidia 21–40 × 3–6·5μ. Cystidia the inferior ones *brownish*, oboval, clavate, or subfusiform, 20–35–80 × 4·5–6–14μ, soon vitrified, central cavity tubular; the upper ones basidia-like. Hyphae rarely distinct, 3μ in diam." Bourd. & Galz. Dead wood, bark, and branches. Jan.—Dec. Common. (*v.v.*)

2352. **P. laevigata** (Fr.) Massee. (= *Corticium laevigatum* Fr.) *Laevigata*, made smooth.

R. 1–3 cm., *ferruginous cinnamon, becoming paler*, effused, very adnate, indeterminate; margin *at length free*. Hymenium *concolorous*, pruinose, finally cracked. Flesh *paler*, firm, thin. Spores white, cylindrical, depressed on the side, 7–9 × 4–5μ; basidia 25–30 × 4–5μ. Cystidia *brownish*, or *yellowish*, fusiform, pointed, or obtuse, 30–50 × 6–9μ, thick walled, slightly incrusted. Hyphae sparse, 2–6μ in diam. Living yew trees, and junipers. Sept.—April. Uncommon. (*v.v.*)

2353. **P. quercina** (Pers.) Cke. (= *Corticium quercinum* (Pers.) Fr.; ? *Peniophora pezizoides* Mass. sec. Wakef. in litt.) Grev. Scot. Crypt. Fl. t. 142, as *Thelephora quercina* Pers. *Quercina*, pertaining to oak.

R. 1–18 cm., *flesh colour, or orange, then lilac, or greyish,* and *finally slate colour*, effused, then cup-shaped and free, *smooth and becoming black beneath*; margin *free, revolute*. Flesh *pale*, or *brownish*, coriaceous, thick, firm. Spores white, sausage-shaped, or cylindrical, often bent, 10–12 × 3–4μ; basidia 30–40 × 5–7μ. Cystidia clavate, or fusiform, 50–70 × 5–12μ, thick walled, smooth, or rugose. Basal hyphae brown, walls more or less thickened, 3–4μ in diam. Fallen branches, especially oak. Jan.—Dec. Common. (*v.v.*)

2354. **P. limitata** (Fr.) Cke. (= *Corticium limitatum* Fr.) Grevillea, VIII, t. 123, no. 7. *Limitata*, marked-off.

R. 1–8 cm., *lurid, becoming pale*, subrotund, tubercular, soon confluent, closely adnate; margin *black*. Hymenium *ochraceous*, minutely velvety, often finally cracked. Flesh grumous, then cartilaginous, rather thick. Spores white, "elliptic-oblong, with a minute basal apiculus, slightly curved, 20–22 × 6μ. Cystidia fusoid, 30–40 × 15–20μ, above the level of the hymenium" Massee. Bark, wood and broom. Rare.

7. Doubtful British species insufficiently described.

2355. **P. phyllophila** Massee. φύλλον, a leaf; φίλος, loving.

R. 2–11 cm., *pallid, or cream colour*, broadly effused; margin fibrillose, often indeterminate. Hymenium *concolorous*, continuous. Flesh membranaceous. Spores white, elliptical, 12 × 6μ. Cystidia fusoid or cylindrical with the apex sometimes thickened, 60–80 × 20–30μ above the level of the hymenium. Dead leaves. Rare.

2356. **P. ochracea** (Fr.) Massee. (=?*Peniophora nuda* (Fr.) Bres. sec. Wakef. in litt.) ὠχρός, pale.

R. 3–13 cm., *ochraceous*, broadly effused, inseparable; margin *white*, byssoid, radiating, soon disappearing. Hymenium *concolorous, sprinkled with golden-glistening atoms*, cracked when dry. Flesh soft, waxy. Spores white, "elliptical, 10 × 5μ. Cystidia fusiform, 40–50 × 20μ, above the level of the hymenium" Massee. Wood and bark. Rare.

P. scotica Massee = **Peniophora velutina** (DC.) Cke. sec. Wakef.

P. rimosa Cke. = **Peniophora hydnoides** Cke. & Massee sec. Wakef.

P. terrestris Massee = **Peniophora hydnoides** Cke. & Massee sec. Wakef.

P. Crosslandii Massee = **Peniophora gigantea** (Fr.) Massee sec. Wakef.

P. pezizoides Mass. = **Peniophora quercina** (Pers.) Cke. sec. Wakef.

P. Chrysanthemi Plowr. = **Corticium Sambuci** (Pers.) Fr. sec. Wakef.

7. Cyphellaceae.

Hymenium covering the whole of the interior of cup-shaped, urceolate, or cylindrical receptacles; smooth or veined.

Cytidia Quél.

(= **Auriculariopsis** R. Maire).

(κύτος, a hollow vessel.)

Receptacle coriaceous-gelatinous, cup-shaped, sessile, scattered, crowded, or confluent. Hymenium smooth, becoming wrinkled, or veined. Spores white, or slightly coloured, boat-shaped, globose, or cylindrical, smooth; basidia elongate, narrow, cylindrical with 4 thin, short sterigmata. Growing on wood.

2357. **C. flocculenta** (Fr.) von Hoehn. & Litsch. (= *Corticium flocculentum* Fr.; *Cyphella ampla* (Lév.) Fr.; *Auriculariopsis ampla* (Lév.) R. Maire.) *Flocculenta*, woolly.

R. 4–12 mm., cup-shaped, hood-shaped and inverted when dry, *externally pale in colour*, and *tomentose*. Hymenium *fawn, or bright brown*; margin *white*, at length wrinkled and veined. Flesh *brownish*, gelatinous, thin. Spores white, boat-shaped, 8–10 × 3–4μ; basidia 30–36 × 4–5μ. Hyphae brown, gelatinous, thick walled, 4–6μ in diam. Twigs, and fallen branches. Oct.—March. Uncommon. (*v.v.*)

2358. **C. rutilans** (Pers.) Quél. (= *Corticium salicinum* Fr.) *Rutilans*, being reddish.

R. 1–10 cm., *blood red*, transparent, cup-shaped, then expanded, or confluent, adfixed by the centre, white villose on the outside, and delicately zoned. Hymenium *red blood colour, or orange*, even, naked, zoned. Flesh gelatinous, then horny, thin. Spores "hyaline, or slightly rosy, globose, 8μ" Quél. *Salix aurita*, more rarely on poplar. Aug.—Sept. Rare.

Cyphella Fr.

(*κύφελλα*, the hollow of the ear.)

Receptacle waxy, membranaceous, or subgelatinous, cup-shaped, or urceolate, stipitate, sessile, or pendulous. Hymenium smooth, rugulose, or veined. Spores white, elliptical, obovate, globose, pruniform, subpyriform, ovate, clavate or pip-shaped, smooth; basidia clavate, with 2-4-sterigmata. Cystidia rarely present. Growing on wood, bark, herbaceous stems, and mosses; scattered, or gregarious.

2359. **C. griseo-pallida** Weinm. *Griseus*, grey; *pallida*, pale.

R. 2–4 mm., *pallid grey*, globose, then campanulate, sessile, *floccose externally*. Hymenium *concolorous*, rugose. Flesh *greyish*, soft, thin. Spores "white, oboval, acuminate at the base, 6–7 × 4–4·5μ; basidia 18–30 × 5–7μ. Hyphae thin walled, 3–8μ in diam., without clamp connections" Bourd. & Galz. Stumps, twigs, elm, and elder bark, and mosses. Dec. Uncommon. (*v.v.*)

2360. **C. fulva** Berk. & Rav. *Fulva*, tawny.

R. 2 mm., *brown tawny*, cup-shaped, *mouth deflexed*, sessile, *externally tomentose* with long, brown, aseptate, thick walled, often curved hairs. Hymenium *concolorous*, even. Flesh thin, membranaceous. Spores white, "elliptical, 16–17 × 8μ" Massee. Scattered, or in little clusters. Dead bark. Jan. Rare.

2361. **C. alboviolascens** (A. & S.) Karst. (= *Cyphella Curreyi* B. & Br.) A. & S. Consp. Fung. t. 8, fig. 4, as *Peziza alboviolascens* A. & S. *Albus*, white; *violascens*, becoming violet.

R. 1–5 mm., *white*, cup-shaped, globose, sessile, or subsessile, often proliferous, *densely white villose*, hairs rough. Hymenium *pallid*, or *violaceous*, smooth. Flesh *whitish*, thin, *firm*. Spores white, broadly elliptical, somewhat inequilateral, 14–15 × 10μ; basidia 60–75 × 6–16μ. Wood, bark, and twigs. Sept.—June. Not uncommon. (*v.v.*)

2362. **C. Bloxamii** B. & Phill.
Rev. A. Bloxam, the well-known mycologist.

R. 1–2 mm., *white*, turbinate, *crenato-lobed*, scattered, floccose. Hymenium *becoming light yellow*. Flesh membranaceous. Spores white, "elliptical, 7–8 × 6μ" Massee. Furze. March—April. Rare.

2363. **C. cyclas** Cke. & Phill. κυκλάς, round.

R. 10–12 mm., *whitish*, conchiform, dimidiate, attached on one side, pendulous, clad with flexuose hairs. Hymenium *very pale rose colour*, even. Dead wood. Rare.

2364. **C. stuppea** B. & Br. στύπη, tow.

R. 1 mm., *brownish*, *becoming white*, erumpent, pezizaeform, sessile, *externally coarsely hispid*. Hymenium *fuscous*. Broom. March. Rare.

2365. **C. brunnea** Phill. *Brunnea*, brown.

R. 8 mm. high, 5 mm. across, *dirty brown*, cupulate, mouth oblique, sessile, scattered, or crowded, clothed near the margin with grey pruina; margin incurved, lacerated. Hymenium *discoloured brown*, smooth. Flesh *paler*, *subgelatinous*. Spores white, globose, 5–6μ. Elder bark, and wood. Rare.

2366. **C. cernua** (Schum.) Massee. Schum. Fl. Dan. t. 1970, fig. 3, as *Peziza cernua* Schum. *Cernua*, looking downwards.

R. 5–6 mm. high, *pale primrose yellow*, obliquely campanulate, *contracted into an elongated equal stem*, glabrous. Hymenium *concolorous*. Flesh thin. Spores white, subglobose, with a basal apiculus, 10 × 8–9μ. Elder bark. Rare.

2367. **C. lacera** (Pers.) Fr. A. & S. Consp. Fung. t. 1, fig. 5, as *Peziza membranacea* A. & S. *Lacera*, torn to pieces.

R. 2–6 mm. high, 2–3 mm. broad, *whitish*, *or yellow*, cup-shaped, *stipitate* from the vertex being extended, pendulous, *then torn into many clefts*, *slightly striate above with dense black fibrils*, *becoming cinereous blackish on the outside*, *and down the stem*. Hymenium *whitish*, *then grey*, slightly wrinkled. Flesh membranaceous, thin. Spores white, "subglobose, 7 × 6μ" Massee, "pruniform, 10–12μ" Quél. Dead twigs. Rare.

2368. **C. capula** (Holmsk.) Fr. Holmsk. Nov. Act. Hafn. I, 286, fig. 7, as *Peziza capula* Holmsk. *Capula*, a small bowl with handle.

R. 4–6 mm. high, 5–8 mm. broad, *whitish*, *becoming greyish and finally blackish*, campanulate, transparent; margin *sinuate*. St. 2 mm., *concolorous*, filiform, flexuose, pubescent and white at the base. Hymenium *whitish*, pruinose, even, then wrinkled. Flesh *whitish*, membranaceous, thin. Spores white, broadly elliptical, 6–7 × 5–6μ, with a large central gutta; basidia 20–30 × 5–7μ, with 2–4-sterigmata,

4–4·5μ long. Hyphae thin walled, 2–8μ cohering. Dead herbaceous stems. Sept.—June. Not uncommon. (*v.v.*)

var. **flavescens** Pat. *Flavescens*, becoming yellow.

Differs from the type in its *yellowish, or brownish colour*. Dead herbaceous stem. Sept.—Nov. Not uncommon. (*v.v.*)

2369. **C. Pimii** Phill. Greenwood Pim, an Irish mycologist.

R. 4 mm. high, 2 mm. broad, *white, or very pale yellow*, cup-shaped, erect, or pendent, *pubescent*; margin somewhat incised. St. *concolorous*, rather slender, crooked, enlarged upwards. Hymenium *concolorous*, smooth. Spores white, subpyriform, 7–10 × 4μ; basidia cylindraceo-clavate, with 2–4-sterigmata. Dead herbaceous stems in water. Feb. Rare.

2370. **C. cuticulosa** (Dicks.) Berk. Dicks. Pl. Crypt. Brit. t. 9, fig. 11, as *Peziza cuticulosa* Dicks. *Cuticulosa*, having a skin.

R. 2–4 mm. high, *white*, diaphanous, at first oblong or digitaliform, then cup-shaped, *elongated into a stem*, smooth externally. Spores white, oval, 6–8 × 4–5μ. Dead grass stems. Oct.—Feb. Rare.

2371. **C. pallida** Rabenh. *Pallida*, pale.

R. ·5–2 mm., *pallid*, cup-shaped, orbicular, *sessile*, sometimes proliferous, *at length irregularly lobed*, plane, tomentose, or slightly hispid. Hymenium *pallid ochraceous*, at length wrinkled. Old stems of *Clematis vitalba*. Nov.—April. Rare.

2372. **C. villosa** (Pers.) Karst. Sow. Eng. Fung. t. 389, fig. 1, as *Peziza sessilis* Sow. *Villosa*, hairy.

R. ·5–1 mm., *white*, globose, *sessile*, gregarious, contracted when dry, externally *white villose*; hairs subfusiform, subulate, pointed, 4–12μ in diam., rough. Hymenium *white*, concave, even. Flesh *white*, membranaceous, thin. Spores white, ovoid, narrower at the apex, broadest at the base, 10–15 × 6–10μ; basidia 40–80 × 7–12μ, with 2–4 straight sterigmata. Stems of herbaceous plants and branches. Oct.—June. Not uncommon. (*v.v.*)

var. **stenospora** Bourd. & Galz. στενός, narrow; σπορά, seed.

Differs from the type in the *narrow oblong spores, attenuated a little obliquely at the base*, 8–10 × 3–4μ, *the smaller basidia* 15–18 × 6–8μ, *and the narrower hairs* 3–4μ *in diam*. Dead fronds of *Lastraea Filix-mas* and *Athyrium Filix-foemina*. Oct.—Nov. Uncommon. (*v.v.*)

2373. **C. dochmiospora** B. & Br. δόχμιος, aslant; σπορά, seed.

R. ·5–1 mm., *snow white*, cup-shaped, *sessile*, minutely villose. Spores white, oblique, ovate, rather acute, 14–17μ. Stems of herbaceous plants. Oct. Rare.

2374. **C. Berkeleyi** Massee. (= *Cyphella griseo-pallida* (Weinm.) Berk.) Rev. Miles Joseph Berkeley, the father of English mycology.

R. 1–2 mm., *reddish grey*, globose, then expanding and becoming campanulate, *sessile*, minutely pilose. Hymenium *concolorous*, even. Spores white, elliptical, 7 × 5μ. Dead *Carex paniculata*. Rare.

2375. **C. Goldbachii** Weinm. Cda. in Sturm, Deutschl. Fl. III, t. 63, as *Chaetocypha variabilis* Cda. Carl Ludwig Goldbach.

R. 2–4 mm. high, 2 mm. broad, *white*, cup-shaped, or campanulate, *sessile, pitcher-shaped-concave*, lobed, externally villose. Hymenium *pallid, or cream colour*, even. Spores white, "globose, 7–8μ" Massee, "broadly elliptical, 4 × 2–3μ" Karst. Dead leaves of *Aira caespitosa* and *Carex paniculata*. Feb. Rare.

2376. **C. lactea** Bres. Bres. Fung. Trid. I, t. 67, fig. 2. *Lactea*, milk white.

R. ·5–1 mm. high and wide, *snow white*, cup-shaped, sessile, tomentose on the outside with *shining, white, clavate hairs*, 5–6μ in diam.; margin entire, *ciliate*. Hymenium *becoming cream colour*, even. Flesh *white*, membranaceous, thin. Spores white, *ovate-clavate*, 9–13 × 3·5–5μ, 3–4-guttulate; basidia 36–45 × 7–10μ, with 2–4 straight sterigmata, 5–6μ long. Dead leaves of *Aira caespitosa*. June. Uncommon. (*v.v.*)

2377. **C. muscigena** (Pers.) Fr. Pers. Myc. Eur. I, t. 7, fig. 6, as *Thelephora vulgaris* Pers. *α. candida* Pers. *Muscus*, moss; *genus*, birth.

R. 3–12 mm., *shining white*, dimidiate, spathulate, becoming plane, *sessile, or stipitate, externally minutely tomentose*. St. *concolorous*, filiform. Hymenium *white*, slightly wrinkled. Flesh *white*, membranaceous, soft. Spores white, pip-shaped, or broadly obovate and apiculate at the attenuated base, 9–10 × 6μ, with a large central gutta. *Polytrichum*, and other large mosses. Sept.—March. Not uncommon. (*v.v.*)

2378. **C. catilla** W. G. Sm. Stevenson, British Fung. II, p. 284, fig. 89. *Catillus*, a small bowl.

R. 18 mm., grey, expanded, often imbricate; margin crisped, undulate. Hymenium *grey, veined*. Flesh submembranaceous. Moss, and dead leaves. Nov. Rare.

Solenia (Hoffm.).

(σωλήν, a pipe.)

Receptacle coriaceous, or membranaceous; tubular, cylindrical, cup-shaped, or pyriform, sessile, seated on a superficial, felt-like, then floccose and fugacious mycelium. Hymenium smooth. Spores white, elliptical, cylindrical, globose, or subglobose; basidia clavate, with 2–4-sterigmata. Growing on wood, gregarious, or fasciculate, rarely solitary.

*White, or whitish.

2379. **S. fasciculata** Pers. Pers. Myc. Eur. I, t. 12, figs. 8–9.
Fasciculata, in small bundles.

R. 2–7 mm. high, *white*, cylindrical, *clavate*, gregarious, and usually fasciculate, externally *minutely silky* and almost smooth. Hymenium *white*, tubular, smooth. Flesh *white*, thin, soft. Spores white, subglobose, pointed at the base, 3·5–5 × 3–4μ, 1-guttulate; basidia 15–20 × 4·5–5μ, with 2–4 straight sterigmata, 4–4·5μ long. Hyphae hyaline, 2–2·75μ in diam., with thin or thick walls, and clamp connections. Rotten wood, and fallen branches; sometimes arising from a white mycelium. Sept.—April. Uncommon. (*v.v.*)

2380. **S. candida** (Hoffm.) Fr. Hoffm. Deutschl. Fl. II, t. 8, fig. 1.
Candida, shining white.

R. 2–3 mm. high, *shining white*, cylindrical, *solitary*, *glabrous*. Hymenium *white*, tubular, smooth. Flesh *white*, *thin*, somewhat diaphanous. Spores white, subglobose, pointed at the base, 4–5 × 3–4μ, with a large central gutta; basidia 12–15 × 4–5μ. Hyphae hyaline, thin, ·5–1μ in diam. Rotten branches of beech, alder, and ash. Oct.—Jan. Uncommon. (*v.v.*)

2381. **S. maxima** Massee. *Maxima*, largest.

R. 2 mm. high, *whitish, or pale buff*, subcylindrical, slightly contracted at the base, gregarious and subfasciculate, externally villose with slender, aseptate hyphae rough with minute particles of lime. Hymenium *concolorous*, tubular, smooth. Flesh *white*, thin. Spores white, elliptical, minutely and obliquely apiculate, 5 × 3μ. Rotten wood. May. Rare.

**Coloured.

2382. **S. anomala** (Pers.) Fr. ἀ, not; ὁμαλός, even.

R. 2–5 mm. high, *dingy ochraceous to ferruginous*, turbinate, or pyriform, usually gregarious, externally villose. Hymenium *pallid*, urceolate; margin incurved. Flesh *brownish*, thin. Spores white, cylindrical, incurved, 7–11 × 3–4μ; basidia 18–30 × 5–6μ. Rotten wood, and fallen branches, especially of alder. Jan.—Dec. Common. (*v.v.*)

var. **ochracea** (Hoffm.) Berk. Hoffm. Deutschl. Fl. II, t. 8, fig. 2.
ὠχρός, pale.

Differs from the type in its *scattered habit, and smaller size*. Rotten wood and bark. Jan.—Dec. Not uncommon. (*v.v.*)

2383. **S. poriaeformis** (DC.) Fr. *Poria*, the genus *Poria*; *forma*, shape.

R. 1–2 mm. high, *grey, cinereous, or brownish*, cup-shaped, hairy, sessile, crowded, seated on a greyish mycelium. Hymenium *pale*, grey, concave. Flesh *brownish*, thin. Spores white, globose, pointed at the base, 5–6μ, with a large central gutta; basidia 18–24 × 5–8μ, with 2–4 conical, straight sterigmata, 5–6μ long. Hyphae hyaline, 1–2μ in diam., with clamp connections. Wood and bark. Oct. Uncommon. (*v.v.*)

Porothelium Fr.

(πόρος, a pore; θηλή, a nipple.)

Receptacles cup-shaped, sessile, more or less crowded, distinct, seated on, or immersed in, a membranaceous, or floccose stroma. Hymenium smooth. Spores white, oblong, elliptical, or linear oblong; smooth; basidia with 2–4-sterigmata. Cystidia none. Growing on wood.

2384. **P. fimbriatum** (Pers.) Fr. Fr. Icon. t. 192, fig. 1, as *Porothelium lacerum* Fr. *Fimbriatum*, fringed.

R. 1–10 cm., *snow white*, effused, membranaceous, firm, separable, minutely tomentose, pruinose, attached to the matrix by a *white, cord-like mycelium*; margin fringed, silky. Pores white, aggregated in places, papillate, then *urceolate, fringed with a pink border*, pubescent. Spores white, "oblong, slightly depressed on one side, 4·5–6 × 3–3·5μ, multi-guttulate; basidia 15–23 × 4·5–6μ, with 2–4-sterigmata, 2–3μ long. Hyphae firm, thick walled, 1–2·5μ in diam., with scattered, small, often oblique clamp connections" Bourd. & Galz. Beech, birch, and hornbeam stumps, and fallen branches. Jan.—Dec. Uncommon. (*v.v.*)

2385. **P. Friesii** Mont. Cke. Handb. fig. 69.
Elias Fries, the illustrious mycologist.

R. 2–7·5 cm., *white, becoming tan colour*, effused, confluent, *flocculoso-membranaceous*, margin simple. Pores *yellowish*, papillate at first, immersed, then open and urceolate. Spores hyaline, elliptical, 5 × 3μ. Pine, and juniper. Sept.—Feb. Uncommon. (*v.v.*)

P. Vaillantii (Fr.) Quél. = **Poria Vaillantii** (DC.) Fr.

P. confusum B. & Br = **Odontia sudans** (A. & S.) Bres.

P. Stevensonii B. & Br. = **Odontia sudans** (A. & S.) Bres.

2386. **P. Keithii** B. & Br. Rev. Dr James Keith, a Scotch mycologist.

R. 2·5–5 cm., *pale umber*, closely adnate, inseparable, thin, at first *subgelatinous*, forming patches; margin very thin, subpulverulent. Pores *pallid*, scattered, papillate, at length collapsing. Spores linear-oblong, 5 × 2μ. Dead fir. April. Rare.

Phaeocyphella Pat.

(φαιός, dusky; κύφελλα, the hollow of the ear.)

Receptacle waxy, fleshy, or membranaceous; cup-shaped, or urceolate, sessile, pendulous. Hymenium smooth, rugulose, or wrinkled. Spores coloured, elliptical, or subglobose; smooth, punctate, verrucose, or echinulate; basidia with 2–4-sterigmata. Growing on wood, or on mosses.

2387. **P. ochroleuca** (B. & Br.) Rea. (= *Cyphella ochroleuca* B. & Br.) ὠχρός, pale; λευκός, white.

R. 2–3 mm., *ochrey white*, cup-shaped, sessile, villose above; margin at *length split*. Hymenium *pale ochre, brighter than the receptacle*, even. Flesh membranaceous. Spores "very pale ochraceous, elliptical, 6 × 4μ" Massee. Dead bramble stems. July—Oct. Rare.

2388. **P. fraxinicola** (B. & Br.) Rea. (= *Cyphella fraxinicola* B. & Br.) *Fraxinus*, ash; *colo*, I inhabit.

R. ·25–·2 mm., *snow-white*, orbicular, sessile, scattered, or gregarious, externally shortly villose. Hymenium *light yellow, becoming fuscous with the spores*, proliferous. Spores "pale olive, elliptical, 6 × 4μ" Massee. Ash bark. Dec.—Feb. Rare.

2389. **P. fuscospora** (Curr. ex Cke.) Rea. (= *Cyphella fuscospora* Curr. ex Cke.) *Fuscus*, dark; σπορά, seed.

R. ·25–·2 mm., *white*, tomentose; margin connivent. Hymenium *becoming yellow*. Spores fuscous, subglobose, punctulate. Bark. Rare.

2390. **P. galeata** (Schum.) Bres. (= *Cyphella galeata* (Schum.) Fr.) Fl. Dan. t. 2027, fig. 1. *Galeata*, covered with a helmet.

R. 2–5 mm., *whitish, or grey when moist, becoming snow white when dry, then rufescent*; *cup-shaped, then dimidiate, helmet-shaped*, sessile, even; margin quite entire. Hymenium *at length rufescent*, slightly wrinkled. Flesh *whitish*, membranaceous, soft. Spores tawny, rough or verrucose, subglobose, 8–10μ; basidia 18–30 × 7–9μ, with 4 curved sterigmata, 5–6 × 2–2·5μ. Hyphae very thin walled, 2–5μ, with clamp connections. Mosses. Nov.—Dec. Uncommon. (*v.v.*)

2391. **P. muscicola** (Fr.) Rea. (= *Cyphella muscicola* Fr.) Fl. Dan. t. 2083, fig. 2. *Muscus*, moss; *colo*, I inhabit.

R. 2–5 mm., *whitish, or cinereous*, persistently *cup-shaped*, sometimes irregular, nodding, *externally slightly fibrilloso-striate*; *margin slightly downy, repand, torn*. Hymenium *white, then grey*, even, then rugulose. Spores pinkish, or pale brown, subglobose, 8–10μ. Mosses. Nov.—May. Uncommon. (*v.v.*)

II. CLAVARIINEAE.

Receptacle erect, dendroid, coralloid, simple, or branched, never pileate. Hymenium more or less amphigenous.

Clavariaceae.

Same characters as the suborder.

Clavaria (Vaill.) Fr.

(*Clava*, a club.)

Receptacle fleshy, or subcoriaceous, erect, branched, or simple and clavate, smooth, or longitudinally striate. Hymenium even, amphigenous, absent in the stem-like portion of the simple clubs. Spores white, or ochraceous, rarely reddish ochre or brownish; elliptical, globose, subglobose, oboval, pip-shaped, pyriform, almond-shaped, reniform, oblong, oblong elliptical, or subfusiform; smooth, punctate, or verrucose; basidia with 2–4-sterigmata. Cystidia none. Putrescent. Growing on the ground, or on wood; solitary, gregarious, caespitose, or caespitoso-connate.

I. Branched.

A. Spores white, slightly coloured in no. 2402; basidia often with 2-sterigmata.

*Growing on the ground.

2392. **C. coralloides** (Linn.) Fr. (? = *Clavaria cristata* (Holmsk.) Fr. sec. Cotton & Wakef.) Sow. Eng. Fung. t. 278.

κοράλλιον, coral; *εἶδος*, like.

R. 5–10 cm. high, *white*, repeatedly and *irregularly much branched*; trunk short, rather thick, *often hollow*. Branches unequal, dilated upwards; branchlets crowded, acute. Flesh *white*, brittle. Spores "white, elliptical, subglobose, 6–8 × 4–5μ" Bourd. & Galz. Edible. Shady deciduous woods. Aug.—Oct. Rare.

2393. **C. cristata** (Holmsk.) Fr. Rolland, Champ. t. 103, no. 230.

Cristata, crested.

R. 2·5–7·5 cm. high, *white, sometimes tinged ochraceous*, often becoming cinereous with age when infected with *Rosellinia Clavariae*, divided into numerous, irregular branches; trunk short, firm, villose. Branches *dilated above*, often flattened, *acute, incised*, crested. Flesh *white*, tough, firm. Spores white, subglobose, 7–8 × 6–7μ, with a large central gutta; "basidia small, 25 × 6–7μ, contents densely granular, with 2-sterigmata. Hyphae loosely interwoven, more or less parallel, fairly regular, frequently septate, segments 35–40 × 5–6μ, in the centre 50–70 × 6–9μ" Cotton & Wakef. Edible. Woods. June—Dec. Common. (*v.v.*)

2394. **C. cinerea** (Bull.) Fr. Rolland, Champ. t. 103, no. 232.

Cinerea, colour of ashes.

R. 2·5–5 cm. high, *cinereous*, very much branched; trunk *whitish, or concolorous*, becoming almost black when infected with *Rosellinia Clavariae*, short, stout, or thin. Branches and branchlets *thickened*, irregularly shaped, somewhat wrinkled, obtuse, often crested and paler. Flesh *white in the trunk, cinereous* upwards, somewhat firm.

Spores white, subglobose, or very broadly elliptical, 7–11 × 7–8μ, with a large central gutta; "basidia long, conspicuous, 35–50(–70) × 6–10μ, contents finely granular, with 2-sterigmata. Hyphae filamentous, loose, 8–10(–12)μ in diam., with occasional inflations, slightly septate, irregular in transverse section" Cotton & Wakef. Edible. Woods. July—Jan. Common. (*v.v.*)

var. **gracilis** Rea. Trans. Brit. Myc. Soc. VI, t. 2. *Gracilis*, slender.

Differs from the type in the *long, slender trunk, and the numerous thin, tapering, acute branches and branchlets.* Spores white, subglobose, with a basal apiculus, 9 × 8μ, with a large central gutta. Bare soil in damp woods. Sept.—Nov. Uncommon. (*v.v.*)

2395. **C. gigaspora** Cotton. γίγας, giant; σπορά, seed.

R. 2–3 cm. high, *greyish with a tinge of yellow*, irregularly branched, sometimes almost palmate; trunk hardly distinct, about 1 cm. long. Branches erect, occasionally forked, often wrinkled, solid, terete, or compressed, much compressed at the acute angles, ultimate branches attenuated, apices blunt. Flesh tough, horny when dry. Spores white, broadly elliptical, slightly oblique, 10–20 × 7–9μ, av. 12–16 × 8μ, guttulate, then granular; basidia 60–70 × 15μ, with 4 short sterigmata, 8–10μ long, contents granular. Hyphae 4–4·5μ in diam., densely packed, forming a firm tough tissue, rather horny when dry. Amongst moss on rocky, heathy slope. Nov. Rare.

2396. **C. amethystina** (Batt.) Fr. Batt. Fung. Arim. Hist. t. 1, fig. C. ἀμέθυστος, amethyst.

R. 5–10 cm. high, *violet*, very much branched; trunk *concolorous*, or *whitish*, thin, equal. Branches round, smooth, or somewhat rugulose, obtuse, often forked at the apex. Flesh *tinged violet, becoming whitish*, rather brittle. Spores white, elliptical, obtuse at both ends, 6–7 × 3–4μ, with a large central gutta; "basidia rather large, 50–60 × 7–10μ, with 2–4-sterigmata. Hyphae densely interwoven, frequently septate, cells 50–100 × 8–12μ, not pseudoparenchymatous in transverse section. Spores globose, with a minute basal apiculus, 5–7μ in diam." Cotton & Wakef. Edible. Woods, and pastures. Aug.—Nov. Uncommon. (*v.v.*)

var. **lilacina** (Fr.) Quél. Schaeff. Icon. t. 172, as *Clavaria purpurea*. *Lilacina*, lilac coloured.

Differs from the type in the *lilac purple colour becoming brownish when dry, in the firmer texture and in the less numerous, dentate, often twisted branches.* Pastures. Rare. (*v.v.*)

2397. **C. rugosa** (Bull.) Fr. Rolland, Champ. t. 103, no. 233. *Rugosa*, wrinkled.

R. 5–10 cm. high, *white, simple,* or *sparingly branched, thickened upwards, wrinkled.* Branches irregular, few, obtuse, rarely crested. Flesh *whitish*, tough, firm. Spores white, subglobose, often apiculate at the base, 8–9 × 6–8μ, with a large central gutta; "basidia long, conspicuous, 60 × 5–6μ, with 2-sterigmata, contents granular. Hyphae somewhat densely interwoven, looser in the centre, 8–10μ in diam., frequently septate" Cotton & Wakef. Edible. Woods, and pastures. Aug.—Dec. Common. (*v.v.*)

var. **fuliginea** (Pers.) Fr. *Fuliginea*, sooty.

Differs from the type in the *dark sooty colour of the clubs and flesh.* Woods. Sept.—Oct. Uncommon. (*v.v.*)

var. **macrospora** Britzl. μακρός, long; σπορά, seed.

Differs from the type in *the larger spores*, 12–14 × 8–10μ. Rare.

2398. **C. grossa** (Pers.) Quél. (= *Clavaria Krombholzii* Fr. sec. Quél.) *Grossa*, thick.

R. 5–7 cm. high, *snow white*, sparingly branched; trunk 3–4 cm. × 3–5 mm., *somewhat bulbous at the base.* Branches compressed, deformed, acute, or incised. Flesh *white*, brittle. Spores white, elliptical, or subglobose, 9–12 × 6–8μ, with a large central gutta. Woods. Aug.—Oct. Uncommon. (*v.v.*)

2399. **C. crassa** Britzl. (? = *Clavaria rugosa* (Bull.) Fr. sec. Cotton & Wakef.) Britzl. Hymen. Südb. v (Clavaria), t. 39. *Crassa*, thick.

R. 5–7 cm. high, *violet*, or *lilac grey*; trunk slender, expanding upwards, and dividing into several obtuse, subcompressed branches. Spores white, subglobose, 8–10 × 8μ. Scattered, or solitary. Woods. Sept. Rare. (*v.v.*)

2400. **C. Krombholzii** Fr. (= *Clavaria Kunzei* Fr. sec. Cotton & Wakef.) Krombh. t. 53, figs. 15–16, as *Clavaria Kunzei* Fr.
J. von Krombholz, author of "Abbildungen der Schwämme."

R. 3–5 cm. high, *white, very densely tufted*, sparingly branched, *squat, compact.* Branches *more or less compressed, obtuse*, blunt. Flesh *white*, somewhat brittle. Spores white, elliptical, 4 × 3μ, with a large central gutta. Edible. Densely caespitose. Pastures. Sept.—Nov. Not uncommon. (*v.v.*)

2401. **C. Kunzei** Fr. Karl Sebastian Heinrich Kunze.

R. 5–12 cm. high, *ivory to cream white*, irregularly and *dichotomously* branched, *loose*, rarely compact; trunk *sometimes pink at the base*, usually distinct, 1–2 cm. × 3–5 mm. Branches erect, or spreading, cylindrical, or slightly compressed, often elongated, 2–5 mm. thick, even, solid, axils lunate; apices blunt, or pointed. Flesh *white, somewhat brittle.* Spores white, globose, often minutely apiculate,

3·5–4·5μ, with a large central gutta; "basidia 30–35 × 5–6μ, with 4-sterigmata. Internal structure pseudoparenchymatous in transverse sections, cells long, 100–300 × 5–8μ" Cotton & Wakef. Edible. Solitary, or gregarious. In long grass in woods, and pastures. Aug.—Nov. Not uncommon. (*v.v.*)

2402. **C. chionea** (Pers.) Quél. (= *Clavaria Kunzei* Fr. sec. Cotton & Wakef.) Trans. Brit. Myc. Soc. VI, t. 2. *χιών*, snow.

R. 5–7 cm. high, *snow white*, trunk thin, 2–3 cm. × 2–3 mm., very much branched. Branches *long, thin*, unequal, *pointed*. Flesh *white, tough*, somewhat elastic. Spores pale yellow in the mass, hyaline under the microscope, subglobose, with a basal apiculus, 4–5 × 4μ, with a large central gutta. Bare soil in woods. Sept.—Oct. Uncommon. (*v.v.*)

2403. **C. subtilis** (Pers.) Fr. Pers. Comm. t. 4, fig. 2. *Subtilis*, slender.

R. 2–2·5 cm. high, *white, becoming yellowish*, trunk 1–1·5 cm. × 1-2 mm., glabrous at the base, *equal, with few branches*. Branches *dichotomously forked*, subfastigiate. Flesh *white*, tough. Spores white, oblong, elliptical, rounded at both ends, 6 × 3μ, 1-guttulate. Amongst grass in woods. Sept.—Oct. Uncommon. (*v.v.*)

2404. **C. corniculata** (Schaeff.) Fr. (= *Clavaria muscoides* Fr. Hym. Eur.) Schaeff. Icon. t. 173. *Corniculata*, having little horns.

R. 2–5 cm. high, *yellow*, generally only twice or thrice branched; trunk *white towards the base*, thin, firm. Branches rounded at the axils, apex *crescent-shaped*, acute. Flesh *yellow*, firm, tough. Spores white, subglobose, often minutely apiculate, 5–6μ, with a large central gutta; "basidia distinct, 50μ long, vacuolate or clear, with 4-sterigmata, 10μ long. Hyphae not interwoven, running parallel to the axis, easily separable, and becoming twisted, 4–8μ in diam., fairly frequently septate, with cells 100–200μ long, not pseudoparenchymatous in transverse section" Cotton & Wakef. Edible. Pastures. Aug.—Dec. Common. (*v.v.*)

var. **pratensis** (Fr.) Cotton & Wakef. (= *Clavaria fastigiata* (Linn.) Fr. Hym. Eur.) *Pratensis*, growing in meadows.

Differs from the type in being *more branched, with short, divaricate, fastigiate branchlets*, and forming a level top. Edible. Pastures. Aug.—March. Common. (*v.v.*)

2405. **C. umbrinella** Sacc. (= *Clavaria umbrina* Berk.) Trans. Brit. Myc. Soc. III, t. 11, fig. E, spores. *Umbrinella*, somewhat umber.

R. 2·5–3·5 cm. high, *pale brown*, irregularly and dichotomously branched, *stem absent, branches distinct to the base*. Branches erect, cylindrical, slender, 1–2 mm. thick, even, solid; apices blunt, or bifid.

Spores white, pip-shaped, laterally apiculate, 4–5 × 4μ, or 5–6 × 3μ, usually guttulate; basidia 35–40 × 6–7μ, contents finely granular, with 4-sterigmata. Hyphae loosely and slightly interwoven, 7–10μ in diam. Taste pleasant. Lawns, and amongst short grass in woods. July—Oct. Uncommon.

2406. **C. curta** Fr. Fr. Icon. t. 199, fig. 2. *Curta*, short.

R. 1–1·5 cm. high, *greenish yellow*, very much branched, pressed close together. Stem *none*. Branches *crowded, obtuse*. Spores white. Amongst grass, and fir leaves. Oct. Rare.

2407. **C. Bizzozeriana** Sacc. (= *Clavaria conchyliata* Allen.) Trans. Brit. Myc. Soc. III, t. 8, as *Clavaria conchyliata* Allen.
J. Bizzozero.

R. 5–10 mm. high, *brilliant violet, becoming paler*, at first simple and toothed at the apex, then dichotomously branched; stem *reddish yellow* (*nankeen*), *pubescent*. Flesh *white*. Spores white, globose, 2·5–3μ; basidia 6–7μ in diam. Gregarious, but scattered. Bare soil under nut trees. Oct. Uncommon. (*v.v.*)

**Growing on wood.

2408. **C. pyxidata** (Pers.) Fr. Pers. Comm. t. 1, fig. 1.
Pyxidata, box-shaped.

R. 2·5–5 cm. high, *pallid, then tan colour, somewhat rufescent*; trunk thin, smooth, branched. Branches and branchlets *all hollowed out in cup-shape at the apex*, cups radiating in a proliferous manner at the margin. Flesh *white*, firm. Spores white, elliptical, 4 × 3μ. Rotten, and buried wood. Oct. Rare. (*v.v.*)

2409. **C. Kewensis** Massee. (= *Clavaria stricta* (Pers.) Fr. sec. Cotton & Wakef.) *Kewensis*, belonging to Kew.

R. 4–7 cm. high, *rusty brown, becoming ochraceous upwards*, base thick, *dividing almost at once* into numerous, subequal, divergent branches. Branches uniform in thickness throughout, often compressed, imperfectly hollow, dividing near the apex into 2–4 short branchlets, axils rounded, tips obtuse, or divided into 2–4 short finger-like processes, pruinose. Spores white, elliptical, apiculate at the base, 5–6 × 3·5–4μ. Smell pleasant like that of *Lentinus cochleatus*. Forming dense tufts, 12–15 cm. across. Rotten wood. Oct. Rare.

B. Spores more or less ochraceous.

*Growing on the ground.

2410. **C. botrytis** (Pers.) Fr. Rolland, Champ. t. 103, no. 231.
βότρυς, a bunch of grapes.

R. 7–10 cm. high, 6–20 cm. broad, *whitish, or ochraceous*; trunk

3–4 × 1·5–6 cm., *white, becoming yellowish,* firm, passing into very numerous, crowded, irregular, thick branches. Branches *ochraceous,* cylindrical, or compressed, dividing into numerous branchlets, apices *red, toothed.* Flesh *white,* compact. Spores ochraceous in the mass, oblong elliptical, attenuated at the base into a slightly curved point, 12–15 × 4–6μ, 1–2-guttulate, finally longitudinally striate; "basidia long, conspicuous, 60–70 × 8–10μ, contents granular, with 2–4-sterigmata. Internal structure of parallel, septate hyphae, cells 100–150 × 10μ in the centre, smaller towards the margin, scarcely parenchymatous in transverse section" Cotton & Wakef. Smell very pleasant, fruity. Taste agreeable. Edible. Woods. Aug.—Nov. Uncommon. (*v.v.*)

2411. **C. flava** (Schaeff.) Fr. Schaeff. Icon. t. 175. *Flava,* yellow.

R. 7–14 cm. high, 10–20 cm. broad, *lemon yellow*; trunk 4–5 × 5–8 cm., *white, becoming blood red when bruised or handled,* divided up into very numerous branches. Branches crowded, compressed, *fastigiate,* apices toothed. Flesh *white, turning slightly blood colour near the cuticle when quite fresh, brittle.* Spores pale ochraceous in the mass, oblong elliptical, 9–12 × 4–5μ, finally minutely warted; "basidia 45 × 10μ, finely granular, with 4-sterigmata. Hyphae interwoven, 8–12μ in diam., subparenchymatous in transverse section" Cotton & Wakef. Taste pleasant. Edible. Woods, Sept.—Oct. Uncommon. (*v.v.*)

2412. **C. aurea** (Schaeff.) Fr. Schaeff. Icon. 287. *Aurea,* golden.

R. 5–10 cm. high, 10–20 cm. broad, *egg yellow, or somewhat tawny*; trunk 2–4 × 1–5 cm., *whitish at the base, concolorous above,* base somewhat bulbous, divided into numerous, erect, *stout* branches. Branches tense, straight, dichotomous, round, flattened; apices *paler,* toothed, or incised. Flesh *white, yellowish near the cuticle,* tough, elastic. Spores ochraceous in the mass, oblong, or subfusiform, often obliquely apiculate, 9–12 × 3–4μ, 1–3-guttulate. Taste pleasant. Edible. Aug.—Oct. Uncommon. (*v.v.*)

2413. **C. rufescens** (Schaeff.) Fr. *Rufescens,* becoming red.

R. 7–9 cm. high, 3–6 cm. broad, *branchlets all blood red, becoming paler*; trunk 4–5 × 2–3 cm., *whitish at the base, becoming yellow upwards,* divided into numerous erect branches. Branches erect, subcompressed; apices dentate, or forked. Flesh *whitish, reddish at the apex of the branches,* tough. Spores ochraceous. Woods. Sept.—Oct. Rare. (*v.v.*)

2414. **C. formosa** (Pers.) Fr. Pers. Icon. et Descr. t. 3, fig. 6. *Formosa,* handsome.

R. 6–9 cm. high, 7–30 cm. broad, *orange rose colour, or pinkish ochraceous*; trunk 3–4 × 2·5–6 cm., *whitish at the base, becoming rosy yellow upwards*, very much branched. Branches erect, *elongate*, crowded; apices *lemon yellow*, forked, simple, or toothed. Flesh *whitish in the trunk, subconcolorous elsewhere*, tough, elastic. Spores ochraceous in the mass, oblong elliptical, with a basal, or lateral apiculus, 9–11 × 5–6μ, usually 1-guttulate, "becoming finally subverrucose" Bourd. & Galz.; "basidia not conspicuous, 30–40 × 6–8μ, with 4 erect sterigmata. Hyphae frequently septate, loosely interwoven in the centre, 6–8(–10)μ wide, occasionally swollen up to 14μ at the septa, more slender, and more closely interwoven towards the margin, subparenchymatous in transverse section. A few latex hyphae present" Cotton & Wakef. Smell pleasant. Taste agreeable. Edible. Woods. July—Oct. Not uncommon. (*v.v.*)

2415. **C. Broomei** Cotton & Wakef.

C. E. Broome, the well-known mycologist.

R. 5–8 cm. high, 2–4 cm. broad, *ochraceous orange, tips darker orange, turning brown easily on bruising*, base *white, or pinkish*; trunk short, not swollen, *white, becoming pinkish on bruising*, somewhat branched, rooting base small. Branches irregular, or subdichotomous, slight below, more frequent above, axils not rounded, cylindrical, or flattened, short, solid, fairly erect, smooth, or the larger branches much wrinkled, tips flattened. Flesh *white, becoming vinous later especially below*, solid. Spores deep ochraceous, or even orange in the mass, fusiform, or pip-shaped, markedly aculeate, 14–20 × 6–8μ (average 15–16 × 6–7μ); basidia not conspicuous, 40–50 × 8–9μ, with 2-sterigmata, contents granular. Internal structure composed of fine filaments, densely packed, slightly interwoven, 3–6μ in diam., with vesicular ends, 10–12μ in diam., not pseudoparenchymatous in transverse section; large crystals in abundance in the tissue. Smell slight, not pleasant. Taste bitter. Woods. Sept.—Oct. Uncommon.

2416. **C. spinulosa** (Pers.) Fr. Pers. Obs. II, t. 3, fig. 1.

Spinulosa, with little spines.

R. 5–8 cm. high, 3–6 cm. broad, *somewhat cinnamon, brownish amber*; trunk 2–4 × 2–3 cm., *pallid, becoming yellowish*, very much branched. Branches elongate, crowded, tense and straight; apices *concolorous*, forked, simple or toothed. Flesh *white*, firm, tough. Spores ochraceous, elliptical, or oboval, incurved and apiculate at the base, 8–9 × 4–5μ. Woods, and heaths. Sept.—Jan. Not uncommon. (*v.v.*)

2417. **C. abietina** (Pers.) Fr. *Abietina*, pertaining to firs.

R. 2·5–7·5 cm. high, 3–6 cm. broad, *ochraceous, becoming greenish when bruised or weathered*; trunk 1–2·5 cm. × 8–15 mm., *white tomen-*

tose, very much branched. Branches erect, crowded, *longitudinally wrinkled when dry*; apices *often becoming tinged with green,* forked, toothed, acute. Flesh *greenish,* especially at the base, firm, tough. Spores ochraceous, pip-shaped, with a basal apiculus, 6–7 × 3–4μ, 1-guttulate, "becoming verrucose" Bourd. & Galz.; "basidia small, 35–40 × 7–8μ, contents uniform, finely granular, with 4 erect sterigmata. Hyphae filamentous, loosely interwoven, 4–10μ (average 5–7μ), slightly septate" Cotton & Wakef. Taste bitter. Edible. Coniferous woods. Aug.—Dec. Common.

2418. **C. Invalii** Cotton & Wakef. Inval near Haslemere, Surrey.

R. 4–5 cm. high, *deep ochre,* forming dense, compact, almost spherical tufts; trunks more or less distinct, short, often woolly, with white, or yellowish rooting strands, irregularly and frequently branched, axils acute. Branches slender, short, uneven, cylindrical, erect, smooth, solid; apices attenuated, pointed. Flesh *white.* Spores yellow, pip-shaped, slightly incurved at the base, echinulate, 7–9 × 4μ (average 8 × 4μ); basidia conspicuous, 30–40 × 7–9μ, contents finely granular, with 4 erect sterigmata. Internal structure of irregular, wavy, filamentous hyphae, 5–10μ in diam., loosely interwoven, and running equally in each direction. Smell slightly pungent. Taste faint, hardly bitter. Amongst leaves in thick plantations of spruce, larch, etc. Uncommon.

2419. **C. flaccida** Fr. Fr. Icon. t. 199, fig. 4. *Flaccida,* flabby.

R. 3–10 cm. high, 3–4 cm. across, *ochraceous, or somewhat tawny*; trunk 2–3 × 1·5–2 cm. or almost none, *concolorous,* very much branched, flaccid. Branches erect, crowded, even, apices acute, simple, or forked, converging. Flesh *whitish, yellowish upwards,* elastic, firm. Spores very pale ochraceous, elliptical, 6–7 × 3μ, 1-guttulate, "minutely verrucose" Bourd. & Galz.; "basidia small, conspicuous, 30 × 7–9μ, contents finely granular, with 4-sterigmata. Hyphae loosely interwoven, slightly septate, 7–10μ in diam., not parenchymatous in transverse section, and more densely arranged towards the periphery" Cotton & Wakef. Woods, and heaths. Sept.—Dec. Uncommon. (*v.v.*)

2420. **C. condensata** Fr. *Condensata,* pressed close together.

R. 5–10 cm. high, 3–4 cm. across, *tan rufescent*; trunk 1–2 cm. × 5–10 mm., *ochraceous cream, or whitish at the base, divided into numerous branches almost to the base.* Branches *yellow, tense and straight, erect, crowded in a parallel manner,* apices fastigiate, twice or thrice toothed. Flesh *white, deep yellow upwards,* firm, tough. Spores ochraceous, elliptic oblong, depressed on the side, 8–9 × 4μ, 2–3-guttulate. Woods, and heaths. Aug.—Oct. Uncommon. (*v.v.*)

2421. **C. palmata** (Pers.) Fr. *Palmata*, palmate.

R. 5–6 cm. high, 3–4 cm. broad, *tan, or tawny, becoming paler*; trunk *thin, very much branched almost from the base*. Branches bifurcate, flattened, thin, erect, *palmately branched*, apices forked, or toothed. Flesh *pale*, firm. Spores pale ochraceous, minutely punctate, oblong elliptical, 6–7 × 3–4μ, usually 1-guttulate. Smell very pleasant. Coniferous woods. Dec. Rare. (*v.v.*)

2422. **C. crocea** (Pers.) Fr. *κρόκος*, saffron.

R. 1–1·5 cm. high, *saffron yellow*; stem pallid, thin, naked. Branches and branchlets similar, somewhat forked. Spores "ochraceous, elliptical, 6–7 × 3–4μ" Massee. Waste ground, downs, gardens, on peat. Sept.—Feb. Rare.

2423. **C. grisea** (Pers.) Fr. (= *Clavaria cinerea* (Bull.) Fr. sec. Cotton.) *Grisea*, grey.

R. 3–7·5 cm. high, *fuliginous cinereous*; trunk 4 × 2·5 cm., *whitish*. Branches 7·5 cm. long, attenuated, *somewhat wrinkled*; branchlets unequal, obtuse. Flesh firm. Spores "reddish ochre in the mass, slightly brownish, and with a yellowish gutta under the microscope, oboval, 10–12 × 7–7·5μ" Bourd. & Galz. Woods. Sept.—Oct. Rare.

2424. **C. fuliginea** Pers. *Fuliginea*, sooty.

R. 5–7·5 cm. high, 5 cm. broad, *cinereous, becoming rufescent*, very much branched; stem thin; larger branches thick, compressed, lateral rather incomplete; branchlets subfastigiate, short, acute. Spores ochraceous, globose, with an apiculus, 10μ. Ground. Rare.

**Growing on wood.

2425. **C. stricta** (Pers.) Fr. Berk. Outl. Brit. Fung. t. 18, fig. 5. *Stricta*, close.

R. 5–10 cm. high, 3–8 cm. across, *pallid yellow, becoming fuscous when bruised*; trunk 1–3 × 1 cm., *whitish at the base, concolorous upwards*, very much branched. Branches tense and straight, crowded, adpressed; apices acute, or toothed. Flesh white, firm, tough. Spores pale ochraceous, pip-shaped, with a basal, or lateral apiculus, 6–9 × 4–5μ; "basidia distinct, 30–40 × 7–9μ, contents granular, with 4 erect sterigmata. Hyphae interwoven, 4–10μ in diam., not parenchymatous in transverse section, central hyphae rather thick walled" Cotton & Wakef. Rotten stumps, and buried wood. Aug.—Jan. Not uncommon. (*v.v.*)

var. **alba** Cotton. *Alba*, white.

Differs from the type in its *creamy white colour*. On the ground, amongst fallen leaves, etc. Rare.

2426. **C. crispula** Fr. *Crispula*, with little curls.

R. 2·5–7·5 high, *tan, then ochraceous,* very much branched; trunk thin, with villose rootlets. Branches *flexuose,* multifid, *divaricate.* Spores "pale yellow, elliptical, 5 × 3μ" Massee. Base of trunks, ash, and elder. Oct.—Dec. Rare.

II. Simple.

A. Tufted, or caespitose at the base.

*Purple, or reddish.

2427. **C. purpurea** (Müller) Fr. Fl. Dan. t. 837, fig. 2. *Purpurea,* purple.

R. 7–12 cm. high, 1–2 mm. thick, *purple, purple grey,* purplish brown, or dark chocolate, *white villose at the base,* fusiform, compressed, hollow ("solid" Cotton), flexuose, acute. Spores white, "oval, 7–8 × 4–5μ, contents granular; basidia small, 25–30 × 7–8, guttulate, with 4 erect sterigmata. Hyphae irregular, cells 50–60 × 3–5μ, or × 7–9μ towards the centre, pseudoparenchymatous in transverse section" Cotton & Wakef. Fragile. Caespitose. Grassy places, and under conifers. Aug.—Oct. Rare.

C. rufa Fl. Dan. = **Clavaria inaequalis** (Müller) Fr. sec. Cotton.

2428. **C. rosea** (Dalman) Fr. *Rosea,* rose colour.

R. 2–5 cm. high, *rose colour, whitish at the attenuated base,* fusiform, apex acute, rarely toothed, or bifid, cylindrical, or somewhat compressed, sometimes becoming yellowish at the apex. Flesh whitish, stuffed, brittle. Spores white, subglobose, with a distinct basal apiculus, 6–6·5 × 5μ; basidia clavate, 22–37 × 7–10μ, with 4-sterigmata. "Hyphae irregular, 7–12μ in diam., frequently septate, semi-parenchymatous in transverse section; crystals sometimes present" Cotton & Wakef. Caespitose. Pastures. Sept.—Oct. Rare. (*v.v.*)

**Yellow.

2429. **C. fusiformis** (Sow.) Fr. Sow. Eng. Fung. t. 234. *Fusus,* a spindle; *forma,* shape.

R. 5–14 cm. high, *yellow,* elongato-fusiform, attenuated at both ends, smooth, often with a central furrow, apex acute, rarely toothed, *becoming hollow.* Flesh *concolorous, paler upwards,* somewhat firm. Spores white, or tinged with yellow, globose, minutely apiculate at the base, 6–8μ, with a large central gutta; "basidia hyaline, 35–40 × 6–8μ, with 4 slightly curved sterigmata. Hyphae more or less interwoven, 4–6μ in diam., walls sometimes rough, occasionally with dark yellow contents" Cotton & Wakef. Taste bitter. Caespitoso-connate. Woods, and pastures. July—Dec. Common. (*v.v.*)

var. **ceranoides** (Pers.) W. G. Sm. Sow. Eng. Fung. t. 235.
κηρός, wax; εἶδος, like.

Differs from the type in the *blunter, often divided apices of the clubs becoming brown*. Woods, and pastures. Sept.—Oct. Not uncommon. (*v.v.*)

2430. **C. luteo-alba** Rea. Trans. Brit. Myc. Soc. II, t. 3
Luteus, orange yellow; *alba*, white.

R. 4–5 cm. high, 2–3 mm. thick, *deep rich yellow, or apricot colour, becoming pale ochraceous when dried*, apex *white*, cylindrical, or slightly compressed, gradually attenuated downwards to the base, blunt, or acute, stuffed. Flesh *orange yellow, or concolorous*, floccose. Spores white, pip-shaped, or ovoid, 5–8 × 3–4μ; basidia 25–30 × 5–7μ, contents slightly granular, with 4-sterigmata. Hyphae 5–6μ in diam., containing orange coloured granules. Taste "like tallow" Cotton. Fasciculate, or scattered. Woods, pastures, and heaths. Aug.—Dec. Not uncommon. (*v.v.*)

2431. **C. inaequalis** (Müller) Quél. Grev. Scot. Crypt. Fl. t. 37, as *Clavaria fragilis*. *Inaequalis*, unequal.

R. 2·5–6 cm., *golden yellow, sulphur coloured at the base*, simple, or forked, apex acute, *stuffed*. Flesh *pale yellow*. Spores white, "ovoid, globose, apiculate, 7–9 × 6–8μ" Bourd. & Galz. Woods, pastures, and heaths. Aug.—Nov. Uncommon. (*v.v.*)

2432. **C. dissipabilis** Britzl. (= *Clavaria similis* Boud. & Pat.; *Clavaria inaequalis* (Müller) Fr. sec. Cotton.) Britzl. Hymen. Südb. fig. 28.
Dissipabilis, dispersed.

R. 3–6 cm. high, 2–3 mm. thick, *yellow, to orange yellow*, fragile, elongato-clavate, tapering downwards, or cylindrical, obtuse, rarely subacute, smooth, or furrowed, stuffed. Flesh white or *yellowish*, floccose. Spores white, or tinged with yellow, *acutely warted*, subglobose, 5–6μ, with a large central gutta; "basidia conspicuous, clavate, yellowish, 30–40 × 6–8μ, with 4 more or less erect sterigmata" Cotton & Wakef. Solitary, or in small clusters. Woods, pastures, and heaths. Aug.—Dec. Common. (*v.v.*)

2433. **C. persimilis** Cotton. *Persimilis*, very like.

R. 3–5 cm. high, 2–3 mm. thick, *orange yellow, to orange, becoming dark orange when dried*, cylindrical, or subcompressed, apex usually acute. Flesh *pale*, floccose. Spores white, subglobose-oblong, with a conspicuous oblique apiculus, 5–6 × 4μ, guttulate; basidia 30–35 × 7–8μ, contents granular, with 4-sterigmata. Hyphae loosely packed, running longitudinally, 3–6μ in diam., not pseudoparenchymatous in transverse section. Isolated, or fasciculate. Amongst short grass. Woods, and pastures. Not uncommon.

2434. **C. argillacea** (Pers.) Fr. (= *Clavaria ericetorum* Pers.) Boud. Icon. t. 175, as *Clavaria ericetorum* Pers.

Argillacea, clay colour.

R. 3–8 cm. high, 4–8 mm. broad, *whitish yellow, or citron yellow*, simple, elongate, attenuated at the base and *shining*, very rarely bifurcate, *compressed, with one or two longitudinal channels, apex obtuse, or truncate*, stuffed, fragile. Flesh *concolorous*, floccose. Spores white, reniform, or oblong and depressed on one side, 10 × 5–6μ, with granular contents; "basidia conspicuous, 70μ long, with 4 sterigmata, contents granular. Internal structure almost pseudoparenchymatous in transverse section even when old, cells regular, 10–14μ in diam., with small, narrow filaments (4–5μ in diam.) between; segments 50–70μ long towards the margin, but up to 200–300μ in the centre" Cotton & Wakef. Solitary, or in tufts of two or three. Heaths, and hillsides. Aug.—Nov. Not uncommon. (*v.v.*)

2435. **C. straminea** Cotton. Trans. Brit. Myc. Soc. III, t. 11, erroneously lettered *Clavaria persimilis* Cotton.

Straminea, straw colour.

R. 3–5 cm. high, 3–4 mm. thick, *straw-coloured, becoming brownish with age or when handled*, simple, cylindrical, or somewhat compressed, smooth, *apex usually acute*. Stem *usually very distinct, cinnamon yellow*, stuffed, brittle. Flesh *somewhat darker than the hymenium*. Spores white, globose, with a minute basal apiculus, granular, 5–7μ; basidia 40–60 × 7–9μ, with 4-sterigmata, contents granular. Internal structure pseudoparenchymatous in transverse section. Isolated, or caespitose. Amongst short grass, lawns, and pastures. Sept.—Oct. Rare. (*v.v.*)

2436. **C. Michelii** Rea. P. A. Micheli, an early illustrator of fungi.

R. 4–7 cm. high, 1–2 mm. thick, *yellow, white at the base, cylindrical*, hollow, apex *acute*, very fragile. Flesh *yellowish, white at the base*. Spores white, subglobose, 3 × 2μ. Fasciculate. Amongst grass under a cherry tree. Sept. Rare. (*v.v.*)

***White, rarely yellow.

2437. **C. vermicularis** Fr. *Vermicularis*, pertaining to worms.

R. 6–12 cm. high, 3–5 mm. broad, *shining white, elongate fusiform*, attenuated at both ends, *compressed, with a longitudinal channel down the middle*, often twisted, *hollow*, apex acute, often becoming brownish when weathered, *very brittle*. Flesh *white*, fragile. Spores white, elliptical, with a basal apiculus, 5–7 × 3–4μ, with a large central gutta; "basidia small, 30 × 6–7μ, with 4-sterigmata. Hyphae parallel, septate, with rather long cells, pseudoparenchymatous in transverse section, central cells 10–15μ in diam., with smaller cells intermixed"

Cotton & Wakef. Taste pleasant, like cheese straws when cooked. Edible. Densely caespitose. Amongst long grass in woods, and pastures. May—Nov. Common. (*v.v.*)

2438. **C. fragilis** (Holmsk.) Fr. (= *Clavaria vermicularis* Fr. sec. Cotton.) *Fragilis*, brittle.

R. 5–6 cm. high, 2–3 mm. thick, *white, or yellow*, simple, *cylindrical*, slightly attenuated downwards, apex subobtuse, or slightly pointed, stuffed, becoming hollow, fragile. Flesh *white, or slightly yellowish*, loose. Spores white, broadly elliptical, 8–9 × 5–6μ, with granular contents. Edible. Fasciculate. Heaths, and pastures, rarely in woods. Aug.—Nov. Not uncommon. (*v.v.*)

****Fuliginous, or blackish.

2439. **C. striata** (Pers.) Fr. Pers. Icon. et Descr. t. 3, fig. 5. *Striata*, furrowed.

R. 3–5 cm. high, 3–4 mm. wide, *subfuliginous*, cylindrical, attenuated at the apex and base, *extreme base white*, somewhat twisted, *here and there longitudinally striate*, compressed, apex subobtuse, stuffed, then hollow. Flesh *white*, loose. Spores white, subglobose, with a basal apiculus, 3–4 × 2–3μ. Amongst grass under beeches. Sept. Rare.

2440. **C. fumosa** (Pers.) Fr. Krombh. t. 53, fig. 18. *Fumosa*, smoky.

R. 5–14 cm. high, 3–7 mm. thick, *fuliginous*, cylindrical, or incrassated at the apex, straight, subcompressed, hollow, somewhat fragile. Flesh *white*, loose. Spores white, elliptical, somewhat pointed towards the base, 6–7 × 4μ, contents granular; "basidia inconspicuous, 35 × 6–7μ, with 2–4 short sterigmata. Hyphae parallel, with short, cylindrical cells, 30–50 × 10–15μ, pseudoparenchymatous in transverse section" Cotton & Wakef. Densely caespitose. Amongst grass in pastures, and on lawns, rarely in woods. Aug.—Oct. Uncommon. (*v.v.*)

B. Clubs distinct at the base.

*Yellow, ochraceous, or tawny, often becoming rufescent.

2441. **C. pistillaris** (Linn.) Fr. Rolland, Champ. t. 104, no. 235. *Pistillaris*, like a pestle.

R. 7–30 cm. high, 2–5 cm. broad, *light yellow, then rufescent, finally dingy brown, obovate-clavate*, obtuse, pruinose, stuffed. Flesh *white*, floccose, firm. Spores white, or tinged yellowish, pip-shaped, or elliptic-oblong with a lateral apiculus, 12–13 × 7–8μ, often 1-guttulate; "basidia about 70μ long, with 2–4-sterigmata" Cotton & Wakef. Edible. Woods. Sept.—Dec. Not uncommon. (*v.v.*)

2442. **C. ligula** (Schaeff.) Fr. *Ligula*, a little tongue.

R. 3–6 cm. high, 1–1·5 cm. broad, *yellowish, then pallid rufescent*, elongato-clavate, obtuse, *base villose, white*, stuffed. Flesh *white*, floccose, loose. Spores white, elliptical, often with a lateral apiculus, 10 × 5–6 μ, contents granular; "basidia conspicuous, 40 × 6–8 μ, with 4-sterigmata" Cotton & Wakef. Edible. Woods. Aug.—Oct. Uncommon. (*v.v.*)

2443. **C. contorta** (Holmsk.) Fr. (= *Clavaria fistulosa* (Holmsk.) Fr. sec. von Hoehn. and Bourd. & Galz.) Boud. Bull. Soc. Myc. Fr. XXXIII, t. 1, fig. 5. *Contorta*, twisted.

R. ·5–2 cm. high, 2–6 mm. thick, *watery yellowish, often tinged with brown, erumpent*, simple, sometimes spathulate, *somewhat twisted, wrinkled*, obtuse, compressed, stuffed. Flesh *yellowish, floccose*, loose. Spores white, almond-shaped, 12–14 × 9 μ, contents granular, "fusiform, 15–18 × 7–9 μ" Boud.; "basidia rather large, very distinct, 50 × 10 μ, contents minutely granular, with 4-sterigmata. Internal structure in longitudinal section of long cells, 10–15 μ in diam., with narrower elements on either side; latex tubes present" Cotton & Wakef. Dead twigs. Oct.—Nov. Uncommon. (*v.v.*)

2444. **C. fistulosa** (Holmsk.) Fr. *Fistulosa*, full of holes.

R. 10–20 cm. high, 2 mm. thick, *yellow, then rufescent*, cylindrical, subobtuse, straight, or curved, often contorted at the base, hollow; root short, *villose*. Flesh *yellowish*, firm. Spores white, verrucose ("smooth" Cotton), pip-shaped, or subfusiform, 12 × 7 μ, with a large central gutta; "basidia conspicuous, about 40 μ long, with 4 erect sterigmata, contents finely granular. Internal structure possessing a system of lacticiferous hyphae, aseptate, frequently branched, 6 μ in diam., contents granular" Cotton & Wakef. Dead branches, and pea sticks. Nov.—Feb. Uncommon. (*v.v.*)

2445. **C. Ardenia** (Sow.) Fr. (= *Clavaria fistulosa* (Holmsk.) Fr. sec. Quél.) Sow. Eng. Fung. t. 215. *Ardenia*, Lady Arden.

R. 12–30 cm. high, 8 mm. broad at the apex, *ferruginous, then date brown*, elongate, *incrassated at the obtuse or acute apex*, hollow, base tomentose, not rooting. Flesh *yellowish*, rigid, firm. Spores white, oblong elliptical, attenuated towards the base, 15 × 6 μ, contents granular. Fallen branches, and mosses. Sept.—Dec. Uncommon. (*v.v.*)

C. tuberosa (Sow.) Berk. = **Calocera tuberosa** (Sow.) Fr.

2446. **C. juncea** (A. & S.) Fr. Boud. Icon. t. 176. *Juncea*, like a rush.

R. 5–15 cm. high, 1–2 mm. thick, *pallid, then rufescent, filiform, flaccid, fistulose*, apex acute, creeping base fibrillose. Flesh *yellowish*.

Spores white, elliptical, sometimes depressed on the side towards the base, 8–9 × 4–5μ, 1-guttulate; "basidia small, 30–35 × 6–7(–8)μ, with 4-sterigmata, slightly vacuolar. Internal structure parenchymatous in transverse section, with occasional large air spaces" Cotton & Wakef. Amongst leaves in woods. Sept.—Nov. Uncommon. (*v.v.*)

**White, or greyish.

2447. **C. canaliculata** Fr. (= *Clavaria rugosa* (Bull.) Fr. sec. Cotton & Wakef.) *Canaliculata*, channelled.

R. 3–20 cm. high, 4–6 mm. thick, *snow white, or greyish*, clavate, or cylindrical, *at length compressed, channelled*, or splitting longitudinally, stuffed, then hollow. Flesh *white*, loose. Spores white, broadly elliptical, 11–12 × 7μ, with a large central gutta. Solitary, rarely in pairs, or threes. Heaths, and pastures. Sept.—Oct. Rare. (*v.v.*)

2448. **C. tenuipes** B. & Br. (= *Pistillaria tenuipes* (B. & Br.) Massee.) B. & Br. Ann. and Mag. Nat. Hist. ser. 2, vol. II, t. 9, fig. 2. *Tenuis*, thin; *pes*, foot.

R. 1·5–6 cm. high, 2–10 mm. wide, *pallid clay colour*, *pale grey* to *drab-coloured*, clavate, or cylindrical, often compressed, smooth, or slightly rugulose, apex blunt, hollow when old. Stem slender, 1–2 cm. × 2–3 mm., more or less sharply marked. Flesh *concolorous*, loose. Spores white, elliptical, or oval, often with a minute basal apiculus, 6–9 × 4–5μ, guttulate, then granular, "basidia 30–40 × 7–9μ, with 4-sterigmata, contents granular. Internal structure consisting of loosely packed, oblong cells, 8–10μ in diam., cells 50–150μ long, hyphae unbranched" Cotton & Wakef. Isolated, or in small groups. Amongst short grass, bare soil, and old charcoal heaps, on heaths, pastures, and in woods. Sept.—Jan. (*v.v.*)

2449. **C. asterospora** Pat. Pat. Tab. Anal. ser. 2 (1886), fig. 568. *ἀστήρ*, a star; *σπορά*, seed.

R. 2–3 cm. high, *pure white*, simple, slender, cylindrical, apex blunt, or pointed, smooth, hollow. St. slender, *greenish at the base*, not markedly distinct from the club. Flesh brittle. Spores hyaline, *with long, scattered spines*, globose, spore body 7–8μ in diam.; basidia clavate, 30–40 × 8μ, with 4-sterigmata. Internal structure pseudoparenchymatous in transverse section. Bare soil. Rare.

2450. **C. acuta** (Sow.) Fr. Sow. Eng. Fung. t. 333. *Acuta*, pointed.

R. 3–7 cm. high, 2–3 mm. thick, *glistening white*, cylindrical, or compressed, smooth, becoming hollow, very brittle, attenuated; apex acute, or obtuse. Stem usually very distinct, 1–2 cm. long. Flesh *white*, loose. Spores white, subglobose, minutely apiculate, 7–10 × 6–9μ, guttulate, then granular; "basidia 30–35 × 7–8μ, with 4-sterigmata, contents granular. Internal structure pseudoparenchymatous in transverse section, cells av. 10μ in diam." Cotton. Isolated, or in twos or threes. Amongst short grass in woods, pastures, shady lawns, and in flower-pots. Sept.—Nov. Not uncommon. (*v.v.*)

2451. **C. Crosslandii** Cotton.
Charles Crossland, the well-known Yorkshire mycologist.

R. 2–3 cm. high, 1–3 mm. thick, *greyish white, or grey,* becoming darker with age, cylindrical, apex usually pointed, brittle. Stem hardly distinct. Flesh *somewhat darker than the hymenium.* Spores white, pip-shaped, 4–5 × 2·5–3μ; basidia 20–25 × 4–5μ, with 4-sterigmata, contents granular. Internal structure pseudoparenchymatous in transverse section, cells 5–8μ in diam. Isolated, or fasciculate. Amongst short grass in woods. Sept. Rare.

2452. **C. tenerrima** Massee & Crossl. (? = *Clavaria acuta* (Sow.) Fr. sec. Cotton.) *Tenerrima,* very delicate.

R. 1·5–4 cm. high, 1–2 mm. thick, *hyaline or shining white,* simple, cylindrical, flexuose, smooth, subattenuated above, but by no means acute, not narrowed at the base, stuffed, firm. Spores hyaline, verrucose, subglobose, apiculate, 8–9μ; basidia with 4-sterigmata. Gregarious but distinct at the base. Amongst short grass.

C. uncialis Grev. = **Pistillaria uncialis** (Grev.) Cost. & Dufour.

***Rosy.

2453. **C. incarnata** Weinm. *Incarnata,* flesh colour.

R. 1–4 cm. high, 2–3 mm. thick, *flesh colour, or rosy,* cylindrical, or compressed, pointed, or blunt, *pruinose,* base *whitish,* hairy, solid. Flesh *purple, sometimes white,* firm. Spores white, broadly elliptical, or pear-shaped, 10 × 6–7μ, 1-guttulate; "basidia 35–40 × 7–8(–10)μ, contents finely granular, with 4 erect sterigmata. Hyphae loosely interwoven, frequently septate, cells 50–100 × 5–10μ, with trumpet-shaped expansions in the subhymenial layer, pseudoparenchymatous in transverse section" Cotton & Wakef. Gregarious. Bare soil in woods. Aug.—Oct. Uncommon. (*v.v.*)

Typhula (Pers.) Fr.

(*Typha,* the reed-mace.)

Receptacle fleshy, waxy, or tough, erect, simple, very rarely branched, cylindrically clavate, with a long, thin stem, often springing from a sclerotium. Hymenium smooth, confined to the clavate portion of the receptacle. Spores white, oblong, ovate, subglobose, pip-shaped, or subcylindrical, smooth; basidia clavate, with 2–4-sterigmata. Cystidia none, or inconspicuous. Growing on dead leaves, herbaceous stems, twigs and wood.

*Springing from a sclerotium at the base.

2454. **T. erythropus** (Bolt.) Fr. Grev. Scot. Crypt. Fl. t. 43, as *Phacoriza erythropus*. ἐρυθρός, red; πούς, foot.

R. 1–2 cm. high, club *white*, 4–6 mm. long, cylindrical, linear, smooth. Stem *reddish black*, 12 mm., filiform, often twisted, springing from an elliptical, brown, or blackish sclerotium (sclerotium sometimes wanting), clothed with concolorous hairs at the base. Spores white, oblong, or subcylindrical, sometimes depressed on the side, 6–9 × 3–4μ. Dead leaves, and sticks, and dead herbaceous stems. Sept.—Dec. Common. (*v.v.*)

2455. **T. phacorrhiza** (Reich.) Fr. Stev. Brit. Fung. II, p. 304, fig. 94. φακός, a lentil; ῥίζα, root.

R. 2·5–7·5 cm. high, brownish, *filiform*, acute, *paler* and tomentose at the base, flexuose, springing from a *compressed*, pallid, then fuscous sclerotium. Flesh *concolorous*. Spores white, subglobose, 7–8 × 7μ, with a large central gutta. Dead leaves, and herbaceous stems. Oct.—Nov. Not uncommon. (*v.v.*)

2456. **T. incarnata** (Lasch) Fr. Grev. Scot. Crypt. Fl. t. 93, as *Phacorhiza filiformis*. *Incarnata*, flesh colour.

R. 5–10 cm. high, *whitish, flesh colour upwards*, filiform, cylindrical, attenuated and *hairy downwards*, springing from a *compressed*, fuscous sclerotium. Spores white, "subglobose, 5 × 4μ" Massee. Dead herbaceous stems. Sept.—Oct. Uncommon.

2457. **T. gyrans** (Batsch) Fr. γῦρος, a circle.

R. 1–4 cm. high, *white*, very slender, subcylindrical, or subfusiform, 3–5 mm. long; stem *hyaline*, diaphanous, filiform, *pubescent*, springing from an elliptical, pallid, then fuscous sclerotium. Spores white, "oblong, 5–6 × 2μ" Pat. Straw, dead stems of grasses, and twigs. Oct. Uncommon.

T. muscicola (Pers.) Fr. = **Eocronartium muscicola** (Pers.) Fitzpat.

**No sclerotium at the base.

2458. **T. Grevillei** Fr.
Dr Robert Kaye Greville, the eminent Scotch mycologist.

R. 5–12 mm. high, *white*, club *elliptical*, 2–3 mm., obtuse. Stem *filiform*, *pilose*. Spores white, pip-shaped, or oblong with a lateral apiculus, 4–5 × 2μ. Dead leaves, especially alder, and poplar. Sept.—Dec. Common. (*v.v.*)

2459. **T. filiformis** (Bull.) Fr. Bull. Hist. Champ. Fr. t. 448, fig. 1, as *Clavaria filiformis*. *Filum*, a thread; *forma*, shape.

R. 2·5–7·5 cm. high, *club white*, cylindrical, fusiform; stem *date brown*, filiform, *somewhat branched, decumbent*, smooth. Spores white, "about 5–4μ" Massee. Amongst dead leaves. Nov. Rare.

2460. **T. gracillima** White ex B. & Br. *Gracillima*, very slender.

R. ·5–2 cm. high, *white*, club elongate; stem very slender, curved, smooth. Various herbaceous plants.

2461. **T. gracilis** Berk. & Desm. B. & Br. Ann. and Mag. Nat. Hist. no. 84, t. 8, fig. 1. *Gracilis*, slender.

R. 2–3 mm. high, club *pallid, simple, or forked*, acute, rough with spores and little prominent bristles; stem short, distinct, smooth, or strigose with "hyaline, flexuose hairs, 100–200 × 1–3μ. Spores white, oblong, 7–9 × 3·5–4μ; basidia 21–28 × 6μ, with 2-sterigmata. Hyphae 3–9μ in diam." Bourd. & Galz. Rotten leaves. Dec. Uncommon.

2462. **T. tenuis** (Sow.) Fr. Sow. Eng. Fung. t. 386, fig. 5, as *Clavaria tenuis.* *Tenuis*, thin.

R. 4–6 mm. high, *fuscous black*, club thickened, smooth; stem filiform. Rotten wood in cellar. Rare.

2463. **T. translucens** B. & Br. (Not a fungus sec. Massee.) *Translucens*, transparent.

R. *white*, minute, pellucid, club irregular, somewhat obovate; stem short, thickened upwards. On the ground. Oct. Rare.

T pusilla Schroet. = **Pistillaria pusilla** (Pers.) Fr.

Pistillaria Fr.

(*Pistillum*, a pestle.)

Receptacle fleshy, or waxy, erect, simple, very rarely forked, club-shaped, with a short, thick, glabrous, or villose stem, rarely springing from a sclerotium. Hymenium smooth, confined to the clavate portion of the receptacle. Spores white, oblong, elliptical, subcylindrical, pruniform oblong, oblong elliptical, or sausage-shaped, smooth; basidia clavate, with 1–2–4-sterigmata. Cystidia none, or inconspicuous. Growing on dead herbaceous stems, and leaves.

*Rosy.

2464. **P. micans** (Pers.) Fr. Boud. Icon. t. 177. *Micans*, sparkling.

R. ·5–1 mm. high, club *glistening rose colour*, obovate, or oblong, obtuse, pruinose; stem *white*, or *concolorous*, cylindrical. Flesh *concolorous*, soft. Spores white, elliptical, or oblong elliptical, 8–12 × 6–7μ; basidia 35–40 × 7–8μ, with 2–4-sterigmata. Dead herbaceous stems, and leaves, especially thistles. July—Feb. Not uncommon. (*v.v.*)

**Tawny.

2465. **P. fulgida** Fr. Sow. Eng. Fung. t. 391, as *Clavaria minuta*. *Fulgida*, shining.

R. 1–3 mm. high, club *tawny*, or *tawny orange*, cylindrical, or lanceolate, subacute, flexuose, smooth; stem *whitish*, or *yellow*, equal, or constricted at the apex. Spores white, "oblong or subcylindrical, obliquely attenuated at the base, 8–10 × 4·5–5μ; basidia 18–25 × 7–9μ, with 2–4-sterigmata" Bourd. & Galz. Stems of *Dipsacus pilosus*, and *Helianthus tuberosus*. Sept. Rare.

***White, or yellowish.

2466. **P. culmigena** Mont. & Fr. Ann. Sc. Nat. (1836), t. 12, fig. 2. *Culmus*, stalk; *gena*, borne.

R. 1·5–3 mm. high, *hyaline pellucid*, club ovate, obtuse, compressed; stem distinct, very short. Flesh *white*, soft, then cartilaginous. Spores white, "elliptical cylindrical, 6–7μ" Quél. Dead grass stems. Oct.—Jan. Not uncommon.

2467. **P. quisquiliaris** Fr. Sow. Eng. Fung. t. 334, fig. 1, as *Clavaria obtusa*. *Quisquiliaris*, pertaining to refuse.

R. 3–8 mm. high, *whitish*, club-shaped, oval, rarely bifid, more or less compressed, attenuated downwards, *smooth*, sometimes springing from a minute sclerotium. Flesh white, soft, then rigid. Spores white, sausage-shaped, or oblong, and depressed on one side, 12–15 × 5–6μ, contents granular. Dead fern stems, especially bracken. April—Dec. Common. (*v.v.*)

2468. **P. ovata** (Pers.) Fr. *Ovata*, ovate.

R. 4–6 mm. high, *white*, club ovate, or obovate, subglobose, pyriform, or turbinate, *hollow*, attenuated downwards into a smooth, diaphanous stem. Spores white, "pruniform oblong, 12μ" Quél. Dead elm, and bramble leaves. Rare.

2469. **P. furcata** W. G. Sm. ex Cke. *Furcata*, forked.

R. 2·5–3 cm. high, *white, or yellowish*, compressed, broad at the apex, attenuated downwards, generally furcate. Flesh waxy, then tough. Caespitose. In greenhouses.

2470. **P. uncialis** (Grev.) Cost. & Dufour. (= *Clavaria uncialis* Grev.; *Ceratella uncialis* (Grev.) Quél.) Grev. Scot. Crypt. Fl. t. 98, as *Clavaria uncialis*. *Uncialis*, a twelfth part.

R. 1–2·5 cm., *white, becoming yellowish with age*, cylindrical, gradually attenuated downwards, obtuse, smooth, naked, flexuose, stuffed.

Flesh *white, tough.* Spores white, "elliptical, 5 × 3μ" Massee. Dead stems of herbaceous plants, especially umbellifers. May—Nov. Uncommon.

2471. **P. puberula** Berk. Sow. Eng. Fung. t. 334, fig. 2, as *Clavaria obtusa.* *Puberula*, somewhat downy.

R. 2–6 mm. high, *white*, obovate, ventricose; stem short, distinct, equal, or attenuated upwards, pellucid, *tomentose.* Flesh *white*, becoming firm. Spores white, "elliptical, 5 × 3μ" Massee. Dead bracken stems. Sept.—Nov. Not uncommon. (*v.v.*)

2472. **P. pusilla** (Pers.) Fr. *Pusilla*, very small.

R. ·5–2 mm. high, *white, linear*, or slightly thickened upwards, nodding when dry, smooth, even, with no distinct stem. Flesh *white*, soft, then firm. Spores white, oblong elliptical, 10 × 4μ; basidia with 2-sterigmata. Dead leaves. Sept.—Feb. Not uncommon. (*v.v.*)

****Purple.

2473. **P. purpurea** W. G. Sm. *Purpurea*, purple.

R. 2 mm. high, *purple*, obovate. St. *whitish.* Spores white, 4 × 3μ. Damp rotten leaves. Rare.

Pterula Fr.

(πτερόν, a feather.)

Receptacle firm, tough, filiform, branched, or simple, branches equal. Hymenium smooth. Spores white, oval, elliptical, or pip-shaped, smooth; basidia with 2–4-sterigmata. Cystidia none, or inconspicuous. Growing on the ground, or on wood.

2474. **P. subulata** Fr. *Subulata*, awl-shaped.

R. 3–4 cm. high, *whitish cinereous*, densely branched, tense and straight, equal, except at the attenuated base, tough. Branches *becoming yellow*, growing into each other, *multifid at the apex*, awl-shaped, smooth, not thicker than a fine thread. Spores white, "oval, 8–10 × 5–7μ" Rabenh. Densely tufted. On wood amongst pines and in gardens. Uncommon.

2475. **P. multifida** Fr. Fr. Icon. t. 200, fig. 2. *Multifida*, many cleft.

R. 2·5–5 cm. high, *pallid whitish, then dirty pale yellowish*, and *finally brownish, almost black when dried, very much branched*, very delicate, flaccid, but slightly tough. Branches tense and straight, not much thicker than a hair, heaped as if swept together, subfastigiate, *apex spear-shaped, or crisped.* Spores white, pip-shaped, or elliptical, depressed on one side with a lateral apiculus, 6 × 3μ, contents granular. Dead branches and running over pine needles. Sept.—Nov. Uncommon. (*v.v.*)

EXOBASIDIINEAE.

**Parasites.

EXOBASIDIALES.

Hymenium effused, rarely consisting of basidia only. Parasitic on leaves, etc. (especially Ericaceae).

Exobasidiaceae.

Same characters as the order.

Exobasidium Woronin.

(ἐξ, out of; *basidium*, a basidium.)

Mycelium vegetating in the interior of the living host, and giving rise, on the exterior, to basidia. Hymenium discontinuous. Spores white, elongate fusiform, or oblong reniform, smooth, simple, or septate; basidia cylindrical, with 4–5–6-sterigmata. Cystidia none. Parasitic on living leaves and stems.

2476. **E. Vaccinii** (Fuck.) Woron. (= *Exobasidium Andromedae* Peck; *Exobasidium Azaleae* Peck; *Exobasidium Rhododendri* Cramer.) Engl. & Prantl, Nat. Pflanzenfam. I, 1**, fig. 65.
Vaccinium, whortle-berry.

Galls 1–2 cm., greyish, innate, effused, hypophyllous, or amphigenous, usually orbicular, or elongated, forming a scurfy or felted hymenium. Spores white, elongate fusiform, often curved, 10–20 × 2·5–5μ, sometimes 1-septate, and becoming 3-septate on germination. Leaves, young shoots, and flowers of *Vaccinium Myrtillus*, *Vaccinium Vitis-idaea*, *Rhododendron ferrugineum*, *Rhododendron hirsutum*, *Rhododendron Wilsonii*, and *Azalea pontica*. May—Oct. Not uncommon. (*v.v.*)

2477. **E. japonicum** Shirai. *Japonica*, Japanese.

Galls 1–3 cm., subcuticular. Spores oblong-reniform, 14·5 × 4μ. Leaves and terminal shoots of *Rhododendron indicum*. Uncommon.

HETEROBASIDIAE.

Basidia longitudinally divided, transversely septate, or simple; spores on germination producing sporidiola, or a mycelium, but the former only in the case of the simple basidia.

AURICULARIALES.

Basidia transversely septate, cylindrical, straight, or curved, consisting of either probasidia, or normal basidia. Spores producing sporidiola, or a mycelium on germination.

1. Parasites, with or without probasidia.

PUCCINIINEAE
COLEOSPORIINEAE } Not dealt with in the present work.
USTILAGINEAE

2. Saprophytes, without probasidia.

AURICULARIINEAE.

Hymenium fully exposed from the first.

AURICULARIACEAE.

Same characters as suborder.

Helicobasidium Pat.

(ἕλιξ, twisted; *basidium*, a basidium.)

Receptacle membranaceous, soft, floccose; effused, incrusting. Hymenium smooth. Basidia cylindrical, more or less incurved, transversely 2–4-septate, with subulate, unilateral sterigmata. Spores white, oval, or pear-shaped, smooth, producing, on germination, either sporidiola, or a mycelium. Growing on humus, or wood.

2478. **H. purpureum** (Tul.) Pat. (= *Corticium lilacinum* Quél.) Trans. Brit. Myc. Soc. III, t. 7. *Purpureum*, purple.

R. 3–6 cm., *dingy reddish purple, margin paler*, broadly effused, incrusting, indeterminate, adnate, inseparable. Hymenium *concolorous, then deep vinous colour*, and covered with a white pruina. Flesh *whitish*, floccose, loose, thin. Spores white, pear-shaped, 10–12 × 6–8μ; basidia cylindrical, incurved, 3–5μ in diam., 3-septate. Basal hyphae pale brown, 4–7μ in diam., septate, without clamp connections. Half buried ash bough amongst leaf debris. March. Rare. (*v.v.*)

Platygloea Schroeter.

(πλατύς, broad; γλοία, glue.)

Receptacle homogeneous, waxy, gelatinous, or coriaceous gelatinous, tubercular, wart-like, or consisting of spreading, or erect, convolute plates. Hymenium smooth, unilateral, or amphigenous; basidia cylindrical, straight, palisade-like, transversely septate, with long sterigmata. Spores white, oval, or elliptical, obtuse, or apiculate, straight, or curved; producing sporidiola on germination. Growing on dead wood.

2479. **P. effusa** Schroet. Trans. Brit. Myc. Soc. VI, text-figs. p. 138. *Effusa*, spread out.

R. *greyish, then whitish*, effused, closely adnate, thin, subgelatinous, firm. Hymenium *concolorous*, pulverulent under a lens. Spores white, smooth, elliptical, or ovate, with an oblique apiculus, 7–8–10 × 4–5μ;

basidia elongate, cylindrical, wavy, apex frequently incurved, and almost circinate, transversely 4-septate, 40–50 × 4–5μ. Subhymenial hyphae thin, guttulate, 1–2μ in diam., arising erect and parallel from a compact, pseudoparenchymatous basal stratum of broader hyphae, 4–5μ in diam. Fallen branches. June. Rare.

Auricularia (Bull.) (= **Hirneola** Fr. p.p.).

(*Auricula*, the ear.)

Receptacle gelatinous-coriaceous, cartilaginous when dry; dimidiate, or cup-shaped, substipitate, or sessile; consisting of three layers, the upper layer thin and compact, very rarely glabrous, generally tomentose with thick, cylindrical, simple, erect or decumbent hairs, the intermediate layer consisting of thin, gelatinous hyphae forming a compact tissue, and the lower layer forming the hymenium. Hymenium smooth, reticulate, or ribbed, fully exposed from the first. Basidia cylindrical, transversely 3-septate, with long, thin sterigmata, and forming a firm, palisade-like layer. Spores white, cylindrical, oblong, or subreniform, producing sporidiola on germination. Growing on wood.

2480. **A. mesenterica** (Dicks.) Fr. Rolland, Champ. t. 106, no. 242.
μέσος, middle; ἔντερον, intestine.

R. 5–30 cm., *fuscous cinereous, grey, or tawny*, resupinate, often cup-shaped, then confluent and reflexed, imbricate, flexuose, *villose, fasciato-zoned.* Hymenium *pale, or greyish, then fuscous violaceous,* costato-plicate, pruinose with the spores. Flesh *fuscous*, gelatinous, then cartilaginous. Spores white, cylindrical, curved, 17–20 × 6–7μ, 1–3-guttulate. Stumps and felled trunks, especially elm. Jan.—Dec. Common. (*v.v.*)

var. **lobata** (Sommerf.) Quél. Berk. Outl. Brit. Fung. t. 18, fig. 1, as *Auricularia lobata.* *Lobata*, lobed.

Differs from the type in the *lobed margin of the p.* Stumps. Jan.—Dec. Not uncommon. (*v.v.*)

2481. **A. auricula-Judae** (Linn.) Schroet. (= *Hirneola auricula-Judae* (Linn.) Berk.) Berk. Outl. Brit. Fung. t. 18, fig. 7, as *Hirneola auricula-Judae.* *Auricula*, ear; *Judae*, of a Jew.

R. 2–8 cm., *grey, then olivaceous, or brownish, and finally black,* cup-shaped, hemispherical, concave, then ear-shaped, flexuose, *plicate*, transparent, *tomentose with subbulbous hairs.* Hymenium *pale, then greyish and finally fuscous*, smooth, *then venoso-plicate.* Flesh *whitish*, gelatinous, then cartilaginous, tough. Spores white, oblong, or cylindrical, curved, 16–20 × 6–9μ, often 2-guttulate. Edible. Common on old elders, more rarely on beech, elm, oak, walnut, willow, holly and *Berberis arcuata.* Jan.—Dec. Common. (*v.v.*)

2482. **A. polytricha** (Mont.) Pat. (= *Hirneola polytricha* Mont.) Jungh. Fl. Crypt. Jav. Ins. t. 13, as *Exidia purpurascens*.
πολύθριξ, with much hair.

R. 2–8 cm., *greyish*, cup-shaped, hemispherical, then expanded and ear-shaped, sessile, or extended behind into a short, oblique stem; margin undulate, *densely covered on the outside with long, obtuse, greyish or tawny hairs*, 400 × 5–9 μ. Hymenium *pale, then brownish purple, and finally becoming blackish, smooth*. Flesh *pale, or brownish*, subgelatinous, then horny, tough, elastic, thick. Spores white, subreniform, "20–22 × 8 μ" Sacc.; basidia cylindrical, 85–90 × 6–7 μ, 3-septate. Wooden fire blocks. April—Nov. Uncommon. (*v.v.*)

Eocronartium Atkinson

(= **Clavaria** p.p.; **Helicobasidium** sec. Pat.).

(*ἠώς*, dawn; *cronartium*, the genus *Cronartium*.)

Receptacle subgelatinous, tough, filamentous, erect, filiform, or subulate. Hymenium smooth. Basidia cylindrical, transversely 3-septate. Spores white, continuous, producing a mycelium on germination. Growing on mosses.

2483. **E. muscicola** (Pers.) Fitzpat. (= *Typhula muscicola* (Pers.) Fr.) Pers. Obs. Myc. II, t. 3, fig. 2, as *Clavaria muscicola* Pers.
Muscus, moss; *colo*, I inhabit.

R. 5–7·5 cm. high, *white*, simple, filiform, clavate, club cylindrical, attenuated into a thin, long, *smooth* st., obtuse, 2–4 mm. thick. Flesh *white*, thin. Spores white, "fusoid, curved or inequilateral, granular, 18–24 × 3·5–5 μ" Atk. Basidia cylindrical, "curved, or more or less sinuous, 25–40 × 6–9 μ, 3–5-septate. Sterigmata flexuous elongate, usually 10–20 × 3–4 μ" Atk. Mosses. Aug.—Oct. Uncommon. (*v.v.*)

Stilbum (Tode) Juel.

(*στίλβω*, I shine.)

Receptacle erect, globose, stipitate. Hymenium consisting of branched threads terminated by a basidium. Basidia short, pear-shaped, transversely 1-septate. Spores white, elliptical. Growing on dead wood.

2484. **S. vulgare** (Tode) Juel. Tode, Fung. Mecklenb. t. 2, fig. 16.
Vulgare, common.

R. 1–2 mm. high, *white, then yellowish*, globose. St. *concolorous*, becoming thinner upwards, fibrous, smooth. Spores white, "elliptical, hyaline, 8 × 5–6 μ" Massee. Rotten wood, oak cupules, etc. Sept.—Nov. Common. (*v.v.*)

ECCHYNINEAE.

Hymenium inclosed within a peridium.

Ecchynaceae.

Same characters as suborder.

Ecchyna Fr.

(ἐκχύνω, I pour out.)

Peridium fibrillose, subglobose, stipitate, or substipitate, thin, fugacious. Gleba threads radiating, branched, flexuose at the ends. Basidia cylindrical, straight, or curved, transversely 3-septate; bearing sessile, or very short pedicellate spores; scattered, or in tufts on the lower portions of the threads. Spores fawn colour, elliptical, or subglobose, smooth, producing sporidiola on germination. Growing on wood.

2485. **E. faginea** (B. & Br.) Fr. (= *Pilacre faginea* B. & Br.; *Pilacre Petersii* Berk. & Curt.[1]) B. & Br. Ann. Nat. Hist. no. 380, t. 11, fig. 5, as *Pilacre faginea*. *Faginea*, pertaining to beech.

P. 3–6 mm. high, 1–3 mm. across, *whitish, or fawn colour*, forming a globose head, smooth, then pulverulent. St. *concolorous*, or *becoming black with age*. Flesh *fawn colour*, floccose, loose. Spores fawn colour, elliptical, or subglobose, depressed or umbilicate on the one side, 5–6 × 4–5μ; basidia cylindrical, 3-*septate*. Hyphae concolorous, 3–4μ in diam., *with numerous clamp connections*. Beech, ash, hornbeam, and holly. Sept.—Jan. Not uncommon.

TREMELLALES.

Basidia subglobose, longitudinally, or vertically, cruciately divided into 2–4 parts. Spores producing sporidiola or a mycelium on germination.

Tremellaceae.

Same characters as the order.

Tremella (Dill.) Fr.

(*Tremo*, I tremble.)

Receptacle gelatinous, or waxy, soft; foliaceous, brain-like, or tubercular. Hymenium spread over the whole surface, very rarely papillate. Basidia amphigenous, superficial, or immersed. Conidia on the same receptacle, preceding, or accompanying the spores. Spores white, rarely yellowish, globose, subglobose, oval, elliptical,

[1] Bayliss-Elliott and Grove hazard the opinion that both these species are only conidial forms of *Roesleria pallida* Sacc., but the transversely septate basidia, and frequent clamp connections of the hyphae seem to disprove this conclusion.

or pyriform; smooth, or punctate; producing sporidiola on germination, or tufts of conidia, that bud in a yeast-like manner. Hyphae filamentous, thin, gelatinous, sometimes inclosing mineral concretions, which form an irregular central nucleus. Growing on wood, rarely on the ground.

I. Foliaceous, divided up into lobes, and variously twisted.

2486. **T. fimbriata** (Pers.) Fr. Bull. Hist. Champ. Fr. t. 272, as *Tremella verticalis*. *Fimbriata*, fringed.

R. 5–7·5 cm. high and broad, *olivaceous, inclining to black, erect, corrugated*; lobes flaccid, incised at the margin, undulato-fimbriate. Flesh gelatinous, very soft. Spores white, "subpyriform" Massee, "5–7μ long" Möll. Caespitose. Darkening water when soaked in it, and staining the fingers black. Dead branches, especially alder. Nov.—Feb. Rare.

2487. **T. frondosa** Fr. Bull. Hist. Champ. Fr. t. 499, fig. T, as *Tremella quercina*. *Frondosa*, leafy.

R. 10–12 cm., *yellow inclining to pale, or pale pinkish yellow*, spathulate, or tongue-shaped, large, *even*, plicate at the base; lobes gyroso-undulated. Flesh *concolorous*, gelatinous, thin, except at the base. Spores white, subglobose, apiculate at the base, 8–10μ, 1-2-guttulate. Caespitose. Stumps, and fallen branches, especially oak. July—Dec. Not uncommon. (*v.v.*)

2488. **T. foliacea** (Pers.) Fr. non Bref. Bres. Fung. Trid. II, t. 209, fig. 1. *Foliacea*, leafy.

R. 3–10 cm., *cinnamon flesh colour, rarely deep brown, or tinged umber violaceous, very much lobed and waved*, segments thin, springing from a plicate base. Flesh *paler*, gelatinous, subdiaphanous. Spores white, minutely punctate, globose, 5–7μ; basidia 13–16 × 10–14μ. Hyphae 1–4μ in diam. Caespitose. Coniferous stumps, and branches, more rarely on deciduous trees. July—Nov. Not uncommon. (*v.v.*)

2489. **T. lutescens** Pers. Pers. Icon. et Descr. t. 8, fig. 9. *Lutescens*, becoming yellow.

R. 1–3 cm., *yellowish*, tubercular, then *undulato-gyrose*, lobes entire, naked. Flesh *yellowish*, gelatinous, *subdeliquescent*, pellucid. Spores white, "oval elliptic, 10–16 × 7–10μ, 1-pluri-guttulate; basidia ovoid, 19–25 × 17–18μ. Hyphae 1–3μ in diam." Bourd. & Galz. Stumps, and fallen branches. Aug.—May. Common. (*v.v.*)

II. Brain-like, with obtuse, and twisted veins.

2490. **T. mesenterica** (Retz.) Fr. Rolland, Champ. t. 105, no. 237. μέσος, middle; ἔντερον, intestine.

R. 1–8 cm., *orange*, variously contorted, brain-like, plicato-undulate, *gyrose*, pruinose with the spores. Flesh *concolorous*, gelatinous, becoming firm, tough. Spores white, broadly elliptical, 13–14 × 7–8μ; "basidia 15–20 × 12–18μ. Conidia ovoid globose, 3–5μ in diam. Hyphae 2–3μ in diam." Bourd. & Galz. Dead branches, sticks, rails, furze, broom, and ivy. Jan.—Dec. Common. (*v.v.*)

2491. **T. vesicaria** Sm. Engl. Bot. t. 2451. *Vesicaria*, bladdery.

R. 5 cm. high and broad, *pallid, or yellowish*, erect, undulated, gyrose. Flesh *pale*, bladdery, very viscid. Spores "11 × 6μ" B. & Br. On the ground. Sept. Rare.

2492. **T. albida** (Huds.) Fr. Engl. Bot. t. 2117. (= *Dacryomyces hyalinus* Pers. sec. Quél.) *Albida*, whitish.

R. 1–4 cm., *whitish, hyaline, becoming brownish*, erumpent, undulated, *somewhat gyrose*, pruinose. Flesh *whitish*, gelatinous, subdiaphanous, fairly thick. Spores white, globose, often with an obtuse apiculus at the base, 9–10μ; "basidia 12–23 × 12–15, with 100μ and more long sterigmata" Bourd. & Galz. Hyphae hyaline, 2–3μ in diam., with a few clamp connections. Fallen branches, sticks and rails. Sept.—May. Common. (*v.v.*)

III. Crustaceous, effused, smooth.

2493. **T. epigaea** B. & Br. (= *Sebacina laciniata* (Bull.) Bres. f. *epigaea* B. & Br. sec. Bourd. & Maire.) Ann. Nat. Hist. ser. 2, II, t. 9, fig. 3. ἐπίγαιος, upon the earth.

R. 3–10 cm., *white*, effused, *gyroso-plicate*, sprinkled with the white spores. Flesh *white*, gelatinous, thin. Spores white, subglobose, 6 × 4μ. Naked soil. Aug.—Sept. Rare.

IV. Tubercular, small, suberumpent.

2494. **T. violacea** (Relh.) Fr. *Violacea*, violet.

R. 3–8 mm., *violaceous, becoming black when dry*, erumpent, subcompressed, *gyrose*. Flesh paler, gelatinous, then firm, 2 mm. thick. Spores "cream citron, punctate, ovoid, 5μ" Quél. Gregarious. Pear trunks. Rare.

2495. **T. indecorata** Sommerf. Fr. Icon. t. 200, fig. 4. *Indecorata*, ugly.

R. 4–30 mm., *dingy, livid, or olivaceous, becoming fuscous*, black *fuscous when dry*, erumpent, convex when moist, then *plicate*. Flesh *dingy*, gelatinous, then firm, opaque. Spores white, globose, 6–9μ, minutely punctate. Birch, oak, willow, and poplar. Oct.—Nov. Rare.

2496. **T. moriformis** Berk. (= *Tremella nigrescens* (Fr.) Quél.) Engl. Bot. t. 2446. *Morus*, mulberry; *forma*, shape.

R. 1–2·5 cm., *mulberry black*, erumpent, *spherical*, sinuous. Flesh *deep violet*, translucent, subgelatinous, firm. Spores "subglobose, apiculate, 5 × 4μ" Massee. Stains paper violet, and gives a purple colour when treated with potash. Dead branches of maple, chestnut, *Robinia*, elm, and plane. June—Feb. Rare.

2497. **T. tubercularia** Berk. *Tuberculum*, a tubercle.

R. 4–8 mm. high, *dingy white*, *nearly black when dry*, erumpent, *stipitate*, *head pileate*. St. short, round. Flesh semi-transparent, horny. Spores "subglobose, with a large, obtuse, lateral apiculus, 5–7 × 5–8μ" Bourd. & Maire. Conidia straight, or slightly curved, 2 × ·5μ. Fallen branches, especially oak. Oct.—Dec. Frequent. (*v.v.*)

T. torta Berk. = **Dacryomyces tortus** (Berk.) Massee.

2498. **T. versicolor** Berk. *Versicolor*, changeable in colour.

R. 2–3 mm., *orange*, *at length brown*, *orbicular*, tear-like, convex. Flesh gelatinous, firm. Spores white, "broadly elliptical, apiculate, 6 × 4μ" Massee. *Peniophora nuda*, ash, and briar. Feb.—May. Uncommon.

2499. **T. atrovirens** Fr. *Ater*, black; *virens*, green.

R. 1–6 mm., *sooty green when moist*, *blackish when dry*, erumpent, pulvinate, gregarious, minutely papillate and rugose. Spores white, "elliptical, 10–12 × 7–9μ; basidia 21–25 × 8–15μ. Hyphae 1–2μ in diam." Bourd. & Galz. Dead broom branches. Oct. Rare.

2500. **T. clavata** (Pers.) Berk. (? = the conidial form of *Coryne sarcoides* (Jacq.) Tul.) Pers. Icon. Pict. t. 10, fig. 2. *Clavata*, club-shaped.

R. 2·5 cm. high, 4 mm. broad, *reddish flesh colour*, *becoming black at the base*, club-shaped, solitary, simple, slightly twisted. Stumps. Rare.

T. sarcoides Sm. = **Coryne sarcoides** (Jacq.) Tul.

T. foliicola Fuck. ? = **Hypocrea rufa** (Pers.) Fr. sec. W. G. Sm.

V. With a firm, hard nucleus.

2501. **T. encephala** (Willd.) Quél. (= *Naematelia encephala* (Willd.) Fr.) Willd. Bot. Mag. I, t. 4, fig. 14, as *Tremella encephala* Willd. ἐν, in; κεφαλή, head.

R. 1–3 cm., *pallid flesh colour*, pulvinate, subsessile, diaphanous, plicato-rugose, white pruinose, rooting. Nucleus *white*, large, hard.

Spores white, "pear-shaped, or subglobose, 12–16 × 10μ, or 9–10μ" Karst. Solitary, or clustered. Conifer branches. Sept.—March. Not uncommon. (*v.v.*)

2502. **T. rubiformis** (Fr.) Quél. (= *Naematelia rubiformis* Fr.) Corda, Icon. I, fig. 299, A. *Rubus*, blackberry; *forma*, shape.

R. 4–5 mm., *yellow*, hemispherical, subsessile, gyróso-tuberculose. Nucleus small. Dead branches, twigs, and moss. Oct. Rare.

2503. **T. virescens** (Schum.) Quél. (= *Naematelia virescens* Cda. sec. Fr.) Corda, Icon. III, t. 6, fig. 90, as *Naematelia virescens* Cda. *Virescens*, becoming green.

R. 5–6 mm., *greenish*, suborbicular, sessile, *depressed*, gyroso-tubercular. Flesh subgelatinous. Spores white, "elliptical, apiculate, 18 × 11μ" Massee. Rotten wood, furze, ivy. Jan.—Dec. Not uncommon. (*v.v.*)

Phaeotremella Rea.

(φαιός, dark; *tremella*, the genus *Tremella*.)

Same characters as *Tremella*, but the spores dark coloured.

2504. **P. pseudofoliacea** Rea. Trans. Brit. Myc. Soc. III, t. 20. ψευδής, false; *foliacea*, the species *T. foliacea*.

R. 4–10 cm., *somewhat cinnamon*, very much lobed, undulate, smooth, plicate at the base. Flesh concolorous, gelatinous, subdiaphanous. Spores *umber*, globose, or broadly obovate, 12 × 9–12μ. Conidia hyaline, elliptical, 9 × 6μ. Stumps, and posts. May—Nov. Not uncommon. (*v.v.*)

Guepinia Fr. (= **Gyrocephalus** Pers.).

(Jean Pierre Guepin, a celebrated botanist.)

Receptacle gelatinous, firm, erect, ear-shaped, spathulate, or infundibuliform, substipitate, or sessile. Hymenium inferior, smooth, or indistinctly veined. Basidia ovoid, vertically cruciately divided into one or two compartments, with long sterigmata. Spores white, oblong, or oval, smooth, producing sporidiola on germination. Growing on the ground, and on wood.

2505. **G. helvelloides** (DC.) Fr. (= *Gyrocephalus rufus* (Jacq.) Bref.) Rolland, Champ. t. 105, no. 240, as *Guepinia rufa*. *Helvella*, a genus of fungi; εἶδος, like.

R. 5–10 cm. high, 4–6 cm. broad, *rosy orange, becoming red, erect*, variable in form, subspathulate, deeply infundibuliform, *like Craterellus cornucopioides*, substipitate, or sessile, diaphanous, entire, or lobed. Hymenium *concolorous*, smooth, or slightly wrinkled, white pruinose. Flesh *reddish,* gelatinous, then cartilaginous, firm. Spores white, oblong, depressed on one side, 10–12 × 4–5μ; "basidia ovoid, or oblong, 16–21 × 10–12μ. Subhymenial hyphae granular, 1·5–3μ in diam." Bourd. & Galz. Under conifers. Sept.—Oct. Uncommon. (*v.v.*)

Exidia Fr.

(ἐξιδίω, I exude.)

Receptacle gelatinous, soft, pellucid; globose, or hemispherical, marginate, substipitate, or sessile, sterile on the upper surface. Hymenium inferior, smooth, reticulately veined, foliaceous, even, or papillose with short, sterile papillae. Basidia deeply immersed in the gelatinous hyphae, and covered by a layer traversed by the sterigmata. Spores white, rarely tinged brownish, allantoid, cylindrical, or oblong, smooth, producing on germination, either strongly curved, or straight and rod-like sporidiola, or bunches of cylindrical conidia. Growing on wood.

I. Spores on germination giving rise to strongly curved sporidiola.

2506. **E. truncata** Fr. *Truncata,* cut off.

R. 1–2 cm. high and broad, *brown bistre, truncato-plane,* hemispherical, soft, *rough with dots,* or granular beneath. St. *concolorous,* very short, becoming cavernous. Hymenium *very black,* shining, *glandular with concolorous papillae.* Flesh *brownish,* gelatinous, not becoming swollen when moistened. Spores "white, very rarely tinged with brownish bistre, cylindrical, curved, 14–20 × 4·5–6μ; basidia ovoid, 13–18 × 11–13μ. Hyphae 1–3μ in diam., septate, with clamp connections" Bourd. & Galz. Lime, and willow branches. Dec. Rare.

2507. **E. intumescens** (Sm.) Rea. (= *Tremella intumescens* Sm.) Engl. Bot. t. 1870, as *Tremella intumescens.*

Intumescens, swollen up.

R. 2–5 cm., *brown, becoming black when dry,* rounded, or conglomerate, somewhat tortuously lobed, resembling the intestines of some animal, *obsoletely punctate.* Flesh *whitish, streaked with fuscous,* gelatinous, shrivelling to a mere skin when dry. Spores white, "oblong, slightly curved, 13 × 4μ" Karst. Fallen beech trunks, branches, and rails. Oct.—Jan. Uncommon. (*v.v.*)

2508. **E. recisa** (Ditm.) Fr. Ditm. in Sturm's Deutschl. Fl. I, t. 13, as *Tremella recisa* Ditm. *Recisa,* cut off.

R. 1–3 cm., *fuscous amber colour, truncato-plane,* hemispherical, or obconical, somewhat repand, rough with dots beneath. St. *concolorous,* short, oblique, *excentric.* Hymenium *brighter coloured,* plane, often veined. Flesh *concolorous,* gelatinous, soft, subdiaphanous. Spores white, oblong, or cylindrical, curved, 15–18 × 4–5μ. Willow, sloe, and poplar branches. Sept.—Dec. Not uncommon. (*v.v.*)

2509. **E. glandulosa** (Bull.) Fr. Bull. Hist. Champ. Fr. t. 420, fig. 1. *Glandulosa*, full of glands.

R. 5–10 cm., *blackish*, globose, or lens-shaped, truncate, or pendulous, *somewhat plane, undulate, cinereous and subtomentose beneath*, feeling like black crape. Hymenium *concolorous, studded with conical papillae*. Flesh *blackish*, gelatinous, *diaphanous*, soft. Spores white, oblong, or cylindrical, curved, 12–15 × 4–5μ; basidia ovoid, 15–21 × 9–11μ. Hyphae 1–3μ in diam., with clamp connections. Dead branches, especially lime. Aug.—May. Common. (*v.v.*)

2510. **E. nucleata** (Schwein.) Rea. (= *Naematelia nucleata* (Schwein.) Fr.; *Tremella gemmata* Lév. sec Quél.; *Exidia gemmata* (Lév.) Bourd. & Maire.) Demid. Exped. t. 4, fig. 1, as *Tremella gemmata* Lév. *Nucleata*, having a kernel.

R. 2–10 mm., *hyaline, or tinged with amethyst or lilac, then opaline and finally brick or flesh colour, date brown when dry*, tubercular, round, then pulvinate and undulato-plicate, finally effused, and confluent, 4–5 cm., either inclosing a whitish separable core of oxalate of lime, or without a core (*Tremella hyalina* Pers.). Spores white, cylindrical, more or less curved, 10–14 × 5–7μ; basidia ovoid, 12–16 × 9–12μ. Hyphae 1·5–3μ in diam. with indistinct clamp connections. Rotten wood, and fallen branches. Jan.—Dec. Not uncommon. (*v.v.*)

2511. **E. Thuretiana** (Lév.) Fr. G. Thuret.

R. 1–4 cm., *opalescent when fresh, sometimes tinged with pink*, effused in thick, undulating, pulvinate, or tuberculate patches, firm, gelatinous. Hymenium *concolorous*, pruinose, finally collapsing into a thin, horny, *yellowish film*. Spores hyaline, cylindrical, curved, 15–20 × 5–7μ; basidia longitudinally septate, 15–21 × 11–15μ. Hyphae 1–2·5μ in diam. Underside of sticks, especially beech. Jan.—May. Not uncommon. (*v.v.*)

2512. **E. viscosa** (Berk.) Rea. (= *Tremella viscosa* Berk.) Boud. Icon. t. 180, as *Tremella viscosa* Berk. *Viscosa*, sticky.

R. 1–4 cm., *white, with a greyish, or violaceous tinge*, flattened, undulate, *smooth*, slightly viscid; margin more or less sinuate. Flesh *whitish*, gelatinous, diaphanous, somewhat thin. Spores white, oblong, or sausage-shaped, often curved, 16–22 × 6–7μ. Rotten wood, and fallen branches. Jan.—Dec. Common. (*v.v.*)

II. Spores on germination giving rise to straight, or rod-like sporidiola.

2513. **E. saccharina** Fr. (= *Ulocolla saccharina* (Fr.) Bref.) *Saccharina*, like sugar.

R. 3–8 cm., *tawny cinnamon, or the colour of crystallised sugar*, effused, tubercular, *gyroso-undulated*, sprinkled with a few minute, obtuse, vanishing papillae. Flesh thick. Spores white, "cylindrical, curved, 12–18 × 4·5–6μ; basidia ovoid, 15–22 × 9–12μ. Hyphae 1–3μ in diam., with clamp connections" Bourd. & Galz. Coniferous trunks, and branches. Nov. Uncommon.

var. **foliacea** (Bref.) Bres. (= *Ulocolla foliacea* Bref.) Bref. Unters. Heft VII, t. VI, fig. 2. *Foliacea*, leafy.

Differs from the type in its *larger size, and lobes*. Coniferous trunks, and branches. Nov. Uncommon.

Tremellodon Pers.

(*Tremo*, I tremble; ὀδών, a tooth.)

Receptacle gelatinous, soft, dimidiate, or spathulate, substipitate, or sessile. Hymenium covering fertile spines or teeth. Basidia globose, or ovoid, longitudinally cruciately septate. Spores white, subglobose, smooth, producing a mycelium on germination. Growing on wood.

2514. **T. gelatinosum** (Scop.) Pers. Boud. Icon. t. 178. *Gelatinosum*, jelly-like.

R. 3–6 cm., *glaucous fuscous, or tawny brownish*, dimidiate, spathulate, or fan-shaped, rounded in front, attenuated behind, sessile or substipitate, *surface papillose*. Spines *white, or glaucous*, 2–4 mm. long, conical, straight, gelatinous. Flesh hyaline, gelatinous, transparent, thick. Spores white, subglobose, 4–7μ, multi-guttulate; basidia globose, longitudinally septate, 14–18 × 10–12μ, with 2–4-sterigmata. Edible. Coniferous stumps and on the ground. Sept.—Dec. Not uncommon. (*v.v.*)

Protodontia von Hoehn.

(πρῶτος, first; ὀδούς, a tooth.)

Like the genus *Odontia* in appearance, but possessing vertically septate basidia. Growing on wood.

2515. **P. uda** von Hoehn. Trans. Brit. Myc. Soc. VI, p. 69, text-figs. *Uda*, moist.

R. 1–2 cm., *pure white and slightly hyaline when fresh, yellowish* when dry, effused, indeterminate, very soft and delicate. Spines *hyaline, then yellowish* and collapsing, slender, acute up to 400μ long, 100–150μ wide at the base. Flesh very thin, sometimes almost wanting, white and mealy when dry. Spores white, elliptical, one side flattened, 6–8(–9) × 3–4μ; basidia globose, immersed, vertically septate, 7–8μ in diam., with 2–4-sterigmata, 8–10 × 1·5μ. Hyphae very fine and closely adherent. Very soft, rotten wood. Dec. Rare.

Sebacina Tul. (= **Thelephora** (Ehrb.) Fr. p.p.).

(*Sebacina*, greasy.)

Receptacle coriaceous, gelatinous, membranaceous, waxy, floccose, or pulverulent, resupinate, effused, adnate, or crustaceous, and with the habit of a *Corticium*. Hymenium smooth, or papillose. Basidia longitudinally, cruciately divided, close together, or scattered, sometimes intermixed with the conidiophores. Spores white, cylindrical, oval, oblong, reniform or globose, smooth; producing sporidiola, or bunches of conidia on germination. Growing on the ground, or on wood.

Subgenus **Eusebacina** Rea.

(εὖ, typical; *Sebacina*, the genus *Sebacina*.)

2516. **S. incrustans** (Pers.) Tul. (= *Thelephora sebacea* (Pers.) Fr.; *Thelephora cristata* (Pers.) Fr.) Berk. Outl. Brit. Fung. t. 17, fig. 6, as *Thelephora sebacea*. *Incrustans*, covering with a coat.

R. 5–10 cm. high, *whitish*, effused, incrusting, very variable in form, tubercular, or resembling stalactites, apices somewhat subulate, or fringed. Hymenium *whitish*, even, flocculoso-pruinose. Flesh *whitish* waxy, then *coriaceous*, firm, loose. Spores white, oblong, curved or flattened on the one side, 11–13 × 4–5μ, often guttulate; "basidia ovoid, 15–20 × 12–15μ. Hyphae firm, 2·5–3μ in diam." Bourd. & Galz. Incrusting grass, twigs, stems, mosses, leaves, etc. Feb.—Nov. Common. (*v.v.*)

2517. **S. calcea** (Pers.) Bres. (= ? *Corticium calceum* (Pers.) Fr.) Bres. Fung. Trid. t. 175. *Calcea*, limy.

R. 3–10 cm., *shining white*, or chalk white *becoming somewhat buff or greyish*, effused, closely adnate, crustaceous, slightly pulverulent; margin mealy, thinner. Hymenium *concolorous*, smooth, or papillose and somewhat tuberculose, at length cracked. Flesh *whitish*, subgelatinous, floccose. Spores white, "reniform, curved, 15–16 × 7–8μ; basidia at first globose, then more elongated and longitudinally, cruciately septate, 18–25 × 12–16μ, with sterigmata 30 × 3–4μ" Wakef. "Paraphyses branched at the apex into very fine branches, loaded with minute granules. Basal hyphae 2μ in diam., the wall gelatinously modified" Burt. Bark and dead wood. Sept.—April. Uncommon. (*v.v.*)

2518. **S. fugacissima** Bourd. & Galz. *Fugacissima*, very fleeting.

R. ·5–2 cm., *greyish*, effused in a very thin, mucous, hyaline film, which disappears completely on drying, or leaves only a slightly glistening trace barely visible under a lens. Spores hyaline, cylindrical, curved, 4·5 × 2·5–4μ; basidia longitudinally septate, 6–7 × 5–6μ. Basal hyphae, thin walled, 2–3μ in diam. Very rotten wood. Feb. Rare.

Subgenus **Heterochaetella** Bourd.

(ἕτερος, different; χαίτη, hair.)

Differs from *Eusebacina* in the hymenium possessing true cystidia.

2519. **S. (Heteroch.) crystallina** Bourd. in Trans. Brit. Myc. Soc. VII, text fig. 2, p. 52. κρυστάλλινος, crystalline.

R. ·5–2 cm., *hyaline*, indeterminate, interruptedly effused, ceraceo-gelatinous, transparent, then collapsed and depressed, very thin, *rough with hyaline setae under a lens*; margin similar, or less continuous and reticulate. Spores hyaline, subglobose, or obovate, more or less apiculate at the base, (4)–4·5–6 × (3)–4·5μ, often 1-guttulate, germinating laterally or at the apiculus and then truly elongate, or conical; basidia obovate, or subglobose, 8–12 × 6–9μ, longitudinally septate, with 2–4 subulate sterigmata, 5–6μ and finally up to 15μ long. Cystidia scattered, or fasciculate, cylindrical, obtuse, 60–180 × 7–12μ, projecting 10–45μ above the hymenium, thin walled. Very rotten pine and juniper wood. Not uncommon. (*v.v.*)

Subgenus **Bourdotia** Bres.

(L'abbé H. Bourdot, the eminent French mycologist.)

Receptacle waxy or pulverulent, entirely resupinate, with the habit of a *Corticium*, possessing tubular, thin walled gloeocystidia, filled with a coloured juice, and rising perpendicularly in the hyphae.

2520. **S. (Bour.) Eyrei** Wakefield. (= *Gloeocystidium croceotingens* Wakef. sec. Bres.)
Rev. W. L. W. Eyre, a former President of the British Mycological Society.

R. 2–6 cm., *hyaline grey, becoming whitish grey when dry*, broadly and irregularly effused, closely adnate. Hymenium *concolorous*, smooth, or here and there tuberculate. Flesh very thin, 40–45 μ thick. Spores white, globose, 4–6μ; basidia at first obovate, continuous, at length longitudinally, cruciately divided into 4 parts, 10–13 × 7–8μ, with 4 curved sterigmata, 7–9 × 1·5μ (at the base). Gloeocystidia very numerous, arising from the base, not or scarcely projecting, cylindrical, or subfusiform, 25–35 × 5–7μ, filled with a granular, yellow olivaceous juice. Vegetative hyphae either thin and hyaline, or red brown, thicker walled, 3–6μ in diam., frequently septate, sometimes almost moniliform. The coloured hyphae often stain the wood a bright orange-red. Decorticated beech log. May—Oct. Rare.

Eichleriella Bres.

(Bogumil Eichler.)

Receptacle coriaceous, waxy, or membranaceous, subgelatinous, cup-shaped, or plano-concave, rarely pendulous; margin free. Hymenium smooth, rugulose, or *Radulum*-like. Basidia globose-ovoid, longitudinally, cruciately divided, with 2–4-sterigmata. Spores white, cylindrical, or oblong, smooth, producing sporidiola on germination. Growing on wood.

2521. **E. spinulosa** (Berk. & Curt.) Burt. (= *Radulum spinulosum* Berk. & Curt.; *Radulum deglubens* B. & Br.; *Eichleriella Kmetii* Bres.; *Stereum rufum* Eng. Auth. non Fr. sec. Wakefield.)

Spinulosa, full of little thorns.

R. 3–6 cm., *flesh colour, then wood brown*, longitudinally and broadly effused, confluent, separable; margin *white*, free, or reflexed, tomentose beneath. Hymenium *concolorous*, pruinose, *becoming red when rubbed, usually bearing tubercles*, distant, simple, or multifid. Flesh *concolorous*, coriaceous, soft, thick. Spores white, oblong, subcylindrical, curved, very obtuse, 15–18 × 6–10μ; basidia longitudinally septate, clavate, then fusiform, 30–45 × 9–12μ, with 2–3-, rarely 4-sterigmata; paraphyses brown at the apex. Dead branches of ash, and poplar. Sept.—March. Uncommon.

TULASNELLALES.

Basidia subglobose, simple, with 2–4 very thick, stout sterigmata. Spores white, producing sporidiola on germination. Hymenium fully exposed from the first.

TULASNELLACEAE.

Same characters as the order.

Tulasnella Schroet.

(L. R. and C. Tulasne, the eminent French mycologists.)

Receptacle fleshy membranaceous, or gelatinous, then cartilaginous, resupinate, effused. Hymenium smooth, or plicate. Basidia globose, with 2–4 very thick, stout sterigmata, at first obtuse, then becoming elongated and filiform, springing from the apex, or the side of the basidia. Spores white, globose, ovoid, elliptical, pyriform, or pip-shaped, smooth, producing conidia or a mycelium on germination. Growing on wood, and humus.

2522. **T. violea** (Quél.) Bourd. & Galz. (= *Hypochnus violeus* Quél.)

Violea, violet.

R. 2–10 cm., *lilac violet, or rosy lilac, becoming rosy greyish or decoloured when dry*, broadly effused; margin *brighter coloured*, byssoid. Hymenium concolorous, membranaceous, tomentose. Flesh membranaceous, very thin. Spores (or sterigmata) white, elliptical, 6 × 4μ, "globose or almost globose, 5–7·5–10 × 4·5–6·5–8μ, basidia oboval, globose, 9–12 × 8–10μ. Hyphae thin walled, 3–6μ" Bourd. & Galz. Dead wood. Sept.—Dec. Uncommon. (*v.v.*)

2523. **T. incarnata** Juel. *Incarnata*, flesh colour.

R. 1–3 cm., *bright pink*, effused, forming irregular, small patches. Hymenium *concolorous*, continuous, membranaceous. Flesh membranaceous, very thin. Spores white, pyriform, or pip-shaped, 8–11 × 5–7μ; basidia obovate or clavate, 9μ in diam. Sterigmata (or sessile spores according to Juel) usually 4, but sometimes 3 or 5, at first obovate, or elliptical, 10–13 × 5–7·5μ, at length with a terminal elongation, 7–17 × 1–2·5μ. Hyphae 3μ in diam. Fallen branches of oak. Sept.—Oct. Rare. (*v.v.*)

2524. **T. tremelloides** Wakef. & Pears. Trans. Brit. Myc. Soc. vi, text-figs., p. 70. *Tremella*, the genus *Tremella*, εἶδος, like.

R. 1–30 cm., *purple, becoming blackish when dry*, broadly effused; margin *concolorous*. Hymenium *concolorous, undulately plicate*. Flesh *pale purplish, gelatinous, then horny*, finally collapsing into a thin film. Spores white, elliptical, depressed on one side, laterally apiculate at the base, 8–10 × 4·5–5·5μ; basidia clavate, 15–18 × 6·5–7μ, with 4-sterigmata; sterigmata at first oblong, granular inside, 15 × 3–3·5μ, at length collapsed, up to 20μ long. Subhymenial hyphae, 6–8μ in diam., basal hyphae pale purplish, 4μ in diam. Pine needles, at the base of a stump. Nov. Rare.

CALOCERALES.

Basidia cylindrical, becoming forked with two long, pointed sterigmata. Hymenium fully exposed from the first. Spores always becoming septate on germination, and producing from each cell either one sporidiolum, or a bunch of conidia.

Caloceraceae.

Same characters as the order.

Dacryomyces Nees.

(δάκρυον, a tear; μύκης, fungus.)

Receptacle gelatinous, or subgelatinous, homogeneous; globose, subglobose, tuberculate, often becoming cup-shaped, and sometimes flattened, sessile, rarely stipitate or substipitate. Hymenium smooth, wrinkled, or folded. Basidia with two long, pointed sterigmata. Spores white, or yellowish, oblong, cylindrical, ovoid, subelliptical, or ovato-triangular; simple, transversely septate, or muriform. Receptacles producing conidia, globose, consisting of radiating, septate, moniliform threads. Growing on wood.

*Rosy.

2525. **D. macrosporus** B. & Br. (= *Dacryomyces fragiformis* (Pers.) Fr. sec. Quél.) B. & Br. Ann. Nat. Hist. no. 1374, t. 7, fig. 1. μακρός, long; σπορά, seed.

R. 6–20 mm., *rose colour*, tuberculate, rounded, irregularly gyrate. Flesh gelatinous, diaphanous, firm. Spores white, oblong, 40–50 × 8–11μ, 3–5-septate. Conidia elliptical, 14μ. Parasitic on old remains of *Diatrype stigma*. Dec.—April. Uncommon.

**Yellow, or orange.

2526. **D. deliquescens** (Bull.) Duby. Bull. Hist. Champ. Fr. 455, fig. 3, as *Tremella deliquescens*. *Deliquescens*, dissolving.

R. 2–12 mm., *yellow, or orange*, somewhat round, convex, then lens-shaped, immarginate, at length twisted, sessile, sometimes stipitate and root-like. Flesh *pale*, gelatinous, *hyaline*. Spores white, cylindrical, curved, 8–22 × 4–7μ, becoming 3-septate, "each compartment producing 1–2 ovoid sporidiola, 3–4 × 2μ; basidia 20–45 × 3–5μ. Hyphae 1–3μ in diam." Bourd. & Galz. Dead wood and fallen branches. Jan.—Dec. Common. (*v.v.*)

var. **hyalinus** (Pers.) Bourd. & Galz. (= *Dacryomyces hyalinus* (Pers.) Quél.; *Tremella albida* Huds. sec. Quél.) *Hyalinus*, transparent.

Differs from the type only in being at the first *entirely hyaline, then opaline*. Birch. Sept.—May. Not uncommon. (*v.v.*)

2527. **D. stillatus** (Nees) Fr. Nees Syst. t. 90. *Stillatus*, distilled.

R. 2–8 mm., *yellow, then orange*, colour persistent, globose, then umbilicate and *Peziza*-like, somewhat confluent, at length plicate, sessile, or substipitate, white villose at the base, often yellow pruinose. Flesh *paler*, gelatinous, then firm. Spores white, or yellowish, "ovoid, or oblong, rarely depressed, 18–25 × 7–10μ, 1-pluri-guttulate, gutta yellow, finally 1-septate; basidia 50–60 × 7–12μ, filled with orange granules. Hyphae 1·5–3μ in diam., with small clamp connections" Bourd. & Galz. Dead pine branches. Jan.—Dec. Common. (*v.v.*)

2528. **D. chrysocomus** (Bull.) Tul. Bull. Hist. Champ. Fr. t. 376, fig. 2, as *Peziza chrysocoma*. χρυσός, gold; κόμη, hair.

R. 2–3 mm., *golden*, orbicular, *spherical when young*, immarginate, *soon collapsing, pezizoid*, at length flattened, persistently *even*. Flesh *paler*, gelatinous, then cartilaginous, firm. Spores yellowish, "oblong oval, or subelliptical, incurved especially towards the base, 12–24 × 6–9μ, becoming larger and 10-septate; basidia 45–85 × 4–6μ. Hyphae 1·5–4μ in diam., with clamp connections" Bourd. & Galz. Coniferous branches. Jan.—Dec. Not uncommon. (*v.v.*)

2529. **D. tortus** (Berk.) Massee. (= *Tremella torta* Berk.) *Tortus*, twisted.

R. 4–8 mm., *yellow, or orange*, rounded, *depressed, gyroso-tuberculate*. Flesh *paler*, gelatinous, firm. Spores white, cylindrical, curved, 12 × 4–5μ, 3-septate. Dead oak branches. Oct.—May. Not uncommon. (*v.v.*)

2530. **D. succineus** Fr. (= *Peziza electrina* Phill. & Plowr.) Boud. Icon. t. 181. *Succineus*, amber colour.

R. ·5 mm., *yellow amber*, punctiform, globose, gregarious, forming patches 3–6 cm.; margin minutely toothed, paler on the outside.

Hymenium *darker*, velvety with the projecting sterigmata. Flesh subgelatinous. Spores white, oblong, slightly curved, 7–15 × 4–5μ; basidia with two long sterigmata. Fallen oak branches and pine leaves. Rare.

***Pallid, or fuscous.

2531. **D. sebaceus** B. & Br. B. & Br. Ann. Nat. Hist. no. 1305, t. 18, fig. 2. *Sebaceus*, like tallow.

R. 4–8 mm., *whitish*, somewhat round, cup-shaped. Flesh white, gelatinous, firm. Spores white, ovato-triangular, 14 × 6–8μ. Hyphae often clavate above. Ash, and maple twigs. Jan.—May. Uncommon. (*v.v.*)

2532. **D. vermiformis** B. & Br. B. & Br. Ann. Nat. Hist. no. 700, t. 3, fig. 1. *Vermis*, a worm; *forma*, shape.

R. 1 mm., *grey*, worm-shaped. Sporophores 14μ, spores 6μ. Rotten wood. April—Sept. Rare.

Ditiola Fr.

(δίς, twice; ἴουλος, down.)

Receptacle gelatinous, subgelatinous, sometimes becoming horny, always firm and becoming indurated in the stem; cup-shaped, tubercular, or globose, sometimes branched, or lobed; stipitate, or substipitate. Hymenium discoid, unilateral, smooth. Basidia cylindrical, with two long sterigmata. Spores white, oblong, cylindrical, elliptical, or elliptic cylindrical, smooth, simple, or 1–3 transversely septate. Growing on wood.

2533. **D. radicata** (A. & S.) Fr. (= *Femsjonia luteo-alba* Fr. sec. Quél.) Eng. & Prantl. Pflanz. Fam. 1**, p. 98, fig. 63, M–Q. *Radicata*, rooted.

R. 2–8 mm., *orange*, tubercular, *then nail-shaped with a distinct stem*, convex, slightly viscid; margin obtuse. St. 3–6 × 2–3 mm., *at first whitish, soon concolorous*, rooting. Flesh *paler*, subgelatinous above, firm below. Spores white, oblong, cylindrical, rounded at both ends, 9–10 × 4μ, becoming 1–3-septate; basidia cylindrical. Hyphae with clamp connections. Dead coniferous branches. Oct.—March. Uncommon. (*v.v.*)

2534. **D. Ulicis** Plowr. (=*Femsjonia luteo-alba* Fr. sec. Lloyd.) Trans. Brit. Myc. Soc. I, t. 2, figs. 2–6. *Ulex*, furze.

R. 1·5–5 mm., *pale lemon yellow, becoming darker*, head globose, then flattened and wrinkled, at first slightly villose with a thin, white, hyaline tomentum. St. ·5–1 mm., sometimes absent, hyaline-villose when young. Spores white, elliptico-cylindrical, 15 × 5μ, 4–5-guttulate, then cylindrical, with an oblique, large apiculus at the base, 15–18 × 5μ, 3-septate. Dead furze stems. Jan. Uncommon.

2535. **D. merulina** (Pers.) Rea. (= *Guepinia merulina* (Pers.) Quél.; *Guepinia peziza* Tul. sec. Pat.) Quél. Jur. et Vosg. I, t. 20, fig. 6, as *Tremella lutescens* Pers.

Merulina, like the genus *Merulius*.

R. 1–3 cm., *yellow amber*, cup-shaped, oblique, often irregular, and with linear folds on the outside; margin thin, waved. St. concolorous, slender. Flesh *yellowish*, diaphanous, gelatinous, then firm. Spores white, "oboval, or oblong, depressed at the side, 9–13 × 5–6μ, 1–3-septate. Conidia rough, subglobose, 9–12μ, formed on the exterior of the receptacle. Hyphae swollen at the ends, 5–6μ in diam., with *Opuntia*-like branches" Bourd. & Galz. Dead branches, and wood, especially beech. Dec.—Feb. Uncommon.

2536. **D. obliqua** (Massee) Rea. *Obliqua*, slanting.

R. 4 mm. high, *orange red*, concave, oblique, often like a rabbit's ear. St. *concolorous*, short, oblique. Flesh gelatinous, then horny. Spores white, elliptical, slightly curved, 12 × 5–6μ. Gregarious. Dead wood. Nov. Rare.

Femsjonia Fr.

(*Femsjonia*, belonging to Femsjo.)

Receptacle gelatinous, or floccose, heterogeneous, erumpent, convex, then plane, obconic, sessile. Hymenium smooth, becoming wrinkled. Basidia cylindrical, with two long, pointed sterigmata. Spores yellowish, boat-shaped, simple, and multi-guttulate, then becoming oblong and multi-septate. Growing on wood.

2537. **F. luteo-alba** Fr. *Luteus*, yellow; *alba*, white.

R. 2–15 cm., *bright golden yellow*, *erumpent*, convex, then plane, obconic, sessile, somewhat rooting, disc becoming wrinkled with age, *white tomentose beneath*. Flesh *yellowish*, subgelatinous near the hymenium, floccose below, firm. Spores yellowish, boat-shaped at first and multi-guttulate, 12–21 × 7–8μ, becoming oblong, and 8–10–or more-septate, 18–22 × 7–8μ. Basidia cylindrical, 75–80 × 5–7μ, with two long sterigmata, 35–40μ. Hyphae with clamp connections. Fallen branches of oak, and birch. Sept.—Nov. Not uncommon. (*v.v.*)

Dacryomitra Tul. (= **Dacryopsis** Massee).

(*δάκρυον*, a tear; *μίτρα*, a turban.)

Receptacle gelatinous, or subgelatinous, firm, erect, cylindrical, apex globose, or elongate, stipitate. Hymenium smooth, or rugosely wrinkled. Basidia cylindrical, with two long sterigmata. Conidiophores present, or absent, accompanying, or preceding the basidia. Spores white, oblong, or elliptical, smooth, simple, or transversely septate. Growing on wood.

*Hymenium with basidia only.

2538. **D. glossoides** (Pers.) Bref. (= *Calocera glossoides* (Pers.) Fr.) *γλῶσσα*, tongue; *εἶδος*, like.

R. 3–12 mm. high, *yellow*, *clavate*, or pear-shaped, *thickened*, obtuse, compressed, slightly viscid. St. *concolorous*, round, white floccose at the base. Flesh *concolorous*, gelatinous, firm. Spores white, broadly oblong, or elliptical, incurved, often attenuated at the base, 13–15 × 5–6μ, 2–3-septate. Dead oak branches. Sept.—Dec. Uncommon. (*v.v.*)

**Hymenium with conidiophores, and basidia.

2539. **D. nuda** (Berk.) Pat. (= *Ditiola nuda* Berk.) Massee, Brit. Fung. Fl. I, p. 56, figs. 5–6, as *Dacryopsis nuda* Massee. *Nuda*, naked.

R 3–4 mm., *reddish orange*, head hemispherical, flattened below. St. 3–4 × 2–3·5 mm., *white, or tinged yellow*, minutely tomentose. Flesh subgelatinous. Spores white, elliptic oblong, with an oblique apiculus, 14 × 5μ, 3-septate; basidia cylindrical, 56–60 × 5–6μ. Conidiophores linear, straight, aseptate, simple, or rarely with 1–3 short branchlets near the apex, 35–40 × 1·5μ; conidia elliptic oblong, 3 × 1μ. Fir stumps. Sept. Rare.

Calocera Fr.

(*καλός*, beautiful; *κέρας*, a horn.)

Receptacle gelatinous coriaceous, cartilaginous when dry; erect cylindrical, simple, or branched. Hymenium smooth, amphigenous. Basidia with two long sterigmata. Spores white, or yellow, elliptical, elliptic oblong, oblong, or comma-shaped, smooth, or punctate, simple, becoming septate on germination. Conidiophores rarely accompanying the basidia. Growing on wood, more rarely amongst leaves.

*Branched.

2540. **C. viscosa** (Pers.) Fr. Rolland, Champ. t. 104, no. 236, as *Calocera flammea* (Schaeff.) Quél. *Viscosa*, sticky.

R. 1–10 cm. high, *golden-egg-yellow, becoming orange when dry*, branched, *long rooted*, viscid, even, linear. Branches *concolorous*, round, or compressed, tense, straight, repeatedly dichotomous, apex generally forked. Flesh *yellow*, gelatinous, then cartilaginous, firm. Spores deep ochraceous, elliptic oblong, often depressed on one side, 10–11 × 4–5μ, 3-guttulate; basidia 40–50 × 5–6μ. Coniferous stumps. July—Jan. Common. (*v.v.*)

2541. **C. palmata** (Schum.) Fr. *Palmata*, palmate.

R. 2 cm. high, *orange, then yellow*, branched, compressed, *dilated upwards*, divided. Branches *concolorous*, somewhat round, *obtuse*.

Flesh *yellow*, gelatinous, tough. Spores "oblong, depressed on one side, 7–12 × 3·5–4·5μ; basidia 22–36 × 4–5μ. Hyphae 2–3μ in diam." Bourd. & Galz Wood. Rare.

**Caespitose.

2542. **C. tuberosa** (Sow.) Fr. Sow. Eng. Fung. t. 199. *Tuberosa*, bulbous.

R. 3–5 cm. high, *yellowish*, linear, simple, subacute, caespitose, two or three springing from a thick, strigose, *subglobose*, *rooting*, *tuberous base*. Flesh gelatinous, tough. Spores white, "elliptical, comma-shaped, 10μ, punctate" Quél., "9–10 × 6μ" Massee. Rotten trunks. Rare.

2543. **C. cornea** (Batsch) Fr. Sow. Eng. Fung. t. 40. *Cornea*, horny.

R. 4–10 mm. high, *yellow orange*, clubs small, *curt*, awl-shaped, connate at the base, rarely with a minute branchlet, viscid, white villose at the base, rooting. Flesh *paler*, gelatinous, soon horny, firm. Spores white, oblong, often curved, 7–9 × 3·5–4μ; "basidia 30–35 × 4–5μ. Hyphae 2–4μ in diam." Bourd. & Galz. Fallen branches, and worked wood. Jan.—Dec. Common. (*v.v.*)

2544. **C. corticalis** Fr. *Corticalis*, pertaining to the bark.

R. 1–2 mm. high, *pallid flesh colour*, clubs awl-shaped, somewhat distinct, erumpent. Flesh pellucid, soft. Caespitose. Dead bark. Dec.—Jan. Rare.

***Simple, distinct.

2545. **C. stricta** Fr. *Stricta*, rigid.

R. 10–25 mm. high, *yellow*, clubs solitary, simple, elongate; base white villose, blunt. Spores "club-shaped, acute downwards, 9–12 × 4–5μ, hyaline, slenderly septate in the middle, not constricted" Sacc. Decorticated wood. Sept.—April. Not uncommon. (*v.v.*)

var. **epiphylla** Fr. ἐπί, upon; φύλλον, a leaf.

Differs from the type in being *longer*, 5–7·5 *cm. high*, *in the base being naked and bluntly rooted*, *and in growing amongst pine needles*. Sept.—Oct. Uncommon.

2546. **C. striata** (Hoffm.) Fr. Hoffm. Fl. Germ. Cr. 2, t. 6, fig. 1, as *Clavaria striata*. *Striata*, furrowed.

R. 1–7 cm. high, *yellow*, clubs solitary, simple, *lanceolate*, acute, *striate when dry*. Flesh tough, translucid. Spores "oblong, depressed on one side, 7–10 × 3–5μ; basidia 28–36 × 4–5μ. Hyphae 2–3μ in diam." Bourd. & Galz. Prostrate trunks. March. Rare.

Apyrenium lignatile Fr. = the conidial condition of **Hypocrea rufa** (Pers.) Fr.

Apyrenium armeniacum B. & Br. = the conidial condition of **Hypocrea gelatinosa** (Tode) Fr.

BIBLIOGRAPHY

Adans.—M. ADANSON, in A. L. de Jussieu's Genera plantarum, 1789.
Afz.—A. AFZELIUS, in Kongliga Vetenskaps-Akademien Nya Handlingar, 1783.
A. & S.—J. B. DE ALBERTINI and L. D. DE SCHWEINITZ, Conspectus Fungorum in Lusatiae superioris agro Niskiensi crescentium, 1805.
Ann. of Bot.—Annals of Botany, 1887–
Ann. Mycol.—Annales Mycologici, 1903–
Ann. Nat. Hist.—Annals and Magazine of Natural History, 1838–1885.
Ann. Sc. Nat.—Annales des Sciences Naturelles, 1824–
Arrh.—J. P. ARRHENIUS, in Fries' Monographia Hymen. Sueciae, II, 1863.
Ass. fr.—Association française pour l'avancement des Sciences, 1872–1907.
Atkinson—G. F. ATKINSON, in Journal of Mycology, vol. VIII, 1902.
Awd.—Auerswald, in Gonnermann and Rabenhorst, 1869–1870.

Badham—C. D. BADHAM, On the Esculent Funguses of England, 1847.
Bamb.—C. VAN BAMBEKE, in Bulletin de la Société Royale de Botanique de Belgique, XLII, XLIII, 1906.
Barbier—M. BARBIER, Description synthétique des Russules de France, in Bull. Soc. d'Hist. Nat. de Châlon-sur-Saône, 1907.
Barl.—J. H. J. B. BARLA, Les Champignons de la province de Nice, 1859; Suite aux Champignons de Nice, 1886; Les Champignons des Alpes-Maritimes, 1888–1892.
Bary, de—H. A. DE BARY, Vergleichende Morphologie und Biologie der Pilze, 1884. (English translation: Comparative morphology and biology of the Fungi, Mycetozoa and Bacteria, 1887.)
Bat.—F. BATAILLE, Flore monographique des Amanites et des Lépiotes, 1902; Flore monographique des Astérosporés, 1908; Les Bolets, 1908; Flore monographique des Cortinaires d'Europe, 1912; Flore monographique des Hygrophores, 1910; Flore analytique des Inocybes d'Europe, 1910; Flore monographique des Marasmes d'Europe, 1919.
Batsch—A. J. G. C. BATSCH, Elenchus fungorum, 1783–1789.
Batt.—A. J. A. BATTARRA, Fungorum agri Ariminensis Historia, 1755.
Beck—G. BECK, Pilzflora Niederösterreichs, in Verh. Zool.-Bot. Ges. Wien, XXX, 1880.
Berk.—M. J. BERKELEY, in J. E. Smith's English Flora, vol. V, pt. 2, 1836; Outlines of British Fungology, 1860; Grevillea, 1872–1876.
B. & Br.—M. J. BERKELEY and C. E. BROOME, Notices of British Fungi, in Annals and Magazine of Natural History, 1838–1889.
B. & Curt.—M. J. BERKELEY and M. A. CURTIS, in Annals and Magazine of Natural History, 1853, 1859; Centuries of North American Fungi.
Berk. & Rav.—M. J. BERKELEY and H. W. RAVENEL, in Notices of North American Fungi, by M. J. Berkeley, in Grevillea, I, 1872.
Big. & Guill.—R. BIGEARD and H. GUILLEMIN, Flore des Champignons supérieurs de France, 1909–1913.
Bisch.—N. BISCHOFF, in Schmidel's Icones plantarum, etc., 1782–1783.
Bolt.—J. BOLTON, An History of Fungusses growing about Halifax, 1788–1791.
Bonord.—H. F. BONORDEN, Handbuch der allgemeinen Mykologie, 1851; in Botanische Zeitung, 1857–1859.
Bot. Magaz.—Botanical Magazine, 1887–1907.

Bot. Zeit.—Botanische Zeitung, 1843–

Boud.—E. BOUDIER, Icones Mycologicae, 1904–1911; in Bulletin de la Société Mycologique de France, 1885–1917.

Bourd. & Galz.—H. BOURDOT and A. GALZIN, Hyménomycètes de France, in Bulletin de la Société Mycologique de France, vols. XXV–XXXVII, 1909–1921.

Bourd. & Maire—H. BOURDOT and L. MAIRE, Notes sur quelques Hyménomycètes &c., in Bulletin de la Société Mycologique de France, vol. XXXVI, 1920.

Bref.—O. BREFELD, Untersuchungen aus dem Gesamtgebiete der Mykologie, 1872–1889.

Bres.—J. BRESADOLA, Fungi Tridentini novi, vel nondum delineati, 1881–1900; I funghi mangerecci e velenosi dell' Europa media, 1899; Schulzeria, nuovo genere d' imenomiceti, 1886; Hymenomycetes Hungarici Kmetiani, in Atti del I. R. Accad. Agiati, III, 1897; Fungi Polonici, in Ann. Myc. I, 1903.

Brig.—V. BRIGANTI, De fungis rarioribus Regni Neapolitani historia, 1824–1851; Historia Fungorum Regni Neapolitani, 1848 (with Fr. Briganti).

Britz.—M. BRITZELMAYR, Die Hymenomyceten Augsburgs und seiner Umgebung, 1879–1881; Hymenomyceten aus Südbayern, 1879–1894.

Brond.—L. DE BRONDEAU, Recueil de Plantes-Cryptogames de l'Agenaise, 1828–1830.

Brot.—F. DE AVELLAR BROTERO, Flora Lusitanica, 1804.

Buckn.—C. BUCKNALL, The fungi of the Bristol District, in Proceedings Bristol Naturalists' Soc. 1880–1886.

Buller—A. H. R. BULLER, Researches on Fungi, 1909.

Bull.—P. BULLIARD, Herbier de la France, 1780–1793; Histoire des Champignons de la France, 1791–1812.

Bull. Soc. Belg.—Bulletin de la Société Royale de Botanique de Belgique, 1862–

Bull. Soc. Bot. Fr.—Bulletin de la Société Botanique de France, 1854–

Bull. Soc. Myc. Fr.—Bulletin de la Société Mycologique de France, 1885–

Burt.—E. A. BURT, Corticiums causing plant diseases—Merulius in North America—The Thelephoraceae of North America, in Annals of the Missouri Botanical Garden, 1914–

Buxb.—J. C. BUXBAUM, Plantarum minus cognitarum centuriae quinque, 1728–1740.

Cat. de S. et L.—F. X. GILLOT et L. LUCAND, Catalogue raisonné des champignons supérieurs (Hyménomycètes) de Saône et Loire, 1891; in Bull. Soc. d'Hist. d'Autun, IV, 1891.

Ces.—V. DE CESATI, in Commentario della Società crittogamica Italiana, 1861.

Chev.—F. F. CHEVALLIER, Flora générale des Environs de Paris, 1826–1836.

Cke.—M. C. COOKE, British Edible Fungi, 1891; Edible and poisonous Mushrooms, 1894; Handbook of Australian Fungi, 1892; Handbook of British Fungi, 1871, 2nd revised edition, 1883–1891; Illustrations of British Fungi, 1881–1891.

Cke. & Quél.—M. C. COOKE and L. QUÉLET, Clavis synoptica Hymenomycetum Europaeorum, 1878.

Clusius—C. CLUSIUS, Rariorum plantarum historia, 1601.

Cda.—A. C. J. CORDA, Icones Fungorum hucusque cognitorum, 1837–1854.

Cordier—F. S. CORDIER, Les Champignons de la France, 1870.

Costantin—J. COSTANTIN, Atlas des Champignons, 1895.

Cost. & Duf.—J. COSTANTIN and L. DUFOUR, Nouvelle flore des Champignons, 3rd edition, 1901.

Crouan—P. L. and M. H. Crouan, Florule du Finistère, 1867.
Curr.—F. Currey, in Transactions Linnean Society, Botany, 1863–1875.
Curt.—W. Curtis, Flora Londinensis, 1817–1828.

Dalm.—J. W. Dalman, Swartz, in Vetensk. Akadem. Nya Handl. 1811.
DC.—A. P. de Candolle, Flore française, 1815; Rapport sur un voyage botanique dans l'ouest de la France, 1807.
Del.—C. J. L. Delastre, Flore analytique et descriptive du Département de la Vienne, 1842; Supplément, in Ann. des Sci. nat.
Desm.—J. B. H. J. Desmazières, Annales d'Histoire naturelle—Plantes cryptogamiques de la France, 1825–1860.
Desv.—N. A. Desvaux, in Desvaux's Journ. Bot. ii, 1809.
Dicks.—J. Dickson, Fasciculus Plantarum cryptogamicarum Britanniae, i–iv, 1785–1801.
Diet.—A. H. Dietrich, Flora regni Borussici, 1833–1844.
Dill.—J. Dillenius, Catalogus plantarum sponte circa Giessam nascentium, 1719.
Ditm.—L. P. F. Ditmar, in Sturm's Deutschlands Flora, 1813–1817.
Duby—J. E. Duby, in A. P. de Candolle's Botanicon Gallicum, 1828–1830.
Duf.—L. Dufour, Atlas des Champignons comestibles et vénéneux, 1891.
Dum.—P. Dumée, Nouvel atlas de poche des Champignons comestibles et vénéneux, 1909.
Dur.—de M. C. Durieu and C. Montagne, Flore d'Algérie, 1846–1849.

Ehrenb.—C. G. Ehrenberg, Sylvae Mycologicae Berolinenses, 1818.
Ehrh.—F. Ehrhart, Beiträge kryptogam. Botanik, 1787–1792.
Ellis—J. B. Ellis, in Bulletin Torrey Botanical Club, vol. lvii, 1881.
Ellis & Cragin—J. B. Ellis and F. W. Cragin, in Bulletin Washburn College Laboratory, Natural History, vol. i, 1885.
Ellis & Everh.—J. B. Ellis and B. M. Everhart, in Journal of Mycology, vol. v, 1889.
Engl. Bot.—English Botany, by Smith and Sowerby, 1790.
Engl. Flo.—English Flora, vol. v, pt. 2, 1837.
Engl. & Prantl—A. Engler and K. Prantl, Die Natürlichen Pflanzenfamilien, I. Teil, Abteilung 1, 1**, 1897–1900.

Falck—R. Falck, Ueber den Hausschwamm, in Zeitschrift f. Hygiene und Infectionskrankheit, lv, 1906.
Fayod—V. Fayod, Prodrome d'une histoire naturelle des Agaricinés, in Ann. Sci. Nat., 7 ser., ix, 1889.
Fitzpat.—H. M. Fitzpatrick, Phytopathology, viii, 1918.
Flora—Flora, 1818–1900.
Fl. Boruss.—Flora Borussica, by C. J. Wulff, 1765.
Fl. Dan.—Flora Danica, 1764–1874.
Fl. Wett.—Gärtner, Meyer and Scherbius, Ökonomische technische Flora der Wetterau, 1799–1802.
Fr.—E. M. Fries, Elenchus Fungorum, 1828; Epicrisis Systematis Mycologici, 1836–1838; Genera Hymenomycetum, 1836; J. A. Walbergii Fungi Natalenses, 1848; Hymenomycetes Europaei, 1874; Icones selectae Hymenomycetum nondum delineatorum, 1867–1884; Monographia Hymenomycetum Sueciae, 1857–1863; Novae symbolae mycologicae, 1851; Observationes mycologicae, 1815–1818; Summa Vegetabilium Scandinaviae, 1846–1849; Sveriges ätliga och giftiga Svampar, 1861; Symbolae Gasteromycorum ad illustrandam Floram Suecicam, 1817–1818; Systema Mycologicum, 1821–1832; Systema orbis vegetabilis, 1825.

Fuck.—L. FUCKEL, Fungi Rhenani exsiccati, 1863–1874; Symbolae mycologicae, 1869–1875.

Genev.—G. GENEVIER, in Bull. Soc. Bot. Fr. XXIII, 1876.
Gill.—C. C. GILLET, Les Champignons de la France, Hyménomycètes, 1874–1893; Tableaux analytiques des Hyménomycètes de France, 1884.
Gmel.—J. F. GMELIN, editor of Linnaeus' Systema Naturae, 13th edition, 1791.
Godey—C. GODEY, in Gillet's Les Champignons de la France, Hyménomycètes, 1874–1877.
Gonn. & Rabenh.—W. GONNERMANN and L. RABENHORST, Mycologia Europaea, 1869–1870.
Grev.—R. K. GREVILLE, Scottish Cryptogamic Flora, 1823–1828.
Grevillea—Grevillea: a quarterly record of Cryptogamic Botany, 1872–1892.

Hall.—A. VAN HALLER, Enumeratio methodica stirpium Helveticae indigenarum, 1742.
Hart.—R. HARTIG, Wichtige Krankheiten der Waldbäume, 1874.
Harz.—C. F. HARZER, Abbildungen der Pilze, 1842–1845.
Hedw.—Hedwigia, 1852–1908.
Henn.—PAUL HENNINGS, Beitr. zur Pilzenflora von Schleswig-Holstein, etc., 1892.
Herpell—G. HERPELL, Beitrag zur Kenntnis der Hutpilze in den Rheinlanden, in Hedwigia, XLIX, 1909.
Heufl.—L. F. VON HOHENBÜHEL-HEUFLER, in Oesterr. bot. Zeitschr. XX, 1870.
Hoehn.—F. VON HÖHNEL, Fragmente zur Mykologie, in Sitz. der Kaiserl. Akad. Wissenschaft in Wien, Mathem.-naturw. Klasse, CXI–CXXIV, 1902–1915.
Hoehn. & Litsch.—F. VON HÖHNEL and V. LITSCHAUER, Beitrag zur Kenntnis der Corticeen, in Sitz. der Kaiserl. Akad. Wissenschaft in Wien, Mathem.-naturw. Klasse, CXV–CXVII, 1906–1908.
Hoffm.—G. F. HOFFMANN, Deutschlands Flora, 1791–1795; Flora Germaniae, 1795; Nomenclator Fungorum, 1789–1790; Vegetabilia Cryptogama, 1787–1790.
Hoffm. H.—H. HOFFMANN, Icones analyticae Fungorum, 1861–1865.
Hollós—L. HOLLÓS, Die Gasteromyceten Ungarns, 1904.
Holmsk.—T. HOLMSKIOLD, Beata ruris otia Fungis Danicis impensa, 1790–1799; Nova Acta Hafn. I.
Hornem.—J. W. HORNEMANN, Flora Danica.
Huds.—W. HUDSON, Flora Anglica, 1778.
Huss.—T. J. HUSSEY, Illustrations of British Mycology, 1847–1855.

Inzenga—G. INZENGA, Funghi Siciliani, 1869–1879.

Jacq.—N. J. DE JACQUIN, Collectanea ad botanicam, et historiam naturalem spectantia, 1786–1796; Flora Austriaca, 1773; Miscellanea Austriaca ad botanicam, 1773–1778.
Johnst.—G. JOHNSTON, Berkeley, in Ann. and Mag. Nat. History, 1842.
Jour. Bot.—Journal de Botanique, 1887–1910.
Jour. of Bot.—Journal of Botany, 1862–1919.
Juel—H. O. JUEL, Die Kernteilungen in den Basidien und die Phylogenie der Basidiomyceten, in Pringsh. Jahrb. f. wiss. Bot. XXXII, 1898; *Stilbum vulgare* Tode, ein bisher verkannter Basidiomycet, in Bihang till Svenska Vet.-Akad. Handlingar, XXIV, 1898.
Jungh.—F. W. JUNGHUHN, in Linnaea, vol. V, 1830.

Kalchbr.—C. KALCHBRENNER, Icones selectae Hymenomycetum Hungariae, 1873–1877.

Karst.—P. A. KARSTEN, Icónes selectae Hymenomycetum Fenniae nondum delineatorum, 1884–1889; Kritisk öfversigt af Finlands Basidsvampar, 1889–1898; Mycologia Fennica, 1871–1879; Rysslands, Finlands och den Skandinaviska Holföns Hattsvampar, 1879–1882; Symbolae ad Mycologiam Fennicam, 1870–1893.

Kauffm.—C. H. KAUFFMAN, Amanita—Clitocybe—Collybia—Cortinarius —Hygrophorus—Lepiota—Russula, in the Agaricaceae of Michigan, Publication Biological Series 5, Michigan Geological and Biological Survey, December, 1918.

Kickx—J. KICKX, Flore cryptogamique des Flandres, 1867.

Klotz.—J. F. KLOTZSCH, Berkeley, in Eng. Fl. v, 1836; Dietrich's Flora regni Borussici, 1838; Flora Borussici, 1833–1841; Linnaea, 1832.

Knapp—J. L. KNAPP, Journal of a Naturalist, 1829.

Krombh.—J. VON KROMBHOLZ, Naturgetreue Abbildungen und Beschreibungen der Essbaren, Schädlichen und Verdächtigen Schwämme, 1831–1846.

Kunt.—O. KUNTZE, Revisio generum Plantarum—Fungi, 1898.

Kunz. & Schm.—G. KUNZE and J. C. SCHMIDT, in Mycologische Hefte, 1817–1823.

La Bill.—J. J. DE LABILLARDIÈRE, Voyage à la recherche de La Pérouse, 1791–1792.

Lam.—J. B. P. A. DE M. DE LAMARCK, Encyclopédie méthodique, 1783–1808; Supplément, 1810–1817; Flora Gallica, 1815.

Lamb.—E. LAMBOTTE, Flore mycologique de la Belgique, 1880.

Lange—J. E. LANGE, Studies in the Agarics of Denmark, Amanita—Collybia—Coprinus—Inocybe—Lepiota—Marasmius—Mycena—Pholiota—Pluteus—Rhodophyllus, in Dansk Botanisk Arkiv, 1914–1921.

Larb.—G. LARBER, Sui Funghi, saggio generale con descrizione dei funghi mangerecci d' Italia, 1829.

Lasch—W. G. LASCH, in Linnaea, 1826–1866.

Lenz—F. A. LENZ, Die Nützlichen und Schädlichen Schwämme, 1840.

Letell.—J. B. LETELLIER, Histoire des Champignons, etc., 1826.

Lév.—J. H. LÉVEILLÉ, in Annales d'histoire nâturelle, 1842–1848.

Leys.—F. W. LEYSSER, Flora Halensis, 1783.

Lib.—M. A. LIBERT, Plantae cryptogamicae quas in Arduenna collegit, 1830–1837.

Lightf.—J. LIGHTFOOT, Flora Scotica, 1777.

Lindgr.—S. J. LINDGREN, Botaniska Notiser, 1845.

Linnaea—Linnaea, 1826–1882.

Linn.—C. VON LINNAEUS, Flora Suecica, 1745–1755; Genera plantarum, 1737; Species plantarum, 1753; Systema Naturae, 1735; Systema Vegetabilium, 1774–1784.

Linn. Trans.—Transactions of the Linnean Society of London, 1791–1818.

Lloyd—C. G. LLOYD, Mycological Notes, 1898–1920; Synopsis of the genus Fomes, 1915; Synopsis of the known Phalloids, 1909; Synopsis of the section Apus of the genus Polyporus, 1915; Synopsis of the section Ovinus of Polyporus, 1911; Synopsis of the Stipitate Polyporoids, 1912; Synopsis of the Stipitate Stereums, 1913; The Geastrae, 1902; The Nidulariaceae, 1906; The Phalloids of Australia, 1907; The Tylostomeae, 1906.

Luc.—L. LUCAND, Figures peintes des Champignons de la France, 1881–1895.

McAlp.—McAlpine, in Annales Mycologici, iv, 1906.
Maire—René Maire, Les bases de la classification dans le genre Russula, 1910; Recherches cytologiques et taxonomiques sur les Basidiomycètes, 1902; Bulletins de la Société Mycologique de France, de la Société d'Histoire Naturelle de l'Afrique du Nord, etc.
Massee—G. Massee, British Fungi and Lichens, 1911; British Fungus-Flora, 1892–1895; European Fungus Flora—Agaricaceae, 1902; Monograph of the British Gastromycetes, in Ann. of Bot. iv (1889–1891); Monograph of the genus Lycoperdon, in Journ. Roy. Micr. Soc., 1887; Monograph of the Thelephoreae, in Linnean Society's Journ.—Bot. xxv, xxvii (1888–1891); Revision of the genus Coprinus, in Ann. of Bot. x (1896).
Massee & Crossl.—G. Massee and C. Crossland, The Fungus Flora of Yorkshire, 1905.
Matt.—H. G. von Mattuschka, Flora Silesiaca, 1776–1777.
Mich.—P. A. Micheli, Nova plantarum genera, 1729.
Möller—A. Möller, Protobasidiomyceten, in A. F. W. Schimper's Botanische Mittheilungen aus den Tropen, pt. viii, 1895.
Mont.—C. Montagne, in Annales des Sciences naturelles, Botanique, 1824–1886; Sylloge generum specierumque Cryptogamarum, 1856.
Moretti—G. Moretti, Il Botanico Italiano, 1826.
Morg.—A. P. Morgan, North American Fungi—The Gastromycetes, in the Journal of the Cincinnati Society of Natural History, 1888–1892.
Moug.—J. B. Mougeot, in Stirpes cryptogamicae Vogeso-rhenanae, 1810–1854.
Müll.—O. F. Müller, Ueber die Schwämme überhaupt, 1776, in Flora Danica, t. 601–900.

Nat.—The Naturalist, 1874–1919.
Nees—C. G. D. Nees von Esenbeck, in Act. Nov. Acad. Leop.-Carol. 1826; Ueberblick des Systems der Pilze und Schwämme, 1816–1817.
Nouel.—Nouel-Malingié, Observations mycologiques, Mém. Soc. Roy. Sci. à Lille, 1829 and 1830 (1831).

Opat.—W. Opatowski, Commentatio historico-naturalis de Familia Fungorum Boletoideorum, 1836.
Osbeck—P. Osbeck, in D. M. and J. Retzius' Dissertatio sistens Supplementum et Emendationes in Editionem secundam Prodromi Florae Scandinaviae, 1805.
Otth—G. H. Otth, in A. P. de Candolle's Prodromus, 1824–1873.
Otto—J. G. Otto, Versuch einer Anordnung der Agaricorum, 1816.
Oud.—C. A. J. A. Oudemans, Revision des Champignons trouvés dans les Pays-Bas, 1893.

Pass.—G. Passerini, Fungi Parmensi, in Nuovo Giorn. Bot. Ital. 1872–1881.
Pat.—N. Patouillard, Catalogue raisonné de Plantes Cellulaires de la Tunisie, 1897; Essai taxonomique sur les familles et les genres des Hyménomycètes, 1900; Les Hyménomycètes d'Europe, 1887; Tabulae analyticae Fungorum, 1883–1889.
Paul.—J. J. Paulet, Traité des Champignons, 1793.
Peck—C. H. Peck, 24th and 27th Reports New York Museum for 1870, 1873, issued 1873, 1875.
Pelt.—E. Peltereau, Etudes et observations sur les Russules, in Bulletin de la Société Mycologique de France, xxiv, 1908.
Pers.—C. H. Persoon, Commentatio de Fungis Clavaeformibus, 1797; Tentamen dispositionis methodicae Fungorum, 1797; Icones et De-

scriptiones Fungorum minus cognitorum, 1798–1800; Icones pictae rariorum Fungorum, 1803–1806; Mycologia Europaea, 1822–1828; Observationes Mycologicae, 1799; Synopsis methodica Fungorum, 1801.

Petch—T. Petch, in Annals of the Royal Botanic Gardens, Peradeniya, 1907–1918.

Phill. & Plowr.—W. Phillips and C. B. Plowright, New and rare British Fungi, in Grevillea.

Pico—V. Pico, Recueil des Mémoires de la Société de Médecine, III, 1780–1781.

Porta—J. B. Porta, Phytognomonica octo libris contenta; in quibus... effertur methodus, qua plantarum...abditas vires assignatur, 1650.

Price—S. Price, Illustrations of the Fungi of our fields and woods, 1864–1865.

Prill. & Del.—E. Prilleux and G. Delacroix, Maladies des plantes agricoles, etc., 1895–1897.

Purton—T. Purton, Midland Flora, 1817–1821.

Quél.—L. Quélet, in Association française pour l'avancement des sciences, 1880–1893; in Bulletin de la Société Botanique de Fr. 1877–1879; Enchiridion Fungorum in Europa media et praesertim in Gallia vigentium, 1886; Flore mycologique de la France et des pays limitrophes, 1888; Les Champignons du Jura et des Vosges, 1869–1875.

Quél. & Bat.—L. Quélet and F. Bataille, Flore monographique des Amanites et des Lépiotes, 1902.

Rabenh.—L. Rabenhorst, Deutschlands Kryptogamenflora, 1844–1853; Fungi Europ. Exsicc. 1824; Kryptogamen-Flora von Deutschland, Oesterreich und der Schweiz, 1884.

Ray—J. Ray, Synopsis methodica stirpium Britanniae, 1696.

Reb.—J. Rebentisch, Prodromus Florae Neomarchicae, 1804.

Reich.—J. J. Reichard, in Schriften Berlin Ges. Naturf.-Freunde, 1780.

Relh.—R. Relhan, Flora Cantabrigiensis, 1785–1793.

Retz.—A. J. Retzius, in Vet. Ak. Handl. 1769; Observationes Botanicae, 1779–1791.

Retz.—D. M. and J. Retzius, Dissertatio sistens Supplementum et Emendationes in Editionem secundam Prodromi Florae Scandinaviae, 1805.

Rev. Myc.—Revue Mycologique, 1879–1906.

Rich. & Roze—Atlas des Champignons comestibles et vénéneux de la France et des pays circonvoisins, 1888.

Rick.—A. Ricken, Die Blätterpilze Deutschlands und der angrenzenden Länder, 1910–1915.

Roll.—L. Rolland, Atlas des Champignons de France, Suisse et Belgique, 1906–1910.

Romell—L. Romell, Hymenomycetes of Lapland, 1911; Observationes mycologicae, I. De genere Russula, 1891; Remarks on some species of the genus Polyporus, 1912.

Roq.—J. Roques, Champignons comestibles et vénéneux, 1832.

Rostk.—J. F. Rostkovius, in Sturm's Deutschlands Flora, 1828–1848.

Roth—A. W. Roth, Catalecta botanica quibus Plantae novae et minus cognitae describuntur atque illustrantur, 1797–1806; Tentamen Florae Germanicae, 1788–1800.

Roze—E. Roze, in Karsten's Rysslands, Finlands och den Skandinaviska Holföns Hattsvampar, 1879–1882.

Sacc.—P. A. SACCARDO, Michelia, 1877–1882; Sylloge Fungorum omnium hucusque cognitorum, 1882–1913; Hymeniales (Fl. Ital. Crypt. Pars I—Fungi), 1915–1916.
Saund. & Sm.—W. W. SAUNDERS and W. G. SMITH, Mycological Illustrations, 1871.
Saut.—A. E. SAUTER, in Flora, vol. XXIV, 1841.
Schaeff.—J. C. SCHAEFFER, Fungorum qui in Bavaria et Palatinatu circa Ratisbonam nascuntur icones, 1762–1774.
Schmid.—C. C. SCHMIDEL, Icones plantarum et analyses partium, 1793–1797.
Schrad.—H. A. SCHRADER, Spicilegium Florae Germanicae, 1794; Nova Genera Plantarum, 1797.
Schrank—F. VON P. VON SCHRANK, Baiersche Flora, 1789.
Schroet.—J. SCHROETER, Die Pilzflora von Schlesien, in Kryptogamen-Flora von Schlesien, 1885–1908.
Schulz.—S. SCHULZER VON MUEGGENBURG, in Kalchbrenner's Icones selectae Hymenomycetum Hungariae, 1873; Mycologische Beiträge, 1876–1885; Verh. Oesterreich. Zool. Bot. Gesell. 1868.
Schum.—C. E. SCHUMACHER, Enumeratio Plantarum Saellandiae, 1801–1803.
Schwein.—L. DE SCHWEINITZ, Synopsis Fungorum Carolinae superioris, 1822.
Scop.—J. A. SCOPOLI, Flora Carniolica, 1772.
Secr.—L. SECRETAN, Mycographie Suisse, 1833.
Sm.—WORTHINGTON G. SMITH, Clavis Agaricinorum, 1870; Guide to Mr Worthington Smith's Drawings of Field and Cultivated Mushrooms and Poisonous or Worthless Fungi often mistaken for Mushrooms, 1910; Guide to Sowerby's Models of British Fungi in the Department of Botany, British Museum (Natural History), 2nd edition, 1908; Supplement to the Outlines of British Fungology by Berkeley, 1891; Synopsis of the British Basidiomycetes, 1908.
Sm. & Sow.—J. E. SMITH and J. SOWERBY, English Botany, 1790.
Sommerf.—C. SOMMERFELT, Supplementum florae Lapponicae, 1826.
Sow.—J. SOWERBY, Coloured Figures of English Fungi, 1797–1809.
Speg.—C. SPEGAZZINI, Fungi Argentini, 1880–1882, 1899.
Sterb.—F. VAN STERBEECK, Theatrum Fungorum, 1675.
Stev.—J. STEVENSON, British Fungi, 1886; Mycologia Scotica, 1879.
Studer—B. STUDER, in Hedwigia, vol. XXXIX, 1900.
Sturm—J. STURM, Deutschlands Flora, 1797–1862.
Sv. Bot.—Svensk Botanisk, 1802, etc.
Swanton—E. W. SWANTON, Fungi and how to know them, 1909.
Swz.—O. SWARTZ, in Nov. Act. Reg. Acad. Sci. Holm. 1808–1810.

Tavel—F. VON TAVEL, Vergleichende Morphologie der Pilze, 1892.
Thore—G. THORE, Essai d'un Chloris du dép. des Landes, 1803.
Tode—H. J. TODE, Fungi Mecklenburgenses selecti, 1790–1791.
Tourn.—J. P. DE TOURNEFORT, Institutiones, 1700.
Trans. Brit. Myc. Soc.—Transactions of the British Mycological Society, 1896–1920.
Trans. Woolh. Nat. Field Club—Transactions of the Woolhope Naturalists' Field Club, 1852–1913.
Trans. Wor. Nats. Club—Transactions of the Worcestershire Naturalists' Club, 1847–1920.
Tratt.—L. TRATTINICK, Die Essbaren Schwämme, 1809; Fungi Austriaci iconibus illustrati cum descriptionibus ac historia naturali completa, 1806.

Trog—J. G. Trog, Verzeichniss der in der Gegend von Thum vorkommenden Schwämme; Über der Wachsthum der Schwämme; Verzeichniss der in der Umgegend von Thum vorkommenden Schwämme, in Flora, xv, xix, xx, xxii, 1832, 1836, 1837, 1839; Verzeichniss schweizerischer Schwämme, in Mitt. d. schweiz. naturf. Ges. in Bern, 1844.

Tul.—L. R. and C. Tulasne, Fungi Hypogaei, 1851; Selecta Fungorum Carpologia, 1861–1865; Monographie des Nidulariées, in Ann. Sc. Nat. 1844; Sur les Fungi Tremellini et leurs alliés, id. 1862. Ann. Sc. Nat. 1842–1872.

Vahl—M. Vahl, in Flora Danica, t. 901–1260.

Vaill.—S. Vaillant, Botanicon Parisiense, 1727.

Vent.—J. Venturi, Miceti dell' agro Bresciano, 1845–1860.

Vill.—D. Villars, Histoire des Plantes du Dauphiné, 1786–1789.

Vitt.—C. Vittadini, Descrizione dei Funghi mangerecci più communi dell' Italia e de' velenosi che possono co' medesimi confondersi, 1835; Monographia Lycoperdineorum, 1842; Monographia Tuberacearum, 1831; Tentamen Mycologicum, 1826.

Viv.—D. Viviani, Funghi d' Italia, 1834.

Wahl.—G. Wahlenberg, Flora Lapponica, 1812; Flora Suecica, 1824–1826; Flora Upsaliensis, 1820.

Wallr.—F. W. Wallroth, Flora Cryptogamica Germaniae, 1831–1833.

Weinm.—C. G. Weinmann, Hymeno- et Gastero-mycetes hucusque in Imperio Rossico observatos, 1836.

Willd.—C. L. Willdenow, Florae Berolinensis prodromus, 1787.

With.—W. Withering, Botanical arrangement of British Plants, 3rd edition, 1796.

Woron.—M. Woronin, in Verhandl. Naturforsch. Ges. zu Freiburg, vol. iv, 1867.

Wulf.—F. X. von Wulfen, in Jacquin's Miscellanea, 1773–1778.

Zoll.—Zollinger, Systematisches Verzeichniss der im indischen Archipel in den Jahren 1842–1848 gesammelten, sowie der aus Japan empfangenen Pflanzen, 1854.

INDEX

Synonyms are in italics

Printed in the United States
By Bookmasters